UPDATE ON THE CHEMICAL THERMODYNAMICS OF URANIUM, NEPTUNIUM, PLUTONIUM, AMERICIUM AND TECHNETIUM

CHEMICAL THERMODYNAMICS

Vol. 1. Chemical Thermodynamics of Uranium (H. Wanner and I. Forest, eds.)
Vol. 2. Chemical Thermodynamics of Americium (R.J. Silva et al.)
Vol. 3. Chemical Thermodynamics of Technetium (M.C.A. Sandino and E. Östhols, eds.)
Vol. 4. Chemical Thermodynamics of Neptunium and
Plutonium (OECD Nuclear Energy Agency, ed.)
Vol. 5. Update on the Chemical Thermodynamics of Uranium, Neptunium, Plutonium, Americium and Technetium (OECD Nuclear Energy Agency, ed.)

CHEMICAL THERMODYNAMICS 5

Update on the Chemical Thermodynamics of Uranium, Neptunium, Plutonium, Americium and Technetium

Robert GUILLAUMONT (Chairman)
Université de Paris XI - Orsay
Laboratoire de Radiochimie - IPN
Orsay, France

Thomas FANGHÄNEL
Forschungszentrum Karlsruhe
Institut für Nukleare Entsorgung (INE)
Karlsruhe, Germany

Volker NECK
Forschungszentrum Karlsruhe
Institut für Nukleare Entsorgung (INE)
Karlsruhe, Germany

Jean FUGER
Institute of Radiochemistry
University of Liège-Sart Tilman
Liège, Belgium

Donald A. PALMER
Oak Ridge National Laboratory
Chemical Sciences Division
Oak Ridge, Tennessee, USA

Ingmar GRENTHE
Royal Institute of Technology
Department of Chemistry
Stockholm, Sweden

Malcolm H. RAND
WintersHill Consultancy
Dry Sandford, Abingdon
United Kingdom

Edited by
Federico J. MOMPEAN, Myriam ILLEMASSENE,
Cristina DOMENECH-ORTI and Katy BEN SAID
OECD Nuclear Energy Agency, Data Bank
Issy-les-Moulineaux (France)

NUCLEAR ENERGY AGENCY
ORGANISATION FOR ECONOMIC CO-OPERATION AND DEVELOPMENT

2003

ELSEVIER

AMSTERDAM • BOSTON • HEIDELBERG • LONDON • NEW YORK • OXFORD
PARIS • SAN DIEGO • SAN FRANCISCO • SINGAPORE • SYDNEY • TOKYO

Published by
ELSEVIER B.V.
Sara Burgerhartstraat 25
P.O. Box 211, 1000 AE Amsterdam, The Netherlands

ISBN: 0-444-51401-5

Copyright © Organisation for Economic Co-Operation and Development, 2003. All rights reserved.
No part of this publication may be reproduced, stored in a retrieval, system or transmitted in any form or by any means, electronic, electrostatic, magnetic tape, mechanical, photocopying, recording or otherwise, without permission in writing from the copyright holders. Queries concerning permission or translation rights should be addressed to: Head of Publications Service, OECD, 2 rue André Pascal, 75775 Paris Cedex 16, France

Special regulations for readers in the U.S.A. - This publication has been registered with the Copyright Clearance Center Inc. (CCC), Salem, Massachusetts. Information can be obtained from the CCC about conditions under which photocopies of parts of this publication may be made in the U.S.A. All other copyright questions, including photocopying outside of the U.S.A., should be referred to the OECD.

No responsibility is assumed by the Publisher or by the OECD for any injury and/or damage to persons or property as a matter of products liability, negligence or otherwise, or from any use or operation of any methods, products, instructions or ideas contained in the material herein.

The opinions expressed and arguments employed in this publication are the sole responsibility of the authors and do not necessarily reflect those of the OECD or of the governments of its member countries. Les idées exprimées et les arguments avancés dans cette publication sont ceux des auteurs et ne reflètent pas nécessairement ceux de l'OCDE ou des gouvernements de ses pays membres.

This volume has been reproduced by the publisher from the formatted pages prepared by the editors.

∞ The paper used in this publication meets the requirements of ANSI/NISO Z39.48-1992 (Permanence of Paper).

Printed in The Netherlands

Preface

This volume is the fifth of the series "Chemical Thermodynamics" edited by the OECD Nuclear Energy Agency (NEA). It is an update of the critical reviews published successively in 1992, as *Chemical Thermodynamics of Uranium*, in 1995 as *Chemical Thermodynamics of Americium*, in 1999 as *Chemical Thermodynamics of Technetium* and in 2001 as *Chemical Thermodynamics of Neptunium and Plutonium*. These previous volumes comprise reviews of the scientific literature in their areas of scope until 1989, 1993, 1995 and 1996, respectively, although later references are, in some isolated cases, also included.

This *Update* was initiated by the Management Board of the NEA Thermochemical Database Project Phase II (NEA TDB II). The first meeting of the U/Am/Tc/Np/Pu Update Review Group was held in October 1998 and six plenary meetings followed at NEA Headquarters at Issy-les-Moulineaux (France) in April 1999, November 1999, March 2000, September 2000 January 2001 and September 2001. The Executive Group of the Management Board provided scientific assistance in the implementation of the NEA TDB Project Guidelines. Jordi Bruno participated in Review Group meetings as the designated member of the Executive Group. At the NEA Data Bank the responsibility for the overall co-ordination of the Project was placed with Eric Östhols (from its initiation in 1998 to February 2000), with Stina Lundberg (from March 2000 to September 2000) and with Federico Mompean (since September 2000). Federico Mompean was in charge of the preparation of the successive drafts, updating the NEA thermodynamic database and editing the book to its present final form, with assistance from Myriam Illemassène, Cristina Domènech-Ortí and Katy Ben Said.

The aim of the members of the Review Group was to review all of the papers published two years prior to the publication dates of the books cited above in order to assure sufficient overlap with previous reviews in collecting and analysing data. The cut-off date was set at roughly the end of 2001 for the papers. The critical selection of thermodynamic data was made with reference to the previous selections. The Review Group focused on literature that contained new information, but in some instances material already analysed was re-examined in order to include all the data referring to a special problem. In all cases the arguments leading to a necessary change in previously selected values were carefully considered.

All the members contributed fully to the main text and the discussions, but the workload was distributed according to the expertise of each member. Malcolm Rand and Jean Fuger reviewed gas and solid-state thermodynamics, while the other members were involved in solution thermodynamics. Volker Neck and Thomas Fanghänel carried out the americium update with the help of Ingmar Grenthe, who mainly reviewed the carbonate and silicate complexes and compounds of uranium and, together with Volker Neck the carbonate complexes and compounds of neptunium and plutonium. Robert Guillaumont reviewed the hydrolysis and complexation of uranium, neptunium and plutonium with the help of Ingmar Grenthe, Volker Neck and Donald Palmer, who was also in charge of reviewing technetium.

The uranium section is the largest due to the amount of material published during the last ten years. This element has been extensively studied, principally because of its role as a nuclear fuel; in addition, it does not present any special radiological risk in laboratory-scale research and is widely present in the natural environment. In the field of solution thermodynamics the classical solubility, potentiometric and spectrophotometric methods to obtain data have been customarily used. A problem identified in all previous compilations is the assignment of uncertainties to the thermodynamic data obtained by these equilibrium analytical methods. During the past ten years scientists have implemented new statistical methods for hypothesis testing that are discussed in this review. However, there are few investigations that have used these methods; uncertainty estimates in this *Update* therefore follow the methods used in prior volumes. Thermodynamic data have also been obtained from new spectroscopic techniques that are applicable in systems with very low total concentrations of metal ions. The spectral resolution of these techniques is often better than in "traditional" spectrophotometry, allowing for a more accurate peak deconvolution. It is gratifying that data obtained by these new techniques in general agrees very well with data obtained by more traditional methods. Quantum chemical methods are also emerging as tools to understand the coordination chemistry of f-elements, thereby providing a valuable tool when discussing speciation.

In general, the sections dealing with the other elements addressed in this *Update* are shorter (except for the section on americium) because there are fewer experimental investigations to be reviewed. This situation is explained by the fact that previous reviews of these elements are more recent and fewer laboratories are currently working on them. The new experimental data concern mostly solution thermodynamics. The section on americium also contains reviews of curium data, the reason for this being the chemical similarity between the two elements and the much higher accuracy of thermodynamic data on Cm(III) that are obtained by fluorescence spectroscopy.

This *Update* does not substantially change the main body of thermodynamic values selected previously by the NEA, but rather supplements it with an appreciable amount of new data.

Palaiseau, France, March 2003   Robert Guillaumont, Chairman

Acknowledgements

For the preparation of this book, the authors have received financial support from the NEA TDB Phase II Project. The following organisations take part in the Project:

ANSTO, Australia
ONDRAF/NIRAS, Belgium
RAWRA, Czech Republic
POSIVA, Finland
ANDRA, France
IPSN (now IRSN), France
FZK, Germany
JNC, Japan
ENRESA, Spain
SKB, Sweden
SKI, Sweden
HSK, Switzerland
NAGRA, Switzerland
PSI, Switzerland
BNFL, UK
NIREX, UK
DoE, USA

The authors would like to express their gratitude to Rod Ewing for providing preprints of several of his publications related to uranium minerals. Comments from the Chairmen of the TDB Phase I Reviews on Americium and Technetium, Robert J. Silva and Joseph A. Rard are also gratefully acknowledged.

Malcolm Rand would like to acknowledge valuable advice on spectroscopic parameters from Rudy Konings, Federico Mompean and a Peer Reviewer, Vladimir Yungman.

At the NEA Data Bank, the bibliographical database was maintained by Linda Furuäng well into 2001. Pierre Nagel and Christian Penon have provided excellent software and advice, which have eased the editorial and database work. Cynthia Picot, Solange Quarmeau and Amanda Costa from NEA Publications have provided considerable help in editing the present series. Their contributions and the support of many NEA staff members are highly appreciated.

The entire manuscript of this book has undergone a peer review by an independent international group of reviewers, according to the procedures in the TDB-6 Guideline, available from the NEA. The peer reviewers have seen and approved the modifications made by the authors in response to their comments. The peer review comment records may be obtained on request from the OECD Nuclear Energy Agency. The peer reviewers were:

Prof. Sue B. Clark, Washington State University, Pullman, Washington, USA

Prof. Diego Ferri, Università Federico II di Napoli, Naples, Italy

Dr. Robert J. Lemire, Atomic Energy of Canada Ltd., Chalk River Laboratories, Ontario, Canada

Dr. Lester R. Morss, Office of Basic Energy Sciences, US Department of Energy, Washington DC, USA

Dr. Dhanpat Rai, Batelle, Pacific Northwest National Laboratory, USA

Dr. Vladimir Yungman, Glushko Thermocenter, Russian Academy of Sciences, Moscow, Russia

Their contributions are gratefully acknowledged.

Note from the Chairman of the NEA TDB Project Phase II

The need to make available a comprehensive, internationally recognised and quality-assured chemical thermodynamic database that meets the modeling requirements for the safety assessment of radioactive waste disposal systems prompted the Radioactive Waste Management Committee (RWMC) of the OECD Nuclear Energy Agency (NEA) to launch in 1984 the Thermochemical Database Project (NEA TDB) and to foster its continuation as a semi-autonomous project known as NEA TDB Phase II in 1998.

The RWMC assigned a high priority to the critical review of relevant chemical thermodynamic data of inorganic species and compounds of the actinides uranium, neptunium, plutonium and americium, as well as the fission product technetium. The first four books in this series on the chemical thermodynamics of uranium, americium, neptunium and plutonium, and technetium originated from this initiative.

The organisation of Phase II of the TDB Project reflects the interest in many OECD/NEA member countries for a timely compilation of the thermochemical data that would meet the specific requirements of their developing national waste disposal programmes.

The NEA TDB Phase II Review Teams, comprising internationally recognised experts in the field of chemical thermodymanics, exercise their scientific judgement in an independent way during the preparation of the review reports. The work of these Review Teams has also been subjected to further independent peer review.

Phase II of the TDB Project consisted of: (i) updating the existing, CODATA-compatible database for inorganic species and compounds of uranium, neptunium, plutonium, americium and technetium, (ii) extending it to include selected data on inorganic species and compounds of nickel, selenium and zirconium, (iii) and further adding data on organic complexes of citrate, oxalate, EDTA and iso-saccharinic acid (ISA) with uranium, neptunium, plutonium, americium, technetium, nickel, selenium, zirconium and some other competing cations.

The NEA TDB Phase II objectives were formulated by the 17 participating organisations coming from the fields of radioactive waste management and nuclear regulation. The TDB Management Board is assisted for technical matters by an Executive Group of experts in chemical thermodynamics. In this second phase of the Project, the NEA acts as coordinator, ensuring the application of the Project Guidelines and liaising with the Review Teams.

The present volume is the first one published within the scope of NEA TDB Phase II and contains an update of the existing databases for inorganic species and compounds of uranium, neptunium, plutonium, americium and technetium. This update was determined by the Project Management Board to be the highest priority task within the established programme of work. We trust that the efforts of the reviewers and the peer reviewers merit the same high recognition from the broader scientific community as received for previous volumes of this series.

Mehdi Askarieh
United Kingdom Nirex limited
Chairman of TDB Project Phase II Management Board
On behalf of the NEA TDB Project Phase II Participating Organisations:

ANSTO, Australia
ONDRAF/NIRAS, Belgium
RAWRA, Czech Republic
POSIVA, Finland
ANDRA, France
IPSN (now IRSN), France
FZK, Germany
JNC, Japan
ENRESA, Spain
SKB, Sweden
SKI, Sweden
HSK, Switzerland
PSI, Switzerland
BNFL, UK
Nirex, UK
DoE, USA

Editor's note

This is the fifth volume of a series of expert reviews of the chemical thermodynamics of key chemical elements in nuclear technology and waste management. This volume is devoted to updating the four previously published reviews on U [92GRE/FUG], Am [95SIL/BID], Tc [99RAR/RAN] and Np and Pu [2001LEM/FUG]. The tables contained in Chapters 3 to 8 list the currently selected thermodynamic values within the NEA TDB Project. The database system developed at the NEA Data Bank, *cf.* Section 2.6, assures consistency among all the selected and auxiliary data sets.

The recommended thermodynamic data are the result of a critical assessment of published information. In many instances (where updating has not been needed) the critical reviews supporting a particular selection will not be found in the present volume, but in the preceding one of this series dealing with the particular element involved (or in the volumes dealing with uranium and technetium for auxiliary data). In order to assist the reader in finding earlier reviewing information, a cross-reference system has been established between the sections of this volume and those of the preceding ones where the same species or group of species has been discussed. For this purpose, the relevant sections of the preceding volumes for a given element are listed in parentheses after the current volume section headings. In this way, for example, "9.4.5.2 Solid uranium iodides (V.4.4.1.3)" refers the reader to section V.4.4.1.3 in volume 1 of the series, [92GRE/FUG] and "10.2.1.2 Neptunium (IV) hydroxide complexes (8.1.4)" to section 8.1.4 in volume 4 of the series, [2001LEM/FUG].

How to contact the NEA TDB Project

Information on the NEA and the TDB Project, on-line access to selected data and computer programs, as well as many documents in electronic format are available at

www.nea.fr.

To contact the TDB project coordinator and the authors of the review reports, send comments on the TDB reviews, or to request further information, please send e-mail to tdb@nea.fr. If this is not possible, write to:

> TDB project coordinator
> OECD Nuclear Energy Agency, Data Bank
> Le Seine-St. Germain
> 12, boulevard des Îles
> F-92130 Issy-les-Moulineaux
> FRANCE

The NEA Data Bank provides a number of services that may be useful to the reader of this book.

- The recommended data can be obtained via internet directly from the NEA Data Bank.

- The NEA Data Bank maintains a library of computer programs in various areas. This includes geochemical codes such as PHREEQE, EQ3/6, MINEQL, MINTEQ and PHRQPITZ, in which chemical thermodynamic data like those presented in this book are required as the basic input data. These computer codes can be obtained on request from the NEA Data Bank.

Contents

Preface v
Acknowledgement vii
Note from the chairman of the NEA TDB Project Phase II ix
Editor's note xi

Part I Introductory material 1

1 INTRODUCTION 3

1.1 Background..3
1.2 Focus of the review..4
1.3 Review procedure and results..5

2 STANDARDS, CONVENTIONS AND CONTENTS OF THE TABLES 9

2.1 Symbols, terminology and nomenclature 9
   2.1.1 Abbreviations 9
   2.1.2 Symbols and terminology...11
   2.1.3 Chemical formulae and nomenclature 13
   2.1.4 Phase designators..13
   2.1.5 Processes 15
   2.1.6 Spectroscopic constants and statistical mechanics calculations for gaseous species 16
   2.1.7 Equilibrium constants...17
      2.1.7.1 Protonation of a ligand 17
      2.1.7.2 Formation of metal complexes 18
      2.1.7.3 Solubility constants 20

	2.1.7.4	Equilibria involving the addition of a gaseous ligand	21
	2.1.7.5	Redox equilibria	21
2.1.8	pH		24
2.1.9	Order of formulae		26
2.1.10	Reference codes		28
2.2	Units and conversion factors		28
2.3	Standard and reference conditions		31
2.3.1	Standard state		31
2.3.2	Standard state pressure		32
2.3.3	Reference temperature		35
2.4	Fundamental physical constants		35
2.5	Uncertainty estimates		36
2.6	The NEA-TDB system		36
2.7	Presentation of the selected data		38

Part II Tables of selected data   41

3   SELECTED DATA FOR URANIUM   43

3.1	General considerations	43
3.2	Precautions to be observed in the use of the tables	44

4   SELECTED DATA FOR NEPTUNIUM   79

4.1	General considerations	79
4.2	Precautions to be observed in the use of the tables	80

5   SELECTED DATA FOR PLUTONIUM   97

5.1	General considerations	97
5.2	Precautions to be observed in the use of the tables	98

6	**SELECTED DATA FOR AMERICIUM**	**113**
6.1	General considerations	113
6.2	Precautions to be observed in the use of the tables	114
7	**SELECTED DATA FOR TECHNETIUM**	**125**
7.1	General considerations	125
7.2	Precautions to be observed in the use of the tables	126
8	**SELECTED AUXILIARY DATA**	**133**

Part III Discussion of new data selection   151

9	**DISCUSSION OF NEW DATA SELECTION FOR URANIUM**	**157**
9.1	Elemental uranium (V.1)	157
9.1.1	Uranium metal (V.1.1)	157
9.1.2	Uranium gas (V.1.2)	157
9.2	Simple uranium aqua ions (V.2)	158
9.2.1	UO_2^{2+} (V.2.1)	159
9.2.2	UO_2^+ (V.2.2)	160
9.2.3	U^{4+} (V.2.3) and U^{3+} (V.2.4)	160
9.3	Uranium oxygen and hydrogen compounds and complexes (V.3)	160
9.3.1	Gaseous uranium oxides and hydroxides	160
9.3.1.1	Gaseous uranium oxides (V.3.1)	160
9.3.1.1.1	UO(g)	160
9.3.1.1.2	UO_3(g)	161
9.3.1.2	Gaseous uranium hydroxides	162
9.3.1.2.1	$UO_2(OH)_2$(g)	162
9.3.2	Aqueous uranium hydroxide complexes (V.3.2)	164

9.3.2.1	U(VI) hydroxide complexes (V.3.2.1)	164
9.3.2.1.1	Major polymeric species at 298.15 K (V.3.2.1.1)	169
9.3.2.1.2	Monomeric cationic and neutral hydrolysis species (V.3.2.1.2)	175
9.3.2.1.3	Anionic hydrolysis species (V.3.2.1.3)	176
9.3.2.1.4	Other polymeric cationic hydrolysis species (V.3.2.1.4)	178
9.3.2.1.5	Relevant solubility studies	178
9.3.2.1.5.1	Schoepite	179
9.3.2.1.5.1.1	Temperature dependence (V.3.2.1.5)	179
9.3.2.1.5.2	Polyuranates	180
9.3.2.1.5.2.1	$CaU_6O_{19} \cdot 11H_2O(cr)$	180
9.3.2.1.5.2.2	$Na_2U_2O_7 \cdot x\ H_2O(cr)$	181
9.3.2.1.5.2.3	$Na_xUO_{(3+x/2)} \cdot H_2O(cr)$	182
9.3.2.1.5.2.4	$K_2U_6O_{19} \cdot 11H_2O(cr)$	182
9.3.2.2	U(IV) hydroxide complexes (V.3.2.3)	182
9.3.2.2.1	Solubility of $UO_2(am)$ (=UO_2(am, hydr.))	184
9.3.2.2.2	Solubility of $UO_2(cr)$	185
9.3.2.2.3	Comparison of the solubilities of $UO_2(am)$ and $UO_2(cr)$	186
9.3.2.2.4	Selection of equilibrium constants	187
9.3.2.3	U(III) hydroxide complexes (V.3.2.4)	188
9.3.3	Crystalline and amorphous uranium oxides and hydroxides (V.3.3)	188
9.3.3.1	U(VI) oxides and hydroxides (V.3.3.1)	188
9.3.3.1.1	Other forms of anhydrous $UO_3(cr)$ (V.3.3.1.2)	188
9.3.3.1.2	U(VI) hydrated oxides and hydroxides (V.3.3.1.5)	188
9.3.3.1.2.1	$UO_3 \cdot 2H_2O$ (cr)	189
9.3.3.1.2.2	$UO_2(OH)_2(cr)$ or $UO_3 \cdot H_2O(cr)$	190
9.3.3.1.2.3	$UO_3 \cdot xH_2O$	191
9.3.3.1.3	U(VI) peroxides (V.3.3.1.4)	192
9.3.3.2	U(IV) oxides (V.3.3.2)	193
9.3.3.2.1	Crystalline UO_2 (uraninite) (V.3.3.2.1)	193
9.3.3.3	Mixed valence oxides (V.3.3.3)	193
9.3.3.3.1	$UO_{2.3333}$ ($\equiv 1/3\ U_3O_7$) (V.3.3.3.2)	193

	9.3.3.3.2 Hydrogen insertion compounds	194
	9.3.3.3.3 Other mixed valence oxides (V.3.3.3.4)	194
9.3.4	Uranium hydrides (V.3.4)	195
9.4	Uranium group 17 (halogen) compounds and complexes (V.4)	195
9.4.1	General	195
	9.4.1.1 Molecular parameters of the gaseous uranium halide species	195
9.4.2	Fluorine compounds and complexes (V.4.1)	199
	9.4.2.1 Gaseous uranium fluorides (V.4.1.1)	199
	9.4.2.1.1 UF(g), UF$_2$(g) and UF$_3$(g) (V.4.1.1.1 and V.4.1.1.2)	199
	9.4.2.1.2 UF$_4$(g) (V.4.1.1.3)	201
	9.4.2.1.3 UF$_5$(g) (V.4.1.1.4)	204
	9.4.2.1.4 UF$_6$(g) (V.4.1.1.5)	206
	9.4.2.1.5 U$_2$F$_{10}$(g) and U$_2$F$_{11}$(g) (V.4.1.1.6)	206
	9.4.2.1.6 UO$_2$F$_2$(g) (V.4.1.1.7)	207
	9.4.2.1.7 UOF$_4$(g) (V.4.1.1.8)	208
	9.4.2.2 Aqueous uranium fluorides (V.4.1.2)	209
	9.4.2.2.1 Aqueous U(VI) fluorides (V.4.1.2.1)	209
	9.4.2.2.1.1 Binary complexes	209
	9.4.2.2.1.2 Ternary U(VI) hydroxide-fluoride complexes	210
	9.4.2.2.2 Aqueous U(IV) fluorides (V.4.1.2.3)	210
	9.4.2.3 Solid uranium fluorides (V.4.1.3)	213
	9.4.2.3.1 Binary uranium fluorides and their hydrates (V.4.1.3.1)	213
	9.4.2.3.1.1 UF$_3$(cr) and UF$_4$(cr)	213
	9.4.2.3.1.1.1 High temperature heat capacity	213
	9.4.2.3.2 Other U(VI) fluorides (V.4.1.3.2)	213
	9.4.2.3.2.1 UO$_2$F$_2$(cr) and its hydrates	213
9.4.3	Chlorine compounds and complexes (V.4.2)	214
	9.4.3.1 Uranium chlorides (V.4.2.1)	214
	9.4.3.1.1 Gaseous uranium chlorides (V.4.2.1.1)	214
	9.4.3.1.1.1 UCl(g), UCl$_2$(g) and UCl$_3$(g)	214
	9.4.3.1.1.2 UCl$_4$(g)	216

9.4.3.1.1.3 $UCl_5(g)$ 217

9.4.3.1.1.4 $UCl_6(g)$ 218

9.4.3.1.1.5 $U_2Cl_8(g)$ 219

9.4.3.1.1.6 $U_2Cl_{10}(g)$ 219

9.4.3.1.1.7 $UO_2Cl_2(g)$ 220

9.4.3.1.2   Aqueous uranium chlorides (V.4.2.1.2.) 221

9.4.3.1.2.1   Aqueous U(VI) chlorides 221

9.4.3.1.2.2   Aqueous U(IV) chlorides 221

9.4.3.1.3   Solid uranium chlorides (V.4.2.1.3) 222

9.4.3.1.3.1 $UCl_3(cr)$ 222

9.4.3.1.3.2 $UCl_4(cr)$ 223

9.4.3.1.3.3 $UCl_6(cr)$ 224

9.4.3.2   Uranium hypochlorites 224

9.4.3.2.1   Aqueous U(VI) hypochlorites 224

9.4.4   Bromine compounds and complexes (V.4.3) 224

9.4.4.1   Gaseous uranium bromides (V.4.3.1.1) 224

9.4.4.1.1 $UBr(g)$, $UBr_2(g)$ and $UBr_3(g)$ 224

9.4.4.1.2 $UBr_4(g)$ 226

9.4.4.1.3 $UBr_5(g)$ 228

9.4.4.2   Solid uranium bromides (V.4.3.1.3) 228

9.4.4.2.1 UBr_4 (cr) 228

9.4.5   Iodides compounds and complexes (V.4.4.1) 228

9.4.5.1   Gaseous uranium iodides (V.4.4.1.1) 228

9.4.5.1.1 $UI(g)$, $UI_2(g)$ and $UI_3(g)$ 229

9.4.5.1.2 $UI_4(g)$ 229

9.4.5.2   Solid uranium iodides (V.4.4.1.3) 230

9.5   Uranium group 16 (chalcogens) compounds and complexes (V.5) 230

9.5.1   Aqueous uranium sulphate (V.5.1.3.1) 230

9.5.1.1   Aqueous U(VI) sulphates 230

9.5.1.1.1   Binary complexes 230

9.5.1.1.2   Ternary hydroxide-sulphate complexes 232

- 9.5.2 Aqueous uranium selenate (V.5.2.3.1) 233
- 9.5.3 Tellurium compounds (V.5.3) 234
  - 9.5.3.1 Uranium tellurides (V.5.3.1) 234
    - 9.5.3.1.1 Binary uranium tellurides (V.5.3.1.1) 234
  - 9.5.3.2 Solid uranium tellurites (V.5.3.2.2) 234
    - 9.5.3.2.1 Uranium(VI) monotellurites 234
    - 9.5.3.2.2 Uranium(VI) polytellurites 236
- 9.6 Uranium group 15 compounds and complexes (V.6) 236
  - 9.6.1 Nitrogen compounds and complexes (V.6.1) 236
    - 9.6.1.1 Uranium nitrides (V.6.1.1) 236
      - 9.6.1.1.1 UN(cr) (V.6.1.1.1) 237
      - 9.6.1.1.2 UN(g) 237
    - 9.6.1.2 Uranium Nitrates (V.6.1.3) 239
      - 9.6.1.2.1 Aqueous U(VI) nitrates 239
  - 9.6.2 Uranium sulphamate 239
    - 9.6.2.1 Aqueous U(VI) sulphamate 239
  - 9.6.3 Uranium phosphorus compounds and complexes (V.6.2) 239
    - 9.6.3.1 Aqueous uranium phosphorus species (V.6.2.1.) 239
      - 9.6.3.1.1 The uranium–phosphoric acid system (V.6.2.1.1) 239
        - 9.6.3.1.1.1 Complex formation in the U(VI)–H_3PO_4 system 239
        - 9.6.3.1.1.2 Complex formation in the U(IV)–H_3PO_4 system 241
    - 9.6.3.2 Solid uranium phosphorus compounds (V.6.2.2) 242
      - 9.6.3.2.1 Uranium orthophosphates (V.6.2.2.5) 242
        - 9.6.3.2.1.1 U(VI) orthophosphates 242
      - 9.6.3.2.2 U(VI) phosphates and vanadates (V.6.2.2.10) 243
      - 9.6.3.2.3 U(VI) fluorophosphates 243
  - 9.6.4 U(VI) arsenates (V.6.3.2.1) 243
    - 9.6.4.1 Aqueous U(VI) arsenates 243
  - 9.6.5 USb(cr) (V.6.4.1) 244
- 9.7 Uranium group 14 compounds and complexes (V.7) 244
  - 9.7.1 Uranium carbides (V.7.1.1) 244

9.7.1.1 UC(cr) (V.7.1.1.1) 244
9.7.2 The aqueous uranium carbonate system (V.7.1.2.1) 244
 9.7.2.1 Uranium(VI) carbonates 244
  9.7.2.1.1 Binary U(VI) carbonate complexes 246
  9.7.2.1.2 Ternary U(VI) hydroxide carbonate complexes 247
  9.7.2.1.3 Ternary U(VI) fluoride carbonate complexes 248
  9.7.2.1.4 Calcium uranium carbonate complex 248
 9.7.2.2 Solid U(VI) carbonates compounds 249
 9.7.2.3 Uranium(V) carbonates 250
9.7.2.3.1 Aqueous U(V) carbonates complexes 250
 9.7.2.4 Uranium(IV) carbonates 250
  9.7.2.4.1 Binary U(IV) carbonate complexes 250
  9.7.2.4.2 Ternary U(IV) hydroxide carbonate complexes 251
9.7.3 Silicon compounds and complexes (V.7.2) 252
 9.7.3.1 Aqueous uranium silicates (V.7.2.1) 252
 9.7.3.2 Solid uranium silicates (V.7.2.2) 254
  9.7.3.2.1 Solid U(IV) silicates 254
  9.7.3.2.2 Solid U(VI) silicates 254
  9.7.3.2.3 $(UO_2)_2SiO_4 \cdot 2H_2O$ (soddyite) (V.7.2.2.1) 254
  9.7.3.2.4 $Ca(UO_2)_2(SiO_3OH)_2 \cdot 5H_2O$ (uranophane), $Na(UO_2)(SiO_3OH) \cdot 2H_2O$ (sodium boltwoodite) and $Na_2(UO_2)_2(Si_2O_5)_3 \cdot 4H_2O$ (sodium weeksite) 256

9.8 Uranium actinide complexes (V.8) 257
 9.8.1 Actinides–actinides interactions (V. 8.1) 257
9.9 Uranium group 2 (alkaline-earth) compounds (V.9) 257
 9.9.1 Magnesium compounds (V.9.2) 257
 9.9.2 Calcium compounds (V.9.3) 259
  9.9.2.1 Calcium uranates (V.9.3.1.) 259
   9.9.2.1.1 $CaUO_4$(cr) (V.9.3.1.1) 259
   9.9.2.1.2 $CaUO_6O_{19} \cdot 11H_2O$(cr) 259
   9.9.2.1.3 Other calcium uranates 259

9.9.3	Strontium compounds (V.9.4)		259
	9.9.3.1	SrUO$_3$(cr)	260
	9.9.3.2	Strontium uranates (VI)	261
		9.9.3.2.1 α–SrUO$_4$ (V.9.4.1.1)	261
		9.9.3.2.2 β–SrUO$_4$ (V.9.4.1.2)	261
		9.9.3.2.3 SrU$_4$O$_{13}$ (V.9.4.2)	262
		9.9.3.2.4 Sr$_2$UO$_5$ (V.9.4.3)	262
		9.9.3.2.5 Sr$_2$U$_3$O$_{11}$ (V.9.4.4)	263
		9.9.3.2.6 Sr3UO6 (V.9.4.5)	263
		9.9.3.2.7 Sr$_3$U$_{11}$O$_{36}$	263
		9.9.3.2.8 Sr$_5$U$_3$O$_{14}$	264
		9.9.3.2.9 Sr$_3$U$_2$O$_9$	264
	9.9.3.3	Strontium uranates (V)	265
		9.9.3.3.1 Sr$_2$UO$_{4.5}$	265
	9.9.3.4	Strontium uranium tellurites	265
9.9.4	Barium compounds (V.9.5)		265
	9.9.4.1	BaUO$_3$(cr) (V.9.5.1.1)	265
9.10	Uranium group 1 (alkali) compounds (V.10)		267
9.10.1	General		267
9.10.2	Lithium compounds (V.10.1)		269
	9.10.2.1	Other uranates	269
9.10.3	Sodium compounds (V.10.2)		271
	9.10.3.1	Sodium monouranates (VI) (V.10.2.1.1)	271
	9.10.3.2	Sodium monouranates (V) (V.10.2.1.2)	271
		9.10.3.2.1 NaUO$_3$(cr)	271
	9.10.3.3	Other sodium uranates	272
	9.10.3.4	Sodium polyuranates (V.10.2.2)	274
		9.10.3.4.1 Na$_2$U$_2$O$_7$(cr) (V.10.2.2.1)	274
		9.10.3.4.2 Na$_6$U$_7$O$_{24}$(cr) (V.10.2.2.3)	274
9.10.4	Potassium compounds (V.10.3)		274
	9.10.4.1	KUO$_3$ (cr) (V.10.3.1)	274

9.10.4.2 $K_2U_2O_7(cr)$ (V.10.3.3) 275

9.10.4.3   K2U4O12(cr), K2U4O13(cr) 275

9.10.4.4 $K_2U_6O_{19} \cdot 11H_2O(cr)$ 275

9.10.5   Rubidium compounds (V.10.4) 275

9.10.5.1 $RbUO_3(cr)$ (V.10.4.1)..275

9.10.5.2 $Rb_2U_3O_{8.5}(cr)$, $Rb_2U_4O_{11}(cr)$, $Rb_2U_4O_{12}(cr)$ and $Rb_2U_4O_{13}(cr)$ 276

9.10.5.3 $Rb_2U(SO_4)_3(cr)$...277

9.10.6   Caesium compounds (V.10.5) 277

9.10.6.1   Caesium monouranate (V.10.5.1)..277

9.10.6.2   Caesium polyuranates (V.10.5.2) 277

9.10.6.2.1 $Cs_2U_2O_7(cr)$ (V.10.5.2.1)..277

9.10.6.2.2 $Cs_4U_5O_{17}(cr)$..278

9.10.6.2.3 $Cs_2U_4O_{12}(cr)$ and $Cs_2U_4O_{13}(cr)$ 278

9.11   Uranium compounds with elements from other groups..............................279

9.11.1   Thallium compounds 279

9.11.1.1   Thallium uranates...279

9.11.2   Zinc compounds. 280

9.11.3   Polyoxometallates...280

9.12   Uranium mineral phases 282

9.12.1   Introduction 282

9.12.2   Review of papers 282

9.12.3   Summary of mineral information 285

10   DISCUSSION OF NEW DATA SELECTION FOR NEPTUNIUM   293

10.1   Neptunium aqua ions (7) 293

10.2   Neptunium oxygen and hydrogen compounds and complexes (8) 294

10.2.1   Aqueous neptunium hydroxide complexes (8.1) 294

10.2.1.1   Neptunium (VI) hydroxide complex (8.1.2)....................................294

10.2.1.2   Neptunium(V) hydroxide complexes (8.1.3)..................................294

10.2.1.3   Neptunium(IV) hydroxide complexes (8.1.4) 294

10.2.2   Crystalline and amorphous neptunium oxides and hydroxides (8.2)...............297

  10.2.2.1 Solubility of neptunium(V) oxides and hydroxides (8.2.4) 297
  10.2.2.2 Neptunium(IV) oxides and hydroxides (8.2.5) 297
   10.2.2.2.1 Solubility of crystalline oxide NpO_2(cr) (8.2.5.1) 297
   10.2.2.2.2 Solubility of amorphous oxide NpO_2(am, hydr.) (8.2.5.2) 298
10.3 Neptunium halide compounds (9.1) 299
 10.3.1 Neptunium fluoride compounds (9.1.2) 299
  10.3.1.1 NpF(g), NpF_2(g), NpF_3(g) (9.1.2.1, 9.1.2.3) 299
10.4 Neptunium group 15 compounds and complexes (11) 300
 10.4.1 Neptunium nitrogen compounds (11.1) 300
  10.4.1.1 NpN(cr) (11.1.1.1) 300
 10.4.2 Neptunium antimony compounds 301
  10.4.2.1 Neptunium antimonides 301
   10.4.2.1.1 Neptunium monoantimonide 301
10.5 Neptunium chloride (9.2.2), nitrate (11.1.4.1) complexes and complexes with other actinides 301
10.6 Neptunium carbon compounds and complexes (12.1) 301
 10.6.1 Aqueous neptunium carbonate complexes (12.1.2.1) 301
  10.6.1.1 Ternary Np(IV) hydroxide-carbonate complexes 302
 10.6.2 Solid neptunium carbonates (12.1.2.2) 305
  10.6.2.1 Solid alkali metal neptunium(V) carbonates hydrates (12.1.2.2.2) 305
   10.6.2.1.1 Sodium neptunium(V) carbonates 305
   10.6.2.1.2 Potassium neptunium(V) carbonates 311

11 DISCUSSION OF NEW DATA SELECTION FOR PLUTONIUM 313

11.1 Plutonium aqua ions (16.1) 313
11.2 Plutonium oxygen and hydrogen compounds and complexes (17) 313
 11.2.1 Aqueous hydroxide complexes (17.1) 313
  11.2.1.1 Plutonium(V) hydroxide complexes (17.1.2) 313
  11.2.1.2 Plutonium(IV) hydroxides complexes (17.1.3) 313
 11.2.2 Solid plutonium oxides and hydroxides (17.2) 315
  11.2.2.1 Plutonium(IV) oxides and hydroxides 315

11.2.2.1.1 Solubility of crystalline oxide PuO$_2$(cr) (17.2.1.2) 316
11.2.2.1.2 Solubility of amorphous oxide and hydroxide PuO$_2$(am, hydr.) 316
11.2.3 Comparison of thermodynamic data for oxides and hydroxide complexes of tetravalent actinides 318
11.2.3.1 Solid oxides and aqueous hydroxide complexes of tetravalent actinides .. 318
11.2.3.2 Crystalline and amorphous oxides 321
11.3 Plutonium group 17 complexes 323
11.3.1 Aqueous plutonium chloride complexes (18.2.2.1) 323
11.3.1.1 Pu(III) chloride complexes 323
11.3.1.2 Pu(VI) chloride complexes 323
11.3.1.3 Cation-cation complexes of Pu(V) 324
11.4 Plutonium group 16 compounds and complexes (19) 324
11.4.1 Plutonium selenium compounds 324
11.4.1.1 Plutonium selenides 324
11.4.1.1.1 Plutonium monoselenide 324
11.4.2 Plutonium tellurium compounds 325
11.4.2.1 Plutonium tellurides 325
11.4.2.1.1 Plutonium monotelluride 325
11.5 Plutonium nitrogen compounds and complexes 325
11.5.1 Plutonium nitrides 325
11.5.1.1 PuN(cr) 325
11.6 Plutonium carbon compounds and complexes (21.1.2) 325
11.6.1 Aqueous plutonium carbonates (21.1.2.1) 325
11.6.1.1 Pu(VI) carbonate complexes (21.1.2.1.1) 325
11.6.1.2 Pu(IV) carbonate complexes (21.1.2.1.3) 327
11.6.2 Solid plutonium carbonate (21.1.2.2) 331
11.6.2.1 Solid Pu(VI) carbonates (21.1.2.2.1) 331
11.7 Plutonium group 2 (alkaline-earth) compounds 332
11.7.1 Plutonium-strontium compounds (22.1) 332
11.7.2 Plutonium-barium compounds (22.1) 332
11.7.2.1 BaPuO$_3$(cr) (22.2.2) 332

12 DISCUSSION OF NEW DATA SELECTION FOR AMERICIUM   333

12.1 Introductory remarks 333
    12.1.1 Estimation of standard entropies (V.4.2.1.2) 333
    12.1.2 Introductory remarks on the inclusion of thermodynamic data of aqueous Cm(III) complexes in this review 334

12.2 Elemental americium (V.1) 335
    12.2.1 Americium ideal monatomic gas (V.1.2) 335
        12.2.1.1 Heat capacity and entropy (V.1.2.1) 335
        12.2.1.2 Enthalpy of formation (V.1.2.2) 336

12.3 Americium oxygen and hydrogen compounds and complexes (V.3) 336
    12.3.1 Aqueous americium hydroxide complexes (V.3.1) 336
        12.3.1.1 Aqueous Am(III) and Cm(III) hydroxide complexes (V.3.1.1) 336
        12.3.1.2 Aqueous Am(V) hydroxide complexes (V.3.1.2) 342
    12.3.2 Solid americium oxides and hydroxides (V.3.2) 343
        12.3.2.1 Americium oxides 343
        12.3.2.2 Solid Am(III) hydroxides (V.3.2.4) 343
        12.3.2.3 Solid Am(V) hydroxides (V.3.2.5) 352

12.4 Americium group 17 (halogen) compounds and complexes (V.4) 354
    12.4.1 Aqueous group 17 (halogen) complexes (V.4.1) 354
        12.4.1.1 Aqueous Am(III) fluorides (V.4.1.1) 354
        12.4.1.2 Aqueous Am(III) chloride complexes (V.4.1.2) 356
        12.4.1.3 Aqueous Am(III) and Cm(III) chlorides (V.4.1.2.1) 356
    12.4.2 Americium halide compounds (V.4.2) 358
        12.4.2.1 Enthalpies of formation 358
        12.4.2.2 Americium fluoride compounds (V.4.2.2) 359
            12.4.2.2.1 Americium trifluoride (V.4.2.2.1) 359
                12.4.2.2.1.1 $AmF_3(cr)$ 359
                12.4.2.2.1.2 $AmF_3(g)$ 360
            12.4.2.2.2 Americium tetrafluoride (V.4.2.2.2) 361
        12.4.2.3 Americium chlorides (V.4.2.3) 362
            12.4.2.3.1 Americium trichloride (V.4.2.3.2) 362

12.4.2.3.1.1 $AmCl_3(cr)$ 362
12.4.2.3.2 Americium oxychloride (V.4.2.3.3) 362
12.4.2.3.3 Quaternary chloride $Cs_2NaAmCl_6(cr)$ (V.4.2.3.4.2) 363
12.4.2.4 Americium bromides (V.4.2.4) 363
12.4.2.4.1 Americium tribromide (V.4.2.4.2) 363
12.4.2.4.1.1 $AmBr_3(cr)$ 363
12.4.2.4.2 Americium oxybromide (V.4.2.4.3) 363
12.4.2.5 Americium iodides (V.4.2.5) 364
12.4.2.5.1 Americium triiodide (V.4.2.5.2) 364
12.4.2.5.1.1 $AmI_3(cr)$ 364
12.5 Americium group 16 (chalcogen) compounds and complexes (V.5) 365
12.5.1 Americium sulphates (V.5.1.2) 365
12.5.1.1 Aqueous Am(III) and Cm(III) sulphate complexes (V.5.1.2.1) 365
12.6 Americium group 14 compounds and complexes (V.7) 369
12.6.1 Carbon compounds and complexes (V.7.1) 369
12.6.1.1 Americium carbonate compounds and complexes (V.7.1.2) 369
12.6.1.1.1 Aqueous Am(III) and Cm(III) carbonate complexes 369
12.6.1.1.1.1 Tetracarbonato complex 374
12.6.1.1.1.2 Am(III) and Cm(III) bicarbonate complexes 375
12.6.1.1.1.3 Mixed Am(III) and Cm(III) hydroxide–carbonate complexes 376
12.6.1.1.2 Am(V) carbonate complexes 376
12.6.1.1.3 Solid americium carbonates (V.7.1.2.2) 378
12.6.1.1.3.1 Americium(III) hydroxycarbonate $AmOHCO_3(s)$ 378
12.6.1.1.3.2 Americium(III) carbonate $Am_2(CO_3)_3(s)$ 381
12.6.1.1.3.3 Sodium americium(III) carbonates 382
12.6.1.1.3.4 Sodium dioxoamericium(V) carbonate $NaAmO_2CO_3(s)$ 384
12.6.2 Aqueous americium silicates (V.7.2.2) 385
12.7 Americium group 6 compounds and complexes (V.10) 386
12.7.1 Americium(III) molybdate compounds and complexes 386
12.7.2 Aqueous complexes with tungstophosphate and tungstosilicate heteropolyanions (V.10.2) 386

LIST OF FIGURES

Figure 12-5: Solubility constants of amorphous Np(V) and Am(V) hydroxides in 0.3 – 5.6 m NaCl solution 353

Figure 12-6: Distribution of Cm species at 298.15 K as a function of the $CaCl_2$ concentration. 357

Figure 12-7: Vapour Pressure of $AmF_3(cr)$ 361

Figure 12-8: Extrapolation to $I = 0$ of experimental data for the formation of Am(III) and Cm(III) sulphate complexes 367

Figure 12-9: Application of the SIT to the stepwise formation constants of Am(III) and Cm(III) carbonate complexes in dilute to concentrated NaCl solutions ...372

Figure A-1: Comparison of solubility measurements for $Am(OH)_3(cr)$ and aged $Am(OH)_3(s)$ in 0.1 M $NaClO_4$ at 25°C. 411

Figure A-2. Solubility of $NaAm(CO_3)_2 \cdot xH_2O$ determined at 25°C and 22°C, respectively, in 5.6 m NaCl solution under an atmosphere of $p_{CO_2} = 10^{-2}$ bar in argon 429

Figure A-3: Speciation diagram of $[U^{4+}] = 10^{-7}$ M and $[CO_3^{2-}] = 1.5 \cdot 10^{-3}$ M 450

Figure A-4: Speciation diagram of $[U^{4+}] = 10^{-7}$ M and $[CO_3^{2-}] = 2.1 \cdot 10^{-1}$ M 450

Figure A-5: Experimental data of [95PAL/NGU] as Z versus – $\log_{10}[H^+]$ 500

Figure A-6: Experimental data Z versus – $\log_{10}[H+]$ from [95PAL/NGU] and the corresponding calculated values from the equilibrium constants obtained by this review using the LETAGROP least-squares program. 502

Figure A-7: Application of the SIT to Np(V) solubility data in K_2CO_3 solutions ≥ 0.17 mol $\cdot$ kg^{-1}, reaction: $KNpO_2(CO_3)(s)+2CO_3^{2-} \rightleftharpoons NpO_2(CO_3)_3^{5-} +K^+$ 572

Figure A-8: Application of the SIT to Np(V) solubility data in K_2CO_3 solutions ≥ 0.17 mol·kg^{-1}, reaction: $K_3NpO_2(CO_3)_2(s)+CO_3^{2-} \rightleftharpoons NpO_2(CO_3)_3^{5-} +3K^+$ 573

Figure A-9: Speciation at zero ionic strength calculated using equilibrium constants given by Meinrath. 599

Figure A-10: Speciation in the U(VI) hydroxide system at zero ionic strength using the equilibrium constants selected in [92GRE/FUG]. 600

Figure A-11: Speciation in the U(VI) hydroxide system in 0.1 M $NaClO_4$ using the equilibrium constants of Meinrath. 601

LIST OF FIGURES

Figure A-12: Speciation in the U(VI) hydroxide system using the equilibrium constants from Meinrath recalculated to 3 M NaClO$_4$. 602

Figure A-13: Speciation in the U(VI) hydroxide system using the equilibrium constants from [92GRE/FUG] recalculated to 3 M NaClO$_4$. 603

Figure A-14: Speciation in the U(VI) hydroxide system using the equilibrium constants from Meinrath recalculated to 3 M NaClO$_4$, but excluding the 7:4 species. 604

Figure A-15: Speciation in the U(VI) hydroxide system using the equilibrium constants from [92GRE/FUG] recalculated to 3 M NaClO$_4$, but excluding the 7:4 species 605

Figure A-16: Speciation diagram of U(VI) in alkaline solutions according to constants given in [2000NGU/PAL] where precipitation of uranyl phases has been suppressed 677

Figure A-17: Experimental Np(IV) solubility data measured upon freshly formed solid particles of Np(OH)$_4$(am) as a function of the H$^+$ or D$^+$ concentration in 0.1 M perchlorate solution. 694

Figure A-18: Np(IV) species distribution in 0.1 M HClO$_4$–NaClO$_4$. 694

Figure A-19: Speciation diagram obtained in perchlorate media assuming that no precipitation occurs 699

Figure A-20: Speciation diagram obtained in nitrate media (no precipitation of uranyl species is assumed) 700

Figure B-1: Plot of $\log_{10}\beta_1 + 4D$ versus I_m for reaction (B.12), at 25°C and 1 bar. 719

Figure D-1: Trace activity coefficients of the H$^+$ ion in NaCl and CsCl solution at 25°C 758

Figure D-2: Trace activity coefficients of the CO_3^{2-} ion in NaCl and NaClO4 solution at 25°C. 761

Figure D-3: Trace activity coefficients of the SO_4^{2-} ion in NaCl and NaTcO$_4$ solution at 25°C. 761

List of Tables

Table 2-1:	Abbreviations for experimental methods	9
Table 2-2:	Symbols and terminology	11
Table 2-3:	Abbreviations used as subscripts of Δ to denote the type of chemical process.	15
Table 2-4:	Unit conversion factors	28
Table 2-5:	Factors ρ for the conversion of molarity, c_B, to molality, m_B, of a substance B, in various media at 298.15 K	30
Table 2-6:	Reference states for some elements at the reference temperature of 298.15 K and standard pressure of 0.1 MPa	31
Table 2-7:	Fundamental physical constants	36
Table 3-1:	Selected thermodynamic data for uranium compounds and complexes	45
Table 3-2:	Selected thermodynamic data for reaction involving uranium compounds and complexes	64
Table 3-3:	Selected temperature coefficients for heat capacities of uranium compounds	76
Table 4-1:	Selected thermodynamic data for neptunium compounds and complexes	81
Table 4-2:	Selected thermodynamic data for reaction involving neptunium compounds and complexes	88
Table 4-3:	Selected temperature coefficients for heat capacities of neptunium compounds	95
Table 5-1:	Selected thermodynamic data for plutonium compounds and complexes	99

Table 5-2:	Selected thermodynamic data for reaction involving plutonium compounds and complexes	105
Table 5-3:	Selected temperature coefficients for heat capacities of plutonium compounds	111
Table 6-1:	Selected thermodynamic data for americium compounds and complexes	115
Table 6-2:	Selected thermodynamic data for reaction involving americium compounds and complexes	119
Table 6-3:	Selected temperature coefficients for heat capacities of americium compounds	123
Table 7-1:	Selected thermodynamic data for technetium compounds and complexes	127
Table 7-2:	Selected thermodynamic data for reaction involving technetium compounds and complexes	129
Table 7-3:	Selected temperature coefficients for heat capacities of technetium compounds	132
Table 8-1:	Selected thermodynamic data for auxiliary compounds and complexes	135
Table 8-2:	Selected thermodynamic data for reactions involving auxiliary compounds and complexes	149
Table 9-1:	Molecular parameters for UO(g)	161
Table 9-2:	Derived values of $\Delta_f H_m^\circ$ (UO$_2$(OH)$_2$, g, 298.15 K)	163
Table 9-3:	Experimental equilibrium constants for the U(VI) hydroxide system for the equilibria:	166
Table 9-4:	Experimental equilibrium constants for the U(VI) hydroxide system	167
Table 9-5:	Literature data for the solubility product of schoepite and Na$_2$U$_2$O$_7$(s)	168
Table 9-6:	Selected values for U(VI) hydrolysis species at 298.15 K	179
Table 9-7:	Equilibrium constants and solubility product [2001NEC/KIM]	187

Table 9-8:	Atomic mass used for the calculation of thermal functions	195
Table 9-9:	Molecular parameters of uranium halide gaseous species	196
Table 9-10:	Enthalpies of the reactions involving UF(g), UF_2(g) and UF_3(g)	199
Table 9-11:	Sums of various enthalpies involving UF(g), UF_2(g) and UF_3(g)	200
Table 9-12:	Optimised enthalpies of formation of UF(g), UF_2(g) and UF_3(g)	200
Table 9-13:	Enthalpy of sublimation of UF_4(cr)	201
Table 9-14:	Enthalpy of formation $\Delta_f H_m^\circ$ (UF_5, g, 298.15 K)	205
Table 9-15:	Enthalpy of sublimation of UO_2F_2(cr)	208
Table 9-16:	Literature data for the formation constants of U(IV) fluoride complexes recalculated to $I = 0$ with the SIT	211
Table 9-17:	Enthalpies of the reactions involving UCl(g), UCl_2(g) and UCl_3(g)	215
Table 9-18:	Sums of various enthalpies involving UCl(g), UCl_2(g) and UCl_3(g)	215
Table 9-19:	Optimised enthalpies of formation of UCl(g), UCl_2(g) and UCl_3(g)	215
Table 9-20:	Vapour pressure data for the vaporisation of UCl_4(cr, l)	216
Table 9-21:	Vapour pressure data for the sublimation of UCl_6(cr)	219
Table 9-22:	Enthalpies of the reactions involving UBr(g), UBr_2(g), UBr_3(g) and UBr_4(g)	225
Table 9-23:	Sums of various enthalpies involving UBr(g), UBr_2(g) and UBr_3(g)	225
Table 9-24:	Optimised enthalpies of formation of UBr(g), UBr_2(g), UBr_3(g) and UBr_4(g)	226
Table 9-25:	$\Delta_{sub} H_m^\circ$ of UBr_4 (cr) from data on UBr_4(cr) only	227
Table 9-26:	Enthalpy of formation of HI(aq),	230
Table 9-27:	Equilibrium constants for binary complexes $UO_2(SO_4)_r^{2-r}$	231
Table 9-28:	Equilibrium constants for ternary complexes, $\log_{10} {}^*\beta_{p,q,r}$, and interaction coefficient $\varepsilon((UO_2)_p(OH)_q(SO_4)_r^{2p-q-2r}, Na^+)$	233
Table 9-29:	Enthalpies of formation of $UTeO_5$(cr) derived from Gibbs energy studies	235

LIST OF TABLES

Table 9-30: Enthalpies of formation of UTe$_3$O$_9$(cr) derived from Gibbs energy studies..236

Table 9-31: Comparison of selected values from [90HAY/THO] and [92GRE/FUG]..237

Table 9-32: Equilibrium constants related to the reaction:
$UO_2^{2+} + r\,H^+ + q\,AsO_4^{3-} \rightleftharpoons UO_2H_r(AsO_4)_q^{2+r-3q}$ 243

Table 9-33: Equilibrium constants in the U(VI) carbonate systems 245

Table 9-34: Equilibrium constants of the $UO_2^{2+} + Si(OH)_4\,(aq) \rightleftharpoons UO_2SiO(OH)_3^+ + H^+$ ...252

Table 9-35: Enthalpies $\Delta_{sol}H_m$ and $\Delta_f H_m$ of SrUO$_{4-x}$ 261

Table 9-36: Derived heat capacity equations...269

Table 9-37: Derived enthalpies of reactions from stoichiometric phases 270

Table 9-38: Derived enthalpies of insertion reactions 270

Table 9-39: Derived enthalpies of reactions from stoichiometric phases 272

Table 9-40: Derived enthalpies of insertion reactions 273

Table 9-41: Conditional equilibrium constants K_1 and K_2 at different concentrations of [H$^+$] 281

Table 9-42: Conditional equilibrium constants of U(VI) with some polyoxometalate anions.. 282

Table 9-43: Minerals of uranium and related solid phases of interest for geochemical modelling..286

Table 10-1: Solubility constants of NaNpO$_2$CO$_3$·3.5H$_2$O(s) and Na$_3$NpO$_2$(CO$_3$)$_2$(s) at 20 – 25°C and conversion to $I = 0$ with the SIT...307

Table 11-1: Summary of solubility and equilibrium constants for tetravalent actinides retained for discussion or selected by this review 318

Table 11-2: Standard Gibbs energy of formation of tetravalent actinides ions and oxides 319

Table 11-3: Comparison of equilibrium constants $\log_{10} K_{s,0}^{\circ}$ derived from thermochemical data and solubility experiments 321

LIST OF TABLES

Table 12-1	Summary of entropy contributions for some Am compounds	334
Table 12-2:	Literature data for Am(III) and Cm(III) hydroxide complexes used in the previous [95SIL/BID] and present reviews for the evaluation of equilibrium constants at $I = 0$	338
Table 12-3:	Solubility constants for Am(III) hydroxides and conversion to $I = 0$, with the SIT coefficients in Appendix B.	344
Table 12-4:	Literature data for the solubility of Am(III) in alkaline solution (pH = 11 to 13).	349
Table 12-5:	Literature values of the formation constants for $AnF_n^{(3-n)}$ and $CmF_n^{(3-n)}$ complexes	355
Table 12-6:	Estimated enthalpies of formation of americium halides at 298.15 K	359
Table 12-7:	Calculations of S_m° (AnF$_4$, cr, 298.15 K)	362
Table 12-8:	Data for Am(III) and Cm(III) carbonate complexes discussed in the present and previous [95SIL/BID] reviews for the evaluation of stepwise formation constants at $I = 0$	369
Table 12-9:	Stepwise formation constants of aqueous Am(V) carbonate complexes from the solubility studies	377
Table 12-10:	Solubility constants reported for Am(III) hydroxycarbonate.	378
Table 12-11:	Solubility constants reported for the reaction: $1/2\, Am_2(CO_3)_3(s) \rightleftharpoons Am^{3+} + 3/2\, CO_3^{2-}$	381
Table 12-12:	Conditional formation constants of Am(III), Cm(III) and Am(IV) complexes with tungstophosphate and tungstosilicate heteropolyanions	386
Table A-1:	Literature data for the solubility of dehydrated schoepite	400
Table A-2:	Derived values of $\Delta_f H_m^\circ$ (UO$_2$(OH)$_2$, g, 298.15 K)	403
Table A-3:	Equilibrium constants for Am(OH)$_3$(s) and Am(III) hydroxide complexes derived from the solubility experiments of Stadler and Kim.	410
Table A-4:	Comparison of selected values from [90HAY/THO] and [92GRE/FUG].	417
Table A-5.	$\log_{10} \beta_n$ values corresponding to an ionic strength of 1.046 m HClO$_4$	421
Table A-6:	Equilibrium constant $\log_{10} \beta_n$ for UO$_2^{2+}$ with Cl$^-$ and NO$_3^-$	432

Table A-7:	Values of X_A, Y_A and W_A for the calculation of enthalpy and Gibbs energy from the Wilcox equation	433
Table A-8:	Values of X_B, Y_B and W_B for the calculation of enthalpy and Gibbs energy from the Wilcox equation	433
Table A-9:	Solubility product and Gibbs energy of formation of soddyite, uranophane, Na–boltwoodite and Na–weeksite:	442
Table A-10:	Equilibrium constants at zero ionic strength for the U(VI) hydroxide system	444
Table A-11:	Heat capacity coefficients and standard heat capacity for $Rb_2U_4O_{12}(cr)$ $Rb_2U_4O_{13}(cr)$, $Cs_2U_4O_{12}(cr)$ and $Cs_2U_4O_{13}(cr)$	445
Table A-12:	Derived values of $\Delta_f H_m^\circ$ ($UO_2(OH)_2$, g, 298.15 K)	455
Table A-13:	Enthalpies of solution and formation of $SrUO_{4-x}$	467
Table A-14:	Equilibrium constants for the dissolution of $NaAmO_2CO_3(s)$.	471
Table A-15:	Experimental and calculated Gibbs energies of sodium uranate reactions	474
Table A-16:	Equilibrium constants derived from the solubility of $AmO_2OH(am)$ and $NpO_2OH(am,$ aged) in 5 M NaCl at 22°C	482
Table A-17:	Equilibrium constants derived in [94RUN/KIM], [96RUN/NEU] from the solubility studies with $NaNpO_2CO_3(s)$ and $NaAmO_2CO_3(s)$ in NaCl solutions at 22°C	483
Table A-18:	Equilibrium constant obtained in various experimental conditions	490
Table A-19:	Equilibrium constant at $I = 0.1$ m and $I = 0$ for $(UO_2)_m(OH)_n^{2m-n}$ species	499
Table A-20:	$-\log_{10}[H^+]$ ranges used in the various titrations.	500
Table A-21:	Tests of various chemical models and refinement of the corresponding equilibrium constants.	501
Table A-22:	Equilibrium constants and lifetime measurements of UO_2^{2+}, $UO_2SO_4(aq)$, $UO_2(SO_4)_2^{2-}$ and UO_2OH^+ and $UO_2(OH)_2(aq)$	509
Table A-23:	Equilibrium constants and lifetime measurements of UO_2^{2+}, $UO_2SO_4(aq)$, $UO_2(SO_4)_2^{2-}$ and $UO_2(SO_4)_3^{4-}$	509

LIST OF TABLES

Table A-24:	Comparison between $\Delta_f G_m^\circ$ calculated from the sum of constituent oxide contributions and $\Delta_f G_m^\circ$ selected in [92GRE/FUG]	519
Table A-25:	Comparison between $\Delta_f G_m^\circ$ calculated from the sum of constituent oxide contribution and $\Delta_f G_m^\circ$ selected in [92GRE/FUG]	535
Table A-26:	Estimated and experimental Gibbs energy of formation for some uranium(VI) minerals.	536
Table A-27:	Main characteristics of fluorescence spectra of some aqueous complexes	537
Table A-28:	Temperature coefficients for $K_2U_4O_{12}(cr)$ and $K_2U_4O_{13}(cr)$ compounds	545
Table A-29:	Temperature coefficient for $Cs_2U_2O_7(cr)$ and $Cs_4U_5O_{17}(cr)$ compounds	546
Table A-30:	Spectroscopic data of some U(VI) complexes	553
Table A-31:	Standard state thermodynamic data	564
Table A-32:	Standard state thermodynamic data	565
Table A-33:	Np(V) carbonate solids observed in [97NOV/ALM], [98ALM/NOV].	570
Table A-34:	Enthalpy of the hydration reaction for trivalent and tetravalent ions M^{n+}	582
Table A-35:	Comparison between the calculated molar standard entropy and selected by TDB review.	582
Table A-36:	Unit cell parameters for schoepite, metaschoepite and dehydrated schoepite crystals	588
Table A-37:	Spectroscopic characteristic of 1:1 and 2:2 species at 25°C	594
Table A-38:	Spectroscopic characterisation of 0:1, 1:1, 2:1, 3:1, 2:2, 5:3 and 7:3 species	613
Table A-39:	Equilibrium constants of U(VI) with polyanions LX^- at pH = 4	620
Table A-40:	Emission wavelengths and lifetimes of emission of UO_2^{2+}, $UO_2H_2PO_4^+$, $UO_2HPO_4(aq)$ and $UO_2PO_4^-$	621
Table A-41:	Experimental and calculated Gibbs energies of potassium uranate reactions	639

LIST OF TABLES

Table A-42:	$\Delta\varepsilon$ values used to calculate the SIT interaction coefficients and $\log_{10} \beta°$ for PuO_2Cl^+, $PuO_2Cl_2(aq)$ and UO_2Cl^+, $UO_2Cl_2(aq)$.	653
Table A-43:	Equilibrium constant $\log_{10} \beta_1°$ and interaction coefficient $\Delta\varepsilon_{(1)}$ for the first chloride complexes of U(VI) and Pu(VI)	654
Table A-44:	Equilibrium constant $\log_{10} \beta_2°$ and interaction coefficient $\Delta\varepsilon_{(2)}$ for the second chloride complexes of U(VI) and Pu(VI)	655
Table A-45:	Equilibrium constants derived from spectroscopic measurements	656
Table A-46:	Values of $\log_{10} K°$ for protonation of arsenic acid at 25°C	657
Table A-47:	$\log_{10} {}^*\beta°_{p,q,r}$ and $\varepsilon(p, q, r, Na^+)$ (kg · mol^{-1})	665
Table A-48:	Values and units of the coefficients of equations (A.127) and (A.128)	667
Table A-49.	Formation constants for $Tc(CO)_3(H_2O)_{3-n}X_n^{(1-n)+}$ complexes	668
Table A-50:	Values of absorption coefficient, ε, for $(UO_2)_n(OH)_m^{2n-m}$ species	674
Table A-51:	Solubility constants for $PuO_2CO_3(s)$ (molal scale in Bold) and conversion to $I = 0$ with the SIT coefficients in Appendix B.	681
Table A-52	Thermodynamic data at zero ionic strength proposed in [2001NEC/KIM]	692
Table A-53:	Values of $\log_{10} {}^*\beta_{n,m}$ for (1,1), (2,2), (4,3), (5,3), (7,4) in 0.1 M KCl, NaClO$_4$ and in 1.0 M KNO$_3$	697
Table A-54:	$\log_{10} {}^*\beta°_{n,m}$ values and $\varepsilon n,m$ for UO_2^{2+}, (1,1), (2,2), (4,3), (5,3) and (7,4) species	698
Table A-55:	Species of U(VI) identified by Raman spectra in basic media according to [2002NGU]	703
Table B-1:	Water activities a_{H_2O} for the most common ionic media at various concentrations applying Pitzer's ion interaction approach and the interaction parameters given in [91PIT]	712
Table B-2:	Debye–Hückel constants as a function of temperature at a pressure of 1 bar below 100°C and at the steam saturated pressure for $t \geq 100°C$.	716
Table B-3:	The preparation of the experimental equilibrium constants for the extrapolation to $I = 0$ with the specific ion interaction method at 25°C and 1 bar, according to reaction (B.12)	718

Table B-4:	Ion interaction coefficients $\varepsilon_{(j,k)}$ (kg·mol^{-1}) for cations j with k = Cl$^-$, ClO$_4^-$ and NO$_3^-$, taken from Ciavatta [80CIA], [88CIA] unless indicated otherwise. 724
Table B-5:	Ion interaction coefficients $\varepsilon_{(j,k)}$ (kg·mol^{-1}) for cations j with k = Li, Na and K, taken from Ciavatta [80CIA], [88CIA] unless indicated otherwise.......731
Table B-6:	Ion interaction coefficients $\varepsilon_{(1,j,k)}$ and $\varepsilon_{(2,j,k)}$ for cations j with k = Cl$^-$, ClO$_4^-$ and NO$_3^-$ (first part), and for anions j with k = Li$^+$, Na$^+$ and K$^+$ (second part), according to the relationship $\varepsilon = \varepsilon_1 + \varepsilon_2 \log_{10}$I/m. The data are taken from Ciavatta [80CIA], [88CIA]. 734
Table C-1:	Details of the calculation of equilibrium constant corrected to I=0, using (C.19) 748

Part I

Introductory material

Chapter 1

Introduction

1.1 Background

The modelling of the behaviour of hazardous materials under environmental conditions is among the most important applications of natural and technical sciences for the protection of the environment. In order to assess, for example, the safety of a waste deposit, it is essential to be able to predict the eventual dispersion of its hazardous components in the environment (geosphere, biosphere). For hazardous materials stored in the ground or in geological formations, the most probable transport medium is the aqueous phase. An important factor is therefore the quantitative prediction of the reactions that are likely to occur between hazardous waste dissolved or suspended in ground water, and the surrounding rock material, in order to estimate the quantities of waste that can be transported in the aqueous phase. It is thus essential to know the relative stabilities of the compounds and complexes that may form under the relevant conditions. This information is often provided by speciation calculations using chemical thermodynamic data. The local conditions, such as ground water and rock composition or temperature, may not be constant along the migration paths of hazardous materials, and fundamental thermodynamic data are the indispensable basis for dynamic modelling of the chemical behaviour of hazardous waste components.

In the field of radioactive waste management, the hazardous material consists to a large extent of actinides and fission products from nuclear reactors. The scientific literature on thermodynamic data, mainly on equilibrium constants and redox potentials in aqueous solution, has been contradictory in a number of cases, especially in actinide chemistry. A critical and comprehensive review of the available literature is necessary in order to establish a reliable thermochemical database that fulfils the requirements for rigorous modelling of the behaviour of the actinide and fission products in the environment.

The International Atomic Energy Agency (IAEA) in Vienna published special issues with compilations of physicochemical properties of compounds and alloys of elements important in reactor technology: Pu, Nb, Ta, Be, Th, Zr, Mo, Hf and Ti between 1966 and 1983. In 1976, IAEA also started the publication of the series "The Chemical Thermodynamics of Actinide Elements and Compounds", oriented towards nuclear engineers and scientists. This international effort has resulted in the publication of several volumes, each concerning the thermodynamic properties of a given type of compounds for the entire actinide series. These reviews cover the literature approximately up to 1984. The latest volume in this series appeared in 1992, under Part 12: The Actinide Aqueous Inorganic Complexes [92FUG/KHO]. Unfortunately, data of importance for radioactive waste management (for example, Part 10: The Actinide Oxides) is lacking in the IAEA series.

The Radioactive Waste Management Committee (RWMC) of the OECD Nuclear Energy Agency recognised the need for an internationally acknowledged, high-quality thermochemical database for application in the safety assessment of radioactive waste disposal, and undertook the development of the NEA Thermochemical Data Base (TDB) project [85MUL], [88WAN], [91WAN]. The RWMC assigned a high priority to the critical review of relevant chemical thermodynamic data of compounds and complexes for this area containing the actinides uranium, neptunium, plutonium and americium, as well as the fission product technetium. The first four books in this series on the chemical thermodynamics of uranium [92GRE/FUG], americium [95SIL/BID], technetium [99RAR/RAN] and neptunium and plutonium [2001LEM/FUG] originated from this initiative. Simultaneously with the NEA's TDB project, other reviews on the physical and chemical properties of actinides appeared, including the book by Cordfunke et al. [90COR/KON2], the series edited by Freeman et al. [84FRE/LAN], [85FRE/LAN], [85FRE/KEL], [86FRE/KEL], [87FRE/LAN], [91FRE/KEL], the two volumes edited by Katz et al. [86KAT/SEA], and Part 12 by Fuger et al. [92FUG/KHO] within the IAEA review series mentioned above.

In 1998, Phase II of the TDB Project (TDB-II) was started to provide for the further needs of the radioactive waste management programs by updating the existing database and applying the TDB review methodology to other elements and to simple organic complexes. In TDB-II the overall objectives are set by a Management Board, integrated by representatives of 17 organisations from the field of radioactive waste management. These participating organisations, together with the NEA, provide financial support for TDB-II. The TDB-II Management Board is assisted in technical matters by a group of experts in chemical thermodynamics (the Executive Group). The NEA acts in this phase as Project Co-ordinator ensuring the implementation of the Project Guidelines and liaising with the Review Teams. The present volume, the fifth in the series, is the first one to be published within this second phase of the TDB Project.

1.2 Focus of the review

This first NEA TDB Update is within the scope and the spirit of previous reviews aimed at helping model the chemical behaviour of actinides and fission products in the near and far field of a radioactive waste repository using consistent data. The present critical review deals with U, Np, Pu, Am (Cm) and Tc. The data discussed and selected in some cases complement those of the previous reviews and in other cases revise them; they cover both solid compounds and soluble species of these elements.

The literature has been surveyed since the last NEA TDB reviews on U, Am, Tc and Np and Pu up to the end of 2001. This survey has revealed that many of the problems pointed out in the previous reviews have been addressed by the scientific community and in some cases have been resolved. This is to a large extent due to the use of new experimental and theoretical methods, in addition to the "traditional" thermodynamic methods such as potentiometric titrations and solubility measurements. The net result in that new data have come available that have been analysed together with the previous data already considered in the previous specific reviews. In addition this review has also used information on the systematics of chemical properties within the actinide series, one example being analogies between Am(III) and Cm(III). The policy of the NEA TDB is concentred on experimental results but it can not be denied that predictive papers present lot of interest. Some have been reviewed and quoted in the discussion of new data selection. Those dealing with thermochemistry are: [92DUC/SAN] (containing [62WIL], [92HIS/BEN] (containing [89LIE/GRE])) and [97ION/MAD] (containing [85BRA/LAG], [86BRA/LAG]).

Although the focus of the review is on actinides it is necessary to use data on a number of other species during the evaluation process that lead to selected data. These auxiliary data are taken both from the publication of CODATA key values [89COX/WAG] and from the evaluation of additional auxiliary data in the series of volumes entitled "Chemical Thermodynamics" [92GRE/FUG], [99RAR/RAN], and their use is recommended by this review. Care has been taken that all the selected thermodynamic data at standard state and conditions (*cf.* section 2.3) and 298.15 K are internally consistent. For this purpose, special software within the NEA TDB database system has been used; *cf.* section 2.6. In order to maintain consistency in the application of the values selected by this review, it is essential to use these auxiliary data when calculating equilibrium constants involving actinide compounds and complexes.

This review does not include any compounds and complexes containing organic ligands.

1.3 Review procedure and results

The objective of the present review is to update the database for the inorganic species of those elements that have been the object of previous NEA TDB reviews. This aim is achieved by an assessment of the new sources of thermodynamic data published since the cut-off dates for the literature searches in the earlier volumes of the series. This assessment is performed in order to decide on the most reliable values that can be recommended. Experimental measurements published in the scientific literature are the main source for the selection of recommended data. Previous reviews are not neglected, but form a valuable source of critical information on the quality of primary publications.

When necessary, experimental source data are re-evaluated by using chemical models that are either found to be more realistic than those used by the original author, or are consistent with side-reactions discussed in another section of the review (for example, data on carbonate complex formation might need to be re-interpreted to take into account consistent values for hydrolysis reactions). Re-evaluation of literature values might be also necessary to correct for known systematic errors (for example, if the junction potentials are neglected in the original publication) or to make extrapolations to standard state conditions ($I = 0$) by using the specific ion interaction (SIT) equations (*cf.* Appendix B). For convenience, these SIT equations are referred to in some places in the text as "the SIT". In order to ensure that consistent procedures are used for the evaluation of primary data, a number of guidelines have been developed. They have been updated and improved since 1987, and their most recent versions are available at the NEA [2000OST/WAN], [2000GRE/WAN], [99WAN/OST], [2000WAN/OST], [99WAN]. Some of these procedures are also outlined in this volume, *cf.* Chapter 2, Appendix B, and Appendix C. Parts of these sections, which were also published in earlier volumes [92GRE/FUG], [95SIL/BID], [99RAR/RAN], [2001LEM/FUG] have been revised in this review. For example, in Chapter 2, the section on "pH" has been revised. Appendix D deals with some limitations encountered in the application of the ionic strength correction procedures. Once the critical review process in the NEA TDB project is completed, the resulting manuscript is reviewed independently by qualified experts nominated by the NEA. The independent peer review is performed according to the procedures outlined in the TDB-6 guideline [99WAN]. The purpose of the additional peer review is to receive an independent view of the judgements and assessments made by the primary reviewers, to verify assumptions, results and conclusions, and to check whether the relevant literature has been exhaustively considered. The independent peer review is performed by persons having technical expertise in the subject matter to be reviewed, to a degree at least equivalent to that needed for the original review. The thermodynamic data selected in the present review (see Chapters 3, 4, 5, 6, 7 and 8) refer to the reference temperature of 298.15 K and to standard conditions, *cf.* section 2.3. For the modelling of real systems it is, in general, necessary to recalculate the standard thermodynamic data to non-standard state conditions. For aqueous species a procedure for the calculation of the activity factors is thus required. This review uses the approximate specific ion interaction method (SIT) for the extrapolation of experimental data to the standard state in the data evaluation process, and in some cases this requires the re-evaluation of original experimental values (solubilities, emf data, *etc.*). For maximum consistency, this method, as described in Appendix B, should always be used in conjunction with the selected data presented in this review. The thermodynamic data selected in this review are provided with uncertainties representing the 95% confidence level. As discussed in Appendix C, there is no unique way to assign uncertainties, and the assignments made in this review are to a large extent based on the subjective choice by the reviewers, supported by their scientific and technical experience in the corresponding area. The quality of thermodynamic models cannot be better than the quality of the data on which they are based. The quality aspect includes both the numerical values of the thermodynamic data used in the model and the "completeness" of the chemical model used, *e.g.*, the inclusion of all the relevant dissolved chemical species and solid phases. For the user it is important to consider that the selected data set presented in this review (Chapters 3 to 8) is certainly not "complete" with respect to all the conceivable systems and conditions; there are gaps in the information. The gaps are pointed out in the various sections of Part III, and this information may be used as a basis for the assignment of research priorities.

Chapter 2

Standards, Conventions and Contents of the Tables

This chapter outlines and lists the symbols, terminology and nomenclature, the units and conversion factors, the order of formulae, the standard conditions, and the fundamental physical constants used in this volume. They are derived from international standards and have been specially adjusted for the TDB publications.

2.1 Symbols, terminology and nomenclature

2.1.1 Abbreviations

Abbreviations are mainly used in tables where space is limited. Abbreviations for methods of measurement are listed in Table 2-1.

Table 2-1: Abbreviations for experimental methods.

AIX	Anion exchange
AES	Atomic Emission Spectroscopy
CAL	Calorimetry
CHR	Chromatography
CIX	Cation exchange
COL	Colorimetry
CON	Conductivity
COU	Coulometry
CRY	Cryoscopy
DIS	Distribution between two phases
DSC	Differential Scanning Calorimetry
DTA	Differential Thermal Analysis
EDS	Energy Dispersive Spectroscopy
EM	Electromigration
EMF	Electromotive force, not specified

(Continued on next page)

Table 2-1: (continued)

EPMA	Electron Probe Micro Analysis
EXAFS	Extended X-ray Absorption Fine Structure
FTIR	Fourier Transform Infra Red
IDMS	Isotope Dilution Mass-Spectroscopy
IR	Infrared
GL	Glass electrode
ISE-X	Ion selective electrode with ion X stated
IX	Ion exchange
KIN	Rate of reaction
LIBD	Laser Induced Breakdown Detection
MVD	Mole Volume Determination
NMR	Nuclear Magnetic Resonance
PAS	Photo Acoustic Spectroscopy
POL	Polarography
POT	Potentiometry
PRX	Proton relaxation
QH	Quinhydrone electrode
RED	Emf with redox electrode
SEM	Scanning Electron Microscopy
SP	Spectrophotometry
SOL	Solubility
TC	Transient Conductivity
TGA	Thermo Gravimetric Analysis
TLS	Thermal Lensing Spectrophotometry
TRLFS	Time Resolved Laser Fluorescence Spectroscopy
UV	Ultraviolet
VLT	Voltammetry
XANES	X-ray Absorption Near Edge Structure
XRD	X-ray Diffraction
?	Method unknown to the reviewers

Other abbreviations may also be used in tables, such as SHE for the standard hydrogen electrode or SCE for the saturated calomel electrode. The abbreviation NHE has been widely used for the "normal hydrogen electrode", which is by definition identical to the SHE. It should nevertheless be noted that NHE customarily refers to a standard state pressure of 1 atm, whereas SHE always refers to a standard state pressure of 0.1 MPa (1 bar) in this review.

2.1.2 Symbols and terminology

The symbols for physical and chemical quantities used in the TDB review follow the recommendations of the International Union of Pure and Applied Chemistry, IUPAC [79WHI], [88MIL/CVI]. They are summarised in Table 2-2.

Table 2-2: Symbols and terminology.

Symbols and terminology	
length	l
height	h
radius	r
diameter	d
volume	V
mass	m
density (mass divided by volume)	ρ
time	t
frequency	ν
wavelength	λ
internal transmittance (transmittance of the medium itself, disregarding boundary or container influence)	T
internal transmission density, (decadic absorbance): $\log_{10}(1/T)$	A
molar (decadic) absorption coefficient: $A/c_B l$	ε
relaxation time	τ
Avogadro constant	N_A
relative molecular mass of a substance[a]	M_r
thermodynamic temperature, absolute temperature	T
Celsius temperature	t
(molar) gas constant	R
Boltzmann constant	k
Faraday constant	F
(molar) entropy	S_m
(molar) heat capacity at constant pressure	$C_{p,m}$
(molar) enthalpy	H_m
(molar) Gibbs energy	G_m
chemical potential of substance B	μ_B
pressure	p
partial pressure of substance B: $x_B p$	p_B
fugacity of substance B	f_B

(Continued next page)

Table 2-2 (continued)

Symbols and terminology	
fugacity coefficient: f_B/p_B	$\gamma_{f,B}$
amount of substance [b]	n
mole fraction of substance B:	x_B
molarity or concentration of a solute substance B (amount of B divided by the volume of the solution) [c]	c_B, [B]
molality of a solute substance B (amount of B divided by the mass of the solvent) [d]	m_B
mean ionic molality [e], $m_{\pm}^{(v_+ + v_-)} = m_+^{v_+} m_-^{v_-}$	$m_{\pm}$
activity of substance B	a_B
activity coefficient, molality basis: a_B / m_B	γ_B
activity coefficient, concentration basis: a_B / c_B	y_B
mean ionic activity [e], $a_{\pm}^{(v_+ + v_-)} = a_B = a_+^{v_+} a_-^{v_-}$	$a_{\pm}$
mean ionic activity coefficient [e], $\gamma_{\pm}^{(v_+ + v_-)} = \gamma_+^{v_+} \gamma_-^{v_-}$	$y_{\pm}$
osmotic coefficient, molality basis	ϕ
ionic strength: $I_m = \frac{1}{2}\sum_i m_i z_i^2$ or $I_c = \frac{1}{2}\sum_i c_i z_i^2$	I
SIT ion interaction coefficient between substance B$_1$ and substance B$_2$, stoichiometric coefficient of substance B (negative for reactants, positive for products)	$\varepsilon(B_1, B_2)$
general equation for a chemical reaction	$0 = \sum_B v_B B$
equilibrium constant [f]	K
charge number of an ion B (positive for cations, negative for anions)	z_B
charge number of a cell reaction	n
electromotive force	E
$pH = -\log_{10}[a_{H^+} /(\text{mol} \cdot \text{kg}^{-1})]$	pH
electrolytic conductivity	κ
superscript for standard state [g]	o

[a] ratio of the average mass per formula unit of a substance to $\frac{1}{12}$ of the mass of an atom of nuclide ^{12}C.

[b] cf. sections 1.2 and 3.6 of the IUPAC manual [79WHI].

[c] This quantity is called "amount-of-substance concentration" in the IUPAC manual [79WHI]. A solution with a concentration equal to $0.1 \text{ mol} \cdot \text{dm}^{-3}$ is called a 0.1 molar solution or a 0.1 M solution.

[d] A solution having a molality equal to $0.1 \text{ mol} \cdot \text{kg}^{-1}$ is called a 0.1 molal solution or a 0.1 m solution.

[e] For an electrolyte $N_{v_+} X_{v_-}$ which dissociates into $v_{\pm}$ $(= v_+ + v_-)$ ions, in an aqueous solution with concentration m, the individual cationic molality and activity coefficient are $m_+ (= v_+ m)$ and $\gamma_+ (= a_+ / m_+)$. A similar definition is used for the anionic symbols. Electrical neutrality requires that $v_+ z_+ = v_- z_-$

[f] Special notations for equilibrium constants are outlined in section 2.1.6. In some cases, K_c is used to indicate a concentration constant in molar units, and K_m a constant in molal units.

[g] See section 2.3.1.

2.1.3 Chemical formulae and nomenclature

This review follows the recommendations made by IUPAC [71JEN], [77FER], [90LEI] on the nomenclature of inorganic compounds and complexes, except for the following items:

- The formulae of coordination compounds and complexes are not enclosed in square brackets [71JEN] (Rule 7.21). Exceptions are made in cases where square brackets are required to distinguish between coordinated and uncoordinated ligands.

- The prefixes "oxy–" and "hydroxy–" are retained if used in a general way, *e.g.*, "gaseous uranium oxyfluorides". For specific formula names, however, the IUPAC recommended citation [71JEN] (Rule 6.42) is used, *e.g.*, "uranium(IV) difluoride oxide" for $UF_2O(cr)$.

An IUPAC rule that is often not followed by many authors [71JEN] (Rules 2.163 and 7.21) is recalled here: the order of arranging ligands in coordination compounds and complexes is the following: central atom first, followed by ionic ligands and then by the neutral ligands. If there is more than one ionic or neutral ligand, the alphabetical order of the symbols of the ligating atoms determines the sequence of the ligands. For example, $(UO_2)_2CO_3(OH)_3^-$ is standard, $(UO_2)_2(OH)_3CO_3^-$ is non-standard and is not used.

Abbreviations of names for organic ligands appear sometimes in formulae. Following the recommendations by IUPAC, lower case letters are used, and if necessary, the ligand abbreviation is enclosed within parentheses. Hydrogen atoms that can be replaced by the metal atom are shown in the abbreviation with an upper case "H", for example: H_3edta^-, $Am(Hedta)(s)$ (where edta stands for ethylenediaminetetraacetate).

2.1.4 Phase designators

Chemical formulae may refer to different chemical species and are often required to be specified more clearly in order to avoid ambiguities. For example, UF_4 occurs as a gas, a solid, and an aqueous complex. The distinction between the different phases is made by phase designators that immediately follow the chemical formula and appear in parentheses. The only formulae that are not provided with a phase designator are aqueous ions. They are the only charged species in this review since charged gases are not considered. The use of the phase designators is described below.

- The designator (l) is used for pure liquid substances, *e.g.*, $H_2O(l)$.

- The designator (aq) is used for undissociated, uncharged aqueous species, *e.g.*, $U(OH)_4(aq)$, $CO_2(aq)$. Since ionic gases are not considered in this review, all ions may be assumed to be aqueous and are not designed with (aq). If a chemical reaction refers to a medium other than H_2O (*e.g.*, D_2O, 90% ethanol/10% H_2O), then (aq) is replaced by a more explicit designator, e.g., "(in D_2O)" or "(sln)". In the case of (sln), the composition of the solution is described in the text.

- The designator (sln) is used for substances in solution without specifying the actual equilibrium composition of the substance in the solution. Note the difference in the designation of H_2O in Eqs.(2.2) and (2.3). $H_2O(l)$ in Reaction (2.2) indicates that H_2O is present as a pure liquid, i.e., no solutes are present, whereas Reaction (2.3) involves an HCl solution, in which the thermodynamic properties of $H_2O(sln)$ may not be the same as those of the pure liquid $H_2O(l)$. In dilute solutions, however, this difference in the thermodynamic properties of H_2O can be neglected, and $H_2O(sln)$ may be regarded as pure $H_2O(l)$.

    Example:

$$UO_2Cl_2(cr) + 2\,HBr(sln) \rightleftharpoons UOBr_2(cr) + 2HCl(sln) \qquad (2.1)$$

$$UO_2Cl_2 \cdot 3H_2O(cr) \rightleftharpoons UO_2Cl_2 \cdot H_2O(cr) + 2\,H_2O(l) \qquad (2.2)$$

$$UO_3(\gamma) + 2\,HCl(sln) \rightleftharpoons UO_2Cl_2(cr) + H_2O(sln) \qquad (2.3)$$

- The designators (cr), (am), (vit), and (s) are used for solid substances. (cr) is used when it is known that the compound is crystalline, (am) when it is known that it is amorphous, and (vit) for glassy substances. Otherwise, (s) is used.

- In some cases, more than one crystalline form of the same chemical composition may exist. In such a case, the different forms are distinguished by separate designators that describe the forms more precisely. If the crystal has a mineral name, the designator (cr) is replaced by the first four characters of the mineral name in parentheses, e.g., SiO_2(quar) for quartz and SiO_2(chal) for chalcedony. If there is no mineral name, the designator (cr) is replaced by a Greek letter preceding the formula and indicating the structural phase, e.g., $\alpha-UF_5$, $\beta-UF_5$.

Phase designators are also used in conjunction with thermodynamic symbols to define the state of aggregation of a compound to which a thermodynamic quantity refers. The notation is in this case the same as outlined above. In an extended notation (cf. [82LAF]) the reference temperature is usually given in addition to the state of aggregation of the composition of a mixture.

Example:

$\Delta_f G_m^\circ (\text{Na}^+, 298.15 \text{ K})$ standard molar Gibbs energy of formation of aqueous Na$^+$ at 298.15 K

$S_m^\circ (\text{UO}_2(\text{SO}_4) \cdot 2.5\text{H}_2\text{O}, \text{cr}, 298.15 \text{ K})$ standard molar entropy of UO$_2$(SO$_4$)·2.5H$_2$O(cr) at 298.15 K

$C_{p,m}^\circ (\text{UO}_3, \alpha, 298.15 \text{ K})$ standard molar heat capacity of $\alpha - \text{UO}_3$ at 298.15 K

$\Delta_f H_m (\text{HF}, \text{sln}, \text{HF} \cdot 7.8\text{H}_2\text{O})$ enthalpy of formation of HF diluted 1:7.8 with water.

2.1.5 Processes

Chemical processes are denoted by the operator Δ, written before the symbol for a property, as recommended by IUPAC [82LAF]. An exception to this rule is the equilibrium constant, *cf.* section 2.1.6. The nature of the process is denoted by annotation of the Δ, e.g., the Gibbs energy of formation, $\Delta_f G_m$, the enthalpy of sublimation, $\Delta_{sub} H_m$, *etc.* The abbreviations of chemical processes are summarised in Table 2-3.

Table 2-3: Abbreviations used as subscripts of Δ to denote the type of chemical process.

Subscript of Δ	Chemical process
at	separation of a substance into its constituent gaseous atoms (atomisation)
dehyd	elimination of water of hydration (dehydration)
dil	dilution of a solution
f	formation of a compound from its constituent elements
fus	melting (fusion) of a solid
hyd	addition of water of hydration to an unhydrated compound
mix	mixing of fluids
r	chemical reaction (general)
sol	process of dissolution
sub	sublimation (evaporation) of a solid
tr	transfer from one solution or liquid phase to another
trs	transition of one solid phase to another
vap	vaporisation (evaporation) of a liquid

The most frequently used symbols for processes are $\Delta_f G$ and $\Delta_f H$, the Gibbs energy and the enthalpy of formation of a compound or complex from the elements in their reference states (*cf.* Table 2-6).

2.1.6 Spectroscopic constants and statistical mechanics calculations for gaseous species

In most cases, the thermal functions for gaseous species have been calculated by well-known statistical-mechanical relations (see for example Chapter 27 of [61LEW/RAN]). The required molecular parameters are given in the current text, see Tables 9-1 and 9-4, for example.

The parameters defining the vibrational and rotational energy levels of the molecule in terms of the rotational (J) and vibrational (v) quantum numbers, and thus many of its thermodynamic properties, are:

- for diatomic molecules (non-rigid rotator, anharmonic oscillator approximation): ω (vibrational frequency in wavenumber units), x (anharmonicity constant), B (rotational constant for equilibrium position), D (centrifugal distortion constant), α (rotational constant correction for excited vibrational states), and σ (symmetry number), where the energy levels with quantum numbers v and J are given by:

$$E_{(v,J)}/hc = \omega(v+1/2) - \omega x(v+1/2)^2 + B J(J+1) \\ - D J^2 (J+1)^2 - \alpha(v+1/2) J (J+1) \quad (2.4)$$

- for linear polyatomic molecules, the parameters are the same as those for diatomic molecules, except that the contributions for anharmonicity are usually neglected.

- for non-linear polyatomic molecules (rigid rotator, harmonic oscillator approximation): I_x I_y I_z, the product of the principal moments of inertia (readily calculated from the geometrical structure of the molecule), $v(i)$, the vibration frequencies and σ, the symmetry number. While the vibrational energy levels for polyatomic molecules are given approximately by the first term of equation (2.4) for each of the normal vibrations, the rotational energy levels cannot be expressed as a simple general formula. However, the required rotational partition function can be expressed with sufficient accuracy simply in terms of the product of the principal moments of inertia. As for linear polyatomic molecules, anharmonic contributions are usually neglected.

In each case, the symmetry number σ, the number of indistinguishable positions into which the molecule can be turned by simple rotations, is required to calculate the correct entropy.

The relations for calculating the thermal functions from the partition function defined by the energy levels are well-known – again, see Chapter 27 of [61LEW/RAN], for a simple description. In each case, the relevant translational and electronic contributions (calculated from the molar mass and the electronic energy levels and degeneracies) must be added. Except where accurate spectroscopic data exist, the geometry and parameters of the excited states are assumed to be the same as those for the ground state.

2.1.7 Equilibrium constants

The IUPAC has not explicitly defined the symbols and terminology for equilibrium constants of reactions in aqueous solution. The NEA has therefore adopted the conventions that have been used in the work *Stability Constants of Metal Ion Complexes* by Sillén and Martell [64SIL/MAR], [71SIL/MAR]. An outline is given in the paragraphs below. Note that, for some simple reactions, there may be different correct ways to index an equilibrium constant. It may sometimes be preferable to indicate the number of the reaction to which the data refer, especially in cases where several ligands are discussed that might be confused. For example, for the equilibrium:

$$m\,M + q\,L \rightleftharpoons M_m L_q \qquad (2.5)$$

both $\beta_{q,m}$ and $\beta\,(2.5)$ would be appropriate, and $\beta_{q,m}\,(2.5)$ is accepted, too. Note that, in general, K is used for the consecutive or stepwise formation constant, and β is used for the cumulative or overall formation constant. In the following outline, charges are only given for actual chemical species, but are omitted for species containing general symbols (M, L).

2.1.7.1 Protonation of a ligand

$$H^+ + H_{r-1}L \rightleftharpoons H_r L \qquad K_{1,r} = \frac{[H_r L]}{[H^+][H_{r-1}L]} \qquad (2.6)$$

$$r\,H^+ + L \rightleftharpoons H_r L \qquad \beta_{1,r} = \frac{[H_r L]}{[H^+]^r [L]} \qquad (2.7)$$

This notation has been proposed and used by Sillén and Martell [64SIL/MAR], but it has been simplified later by the same authors [71SIL/MAR] from $K_{1,r}$ to K_r. This review retains, for the sake of consistency, *cf.* Eqs.(2.8) and (2.9), the older formulation of $K_{1,r}$.

For the addition of a ligand, the notation shown in Eq.(2.8) is used.

$$HL_{q-1} + L \rightleftharpoons HL_q \qquad K_q = \frac{[HL_q]}{[HL_{q-1}][L]} \qquad (2.8).$$

Eq.(2.9) refers to the overall formation constant of the species $H_r L_q$.

$$r\,H^+ + q\,L \rightleftharpoons H_r L_q \qquad \beta_{q,r} = \frac{[H_r L_q]}{[H^+]^r [L]^q} \qquad (2.9).$$

In Eqs.(2.6), (2.7) and (2.9), the second subscript r can be omitted if $r = 1$, as shown in Eq.(2.8).

Example:

$$H^+ + PO_4^{3-} \rightleftharpoons HPO_4^{2-} \qquad \beta_{1,1} = \beta_1 = \frac{[HPO_4^{2-}]}{[H^+][PO_4^{3-}]}$$

$$2 H^+ + PO_4^{3-} \rightleftharpoons H_2PO_4^- \qquad \beta_{1,2} = \frac{[H_2PO_4^-]}{[H^+]^2[PO_4^{3-}]}$$

2.1.7.2 Formation of metal complexes

$$ML_{q-1} + L \rightleftharpoons ML_q \qquad K_q = \frac{[ML_q]}{[ML_{q-1}][L]} \qquad (2.10)$$

$$M + qL \rightleftharpoons ML_q \qquad \beta_q = \frac{[ML_q]}{[M][L]^q} \qquad (2.11)$$

For the addition of a metal ion, *i.e.*, the formation of polynuclear complexes, the following notation is used, analogous to Eq.(2.6):

$$M + M_{m-1}L \rightleftharpoons M_mL \qquad K_{1,m} = \frac{[M_mL]}{[M][M_{m-1}L]} \qquad (2.12).$$

Eq.(2.13) refers to the overall formation constant of a complex M_mL_q.

$$m M + q L \rightleftharpoons M_mL_q \qquad \beta_{q,m} = \frac{[M_mL_q]}{[M]^m[L]^q} \qquad (2.13)$$

The second index can be omitted if it is equal to 1, *i.e.*, $\beta_{q,m}$ becomes β_q if $m = 1$. The formation constants of mixed ligand complexes are not indexed. In this case, it is necessary to list the chemical reactions considered and to refer the constants to the corresponding reaction numbers.

It has sometimes been customary to use negative values for the indices of the protons to indicate complexation with hydroxide ions, OH^-. This practice is not adopted in this review. If OH^- occurs as a reactant in the notation of the equilibrium, it is treated like a normal ligand L, but in general formulae the index variable n is used instead of q. If H_2O occurs as a reactant to form hydroxide complexes, H_2O is considered as a protonated ligand, HL, so that the reaction is treated as described below in

Eqs.(2.14) to (2.16) using n as the index variable. For convenience, no general form is used for the stepwise constants for the formation of the complex $M_mL_qH_r$. In many experiments, the formation constants of metal ion complexes are determined by adding a ligand in its protonated form to a metal ion solution. The complex formation reactions thus involve a deprotonation reaction of the ligand. If this is the case, the equilibrium constant is supplied with an asterisk, as shown in Eqs.(2.14) and (2.15) for mononuclear and in Eq.(2.16) for polynuclear complexes.

$$ML_{q-1} + HL \rightleftharpoons ML_q + H^+ \qquad {}^*K_q = \frac{[ML_q][H^+]}{[ML_{q-1}][HL]} \qquad (2.14)$$

$$M + q\,HL \rightleftharpoons ML_q + qH^+ \qquad {}^*\beta_q = \frac{[ML_q][H^+]^q}{[M][HL]^q} \qquad (2.15)$$

$$m\,M + q\,HL \rightleftharpoons M_mL_q + qH^+ \qquad {}^*\beta_{q,m} = \frac{[M_mL_q][H^+]^q}{[M]^m[HL]^q} \qquad (2.16)$$

Example:

$$UO_2^{2+} + HF(aq) \rightleftharpoons UO_2F^+ + H^+ \qquad {}^*K_1 = {}^*\beta_1 = \frac{[UO_2F^+][H^+]}{[UO_2^{2+}][HF(aq)]}$$

$$3\,UO_2^{2+} + 5\,H_2O(l) \rightleftharpoons (UO_2)_3(OH)_5^+ + 5\,H^+ \qquad {}^*\beta_{5,3} = \frac{[(UO_2)_3(OH)_5^+][H^+]^5}{[UO_2^{2+}]^3}$$

Note that an asterisk is only assigned to the formation constant if the protonated ligand that is added is deprotonated during the reaction. If a protonated ligand is added and coordinated as such to the metal ion, the asterisk is to be omitted, as shown in Eq.(2.17).

$$M + q\,H_rL \rightleftharpoons M(H_rL)_q \qquad \beta_q = \frac{[M(H_rL)_q]}{[M][H_rL]^q} \qquad (2.17)$$

Example:

$$UO_2^{2+} + 3\,H_2PO_4^- \rightleftharpoons UO_2(H_2PO_4)_3^- \qquad \beta_3 = \frac{[UO_2(H_2PO_4)_3^-]}{[UO_2^{2+}][H_2PO_4^-]^3}$$

$$E^\circ = -\frac{1}{nF}\Delta_r G_m^\circ = \frac{RT}{nF}\ln K^\circ \qquad (2.22)$$

and the potential, E, is related to E° by:

$$E = E^\circ - (RT/nF)\sum v_i \ln a_i \qquad (2.23).$$

For example, for the hypothetical galvanic cell:

$$\text{Pt} \mid H_2(g, p = 1 \text{ bar}) \mid HCl(aq, a_\pm = 1, f_{H_2} = 1) \mid \begin{array}{c} Fe(ClO_4)_2 \,(aq, a_{Fe^{2+}} = 1) \\ Fe(ClO_4)_3 \,(aq, a_{Fe^{3+}} = 1) \end{array} \mid \text{Pt} \qquad (2.24)$$

where $\vdots$ denotes a liquid junction and $\mid$ a phase boundary, the reaction is:

$$Fe^{3+} + \frac{1}{2}H_2(g) \rightleftharpoons Fe^{2+} + H^+ \qquad (2.25)$$

For convenience Reaction (2.25) can be represented by half cell reactions, each involving an equal number of "electrons", (designated "e^-"), as shown in the following equations:

$$Fe^{3+} + e^- \rightleftharpoons Fe^{2+} \qquad (2.26)$$

$$\frac{1}{2}H_2(g) \rightleftharpoons H^+ + e^-. \qquad (2.27)$$

The terminology is useful, although it must be emphasised "e^-" here does not represent the hydrated electron.

Equilibrium (2.27) and Nernst law can be used to introduce a_{e^-}:

$$E = E^\circ (2.27) + \frac{RT}{F}\ln(\sqrt{f_{H_2}}/(a_{H^+}a_{e^-})) \qquad (2.28)$$

According to the SHE convention $E^\circ (2.27) = 0$, $f_{H_2} = 1$, $a_{H^+} = 1$, hence

$$E = -\frac{RT}{F}\ln a_{e^-} \qquad (2.29).$$

This equation is used to calculate a numerical value of a_{e^-} from emf measurements vs. the SHE; hence, as for the value of E (V vs. the SHE), the numerical value of a_{e^-} depends on the SHE convention. Equilibrium constants may be written for these half cell reactions in the following way:

$$K^\circ (2.26) = \frac{a_{Fe^{2+}}}{a_{Fe^{3+}} \cdot a_{e^-}} \qquad (2.30)$$

$$K^\circ (2.27) = \frac{a_{H^+} \cdot a_{e^-}}{\sqrt{f_{H_2}}} = 1 \quad \text{(by definition)} \qquad (2.31)$$

Eqs.(2.14) to (2.16) using n as the index variable. For convenience, no general form is used for the stepwise constants for the formation of the complex $M_mL_qH_r$. In many experiments, the formation constants of metal ion complexes are determined by adding a ligand in its protonated form to a metal ion solution. The complex formation reactions thus involve a deprotonation reaction of the ligand. If this is the case, the equilibrium constant is supplied with an asterisk, as shown in Eqs.(2.14) and (2.15) for mononuclear and in Eq.(2.16) for polynuclear complexes.

$$ML_{q-1} + HL \rightleftharpoons ML_q + H^+ \qquad {}^*K_q = \frac{[ML_q][H^+]}{[ML_{q-1}][HL]} \qquad (2.14)$$

$$M + q\,HL \rightleftharpoons ML_q + qH^+ \qquad {}^*\beta_q = \frac{[ML_q][H^+]^q}{[M][HL]^q} \qquad (2.15)$$

$$m\,M + q\,HL \rightleftharpoons M_mL_q + qH^+ \qquad {}^*\beta_{q,m} = \frac{[M_mL_q][H^+]^q}{[M]^m[HL]^q} \qquad (2.16)$$

Example:

$$UO_2^{2+} + HF(aq) \rightleftharpoons UO_2F^+ + H^+ \qquad {}^*K_1 = {}^*\beta_1 = \frac{[UO_2F^+][H^+]}{[UO_2^{2+}][HF(aq)]}$$

$$3\,UO_2^{2+} + 5\,H_2O(l) \rightleftharpoons (UO_2)_3(OH)_5^+ + 5\,H^+ \qquad {}^*\beta_{5,3} = \frac{[(UO_2)_3(OH)_5^+][H^+]^5}{[UO_2^{2+}]^3}$$

Note that an asterisk is only assigned to the formation constant if the protonated ligand that is added is deprotonated during the reaction. If a protonated ligand is added and coordinated as such to the metal ion, the asterisk is to be omitted, as shown in Eq.(2.17).

$$M + q\,H_rL \rightleftharpoons M(H_rL)_q \qquad \beta_q = \frac{[M(H_rL)_q]}{[M][H_rL]^q} \qquad (2.17)$$

Example:

$$UO_2^{2+} + 3\,H_2PO_4^- \rightleftharpoons UO_2(H_2PO_4)_3^- \qquad \beta_3 = \frac{[UO_2(H_2PO_4)_3^-]}{[UO_2^{2+}][H_2PO_4^-]^3}$$

2.1.7.3 Solubility constants

Conventionally, equilibrium constants involving a solid compound are denoted as "solubility constants" rather than as formation constants of the solid. An index "s" to the equilibrium constant indicates that the constant refers to a solubility process, as shown in Eqs.(2.18) to (2.20).

$$M_aL_b(s) \rightleftharpoons a M + b L \qquad K_{s,0} = [M]^a [L]^b \qquad (2.18).$$

$K_{s,0}$ is the conventional solubility product[1], and the subscript "0" indicates that the equilibrium reaction involves only uncomplexed aqueous species. If the solubility constant includes the formation of aqueous complexes, a notation analogous to that of Eq.(2.13) is used:

$$\frac{m}{a}M_aL_b(s) \rightleftharpoons M_mL_q + \left(\frac{mb}{a} - q\right)L \qquad K_{s,q,m} = [M_mL_q][L]^{(\frac{mb}{a}-q)} \qquad (2.19).$$

Example:

$$UO_2F_2(cr) \rightleftharpoons UO_2F^+ + F^- \qquad K_{s,1,1} = K_{s,1} = [UO_2F^+][F^-].$$

Similarly, an asterisk is added to the solubility constant if it simultaneously involves a protonation equilibrium:

$$\frac{m}{a}M_aL_b(s) + \left(\frac{mb}{a} - q\right)H^+ \rightleftharpoons M_mL_q + \left(\frac{mb}{a} - q\right)HL$$

$$*K_{s,q,m} = \frac{[M_mL_q][HL]^{(\frac{mb}{a}-q)}}{[H^+]^{(\frac{mb}{a}-q)}} \qquad (2.20)$$

Example:

$$U(HPO_4)_2 \cdot 4H_2O(cr) + H^+ \rightleftharpoons UHPO_4^{2+} + H_2PO_4^- + 4 H_2O(l)$$

$$*K_{s,1,1} = *K_{s,1} = \frac{[UHPO_4^{2+}][H_2PO_4^-]}{[H^+]}.$$

---

[1] In some cases, most noticeably when dealing with the solubility of actinide oxides, the $K_{s,0}$ notation is customarily applied in NEA TDB reviews to denote the constants for equilibria such as $AnO_2(s) + 2 H_2O(l) \rightleftharpoons An^{4+} + 4OH^-$. A reference to the appropriate chemical equation is attached to the symbol for the equilibrium constant when there is a risk of confusion.

2.1.7.4 Equilibria involving the addition of a gaseous ligand

A special notation is used for constants describing equilibria that involve the addition of a gaseous ligand, as outlined in Eq.(2.21).

$$ML_{q-1} + L(g) \rightleftharpoons ML_q \qquad K_{p,q} = \frac{[ML_q]}{[ML_{q-1}]p_L} \qquad (2.21)$$

The subscript "p" can be combined with any other notations given above.

Example:

$$CO_2(g) \rightleftharpoons CO_2(aq) \qquad K_p = \frac{[CO_2(aq)]}{p_{CO_2}}$$

$$3\,UO_2^{2+} + 6\,CO_2(g) + 6\,H_2O(l) \rightleftharpoons (UO_2)_3(CO_3)_6^{6-} + 12\,H^+$$

$$^*\beta_{p,6,3} = \frac{\left[(UO_2)_3(CO_3)_6^{6-}\right]\left[H^+\right]^{12}}{\left[UO_2^{2+}\right]^3 p_{CO_2}^6}$$

$$UO_2CO_3(cr) + CO_2(g) + H_2O(l) \rightleftharpoons UO_2(CO_3)_2^{2-} + 2\,H^+$$

$$^*K_{p,s,2} = \frac{\left[UO_2(CO_3)_2^{2-}\right]\left[H^+\right]^2}{p_{CO_2}}$$

In cases where the subscripts become complicated, it is recommended that K or β be used with or without subscripts, but always followed by the equation number of the equilibrium to which it refers.

2.1.7.5 Redox equilibria

Redox reactions are usually quantified in terms of their electrode (half cell) potential, E, which is identical to the electromotive force (emf) of a galvanic cell in which the electrode on the left is the standard hydrogen electrode, SHE[1], in accordance with the "1953 Stockholm Convention" [88MIL/CVI]. Therefore, electrode potentials are given as reduction potentials relative to the standard hydrogen electrode, which acts as an electron donor. In the standard hydrogen electrode, $H_2(g)$ is at unit fugacity (an ideal gas at unit pressure, 0.1 MPa), and H^+ is at unit activity. The sign of the electrode potential, E, is that of the observed sign of its polarity when coupled with the standard hydrogen electrode. The standard electrode potential, $E°$, i.e., the potential of a standard galvanic cell relative to the standard hydrogen electrode (all components in their standard state, cf. section 2.3.1, and with no liquid junction potential) is related to the standard Gibbs energy change $\Delta_r G_m°$ and the standard (or thermodynamic) equilibrium constant $K°$ as outlined in Eq.(2.22).

---

[1] The definitions of SHE and NHE are given in section 2.1.1.

$$E^\circ = -\frac{1}{nF}\Delta_r G_m^\circ = \frac{RT}{nF}\ln K^\circ \qquad (2.22)$$

and the potential, E, is related to E° by:

$$E = E^\circ - (RT/nF)\sum v_i \ln a_i \qquad (2.23).$$

For example, for the hypothetical galvanic cell:

$$\text{Pt} \mid H_2(g, p=1\text{ bar}) \mid HCl(aq, a_\pm = 1, f_{H_2} = 1) \mid \begin{array}{l} \text{Fe(ClO}_4)_2 \text{ (aq, } a_{Fe^{2+}} = 1) \\ \text{Fe(ClO}_4)_3 \text{ (aq, } a_{Fe^{3+}} = 1) \end{array} \mid \text{Pt} \qquad (2.24)$$

where ⋮ denotes a liquid junction and │ a phase boundary, the reaction is:

$$Fe^{3+} + \frac{1}{2}H_2(g) \rightleftharpoons Fe^{2+} + H^+ \qquad (2.25)$$

For convenience Reaction (2.25) can be represented by half cell reactions, each involving an equal number of "electrons", (designated "e^-"), as shown in the following equations:

$$Fe^{3+} + e^- \rightleftharpoons Fe^{2+} \qquad (2.26)$$

$$\frac{1}{2}H_2(g) \rightleftharpoons H^+ + e^-. \qquad (2.27)$$

The terminology is useful, although it must be emphasised "e^-" here does not represent the hydrated electron.

Equilibrium (2.27) and Nernst law can be used to introduce a_{e^-}:

$$E = E^\circ (2.27) + \frac{RT}{F}\ln(\sqrt{f_{H_2}}/(a_{H^+}\cdot a_{e^-})) \qquad (2.28)$$

According to the SHE convention $E^\circ (2.27) = 0$, $f_{H_2} = 1$, $a_{H^+} = 1$, hence

$$E = -\frac{RT}{F}\ln a_{e^-} \qquad (2.29).$$

This equation is used to calculate a numerical value of a_{e^-} from emf measurements vs. the SHE; hence, as for the value of E (V vs. the SHE), the numerical value of a_{e^-} depends on the SHE convention. Equilibrium constants may be written for these half cell reactions in the following way:

$$K^\circ (2.26) = \frac{a_{Fe^{2+}}}{a_{Fe^{3+}}\cdot a_{e^-}} \qquad (2.30)$$

$$K^\circ (2.27) = \frac{a_{H^+}\cdot a_{e^-}}{\sqrt{f_{H_2}}} = 1 \text{ (by definition)} \qquad (2.31)$$

In addition, $\Delta_r G_m^\circ$ (2.27) = 0, $\Delta_r H_m^\circ$ (2.27) = 0, $\Delta_r S_m^\circ$ (2.27) = 0 by definition, at all temperatures, and therefore $\Delta_r G_m^\circ$ (2.26) = $\Delta_r G_m^\circ$ (2.25). From $\Delta_r G_m^\circ$ (2.27) and the values given at 298.15 K in Table 8.1 for $H_2(g)$ and H^+, the corresponding values for e^- can be calculated to be used in thermodynamic cycles involving half cell reactions. The following equations describe the change in the redox potential of Reaction (2.25), if p_{H_2} and a_{H^+} are equal to unity (cf. Eq.(2.23)):

$$E(2.25) = E^\circ (2.25) - RT \ln\left(\frac{a_{Fe^{2+}}}{a_{Fe^{3+}}}\right) \qquad (2.32)$$

For the standard hydrogen electrode $a_{e^-} = 1$ (by the convention expressed in Eq.(2.31)), while rearrangement of Eq.(2.30) for the half cell containing the iron perchlorates in cell (2.24) gives:

$$-\log_{10} a_{e^-} = \log_{10} K^\circ (2.26) - \log_{10}\left(\frac{a_{Fe^{2+}}}{a_{Fe^{3+}}}\right)$$

and from Eq.(2.28):

$$-\log_{10} a_{e^-} = \log_{10} K^\circ (2.25) - \log_{10}\left(\frac{a_{Fe^{2+}}}{a_{Fe^{3+}}}\right) \qquad (2.33)$$

and

$$-\log_{10} a_{e^-} = \frac{F}{RT\ln(10)} E (2.25) \qquad (2.34)$$

which is a specific case of the general equation (2.29).

The splitting of redox reactions into two half cell reactions by introducing the symbol "e^-", which according to Eq.(2.28) is related to the standard electrode potential, is arbitrary, but useful (this e^- notation does not in any way refer to solvated electrons). When calculating the equilibrium composition of a chemical system, both "e^-", and H^+ can be chosen as components and they can be treated numerically in a similar way: equilibrium constants, mass balance, etc. may be defined for both. However, while H^+ represents the hydrated proton in aqueous solution, the above equations use only the activity of "e^-", and never the concentration of "e^-". Concentration to activity conversions (or activity coefficients) are never needed for the electron (cf. Appendix B, Example B.3).

In the literature on geochemical modelling of natural waters, it is customary to represent the "electron activity" of an aqueous solution with the symbol "pe" or "pε"($= -\log_{10} a_{e^-}$) by analogy with pH ($= -\log_{10} a_{H^+}$), and the redox potential of an aqueous solution relative to the standard hydrogen electrode is usually denoted by either "Eh" or "E_H" (see for example [81STU/MOR], [82DRE], [84HOS], [86NOR/MUN]).

In this review, the symbol E'° is used to denote the so called "formal potential" [74PAR]. The formal (or "conditional") potential can be regarded as a standard potential for a particular medium in which the activity coefficients are independent (or approximately so) of the reactant concentrations [85BAR/PAR] (the definition of E'° parallels that of "concentration quotients" for equilibria). Therefore, from

$$E = E'^\circ - \frac{RT}{nF}\sum \upsilon_i \ln c_i \qquad (2.35)$$

E'° is the potential E for a cell when the ratio of the *concentrations* (not the activities) on the right–hand side and the left–hand side of the cell reaction is equal to unity, and

$$E'^\circ = E^\circ - \frac{RT}{nF}\sum \upsilon_i \ln \rho \gamma_i = -\frac{\Delta_r G_m}{nF} \qquad (2.36)$$

where the γ_i are the molality activity coefficients and ρ is (m_i/c_i), the ratio of molality to molarity (*cf.* section 2.2). The medium must be specified.

2.1.8 pH

Because of the importance that potentiometric methods have in the determination of equilibrium constants in aqueous solutions, a short discussion on the definition of "pH" and a simplified description of the experimental techniques used to measure pH will be given here.

The acidity of aqueous solutions is often expressed in a logarithmic scale of the hydrogen ion activity. The definition of pH as:

$$\mathrm{pH} = -\log_{10} a_{H^+} = -\log_{10}(m_{H^+} \gamma_{H^+})$$

can only be strictly used in the limiting range of the Debye–Hückel equation (that is, in extremely dilute solutions). In practice the use of pH values requires extra assumptions as to the values for single ion activities. In this review values of pH are used to describe qualitatively the ranges of acidity of experimental studies, and the assumptions described in Appendix B are used to calculate single ion activity coefficients.

The determination of pH is often performed by emf measurements of galvanic cells involving liquid junctions [69ROS], [73BAT]. A common setup is a cell made up of a reference half cell (*e.g.* Ag(s)/AgCl(s) in a solution of constant chloride concentration), a salt bridge, the test solution, and a glass electrode (which encloses a solution of constant acidity and an internal reference half cell):

Pt(s)	Ag(s)	AgCl(s)	KCl(aq)	salt bridge	test solution		KCl(aq)	AgCl(s)	Ag(s)	Pt(s)
				a	b					(2.37)

where ⟨ stands for a glass membrane (permeable to hydrogen ions).

2.1 Symbols, terminology and nomenclature

The emf of such a cell is given by:

$$E = E^* - \frac{RT}{nF} \ln a_{H^+} + E_j$$

where E^* is a constant, and E_j is the liquid junction potential. The purpose of the salt bridge is to minimise the junction potential in junction "b", while keeping constant the junction potential for junction "a". Two methods are most often used to reduce and control the value of E_j. An electrolyte solution of *high* concentration (the "salt bridge") is a requirement of both methods. In the first method, the salt bridge is a saturated (or nearly saturated) solution of potassium chloride. A problem with a bridge of high potassium concentration is that potassium perchlorate might precipitate[1] inside the liquid junction when the test solution contains a high concentration of perchlorate ions.

In the other method the salt bridge contains the same *high* concentration of the same inert electrolyte as the test solution (for example, 3 M NaClO$_4$). However, if the concentration of the background electrolyte in the salt bridge and test solutions is reduced, the values of E_j are dramatically increased. For example, if both the bridge and the test solution have [ClO$_4^-$] = 0.1 M as background electrolyte, the dependence of the liquid junction at "b" on acidity is $E_j \approx -440 \times$ [H$^+$] mV·dm^3 mol^{-1} at 25°C [69ROS] (p.110), which corresponds to an error at pH = 2 of ≥ 0.07 pH units.

Because of the problems in eliminating the liquid junction potentials and in defining individual ionic activity coefficients, an "operational" definition of pH is given by IUPAC [88MIL/CVI]. This definition involves the measurement of pH differences between the test solution and standard solutions of known pH and similar ionic strength (in this way similar values of γ_{H^+} and E_j cancel each other when emf values are substracted).

The measurement and use of pH in equilibrium analytical investigations creates many problems that have not always been taken into account by the investigators, as discussed in many reviews in Appendix A. In order to deduce the stoichiometry and equilibrium constants of complex formation reactions and other equilibria, it is necessary to vary the concentrations of reactants and products over fairly large concentration ranges under conditions where the activity coefficients of the species are either known, or constant. Only in this manner is it possible to use the mass balance equations for the various components together with the measurement of one or more free concentrations to obtain the information desired [61ROS/ROS], [90BEC/NAG], [97ALL/BAN], p. 326-327. For equilibria involving hydrogen ions, it is necessary to use concentration units, rather than hydrogen ion activity. For experiments in an ionic medium, where the concentration of an "inert" electrolyte is much larger than the concentration of reactants and products we can ensure that, as a first approximation, their trace activity coeffi-

---

[1] KClO$_4$(cr) has a solubility of ≈ 0.15 M in pure water at 25°C

cients remain constant even for moderate variations of the corresponding total concentrations. Under these conditions of fixed ionic strength the free proton concentration may be measured directly, thereby defining it in terms of $-\log_{10}[H^+]$ rather than on the activity scale as pH, and the value of $-\log_{10}[H^+]$ and pH will differ by a constant term, *i.e.*, $\log_{10}\gamma_{H^+}$. Equilibrium constants deduced from measurements in such ionic media are therefore *conditional* constants, because they refer to the given medium, not to the standard state. In order to compare the magnitude of equilibrium constants obtained in different ionic media it is necessary to have a method for estimating activity coefficients of ionic species in mixed electrolyte systems to a *common* standard state. Such procedures are discussed in Appendix B.

Note that the precision of the measurement of $-\log_{10}[H^+]$ and pH is virtually the same, in very good experiments, ± 0.001. However, the accuracy is generally considerably poorer, depending in the case of glass electrodes largely on the response of the electrode (linearity, age, pH range, *etc.*), and to a lesser extent on the calibration method employed, although the stoichiometric $-\log_{10}[H^+]$ calibration standards can be prepared far more accurately than the commercial pH standards.

2.1.9 Order of formulae

To be consistent with CODATA, the data tables are given in "*Standard Order of Arrangement*" [82WAG/EVA]. This scheme is presented in Figure 2-1 below, and shows the sequence of the ranks of the elements in this convention. The order follows the ranks of the elements.

For example, for uranium, this means that, after elemental uranium and its monoatomic ions (*e.g.*, U^{4+}), the uranium compounds and complexes with oxygen would be listed, then those with hydrogen, then those with oxygen and hydrogen, and so on, with decreasing rank of the element and combinations of the elements. Within a class, increasing coefficients of the higher rank elements go before increasing coefficients of the lower rank elements. For example, in the U–O–F class of compounds and complexes, a typical sequence would be $UOF_2(cr)$, $UOF_4(cr)$, $UOF_4(g)$, $UO_2F(aq)$, UO_2F^+, $UO_2F_2(aq)$, $UO_2F_2(cr)$, $UO_2F_2(g)$, $UO_2F_3^-$, $UO_2F_4^{2-}$, $U_2O_3F_6(cr)$, *etc.* [92GRE/FUG]. Formulae with identical stoichiometry are in alphabetical order of their designators.

2.1 Symbols, terminology and nomenclature

Figure 2-1: Standard order of arrangement of the elements and compounds based on the periodic classification of the elements (from [82WAG/EVA]).

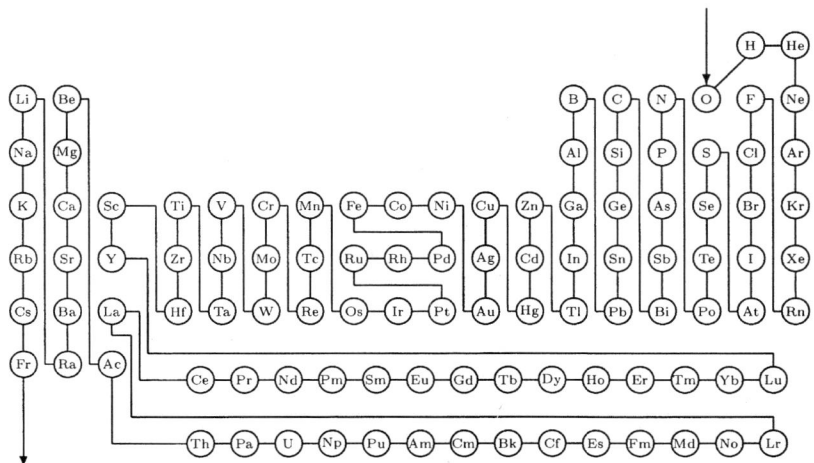

2.1.10 Reference codes

The references cited in the review are ordered chronologically and alphabetically by the first two authors within each year, as described by CODATA [87GAR/PAR]. A reference code is made up of the final two digits of the year of appearance (if the publication is not from the 20th century, the year will be put in full). The year is followed by the first three letters of the surnames of the first two authors, separated by a slash.

If there are multiple reference codes, a "2" will be added to the second one, a "3" to the third one, and so forth. Reference codes are always enclosed in square brackets.

2.2 Units and conversion factors

Thermodynamic data are given according to the *Système International d'unités* (SI units). The unit of energy is the joule. Some basic conversion factors, also for non-thermodynamic units, are given in Table 2-4.

Table 2-4: Unit conversion factors.

To convert from (non-SI unit symbol)	to (SI unit symbol)	multiply by
ångström (Å)	metre (m)	1×10^{-10} (exactly)
standard atmosphere (atm)	pascal (Pa)	1.01325×10^5 (exactly)
bar (bar)	pascal (Pa)	1×10^5 (exactly)
thermochemical calorie (cal)	joule (J)	4.184 (exactly)
entropy unit e.u. $\triangleq$ cal·K^{-1}·mol^{-1}	J·K^{-1}·mol^{-1}	4.184 (exactly)

Since a large part of the NEA TDB project deals with the thermodynamics of aqueous solutions, the units describing the amount of dissolved substance are used very frequently. For convenience, this review uses "M" as an abbreviation of "mol·dm^{-3}" for molarity, c, and, in Appendices B and C, "m" as an abbreviation of "mol·kg^{-1}" for molality, m. It is often necessary to convert concentration data from molarity to molality and vice versa. This conversion is used for the correction and extrapolation of equilibrium data to zero ionic strength by the specific ion interaction theory, which works in molality units (*cf.* Appendix B). This conversion is made in the following way. Molality is defined as m_B moles of substance B dissolved in 1000 grams of pure water. Molarity is defined as c_B moles of substance B dissolved in $(1000 \rho - c_B M)$ grams of pure water, where ρ is the density of the solution and M the molar weight of the solute.

From this it follows that:

$$m_B = \frac{1000 c_B}{1000 \rho - c_B M}.$$

Baes and Mesmer [76BAE/MES], (*p.*439) give a table with conversion factors (from molarity to molality) for nine electrolytes and various ionic strengths. Conversion factors at 298.15 K for twenty one electrolytes, calculated using the density equations reported by Söhnel and Novotný [85SOH/NOV], are reported in Table 2-5.

Example:

$$1.00 \text{ M NaClO}_4 \triangleq 1.05 \text{ m NaClO}_4$$
$$1.00 \text{ M NaCl} \triangleq 1.02 \text{ m NaCl}$$
$$4.00 \text{ M NaClO}_4 \triangleq 4.95 \text{ m NaClO}_4$$
$$6.00 \text{ M NaNO}_3 \triangleq 7.55 \text{ m NaNO}_3$$

It should be noted that equilibrium constants need also to be converted if the concentration scale is changed from molarity to molality or vice versa. For a general equilibrium reaction, $0 = \sum_B \nu_B B$, the equilibrium constants can be expressed either in molarity or molality units, K_c or K_m, respectively:

$$\log_{10} K_c = \sum_B \nu_B \log_{10} c_B$$
$$\log_{10} K_m = \sum_B \nu_B \log_{10} m_B$$

With $(m_B / c_B) = \rho$, or $(\log_{10} m_B - \log_{10} c_B) = \log_{10} \rho$, the relationship between K_c and K_m becomes very simple, as shown in Eq.(2.38).

$$\log_{10} K_m = \log_{10} K_c + \sum_B \nu_B \log_{10} \rho \qquad (2.38)$$

$\sum_B \nu_B$ is the sum of the stoichiometric coefficients of the reaction, cf. Eq. (2.54) and the values of ρ are the factors for the conversion of molarity to molality as tabulated in Table 2-5 for several electrolyte media at 298.15 K. The differences between the values in Table 2-5 and the values listed in the uranium NEA TDB review [92GRE/FUG] (p.23) are found at the highest concentrations, and are no larger than ± 0.003 dm³·kg⁻¹, reflecting the accuracy expected in this type of conversion. The uncertainty introduced by the use of Eq.(2.38) in the values of $\log_{10} K_m$ will be no larger than $\pm 0.001 \sum_B \nu_B$.

Table 2-5: Factors ρ for the conversion of molarity, c_B, to molality, m_B, of a substance B, in various media at 298.15 K (calculated from densities in [85SOH/NOV]).

$\rho = m_B / c_B$ (dm^3 of solution per kg of H$_2$O)

c (M)	HClO$_4$	NaClO$_4$	LiClO$_4$	NH$_4$ClO$_4$	Ba(ClO$_4$)$_2$	HCl	NaCl	LiCl
0.10	1.0077	1.0075	1.0074	1.0091	1.0108	1.0048	1.0046	1.0049
0.25	1.0147	1.0145	1.0141	1.0186	1.0231	1.0076	1.0072	1.0078
0.50	1.0266	1.0265	1.0256	1.0351	1.0450	1.0123	1.0118	1.0127
0.75	1.0386	1.0388	1.0374	1.0523	1.0685	1.0172	1.0165	1.0177
1.00	1.0508	1.0515	1.0496	1.0703	1.0936	1.0222	1.0215	1.0228
1.50	1.0759	1.0780	1.0750	1.1086	1.1491	1.0324	1.0319	1.0333
2.00	1.1019	1.1062	1.1019		1.2125	1.0430	1.0429	1.0441
3.00	1.1571	1.1678	1.1605		1.3689	1.0654	1.0668	1.0666
4.00	1.2171	1.2374	1.2264			1.0893	1.0930	1.0904
5.00	1.2826	1.3167				1.1147	1.1218	1.1156
6.00	1.3547	1.4077				1.1418		1.1423

c (M)	KCl	NH$_4$Cl	MgCl$_2$	CaCl$_2$	NaBr	HNO$_3$	NaNO$_3$	LiNO$_3$
0.10	1.0057	1.0066	1.0049	1.0044	1.0054	1.0056	1.0058	1.0059
0.25	1.0099	1.0123	1.0080	1.0069	1.0090	1.0097	1.0102	1.0103
0.50	1.0172	1.0219	1.0135	1.0119	1.0154	1.0169	1.0177	1.0178
0.75	1.0248	1.0318	1.0195	1.0176	1.0220	1.0242	1.0256	1.0256
1.00	1.0326	1.0420	1.0258	1.0239	1.0287	1.0319	1.0338	1.0335
1.50	1.0489	1.0632	1.0393	1.0382	1.0428	1.0478	1.0510	1.0497
2.00	1.0662	1.0855	1.0540	1.0546	1.0576	1.0647	1.0692	1.0667
3.00	1.1037	1.1339	1.0867	1.0934	1.0893	1.1012	1.1090	1.1028
4.00	1.1453	1.1877	1.1241	1.1406	1.1240	1.1417	1.1534	1.1420
5.00		1.2477		1.1974	1.1619	1.1865	1.2030	1.1846
6.00					1.2033	1.2361	1.2585	1.2309

c (M)	NH$_4$NO$_3$	H$_2$SO$_4$	Na$_2$SO$_4$	(NH$_4$)$_2$SO$_4$	H$_3$PO$_4$	Na$_2$CO$_3$	K$_2$CO$_3$	NaSCN
0.10	1.0077	1.0064	1.0044	1.0082	1.0074	1.0027	1.0042	1.0069
0.25	1.0151	1.0116	1.0071	1.0166	1.0143	1.0030	1.0068	1.0130
0.50	1.0276	1.0209	1.0127	1.0319	1.0261	1.0043	1.0121	1.0234
0.75	1.0405	1.0305	1.0194	1.0486	1.0383	1.0065	1.0185	1.0342
1.00	1.0539	1.0406	1.0268	1.0665	1.0509	1.0094	1.0259	1.0453
1.50	1.0818	1.0619	1.0441	1.1062	1.0773	1.0170	1.0430	1.0686
2.00	1.1116	1.0848		1.1514	1.1055	1.0268	1.0632	1.0934
3.00	1.1769	1.1355		1.2610	1.1675		1.1130	1.1474
4.00	1.2512	1.1935		1.4037	1.2383		1.1764	1.2083
5.00	1.3365	1.2600			1.3194		1.2560	1.2773
6.00	1.4351	1.3365			1.4131			1.3557

2.3 Standard and reference conditions

2.3.1 Standard state

A precise definition of the term "standard state" has been given by IUPAC [82LAF]. The fact that only changes in thermodynamic parameters, but not their absolute values, can be determined experimentally, makes it important to have a well-defined standard state that forms a base line to which the effect of variations can be referred. The IUPAC [82LAF] definition of the standard state has been adopted in the NEA–TDB project. The standard state pressure, $p^\circ = 0.1$ MPa (1 bar), has therefore also been adopted, *cf.* section 2.3.2. The application of the standard state principle to pure substances and mixtures is summarised below. It should be noted that the standard state is always linked to a reference temperature, *cf.* section 2.3.3.

- The standard state for a gaseous substance, whether pure or in a gaseous mixture, is the pure substance at the standard state pressure and in a (hypothetical) state in which it exhibits ideal gas behaviour.

- The standard state for a pure liquid substance is (ordinarily) the pure liquid at the standard state pressure.

- The standard state for a pure solid substance is (ordinarily) the pure solid at the standard state pressure.

- The standard state for a solute B in a solution is a hypothetical liquid solution, at the standard state pressure, in which $m_B = m^\circ = 1$ mol $\cdot$ kg^{-1}, and in which the activity coefficient γ_B is unity.

It should be emphasised that the use of superscript, $^\circ$, *e.g.*, in $\Delta_f H_m^\circ$, implies that the compound in question is in the standard state and that the elements are in their reference states. The reference states of the elements at the reference temperature (*cf.* section 2.3.3) are listed in Table 2-6.

Table 2-6: Reference states for some elements at the reference temperature of 298.15 K and standard pressure of 0.1 MPa [82WAG/EVA], [89COX/WAG], [91DIN].

O_2	gaseous		Al	crystalline, cubic
H_2	gaseous		Zn	crystalline, hexagonal
He	gaseous		Cd	crystalline, hexagonal
Ne	gaseous		Hg	liquid
Ar	gaseous		Cu	crystalline, cubic
Kr	gaseous		Ag	crystalline, cubic
Xe	gaseous		Fe	crystalline, cubic, bcc
F_2	gaseous		Tc	crystalline, hexagonal
Cl_2	gaseous		V	crystalline, cubic

(Continued on next page)

Table 2-6: (continued)

Br_2	liquid	Ti	crystalline, hexagonal
I_2	crystalline, orthorhombic	Am	crystalline, dhcp
S	crystalline, orthorhombic	Pu	crystalline, monoclinic
Se	crystalline, hexagonal ("black")	Np	crystalline, orthorhombic
Te	crystalline, hexagonal	U	crystalline, orthorhombic
N_2	gaseous	Th	crystalline, cubic
P	crystalline, cubic ("white")	Be	crystalline, hexagonal
As	crystalline, rhombohedral ("grey")	Mg	crystalline, hexagonal
Sb	crystalline, rhombohedral	Ca	crystalline, cubic, fcc
Bi	crystalline, rhombohedral	Sr	crystalline, cubic, fcc
C	crystalline, hexagonal (graphite)	Ba	crystalline, cubic
Si	crystalline, cubic	Li	crystalline, cubic
Ge	crystalline, cubic	Na	crystalline, cubic
Sn	crystalline, tetragonal ("white")	K	crystalline, cubic
Pb	crystalline, cubic	Rb	crystalline, cubic
B	β, crystalline, rhombohedral	Cs	crystalline, cubic

2.3.2 Standard state pressure

The standard state pressure chosen for all selected data is 0.1 MPa (1 bar) as recommended by the International Union of Pure and Applied Chemistry IUPAC [82LAF].

However, the majority of the thermodynamic data published in the scientific literature and used for the evaluations in this review, refer to the old standard state pressure of 1 "standard atmosphere" (= 0.101325 MPa). The difference between the thermodynamic data for the two standard state pressures is not large and lies in most cases within the uncertainty limits. It is nevertheless essential to make the corrections for the change in the standard state pressure in order to avoid inconsistencies and propagation of errors. In practice the parameters affected by the change between these two standard state pressures are the Gibbs energy and entropy changes of all processes that involve gaseous species. Consequently, changes occur also in the Gibbs energies of formation of species that consist of elements whose reference state is gaseous (H, O, F, Cl, N, and the noble gases). No other thermodynamic quantities are affected significantly. A large part of the following discussion has been taken from the NBS tables of chemical thermodynamic properties [82WAG/EVA], see also Freeman [84FRE].

The following expressions define the effect of pressure on the properties of all substances:

$$\left(\frac{\partial H}{\partial p}\right)_T = V - T\left(\frac{\partial V}{\partial T}\right)_p = V(1-\alpha T) \tag{2.39}$$

2.3 Standard and reference conditions

$$\left(\frac{\partial C_p}{\partial p}\right)_T = -T\left(\frac{\partial^2 V}{\partial T^2}\right)_p \tag{2.40}$$

$$\left(\frac{\partial S}{\partial p}\right)_T = -V\alpha = -\left(\frac{\partial V}{\partial T}\right)_p \tag{2.41}$$

$$\left(\frac{\partial G}{\partial p}\right)_T = V \tag{2.42}$$

$$\text{where} \quad \alpha \equiv \frac{1}{V}\left(\frac{\partial V}{\partial T}\right)_p \tag{2.43}$$

For ideal gases, $V = \frac{RT}{p}$ and $\alpha = \frac{R}{pV} = \frac{1}{T}$. The conversion equations listed below (Eqs. (2.44) to (2.51)) apply to the small pressure change from 1 atm to 1 bar (0.1 MPa). The quantities that refer to the old standard state pressure of 1 atm are assigned the superscript $^{(atm)}$, and those that refer to the new standard state pressure of 1 bar are assigned the superscript $^{(bar)}$.

For all substances the changes in the enthalpy of formation and heat capacity are much smaller than the experimental accuracy and can be disregarded. This is exactly true for ideal gases.

$$\Delta_f H^{(bar)}(T) - \Delta_f H^{(atm)}(T) = 0 \tag{2.44}$$

$$C_p^{(bar)}(T) - C_p^{(atm)}(T) = 0 \tag{2.45}$$

For gaseous substances, the entropy difference is:

$$S^{(bar)}(T) - S^{(atm)}(T) = R \ln\left(\frac{p^{(atm)}}{p^{(bar)}}\right) = R \ln 1.01325$$

$$= 0.1094 \text{ J} \cdot \text{K}^{-1} \cdot \text{mol}^{-1} \tag{2.46}$$

This is exactly true for ideal gases, as follows from Eq.(2.41) with $\alpha = \frac{R}{pV}$. The entropy change of a reaction or process is thus dependent on the number of moles of gases involved:

$$\Delta_r S^{(bar)} - \Delta_r S^{(atm)} = \delta \cdot R \ln\left(\frac{p^{(atm)}}{p^{(bar)}}\right)$$

$$= \delta \times 0.1094 \text{ J} \cdot \text{K}^{-1} \cdot \text{mol}^{-1} \tag{2.47}$$

where δ is the net increase in moles of gas in the process.

Similarly, the change in the Gibbs energy of a process between the two standard state pressures is:

$$\Delta_r G^{(\text{bar})} - \Delta_r G^{(\text{atm})} = -\delta \cdot RT \ln\left(\frac{p^{(\text{atm})}}{p^{(\text{bar})}}\right)$$

$$= -\delta \cdot 0.03263 \text{ kJ} \cdot \text{mol}^{-1} \text{ at } 298.15 \text{ K}. \tag{2.48}$$

Eq.(2.48) applies also to $\Delta_f G^{(\text{bar})} - \Delta_f G^{(\text{atm})}$, since the Gibbs energy of formation describes the formation process of a compound or complex from the reference states of the elements involved:

$$\Delta_f G^{(\text{bar})} - \Delta_f G^{(\text{atm})} = -\delta \times 0.03263 \text{ kJ} \cdot \text{mol}^{-1} \text{ at } 298.15 \text{ K}. \tag{2.49}$$

The changes in the equilibrium constants and cell potentials with the change in the standard state pressure follows from the expression for Gibbs energy changes, Eq.(2.48)

$$\log_{10} K^{(\text{bar})} - \log_{10} K^{(\text{atm})} = -\frac{\Delta_r G^{(\text{bar})} - \Delta_r G^{(\text{atm})}}{RT \ln 10}$$

$$= \delta \cdot \frac{\ln\left(\frac{p^{(\text{atm})}}{p^{(\text{bar})}}\right)}{\ln 10} = \delta \cdot \log_{10}\left(\frac{p^{(\text{atm})}}{p^{(\text{bar})}}\right)$$

$$= \delta \cdot 0.005717 \tag{2.50}$$

$$E^{(\text{bar})} - E^{(\text{atm})} = -\frac{\Delta_r G^{(\text{bar})} - \Delta_r G^{(\text{atm})}}{nF}$$

$$= \delta \cdot \frac{RT \ln\left(\frac{p^{(\text{atm})}}{p^{(\text{bar})}}\right)}{nF}$$

$$= \delta \cdot \frac{0.0003382}{n} \text{ V at } 298.15 \text{ K} \tag{2.51}.$$

It should be noted that the standard potential of the hydrogen electrode is equal to 0.00 V exactly, by definition.

$$H^+ + e^- \rightleftharpoons \frac{1}{2} H_2(g) \qquad E^\circ \stackrel{\text{def}}{=} 0.00 \text{V} \tag{2.52}.$$

This definition will not be changed, although a gaseous substance, $H_2(g)$, is involved in the process. The change in the potential with pressure for an electrode potential conventionally written as:

$$Ag^+ + e^- \rightleftharpoons Ag\,(cr)$$

should thus be calculated from the balanced reaction that includes the hydrogen electrode,

$$Ag^+ + \frac{1}{2}H_2\,(g) \rightleftharpoons Ag\,(cr) + H^+$$

Here $\delta = -0.5$. Hence, the contribution to δ from an electron in a half cell reaction is the same as the contribution of a gas molecule with the stoichiometric coefficient of 0.5. This leads to the same value of δ as the combination with the hydrogen half cell.

Example:

$Fe(cr) + 2\,H^+ \rightleftharpoons Fe^{2+} + H_2\,(g)$ $\delta = 1$ $E^{(bar)} - E^{(atm)} = 0.00017$ V

$CO_2(g) \rightleftharpoons CO_2(aq)$ $\delta = -1$ $\log_{10} K^{(bar)} - \log_{10} K^{(bar)} = -0.0057$

$NH_3(g) + \frac{5}{4}O_2 \rightleftharpoons NO(g) + \frac{3}{2}H_2O(g)$ $\delta = 0.25$ $\Delta_r G^{(bar)} - \Delta_r G^{(atm)} = -0.008$ kJ·mol^{-1}

$\frac{1}{2}Cl_2(g) + 2\,O_2(g) + e^- \rightleftharpoons ClO_4^-$ $\delta = -3$ $\Delta_r G^{(bar)} - \Delta_r G^{(atm)} = 0.098$ kJ·mol^{-1}

2.3.3 Reference temperature

The definitions of standard states given in section 2.3 make no reference to fixed temperature. Hence, it is theoretically possible to have an infinite number of standard states of a substance as the temperature varies. It is, however, convenient to complete the definition of the standard state in a particular context by choosing a reference temperature. As recommended by IUPAC [82LAF], the reference temperature chosen in the NEA–TDB project is $T = 298.15$ K or $t = 25.00°C$. Where necessary for the discussion, values of experimentally measured temperatures are reported after conversion to the IPTS–68 [69COM]. The relation between the absolute temperature T (K, kelvin) and the Celsius temperature t (°C) is defined by $t = (T - T_o)$ where $T_o = 273.15$ K.

2.4 Fundamental physical constants

The fundamental physical constants are taken from a publication by CODATA [86COD]. Those relevant to this review are listed in Table 2-7.

Table 2-7: Fundamental physical constants. These values have been taken from CODATA [86COD]. The digits in parentheses are the one–standard–deviation uncertainty in the last digits of the given value.

Quantity	Symbol	Value	Units
speed of light in vacuum	c	299 792 458	$m \cdot s^{-1}$
permeability of vacuum	μ_0	$4\pi \times 10^{-7} = 12.566\,370\,614...$	$10^{-7} N \cdot A^{-2}$
permittivity of vacuum	ϵ_0	$1/\mu_0 c^2 = 8.854\,187\,817...$	$10^{-12} C^2 \cdot J^{-1} \cdot m^{-1}$
Planck constant	h	6.626 0755(40)	$10^{-34} J \cdot s$
elementary charge	e	1.602 177 33(49)	$10^{-19} C$
Avogadro constant	N_A	6.022 1367(36)	$10^{23} mol^{-1}$
Faraday constant	F	96 485.309(29)	$C \cdot mol^{-1}$
molar gas constant	R	8.314 510(70)	$J \cdot K^{-1} \cdot mol^{-1}$
Boltzmann constant, R/N_A	k	1.380 658(12)	$10^{-23} J \cdot K^{-1}$
Non-SI units used with SI:			
electron volt, (e/C) J	eV	1.602 177 33(49)	$10^{-19} J$
atomic mass unit, $1u = m_u = \frac{1}{12} m(^{12}C)$	u	1.660 5402(10)	$10^{-27} kg$

2.5 Uncertainty estimates

One of the principal objectives of the NEA TDB development effort is to provide an idea of the uncertainties associated with the data selected in the reviews. In general the uncertainties should define the range within which the corresponding data can be reproduced with a probability of 95%. In many cases, a full statistical treatment is limited or impossible due to the availability of only one or a few data points. Appendix C describes in detail the procedures used for the assignment and treatment of uncertainties, as well as the propagation of errors and the standard rules for rounding.

2.6 The NEA-TDB system

A database system has been developed at the NEA Data Bank that allows the storage of thermodynamic parameters for individual species as well as for reactions. The structure of the database system allows consistent derivation of thermodynamic data for individual species from reaction data at standard conditions, as well as internal recalculations of data at standard conditions. If a selected value is changed, all the dependent values will be recalculated consistently. The maintenance of consistency of all the selected data, including their uncertainties (cf. Appendix C), is ensured by the software developed for this purpose at the NEA Data Bank. The literature sources of the data are also stored in the database.

2.6 The NEA-TDB system

The following thermodynamic parameters, valid at the reference temperature of 298.15 K and at the standard pressure of 1 bar, are stored in the database:

$\Delta_f G_m^\circ$ the standard molar Gibbs energy of formation from the elements in their reference state ($kJ \cdot mol^{-1}$)

$\Delta_f H_m^\circ$ the standard molar enthalpy of formation from the elements in their reference state ($kJ \cdot mol^{-1}$)

S_m° the standard molar entropy ($J \cdot K^{-1} \cdot mol^{-1}$)

$C_{p,m}^\circ$ the standard molar heat capacity ($J \cdot K^{-1} \cdot mol^{-1}$).

For aqueous neutral species and ions, the values of $\Delta_f G_m^\circ$, $\Delta_f H_m^\circ$, S_m° and $C_{p,m}^\circ$ correspond to the standard partial molar quantities, and for individual aqueous ions they are relative quantities, defined with respect to the aqueous hydrogen ion, according to the convention [89COX/WAG] that $\Delta_f H_m^\circ (H^+, aq, T) = 0$ and that $S_m^\circ (H^+, aq, T) = 0$. Furthermore, for an *ionised solute* B containing any number of different cations and anions:

$$\Delta_f H_m^\circ (B_\pm, aq) = \sum_+ \upsilon_+ \Delta_f H_m^\circ (\text{cation, aq}) + \sum_- \upsilon_- \Delta_f H_m^\circ (\text{anion, aq})$$

$$S_m^\circ (B_\pm, aq) = \sum_+ \upsilon_+ S_m^\circ (\text{cation, aq}) + \sum_- \upsilon_- S_m^\circ (\text{anion, aq})$$

As the thermodynamic parameters vary as a function of temperature, provision is made for including the compilation of the coefficients of empirical temperature functions for these data, as well as the temperature ranges over which they are valid. In many cases the thermodynamic data measured or calculated at several temperatures were published for a particular species, rather than the deduced temperature functions. In these cases, a linear regression method is used in this review to obtain the most significant coefficients of the following empirical function for a thermodynamic parameter, X:

$$X(T) = a_X + b_X \cdot T + c_X \cdot T^2 + d_X \cdot T^{-1} + e_X \cdot T^{-2} + f_X \cdot \ln T + g_X \cdot T \ln T \quad (2.53)$$
$$+ h_X \cdot \sqrt{T} + \frac{i_X}{\sqrt{T}} + j_X \cdot T^3 + k_X \cdot T^{-3}.$$

Most temperature variations can be described with three or four parameters. In the present review, only $C_{p,m}(T)$, *i.e.*, the thermal functions of the heat capacities of individual species are considered and stored in the database. They refer to the relation:

$$C_{p,m}(T) = a + b \cdot T + c \cdot T^2 + d \cdot T^{-1} + e \cdot T^{-2}$$

(where the subindices for the coefficients have been dropped) and are listed in Tables 3.3, 4.3, 5.3, 6.3 and 7.3.

The pressure dependence of thermodynamic data has not been the subject of critical analysis in the present compilation. The reader interested in higher temperatures and pressures, or the pressure dependency of thermodynamic functions for geochemical applications, is referred to the specialised literature in this area, *e.g.*, [82HAM],

[84MAR/MES], [88SHO/HEL], [88TAN/HEL], [89SHO/HEL], [89SHO/HEL2], [90MON], [91AND/CAS].

Selected standard thermodynamic data referring to chemical reactions are also compiled in the database. A chemical reaction "r", involving reactants and products 'B', can be abbreviated as:

$$0 = \sum_B \nu'_B \, B \tag{2.54}$$

where the stoichiometric coefficients ν'_B are positive for products, and negative for reactants. The reaction parameters considered in the NEA TDB system include:

$\log_{10} K_r^\circ$ the equilibrium constant of the reaction, logarithmic

$\Delta_r G_m^\circ$ the molar Gibbs energy of reaction ($kJ \cdot mol^{-1}$)

$\Delta_r H_m^\circ$ the molar enthalpy of reaction ($kJ \cdot mol^{-1}$)

$\Delta_r S_m^\circ$ the molar entropy of reaction ($J \cdot K^{-1} \cdot mol^{-1}$)

$\Delta_r C_{p,m}^\circ$ the molar heat capacity of reaction ($J \cdot K^{-1} \cdot mol^{-1}$)

The temperature functions of these data, if available, are stored according to Eq.(2.53).

The equilibrium constant, K_r°, is related to $\Delta_r G_m^\circ$ according to the following relation:

$$\log_{10} K_r^\circ = -\frac{\Delta_r G_m^\circ}{RT \ln(10)}$$

and can be calculated from the individual values of $\Delta_f G_m^\circ(B)$ (for example, those given in Tables 3.1 and 4.1), according to:

$$\log_{10} K_r^\circ = -\frac{1}{RT \ln(10)} \sum_B \nu'_B \, \Delta_f G_m^\circ(B) \tag{2.55}$$

2.7 Presentation of the selected data

The selected data are presented in Chapters 3, 4, 5, 6 and 7. Unless otherwise indicated, they refer to standard conditions (cf. section 2.3) and 298.15 K (25.00°C) and are provided with an uncertainty which should correspond to the 95% confidence level (see Appendix C).

Chapters 3, 4, 5, 6 and 7 contain tables of selected thermodynamic data for individual compounds and complexes of uranium, neptunium, plutonium, americium and technetium (Tables 3.1, 4.1, 5.1, 6.1 and 7.1, respectively), tables of selected reaction data (Tables 3.2, 4.2, 5.2, 6.2 and 7.2) for reactions concerning uranium, neptunium, plutonium, americium and technetium species, respectively, and tables containing selected thermal functions of the heat capacities of individual species of uranium, neptunium, plutonium, americium and technetium (Tables 3.3, 4.3, 5.3, 6.3 and 7.3, respec-

2.7 Presentation of the selected data

tively). The selection of these data is discussed in Part III. The fitted heat capacity coefficients for the gaseous species in Tables 3.3, 4.3, 5.3, 6.3 and 7.3 are valid only up to the maximum temperatures given. These temperatures vary in order to retain only small differences (generally < 0.25 J · K^{-1} · mol^{-1}) between the fitted and calculated heat capacities.

Chapter 14 contains, for auxiliary compounds and complexes that do not contain uranium, neptunium, plutonium, americium or technetium, a table of the thermodynamic data for individual species (Table 8.1) and a table of reaction data (Table 8.2). Most of these values are the CODATA Key Values [89COX/WAG]. The selection of the remaining auxiliary data is discussed in [92GRE/FUG], [99RAR/RAN] and [2001LEM/FUG].

All the selected data presented in Tables 3.1, 3.2, 4.1, 4.2, 5.1, 5.2, 6.1, 6.2, 7.1 and 7.2 are internally consistent. This consistency is maintained by the internal consistency verification and recalculation software developed at the NEA Data Bank in conjunction with the NEA TDB database system, *cf.* section 2.6. Therefore, when using the selected data for uranium, neptunium, plutonium, americium or technetium species, the auxiliary data in Chapter 8 must be used together with the data in Chapter 3, 4, 5, 6 and 7 to ensure internal consistency of the data set.

It is important to note that Tables 3.2, 4.2, 5.2, 6.2, 7.2 and 8.2 include only those species for which the primary selected data are reaction data. The formation data derived there from and listed in Tables 3.1, 4.1, 5.1, 6.1 and 7.1 are obtained using auxiliary data, and their uncertainties are propagated accordingly. In order to maintain the uncertainties originally assigned to the selected data in this review, the user is advised to make direct use of the reaction data presented in Tables 3.2, 4.2, 5.2, 6.2, 7.2 and 8.2, rather than taking the derived values in Tables 3.1, 4.1, 5.1, 6.1, 7.1 and 8.1 to calculate the reaction data with Eq.(2.55). The latter approach would imply a twofold propagation of the uncertainties and result in reaction data whose uncertainties would be considerably larger than those originally assigned.

The thermodynamic data in the selected set refer to a temperature of 298.15 K (25.00°C), but they can be recalculated to other temperatures if the corresponding data (enthalpies, entropies, heat capacities) are available [97PUI/RAR]. For example, the temperature dependence of the standard reaction Gibbs energy as a function of the standard reaction entropy at the reference temperature ($T_0 = 298.15$ K), and of the heat capacity function is:

$$\Delta_r G_m^\circ(T) = \Delta_r H_m^\circ(T_0) + \int_{T_0}^{T} \Delta_r C_{p,m}^\circ(T) \, dT \\ - T \left(\Delta_r S_m^\circ(T_0) + \int_{T_0}^{T} \frac{\Delta_r C_{p,m}^\circ(T)}{T} dT \right),$$

and the temperature dependence of the standard equilibrium constant as a function of the standard reaction enthalpy and heat capacity is:

$$\log_{10} K°(T) = \log_{10} K°(T_0) - \frac{\Delta_r H_m°(T_0)}{R \ln(10)} \left(\frac{1}{T} - \frac{1}{T_0} \right)$$

$$- \frac{1}{RT \ln(10)} \int_{T_0}^{T} \Delta_r C_{p,m}°(T) \, dT + \frac{1}{R \ln(10)} \int_{T_0}^{T} \frac{\Delta_r C_{p,m}°(T)}{T} dT,$$

where R is the gas constant (cf. Table 2-7).

In the case of aqueous species, for which enthalpies of reaction are selected or can be calculated from the selected enthalpies of formation, but for which there are no selected heat capacities, it is in most cases possible to recalculate equilibrium constants to temperatures up to 100 to 150°C, with an additional uncertainty of perhaps about 1 to 2 logarithmic units, due to neglecting the heat capacity contributions to the temperature correction. However, it is important to observe that "new" aqueous species, i.e., species not present in significant amounts at 25°C and therefore not detected, may be significant at higher temperatures, see for example the work by Ciavatta et al. [87CIA/IUL]. Additional high–temperature experiments may therefore be needed in order to ascertain that proper chemical models are used in the modelling of hydrothermal systems. For many species, experimental thermodynamic data are not available to allow a selection of parameters describing the temperature dependence of equilibrium constants and Gibbs energies of formation. The user may find information on various procedures to estimate the temperature dependence of these thermodynamic parameters in [97PUI/RAR]. The thermodynamic data in the selected set refer to infinite dilution for soluble species. Extrapolation of an equilibrium constant K, usually measured at high ionic strength, to $K°$ at $I = 0$ using activity coefficients γ, is explained in Appendix B. The corresponding Gibbs energy of dilution is:

$$\Delta_{dil} G_m = \Delta_r G_m° - \Delta_r G_m \tag{2.56}$$

$$= - RT \Delta_r \ln \gamma_\pm \tag{2.57}$$

Similarly $\Delta_{dil} S_m$ can be calculated from $\ln \gamma_\pm$ and its variations with T, while:

$$\Delta_{dil} H_m = RT^2 \frac{\partial}{\partial T} (\Delta_r \ln \gamma_\pm)_p \tag{2.58}$$

depends only on the variation of γ with T, which is neglected in this review, when no data on the temperature dependence of γ's are available. In this case the Gibbs energy of dilution $\Delta_{dil} G_m$ is entirely assigned to the entropy difference. This entropy of reaction is calculated using the Gibbs-Helmholtz equation, the above assumption $\Delta_{dil} H_m = 0$, and $\Delta_{dil} G_m$.

Part II

Tables of selected data

Chapter 3

Selected Data for Uranium

3.1 General considerations

This chapter presents updated chemical thermodynamic data for uranium species, as described in detail in Chapter 9 of this volume. The newly selected data represent revisions to those chosen in the previous NEA TDB review [92GRE/FUG]. In this respect, it will be found that while new species appear in the Tables, some others have been removed from them. Table 3–1 contains the recommended thermodynamic data of the uranium compounds and complexes, Table 3–2 the recommended thermodynamic data of chemical equilibrium reactions by which the uranium compounds and complexes are formed, and Table 3–3 the temperature coefficients of the heat capacity data of Table 3–1 where available.

The species and reactions in Table 3–1, Table 3–2 and Table 3–3 appear in standard order of arrangement (cf. Figure 2.1). Table 3–2 contains information only on those reactions for which primary data selections are made in this review. These selected reaction data are used, together with data for key uranium species (for example U^{4+}) and auxiliary data listed in Table 8–1, to derive the corresponding formation quantities in Table 3–1. The uncertainties associated with values for the key uranium species and for some of the auxiliary data are substantial, leading to comparatively large uncertainties in the formation quantities derived in this manner. The inclusion of a table for reaction data (Table 3–2) in this report allows the use of equilibrium constants with total uncertainties that are directly based on the experimental accuracy. This is the main reason for including both the table for reaction data (Table 3–2) and the table of $\Delta_f G_m^\circ$, $\Delta_f H_m^\circ$, S_m° and $C_{p,m}^\circ$ values (Table 3–1). In a few cases, the correlation of small uncertainties in values for ligands has been neglected in calculations of uncertainty values for species in Table 3–1 from uncertainty values in Table 3–2. However, for those species the effects are less than 2% of the stated uncertainties.

The selected thermal functions of the heat capacities, listed in Table 3–3, refer to the relation:

$$C_{p,m}(T) = a + b \cdot T + c \cdot T^2 + d \cdot T^{-1} + e \cdot T^{-2} \tag{3.1}$$

No references are given in these tables since the selected data are generally not directly attributable to a specific published source. A detailed discussion of the selection procedure is presented in [92GRE/FUG] and in Chapter 9 of this volume.

3.2 Precautions to be observed in the use of the tables

Geochemical modelling in aquatic systems requires the careful use of the data selected in the NEA TDB reviews. The selected data in the tables must not be adopted without taking into account the chemical background information discussed in the corresponding sections of this book. In particular the following precautions should be observed when using data from the Tables.

- The addition of any aqueous species and its data to this internally consistent database can result in a modified data set which is no longer rigorous and can lead to erroneous results. The situation is similar, to a lesser degree, with the addition of gases and solids. It should also be noted that the data set presented in this chapter may not be "complete" for all the conceivable systems and conditions. Gaps are pointed out in both the previous NEA TDB review and the present update.

- Solubility data for crystalline phases are well defined in the initial state, but not necessarily in the final state after "equilibrium" has been attained. Hence, the solubility calculated from these phases may be very misleading, as discussed for the solubility data of $MO_2(cr)$, M = U, Np, Pu.

- The selected thermodynamic data in [92GRE/FUG], [95SIL/BID] and [2001LEM/FUG] contain thermodynamic data for amorphous phases. Most of these refer to Gibbs energy of formation deduced from solubility measurements. In the present review all such data have been removed. However, the corresponding $\log_{10}K$ values have been retained. The reasons for these changes are:
    o Thermodynamic data are only meaningful if they refer to well–defined systems; this is not the case for amorphous phases. It is well known that their solubility may change with time, due to re–crystallisation with a resulting change in water content and surface area/crystal size; the time scale for these changes can vary widely. These kinetic phenomena are different from those encountered in systems where there is a very high activation barrier for the reaction, e.g., electron exchange between sulphate and sulphide.

    o The solubility of amorphous phases provides useful information for users of the database when modelling the behaviour of complex systems. Therefore, the solubility products have been given with the proviso that they are not thermodynamic quantities.

Table 3–1: Selected thermodynamic data for uranium compounds and complexes. All ionic species listed in this table are aqueous species. Unless noted otherwise, all data refer to the reference temperature of 298.15 K and to the standard state, *i.e.*, a pressure of 0.1 MPa and, for aqueous species, infinite dilution ($I = 0$). The uncertainties listed below each value represent total uncertainties and correspond in principle to the statistically defined 95% confidence interval. Values in bold typeface are CODATA Key Values and are taken directly from reference [89COX/WAG] without further evaluation. Values obtained from internal calculation, *cf.* footnotes (a) and (b), are rounded at the third digit after the decimal point and may therefore not be exactly identical to those given in Part III. Systematically, all the values are presented with three digits after the decimal point, regardless of the significance of these digits. The data presented in this table are available on computer media from the OECD Nuclear Energy Agency.

Compound	$\Delta_f G_m^\circ$ (kJ · mol^{-1})	$\Delta_f H_m^\circ$ (kJ · mol^{-1})	S_m° (J · K^{-1} · mol^{-1})	$C_{p,m}^\circ$ (J · K^{-1} · mol^{-1})
U(cr)	0.000	0.000	**50.200** ±**0.200**	27.660 [c] ±0.050
U(g)	488.400 [a] ±8.000	**533.000** ±**8.000**	**199.790** ±**0.030**	23.690 [c] ±0.030
U^{3+}	−476.473 [b] ±1.810	−489.100 [b] ±3.712	−188.172 [b] ±13.853	−150.000 ±50.000
U^{4+}	−529.860 [b] ±1.765	−591.200 ±3.300	−416.895 [a] ±12.553	−220.000 ±50.000
UO(g)	5.327 [a] ±10.018	35.000 ±10.000	252.300 ±2.000	39.600 [c] ±2.000
UO$_2$(cr)	−1031.833 [a] ±1.004	**−1085.000** ±**1.000**	**77.030** ±**0.200**	63.600 [c] ±0.080
UO$_2$(g)	−481.064 [a] ±20.036	−477.800 ±20.000	266.300 ±4.000	59.500 [c] ±2.000
UO$_2^+$	−961.021 [b] ±1.752	−1025.127 [a] ±2.960	−25.000 ±8.000	
UO$_2^{2+}$	**−952.551** [a] ±**1.747**	**−1019.000** ±**1.500**	**−98.200** ±**3.000**	42.400 [c] ±3.000
β-UO$_{2.25}$ [d]	−1069.083 [a] ±1.702	−1127.400 ±1.700	85.400 ±0.200	[c]
UO$_{2.25}$(cr) [e]	−1069.125 [a] ±1.702	−1128.000 ±1.700	83.530 ±0.170	73.340 [c] ±0.150
α-UO$_{2.3333}$			82.170 ±0.500	71.420 [c] ±0.300
β-UO$_{2.3333}$	−1079.572 [a] ±2.002	−1141.000 ±2.000	83.510 ±0.200	71.840 [c] ±0.140

(Continued on next page)

Table 3–1: (continued)

Compound	$\Delta_f G_m^\circ$ (kJ·mol^{-1})	$\Delta_f H_m^\circ$ (kJ·mol^{-1})	S_m° (J·K^{-1}·mol^{-1})	$C_{p,m}^\circ$ (J·K^{-1}·mol^{-1})
UO$_{2.6667}$(cr)	−1123.157 [a] ±0.804	−1191.600 ±0.800	94.180 ±0.170	79.310 [c] ±0.160
UO$_{2.86}$· 0.5 H$_2$O(cr)		−1367.000 ±10.000		
UO$_{2.86}$· 1.5 H$_2$O(cr)		−1666.000 ±10.000		
α–UO$_{2.95}$		−1211.280 [b] ±1.284		
α–UO$_3$	−1135.330 [a] ±1.482	−1212.410 ±1.450	99.400 ±1.000	81.840 ±0.300
β–UO$_3$	−1142.301 [a] ±1.307	−1220.300 ±1.300	96.320 ±0.400	81.340 [c] ±0.160
γ–UO$_3$	**−1145.739** [a] **±1.207**	**−1223.800** **±1.200**	**96.110** **±0.400**	**81.670** [c] **±0.160**
δ–UO$_3$		−1213.730 ±1.440		
ε–UO$_3$		−1217.200 ±1.300		
UO$_3$(g)	−784.761 [a] ±15.012	−799.200 ±15.000	309.500 ±2.000	64.500 [c] ±2.000
β–UH$_3$	−72.556 [a] ±0.148	−126.980 ±0.130	63.680 ±0.130	49.290 ±0.080
UOH^{3+}	−763.918 [b] ±1.798	−830.120 [b] ±9.540	−199.946 [b] ±32.521	
UO$_2$OH$^+$	−1159.724 [b] ±2.221	−1261.371 [a] ±15.072	17.000 ±50.000	
UO$_3$· 0.393 H$_2$O(cr)		−1347.800 ±1.300		
UO$_3$· 0.648 H$_2$O(cr)		−1424.600 ±1.300		
α–UO$_3$· 0.85 H$_2$O		−1491.900 ±1.300		
α–UO$_3$· 0.9 H$_2$O	−1374.560 [a] ±2.460	−1506.300 ±1.300	126.000 ±7.000	140.000 ±30.000
δ–UO$_3$H$_{0.83}$		−1285.140 [b] ±2.020		

(Continued on next page)

Table 3–1: (continued)

Compound	$\Delta_f G_m^o$ (kJ · mol^{-1})	$\Delta_f H_m^o$ (kJ · mol^{-1})	S_m^o (J · K^{-1} · mol^{-1})	$C_{p,m}^o$ (J · K^{-1} · mol^{-1})
UO$_2$(OH)$_2$(aq)	−1357.479 (b) ±1.794			
β−UO$_2$(OH)$_2$	−1398.683 (a) ±1.765	−1533.800 ±1.300	138.000 ±4.000	141.000 (c) ±15.000
γ−UO$_2$(OH)$_2$		−1531.400 ±1.300		
U(OH)$_4$(aq)	−1421.309 (b) ±8.189	−1624.607 (a) ±11.073	40.000 ±25.000	205.000 ±80.000
UO$_2$(OH)$_3^-$	−1548.384 (b) ±2.969			
UO$_3$· 2 H$_2$O(cr)	−1636.506 (a) ±1.705	−1826.100 ±1.700	188.540 ±0.380	172.070 (c) ±0.340
UO$_2$(OH)$_4^{2-}$	−1716.171 (b) ±4.260			
UO$_4$· 2 H$_2$O(cr)		−1784.000 ±4.200		
UO$_4$· 4 H$_2$O(cr)		−2384.700 ±2.100		
(UO$_2$)$_2$OH^{3+}	−2126.830 (b) ±6.693			
(UO$_2$)$_2$(OH)$_2^{2+}$	−2347.303 (b) ±3.503	−2572.065 (a) ±5.682	−38.000 ±15.000	
(UO$_2$)$_3$(OH)$_4^{2+}$	−3738.288 (b) ±5.517			
(UO$_2$)$_3$(OH)$_5^+$	−3954.594 (b) ±5.291	−4389.086 (a) ±10.394	83.000 ±30.000	
(UO$_2$)$_3$(OH)$_7^-$	−4333.835 (b) ±6.958			
(UO$_2$)$_4$(OH)$_7^+$	−5345.179 (b) ±9.029			
UF(g)	−76.876 (a) ±20.020	−47.000 ±20.000	251.800 ±3.000	37.900 (c) ±3.000
UF^{3+}	−865.153 (b) ±3.474	−932.150 (b) ±3.400	−269.133 (b) ±16.012	
UF$_2$(g)	−558.697 (a) ±25.177	−540.000 ±25.000	315.700 ±10.000	56.200 (c) ±5.000

(Continued on next page)

Table 3–1: (continued)

Compound	$\Delta_f G_m^\circ$ (kJ · mol^{-1})	$\Delta_f H_m^\circ$ (kJ · mol^{-1})	S_m° (J · K^{-1} · mol^{-1})	$C_{p,m}^\circ$ (J · K^{-1} · mol^{-1})
UF_2^{2+}	−1187.431 [b] ±4.632	−1265.400 [b] ±3.597	−139.195 [b] ±18.681	
UF_3(cr)	−1432.531 [a] ±4.702	−1501.400 ±4.700	123.400 ±0.400	95.100 [c] ±0.400
UF_3(g)	−1062.947 [a] ±20.221	−1065.000 ±20.000	347.500 ±10.000	76.200 [c] ±5.000
UF_3^+	−1499.378 [b] ±5.466	−1596.750 [b] ±5.540	−37.537 [b] ±24.410	
UF_4(aq)	−1806.302 [b] ±6.388	−1941.030 [b] ±8.565	3.905 [b] ±21.374	
UF_4(cr)	−1823.538 [a] ±4.201	−1914.200 ±4.200	151.700 ±0.200	116.000 [c] ±0.100
UF_4(g)	−1576.851 [a] ±6.698	−1605.200 [b] ±6.530	360.700 ±5.000	95.100 [c] ±3.000
α–UF_5	−1968.688 [a] ±6.995	−2075.300 ±5.900	199.600 ±12.600	132.200 [c] ±4.200
β–UF_5	−1970.595 [a] ±5.635	−2083.200 ±4.200	179.500 ±12.600	132.200 [c] ±12.000
UF_5(g)	−1862.083 [a] ±15.294	−1913.000 ±15.000	386.400 ±10.000	110.600 [c] ±5.000
UF_5^-	−2095.760 [b] ±5.739			
UF_6(cr)	−2069.205 [a] ±1.842	−2197.700 ±1.800	227.600 ±1.300	166.800 [c] ±0.200
UF_6(g)	−2064.440 [a] ±1.893	−2148.600 [b] ±1.868	376.300 ±1.000	129.400 [c] ±0.500
UF_6^{2-}	−2389.098 [b] ±6.028			
U_2F_9(cr)	−3812.000 ±17.000	−4015.923 [a] ±18.016	329.000 ±20.000	251.000 [c] ±16.700
U_2F_{10}(g)		−3993.200 [b] ±30.366		
U_4F_{17}(cr)	−7464.000 ±30.000	−7849.665 [a] ±32.284	631.000 ±40.000	485.300 [c] ±33.000
UOF_2(cr)	−1434.127 [a] ±6.424	−1504.600 ±6.300	119.200 ±4.200	

(Continued on next page)

Table 3–1: (continued)

Compound	$\Delta_f G_m^\circ$ (kJ·mol⁻¹)	$\Delta_f H_m^\circ$ (kJ·mol⁻¹)	S_m° (J·K⁻¹·mol⁻¹)	$C_{p,m}^\circ$ (J·K⁻¹·mol⁻¹)
UOF₄(cr)	−1816.264 (a) ±4.269	−1924.600 ±4.000	195.000 ±5.000	
UOF₄(g)	−1704.814 (a) ±20.142	−1763.000 ±20.000	363.200 ±8.000	108.100 (c) ±5.000
UO₂F⁺	−1263.527 (b) ±1.911	−1352.650 (b) ±1.637	−7.511 (b) ±3.321	
UO₂F₂(aq)	−1565.999 (b) ±2.276	−1687.600 (b) ±1.994	50.293 (b) ±3.783	
UO₂F₂(cr)	−1557.322 (a) ±1.307	−1653.500 ±1.300	135.560 ±0.420	103.220 (c) ±0.420
UO₂F₂(g)	−1318.081 (a) ±10.242	−1352.500 (b) ±10.084	342.700 ±6.000	86.400 (c) ±3.000
UO₂F₃⁻	−1859.338 (b) ±2.774	−2022.700 (b) ±2.480	76.961 (b) ±4.417	
UO₂F₄²⁻	−2146.226 (b) ±3.334	−2360.110 (b) ±3.038	74.248 (b) ±5.115	
U₂O₃F₆(cr)	−3372.730 (a) ±14.801	−3579.200 (b) ±13.546	324.000 ±20.000	
U₃O₅F₈(cr)	−4890.135 (a) ±9.771	−5192.950 (b) ±3.929	459.000 ±30.000	304.100 ±4.143
H₃OUF₆(cr)		−2641.400 (b) ±3.157		
UOFOH(cr)	−1336.930 (a) ±12.948	−1426.700 ±12.600	121.000 ±10.000	
UOFOH·0.5 H₂O(cr)	−1458.117 (a) ±6.970	−1576.100 ±6.300	143.000 ±10.000	
UOF₂·H₂O(cr)	−1674.474 (a) ±4.143	−1802.000 ±3.300	161.100 ±8.400	
UF₄·2.5 H₂O(cr)	−2440.282 (a) ±6.188	−2671.475 ±4.277	263.500 ±15.000	263.700 ±15.000
UO₂FOH·H₂O(cr)	−1721.700 ±7.500	−1894.500 ±8.400	178.345 (a) ±37.770	
UO₂FOH·2 H₂O(cr)	−1961.032 (b) ±8.408	−2190.010 (b) ±9.392	223.180 (b) ±38.244	
UO₂F₂·3 H₂O(cr)	−2269.658 (a) ±6.939	−2534.390 (b) ±4.398	270.000 ±18.000	

(Continued on next page)

Table 3–1: (continued)

Compound	$\Delta_f G_m^°$ (kJ · mol^{-1})	$\Delta_f H_m^°$ (kJ · mol^{-1})	$S_m^°$ (J · K^{-1} · mol^{-1})	$C_{p,m}^°$ (J · K^{-1} · mol^{-1})
UCl(g)	155.945 [a] ±20.020	187.000 ±20.000	265.900 ±3.000	43.200 [c] ±3.000
UCl^{3+}	−670.895 [b] ±1.918	−777.280 [b] ±9.586	−391.093 [b] ±32.788	
UCl$_2$(g)	−174.624 [a] ±20.221	−155.000 ±20.000	339.100 ±10.000	59.900 [c] ±5.000
UCl$_3$(cr)	−796.103 [a] ±2.006	−863.700 ±2.000	158.100 ±0.500	102.520 [c] ±0.500
UCl$_3$(g)	−521.652 [a] ±20.221	−523.000 ±20.000	380.300 ±10.000	82.400 [c] ±5.000
UCl$_4$(cr)	−929.605 [a] ±2.512	−1018.800 ±2.500	197.200 ±0.800	121.800 [c] ±0.400
UCl$_4$(g)	−789.442 [a] ±4.947	−815.400 [b] ±4.717	409.300 ±5.000	103.500 [c] ±3.000
UCl$_5$(cr)	−930.115 [a] ±3.908	−1039.000 ±3.000	242.700 ±8.400	150.600 [c] ±8.400
UCl$_5$(g)	−849.552 [a] ±15.074	−900.000 [b] ±15.000	438.700 ±10.000	123.600 [c] ±5.000
UCl$_6$(cr)	−937.120 [a] ±3.043	−1066.500 ±3.000	285.500 ±1.700	175.700 [c] ±4.200
UCl$_6$(g)	−901.588 [a] ±5.218	−985.500 [b] ±5.000	438.000 ±5.000	147.200 [c] ±3.000
U$_2$Cl$_{10}$(g)		−1965.300 [b] ±10.208		
UOCl(cr)	−785.654 [a] ±4.890	−833.900 ±4.200	102.500 ±8.400	71.000 [c] ±5.000
UOCl$_2$(cr)	−998.478 [a] ±2.701	−1069.300 ±2.700	138.320 ±0.210	95.060 [c] ±0.420
UOCl$_3$(cr)	−1045.576 [a] ±8.383	−1140.000 ±8.000	170.700 ±8.400	117.200 [c] ±4.200
UO$_2$Cl(cr)	−1095.253 [a] ±8.383	−1171.100 ±8.000	112.500 ±8.400	88.000 [c] ±5.000
UO$_2$Cl$^+$	−1084.738 [b] ±1.755	−1178.080 [b] ±2.502	−11.513 [b] ±7.361	
UO$_2$Cl$_2$(aq)	−1208.707 [b] ±2.885	−1338.160 [b] ±6.188	44.251 [b] ±21.744	

(Continued on next page)

3 Selected uranium data

Table 3–1: (continued)

Compound	$\Delta_f G_m^\circ$ (kJ · mol^{-1})	$\Delta_f H_m^\circ$ (kJ · mol^{-1})	S_m° (J · K^{-1} · mol^{-1})	$C_{p,m}^\circ$ (J · K^{-1} · mol^{-1})
UO$_2$Cl$_2$(cr)	−1145.838 (a) ±1.303	−1243.600 ±1.300	150.540 ±0.210	107.860 (c) ±0.170
UO$_2$Cl$_2$(g)	−938.984 (a) ±15.106	−970.300 ±15.000	373.400 ±6.000	92.600 (c) ±3.000
U$_2$O$_2$Cl$_5$(cr)	−2037.306 (a) ±4.891	−2197.400 ±4.200	326.300 ±8.400	219.400 (c) ±5.000
(UO$_2$)$_2$Cl$_3$(cr)	−2234.755 (a) ±2.931	−2404.500 ±1.700	276.000 ±8.000	203.600 (c) ±5.000
UO$_2$ClO$_3^+$	−963.308 (b) ±2.239	−1126.900 (b) ±1.828	60.592 (b) ±4.562	
U$_5$O$_{12}$Cl(cr)	−5517.951 (a) ±12.412	−5854.400 ±8.600	465.000 ±30.000	
UO$_2$Cl$_2$· H$_2$O(cr)	−1405.003 (a) ±3.269	−1559.800 ±2.100	192.500 ±8.400	
UO$_2$ClOH· 2 H$_2$O(cr)	−1782.219 (a) ±4.507	−2010.400 ±1.700	236.000 ±14.000	
UO$_2$Cl$_2$· 3 H$_2$O(cr)	−1894.616 (a) ±3.028	−2164.800 ±1.700	272.000 ±8.400	
UCl$_3$F(cr)	−1146.573 (a) ±5.155	−1243.000 ±5.000	162.800 ±4.200	118.800 ±4.200
UCl$_2$F$_2$(cr)	−1375.967 (a) ±5.592	−1466.000 ±5.000	174.100 ±8.400	119.700 ±4.200
UClF$_3$(cr)	−1606.360 (a) ±5.155	−1690.000 ±5.000	185.400 ±4.200	120.900 ±4.200
UBr(g)	199.623 (a) ±15.027	245.000 ±15.000	278.500 ±3.000	44.500 (c) ±3.000
UBr^{3+}	−642.044 (b) ±2.109			
UBr$_2$(g)	−86.896 (a) ±15.294	−40.000 ±15.000	359.700 ±10.000	61.400 (c) ±5.000
UBr$_3$(cr)	−673.198 (a) ±4.205	−698.700 ±4.200	192.980 ±0.500	105.830 (c) ±0.500
UBr$_3$(g)	−408.115 (a) ±20.494	−371.000 ±20.000	403.000 ±15.000	85.200 (c) ±5.000
UBr$_4$(cr)	−767.479 (a) ±3.544	−802.100 ±2.500	238.500 ±8.400	128.000 (c) ±4.200

(Continued on next page)

Table 3–1: (continued)

Compound	$\Delta_f G_m^\circ$ (kJ · mol^{-1})	$\Delta_f H_m^\circ$ (kJ · mol^{-1})	S_m° (J · K^{-1} · mol^{-1})	$C_{p,m}^\circ$ (J · K^{-1} · mol^{-1})
UBr$_4$(g)	−634.604 (a) ±4.951	−605.600 (b) ±4.717	451.900 ±5.000	106.900 (c) ±3.000
UBr$_5$(cr)	−769.308 (a) ±9.205	−810.400 ±8.400	292.900 ±12.600	160.700 (c) ±8.000
UBr$_5$(g)	−668.212 (a) ±15.295	−647.945 (b) ±15.000	498.700 ±10.000	129.000 (c) ±5.000
UOBr$_2$(cr)	−929.648 (a) ±8.401	−973.600 ±8.400	157.570 ±0.290	98.000 (c) ±0.400
UOBr$_3$(cr)	−901.498 (a) ±21.334	−954.000 ±21.000	205.000 ±12.600	120.900 (c) ±4.200
UO$_2$Br$^+$	−1057.657 (b) ±1.759			
UO$_2$Br$_2$(cr)	−1066.422 (a) ±1.808	−1137.400 ±1.300	169.500 ±4.200	116.000 (c) ±8.000
UO$_2$BrO$_3^+$	−937.077 (b) ±1.914	−1085.600 (b) ±1.609	75.697 (b) ±3.748	
UO$_2$Br$_2$· H$_2$O(cr)	−1328.644 (a) ±2.515	−1455.900 ±1.400	214.000 ±7.000	
UO$_2$BrOH· 2 H$_2$O(cr)	−1744.162 (a) ±4.372	−1958.200 ±1.300	248.000 ±14.000	
UO$_2$Br$_2$· 3 H$_2$O(cr)	−1818.486 (a) ±5.573	−2058.000 ±1.500	304.000 ±18.000	
UBr$_2$Cl(cr)	−714.389 (a) ±9.765	−750.600 ±8.400	192.500 ±16.700	
UBr$_3$Cl(cr)	−807.114 (a) ±9.766	−852.300 ±8.400	238.500 ±16.700	
UBrCl$_2$(cr)	−760.315 (a) ±9.765	−812.100 ±8.400	175.700 ±16.700	
UBr$_2$Cl$_2$(cr)	−850.896 (a) ±9.765	−907.900 ±8.400	234.300 ±16.700	
UBrCl$_3$(cr)	−893.500 (a) ±9.202	−967.300 ±8.400	213.400 ±12.600	
UI(g)	288.861 (a) ±25.045	342.000 ±25.000	286.500 ±5.000	44.800 (c) ±5.000
UI^{3+}	−588.719 (b) ±2.462			

(Continued on next page)

Table 3–1: (continued)

Compound	$\Delta_f G_m^o$ (kJ · mol^{-1})	$\Delta_f H_m^o$ (kJ · mol^{-1})	S_m^o (J · K^{-1} · mol^{-1})	$C_{p,m}^o$ (J · K^{-1} · mol^{-1})
UI$_2$(g)	40.341 (a) ±25.177	103.000 ±25.000	376.500 ±10.000	61.900 (c) ±5.000
UI$_3$(cr)	−466.122 (a) ±4.892	−466.900 ±4.200	221.800 ±8.400	112.100 (c) ±6.000
UI$_3$(g)	−198.654 (a) ±25.178	−137.000 ±25.000	431.200 ±10.000	86.000 (c) ±5.000
UI$_4$(cr)	−512.671 (a) ±3.761	−518.300 ±2.800	263.600 ±8.400	126.400 (c) ±4.200
UI$_4$(g)	−369.585 (a) ±6.210	−305.000 (b) ±5.731	499.100 ±8.000	108.800 (c) ±4.000
UO$_2$IO$_3^+$	−1090.305 (b) ±1.917	−1228.900 (b) ±1.819	90.959 (b) ±4.718	
UO$_2$(IO$_3$)$_2$(aq)	−1225.718 (b) ±2.493			
UO$_2$(IO$_3$)$_2$(cr)	−1250.206 (b) ±2.410	−1461.281 (a) ±3.609	279.000 ±9.000	
UClI$_3$(cr)	−615.789 (a) ±11.350	−643.800 ±10.000	242.000 ±18.000	
UCl$_2$I$_2$(cr)	−723.356 (a) ±11.350	−768.800 ±10.000	237.000 ±18.000	
UCl$_3$I(cr)	−829.877 (a) ±8.766	−898.300 ±8.400	213.400 ±8.400	
UBrI$_3$(cr)		−589.600 ±10.000		
UBr$_2$I$_2$(cr)		−660.400 ±10.000		
UBr$_3$I(cr)		−727.600 ±8.400		
US(cr)	−320.929 (a) ±12.600	−322.200 ±12.600	77.990 ±0.210	50.540 (c) ±0.080
US$_{1.90}$(cr)	−509.470 (a) ±20.900	−509.900 ±20.900	109.660 ±0.210	73.970 (c) ±0.130
US$_2$(cr)	−519.241 (a) ±8.001	−520.400 ±8.000	110.420 ±0.210	74.640 (c) ±0.130
US$_3$(cr)	−537.253 (a) ±12.600	−539.600 ±12.600	138.490 ±0.210	95.600 (c) ±0.250

(Continued on next page)

Table 3–1: (continued)

Compound	$\Delta_f G_m^\circ$ (kJ · mol^{-1})	$\Delta_f H_m^\circ$ (kJ · mol^{-1})	S_m° (J · K^{-1} · mol^{-1})	$C_{p,m}^\circ$ (J · K^{-1} · mol^{-1})
$U_2S_3(cr)$	−879.787 (a) ±67.002	−879.000 ±67.000	199.200 ±1.700	133.700 (c) ±0.800
$U_2S_5(cr)$			243.000 ±25.000	
$U_3S_5(cr)$	−1425.076 (a) ±100.278	−1431.000 ±100.000	291.000 ±25.000	
USO_4^{2+}	−1311.423 (b) ±2.113	−1492.540 (b) ±4.283	−245.591 (b) ±15.906	
$UO_2SO_3(aq)$	−1477.696 (b) ±5.563			
$UO_2SO_3(cr)$	−1530.370 (a) ±12.727	−1661.000 ±12.600	157.000 ±6.000	
$UO_2S_2O_3(aq)$	−1487.824 (b) ±11.606			
$UO_2SO_4(aq)$	−1714.535 (b) ±1.800	−1908.840 (b) ±2.229	46.010 (b) ±6.173	
$UO_2SO_4(cr)$	−1685.775 (a) ±2.642	−1845.140 ±0.840	163.200 ±8.400	145.000 (c) ±3.000
$U(SO_3)_2(cr)$	−1712.826 (a) ±21.171	−1883.000 ±21.000	159.000 ±9.000	
$U(SO_4)_2(aq)$	−2077.860 (b) ±2.262	−2377.180 (b) ±4.401	−69.007 (b) ±16.158	
$U(SO_4)_2(cr)$	−2084.521 (a) ±14.070	−2309.600 ±12.600	180.000 ±21.000	
$UO_2(SO_4)_2^{2-}$	−2464.190 (b) ±1.978	−2802.580 (b) ±1.972	135.786 (b) ±4.763	
$UO_2(SO_4)_3^{4-}$	−3201.801 (b) ±3.054			
$U(OH)_2SO_4(cr)$	−1766.223 (b) ±3.385			
$UO_2SO_4 \cdot 2.5\ H_2O(cr)$	−2298.475 (b) ±1.803	−2607.000 ±0.900	246.053 (a) ±6.762	
$UO_2SO_4 \cdot 3\ H_2O(cr)$	−2416.561 (b) ±1.811	−2751.500 ±4.600	274.087 (a) ±16.582	
$UO_2SO_4 \cdot 3.5\ H_2O(cr)$	−2535.595 (b) ±1.806	−2901.600 ±0.800	286.520 (a) ±6.628	

(Continued on next page)

Table 3–1: (continued)

Compound	$\Delta_f G_m^\circ$ (kJ·mol^{-1})	$\Delta_f H_m^\circ$ (kJ·mol^{-1})	S_m° (J·K^{-1}·mol^{-1})	$C_{p,m}^\circ$ (J·K^{-1}·mol^{-1})
U(SO$_4$)$_2$·4 H$_2$O(cr)	−3033.310 [a] ±11.433	−3483.200 ±6.300	359.000 ±32.000	
U(SO$_4$)$_2$·8 H$_2$O(cr)	−3987.898 [a] ±16.735	−4662.600 ±6.300	538.000 ±52.000	
USe(cr)	−276.908 [a] ±14.600	−275.700 ±14.600	96.520 ±0.210	54.810 [c] ±0.170
α–USe$_2$	−427.072 [a] ±42.000	−427.000 ±42.000	134.980 ±0.250	79.160 [c] ±0.170
β–USe$_2$	−427.972 [a] ±42.179	−427.000 ±42.000	138.000 ±13.000	
USe$_3$(cr)	−451.997 [a] ±42.305	−452.000 ±42.000	177.000 ±17.000	
U$_2$Se$_3$(cr)	−721.194 [a] ±75.002	−711.000 ±75.000	261.400 ±1.700	
U$_3$Se$_4$(cr)	−988.760 [a] ±85.752	−983.000 ±85.000	339.000 ±38.000	
U$_3$Se$_5$(cr)	−1130.611 [a] ±113.567	−1130.000 ±113.000	364.000 ±38.000	
UO$_2$SeO$_3$(cr)		−1522.000 ±2.300		
UO$_2$SeO$_4$(cr)		−1539.300 ±3.300		
UOTe(cr)			111.700 ±1.300	80.600 ±0.800
UTeO$_5$(cr)		−1603.100 ±2.800		
UTe$_3$O$_9$(cr)		−2275.800 ±8.000		
UN(cr)	−265.082 [a] ±3.001	−290.000 ±3.000	62.430 ±0.220	47.570 [c] ±0.400
β–UN$_{1.466}$		−362.200 ±2.300		
α–UN$_{1.59}$	−338.202 [a] ±5.201	−379.200 ±5.200	65.020 ±0.300	
α–UN$_{1.606}$		−381.400 ±5.000		

(Continued on next page)

Table 3–1: (continued)

Compound	$\Delta_f G_m^\circ$ (kJ·mol^{-1})	$\Delta_f H_m^\circ$ (kJ·mol^{-1})	S_m° (J·K^{-1}·mol^{-1})	$C_{p,m}^\circ$ (J·K^{-1}·mol^{-1})
α–UN$_{1.674}$		−390.800 ±5.000		
α–UN$_{1.73}$	−353.753 (a) ±7.501	−398.500 ±7.500	65.860 ±0.300	
UO$_2$N$_3^+$	−619.077 (b) ±2.705			
UO$_2$(N$_3$)$_2$(aq)	−280.867 (b) ±4.558			
UO$_2$(N$_3$)$_3^-$	59.285 (b) ±6.374			
UO$_2$(N$_3$)$_4^{2-}$	412.166 (b) ±8.302			
UNO$_3^{3+}$	−649.045 (b) ±1.960			
UO$_2$NO$_3^+$	−1065.057 (b) ±1.990			
U(NO$_3$)$_2^{2+}$	−764.576 (b) ±2.793			
UO$_2$(NO$_3$)$_2$(cr)	−1106.094 (a) ±5.675	−1351.000 ±5.000	241.000 ±9.000	
UO$_2$(NO$_3$)$_2$·H$_2$O(cr)	−1362.965 (a) ±10.524	−1664.000 ±10.000	286.000 ±11.000	
UO$_2$(NO$_3$)$_2$·2H$_2$O(cr)	−1620.500 ±2.000	−1978.700 ±1.700	327.520 (a) ±8.806	278.000 ±4.000
UO$_2$(NO$_3$)$_2$·3H$_2$O(cr)	−1864.694 (a) ±1.965	−2280.400 ±1.700	367.900 ±3.300	320.100 ±1.700
UO$_2$(NO$_3$)$_2$·6H$_2$O(cr)	−2584.213 (a) ±1.615	−3167.500 ±1.500	505.600 ±2.000	468.000 ±2.500
UP(cr)	−265.921 (a) ±11.101	−269.800 ±11.100	78.280 ±0.420	50.290 ±0.500
UP$_2$(cr)	−294.555 (a) ±15.031	−304.000 ±15.000	100.700 ±3.200	78.800 ±3.500
U$_3$P$_4$(cr)	−826.435 (a) ±26.014	−843.000 ±26.000	259.400 ±2.600	176.900 ±3.600
UPO$_5$(cr)	−1924.713 (a) ±4.990	−2064.000 ±4.000	137.000 ±10.000	124.000 (c) ±12.000

(Continued on next page)

Table 3–1: (continued)

Compound	$\Delta_f G_m^\circ$ (kJ·mol^{-1})	$\Delta_f H_m^\circ$ (kJ·mol^{-1})	S_m° (J·K^{-1}·mol^{-1})	$C_{p,m}^\circ$ (J·K^{-1}·mol^{-1})
$UO_2PO_4^-$	−2053.559 [b] ±2.504			
$UP_2O_7(cr)$	−2659.272 [a] ±5.369	−2852.000 ±4.000	204.000 ±12.000	184.000 ±18.000
$(UO_2)_2P_2O_7(cr)$	−3930.002 [a] ±6.861	−4232.600 ±2.800	296.000 ±21.000	258.000 [c] ±26.000
$(UO_2)_3(PO_4)_2(cr)$	−5115.975 [a] ±5.452	−5491.300 ±3.500	410.000 ±14.000	339.000 [c] ±17.000
$UO_2HPO_4(aq)$	−2089.862 [b] ±2.777			
$UO_2H_2PO_4^+$	−2108.311 [b] ±2.378			
$UO_2H_3PO_4^{2+}$	−2106.256 [b] ±2.504			
$UO_2(H_2PO_4)_2(aq)$	−3254.938 [b] ±3.659			
$UO_2(H_2PO_4)(H_3PO_4)^+$	−3260.703 [b] ±3.659			
$UO_2HPO_4 \cdot 4H_2O(cr)$	−3064.749 [b] ±2.414	−3469.968 [a] ±7.836	346.000 ±25.000	
$U(HPO_4)_2 \cdot 4H_2O(cr)$	−3844.453 [b] ±3.717	−4334.818 [a] ±8.598	372.000 ±26.000	460.000 ±50.000
$(UO_2)_3(PO_4)_2 \cdot 4H_2O(cr)$	−6138.968 [b] ±6.355	−6739.105 [a] ±9.136	589.000 ±22.000	
$(UO_2)_3(PO_4)_2 \cdot 6H_2O(cr)$	−6613.025 [a] ±13.392	−7328.400 ±10.700	669.000 ±27.000	
$UAs(cr)$	−237.908 [a] ±8.024	−234.300 ±8.000	97.400 ±2.000	57.900 ±1.200
$UAs_2(cr)$	−252.790 [a] ±13.005	−252.000 ±13.000	123.050 ±0.200	79.960 ±0.100
$U_3As_4(cr)$	−725.388 [a] ±18.016	−720.000 ±18.000	309.070 ±0.600	187.530 ±0.200
$UAsO_5(cr)$				138.000 [c] ±14.000
$UO_2(AsO_3)_2(cr)$	−1944.911 [a] ±12.006	−2156.600 ±8.000	231.000 ±30.000	201.000 [c] ±20.000

(Continued on next page)

Table 3–1: (continued)

Compound	$\Delta_f G_m^\circ$ (kJ·mol^{-1})	$\Delta_f H_m^\circ$ (kJ·mol^{-1})	S_m° (J·K^{-1}·mol^{-1})	$C_{p,m}^\circ$ (J·K^{-1}·mol^{-1})
(UO$_2$)$_2$As$_2$O$_7$(cr)	−3130.254 [a] ±12.006	−3426.000 ±8.000	307.000 ±30.000	273.000 [c] ±27.000
(UO$_2$)$_3$(AsO$_4$)$_2$(cr)	−4310.789 [a] ±12.007	−4689.400 ±8.000	387.000 ±30.000	364.000 [c] ±36.000
UO$_2$HAsO$_4$(aq)	−1707.994 [b] ±4.717			
UO$_2$H$_2$AsO$_4^+$	−1726.260 [b] ±4.582			
UO$_2$(H$_2$AsO$_4$)$_2$(aq)	−2486.326 [b] ±8.283			
UAsS(cr)			114.500 ±3.400	80.900 ±2.400
UAsSe(cr)			131.900 ±4.000	83.500 ±2.500
UAsTe(cr)			139.900 ±4.200	81.300 ±2.400
USb(cr)	−140.969 [a] ±7.711	−138.500 ±7.500	104.000 ±6.000	75.000 ±7.000
USb$_2$(cr)	−173.666 [a] ±10.901	−173.600 ±10.900	141.460 ±0.130	80.200 ±0.100
U$_4$Sb$_3$(cr)			379.000 ±20.000	
U$_3$Sb$_4$(cr)	−457.004 [a] ±22.602	−451.900 ±22.600	349.800 ±0.400	188.200 ±0.200
UC(cr)	−98.900 ±3.000	−97.900 ±4.000	59.294 [a] ±16.772	50.100 [c] ±1.000
α−UC$_{1.94}$	−87.400 ±2.100	−85.324 [a] ±2.185	68.300 ±2.000	60.800 [c] ±1.500
U$_2$C$_3$(cr)	−189.317 [a] ±10.002	−183.300 ±10.000	137.800 ±0.300	107.400 [c] ±2.000
UO$_2$CO$_3$(aq)	−1537.188 [b] ±1.799	−1689.230 [b] ±2.512	58.870 [b] ±7.438	
UO$_2$CO$_3$(cr)	−1564.701 [b] ±1.794	−1691.302 [a] ±1.798	144.200 ±0.300	120.100 ±0.100
UO$_2$(CO$_3$)$_2^{2-}$	−2103.161 [b] ±1.982	−2350.960 [b] ±4.301	181.846 [b] ±13.999	

(Continued on next page)

Table 3–1: (continued)

Compound	$\Delta_f G_m^\circ$ (kJ·mol^{-1})	$\Delta_f H_m^\circ$ (kJ·mol^{-1})	S_m° (J·K^{-1}·mol^{-1})	$C_{p,m}^\circ$ (J·K^{-1}·mol^{-1})
$UO_2(CO_3)_3^{4-}$	−2660.914 (b) ±2.116	−3083.890 (b) ±4.430	38.446 (b) ±14.411	
$UO_2(CO_3)_3^{5-}$	−2584.392 (b) ±2.943			
$U(CO_3)_4^{4-}$	−2841.926 (b) ±5.958			
$U(CO_3)_5^{6-}$	−3363.432 (b) ±5.772	−3987.350 (b) ±5.334	−83.051 (b) ±25.680	
$(UO_2)_3(CO_3)_6^{6-}$	−6333.285 (b) ±8.096	−7171.080 (b) ±5.316	228.926 (b) ±23.416	
$(UO_2)_2CO_3(OH)_3^-$	−3139.526 (b) ±4.517			
$(UO_2)_3O(OH)_2(HCO_3)^+$	−4100.695 (b) ±5.973			
$(UO_2)_{11}(CO_3)_6(OH)_{12}^{2-}$	−16698.986 (b) ±22.383			
$UO_2CO_3F^-$	−1840.459 (b) ±1.987			
$UO_2CO_3F_2^{2-}$	−2132.371 (b) ±2.400			
$UO_2CO_3F_3^{3-}$	−2418.518 (b) ±2.813			
$USCN^{3+}$	−454.113 (b) ±4.385	−541.800 (b) ±9.534	−306.326 (b) ±35.198	
$U(SCN)_2^{2+}$	−368.776 (b) ±8.257	−456.400 (b) ±9.534	−107.174 (b) ±42.302	
UO_2SCN^+	−867.842 (b) ±4.558	−939.380 (b) ±4.272	83.671 (b) ±19.708	
$UO_2(SCN)_2(aq)$	−774.229 (b) ±8.770	−857.300 (b) ±8.161	243.927 (b) ±39.546	
$UO_2(SCN)_3^-$	−686.438 (b) ±12.458	−783.800 (b) ±12.153	394.934 (b) ±57.938	
$USiO_4(cr)$	−1883.600 ±4.000	−1991.326 (a) ±5.367	118.000 ±12.000	
$UO_2SiO(OH)_3^+$	−2249.783 (b) ±2.172			

(Continued on next page)

Table 3–1: (continued)

Compound	$\Delta_f G_m^\circ$ (kJ·mol^{-1})	$\Delta_f H_m^\circ$ (kJ·mol^{-1})	S_m° (J·K^{-1}·mol^{-1})	$C_{p,m}^\circ$ (J·K^{-1}·mol^{-1})
Tl$_2$U$_4$O$_{11}$(cr)				360.200 [c] ±20.000
Zn$_{0.12}$UO$_{2.95}$(cr)		−1938.070 ±1.060		
(UO$_2$)$_2$(PuO$_2$)(CO$_3$)$_6^{6-}$	−6140.119 [b] ±9.582			
(UO$_2$)$_2$(NpO$_2$)(CO$_3$)$_6^{6-}$	−6176.973 [b] ±16.977			
Be$_{13}$U(cr)	−165.508 [a] ±17.530	−163.600 ±17.500	180.100 ±3.300	242.300 ±4.200
α–Mg$_{0.17}$UO$_{2.95}$		−1291.440 ±0.990		
MgUO$_4$(cr)	−1749.601 [a] ±1.502	−1857.300 ±1.500	131.950 ±0.170	128.100 ±0.300
MgU$_3$O$_{10}$(cr)			338.600 ±1.000	305.600 ±5.000
β–CaUO$_4$				124.700 ±2.700
CaUO$_4$(cr)	−1888.706 [a] ±2.421	−2002.300 ±2.300	121.100 ±2.500	123.800 ±2.500
Ca$_3$UO$_6$(cr)		−3305.400 ±4.100		
CaU$_6$O$_{19}$·11 H$_2$O(cr)	−10305.460 [b] ±13.964			
SrUO$_3$(cr)		−1672.600 ±8.600		
α–SrUO$_4$	−1881.355 [a] ±2.802	−1989.600 ±2.800	153.150 ±0.170	130.620 ±0.200
β–SrUO$_4$		−1988.400 ±5.400		
Sr$_2$UO$_{4.5}$(cr)		−2494.000 ±2.800		
Sr$_2$UO$_5$(cr)		−2632.900 ±1.900		
Sr$_3$UO$_6$(cr)		−3263.500 ±3.000		
Sr$_3$U$_2$O$_9$(cr)		−4620.000 ±8.000		301.800 [c] ±3.000

(Continued on next page)

Table 3–1: (continued)

Compound	$\Delta_f G_m^\circ$ (kJ · mol^{-1})	$\Delta_f H_m^\circ$ (kJ · mol^{-1})	S_m° (J · K^{-1} · mol^{-1})	$C_{p,m}^\circ$ (J · K^{-1} · mol^{-1})
Sr$_2$U$_3$O$_{11}$(cr)		−5243.700 ±5.000		
SrU$_4$O$_{13}$(cr)		−5920.000 ±20.000		
Sr$_5$U$_3$O$_{14}$(cr)		−7248.600 ±7.500		
Sr$_3$U$_{11}$O$_{36}$(cr)		−15903.800 ±16.500		1064.200 [c] ±10.600
BaUO$_3$(cr)		−1690.000 ±10.000		
BaUO$_4$(cr)	−1883.805 [a] ±3.393	−1993.800 ±3.300	154.000 ±2.500	125.300 ±2.500
Ba$_3$UO$_6$(cr)	−3044.951 [a] ±9.196	−3210.400 ±8.000	298.000 ±15.000	
BaU$_2$O$_7$(cr)	−3052.093 [a] ±6.714	−3237.200 ±5.000	260.000 ±15.000	
Ba$_2$U$_2$O$_7$(cr)	−3547.015 [a] ±7.743	−3740.000 ±6.300	296.000 ±15.000	
Ba$_2$MgUO$_6$(cr)		−3245.900 ±6.500		
Ba$_2$CaUO$_6$(cr)		−3295.800 ±5.900		
Ba$_2$SrUO$_6$(cr)		−3257.300 ±5.700		
Li$_{0.12}$UO$_{2.95}$(cr)		−1254.970 [b] ±1.225		
γ–Li$_{0.55}$UO$_3$		−1394.535 [b] ±2.779		
δ–Li$_{0.69}$UO$_3$		−1434.525 [b] ±2.367		
LiUO$_3$(cr)		−1522.300 ±1.800		
Li$_2$UO$_4$(cr)	−1853.190 [a] ±2.215	−1968.200 ±1.300	133.000 ±6.000	
Li$_4$UO$_5$(cr)		−2639.400 ±1.700		

(Continued on next page)

Table 3–1: (continued)

Compound	$\Delta_f G_m^\circ$ (kJ·mol^{-1})	$\Delta_f H_m^\circ$ (kJ·mol^{-1})	S_m° (J·K^{-1}·mol^{-1})	$C_{p,m}^\circ$ (J·K^{-1}·mol^{-1})
Li$_2$U$_2$O$_7$(cr)		−3213.600 ±5.300		
Li$_{0.19}$U$_3$O$_8$(cr)		−3632.955 [b] ±3.113		
Li$_{0.88}$U$_3$O$_8$(cr)		−3837.090 [b] ±3.748		
Li$_2$U$_3$O$_{10}$(cr)		−4437.400 ±4.100		
Na$_{0.12}$UO$_{2.95}$(cr)		−1247.372 [b] ±1.877		
α–Na$_{0.14}$UO$_3$		−1267.234 [b] ±1.804		
δ–Na$_{0.54}$UO$_3$		−1376.954 [b] ±5.482		
NaUO$_3$(cr)	−1412.495 [a] ±10.001	−1494.900 ±10.000	132.840 ±0.400	108.870 [c] ±0.400
α–Na$_2$UO$_4$	−1779.303 [a] ±3.506	−1897.700 ±3.500	166.000 ±0.500	146.700 [c] ±0.500
β–Na$_2$UO$_4$		−1884.600 ±3.600		
Na$_3$UO$_4$(cr)	−1899.909 [a] ±8.003	−2024.000 ±8.000	198.200 ±0.400	173.000 [c] ±0.400
Na$_4$UO$_5$(cr)		−2457.000 ±2.200		
Na$_2$U$_2$O$_7$(cr)	−3011.454 [a] ±4.015	−3203.800 ±4.000	275.900 ±1.000	227.300 [c] ±1.000
Na$_{0.20}$U$_3$O$_8$(cr)		−3626.450 [b] ±3.252		
Na$_6$U$_7$O$_{24}$(cr)		−10841.700 ±10.000		
Na$_4$UO$_2$(CO$_3$)$_3$(cr)	−3737.836 [b] ±2.342			
KUO$_3$(cr)		−1522.900 ±1.700		137.000 [c] ±10.000
K$_2$UO$_4$(cr)	−1798.499 [a] ±3.248	−1920.700 ±2.200	180.000 ±8.000	

(Continued on next page)

Table 3–1: (continued)

Compound	$\Delta_f G_m^\circ$ (kJ · mol^{-1})	$\Delta_f H_m^\circ$ (kJ · mol^{-1})	S_m° (J · K^{-1} · mol^{-1})	$C_{p,m}^\circ$ (J · K^{-1} · mol^{-1})
K$_2$U$_2$O$_7$(cr)		−3250.500 ±4.500		229.000 (c) ±20.000
K$_2$U$_4$O$_{13}$(cr)				425.000 (c) ±50.000
K$_2$U$_6$O$_{19}$· 11 H$_2$O(cr)	−10337.081 (b) ±10.956			
RbUO$_3$(cr)		−1520.900 ±1.800		
Rb$_2$UO$_4$(cr)	−1800.141 (a) ±3.250	−1922.700 ±2.200	203.000 ±8.000	
Rb$_2$U$_2$O$_7$(cr)		−3232.000 ±4.300		
Rb$_2$U$_4$O$_{11}$(cr)				365.600 (c) ±25.000
Rb$_2$U$_4$O$_{13}$(cr)				420.000 (c) ±50.000
Rb$_2$U(SO$_4$)$_3$(cr)				348.000 (c) ±30.000
Cs$_2$UO$_4$(cr)	−1805.370 (a) ±1.232	−1928.000 ±1.200	219.660 ±0.440	152.760 (c) ±0.310
Cs$_2$U$_2$O$_7$(cr)	−3022.881 (a) ±10.005	−3220.000 ±10.000	327.750 ±0.660	231.200 (c) ±0.500
Cs$_2$U$_4$O$_{12}$(cr)	−5251.058 (a) ±3.628	−5571.800 ±3.600	526.400 ±1.000	384.000 (c) ±1.000
Cs$_4$U$_5$O$_{17}$(cr)				750.000 (c) ±50.000

(a) Value calculated internally with the Gibbs–Helmholtz equation, $\Delta_f G_m^\circ = \Delta_f H_m^\circ - T \sum_i S_{m,i}^\circ$.
(b) Value calculated internally from reaction data (see Table 3–2).
(c) Temperature coefficients of this function are listed in Table 3–3.
(d) Stable phase of UO$_{2.25}$ ($\equiv$ U$_4$O$_9$) above 348 K. The thermodynamic parameters, however, refer to 298.15 K.
(e) Stable phase of UO$_{2.25}$ ($\equiv$ U$_4$O$_9$) below 348 K.

Table 3–2: Selected thermodynamic data for reactions involving uranium compounds and complexes. All ionic species listed in this table are aqueous species. Unless noted otherwise, all data refer to the reference temperature of 298.15 K and to the standard state, *i.e.*, a pressure of 0.1 MPa and, for aqueous species, infinite dilution ($I = 0$). The uncertainties listed below each value represent total uncertainties and correspond in principle to the statistically defined 95% confidence interval. Values obtained from internal calculation, *cf.* footnote (a), are rounded at the third digit after the decimal point and may therefore not be exactly identical to those given in Part III. Systematically, all the values are presented with three digits after the decimal point, regardless of the significance of these digits. The data presented in this table are available on computer media from the OECD Nuclear Energy Agency.

Species	Reaction	$\log_{10} K°$	$\Delta_r G_m°$ (kJ·mol^{-1})	$\Delta_r H_m°$ (kJ·mol^{-1})	$\Delta_r S_m°$ (J·K^{-1}·mol^{-1})
U^{3+}	$U^{4+} + e^- \rightleftharpoons U^{3+}$	−9.353 (b) ±0.070	53.387 ±0.400	102.100 ±1.700	163.383 (a) ±5.857
U^{4+}	$4H^+ + UO_2^{2+} + 2e^- \rightleftharpoons 2H_2O(l) + U^{4+}$	9.038 ±0.041	−51.589 ±0.234		
UO_2(am, hyd)	$U^{4+} + 4OH^- \rightleftharpoons 2H_2O(l) + UO_2$(am, hyd)	54.500 ±1.000	−311.088 ±5.708		
UO_2^+	$UO_2^{2+} + e^- \rightleftharpoons UO_2^+$	1.484 (b) ±0.022	−8.471 ±0.126		
α–UO$_{2.95}$	$0.15 UO_{2.6667}(cr) + 0.85\, \gamma$–UO$_3 \rightleftharpoons \alpha$–UO$_{2.95}$			7.690 ±0.770	
UOH^{3+}	$H_2O(l) + U^{4+} \rightleftharpoons H^+ + UOH^{3+}$	−0.540 ±0.060	3.082 ±0.342	46.910 (a) ±8.951	147.000 ±30.000
UO$_2$OH$^+$	$H_2O(l) + UO_2^{2+} \rightleftharpoons H^+ + UO_2OH^+$	−5.250 ±0.240	29.967 ±1.370		
δ–UO$_3$H$_{0.83}$	$0.415 H_2(g) + \delta$–UO$_3 \rightleftharpoons \delta$–UO$_3H_{0.83}$			−71.410 ±2.480	

(Continued on next page)

Table 3–2: (continued)

Species	Reaction	$\log_{10} K°$	$\Delta_r G_m^\circ$ (kJ·mol^{-1})	$\Delta_r H_m^\circ$ (kJ·mol^{-1})	$\Delta_r S_m^\circ$ (J·K^{-1}·mol^{-1})
$UO_2(OH)_2(aq)$	$2H_2O(l) + UO_2^{2+} \rightleftharpoons 2H^+ + UO_2(OH)_2(aq)$	-12.150 ± 0.070	69.353 ± 0.400		
$U(OH)_4(aq)$	$4OH^- + U^{4+} \rightleftharpoons U(OH)_4(aq)$	46.000 ± 1.400	-262.570 ± 7.991		
$UO_2(OH)_3^-$	$3H_2O(l) + UO_2^{2+} \rightleftharpoons 3H^+ + UO_2(OH)_3^-$	-20.250 ± 0.420	115.588 ± 2.397		
$UO_2(OH)_4^{2-}$	$4H_2O(l) + UO_2^{2+} \rightleftharpoons 4H^+ + UO_2(OH)_4^{2-}$	-32.400 ± 0.680	184.941 ± 3.881		
$(UO_2)_2OH^{3+}$	$H_2O(l) + 2UO_2^{2+} \rightleftharpoons (UO_2)_2OH^{3+} + H^+$	-2.700 ± 1.000	15.412 ± 5.708		
$(UO_2)_2(OH)_2^{2+}$	$2H_2O(l) + 2UO_2^{2+} \rightleftharpoons (UO_2)_2(OH)_2^{2+} + 2H^+$	-5.620 ± 0.040	32.079 ± 0.228		
$(UO_2)_3(OH)_4^{2+}$	$4H_2O(l) + 3UO_2^{2+} \rightleftharpoons (UO_2)_3(OH)_4^{2+} + 4H^+$	-11.900 ± 0.300	67.926 ± 1.712		
$(UO_2)_3(OH)_5^+$	$5H_2O(l) + 3UO_2^{2+} \rightleftharpoons (UO_2)_3(OH)_5^+ + 5H^+$	-15.550 ± 0.120	88.760 ± 0.685		
$(UO_2)_3(OH)_7^-$	$7H_2O(l) + 3UO_2^{2+} \rightleftharpoons (UO_2)_3(OH)_7^- + 7H^+$	-32.200 ± 0.800	183.799 ± 4.566		
$(UO_2)_4(OH)_7^+$	$7H_2O(l) + 4UO_2^{2+} \rightleftharpoons (UO_2)_4(OH)_7^+ + 7H^+$	-21.900 ± 1.000	125.010 ± 5.708		
UF^{3+}	$F^- + U^{4+} \rightleftharpoons UF^{3+}$	9.420 ± 0.510	-53.770 ± 2.911	-5.600 ± 0.500	161.562 [a] ± 9.907

(Continued on next page)

Table 3–2: (continued)

Species	Reaction	$\log_{10} K^\circ$	$\Delta_r G_m^\circ$ (kJ·mol^{-1})	$\Delta_r H_m^\circ$ (kJ·mol^{-1})	$\Delta_r S_m^\circ$ (J·K^{-1}·mol^{-1})
UF_2^{2+}	$2F^- + U^{4+} \rightleftharpoons UF_2^{2+}$	16.560 ±0.710	−94.525 ±4.053	−3.500 ±0.600	305.300 [a] ±13.741
UF_3^+	$3F^- + U^{4+} \rightleftharpoons UF_3^+$	21.890 ±0.830	−124.949 ±4.738	0.500 ±4.000	420.758 [a] ±20.796
$UF_4(aq)$	$4F^- + U^{4+} \rightleftharpoons UF_4(aq)$	26.340 ±0.960	−150.350 ±5.480	−8.430 [a] ±7.464	476.000 ±17.000
$UF_4(g)$	$UF_4(cr) \rightleftharpoons UF_4(g)$			309.000 ±5.000	
UF_5^-	$5F^- + U^{4+} \rightleftharpoons UF_5^-$	27.730 ±0.740	−158.284 ±4.224		
$UF_6(g)$	$UF_6(cr) \rightleftharpoons UF_6(g)$			49.100 ±0.500	
UF_6^{2-}	$6F^- + U^{4+} \rightleftharpoons UF_6^{2-}$	29.800 ±0.700	−170.100 ±3.996		
$U_2F_{10}(g)$	$2UF_5(g) \rightleftharpoons U_2F_{10}(g)$			−167.200 ±4.700	
UO_2F^+	$F^- + UO_2^{2+} \rightleftharpoons UO_2F^+$	5.160 ±0.060	−29.453 ±0.342	1.700 ±0.080	104.489 [a] ±1.180
$UO_2F_2(aq)$	$2F^- + UO_2^{2+} \rightleftharpoons UO_2F_2(aq)$	8.830 ±0.080	−50.402 ±0.457	2.100 ±0.190	176.093 [a] ±1.659
$UO_2F_2(g)$	$UO_2F_2(cr) \rightleftharpoons UO_2F_2(g)$			301.000 ±10.000	

(Continued on next page)

Table 3–2: (continued)

Species	Reaction			
	$\log_{10} K°$	$\Delta_r G_m°$ (kJ·mol^{-1})	$\Delta_r H_m°$ (kJ·mol^{-1})	$\Delta_r S_m°$ (J·K^{-1}·mol^{-1})
$UO_2F_3^-$	$3F^- + UO_2^{2+} \rightleftharpoons UO_2F_3^-$			
	10.900 ±0.100	−62.218 ±0.571	2.350 ±0.310	216.561 (a) ±2.179
$UO_2F_4^{2-}$	$4F^- + UO_2^{2+} \rightleftharpoons UO_2F_4^{2-}$			
	11.840 ±0.110	−67.583 ±0.628	0.290 ±0.470	227.648 (a) ±2.631
$U_2O_3F_6(cr)$	$3UOF_4(cr) \rightleftharpoons U_2O_3F_6(cr) + UF_6(g)$			
			46.000 ±6.000	
$U_3O_5F_8(cr)$	$0.5UF_6(g) + 2.5UO_2F_2(cr) \rightleftharpoons U_3O_5F_8(cr)$			
			15.100 (d) ±2.000	
$H_3OUF_6(cr)$	$6HF(g) + UF_4(cr) + UO_2F_2(cr) \rightleftharpoons 2H_3OUF_6(cr)$			
			−75.300 ±1.700	
$UF_4 \cdot 2.5H_2O(cr)$	$2.5H_2O(l) + UF_4(cr) \rightleftharpoons UF_4 \cdot 2.5H_2O(cr)$			
			−42.700 ±0.800	
$UO_2FOH \cdot 2H_2O(cr)$	$H_2O(g) + UO_2FOH \cdot H_2O(cr) \rightleftharpoons UO_2FOH \cdot 2H_2O(cr)$			
	1.883 (c) ±0.666	−10.750 ±3.800	−53.684 (a) ±4.200	−144.000 ±6.000
$UO_2F_2 \cdot 3H_2O(cr)$	$3H_2O(l) + UO_2F_2(cr) \rightleftharpoons UO_2F_2 \cdot 3H_2O(cr)$			
			−23.400 ±4.200	
UCl^{3+}	$Cl^- + U^{4+} \rightleftharpoons UCl^{3+}$			
	1.720 ±0.130	−9.818 ±0.742	−19.000 ±9.000	−30.797 (a) ±30.289
$UCl_4(g)$	$UCl_4(cr) \rightleftharpoons UCl_4(g)$			
			203.400 ±4.000	
$UCl_6(g)$	$UCl_6(cr) \rightleftharpoons UCl_6(g)$			
			81.000 ±4.000	

(Continued on next page)

3 Selected uranium data

Table 3–2: (continued)

Species	Reaction	$\log_{10} K°$	$\Delta_r G_m°$ (kJ·mol⁻¹)	$\Delta_r H_m°$ (kJ·mol⁻¹)	$\Delta_r S_m°$ (J·K⁻¹·mol⁻¹)
$U_2Cl_{10}(g)$	$Cl_2(g) + 2UCl_4(cr) \rightleftharpoons U_2Cl_{10}(g)$			72.300 ±8.900	
UO_2Cl^+	$Cl^- + UO_2^{2+} \rightleftharpoons UO_2Cl^+$	0.170 ±0.020	−0.970 ±0.114	8.000 ±2.000	30.087 [a] ±6.719
$UO_2Cl_2(aq)$	$2Cl^- + UO_2^{2+} \rightleftharpoons UO_2Cl_2(aq)$	−1.100 ±0.400	6.279 ±2.283	15.000 ±6.000	29.251 [a] ±21.532
$UO_2ClO_3^+$	$ClO_3^- + UO_2^{2+} \rightleftharpoons UO_2ClO_3^+$	0.500 ±0.070	−2.854 ±0.400	−3.900 ±0.300	−3.508 [a] ±1.676
UBr^{3+}	$Br^- + U^{4+} \rightleftharpoons UBr^{3+}$	1.460 ±0.200	−8.334 ±1.142		
$UBr_4(g)$	$UBr_4(cr) \rightleftharpoons UBr_4(g)$			196.500 ±4.000	
$UBr_5(g)$	$0.5Br_2(g) + UBr_4(g) \rightleftharpoons UBr_5(g)$			−57.800 ±2.700	
UO_2Br^+	$Br^- + UO_2^{2+} \rightleftharpoons UO_2Br^+$	0.220 ±0.020	−1.256 ±0.114		
$UO_2BrO_3^+$	$BrO_3^- + UO_2^{2+} \rightleftharpoons UO_2BrO_3^+$	0.630 ±0.080	−3.596 ±0.457	0.100 ±0.300	12.397 [a] ±1.833
UI^{3+}	$I^- + U^{4+} \rightleftharpoons UI^{3+}$	1.250 ±0.300	−7.135 ±1.712		
$UI_4(g)$	$UI_4(cr) \rightleftharpoons UI_4(g)$			213.300 ±5.000	

(Continued on next page)

Table 3–2: (continued)

Species	Reaction			
	$\log_{10} K°$	$\Delta_r G_m^°$ (kJ·mol^{-1})	$\Delta_r H_m^°$ (kJ·mol^{-1})	$\Delta_r S_m^°$ (J·K^{-1}·mol^{-1})
$UO_2IO_3^+$	$IO_3^- + UO_2^{2+} \rightleftharpoons UO_2IO_3^+$			
	2.000 ±0.020	−11.416 ±0.114	9.800 ±0.900	71.159 [a] ±3.043
$UO_2(IO_3)_2(aq)$	$2IO_3^- + UO_2^{2+} \rightleftharpoons UO_2(IO_3)_2(aq)$			
	3.590 ±0.150	−20.492 ±0.856		
$UO_2(IO_3)_2(cr)$	$2IO_3^- + UO_2^{2+} \rightleftharpoons UO_2(IO_3)_2(cr)$			
	7.880 ±0.100	−44.979 ±0.571		
USO_4^{2+}	$SO_4^{2-} + U^{4+} \rightleftharpoons USO_4^{2+}$			
	6.580 ±0.190	−37.559 ±1.085	8.000 ±2.700	152.805 [a] ±9.759
$UO_2SO_3(aq)$	$SO_3^{2-} + UO_2^{2+} \rightleftharpoons UO_2SO_3(aq)$			
	6.600 ±0.600	−37.673 ±3.425		
$UO_2S_2O_3(aq)$	$S_2O_3^{2-} + UO_2^{2+} \rightleftharpoons UO_2S_2O_3(aq)$			
	2.800 ±0.300	−15.983 ±1.712		
$UO_2SO_4(aq)$	$SO_4^{2-} + UO_2^{2+} \rightleftharpoons UO_2SO_4(aq)$			
	3.150 ±0.020	−17.980 ±0.114	19.500 ±1.600	125.710 [a] ±5.380
$U(SO_4)_2(aq)$	$2SO_4^{2-} + U^{4+} \rightleftharpoons U(SO_4)_2(aq)$			
	10.510 ±0.200	−59.992 ±1.142	32.700 ±2.800	310.889 [a] ±10.142
$UO_2(SO_4)_2^{2-}$	$2SO_4^{2-} + UO_2^{2+} \rightleftharpoons UO_2(SO_4)_2^{2-}$			
	4.140 ±0.070	−23.631 ±0.400	35.100 ±1.000	196.986 [a] ±3.612
$UO_2(SO_4)_3^{4-}$	$3SO_4^{2-} + UO_2^{2+} \rightleftharpoons UO_2(SO_4)_3^{4-}$			
	3.020 ±0.380	−17.238 ±2.169		
$U(OH)_2SO_4(cr)$	$2OH^- + SO_4^{2-} + U^{4+} \rightleftharpoons U(OH)_2SO_4(cr)$			
	31.170 ±0.500	−177.920 ±2.854		

(Continued on next page)

Table 3–2: (continued)

Species	Reaction			
	$\log_{10} K°$	$\Delta_r G_m°$ (kJ · mol^{-1})	$\Delta_r H_m°$ (kJ · mol^{-1})	$\Delta_r S_m°$ (J · K^{-1} · mol^{-1})
$UO_2SO_4 \cdot 2.5H_2O(cr)$	$2.5H_2O(l) + SO_4^{2-} + UO_2^{2+} \rightleftharpoons UO_2SO_4 \cdot 2.5H_2O(cr)$			
		1.589 (c) ±0.019	−9.070 ±0.110	
$UO_2SO_4 \cdot 3H_2O(cr)$	$UO_2SO_4 \cdot 3.5H_2O(cr) \rightleftharpoons 0.5H_2O(g) + UO_2SO_4 3H_2O(cr)$			
		−0.831 ±0.023	4.743 ±0.131	
$UO_2SO_4 \cdot 3.5H_2O(cr)$	$3.5H_2O(l) + SO_4^{2-} + UO_2^{2+} \rightleftharpoons UO_2SO_4 \cdot 3.5H_2O(cr)$			
		1.585 (c) ±0.019	−9.050 ±0.110	
$UO_2N_3^+$	$N_3^- + UO_2^{2+} \rightleftharpoons UO_2N_3^+$			
		2.580 ±0.090	−14.727 ±0.514	
$UO_2(N_3)_2(aq)$	$2N_3^- + UO_2^{2+} \rightleftharpoons UO_2(N_3)_2(aq)$			
		4.330 ±0.230	−24.716 ±1.313	
$UO_2(N_3)_3^-$	$3N_3^- + UO_2^{2+} \rightleftharpoons UO_2(N_3)_3^-$			
		5.740 ±0.220	−32.764 ±1.256	
$UO_2(N_3)_4^{2-}$	$4N_3^- + UO_2^{2+} \rightleftharpoons UO_2(N_3)_4^{2-}$			
		4.920 ±0.240	−28.084 ±1.370	
UNO_3^{3+}	$NO_3^- + U^{4+} \rightleftharpoons UNO_3^{3+}$			
		1.470 ±0.130	−8.391 ±0.742	
$UO_2NO_3^+$	$NO_3^- + UO_2^{2+} \rightleftharpoons UO_2NO_3^+$			
		0.300 ±0.150	−1.712 ±0.856	
$U(NO_3)_2^{2+}$	$2NO_3^- + U^{4+} \rightleftharpoons U(NO_3)_2^{2+}$			
		2.300 ±0.350	−13.128 ±1.998	
$UO_2PO_4^-$	$PO_4^{3-} + UO_2^{2+} \rightleftharpoons UO_2PO_4^-$			
		13.230 ±0.150	−75.517 ±0.856	

(Continued on next page)

Table 3–2: (continued)

Species	Reaction	$\log_{10} K°$	$\Delta_r G_m^\circ$ (kJ·mol⁻¹)	$\Delta_r H_m^\circ$ (kJ·mol⁻¹)	$\Delta_r S_m^\circ$ (J·K⁻¹·mol⁻¹)
$UO_2HPO_4(aq)$	$HPO_4^{2-} + UO_2^{2+} \rightleftharpoons UO_2HPO_4(aq)$	7.240 ±0.260	−41.326 ±1.484		
$UO_2H_2PO_4^+$	$H_3PO_4(aq) + UO_2^{2+} \rightleftharpoons H^+ + UO_2H_2PO_4^+$	1.120 ±0.060	−6.393 ±0.342		
$UO_2H_3PO_4^{2+}$	$H_3PO_4(aq) + UO_2^{2+} \rightleftharpoons UO_2H_3PO_4^{2+}$	0.760 ±0.150	−4.338 ±0.856		
$UO_2(H_2PO_4)_2(aq)$	$2H_3PO_4(aq) + UO_2^{2+} \rightleftharpoons 2H^+ + UO_2(H_2PO_4)_2(aq)$	0.640 ±0.110	−3.653 ±0.628		
$UO_2(H_2PO_4)(H_3PO_4)^+$	$2H_3PO_4(aq) + UO_2^{2+} \rightleftharpoons H^+ + UO_2(H_2PO_4)(H_3PO_4)^+$	1.650 ±0.110	−9.418 ±0.628		
$UO_2HPO_4 \cdot 4H_2O(cr)$	$4H_2O(l) + H_3PO_4(aq) + UO_2^{2+} \rightleftharpoons 2H^+ + UO_2HPO_4 \cdot 4H_2O(cr)$	2.500 ±0.090	−14.270 ±0.514		
$U(HPO_4)_2 \cdot 4H_2O(cr)$	$4H_2O(l) + 2H_3PO_4(aq) + U^{4+} \rightleftharpoons 4H^+ + U(HPO_4)_2 \cdot 4H_2O(cr)$	11.790 ±0.150	−67.298 ±0.856		
$(UO_2)_3(PO_4)_2 \cdot 4H_2O(cr)$	$4H_2O(l) + 2H_3PO_4(aq) + 3UO_2^{2+} \rightleftharpoons (UO_2)_3(PO_4)_2 \cdot 4H_2O(cr) + 6H^+$	5.960 ±0.300	−34.020 ±1.712		
$UO_2HAsO_4(aq)$	$AsO_4^{3-} + H^+ + UO_2^{2+} \rightleftharpoons UO_2HAsO_4(aq)$	18.760 ±0.310	−107.083 ±1.769		
$UO_2H_2AsO_4^+$	$AsO_4^{3-} + 2H^+ + UO_2^{2+} \rightleftharpoons UO_2H_2AsO_4^+$	21.960 ±0.240	−125.349 ±1.370		
$UO_2(H_2AsO_4)_2(aq)$	$2AsO_4^{3-} + 4H^+ + UO_2^{2+} \rightleftharpoons UO_2(H_2AsO_4)_2(aq)$	41.530 ±0.200	−237.055 ±1.142		

(Continued on next page)

Table 3–2: (continued)

Species	Reaction	$\log_{10} K°$	$\Delta_r G_m°$ (kJ·mol⁻¹)	$\Delta_r H_m°$ (kJ·mol⁻¹)	$\Delta_r S_m°$ (J·K⁻¹·mol⁻¹)
$UO_2CO_3(aq)$	$CO_3^{2-} + UO_2^{2+} \rightleftharpoons UO_2CO_3(aq)$	9.940 ±0.030	−56.738 ±0.171	5.000 ±2.000	207.070 [a] ±6.733
$UO_2CO_3(cr)$	$CO_3^{2-} + UO_2^{2+} \rightleftharpoons UO_2CO_3(cr)$	14.760 ±0.020	−84.251 ±0.114		
$UO_2(CO_3)_2^{2-}$	$2CO_3^{2-} + UO_2^{2+} \rightleftharpoons UO_2(CO_3)_2^{2-}$	16.610 ±0.090	−94.811 ±0.514	18.500 ±4.000	380.046 [a] ±13.526
$UO_2(CO_3)_3^{4-}$	$3CO_3^{2-} + UO_2^{2+} \rightleftharpoons UO_2(CO_3)_3^{4-}$	21.840 ±0.040	−124.664 ±0.228	−39.200 ±4.100	286.646 [a] ±13.773
$UO_2(CO_3)_3^{5-}$	$3CO_3^{2-} + UO_2^+ \rightleftharpoons UO_2(CO_3)_3^{5-}$	6.950 ±0.360	−39.671 ±2.055		
$U(CO_3)_4^{4-}$	$U(CO_3)_5^{6-} \rightleftharpoons CO_3^{2-} + U(CO_3)_4^{4-}$	1.120 ±0.250	−6.393 ±1.427		
$U(CO_3)_5^{6-}$	$5CO_3^{2-} + U^{4+} \rightleftharpoons U(CO_3)_5^{6-}$	34.000 ±0.900	−194.073 ±5.137	−20.000 ±4.000	583.845 [a] ±21.838
$(UO_2)_3(CO_3)_6^{6-}$	$6CO_3^{2-} + 3UO_2^{2+} \rightleftharpoons (UO_2)_3(CO_3)_6^{6-}$	54.000 ±1.000	−308.234 ±5.708	−62.700 ±2.400	823.526 [a] ±20.768
$(UO_2)_2CO_3(OH)_3^-$	$CO_2(g) + 4H_2O(l) + 2UO_2^{2+} \rightleftharpoons (UO_2)_2CO_3(OH)_3^- + 5H^+$	−19.010 ±0.500	108.510 ±2.854		
$(UO_2)_3O(OH)_2(HCO_3)^+$	$CO_2(g) + 4H_2O(l) + 3UO_2^{2+} \rightleftharpoons (UO_2)_3O(OH)_2(HCO_3)^+ + 5H^+$	−17.500 ±0.500	99.891 ±2.854		
$(UO_2)_{11}(CO_3)_6(OH)_{12}^{2-}$	$6CO_2(g) + 18H_2O(l) + 11UO_2^{2+} \rightleftharpoons (UO_2)_{11}(CO_3)_6(OH)_{12}^{2-} + 24H^+$	−72.500 ±2.000	413.833 ±11.416		

(Continued on next page)

Table 3–2: (continued)

Species	Reaction			
	$\log_{10} K°$	$\Delta_r G_m^\circ$ (kJ·mol^{-1})	$\Delta_r H_m^\circ$ (kJ·mol^{-1})	$\Delta_r S_m^\circ$ (J·K^{-1}·mol^{-1})
$UO_2CO_3F^-$	$CO_3^{2-} + F^- + UO_2^{2+} \rightleftharpoons UO_2CO_3F^-$			
	13.750 ±0.090	−78.486 ±0.514		
$UO_2CO_3F_2^{2-}$	$CO_3^{2-} + 2F^- + UO_2^{2+} \rightleftharpoons UO_2CO_3F_2^{2-}$			
	15.570 ±0.140	−88.874 ±0.799		
$UO_2CO_3F_3^{3-}$	$CO_3^{2-} + 3F^- + UO_2^{2+} \rightleftharpoons UO_2CO_3F_3^{3-}$			
	16.380 ±0.110	−93.498 ±0.628		
$USCN^{3+}$	$SCN^- + U^{4+} \rightleftharpoons USCN^{3+}$			
	2.970 ±0.060	−16.953 ±0.342	−27.000 ±8.000	−33.698 [a] ±26.857
$U(SCN)_2^{2+}$	$2SCN^- + U^{4+} \rightleftharpoons U(SCN)_2^{2+}$			
	4.260 ±0.180	−24.316 ±1.027	−18.000 ±4.000	21.185 [a] ±13.852
UO_2SCN^+	$SCN^- + UO_2^{2+} \rightleftharpoons UO_2SCN^+$			
	1.400 ±0.230	−7.991 ±1.313	3.220 ±0.060	37.603 [a] ±4.408
$UO_2(SCN)_2(aq)$	$2SCN^- + UO_2^{2+} \rightleftharpoons UO_2(SCN)_2(aq)$			
	1.240 ±0.550	−7.078 ±3.139	8.900 ±0.600	53.590 [a] ±10.720
$UO_2(SCN)_3^-$	$3SCN^- + UO_2^{2+} \rightleftharpoons UO_2(SCN)_3^-$			
	2.100 ±0.500	−11.987 ±2.854	6.000 ±1.200	60.328 [a] ±10.384
$UO_2SiO(OH)_3^+$	$Si(OH)_4(aq) + UO_2^{2+} \rightleftharpoons H^+ + UO_2SiO(OH)_3^+$			
	−1.840 ±0.100	10.503 ±0.571		
$(UO_2)_2(PuO_2)(CO_3)_6^{6-}$	$PuO_2(CO_3)_3^{4-} + 2UO_2(CO_3)_3^{4-} \rightleftharpoons (UO_2)_2(PuO_2)(CO_3)_6^{6-} + 3CO_3^{2-}$			
	−8.200 ±1.300	46.806 ±7.420		
$(UO_2)_2(NpO_2)(CO_3)_6^{6-}$	$NpO_2(CO_3)_3^{4-} + 2UO_2(CO_3)_3^{4-} \rightleftharpoons (UO_2)_2(NpO_2)(CO_3)_6^{6-} + 3CO_3^{2-}$			
	−8.998 ±2.690	51.364 ±15.355		

(Continued on next page)

Table 3–2: (continued)

Species	Reaction	$\log_{10} K°$	$\Delta_r G_m°$ (kJ·mol^{-1})	$\Delta_r H_m°$ (kJ·mol^{-1})	$\Delta_r S_m°$ (J·K^{-1}·mol^{-1})
$CaU_6O_{19}·11H_2O(cr)$	$Ca^{2+} + 18H_2O(l) + 6UO_2^{2+} \rightleftharpoons CaU_6O_{19}·11H_2O(cr) + 14H^+$		-40.500 ±1.600	231.176 ±9.133	
$Li_{0.12}UO_{2.95}(cr)$	$0.12LiUO_3(cr) + 0.15UO_{2.6667}(cr) + 0.73\ \gamma-UO_3 \rightleftharpoons Li_{0.12}UO_{2.95}(cr)$			-0.180 ±0.820	
$\gamma-Li_{0.55}UO_3$	$0.55LiUO_3(cr) + 0.45\ \gamma-UO_3 \rightleftharpoons \gamma-Li_{0.55}UO_3$			-6.560 ±2.540	
$\delta-Li_{0.69}UO_3$	$0.69LiUO_3(cr) + 0.31\ \gamma-UO_3 \rightleftharpoons \delta-Li_{0.69}UO_3$			-4.760 ±1.980	
$Li_{0.19}U_3O_8(cr)$	$0.19LiUO_3(cr) + 3UO_{2.6667}(cr) \rightleftharpoons Li_{0.19}U_3O_8(cr) + 0.19\ \gamma-UO_3$			-1.440 ±1.940	
$Li_{0.88}U_3O_8(cr)$	$0.88LiUO_3(cr) + 3UO_{2.6667}(cr) \rightleftharpoons Li_{0.88}U_3O_8(cr) + 0.88\ \gamma-UO_3$			0.390 ±2.160	
$Na_{0.12}UO_{2.95}(cr)$	$0.12NaUO_3(cr) + 0.15UO_{2.6667}(cr) + 0.73\ \gamma-UO_3 \rightleftharpoons Na_{0.12}UO_{2.95}(cr)$			4.130 ±1.140	
$\alpha-Na_{0.14}UO_3$	$0.14NaUO_3(cr) + 0.86\ \gamma-UO_3 \rightleftharpoons \alpha-Na_{0.14}UO_3$			-5.480 ±0.480	
$\delta-Na_{0.54}UO_3$	$0.54NaUO_3(cr) + 0.46\ \gamma-UO_3 \rightleftharpoons \delta-Na_{0.54}UO_3$			-6.760 ±0.770	
$Na_{0.2}U_3O_8(cr)$	$0.2NaUO_3(cr) + 3UO_{2.6667}(cr) \rightleftharpoons Na_{0.2}U_3O_8(cr) + 0.2\ \gamma-UO_3$			2.570 ±0.870	

(Continued on next page)

Table 3–2: (continued)

Species	Reaction			
	$\log_{10} K°$	$\Delta_r G_m°$ (kJ · mol^{-1})	$\Delta_r H_m°$ (kJ · mol^{-1})	$\Delta_r S_m°$ (J · K^{-1} · mol^{-1})
Na$_4$UO$_2$(CO$_3$)$_3$(cr)	$4Na^+ + UO_2(CO_3)_3^{4-} \rightleftharpoons Na_4UO_2(CO_3)_3(cr)$			
	5.340 ±0.160	−30.481 ±0.913		
K$_2$U$_6$O$_{19}$·11H$_2$O(cr)	$18H_2O(l) + 2K^+ + 6UO_2^{2+} \rightleftharpoons 14H^+ + K_2U_6O_{19} \cdot 11H_2O(cr)$			
	−37.100 ±0.540	211.768 ±3.082		

(a) Value calculated internally with the Gibbs–Helmholtz equation, $\Delta_r G_m° = \Delta_r H_m° - T \Delta_r S_m°$.

(b) Value calculated from a selected standard potential.

(c) Value of $\log_{10} K°$ calculated internally from $\Delta_r G_m°$.

(d) For the reaction $0.5UF_6(g) + 2.5\ UO_2F_2(cr) \rightleftharpoons U_3O_5F_8(cr)$, a heat capacity of $\Delta_r C_{p,m}° (298.15\ K) = -(18.7 \pm 4.0)$ J·K^{-1}·mol^{-1} is selected.

Table 3–3: Selected temperature coefficients for heat capacities marked with [(c)] in Table 3–1, according to the form $C_{p,m}(T) = a + bT + cT^2 + eT^{-2}$. The functions are valid between the temperatures T_{min} and T_{max} (in K). The notation E±nn indicates the power of 10. $C_{p,m}$ units are $J \cdot K^{-1} \cdot mol^{-1}$.

Compound	a	b	c	e	T_{min}	T_{max}
U(cr)	2.69200E+01	−2.50200E−03	2.65560E−05	−7.69860E+04	298	941
U(g)	3.25160E+01	−2.25394E−02	1.46050E−05	−3.02280E+05	298	900
UO(g)	3.00540E+01	5.28803E−02	−4.22663E−05	−2.21060E+05	298	500
UO_2(cr)	6.27740E+01	3.17400E−02	0	−7.69300E+05	250	600
UO_2(g)	5.78390E+01	9.01050E−03	−9.36370E−07	−7.83110E+04	298	1000
UO_2^{2+}	(a)				283	328
β–$UO_{2.25}$	7.90890E+01	1.36500E−02	0	−1.03800E+06	348	600
$UO_{2.25}$(cr)	1.48760E+03	−6.97370E+00	9.73600E−03	−1.78600E+07	250	348
α–$UO_{2.3333}$	6.41490E+01	4.91400E−02	0	−6.72000E+05	237	347
β–$UO_{2.3333}$	6.43380E+01	4.97900E−02	0	−6.55000E+05	232	346
$UO_{2.6667}$(cr)	8.72760E+01	2.21600E−02	0	−1.24400E+06	233	600
β–UO_3	8.61700E+01	2.49840E−02	0	−1.09150E+06	298	678
γ–UO_3	8.81030E+01	1.66400E−02	0	−1.01280E+06	298	850
UO_3(g)	6.44290E+01	3.17330E−02	−1.51980E−05	−7.14580E+05	298	1000
β–$UO_2(OH)_2$	4.18000E+01	2.00000E−01	0	3.53000E+06	298	473
$UO_3 \cdot 2H_2O$(cr)	8.42380E+01	2.94590E−01	0	0	298	400
UF(g)	3.68030E+01	3.76720E−02	−3.33230E−05	−6.33380E+05	298	500
UF_2(g)	4.30910E+01	4.23483E−02	−1.97810E−05	1.96710E+05	298	900
UF_3(cr)	1.06539E+02	7.05000E−04	0	−1.03550E+06	298	1768
UF_3(g)	8.13270E+01	−4.30000E−06	2.42700E−06	−4.76300E+05	298	1800
UF_4(cr)	1.38865E+02	−3.20680E−02	2.79880E−05	−1.40200E+06	298	1309
UF_4(g)	1.03826E+02	9.54900E−03	−1.45100E−06	−1.02132E+06	298	3000
α–UF_5	1.25159E+02	3.02080E−02	0	−1.92500E+05	298	1000
β–UF_5	1.25159E+02	3.02080E−02	0	−1.92500E+05	298	1000
UF_5(g)	1.16738E+02	3.13041E−02	−1.25380E−05	−1.27300E+06	298	1100
UF_6(cr)	5.23180E+01	3.83798E−01	0	0	298	337
UF_6(g)	1.37373E+02	3.96050E−02	−2.17880E−05	−1.58687E+06	298	700
U_2F_9(cr)	2.35978E+02	5.99149E−02	0	−2.17568E+05	298	600
U_4F_{17}(cr)	4.53546E+02	1.18491E−01	0	−2.67776E+05	298	600
UOF_4(g)	1.16407E+02	2.85419E−02	−1.38740E−05	−1.38364E+06	298	900
UO_2F_2(cr)	1.06238E+02	2.83260E−02	0	−1.02080E+06	298	2000
UO_2F_2(g)	8.85810E+01	3.46640E−02	−1.76390E−05	−9.74480E+05	298	800
UCl(g)	6.30000E+01	−4.14373E−02	2.89720E−05	−8.93400E+05	298	700
UCl_2(g)	5.78070E+01	3.46400E−04	2.38800E−06	1.58480E+05	298	1100
UCl_3(cr)	1.06967E+02	−2.08595E−02	3.63890E−05	−1.29994E+05	298	1115
UCl_3(g)	8.40180E+01	−3.47320E−03	3.61300E−06	−7.98900E+04	298	1700
UCl_4(cr)	1.16320E+02	3.10837E−02	0	−3.40402E+05	298	863
UCl_4(g)	1.10634E+02	3.23750E−03	−3.12000E−07	−7.15600E+05	298	3000
UCl_5(cr)	1.40164E+02	3.55640E−02	0	0	298	600
UCl_5(g)	1.28655E+02	1.06600E−02	−2.66100E−06	−7.10000E+05	298	1900
UCl_6(cr)	1.73427E+02	3.50619E−02	0	−7.40568E+05	298	452
UCl_6(g)	1.57768E+02	9.73000E−05	−1.10000E−08	−9.46160E+05	298	3000
UOCl(cr)	7.58140E+01	1.43510E−02	0	−8.28430E+05	298	900
$UOCl_2$(cr)	9.88120E+01	2.22100E−02	0	−9.22160E+05	298	700
$UOCl_3$(cr)	1.05270E+02	3.99150E−02	0	−2.09200E+04	298	900
UO_2Cl(cr)	9.01230E+01	2.22590E−02	0	−7.74040E+05	298	1000
UO_2Cl_2(cr)	1.15000E+02	1.82230E−02	0	−1.14180E+06	298	650
UO_2Cl_2(g)	9.59130E+01	1.96284E−02	−8.92180E−06	−7.46590E+05	298	1000
$U_2O_2Cl_5$(cr)	2.34300E+02	3.55640E−02	0	−2.26770E+06	298	700
$(UO_2)_2Cl_3$(cr)	2.25940E+02	3.55640E−02	0	−2.92880E+06	298	900

(Continued on next page)

Table 3–3 :(continued)

Compound	a	b	c	e	T_{min}	T_{max}
UBr(g)	6.34750E+01	−4.16971E−02	2.94940E−05	−8.16640E+05	298	700
UBr$_2$(g)	5.81290E+01	−1.74200E−04	2.62000E−06	2.70820E+05	298	1100
UBr$_3$(cr)	9.79710E+01	2.63600E−02	0	0	298	600
UBr$_3$(g)	8.44060E+01	−3.95240E−03	3.76800E−06	1.41230E+05	298	1700
UBr$_4$(cr)	1.19244E+02	2.97064E−02	0	0	298	792
UBr$_4$(g)	1.10817E+02	3.13740E−03	−3.02000E−07	−4.30880E+05	298	3100
UBr$_5$(cr)	1.50624E+02	3.34720E−02	0	0	298	400
UBr$_5$(g)	1.27620E+02	1.32749E−02	−4.04900E−06	−2.01280E+05	298	1400
UOBr$_2$(cr)	1.10580E+02	1.36820E−02	0	−1.47900E+06	298	900
UOBr$_3$(cr)	1.30540E+02	2.05020E−02	0	−1.38070E+06	298	1100
UO$_2$Br$_2$(cr)	1.04270E+02	3.79380E−02	0	0	298	500
UI(g)	6.35060E+01	−4.15266E−02	2.95180E−05	−7.94920E+05	298	700
UI$_2$(g)	5.81850E+01	−2.65300E−04	2.66100E−06	3.17020E+05	298	1100
UI$_3$(cr)	1.05018E+02	2.42672E−02	0	0	298	800
UI$_3$(g)	8.44400E+01	−3.99330E−03	3.78200E−06	2.11250E+05	298	1700
UI$_4$(cr)	1.45603E+02	9.95792E−03	0	−1.97486E+06	298	720
UI$_4$(g)	1.10978E+02	2.95700E−03	−2.55000E−07	−2.70780E+05	298	3000
US(cr)	5.28560E+01	6.51570E−03	0	−3.78290E+05	298	2000
US$_{1.90}$(cr)	7.80570E+01	6.63580E−03	0	−5.38900E+05	298	1900
US$_2$(cr)	7.56630E+01	8.93700E−03	0	−3.27780E+05	298	1800
US$_3$(cr)	1.00670E+02	1.12130E−02	0	−7.44750E+05	298	1100
U$_2$S$_3$(cr)	1.29450E+02	1.46860E−02	0	0	298	2000
UO$_2$SO$_4$(cr)	1.12470E+02	1.08780E−01	0	0	298	820
USe(cr)	5.41470E+01	7.96240E−03	0	−1.52180E+05	298	800
α–USe$_2$	7.96930E+01	8.64870E−03	0	−2.64830E+05	298	800
UN(cr)	5.05400E+01	1.06600E−02	0	−5.23800E+05	298	1000
UPO$_5$(cr)	1.10430E+02	8.52280E−02	0	−1.03790E+06	298	600
(UO$_2$)$_2$P$_2$O$_7$(cr)	2.50670E+02	1.54300E−01	0	−3.40030E+06	298	600
(UO$_2$)$_3$(PO$_4$)$_2$(cr)	3.26370E+02	1.96440E−01	0	−4.08400E+06	298	600
UAsO$_5$(cr)	1.30700E+02	6.42580E−02	0	−1.05420E+06	298	800
UO$_2$(AsO$_3$)$_2$(cr)	2.08460E+02	8.89100E−02	0	−3.03460E+06	298	750
(UO$_2$)$_2$As$_2$O$_7$(cr)	3.07410E+02	1.05820E−01	0	−5.85130E+06	298	850
(UO$_2$)$_3$(AsO$_4$)$_2$(cr)	3.31150E+02	2.11360E−01	0	−2.68180E+06	298	850
UC(cr)	5.55710E+01	9.35940E−03	0	−7.32310E+05	298	600
α–UC$_{1.94}$	6.57750E+01	2.03740E−02	0	−9.86520E+05	298	600
U$_2$C$_3$(cr)	1.14820E+02	3.10060E−02	0	−1.48450E+06	298	600
Tl$_2$U$_4$O$_{11}$(cr)	3.68200E+02	2.48860E−02	0	−1.37500E+06	298	673
Sr$_3$U$_2$O$_9$(cr)	3.19180E+02	1.16020E−01	0	−4.62010E+06	298	1000
Sr$_3$U$_{11}$O$_{36}$(cr)	9.62720E+02	3.55260E−01	0	−3.95400E+05	298	1000
NaUO$_3$(cr)	1.15490E+02	1.91670E−02	0	−1.09660E+06	415	931
α–Na$_2$UO$_4$	1.62540E+02	2.58860E−02	0	−2.09660E+06	618	1165
Na$_3$UO$_4$(cr)	1.88900E+02	2.51790E−02	0	−2.08010E+06	523	1212
Na$_2$U$_2$O$_7$(cr)	2.62830E+02	1.46530E−02	0	−3.54900E+06	390	540
KUO$_3$(cr)	1.33258E+02	1.25580E−02	0	0	298	714
K$_2$U$_2$O$_7$(cr)	1.49084E+02	2.69500E−01	0	0	391	683
K$_2$U$_4$O$_{13}$(cr)	4.71068E+02	−4.68900E−02	0	−2.87540E+06	411	888
Rb$_2$U$_4$O$_{11}$(cr)	3.30400E+02	1.41340E−01	0	−6.19800E+05	396	735
Rb$_2$U$_4$O$_{13}$(cr)	4.12560E+02	2.50000E−02	0	0	325	805
Rb$_2$U(SO$_4$)$_3$(cr)	3.86672E+02	6.88000E−02	0	−5.25917E+06	298	628
Cs$_2$UO$_4$(cr)	1.64880E+02	1.70230E−02	0	−1.52850E+06	298	1061
Cs$_2$U$_2$O$_7$(cr)	3.55325E+02	8.15759E−02	0	−1.31956E+07	298	852
Cs$_2$U$_4$O$_{12}$(cr)	4.23726E+02	7.19405E−02	0	−5.43750E+06	361	719
Cs$_4$U$_5$O$_{17}$(cr)	6.99211E+02	1.71990E−01	0	0	368	906

(a) The thermal function is $C_{p,m}^{o}(UO_2^{2+},T) = (350.5 - 0.8722T - \frac{5308}{T-190})$ J·K^{-1}·mol^{-1} for $283\,K \le T \le 328\,K$.

Chapter 4

Selected Data for Neptunium

4.1 General considerations

This chapter presents updated chemical thermodynamic data for neptunium species, as described in detail in Chapter 10 of this volume. The newly selected data represent revisions to those chosen in the previous NEA TDB review [2001LEM/FUG]. In this respect, it will be found that while new species appear in the Tables, some others have been removed from them. Table 4–1 contains the recommended thermodynamic data of the neptunium compounds and complexes, Table 4–2 the recommended thermodynamic data of chemical equilibrium reactions by which the neptunium compounds and complexes are formed, and Table 4–3 the temperature coefficients of the heat capacity data of Table 4–1 where available.

The species and reactions in Table 4–1, Table 4–2 and Table 4–3 appear in standard order of arrangement (cf. Figure 2.1). Table 4–2 contains information only on those reactions for which primary data selections are made in this review. These selected reaction data are used, together with data for key neptunium species (for example Np^{4+}) and auxiliary data listed in Table 8–1, to derive the corresponding formation quantities in Table 4–1. The uncertainties associated with values for the key neptunium species and for some of the auxiliary data are substantial, leading to comparatively large uncertainties in the formation quantities derived in this manner. The inclusion of a table for reaction data (Table 4–2) in this report allows the use of equilibrium constants with total uncertainties that are directly based on the experimental accuracies. This is the main reason for including both the table for reaction data (Table 4–2) and the table of $\Delta_f G_m^\circ$, $\Delta_f H_m^\circ$, S_m° and $C_{p,m}^\circ$ values (Table 4–1). In a few cases, the correlation of small uncertainties in values for ligands has been neglected in calculations of uncertainty values for species in Table 4–1 from uncertainty values in Table 4–2. However, for those species the effects are less than 2% of the stated uncertainties.

The selected thermal functions of the heat capacities, listed in Table 4–3, refer to the relation:

$$C_{p,m}(T) = a + b \cdot T + c \cdot T^2 + d \cdot T^{-1} + e \cdot T^{-2} \tag{4.1}$$

No references are given in these tables since the selected data are generally not directly attributable to a specific published source. A detailed discussion of the selection procedure is presented in [2001LEM/FUG] and in Chapter 10 of this volume.

4.2 Precautions to be observed in the use of the tables

Geochemical modelling in aquatic systems requires the careful use of the data selected in the NEA TDB reviews. The data in the tables of selected data must not be adopted without taking into account the chemical background information discussed in the corresponding sections of this book. In particular the following precautions should be observed when using data from the Tables.

- The addition of any aqueous species and its data to this internally consistent database can result in a modified data set which is no longer rigorous and can lead to erroneous results. The situation is similar, to a lesser degree, with the addition of gases and solids. It should also be noted that the data set presented in this chapter may not be "complete" for all the conceivable systems and conditions. Gaps are pointed out in both the previous NEA TDB review and the present update.

- Solubility data for crystalline phases are well defined in the initial state, but not necessarily in the final state after "equilibrium" has been attained. Hence, the solubility calculated from these phases may be very misleading, as discussed for the solubility data of $MO_2(cr)$, M = U, Np, Pu.

- The selected thermodynamic data in [92GRE/FUG], [95SIL/BID] and [2001LEM/FUG] contain thermodynamic data for amorphous phases. Most of these refer to Gibbs energy of formation deduced from solubility measurements. In the present review all such data have been removed. However, the corresponding $\log_{10}K$ values have been retained. The reasons for these changes are:

    o Thermodynamic data are only meaningful if they refer to well–defined systems; this is not the case for amorphous phases. It is well known that their solubility may change with time, due to re–crystallisation with a resulting change in water content and surface area/crystal size; the time scale for these changes can vary widely. These kinetic phenomena are different from those encountered in systems where there is a very high activation barrier for the reaction, *e.g.*, electron exchange between sulphate and sulphide.

    o The solubility of amorphous phases provides useful information for users of the database when modelling the behaviour of complex systems. Therefore, the solubility products have been given with the proviso that they are not thermodynamic quantities.

Table 4–1: Selected thermodynamic data for neptunium compounds and complexes. All ionic species listed in this table are aqueous species. Unless noted otherwise, all data refer to the reference temperature of 298.15 K and to the standard state, *i.e.*, a pressure of 0.1 MPa and, for aqueous species, infinite dilution ($I = 0$). The uncertainties listed below each value represent total uncertainties and correspond in principle to the statistically defined 95% confidence interval. Values obtained from internal calculation, *cf.* footnotes (a) and (b), are rounded at the third digit after the decimal point and may therefore not be exactly identical to those given in Part III. Systematically, all the values are presented with three digits after the decimal point, regardless of the significance of these digits. The data presented in this table are available on computer media from the OECD Nuclear Energy Agency.

Compound	$\Delta_f G_m^\circ$ (kJ·mol^{-1})	$\Delta_f H_m^\circ$ (kJ·mol^{-1})	S_m° (J·K^{-1}·mol^{-1})	$C_{p,m}^\circ$ (J·K^{-1}·mol^{-1})
Np(cr)	0.000	0.000	50.460 ±0.800	29.620 [c] ±0.800
β–Np				[d]
γ–Np				[d]
Np(g)	421.195 [a] ±3.009	465.100 [b] ±3.000	197.719 ±0.005	20.824 [c] ±0.020
Np^{3+}	−512.866 [b] ±5.669	−527.184 ±2.092	−193.584 ±20.253	
Np^{4+}	−491.774 [a] ±5.586	−556.022 ±4.185	−426.390 [b] ±12.386	
NpO$_2$(cr)	−1021.731 [a] ±2.514	−1074.000 ±2.500	80.300 ±0.400	66.200 [c] ±0.500
NpO$_2^+$	−907.765 [a] ±5.628	−978.181 ±4.629	−45.904 ±10.706	−4.000 ±25.000
NpO$_2^{2+}$	−795.939 ±5.615	−860.733 ±4.662	−92.387 [b] ±10.464	
Np$_2$O$_5$(cr)	−2031.574 [a] ±11.227	−2162.700 ±9.500	174.000 ±20.000	128.600 [c] ±5.000
NpOH^{2+}	−711.191 [b] ±5.922			
NpOH^{3+}	−732.053 [b] ±5.702			
Np(OH)$_2^{2+}$	−968.052 [b] ±5.844			
NpO$_2$OH(aq)	−1080.405 [b] ±6.902	−1199.226 [a] ±19.176	25.000 ±60.000	

(Continued on next page)

Table 4–1: (continued)

Compound	$\Delta_f G_m^\circ$ (kJ·mol⁻¹)	$\Delta_f H_m^\circ$ (kJ·mol⁻¹)	S_m° (J·K⁻¹·mol⁻¹)	$C_{p,m}^\circ$ (J·K⁻¹·mol⁻¹)
NpO_2OH^+	−1003.968 (b) ±6.062			
$NpO_2(OH)_2(cr)$	−1239.000 ±6.400	−1377.000 ±5.000	128.590 (a) ±27.252	120.000 ±20.000
$NpO_2(OH)_2^-$	−1247.336 (b) ±6.311	−1431.230 (a) ±30.476	40.000 ±100.000	
$NpO_3 \cdot H_2O(cr)$	−1238.997 (b) ±6.062			
$Np(OH)_4(aq)$	−1392.927 (b) ±8.409			
$(NpO_2)_2(OH)_2^{2+}$	−2030.369 (b) ±11.294			
$(NpO_2)_3(OH)_5^+$	−3475.795 (b) ±16.893			
$NpF(g)$	−109.560 (a) ±25.046	−80.000 ±25.000	251.000 ±5.000	33.800 (c) ±3.000
NpF^{3+}	−824.441 (b) ±5.686	−889.872 (b) ±4.684	−263.621 (b) ±14.361	
$NpF_2(g)$	−590.131 (a) ±30.149	−575.000 ±30.000	304.000 ±10.000	55.900 (c) ±5.000
NpF_2^{2+}	−1144.436 (b) ±6.005			
$NpF_3(cr)$	−1460.501 (a) ±8.325	−1529.000 ±8.300	124.900 ±2.000	94.200 (c) ±3.000
$NpF_3(g)$	−1107.801 (a) ±25.178	−1115.000 ±25.000	330.500 ±10.000	72.200 (c) ±5.000
$NpF_4(cr)$	−1783.797 (a) ±16.046	−1874.000 ±16.000	153.500 ±4.000	116.100 (c) ±4.000
$NpF_4(g)$	−1535.287 (a) ±22.202	−1561.000 (b) ±22.000	369.800 ±10.000	95.300 (c) ±5.000
$NpF_5(cr)$	−1834.430 (a) ±25.398	−1941.000 ±25.000	200.000 ±15.000	132.800 (c) ±8.000
$NpF_6(cr)$	−1841.872 (a) ±20.002	−1970.000 ±20.000	229.090 ±0.500	167.440 (c) ±0.400
$NpF_6(g)$	−1837.525 (a) ±20.002	−1921.660 (b) ±20.000	376.643 ±0.500	129.072 (c) ±1.000
$NpF_6(l)$				(c)

(Continued on next page)

4 Selected neptunium data

Table 4–1: (continued)

Compound	$\Delta_f G_m^\circ$ (kJ·mol^{-1})	$\Delta_f H_m^\circ$ (kJ·mol^{-1})	S_m° (J·K^{-1}·mol^{-1})	$C_{p,m}^\circ$ (J·K^{-1}·mol^{-1})
NpO$_2$F(aq)	−1196.138 [b] ±5.923			
NpO$_2$F$^+$	−1103.548 [b] ±5.672			
NpO$_2$F$_2$(aq)	−1402.366 [b] ±5.801			
NpO$_2$F$_2$(cr)				[c]
NpCl^{3+}	−631.553 [b] ±5.844			
NpCl$_3$(cr)	−829.811 [a] ±3.237	−896.800 ±3.000	160.400 ±4.000	101.850 [c] ±4.000
NpCl$_3$(g)	−582.357 [a] ±10.822	−589.000 [b] ±10.400	362.800 ±10.000	78.500 [c] ±5.000
NpCl$_4$(cr)	−895.562 [a] ±2.998	−984.000 ±1.800	200.000 ±8.000	122.000 [c] ±6.000
NpCl$_4$(g)	−765.050 [a] ±5.487	−787.000 [b] ±4.600	423.000 ±10.000	105.000 [c] ±5.000
NpCl$_4$(l)				[c]
NpOCl$_2$(cr)	−960.645 [a] ±8.141	−1030.000 ±8.000	143.500 ±5.000	95.000 [c] ±4.000
NpO$_2$Cl$^+$	−929.440 [b] ±5.699			
NpO$_2$ClO$_4$(aq)				−32.000 [c] ±25.000
NpBr$_3$(cr)	−705.521 [a] ±3.765	−730.200 ±2.900	196.000 ±8.000	103.800 [c] ±6.000
NpBr$_4$(cr)	−737.843 [a] ±3.495	−771.200 ±1.800	243.000 ±10.000	128.000 [c] ±4.000
NpOBr$_2$(cr)	−906.933 [a] ±11.067	−950.000 ±11.000	160.800 ±4.000	98.200 [c] ±4.000
NpI^{3+}	−552.059 [b] ±6.036			
NpI$_3$(cr)	−512.498 [a] ±3.715	−512.400 ±2.200	225.000 ±10.000	110.000 [c] ±8.000

(Continued on next page)

Table 4–1: (continued)

Compound	$\Delta_f G_m^\circ$ (kJ · mol^{-1})	$\Delta_f H_m^\circ$ (kJ · mol^{-1})	S_m° (J · K^{-1} · mol^{-1})	$C_{p,m}^\circ$ (J · K^{-1} · mol^{-1})
NpO$_2$IO$_3$(aq)	−1036.957 [b] ±5.934			
NpO$_2$IO$_3^+$	−929.126 [b] ±5.922			
NpSO$_4^{2+}$	−1274.887 [b] ±5.809	−1435.522 [b] ±9.796	−176.635 [b] ±32.277	
NpO$_2$SO$_4$(aq)	−1558.666 [b] ±5.641	−1753.373 [b] ±4.706	44.920 [b] ±10.667	
NpO$_2$SO$_4^-$	−1654.281 [b] ±5.850	−1864.321 [b] ±8.569	58.833 [b] ±26.920	
Np(SO$_4$)$_2$(aq)	−2042.873 [b] ±6.360	−2319.322 [b] ±5.871	7.964 [b] ±18.924	
NpO$_2$(SO$_4$)$_2^{2-}$	−2310.775 [b] ±5.705	−2653.413 [b] ±4.880	121.798 [b] ±11.402	
NpN(cr)	−280.443 [a] ±10.013	−305.000 ±10.000	63.900 ±1.500	48.700 [c] ±0.900
NpNO$_3^{3+}$	−613.413 [b] ±5.667			
NpO$_2$(NO$_3$)$_2$· 6 H$_2$O(s)	−2428.069 [a] ±5.565	−3008.241 ±5.022	516.306 ±8.000	
NpO$_2$HPO$_4$(aq)	−1927.314 [b] ±7.067			
NpO$_2$HPO$_4^-$	−2020.589 [b] ±5.870			
NpO$_2$H$_2$PO$_4^+$	−1952.042 [b] ±6.491			
NpO$_2$(HPO$_4$)$_2^{2-}$	−3042.135 [b] ±8.598			
NpSb(cr)			101.400 ±6.100	48.900 ±2.900
NpC$_{0.91}$(cr)	−76.024 [a] ±10.028	−71.100 ±10.000	72.200 ±2.400	50.000 [c] ±1.000
Np$_2$C$_3$(cr)	−192.427 ±19.436	−187.400 ±19.200	135.000 ±10.000	110.000 ±8.000
NpO$_2$CO$_3$(aq)	−1377.040 [b] ±6.617			

(Continued on next page)

Table 4–1: (continued)

Compound	$\Delta_f G_m^\circ$ (kJ·mol^{-1})	$\Delta_f H_m^\circ$ (kJ·mol^{-1})	S_m° (J·K^{-1}·mol^{-1})	$C_{p,m}^\circ$ (J·K^{-1}·mol^{-1})
NpO$_2$CO$_3$(s)	−1407.156 (b) ±6.233			
NpO$_2$CO$_3^-$	−1463.988 (b) ±5.652			
NpO$_2$(CO$_3$)$_2^{2-}$	−1946.015 (b) ±7.033			
NpO$_2$(CO$_3$)$_2^{3-}$	−2000.861 (b) ±5.685			
Np(CO$_3$)$_3^{3-}$	−2185.949 (b) ±15.451			
NpO$_2$(CO$_3$)$_3^{4-}$	−2490.208 (b) ±5.759	−2928.323 ±6.254	−12.070 ±17.917	
NpO$_2$(CO$_3$)$_3^{5-}$	−2522.859 (b) ±5.733	−3017.120 ±6.893	−135.050 ±20.467	
Np(CO$_3$)$_4^{4-}$	−2812.775 (b) ±8.240			
Np(CO$_3$)$_5^{6-}$	−3334.567 (b) ±8.425			
(NpO$_2$)$_3$(CO$_3$)$_6^{6-}$	−5839.709 (b) ±19.185			
NpO$_2$(CO$_3$)$_2$OH^{4-}	−2170.417 (b) ±8.785			
(NpO$_2$)$_2$CO$_3$(OH)$_3^-$	−2814.914 (b) ±14.665			
(NH$_4$)$_4$NpO$_2$(CO$_3$)$_3$(s)	−2850.284 (b) ±6.106			
Np(SCN)$^{3+}$	−416.198 (b) ±7.081	−486.622 (b) ±6.520	−248.165 (b) ±25.449	
Np(SCN)$_2^{2+}$	−329.777 (b) ±10.166	−412.222 (b) ±12.748	−89.545 (b) ±50.953	
Np(SCN)$_3^+$	−241.072 (b) ±13.541	−339.822 (b) ±15.573	54.707 (b) ±66.304	
Sr$_3$NpO$_6$(cr)		−3125.800 ±5.900		
Ba$_3$NpO$_6$(cr)		−3085.600 ±9.600		

(Continued on next page)

Table 4–1: (continued)

Compound	$\Delta_f G_m^\circ$ (kJ·mol⁻¹)	$\Delta_f H_m^\circ$ (kJ·mol⁻¹)	S_m° (J·K⁻¹·mol⁻¹)	$C_{p,m}^\circ$ (J·K⁻¹·mol⁻¹)
Ba₂MgNpO₆(cr)		−3096.900 ±8.200		
Ba₂CaNpO₆(cr)		−3159.300 ±7.900		
Ba₂SrNpO₆(cr)		−3122.500 ±7.800		
Li₂NpO₄(cr)		−1828.200 ±5.800		
α–Na₂NpO₄		−1763.800 ±5.700		
β–Na₂NpO₄		−1748.500 ±6.100		
β–Na₄NpO₅		−2315.400 ±5.700		
Na₂Np₂O₇(cr)		−2894.000 ±11.000		
Na₃NpF₈(cr)	−3521.239 [a] ±21.305	−3714.000 [b] ±21.000	369.000 [b] ±12.000	272.250 [c] ±12.000
Na₃NpO₂(CO₃)₂(cr)	−2830.592 [b] ±6.365			
NaNpO₂CO₃·3.5 H₂O(cr)	−2590.397 [b] ±5.808			
K₂NpO₄(cr)		−1784.300 ±6.400		
K₂Np₂O₇(cr)		−2932.000 ±11.000		
KNpO₂CO₃(s)	−1793.235 [b] ±5.746			
K₃NpO₂(CO₃)₂(s)	−2899.340 [b] ±5.765			
K₄NpO₂(CO₃)₃(s)	−3660.395 [b] ±7.641			
Rb₂Np₂O₇(cr)		−2914.000 ±12.000		
Cs₂NpO₄(cr)		−1788.100 ±5.700		

(Continued on next page)

Table 4–1: (continued)

Compound	$\Delta_f G_m^\circ$ (kJ·mol^{-1})	$\Delta_f H_m^\circ$ (kJ·mol^{-1})	S_m° (J·K^{-1}·mol^{-1})	$C_{p,m}^\circ$ (J·K^{-1}·mol^{-1})
Cs$_2$NpCl$_6$(cr)	−1833.039 (a) ±4.871	−1976.200 ±1.900	410.000 ±15.000	
Cs$_2$NpO$_2$Cl$_4$(cr)		−2056.100 ±5.400		
Cs$_3$NpO$_2$Cl$_4$(cr)		−2449.100 ±4.800		
Cs$_2$NpBr$_6$(cr)	−1620.121 (a) ±3.616	−1682.300 ±2.000	469.000 ±10.000	
Cs$_2$NaNpCl$_6$(cr)		−2217.200 ±3.100		

(a) Value calculated internally with the Gibbs–Helmholtz equation, $\Delta_f G_m^\circ = \Delta_f H_m^\circ - T \sum_i S_{m,i}^\circ$.
(b) Value calculated internally from reaction data (see Table 4–2).
(c) Temperature coefficients of this function are listed in Table 4–3.
(d) A temperature function for the heat capacity is given in Table 4–3.

Table 4–2: Selected thermodynamic data for reactions involving neptunium compounds and complexes. All ionic species listed in this table are aqueous species. Unless noted otherwise, all data refer to the reference temperature of 298.15 K and to the standard state, *i.e.*, a pressure of 0.1 MPa and, for aqueous species, infinite dilution ($I = 0$). The uncertainties listed below each value represent total uncertainties and correspond in principle to the statistically defined 95% confidence interval. Values obtained from internal calculation, *cf.* footnote (a), are rounded at the third digit after the decimal point and may therefore not be exactly identical to those given in Part III. Systematically, all the values are presented with three digits after the decimal point, regardless of the significance of these digits. The data presented in this table are available on computer media from the OECD Nuclear Energy Agency.

Species	Reaction	$\log_{10} K°$	$\Delta_r G_m°$ (kJ · mol^{-1})	$\Delta_r H_m°$ (kJ · mol^{-1})	$\Delta_r S_m°$ (J · K^{-1} · mol^{-1})
Np(g)	Np(cr) $\rightleftharpoons$ Np(g)			465.100 ±3.000	
Np^{3+}	$0.5H_2(g) + Np^{4+} \rightleftharpoons H^+ + Np^{3+}$	3.695 (c) ±0.169		−21.092 ±0.965	
Np^{4+}	$3H^+ + 0.5H_2(g) + NpO_2^+ \rightleftharpoons 2H_2O(l) + Np^{4+}$				−305.930 ±6.228
NpO$_2$(am, hyd)	$Np^{4+} + 4OH^- \rightleftharpoons 2H_2O(l) + NpO_2(am, hyd)$	56.700 ±0.500		−323.646 ±2.854	
NpO$_2^+$	$0.5H_2(g) + NpO_2^{2+} \rightleftharpoons H^+ + NpO_2^+$				−18.857 ±2.264
NpO$_2^{2+}$	$NpO_2(NO_3)_2 \cdot 6H_2O(s) \rightleftharpoons 6H_2O(l) + 2NO_3^- + NpO_2^{2+}$				104.410 ±6.695
NpOH^{2+}	$H_2O(l) + Np^{3+} \rightleftharpoons H^+ + NpOH^{2+}$	−6.800 ±0.300		38.815 ±1.712	

(Continued on next page)

Table 4–2: (continued)

Species	Reaction			
	$\log_{10} K°$	$\Delta_r G_m°$ (kJ·mol^{-1})	$\Delta_r H_m°$ (kJ·mol^{-1})	$\Delta_r S_m°$ (J·K^{-1}·mol^{-1})
NpOH^{3+}	$H_2O(l) + Np^{4+} \rightleftharpoons H^+ + NpOH^{3+}$			
	0.550 ±0.200	−3.139 ±1.142		
Np(OH)$_2^{2+}$	$2H_2O(l) + Np^{4+} \rightleftharpoons 2H^+ + Np(OH)_2^{2+}$			
	0.350 ±0.300	−1.998 ±1.712		
NpO$_2$OH(am, aged)	$H_2O(l) + NpO_2^+ \rightleftharpoons H^+ + NpO_2OH(am, aged)$			
	−4.700 ±0.500	26.828 ±2.854		
NpO$_2$OH(am, fresh)	$H_2O(l) + NpO_2^+ \rightleftharpoons H^+ + NpO_2OH(am, fresh)$			
	−5.300 ±0.200	30.253 ±1.142		
NpO$_2$OH(aq)	$H_2O(l) + NpO_2^+ \rightleftharpoons H^+ + NpO_2OH(aq)$			
	−11.300 ±0.700	64.501 ±3.996		
NpO$_2$OH$^+$	$H_2O(l) + NpO_2^{2+} \rightleftharpoons H^+ + NpO_2OH^+$			
	−5.100 ±0.400	29.111 ±2.283		
NpO$_2$(OH)$_2^-$	$2H_2O(l) + NpO_2^+ \rightleftharpoons 2H^+ + NpO_2(OH)_2^-$			
	−23.600 ±0.500	134.710 ±2.854		
NpO$_3$·H$_2$O(cr)	$2H_2O(l) + NpO_2^{2+} \rightleftharpoons 2H^+ + NpO_3 \cdot H_2O(cr)$			
	−5.470 ±0.400	31.223 ±2.283		
Np(OH)$_4$(aq)	$4H_2O(l) + Np^{4+} \rightleftharpoons 4H^+ + Np(OH)_4(aq)$			
	−8.300 ±1.100	47.377 ±6.279		
(NpO$_2$)$_2$(OH)$_2^{2+}$	$2H_2O(l) + 2NpO_2^{2+} \rightleftharpoons (NpO_2)_2(OH)_2^{2+} + 2H^+$			
	−6.270 ±0.210	35.789 ±1.199		
(NpO$_2$)$_3$(OH)$_5^+$	$5H_2O(l) + 3NpO_2^{2+} \rightleftharpoons (NpO_2)_3(OH)_5^+ + 5H^+$			
	−17.120 ±0.220	97.722 ±1.256		

(Continued on next page)

Table 4–2: (continued)

Species	Reaction	$\log_{10} K°$	$\Delta_r G_m°$ (kJ·mol^{-1})	$\Delta_r H_m°$ (kJ·mol^{-1})	$\Delta_r S_m°$ (J·K^{-1}·mol^{-1})
NpF^{3+}	F$^-$ + Np^{4+} ⇌ NpF^{3+}	8.960 ±0.140	−51.144 ±0.799	1.500 ±2.000	176.570 [a] ±7.224
NpF$_2^{2+}$	2F$^-$ + Np^{4+} ⇌ NpF$_2^{2+}$	15.700 ±0.300	−89.616 ±1.712		
NpF$_4$(g)	NpF$_4$(cr) ⇌ NpF$_4$(g)			313.000 ±15.000	
NpF$_6$(g)	NpF$_6$(cr) ⇌ NpF$_6$(g)			48.340 ±0.070	
NpO$_2$F(aq)	F$^-$ + NpO$_2^+$ ⇌ NpO$_2$F(aq)	1.200 ±0.300	−6.850 ±1.712		
NpO$_2$F$^+$	F$^-$ + NpO$_2^{2+}$ ⇌ NpO$_2$F$^+$	4.570 ±0.070	−26.086 ±0.400		
NpO$_2$F$_2$(aq)	2F$^-$ + NpO$_2^{2+}$ ⇌ NpO$_2$F$_2$(aq)	7.600 ±0.080	−43.381 ±0.457		
NpCl^{3+}	Cl$^-$ + Np^{4+} ⇌ NpCl^{3+}	1.500 ±0.300	−8.562 ±1.712		
NpCl$_3$(g)	NpCl$_3$(cr) ⇌ NpCl$_3$(g)			307.800 ±10.000	
NpCl$_4$(g)	NpCl$_4$(cr) ⇌ NpCl$_4$(g)			197.000 ±3.000	
NpO$_2$Cl$^+$	Cl$^-$ + NpO$_2^{2+}$ ⇌ NpO$_2$Cl$^+$	0.400 ±0.170	−2.283 ±0.970		

(Continued on next page)

Table 4–2: (continued)

Species	Reaction			
	$\log_{10} K°$	$\Delta_r G_m°$ (kJ·mol^{-1})	$\Delta_r H_m°$ (kJ·mol^{-1})	$\Delta_r S_m°$ (J·K^{-1}·mol^{-1})
NpI^{3+}	I$^-$ + Np^{4+} ⇌ NpI^{3+}			
	1.500 ±0.400	−8.562 ±2.283		
NpO$_2$IO$_3$(aq)	IO$_3^-$ + NpO$_2^+$ ⇌ NpO$_2$IO$_3$(aq)			
	0.500 ±0.300	−2.854 ±1.712		
NpO$_2$IO$_3^+$	IO$_3^-$ + NpO$_2^{2+}$ ⇌ NpO$_2$IO$_3^+$			
	1.200 ±0.300	−6.850 ±1.712		
NpSO$_4^{2+}$	HSO$_4^-$ + Np^{4+} ⇌ H$^+$ + NpSO$_4^{2+}$			
	4.870 ±0.150	−27.798 ±0.856	7.400 ±8.800	118.055 [a] ±29.655
NpO$_2$SO$_4$(aq)	NpO$_2^{2+}$ + SO$_4^{2-}$ ⇌ NpO$_2$SO$_4$(aq)			
	3.280 ±0.060	−18.722 ±0.342	16.700 ±0.500	118.807 [a] ±2.033
NpO$_2$SO$_4^-$	NpO$_2^+$ + SO$_4^{2-}$ ⇌ NpO$_2$SO$_4^-$			
	0.440 ±0.270	−2.512 ±1.541	23.200 ±7.200	86.237 [a] ±24.696
Np(SO$_4$)$_2$(aq)	2HSO$_4^-$ + Np^{4+} ⇌ 2H$^+$ + Np(SO$_4$)$_2$(aq)			
	7.090 ±0.250	−40.470 ±1.427	10.500 ±3.600	170.954 [a] ±12.988
NpO$_2$(SO$_4$)$_2^{2-}$	NpO$_2^{2+}$ + 2SO$_4^{2-}$ ⇌ NpO$_2$(SO$_4$)$_2^{2-}$			
	4.700 ±0.100	−26.828 ±0.571	26.000 ±1.200	177.185 [a] ±4.457
NpNO$_3^{3+}$	NO$_3^-$ + Np^{4+} ⇌ NpNO$_3^{3+}$			
	1.900 ±0.150	−10.845 ±0.856		
NpO$_2$HPO$_4$(aq)	HPO$_4^{2-}$ + NpO$_2^{2+}$ ⇌ NpO$_2$HPO$_4$(aq)			
	6.200 ±0.700	−35.390 ±3.996		
NpO$_2$HPO$_4^-$	HPO$_4^{2-}$ + NpO$_2^+$ ⇌ NpO$_2$HPO$_4^-$			
	2.950 ±0.100	−16.839 ±0.571		

(Continued on next page)

Table 4–2: (continued)

Species	Reaction			
	$\log_{10} K°$	$\Delta_r G_m^°$ (kJ · mol^{-1})	$\Delta_r H_m^°$ (kJ · mol^{-1})	$\Delta_r S_m^°$ (J · K^{-1} · mol^{-1})
NpO$_2$H$_2$PO$_4^+$	H$_2$PO$_4^-$ + NpO$_2^{2+}$ $\rightleftharpoons$ NpO$_2$H$_2$PO$_4^+$			
	3.320 ±0.500	−18.951 ±2.854		
NpO$_2$(HPO$_4$)$_2^{2-}$	2HPO$_4^{2-}$ + NpO$_2^{2+}$ $\rightleftharpoons$ NpO$_2$(HPO$_4$)$_2^{2-}$			
	9.500 ±1.000	−54.226 ±5.708		
NpO$_2$CO$_3$(aq)	CO$_3^{2-}$ + NpO$_2^{2+}$ $\rightleftharpoons$ NpO$_2$CO$_3$(aq)			
	9.320 (c) ±0.610	−53.201 ±3.480		
NpO$_2$CO$_3$(s)	CO$_3^{2-}$ + NpO$_2^{2+}$ $\rightleftharpoons$ NpO$_2$CO$_3$(s)			
	14.596 (c) ±0.469	−83.317 ±2.678		
NpO$_2$CO$_3^-$	CO$_3^{2-}$ + NpO$_2^+$ $\rightleftharpoons$ NpO$_2$CO$_3^-$			
	4.962 ±0.061	−28.323 ±0.348		
NpO$_2$(CO$_3$)$_2^{2-}$	2CO$_3^{2-}$ + NpO$_2^{2+}$ $\rightleftharpoons$ NpO$_2$(CO$_3$)$_2^{2-}$			
	16.516 (c) ±0.729	−94.276 ±4.162		
NpO$_2$(CO$_3$)$_2^{3-}$	CO$_3^{2-}$ + NpO$_2$CO$_3^-$ $\rightleftharpoons$ NpO$_2$(CO$_3$)$_2^{3-}$			
	1.572 ±0.083	−8.973 ±0.474		
Np(CO$_3$)$_3^{3-}$	Np(CO$_3$)$_5^{6-}$ + e$^-$ $\rightleftharpoons$ 2CO$_3^{2-}$ + Np(CO$_3$)$_3^{3-}$			
	−16.261 ±2.265	92.818 ±12.929		
NpO$_2$(CO$_3$)$_3^{4-}$	NpO$_2$(CO$_3$)$_3^{5-}$ $\rightleftharpoons$ NpO$_2$(CO$_3$)$_3^{4-}$ + e$^-$			
	−5.720 (c) ±0.095	32.651 ±0.540		
NpO$_2$(CO$_3$)$_3^{4-}$	3CO$_3^{2-}$ + NpO$_2^{2+}$ $\rightleftharpoons$ NpO$_2$(CO$_3$)$_3^{4-}$			
			−41.900 ±4.100	
NpO$_2$(CO$_3$)$_3^{5-}$	CO$_3^{2-}$ + NpO$_2$(CO$_3$)$_2^{3-}$ $\rightleftharpoons$ NpO$_2$(CO$_3$)$_3^{5-}$			
	−1.034 ±0.110	5.902 ±0.628		

(Continued on next page)

Table 4–2: (continued)

Species	Reaction	$\log_{10} K°$	$\Delta_r G_m^°$ (kJ·mol^{-1})	$\Delta_r H_m^°$ (kJ·mol^{-1})	$\Delta_r S_m^°$ (J·K^{-1}·mol^{-1})
$NpO_2(CO_3)_3^{5-}$	$NpO_2(CO_3)_3^{4-} + e^- \rightleftharpoons NpO_2(CO_3)_3^{5-}$			−88.800 ±2.900	
$Np(CO_3)_4^{4-}$	$4CO_3^{2-} + 2H_2O(l) + NpO_2(am, hyd) \rightleftharpoons Np(CO_3)_4^{4-} + 4OH^-$	−17.790 ±0.220	101.550 ±1.256		
$Np(CO_3)_5^{6-}$	$CO_3^{2-} + Np(CO_3)_4^{4-} \rightleftharpoons Np(CO_3)_5^{6-}$	−1.070 ±0.300	6.108 ±1.712		
$(NpO_2)_3(CO_3)_6^{6-}$	$3NpO_2(CO_3)_3^{4-} \rightleftharpoons (NpO_2)_3(CO_3)_6^{6-} + 3CO_3^{2-}$	−8.272 (c) ±1.447	47.215 ±8.260		
$NpO_2(CO_3)_2OH^{4-}$	$NpO_2(CO_3)_3^{5-} + OH^- \rightleftharpoons CO_3^{2-} + NpO_2(CO_3)_2OH^{4-}$	3.195 (c) ±1.164	−18.238 ±6.644		
$(NpO_2)_2CO_3(OH)_3^-$	$7H^+ + 2NpO_2(CO_3)_3^{4-} \rightleftharpoons (NpO_2)_2CO_3(OH)_3^- + 5CO_2(g) + 2H_2O(l)$	49.166 (c) ±1.586	−280.640 ±9.053		
$(NH_4)_4NpO_2(CO_3)_3(s)$	$4NH_4^+ + NpO_2(CO_3)_3^{4-} \rightleftharpoons (NH_4)_4NpO_2(CO_3)_3(s)$	7.443 (c) ±0.297	−42.485 ±1.698		
$Np(SCN)^{3+}$	$Np^{4+} + SCN^- \rightleftharpoons Np(SCN)^{3+}$	3.000 ±0.300	−17.124 ±1.712	−7.000 ±3.000	33.956 (a) ±11.586
$Np(SCN)_2^{2+}$	$Np^{4+} + 2SCN^- \rightleftharpoons Np(SCN)_2^{2+}$	4.100 ±0.500	−23.403 ±2.854	−9.000 ±9.000	48.308 (a) ±31.668
$Np(SCN)_3^+$	$Np^{4+} + 3SCN^- \rightleftharpoons Np(SCN)_3^+$	4.800 ±0.500	−27.399 ±2.854	−13.000 ±9.000	48.293 (a) ±31.668
$Na_3NpF_8(cr)$	$3NaF(cr) + NpF_6(g) \rightleftharpoons 0.5F_2(g) + Na_3NpF_8(cr)$	7.876 (c) ±1.350	−44.954 (a) ±7.705	−62.678 ±6.900	−59.447 ±11.500

(Continued on next page)

Table 4–2: (continued)

Species	Reaction			
	$\log_{10} K°$	$\Delta_r G_m°$ (kJ·mol^{-1})	$\Delta_r H_m°$ (kJ·mol^{-1})	$\Delta_r S_m°$ (J·K^{-1}·mol^{-1})
Na$_3$NpO$_2$(CO$_3$)$_2$(cr)	$2CO_3^{2-} + 3Na^+ + NpO_2^+ \rightleftharpoons$ Na$_3$NpO$_2$(CO$_3$)$_2$(cr)			
	14.220 ±0.500		−81.168 ±2.854	
NaNpO$_2$CO$_3$·3.5H$_2$O(cr)	$CO_3^{2-} + 3.5H_2O(l) + Na^+ + NpO_2^+ \rightleftharpoons$ NaNpO$_2$CO$_3$·3.5H$_2$O(cr)			
	11.000 ±0.240		−62.788 ±1.370	
KNpO$_2$CO$_3$(s)	$CO_3^{2-} + K^+ + NpO_2^+ \rightleftharpoons$ KNpO$_2$CO$_3$(s)			
	13.150 ±0.190		−75.061 ±1.085	
K$_3$NpO$_2$(CO$_3$)$_2$(s)	$2CO_3^{2-} + 3K^+ + NpO_2^+ \rightleftharpoons$ K$_3$NpO$_2$(CO$_3$)$_2$(s)			
	15.460 ±0.160		−88.246 ±0.913	
K$_4$NpO$_2$(CO$_3$)$_3$(s)	$4K^+ + NpO_2(CO_3)_3^{4-} \rightleftharpoons$ K$_4$NpO$_2$(CO$_3$)$_3$(s)			
	7.033 (c) ±0.876		−40.147 ±5.001	

(a) Value calculated internally with the Gibbs–Helmholtz equation, $\Delta_r G_m° = \Delta_r H_m° - T \Delta_r S_m°$.
(b) Value calculated from a selected standard potential.
(c) Value of $\log_{10} K°$ calculated internally from $\Delta_r G_m°$.

Table 4–3: Selected temperature coefficients for heat capacities marked with [c] in Table 4–1, according to the form $C_{p,m}(T) = a + bT + cT^2 + dT^{-1} + eT^{-2}$. The functions are valid between the temperatures T_{min} and T_{max} (in K). The notation E±nn indicates the power of 10. $C_{p,m}$ units are $J \cdot K^{-1} \cdot mol^{-1}$.

Compound	a	b	c	d	e	T_{min}	T_{max}
Np(cr)	−4.05430E+00	8.25540E−02	0	0	8.05710E+05	298	553
β–Np	3.93300E+01	0	0	0	0	553	849
γ–Np	3.64010E+01	0	0	0	0	849	912
Np(g)	1.77820E+01	3.35250E−03	6.68070E−06	0	1.28800E+05	298	800
	1.41320E+01	1.55670E−02	−1.59546E−06	0	−3.98870E+05	800	2000
NpO$_2$(cr)	6.75110E+01	2.65990E−02	0	0	−8.19000E+05	220	800
Np$_2$O$_5$(cr)	9.92000E+01	9.86000E−02	0	0	0	298	750
NpF(g)	3.59880E+01	2.30310E−03	−7.30170E−07	0	−2.50010E+05	298	1400
NpF$_2$(g)	6.74740E+01	5.56480E−03	−5.95160E−06	0	−1.12930E+06	298	600
NpF$_3$(cr)	1.05200E+02	8.12000E−04	0	0	−1.00000E+06	298	1735
NpF$_3$(g)	8.16540E+01	1.20840E−03	−2.42780E−07	0	−8.67160E+05	298	3000
NpF$_4$(cr)	1.22640E+02	9.68400E−03	0	0	−8.36470E+05	298	1305
NpF$_4$(g)	1.05840E+02	8.61400E−04	1.57430E−06	0	−9.75640E+05	298	1500
NpF$_5$(cr)	1.26000E+02	3.00000E−02	0	0	−1.90000E+05	298	600
NpF$_6$(cr)	6.23330E+01	3.52550E−01	0	0	0	298	328
NpF$_6$(g)	1.43240E+02	2.44160E−02	−1.13120E−05	0	−1.81740E+06	298	1000
NpF$_6$(l)	1.50340E+02	1.10080E−01	0	0	0	328	350
NpO$_2$F$_2$(cr)	1.06240E+02	2.83260E−02	0	0	−1.02080E+06	298	1000
NpCl$_3$(cr)	8.96000E+01	2.75000E−02	0	0	3.60000E+05	298	1075
NpCl$_3$(g)	7.19890E+01	2.47410E−02	−8.48420E−06	0	−1.15000E+04	298	1000
NpCl$_4$(cr)	1.12500E+02	3.60000E−02	0	0	−1.10000E+05	298	811
NpCl$_4$(g)	1.08770E+02	−2.86580E−03	3.15880E−06	0	−2.80770E+05	298	1800
NpCl$_4$(l)	1.08000E+02	6.00000E−02	0	0	0	811	1000
NpOCl$_2$(cr)	9.88000E+01	2.20000E−02	0	0	−9.20000E+05	298	1000
NpO$_2$ClO$_4$(aq)[a]	3.56770E+03	−4.95930E+00	0	−6.32340E+05	0	291	398
NpBr$_3$(cr, hex)	1.01230E+02	2.06800E−02	0	0	−3.20000E+05	298	975
NpBr$_4$(cr)	1.19000E+02	3.00000E−02	0	0	0	298	800
NpOBr$_2$(cr)	1.11000E+02	1.37000E−02	0	0	−1.50000E+06	298	800
NpI$_3$(cr)	1.04000E+02	2.00000E−02	0	0	0	298	975
NpN(cr)	4.76700E+01	1.31740E−02	0	0	−2.57620E+05	298	2000
NpC$_{0.91}$(cr)	6.12500E+01	−3.52750E−02	3.62830E−05	0	−3.48420E+05	298	1000
Na$_3$NpF$_8$(cr)	2.70000E+02	5.66000E−02	0	0	−1.30000E+06	298	800

(a) partial molar heat capacity of solute.

Chapter 5

Selected Data for Plutonium

5.1 General considerations

This chapter presents updated chemical thermodynamic data for plutonium species, as described in detail in Chapter 11 of this volume. The newly selected data represent revisions to those chosen in the previous NEA TDB review [2001LEM/FUG]. In this respect, it will be found that while new species appear in the Tables, some others have been removed from them. Table 5–1 contains the recommended thermodynamic data of the plutonium compounds and complexes, Table 5–2 the recommended thermodynamic data of chemical equilibrium reactions by which the plutonium compounds and complexes are formed, and Table 5–3 the temperature coefficients of the heat capacity data of Table 5–1 where available.

The species and reactions in Table 5–1, Table 5–2 and Table 5–3 appear in standard order of arrangement (cf. Figure 2.1). Table 5–2 contains information only on those reactions for which primary data selections are made in this review. These selected reaction data are used, together with data for key plutonium species (for example Pu^{4+}) and auxiliary data listed in Table 8–1, to derive the corresponding formation quantities in Table 5–1. The uncertainties associated with values for the key plutonium species and for some of the auxiliary data are substantial, leading to comparatively large uncertainties in the formation quantities derived in this manner. The inclusion of a table for reaction data (Table 5–2) in this report allows the use of equilibrium constants with total uncertainties that are directly based on the experimental accuracies. This is the main reason for including both the table for reaction data (Table 5–2) and the table of $\Delta_f G_m^\circ$, $\Delta_f H_m^\circ$, S_m° and $C_{p,m}^\circ$ values (Table 5–1). In a few cases, the correlation of small uncertainties in values for ligands has been neglected in calculations of uncertainty values for species in Table 5–1 from uncertainty values in Table 5–2. However, for those species the effects are less than 2% of the stated uncertainties.

The selected thermal functions of the heat capacities, listed in Table 5–3, refer to the relation:

$$C_{p,m}(T) = a + b \cdot T + c \cdot T^2 + d \cdot T^{-1} + e \cdot T^{-2} \tag{5.1}$$

No references are given in these tables since the selected data are generally not directly attributable to a specific published source. A detailed discussion of the selection procedure is presented in [2001LEM/FUG] and in Chapter 11 of this volume.

5.2 Precautions to be observed in the use of the tables

Geochemical modelling in aquatic systems requires the careful use of the data selected in the NEA TDB reviews. The data in the tables of selected data must not be adopted without taking into account the chemical background information discussed in the corresponding sections of this book. In particular the following precautions should be observed when using data from the Tables.

- The addition of any aqueous species and its data to this internally consistent database can result in a modified data set which is no longer rigorous and can lead to erroneous results. The situation is similar, to a lesser degree, with the addition of gases and solids. It should also be noted that the data set presented in this chapter may not be "complete" for all the conceivable systems and conditions. Gaps are pointed out in both the previous NEA TDB review and the present update.

- Solubility data for crystalline phases are well defined in the initial state, but not necessarily in the final state after "equilibrium" has been attained. Hence, the solubility calculated from these phases may be very misleading, as discussed for the solubility data of $MO_2(cr)$, M = U, Np, Pu.

- The selected thermodynamic data in [92GRE/FUG], [95SIL/BID] and [2001LEM/FUG] contain thermodynamic data for amorphous phases. Most of these refer to Gibbs energy of formation deduced from solubility measurements. In the present review all such data have been removed. However, the corresponding $\log_{10} K$ values have been retained. The reasons for these changes are:

    o Thermodynamic data are only meaningful if they refer to well–defined systems; this is not the case for amorphous phases. It is well known that their solubility may change with time, due to re–crystallisation with a resulting change in water content and surface area/crystal size; the time scale for these changes can vary widely. These kinetic phenomena are different from those encountered in systems where there is a very high activation barrier for the reaction, *e.g.*, electron exchange between sulphate and sulphide.

    o The solubility of amorphous phases provides useful information for users of the database when modelling the behaviour of complex systems. Therefore, the solubility products have been given with the proviso that they are not thermodynamic quantities.

5 Selected plutonium data

Table 5–1: Selected thermodynamic data for plutonium compounds and complexes. All ionic species listed in this table are aqueous species. Unless noted otherwise, all data refer to the reference temperature of 298.15 K and to the standard state, *i.e.*, a pressure of 0.1 MPa and, for aqueous species, infinite dilution ($I = 0$). The uncertainties listed below each value represent total uncertainties and correspond in principle to the statistically defined 95% confidence interval. Values obtained from internal calculation, *cf.* footnotes (a) and (b), are rounded at the third digit after the decimal point and may therefore not be exactly identical to those given in Part III. Systematically, all the values are presented with three digits after the decimal point, regardless of the significance of these digits. The data presented in this table are available on computer media from the OECD Nuclear Energy Agency.

Compound	$\Delta_f G_m^\circ$ (kJ · mol^{-1})	$\Delta_f H_m^\circ$ (kJ · mol^{-1})	S_m° (J · K^{-1} · mol^{-1})	$C_{p,m}^\circ$ (J · K^{-1} · mol^{-1})
Pu(cr)	0.000	0.000	54.460 ±0.800	31.490 [c] ±0.400
β–Pu				[d]
δ–Pu				[d]
δ'–Pu				[d]
γ–Pu				[d]
ε–Pu				[d]
Pu(g)	312.415 [a] ±3.009	349.000 [b] ±3.000	177.167 ±0.005	20.854 [c] ±0.010
Pu^{3+}	−578.984 ±2.688	−591.790 ±1.964	−184.510 [b] ±6.154	
Pu^{4+}	−477.988 ±2.705	−539.895 ±3.103	−414.535 ±10.192	
PuO$_{1.61}$(bcc)	−834.771 [a] ±10.113	−875.500 ±10.000	83.000 ±5.000	61.200 [c] ±5.000
PuO$_2$(cr)	−998.113 [a] ±1.031	−1055.800 ±1.000	66.130 ±0.260	66.250 [c] ±0.260
PuO$_2^+$	−852.646 [b] ±2.868	−910.127 [a] ±8.920	1.480 [b] ±30.013	
PuO$_2^{2+}$	−762.353 ±2.821	−822.036 ±6.577	−71.246 ±22.120	

(Continued on next page)

Table 5–1: (continued)

Compound	$\Delta_f G_m^\circ$ (kJ·mol^{-1})	$\Delta_f H_m^\circ$ (kJ·mol^{-1})	S_m° (J·K^{-1}·mol^{-1})	$C_{p,m}^\circ$ (J·K^{-1}·mol^{-1})
Pu$_2$O$_3$(cr)	−1580.375 (a) ±10.013	−1656.000 ±10.000	163.000 ±0.600	117.000 (c) ±0.500
PuOH^{2+}	−776.739 (b) ±3.187			
PuOH^{3+}	−718.553 (b) ±2.936			
Pu(OH)$_2^{2+}$	−955.693 (b) ±3.203			
Pu(OH)$_3$(cr)	−1200.218 (b) ±8.975			
Pu(OH)$_3^+$	−1176.280 (b) ±3.542			
PuO$_2$OH(aq)	≥ −1034.247 (b)			
PuO$_2$OH$^+$	−968.099 (b) ±4.013	−1079.866 (b) ±16.378	−12.680 (b) ±55.786	
PuO$_2$(OH)$_2$(aq)	−1161.287 (b) ±9.015			
Pu(OH)$_4$(aq)	−1378.031 (b) ±3.936			
PuO$_2$(OH)$_2$·H$_2$O(cr)	−1442.380 (b) ±6.368	−1632.809 (a) ±13.522	190.000 ±40.000	170.000 ±20.000
(PuO$_2$)$_2$(OH)$_2^{2+}$	−1956.176 (b) ±8.026			
PuF(g)	−140.967 (a) ±10.113	−112.600 ±10.000	251.000 ±5.000	33.500 (c) ±3.000
PuF^{3+}	−809.970 (b) ±2.850	−866.145 (b) ±3.859	−228.573 (b) ±12.753	
PuF$_2$(g)	−626.151 (a) ±6.704	−614.300 ±6.000	297.000 ±10.000	51.500 (c) ±5.000
PuF$_2^{2+}$	−1130.651 (b) ±3.246	−1199.595 (b) ±6.026	−104.666 (b) ±20.058	
PuF$_3$(cr)	−1517.369 (a) ±3.709	−1586.700 ±3.700	126.110 ±0.360	92.640 (c) ±0.280
PuF$_3$(g)	−1161.081 (a) ±4.758	−1167.800 (b) ±3.700	336.110 ±10.000	72.240 (c) ±5.000

(Continued on next page)

Table 5–1: (continued)

Compound	$\Delta_f G_m^\circ$ (kJ·mol^{-1})	$\Delta_f H_m^\circ$ (kJ·mol^{-1})	S_m° (J·K^{-1}·mol^{-1})	$C_{p,m}^\circ$ (J·K^{-1}·mol^{-1})
PuF$_4$(cr)	−1756.741 [a] ±20.002	−1850.000 ±20.000	147.250 ±0.370	116.190 ±0.290
PuF$_4$(g)	−1517.874 [a] ±22.202	−1548.000 [b] ±22.000	359.000 ±10.000	92.400 [c] ±5.000
PuF$_6$(cr)	−1729.856 [a] ±20.174	−1861.350 [b] ±20.170	221.800 ±1.100	168.100 [c] ±2.000
PuF$_6$(g)	−1725.064 [a] ±20.104	−1812.700 [b] ±20.100	368.900 ±1.000	129.320 [c] ±1.000
PuOF(cr)	−1091.571 [a] ±20.222	−1140.000 ±20.000	96.000 ±10.000	69.400 [c] ±10.000
PuO$_2$F$^+$	−1069.905 [b] ±3.121			
PuO$_2$F$_2$(aq)	−1366.783 [b] ±4.059			
PuCl^{2+}	−717.051 [b] ±2.923			
PuCl^{3+}	−619.480 [b] ±3.204			
PuCl$_3$(cr)	−891.806 [a] ±2.024	−959.600 ±1.800	161.700 ±3.000	101.200 [c] ±4.000
PuCl$_3$(g)	−641.299 [a] ±3.598	−647.400 [b] ±1.868	368.620 ±10.000	78.470 [c] ±5.000
PuCl$_4$(cr)	−879.368 [a] ±5.826	−968.700 ±5.000	201.000 ±10.000	121.400 ±4.000
PuCl$_4$(g)	−764.683 [a] ±10.438	−792.000 [b] ±10.000	409.000 ±10.000	103.400 [c] ±5.000
PuOCl(cr)	−882.409 [a] ±1.936	−931.000 ±1.700	105.600 [b] ±3.000	71.600 [c] ±4.000
PuO$_2$Cl$^+$	−894.883 [b] ±2.829			
PuO$_2$Cl$_2$(aq)	−1018.224 [b] ±3.308			
PuCl$_3$·6H$_2$O(cr)	−2365.347 [a] ±2.586	−2773.400 ±2.100	420.000 ±5.000	
PuBr^{3+}	−590.971 [b] ±3.206			

(Continued on next page)

Table 5–1: (continued)

Compound	$\Delta_f G_m^\circ$ (kJ·mol^{-1})	$\Delta_f H_m^\circ$ (kJ·mol^{-1})	S_m° (J·K^{-1}·mol^{-1})	$C_{p,m}^\circ$ (J·K^{-1}·mol^{-1})
PuBr$_3$(cr)	−767.324 [a] ±2.697	−792.600 ±2.000	198.000 ±6.000	101.800 [c] ±6.000
PuBr$_3$(g)	−529.808 [a] ±15.655	−488.000 [b] ±15.000	423.000 [b] ±15.000	81.600 [c] ±10.000
PuOBr(cr)	−838.354 [a] ±8.541	−870.000 [b] ±8.000	127.000 [b] ±10.000	73.000 [c] ±8.000
PuI^{2+}	−636.987 [b] ±3.529			
PuI$_3$(cr)	−579.000 [a] ±4.551	−579.200 ±2.800	228.000 ±12.000	110.000 [c] ±8.000
PuI$_3$(g)	−366.517 [a] ±15.655	−305.000 ±15.000	435.000 ±15.000	82.000 [c] ±5.000
PuOI(cr)	−776.626 [a] ±20.495	−802.000 ±20.000	130.000 ±15.000	75.600 [c] ±10.000
PuSO$_4^+$	−1345.315 [b] ±4.599	−1483.890 [b] ±2.976	−33.301 [b] ±15.108	
PuSO$_4^{2+}$	−1261.329 [b] ±3.270			
PuO$_2$SO$_4$(aq)	−1525.650 [b] ±3.072	−1715.276 [b] ±6.616	65.963 [b] ±22.543	
Pu(SO$_4$)$_2$(aq)	−2029.601 [b] ±4.225			
Pu(SO$_4$)$_2^-$	−2099.545 [b] ±5.766	−2398.590 [b] ±16.244	1.520 [b] ±56.262	
PuO$_2$(SO$_4$)$_2^{2-}$	−2275.477 [b] ±3.156	−2597.716 [b] ±11.176	194.214 [b] ±37.627	
PuSe(cr)			92.100 ±1.800	59.700 ±1.200
PuTe(cr)			107.900 ±4.300	73.100 ±2.900
PuN(cr)	−273.719 [a] ±2.551	−299.200 ±2.500	64.800 ±1.500	49.600 ±1.000
PuNO$_3^{3+}$	−599.913 [b] ±2.868			
PuO$_2$(NO$_3$)$_2$·6H$_2$O(cr)	−2393.300 ±3.200			

(Continued on next page)

Table 5–1: (continued)

Compound	$\Delta_f G_m^\circ$ (kJ · mol^{-1})	$\Delta_f H_m^\circ$ (kJ · mol^{-1})	S_m° (J · K^{-1} · mol^{-1})	$C_{p,m}^\circ$ (J · K^{-1} · mol^{-1})
PuP(cr)	−313.757 [a] ±21.078	−318.000 ±21.000	81.320 ±6.000	50.200 ±4.000
PuPO$_4$(s, hyd)	−1744.893 [b] ±5.528			
PuH$_3$PO$_4^{4+}$	−1641.055 [b] ±3.569			
PuAs(cr)	−241.413 [a] ±20.111	−240.000 ±20.000	94.300 ±7.000	51.600 ±4.000
PuSb(cr)	−152.063 [a] ±20.126	−150.000 ±20.000	106.900 ±7.500	52.800 ±3.500
PuBi(cr)	−119.624 [a] ±20.223	−117.000 ±20.000	120.000 ±10.000	
PuBi$_2$(cr)	−124.527 [a] ±22.395	−126.000 ±22.000	163.000 ±14.000	
PuC$_{0.84}$(cr)	−49.827 [a] ±8.028	−45.200 ±8.000	74.800 ±2.100	47.100 [c] ±1.000
Pu$_3$C$_2$(cr)	−123.477 [a] ±30.046	−113.000 ±30.000	210.000 ±5.000	136.800 [c] ±2.500
Pu$_2$C$_3$(cr)	−156.514 [a] ±16.729	−149.400 ±16.700	150.000 ±2.900	114.000 [c] ±0.400
PuO$_2$CO$_3$(aq)	−1344.479 [b] ±4.032			
PuO$_2$CO$_3$(s)	−1373.876 [b] ±3.912			
PuO$_2$CO$_3^-$	−1409.771 [b] ±3.002			
PuO$_2$(CO$_3$)$_2^{2-}$	−1902.061 [b] ±4.088	−2199.496 [b] ±7.714	19.625 [b] ±27.657	
PuO$_2$(CO$_3$)$_3^{4-}$	−2448.797 [b] ±4.180	−2886.326 [b] ±6.915	−6.103 [b] ±25.198	
PuO$_2$(CO$_3$)$_3^{5-}$	−2465.031 [b] ±6.096	−2954.927 [b] ±12.344	−116.406 [b] ±45.084	
Pu(CO$_3$)$_4^{4-}$	−2800.785 [b] ±7.013			
Pu(CO$_3$)$_5^{6-}$	−3320.979 [b] ±7.261			

(Continued on next page)

Table 5–1: (continued)

Compound	$\Delta_f G_m^\circ$ (kJ · mol^{-1})	$\Delta_f H_m^\circ$ (kJ · mol^{-1})	S_m° (J · K^{-1} · mol^{-1})	$C_{p,m}^\circ$ (J · K^{-1} · mol^{-1})
PuSCN^{2+}	−493.704 [b] ±5.333	−515.390 [b] ±5.988	−15.355 [a] ±26.906	
Sr$_3$PuO$_6$(cr)		−3042.100 ±7.900		
BaPuO$_3$(cr)		−1654.200 ±8.300		
Ba$_3$PuO$_6$(cr)		−2997.000 ±10.000		
Ba$_2$MgPuO$_6$(cr)		−2995.800 ±8.800		
Ba$_2$CaPuO$_6$(cr)		−3067.500 ±8.900		
Ba$_2$SrPuO$_6$(cr)		−3023.300 ±9.000		
Cs$_2$PuCl$_6$(cr)	−1838.243 [a] ±6.717	−1982.000 ±5.000	412.000 ±15.000	
Cs$_3$PuCl$_6$(cr)	−2208.045 [b] ±9.491	−2364.415 [b] ±9.040	454.925 [b] ±10.959	258.600 [c] ±10.000
CsPu$_2$Cl$_7$(cr)	−2235.119 [b] ±5.284	−2399.380 [b] ±5.734	424.000 [b] ±7.281	254.900 [c] ±10.000
Cs$_2$PuBr$_6$(cr)	−1634.326 [a] ±6.150	−1697.400 ±4.200	470.000 ±15.000	
Cs$_2$NaPuCl$_6$(cr)	−2143.496 [a] ±5.184	−2294.200 ±2.600	440.000 ±15.000	

(a) Value calculated internally with the Gibbs–Helmholtz equation, $\Delta_f G_m^\circ = \Delta_f H_m^\circ - T \sum_i S_{m,i}^\circ$.
(b) Value calculated internally from reaction data (see Table 5–2).
(c) Temperature coefficients of this function are listed in Table 5–3.
(d) A temperature function for the heat capacity is given in Table 5–3.

Table 5–2: Selected thermodynamic data for reactions involving plutonium compounds and complexes. All ionic species listed in this table are aqueous species. Unless noted otherwise, all data refer to the reference temperature of 298.15 K and to the standard state, *i.e.*, a pressure of 0.1 MPa and, for aqueous species, infinite dilution ($I = 0$). The uncertainties listed below each value represent total uncertainties and correspond in principle to the statistically defined 95% confidence interval. Values obtained from internal calculation, *cf.* footnote (a), are rounded at the third digit after the decimal point and may therefore not be exactly identical to those given in Part III. Systematically, all the values are presented with three digits after the decimal point, regardless of the significance of these digits. The data presented in this table are available on computer media from the OECD Nuclear Energy Agency.

Species	Reaction	$\log_{10} K^\circ$	$\Delta_r G_m^\circ$ (kJ · mol^{-1})	$\Delta_r H_m^\circ$ (kJ · mol^{-1})	$\Delta_r S_m^\circ$ (J · K^{-1} · mol^{-1})
Pu(g)	Pu(cr) ⇌ Pu(g)			349.000 ±3.000	
Pu^{3+}	PuCl$_3$· 6H$_2$O(cr) ⇌ 3Cl$^-$ + 6H$_2$O(l) + Pu^{3+}			−15.010 ±3.533	
Pu^{4+}	H$^+$ + Pu^{3+} ⇌ 0.5H$_2$(g) + Pu^{4+}			−164.680 ±8.124	
PuO$_2$(am, hyd)	4OH$^-$ + Pu^{4+} ⇌ 2H$_2$O(l) + PuO$_2$(am, hyd)	58.330 ±0.520	−332.950 ±2.968		
PuO$_2^+$	0.5H$_2$(g) + PuO$_2^{2+}$ ⇌ H$^+$ + PuO$_2^+$	15.819 (c) ±0.090	−90.293 ±0.515	−88.091 ±6.026	7.386 (a) ±20.285
PuOH^{2+}	H$_2$O(l) + Pu^{3+} ⇌ H$^+$ + PuOH^{2+}	−6.900 ±0.300	39.385 ±1.712		
PuOH^{3+}	H$_2$O(l) + Pu^{4+} ⇌ H$^+$ + PuOH^{3+}	0.600 ±0.200	−3.425 ±1.142		

(Continued on next page)

Table 5–2: (continued)

Species	Reaction	$\log_{10} K°$	$\Delta_r G_m°$ (kJ · mol^{-1})	$\Delta_r H_m°$ (kJ · mol^{-1})	$\Delta_r S_m°$ (J · K^{-1} · mol^{-1})
$Pu(OH)_2^{2+}$	$2H_2O(l) + Pu^{4+} \rightleftharpoons 2H^+ + Pu(OH)_2^{2+}$	0.600 ±0.300	−3.425 ±1.712		
$Pu(OH)_3(cr)$	$3H_2O(l) + Pu^{3+} \rightleftharpoons 3H^+ + Pu(OH)_3(cr)$	−15.800 ±1.500	90.187 ±8.562		
$Pu(OH)_3^+$	$3H_2O(l) + Pu^{4+} \rightleftharpoons 3H^+ + Pu(OH)_3^+$	−2.300 ±0.400	13.128 ±2.283		
$PuO_2OH(am)$	$H_2O(l) + PuO_2^+ \rightleftharpoons H^+ + PuO_2OH(am)$	−5.000 ±0.500	28.540 ±2.854		
$PuO_2OH(aq)$	$H_2O(l) + PuO_2^+ \rightleftharpoons H^+ + PuO_2OH(aq)$	≤ −9.730	≥ 55.539		
PuO_2OH^+	$H_2O(l) + PuO_2^{2+} \rightleftharpoons H^+ + PuO_2OH^+$	−5.500 ±0.500	31.394 ±2.854	28.000 ±15.000	−11.384 [a] ±51.213
$PuO_2(OH)_2(aq)$	$2H_2O(l) + PuO_2^{2+} \rightleftharpoons 2H^+ + PuO_2(OH)_2(aq)$	−13.200 ±1.500	75.346 ±8.562		
$Pu(OH)_4(aq)$	$4H_2O(l) + Pu^{4+} \rightleftharpoons 4H^+ + Pu(OH)_4(aq)$	−8.500 ±0.500	48.518 ±2.854		
$PuO_2(OH)_2 \cdot H_2O(cr)$	$3H_2O(l) + PuO_2^{2+} \rightleftharpoons 2H^+ + PuO_2(OH)_2 \cdot H_2O(cr)$	−5.500 ±1.000	31.394 ±5.708		
$(PuO_2)_2(OH)_2^{2+}$	$2H_2O(l) + 2PuO_2^{2+} \rightleftharpoons (PuO_2)_2(OH)_2^{2+} + 2H^+$	−7.500 ±1.000	42.810 ±5.708		
PuF^{3+}	$F^- + Pu^{4+} \rightleftharpoons PuF^{3+}$	8.840 ±0.100	−50.459 ±0.571	9.100 ±2.200	199.762 [a] ±7.623

(Continued on next page)

Table 5-2: (continued)

Species	Reaction	$\log_{10} K°$	$\Delta_r G_m°$ (kJ·mol^{-1})	$\Delta_r H_m°$ (kJ·mol^{-1})	$\Delta_r S_m°$ (J·K^{-1}·mol^{-1})
PuF$_2^{2+}$	2F$^-$ + Pu^{4+} ⇌ PuF$_2^{2+}$	15.700 ±0.200	−89.616 ±1.142	11.000 ±5.000	337.469 [a] ±17.202
PuF$_3$(g)	PuF$_3$(cr) ⇌ PuF$_3$(g)			418.900 ±0.500	
PuF$_4$(g)	PuF$_4$(cr) ⇌ PuF$_4$(g)			301.800 ±3.600	
PuF$_6$(cr)	PuF$_6$(g) ⇌ PuF$_6$(cr)			−48.650 ±1.000	
PuF$_6$(g)	F$_2$(g) + PuF$_4$(cr) ⇌ PuF$_6$(g)			37.300 ±2.400	
PuO$_2$F$^+$	F$^-$ + PuO$_2^{2+}$ ⇌ PuO$_2$F$^+$	4.560 ±0.200	−26.029 ±1.142		
PuO$_2$F$_2$(aq)	2F$^-$ + PuO$_2^{2+}$ ⇌ PuO$_2$F$_2$(aq)	7.250 ±0.450	−41.383 ±2.569		
PuCl^{3+}	Cl$^-$ + Pu^{4+} ⇌ PuCl^{3+}	1.800 ±0.300	−10.274 ±1.712		
PuCl$_3$(g)	PuCl$_3$(cr) ⇌ PuCl$_3$(g)			312.200 ±0.500	
PuCl$_4$(g)	0.5Cl$_2$(g) + PuCl$_3$(cr) ⇌ PuCl$_4$(g)			167.600 ±1.000	
PuOCl(cr)	H$_2$O(g) + PuCl$_3$(cr) ⇌ 2HCl(g) + PuOCl(cr)			128.710 ±0.490	

(Continued on next page)

Table 5–2: (continued)

Species	Reaction	$\log_{10} K^\circ$	$\Delta_r G_m^\circ$ (kJ·mol^{-1})	$\Delta_r H_m^\circ$ (kJ·mol^{-1})	$\Delta_r S_m^\circ$ (J·K^{-1}·mol^{-1})
PuO_2Cl^+	$Cl^- + PuO_2^{2+} \rightleftharpoons PuO_2Cl^+$	0.230 ±0.030	−1.313 ±0.171		
$PuO_2Cl_2(aq)$	$2Cl^- + PuO_2^{2+} \rightleftharpoons PuO_2Cl_2(aq)$	−1.150 ±0.300	6.564 ±1.712		
$PuBr^{3+}$	$Br^- + Pu^{4+} \rightleftharpoons PuBr^{3+}$	1.600 ±0.300	−9.133 ±1.712		
$PuBr_3(g)$	$PuBr_3(cr) \rightleftharpoons PuBr_3(g)$	−41.565 [c] ±2.728	237.257 [a] ±15.570	304.400 ±15.000	225.200 ±14.000
$PuOBr(cr)$	$H_2O(g) + PuBr_3(cr) \rightleftharpoons 2HBr(g) + PuOBr(cr)$	−8.893 [c] ±0.914	50.764 [a] ±5.218	91.700 ±5.000	137.300 ±5.000
PuI^{2+}	$I^- + Pu^{3+} \rightleftharpoons PuI^{2+}$	1.100 ±0.400	−6.279 ±2.283		
$PuSO_4^+$	$HSO_4^- + Pu^{3+} \rightleftharpoons H^+ + PuSO_4^+$	1.930 ±0.610	−11.017 ±3.482	−5.200 ±2.000	19.509 [a] ±13.468
$PuSO_4^{2+}$	$HSO_4^- + Pu^{4+} \rightleftharpoons H^+ + PuSO_4^{2+}$	4.910 ±0.220	−28.026 ±1.256		
$PuO_2SO_4(aq)$	$PuO_2^{2+} + SO_4^{2-} \rightleftharpoons PuO_2SO_4(aq)$	3.380 ±0.200	−19.293 ±1.142	16.100 ±0.600	118.709 [a] ±4.326
$Pu(SO_4)_2(aq)$	$2HSO_4^- + Pu^{4+} \rightleftharpoons 2H^+ + Pu(SO_4)_2(aq)$	7.180 ±0.320	−40.984 ±1.827		
$Pu(SO_4)_2^-$	$2HSO_4^- + Pu^{3+} \rightleftharpoons 2H^+ + Pu(SO_4)_2^-$	1.740 ±0.760	−9.932 ±4.338	−33.000 ±16.000	−77.370 [a] ±55.602

(Continued on next page)

Table 5–2: (continued)

Species	Reaction			
	$\log_{10} K°$	$\Delta_r G_m°$ (kJ·mol⁻¹)	$\Delta_r H_m°$ (kJ·mol⁻¹)	$\Delta_r S_m°$ (J·K⁻¹·mol⁻¹)
$PuO_2(SO_4)_2^{2-}$	$PuO_2^{2+} + 2SO_4^{2-} \rightleftharpoons PuO_2(SO_4)_2^{2-}$			
	4.400 ±0.200	−25.115 ±1.142	43.000 ±9.000	228.460 [a] ±30.428
$PuNO_3^{3+}$	$NO_3^- + Pu^{4+} \rightleftharpoons PuNO_3^{3+}$			
	1.950 ±0.150	−11.131 ±0.856		
$PuPO_4(s, hyd)$	$PO_4^{3-} + Pu^{3+} \rightleftharpoons PuPO_4(s, hyd)$			
	24.600 ±0.800	−140.418 ±4.566		
$PuH_3PO_4^{4+}$	$H_3PO_4(aq) + Pu^{4+} \rightleftharpoons PuH_3PO_4^{4+}$			
	2.400 ±0.300	−13.699 ±1.712		
$Pu(HPO_4)_2(am, hyd)$	$2HPO_4^{2-} + Pu^{4+} \rightleftharpoons Pu(HPO_4)_2(am, hyd)$			
	30.450 ±0.510	−173.810 ±2.911		
$PuO_2CO_3(aq)$	$CO_3^{2-} + PuO_2^{2+} \rightleftharpoons PuO_2CO_3(aq)$			
	9.500 ±0.500	−54.226 ±2.854		
$PuO_2CO_3(s)$	$CO_3^{2-} + PuO_2^{2+} \rightleftharpoons PuO_2CO_3(s)$			
	14.650 ±0.470	−83.623 ±2.683		
$PuO_2CO_3^-$	$CO_3^{2-} + PuO_2^+ \rightleftharpoons PuO_2CO_3^-$			
	5.120 ±0.140	−29.225 ±0.799		
$PuO_2(CO_3)_2^{2-}$	$2CO_3^{2-} + PuO_2^{2+} \rightleftharpoons PuO_2(CO_3)_2^{2-}$			
	14.700 ±0.500	−83.908 ±2.854	−27.000 ±4.000	190.871 [a] ±16.481
$PuO_2(CO_3)_3^{4-}$	$3CO_3^{2-} + PuO_2^{2+} \rightleftharpoons PuO_2(CO_3)_3^{4-}$			
	18.000 ±0.500	−102.745 ±2.854	−38.600 ±2.000	215.143 [a] ±11.689
$PuO_2(CO_3)_3^{5-}$	$3CO_3^{2-} + PuO_2^+ \rightleftharpoons PuO_2(CO_3)_3^{5-}$			
	5.025 [c] ±0.920	−28.685 ±5.250	−19.110 ±8.500	32.115 [a] ±33.509

(Continued on next page)

Table 5–2: (continued)

Species	Reaction $\log_{10} K°$	$\Delta_r G_m^\circ$ (kJ·mol^{-1})	$\Delta_r H_m^\circ$ (kJ·mol^{-1})	$\Delta_r S_m^\circ$ (J·K^{-1}·mol^{-1})
Pu(CO$_3$)$_4^{4-}$	$4CO_3^{2-} + Pu^{4+} \rightleftharpoons Pu(CO_3)_4^{4-}$			
	37.000 ±1.100	−211.198 ±6.279		
Pu(CO$_3$)$_5^{6-}$	$5CO_3^{2-} + Pu^{4+} \rightleftharpoons Pu(CO_3)_5^{6-}$			
	35.650 ±1.130	−203.492 ±6.450		
PuSCN^{2+}	$Pu^{3+} + SCN^- \rightleftharpoons PuSCN^{2+}$			
	1.300 ±0.400	−7.420 ±2.283	0.000 ±4.000	24.888 (a) ±15.448
Cs$_3$PuCl$_6$(cr)	CsCl(cr) + 0.2Cs$_2$Pu$_2$Cl$_7$(cr) $\rightleftharpoons$ 0.4Cs$_3$PuCl$_6$(cr)			
	3.922 (c) ±0.638	−22.387 (a) ±3.640	−23.580 ±3.426	−4.000 ±4.130
CsPu$_2$Cl$_7$(cr)	CsCl(cr) + 2PuCl$_3$(cr) $\rightleftharpoons$ CsPu$_2$Cl$_7$(cr)			
	6.605 (c) ±0.594	−37.700 (a) ±3.390	−37.870 ±3.160	−0.570 ±4.120

(a) Value calculated internally with the Gibbs–Helmholtz equation, $\Delta_r G_m^\circ = \Delta_r H_m^\circ - T \Delta_r S_m^\circ$.
(b) Value calculated from a selected standard potential.
(c) Value of $\log_{10} K°$ calculated internally from $\Delta_r G_m^\circ$.

Table 5–3: Selected temperature coefficients for heat capacities marked with [(c)] in Table 5–1, according to the form $C_{p,m}(T) = a + bT + cT^2 + eT^{-2}$. The functions are valid between the temperatures T_{min} and T_{max} (in K). The notation E±nn indicates the power of 10. for $C_{p,m}$ units are $J \cdot K^{-1} \cdot mol^{-1}$.

Compound	a	b	c	e	T_{min}	T_{max}
Pu(cr)	1.81260E+01	4.48200E–02	0	0	298	398
β–Pu	2.74160E+01	1.30600E–02	0	0	398	488
δ–Pu	2.84780E+01	1.08070E–02	0	0	593	736
δ'–Pu	3.55600E+01	0	0	0	736	756
γ–Pu	2.20230E+01	2.29590E–02	0	0	488	593
ε–Pu	3.37200E+01	0	0	0	756	913
Pu(g)	2.05200E+01	–9.39140E–03	2.57550E–05	7.51200E+04	298	500
	4.59200E+00	3.31521E–02	–5.64466E–06	7.01590E+05	500	1100
	–9.58800E+00	4.60885E–02	–8.63248E–06	5.01446E+05	1100	2000
PuO$_{1.61}$(bcc)	6.59100E+01	1.38500E–02	0	–8.75700E+05	298	2300
PuO$_2$(cr)	8.44950E+01	1.06390E–02	–6.11360E–07	–1.90060E+06	298	2500
Pu$_2$O$_3$(cr)	1.69470E+02	–7.99800E–02	0	–2.54590E+06	298	350
	1.22953E+02	2.85480E–02	0	–1.50120E+05	350	2358
PuF(g)	3.66410E+01	9.20600E–04	–1.44430E–07	–3.02540E+05	298	3000
PuF$_2$(g)	5.73100E+01	7.27000E–04	–1.47210E–07	–5.38960E+05	298	3000
PuF$_3$(cr)	1.04080E+02	7.07000E–04	0	–1.03550E+06	298	1700
PuF$_3$(g)	7.92670E+01	5.60920E–03	–2.20240E–06	–7.55940E+05	298	1300
PuF$_4$(g)	1.05110E+02	2.84120E–03	–6.88730E–07	–1.19680E+06	298	2400
PuF$_6$(cr)	7.23480E+01	3.21300E–01	0	0	298	325
PuF$_6$(g)	1.43990E+02	2.32110E–02	–1.07640E–05	–1.83430E+06	298	1000
PuOF(cr)	7.20000E+01	1.60000E–02	–3.30000E–06	–6.20000E+05	298	1500
PuCl$_3$(cr)	9.13500E+01	2.40000E–02	0	2.40000E+05	298	1041
PuCl$_3$(g)	7.71030E+01	1.29970E–02	–4.31250E–06	–1.88730E+05	298	1100
PuCl$_4$(g)	1.10430E+02	4.08180E–03	–9.76160E–07	–7.23430E+05	298	3000
PuOCl(cr)	7.30300E+01	1.71000E–02	0	–5.83000E+05	298	1100
PuBr$_3$(cr)	1.04500E+02	1.50000E–02	0	–6.38000E+05	298	935
PuBr$_3$(g)	8.31350E+01	3.90000E–06	0	–1.38320E+05	298	1500
PuOBr(cr)	7.37000E+01	1.70000E–02	0	–5.15000E+05	298	1100
PuI$_3$(cr)	1.04000E+02	2.00000E–02	0	0	298	930
PuI$_3$(g)	8.31500E+01	1.40000E–06	0	–1.00000E+05	298	1500
PuOI(cr)	6.70000E+01	3.57000E–02	–1.20000E–05	–9.00000E+04	298	1000
PuC$_{0.84}$(cr)	7.15910E+01	–5.95040E–02	4.94350E–05	–9.93200E+05	298	1875
Pu$_3$C$_2$(cr)	1.20670E+02	4.68600E–02	0	1.94560E+05	298	850
Pu$_2$C$_3$(cr)	1.56000E+02	–7.98730E–02	7.04170E–05	–2.17570E+06	298	2285
Cs$_3$PuCl$_6$(cr)	2.56600E+02	3.46000E–02	0	–7.40000E+05	298	900
CsPu$_2$Cl$_7$(cr)	2.37800E+02	5.15000E–02	0	1.55000E+05	298	900

Chapter 6

Selected Data for Americium

6.1 General considerations

This chapter presents updated chemical thermodynamic data for americium species, as described in detail in Chapter 12 of this volume. The newly selected data represent revisions to those chosen in the previous NEA TDB review [95SIL/BID]. In this respect, it will be found that while new species appear in the Tables, some others have been removed from them. Table 6–1 contains the recommended thermodynamic data of the americium compounds and complexes, Table 6–2 the recommended thermodynamic data of chemical equilibrium reactions by which the americium compounds and complexes are formed, and Table 6–3 the temperature coefficients of the heat capacity data of Table 6–1 where available.

The species and reactions in Table 6–1, Table 6–2 and Table 6–3 appear in standard order of arrangement (*cf.* Figure 2.1). Table 6–2 contains information only on those reactions for which primary data selections are made in this review. These selected reaction data are used, together with data for key americium species (for example Am^{3+}) and auxiliary data listed in Table 8–1, to derive the corresponding formation quantities in Table 6–1. The uncertainties associated with values for the key americium species and for some of the auxiliary data are substantial, leading to comparatively large uncertainties in the formation quantities derived in this manner. The inclusion of a table for reaction data (Table 6–2) in this report allows the use of equilibrium constants with total uncertainties that are directly based on the experimental accuracies. This is the main reason for including both the table for reaction data (Table 6–2) and the table of $\Delta_f G_m^\circ, \Delta_f H_m^\circ, S_m^\circ$ and $C_{p,m}^\circ$ values (Table 6–1). In a few cases, the correlation of small uncertainties in values for ligands has been neglected in calculations of uncertainty values for species in Table 6–1 from uncertainty values in Table 6–2. However, for those species the effects are less than 2% of the stated uncertainties.

The selected thermal functions of the heat capacities, listed in Table 6–3, refer to the relation:

$$C_{p,m}(T) = a + b \cdot T + c \cdot T^2 + d \cdot T^{-1} + e \cdot T^{-2} \qquad (6.1)$$

No references are given in these tables since the selected data are generally not directly attributable to a specific published source. A detailed discussion of the selection procedure is presented in [95SIL/BID] and in Chapter 12 of this volume.

6.2 Precautions to be observed in the use of the tables

Geochemical modelling in aquatic systems requires the careful use of the data selected in the NEA TDB reviews. The data in the tables of selected data must not be adopted without taking into account the chemical background information discussed in the corresponding sections of this book. In particular the following precautions should be observed when using data from the Tables.

- The addition of any aqueous species and its data to this internally consistent database can result in a modified data set which is no longer rigorous and can lead to erroneous results. The situation is similar, to a lesser degree, with the addition of gases and solids. It should also be noted that the data set presented in this chapter may not be "complete" for all the conceivable systems and conditions. Gaps are pointed out in both the previous NEA TDB review and the present update.

- Solubility data for crystalline phases are well defined in the initial state, but not necessarily in the final state after "equilibrium" has been attained. Hence, the solubility calculated from these phases may be very misleading, as discussed for the solubility data of $MO_2(cr)$, M = U, Np, Pu.

- The selected thermodynamic data in [92GRE/FUG], [95SIL/BID] and [2001LEM/FUG] contain thermodynamic data for amorphous phases. Most of these refer to Gibbs energy of formation deduced from solubility measurements. In the present review all such data have been removed. However, the corresponding $\log_{10}K$ values have been retained. The reasons for these changes are:

    o Thermodynamic data are only meaningful if they refer to well–defined systems; this is not the case for amorphous phases. It is well known that their solubility may change with time, due to re–crystallisation with a resulting change in water content and surface area/crystal size; the time scale for these changes can vary widely. These kinetic phenomena are different from those encountered in systems where there is a very high activation barrier for the reaction, e.g., electron exchange between sulphate and sulphide.

    o The solubility of amorphous phases provides useful information for users of the database when modelling the behaviour of complex systems. Therefore the solubility products have been given, however with the proviso that they are not thermodynamic quantities.

6 Selected americium data

Table 6–1: Selected thermodynamic data for americium compounds and complexes. All ionic species listed in this table are aqueous species. Unless noted otherwise, all data refer to the reference temperature of 298.15 K and to the standard state, *i.e.*, a pressure of 0.1 MPa and, for aqueous species, infinite dilution ($I = 0$). The uncertainties listed below each value represent total uncertainties and correspond in principle to the statistically defined 95% confidence interval. Values obtained from internal calculation, *cf.* footnotes (a) and (b), are rounded at the third digit after the decimal point and may therefore not be exactly identical to those given in Part III. Systematically, all the values are presented with three digits after the decimal point, regardless of the significance of these digits. The data presented in this table are available on computer media from the OECD Nuclear Energy Agency.

Compound	$\Delta_f G_m^\circ$ (kJ · mol^{-1})	$\Delta_f H_m^\circ$ (kJ · mol^{-1})	S_m° (J · K^{-1} · mol^{-1})	$C_{p,m}^\circ$ (J · K^{-1} · mol^{-1})
Am(cr)	0.000	0.000	55.400 ±2.000	25.500 [c] ±1.500
β–Am				[d]
γ–Am				[d]
Am(l)				[d]
Am(g)	242.312 [a] ±1.614	283.800 ±1.500	194.550 ±0.050	20.786 [c] ±0.010
Am^{2+}	−376.780 [b] ±15.236	−354.633 [a] ±15.890	−1.000 ±15.000	
Am^{3+}	−598.698 [a] ±4.755	−616.700 ±1.500	−201.000 ±15.000	
Am^{4+}	−346.358 [a] ±8.692	−406.000 ±6.000	−406.000 ±21.000	
AmO$_2$(cr)	−877.683 [a] ±4.271	−932.200 ±3.000	77.700 ±10.000	66.170 [c] ±10.000
AmO$_2^+$	−739.796 [a] ±6.208	−804.260 [b] ±5.413	−21.000 ±10.000	
AmO$_2^{2+}$	−585.801 [a] ±5.715	−650.760 [b] ±4.839	−88.000 ±10.000	
Am$_2$O$_3$(cr)	−1605.449 [a] ±8.284	−1690.400 ±8.000	133.600 ±6.000	117.500 [c] ±15.000
AmH$_2$(cr)	−134.661 [a] ±15.055	−175.800 ±15.000	48.100 ±3.800	38.200 [c] ±2.500

(Continued on next page)

Table 6–1: (continued)

Compound	$\Delta_f G_m^\circ$ (kJ·mol⁻¹)	$\Delta_f H_m^\circ$ (kJ·mol⁻¹)	S_m° (J·K⁻¹·mol⁻¹)	$C_{p,m}^\circ$ (J·K⁻¹·mol⁻¹)
AmOH²⁺	−794.740 [b] ±5.546			
Am(OH)₂⁺	−986.787 [b] ±6.211			
Am(OH)₃(aq)	−1160.568 [b] ±5.547			
Am(OH)₃(cr)	−1221.073 [b] ±5.861	−1353.198 [a] ±6.356	116.000 ±8.000	
AmF²⁺	−899.628 [b] ±5.320			
AmF₂⁺	−1194.851 [b] ±5.082			
AmF₃(cr)	−1519.765 [a] ±14.126	−1594.000 ±14.000	110.600 ±6.000	
AmF₃(g)	−1147.798 [a] ±16.771	−1156.500 [b] ±16.589	330.400 ±8.000	72.200 ±5.000
AmF₄(cr)	−1632.503 [a] ±17.177	−1724.000 ±17.000	154.100 ±8.000	
AmCl²⁺	−731.285 [b] ±4.759			
AmCl₂⁺	−856.908 [b] ±4.769			
AmCl₃(cr)	−905.105 [a] ±2.290	−977.800 ±1.300	146.200 ±6.000	103.000 ±10.000
AmOCl(cr)	−897.052 [a] ±6.726	−949.800 [b] ±6.000	92.600 [b] ±10.000	70.400 [c] ±10.000
AmBr₃(cr)	−773.674 [a] ±6.728	−804.000 ±6.000	182.000 ±10.000	
AmOBr(cr)	−848.485 [a] ±9.794	−887.000 [b] ±9.000	104.900 [b] ±12.800	
AmI₃(cr)	−609.451 [a] ±10.068	−615.000 ±9.000	211.000 ±15.000	
AmS(cr)			92.000 ±12.000	
AmSO₄⁺	−1361.538 [b] ±4.849			

(Continued on next page)

Table 6–1: (continued)

Compound	$\Delta_f G_m^\circ$ (kJ·mol⁻¹)	$\Delta_f H_m^\circ$ (kJ·mol⁻¹)	S_m° (J·K⁻¹·mol⁻¹)	$C_{p,m}^\circ$ (J·K⁻¹·mol⁻¹)
$Am(SO_4)_2^-$	−2107.826 [b] ±4.903			
$AmSe(cr)$			109.000 ±12.000	
$AmTe(cr)$			121.000 ±12.000	
AmN_3^{2+}	−260.030 [b] ±5.190			
$AmNO_3^{2+}$	−717.083 [b] ±4.908			
$AmH_2PO_4^{2+}$	−1752.974 [b] ±5.763			
$Am_2C_3(cr)$	−156.063 [a] ±42.438	−151.000 ±42.000	145.000 ±20.000	
$AmCO_3^+$	−1172.262 [b] ±5.289			
$AmO_2CO_3^-$	−1296.807 [b] ±6.844			
$Am(CO_3)_2^-$	−1728.131 [b] ±5.911			
$AmO_2(CO_3)_2^{3-}$	−1833.840 [b] ±7.746			
$Am(CO_3)_3^{3-}$	−2268.018 [b] ±7.521			
$AmO_2(CO_3)_3^{5-}$	−2352.607 [b] ±8.514			
$Am(CO_3)_5^{6-}$	−3210.227 [b] ±7.919			
$AmHCO_3^{2+}$	−1203.238 [b] ±5.060			
$AmCO_3OH \cdot 0.5\,H_2O(cr)$	−1530.248 [b] ±5.560	−1682.900 ±2.600	141.413 [a] ±20.683	
$AmSiO(OH)_3^{2+}$	−1896.844 [b] ±5.000			
$AmSCN^{2+}$	−513.418 [b] ±6.445			

(Continued on next page)

Table 6–1: (continued)

Compound	$\Delta_f G_m^\circ$ (kJ·mol^{-1})	$\Delta_f H_m^\circ$ (kJ·mol^{-1})	S_m° (J·K^{-1}·mol^{-1})	$C_{p,m}^\circ$ (J·K^{-1}·mol^{-1})
SrAmO$_3$(cr)		−1539.000 ±4.100		
BaAmO$_3$(cr)		−1544.600 ±3.400		
NaAmO$_2$CO$_3$(s)	−1591.867 (b) ±6.627			
NaAm(CO$_3$)$_2$·5H$_2$O(cr)	−3222.021 (b) ±5.605			
Cs$_2$NaAmCl$_6$(cr)	−2159.151 (a) ±4.864	−2315.800 ±1.800	421.000 ±15.000	260.000 ±15.000

(a) Value calculated internally with the Gibbs–Helmholtz equation, $\Delta_f G_m^\circ = \Delta_f H_m^\circ - T\sum_i S_{m,i}^\circ$.
(b) Value calculated internally from reaction data (see Table 6–2).
(c) Temperature coefficients of this function are listed in Table 6–3.
(d) A temperature function for the heat capacity is given in Table 6–3.

Table 6–2: Selected thermodynamic data for reactions involving americium compounds and complexes. All ionic species listed in this table are aqueous species. Unless noted otherwise, all data refer to the reference temperature of 298.15 K and to the standard state, i.e., a pressure of 0.1 MPa and, for aqueous species, infinite dilution ($I = 0$). The uncertainties listed below each value represent total uncertainties and correspond in principle to the statistically defined 95% confidence interval. Values obtained from internal calculation, cf. footnote (a), are rounded at the third digit after the decimal point and may therefore not be exactly identical to those given in Part III. Systematically, all the values are presented with three digits after the decimal point, regardless of the significance of these digits. The data presented in this table are available on computer media from the OECD Nuclear Energy Agency.

Species	Reaction	$\log_{10} K°$	$\Delta_r G_m°$ (kJ · mol^{-1})	$\Delta_r H_m°$ (kJ · mol^{-1})	$\Delta_r S_m°$ (J · K^{-1} · mol^{-1})
Am(g)	Am(cr) $\rightleftharpoons$ Am(g)			283.800 ±1.500	
Am^{2+}	Am^{3+} + e$^-$ $\rightleftharpoons$ Am^{2+}	−38.878 (b) ±2.536	221.920 ±14.476		
AmO$_2^+$	Am^{3+} + 2H$_2$O(l) $\rightleftharpoons$ AmO$_2^+$ + 4H$^+$ + 2 e$^-$			384.100 ±5.200	
AmO$_2^{2+}$	Am^{3+} + 2H$_2$O(l) $\rightleftharpoons$ AmO$_2^{2+}$ + 4H$^+$ + 3 e$^-$			537.600 ±4.600	
AmOH^{2+}	Am^{3+} + H$_2$O(l) $\rightleftharpoons$ AmOH^{2+} + H$^+$	−7.200 ±0.500	41.098 ±2.854		
Am(OH)$_2^+$	Am^{3+} + 2H$_2$O(l) $\rightleftharpoons$ Am(OH)$_2^+$ + 2H$^+$	−15.100 ±0.700	86.191 ±3.996		
AmO$_2$OH(am)	AmO$_2^+$ + H$_2$O(l) $\rightleftharpoons$ AmO$_2$OH(am) + H$^+$	−5.300 ±0.500	30.253 ±2.854		
Am(OH)$_3$(cr)	Am^{3+} + 3H$_2$O(l) $\rightleftharpoons$ Am(OH)$_3$(cr) + 3H$^+$	−15.600 ±0.600	89.045 ±3.425		

(Continued on next page)

Table 6–2: (continued)

Species	Reaction	$\log_{10} K°$	$\Delta_r G_m^o$ (kJ·mol^{-1})	$\Delta_r H_m^o$ (kJ·mol^{-1})	$\Delta_r S_m^o$ (J·K^{-1}·mol^{-1})
Am(OH)$_3$(am)	Am^{3+} + 3H$_2$O(l) $\rightleftharpoons$ Am(OH)$_3$(am) + 3H$^+$	−16.900 ±0.800	96.466 ±4.566		
Am(OH)$_3$(aq)	Am^{3+} + 3H$_2$O(l) $\rightleftharpoons$ Am(OH)$_3$(aq) + 3H$^+$	−26.200 ±0.500	149.551 ±2.854		
AmF^{2+}	Am^{3+} + F$^-$ $\rightleftharpoons$ AmF^{2+}	3.400 ±0.400	−19.407 ±2.283		
AmF$_2^+$	Am^{3+} + 2F$^-$ $\rightleftharpoons$ AmF$_2^+$	5.800 ±0.200	−33.107 ±1.142		
AmF$_3$(g)	AmF$_3$(cr) $\rightleftharpoons$ AmF$_3$(g)			437.500 ±8.900	
AmCl^{2+}	Am^{3+} + Cl$^-$ $\rightleftharpoons$ AmCl^{2+}	0.240 ±0.030	−1.370 ±0.171		
AmCl$_2^+$	Am^{3+} + 2Cl$^-$ $\rightleftharpoons$ AmCl$_2^+$	−0.740 ±0.050	4.224 ±0.285		
AmOCl(cr)	AmCl$_3$(cr) + H$_2$O(g) $\rightleftharpoons$ AmOCl(cr) + 2HCl(g)	−8.066 [c] ±1.115	46.042 [a] ±6.364	85.213 ±5.900	131.380 ±8.000
AmOBr(cr)	AmBr$_3$(cr) + H$_2$O(g) $\rightleftharpoons$ AmOBr(cr) + 2HBr(g)	−8.246 [c] ±2.661	47.070 [a] ±15.188	86.256 ±15.000	131.430 ±8.000
AmSO$_4^+$	Am^{3+} + SO$_4^{2-}$ $\rightleftharpoons$ AmSO$_4^+$	3.300 ±0.150	−18.837 ±0.856		
Am(SO$_4$)$_2^-$	Am^{3+} + 2SO$_4^{2-}$ $\rightleftharpoons$ Am(SO$_4$)$_2^-$	3.700 ±0.150	−21.120 ±0.856		

(Continued on next page)

Table 6–2: (continued)

Species	Reaction			
	$\log_{10} K°$	$\Delta_r G_m°$ (kJ·mol^{-1})	$\Delta_r H_m°$ (kJ·mol^{-1})	$\Delta_r S_m°$ (J·K^{-1}·mol^{-1})
AmN_3^{2+}	$Am^{3+} + N_3^- \rightleftharpoons AmN_3^{2+}$			
	1.670 ±0.100	−9.532 ±0.571		
$AmNO_2^{2+}$	$Am^{3+} + NO_2^- \rightleftharpoons AmNO_2^{2+}$			
	2.100 ±0.200	−11.987 ±1.142		
$AmNO_3^{2+}$	$Am^{3+} + NO_3^- \rightleftharpoons AmNO_3^{2+}$			
	1.330 ±0.200	−7.592 ±1.142		
$AmPO_4$(am, hyd)	$Am^{3+} + PO_4^{3-} \rightleftharpoons AmPO_4$(am, hyd)			
	24.790 ±0.600	−141.500 ±3.425		
$AmH_2PO_4^{2+}$	$Am^{3+} + H_2PO_4^- \rightleftharpoons AmH_2PO_4^{2+}$			
	3.000 ±0.500	−17.124 ±2.854		
$AmCO_3^+$	$Am^{3+} + CO_3^{2-} \rightleftharpoons AmCO_3^+$			
	8.000 ±0.400	−45.664 ±2.283		
$AmO_2CO_3^-$	$AmO_2^+ + CO_3^{2-} \rightleftharpoons AmO_2CO_3^-$			
	5.100 ±0.500	−29.111 ±2.854		
$Am(CO_3)_2^-$	$Am^{3+} + 2CO_3^{2-} \rightleftharpoons Am(CO_3)_2^-$			
	12.900 ±0.600	−73.634 ±3.425		
$AmO_2(CO_3)_2^{3-}$	$AmO_2^+ + 2CO_3^{2-} \rightleftharpoons AmO_2(CO_3)_2^{3-}$			
	6.700 ±0.800	−38.244 ±4.566		
$Am(CO_3)_3^{3-}$	$Am^{3+} + 3CO_3^{2-} \rightleftharpoons Am(CO_3)_3^{3-}$			
	15.000 ±1.000	−85.621 ±5.708		
$Am_2(CO_3)_3$(am)	$2Am^{3+} + 3CO_3^{2-} \rightleftharpoons Am_2(CO_3)_3$(am)			
	16.700 ±1.100	−95.324 ±6.279		

(Continued on next page)

Table 6–2: (continued)

Species	Reaction	$\log_{10} K°$	$\Delta_r G_m°$ (kJ·mol^{-1})	$\Delta_r H_m°$ (kJ·mol^{-1})	$\Delta_r S_m°$ (J·K^{-1}·mol^{-1})
$AmO_2(CO_3)_3^{5-}$	$AmO_2^+ + 3CO_3^{2-} \rightleftharpoons AmO_2(CO_3)_3^{5-}$	5.100 ±1.000	−29.111 ±5.708		
$Am(CO_3)_5^{6-}$	$Am(CO_3)_3^{3-} + 2CO_3^{2-} \rightleftharpoons Am(CO_3)_5^{6-} + e^-$	−20.100 (b) ±0.900	114.730 ±5.137		
$AmHCO_3^{2+}$	$Am^{3+} + HCO_3^- \rightleftharpoons AmHCO_3^{2+}$	3.100 ±0.300	−17.695 ±1.712		
$AmCO_3OH \cdot 0.5H_2O(cr)$	$Am^{3+} + CO_3^{2-} + 0.5H_2O(l) + OH^- \rightleftharpoons AmCO_3OH \cdot 0.5H_2O(cr)$	22.400 ±0.500	−127.860 ±2.854		
$AmCO_3OH(am, hyd)$	$Am^{3+} + CO_3^{2-} + OH^- \rightleftharpoons AmCO_3OH(am, hyd)$	20.200 ±1.000	−115.302 ±5.708		
$AmSiO(OH)_3^{2+}$	$Am^{3+} + Si(OH)_4(aq) \rightleftharpoons AmSiO(OH)_3^{2+} + H^+$	−1.680 ±0.180	9.590 ±1.027		
$AmSCN^{2+}$	$Am^{3+} + SCN^- \rightleftharpoons AmSCN^{2+}$	1.300 ±0.300	−7.420 ±1.712		
$NaAmO_2CO_3(s)$	$AmO_2^+ + CO_3^{2-} + Na^+ \rightleftharpoons NaAmO_2CO_3(s)$	10.900 ±0.400	−62.218 ±2.283		
$NaAm(CO_3)_2 \cdot 5H_2O(cr)$	$Am^{3+} + 2CO_3^{2-} + 5H_2O(l) + Na^+ \rightleftharpoons NaAm(CO_3)_2 \cdot 5H_2O(cr)$	21.000 ±0.500	−119.869 ±2.854		

(a) Value calculated internally with the Gibbs–Helmholtz equation, $\Delta_r G_m° = \Delta_r H_m° - T \Delta_r S_m°$.
(b) Value calculated from a selected standard potential.
(c) Value of $\log_{10} K°$ calculated internally from $\Delta_r G_m°$.

Table 6–3: Selected temperature coefficients for heat capacities marked with [c] in Table 6–1, according to the form $C_{p,m}(T) = a + bT + cT^2 + eT^{-2}$. The functions are valid between the temperatures T_{min} and T_{max} (in K). The notation E±nn indicates the power of 10. $C_{p,m}$ units are $J \cdot K^{-1} \cdot mol^{-1}$.

Compound	a	b	c	e	T_{min}	T_{max}
Am(cr)	2.11870E+01	1.11990E–02	3.24620E–06	6.08500E+04	298	1042
β–Am	1.94410E+01	1.08360E–02	2.25140E–06	5.20870E+05	1042	1350
γ–Am	3.97480E+01	0	0	0	1350	1449
Am(l)	4.18400E+01	0	0	0	1449	3000
Am(g)	2.07860E+01	0	0	0	298	1100
AmO_2(cr)	8.47390E+01	1.07200E–02	–8.15900E–07	–1.92580E+06	298	2000
Am_2O_3(cr)	1.13930E+02	5.93700E–02	–2.30100E–05	–1.07100E+06	298	1000
AmH_2(cr)	2.48000E+01	4.50000E–02	0	0	298	1200
AmF_3(g)	8.16540E+01	1.20840E–03	–2.42780E–07	–8.67160E+05	298	3000
AmOCl(cr)	6.12840E+01	4.58930E–02	–1.73070E–05	–2.69380E+05	298	1100

Chapter 7

Selected Data for Technetium

7.1 General considerations

This chapter presents updated chemical thermodynamic data for technetium species, as described in detail in Chapter 13 of this volume. The newly selected data represent revisions to those chosen in the previous NEA TDB review [99RAR/RAN]. According to the discussion in Chapter 13 of this volume, this update has not resulted in any changes to the values selected there. However it should be noted that several typographical errors slipped into the selected value tables published in [99RAR/RAN]. For this reason, the tables reproduced here have been recalculated and all the entries in them are now in agreement with the discussions presented in [99RAR/RAN] and in this volume. Table 7–1 contains the recommended thermodynamic data of the technetium compounds and complexes, Table 7–2 the recommended thermodynamic data of chemical equilibrium reactions by which the technetium compounds and complexes are formed, and Table 7–3 the temperature coefficients of the heat capacity data of Table 7–1 where available.

The species and reactions in Table 7–1, Table 7–2 and Table 7–3 appear in standard order of arrangement (*cf.* Figure 2.1). Table 7–2 contains information only on those reactions for which primary data selections are made in this review. These selected reaction data are used, together with data for key technetium species (for example TcO_4^-) and auxiliary data listed in Table 8–1, to derive the corresponding formation quantities in Table 7–1. The uncertainties associated with values for the key technetium species and for some of the auxiliary data are substantial, leading to comparatively large uncertainties in the formation quantities derived in this manner. The inclusion of a table for reaction data (Table 7–2) in this report allows the use of equilibrium constants with total uncertainties that are directly based on the experimental accuracies. This is the main reason for including both the table for reaction data (Table 7–2) and the table of $\Delta_f G_m^\circ$, $\Delta_f H_m^\circ$, S_m° and $C_{p,m}^\circ$ values (Table 7–1). In a few cases, the correlation of small uncertainties in values for ligands has been neglected in calculations of uncertainty values for species in Table 7–3 from uncertainty values in Table 7–2. However, for those species the effects are less than 2% of the stated uncertainties.

The selected thermal functions of the heat capacities, listed in Table 7–3, refer to the relation:

$$C_{p,m}(T) = a + b \cdot T + c \cdot T^2 + d \cdot T^{-1} + e \cdot T^{-2} \qquad (7.1)$$

No references are given in these tables since the selected data are generally not directly attributable to a specific published source. A detailed discussion of the selection procedure is presented in [99RAR/RAN] and in Chapter 13 of this volume.

7.2 Precautions to be observed in the use of the tables

Geochemical modelling in aquatic systems requires the careful use of the data selected in the NEA TDB reviews. The data in the tables of selected data must not be adopted without taking into account the chemical background information discussed in the corresponding sections of this book. In particular the following precautions should be observed when using data from the Tables.

- The addition of any aqueous species and its data to this internally consistent database can result in a modified data set which is no longer rigorous and can lead to erroneous results. The situation is similar, to a lesser degree, with the addition of gases and solids. It should also be noted that the data set presented in this chapter may not be "complete" for all the conceivable systems and conditions. Gaps are pointed out in both the previous NEA TDB review and the present update.

- Solubility data for crystalline phases are well defined in the initial state, but not necessarily in the final state after "equilibrium" has been attained. Hence, the solubility calculated from these phases may be very misleading, as discussed for the solubility data of $MO_2(cr)$, M = U, Np, Pu.

Table 7–1: Selected thermodynamic data for technetium compounds and complexes. Selected thermodynamic functions for some Tc species are also given in Appendix B. All ionic species listed in this table are aqueous species. Unless noted otherwise, all data refer to 298.15 K and a pressure of 0.1 MPa and, for aqueous species, infinite dilution ($I = 0$). The uncertainties listed below each value represent total uncertainties and correspond in principle to the statistically defined 95% confidence interval. Values obtained from internal calculation, cf. footnotes (a) and (b), are rounded at the third digit after the decimal point and may therefore not be exactly identical to those given in Chapter V. Systematically, all the values are presented with three digits after the decimal point, regardless of the significance of these digits. It should be noted that insufficient auxiliary data are available in a number of cases to derive formation data from the reactions listed in Table 7–2. The data presented in this table are available on computer media from the OECD Nuclear Energy Agency.

Compound	$\Delta_f G_m^\circ$ (kJ·mol^{-1})	$\Delta_f H_m^\circ$ (kJ·mol^{-1})	S_m° (J·K^{-1}·mol^{-1})	$C_{p,m}^\circ$ (J·K^{-1}·mol^{-1})
Tc(cr)	0.000	0.000	32.506 ±0.700	24.879 [c,d] ±1.000
Tc(g)	630.711 [a] ±25.001	675.000 [b] ±25.000	181.052 ±0.010	20.795 [c,d] ±0.010
TcO(g)	357.494 [a] ±57.001	390.000 ±57.000	244.109 ±0.600	31.256 [d] ±0.750
TcO^{2+}	> −116.799 [b]			
TcO$_2$(cr)	−401.850 [a] ±11.762	−457.800 ±11.700	50.000 ±4.000	
TcO$_4^-$	−637.406 [a] ±7.616	−729.400 ±7.600	199.600 ±1.500	−15.000 ±8.000
TcO$_4^{2-}$	−575.759 [b] ±8.133			
Tc$_2$O$_7$(cr)	−950.280 [a] ±15.562	−1126.500 ±14.900	192.000 ±15.000	160.400 [c,d] ±15.000
Tc$_2$O$_7$(g)	−904.820 [a] ±16.462	−1008.100 [b] ±16.063	436.641 ±12.000	146.736 [d] ±5.000
TcO(OH)$^+$	−345.377 [b] ±9.009			
TcO(OH)$_2$(aq)	−568.247 [b] ±8.845			
TcO$_2$·1.6 H$_2$O(s)	−758.479 [b] ±8.372			

(Continued on next page)

Table 7–1: (continued)

Compound	$\Delta_f G_m^\circ$ (kJ · mol^{-1})	$\Delta_f H_m^\circ$ (kJ · mol^{-1})	S_m° (J · K^{-1} · mol^{-1})	$C_{p,m}^\circ$ (J · K^{-1} · mol^{-1})
TcO(OH)$_3^-$	−743.170 (b) ±9.135			
Tc$_2$O$_7$· H$_2$O(s)	−1194.300 ±15.500	−1414.146 (b) ±14.905	278.933 (a) ±72.138	
TcF$_6$(cr, cubic)			253.520 ±0.510	157.840 ±0.320
TcF$_6$(g)			359.136 ±4.500	120.703 ±2.629
TcO$_3$F(g)			306.879 ±0.677	77.261 (c) ±0.308
TcO$_3$Cl(g)			317.636 ±0.797	80.365 (c) ±0.308
TcS(g)	491.925 (a) ±65.001	549.000 ±65.000	255.990 ±1.000	34.474 (d) ±1.000
NH$_4$TcO$_4$(cr)	−721.998 (b) ±7.632			
TcC(g)	765.602 (a) ±40.250	826.500 ±40.000	242.500 ±15.000	
TcCO$_3$(OH)$_2$(aq)	−968.899 (b) ±9.010			
TcCO$_3$(OH)$_3^-$	−1158.662 (b) ±9.486			
TlTcO$_4$(cr)	−700.173 (b) ±7.653			
AgTcO$_4$(cr)	−578.975 (b) ±7.654			
NaTcO$_4$· 4 H$_2$O(s)	−1843.411 (b) ±7.622			
KTcO$_4$(cr)	−932.921 (a) ±7.604	−1035.100 ±7.600	164.780 ±0.330	123.300 ±0.250
CsTcO$_4$(cr)	−949.700 (b) ±7.700			

(a) Value calculated internally with the Gibbs–Helmholtz equation, $\Delta_f G_m^\circ = \Delta_f H_m^\circ - T \sum_i S_{m,i}^\circ$.
(b) Value calculated internally from reaction data.
(c) Temperature coefficients of this function are listed in Table 7–3.
(d) Tables giving the temperature dependence of thermodynamic functions for this substance or compound can be found in appendix B of the NEA TDB review on Tc [99RAR/RAN]

Table 7-2: Selected thermodynamic data for reactions involving technetium compounds and complexes. All ionic species listed in this table are aqueous species. Unless noted otherwise, all data refer to 298.15 K and a pressure of 0.1 MPa and, for aqueous species, infinite dilution ($I = 0$). The uncertainties listed below each value represent total uncertainties and correspond in principle to the statistically defined 95% confidence interval. Values obtained from internal calculation, cf. footnote (a), are rounded at the third digit after the decimal point and may therefore not be exactly identical to those given in Chapter 13. Systematically, all the values are presented with three digits after the decimal point, regardless of the significance of these digits. The data presented in this table are available on computer media from the OECD Nuclear Energy Agency.

Species	Reaction			
	$\log_{10} K^\circ$	$\Delta_r G_m^\circ$ (kJ·mol^{-1})	$\Delta_r H_m^\circ$ (kJ·mol^{-1})	$\Delta_r S_m^\circ$ (J·K^{-1}·mol^{-1})
Tc(g)	Tc(cr) $\rightleftharpoons$ Tc(g)			
			675.000	
			±25.000	
TcO^{2+}	2H$^+$ + TcO(OH)$_2$(aq) $\rightleftharpoons$ 2H$_2$O(l) + TcO^{2+}			
	< 4.000	> −22.832		
TcO$_4^{2-}$	TcO$_4^-$ + e$^-$ $\rightleftharpoons$ TcO$_4^{2-}$			
	−10.800 (b)		61.647	
	±0.500		±2.854	
Tc$_2$O$_7$(g)	Tc$_2$O$_7$(cr) $\rightleftharpoons$ Tc$_2$O$_7$(g)			
			118.400	
			±6.000	
TcO(OH)$^+$	H$^+$ + TcO(OH)$_2$(aq) $\rightleftharpoons$ H$_2$O(l) + TcO(OH)$^+$			
	2.500		−14.270	
	±0.300		±1.712	
TcO(OH)$_2$(aq)	TcO$_2$·1.6H$_2$O(s) $\rightleftharpoons$ 0.6H$_2$O(l) + TcO(OH)$_2$(aq)			
	−8.400		47.948	
	±0.500		±2.854	
TcO$_2$·1.6H$_2$O(s)	4H$^+$ + TcO$_4^-$ + 3 e$^-$ $\rightleftharpoons$ 0.4H$_2$O(l) + TcO$_2$·1.6H$_2$O(s)			
	37.829 (b)		−215.930	
	±0.609		±3.476	
TcO(OH)$_3^-$	H$_2$O(l) + TcO(OH)$_2$(aq) $\rightleftharpoons$ H$^+$ + TcO(OH)$_3^-$			
	−10.900		62.218	
	±0.400		±2.283	

(Continued on next page)

Table 6–2: (continued)

Species	Reaction $\log_{10} K°$	$\Delta_r G_m°$ (kJ·mol^{-1})	$\Delta_r H_m°$ (kJ·mol^{-1})	$\Delta_r S_m°$ (J·K^{-1}·mol^{-1})
$Tc_2O_7 \cdot H_2O(s)$	$H_2O(g) + Tc_2O_7(cr) \rightleftharpoons Tc_2O_7 \cdot H_2O(s)$		-45.820 ± 0.400	
$TcF_6(g)$	$TcF_6(cr, cubic) \rightleftharpoons TcF_6(g)$ -0.535 [c] ± 0.327	3.055 [a] ± 1.868	34.540 ± 1.300	105.600 ± 4.500
$TcO_2Cl_4^{3-}$	$H_2O(l) + TcOCl_5^{2-} \rightleftharpoons Cl^- + 2H^+ + TcO_2Cl_4^{3-}$ -2.950 ± 0.150		16.839 ± 0.856	
$NH_4TcO_4(cr)$	$NH_4^+ + TcO_4^- \rightleftharpoons NH_4TcO_4(cr)$ 0.910 ± 0.070		-5.194 ± 0.400	
$(NH_4)_2TcCl_6(cr)$	$2NH_4^+ + TcCl_6^{2-} \rightleftharpoons (NH_4)_2TcCl_6(cr)$ 7.988 ± 1.000		-45.596 ± 5.708	
$(NH_4)_2TcBr_6(cr)$	$2NH_4^+ + TcBr_6^{2-} \rightleftharpoons (NH_4)_2TcBr_6(cr)$ 6.680 ± 1.000		-38.130 ± 5.708	
$TcCO_3(OH)_2(aq)$	$CO_2(g) + TcO(OH)_2(aq) \rightleftharpoons TcCO_3(OH)_2(aq)$ 1.100 ± 0.300		-6.279 ± 1.712	
$TcCO_3(OH)_3^-$	$CO_2(g) + H_2O(l) + TcO(OH)_2(aq) \rightleftharpoons H^+ + TcCO_3(OH)_3^-$ -7.200 ± 0.600		41.098 ± 3.425	
$TlTcO_4(cr)$	$TcO_4^- + Tl^+ \rightleftharpoons TlTcO_4(cr)$ 5.320 ± 0.120		-30.367 ± 0.685	
$AgTcO_4(cr)$	$Ag^+ + TcO_4^- \rightleftharpoons AgTcO_4(cr)$ 3.270 ± 0.130		-18.665 ± 0.742	
$NaTcO_4 \cdot 4H_2O(s)$	$4H_2O(l) + Na^+ + TcO_4^- \rightleftharpoons NaTcO_4 \cdot 4H_2O(s)$ -0.790 ± 0.040		4.509 ± 0.228	

(Continued on next page)

Table 6–2: (continued)

Species	Reaction			
	$\log_{10} K°$	$\Delta_r G_m^°$ (kJ·mol^{-1})	$\Delta_r H_m^°$ (kJ·mol^{-1})	$\Delta_r S_m^°$ (J·K^{-1}·mol^{-1})
KTcO$_4$(cr)	K$^+$ + TcO$_4^-$ $\rightleftharpoons$ KTcO$_4$(cr)			
	2.288 ±0.026	−13.060 ±0.148	−53.620 ±0.420	−136.039 [a] ±1.494
K$_2$TcCl$_6$(cr)	2K$^+$ + TcCl$_6^{2-}$ $\rightleftharpoons$ K$_2$TcCl$_6$(cr)			
	9.610 ±1.000	−54.854 ±5.708		
K$_2$TcBr$_6$(cr)	2K$^+$ + TcBr$_6^{2-}$ $\rightleftharpoons$ K$_2$TcBr$_6$(cr)			
	6.920 ±1.000	−39.500 ±5.708		
Rb$_2$TcCl$_6$(cr)	2Rb$^+$ + TcCl$_6^{2-}$ $\rightleftharpoons$ Rb$_2$TcCl$_6$(cr)			
	11.120 ±1.000	−63.473 ±5.708		
Rb$_2$TcBr$_6$(cr)	2Rb$^+$ + TcBr$_6^{2-}$ $\rightleftharpoons$ Rb$_2$TcBr$_6$(cr)			
	9.470 ±1.000	−54.055 ±5.708		
CsTcO$_4$(cr)	Cs$^+$ + TcO$_4^-$ $\rightleftharpoons$ CsTcO$_4$(cr)			
	3.617 ±0.047	−20.646 ±0.268		
Cs$_2$TcCl$_6$(cr)	2Cs$^+$ + TcCl$_6^{2-}$ $\rightleftharpoons$ Cs$_2$TcCl$_6$(cr)			
	11.430 ±1.000	−65.243 ±5.708		
Cs$_2$TcBr$_6$(cr)	2Cs$^+$ + TcBr$_6^{2-}$ $\rightleftharpoons$ Cs$_2$TcBr$_6$(cr)			
	11.240 ±1.000	−64.158 ±5.708		

(a) Value calculated internally with the Gibbs–Helmholtz equation, $\Delta_r G_m^° = \Delta_r H_m^° - T\Delta_r S_m^°$.
(b) Value calculated from a selected standard potential.
(c) Value of $\log_{10} K°$ calculated internally from $\Delta_r G_m^°$.

Table 7–3: Selected temperature coefficients for heat capacities marked with $^{(c)}$ in Table 7–1, according to the form $C_{p,m}(T) = a + bT + cT^2 + eT^{-2}$. The functions are valid between the temperatures T_{min} and T_{max} (in K). The notation E±nn indicates the power of 10. $C_{p,m}$ units are $J \cdot K^{-1} \cdot mol^{-1}$.

Compound	a	b	c	e	T_{min}	T_{max}
Tc(cr)	2.50940E+01	4.31450E–03	–2.75460E–07	–1.31300E+05	298	2430
Tc(g)	2.49130E+01	–1.91830E–02	2.51280E–05	–5.61800E+04	298	600
Tc$_2$O$_7$(cr)	1.55000E+02	8.60000E–02	0	–1.80000E+06	298	392
TcF$_6$(cr, cubic)	7.20810E+01	2.87666E–01	0	0	268	311
TcO$_3$F(g)	7.61200E+01	5.34274E–02	–2.53393E–05	–1.11446E+06	298	1000
TcO$_3$Cl(g)	8.03360E+01	4.63950E–02	–2.20086E–05	–1.05321E+06	298	1000

Chapter 8

Selected auxiliary data

This chapter presents the chemical thermodynamic data for auxiliary compounds and complexes which are used within the NEA TDB project. Most of these auxiliary species are used in the evaluation of the recommended uranium, neptunium, plutonium, americium and technetium data in Table 3–1, Table 3–2, Table 4–1, Table 4–2, Table 5–1, Table 5–2, Table 6–1, Table 6–2, Table 7–1 andTable 7–2. It is therefore essential to always use these auxiliary data in conjunction with the selected data for uranium, neptunium, plutonium, americium and technetium. The use of other auxiliary data can lead to inconsistencies and erroneous results.

The values in the tables of this chapter are either CODATA Key Values, taken from [89COX/WAG], or were evaluated within the NEA TDB project, as described in Chapter VI of [92GRE/FUG] and [99RAR/RAN], and in Chapter 14 of this review.

Table 8–1 contains the selected thermodynamic data of the auxiliary species and Table 8–2 the selected thermodynamic data of chemical reactions involving auxiliary species. The reason for listing both reaction data and entropies, enthalpies and Gibbs energies of formation is, as described in Chapters 3, 4, 5, 6 and 7, that uncertainties in reaction data are often smaller than those derived for S_m°, $\Delta_f H_m^\circ$ and $\Delta_f G_m^\circ$, due to uncertainty accumulation during the calculations.

All data in Table 8–1 and Table 8–2 refer to a temperature of 298.15 K, the standard state pressure of 0.1 MPa and, for aqueous species and reactions, to the infinite dilution reference state ($I = 0$).

The uncertainties listed below each reaction value in Table 8–2 are total uncertainties, and correspond mainly to the statistically defined 95% confidence interval. The uncertainties listed below each value in Table 8–1 have the following significance:

- for CODATA values from [89COX/WAG], the ± terms have the meaning: "it is probable, but not at all certain, that the true values of the thermodynamic quantities differ from the recommended values given in this report by no more than twice the ± terms attached to the recommended values".

- for values from [92GRE/FUG] or [99RAR/RAN], the ± terms are derived from the total uncertainties in the corresponding equilibrium constant of reaction (*cf.* Table 8–2), and from the ± terms listed for the necessary CODATA key values.

CODATA [89COX/WAG] values are available for $CO_2(g)$, HCO_3^-, CO_3^{2-}, $H_2PO_4^-$ and HPO_4^{2-}. From the values given for $\Delta_f H_m^\circ$ and S_m°, the values of $\Delta_f G_m^\circ$ and, consequently, all the relevant equilibrium constants and enthalpy changes can be calculated. The propagation of errors during this procedure, however, leads to uncertainties in the resulting equilibrium constants that are significantly higher than those obtained from the experimental determination of the constants. Therefore, reaction data for $CO_2(g)$, HCO_3^-, CO_3^{2-}, which were originally absent from the corresponding Table 8–2 in [92GRE/FUG], are included in this table to provide the user of selected data for species of uranium, neptunium, plutonium, americium and technetium (*cf.* Chapters 3, 4, 5, 6 and 7) with the data needed to obtain the lowest possible uncertainties in the reaction properties.

Note that the values in Table 8–1 and Table 8–2 may contain more digits than those listed in either [89COX/WAG] or in Chapter VI of [92GRE/FUG] and [99RAR/RAN], because the data in the present chapter are retrieved directly from the computerised database and rounded to three digits after the decimal point throughout.

Table 8–1: Selected thermodynamic data for auxiliary compounds and complexes, including the CODATA Key Values [89COX/WAG] of species not containing uranium, neptunium, plutonium, americium or technetium as well as other data that were evaluated in Chapter VI of [92GRE/FUG] and [99RAR/RAN]. All ionic species listed in this table are aqueous species. Unless noted otherwise, all data refer to 298.15 K and a pressure of 0.1 MPa and, for aqueous species, a reference state or standard state of infinite dilution ($I = 0$). The uncertainties listed below each value represent total uncertainties and correspond in principle to the statistically defined 95% confidence interval. Values in bold typeface are CODATA Key Values and are taken directly from reference [89COX/WAG] without further evaluation. Values obtained from internal calculation, *cf.* footnotes (a) and (b), are rounded at the third digit after the decimal point and may therefore not be exactly identical to those given in Chapter VI of Ref. [92GRE/FUG] or [99RAR/RAN]. Systematically, all the values are presented with three digits after the decimal point, regardless of the significance of these digits. The data presented in this table are available on computer media from the OECD Nuclear Energy Agency.

Compound	$\Delta_f G_m^\circ$ (kJ·mol^{-1})	$\Delta_f H_m^\circ$ (kJ·mol^{-1})	S_m° (J·K^{-1}·mol^{-1})	$C_{p,m}^\circ$ (J·K^{-1}·mol^{-1})
O(g)	231.743 (a) ±0.100	249.180 ±0.100	161.059 ±0.003	21.912 ±0.001
O$_2$(g)	0.000	0.000	205.152 ±0.005	29.378 ±0.003
H(g)	203.276 (a) ±0.006	217.998 ±0.006	114.717 ±0.002	20.786 ±0.001
H$^+$	0.000	0.000	0.000	0.000
H$_2$(g)	0.000	0.000	130.680 ±0.003	28.836 ±0.002
OH$^-$	−157.220 (a) ±0.072	−230.015 ±0.040	−10.900 ±0.200	
H$_2$O(g)	−228.582 (a) ±0.040	−241.826 ±0.040	188.835 ±0.010	33.609 ±0.030
H$_2$O(l)	−237.140 (a) ±0.041	−285.830 ±0.040	69.950 ±0.030	75.351 ±0.080
H$_2$O$_2$(aq)		−191.170 (c) ±0.100		
He(g)	0.000	0.000	126.153 ±0.002	20.786 ±0.001
Ne(g)	0.000	0.000	146.328 ±0.003	20.786 ±0.001
Ar(g)	0.000	0.000	154.846 ±0.003	20.786 ±0.001

(Continued on next page)

Table 8–1: (continued)

Compound	$\Delta_f G_m^\circ$ (kJ·mol^{-1})	$\Delta_f H_m^\circ$ (kJ·mol^{-1})	S_m° (J·K^{-1}·mol^{-1})	$C_{p,m}^\circ$ (J·K^{-1}·mol^{-1})
Kr(g)	0.000	0.000	164.085 ±0.003	20.786 ±0.001
Xe(g)	0.000	0.000	169.685 ±0.003	20.786 ±0.001
F(g)	62.280 [a] ±0.300	79.380 ±0.300	158.751 ±0.004	22.746 ±0.002
F$^-$	−281.523 [a] ±0.692	−335.350 ±0.650	−13.800 ±0.800	
F$_2$(g)	0.000	0.000	202.791 ±0.005	31.304 ±0.002
HF(aq)	−299.675 [b] ±0.702	−323.150 [b] ±0.716	88.000 [a] ±3.362	
HF(g)	−275.400 [a] ±0.700	−273.300 ±0.700	173.779 ±0.003	29.137 ±0.002
HF$_2^-$	−583.709 [b] ±1.200	−655.500 [b] ±2.221	92.683 [a] ±8.469	
Cl(g)	105.305 [a] ±0.008	121.301 ±0.008	165.190 ±0.004	21.838 ±0.001
Cl$^-$	−131.217 [a] ±0.117	−167.080 ±0.100	56.600 ±0.200	
Cl$_2$(g)	0.000	0.000	223.081 ±0.010	33.949 ±0.002
ClO$^-$	−37.669 [b] ±0.962			
ClO$_2^-$	10.250 [b] ±4.044			
ClO$_3^-$	−7.903 [a] ±1.342	−104.000 ±1.000	162.300 ±3.000	
ClO$_4^-$	−7.890 [a] ±0.600	−128.100 ±0.400	184.000 ±1.500	
HCl(g)	−95.298 [a] ±0.100	−92.310 ±0.100	186.902 ±0.005	29.136 ±0.002
HClO(aq)	−80.023 [b] ±0.613			
HClO$_2$(aq)	−0.938 [b] ±4.043			

(Continued on next page)

8 Selected auxiliary data

Table 8–1: (continued)

Compound	$\Delta_f G_m^\circ$ (kJ·mol^{-1})	$\Delta_f H_m^\circ$ (kJ·mol^{-1})	S_m° (J·K^{-1}·mol^{-1})	$C_{p,m}^\circ$ (J·K^{-1}·mol^{-1})
Br(g)	82.379 [a] ±0.128	111.870 ±0.120	175.018 ±0.004	20.786 ±0.001
Br$^-$	−103.850 [a] ±0.167	−121.410 ±0.150	82.550 ±0.200	
Br$_2$(aq)	4.900 ±1.000			
Br$_2$(g)	3.105 [a] ±0.142	30.910 ±0.110	245.468 ±0.005	36.057 ±0.002
Br$_2$(l)	0.000	0.000	152.210 ±0.300	
BrO$^-$	−32.095 ±1.537			
BrO$_3^-$	19.070 [a] ±0.634	−66.700 ±0.500	161.500 ±1.300	
HBr(g)	−53.361 [a] ±0.166	−36.290 ±0.160	198.700 ±0.004	29.141 ±0.003
HBrO(aq)	−81.356 [b] ±1.527			
I(g)	70.172 [a] ±0.060	106.760 ±0.040	180.787 ±0.004	20.786 ±0.001
I$^-$	−51.724 [a] ±0.112	−56.780 ±0.050	106.450 ±0.300	
I$_2$(cr)	0.000	0.000	116.140 ±0.300	
I$_2$(g)	19.323 [a] ±0.120	62.420 ±0.080	260.687 ±0.005	36.888 ±0.002
IO$_3^-$	−126.338 [a] ±0.779	−219.700 ±0.500	118.000 ±2.000	
HI(g)	1.700 [a] ±0.110	26.500 ±0.100	206.590 ±0.004	29.157 ±0.003
HIO$_3$(aq)	−130.836 [b] ±0.797			
S(cr)[d]	0.000	0.000	32.054 ±0.050	22.750 ±0.050
S(g)	236.689 [a] ±0.151	277.170 ±0.150	167.829 ±0.006	23.674 ±0.001

(Continued on next page)

Table 8–1: (continued)

Compound	$\Delta_f G_m^\circ$ (kJ·mol^{-1})	$\Delta_f H_m^\circ$ (kJ·mol^{-1})	S_m° (J·K^{-1}·mol^{-1})	$C_{p,m}^\circ$ (J·K^{-1}·mol^{-1})
S^{2-}	120.695 [b] ±11.610			
$S_2(g)$	79.686 [a] ±0.301	128.600 ±0.300	228.167 ±0.010	32.505 ±0.010
$SO_2(g)$	−300.095 [a] ±0.201	−296.810 ±0.200	248.223 ±0.050	39.842 ±0.020
SO_3^{2-}	−487.472 [b] ±4.020			
$S_2O_3^{2-}$	−519.291 [b] ±11.345			
SO_4^{2-}	−744.004 [a] ±0.418	−909.340 ±0.400	18.500 ±0.400	
HS^-	12.243 [a] ±2.115	−16.300 ±1.500	67.000 ±5.000	
$H_2S(aq)$	−27.648 [a] ±2.115	−38.600 ±1.500	126.000 ±5.000	
$H_2S(g)$	−33.443 [a] ±0.500	−20.600 ±0.500	205.810 ±0.050	34.248 ±0.010
HSO_3^-	−528.684 [b] ±4.046			
$HS_2O_3^-$	−528.366 [b] ±11.377			
$H_2SO_3(aq)$	−539.187 [b] ±4.072			
HSO_4^-	−755.315 [a] ±1.342	−886.900 ±1.000	131.700 ±3.000	
$Se(cr)$	0.000	0.000	42.270 ±0.050	25.030 ±0.050
$SeO_2(cr)$		−225.100 ±2.100		
SeO_3^{2-}	−361.597 [b] ±1.473			
$HSeO_3^-$	−409.544 [b] ±1.358			

(Continued on next page)

Table 8–1: (continued)

Compound	$\Delta_f G_m^\circ$ (kJ·mol^{-1})	$\Delta_f H_m^\circ$ (kJ·mol^{-1})	S_m° (J·K^{-1}·mol^{-1})	$C_{p,m}^\circ$ (J·K^{-1}·mol^{-1})
H$_2$SeO$_3$(aq)	−425.527 [b] ±0.736			
Te(cr)	0.000	0.000	49.221 ±0.050	25.550 ±0.100
TeO$_2$(cr)	−265.996 [a] ±2.500	−321.000 ±2.500	69.890 ±0.150	60.670 [q] ±0.150
N(g)	**455.537** [a] **±0.400**	**472.680** **±0.400**	**153.301** **±0.003**	**20.786** **±0.001**
N$_2$(g)	0.000	0.000	191.609 ±0.004	29.124 ±0.001
N$_3^-$	348.200 ±2.000	275.140 ±1.000	107.710 [a] ±7.500	
NO$_3^-$	−110.794 [a] ±0.417	−206.850 ±0.400	146.700 ±0.400	
HN$_3$(aq)	321.372 [b] ±2.051	260.140 [b] ±10.050	147.381 [b] ±34.403	
NH$_3$(aq)	−26.673 [b] ±0.305	−81.170 [b] ±0.326	109.040 [b] ±0.913	
NH$_3$(g)	−16.407 [a] ±0.350	−45.940 ±0.350	192.770 ±0.050	35.630 ±0.005
NH$_4^+$	−79.398 [a] ±0.278	−133.260 ±0.250	111.170 ±0.400	
P(am)[e]		−7.500 ±2.000		
P(cr)[e]	0.000	0.000	41.090 ±0.250	23.824 ±0.200
P(g)	280.093 [a] ±1.003	316.500 ±1.000	163.199 ±0.003	20.786 ±0.001
P$_2$(g)	103.469 [a] ±2.006	144.000 ±2.000	218.123 ±0.004	32.032 ±0.002
P$_4$(g)	24.419 [a] ±0.448	58.900 ±0.300	280.010 ±0.500	67.081 ±1.500
PO$_4^{3-}$	−1025.491 [b] ±1.576	−1284.400 [b] ±4.085	−220.970 [b] ±12.846	
P$_2$O$_7^{4-}$	−1935.503 [b] ±4.563			

(Continued on next page)

Table 8–1: (continued)

Compound	$\Delta_f G_m^\circ$ (kJ · mol^{-1})	$\Delta_f H_m^\circ$ (kJ · mol^{-1})	S_m° (J · K^{-1} · mol^{-1})	$C_{p,m}^\circ$ (J · K^{-1} · mol^{-1})
HPO_4^{2-}	−1095.985 [a] ±1.567	−1299.000 ±1.500	−33.500 ±1.500	
$H_2PO_4^-$	−1137.152 [a] ±1.567	−1302.600 ±1.500	92.500 ±1.500	
$H_3PO_4(aq)$	−1149.367 [b] ±1.576	−1294.120 [b] ±1.616	161.912 [b] ±2.575	
$HP_2O_7^{3-}$	−1989.158 [b] ±4.482			
$H_2P_2O_7^{2-}$	−2027.117 [b] ±4.445			
$H_3P_2O_7^-$	−2039.960 [b] ±4.362			
$H_4P_2O_7(aq)$	−2045.668 [b] ±3.299	−2280.210 [b] ±3.383	274.919 [b] ±6.954	
As(cr)	0.000	0.000	35.100 ±0.600	24.640 ±0.500
AsO_2^-	−350.022 [a] ±4.008	−429.030 ±4.000	40.600 ±0.600	
AsO_4^{3-}	−648.360 [a] ±4.008	−888.140 ±4.000	−162.800 ±0.600	
$As_2O_5(cr)$	−782.449 [a] ±8.016	−924.870 ±8.000	105.400 ±1.200	116.520 ±0.800
$As_4O_6(cubic)$[f]	−1152.445 [a] ±16.032	−1313.940 ±16.000	214.200 ±2.400	191.290 ±0.800
$As_4O_6(mono)$[g]	−1154.009 [a] ±16.041	−1309.600 ±16.000	234.000 ±3.000	
$HAsO_2(aq)$	−402.925 [a] ±4.008	−456.500 ±4.000	125.900 ±0.600	
$H_2AsO_3^-$	−587.078 [a] ±4.008	−714.790 ±4.000	110.500 ±0.600	
$H_3AsO_3(aq)$	−639.681 [a] ±4.015	−742.200 ±4.000	195.000 ±1.000	
$HAsO_4^{2-}$	−714.592 [a] ±4.008	−906.340 ±4.000	−1.700 ±0.600	
$H_2AsO_4^-$	−753.203 [a] ±4.015	−909.560 ±4.000	117.000 ±1.000	

(Continued on next page)

Table 8–1: (continued)

Compound	$\Delta_f G_m^\circ$ (kJ · mol^{-1})	$\Delta_f H_m^\circ$ (kJ · mol^{-1})	S_m° (J · K^{-1} · mol^{-1})	$C_{p,m}^\circ$ (J · K^{-1} · mol^{-1})
H$_3$AsO$_4$(aq)	−766.119 [a] ±4.015	−902.500 ±4.000	184.000 ±1.000	
(As$_2$O$_5$)$_3$· 5 H$_2$O(cr)		−4248.400 ±24.000		
Sb(cr)	0.000	0.000	45.520 ±0.210	25.260 ±0.200
Bi(cr)[h]	0.000	0.000	56.740 ±0.420	25.410 ±0.200
C(cr)	0.000	0.000	5.740 ±0.100	8.517 ±0.080
C(g)	671.254 [a] ±0.451	716.680 ±0.450	158.100 ±0.003	20.839 ±0.001
CO(g)	−137.168 [a] ±0.173	−110.530 ±0.170	197.660 ±0.004	29.141 ±0.002
CO$_2$(aq)	−385.970 [a] ±0.270	−413.260 ±0.200	119.360 ±0.600	
CO$_2$(g)	−394.373 [a] ±0.133	−393.510 ±0.130	213.785 ±0.010	37.135 ±0.002
CO$_3^{2-}$	−527.900 [a] ±0.390	−675.230 ±0.250	−50.000 ±1.000	
HCO$_3^-$	−586.845 [a] ±0.251	−689.930 ±0.200	98.400 ±0.500	
SCN$^-$	92.700 ±4.000	76.400 ±4.000	144.268 [a] ±18.974	
Si(cr)	0.000	0.000	18.810 ±0.080	19.789 ±0.030
Si(g)	405.525 [a] ±8.000	450.000 ±8.000	167.981 ±0.004	22.251 ±0.001
SiO$_2$(quar)[i]	−856.287 [a] ±1.002	−910.700 ±1.000	41.460 ±0.200	44.602 ±0.300
SiO$_2$(OH)$_2^{2-}$	−1175.651 [b] ±1.265	−1381.960 [b] ±15.330	−1.488 [b] ±51.592	
SiO(OH)$_3^-$	−1251.740 [b] ±1.162	−1431.360 [b] ±3.743	88.024 [b] ±13.144	
Si(OH)$_4$(aq)	−1307.735 [b] ±1.156	−1456.960 [b] ±3.163	189.973 [b] ±10.245	

(Continued on next page)

Table 8–1: (continued)

Compound	$\Delta_f G_m^\circ$ (kJ·mol^{-1})	$\Delta_f H_m^\circ$ (kJ·mol^{-1})	S_m° (J·K^{-1}·mol^{-1})	$C_{p,m}^\circ$ (J·K^{-1}·mol^{-1})
Si$_2$O$_3$(OH)$_4^{2-}$	−2269.878 (b) ±2.878			
Si$_2$O$_2$(OH)$_5^-$	−2332.096 (b) ±2.878			
Si$_3$O$_6$(OH)$_3^{3-}$	−3048.536 (b) ±3.870			
Si$_3$O$_5$(OH)$_5^{3-}$	−3291.955 (b) ±3.869			
Si$_4$O$_8$(OH)$_4^{4-}$	−4075.179 (b) ±5.437			
Si$_4$O$_7$(OH)$_5^{3-}$	−4136.826 (b) ±4.934			
SiF$_4$(g)	−1572.773 (a) ±0.814	−1615.000 ±0.800	282.760 ±0.500	73.622 ±0.500
Ge(cr)	0.000	0.000	31.090 ±0.150	23.222 ±0.100
Ge(g)	331.209 (a) ±3.000	372.000 ±3.000	167.904 ±0.005	30.733 ±0.001
GeO$_2$(tetr)$^{(j)}$	−521.404 (a) ±1.002	−580.000 ±1.000	39.710 ±0.150	50.166 ±0.300
GeF$_4$(g)	−1150.018 (a) ±0.584	−1190.200 ±0.500	301.900 ±1.000	81.602 ±1.000
Sn(cr)	0.000	0.000	51.180 ±0.080	27.112 ±0.030
Sn(g)	266.223 (a) ±1.500	301.200 ±1.500	168.492 ±0.004	21.259 ±0.001
Sn^{2+}	−27.624 (a) ±1.557	−8.900 ±1.000	−16.700 ±4.000	
SnO(tetr)$^{(j)}$	−251.913 (a) ±0.220	−280.710 ±0.200	57.170 ±0.300	47.783 ±0.300
SnO$_2$(cass)$^{(k)}$	−515.826 (a) ±0.204	−577.630 ±0.200	49.040 ±0.100	53.219 ±0.200
Pb(cr)	0.000	0.000	64.800 ±0.300	26.650 ±0.100
Pb(g)	162.232 (a) ±0.805	195.200 ±0.800	175.375 ±0.005	20.786 ±0.001

(Continued on next page)

Table 8–1: (continued)

Compound	$\Delta_f G_m^\circ$ (kJ·mol^{-1})	$\Delta_f H_m^\circ$ (kJ·mol^{-1})	S_m° (J·K^{-1}·mol^{-1})	$C_{p,m}^\circ$ (J·K^{-1}·mol^{-1})
Pb^{2+}	−24.238 [a] ±0.399	0.920 ±0.250	18.500 ±1.000	
PbSO$_4$(cr)	−813.036 [a] ±0.447	−919.970 ±0.400	148.500 ±0.600	
B(cr)	0.000	0.000	5.900 ±0.080	11.087 ±0.100
B(g)	521.012 [a] ±5.000	565.000 ±5.000	153.436 ±0.015	20.796 ±0.005
B$_2$O$_3$(cr)	−1194.324 [a] ±1.404	−1273.500 ±1.400	53.970 ±0.300	62.761 ±0.300
B(OH)$_3$(aq)	−969.268 [a] ±0.820	−1072.800 ±0.800	162.400 ±0.600	
B(OH)$_3$(cr)	−969.667 [a] ±0.820	−1094.800 ±0.800	89.950 ±0.600	86.060 ±0.400
BF$_3$(g)	−1119.403 [a] ±0.803	−1136.000 ±0.800	254.420 ±0.200	50.463 ±0.100
Al(cr)	0.000	0.000	28.300 ±0.100	24.200 ±0.070
Al(g)	289.376 [a] ±4.000	330.000 ±4.000	164.554 ±0.004	21.391 ±0.001
Al^{3+}	−491.507 [a] ±3.338	−538.400 ±1.500	−325.000 ±10.000	
Al$_2$O$_3$(coru)$^{(l)}$	−1582.257 [a] ±1.302	−1675.700 ±1.300	50.920 ±0.100	79.033 ±0.200
AlF$_3$(cr)	−1431.096 [a] ±1.309	−1510.400 ±1.300	66.500 ±0.500	75.122 ±0.400
Tl$^+$	−32.400 ±0.300			
Zn(cr)	0.000	0.000	41.630 ±0.150	25.390 ±0.040
Zn(g)	94.813 [a] ±0.402	130.400 ±0.400	160.990 ±0.004	20.786 ±0.001
Zn^{2+}	−147.203 [a] ±0.254	−153.390 ±0.200	−109.800 ±0.500	
ZnO(cr)	−320.479 [a] ±0.299	−350.460 ±0.270	43.650 ±0.400	

(Continued on next page)

Table 8–1: (continued)

Compound	$\Delta_f G_m^\circ$ (kJ·mol^{-1})	$\Delta_f H_m^\circ$ (kJ·mol^{-1})	S_m° (J·K^{-1}·mol^{-1})	$C_{p,m}^\circ$ (J·K^{-1}·mol^{-1})
Cd(cr)	0.000	0.000	51.800 ±0.150	26.020 ±0.040
Cd(g)	77.230 [a] ±0.205	111.800 ±0.200	167.749 ±0.004	20.786 ±0.001
Cd^{2+}	−77.733 [a] ±0.750	−75.920 ±0.600	−72.800 ±1.500	
CdO(cr)	−228.661 [a] ±0.602	−258.350 ±0.400	54.800 ±1.500	
CdSO$_4$· 2.667 H$_2$O(cr)	−1464.959 [a] ±0.810	−1729.300 ±0.800	229.650 ±0.400	
Hg(g)	31.842 [a] ±0.054	61.380 ±0.040	174.971 ±0.005	20.786 ±0.001
Hg(l)	0.000	0.000	75.900 ±0.120	
Hg^{2+}	164.667 [a] ±0.313	170.210 ±0.200	−36.190 ±0.800	
Hg$_2^{2+}$	153.567 [a] ±0.559	166.870 ±0.500	65.740 ±0.800	
HgO(mont)[m]	−58.523 [a] ±0.154	−90.790 ±0.120	70.250 ±0.300	
Hg$_2$Cl$_2$(cr)	−210.725 [a] ±0.471	−265.370 ±0.400	191.600 ±0.800	
Hg$_2$SO$_4$(cr)	−625.780 [a] ±0.411	−743.090 ±0.400	200.700 ±0.200	
Cu(cr)	0.000	0.000	33.150 ±0.080	24.440 ±0.050
Cu(g)	297.672 [a] ±1.200	337.400 ±1.200	166.398 ±0.004	20.786 ±0.001
Cu^{2+}	65.040 [a] ±1.557	64.900 ±1.000	−98.000 ±4.000	
CuCl(g)		77.000 ±10.000		
CuSO$_4$(cr)	−662.185 [a] ±1.206	−771.400 ±1.200	109.200 ±0.400	
Ag(cr)	0.000	0.000	42.550 ±0.200	25.350 ±0.100

(Continued on next page)

Table 8–1: (continued)

Compound	$\Delta_f G_m^\circ$ (kJ·mol^{-1})	$\Delta_f H_m^\circ$ (kJ·mol^{-1})	S_m° (J·K^{-1}·mol^{-1})	$C_{p,m}^\circ$ (J·K^{-1}·mol^{-1})
Ag(g)	246.007 [a] ±0.802	284.900 ±0.800	172.997 ±0.004	20.786 ±0.001
Ag$^+$	77.096 [a] ±0.156	105.790 ±0.080	73.450 ±0.400	
AgCl(cr)	−109.765 [a] ±0.098	−127.010 ±0.050	96.250 ±0.200	
Ti(cr)	0.000	0.000	30.720 ±0.100	25.060 ±0.080
Ti(g)	428.403 [a] ±3.000	473.000 ±3.000	180.298 ±0.010	24.430 ±0.030
TiO$_2$(ruti)[n]	−888.767 [a] ±0.806	−944.000 ±0.800	50.620 ±0.300	55.080 ±0.300
TiCl$_4$(g)	−726.324 [a] ±3.229	−763.200 ±3.000	353.200 ±4.000	95.408 ±1.000
Th(cr)	0.000	0.000	51.800 ±0.500	26.230 ±0.050
Th(g)	560.745 [a] ±6.002	602.000 ±6.000	190.170 ±0.050	20.789 ±0.100
ThO$_2$(cr)	−1169.238 [a] ±3.504	−1226.400 ±3.500	65.230 ±0.200	
Be(cr)	0.000	0.000	9.500 ±0.080	16.443 ±0.060
Be(g)	286.202 [a] ±5.000	324.000 ±5.000	136.275 ±0.003	20.786 ±0.001
BeO(brom)[o]	−580.090 [a] ±2.500	−609.400 ±2.500	13.770 ±0.040	25.565 ±0.100
Mg(cr)	0.000	0.000	32.670 ±0.100	24.869 ±0.020
Mg(g)	112.521 [a] ±0.801	147.100 ±0.800	148.648 ±0.003	20.786 ±0.001
Mg^{2+}	−455.375 [a] ±1.335	−467.000 ±0.600	−137.000 ±4.000	
MgO(cr)	−569.312 [a] ±0.305	−601.600 ±0.300	26.950 ±0.150	37.237 ±0.200
MgF$_2$(cr)	−1071.051 [a] ±1.210	−1124.200 ±1.200	57.200 ±0.500	61.512 ±0.300

(Continued on next page)

Table 8–1: (continued)

Compound	$\Delta_f G_m^\circ$ (kJ·mol⁻¹)	$\Delta_f H_m^\circ$ (kJ·mol⁻¹)	S_m° (J·K⁻¹·mol⁻¹)	$C_{p,m}^\circ$ (J·K⁻¹·mol⁻¹)
Ca(cr)	0.000	0.000	41.590 ±0.400	25.929 ±0.300
Ca(g)	144.021 [a] ±0.809	177.800 ±0.800	154.887 ±0.004	20.786 ±0.001
Ca²⁺	−552.806 [a] ±1.050	−543.000 ±1.000	−56.200 ±1.000	
CaO(cr)	−603.296 [a] ±0.916	−634.920 ±0.900	38.100 ±0.400	42.049 ±0.400
CaF(g)	−302.118 ±5.104	−276.404 ±5.100	229.244 ±0.500	33.671 ±0.500
CaCl(g)	−129.787 ±5.001	−103.400 ±5.000	241.634 ±0.300	35.687 ±0.010
Sr(cr)	0.000	0.000	55.700 ±0.210	
Sr²⁺	−563.864 [a] ±0.781	−550.900 ±0.500	−31.500 ±2.000	
SrO(cr)	−559.939 [a] ±0.914	−590.600 ±0.900	55.440 ±0.500	
SrCl₂(cr)	−784.974 [a] ±0.714	−833.850 ±0.700	114.850 ±0.420	
Sr(NO₃)₂(cr)	−783.146 [a] ±1.018	−982.360 ±0.800	194.600 ±2.100	
Ba(cr)	0.000	0.000	62.420 ±0.840	
Ba(g)	152.852 ±5.006	185.000 ±5.000	170.245 ±0.010	20.786 ±0.001
Ba²⁺	−557.656 [a] ±2.582	−534.800 ±2.500	8.400 ±2.000	
BaO(cr)	−520.394 [a] ±2.515	−548.100 ±2.500	72.070 ±0.380	
BaF(g)	−349.569 ±6.705	−324.992 ±6.700	246.219 ±0.210	34.747 ±0.300
BaCl₂(cr)	−806.953 [a] ±2.514	−855.200 ±2.500	123.680 ±0.250	

(Continued on next page)

Table 8–1: (continued)

Compound	$\Delta_f G_m^o$ (kJ·mol^{-1})	$\Delta_f H_m^o$ (kJ·mol^{-1})	S_m^o (J·K^{-1}·mol^{-1})	$C_{p,m}^o$ (J·K^{-1}·mol^{-1})
Li(cr)	0.000	0.000	29.120 ±0.200	24.860 ±0.200
Li(g)	126.604 [a] ±1.002	159.300 ±1.000	138.782 ±0.010	20.786 ±0.001
Li$^+$	−292.918 [a] ±0.109	−278.470 ±0.080	12.240 ±0.150	
Na(cr)	0.000	0.000	51.300 ±0.200	28.230 ±0.200
Na$^+$	−261.953 [a] ±0.096	−240.340 ±0.060	58.450 ±0.150	
NaF(cr)[p]	−546.327 [a] ±0.704	−576.600 ±0.700	51.160 ±0.150	46.820
NaCl(cr)[p]	−384.221 ±0.147	−411.260 ±0.120	72.150 ±0.200	50.500
NaNO$_3$(cr)		−467.580 ±0.410		
K(cr)	0.000	0.000	64.680 ±0.200	29.600 ±0.100
K(g)	60.479 [a] ±0.802	89.000 ±0.800	160.341 ±0.003	20.786 ±0.001
K$^+$	−282.510 [a] ±0.116	−252.140 ±0.080	101.200 ±0.200	
Rb(cr)	0.000	0.000	76.780 ±0.300	31.060 ±0.100
Rb(g)	53.078 [a] ±0.805	80.900 ±0.800	170.094 ±0.003	20.786 ±0.001
Rb$^+$	−284.009 [a] ±0.153	−251.120 ±0.100	121.750 ±0.250	

(Continued on next page)

Table 8–1: (continued)

Compound	$\Delta_f G_m^\circ$ (kJ · mol^{-1})	$\Delta_f H_m^\circ$ (kJ · mol^{-1})	S_m° (J · K^{-1} · mol^{-1})	$C_{p,m}^\circ$ (J · K^{-1} · mol^{-1})
Cs(cr)	0.000	0.000	85.230 ±0.400	32.210 ±0.200
Cs(g)	49.556 [a] ±1.007	76.500 ±1.000	175.601 ±0.003	20.786 ±0.001
Cs$^+$	−291.456 [a] ±0.535	−258.000 ±0.500	132.100 ±0.500	
CsCl(cr)	−413.807 [a] ±0.208	−442.310 ±0.160	101.170 ±0.200	52.470
CsBr(cr)[p]	−391.171 ±0.305	−405.600 ±0.250	112.940 ±0.400	52.930

(a) Value calculated internally with the Gibbs–Helmholtz equation, $\Delta_f G_m^\circ = \Delta_f H_m^\circ - T \sum_i S_{m,i}^\circ$.
(b) Value calculated internally from reaction data (see Table 8–2).
(c) From [82WAG/EVA], uncertainty estimated in the uranium review [92GRE/FUG].
(d) Orthorhombic.
(e) P(cr) refers to white, crystalline (cubic) phosphorus and is the reference state for the element phosphorus. P(am) refers to red, amorphous phosphorus.
(f) Cubic.
(g) Monoclinic.
(h) Data from [82WAG/EVA], [73HUL/DES], with the uncertainty in S° from the latter
(i) α-Quartz.
(j) Tetragonal.
(k) Cassiterite, tetragonal.
(l) Corundum.
(m) Montroydite, red.
(n) Rutile.
(o) Bromellite.
(p) Data from [82GLU/GUR], compatible with [89COX/WAG].
(q) Temperature coefficients of this function are given in Section 14.1.1.1.

Table 8–2: Selected thermodynamic data for reactions involving auxiliary compounds and complexes used in the evaluation of thermodynamic data for the NEA TDB project data. All ionic species listed in this table are aqueous species. The selection of these data is described in Chapter VI of the uranium review [92GRE/FUG]. Unless noted otherwise, all data refer to 298.15 K and a pressure of 0.1 MPa and, for aqueous species, a reference state or standard state of infinite dilution ($I = 0$). The uncertainties listed below each value represent total uncertainties and correspond in principle to the statistically defined 95% confidence interval. Systematically, all the values are presented with three digits after the decimal point, regardless of the significance of these digits. The data presented in this table are available on computer media from the OECD Nuclear Energy Agency.

Species	Reaction	$\log_{10} K°$	$\Delta_r G_m°$ (kJ·mol^{-1})	$\Delta_r H_m°$ (kJ·mol^{-1})	$\Delta_r S_m°$ (J·K^{-1}·mol^{-1})
HF(aq)	F$^-$ + H$^+$ ⇌ HF(aq)	3.180 ±0.020	−18.152 ±0.114	12.200 ±0.300	101.800 [a] ±1.077
HF$_2^-$	F$^-$ + HF(aq) ⇌ HF$_2^-$	0.440 ±0.120	−2.511 ±0.685	3.000 ±2.000	18.486 [a] ±7.090
ClO$^-$	HClO(aq) ⇌ ClO$^-$ + H$^+$	−7.420 ±0.130	42.354 ±0.742	19.000 ±9.000	−78.329 [a] ±30.289
ClO$_2^-$	HClO$_2$(aq) ⇌ ClO$_2^-$ + H$^+$	−1.960 ±0.020	11.188 ±0.114		
HClO(aq)	Cl$_2$(g) + H$_2$O(l) ⇌ Cl$^-$ + H$^+$ + HClO(aq)	−4.537 ±0.105	25.900 ±0.600		
HClO$_2$(aq)	H$_2$O(l) + HClO(aq) ⇌ 2H$^+$ + HClO$_2$(aq) + 2 e$^-$	−55.400 [b] ±0.700	316.230 ±3.996		
BrO$^-$	HBrO(aq) ⇌ BrO$^-$ + H$^+$	−8.630 ±0.030	49.260 ±0.171	30.000 ±3.000	−64.600 [a] ±10.078
HBrO(aq)	Br$_2$(aq) + H$_2$O(l) ⇌ Br$^-$ + H$^+$ + HBrO(aq)	−8.240 ±0.200	47.034 ±1.142		

(Continued on next page)

Table 8–2: (continued)

Species	Reaction	$\log_{10} K^\circ$	$\Delta_r G_m^\circ$ (kJ·mol^{-1})	$\Delta_r H_m^\circ$ (kJ·mol^{-1})	$\Delta_r S_m^\circ$ (J·K^{-1}·mol^{-1})
HIO$_3$(aq)	H$^+$ + IO$_3^-$ ⇌ HIO$_3$(aq)				
		0.788 ±0.029	−4.498 ±0.166		
S^{2-}	HS$^-$ ⇌ H$^+$ + S^{2-}				
		−19.000 ±2.000	108.450 ±11.416		
SO$_3^{2-}$	H$_2$O(l) + SO$_4^{2-}$ + 2 e$^-$ ⇌ 2OH$^-$ + SO$_3^{2-}$				
		−31.400 [b] ±0.700	179.230 ±3.996		
S$_2$O$_3^{2-}$	3H$_2$O(l) + 2SO$_3^{2-}$ + 4 e$^-$ ⇌ 6OH$^-$ + S$_2$O$_3^{2-}$				
		−39.200 [b] ±1.400	223.760 ±7.991		
H$_2$S(aq)	H$_2$S(aq) ⇌ H$^+$ + HS$^-$				
		−6.990 ±0.170	39.899 ±0.970		
HSO$_3^-$	H$^+$ + SO$_3^{2-}$ ⇌ HSO$_3^-$				
		7.220 ±0.080	−41.212 ±0.457	66.000 ±30.000	359.590 [a] ±100.630
HS$_2$O$_3^-$	H$^+$ + S$_2$O$_3^{2-}$ ⇌ HS$_2$O$_3^-$				
		1.590 ±0.150	−9.076 ±0.856		
H$_2$SO$_3$(aq)	H$^+$ + HSO$_3^-$ ⇌ H$_2$SO$_3$(aq)				
		1.840 ±0.080	−10.503 ±0.457	16.000 ±5.000	88.891 [a] ±16.840
HSO$_4^-$	H$^+$ + SO$_4^{2-}$ ⇌ HSO$_4^-$				
		1.980 ±0.050	−11.302 ±0.285		
SeO$_3^{2-}$	HSeO$_3^-$ ⇌ H$^+$ + SeO$_3^{2-}$				
		−8.400 ±0.100	47.948 ±0.571	−5.020 ±0.500	−177.650 [a] ±2.545
H$_2$Se(aq)	H$^+$ + HSe$^-$ ⇌ H$_2$Se(aq)				
		3.800 ±0.300	−21.691 ±1.712		

(Continued on next page)

Table 8–2: (continued)

Species	Reaction			
	$\log_{10} K°$	$\Delta_r G_m°$ (kJ·mol^{-1})	$\Delta_r H_m°$ (kJ·mol^{-1})	$\Delta_r S_m°$ (J·K^{-1}·mol^{-1})
$HSeO_3^-$	$H_2SeO_3(aq) \rightleftharpoons H^+ + HSeO_3^-$			
	−2.800 ±0.200	15.983 ±1.142	−7.070 ±0.500	−77.319 [a] ±4.180
$H_2SeO_3(aq)$	$3H_2O(l) + 2I_2(cr) + Se(cr) \rightleftharpoons 4H^+ + H_2SeO_3(aq) + 4I^-$			
	−13.840 ±0.100	78.999 ±0.571		
$HSeO_4^-$	$H^+ + SeO_4^{2-} \rightleftharpoons HSeO_4^-$			
	1.800 ±0.140	−10.274 ±0.799	23.800 ±5.000	114.290 [a] ±16.983
$HN_3(aq)$	$H^+ + N_3^- \rightleftharpoons HN_3(aq)$			
	4.700 ±0.080	−26.828 ±0.457	−15.000 ±10.000	39.671 [a] ±33.575
$NH_3(aq)$	$NH_4^+ \rightleftharpoons H^+ + NH_3(aq)$			
	−9.237 ±0.022	52.725 ±0.126	52.090 ±0.210	−2.130 [a] ±0.821
$HNO_2(aq)$	$H^+ + NO_2^- \rightleftharpoons HNO_2(aq)$			
	3.210 ±0.160	−18.323 ±0.913	−11.400 ±3.000	23.219 [a] ±10.518
PO_4^{3-}	$HPO_4^{2-} \rightleftharpoons H^+ + PO_4^{3-}$			
	−12.350 ±0.030	70.494 ±0.171	14.600 ±3.800	−187.470 [a] ±12.758
$P_2O_7^{4-}$	$HP_2O_7^{3-} \rightleftharpoons H^+ + P_2O_7^{4-}$			
	−9.400 ±0.150	53.656 ±0.856		
$H_2PO_4^-$	$H^+ + HPO_4^{2-} \rightleftharpoons H_2PO_4^-$			
	7.212 ±0.013	−41.166 ±0.074	−3.600 ±1.000	126.000 [a] ±3.363
$H_3PO_4(aq)$	$H^+ + H_2PO_4^- \rightleftharpoons H_3PO_4(aq)$			
	2.140 ±0.030	−12.215 ±0.171	8.480 ±0.600	69.412 [a] ±2.093
$HP_2O_7^{3-}$	$H_2P_2O_7^{2-} \rightleftharpoons H^+ + HP_2O_7^{3-}$			
	−6.650 ±0.100	37.958 ±0.571		

(Continued on next page)

Table 8–2: (continued)

Species	Reaction			
	$\log_{10} K°$	$\Delta_r G_m°$ (kJ·mol^{-1})	$\Delta_r H_m°$ (kJ·mol^{-1})	$\Delta_r S_m°$ (J·K^{-1}·mol^{-1})
$H_2P_2O_7^{2-}$	$H_3P_2O_7^- \rightleftharpoons H^+ + H_2P_2O_7^{2-}$			
	−2.250 ±0.150	12.843 ±0.856		
$H_3P_2O_7^-$	$H_4P_2O_7(aq) \rightleftharpoons H^+ + H_3P_2O_7^-$			
	−1.000 ±0.500	5.708 ±2.854		
$H_4P_2O_7(aq)$	$2H_3PO_4(aq) \rightleftharpoons H_2O(l) + H_4P_2O_7(aq)$			
	−2.790 ±0.170	15.925 ±0.970	22.200 ±1.000	21.045 [a] ±4.673
$CO_2(aq)$	$H^+ + HCO_3^- \rightleftharpoons CO_2(aq) + H_2O(l)$			
	6.354 ±0.020	−36.269 ±0.114		
$CO_2(g)$	$CO_2(aq) \rightleftharpoons CO_2(g)$			
	1.472 ±0.020	−8.402 ±0.114		
HCO_3^-	$CO_3^{2-} + H^+ \rightleftharpoons HCO_3^-$			
	10.329 ±0.020	−58.958 ±0.114		
$SiO_2(OH)_2^{2-}$	$Si(OH)_4(aq) \rightleftharpoons 2H^+ + SiO_2(OH)_2^{2-}$			
	−23.140 ±0.090	132.080 ±0.514	75.000 ±15.000	−191.460 [a] ±50.340
$SiO(OH)_3^-$	$Si(OH)_4(aq) \rightleftharpoons H^+ + SiO(OH)_3^-$			
	−9.810 ±0.020	55.996 ±0.114	25.600 ±2.000	−101.950 [a] ±6.719
$Si(OH)_4(aq)$	$2H_2O(l) + SiO_2(quar) \rightleftharpoons Si(OH)_4(aq)$			
	−4.000 ±0.100	22.832 ±0.571	25.400 ±3.000	8.613 [a] ±10.243
$Si_2O_3(OH)_4^{2-}$	$2Si(OH)_4(aq) \rightleftharpoons 2H^+ + H_2O(l) + Si_2O_3(OH)_4^{2-}$			
	−19.000 ±0.300	108.450 ±1.712		
$Si_2O_2(OH)_5^-$	$2Si(OH)_4(aq) \rightleftharpoons H^+ + H_2O(l) + Si_2O_2(OH)_5^-$			
	−8.100 ±0.300	46.235 ±1.712		

(Continued on next page)

Table 8–2: (continued)

Species	Reaction			
	$\log_{10} K°$	$\Delta_r G_m°$ (kJ·mol^{-1})	$\Delta_r H_m°$ (kJ·mol^{-1})	$\Delta_r S_m°$ (J·K^{-1}·mol^{-1})
$Si_3O_6(OH)_3^{3-}$	$3Si(OH)_4(aq) \rightleftharpoons 3H^+ + 3H_2O(l) + Si_3O_6(OH)_3^{3-}$			
	−28.600 ±0.300		163.250 ±1.712	
$Si_3O_5(OH)_5^{3-}$	$3Si(OH)_4(aq) \rightleftharpoons 3H^+ + 2H_2O(l) + Si_3O_5(OH)_5^{3-}$			
	−27.500 ±0.300		156.970 ±1.712	
$Si_4O_8(OH)_4^{4-}$	$4Si(OH)_4(aq) \rightleftharpoons 4H^+ + 4H_2O(l) + Si_4O_8(OH)_4^{4-}$			
	−36.300 ±0.500		207.200 ±2.854	
$Si_4O_7(OH)_5^{3-}$	$4Si(OH)_4(aq) \rightleftharpoons 3H^+ + 4H_2O(l) + Si_4O_7(OH)_5^{3-}$			
	−25.500 ±0.300		145.560 ±1.712	

(a) Value calculated internally with the Gibbs–Helmholtz equation, $\Delta_r G_m° = \Delta_r H_m° - T \Delta_r S_m°$.

(b) Value calculated from a selected standard potential.

(c) Value of $\log_{10} K°$ calculated internally from $\Delta_r G_m°$.

Part III

Discussion of new data selection

Chapter 9

Discussion of new data selection for Uranium

9.1 Elemental uranium (V.1)

9.1.1 Uranium metal (V.1.1)

The only new data on condensed uranium are those by Sheldon and Mulford [91SHE/MUL]. These authors have re-measured the emissivity of U(l), and calculated the concomitant corrections (which are appreciable) to the earlier measurements of $H_m(T) - H_m^o(298.15\,\text{K})$ for U(l) from 2357 to 5340 K by Mulford and Sheldon [88MUL/SHE]. Unfortunately, this correction removes the previous good agreement with the enthalpy data of Stephens [74STE] from levitation calorimetry and corresponds to a higher emissivity at the melting point. However Stephens' data at the lower temperatures, relevant to the vapour pressure measurements, are in good accord with the conventional enthalpy drop calorimetric measurements by Levinson [64LEV]. Thus, the current values of the heat capacity of U(l) used in the derivation of $\Delta_{sub}H_m$ (U, cr, 298.15 K), namely those given by [82GLU/GUR], have been retained until the discrepancy in the emissivity is resolved.

Yoo *et al.* [98YOO/CYN] have reported X–ray diffraction measurements on uranium up to temperatures of 4300 K and pressures of 100 GPa. Details of the results are given in Appendix A. The data for $U(\alpha, \beta, \gamma, l)$ assessed by McBride and Gordon [93MCB/GOR] are essentially identical to those given by [82GLU/GUR], which form the basis of the CODATA Key Values selection used in [92GRE/FUG], which are thus retained.

Boivineau *et al.* [93BOI/ARL] have also measured the thermophysical properties of uranium at high temperatures, but this study does not provide any new data relevant to this review.

9.1.2 Uranium gas (V.1.2)

The latest listing of the energy levels of U(g) [92BLA/WYA], required for statistical–mechanical calculations, gives a total of 2252 levels with a total statistical weight of 26050, up to *ca.* 50 cm^{-1} below the ionisation limit. This is a noticeable increase in the number of levels used in previous compilations (1133 by Oetting *et al.* [76OET/RAN] and 1596 by Glushko et al. [82GLU/GUR]). However, most of the new energy levels are above 20000 cm^{-1}, so the calculated values of the thermodynamic functions at 298.15 K,

$$S_m^\circ (U, g, 298.15 K) = (199.79 \pm 0.03) \text{ J} \cdot \text{K}^{-1} \cdot \text{mol}^{-1},$$

$$C_{p,m}^\circ (U, g, 298.15 K) = (23.69 \pm 0.03) \text{ J} \cdot \text{K}^{-1} \cdot \text{mol}^{-1},$$

are the same as those given in [92GRE/FUG]; the differences at higher temperatures remain small – see below.

The enthalpy of formation selected by [92GRE/FUG] is from the CODATA Key Values [89COX/WAG], which is based on four reasonably consistent measurements of the vapour pressure from ca. 1900 to 2400 K. Over this temperature range, the revised values of $(G_m(T) - H_m^\circ(298.15 K))/T$ for U(g) differ from those used in the CODATA assessment by 0.003 to 0.004 J · K^{-1} · mol^{-1}, so the derived value of the enthalpy of sublimation using the revised thermal functions will differ from this value by less than 0.01 kJ · mol^{-1}. The value selected in [92GRE/FUG] is therefore retained:

$$\Delta_f H_m^\circ (U, g, 298.15 K) = (533.0 \pm 8.0) \text{ kJ} \cdot \text{mol}^{-1}.$$

9.2 Simple uranium aqua ions (V.2)

No new experimental thermodynamic data related to this section appeared since the previous review in this series [92GRE/FUG]. However, a number of points call for attention.

This is particularly the case of the evaluation of the partial molar heat capacities of the U^{3+} and U^{4+} ions. In [92GRE/FUG], the estimates of Lemire and Tremaine [80LEM/TRE], were based on the Criss-Cobble relationships [64CRI/COB], [64CRI/COB2], giving:

$$C_{p,m} (U^{3+}) = - (64 \pm 22) \text{ J} \cdot \text{K}^{-1} \cdot \text{mol}^{-1} \quad (T = 298 \text{ to } 473 \text{ K}),$$

$$C_{p,m} (U^{4+}) = - (48 \pm 15) \text{ J} \cdot \text{K}^{-1} \cdot \text{mol}^{-1} \quad (T = 298 \text{ to } 473 \text{ K}).$$

These are average values for the temperature range 298.15 to 473 K. However, in the last decade or so, several authors have pointed out the severe limitations of the Criss-Cobble relationships when applied to tri- and tetravalent aqueous ions. This was the case of Shock and Helgeson [88SHO/HEL] who used estimation methods based on revised Helgeson–Kirkham–Flowers (HKF) equations of state for aqueous ions. Along this line, Shock et al. [97SHO/SAS2] (as discussed in Appendix A) estimated $C_{p,m}^\circ$ (U^{3+}, 298.15 K) = − 152.3 J · K^{-1} · mol^{-1}, thus ca. 90 J · K^{-1} · mol^{-1} more negative than the value accepted in [92GRE/FUG], and $C_{p,m}^\circ$ (U^{4+}, 298.15 K) = 0.8 J · K^{-1} · mol^{-1}. The latter value was based on the experimental result $C_{p,m}^\circ$ (Th^{4+}, 298.15 K) = − (1 ± 11) J · K^{-1} · mol^{-1} by Morss and McCue [76MOR/MCC] (see Appendix A).

More recently, Hovey [97HOV] (see Appendix A) determined the value $C_{p,m}^{o}$(Th^{4+}, 298.15 K) = $-(224 \pm 5)$ J·K^{-1}·mol^{-1}. Using a more recent value for the heat capacity of the nitrate ion, this author also recalculated the results of Morss and McCue, obtaining $-(60 \pm 11)$ J·K^{-1}·mol^{-1}. Thus, the difference between the two sets amounts to *ca.* 140 – 160 J·K^{-1}·mol^{-1}. The experimental conditions used by [97HOV], especially in terms of minimising thorium hydrolysis and complexation, were preferable to those selected by [76MOR/MCC]. Reasons for the discrepancy of the value of $C_{p,m}^{o}$(Th^{4+}) = 111 J·K^{-1}·mol^{-1} obtained by [75APE/SAH] using the bulk heat capacity of thorium nitrates solutions at 303 K are briefly discussed in the comments on [97HOV] (see Appendix A).

In an earlier study of Al^{3+}, Hovey and Tremaine [86HOV/TRE] also determined $C_{p,m}^{o}$(Al^{3+}, 298.15 K) = $-$ 119 J·K^{-1}·mol^{-1}, while the Criss-Cobble relations led to a value of 16 J·K^{-1}·mol^{-1}, thus giving a similar difference of 135 J·K^{-1}·mol^{-1}. In view of the above, this review adopts:

$$C_{p,m}^{o}(U^{3+}, 298.15 \text{ K}) = -(150 \pm 50) \text{ J·K}^{-1}\text{·mol}^{-1},$$

$$C_{p,m}^{o}(U^{4+}, 298.15 \text{ K}) = -(220 \pm 50) \text{ J·K}^{-1}\text{·mol}^{-1}.$$

The large uncertainties are due, in part, to the fact that the comparison between average values for the temperature range 298.15 to 473 K with values at 298.15 K is not strictly correct. The experimental basis for the selected values is not yet satisfactory and revisions might occur as a result of new studies. The selected heat capacity values should not be used at temperatures above 373 K.

The value $C_{p,m}^{o}$(UO$_2^{2+}$, 298.15 K) = (42.4 ± 3.0) J·K^{-1}·mol^{-1} adopted by [92GRE/FUG] on the basis of the experimental results of [89HOV/NGU] is maintained.

The reduction behaviour of U(VI) to U(IV) in aqueous solutions has been studied by quantum chemical calculations [99VAL/MAR]. As discussed in Appendix A, this approach might soon provide new support when selecting thermodynamic data. Theoretical predictions of redox properties are also given in [2000HAY/MAR] (see Appendix A).

Finally, with regard to simple aqueous uranium ions, semi–theoretical data of entropies, Gibbs energy and enthalpy of hydration of monatomic aqueous ions (U^{3+}, U^{4+}) are discussed in [98DAV/FOU] (see Appendix A). The stability of the oxidation states is discussed in [99MIK/RUM].

9.2.1 UO$_2^{2+}$ (V.2.1)

The entropy value, S_m^{o}(UO$_2^{2+}$, 298.15 K) = $-(98.2 \pm 3.0)$ J·K^{-1}·mol^{-1} selected in [92GRE/FUG] has been questioned in [94SER/DEV] and [97GUR/SER]. According to Gurevich *et al.* this value is too low by about 15 to 20 J·K^{-1}·mol^{-1}. As discussed in Appendix A, this review retains the value selected previously.

The chemical state of UO_2^{2+} in highly acidic non–complexing media (perchloric and triflic acids) has been discussed on the basis of laser fluorescence spectroscopic data [97GEI/RUT], [99BOU/BIL] and [2001BIL/RUS]. The question of the possible existence of U(VI) complexes is discussed with the conclusion that there is no evidence that they exist. EXAFS measurements of Semon *et al.* [2001SEM/BOE] show definitely that ClO_4^- is not coordinated to UO_2^{2+}, even in 10 M $HClO_4$, whereas in 10 M CF_3SO_3H the triflate ion forms an inner-sphere complex (see Appendix A).

Recent EXAFS and NMR studies have confirmed the structure of $UO_2(H_2O)_5^{2+}$ and quantum chemical calculations have shown that it has the lowest energy among other possibilities of hydration and structure [98BAR/RUB], [99WAH/MOL] and [2001SEM/BOE] (see Appendix A).

9.2.2 UO_2^+ (V.2.2)

There are no new experimental studies relevant to the thermodynamics of UO_2^+. Most of the studies involving U(VI)/U(V) species refer to carbonate solutions [93MIZ/PAR], [99DOC/MOS] (see section 9.7.2.3).

9.2.3 U^{4+}(V.2.3) and U^{3+}(V.2.4)

The new values adopted for the heat capacity of U^{4+} and U^{3+} at 298.15 K are discussed at the beginning of section 9.2.

The coordination number of the aqueous ion U^{4+} has been reported as (10 ± 1) [99MOL/DEN]. Giridhar and Langmuir [91GIR/LAN] repeated the study of [85BRU/GRE] on the reduction of UO_2^{2+} by Cu(cr). As reported in Appendix A the final computed value of $E°$ for the couple UO_2^{2+}/U^{4+} is (0.263 ± 0.004) V, which is within the combined uncertainties of the [92GRE/FUG] recommended value of (0.2673 ± 0.0012) V. This review keeps the $\Delta_f G_m°$ (U^{4+}) value of [92GRE/FUG] despite the proposal of Langmuir [97LAN] to increase it by 0.8 kJ · mol^{-1} (see Appendix A).

Results reported in [94AHO/ERV] confirm that the data selected in [92GRE/FUG] for U^{4+}/UO_2^{2+}, U(OH)$_4$(aq), di- and tri-dioxouranium(VI)carbonate apply to a natural environment (see Appendix A).

The oxidation of U^{4+} is discussed in [92HAS2] (see Appendix A).

9.3 Uranium oxygen and hydrogen compounds and complexes (V.3)

9.3.1 Gaseous uranium oxides and hydroxides

9.3.1.1 Gaseous uranium oxides (V.3.1)

9.3.1.1.1 UO(g)

There is new information on the energy levels of UO(g) which allows a better calculation of the thermal functions.

Earlier data derived from matrix isolation studies have been amplified by the recent detailed spectroscopic studies of Kaledin et al. [94KAL/MCC] and Kaledin and Heaven [97KAL/HEA]. They have derived the electronic levels of 16 low-lying electronic states of UO(g), and estimated the distribution of missing energy levels up to 40000 cm^{-1}. Although they have given partial vibrational-rotational data for three of the low-lying levels, we have preferred to use the better-established parameters for the ground state for all the excited levels, as given in Table 9-1, to calculate the thermal functions, using the usual non-rigid rotator, anharmonic oscillator approximation, plus electronic contributions. The derived values at 298.15 K are:

$$S_m^\circ \text{ (UO, g, 298.15 K)} = (252.3 \pm 2.0) \text{ J} \cdot \text{K}^{-1} \cdot \text{mol}^{-1},$$
$$C_{p,m}^\circ \text{ (UO, g, 298.15 K)} = (39.6 \pm 2.0) \text{ J} \cdot \text{K}^{-1} \cdot \text{mol}^{-1}.$$

Table 9-1: Molecular parameters for UO(g) *.

r (Å)	ω (cm^{-1})	ωx (cm^{-1})	B (cm^{-1})	α (cm^{-1})	D (cm^{-1})	Symmetry number
1.838	846.5	2.3	0.3333	3.24 x 10^{-3}	3.43 x 10^{-7}	1

Electronic levels (multiplicities) cm^{-1}

0 (2)   294.1 (2)   651.1 (2)   958.7 (2)   1043.0 (2)   1181.3 (2)   1253.7 (1)
1493.3 (2)   1574.1 (2)   1941.5 (2)   2118.8 (2)   2235.0 (2)   2272.0 (1)   2276.5 (2)   2412.6 (1)
2461.9 (2)   5000 (96)   10000 (313)   15000 (507)   20000 (879)   25000 (1271)   30000 (3576)
35000 (6999)   40000 (12810)

*: See section 2.1.6 for notations

We have recalculated the enthalpy of formation from the seven Gibbs energy studies discussed in [82GLU/GUR], giving considerably more weight to mass-loss effusion measurements (under mass-spectrometric monitoring) than to the purely mass-spectroscopic data, since the more recent paper by Storms [85STO2] suggests that the ionisation cross-sections of the gaseous uranium oxides are far from certain. The more reliable data [66DRO/PAT], [68PAT/DRO], [69ACK/RAU] give values of $\Delta_f H_m^\circ$ (UO, g, 298.15 K) from 31.5 to 42.4 kJ · mol^{-1}, from third-law analyses.

The selected value is:

$$\Delta_f H_m^\circ \text{ (UO, g, 298.15 K)} = (35.0 \pm 10.0) \text{ kJ} \cdot \text{mol}^{-1},$$

where the uncertainty is increased due to the appreciable discrepancy with the second-law enthalpy values.

This value is just consistent with the value of (7.81 ± 0.1) eV for the dissociation energy of UO(g) derived from appearance potentials given by Capone et al. [99CAP/COL], which corresponds to $\Delta_f H_m$ (UO, g) = (25.4 ± 10.0) kJ · mol^{-1}, if their value is taken to refer to 0 K.

9.3.1.1.2 UO$_3$(g)

Ebbinghaus [95EBB] has repeated the calculations of the thermal functions of UO$_3$(g), using essentially the same molecular parameters as those used by Glushko et al. [82GLU/GUR], which were adopted by [92GRE/FUG]. Krikorian et al. [93KRI/EBB] have reported transpiration data on the pressure of UO$_3$(g) in equilibrium with U$_3$O$_8$(cr) and oxygen gas from 1273 to 1573 K. Their pressures of UO$_3$(g) from the reaction:

$$\tfrac{1}{3} U_3O_8(cr) + \tfrac{1}{6} O_2(g) \rightleftharpoons UO_3(g)$$

are in good agreement with the similar data by [74DHA/TRI], and other literature data which form the basis of the choice of $\Delta_f G_m^\circ$(UO$_3$, g) in [82GLU/GUR] and [92GRE/FUG]. The [92GRE/FUG] data for UO$_3$(g) are therefore retained. It is difficult to derive any substantial conclusions from the work of Guido and Balducci [91GUI/BAL], who report values of the equilibrium constant for the reaction UO(g) + UO$_3$(g) $\rightleftharpoons$ 2 UO$_2$(g) at three temperatures from 2304 to 2414 K, measured mass-spectrometrically. The derived third-law value of $\Delta_f H_m^\circ$ (298.15 K) for this reaction is -197.3 kJ $\cdot$ mol^{-1}, with a large uncertainty due to the uncertain cross-sections and corrections for fragmentation. This is in good agreement with that derived from the selected values, $-(191.4 \pm 33.5)$ kJ $\cdot$ mol^{-1}. On the other hand the second-law value from the measurements of [91GUI/BAL] is considerably different, at $-(138.6 \pm 2.9)$ kJ $\cdot$ mol^{-1}, perhaps due to the quite small temperature range. The most can be said is that these data are not inconsistent with the selected values

9.3.1.2 Gaseous uranium hydroxides

9.3.1.2.1 UO$_2$(OH)$_2$(g)

There are a number of studies relating to gaseous dioxouranium(VI) hydroxide. Dharwadkar et al. [74DHA/TRI] were the first to show that the volatility of the higher uranium oxides above 1300 K, is substantially enhanced in the presence of water vapour, due to the formation of UO$_2$(OH)$_2$(g). Krikorian et al. [93KRI/EBB] have repeated these measurements, but find equilibrium constants for the reaction:

$$\tfrac{1}{3} U_3O_8(cr) + \tfrac{1}{6} O_2(g) + H_2O(g) \rightleftharpoons UO_2(OH)_2(g),$$

smaller by a factor of more than 300. More recently, Krikorian et al. [97KRI/FON] have studied the release of uranium-containing species from a ^{238}PuO$_2$ sample containing 3.3 mol% of ^{234}UO$_2$ under similar conditions. Despite some uncertainties in the calculation (particularly of the activity of UO$_2$ in the (U, Pu)O$_{2+x}$ solid solution presumably formed), the releases were in general agreement with those predicted from the data of [93KRI/EBB].

Both Ebbinghaus [95EBB] and Gorokhov and Sidorova [98GOR/SID] have calculated the thermal functions of this species by statistical mechanical calculations, using estimated molecular parameters. However, as noted in Appendix A, the different estimates of the molecular parameters, principally the vibration frequencies, lead to calculated standard entropies, which differ by ca. 22.5 J · K^{-1} · mol^{-1}.

The transpiration measurements of Dharwadkar et al. [74DHA/TRI] and Krikorian et al. [93KRI/EBB] have been analysed with both sets of calculated thermal functions of UO$_2$(OH)$_2$(g) in the respective entries in Appendix A. The results are summarised in the first four rows of Table 9-2.

Alexander and Ogden [87ALE/OGD] have also studied reactions involving UO$_2$(OH)$_2$(g). They report, with no further details, the Gibbs energy of the gaseous reaction:

$$UO_3(g) + H_2O(g) \rightleftharpoons UO_2(OH)_2(g) \qquad (9.1)$$

to be $\Delta_r G_m$ (9.1) = − 333500 + 156.9 T, J · mol^{-1} over an unspecified temperature range. Although the scantily-reported results of [87ALE/OGD] could never be used as a basis to select data, given the discrepant data from the two other studies of gaseous dioxouranium(VI) hydroxide, we have processed the data by second- and third-law analyses. The derived standard enthalpies of formation are included in Table 9-2.

Table 9-2: Derived values of $\Delta_f H_m^\circ$ (UO$_2$(OH)$_2$, g, 298.15 K), kJ · mol^{-1}.

Reference	Method	Thermal functions	
		[95EBB]	[98GOR/SID]
[74DHA/TRI]	Second-law	− (1212.9 ± 21.5)	− (1218.2 ± 21.5)
[74DHA/TRI]	Third-law	− (1262.6 ± 7.8)	− (1291.8 ± 10.4)
[93KRI/EBB]	Second-law	− (1153.6 ± 58.1)	− (1157.6 ± 58.1)
[93KRI/EBB]	Third-law	− (1197.2 ± 12.5)	− (1224.5 ± 13.3)
[87ALE/OGD]	Second-law	− 1388.9	− 1397.8
[87ALE/OGD]	Third-law	− 1301.3	− 1339.6

As noted in Appendix A, Hashizume et al. [99HAS/WAN] have also studied this system. The principal aims were to clarify the kinetics and mechanism of the volatilisation process and to assess the validity of selected thermodynamic data needed to interpret the experimental data. However, no new thermodynamic data can be derived from this paper and the authors used the data proposed in the related paper by Olander [99OLA], based principally on the data of Krikorian et al. [93KRI/EBB] and Ebbinghaus [95EBB], to interpret their results.

As can be seen from the Table 9-2, there is a wide variation in the values for the derived enthalpy of formation of UO$_2$(OH)$_2$(g). These big differences, especially the unexplained large discrepancy between the results of the similar studies by [74DHA/TRI] and [93KRI/EBB], plus the appreciably different thermal functions proposed by [95EBB], [98GOR/SID] and the substantial differences in the second- and third-law analyses shown in Table 9-2, mean that no reliable data for this species can be selected in this review.

9.3.2 Aqueous uranium hydroxide complexes (V.3.2)

All the experimental data since 1991 have confirmed the chemical models selected in [92GRE/FUG] and in most cases also the numerical values of the corresponding equilibrium constants. For U(VI), no new species nor the composition of those selected in [92GRE/FUG] have been added or changed. Only some selected equilibrium constants have been revised. For U(IV), the hydrolysis of U^{4+} has been reviewed according to the behaviour of the other aqueous actinide (IV) ions. New equilibrium constants have been selected.

Different experimental methods have been used with spectroscopy evolving into a very powerful tool for the determination of speciation and calculation of equilibrium constants, in addition to conventional potentiometric and solubility measurements. Spectroscopy provides a direct measure of the concentration of the U(VI) complexes formed, as compared to standard potentiometry where these quantities must be calculated from known total concentrations of the reactants and the free hydrogen-ion concentration. The concordant results of these two very different experimental methods provide very strong support for the selected stoichiometry and equilibrium constants. There are in general no serious systematic discrepancies between the results gained from the different approaches. New theoretical developments on both U(VI) and U(IV) have appeared recently and these can be used as a guide for the selection of species.

The reader should note that the name schoepite is commonly applied to a mineral or synthetic preparation with a formula close to $UO_3·2H_2O$. It should be named metaschoepite [98FIN/HAW] (see Appendix A and section 9.3.3.1.2). However, throughout this review the name 'schoepite', commonly used by chemists, is retained.

9.3.2.1 U(VI) hydroxide complexes (V.3.2.1)

A number of new papers dealing with U(VI) hydrolysis have appeared since the previous review, [92GRE/FUG]. Those that contain thermodynamic data of dioxouranium(VI) hydrolysis species fall into three categories.

The first consists of papers that give new independent data with regard to the constants selected in [92GRE/FUG]. They are evaluated and compared with those in [92GRE/FUG], and provide the basis for new selected values and/or uncertainty ranges. The equilibrium constants for the formation of complexes are collected in Table 9-3 and those for the solubility products of solid phases in Table 9-5. These papers are discussed in sections 9.3.2.1.1 to 9.3.2.1.5 and reviewed in Appendix A.

The second category comprises papers that confirm the [92GRE/FUG] selections on either a quantitative or qualitative basis, where the qualitative support is obtained by comparing chemical modelling based on the [92GRE/FUG] data with observations of field or laboratory data from "real" systems. The data are briefly discussed in the appropriate sections 9.3.2.1.1 to 9.3.2.1.5, (and reviewed in Appendix the appropriate sections 9.3.2.1.1 to 9.3.2.1.5, (and reviewed in Appendix A) with a special emphasis on the possibility of estimating the uncertainties in the previous selected data.

The third category of papers deals with the aqueous chemistry of U(VI). It provides thermodynamic information only through the use of the constants selected in [92GRE/FUG]: behaviour of uranium in environmental solutions [97BRU/CAS], [97ELL/ARM], [97GEI/BER], [97MUR], [98CHA/TRI], [98KAP/GER], [99MEI/VOL]; dissolution of spent fuel [96PAR/PYO], [98SER/RON], [98WER/SPA], [2000BRU/CER], [2000BUR/OLS]; spectroscopic identification of species [95ELI/BID], [97SCA/ANS], [98KIT/YAM], [98MOU/LAS] and the calculation of thermodynamic quantities as functions of temperature and pressure [97SHO/SAS]. Some of these papers are reviewed in Appendix A.

There are also papers from which no thermodynamic data can be derived at all, because they lack reliable solubility data [96DIA/GAR], information on the solid phase [97VAL/RAG] or sufficient experimental information [89SER/SAV], [96RAK/TSY] and [97RED].

Experimentally derived equilibrium data reported in the original publications for the equilibrium:

$$m\ UO_2^{2+} + n\ H_2O(l) \rightleftharpoons (UO_2)_m(OH)_n^{2m-n} + n\ H^+, \qquad (9.2)$$

are given in Table 9-3 as $\log_{10} {}^*\beta_{n,m}$. The experimental studies of [92SAN/BRU], [93MEI/KAT], [96MEI/KAT], [98MEI2], [98MEI3], [98YAM/KIT] and [2002RAI/FEL] are based on solubility data and those of [92KRA/BIS], [93FER/SAL], [95PAL/NGU] and [2002BRO] on potentiometry. Most of the data refer to 0.1 M ionic strength, but a wide range of ionic strength is covered. They are useful to ascertain that the chemical models do not vary with ionic strength or ionic medium.

Table 9-3: Experimental equilibrium constants for the U(VI) hydroxide system for the equilibria: $m\,UO_2^{2+} + n\,H_2O(l) \rightleftharpoons (UO_2)_m(OH)_n^{2m-n} + n\,H^+$.

n:m	t (°C)	Method	I (medium)	$\log_{10} {}^*\beta_{n,m}$	Reference
1:1	25	dis	0.1 M (NaClO$_4$)	$-(5.91 \pm 0.08)$	[91CHO/MAT]
			1 M (NaClO$_4$)	$-(5.75 \pm 0.07)$	
2:1	25	dis	0.1 M (NaClO$_4$)	$-(12.43 \pm 0.09)$	[91CHO/MAT]
			1 M (NaClO$_4$)	$-(12.29 \pm 0.09)$	
2:2	25	pot	0.1 M (NaClO$_4$)	$-(5.40 \pm 0.03)^{(a)}$	[92KRA/BIS]
5:3	25	pot	0.1 M (NaClO$_4$)	$-(15.85 \pm 0.04)^{(a)}$	[92KRA/BIS]
5:3	25	pot	0.1 M (NaClO$_4$)	$-(17.04 \pm 0.18)^{(b)}$	[92KRA/BIS]
3:1	25	sol	0.5 M (NaClO$_4$)	$-(19.83 \pm 0.34)$	[92SAN/BRU]
				$-(20.18 \pm 0.19)$	
				$-(19.67 \pm 0.17)$	
7:3	25	sol	0.5 M (NaClO$_4$)	$-(32.00 \pm 0.17)$	[92SAN/BRU]
				$-(33.32 \pm 0.22)$	
2:2	25	pot	3 M (NaClO$_4$)	$-(5.98 \pm 0.02)$	[93FER/SAL]
5:3	25	pot	3 M (NaClO$_4$)	$-(16.23 \pm 0.05)$	[93FER/SAL]
2:2	24	sol, sp	0.1 M (NaClO$_4$)	$-(5.89 \pm 0.12)$	[93MEI/KAT]
2:2	24	sp	0.1 M (NaClO$_4$)	$-(5.97 \pm 0.06)$	[93MEI/KAT]
2:2	25	pot	0.1 m (tmatfms)*	$-(5.77 \pm 0.01)$	[95PAL/NGU]
				$-(5.70 \pm 0.01)^{(c)}$	
5:3	25	pot	0.1 m (tmatfms)	$-(16.10 \pm 0.01)$	[95PAL/NGU]
				$-(16.18 \pm 0.01)^{(c)}$	
7:3	25	pot	0.1 m (tmatfms)	$-(28.80 \pm 0.04)$	[95PAL/NGU]
				$-(28.25 \pm 0.04)^{(c)}$	
8:3	25	pot	0.1 m (tmatfms)	$-(37.62 \pm 0.07)$	[95PAL/NGU]
10:3	25	pot	0.1 m (tmatfms)	$-(60.53 \pm 0.08)$	[95PAL/NGU]
2:2	25	pot	3 M (NaClO$_4$)	$-(6.24 \pm 0.02)$	[97LUB/HAV2]
5:3	25	pot	3 M (NaClO$_4$)	$-(16.80 \pm 0.04)$	[97LUB/HAV2]
4:3	25	pot	3 M (NaClO$_4$)	$-(12.8 \pm 0.1)$	[97LUB/HAV2]
2:2	25	sp	3 M (NaClO$_4$)	$-(6.13 \pm 0.02)$	[97LUB/HAV2]
5:3	25	sp	3 M (NaClO$_4$)	$-(16.81 \pm 0.02)$	[97LUB/HAV2]
4:3	25	sp	3 M (NaClO$_4$)	$-(12.57 \pm 0.02)$	[97LUB/HAV2]
1:2	25	RAMAN	var 0.1 to 2.3 M (tmatfms)	(2.1 ± 0.2) to (2.7 ± 0.3)	[2000NGU/PAL]
11:3	25	RAMAN	var 0.02 to 0.56 M (tmatfms)	$-(78 \pm 8)$	[2000NGU/PAL]
5:3	25	sol, sp	0.1 M (NaClO$_4$)	$-(17.14 \pm 0.13)$	[96MEI/KAT]
2:2	25	sp	0.1 M (NaClO$_4$)	$-(6.14 \pm 0.08)$	[96MEI/SCH], [97MEI/SCH]
2:2	25	sp	0.1 M (NaClO$_4$)	$-(6.145 \pm 0.088)$	[97MEI]
5:3	25	sp	0.1 M (NaClO$_4$)	$-(17.142 \pm 0.138)$	[97MEI]
2:2	25	sp	0.1 M (NaClO$_4$)	$-(6.237 \pm 0.103)$	[98MEI]
5:3	25	sp	0.1 M (NaClO$_4$)	$-(17.203 \pm 0.157)$	[98MEI]
5:3	25	sp	0.1 M (NaClO$_4$)	$-(17.00 \pm 0.17)$	[98MEI2]
5:3	25	sol	0.1 M (NaClO$_4$)	$-(17.16 \pm 0.18)$	[98MEI2]
2:2	25	sol	0.1 M (NaClO$_4$)	$-(6.168 \pm 0.056)^{(d)}$	[98MEI3]
5:3	25	sol	0.1 M (NaClO$_4$)	$-(17.123 \pm 0.069)^{(d)}$	[98MEI3]
2:2	25	sol	0.1 M (NaClO$_4$)	$-(6.145 \pm 0.088)$	[98MEI3]
5:3	25	sol	0.1 M (NaClO$_4$)	$-(17.142 \pm 0.014)$	[98MEI3]

(Continued on next page)

Table 9-3 (continued)

n:m	t (°C)	Method	I (medium)	$\log_{10} {}^*\beta_{n,m}$	Reference
1:1	25	pot	0.1 M (NaClO$_4$)	$-(5.01 \pm 0.03)$	[2002BRO]
2:2	25	pot	0.1 M (NaClO$_4$)	$-(5.98 \pm 0.04)$	[2002BRO]
4:3	25	pot	0.1 M (NaClO$_4$)	$-(12.39 \pm 0.05)$	[2002BRO]
5:3	25	pot	0.1 M (NaClO$_4$)	$-(16.36 \pm 0.05)$	[2002BRO]
1:1	25	pot	0.1 M (KCl)	$-(5.17 \pm 0.03)$	[2002BRO]
2:2	25	pot	0.1 M (KCl)	$-(5.86 \pm 0.04)$	[2002BRO]
4:3	25	pot	0.1 M (KCl)	$-(12.00 \pm 0.06)$	[2002BRO]
5:3	25	pot	0.1 M (KCl)	$-(16.09 \pm 0.06)$	[2002BRO]
1:1	25	pot	1.0 M (KNO$_3$)	$-(5.85 \pm 0.03)$	[2002BRO]
4:3	25	pot	1.0 M (KNO$_3$)	$-(11.95 \pm 0.05)$	[2002BRO]
5:3	25	pot	1.0 M (KNO$_3$)	$-(16.40 \pm 0.06)$	[2002BRO]
7:4	25	pot	1.0 M (KNO$_3$)	$-(21.79 \pm 0.06)$	[2002BRO]

(*) Tetramethylammonium trifluoromethanesulphonate.
(a) Calculated from $\log_{10} \beta_{n,m}$ with $pK_w = (13.78 \pm 0.01)$.
(b) Reinterpretation by [98MEI2].
(c) Fixing $\log_{10} {}^*\beta_{1,1} = -5.50$, $\log_{10} {}^*\beta_{7,4} = -22.67$.
(d) Uncertainty in pH not taken into account.

Table 9-4 gives the experimental values of $\log_{10} \beta_{n,1}$ for the equilibria:

$$0.5\,Na_2U_2O_7 \cdot xH_2O + (n-3)OH^- \rightleftharpoons UO_2(OH)_n^{2-n} + Na^+ + (x-1.5)H_2O(l) \qquad (9.3)$$
($x = 3$ to 5).

Table 9-4: Experimental equilibrium constants for the U(VI) hydroxide system for the equilibria (9.3).

n:m	t (°C)	Method	I (medium)	$\log_{10} {}^*\beta_{n,m}$	Reference
3:1	25	sol	0.5 M (NaClO$_4$)	$-(7.28 \pm 0.37)$	[98YAM/KIT]
			1 M (NaClO$_4$)	$-(7.45 \pm 0.38)$	[98YAM/KIT]
			2 M (NaClO$_4$)	$-(6.04 \pm 0.11)$	[98YAM/KIT]
4:1	25	sol	0.5 M (NaClO$_4$)	$-(5.05 \pm 0.06)$	[98YAM/KIT]
			1 M (NaClO$_4$)	$-(4.87 \pm 0.04)$	[98YAM/KIT]
			2 M (NaClO$_4$)	$-(4.50 \pm 0.14)$	[98YAM/KIT]

Table 9-5 gives the solubility product of the solid phases used to deduce equilibrium constants for U(VI) hydrolysis. The solubility product for UO$_3$·2 H$_2$O, $\log_{10} K_{s,0}$ refers to the reaction:

$$UO_3 \cdot 2H_2O(cr) \rightleftharpoons UO_2^{2+} + 2\,OH^- + H_2O(l). \qquad (9.4)$$

The solubility product for sodium uranate refers to the reaction:

$$1/2\,Na_2U_2O_7(s) + 3/2\,H_2O(l) \rightleftharpoons Na^+ + UO_2^{2+} + 3\,OH^-, \qquad (9.5)$$

where the values at zero ionic strength were calculated with the SIT by [92SAN/BRU], [98YAM/KIT] or by this review. In selecting values for equilibrium constants, this review has considered the relative importance of systematic and random errors. These

Table 9-5: Literature data for the solubility product of $UO_3 \cdot 2H_2O$ and $Na_2U_2O_7(s)$ according to the reactions (9.4) and (9.5), ($t = 25°C$).

compound	I (medium)	$\log_{10} {}^*K_{s,0}$	$\log_{10} {}^*K^\circ_{s,0}$	Reference
$UO_3 \cdot 2H_2O$(cr)	$I = 0$		$-(23.19 \pm 0.43)^{(a)}$	[92GRE/FUG]
$UO_3 \cdot 2H_2O$(s)	0.1 M (NaClO$_4$)	$-(22.21 \pm 0.01)$	$-(22.81 \pm 0.01)$	[92KRA/BIS]
		$-(21.77 \pm 0.03)^{(b)}$	$-(22.37 \pm 0.03)$	[92KRA/BIS]
$UO_3 \cdot 2H_2O$(cr)	0.5 M (NaClO$_4$)	$-(21.25 \pm 0.14)$	$-(22.03 \pm 0.14)$	[92SAN/BRU]
$UO_3 \cdot 2H_2O$(am)	0.5 M (NaClO$_4$)	$-(20.89 \pm 0.14)$	$-(21.67 \pm 0.14)$	[92SAN/BRU]
$UO_3 \cdot 2H_2O$(s)	0.1 M (NaClO$_4$)	$-(22.28 \pm 0.19)$	$-(22.88 \pm 0.19)$	[93MEI/KIM2]
$UO_3 \cdot 2H_2O$(s)	0.1 M (NaClO$_4$)	$-(22.30 \pm 0.07)$	$-(22.90 \pm 0.07)$	[96MEI/KAT]
		$-(21.84 \pm 0.02)^{(d)}$	$-(22.44 \pm 0.02)$	[96MEI/KAT]
$UO_3 \cdot 2H_2O$(s)	0.1 M (NaClO$_4$)	$-(22.15 \pm 0.06)$	$-(22.75 \pm 0.06)$	[96KAT/KIM]
$UO_3 \cdot 2H_2O$(s)	0.1 M (NaClO$_4$)	$-(21.81 \pm 0.03)$	$-(22.41 \pm 0.03)$	[98MEI2]
$UO_3 \cdot 2H_2O$(s)	1 M (NaCl)	$-(21.94 \pm 0.08)^{(d)}$	$-(22.62 \pm 0.20)$	[94TOR/CAS]
$UO_3 \cdot 2H_2O$(s)	0.5 m (NaClO$_4$)	$-22.35^{(d,f)}$	$-23.14^{(f)}$	[98DIA/GRA]
$Na_2U_2O_7$(cr)	$I = 0$		$-(30.7 \pm 0.5)^{(a)}$	[92GRE/FUG]
$Na_2U_2O_7 \cdot x\,H_2O$ $x = 3-5$	0.5, 1.0 and 2.0 M (NaClO$_4$)			[98YAM/KIT]
	$I = 0$		$-(29.45 \pm 1.04)^{(c)}$	[98YAM/KIT]
$Na_2U_2O_7 \cdot x\,H_2O$	0.1 M (NaClO$_4$)	$-(29.2 \pm 0.1)^{(e)}$	$-(29.7 \pm 0.1)^{(e)}$	[98MEI/FIS]

(a) Calculated from thermochemical data for crystalline solids.
(b) Reinterpretation by [98MEI2].
(c) Calculated using the formation constant of tricarbonato dioxouranium(VI) of [92GRE/FUG] (see 9.3.2.1.5.2.2).
(d) Value derived using mainly hydrolysis constants selected in [92GRE/FUG].
(e) Calculated from $\log_{10} {}^*K_{s,0} = (24.2 \pm 0.2)$ in 0.1 M NaClO$_4$ and $\log_{10} {}^*K^\circ_{s,0} = (24.6 \pm 0.2)$.
(f) Given without uncertainty.

9.3.2.1.1 Major polymeric species at 298.15 K (V.3.2.1.1)

The solubility of $UO_2(OH)_2 \cdot H_2O$ and UO_2CO_3 were studied by Kramer–Schnabel et al. [92KRA/BIS] at 298.15 K in 0.1 M NaClO$_4$ in the absence and the presence of carbonate. The pH range was generally 4.5 to 5.5 for the study with no added carbonate. Spectrophotometric measurements showed the presence of the 2:2, $(UO_2)_2(OH)_2^{2+}$ and 5:3, $(UO_2)_3(OH)_5^+$, species. The solubility data were then used to obtain $\log_{10} \beta_{2,2} = (22.16 \pm 0.03)$, $\log_{10} \beta_{5,3} = (53.05 \pm 0.04)$ and $\log_{10} K_{s,0}$ ($UO_2 \cdot 2H_2O$) $= -(22.21 \pm 0.01)$. This review has recalculated these data to obtain equilibrium constants for the corresponding reactions (9.2) using $pK_w = (13.78 \pm 0.01)$ at $I = 0.1$ M, and obtains $\log_{10} {}^*\beta_{2,2} = -(5.40 \pm 0.03)$ and $\log_{10} {}^*\beta_{5,3} = -(15.85 \pm 0.04)$. Using the interaction coefficients in [92GRE/FUG] this review finds at zero ionic strength:

$$\log_{10} {}^*\beta^\circ_{2,2} = -(5.19 \pm 0.09),$$
$$\log_{10} {}^*\beta^\circ_{5,3} = -(15.22 \pm 0.17),$$
$$\log_{10} K^\circ_{s,0} (UO_3 \cdot 2H_2O) = -(22.81 \pm 0.01),$$

where $\log_{10} {}^*\beta^\circ_{2,2}$ is significantly different from that of [92GRE/FUG], where the selected values are:

$$\log_{10} {}^*\beta^\circ_{2,2} = -(5.62 \pm 0.04),$$
$$\log_{10} {}^*\beta^\circ_{5,3} = -(15.55 \pm 0.12),$$
$$\log_{10} K^\circ_{s,0} (UO_3 \cdot 2H_2O) = -(23.19 \pm 0.43).$$

The solubility product of schoepite in [92GRE/FUG] is based on the value $\Delta_r G^\circ_m = -(27.47 \pm 2.44)$ kJ · mol^{-1} for the reaction:

$$UO_3 \cdot 2H_2O(cr) + 2H^+ \rightleftharpoons UO_2^{2+} + 3H_2O(l),$$

$\log_{10} {}^*K^\circ_{s,0} = (4.81 \pm 0.43)$). $\Delta_r G^\circ_m$ is obtained from the appropriate $\Delta_f G^\circ_m$ values, with $\Delta_f G^\circ_m (UO_3 \cdot 2H_2O, cr)$ coming from the Gibbs energy of hydration of γ–UO$_3$(cr) and from $\Delta_f G^\circ_m (UO_2^{2+})$ (in [92GRE/FUG]). Part of the data of [92KRA/BIS] has been reinterpreted in [98MEI2] (see below).

Potentiometric titrations by Ferri *et al.* [93FER/SAL] in 3.0 M NaClO$_4$ at 298.15 K were interpreted in terms of hydrolysis with the equilibrium constants $\log_{10} {}^*\beta_{2,2} = -(5.98 \pm 0.02)$ and $\log_{10} {}^*\beta_{5,3} = -(16.23 \pm 0.05)$. The respective values at zero ionic strength using SIT values from [92GRE/FUG] calculated by this review are $-(5.54 \pm 0.04)$ and $-(15.08 \pm 0.70)$, which are in good agreement with those of [92GRE/FUG].

Palmer and Nguyen-Trung [95PAL/NGU] used potentiometric titrations of U(VI) in a 0.1 M tetramethylammonium trifluoromethanesulphonate (tmatfms) medium (298.15 K), over the pH range 2.4 to 11.9, to obtain equilibrium constants for the cationic U(VI) hydroxide species 2:2 and 5:3, $\log_{10} {}^*\beta_{2,2} = -(5.77 \pm 0.01)$ and $\log_{10} {}^*\beta_{5,3} = -(16.10 \pm 0.01)$, and the anionic 7:3 complex, $(UO_2)_3(OH)_7^-$, $\log_{10} {}^*\beta_{7,3} = -(28.80 \pm 0.04)$. If two additional complexes UO_2OH^+ and $(UO_2)_4(OH)_7^+$ are added to the previous chemical model slightly different values of the equilibrium constants are obtained, despite the fact that these species are present in only small amounts. The equilibrium constants for the 1:1, UO_2OH^+, and 7:4, $(UO_2)_4(OH)_7^+$, complexes could not be determined from the data of [95PAL/NGU], instead the values from [79SYL/DAV] were used. The values $\log_{10} {}^*\beta_{1,1} = -5.50$ and $\log_{10} {}^*\beta_{7,4} = -22.76$, were used as fixed parameters in a least-squares fit that gave the following values at $I = 0.1$ M:

$$\log_{10} {}^*\beta_{2,2} = -(5.70 \pm 0.01),$$
$$\log_{10} {}^*\beta_{5,3} = -(16.18 \pm 0.01),$$
$$\log_{10} {}^*\beta_{7,3} = -(28.25 \pm 0.04).$$

These values compare well with other data in Table V.5 (page 99) of [92GRE/FUG]. A reinterpretation made by this review using the primary experimental data obtained from the authors, using the LETAGROP least-squares program, gives similar values as shown in the discussion of the paper in Appendix A, and hence, the values at zero ionic strength are also in agreement.

The formation constants of other trimeric species (8:3, 10:3 and 11:3) reported in [95PAL/NGU] are not accepted and not retained by this review as discussed in section 9.3.2.1.3 below and in the Appendix A.

Meinrath and co–workers have devoted a number of papers to the determination of $\log_{10} {}^*\beta_{2,2}$ and $\log_{10} {}^*\beta_{5,3}$ and the solubility product of schoepite, [93MEI/KAT], [93MEI/KIM2], [96MEI/KAT], [96MEI/SCH], [97MEI/SCH], [97MEI], [98MEI], [98MEI2] and [98MEI3], using spectroscopic and solubility measurements. These studies have been made in 0.1 M NaClO$_4$, at a pH less than 6–7 and temperatures between 297.15 to 298.15 K, under variable partial pressures of CO$_2$, p_{CO_2}. It is not always clear if the authors consider new experimental data in each paper or use already published data for their reinterpretation.

Values of $\log_{10} {}^*\beta_{n,m}$ are very sensitive to random and systematic errors in the measurement of [H$^+$]. Meinrath et al. have paid attention to this. They have made multiple calibration points of the glass electrode (mV vs. pH of NIST reference solutions). These data were used in a least-squares analysis to determine a calibration curve. This calibration gives the statistical ΔpH (random error) for a given potential. For instance, from a 5 point pH calibration (1.7 to 10) they derived the following relation between the pH and the measured emf, E, where E is in mV:

$$pH = (7.078 \pm 0.047) - (0.017029 \pm 0.000339) E.$$

It is unfortunate that Meinrath et al. have not calibrated their electrode directly in concentration units to obtain the free hydrogen ion concentration. They instead assume $\log_{10} \gamma_{H^+} = -0.09$, which is in reasonable agreement with the value -0.1092 deduced from the SIT method [97MEI], [98MEI3]. It can be noted that hydrogen ion concentration as $-\log_{10}$ [H$^+$] can be measured more precisely than the ± 0.047 given in the previous calibration equation. A general comment from this review is that more attention should be given to electrode calibrations in potentiometric determinations of equilibrium constants and where practical, they should also be made in concentration units.

The two first papers of the series [93MEI/KAT] and [93MEI/KIM2] complement each other and give:

- A clear identification of the 2:2 species, and a crude estimation of $\log_{10} {}^*\beta_{2,2}$ from saturated $-(5.89 \pm 0.12)$ and unsaturated solutions $-(5.96 \pm 0.06)$ [93MEI/KAT].

- The identification of rutherfordine and schoepite (in fact metaschoepite) as the solubility limiting phases, with the solubility products $\log_{10} K_{s,0} (UO_2CO_3) = -(13.89 \pm 0.11)$, and $\log_{10} K_{s,0} (UO_3 \cdot 2H_2O) = -(22.28 \pm 0.19)$, respectively [93MEI/KIM2]. The solubility product for schoepite has been obtained by assuming a value of $\log_{10} {}^*\beta_{2,2} = -(5.97 \pm 0.16)$ [93MEI/KIM] because the solubility product and $\log_{10} {}^*\beta_{2,2}$ are strongly correlated and cannot be determined separately in these experiments. These data provide a basis for the subsequent studies of Meinrath and co–workers, where a more detailed statistical analysis of the data is often used as a tool to assign better estimates of the uncertainty in experiments of this type.

The following studies, [96MEI/SCH], [97MEI/SCH], give additional data from unsaturated solutions, leading to a slightly different value of $\log_{10} {}^*\beta_{2,2} = -(6.14 \pm 0.08)$.

Two sets of data in [96MEI/KAT] deal with hydrolysis of U(VI).

The first set consists of solubility measurements of schoepite, (under 0.3 % CO_2, pH = 3.4 to 4.8) and leads to the solubility product $\log_{10} K_{s,0} (UO_3 \cdot 2H_2O) = -(22.30 \pm 0.05)$ (using $\log_{10} {}^*\beta_{2,2} = -(6.00 \pm 0.06)$, as discussed above). However, the method used to calculate the solubility product of schoepite is not clearly outlined (see Appendix A) and therefore this review does not accept the proposed value to select new data.

The second data set consists of solubility measurements of schoepite in equilibrium with air, $p_{CO_2} = 10^{-3.5}$ atm and pH varying from 3.8 to slightly below 7 to avoid phase transformation to sodium uranate. These data also lead to the solubility product $\log_{10} K_{s,0} (UO_3 \cdot 2H_2O)$. The solubility product of schoepite is determined assuming the formation of $(UO_2)_2(OH)_2^{2+}$ and $(UO_2)_3(OH)_5^+$ in addition to UO_2^{2+} and $UO_2CO_3(aq)$, the latter being identified by TRLFS, [96KAT/KIM]. Several chemical models were tested. The best fit is obtained by fixing: $\log_{10} {}^*\beta_{2,2} = -(6.00 \pm 0.06)$, $\log_{10} {}^*\beta (UO_2CO_3, aq) = -(9.23 \pm 0.03)$, using the value from [93MEI/KIM] and $-6.08 < \log_{10} {}^*\beta_{1,1} < -5.6$. The resulting values of $\log_{10} {}^*K_{s,0}$ and $\log_{10} {}^*\beta_{5,3}$ are: $5.72 < \log_{10} {}^*K_{s,0} (UO_3 \cdot 2H_2O) < 5.74$ and $-17.13 < \log_{10} {}^*\beta_{5,3} < -17.74$. Meinrath et al. retained the average value $\log_{10} {}^*K_{s,0} (UO_3 \cdot 2H_2O) = (5.73 \pm 0.01)$ and calculated the value of $\log_{10} K_{s,0} = -(21.83 \pm 0.02)$ using $\log_{10} K_w = (13.78 \pm 0.01)$.

In [97MEI], Meinrath reports the equilibrium constants $\log_{10} {}^*\beta_{2,2}$ and $\log_{10} {}^*\beta_{5,3}$ deduced from spectroscopic data in the pH–ranges (3.504 ± 0.032) to (4.718 ± 0.019) and (3.939 ± 0.029) to (4.776 ± 0.014), using a chemical model with three species. The weighted average values and standard deviations reported are respectively $\log_{10} {}^*\beta_{2,2} = -(6.145 \pm 0.088)$ and $\log_{10} {}^*\beta_{5,3} = -(17.142 \pm 0.138)$. The modelling and least-squares analysis relies on statistical criteria using a multivariate technique for analysing the optical density versus pH data. Part of these data (pH = 3.939 to 4.776) is refined in [98MEI] and results in slightly different values of $\log_{10} {}^*\beta_{2,2} = -(6.237 \pm 0.103)$ and $\log_{10} {}^*\beta_{5,3} = -(17.203 \pm 0.157)$.

In [98MEI2], Meinrath combines new solubility data for schoepite under atmospheric pressure (pH = 4 to 5.5), with previous data from [93MEI/KIM], [92KRA/BIS], [96MEI/KAT] and [96KAT/KIM], which relate to the same system and analyses them by using chemometric methods. As the direct determination of the free concentration, $[UO_2^{2+}]$, from the spectroscopic data is very uncertain, $\log_{10} {}^*\beta_{5,3}$ was calculated from the measured concentrations of $[H^+]$ and $(UO_2)_2(OH)_2^{2+}$ by taking $\log_{10} {}^*\beta_{2,2} = -(6.145 \pm 0.088)$ [97MEI]. From the concentrations of $[H^+]$ and $(UO_2)_3(OH)_5^+$ one then obtains $\log_{10} {}^*\beta_{5,3} = -(17.00 \pm 0.17)$. By modelling the solubility data with the same value of $\log_{10} {}^*\beta_{2,2}$ Meinrath finds $\log_{10} K_{s,0} (UO_3 \cdot 2H_2O) = -(21.81 \pm 0.03)$ and $\log_{10} {}^*\beta_{5,3} = -(17.16 \pm 0.18)$ where he considers the latter value more precise than the one obtained by spectroscopy. According to Meinrath et al. the species 5:3 dominates only in the narrow region close to the solubility limit of schoepite, [98MEI]. This is the optimal region for the determination of ${}^*\beta_{5,3}$.

Reinterpretation of [92KRA/BIS] data by [98MEI] with a fixed value of ${}^*\beta_{2,2}$, gives $\log_{10} K_{s,0} (UO_3 \cdot 2H_2O) = -(21.77 \pm 0.03)$ and $\log_{10} {}^*\beta_{5,3} = -(17.04 \pm 0.18)$.

The paper [98MEI3] gives some additional spectroscopic and solubility data, which do not change their previous values, only the uncertainties, which seem to depend on the calculation methods employed (see Table 9-3).

The investigations of Meinrath and co-workers clearly show the difficulty of finding a set of data fitting both spectroscopic measurements and solubility data. They emphasise the importance of the 2:2 species in U(VI) hydrolysis because its concentration can be determined spectroscopically.

This review considers the most recent equilibrium constants reported by Meinrath [98MEI3] as the most reliable, $\log_{10} {}^*\beta_{2,2} = -(6.145 \pm 0.088)$, $\log_{10} {}^*\beta_{5,3} = -(17.16 \pm 0.18)$ and $\log_{10} K_{s,0} (UO_3 \cdot 2H_2O) = -(21.81 \pm 0.03)$, because they were established from the more precise and cross-checked experimental data that the author obtained (see Appendix A). This review also notes that the values of $\log_{10} {}^*\beta_{2,2}$ proposed by Meinrath have decreased in magnitude and increased in accuracy from 1993 to 1998 and that the uncertainty in $\log_{10} K_{s,0} (UO_3 \cdot 2H_2O)$ seems very low. Using the auxiliary data in [92GRE/FUG] to derive values at zero ionic strength from 0.1 M NaClO$_4$ gives:

$$\log_{10} {}^*\beta_{2,2}^\circ = -(5.93 \pm 0.09),$$
$$\log_{10} {}^*\beta_{5,3}^\circ = -(16.51 \pm 0.18),$$
$$\log_{10} K_{s,0}^\circ (UO_3 \cdot 2H_2O) = -(22.41 \pm 0.03).$$

9.3 Uranium oxygen and hydrogen compounds and complexes (V.3)

As mentioned above, the selected values in [92GRE/FUG] are:

$\log_{10} {}^*\beta^o_{2,2} = -(5.62 \pm 0.04)$,

$\log_{10} {}^*\beta^o_{5,3} = -(15.55 \pm 0.12)$,

$\log_{10} K^o_{s,0} (UO_3 \cdot 2H_2O) = -(23.19 \pm 0.43)$.

There are obvious systematic deviations between all of the values from Meinrath et al. and those selected by [92GRE/FUG]. The solubility product is 0.8 orders of magnitude higher, ${}^*\beta^o_{2,2}$ is 0.3 orders of magnitude lower and ${}^*\beta^o_{5,3}$ is one order of magnitude lower. Selected values by [92GRE/FUG] are based on the analysis of numerous compatible experimental values obtained by different groups and different experimental methods. A plausible explanation for the observed discrepancy could be either a systematic error in the determination of $\log_{10} [H^+]$ or a real chemical effect caused by the difference in ionic strength and perchlorate concentration between the experiments of Meinrath et al. (0.1 M NaClO$_4$) and those of most other investigators who in general have worked at much higher ionic strength. For example, an error of 0.15 in $\log_{10}[H^+]$ would result in $\log_{10} {}^*\beta^o_{2,2} = -5.62$ and $\log_{10} {}^*\beta^o_{5,3} = -15.75$. Small differences in the crystallinity of schoepite (in fact metaschoepite) could be also a reason for the differences in the values of $\log_{10} {}^*K^o_{s,0}$.

This review has compared speciation diagrams in 0.1 and 3 M NaClO$_4$ for the U(VI) hydroxide system using the equilibrium constants proposed by Meinrath and co-workers for species 2:2 and 5:3 and those selected in [92GRE/FUG] for zero ionic strength and recalculated to 0.1 and 3 M NaClO$_4$ (see [98MEI] in Appendix A). The $-\log_{10}[H^+]$ ranged from 2.5 to 5 and the concentration of uranium was chosen as 10^{-2} M, which optimises the chances of finding the 5:3 species. This concentration is within the range suitable for potentiometric studies. The data selected in [92GRE/FUG] result in much higher concentrations of $(UO_2)_3(OH)_5^+$ than found when using the data of Meinrath et al. As a matter of fact it would be very difficult to identify this species from potentiometric data if the constant proposed by Meinrath and recalculated to high ionic strength is correct.

This review does not select the Meinrath values despite the fact that his experimental and modelling work are well done. There are no arguments to reject the previous experimental values of ${}^*\beta^o_{2,2}$, ${}^*\beta^o_{5,3}$ and $K^o_{s,0}$ which are supported by other recent studies. On the other hand, a weighted average of all the $\log_{10} {}^*\beta^o_{2,2}$, $\log_{10} {}^*\beta^o_{5,3}$ and $\log_{10} K^o_{s,0}$ values would not change significantly the values selected by [92GRE/FUG] due to the large number of other determinations on which they are based.

The recent potentiometric data of Brown [2002BRO] in 0.1 M NaClO$_4$, 0.1 M KCl and 1.0 M KNO$_3$ (pH = 3 to 6), (see Appendix A) support well the values $\log_{10} {}^*\beta^o_{2,2}$ and $\log_{10} {}^*\beta^o_{5,3}$ selected by [92GRE/FUG]. This review has calculated from the data obtained in perchlorate medium, $\log_{10} {}^*\beta^o_{2,2} = -(5.76 \pm 0.04)$ and $\log_{10} {}^*\beta^o_{5,3} = -(15.77 \pm 0.05)$.

In conclusion, considering the new data with regard to the determination of $\log_{10} {}^*\beta_{2,2}^\circ$ and $\log_{10} {}^*\beta_{5,3}^\circ$, this review accepts the equilibrium constants values selected by [92GRE/FUG] (see Table 9-6).

From the data of Brown [2002BRO], this review has also calculated $\log_{10} {}^*\beta_{4,3}^\circ = -(11.93 \pm 0.06)$ which agrees very well with the value selected by [92GRE/FUG]. Therefore, the value of [92GRE/FUG] is retained.

Papers that give a quantitative confirmation of the data in [92GRE/FUG] focus on the 2:2 and 5:3 species. Only one, [95PAL/NGU], covers a broad concentration – $\log_{10}[H^+]$ range that involves other species as well. One solubility experiment by Diaz et al. [98DIA/GRA] at 298.15 K in 0.5 molal $NaClO_4$ ($-\log_{10}[H^+] = 4.70$), in which crystalline schoepite was characterised as the solid phase, yielded $\log_{10} {}^*K_{s,0} = 5.14$, which by applying SIT reduces to $\log_{10} {}^*K_{s,0}^\circ (UO_3 \cdot 2H_2O) = 4.86$ and gives $\log_{10} K_{s,0}^\circ (UO_3 \cdot 2H_2O) = -23.14$, which is close to the value selected in [92GRE/FUG] (see Appendix A).

Rizkalla et al. [94RIZ/RAO] conducted calorimetric titrations of solutions containing the 2:2 species ($I = 1$ M, CO_2 free tetramethylammonium chloride) and found $\Delta_r H_m^\circ (2:2) = (44.4 \pm 1.9)$ kJ $\cdot$ mol^{-1} and $\Delta_r S_m^\circ (2:2) = -(36 \pm 6)$ J $\cdot$ K$^{-1} \cdot$ mol^{-1} values close to the ones reported in [92GRE/FUG], p.118, Table V. 8. Dai et al. [98DAI/BUR] reported the determination of the enthalpy of reaction from spectroscopic measurements in 1 M CF_3SO_3H: $\Delta_r H_m^\circ (2:2) = 45.5$ kJ $\cdot$ mol^{-1}. This value is also in agreement with the values reported in [92GRE/FUG] (Table V.8, p.118).

Sergeyeva et al. [94SER/DEV] analysed the same literature data as [92GRE/FUG] and came to slightly different thermodynamic values at zero ionic strength (298.15 K, 1 bar), due probably to a different appreciation of the validity of experimental data and the method used for extrapolation to $I = 0$, as discussed in Appendix A.

The solubility study of $UO_3 \cdot 2 H_2O$ in 0.1 M $NaClO_4$ (298.15 K) made by Kato et al. [96KAT/KIM] gives a solubility product of this hydrated oxide which, recalculated to zero ionic strength, is $\log_{10} K_{s,0}^\circ (UO_3 \cdot 2H_2O) = -(22.75 \pm 0.06)$, lower than the value obtained by [92SAN/BRU] for crystalline $UO_3 \cdot 2H_2O$, $\log_{10} K_{s,0}^\circ = -(22.03 \pm 0.14)$ and within the uncertainty bands of the value selected in [92GRE/FUG], $-(23.19 \pm 0.43)$. The observed difference might be due to differences in the crystallinity of the samples used.

Wruck et al. [97WRU/PAL] used laser induced photoacoustic spectroscopy (LIPAS) to investigate the hydrolysis of U(VI) in 0.1 m $NaClO_4$ solutions and a temperature of (298.15 ± 0.5) K in the pH range 3 to 5. Modelling of these data included three species: UO_2^{2+}, UO_2OH^+ and $(UO_2)_2(OH)_2^{2+}$ with fixed $\log_{10} {}^*\beta_{1,1} = -5.8$, and resulted in an equilibrium constant $\log_{10} {}^*\beta_{2,2} = -(5.45 \pm 0.05)$. This value is not very sensitive to the choice of $\log_{10} {}^*\beta_{1,1}$. Extrapolation by this review to zero ionic strength using the Davies equation gives $\log_{10} {}^*\beta_{2,2}^\circ = -(5.23 \pm 0.05)$ and $\log_{10} {}^*\beta_{2,2}^\circ = -(5.24 \pm 0.05)$ using the SIT approach. These values are considerably different from the value selected by [92GRE/FUG], (see Appendix A).

Raman spectroscopic data of Nguyen-Trung et al. [2000NGU/PAL] confirmed the existence of the species 2:2, 5:3 and 1:2 in acidic medium (pH < 5.6) of variable ionic strengths as discussed in Appendix A. The corresponding values of the formation constants agree with those of [95PAL/NGU] and [92GRE/FUG] (see above). In the same way, Raman and infrared spectra obtained by Quilès et al. [2000QUI/BUR] confirmed the presence of the species 2:2 and 5:3 at pH < 4.2, but the values of $\log_{10} {}^*\beta_{2,2}^\circ$ and of $\log_{10} {}^*\beta_{5,3}^\circ$ are higher than those selected by [92GRE/FUG] (see Appendix A).

Solubility studies in 1 M NaCl (298.15 K) involving dissolution of synthetic amorphous schoepite and precipitation of schoepite on UO_2 seeds are described, but no experimental results other than figures are made available [94TOR/CAS]. The authors present only $\log_{10} {}^*K_{s,0}$ values which were derived from a speciation model using the data of [92GRE/FUG], and find $\log_{10} {}^*K_{s,0}^\circ = (5.57 \pm 0.08)$ and (5.92 ± 0.08), respectively, for the amorphous and crystalline phases. Further refinement yielded $\log_{10} {}^*K_{s,0}^\circ = (5.38 \pm 0.20)$ and (5.73 ± 0.28). The second of these values is considerably higher than that recommended by [92GRE/FUG], $\log_{10} {}^*K_{s,0}^\circ = (4.81 \pm 0.43)$. The reason for this is not clear due to lack of experimental detail in the paper.

For reasons discussed in Appendix A the values of $\log_{10} {}^*\beta_{2,2}^\circ = -(5.80 \pm 0.04)$ or $-(5.69 \pm 0.04)$, $\log_{10} {}^*\beta_{5,3}^\circ = -(15.65 \pm 0.70)$ or $-(15.66 \pm 0.70)$ and $\log_{10} {}^*\beta_{4,3}^\circ = -(15.00 \pm 0.90)$ or $-(14.78 \pm 0.90)$ derived from the study of Lubal and Havel [97LUB/HAV2], are not considered by this review despite the fact that the two first sets confirm the selected data.

All the reported data, which use the [92GRE/FUG] selected data as input parameters in modelling, do not reduce the uncertainties in the thermodynamic values.

Data reported in [92FUG] which includes those of [92FUG/KHO] are considered in [92GRE/FUG].

9.3.2.1.2 Monomeric cationic and neutral hydrolysis species (V.3.2.1.2)

The first and second hydrolysis constants for the mononuclear hydroxo species 1:1 and 2:1, $UO_2(OH)_2(aq)$, were determined by Choppin and Mathur [91CHO/MAT] at 298.15 K using a solvent extraction technique, in aqueous solutions with the ionic strength 0.1 and 1.0 m $NaClO_4$. The solutions were in the pH range 5 to 7.5 and were open to the atmosphere. The resulting hydrolysis constants are $\log_{10} {}^*\beta_{1,1} = -(5.91 \pm 0.08)$ ($I = 0.1$ M) and $\log_{10} {}^*\beta_{1,1} = -(5.75 \pm 0.07)$ ($I = 1.0$ M), and $\log_{10} {}^*\beta_{2,1} = -(12.43 \pm 0.09)$ and $-(12.29 \pm 0.09)$, respectively. The latter value is consistent with that reported for PuO_2^{2+} in [2001LEM/FUG] based on the data from [95PAS/KIM], $\log_{10} {}^*\beta_{2,1} = -13.15$ and gives confidence that the data of Choppin and Mathur are not affected by the presence of carbonate complexes.

Based on the SIT model and using the values of the ion interaction parameters of [92GRE/FUG], the infinite dilution hydrolysis constants are:

$\log_{10} {}^*\beta_{1,1}^\circ = -(5.74 \pm 0.88)$ (from $I = 1.05$ m),

$\log_{10} {}^*\beta_{1,1}^\circ = -(5.73 \pm 0.38)$ (from $I = 0.10$ m),

$\log_{10} {}^*\beta_{2,1}^\circ = -(12.07 \pm 0.10)$ (from $I = 1.05$ m),

$\log_{10} {}^*\beta_{2,1}^\circ = -(12.23 \pm 0.09)$ (from $I = 0.10$ m).

The selected value for $\log_{10} {}^*\beta_{1,1}$ from Table V.7 p.107 of [92GRE/FUG] is $-(5.2 \pm 0.3)$, while for $\log_{10} {}^*\beta_{2,1}$ only an upper limit -10.3 is given.

From the potentiometric data of Brown [2002BRO], this review has calculated $\log_{10} {}^*\beta_{1,1}^\circ = -(4.72 \pm 0.37)$ which is compatible with the value selected by [92GRE/FUG] within the large uncertainty limit.

Based on the new data this review selects the weighted average value of $\log_{10} {}^*\beta_{1,1}^\circ$ from [92GRE/FUG], [91CHO/MAT] and [2002BRO], $\log_{10} {}^*\beta_{1,1}^\circ = -(5.25 \pm 0.24)$ and $\log_{10} {}^*\beta_{2,1}^\circ = -(12.15 \pm 0.07)$ which is the weighted average value of the constants obtained by [91CHO/MAT]. The value $\log_{10} {}^*\beta_{2,1}^\circ = -(11.3 \pm 1.0)$ obtained by Rai et al. [2002RAI/FEL] from a solubility study of becquerelite has not been considered for the selection for reasons discussed in Appendix A.

9.3.2.1.3 Anionic hydrolysis species (V.3.2.1.3)

The stoichiometry and equilibrium constant of the limiting complex formed in strong alkaline solutions were left open in [92GRE/FUG]. A number of publications dealing with this problem have appeared since 1991.

The data of Sandino [91SAN] reported in [92GRE/FUG] have been reinterpreted [92SAN/BRU] leading to $\log_{10} {}^*\beta_{7,3}^\circ = -(32.2 \pm 0.8)$ and $\log_{10} {}^*\beta_{3,1}^\circ = -(20.1 \pm 0.5)$. These are the unweighted averages derived from solubility measurements of crystalline and amorphous schoepite for which solubility products are, respectively, $\log_{10} K_{s,0}^\circ = -(22.03 \pm 0.14)$ and $-(21.67 \pm 0.14)$. These values are higher than the solubility product deduced from calorimetric data for crystalline schoepite, $\log_{10} K_{s,0}^\circ = -(23.19 \pm 0.43)$ [92GRE/FUG]. The weighted averaged values of $\log_{10} {}^*\beta_{3,1}^\circ$ and $\log_{10} {}^*\beta_{7,3}^\circ$ obtained by [92SAN/BRU] are slightly different but the uncertainties are reduced (see Appendix A).

The study of Palmer and Nguyen-Trung [95PAL/NGU] has been discussed previously. They interpret their data using a set of trimeric species 7:3 $(UO_2)_3(OH)_7^-$, 8:3 $(UO_2)_3(OH)_8^{2-}$, and 10:3 $(UO_2)_3(OH)_{10}^{4-}$, which are formed in the pH range 6 to 10. These data are discussed in Appendix A. Based on the review in Appendix A, only the identification of the 7:3 species with the value $\log_{10} {}^*\beta_{7,3} = -(28.80 \pm 0.04)$ is accepted. Extrapolation to zero ionic strength using SIT with interaction parameters valid in perchlorate media gives $\log_{10} {}^*\beta_{7,3}^\circ = -(28.40 \pm 0.04)$ which is far from the value of Sandino and Bruno [92SAN/BRU].

This review selects the value from Sandino and Bruno: $\log_{10} {}^*\beta_{7,3}^\circ = -(32.2 \pm 0.8)$, which is based on consistent solubility data and do not assume the formation of the 8:3 and 10:3 species, that are not accepted by this review.

The existence of the species 8:3 and 10:3 is not sufficiently established by the potentiometric titrations, cf. Appendix A. The Raman spectroscopic data of Nguyen et al. [2000NGU/PAL] do not allow a determination of the stoichiometry of the anionic species, as discussed in Appendix A, except for the complex $UO_2(OH)_4^{2-}$. The identification of trimeric species in the pH range 7 to 13 at uranium concentrations above 10^{-4} M, which have also been postulated by [49SUT], [72MUS], remains a problem to be solved.

Yamamura et al. [98YAM/KIT] have determined the equilibrium constant for the species 3:1, $UO_2(OH)_3^-$ and 4:1 $UO_2(OH)_4^{2-}$ for reactions:

$$UO_2^{2+} + 3 H_2O(l) \rightleftharpoons UO_2(OH)_3^- + 3 H^+,$$
$$UO_2^{2+} + 4 H_2O(l) \rightleftharpoons UO_2(OH)_4^{2-} + 4 H^+.$$

The solid-state structure of the latter complex was determined by Clark et al. [99CLA/CON]. The structure of the limiting complex at high hydroxide concentrations using EXAFS has also been determined by Clark et al. [99CLA/CON] and by Wahlgren et al. [99WAH/MOL]. The two groups give different interpretations of the experimental findings. Clark et al. propose the species 5:1, $UO_2(OH)_5^{3-}$, as the limiting complex based on the co–ordination number determined by EXAFS. Wahlgren et al. find the same co–ordination number and bond distances as Clark et al., but suggest the tetra–hydroxide as the limiting complex, based on a comparison of the experimental U(VI)–OH bond length in the solid and solution at high pH. In addition a comparison is made between these experimental data and the bond lengths calculated for 4:1 and 5:1 complexes $UO_2(OH)_4^{2-}$ and $UO_2(OH)_5^{3-}$ using ab initio methods of the theoretical study of Vallet et al. [2001VAL/WAH]. The problem has recently been discussed by Moll et al. [2000MOL/REI] and their conclusion is that $UO_2(OH)_4^{2-}$ has a very broad range of existence at high pH, but with evidence for the formation of small amounts of $UO_2(OH)_5^{3-}$ at pH > 14 (see Appendix A).

Nguyen et al. [2000NGU/PAL] concluded from their data that there is no evidence for the existence of the species 5:1, and they propose $\log_{10} {}^*\beta_{4,1}^\circ = -(32.2 \pm 1.6)$, close to the value of [92GRE/FUG]. The solubility experiments of Yamamura et al. [98YAM/KIT] are also in agreement with this interpretation and give more precise values of the equilibrium constants than those of Nguyen et al. This review includes the data of [98YAM/KIT] recalculated to zero ionic strength $\log_{10} {}^*\beta_{2,1}^\circ = -(20.86 \pm 0.79)$ and $\log_{10} {}^*\beta_{4,1}^\circ = -(32.40 \pm 0.68)$ in the set used to deduce the selected equilibrium constants.

All these data have been used to revise the corresponding equilibrium constants given in [92GRE/FUG]. The new selected value of $\log_{10} {}^*\beta^\circ_{3,1} = -(20.25 \pm 0.42)$ is based on the weighted average of values given by [92SAN/BRU] ($\log_{10} {}^*\beta^\circ_{3,1} = -(20.1 \pm 0.5)$) and [98YAM/KIT] ($\log_{10} {}^*\beta^\circ_{3,1} = -(20.86 \pm 0.79)$). The value for $\log_{10} {}^*\beta^\circ_{3,1} \leq -21.5$ [2002RAI/FEL] has not been considered in this selection (see Appendix A). The selected value $\log_{10} {}^*\beta^\circ_{4,1} = -(32.40 \pm 0.68)$ is taken from [98YAM/KIT], (see Appendix A) (Table 9-6). The estimated uncertainty is large in all experimental determinations and more precise new information is desirable.

The recent potentiometric study of Brown [2002BRO] in 1.0 M KNO$_3$ medium (pH = 3 to 6) has shown the formation of $(UO_2)_4(OH)_7^-$ with $\log_{10} {}^*\beta^\circ_{7,4} = -(21.79 \pm 0.06)$ (see Appendix A). This value is close to that selected by [92GRE/FUG] which is therefore retained by this review.

The behaviour of U(VI) in alkaline conditions is still under investigation [2000KON/CLA], [2000NGU/BUR], [2002NGU] and has been the subject of previous investigations [95HOB/KAR] (see Appendix A).

9.3.2.1.4 Other polymeric cationic hydrolysis species (V.3.2.1.4)

Some potentiometric titrations (298.15 K) in sulphate medium (pH less than 6.8) [93GRE/LAG], [2000COM/BRO], [2000MOL/REI] and EXAFS and NMR measurements, provide evidence for the existence of ternary polynuclear U(VI) complexes containing sulphate and hydroxide/oxide of the type $(UO_2)_m(OH)_n(SO_4)_r^{2m-n-2r}$ with $m = 2, 3$ and 4; $n = 2, 4, 5$ and 7, while the value of r varies between 2 and 4 [2000COM/BRO], [93GRE/LAG].

The EXAFS data indicate that the sulphate ligand is bonded as a chelate to U(VI), with no evidence of sulphate bridges. The conditional equilibrium constants, cf. Appendix A, are precise, while the values of $\log_{10} \beta^\circ_{m,n,r}$ depend strongly on the method used to correct the activity coefficient variations in the test solutions, cf. Appendix A and are much less precise.

No values are therefore selected by this review.

9.3.2.1.5 Relevant solubility studies

This section deals with the selection of equilibrium constants for the dissolution of some U(VI) compounds including: $UO_3 \cdot 2H_2O$, named "schoepite". The section 9.3.3.1.2 deals with thermodynamic data for U(VI) oxide hydrates, $UO_3 \cdot xH_2O$ and U(VI) hydroxide hydrate, $UO_2(OH)_2$.

9.3.2.1.5.1 Schoepite

The name schoepite is commonly applied to a mineral or synthetic preparation with a formula close to $UO_3 \cdot 2H_2O$. It should be named metaschoepite [98FIN/HAW] (see Appendix A and section 9.3.3.1.2). In the present section and throughout the review, the name "schoepite", commonly used by chemists, is however retained.

Determination of the solubility product of schoepite by solubility measurements requires consideration of the hydrolysis of UO_2^{2+}. The values in Table 9-5 with the exception of those of [94TOR/CAS] and [98DIA/GRA] are systematically larger than that selected by [92GRE/FUG] based on calorimetric data, $\log_{10} K_{s,0}^\circ = -(23.19 \pm 0.43)$. The varying solubility products found in many papers may be a result of differences in the structure of the solid phase. The weighted average value of the 10 values reported in Table 9-5 is $\log_{10} K_{s,0}^\circ = -(22.71 \pm 0.01)$ in good agreement with the value selected by [92GRE/FUG].

Based on the previous detailed discussions of the selection of the hydrolysis constants, this review does not find compelling reasons to change the previously selected value. Hence, this review keeps the equilibrium constant for the reaction:

$$UO_3 \cdot 2H_2O(cr) \rightleftharpoons UO_2^{2+} + 2\,OH^- + H_2O(l),$$
$$\log_{10} K_{s,0}^\circ = -(23.19 \pm 0.43).$$

Table 9-6: Selected values for U(VI) hydrolysis species at 298.15 K.

Species $n{:}m$	$\log_{10} {}^*\beta_{n,m}^\circ ((UO_2)_m(OH)_n^{2m-n}, 298.15\,K)$
1:1	$-(5.25 \pm 0.24)$*
2:1	$-(12.15 \pm 0.07)$*
3:1	$-(20.25 \pm 0.42)$*
4:1	$-(32.40 \pm 0.68)$*
1:2	$-(2.7 \pm 1.0)$
2:2	$-(5.62 \pm 0.04)$
4:3	$-(11.9 \pm 0.3)$
5:3	$-(15.55 \pm 0.12)$
7:3	$-(32.2 \pm 0.8)$*
7:4	$-(21.9 \pm 1.0)$

* Selected by this review, otherwise selected by [92GRE/FUG]

9.3.2.1.5.1.1   Temperature dependence (V.3.2.1.5)

Few reliable data can be extracted from the paper [89SER/SAV] dealing with the hydrolysis of U(VI) as a function of temperature.

The results of the investigation of Redkin [97RED] on the solubility of $UO_3 \cdot H_2O$ in water in the presence of Cu_2O/CuO between 473 and 773 K show that there is a progressive dehydration of the oxide to $UO_3 \cdot 0.33H_2O$ (423 < T < 573 K) followed by the precipitation of U(VI) hydroxides. The molal solubility of uranium, $m(U)$, in the presence of the two copper oxides is given by:

$$\log_{10} m(U) = 15375.3\ T^{-1} + 0.0647\ T - 67.05 \qquad (T = 473\text{–}573\ K)$$
$$\log_{10} m(U) = -221.7\ T^{-1} - 2.71 \qquad (T = 573\text{–}773\ K)$$

with $(UO_2)_3(OH)_7^-$ postulated to be the predominant species. No additional quantitative thermodynamic information can be extracted from this paper. The solubility of $UO_3 \cdot H_2O$ as a function of T is further discussed in section V.3.3.1 in [92GRE/FUG], p.133.

The results of [98DAI/BUR] mentioned in section 9.3.2.1 are discussed in Appendix A.

9.3.2.1.5.2   Polyuranates

9.3.2.1.5.2.1 $CaU_6O_{19} \cdot 11H_2O(cr)$

Solubility measurements of the solid phase becquerelite $CaU_6O_{19} \cdot 11H_2O(cr)$ were performed by Sandino and Grambow [94SAN/GRA] at 298.15 K in 1 molal $CaCl_2$ at several pH values, following the equilibrium:

$$CaU_6O_{19} \cdot 11H_2O\ (cr) + 14\,H^+ \rightleftharpoons Ca^{2+} + 6\,UO_2^{2+} + 18\,H_2O(l).$$

The mean values for $\log_{10}{}^*K_{s,0}$ were calculated by this review as $\log_{10}{}^*K_{s,0}(CaU_6O_{19} \cdot 11H_2O) = (44.4 \pm 0.9)$, which yielded $\log_{10}{}^*K_{s,0}^\circ$ values of (39.5 ± 1.0) using SIT. This infinite dilution constant disagrees for unknown reasons substantially from those tabulated by the authors. Rai et al. [2002RAI/FEL] have made a very careful study of the solubility product of a synthetic becquerelite in $2 \cdot 10^{-2}$, 0.1 and 0.5 M $CaCl_2$ at (296 ± 2) K. In the pH range 4.4 to 9, the data refined using hydrolysis data for U(VI) selected by [92GRE/FUG] and a Pitzer approach give $\log_{10}{}^*K_{s,0}^\circ = (41.4 \pm 0.2)$. For reasons discussed in Appendix A, this review increases the uncertainty and retains $\log_{10}{}^*K_{s,0}^\circ = (41.4 \pm 1.2)$ for the selection of the solubility product of becquerelite. The selected value is the average of the values (39.5 ± 1.0) and (41.4 ± 1.2):

$$\log_{10}{}^*K_{s,0}^\circ = (40.5 \pm 1.6).$$

The value given by Vochten et al. [90VOC/HAV], $\log_{10}{}^*K_{s,0} = (43.6 \pm 0.3)$ at ionic strength 10^{-2} M is not considered by this review for reasons discussed in Appendix A. The Gibbs energy of formation calculated from $\log_{10}{}^*K_{s,0}^\circ$ is:

$\Delta_f G_m^\circ$ (CaU$_6$O$_{19}$·11H$_2$O, cr, 298.15 K) = – (10305.5 ± 14.0) kJ·mol^{-1}.

Using natural well characterised becquerelite, Casas et al. [97CAS/BRU] derived from solubility measurements at low ionic strength, $\log_{10} {}^*K_{s,0}^\circ$ = (29 ± 1) (see Appendix A). The discrepancy with the [94SAN/GRA] results is attributed to the differences in experimental conditions with regard to aqueous phases and size of the crystallites.

9.3.2.1.5.2.2 Na$_2$U$_2$O$_7$ · x H$_2$O(cr)

Yamamura et al. [98YAM/KIT] have determined the solubility of a solid phase claimed to contain 3–5 waters of crystallisation. A discussion of the data is given in Appendix A. The experimental solubility data refer to the reaction:

$$0.5\,\text{Na}_2\text{U}_2\text{O}_7(s) + 3\,\text{CO}_3^{2-} + 1.5\,\text{H}_2\text{O(l)} \rightleftharpoons \text{UO}_2(\text{CO}_3)_3^{4-} + 3\,\text{OH}^- + \text{Na}^+ \qquad (9.6)$$

By combining this equilibrium constant with that taken from [92GRE/FUG] for the reaction:

$$\text{UO}_2^{2+} + 3\,\text{CO}_3^{2-} \rightleftharpoons \text{UO}_2(\text{CO}_3)_3^{4-}, \qquad (9.7)$$

they calculate the solubility product $\log_{10} K_{s,0}^\circ$ (9.8) = – (29.45 ± 1.04) and $\log_{10} {}^*K_{s,0}^\circ$ (9.9) = (25.1 ± 2.1) for the reactions:

$$0.5\,\text{Na}_2\text{U}_2\text{O}_7(s) + 1.5\,\text{H}_2\text{O(l)} \rightleftharpoons \text{UO}_2^{2+} + 3\,\text{OH}^- + \text{Na}^+, \qquad (9.8)$$

$$\text{Na}_2\text{U}_2\text{O}_7(s) + 6\,\text{H}^+ \rightleftharpoons 2\,\text{UO}_2^{2+} + 3\,\text{H}_2\text{O(l)} + 2\,\text{Na}^+. \qquad (9.9)$$

The value of $\log_{10} {}^*K_{s,0}^\circ$ (9.9) = (25.1 ± 2.1) is consistent with, but less precise than, the value calculated using the data in [92GRE/FUG], $\log_{10} {}^*K_{s,0}^\circ$ (9.9) = (22.6 ± 1.0). The respective values for equilibrium (9.8) are $\log_{10} K_{s,0}^\circ$ (9.8) = – (29.45 ± 1.04) [98YAM/KIT] and $\log_{10} K_{s,0}^\circ$ (9.8) = – (30.7 ± 0.5) [92GRE/FUG]. The Gibbs energy of formation of Na$_2$U$_2$O$_7$(cr) in [92GRE/FUG] refers to a crystalline anhydrous phase. The good agreement between the two values indicates either that the Gibbs energy contribution of structural water is not significantly different from that of free liquid water, or that the solid phase used in [98YAM/KIT] is anhydrous.

The solubility product for reaction (9.9) has also been calculated by Meinrath et al. [98MEI/FIS], $\log_{10} {}^*K_{s,0}^\circ$ = (24.6 ± 0.2), but as indicated in Appendix A, this value is based on assumptions of the composition of the solid phase that are not justified, so even if these data confirm the values given above, they are not used to calculate the solubility product for equation (9.9).

This review keeps the value selected in [92GRE/FUG].

9.3.2.1.5.2.3 Na$_x$UO$_{(3+x/2)}$·H$_2$O(cr)

Díaz Arocas and Grambow [98DIA/GRA] studied the precipitation of U-containing phases and found compositions of hydrated Na$_x$UO$_{(3+x/2)}$ phases with values of x from 0 to 1, in NaCl, as the ratio Na^+/H^+ is increased. For x = 1/3 the solubility product for the reaction:

$$Na_{0.33}UO_{3.165} \cdot 2H_2O(s) + 2.33\, H^+ \rightleftharpoons UO_2^{2+} + 3.165\, H_2O(l) + 0.33\, Na^+$$

is $\log_{10}{}^*K_{s,0}$ = (7.95 ± 0.15), at 298.15 K, and I = 3 m NaCl. The authors give $\log_{10}{}^*K_{s,0}^\circ$ = (7.13 ± 0.15) (see Appendix A). No other reliable solubility product for this phase has been reported and this review does not select a value for the solubility product for $Na_2U_6O_{19} \cdot 12\, H_2O$.

9.3.2.1.5.2.4 $K_2U_6O_{19} \cdot 11H_2O(cr)$

Solubility measurements of the compreignacite, $K_2U_6O_{19} \cdot 11H_2O(cr)$, were performed by Sandino and Grambow [94SAN/GRA] at 298.15 K in 1 molal KCl at several pH values, following the equilibrium:

$$K_2U_6O_{19} \cdot 11H_2O\,(cr) + 14\, H^+ \rightleftharpoons 2\, K^+ + 6\, UO_2^{2+} + 18\, H_2O(l).$$

The mean values for $\log_{10}{}^*K_{s,0}$ was calculated by this review as $\log_{10}{}^*K_{s,0}(K_2U_6O_{19} \cdot 11H_2O)$ = (40.2 ± 1.2), which yielded a $\log_{10}{}^*K_{s,0}^\circ$ value of (37.1 ± 0.5) using SIT. This infinite dilution constant disagrees for unknown reasons substantially from those tabulated by the authors. This review retains the recalculated solubility products. The Gibbs energies of formation calculated from $\log_{10}{}^*K_{s,0}^\circ$ is:

$$\Delta_f G_m^\circ (K_2U_6O_{19} \cdot 11H_2O,\, cr,\, 298.15\, K) = -(10337.1 \pm 11.0)\, kJ \cdot mol^{-1}.$$

9.3.2.2   U(IV) hydroxide complexes (V.3.2.3)

The hydrolysis behaviour of U^{4+} aqueous ions is difficult to ascertain, because of the easy oxidation of U(IV) to U(VI) and the precipitation of hydrous oxide phases, with poor crystallinity and small particle size, which are commonly denoted $UO_2(am)$ and microscrystalline $UO_2(cr)$. The problem encountered in these systems has been extensively discussed in [92GRE/FUG] where it was concluded that it was only possible to select the following thermodynamic quantities:

$\log_{10}{}^*\beta_1^\circ (UOH^{3+}) = -(0.54 \pm 0.06)$ (page 123),

$\Delta_f G_m^\circ (UO_2,\, cr,\, 298.15\, K) = -(1031.8 \pm 1.0)\, kJ \cdot mol^{-1}$ (page 140),

$\Delta_f G_m^\circ (U(OH)_4,\, aq,\, 298.15\, K) = -(1452.5 \pm 8.0)\, kJ \cdot mol^{-1}$ (page 126).

To clarify the following discussion in this section, we have used auxiliary data of [92GRE/FUG] to calculate the equilibrium constants for the following reactions:

$UO_2(cr) + 4\, H^+ \rightleftharpoons U^{4+} + 2\, H_2O(l)$ $\log_{10}{}^*K_{s,0}^\circ = -(4.85 \pm 0.33)$,

$UO_2(cr) + 2\, H_2O(l) \rightleftharpoons U^{4+} + 4\, OH^-$ $\log_{10} K_{s,0}^\circ = -(60.86 \pm 0.36)$,

$UO_2(cr) + 2\, H_2O(l) \rightleftharpoons U(OH)_4(aq)$ $\log_{10} K_{s,4}^\circ = -(9.5 \pm 1.0)$,

which are consistent with a value of $\log_{10} \beta_4^\circ = (51.36 \pm 1.06)$.

9.3 Uranium oxygen and hydrogen compounds and complexes (V.3)

In [92GRE/FUG] no thermodynamic quantities are selected for UO_2(am), because thermodynamic constants are not defined for an amorphous phase and will vary with the "ripening" of this phase. However, thermodynamic data were given for some other amorphous phases in [92GRE/FUG]. The present review prefers, for the reason given, not to select Gibbs energies of formation for amorphous phases. A solubility product for UO_2(am) is given in Table 9-7 with a remark that it is selected as the value to be used to calculate the U(IV) concentration in the presence of UO_2(am). The literature data for $\Delta_f G_m^\circ$(UO_2, am, 298.15 K) range from -977 to -1003.6 kJ·mol^{-1} [92GRE/FUG].

Some additional information can be extracted from the papers, which have appeared either since 1991, dealing with U(VI) hydrolysis [94SER/DEV], [95RAI/FEL], [95YAJ/KAW], [96RAK/TSY], [97LAN], [97RED], [97TOR/BAR], [98CAS/PAB], [99NEC/KIM] and [2001NEC/KIM] or before 1991, [90PHI].

The report of Philips [90PHI] is not explicitly mentioned in [92GRE/FUG], but his selected values for U(IV) hydrolysis, are either estimated or less critically evaluated than those selected in [92GRE/FUG], as discussed in Appendix A. The same is the case for the paper of Sergeyeva et al. [94SER/DEV]. Analysis of the same data in [94SER/DEV] and [92GRE/FUG] leads to slight differences at zero ionic strength (298.15 K), probably due to different conclusions as to the accuracy of experimental data and the method used for extrapolation to $I = 0$. For instance for the reaction:

$$U^{4+} + H_2O(l) \rightleftharpoons UOH^{3+} + H^+$$

the value recommended by Sergeyeva et al. is $\log_{10} {}^*\beta_1^\circ = -(0.4 \pm 0.2)$, while the value selected in [92GRE/FUG] is $\log_{10} {}^*\beta_1^\circ = -(0.54 \pm 0.06)$.

Langmuir [97LAN] reviewed the literature data including those considered in [92GRE/FUG] and questioned the selected values concerning U(OH)$_4$(aq) and UO_2(am). This review gives comments on the remarks of Langmuir in Appendix A.

In a critical review of the hydrolysis data of all the tetravalent actinides and from theoretical considerations, Neck and Kim [99NEC/KIM] calculated $\log_{10} \beta_1^\circ = 13.8$, $\log_{10} \beta_2^\circ = 27.5$, $\log_{10} \beta_3^\circ = 36.2$ and $\log_{10} \beta_4^\circ = 45.7$, by assuming as an input experimental parameter $\log_{10} \beta_1^\circ = (13.6 \pm 0.2)$ (see Appendix A). This last value is different from that selected in [92GRE/FUG]. Neck and Kim take an averaged value of the values selected by [92GRE/FUG] $\log_{10} {}^*\beta_1^\circ = -(0.54 \pm 0.06)$ and by [92FUG/KHO], $\log_{10} {}^*\beta_1^\circ = -(0.34 \pm 0.20)$. The reason is that the selection of [92FUG/KHO] is not explicitly discussed in [92GRE/FUG]. The value taken by Neck and Kim, $\log_{10} {}^*\beta_1^\circ = -(0.4 \pm 0.2)$ is the same as the recommended value of [94SER/DEV]. Refined estimated values are given by Neck and Kim in [2001NEC/KIM] and are considered by this review; $\log_{10} \beta_2^\circ = (26.9 \pm 1.0)$ and $\log_{10} \beta_3^\circ = (37.3 \pm 1.0)$, but are not retained (see 9.3.2.2.4) for the selection of equilibrium constants. The $\log_{10} \beta_{4,1}^\circ$ value derived from solubility data is discussed in section 9.3.2.2.3.

No evidence for the formation of the anionic complex $U(OH)_5^-$ has been given since the estimate of $\Delta_f G_m^\circ (U(OH)_5^-, 298.15 \text{ K})$ by [92GRE/FUG]. This review does not select any value for $\log_{10}{}^*\beta_5^\circ$.

The following discussion concerns the dissolution of, first $UO_2(am)$, and then $UO_2(cr)$.

Grenthe et al. [92GRE/FUG] reported a formation constant for the polynuclear U(IV) hydroxide complex $U_6(OH)_{15}^{9+}$, proposed by Baes and Mesmer [76BAE/MES] from a reanalysis of the potentiometric titration study of Hietanen [56HIE]. However, it is not possible to extrapolate the value of $\log_{10}{}^*\beta_{15,6}$ (3 M NaClO$_4$) = $-(16.9 \pm 0.6)$ to $I = 0$. For a complex of such high charge there are no SIT coefficients available nor can they be estimated by analogies.

9.3.2.2.1 Solubility of $UO_2(am)$ ($\equiv UO_2(am, hydr.)$)

In the previous review [92GRE/FUG] no thermodynamic data have been selected for $UO_2(am)$. The equilibrium constants reported for the reactions:

$$UO_2(am) + 4H^+ \rightleftharpoons U^{4+} + 2H_2O(l) \qquad (9.10)$$

are largely discrepant. They range from $\log_{10}{}^* K_{s,0}^\circ (9.10) = -(3.2 \pm 0.2)$, [78NIK/PIR] to an upper limit of $\log_{10}{}^* K_{s,0}^\circ (9.10) < 4.3$ [60STE/GAL] for an active form of hydrated oxide. The values determined by Bruno et al. [86BRU/FER] in 3 M NaClO$_4$, $\log_{10}{}^* K_{s,0} (9.10) = (0.5 \pm 0.3)$ for an amorphous phase and $\log_{10}{}^* K_{s,0} (9.10) = -(1.2 \pm 0.1)$ for the "ripened" phase, show the effect of differences in crystallinity. The data of Rai et al. [98RAI/FEL] on the solubility of $UO_2(am)$ in carbonate media, corrected by taking into account the formation constant of $U(CO_3)_5^{6-}$, give $\log_{10}{}^* K_{s,0}^\circ (9.10) = -0.2$ (see Appendix A).

The solubility of freshly precipitated amorphous $UO_2(am, hydr.)$ was measured by Rai et al. [97RAI/FEL] at various ionic strengths (0.03 to 6 m NaCl, 1 to 3 m MgCl$_2$) in the pH range 2 to 4.5 and under conditions designed to prevent U(IV) oxidation. The data were processed by the Pitzer ion-interaction approach to give $\Delta_f G_m^\circ$ of U^{4+}, UOH^{3+} and $UO_2(am)$. The two first values are very close to the recommended values of [92GRE/FUG] as discussed in Appendix A. The reported value of $\Delta_f G_m^\circ (UO_2, am, 298.15 \text{ K}) = -990.77$ kJ·mol^{-1} by [97RAI/FEL] is within the range of values given in [92GRE/FUG] (see section 9.3.2.2). The corresponding value of the equilibrium constant for

$$UO_2(am, hydr.) + 2H_2O(l) \rightleftharpoons U^{4+} + 4OH^- \qquad (9.11)$$

is $\log_{10} K_{s,0}^\circ (UO_2, am, hydr., (9.11)) = -53.45$. The data of [97RAI/FEL] are considered in the section 9.3.2.2.3 below to derive a selected value of $\log_{10} K_{s,0}^\circ (UO_2, am, hydr., (9.11))$.

9.3.2.2.2 Solubility of $UO_2(cr)$

The solubility of $UO_2(cr)$ is discussed in a short paper by Redkin [97RED]. He gives a value of $2 \cdot 10^{-9}$ molal for the solubility of $UO_2(cr)$ over the temperature range 573.15–873.15 K. The fugacity of oxygen was controlled by Ni–NiO and Fe_2O_3–Fe_3O_4 couples and the pressure was 1 kbar. This confirms earlier results of the negligible temperature dependence [87DUB/RAM], [89RED/SAV] of the solubility of UO_2, also discussed in [92GRE/FUG]. For the dissolution reaction:

$$UO_2(cr) + 2 H_2O(l) \rightleftharpoons U(OH)_4(aq), \qquad (9.12)$$

Redkin gives $\log_{10} K^\circ_{s,4} = -8.7$ (pH = 7, 1 kbar) compared to an average value of $\log_{10} K^\circ_{s,4} = -(9.47 \pm 0.56)$ reported by [88PAR/POH] from data at pH > 4 over the temperature range 373 – 573 K (0.5 kbar of H_2). This paper also gives other data on the solubility of $UO_2(cr)$ in acidic and basic solutions at various redox potentials from which little quantitative thermodynamic information can be extracted.

Rakitskaya et al. [96RAK/TSY] showed that the solubility of non–stoichiometric $UO_2(cr)$ (O/U = 2.005 to 1.985) in water (pH = 7) at 300 K, after one year of equilibration, is higher for the oxide which contains the highest content of U(VI). This behaviour is not surprising. According to these authors, the solubility of UO_2 is dependent on oxide ion diffusion. On the other hand, the work of Torrero et al. [97TOR/BAR] shows that the dissolution rate of $UO_2(cr)$ in $NaClO_4$ is pH and O_2 dependent with a fractional rate order that indicates that adsorbed oxygen acts as an oxidant (a layer of UO_{2+x} is formed with, according to the authors, O/U = 2.25 at pH > 6.7).

The solubilities of crystalline and microcrystalline $UO_2(cr)$ were measured by Yajima et al. [95YAJ/KAW] in 0.1 M $NaClO_4$ (pH = 2 to 12, T = 298.15 K), where the approach to equilibrium was both from under- and over-saturation. In general, lower solubilities were obtained than in [90RAI/FEL]. Below pH = 2, the slope of $\log_{10}m(U)$ versus pH, was approximately – 4, whereas above pH = 4 the uranium concentration was independent of pH. The constants derived from the experimental results are scattered with $\log_{10} {}^*K^\circ_{s,0} = (0.34 \pm 0.4)$ and $\log_{10} K^\circ_{s,4} = -(8.7 \pm 0.4)$, respectively, for the reactions:

$$UO_2(cr) + 4 H^+ \rightleftharpoons U^{4+} + 2 H_2O(l), \qquad (9.13)$$

$$UO_2(cr) + 2 H_2O(l) \rightleftharpoons U(OH)_4(aq). \qquad (9.14)$$

The solubility constant for the reaction,

$$UO_2(cr) + 2 H_2O \rightleftharpoons U^{4+} + 4 OH^-, \qquad (9.15)$$

is more than four orders of magnitude larger than $\log_{10} K^\circ_{s,0}$ (UO_2, cr, 298.15 K, (9.15)) $= -(60.86 \pm 0.36)$ calculated from the $\Delta_f G^\circ_m$ values of $UO_2(cr)$ and U^{4+} and auxiliary data selected in [92GRE/FUG]. This could be due to either poor control on the redox potential or formation of $UO_2(am)$[1]. The value of $\log_{10} {}^*\beta_4^\circ = -(9.0 \pm 0.5)$ or $\log_{10} \beta_4^\circ = (47.0 \pm 0.5)$ is in reasonable agreement with [2001NEC/KIM]. This value is considered in the next section for the selection of thermodynamic data.

9.3.2.2.3 Comparison of the solubilities of $UO_2(am)$ and $UO_2(cr)$

All the recent [95YAJ/KAW], [97RAI/FEL] and earlier [83RYA/RAI], [90RAI/FEL] experimental data on the solubility of $UO_2(am)$ and $UO_2(cr)$ as a function of $-\log_{10}[H^+]$, have been reinterpreted by Neck and Kim [99NEC/KIM], [2001NEC/KIM], taking into account the literature data on other tetravalent actinides (see Appendix A).

What appears new in these studies is that the U(IV) concentrations in equilibrium with $UO_2(cr)$ and $UO_2(am)$ are the same at pH above 6, such that $UO_2(am)$ is the solubility limiting solid phase of U(IV) in neutral and alkaline aqueous solutions[2]. According to [2001NEC/KIM], the experimental data determined with $UO_2(cr)$ must be ascribed to the dissolution of an amorphous surface layer according to :

$$UO_2(am) + 2H_2O(l) \rightleftharpoons U(OH)_4(aq). \tag{9.16}$$

Consequently the $\Delta_f G_m^\circ (U(OH)_4, aq, 298.15\ K)$ value must be revised. Indeed this value comes from studies which assume that $UO_2(cr)$ is the equilibrium phase [92GRE/FUG].

For $UO_2(am)$ [2001NEC/KIM] provides the best fit of solubility data and gives $\log_{10} K_{s,0}^\circ (UO_2, am, hydr., (9.11)) = -(54.5 \pm 1.0)$ and $\log_{10} \beta_4^\circ = (46.0 \pm 1.4)$, deduced according to the relationship:

$$\log_{10} [U(OH)_4(aq)] = \log_{10} K_{s,0}^\circ (UO_2, am, (9.11)) + \log_{10} \beta_4^\circ = -(8.5 \pm 1).$$

The constants $K_{s,0}^\circ$ (9.11) for $UO_2(am)$ and β_4° derived by Neck and Kim come from experimental data and values of β_2° and β_3° are estimated on the basis of selected values of β_1° as discussed above. For U(IV) this is currently the best manner in which to derive the thermodynamic values $\log_{10} \beta_2^\circ$ and $\log_{10} \beta_3^\circ$.

Using the relationship:

$$\log_{10} K_{s,4}^\circ (UO_2, cr) = \log_{10} K_{s,0}^\circ (UO_2, cr, (9.15)) + \log_{10} \beta_4^\circ,$$

this review has calculated for the reaction:

$$UO_2(cr) + 2H_2O(l) \rightleftharpoons U(OH)_4(aq),$$

---

[1] During the Peer Review process the value $\log_{10} K_{s,0}^\circ = -(60.20 \pm 0.24)$ for the equilibrium, $UO_2(cr) + 2H_2O \rightleftharpoons U^{4+} + 4OH^-$, was made available [2003RAI/YUI]. This latter value is very close to that calculated from calorimetric data. The value $\log_{10} {}^*\beta_4^\circ < -11.6$ (or $\log_{10} \beta_4^\circ < 44.4$) was also made available.

[2] During the Peer Review process additional data was made available [2003RAI/YUI] supporting this conclusion.

$\log_{10} K^\circ_{s,4}$ (UO$_2$, cr) = $-$ (14.86 $\pm$ 1.44), which using auxiliary data gives $\Delta_f G^\circ_m$ (U(OH)$_4$, aq, 298.15 K) = $-$ (1421.30 $\pm$ 8.18) kJ$\cdot$mol^{-1}. This review does not select any value of $\Delta_f G^\circ_m$ (UO$_2$, am, 298.15 K) from $\log_{10} K^\circ_{s,0}$ (UO$_2$, am, hydr., (9.11)) and auxiliary data following [92GRE/FUG] for the reasons discussed above (section 9.3.2.2).

9.3.2.2.4 Selection of equilibrium constants

The review retains as new selected equilibrium constants for U(IV), the value $\log_{10} \beta^\circ_4$ = (46.0 $\pm$ 1.4) given by Neck and Kim [2001NEC/KIM]. The values $\log_{10} \beta^\circ_2$ = (26.9 $\pm$ 1.0) and $\log_{10} \beta^\circ_3$ = (37.3 $\pm$ 1.0) also given by Neck and Kim [2001NEC/KIM], which are estimated, cannot be selected according to the NEA TDB Guidelines because only data supported by experiment can be selected. The estimation of $\log_{10} \beta^\circ_2$ and $\log_{10} \beta^\circ_3$ is well founded. It relies on general principles of actinide complexation chemistry and correlation among different actinides and oxidation states. There is some evidence for U(OH)$_2^{2+}$ in [89GRE/BID] and for U(OH)$_3^+$ in [87BRU/CAS] (see [92GRE/FUG]). As no new experimental data have appeared since the previous review on the formation of the species UOH^{3+}, this review keeps the value selected by [92GRE/FUG], $\log_{10} \beta^\circ_1$ = (13.46 $\pm$ 0.06), ($\log_{10} {^*\beta^\circ_1}$ = $-$ (0.54 $\pm$ 0.06)). The value $\log_{10} K^\circ_{s,0}$ (UO$_2$, am, hydr., (9.11)) = $-$ (54.5 $\pm$ 1.0) is selected as a constant to be used to calculate a concentration of U(IV) in aqueous solutions equilibrated with UO$_2$(am). Table 9-7 summarises the selected equilibrium constants by this review and [92GRE/FUG] for U(IV) hydroxide complexes and compounds and a value for $\log_{10} K^\circ_{s,0}$ (UO$_2$, am, hydr., (9.11)).

Table 9-7: Solubility products and overall formation constants [2001NEC/KIM].

Species or compound	$\log_{10} K^\circ_{s,0}$	$\log_{10} \beta^\circ_m$
UO$_2$(am, hydr.)	$-$ (54.5 $\pm$ 1.0) *#	
UO$_2$(cr)	$-$ (60.86 $\pm$ 0.36)**	
U(OH)$^{3+}$		(13.46 $\pm$ 0.06)**, 13.8$^{(a)}$, (13.6 $\pm$ 0.2)$^{(b)}$
U(OH)$_2^{2+}$		(26.9 $\pm$ 1.0)$^{(a)}$
U(OH)$_3^+$		(37.3 $\pm$ 1.0)$^{(a)}$
U(OH)$_4$ (aq)		(46.0 $\pm$ 1.4)*

* selected by this review, (a) estimated.
** from [92GRE/FUG]. (b) averaged selected constants (see text).
\# to calculate U(IV) concentration in equilibrium with UO$_2$(am).

Casas et al. [98CAS/PAB] used all the selected values by [92GRE/FUG] for U(IV) hydrolysis for modelling the solubility of UO$_2$(cr) and uraninite UO$_{2+x}$(cr) in anoxic (and oxic) conditions. As discussed in Appendix A, discrepancies between predicted and measured uranium concentrations could be due to the size of microcrystallites of uranium dioxide.

9.3.2.3 U(III) hydroxide complexes (V.3.2.4)

No experimental data have been reported but thermodynamic estimations can be found in [98SAV]. They have not been retained for reasons given in Appendix A.

9.3.3 Crystalline and amorphous uranium oxides and hydroxides (V.3.3)

9.3.3.1 U(VI) oxides and hydroxides (V.3.3.1)

9.3.3.1.1 Other forms of anhydrous UO$_3$(cr) (V.3.3.1.2)

The value selected in [92GRE/FUG] for the enthalpy of formation of δ–UO$_3$, $-(1209 \pm 5)$ kJ · mol^{-1}, was based on an early determination by Cordfunke [64COR] who used a sample which was a mixture of α–UO$_3$ and δ–UO$_3$. More recently, in a thesis, [90POW] reported a new value, using a sample of δ–UO$_3$ prepared by thermal decomposition, at 648 K, of β–UO$_3$·H$_2$O. X-ray powder diffraction indicated that no phases other than δ–UO$_3$ were present. The medium used for the dissolution of the sample was the same as that used by Cordfunke and Ouweltjes [77COR/OUW2], 0.0350 M Ce(SO$_4$)$_2$ + 1.505 M H$_2$SO$_4$. The author [90POW] reported $\Delta_{sol}H_m$ (UO$_3$, δ) = $-(94.71 \pm 0.72)$ kJ · mol^{-1} (the uncertainty limits being ours). These results are also reported with less detail by Dickens et al. [89DIC/LAW]. The use of the enthalpy of solution of γ–UO$_3$ from [77COR/OUW2] ($-(84.64 \pm 0.38)$ kJ · mol^{-1}), together with the [92GRE/FUG] value for the standard enthalpy of formation of γ–UO$_3$, leads to:

$$\Delta_f H_m^\circ (UO_3, \delta, 298.15 \text{ K}) = -(1213.73 \pm 1.44) \text{ kJ} \cdot \text{mol}^{-1}$$

which is the accepted value, consistent with, but more precise than, that selected in [92GRE/FUG].

Using the same medium, Dickens et al. [88DIC/POW] report $\Delta_{sol}H_m$ (UO$_3$, α) $= -(96.03 \pm 0.72)$ kJ · mol^{-1}. This value leads to:

$$\Delta_f H_m^\circ (UO_3, \alpha, 298.15 \text{ K}) = -(1212.41 \pm 1.45) \text{ kJ} \cdot \text{mol}^{-1},$$

which we accept, being more precise than that selected in [92GRE/FUG] ($\Delta_f H_m^\circ$ (UO$_3$, α, 298.15 K) $= -(1217.5 \pm 3.0)$ kJ · mol^{-1}) and consistent with the selection for the δ phase.

It is worth recalling here that α–UO$_3$ has a vacancy structure [66LOO/COR] and that small variations in stoichiometry may be responsible for variations in the enthalpies of solution.

9.3.3.1.2 U(VI) hydrated oxides and hydroxides (V.3.3.1.5)

Structural and thermodynamic data on synthetic or natural uranium trioxide hydrates and dioxouranium hydroxide are discussed in [92GRE/FUG]. The inter-relationships between all of them is complicated and it is generally difficult to identify, at the conditions reported in papers dealing with solubility studies, which solid phase is really present. Since the previous review, Finch et al. [92FIN/MIL] focus on the synthetic and naturally occurring trioxide UO$_3\cdot$2H$_2$O. Gurevich et al. [97GUR/SER] discuss some transformations among dioxouranium hydroxide and the uranium trioxide hydrates. Finch et al. [98FIN/HAW], [96FIN/COO] and Sowder et al. [99SOW/CLA] clarified the situation among U(VI) trioxide hydrates with the identification of schoepite as (UO$_2$)$_8$O$_2$(OH)$_{12}\cdot$12H$_2$O (formally UO$_3\cdot$2.25H$_2$O), metaschoepite as (UO$_2$)$_8$O$_2$(OH)$_{12}\cdot$10H$_2$O (formally UO$_3\cdot$2H$_2$O) and dehydrated schoepite as UO$_2$O$_{(0.25-x)}$(OH)$_{(1.5+2x)}$ $0 < x < 0.25$ (formally UO$_3\cdot$0.75H$_2$O to UO$_3\cdot$H$_2$O). Schoepite and metaschoepite are orthorhombic and dehydrated schoepite is isostructural with α-UO$_2$(OH)$_2$ but with anion vacancies. All forms of schoepite are structurally and chemically distinct from the three dioxouranium hydrates, UO$_2$(OH)$_2$.

To summarise, for UO$_3\cdot$xH$_2$O three polymorphs exist (α-orthorhombic: schoepite, x = 2.25; β-orthorhombic: metaschoepite, x = 2 and γ-orthorhombic: dehydrated schoepite, x < 0.75) and for UO$_2$(OH)$_2$ or UO$_3\cdot$H$_2$O three polymorphs exist (α-orthorhombic, β-orthorhombic and γ-monoclinic) (see [98FIN/HAW] in Appendix A). Numerous compositions for x = 0.9 to 0.333 have been identified, but it is not clear to which series they belong.

9.3.3.1.2.1   UO$_3\cdot$2H$_2$O (cr)

The value $\Delta_f H_m^\circ$(UO$_3\cdot$2H$_2$O, 298.15 K) = $-$ (1826.1 ± 1.7) kJ·mol^{-1} was selected in [92GRE/FUG] on the basis of an assessed value of $-$ (30.6 ± 1.2) kJ·mol^{-1} for the difference of dissolution in dilute hydrofluoric acid of UO$_3\cdot$2H$_2$O(cr) and γ–UO$_3$. The following thermodynamic quantities were also selected in [92GRE/FUG]:

$$S_m^\circ(\text{UO}_3 \cdot 2\text{H}_2\text{O, cr, 298.15 K}) = (188.54 \pm 0.38) \text{ J} \cdot \text{K}^{-1} \cdot \text{mol}^{-1},$$
$$C_{p,m}^\circ(\text{UO}_3 \cdot 2\text{H}_2\text{O, cr, 298.15 K}) = 84.238 + 0.294592\ T\ \text{ J} \cdot \text{K}^{-1} \cdot \text{mol}^{-1}$$

which gives: $C_{p,m}^\circ$(UO$_3 \cdot$ 2H$_2$O, cr, 298.15 K) = (172.070 ± 0.340) J·K^{-1}·mol^{-1}. Finally, $\Delta_f G_m^\circ$(UO$_3 \cdot$ 2H$_2$O, cr, 298.15 K) = $-$ (1636.5 ± 1.7) kJ·mol^{-1}.

These selected values come mainly from [88TAS/OHA] (see Appendix A).

There were no more recent determinations on these compounds, which are nevertheless mentioned here for the convenience of the following discussions.

The Gibbs enthalpy of dissolution of UO$_2$(OH)$_2$(aq) is calculated from the value selected by this review, of $\log_{10} {}^*\beta_{2,1}^\circ = -$ (12.15 ± 0.07) (see Table 9-6). For the reaction:

$$\text{UO}_3 \cdot 2\text{H}_2\text{O (cr)} \rightleftharpoons \text{UO}_2(\text{OH})_2(\text{aq}) + \text{H}_2\text{O(l)},$$

the value $\Delta_r G_m^\circ = $ (41.9 ± 2.5) kJ·mol^{-1} is obtained using auxiliary data.

9.3.3.1.2.2   UO$_2$(OH)$_2$(cr) or UO$_3\cdot$H$_2$O(cr)

The values of $C_{p,m}^\circ$, S_m° and $\Delta_f G_m^\circ$ for β–orthorhombic UO$_3\cdot$H$_2$O(cr) at 298.15 K have been selected in [92GRE/FUG].

Recently, Gurevich et al. [97GUR/SER] reported heat capacity measurements on $UO_2(OH)_2(\alpha,\text{ orth})$ in the temperature range 14 – 316 K. It should be noted that throughout the paper the abbreviation "rhomb β" was mistakenly given for "orth β". The experimental results were fitted by a polynomial of the type:

$$\sum_x A_x [1 - \exp(-0.001\ T)]^x, \quad x = 1 \text{ to } 5.$$

At room temperature, the following values were reported:

$C_{p,m}^\circ (UO_3 \cdot H_2O, \alpha, \text{orth}, 298.15\,\text{K}) = (113.96 \pm 0.12)\ \text{J} \cdot \text{K}^{-1} \cdot \text{mol}^{-1}$,
$S_m^\circ (UO_3 \cdot H_2O, \alpha, \text{orth}, 298.15\,\text{K}) = (128.10 \pm 0.20)\ \text{J} \cdot \text{K}^{-1} \cdot \text{mol}^{-1}$,
$H_m^\circ(298.15\,\text{K}) - H_m(0) = 19.703\ \text{J} \cdot \text{K}^{-1} \cdot \text{mol}^{-1}$.

These values do not appear incompatible with those selected in [92GRE/FUG] for the β–phase, which is also orthorhombic:

$C_{p,m}^\circ (UO_3 \cdot H_2O, \beta, \text{orth}, 298.15\,\text{K}) = (141 \pm 15)\ \text{J} \cdot \text{K}^{-1} \cdot \text{mol}^{-1}$,
$S_m^\circ (UO_3 \cdot H_2O, \beta, \text{orth}, 298.15\,\text{K}) = (138 \pm 4)\ \text{J} \cdot \text{K}^{-1} \cdot \text{mol}^{-1}$,

nor with the selected values for the heat capacity and entropy of $UO_3 \cdot 2H_2O(cr)$, given the difference of one H_2O.

Since the selections by [92GRE/FUG] rest partially on reaction enthalpies by [73HOE/SIE], it is appropriate to note here that the lattice parameters given by [97GUR/SER] for the α–phase are in reasonable agreement with those given for the same phase by [73HOE/SIE].

Gurevich et al. derived from their data, as discussed in Appendix A,

$\Delta_f H_m^\circ (UO_3 \cdot H_2O, \alpha\text{ orth}, 298.15\,\text{K}) = -(1536.87 \pm 1.30)\ \text{kJ} \cdot \text{mol}^{-1}$,

$\Delta_f H_m^\circ (UO_3 \cdot H_2O, \beta\text{ orth}, 298.15\,\text{K}) = -(1533.87 \pm 1.30)\ \text{kJ} \cdot \text{mol}^{-1}$.

The second value agrees with the selected value by [92GRE/FUG], $-(1533.8 \pm 1.3)\ \text{kJ} \cdot \text{mol}^{-1}$.

The first value was obtained by extrapolation of literature values to $x = 1$, as $-(27.4 \pm 0.2)\ \text{kJ} \cdot \text{mol}^{-1}$, for $\Delta_r H_m^\circ (298.15\,\text{K})$ for the reaction:

$$\gamma - UO_3 + x\ H_2O(l) \rightleftharpoons UO_3 \cdot xH_2O(\alpha, \text{orth}),$$

using the results of [72SAN/VID] for $x = 0.393$ and 0.648, [64COR] for $x = 0.85$ and [88OHA/LEW] for $x = 0.9$.

As discussed in Appendix A, a new extrapolation taking equally into account all experimental values leads to $-(26.8 \pm 1.5)\ \text{kJ} \cdot \text{mol}^{-1}$, which is nearly the same value as that obtained by [97GUR/SER] with greatly increased uncertainty limits. Keeping in mind the precarious character of such an extrapolation, we calculate:

$\Delta_f H_m^\circ (UO_3 \cdot H_2O, \alpha\text{ orth}, 298.15\,\text{K}) = -(1536.4 \pm 1.9)\ \text{kJ} \cdot \text{mol}^{-1}$.

In order to obtain another cycle in the evaluation of the stability of $UO_3 \cdot H_2O(\alpha, \text{orth})$ we have used (as done initially by [97GUR/SER]) the results of [71NIK/PIR] who measured potentiometrically the solubility of the compound in the temperature range 295 – 423 K.

For the reaction:

$$UO_3 \cdot H_2O\ (\alpha, \text{orth}) + 2\ H^+ \rightleftharpoons UO_2^{2+} + 2\ H_2O(l) \tag{9.17}$$

we recalculate, at infinite dilution, a value of $\log_{10}{}^*K^\circ_{s,0}((9.17), 298.15\ \text{K}) = -(5.80 \pm 0.10)$ leading to $\Delta_f G^\circ_m\ (UO_3 \cdot H_2O, \alpha, \text{orth}, 298.15\text{K}) = -(1393.72 \pm 1.84)\ \text{kJ} \cdot \text{mol}^{-1}$. The values for $\log_{10} K^\circ_{s,0}$ (9.17) at temperatures up to 423 K have also been recalculated, but, as discussed in Appendix A, they are given for information only.

Using $S^\circ_m\ (UO_3 \cdot H_2O, \alpha, \text{orth}, 298.15\text{K}) = (128.10 \pm 0.20)\ \text{J} \cdot \text{K}^{-1} \cdot \text{mol}^{-1}$ determined by [97GUR/SER] and other auxiliary values accepted by this review, we obtain from the above Gibbs energy value, $\Delta_f H^\circ_m\ (UO_3 \cdot H_2O, \alpha, \text{orth}, 298.15\text{K}) = -(1531.79 \pm 1.84)\ \text{kJ} \cdot \text{mol}^{-1}$, which is slightly different from the value obtained by the extrapolation above. Taking the average of the two values yields, $\Delta_f H^\circ_m\ (UO_3 \cdot H_2O, \alpha, \text{orth}, 298.15\text{K}) = -(1534.1 \pm 4.1)\ \text{kJ} \cdot \text{mol}^{-1}$. Consequently with the value of the entropy given by [97GUR/SER], $\Delta_f G^\circ_m\ (UO_3 \cdot H_2O, \alpha, \text{orth}, 298.15\text{K}) = -(1396.0 \pm 4.2)\ \text{kJ} \cdot \text{mol}^{-1}$.

From this result and $\Delta_f G^\circ_m\ (UO_3 \cdot H_2O, \beta, \text{orth}, 298.15\text{K}) = -(1398.7 \pm 1.8)\ \text{kJ} \cdot \text{mol}^{-1}$ selected by [92GRE/FUG], it is impossible to decide which of the two forms is thermodynamically more stable. The claim of [97GUR/SER] as to the metastability of the β–phase toward the α–phase cannot be supported or rejected. Indeed these authors pointed out that the β– to α–phase transformation could be due to mostly uncontrollable factors.

The value $\Delta_f H^\circ_m\ (UO_3 \cdot H_2O, \gamma, \text{mono}, 298.15\ \text{K}) = -(1531.4 \pm 1.3)\ \text{kJ} \cdot \text{mol}^{-1}$ was also selected in [92GRE/FUG], with the indication that this species could correspond to a metastable form. No further information has appeared on this compound.

9.3.3.1.2.3 $UO_3 \cdot xH_2O$

The values of $\Delta_f H^\circ_m\ (UO_3 \cdot xH_2O)$ for x = 0.9, 0.85, 0.648, and 0.393 are given by [92GRE/FUG], and the variations of $\Delta_f H^\circ_m$, S°_m and $\Delta_f G^\circ_m$ as a function of x are discussed by [88OHA/LEW].

According to [72TAY/KEL], the transformation reaction of the α–$UO_3 \cdot 0.85\ H_2O$(cr) to the β orthorhombic $UO_3 \cdot H_2O$ phase:

$$UO_3 \cdot 0.85\ H_2O(cr) + 0.15\ H_2O(l) \rightleftharpoons UO_3 \cdot H_2O(cr)$$

is accompanied by an enthalpy change of $-5.69\ \text{kJ} \cdot \text{mol}^{-1}$ at 298.15 K. As noted by [92GRE/FUG], this effect is not consistent with the values adopted in this review for the enthalpies of formation of these two hydrates. It is also not consistent with the value adopted above for $UO_3 \cdot H_2O$ (α, orth).

Gurevich et al. [97GUR/SER] give estimates of the standard entropy and enthalpy of formation of the species $UO_3 \cdot 0.5H_2O(cr)$. These values are mentioned in Appendix A but are not selected here.

9.3.3.1.3 U(VI) peroxides (V.3.3.1.4)

Three misprints in the [92GRE/FUG] text of this section were pointed out by [95GRE/PUI]. The revised text below incorporates these corrections and includes a discussion of an earlier relevant reference.

Cordfunke [66COR] measured the enthalpy of precipitation of $UO_4 \cdot 4H_2O$ (cr), when adding a 0.5% solution of H_2O_2 to a solution of dioxouranium(VI) nitrate in water, to be $-(31.21 \pm 0.29)$ kJ · mol^{-1}. The data in [82WAG/EVA] indicate that the enthalpy of formation of H_2O_2 in the 0.5% solution will be negligibly different from that at infinite dilution, $\Delta_f H_m^\circ$ (H_2O_2, aq, 298.15 K) = $-(191.17 \pm 0.10)$ kJ · mol^{-1}. Thus, the value $\Delta_f H_m^\circ$ ($UO_4 \cdot 4H_2O$, cr, 298.15 K) = $-(2384.7 \pm 2.1)$ kJ · mol^{-1} is calculated. The enthalpy of dehydration of the tetrahydrate to form the dihydrate,

$$UO_4 \cdot 4H_2O(cr) \rightleftharpoons UO_4 \cdot 2H_2O(cr) + 2H_2O(g) \tag{9.18}$$

was calculated to be (117.02 ± 1.00) kJ · mol^{-1} from vapour pressure measurements of [63COR/ALI] (312.1 to 331.8 K), corrected by [66COR] to take account of the fact that the mercury in the manometer was at a variable temperature, different from 0°C. The correction of this enthalpy from the mean temperature to 298.15 K is estimated to be an order of magnitude smaller than the final uncertainty, so this value of $\Delta_f H_m^\circ$ (9.18) = (117.02 ± 1.00) kJ · mol^{-1} is assumed to apply at 298.15 K, and hence $\Delta_f H_m^\circ$ ($UO_4 \cdot 2H_2O$, cr, 298.15 K) = $-(1784.0 \pm 4.2)$ kJ · mol^{-1} is calculated. The uncertainties for the enthalpies of formation given above are those proposed by Cordfunke and O'Hare [78COR/OHA].

The enthalpy of formation of $UO_4 \cdot 2H_2O(cr)$ can also be determined using the early work of Pissarjewsky [00PIS], who measured the enthalpy of dissolution of an analysed sample of this compound and that of $UO_3 \cdot H_2O(s)$ in 1 M sulphuric acid. As described in Appendix A, this review calculates from these results, $\Delta_f H_m^\circ$ ($UO_4 \cdot 2H_2O$, cr, 298.15 K) = $-(1789.7 \pm 4.1)$ kJ · mol^{-1}, in good agreement with the results discussed immediately above. This agreement had been implicitly noted by [66COR], who compared his results with those given in the NBS Circular 500 Tables [52ROS/WAG], which were based on the data of Pissarjewsky [00PIS].

This review maintains the selections made on the basis of the results of [66COR] and [63COR/ALI]:

$$\Delta_f H_m^\circ (UO_4 \cdot 4H_2O, \text{cr}, 298.15 \text{ K}) = -(2384.7 \pm 2.1) \text{ kJ} \cdot \text{mol}^{-1},$$

$$\Delta_f H_m^\circ (UO_4 \cdot 2H_2O, \text{cr}, 298.15\text{ K}) = -(1784.0 \pm 4.2) \text{ kJ} \cdot \text{mol}^{-1}.$$

No entropy or Gibbs energy data are available for these peroxide hydrates.

9.3.3.2 U(IV) oxides (V.3.3.2)

9.3.3.2.1 Crystalline UO_2 (uraninite) (V.3.3.2.1)

Fink [2000FIN] has updated her earlier assessment [81FIN/CHA] of the thermodynamic and transport properties of UO_2(cr, l), to incorporate an appreciable amount of new data, principally on the liquid. Fink gives two fitting equations for the enthalpy increment of UO_2(cr), one a polynomial, and one containing terms representing the individual contributions to the enthalpy and heat capacity. These equations give very similar values for both $H_m(T) - H_m^\circ(298.15\text{K})$ and $C_{p,m}(T)$, and either (see Appendix A) can be used to represent the data for UO_2(cr) up to its melting point.

The equation for $C_{p,m}(UO_2, \text{cr})$ in [92GRE/FUG], valid only to 600 K, gives values which are close to those derived from the revised assessment by Fink, [2000FIN], so the selected data [92GRE/FUG] will not be adjusted. Takahashi [97TAK] has also reported briefly on additional enthalpy increment measurements on UO_2(cr) from *ca.* 300 to 1500 K consistent with the earlier measurements.

Fink also gives recommended data for the heat capacity of the liquid, which decreases sharply from *ca.* 136 J · K^{-1} · mol^{-1} at the melting point (3120 K) to *ca.* 66 J · K^{-1} · mol^{-1} (at 4500 K) and the enthalpy of fusion (70 ± 4) kJ · mol^{-1}.

9.3.3.3 Mixed valence oxides (V.3.3.3)

9.3.3.3.1 $UO_{2.3333}$ (≡ 1/3 U_3O_7) (V.3.3.2)

The data selected by [92GRE/FUG] for β–$UO_{2.3333}$ make this phase stable with respect to U_4O_9 and U_3O_8 from 298.15 to *ca.* 400 K, for which there is no experimental evidence. We therefore select a value of $\Delta_f H_m^\circ (UO_{2.3333})$ which is more positive by 1 kJ · mol^{-1}, within the [92GRE/FUG] uncertainty, thereby removing the problem, so that the new selected value is:

$$\Delta_f H_m^\circ (UO_{2.3333}, \beta, 298.15\text{ K}) = -(1141.0 \pm 2.0) \text{ kJ} \cdot \text{mol}^{-1}.$$

9.3.3.3.2 Hydrogen insertion compounds

Powell [90POW] reported the enthalpy of formation of the insertion species δ–$H_{0.83}UO_3$, obtained by hydrogen spill-over and characterised by X-ray diffraction and redox titration (mean oxidation state of uranium). The determination rests on the enthalpy of solution of the constituents of the equation:

$$0.415\ \alpha - U_3O_8 + 0.415\ H_2O(l) \rightleftharpoons \delta - H_{0.83}UO_3 + 0.245\ \gamma - UO_3 \quad (9.19)$$

The medium was that used by [77COR/OUW2], [81COR/OUW] (see section 9.10.2.1.) and the values for the dissolution of the binary oxides and the transfer of H_2O

are taken from these authors. The enthalpy of solution of the compound is reported as $-(142.25 \pm 1.64)$ kJ·mol^{-1}. This leads to $\Delta_r H_m^\circ ((9.19), 298.15\text{ K}) = -(17.19 \pm 1.73)$ kJ·mol^{-1}. With NEA selected values, we obtain:

$$\Delta_f H_m^\circ (\text{H}_{0.83}\text{UO}_3, \delta, 298.15\text{ K}) = -(1285.14 \pm 2.02) \text{ kJ·mol}^{-1}.$$

For the insertion reaction:

$$0.415 \text{ H}_2(\text{g}) + \delta - \text{UO}_3 \rightleftharpoons \delta - \text{H}_{0.83}\text{UO}_3 \tag{9.20}$$

using the revised value for the standard enthalpy of formation of δ–UO$_3$ (see section 9.3.3.1.1), we obtain:

$$\Delta_r H_m^\circ ((9.20), 298.15\text{ K}) = -(71.41 \pm 2.48) \text{ kJ·mol}^{-1}.$$

Powell also quotes without any reference earlier results on α–H$_{1.08}$UO$_3$, which we shall not consider further here.

9.3.3.3.3   Other mixed valence oxides (V.3.3.3.4)

[93PAT/DUE] have reported the standard enthalpy of formation of hexagonal α–UO$_{2.95}$ using an oxide sample characterised by powder X-ray diffraction. The O/U ratio was accurately determined by potentiometric titration (dissolution in potassium dichromate and back titration using a Fe(II) standard solution).

The calorimetric determination was based on the enthalpy of solution of the constituents of the reaction:

$$0.05 \, \alpha - \text{U}_3\text{O}_8 + 0.85 \, \gamma - \text{UO}_3 \rightleftharpoons \alpha - \text{UO}_{2.95}. \tag{9.21}$$

The calorimetric reagent was 0.274 M Ce(SO$_4$)$_2$ + 0.484 M H$_3$BO$_3$ + 0.93 M H$_2$SO$_4$, the same as that used by [81COR/OUW2] for the dissolution of γ–UO$_3$ and α–U$_3$O$_8$. In fact, the values for the enthalpy of dissolution of the last two reagents were taken from these authors ($-(79.94 \pm 0.48)$ kJ·mol^{-1} and $-(354.61 \pm 1.70)$ kJ·mol^{-1}, respectively). $\Delta_{sol} H_m$ (UO$_{2.95}$, α) was determined to be $-(93.37 \pm 0.65)$ kJ·mol^{-1}, the reported uncertainty limits being for the 1.96σ interval. These values lead to $\Delta_r H_m^\circ ((9.21), 298.15\text{ K}) = (7.69 \pm 0.77)$ kJ·mol^{-1}. Using NEA accepted values for the standard enthalpy of formation of γ–UO$_3$ and α–U$_3$O$_8$, we recalculate:

$$\Delta_f H_m^\circ (\text{UO}_{2.95}, \alpha, 298.15\text{K}) = -(1211.28 \pm 1.28) \text{ kJ·mol}^{-1},$$

identical, except for the uncertainty limits, to that reported by [93PAT/DUE]. This value, accepted here, is consistent with: $\Delta_f H_m^\circ (\text{UO}_3, \alpha, 298.15\text{K}) = -(1212.41 \pm 1.45)$ kJ·mol^{-1} selected above.

9.3.4 Uranium hydrides (V.3.4)

The only relevant new study on uranium hydrides is that by Ito et al. [98ITO/YAM] who give pressure-composition isotherms for the hydrogenation of U$_6$Mn(s) and U$_6$Ni(s) at 573 and 673 K, and compare them to those for the binary U–H system. Ternary hydrides of approximate composition U_6MnH_{18}(s) and U_6NiH_{14}(s) are probably formed and the amount of hydrogen in the alloys in equilibrium with these phases is much higher than for unalloyed uranium. The preparation of high density essentially stoichiometric UH_3 (s) and UD_3(s) has been reported by [90COS/LAK].

9.4 Uranium group 17 (halogen) compounds and complexes (V.4)

9.4.1 General

The data for the condensed phases of the uranium halides have not been changed, except to correct a few minor errors in [92GRE/FUG] and to modify slightly some heat capacity equations in order to extend their range of applicability.

However, the data for nearly all the uranium halide gaseous species have been changed, for reasons noted in section 9.4.2.1.2. For convenience, the molecular parameters of all the relevant gaseous species used are summarised in this section.

9.4.1.1 Molecular parameters of the gaseous uranium halide species

The thermal functions have been calculated with the natural atomic masses [92IUP] for non-actinide elements and uranium and the most common isotopes of the other actinides [99FIR], as given in Table 9-8.

Table 9-8: Atomic mass used for the calculation of thermal functions.

Element	Isotope	Molar mass (g · mol^{-1})
Americium	241	241.057
Bromine		79.904
Chlorine		35.4527
Fluorine		18.99840
Iodine		126.90447
Neptunium	237	237.048
Oxygen		15.9994
Plutonium	239	239.052
Uranium	238	238.0289

Most of the thermal functions for the gaseous species were calculated from the structural parameters and energy levels of the molecules. For convenience, these data are summarised here in Table 9-9, but the sources of these data are discussed, where necessary, under each species. In fact, most of these molecular parameters are the same as, or based on, those estimated by [82GLU/GUR] for the fluorides, and [85HIL/GUR] for the Cl, Br and I species. The principal exceptions are for the four tetrahalides, where, as noted in the text, there are experimental measurements for UF_4(g) and UCl_4(g).

The estimated thermal functions calculated from these data are generally valid for temperatures up to 3000 K; electronic energy levels which contribute only above this temperature, have, nevertheless, been included for completeness.

The uncertainties in S_m° and $C_{p,m}^\circ$ at 298.15 K arise mainly from the estimated electronic levels and vibration frequencies. They are fairly conservative, and are selected to encompass the likely results of any subsequent measurements or calculations.

Some uncertainties have been revised from those given in [92GRE/FUG]. In particular, those for the UX(g) species, with only one estimated vibration frequency, have generally been reduced, while those for a few higher halides, with many frequencies, have been increased.

The estimates in Table 9-9 are mostly taken from the assessments of Glushko et al., [82GLU/GUR] and Hildenbrand et al., [85HIL/GUR]. While this review was being peer-reviewed, Joubert and Maldivi published details of theoretical predictions of the structural and vibrational parameters of the uranium trihalide gases [2001JOU/MAL], using seven different approximations. All the calculated parameters depend appreciably on the approximation used, but the general results for the interatomic distances and angles are in rather good agreement with the estimates in Table 9-9, except that interatomic distances in $UI_3(g)$ might be slightly too large. There are also indications that the low frequency vibrations in $UBr_3(g)$ and $UI_3(g)$ have been over–estimated. Theoretical calculations of this type [2001JOU/MAL] will clearly become very valuable when they become more precise, but for the current review we have retained the estimated parameters, which form a consistent set for all the uranium halides. The associated uncertainties of the thermal functions do, in fact, cover those calculated from typical calculated parameters in [2001JOU/MAL].

Table 9-9: Molecular parameters of uranium halide gaseous species[*].
Part A: diatomic species.

	r Å	ω cm^{-1}	ωx cm^{-1}	B cm^{-1}	α cm^{-1}	D cm^{-1}	Symmetry number
UF	2.03	600	1.7	0.230	1.00 10^{-3}	1.40 10^{-7}	1
UCl	2.50	350	0.8	0.0875	4.00 10^{-4}	2.00 10^{-7}	1
UBr	2.65	200	0.8	0.0401	4.00·10^{-4}	2.00·10^{-7}	1
UI	2.85	150	0.4	0.0251	4.00 10^{-4}	2.00 10^{-7}	1

Electronic levels (multiplicities) cm^{-1}	
UF	0(2) 1000(6) 3000(12) 5000(20) 1000(120) 20000(300) 30000(500) 40000(500)
UCl	0(2) 500(2) 1000(6) 3000(20) 5000(30) 10000(100) 20000(250) 30000(300)
UBr	0(2) 500(2) 1000(6) 3000(20) 5000(30) 10000(100) 20000(250) 30000(300)
UI	0(2) 500(2) 1000(6) 3000(20) 5000(30) 10000(100) 20000(250) 30000(300)

(Continued on next page)

Table 9-9: (continued):
Part B: polyatomic species.

Species	Point Group (Symmetry number)	Interatomic Distance Å	Angle deg	Vibration frequencies, (multiplicities) cm^{-1}	Product of moments of inertia (g^3·cm^{-6})
UF$_2$	C$_{2v}$ (2)	2.10	110	530 115 510	3.92 10^{-114}
UF$_3$	C$_{3v}$ (3)	2.10	110	570 80 520(2) 130(2)	1.84 10^{-113}
UF$_4$	T$_d$ (12)	2.059	109.5	625 123(2) 539(3) 114(3)	4.54 10^{-113}
UF$_5$	C$_{2v}$ (2)	2.02	100	649 572 130 515 60 200 593(2) 200(2) 180(2)	7.72 10^{-113}
UF$_6$	O$_h$ (24)	1.9962	90	668.2 534.5(2) 627.7(3) 187.5(3) 201.0(3)143.0(3)	1.27 10^{-112}
UO$_2$F$_2$	C$_{2v}$ (2)	U–O = 1.80 U–F = 2.00	O–U–O = 100 F–U–F = 120	830 550 200 150 150 850 180 600 200	2.21 10^{-113}
UOF$_4$	C$_{4v}$ (4)	U–O = 1.80 U–F = 2.00	O–U–F = 105	830 650 150 500 200 180 600(2) 200(2) 170(2)	5.92 10^{-113}
UCl$_2$	C$_{2v}$ (2)	2.50	100	310 75 295	6.74 10^{-113}
UCl$_3$	C$_{3v}$ (3)	2.55	110	325 55 315(2) 90(2)	3.26 10^{-112}
UCl$_4$	T$_d$ (12)	2.506	109.5	327 62(2) 337(3) 72(3)	9.58 10^{-112}
UCl$_5$	C$_{2v}$ (2)	2.50	100	350 325 90 295 70 135 360(2) 125(2) 105(2)	1.80 10^{-111}
UCl$_6$	O$_h$ (24)	2.52	90	350 275(2) 320(3) 130(3) 150(3) 110(3)	3.34 10^{-111}
UO$_2$Cl$_2$	C$_{2v}$ (2)	U–O = 1.80 U–Cl = 2.50	O–U–O = 100 Cl–U–Cl = 120	810 290 180 80 100 850 130 300 150	1.33 10^{-112}
UBr$_2$	C$_{2v}$ (2)	2.65	100	210 65 190	7.83 10^{-112}
UBr$_3$	C$_{3v}$ (3)	2.65	110	200 100 170(2) 90(2)	4.88 10^{-111}
UBr$_4$	T$_d$ (12)	2.693	109.5	220 50(2) 233(3) 45(3)	1.69 10^{-110}
UBr$_5$	C$_{2v}$ (2)	2.66	100	240 210 60 200 60 70 220(2) 70(2) 65(2)	2.95 10^{-110}
UI$_2$	C$_{2v}$ (2)	2.85	100	170 45 130	3.70 10^{-111}
UI$_3$	C$_{3v}$ (3)	2.85	110	150 60 140(2) 55(2)	2.88 10^{-110}
UI$_4$	T$_d$ (12)	2.85	109.5	150 30(2) 150 25(3)	9.51 10^{-110}

(Continued on next page)

Table 9-9: (Part B: polyatomic species continued)

	Electronic levels (multiplicities) cm^{-1}
UF$_2$	0(1) 200(6) 500(6) 2000(25) 4000(25) 6000(25) 8000(25) 10000(25) 15000(50) 20000(50)
UF$_3$	0(2) 200(4) 400(4) 4500(12) 7000(20) 10000(35) 15000(35) 20000(35)
UF$_4$	0(3) 500(2) 2000(3) 3000(1) 4000(5) 6000(8) 7000(3) 9000(14) 11000(11) 12000(4) 14000(5) 17000(10) 19000(15) 22000(6)
UF$_5$	0(2) 2000(2) 5000(2) 7000(2) 9000(2) 11000(2) 14000(2)
UF$_6$	0(1)
UO$_2$F$_2$	0(1)
UOF$_4$	0(1)
UCl$_2$	0(1) 200(6) 500(6) 3000(10) 5000(10) 8000(20) 10000(15) 15000(30) 20000(30)
UCl$_3$	0(2) 200(4) 400(4) 4500(12) 7000(20) 10000(35) 15000(35) 20000(35)
UCl$_4$	0(3) 710(2) 1700(3) 2500(1) 4200(5) 6200(8) 7300(3) 9100(4) 11200(11) 12100(4) 14200(5) 16500(10) 19300(15) 22000(6)
UCl$_5$	0(2) 2000(2) 4500(2) 6500(2) 9000(2) 10000(2) 12000(2)
UCl$_6$	0(1)
UBr$_2$	0(1) 200(6) 500(6) 3000(10) 5000(10) 8000(20) 10000(15) 15000(30) 20000(30)
UBr$_3$	0(2) 200(4) 400(4) 4500(12) 7000(20) 10000(35) 15000(35) 20000(35)
UBr$_4$	0(3) 710(2) 1700(3) 2500(1) 4200(5) 6200(8) 7300(3) 9100(4) 11200(11) 12100(4) 14200(5) 16500(10) 19300(15) 22000(6)
UBr$_5$	0(2) 2000(2) 4500(2) 6500(2) 9000(2) 10000(2) 12000(2)
UI$_2$	0(1) 200(6) 500(6) 3000(10) 5000(10) 8000(20) 10000(15) 15000(30) 20000(30)
UI$_3$	0(2) 200(4) 400(4) 4500(12) 7000(20) 10000(35) 15000(35) 20000(35)
UI$_4$	0(3) 710(2) 1700(3) 2500(1) 4200(5) 6200(8) 7300(3) 9100(4) 11200(11) 12100(4) 14200(5) 16500(10) 19300(15) 22000(6)

*: See section 2.1.6 for notations

9.4.2 Fluorine compounds and complexes (V.4.1)

9.4.2.1 Gaseous uranium fluorides (V.4.1.1)

9.4.2.1.1 UF(g), UF$_2$(g) and UF$_3$(g) (V.4.1.1.1 and V.4.1.1.2)

These species are considered together since their data are intimately linked by the various experimental studies, mainly involving mass-spectrometry, and thus involving some uncertainties in absolute pressures.

The thermal functions of UF(g), UF$_2$(g) and UF$_3$(g) were calculated assuming the molecular parameters shown in Table 9-9, which are essentially those estimated by Glushko et al. [82GLU/GUR]. These assumptions give:

$$S_m^o (UF, g, 298.15 K) = (251.8 \pm 3.0) \text{ J} \cdot \text{K}^{-1} \cdot \text{mol}^{-1}$$
$$C_{p,m}^o (UF, g, 298.15 K) = (37.9 \pm 3.0) \text{ J} \cdot \text{K}^{-1} \cdot \text{mol}^{-1},$$
$$S_m^o (UF_2, g, 298.15 K) = (315.7 \pm 10.0) \text{ J} \cdot \text{K}^{-1} \cdot \text{mol}^{-1}$$
$$C_{p,m}^o (UF_2, g, 298.15 K) = (56.2 \pm 5.0) \text{ J} \cdot \text{K}^{-1} \cdot \text{mol}^{-1},$$

$S_m^°$ (UF$_3$, g, 298.15 K) = (347.5 ± 10.0) J · K^{-1} · mol^{-1}

$C_{p,m}^°$ (UF$_3$, g, 298.15 K) = (76.2 ± 5.0) J · K^{-1} · mol^{-1},

which after rounding are the same as those in [92GRE/FUG].

The nine studies involving equilibria between UF(g), UF$_2$(g), UF$_3$(g) and UF$_4$(g) are summarised in Table 9-10, where the enthalpies of the given reactions at 298.15 K derived from both second- and third-law analyses are given. All the auxiliary data for the calcium and barium gaseous species are taken from the assessment by Glushko et al. [81GLU/GUR]. The study by Roy et al. [82ROY/PRA] of the reaction:

$$UF_4(g) + 0.5\ H_2(g) \rightleftharpoons UF_3(g) + HF(g)$$

from 1229 to 1367 K using a transpiration method, gives a much more negative enthalpy of formation of UF$_3$(g), and as in [92GRE/FUG] has been rejected, owing to the uncertainties in the composition of the vapour and the solid collected.

In all of the studies considered, the pressures were derived from mass-spectrometric intensities, with the well-known problems of fragmentation and the necessity to estimate ionisation cross-sections. Second-law calculations were not always possible, because the temperature ranges of measurement were too small, or the original data were not given in full. It will be seen that the agreement between the second- and third-law analyses is mostly rather poor.

Table 9-10: Enthalpies of the reactions involving UF(g), UF$_2$(g) and UF$_3$(g).

Reference	Reaction	Number of points	$\Delta_r H_m^°$ (298.15 K) kJ · mol^{-1}	
			Second-law	Third-law
[69ZMB]	UF$_3$(g) + CaF(g) $\rightleftharpoons$ UF$_4$(g) + Ca(g)	13	– 108.9	– 86.4
[82LAU/HIL]	UF$_3$(g) + CaF(g) $\rightleftharpoons$ UF$_4$(g) + Ca(g)	9	– 95.2	– 66.6
[82LAU/HIL]	U(g) + BaF(g) $\rightleftharpoons$ UF(g) + Ba(g)	9	– 85.1	– 87.6
[82LAU/HIL]	UF(g) + BaF(g) $\rightleftharpoons$ UF$_2$(g) + Ba(g)	15	15.6	– 8.6
[82LAU/HIL]	UF$_2$(g) + BaF(g) $\rightleftharpoons$ UF$_3$(g) + Ba(g)	15	– 46.0	– 35.5
[84GOR/SMI]	UF$_3$(cr) $\rightleftharpoons$ UF$_3$(g)	22	412.7	447.3
[84GOR/SMI]	2UF$_3$(g) $\rightleftharpoons$ UF$_2$(g) + UF$_4$(g)	21	– 16.6	– 29.1
[84GOR/SMI]	2UF$_2$(g) $\rightleftharpoons$ UF(g) + UF$_3$(g)	2	—	– 29.5
[91HIL/LAU]	UF$_2$(g) + U(g) $\rightleftharpoons$ 2UF(g)	8	– 93.6	– 79.6

With these data, we have eight or nine essentially independent measurements to define the three enthalpies of formation of UF(g), UF$_2$(g), UF$_3$(g), as shown in Table 9-11. We have used a least-squares analysis to find the optimal solutions to the two over-determined sets of linear equations, using either the second-law or the third-law enthalpies, with the results shown in Table 9-12.

Table 9-11: Sums of various enthalpies involving $UF(g)$, $UF_2(g)$ and $UF_3(g)$.

Reference	Expression	$\Delta_f H_m^\circ$ (298.15 K) kJ·mol^{-1}	
		Second-law	Third-law
[69ZMB]	$UF_3(g)$	−1042.1	−1064.8
[82LAU/HIL]	$UF_3(g)$	−1055.7	−1084.3
[82LAU/HIL]	$UF(g)$	−62.1	−64.6
[82LAU/HIL]	$-UF(g) + UF_2(g)$	−494.4	−518.7
[82LAU/HIL]	$-UF_2(g) + UF_3(g)$	−556.0	−545.5
[84GOR/SMI]	$UF_3(g)$	−1088.7	−1054.1
[84GOR/SMI]	$-UF_2(g) + 2UF_3(g)$	−1588.7	−1576.0
[84GOR/SMI]	$UF(g) - 2UF_2(g) + UF_3(g)$	—	−29.5
[91HIL/LAU]	$2UF(g) - UF_2(g)$	439.4	453.4

Table 9-12: Optimised enthalpies of formation, $\Delta_f H_m^\circ$ (298.15 K) kJ·mol^{-1}.

Species	Second-law	Third-law	Selected	[92GRE/FUG]
$UF(g)$	−46.3	−47.3	−(47 ± 20)	−(52 ± 20)
$UF_2(g)$	−529.6	−546.1	−(540 ± 25)	−(530 ± 30)
$UF_3(g)$	−1063.7	−1068.2	−(1065 ± 20)	−(1054.2 ± 15.7)

The selected values are rounded from the mean of the second- and third-law enthalpies of formation, which are surprisingly close, considering the appreciable discrepancies in the values for the individual reactions in Table 9-10.

$$\Delta_f H_m^\circ (UF, g, 298.15 K) = -(47 \pm 20) \text{ kJ} \cdot \text{mol}^{-1},$$
$$\Delta_f H_m^\circ (UF_2, g, 298.15 K) = -(540 \pm 25) \text{ kJ} \cdot \text{mol}^{-1},$$
$$\Delta_f H_m^\circ (UF_3, g, 298.15 K) = -(1065 \pm 20) \text{ kJ} \cdot \text{mol}^{-1}.$$

The uncertainties have been increased substantially from the purely statistical values to allow for the uncertainties in the thermal functions and the conversion from mass-spectrometric intensities to pressures.

It will be seen that the new analysis of the experimental data for these species yields somewhat different values for the enthalpies of formation of $UF_2(g)$ and $UF_3(g)$ from those selected by Grenthe et al. [92GRE/FUG] although well within the uncertainty limits.

9.4.2.1.2 UF$_4$(g) (V.4.1.1.3)

Konings *et al.* [96KON/BOO] have recently studied the infrared spectrum of UF$_4$ vapour between 1300 and 1370 K. Based on this, and a re-analysis of the previously determined gas electron diffraction data, they have demonstrated that the UF$_4$(g) molecule almost certainly has tetrahedral symmetry. A further paper by Konings and Hildenbrand [98KON/HIL] has extended the discussion to other actinide tetrahalide molecules. The molecular parameters for UF$_4$(g) reported in [98KON/HIL] have thus been adopted for the UF$_4$(g) molecule, with r(U–F) distance = 2.059 Å. The ground-state energy level was assumed to have a statistical weight of three, and the higher electronic levels were taken from Glushko *et al.* [82GLU/GUR], except that, following [98KON/HIL], the statistical weight of the first excited level at 500 cm^{-1} is taken to be two rather than four. The calculated values for the entropy and heat capacity of UF$_4$(g) at 298.15 K are:

$$S_m^o (UF_4, g, 298.15 \text{ K}) = (360.7 \pm 5.0) \text{ J} \cdot \text{K}^{-1} \cdot \text{mol}^{-1},$$
$$C_{p,m}^o (UF_4, g, 298.15 \text{ K}) = (95.1 \pm 3.0) \text{ J} \cdot \text{K}^{-1} \cdot \text{mol}^{-1}$$

and these are the selected values.

There are numerous investigations of the vapour pressures of UF$_4$(cr) and UF$_4$(l), as summarised in Table 9-13.

There are two other studies which should be mentioned. The early results of Altman [43ALT] give very much lower pressures and have been ignored, while the unpublished data by Khodeev *et al.*, quoted by Glushko *et al.* [82GLU/GUR], as judged by the analysis in [82GLU/GUR], seem to give slightly higher pressures than those of Chudinov and Choporov [70CHU/CHO4].

Table 9-13: Enthalpy of sublimation of UF$_4$(cr).

Reference	Method	T range (K)	$\Delta_{sub}H_m^o$ (298.15 K) kJ · mol^{-1}
[47JOH]	Diaphragm gauge	1120–1275	(301.6 ± 5.8) *
[47RYO/TWI]	Effusion; boiling point	1013–1457	(311.2 ± 4.0)
[59POP/KOS]	Transpiration	1148–1273	(312.2 ± 2.0)
[60LAN/BLA]	Gauge; boiling point	1291–1575	(307.8 ± 0.75)
[61AKI/KHO]	Mass-spectrometric effusion	917–1041	302.0 (from v.p. equation)*
[70CHU/CHO4]	Effusion	828–1280	(303.9 ± 4.1) *
[77HIL]	Torsion–effusion	980–1130	(309.8 ± 1.0)
[80NAG/BHU]	Transpiration; boiling point	1169–1427	(309.8 ± 1.4)

* Data not included in the mean

Since these measurements, with the exception of those by [61AKI/KHO], give absolute values of the pressure, we have analysed the data by the third-law method, using the thermal functions for UF$_4$ discussed above, to give the results shown in Table 9-13. There is quite good agreement between all the studies in the table, which are plotted in and Figure 9-2. However, it is clear from these figures that the data of the three studies marked with an asterisk in the Table 9-13 are appreciably more discrepant from the mean than those of the other five studies. The selected enthalpy of sublimation is therefore based on the weighted mean of the studies shown in .

$$\Delta_{sub}H_m^\circ (UF_4, 298.15\ K) = (309.0 \pm 5.0)\ kJ \cdot mol^{-1}$$

where the uncertainty has been increased substantially from the statistical 1.96 σ value (± 0.5 kJ·mol^{-1}) to allow for uncertainties in the thermal functions of UF$_4$(g). This value is also in good accord with that (307.4 ± 10.7) kJ·mol^{-1} given by a second-law analysis of the more extensive mass-spectrometric data given by [61AKI/KHO] for which no absolute pressures are reported.

The derived enthalpy of formation of UF$_4$(g) is thus:

$$\Delta_f H_m^\circ (UF_4, g, 298.15K) = -(1605.2 \pm 6.5)\ kJ \cdot mol^{-1}.$$

The vapour pressures for the liquid are calculated with an enthalpy of fusion of 47 kJ·mol^{-1} based on the enthalpy drop measurements of Dworkin [72DWO]. The vapour pressure data would actually be best fitted with an enthalpy of fusion as low as 36 kJ·mol^{-1}, but Dworkin's data seem quite reliable, and the higher value has been preferred.

Figure 9-1: Vapour pressure of UF$_4$(cr, l), included data.

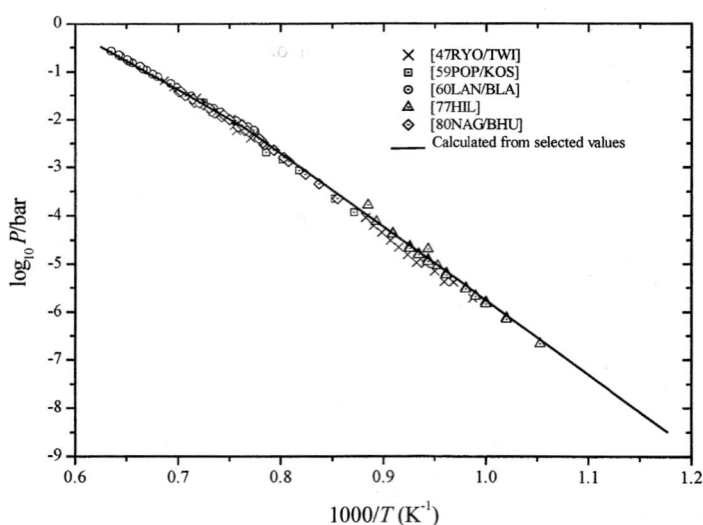

Figure 9-2: Vapour pressure of UF$_4$(cr, l), excluded data.

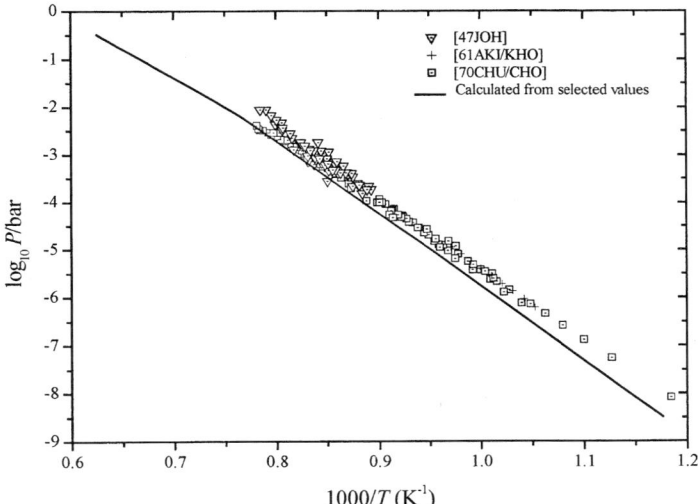

9.4.2.1.3 UF$_5$(g) (V.4.1.1.4)

The data for UF$_5$(g) are not of high precision, for a number of reasons, apart from its high reactivity. The region of the U–F system between UF$_4$ and UF$_6$ is quite complex, both in the solid and gaseous states. Thus most of the data are derived from mass-spectrometric intensities, with the well-known problems of fragmentation and the necessity to estimate cross-sections. At higher pressures the dimer U$_2$F$_{10}$(g) is by far the dominant species, so that substantial corrections to vapour pressure data are required.

The thermal functions of UF$_5$(g) were calculated assuming the molecular parameters shown in Table 9-9, which are those estimated by Glushko *et al.* [82GLU/GUR]. The molecule is assumed to be a square pyramid with a slight distortion to reduce the symmetry to C$_{2v}$. These assumptions give:

$$S_m^o \text{ (UF}_5\text{, g, 298.15 K)} = (386.4 \pm 10.0) \text{ J} \cdot \text{K}^{-1} \cdot \text{mol}^{-1},$$
$$C_{p,m}^o \text{ (UF}_5\text{, g, 298.15 K)} = (110.6 \pm 5.0) \text{ J} \cdot \text{K}^{-1} \cdot \text{mol}^{-1},$$

which are not significantly different from those in [92GRE/FUG].

The reactions used to calculate the stability of UF$_5$(g) are summarised in Table 9-14 and are now discussed in detail.

Leitnaker [80LEI] has re-analysed the transpiration measurements of the vapour pressure of UF$_5$(cr, l) by Wolf *et al.* [65WOL/POS] to take account of the dimerisation of UF$_5$(g) and the probable formation of U$_2$F$_{11}$(g) in the presence of the transporting gas UF$_6$(g). There is a minor flaw in Leitnaker's re-analysis; it was difficult to impose a constraint of equal vapour pressure at the known melting point (621 K) when refitting the vapour equations for the solid and liquid. However, in our final analysis, this is not important, since we have only utilised the data over the solid, which are appreciably more consistent than those over the liquid. For each data point given by Wolf et al. we have used Leitnaker's extrapolation of the effective pressure $p_{eff} = p_{UF_5(g)} + 2 p_{U_2F_{10}(g)}$ to $p_{UF_6(g)} = 0$, and calculated the individual pressures of $UF_5(g)$ and $U_2F_{10}(g)$, using the equilibrium constant given by Kleinschmidt and Hildenbrand [79KLE/HIL].

Gorokhov et al. [84GOR/SMI] have used three reactions to obtain the stability of $UF_5(g)$:

$$UF_4(g) + UF_6(g) \rightleftharpoons 2\,UF_5(g),$$

$$\frac{1}{3}U_4F_{17}(cr) + UF_6(g) \rightleftharpoons \frac{7}{3}UF_5(g),$$

$$UF_4(g) + CuF_2(g) \rightleftharpoons UF_5(g) + CuF(g).$$

We have included the first two reactions in the analysis, but not the third. The stabilities of both $CuF(g)$ and $CuF_2(g)$ are quite poorly defined - there is a difference of 11.3 kJ · mol^{-1} in the relevant $\Delta_f G_m$ ($CuF(g)$–$CuF_2(g)$) at 1000 K, (near the mid point of the experimental temperature range of [84GOR/SMI]) between two current authoritative assessments [98CHA], [2000LAN].

Smirnov and Gorokhov [84SMI/GOR] and Lau et al. [85LAU/BRI] have both used the decomposition of $UO_2F_2(cr)$ to investigate the stability of $UF_5(g)$. They both find that the principal gaseous species over $UO_2F_2(cr)$ are $UO_2F_2(g)$, $UF_5(g)$, $O_2(g)$, $UF_4(g)$ and $UOF_4(g)$. These authors choose to use the equilibrium constants of two different reactions to derive $\Delta_f G_m^\circ$ ($UF_5(g)$):

[84SMI/GOR]: $UO_2F_2(g) + 2\,UF_5(g) \rightleftharpoons 3\,UF_4(g) + O_2(g)$ (1397-1539 K)

[85LAU/BRI]: $\frac{5}{2}UO_2F_2(cr) \rightleftharpoons UF_5(g) + \frac{1}{2}O_2(g) + \frac{3}{2}UO_{2.6667}(cr)$ (896-1036 K).

We have utilised the results of [85LAU/BRI] (see Table 9-14), but we cannot use the data of [84SMI/GOR], as they report only the derived $\Delta_f H_m^\circ$ (UF_5, g, 298.15 K) value, $-(1908 \pm 13)$ kJ · mol^{-1}, using, presumably, the data of [82GLU/GUR] for the auxiliary data.

Table 9-14: Enthalpy of formation $\Delta_f H_m^\circ$ (UF$_5$, g, 298.15 K).

Reference	Reaction	T range (K)	$\Delta_f H_m^\circ$ (kJ·mol^{-1}) Third-law
[80LEI]	UF$_5$(cr) $\rightleftharpoons$ UF$_5$(g)	556–615	$-(1918.0 \pm 4.2)$
[84GOR/SMI]	UF$_4$(cr) + UF$_6$(g) $\rightleftharpoons$ 2 UF$_5$(g)	440–452	$-(1911.7 \pm 4.8)$
[84GOR/SMI]	1/3 U$_4$F$_{17}$(cr) + UF$_6$(g) $\rightleftharpoons$ 7/3 UF$_5$(g)	440–452	$-(1927.9 \pm 4.9)$
[85LAU/BRI]	5/2UO$_2$F$_2$(cr) $\rightleftharpoons$ UF$_5$(g) +1/2O$_2$(g) +3/2 UO$_{2.6667}$(cr)	896–1036	$-(1894.1 \pm 24.3)$*
[87BON/KOR]	UF$_4$(cr) + UF$_6$(g) $\rightleftharpoons$ 2 UF$_5$(g)	663–809	$-(1908.1 \pm 3.4)$
[93NIK/TSI]	UF$_4$(g) +1/2 NiF$_2$(cr) $\rightleftharpoons$ UF$_5$(g) + 1/2 Ni(cr)	1017–1109	$-(1914.3 \pm 4.4)$

* Second-law value

Bondarenko *et al.* [87BON/KOR] have made mass-spectrometric measurements of the reaction:

$$UF_4(g) + UF_6(g) \rightleftharpoons 2\ UF_5(g),$$

with results that are quite similar to those of [84GOR/SMI] for the same reaction.

Nikitin and Tsirel'nikov [93NIK/TSI] have measured the ratio of UF$_5$/UF$_4$ in equilibrium with the Ni/NiF$_2$ couple from 1017 to 1090 K – see Table 9-14.

The mass-spectrometric studies by Hildenbrand [77HIL] and Hildenbrand and Lau [91HIL/LAU2] of the reaction:

$$UF_5(g) + Ag(g) \rightleftharpoons UF_4(g) + AgF(g)$$

give stabilities markedly different from the assembled studies, as noted by [92GRE/FUG] and by Nikitin and Tsirel'nikov [93NIK/TSI]. However, as for the copper fluorides used by [84GOR/SMI], the thermodynamic data for AgF(g) are far from well-established, and these studies have not been included in the current analysis.

The selected value is the mean of the four most consistent results, and is also close to the mean of the six studies:

$$\Delta_f H_m^\circ (UF_5, g, 298.15\ K) = -(1913 \pm 15)\ kJ \cdot mol^{-1},$$

the uncertainty having been increased to allow for the uncertainties in ionisation cross-sections, *etc.*

9.4.2.1.4 UF$_6$(g) (V.4.1.1.5)

Uranium hexafluoride gas is one of the few uranium fluorides whose data are not dependent in some way on those for UF$_4$(g), and since there have been no new experimental studies since the publication of Grenthe *et al.* [92GRE/FUG], the data proposed there have been accepted. For completeness, they are summarised below.

The thermal functions of $UF_6(g)$ ideal gas were calculated assuming the molecular parameters given in Table 9-9, using the rigid–rotator, harmonic-oscillator approximation. The molecule is taken to be octahedral with O_h symmetry. All the molecular parameters are taken from the comprehensive paper by Aldridge et al. [85ALD/BRO].

$$S_m^\circ (UF_6, g, 298.15\ K) = (376.3 \pm 1.0)\ J \cdot K^{-1} \cdot mol^{-1},$$
$$C_{p,m}^\circ (UF_6, g, 298.15\ K) = (129.4 \pm 0.5)\ J \cdot K^{-1} \cdot mol^{-1}.$$

These are both slightly smaller than the values selected in [92GRE/FUG].

However, with these data, the enthalpy of sublimation derived from the eight consistent determinations of the vapour pressure noted in [92GRE/FUG] remains unchanged at $\Delta_{sub}H_m^\circ (UF_6, g, 298.15\ K) = (49.1 \pm 0.5)\ kJ \cdot mol^{-1}$, corresponding to the unchanged selected value of the enthalpy of formation:

$$\Delta_f H_m^\circ (UF_6, g, 298.15\ K) = - (2148.6 \pm 1.9)\ kJ \cdot mol^{-1}.$$

Calculations of the populations of the rotational and vibrational states and band contours of $UF_6(g)$ around 300 K have been reported by Oda [93ODA]. This reference is given for information only, as it does not lead to changes in the selections reported above.

9.4.2.1.5 $U_2F_{10}(g)$ and $U_2F_{11}(g)$ (V.4.1.1.6)

Kleinschmidt and Hildenbrand [79KLE/HIL] have measured by mass-spectrometry the dimerisation constants for the reaction:

$$2\ UF_5(g) \rightleftharpoons U_2F_{10}(g), \tag{9.22}$$

$$\log_{10} (K\ /bar) = \frac{8523}{T} - 8.744, \qquad (608\ to\ 866\ K)$$

corresponding to:

$$\Delta_r G_m^\circ (9.22) = - 163170 + 167.40\ T\quad J \cdot mol^{-1}.$$

We have not estimated any thermal functions for $U_2F_{10}(g)$, but instead used the following procedure: the enthalpy of formation at the mid-range temperature (737 K) has been recalculated to 298.15 K by assuming that the value of $(H_m(737\ K) - H_m^\circ(298.15\ K))$ for reaction (9.22) is $4000\ J \cdot mol^{-1}$, slightly larger than that for the similar reaction involving Mo, $(3489\ J \cdot mol^{-1})$, the data for $MoF_5(g)$ and $Mo_2F_{10}(g)$ being taken from the NIST–JANAF database [98CHA].

Thus, $\Delta_r H_m^\circ ((9.22), 298.15\ K) = - (167.2 \pm 4.7)\ kJ \cdot mol^{-1}$, giving finally:

$$\Delta_f H_m^\circ (U_2F_{10}(g), 298.15\ K) = - (3993 \pm 30)\ kJ \cdot mol^{-1},$$

which is the selected value. This is appreciably less negative than that selected in [92GRE/FUG], $- (4021 \pm 30)\ kJ \cdot mol^{-1}$, since they erroneously included in their average a value of $\Delta_f H_m^\circ (U_2F_{10}(g), 298.15\ K)$ from Hildenbrand et al. [85HIL/GUR], derived from the same dimerisation data, but based on their own [85HIL/GUR] much more negative enthalpy of formation of $UF_5(g)$.

In their dimerisation study, Kleinschmidt and Hildenbrand [79KLE/HIL] generated the vapour phase by reacting $UF_4(cr)$ with $UF_6(g)$, and observed a small signal at a mass corresponding to U_2F_{11}. They attributed this to a secondary ion reaction between UF_5^+ and UF_6. However, Leitnaker [80LEI] pointed out that in the transpiration experiments of Wolf et al. [65WOL/POS], the measured total pressure increased systematically with the pressure of the transporting $UF_6(g)$, attributing this to the formation of some $U_2F_{11}(g)$ under these circumstances. However, the data are neither conclusive, nor precise enough, to derive any reliable thermodynamic data for this species.

9.4.2.1.6 $UO_2F_2(g)$ (V.4.1.1.7)

The thermal functions of $UO_2F_2(g)$ were calculated assuming the molecular parameters shown in Table 9-9, which are those estimated by Glushko et al. [82GLU/GUR]. These assumptions give:

$$S_m^o (UO_2F_2, g, 298.15 \text{ K}) = (342.7 \pm 6.0) \text{ J} \cdot \text{K}^{-1} \cdot \text{mol}^{-1},$$
$$C_{p,m}^o (UO_2F_2, g, 298.15 \text{ K}) = (86.4 \pm 3.0) \text{ J} \cdot \text{K}^{-1} \cdot \text{mol}^{-1},$$

which give entropy and heat capacity values that are negligibly different from those in [92GRE/FUG].

Ebbinghaus [95EBB] has made somewhat different estimates of the molecular parameters (particularly for some of the vibration frequencies), which lead to appreciably lower values of the entropy and heat capacity:

$$S_m^o (298.15 \text{ K}) = 327.9 \text{ J} \cdot \text{K}^{-1} \cdot \text{mol}^{-1}, \; C_{p,m}^o (298.15 \text{ K}) = 83.1 \text{ J} \cdot \text{K}^{-1} \cdot \text{mol}^{-1}.$$

However for consistency with the molecular parameters of the other gaseous uranium fluorides, we have preferred to retain the estimates of Glushko et al. [82GLU/GUR].

The vaporisation behaviour of UO_2F_2 is complex. The mass-spectrometric studies of Smirnov et al. [84SMI/GOR] and Lau et al. [85LAU/BRI] are in good agreement. They both find that the principal gaseous species over $UO_2F_2(cr)$ are $UO_2F_2(g)$, $UF_5(g)$, $O_2(g)$, $UF_4(g)$ and $UOF_4(g)$ (there was no evidence for $UF_6(g)$, contrary to the assumption of Knacke et al. [69KNA/LOS]). The latter authors measured the total vapour pressure over UO_2F_2 and subtracted that attributed to $UF_6(g)$ from the mass of U_3O_8 formed in the residue. This calculation will now be slightly in error, but the calculated pressures of $UO_2F_2(g)$ are quite scattered (and also up to the limit of validity of the Knudsen equation for effusion). In fact we have excluded their measurement at their highest temperature (1132 K), since the pressure was ca. $7 \cdot 10^{-4}$ bar.

Table 9-15: Enthalpy of sublimation of UO_2F_2(cr), $\Delta_{sub}H_m^o$ (UO_2F_2, 298.15 K).

Reference	Method	T range (K)	$\Delta_{sub}H_m^o$ (kJ · mol^{-1})	
			Second-law	Third-law
[69KNA/LOS]	Effusion	1033–1099	–	(297.3 ± 9.8)
[84SMI/GOR]	Mass-spectrometric effusion	929–1094	(290.3 ± 12.8)	(302.1 ± 1.2)
[85LAU/BRI]	Torsion effusion + mass-spectrometry	957–1000	(303.1 ± 3.0)	(301.6 ± 2.7)

The selected value is based on the third-law values $\Delta_{sub}H_m^o$(UO_2F_2, 298.15 K) = (301 ± 10) kJ · mol^{-1} giving finally:

$$\Delta_f H_m^o (UO_2F_2, g, 298.15 K) = -(1352.5 \pm 10.1) \text{ kJ} \cdot \text{mol}^{-1},$$

which is the same value as in [92GRE/FUG], but with a smaller uncertainty.

9.4.2.1.7 UOF_4(g) (V.4.1.1.8)

The thermal functions of UOF_4(g) were calculated assuming the molecular parameters shown in Table 9-9, which are those estimated by Glushko et al. [82GLU/GUR]; the molecule is assumed to be a square pyramid. These assumptions give:

$$S_m^o (UOF_4, g, 298.15 K) = (363.2 \pm 8.0) \text{ J} \cdot \text{K}^{-1} \cdot \text{mol}^{-1},$$
$$C_{p,m}^o (UOF_4, g, 298.15 K) = (108.1 \pm 5.0) \text{ J} \cdot \text{K}^{-1} \cdot \text{mol}^{-1}.$$

These values are negligibly different from those in [92GRE/FUG], although the uncertainty in the entropy has been decreased slightly.

The mass-spectrometric studies of Smirnov and Gorokhov [84SMI/GOR] and Lau et al. [85LAU/BRI] both indicate that there is a small amount of UOF_4 in the vapour over UO_2F_2(cr). They choose different reactions to calculate the stability of UOF_4(g):

[84SMI/GOR]: $UO_2F_2(g) + 2 UF_5(g) + \frac{1}{2} O_2(g) \rightleftharpoons 3 UOF_4(g)$ (1028 – 1100 K)

[85LAU/BRI]: $2 UO_2F_2(cr) \rightleftharpoons UOF_4(g) + \frac{1}{6} O_2(g) + UO_{2.6667}(cr)$ (913 – 1036 K)

although Lau et al. give sufficient data to calculate the equilibrium constant only for 1000 K. Third-law analyses of the data give enthalpies of reaction of $\Delta_r H_m^o$(UOF_4, g, 298.15 K) = – (121.6 ± 5.9) kJ·mol^{-1} and 355.4 kJ·mol^{-1} for the respective reactions and thus $\Delta_f H_m^o$(UOF_4, g, 298.15 K) = –(1766.7 ± 11.2) kJ·mol^{-1} and – (1760.0 ± 2.7) kJ·mol^{-1}, respectively. The selected value is:

$$\Delta_f H_m^o (UOF_4, g, 298.15 K) = -(1763 \pm 20) \text{ kJ} \cdot \text{mol}^{-1},$$

where the uncertainty has been increased to allow for the uncertainties in ionisation cross-sections and the difficulty of measuring oxygen pressures accurately in a mass-spectrometer.

9.4.2.2 Aqueous uranium fluorides (V.4.1.2)

9.4.2.2.1 Aqueous U(VI) fluorides (V.4.1.2.1)

9.4.2.2.1.1 Binary complexes

The only new study providing information on equilibrium constants is the study of Ferri et al. [93FER/SAL] who performed potentiometric titrations of U(VI) solutions at $-\log_{10}[\text{H}^+] = 3.50$, the ionic strength being 3 M (NaClO$_4$) (298.15 K). As discussed in Appendix A the formation constants for the binary complexes $\text{UO}_2\text{F}_q^{2-q}$, $\log_{10} \beta_1 = (4.86 \pm 0.02)$, $\log_{10} \beta_2 = (8.62 \pm 0.04)$, $\log_{10} \beta_3 = (11.71 \pm 0.06)$ and $\log_{10} \beta_4 = (13.78 \pm 0.08)$ are retained by this review. As the average number of coordinated fluoride ligands per U(VI) reaches 2.35, the uncertainty in $\log_{10} \beta_4$ seems surprisingly low. Ferri et al. reported an equilibrium constant $K_5 = 3.0$ M^{-1} at 269.4 K for the reaction:

$$\text{UO}_2\text{F}_4^{2-} + \text{F}^- \rightleftharpoons \text{UO}_2\text{F}_5^{3-}$$

obtained by ^{19}F NMR data. However, more recent NMR investigations [2002VAL/WAH] indicate that this value is too large and that $K_5 = (0.60 \pm 0.05)$ M^{-1} is a better estimate. The $\text{UO}_2\text{F}_5^{3-}$ complex is weak and is therefore only formed in very concentrated fluoride solutions, where its presence is clearly demonstrated by the ^{19}F NMR data in both studies. No equilibrium constant has been selected by this review for $\text{UO}_2\text{F}_5^{3-}$.

By combining these data with those of [71AHR/KUL] at 1 M ionic strength, the following values of the ion interaction parameters are given by Ferri et al. [93FER/SAL]: $\varepsilon(\text{UO}_2\text{F}^+, \text{ClO}_4^-) = (0.28 \pm 0.04)$ kg·mol^{-1}, $\varepsilon(\text{UO}_2\text{F}_3^-, \text{Na}^+) = -(0.14 \pm 0.05)$ kg·mol^{-1} and $\varepsilon(\text{UO}_2\text{F}_4^{2-}, \text{Na}^+) = -(0.30 \pm 0.06)$ kg·mol^{-1}. They also give $\varepsilon(\text{UO}_2\text{F}_2(\text{aq}), \text{Na}^+ \text{ or ClO}_4^-) = (0.13 \pm 0.05)$ kg·mol^{-1} (see Appendix A). The first value agrees well with the selected value (0.29 ± 0.05) kg·mol^{-1} in [92GRE/FUG], whereas the other values are new.

Extrapolation to zero ionic strength gives:

$$\log_{10} \beta_1^\circ = (5.16 \pm 0.06),$$

which agrees with the value (5.09 ± 0.13) previously selected [92GRE/FUG],

$$\log_{10} \beta_2^\circ = (8.83 \pm 0.08),$$

$$\log_{10} \beta_3^\circ = (10.90 \pm 0.10)$$

$$\log_{10} \beta_4^\circ = (11.84 \pm 0.11),$$

which differ only slightly, even for $\log_{10} \beta_4^\circ$, from the values of [92GRE/FUG] ($\log_{10} \beta_2^\circ = (8.62 \pm 0.04)$, $\log_{10} \beta_3^\circ = (10.9 \pm 0.4)$ and $\log_{10} \beta_4^\circ = (11.7 \pm 0.7)$). As the selected values in [92GRE/FUG] are based on estimated ε values for the fluoride complexes, this review selects the new results of [93FER/SAL] for the $\text{UO}_2\text{F}_n^{2-n}$ complexes, $n = 1$ to 4.

Fazekas et al. [98FAZ/YAM] have studied the U(VI) – fluoride system using fluorescence spectroscopy. This study does not give new thermodynamic information.

9.4.2.2.1.2 Ternary U(VI) hydroxide-fluoride complexes

Upper limits for the equilibrium constants of possible ternary hydroxo complexes are reported in [93FER/SAL] in 3.00 M NaClO$_4$ and 298.15 K. Upper limits for the equilibrium constants of these complexes are, respectively, $\log_{10} {}^*\beta_{2,2,1} < -2.2$ and $\log_{10} {}^*\beta_{5,3,1} < -12.7$ for the reactions:

$$2\,UO_2^{2+} + 2\,H_2O + F^- \rightleftharpoons (UO_2)_2(OH)_2 F^+ + 2\,H^+$$

and

$$3\,UO_2^{2+} + 5\,H_2O + F^- \rightleftharpoons (UO_2)_3(OH)_5 F(aq) + 5\,H^+.$$

9.4.2.2.2 Aqueous U(IV) fluorides (V.4.1.2.3)

The work of Sawant and Chaudhuri [90SAW/CHA2] was not analysed in [92GRE/FUG]. They report equilibrium constants $\log_{10} \beta_q^\circ$, $q = 1$ to 4, for the reactions:

$$U^{4+} + q\,F^- \rightleftharpoons UF_q^{4-q},$$

at $I = 1$ M (HClO$_4$, NaF) at 296.15 K (see Appendix A). The equilibrium constants $\log_{10} \beta_q^\circ$ for the reactions were calculated using the auxiliary data in [92GRE/FUG] by this review and are reported in Table 9-16 together with the data used in [92GRE/FUG] (page 164-167) for the previous data selection. The values reported by [90SAW/CHA2] are generally somewhat higher, but not better than the results in the papers [69GRE/VAR], [69NOR], [74KAK/ISH], [76CHO/UNR] used previously. This review selects the unweighted average values of the $\log_{10} \beta_n^\circ$ ($n = 1$-4) values reported in Table 9-16 as:

$$\log_{10} \beta_1^\circ = (9.42 \pm 0.51),$$
$$\log_{10} \beta_2^\circ = (16.56 \pm 0.71),$$
$$\log_{10} \beta_3^\circ = (21.89 \pm 0.83),$$
$$\log_{10} \beta_4^\circ = (26.34 \pm 0.96).$$

Table 9-16: Literature data for the formation constants of U(IV) fluoride complexes recalculated to $I = 0$ with the SIT and auxiliary data from [92GRE/FUG].

Complex		$\log_{10} \beta_n^\circ$	Reference
UF^{3+}		(9.42 ± 0.25)	[69GRE/VAR]
		(9.54 ± 0.25)	[69NOR]
		(9.28 ± 0.15)	[74KAK/ISH]
		(9.09 ± 0.17)	[76CHO/UNR]
		(9.78 ± 0.12)[a]	[90SAW/CHA2]
	Average	(9.42 ± 0.51)[*]	This review
		(9.28 ± 0.09)	[92GRE/FUG][b]
UF$_2^{2+}$		(16.37 ± 0.50)	[69GRE/VAR]
		(16.72 ± 0.50)	[69NOR]
		(16.16 ± 0.16)	[74KAK/ISH]
		(16.97 ± 0.13)[a]	[90SAW/CHA2]
	Average	(16.56 ± 0.71)[*]	This review
		(16.23 ± 0.15)	[92GRE/FUG][b]
UF$_3^+$		(22.06 ± 0.50)	[69NOR]
		(21.23 ± 0.17)	[74KAK/ISH]
		(22.38 ± 0.14)[a]	[90SAW/CHA2]
	Average	(21.89 ± 0.83)[**]	This review
		(21.6 ± 1.0)	[92GRE/FUG][b]
UF$_4$(aq)		(25.61 ± 0.23)	[74KAK/ISH]
		(27.06 ± 0.12)[a]	[90SAW/CHA2]
	Average	(26.34 ± 0.96)[**]	This review
		(25.6 ± 1.0)	[92GRE/FUG][b]

(a) Calculated by this review.
(b) The values selected by [92GRE/FUG] are based on the same experimental data, except those in [90SAW/CHA2], which were not discussed.
(*) Uncertainty: ± 1.96σ.
(**) The uncertainty is chosen to cover the uncertainty ranges of the discrepant individual values.

The new selected values are only slightly different from those selected by [92GRE/FUG] but with uncertainties much greater than those proposed there.

To preserve the consistency of selected data, the value of $\log_{10} K_{s,0}^\circ$ (UF$_4$·2.5H$_2$O(cr)) from [60SAV/BRO] must be recalculated and new values of $\log_{10} \beta_5^\circ$ and $\log_{10} \beta_6^\circ$ must be selected. Indeed, [92GRE/FUG], in a first step used their selected $\log_{10} \beta_2^\circ$ to $\log_{10} \beta_4^\circ$ values and $\log_{10} K_{s,2}^\circ$, $\log_{10} K_{s,3}^\circ$, $\log_{10} K_{s,4}^\circ$ values from solubility data in [60SAV/BRO] to calculate $\log_{10} K_{s,0}^\circ$, which therefore depends directly on the selected $\log_{10} \beta_n^\circ$ (n = 2-4) values. In a second step [92GRE/FUG] used the thus evaluated $\log_{10} K_{s,0}^\circ$ value together with $\log_{10} K_{s,5}^\circ$ and $\log_{10} K_{s,6}^\circ$ also derived from the solubility data in [60SAV/BRO] to calculate $\log_{10} \beta_5^\circ$ and $\log_{10} \beta_6^\circ$, which therefore also depend on the selected $\log_{10} \beta_n^\circ$ ($n = 2$-4).

Following the procedure applied in [92GRE/FUG] (Appendix A, review of [60SAV/BRO]), the $\log_{10} K_{s,n}^\circ$ values calculated in [92GRE/FUG] from the solubility data for $UF_4 \cdot 2.5\ H_2O(cr)$ reported by Savage and Browne [60SAV/BRO] lead to:

$$\log_{10} K_{s,2}^\circ = -(13.15 \pm 0.04),$$

$$\log_{10} K_{s,3}^\circ = -(8.47 \pm 0.09),$$

$$\log_{10} K_{s,4}^\circ = -(3.96 \pm 0.03),$$

which give, respectively:

$$\log_{10} K_{s,0}^\circ = -(29.71 \pm 0.71),$$

$$\log_{10} K_{s,0}^\circ = -(30.36 \pm 0.83),$$

$$\log_{10} K_{s,0}^\circ = -(30.30 \pm 0.96),$$

and an unweighted average value of $\log_{10} K_{s,0}^\circ$ ($UF_4 \cdot 2.5\ H_2O(cr)$) = $-(30.12 \pm 0.70)$.

Using this solubility constant and $\log_{10} K_{s,5}^\circ = -(2.39 \pm 0.25)$ and $\log_{10} K_{s,6}^\circ = -(0.32 \pm 0.08)$ as evaluated by [92GRE/FUG] from the experimental data of [60SAV/BRO], the formation constants of the penta- and hexa-fluoro complexes are calculated to be:

$$\log_{10} \beta_5^\circ = (27.73 \pm 0.74),$$

$$\log_{10} \beta_6^\circ = (29.80 \pm 0.70),$$

which are selected by this review. These are somewhat different from those selected in [92GRE/FUG], $\log_{10} \beta_5^\circ = (27.01 \pm 0.30)$ and $\log_{10} \beta_6^\circ = (29.08 \pm 0.18)$ although they overlap within the combined uncertainties.

The solubility study of Savage and Browne [60SAV/BRO] was not used by [92GRE/FUG] to select thermodynamic data for $UF_4 \cdot 2.5\ H_2O(cr)$, because this value is considerably larger than $\log_{10} K_{s,0}^\circ$ ($UF_4 \cdot 2.5\ H_2O(cr)$) = $-(33.5 \pm 1.2)$ as calculated from the selected thermochemical data.

The solubility measurements of $UO_2(cr)$ at 773.15 K, 1 kbar, in HF solutions reported by Redkin [97RED] give no reliable new thermodynamic data.

9.4.2.3 Solid uranium fluorides (V.4.1.3)

9.4.2.3.1 Binary uranium fluorides and their hydrates (V.4.1.3.1)

9.4.2.3.1.1 UF$_3$(cr) and UF$_4$(cr)

9.4.2.3.1.1.1 High temperature heat capacity

The $C_{p,m}$ equation for UF$_3$(cr) in [92GRE/FUG] is a refit to 800 K of the data given by [82GLU/GUR]. Since the experimental data involving uranium trifluoride extend above 800 K, it has been replaced by the actual equation from [82GLU/GUR]:

$$C_{p,m}(\text{UF}_3, \text{cr}, T) = 106.539 + 7.05 \times 10^{-4}\, T - 1.0355 \times 10^6\, T^{-2} \quad \text{J} \cdot \text{K}^{-1} \cdot \text{mol}^{-1}$$
(298.15 to 1768 K)

The value of $C_{p,m}^\circ$ (298.15 K) is unchanged:

$$C_{p,m}^\circ(\text{UF}_3, \text{cr}, 298.15\, \text{K}) = (95.1 \pm 0.4)\, \text{J} \cdot \text{K}^{-1} \cdot \text{mol}^{-1}$$

Similarly, the equation for $C_{p,m}(T)$ in [92GRE/FUG] for UF$_4$(cr) (a refit to 1200 K) of the data given by [82GLU/GUR] has been replaced by the actual equation from [82GLU/GUR]:

$$C_{p,m}(\text{UF}_4, \text{cr}, T) = 138.865 - 3.2068 \cdot 10^{-2}\, T + 2.7988 \cdot 10^{-5}\, T^2$$
$$- 1.4020 \cdot 10^6\, T^{-2} \quad \text{J} \cdot \text{K}^{-1} \cdot \text{mol}^{-1}.$$
(298.15 to 1309 K)

As noted in section 9.4.2.1.2 the vapour pressure data for UF$_4$(cr, l) are more consistent with a somewhat smaller value of the enthalpy of fusion than the experimental value [72DWO], but the latter seems well-defined and has been retained, together with the heat capacity of the liquid from [82GLU/GUR]:

$$T_{\text{fus}} = (1305 \pm 30)\, \text{K},$$
$$\Delta_{\text{fus}} H_m^\circ(\text{UF}_4, \text{cr}, 1305\, \text{K}) = (47 \pm 5)\, \text{kJ} \cdot \text{mol}^{-1},$$
$$C_{p,m}^\circ(\text{UF}_4, \text{l}) = (167 \pm 15)\, \text{J} \cdot \text{K}^{-1} \cdot \text{mol}^{-1}.$$

9.4.2.3.2 Other U(VI) fluorides (V.4.1.3.2)

9.4.2.3.2.1 UO$_2$F$_2$(cr) and its hydrates

The heat capacity equation for UO$_2$F$_2$(cr) given in Table III.3 of [92GRE/FUG] is valid only to 400 K. Since heat capacity values at higher temperatures were required for vapour pressure calculations (section 9.4.2.1.6), this expression has been replaced by the equation given by Glushko et al.[82GLU/GUR]:

$$C_{p,m}(\text{UO}_2\text{F}_2, \text{cr}, T) = 106.238 + 2.8326 \times 10^{-2}\, T - 1.0208 \times 10^6\, T^{-2} \quad \text{J} \cdot \text{K}^{-1} \cdot \text{mol}^{-1}$$
(298.15 to 2100 K).

The value at 298.15 K is unchanged:

$$C_{p,m}^\circ(\text{UO}_2\text{F}_2, \text{cr}, 298.15\, \text{K}) = (103.22 \pm 0.42)\, \text{J} \cdot \text{K}^{-1} \cdot \text{mol}^{-1}.$$

9.4.3 Chlorine compounds and complexes (V.4.2)

9.4.3.1 Uranium chlorides (V.4.2.1)

9.4.3.1.1 Gaseous uranium chlorides (V.4.2.1.1)

9.4.3.1.1.1 UCl(g), UCl$_2$(g) and UCl$_3$(g)

These species are considered together since their data are linked by the four mass-spectrometic studies, which thus involve some uncertainties in absolute pressures.

The thermal functions of UCl(g), UCl$_2$(g) and UCl$_3$(g) were calculated assuming the molecular parameters shown in Table 9-9, which are essentially those estimated by Hildenbrand et al. [85HIL/GUR]. These assumptions give:

$$S_m^\circ \text{(UCl, g, 298.15 K)} = (265.9 \pm 3.0) \text{ J} \cdot \text{K}^{-1} \cdot \text{mol}^{-1}$$
$$C_{p,m}^\circ \text{(UCl, g, 298.15 K)} = (43.2 \pm 3.0) \text{ J} \cdot \text{K}^{-1} \cdot \text{mol}^{-1},$$
$$S_m^\circ \text{(UCl}_2\text{, g, 298.15 K)} = (339.1 \pm 10.0) \text{ J} \cdot \text{K}^{-1} \cdot \text{mol}^{-1}$$
$$C_{p,m}^\circ \text{(UCl}_2\text{, g, 298.15 K)} = (59.9 \pm 5.0) \text{ J} \cdot \text{K}^{-1} \cdot \text{mol}^{-1},$$
$$S_m^\circ \text{(UCl}_3\text{, g, 298.15 K)} = (380.3 \pm 10.0) \text{ J} \cdot \text{K}^{-1} \cdot \text{mol}^{-1}$$
$$C_{p,m}^\circ \text{(UCl}_3\text{, g, 298.15 K)} = (82.4 \pm 5.0) \text{ J} \cdot \text{K}^{-1} \cdot \text{mol}^{-1},$$

which after rounding are the same as those in [92GRE/FUG].

Lau and Hildenbrand [84LAU/HIL] have studied four equilibria between UCl(g), UCl$_2$(g), UCl$_3$(g) and UCl$_4$(g) which are summarised in Table 9-17, where the enthalpies of the given reactions at 298.15 K derived from both second- and third-law analyses are given. The auxiliary data for Ca(g) and CaCl(g) are taken from the CODATA assessments [87GAR/PAR], [89COX/WAG]; the data for CuCl(g) are discussed in chapter 14. In addition there were "tentative" measurements of the vapour pressure of UCl$_3$(cr, l) by Altman [43ALT]. However, such studies are not straightforward, since UCl$_3$ is known to disproportionate to U(cr, l) and UCl$_4$(g) (and possibly UCl$_5$(g)) – see discussions in [83FUG/PAR], section 8.5.B.1.5 and [51KAT/RAB], pages 458-459. This presumably accounts for the fact that the pressures measured by Altman are appreciably higher than those calculated for the simple vaporisation of UCl$_3$(cr) with the selected data. The measurements of Altman [43ALT] are therefore not included in the analysis.

Table 9-17: Enthalpies of the reactions involving UCl(g), UCl$_2$(g) and UCl$_3$(g).

Reference	Reaction	Number of points	$\Delta_r H_m^\circ$ (298.15 K) kJ · mol^{-1}	
			Second-law	Third-law
[84LAU/HIL]	UCl$_2$(g) + CaCl(g) $\rightleftharpoons$ UCl$_3$(g) +Ca(g)	14	– 92.3	– 89.9
[84LAU/HIL]	UCl(g) + CaCl(g) $\rightleftharpoons$ UCl$_2$(g)+Ca(g)	6	– 78.3	– 68.8
[84LAU/HIL]	UCl$_2$(g) + U(g) $\rightleftharpoons$ 2UCl(g)	9	17.5	6.4
[84LAU/HIL]	UCl$_3$(g)+ CuCl(g) $\rightleftharpoons$ UCl$_4$(g)+Cu(g)	10	– 51.3	– 34.9

Of necessity, the pressures for most of these reactions were derived from mass-spectrometric intensities, leading to uncertainties in the absolute pressures. It will be seen that the agreement between the second- and third-law analyses is mostly quite poor.

Table 9-18: Sums of various enthalpies involving UCl(g), UCl$_2$(g) and UCl$_3$(g).

Reference	Expression	$\Delta_f H_m^\circ$ (298.15 K) kJ · mol^{-1}	
		Second-law	Third-law
[84LAU/HIL]	– UCl$_2$(g) + UCl$_3$(g)	– 373.5	– 371.1
[84LAU/HIL]	– UCl(g) + UCl$_2$(g)	– 359.5	– 350.0
[84LAU/HIL]	–2 UCl(g) + UCl$_2$(g)	– 515.5	– 526.6
[84LAU/HIL]	UCl$_3$(g)	– 503.7	– 520.1

With these data, we have four quasi–independent measurements to define the three enthalpies of formation of UCl(g), UCl$_2$(g), UCl$_3$(g), as shown in Table 9-18. We have used a least-squares analysis to find the optimal solutions to the two over–determined sets of linear equations, using either the second-law or the third-law enthalpies, with the results shown in Table 9-19.

Table 9-19: Optimised enthalpies of formation, $\Delta_f H_m^\circ$ (298.15 K), kJ · mol^{-1}.

Species	Second-law	Third-law	Selected	[92GRE/FUG]
UCl(g)	187.4	187.1	(187 ± 20)	(188.2 ± 20.0)
UCl$_2$(g)	– 151.1	– 156.0	– (155 ± 20)	– (163.0 ± 22.3)
UCl$_3$(g)	– 514.2	– 523.6	– (523 ± 20)	– (537.1 ± 15.8)

The noticeable differences from the values selected in [92GRE/FUG] for UCl$_2$(g) and UCl$_3$(g) are due principally to the difference in the auxiliary data. The selected values are thus:

$$\Delta_f H_m^\circ (\text{UCl, g, 298.15 K}) = (187 \pm 20) \text{ kJ} \cdot \text{mol}^{-1},$$
$$\Delta_f H_m^\circ (\text{UCl}_2\text{, g, 298.15 K}) = - (155 \pm 20) \text{ kJ} \cdot \text{mol}^{-1},$$
$$\Delta_f H_m^\circ (\text{UCl}_3\text{, g, 298.15 K}) = - (523 \pm 20) \text{ kJ} \cdot \text{mol}^{-1}.$$

The uncertainties have been increased substantially from the purely statistical values to allow for the uncertainties in the thermal functions and the conversion from mass-spectrometric intensities to pressures.

9.4.3.1.1.2   UCl$_4$(g)

Haarland et al. [95HAA/MAR] have studied the electron diffraction and infrared spectroscopy of UCl$_4$(g) and have concluded that the molecule is an undistorted tetrahedron, with a U–Cl distance of 2.503 to 2.51 Å; we have taken the value of 2.506 Å suggested by Konings and Hildenbrand [98KON/HIL]. We also accept the vibration frequencies given by these authors [95HAA/MAR], [98KON/HIL]; the electronic states were taken to be those given by Hildenbrand et al. [85HIL/GUR], which are a simplification of the levels deduced by Gruber and Hecht [73GRU/HEC] from their spectroscopic measurements. The full data used are given in Table 9-9.

The calculated values for the entropy and heat capacity of UCl$_4$(g) at 298.15 K are:

$$S_m^\circ \text{(UCl}_4\text{, g, 298.15 K)} = (409.3 \pm 5.0) \text{ J} \cdot \text{K}^{-1} \cdot \text{mol}^{-1},$$
$$C_{p,m}^\circ \text{(UCl}_4\text{, g, 298.15 K)} = (103.5 \pm 3.0) \text{ J} \cdot \text{K}^{-1} \cdot \text{mol}^{-1},$$

and these are the selected values.

The vapour pressures of UCl$_4$(cr) and UCl$_4$(l) have been measured by a number of investigators, as summarised in Table 9-20.

Table 9-20: Vapour pressure data for the vaporisation of UCl$_4$(cr, l).

Reference	Method	T range (K)	$\Delta_{sub}H_m^\circ$ (298.15 K) kJ · mol^{-1}
[42JEN/AND]	Boiling point	873–1103	(202.4 ± 4.0)
[45DAV]	Tube flow	710–775	(204.2 ± 1.9)
[45DAV/STR]	Effusion	640–732	(204.5 ± 1.8)
[45WAG/GRA]	Boiling point	864–1064	(202.1 ± 0.8)
[46GRE2]	Diaphragm gauge	823–1044	(202.1 ± 2.4)
[48MUE]	Effusion	630–763	(205.1 ± 2.2)
[48THO/SCH]	Effusion	603–783	(203.8 ± 2.4)
[56SHC/VAS2]	Effusion	631–708	(204.0 ± 1.1)
[58JOH/BUT]	Transpiration	723–948	(196.9 ± 2.3) *
[58YOU/GRA]	Boiling point	864–1064	(202.0 ± 0.6)
[68CHO/CHU]	Effusion	648–748	(202.1 ± 1.5)
[75HIL/CUB]	Mass-spectrometric effusion	575–654	(203.6 ± 3.7)
[78SIN/PRA]	Transpiration	763–971	(202.7 ± 1.1)
[91COR/KON]	Transpiration	699–842	(205.5 ± 1.3)
[91HIL/LAU]	Torsion–effusion	588–674	(204.0 ± 0.9)

(*) Data not included in the mean

All these data have been analysed by the third-law method to give the results shown in the Table 9-20. There is good agreement between all the studies, which are plotted in Figure 4-3, with the exception of the transpiration data of Johnson *et al.* [58JOH/BUT]. The selected enthalpy of sublimation is therefore based on the mean of the concordant studies:

$$\Delta_{sub}H_m^\circ (UCl_4, 298.15\ K) = (203.4 \pm 4.0)\ kJ \cdot mol^{-1},$$

where the uncertainty has been increased substantially from the statistical 1.96σ value ($\pm 0.3\ kJ \cdot mol^{-1}$) to allow for uncertainties in the thermal functions of $UCl_4(g)$.

The derived enthalpy of formation of $UCl_4(g)$ is thus:

$$\Delta_f H_m^\circ (UCl_4, g, 298.15\ K) = -(815.4 \pm 4.7)\ kJ \cdot mol^{-1}.$$

As noted in section 9.4.3.1.3, Popov *et al.* [59POP/GAL] found a number of enthalpy effects just below the melting point (possibly related to reaction with the silica container), but gave an integrated enthalpy of fusion of *ca.* $49.8\ kJ \cdot mol^{-1}$, which should be regarded as the upper limit. The calculated vapour pressure of $UCl_4(l)$ using an enthalpy of fusion of $45\ kJ \cdot mol^{-1}$ is shown in Figure 9-3. While it is clear that a somewhat smaller enthalpy of fusion would give a slightly better fit to the liquid data, it is possible that there is a small contribution of dimer molecules to the total pressure near the boiling point, so we have preferred to use this value as the selected value of the enthalpy of fusion:

$$\Delta_{fus}H_m^\circ (UCl_4, cr, 863\ K) = (45.0 \pm 8.0)\ kJ \cdot mol^{-1}.$$

9.4.3.1.1.3 $UCl_5(g)$

The thermal functions are calculated from the molecular parameters given in Table 9-9 based on a square pyramid with a slight distortion to C_{2v} symmetry, as in Hildenbrand *et al.* [85HIL/GUR].

$$S_m^\circ (UCl_5, g, 298.15\ K) = (438.7 \pm 10.0)\ J \cdot K^{-1} \cdot mol^{-1},$$
$$C_{p,m}^\circ (UCl_5, g, 298.15\ K) = (123.6 \pm 5.0)\ J \cdot K^{-1} \cdot mol^{-1}.$$

There is only one study leading to the stability of the pentachloride, by Lau and Hildenbrand [84LAU/HIL], who measured mass-spectrometrically the equilibrium constant of the reaction from 613 to 920 K:

$$UCl_4(g) + \frac{1}{2} Cl_2(g) \rightleftharpoons UCl_5(g) \quad (9.23)$$

Second- and third-law analyses of these data give respectively:

$$\Delta_r H_m (9.23) = -(86.4 \pm 1.5)\ \text{and}\ -(72.6 \pm 3.9)\ kJ \cdot mol^{-1},$$

resulting in $\Delta_f H_m^\circ (UCl_5, 298.15\ K) = -(901.8 \pm 4.9)$ and $-(888.0 \pm 6.1)\ kJ \cdot mol^{-1}$. The selected value is:

$$\Delta_f H_m^\circ (UCl_5, g, 298.15\ K) = -(900 \pm 15)\ kJ \cdot mol^{-1}.$$

Figure 9-3: Vapour pressure of UCl$_4$(cr, l).

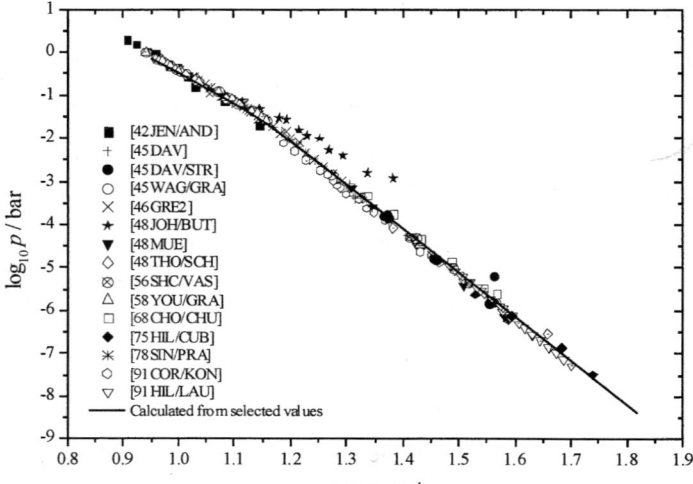

9.4.3.1.1.4 UCl$_6$(g)

Although the data for UCl$_6$(g) are independent of those for other gaseous uranium halides there are new experimental data on the vapour pressure to supplement the (early) studies considered by Grenthe et al. [92GRE/FUG].

The thermal functions of UCl$_6$(g) ideal gas were calculated from the estimated molecular parameters given in Table 9-9, using the rigid–rotator, harmonic oscillator approximation. The molecule is taken to be octahedral with O$_h$ symmetry.

$$S_m^\circ (\text{UCl}_6, \text{g}, 298.15\text{K}) = (438.0 \pm 5.0) \text{ J} \cdot \text{K}^{-1} \cdot \text{mol}^{-1},$$
$$C_{p,m}^\circ (\text{UCl}_6, \text{g}, 298.15\text{K}) = (147.2 \pm 3.0) \text{ J} \cdot \text{K}^{-1} \cdot \text{mol}^{-1}.$$

Sevast'yanov et al. [90SEV/ALI] have used slightly different molecular parameters for similar calculations to obtain values of S_m (UCl$_6$, g) at 285, 320 and 353 K which differ from those derived from the selected values by ca. 2.4 J · K^{-1} · mol^{-1}.

The heat capacities of the vapour have been refitted to the expression:

$$C_{p,m}(\text{UCl}_6, \text{g}, T) = 157.768 + 9.730 \cdot 10^{-5} \, T - 1.059 \cdot 10^{-8} \, T^2 - 9.4616 \cdot 10^5 \, T^{-2}$$
$$(\text{in J} \cdot \text{K}^{-1} \cdot \text{mol}^{-1}, 298.15 \text{ to } 1000 \text{ K}).$$

Table 9-21 shows the results of third-law analyses of the three measurements of the vapour pressure of UCl$_6$(cr) by Altman et al. [43ALT/LIP] (clicker gauge) and by Johnson et al. [58JOH/BUT] by transpiration (work done in 1944) and the more recent study by Knudsen effusion by Sevast'yanov et al. [90SEV/ALI].

Table 9-21: Vapour pressure data for the sublimation of $UCl_6(cr)$.

Reference	Method	T range (K)	$\Delta_{sub}H_m^\circ$ (298.15 K) kJ·mol^{-1}
[43ALT/LIP]	Clicker gauge	327–417	(76.8 ± 3.0)
[58JOH/BUT]	Transpiration	349–411	(81.4 ± 1.6)
[90SEV/ALI]	Mass-spectrometric effusion	285–353	(81.1 ± 2.1)

The higher vapour pressures measured by Altman *et al.* [43ALT/LIP] probably result from partial decomposition to UCl_5, which becomes noticeable around 420 K. The selected value for the enthalpy of sublimation is $\Delta_{sub}H_m^\circ$ (298.15 K) = (81.0 ± 4.0) kJ·mol^{-1}, where the uncertainty includes that in the thermal functions and other uncertainties. The corresponding enthalpy of formation is:

$$\Delta_f H_m^\circ (UCl_6, g, 298.15 K) = - (985.5 \pm 5.0) \text{ kJ·mol}^{-1},$$

which is the selected value.

9.4.3.1.1.5 $U_2Cl_8(g)$

We have not attempted to estimate the molecular parameters of $U_2Cl_8(g)$ in order to calculate the thermal functions. In [92GRE/FUG], values of the thermal functions from [84COR/KUB] were adopted, based on the procedure suggested by Kubaschewski [84KUB] for gaseous dimers from the monomer values. However, the monomer values (of $UCl_4(g)$) used were different from the selected values of [92GRE/FUG], which must lead to an inconsistency between the experimental and calculated equilibrium constants for the dimerisation reaction. We shall defer selection of any thermal function values at this time.

Binnewies and Schäfer [74BIN/SCH] give mass-spectrometric data from which the equilibrium constant of the dimerisation reaction:

$$2 \, UCl_4(g) \rightleftharpoons U_2Cl_8(g)$$

is calculated to be 4.2 bar^{-1} at 710 K and 1.25 bar^{-1} at 737 K. Given the very short temperature range, and the uncertainties in the ionisation cross-sections of $UCl_4(g)$ and $U_2Cl_8(g)$, we have not attempted to use these values to provide data for $U_2Cl_8(g)$ at 298.15 K.

9.4.3.1.1.6 $U_2Cl_{10}(g)$

As with $U_2Cl_8(g)$, we have not attempted to estimate the molecular parameters of $U_2Cl_8(g)$ in order to calculate the thermal functions.

Gruen and McBeth [69GRU/MCB] studied spectrometrically the reaction of $UCl_4(cr)$ with $Cl_2(g)$ (at pressures of *ca.* 3 bar) from 460 to 660 K. They were not able to define unambiguously the vapour species formed, but on the basis of a much greater consistency for the calculated equilibrium constants, suggested that the principal reaction was:

$$2\,UCl_4(cr) + Cl_2(g) \rightleftharpoons U_2Cl_{10}(g), \tag{9.24}$$

rather than the formation of monomeric $UCl_5(g)$ or higher chlorides. This conclusion is confirmed by calculations with our selected data, for which the pressures of $UCl_5(g)$ are less than 1% of the measured total pressures.

The equilibrium constants for the formation of $U_2Cl_{10}(g)$ correspond to $\Delta_r H_m^\circ((9.24), 560\,K) = (63.3 \pm 8.0)\,kJ \cdot mol^{-1}$. In order to reduce this to 298.15 K, we assume that $(H_m(560K) - H_m(298.15K))$ for reaction (9.24) is $-(9.0 \pm 4.0)\,kJ \cdot mol^{-1}$, slightly greater than that of the corresponding reaction with molybdenum $(-9.9\,kJ \cdot mol^{-1})$ [98CHA]. Thus $\Delta_r H_m^\circ((9.24), 298.15\,K) = (72.3 \pm 8.9)\,kJ \cdot mol^{-1}$, giving the selected value:

$$\Delta_f H_m^\circ(U_2Cl_{10}, g, 298.15\,K) = -(1965.3 \pm 10.2)\,kJ \cdot mol^{-1}.$$

9.4.3.1.1.7 $UO_2Cl_2(g)$

No new data have been reported for this species, but the data in [92GRE/FUG] have been slightly modified to take account of the recent spectroscopic results of $UCl_4(g)$.

The thermal functions were calculated from estimated data given in Table 9-9; the derived values of the standard entropy and heat capacity are:

$$S_m^\circ(UO_2Cl_2, g, 298.15\,K) = (373.4 \pm 6.0)\,J \cdot K^{-1} \cdot mol^{-1},$$

$$C_{p,m}^\circ(UO_2Cl_2, g, 298.15\,K) = (92.6 \pm 3.0)\,J \cdot K^{-1} \cdot mol^{-1}.$$

The enthalpy of formation was calculated from the investigation of the reaction:

$$U_3O_8(cr) + 3\,Cl_2(g) \rightleftharpoons 3\,UO_2Cl_2(g) + O_2(g) \tag{9.25}$$

by Cordfunke and Prins [74COR/PRI] from 1135 to 1327 K. Second- and third-law treatments of their results give $\Delta_r H_m(9.25) = (648.6 \pm 15.0)\,kJ \cdot mol^{-1}$ and $(665.0 \pm 3.1)\,kJ \cdot mol^{-1}$, giving for $\Delta_f H_m^\circ(UO_2Cl_2, g, 298.15\,K) = -(975.4 \pm 5.1)\,kJ \cdot mol^{-1}$ and $-(969.0 \pm 1.3)\,kJ \cdot mol^{-1}$, respectively. However, the uncertainty in the selected value has been increased substantially, since the reaction is almost certainly more complex than assumed, and $U_3O_8(cr)$ will certainly be sub-stoichiometric under the conditions of the experimentation. The selected value is the weighted mean:

$$\Delta_f H_m^\circ(UO_2Cl_2, g, 298.15\,K) = -(970.3 \pm 15.0)\,kJ \cdot mol^{-1}.$$

Kangro [63KAN] has also studied reaction (9.25) from 1273 to 1373 K, with quite different results from those of Cordunke and Prins [74COR/PRI]. However, the entropy change calculated from his results (ca. 91 $J \cdot K^{-1} \cdot mol^{-1}$) is too small to correspond to a reaction in which there is a net formation of one mole of gas, and those results have been discounted.

9.4.3.1.2 Aqueous uranium chlorides (V.4.2.1.2.)

9.4.3.1.2.1 Aqueous U(VI) chlorides

The chloride complexes of U(VI) are weak and it is therefore difficult to distinguish between complex formation and activity variations caused by the often large changes in the ligand concentration. However, Allen *et al.* [97ALL/BUC] showed in an EXAFS study that U(VI) chloride complexes were formed at very high chloride concentrations.

Choppin and Du [92CHO/DU] studied by solvent extraction the chloride complexation of U(VI) in acidic medium of high ionic strength (3.5 to 14.1 m NaClO$_4$) and reported equilibrium constants for the formation of UO$_2$Cl$^+$ and UO$_2$Cl$_2$(aq). These data are not retained by this review as discussed in Appendix A.

9.4.3.1.2.2 Aqueous U(IV) chlorides

Rai *et al.* [97RAI/FEL] interpreted solubility measurements of UO$_2$(am) in NaCl and MgCl$_2$ solutions using the Pitzer ion interaction model without taking complex formation into account. It is well known that it is virtually impossible to distinguish between ionic medium effects and complex formation in systems where weak complexes are formed (as in the case here) using thermodynamic data alone. Runde *et al.* [97RUN/NEU] made a spectroscopic study of Np(IV) chloride complexes in 5 M NaCl and found no evidence for the formation of inner-phere chloride complexes. On the other hand, Allen *et al.* [97ALL/BUC] found clear evidence for the formation of such complexes using EXAFS. This review finds the EXAFS evidence for the formation of chloride complexes of Np(IV) convincing. The new information on chloride complex formation of M^{4+} actinide ions is contradictory. It is quite clear that the chloride complexes are weak and that they therefore only can be identified in concentrated chloride solutions. In view of the chemical similarity between U(IV) and Np(IV), there are good reasons to assume that chloride complexes are formed also for U(IV).

The equilibrium constant for the complex UCl^{3+} given in [92GRE/FUG] is $\log_{10} \beta_1^\circ = (1.72 \pm 0.13)$. Allen *et al.* [97ALL/BUC] estimated $\log_{10} \beta_1 = -0.78$ for NpCl^{3+} at [Cl$^-$] = 6 M, [H$^+$] = 5·10^{-3} M.

This review keeps the value of $\log_{10} \beta_1^\circ$ (UCl^{3+}) selected by [92GRE/FUG] but points out that it could be too high. The ion interaction coefficient ε(UCl^{3+}, ClO$_4^-$) reported as (0.59 ± 0.10) kg·mol^{-1} on page 197 in [92GRE/FUG] is erroneous. Using the linear regression in Figure V.12 on page 199 in [92GRE/FUG] and auxiliary interaction coefficients, one obtains ε(UCl^{3+}, ClO$_4^-$) = (0.50 ± 0.10) kg·mol^{-1}. This value is very similar to that of other +3 ions; the interaction coefficients reported in Appendix B, Table B.4 has been changed accordingly.

The existence of a U(IV) oxychloride phase limiting solubility has been suggested by [91AGU/CAS] (see Appendix A).

9.4.3.1.3   Solid uranium chlorides (V.4.2.1.3)

9.4.3.1.3.1   UCl₃(cr)

There are a number of errors in the heat capacity data in the tables and text relating to UCl$_3$(cr) in [92GRE/FUG] as noted in [95GRE/PUI], so the original data have been re-examined and refitted.

The two sets of data on the low temperature heat capacities [44FER/PRA] (from the citation in [58MAC]), [89COR/KON] agree excellently. Although [89COR/KON] only give their data as a small graph, plus the actual values of $C_{p,m}^\circ$ and S_m° at 298.15 K, these have been preferred, since they almost certainly pertain to a purer sample. The data of [44FER/PRA] at 298.15 K are 0.5 J · K^{-1} · mol^{-1} higher than the preferred values, so their $C_{p,m}$ values from 270 – 320 K have been decreased by the same amount in the fitting. These data (four points) and the enthalpy drop measurements of Ginnings and Corruccini [47GIN/COR] (nine values of ($H_m(T) - H_m(273.15\ K)$) from 323.5 to 997.0 K) have been fitted simultaneously with a constraint of $C_{p,m}^\circ$ (298.15 K) = 102.52 J · K^{-1} · mol^{-1} [89COR/KON]. A four-term $C_{p,m}^\circ$ expression was preferred, since this avoids a shallow minimum in $C_{p,m}^\circ$ just above 300 K.

The final fit gives:

$$C_{p,m}(UCl_3, cr, T) = 106.967 - 2.0859 \cdot 10^{-2}\ T + 3.63895 \cdot 10^{-5}\ T^2$$
$$- 1.29994 \cdot 10^{5}\ T^{-2}\ \text{J} \cdot \text{K}^{-1} \cdot \text{mol}^{-1} \quad (298.15\ \text{to}\ 1115\ \text{K}).$$

Note that the value of the heat capacity at 298.15 K is incorrectly quoted in [92GRE/FUG]. The correct value is:

$$C_{p,m}^\circ (UCl_3, cr, 298.15\ K) = (102.52 \pm 0.50)\ \text{J} \cdot \text{K}^{-1} \cdot \text{mol}^{-1}.$$

The selected melting point and enthalpy of fusion are those measured by differential thermal analysis by Kovács et al. [96KOV/BOO]; the melting point agrees with those (1108 – 1115 K) found in work in the Manhattan Project (see references quoted in [96KOV/BOO]):

$$T_{fus} = (1115 \pm 2)\ \text{K},$$
$$\Delta_{fus} H_m^\circ (1115\ K) = (49.0 \pm 2.0)\ \text{kJ} \cdot \text{mol}^{-1}.$$

The selected heat capacity of UCl$_3$(l) in the temperature range from the melting point to 1200 K, is that suggested by [83FUG/PAR], although the source of this is not clear:

$$C_{p,m}(UCl_3, l, T) = (129.7 \pm 10.0)\ \text{J} \cdot \text{K}^{-1} \cdot \text{mol}^{-1}.$$
$$(1115\ \text{to}\ 1200\ \text{K})$$

9.4.3.1.3.2 UCl$_4$(cr)

The data for the heat capacity of UCl$_4$(cr) in [92GRE/FUG] are also unsatisfactory, in that the equation given in their Table III.3 does not reproduce the correct $C_{p,m}^\circ$ at 298.15 K, so as for UCl$_3$, we have re-examined and refitted the original data. There are three determinations: Ferguson *et al.* [44FER/PRA2], reported in detail in [51KAT/RAB], measured the heat capacity from 15 to 355 K, and Ginnings and Corruccini [47GIN/COR] gave enthalpy increments ($H_m(T) - H_m(273.15 \text{ K})$) from 323.4 to 699.9 K for a (probably) purer sample contained in a Nichrome capsule.

Later, Popov *et al.* [59POP/GAL] reported $C_{p,m}$ measurements from 381 to 920 K (thus extending into the liquid range). Since these experiments were apparently carried out using a silica vessel, it is not surprising that the resultant data are rather scattered. In particular, the data at temperatures close to the melting point showed irregularities, and the reported $C_{p,m}$ values below 450 K are appreciably lower than those from the other two studies (which agree well, but not perfectly). In fitting the data, we have given double weight to the most accurate data of [47GIN/COR], ignored the data of [59POP/GAL] below 450 K and imposed a constraint of $C_{p,m}^\circ$ (298.15 K) = (121.8 ± 0.4) J · K^{-1} · mol^{-1} [44FER/PRA2].

$$C_{p,m}(\text{UCl}_4, \text{cr}, T) = 116.362 + 3.10837 \cdot 10^{-2} \, T - 3.40402 \cdot 10^5 \, T^{-2} \text{ J} \cdot \text{K}^{-1} \cdot \text{mol}^{-1}$$
(298.15 to 863 K).

The resulting $C_{p,m}^\circ$ for UCl$_4$(cr) fits all the data used within their uncertainties.

Note that the selected value of $C_{p,m}^\circ$ (298.15 K) = 121.8 J · K^{-1} · mol^{-1} as given in [51KAT/RAB], is slightly lower than that suggested by [83FUG/PAR] and hence [92GRE/FUG]. This latter value (122.0 J · K^{-1} · mol^{-1}) seems to have suffered from a "J to cal to J" rounding error during conversion of the original data of [44FER/PRA2], [51KAT/RAB], which were given in joules. Similarly we select the original [44FER/PRA2] value of the standard entropy:

$$S_m^\circ (\text{UCl}_4, \text{cr}, 298.15 \text{ K}) = (197.2 \pm 0.8) \text{ J} \cdot \text{K}^{-1} \cdot \text{mol}^{-1},$$

rather than the perturbed value of 197.1 J · K^{-1} · mol^{-1} used in [92GRE/FUG].

The melting point of UCl$_4$(cr) is taken to be (863 ± 1) K, as given in early work summarised by Katz and Rabinowitch [51KAT/RAB] in agreement with the less precise value from the thermal study by Popov *et al.* [59POP/GAL]. The latter authors found a number of enthalpy effects just below the melting point (possibly related to reaction with the silica container), but gave an integrated enthalpy of fusion of *ca.* 49.8 kJ · mol^{-1}. However, the vapour pressure data suggest an appreciably smaller fusion enthalpy (see section 9.4.3.1.1). The selected value is:

$$\Delta_{fus}H_m^\circ (\text{UCl}_4, \text{cr}, 863 \text{ K}) = (45.0 \pm 8.0) \text{ kJ} \cdot \text{mol}^{-1},$$

which still gives excellent, if not optimal, agreement with the vapour pressure data - as shown in Figure 9-3 (section 9.4.3.1.1).

9.4.3.1.3.3   UCl$_6$(cr)

The value for S_m° (UCl$_6$, cr, 298.15 K) = (285.8 ± 1.7) J · K^{-1} · mol^{-1} selected by Grenthe et al. [92GRE/FUG] was based on the value of 68.3 cal·K^{-1} · mol^{-1} selected by Fuger et al. [83FUG/PAR], in turn based on the original low temperature heat capacity measurements by Ferguson and Rand [45FER/RAN]. However, the latter were actually reported as 285.54 J · K^{-1} · mol^{-1}.

We therefore select:

$$S_m^\circ (UCl_6, cr, 298.15 K) = (285.5 \pm 1.7) \text{ J} \cdot \text{K}^{-1} \cdot \text{mol}^{-1},$$

for the standard entropy of UCl$_6$(cr), where the relatively high uncertainty arises from the presence of 6.3% of UCl$_4$(cr) in the calorimetric sample.

The heat capacity at 298.15 K has not suffered from the same "J to cal to J" conversions and remains unchanged:

$$C_{p,m}^\circ (UCl_6, cr, 298.15 K) = (175.7 \pm 4.2) \text{ J} \cdot \text{K}^{-1} \cdot \text{mol}^{-1}.$$

9.4.3.2   Uranium hypochlorites

9.4.3.2.1   Aqueous U(VI) hypochlorites

Kim et al. [94KIM/CHO] reported an increase (up to twofold) in the solubility of schoepite with increasing ClO$^-$ concentration in 0.1 NaCl (pH = 5 to 10), but were unable to decide if this was a result of the formation of hypochlorite complex(es) formed from the hydrolysed species, $(UO_2)_2(OH)_2^{2+}$ or $(UO_2)_3(OH)_5^+$, or not.

9.4.4   Bromine compounds and complexes (V.4.3)

9.4.4.1   Gaseous uranium bromides (V.4.3.1.1)

9.4.4.1.1   UBr(g), UBr$_2$(g) and UBr$_3$(g)

These species are considered together since their data are linked by six experimental studies, mainly involving mass-spectrometry, and thus involving some uncertainties in absolute pressures.

The thermal functions of UBr(g), UBr$_2$(g) and UBr$_3$(g) were calculated assuming the molecular parameters shown in Table 9-9, which are those estimated by Hildenbrand et al. [85HIL/GUR], except that higher electronic energy levels (assumed to be the same as for the corresponding chlorides) have been included. These calculations give:

$$S_m^\circ (UBr, g, 298.15 K) = (278.5 \pm 3.0) \text{ J} \cdot \text{K}^{-1} \cdot \text{mol}^{-1}$$
$$C_{p,m}^\circ (UBr, g, 298.15 K) = (44.5 \pm 3.0) \text{ J} \cdot \text{K}^{-1} \cdot \text{mol}^{-1},$$

$$S_m^\circ (UBr_2, g, 298.15 K) = (359.7 \pm 10.0) \text{ J} \cdot \text{K}^{-1} \cdot \text{mol}^{-1}$$
$$C_{p,m}^\circ (UBr_2, g, 298.15 K) = (61.4 \pm 5.0) \text{ J} \cdot \text{K}^{-1} \cdot \text{mol}^{-1},$$

S_m° (UBr$_3$, g, 298.15 K) = (403.0 ± 15.0) J · K^{-1} · mol^{-1}
$C_{p,m}^\circ$ (UBr$_3$, g, 298.15 K) = (85.2 ± 5.0) J · K^{-1} · mol^{-1}.

Lau and Hildenbrand [87LAU/HIL] have studied four equilibria between UBr(g), UBr$_2$(g), UBr$_3$(g) and UBr$_4$(g), which are summarised in Table 9-22, where the enthalpies of the given reactions at 298.15 K derived from both second- and third-law analyses are given. In addition, Altman [43ALT] and Webster [43WEB] measured the vapour pressure of UBr$_3$(cr, l) using, respectively, effusion and transpiration techniques. However, such studies cannot give reliable data on the pressure of UBr$_3$(g), since on heating, UBr$_3$(cr) probably disproportionates to U(cr) and UBr$_4$(g). [51KAT/RAB] review early studies which indicate the formation of UBr$_4$(g) on heating UBr$_3$(cr) above 1173 K, with a concomitant increase of the U/Br ratio in the solid, suggesting that UBr$_3$(cr, l) disproportionates to U(cr) and UBr$_4$(g). This accounts for the fact that the pressures measured by these authors were appreciably higher than those calculated for the simple vaporisation of UBr$_3$(cr, l) with the selected data.

Table 9-22: Enthalpies of the reactions involving UBr(g), UBr$_2$(g), UBr$_3$(g) and UBr$_4$(g).

Reference	Reaction	Number of points	$\Delta_r H_m^\circ$ (298.15 K) kJ · mol^{-1}	
			Second-law	Third-law
[87LAU/HIL]	U(g) + Br(g) ⇌ UBr(g)	15	− 387.7	− 395.5
[87LAU/HIL]	UBr$_2$(g) + U(g) ⇌ 2UBr(g)	7	32.0	− 6.4
[87LAU/HIL]	UBr(g) + UBr$_3$(g) ⇌ 2 UBr$_2$(g)	10	10.3	44.7
[87LAU/HIL]	UBr$_2$(g) + UBr$_4$(g) ⇌ 2 UBr$_3$(g)	8	− 76.1	− 98.4

Of necessity, the pressures for most of these reactions were derived from mass-spectrometric intensities, leading to uncertainties in the absolute pressures. However, the agreement between the second- and third-law analyses is rather poor.

With these data, we have four quasi–independent measurements to define the three enthalpies of formation of UBr(g), UBr$_2$(g), UBr$_3$(g), as shown in Table 9-23. We have used a least-squares analysis to find the optimal solutions to the two over-determined sets of linear equations, using either the second-law or the third-law enthalpies, with the results shown in Table 9-24.

Table 9-23: Sums of various enthalpies involving UBr(g), UBr$_2$(g) and UBr$_3$(g).

Reference	Expression	$\Delta_f H_m^\circ$ (298.15 K) kJ · mol^{-1}	
		Second-law	Third-law
[87LAU/HIL]	UBr(g)	257.1	249.4
[87LAU/HIL]	2 UBr(g) − UBr$_2$(g)	565.0	526.6
[87LAU/HIL]	− UBr(g) + 2 UBr$_2$(g) − UBr$_3$(g)	10.3	44.7
[84GOR/SMI]	− UBr$_2$(g) + 2 UBr$_3$(g)	− 681.7	− 704.0

The selected values are:

$$\Delta_f H_m^\circ \text{ (UBr, g, 298.15 K)} = (245 \pm 15) \text{ kJ} \cdot \text{mol}^{-1},$$
$$\Delta_f H_m^\circ \text{ (UBr}_2\text{, g, 298.15 K)} = -(40 \pm 15) \text{ kJ} \cdot \text{mol}^{-1},$$
$$\Delta_f H_m^\circ \text{ (UBr}_3\text{, g, 298.15 K)} = -(371 \pm 20) \text{ kJ} \cdot \text{mol}^{-1},$$

where the uncertainties have been estimated here. It will be seen that the new analysis of the experimental data for these species yields somewhat different values for the enthalpies of formation of UBr$_2$(g) and UBr$_3$(g) from those selected by Grenthe et al. [92GRE/FUG], although well within the substantial uncertainty limits.

Table 9-24: Optimised enthalpies of formation, $\Delta_f H_m^\circ$ (298.15 K) kJ·mol^{-1}.

Species	Second-law	Third-law	Selected	[92GRE/FUG]
UBr(g)	257.8	245.1	(245 ± 15)	(247 ± 17)
UBr$_2$(g)	− 48.8	− 39.7	− (40 ± 15)	− (31 ± 25)
UBr$_3$(g)	− 365.3	− 371.3	− (371 ± 20)	− (364 ± 37)

9.4.4.1.2 UBr$_4$(g)

The UBr$_4$(g) molecule is assumed to have T$_d$ symmetry, as for UF$_4$(g) and UCl$_4$(g), with a U–Br distance of 2.693 Å (see Konings and Hildenbrand [98KON/HIL]). We also accept the vibration frequencies given by these authors; the electronic states were assumed to be the same as those given by Hildenbrand et al. [85HIL/GUR] for UCl$_4$(g).

The calculated values for the entropy and heat capacity of UBr$_4$(g) at 298.15 K are:

$$S_m^\circ \text{ (UBr}_4\text{, g, 298.15 K)} = (451.9 \pm 5.0) \text{ J} \cdot \text{K}^{-1} \cdot \text{mol}^{-1},$$
$$C_{p,m}^\circ \text{ (UBr}_4\text{, g, 298.15 K)} = (106.9 \pm 3.0) \text{ J} \cdot \text{K}^{-1} \cdot \text{mol}^{-1},$$

and these are the selected values.

There are several measurements of the vapour pressures of UBr$_4$(cr) and UBr$_4$(l), with excellent agreement. The early data of Thompson and Schelberg, and Nottorf and Powell are summarised by Katz and Rabinowitch [51KAT/RAB], and those by Gregory [46GRE3] are given in the report by Mueller [48MUE]. Subsequently Prasad et al. [79PRA/NAG] studied the vaporisation of (principally) the liquid, while Hildenbrand and Lau [91HIL/LAU2] used a torsion effusion technique to study the sublimation. The latter study is particularly valuable, since it shows that the molar mass of the vaporising species corresponds closely to that of UBr$_4$ in the range of their experiments. Since the enthalpy of fusion of UBr$_4$(cr) is not well-defined, only the vapour pressures of the solid have been used to derive, by a third-law analysis, the enthalpy of sublimation (Table 9-25) .

Table 9-25: $\Delta_{sub}H_m^o$ (298.15K) of UBr$_4$ (cr) from data on UBr$_4$(cr) only.

Reference	Method	T range (K)	$\Delta_{sub}H_m^o$ (298.15 K), kJ·mol^{-1}
[42THO/SCH]	Effusion	573–723	(194.8 ± 3.1)
[44NOT/POW]	Transpiration	723–792	(198.7 ± 1.6)
[79PRA/NAG]	Transpiration	759–792	(196.1 ± 0.4) (from v.p equation)
[91HIL/LAU]	Torsion–effusion	579–693	(196.8 ± 0.3)

The selected enthalpy of sublimation is:

$$\Delta_{sub}H_m^o(\text{UBr}_4, 298.15\text{ K}) = (196.5 \pm 4.0) \text{ kJ} \cdot \text{mol}^{-1},$$

where the uncertainty has been increased substantially to allow for uncertainties in the thermal functions of UBr$_4$(g).

Figure 9-4: Vapour pressure of UBr$_4$(cr, l).

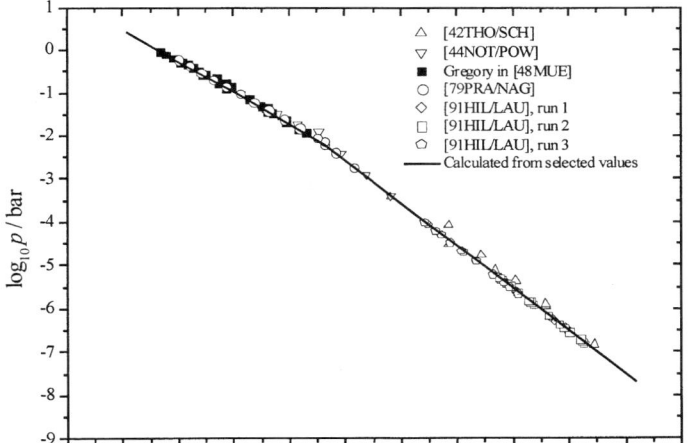

The derived enthalpy of formation of UBr$_4$(g) is thus:

$$\Delta_f H_m^o(\text{UBr}_4, \text{g}, 298.15\text{ K}) = -(605.6 \pm 4.7) \text{ kJ} \cdot \text{mol}^{-1}.$$

The enthalpy of fusion of UBr$_4$(cr) can then be derived by fitting the vapour pressure of the liquid with a constrained T_{fus} = 792 K. The best value to fit the vapour pressure of the liquid (see Figure 9-4) gives:

$$\Delta_{fus}H_m^o(\text{UBr}_4, 792\text{ K}) = (36.0 \pm 5.0) \text{ kJ} \cdot \text{mol}^{-1},$$

where the uncertainty includes the possibility of a small contribution from dimer molecules at higher pressures.

9.4.4.1.3   UBr$_5$(g)

The thermal functions are calculated from the molecular parameters given in Table 9-9, based on a square pyramidal structure with a slight distortion to C$_{2v}$ symmetry and a r(U-Br) distance slightly less than that in UBr$_4$(g). The vibration frequencies are those suggested by Lau and Hildenbrand [87LAU/HIL]. As for the other gaseous bromides, the electronic levels have been assumed to be the same as those for the corresponding chloride. The selected values are thus:

$$S_m^\circ (\text{UBr}_5, \text{g}, 298.15 \text{ K}) = (498.7 \pm 10.0) \text{ J} \cdot \text{K}^{-1} \cdot \text{mol}^{-1},$$
$$C_{p,m}^\circ (\text{UBr}_5, \text{g}, 298.15 \text{ K}) = (129.0 \pm 5.0) \text{ J} \cdot \text{K}^{-1} \cdot \text{mol}^{-1}.$$

There is only one study leading to the stability of the pentabromide, by Lau and Hildenbrand [87LAU/HIL], who measured mass-spectrometrically the equilibrium constant of the reaction:

$$\text{UBr}_4 (\text{g}) + \tfrac{1}{2}\text{Br}_2 (\text{g}) \rightleftharpoons \text{UBr}_5 (\text{g}) \tag{9.26}$$

from 613 to 810 K. Second- and third-law analyses of these data give respectively $\Delta_r H_m^\circ ((9.26), 298.15 \text{ K}) = -(69.6 \pm 1.5)$ and $-(57.8 \pm 2.7)$ kJ·mol^{-1}. In order to maintain consistency with the thermal functions used, the third-law value is selected giving $\Delta_f H_m^\circ (\text{UBr}_5, \text{g}, 298.15 \text{ K}) = -(647.9 \pm 8.9)$ kJ·mol^{-1}. This, after rounding and increasing the uncertainty gives the selected value:

$$\Delta_f H_m^\circ (\text{UBr}_5, \text{g}, 298.15 \text{ K}) = -(648 \pm 15) \text{ kJ} \cdot \text{mol}^{-1}.$$

9.4.4.2   Solid uranium bromides (V.4.3.1.3)

9.4.4.2.1   UBr$_4$ (cr)

The re-assessed data for the vapour pressure of UBr$_4$(cr, l) have been used to provide a better value of the enthalpy of fusion (see section 9.4.4.1.2):

$$\Delta_{\text{fus}} H_m^\circ (\text{UBr}_4, 792 \text{ K}) = (36.0 \pm 5.0) \text{ kJ} \cdot \text{mol}^{-1}.$$

9.4.5   Iodides compounds and complexes (V.4.4.1)

9.4.5.1   Gaseous uranium iodides (V.4.4.1.1)

There are few reliable data for the thermodynamic properties of the gaseous iodides.

9.4.5.1.1   UI(g), UI$_2$(g) and UI$_3$(g)

The thermal functions are calculated from the data in Table 9-26. These are essentially those selected by Hildenbrand et al. [85HIL/GUR], with the addition of an anharmonicity term for UI(g), and the inclusion of higher electronic levels (assumed to be the same as those for corresponding chlorides). The calculated values at 298.15 K are:

$S_m^°$ (UI, g, 298.15 K) = (286.5 ± 5.0) J·K^{-1}·mol^{-1}
$C_{p,m}^°$ (UI, g, 298.15 K) = (44.8 ± 5.0) J·K^{-1}·mol^{-1},

$S_m^°$ (UI$_2$, g, 298.15 K) = (376.5 ± 10.0) J·K^{-1}·mol^{-1}
$C_{p,m}^°$ (UI$_2$, g, 298.15 K) = (61.9 ± 5.0) J·K^{-1}·mol^{-1},

$S_m^°$ (UI$_3$, g, 298.15 K) = (431.2 ± 10.0) J·K^{-1}·mol^{-1}
$C_{p,m}^°$ (UI$_3$, g, 298.15 K) = (86.0 ± 5.0) J·K^{-1}·mol^{-1}.

As noted by [92GRE/FUG], all the data for these species are estimated by Hildenbrand et al. [85HIL/GUR] from $\Delta_f H_m^°$ (UI$_4$, g, 298.15 K), assuming the bond energies of these lower iodides follow the same pattern as the chlorides and bromides [84LAU/HIL]. We have rescaled this calculation, using a revised value for $\Delta_f H_m^°$ (UI$_4$, g, 298.15 K) from the next section. The revised values for the enthalpies of formation are:

$\Delta_f H_m^°$ (UI, g, 298.15 K) = (342 ± 25) kJ·mol^{-1},

$\Delta_f H_m^°$ (UI$_2$, g, 298.15 K) = (103 ± 25) kJ·mol^{-1},

$\Delta_f H_m^°$ (UI$_3$, g, 298.15 K) = − (137 ± 25) kJ·mol^{-1},

where the uncertainties are those selected by [92GRE/FUG].

9.4.5.1.2 UI$_4$(g)

The UI$_4$(g) molecule is assumed to have T$_d$ symmetry, as for UF$_4$(g) and UCl$_4$(g) with a U–I distance of 2.85 Å (see Konings and Hildenbrand [98KON/HIL]). We also accept the vibration frequencies given by these authors; the electronic states were assumed to be the same as those given by Hildenbrand et al. [85HIL/GUR] for UCl$_4$(g).

The calculated values for the entropy and heat capacity of UI$_4$(g) at 298.15 K are:

$S_m^°$ (UI$_4$, g, 298.15 K) = (499.1 ± 8.0) J·K^{-1}·mol^{-1},
$C_{p,m}^°$ (UI$_4$, g, 298.15 K) = (108.8 ± 4.0) J·K^{-1}·mol^{-1},

and these are the selected values.

Following [92GRE/FUG] we prefer the effusion data of Schelberg and Thompson [42SCH/THO] for the vapour pressure of UI$_4$(cr), since these were measured at lower temperatures (573 to 683 K) than those of Gregory [46GRE] for the liquid (823 to 837 K), and will thus suffer less from the problems of dissociation to UI$_3$ and I$_2$(g). A third-law analysis of their data gives: $\Delta_{sub} H_m^°$ (UI$_4$, g, 298.15 K) = (213.3 ± 5.0) kJ·mol^{-1}, and thus the selected value for the enthalpy of formation:

$\Delta_f H_m^°$ (UI$_4$, g, 298.15 K) = − (305.0 ± 5.7) kJ·mol^{-1},

where the uncertainty has been increased from the statistical value (3.0 kJ·mol^{-1}) to allow for the uncertainty in the thermal functions of UI$_4$(g) and possible dissociation.

9.4.5.2 Solid uranium iodides (V.4.4.1.3)

There is a small error in $\Delta_f H_m^\circ$ (UI$_3$, cr, 298.15 K) and $\Delta_f H_m^\circ$ (UI$_4$, cr, 298.15 K), in [92GRE/FUG] who adopted the values unchanged from [83FUG/PAR]. However, these were based on the 1977 CODATA Key Value for $\Delta_f H_m^\circ$ (I$^-$, 298.15 K), which was 0.12 kJ·mol^{-1} more negative than the final value [89COX/WAG] used in the [92GRE/FUG] assessment (and the current revision). The corresponding values for the enthalpy of formation of HI(aq) in 1.0 and 6.0 M HCl solutions are given in Table 9-26.

Table 9-26: Enthalpy of formation of HI(aq).

Reference	Species	Property	kJ·mol^{-1}
[89COX/WAG]	HI (aq)	$\Delta_f H_m^\circ$	− (56.780 ± 0.050)
interpolated	HI (1.0 M HCl)	$\Delta_f H_m^\circ$ (partial)	− (55.379 ± 0.100)
interpolated	HI (6.0 M HCl)	$\Delta_f H_m^\circ$ (partial)	− (46.75 ± 0.15)

The revised values of the enthalpy of formation of UI$_3$(cr) and UI$_4$(cr) are:

$\Delta_f H_m^\circ$ (UI$_3$, cr, 298.15 K) = − (466.9 ± 4.2) kJ·mol^{-1}
$\Delta_f H_m^\circ$ (UI$_4$, cr, 298.15 K) = − (518.3 ± 2.8) kJ·mol^{-1}.

These are both more positive than the corresponding values in [92GRE/FUG] by 0.5 kJ·mol^{-1}, since $\Delta_f H_m^\circ$ (UI$_3$, cr, 298.15 K) is based predominantly on that of UI$_4$(cr), using the difference in the enthalpies of solution of the two compounds in 12 M HCl, as discussed in [83FUG/PAR].

9.5 Uranium group 16 (chalcogens) compounds and complexes (V.5)

9.5.1 Aqueous uranium sulphate (V.5.1.3.1)

9.5.1.1 Aqueous U(VI) sulphates

9.5.1.1.1 Binary complexes

Geipel et al. [96GEI/BRA] performed fluorescence spectroscopic measurements on solutions with I = 0.2, 0.5 and 1 M (NaClO$_4$ + HClO$_4$) of different pH and U concentrations and identified the species UO$_2$SO$_4$(aq), UO$_2$(SO$_4$)$_2^{2-}$ and UO$_2$(SO$_4$)$_3^{4-}$. They derived equilibrium constants from the spectroscopic data, which are given in Table 9-27. Comarmond and Brown [2000COM/BRO] determined equilibrium constants for the two first binary U(VI)–sulphate complexes using potentiometry and the values of $\log_{10} \beta_1$ and $\log_{10} \beta_2$ are also given in Table 9-27. These values extrapolated to zero ionic strength by this review are: $\log_{10} \beta_1^\circ$ = (2.81 ± 0.06) and $\log_{10} \beta_2^\circ$ = (3.78 ± 0.13). They are only in fair agreement with the selected values in [92GRE/FUG], $\log_{10} \beta_1^\circ$ = (3.15 ± 0.02) and $\log_{10} \beta_2^\circ$ = (4.14 ± 0.07), respectively, and for reasons detailed in Appendix A, they have not been used to select the recommended equilibrium constants.

The $\log_{10} \beta_r^\circ$ values at zero ionic strength are calculated by Geipel et al. according to SIT theory using $\varepsilon(UO_2(SO_4)_3^{4-}, Na^+) = -0.24$ kg·mol^{-1} estimated from $P_2O_7^{4-}$, Na$^+$ as an analogue. The two first values are somewhat higher than those selected in [92GRE/FUG] (see above) with larger uncertainty. The weighted averaged values of the constants given by [92GRE/FUG] and [96GEI/BRA] are $\log_{10} \beta_1^\circ = (3.15 \pm 0.02)$ and $\log_{10} \beta_2^\circ = (4.15 \pm 0.06)$. These values do not cause noticeable changes in the values selected by [92GRE/FUG] which are retained by this review.

This review selects,

$$\log_{10} \beta_3^\circ = (3.02 \pm 0.38),$$

which is the only value available and recalculated in Appendix A, review of [96GEI/BRA].

Some indications on saturated solutions of dioxouranium(VI) sulphate are given in [97DEM/SER].

Table 9-27: Equilibrium constants[1] for binary complexes $UO_2(SO_4)_r^{2-2r}$
- from spectroscopic measurements [96GEI/BRA].

Species	$\log_{10} \beta_r$, $I = 0.2$ M	$\log_{10} \beta_r^{\circ}$ [a]
UO_2SO_4 (aq)	(2.42 ± 0.14)	(3.35 ± 0.15)
$UO_2(SO_4)_2^{2-}$	(3.30 ± 0.17)	(4.21 ± 0.17)
	$\log_{10} \beta_r$, $I = 1$ M	$\log_{10} \beta_r^\circ$
UO_2SO_4 (aq)	(1.88 ± 0.27)	(3.33 ± 0.29)
$UO_2(SO_4)_2^{2-}$	(2.9 ± 0.4)	(4.29 ± 0.45)
$UO_2(SO_4)_3^{4-}$	(3.2 ± 0.25)	(2.73 ± 0.33)

- from potentiometric measurements [2000COM/BRO].

Species	$\log_{10} \beta_r$, Sulphate concentration, mol·kg^{-1}			
	0	0.1004	1.027	1.566
UO_2SO_4 (aq)	(2.81 ± 0.06)	(1.92 ± 0.03)	(1.06 ± 0.19)	(1.9 ± 0.19)[b]
$UO_2(SO_4)_2^{2-}$	(3.78 ± 0.13)	(2.90 ± 0.08)	(2.06 ± 0.32)	(2.29 ± 0.24)[b]

(a) r is the number of ligands in the complex $UO_2(SO_4)_r^{2-2r}$.
(b) calculated by [2000COM/BRO] from [2000MOL/REI2] and [93GRE/LAG].

---

[1] During the Peer Review process, the following sets of unpublished data by L. Ciavatta et al. (submitted to Ann. Chim. (Rome)), related to 3 M NaClO$_4$ at 298.15 K were given to the reviewers: $\log_{10} \beta_1 = (1.77 \pm 0.01)$ and $\log_{10} \beta_2 = (2.99 \pm 0.02)$, $\log_{10} \beta_1 = (1.83 \pm 0.06)$ and $\log_{10} \beta_2 = (2.83 \pm 0.10)$, $\log_{10} \beta_1 = (1.82 \pm 0.07)$ and $\log_{10} \beta_2 = (2.95 \pm 0.05)$, obtained from potentiometry, spectrophotometry and solubility of UO$_2$(IO$_3$)$_2$, respectively, which are consistent with those listed in this Table.

9.5.1.1.2 Ternary hydroxide-sulphate complexes

Ternary hydroxo sulphato complexes, $(UO_2)_m(OH)_n(SO_4)_r^{2m-n-2r}$, have been identified by Grenthe and Lagerman [93GRE/LAG] by combining new potentiometric titrations in a 0.5 M Na_2SO_4 + 2 M $NaClO_4$ with previous data in 1.5 M Na_2SO_4 [61PET] at 298.15 K. The first step in the analysis is a determination of the conditional equilibrium constants in the two media that provides information on the stoichiometric coefficients m and n. In the second step the stoichiometric coefficient r was determined from the sulphate concentration dependence of these constants. This is not straightforward and requires estimates of the activity coefficients of reactants and products in the two media investigated. Grenthe and Lagerman [93GRE/LAG] used the SIT model for this purpose, but the resulting equilibrium constants for the reactions:

$$m\, UO_2^{2+} + n\, H_2O + r\, SO_4^{2-} \rightleftharpoons (UO_2)_m(OH)_n(SO_4)_r^{2m-n-2r} + n\, H^+$$

are uncertain because of the approximate nature of the method. Grenthe and Lagerman [93GRE/LAG] propose the following m:n:r set for the ternary complexes: 2:2:2, 3:4:3, 3:4:4 and 5:8:6 (see Appendix A).

Moll et al. [2000MOL/REI2] repeated the potentiometric measurements at the same conditions as in [93GRE/LAG] and [61PET], but used data that covered a larger range of $-\log_{10}[H^+]$ and total concentration of U(VI). These data confirm the previous conditional equilibrium constants. Moll et al. tested the chemical model proposed by Comarmond and Brown [2000COM/BRO] and found that inclusion of the species with $m = 4$, $n = 7$, as suggested by Comarmond and Brown, gave a slightly better fit of the experimental data. From the EXAFS data Moll et al. [2000MOL/REI2] conclude that the sulphate ions form bidentate complexes both in the binary U(VI)-sulphate complexes and in the ternary complexes. There is no evidence for bridging sulphate in the systems as suggested in [93GRE/LAG]. Grenthe and Lagerman give tentative SIT interaction coefficients for the media considered, but they indicate that these coefficients should be used with caution.

Comarmond and Brown [2000COM/BRO] added new additional potentiometric data in 0.1 and 1 M Na_2SO_4 to those from the two previous studies. From the experimental data they proposed that species with m:n:r 1:0:1, 1:0:2, 2:2:2, 3:4:3, 4:7:4 and 5:8:4 were formed. The $\beta_{m,n,r}$ and interaction coefficients, ε, and the extrapolated values $\beta^{\circ}_{m,n,r}$ from SIT, are given in Table 9-28 together with the data from [93GRE/LAG] and [61PET]. The SIT analysis in [2000COM/BRO] is not identical with the one in [93GRE/LAG]. This difference is reflected both in the different sulphate stoichiometry proposed for the (8,5) complex and in the equilibrium constants. This difference may be used as an estimate of the model/extrapolation uncertainty. Because of this, no equilibrium constants have been retained in the present review. However, the stoichiometry and equilibrium constants proposed in the experimental studies discussed here may be used as guidance.

Table 9-28: Equilibrium constants[1] for ternary complexes, $\log_{10} {}^*\beta_{m,n,r}$, and interaction coefficient $\varepsilon((UO_2)_m(OH)_n(SO_4)_r^{2m-n-2r}, Na^+)$ in kg·mol^{-1}.

Species	ε	$\log_{10} {}^*\beta_{m,n,r}$		
	kg·mol^{-1}	mol·kg^{-1} [a]		
	0	0.1004	1.027	1.566
$(UO_2)_2(OH)_2(SO_4)_2^{2-}$	$-(0.14 \pm 0.22)$	$-(0.64 \pm 0.01)$	$-(2.17 \pm 0.15)$	$-(3.02 \pm 0.68)$ $-(3.20 \pm 0.82)$[c]
$(UO_2)_3(OH)_4(SO_4)_3^{4-}$	(0.6 ± 0.6)	$-(5.9 \pm 0.2)$	$-(6.60 \pm 0.17)$	$-(7.18 \pm 0.70)$ $-(9.01 \pm 0.87)$[c]
$(UO_2)_4(OH)_7(SO_4)_4^{7-}$	(2.8 ± 0.7)	$-(18.9 \pm 0.2)$	$-(15.85 \pm 0.28)$	$-(18.4 \pm 1.3)$ $-(22.6 \pm 1.2)$[c]
$(UO_2)_5(OH)_8(SO_4)_4^{6-}$	(1.1 ± 0.5)	$-(18.7 \pm 0.1)$	$-(17.69 \pm 0.20)$	$-(19.61 \pm 0.73)$ $-(20.14 \pm 0.88)$

Species	Sulphate concentration, M [b]	
	1.500 M Na$_2$SO$_4$	0.500 M Na$_2$SO$_4$ + 2.00 M NaClO$_4$
$(UO_2)_2(OH)_2(SO_4)_2^{2-}$	$-(3.26 \pm 0.46)$	$-(2.73 \pm 0.09)$
$(UO_2)_3(OH)_4(SO_4)_3^{4-}$	$-(8.64 \pm 0.69)$	$-(8.15 \pm 0.17)$
$(UO_2)_3(OH)_4(SO_4)_4^{6-}$	$-(8.81 \pm 0.69)$	$-(7.84 \pm 0.17)$
$(UO_2)_5(OH)_8(SO_4)_6^{10-}$	$-(19.79 \pm 1.20)$	$-(18.53 \pm 0.25)$

(a) from [2000COM/BRO].
(b) from [93GRE/LAG].
(c) calculated by [2000COM/BRO] from [2000MOL/REI2] and [93GRE/LAG].

9.5.2 Aqueous uranium selenate (V.5.2.3.1)

The data reported by Kumok and Batyreva [90KUM/BAT], and Serezhkina and Serezhkin [94SER/SER] on the solubility of $UO_2SeO_4 \cdot 4H_2O$ do not allow derivation of any reliable thermodynamic data. It is the same for the solubility data reported by Tatarinova *et al.* [89TAT/SER] for some seleniate compounds (see Appendix A). Lubal and Havel [97LUB/HAV] identified the species UO_2SeO_4 (aq) and $UO_2(SeO_4)_2^{2-}$ by spectrophotometry in solutions at $I = 3$ M (NaClO$_4$ + NaSeO$_4$) at 298.15 K and reported $\log_{10} \beta_1 = (1.576 \pm 0.016)$ and $\log_{10} \beta_2 = (2.423 \pm 0.013)$. They also carried out potentiometric measurements at variable low ionic strengths, less than 10^{-2} M, from which $\log_{10} \beta_1 = (2.64 \pm 0.01)$ can be calculated. If it is assumed that this value is equal to $\log_{10} \beta_1^\circ$ it is possible to calculate a SIT interaction coefficient, $\varepsilon(SeO_4^{2-}, Na^+) = -0.21$ kg·mol^{-1}, in fair agreement with the corresponding interaction coefficient for sulphate

---

[1] During the Peer Review process, the following sets of unpublished data by L. Ciavatta *et al.* (submitted to *Ann. Chim.* (Rome)) related to 3 M NaClO$_4$ at 298.15 K were given to the reviewers: $\log_{10} {}^*\beta_{2,2,2} = -(2.94 \pm 0.03)$, $\log_{10} {}^*\beta_{3,4,1} = -(9.82 \pm 0.06)$, $\log_{10} {}^*\beta_{2,1,1} = -(0.30 \pm 0.09)$, $\log_{10} {}^*\beta_{2,1,2} = (1.09 \pm 0.09)$, $\log_{10} {}^*\beta_{3,5,1} = -(15.04 \pm 0.09)$, $\log_{10} {}^*\beta_{4,6,2} = -(14.40 \pm 0.06)$. Only the constant $\log_{10} \beta_{2,2,2}$ is consistent with those listed in this table. The stoichiometry of the other complexes in the study of Ciavatta *et al.* are not consistent with those in the table, possibly because of different sulphate concentrations.

as discussed in Appendix A. For reasons given in Appendix A this review does not select any equilibrium constants for selenate complexes.

9.5.3 Tellurium compounds (V.5.3)

9.5.3.1 Uranium tellurides (V.5.3.1)

9.5.3.1.1 Binary uranium tellurides (V.5.3.1.1)

No data for the uranium tellurides are selected in [92GRE/FUG], nor here, but the cautious use of the data assessed by [84GRO/DRO] is recommended. Similar caution probably applies to the use of the heat capacity value of:

$$C_{p,m}^\circ (\text{UTe, cr, 298.15 K}) = (84 \pm 8) \text{ J} \cdot \text{K}^{-1} \cdot \text{mol}^{-1}$$

derived from a small graph of the low–temperature heat capacity measurements given by Ochiai et al. [94OCH/SUZ]. The sample, characterised only by X–ray diffraction, showed a pronounced ferromagnetic peak in $C_{p,m}$, at a temperature different by several K from earlier studies.

9.5.3.2 Solid uranium tellurites (V.5.3.2.2)

There are consistent thermodynamic data for two U(VI) tellurites, UO_2TeO_3(cr), (schmitterite) referred to herein as $UTeO_5$(cr), and UTe_3O_9(cr).

9.5.3.2.1 Uranium(VI) monotellurites

Brandenburg [78BRA] measured the enthalpies of dissolution of γ–UO_3, TeO_2(cr) and $UTeO_5$(cr) in ca. 11 mol·dm^{-3} HF solution. The enthalpy of formation given in [92GRE/FUG] as $\Delta_f H_m^\circ (UO_2TeO_3, \text{cr, 298.15 K}) = - (1605.4 \pm 1.3)$ kJ·mol^{-1}, as reported by [78BRA] was based on $\Delta_f H_m^\circ (TeO_2, \text{cr, 298.15 K}) = - 322.6$ kJ·mol^{-1} without specified uncertainty limits. As noted in the auxiliary section (chapter 14), we now prefer a slightly more positive value of $\Delta_f H_m^\circ (TeO_2, \text{cr, 298.15 K}) = - (321.0 \pm 2.5)$ kJ·mol^{-1}, based on more recent data and thus obtain $\Delta_f H_m^\circ (UTeO_5, \text{cr, 298.15 K}) = - (1603.7 \pm 2.9)$ kJ·mol^{-1}.

Basu et al. [99BAS/MIS] have recently made similar measurements in 11 mol·dm^{-3} HCl solution, but with β–UO_3 rather than γ–UO_3. With $\Delta_f H_m^\circ (TeO_2, \text{cr, 298.15 K}) = - (321.0 \pm 2.5)$ kJ·mol^{-1}, these data give $\Delta_f H_m^\circ (UTeO_5, \text{cr, 298.15 K}) = - (1602.5 \pm 2.8)$ kJ·mol^{-1}, in excellent agreement with the earlier data [78BRA].

In addition to these calorimetric data, there are three measurements which lead to the Gibbs energy of $UTeO_5$(cr). Mishra et al. [98MIS/NAM] have studied the partial pressure of TeO_2(g) in the decomposition of $UTeO_5$(cr) in pure oxygen according to the reaction:

$$3 \text{ UTeO}_5(\text{cr}) \rightleftharpoons U_3O_8(\text{cr}) + \tfrac{1}{2} O_2(\text{g}) + 3 \text{ TeO}_2(\text{g}) \quad (9.27)$$

and Krishnan et al. [97KRI/RAM], [98KRI/RAM] have made similar measurements by Knudsen effusion but at the self-generated oxygen pressure of the reaction.

As discussed in the relevant entries in Appendix A, we have processed these data, with estimated data for:

$$S_m^\circ(\text{UTeO}_5, \text{cr}, 298.15 \text{ K}) = (166 \pm 10) \text{ J} \cdot \text{K}^{-1} \cdot \text{mol}^{-1}$$

$$C_{p,m}^\circ(\text{TeO}_2, \text{cr}, T) = 151.845 + 3.6552 \cdot 10^{-2} \, T - 1.8129 \cdot 10^6 \, T^{-2} \text{ J} \cdot \text{K}^{-1} \cdot \text{mol}^{-1}.$$

The resulting standard enthalpies of formation of UTeO$_5$(cr) are given in Table 9-29.

In a further study Singh *et al.* [99SIN/DAS] have measured the Gibbs energy of a reaction similar to (9.27):

$$3 \text{ UTeO}_5(\text{cr}) \rightleftharpoons \text{U}_3\text{O}_8(\text{cr}) + \frac{1}{2} \text{O}_2(\text{g}) + 3 \text{ TeO}_2(\text{cr}) \tag{9.28}$$

by a solid-state emf cell. As discussed in Appendix A, the derived Gibbs energy of reaction corresponds to a surprisingly large entropy of reaction. The derived third- and second-law enthalpies of formation are included in the Table 9-29.

Table 9-29: Enthalpies of formation of UTeO$_5$(cr) derived from Gibbs energy studies.

Reference	Technique	T Range (K)	$\Delta_f H_m^\circ$ (UTeO$_5$, cr, 298.15 K) kJ · mol^{-1}	
			Second-law	Third-law
[98KRI/RAM]	Effusion	1063 – 1155	– 1532.0	– (1607.8 ± 5.2)
[98MIS/NAM]	Transpiration	1107 – 1207	– 1611.8	– (1604.4 ± 3.0)
[99SIN/DAS]	emf	821 – 994	– 1640.1	– (1608.8 ± 6.1)

These consistent Gibbs energy data thus give good support to the calorimetric determinations, and the selected value is the mean of the calorimetric values by [78BRA] and [99BAS/MIS]:

$$\Delta_f H_m^\circ(\text{UTeO}_5, \text{cr}, 298.15 \text{ K}) = - (1603.1 \pm 2.8) \text{ kJ} \cdot \text{mol}^{-1}.$$

This corresponds to an enthalpy of formation from the binary oxides (γ–UO$_3$ and TeO$_2$(cr)) of – 58.3 kJ · mol^{-1}.

No value is selected for the standard entropy, but the good agreement of the calorimetric and Gibbs energy data using S_m°(UTeO$_5$, cr, 298.15 K) = (166 ± 10) J · K^{-1} · mol^{-1} indicates that this is a reasonable estimate.

9.5.3.2.2 Uranium(VI) polytellurites

Basu *et al.* [99BAS/MIS] also measured the enthalpy of dissolution of UTe$_3$O$_9$(cr) in *ca.* 11 mol · dm^{-3} HCl solution. When combined with the enthalpies of solution of the individual oxides in the same solvent, the derived enthalpy of formation becomes $\Delta_f H_m^\circ$(UTe$_3$O$_9$ cr, 298.15 K) = – (2275.8 ± 8.0) kJ · mol^{-1}, where the major uncertainty arises from that in the enthalpy of formation of TeO$_2$(cr).

In their studies of the decomposition reactions of dioxouranium(VI) tellurites noted above, both Mishra et al. [98MIS/NAM] and Krishnan et al. [98KRI/RAM] also measured the partial pressure of $TeO_2(g)$ in the decomposition reaction:

$$UTe_3O_9(cr) \rightleftharpoons UTeO_5(cr) + 2\ TeO_2(g). \tag{9.29}$$

Table 9-30: Enthalpies of formation of $UTe_3O_9(cr)$ derived from Gibbs energy studies.

Reference	Technique	Temperature range K	$\Delta_f H_m^\circ$ (UTe_3O_9, cr, 298.15K) kJ·mol^{-1}	
			Second-law	Third-law
[98KRI/RAM]	Effusion	1063 – 1155	– 2274.9	– (2249.0 ± 6.2)
[98MIS/NAM]	Transpiration	947 – 1011	– 2270.8	– (2252.1 ± 6.4)

As noted in the Appendix A reviews, our processing of the data, with estimated data for the entropy and heat capacity of $UTe_3O_9(cr)$, gives the second- and third-law enthalpies of formation $\Delta_f H_m^\circ$ (UTe_3O_9, cr, 298.15 K) given in Table 9-30. These consistent Gibbs energy data thus give support to the calorimetric value for the enthalpy of formation, which is the selected value:

$$\Delta_f H_m^\circ (UTe_3O_9, cr, 298.15\ K) = - (2275.8 \pm 8.0)\ kJ \cdot mol^{-1}.$$

This corresponds to an enthalpy of formation from the binary oxides (γ–UO_3 and $TeO_2(cr)$) of – 87.5 kJ · mol^{-1}.

9.6 Uranium group 15 compounds and complexes (V.6)

9.6.1 Nitrogen compounds and complexes (V.6.1)

9.6.1.1 Uranium nitrides (V.6.1.1)

Following the position adopted in [92GRE/FUG], this review does not include any detailed treatment of the complex U-N system. However, for interest it may be noted that Nakagawa et al. [98NAK/NIS] have studied the nitrogen pressures in the α–U_2N_{3+x} single-phase region for $0.26 < x < 0.52$, as noted in Appendix A.

9.6.1.1.1 UN(cr) (V.6.1.1.1)

Hayes et al. [90HAY/THO] have made a comprehensive review/assessment of the thermodynamics and vaporisation of UN(cr), which is mentioned only briefly in [92GRE/FUG]. Table 9-31 is a comparison of the two sets of selected values for the heat capacity and entropy at 298.15 K.

Table 9-31: Comparison of selected values from [90HAY/THO] and [92GRE/FUG].

Property	[90HAY/THO]	[92GRE/FUG]
$C_{p,m}^\circ$ (298.15 K) (J · mol^{-1}·K^{-1})	47.96	(47.57 ± 0.40)
S_m° (298.15 K) (J · mol^{-1}·K^{-1})	62.68	(62.43 ± 0.22)

The two assessments overlap within the combined uncertainties and we shall retain the values selected in [92GRE/FUG].

Suzuki and Arai [98SUZ/ARA] have reviewed the properties of the actinide mononitrides, including the vaporisation behaviour, but have not presented any new data that are relevant to this assessment.

Ogawa et al. [98OGA/KOB] have discussed the discrepancy in the Gibbs energies of formation of UN(cr) derived from calorimetric and vaporisation data, as have [92GRE/FUG] and other authors. They [98OGA/KOB] prefer a Gibbs energy expression derived from an early assessment which predates many of the experimental data, and the proposed equation for $\Delta_f G_m$ (UN, T) has an entropy of formation near 2000 K about 4.3 J · K^{-1} · mol^{-1} different from the well-defined value calculated from the calorimetric data. We therefore prefer to retain the data selected in [92GRE/FUG].

Ogawa [93OGA] has presented an interesting discussion on modelling of substoichiometric (U, Pu) nitrides, which is beyond the scope of the present review. Ogorodinkov and Rogovoi [93OGO/ROG] have attempted to establish regularities between six thermodynamic parameters at 298.15 K for the mononitrides of 5d transition elements plus Th, U, and Pu. From their relations, they predict missing values, including the specific heat capacity (C_V). As briefly discussed in Appendix A, their value for UN(cr) seems reasonable, but that for PuN(cr) is probably too small.

9.6.1.1.2 UN(g)

Venugopal et al. [92VEN/KUL] have added to the studies on vaporisation of UN(cr), using Knudsen effusion mass-spectrometry from 1757 to 2400 K. The principal reaction is the loss of N$_2$(g) to form a nitrogen–saturated U(l), but U(g) and UN(g) are also present in the vapour. The pressures of U(g) were measured from 1757 to 2396 K and the lower pressures of UN(g) from 2190 to 2400 K.

As noted in [92GRE/FUG], such studies relate to vaporisation from a uranium liquid saturated with nitrogen. Venugopal et al. used a tantalum effusion cell, so in this instance the relevant phase will also contain some tantalum, since this metal is appreciably soluble in U(l).

The uranium ion intensity became steady state after an initial period (perhaps due to the time needed to establish the UN(cr) + U(l) phase equilibrium), and corresponded to the equation:

$$\log_{10}(p_U/\text{bar}) = 5.59 - \frac{26857}{T} \qquad (1757 \text{ to } 2396 \text{ K}) \qquad$$

with very good consistency among six different runs. The reported pressure at 2000 K is at the lower end of the range of pressures given by five previous studies (quoted in [92GRE/FUG]).

As noted in Appendix A, it is therefore not surprising that the derived enthalpies of sublimation of U(cr) are appreciably more positive than the CODATA value for pure uranium adopted by [92GRE/FUG]. In the absence of any measurements of the co-existing nitrogen pressures, no data pertaining to UN(cr) can be derived from this study.

The pressures for UN(g), corresponding to the sublimation:

$$\text{UN(cr)} \rightleftharpoons \text{UN(g)}, \qquad (9.30)$$

(although the solid will in fact be slightly hypostoichiometric) were fitted to the equation:

$$\log_{10}(p_{UN}/\text{bar}) = 7.19 - \frac{37347}{T} \qquad (2190 \text{ to } 2400 \text{ K}).$$

Thus,

$$\Delta_r G_m (9.30) = 715003 - 137.65 \, T \qquad (\text{J} \cdot \text{mol}^{-1}).$$

The authors combine this equation with values of $\Delta_f G_m$(UN, cr) derived from the assessment by Matsui and Ohse [87MAT/OHS], $\Delta_f G_m$(UN, cr, T) = $-304890 + 88.2 \, T$ (J·mol^{-1}) to define $\Delta_f G_m$(UN, g, T) (but seem to have made a numerical error in their first term). We have changed the equation for $\Delta_f G_m$(UN, cr, T) to be consistent with the [92GRE/FUG] selected value for $\Delta_f H_m^\circ$(UN, cr, 298.15 K) and to relate to the mid–temperature of the measurements involving UN(g), 2300 K. The NEA TDB compatible equation from 2200 to 2400 K is then:

$$\Delta_f G_m (\text{UN, cr}, T) = -297596 + 87.53 \, T \quad \text{J} \cdot \text{mol}^{-1},$$

from which we derive:

$$\Delta_f G_m (\text{UN, g}, T) = 417407 - 50.12 \, T \quad \text{J} \cdot \text{mol}^{-1} \qquad (2200 - 2400 \text{ K}).$$

This is the first significant experimental determination of the stability of UN(g). However, since no thermal functions are available for UN(g), these Gibbs energy values cannot be reliably converted to provide standard state data at 298.15 K.

9.6.1.2 Uranium Nitrates (V.6.1.3)

9.6.1.2.1 Aqueous U(VI) nitrates

Cohen-Adad *et al.* [95COH/LOR] (see Appendix A) have analysed the solubility of $UO_2(NO_3)_2 \cdot 6H_2O$ as a function of temperature, and give at T = 298.15 K, m_{sat} = 3.21 mol·kg^{-1}. However since this value is smaller than all except one of the twelve experimental values at 298.15 K, we have retained the value of m_{sat} = 3.24 mol·kg^{-1} selected by Cox *et al.* [89COX/WAG]. Apelblat and Korin

[98APE/KOR] have measured the vapour pressure of aqueous solutions saturated with - from 278 to 323 K. They derived the enthalpy of dissolution of dioxouranium(VI) nitrate hexahydrate in saturated solutions as $\Delta_{sol}H_m^\circ$ (298.15 K, m_{sat} = 3.323 mol·kg^{-1}) = 43.4 kJ·mol^{-1}. (As noted in Appendix A, this saturation molality is slightly higher than one selected by Cox *et al.* [89COX/WAG], 3.24 mol·kg^{-1} and retained here). From the values given in [92GRE/FUG], this review calculates (Appendix A) the enthalpy of dilution from m = 3.323 mol·kg^{-1} to the standard state at infinite dilution, $\Delta_{dil}H_m^\circ$ (298.15 K) = − 23.6 kJ·mol^{-1} with an unknown uncertainty.

It may be noted that the isopiestic measurements of Robinson and Lim [51ROB/LIM] extend up to a concentration of 5.511 mol·kg^{-1} UO$_2$(NO$_3$)$_2$, considerably above the saturation limit. The occurrence of supersaturation in isopiestic measurements is not uncommon, as noted in the Appendix A entry for [98APE/KOR].

As previously noted in the U(VI) chloride section 9.4.3.1.2.1, the paper of Choppin *et al.* [92CHO/DU], which deals also with complexing of UO$_2^{2+}$ by nitrate at high ionic strengths (3 to 7 M), does not contribute any additional data to those in [92GRE/FUG].

9.6.2 Uranium sulphamate

9.6.2.1 Aqueous U(VI) sulphamate

Standritchuk *et al.* [89STA/MAK] give estimated data on the solubility of U(VI) sulphamate UO$_2$(SO$_3$NH$_2$)$_2$ (see Appendix A).

9.6.3 Uranium phosphorus compounds and complexes (V.6.2)

9.6.3.1 Aqueous uranium phosphorus species (V.6.2.1.)

9.6.3.1.1 The uranium–phosphoric acid system (V.6.2.1.1)

9.6.3.1.1.1 Complex formation in the U(VI)–H$_3$PO$_4$ system

Few additional data have appeared in the literature since the last review [92GRE/FUG] where a detailed discussion on the dominant species in the system U(VI)–H$_3$PO$_4$ is given.

Scapolan *et al.* [98SCA/ANS] have used fluorescence spectroscopy to identify the species UO$_2$(H$_2$PO$_4$)$^+$, UO$_2$(HPO$_4$)(aq) and UO$_2$(PO$_4$)$^-$ in phosphoric acid (10^{-4} to 10^{-3} M) between pH = 1.5 and 7.5. These data provide an independent confirmation of the identity of these species and their formation constants as reported in [92GRE/FUG].

Brendler *et al.* [96BRE/GEI] have also used spectroscopy to identify the two first complexes in dilute phosphoric acid near pH = 3.5. In addition, they have carried out potentiometric titrations of phosphoric solutions at very low variable ionic strengths (at ca. 10^{-2} M). The latter data are less reliable for reasons discussed in Appendix A. Brendler et al. reported the following values at zero ionic strength using the Davies equation for the reactions:

$$UO_2^{2+} + r\ H^+ + q\ PO_4^{3-} \rightleftharpoons UO_2H_r(PO_4)_q^{2+r-3q}.$$

From potentiometry they obtain:

$$\log_{10} \beta°(UO_2HPO_4(aq)) = (19.87 \pm 0.29),$$

$$\log_{10} \beta°(UO_2H_2(PO_4^+)) = (22.58 \pm 0.17),$$

$$\log_{10} \beta°(UO_2(H_2PO_4)_2(aq)) = (46.90 \pm 0.22)$$

and from spectroscopy, the values are:

$$\log_{10} \beta°(UO_2HPO_4(aq)) = (19.53 \pm 0.14),$$

$$\log_{10} \beta°(UO_2H_2PO_4^+) = (22.31 \pm 0.16).$$

The $\log_{10} K°$ values selected by [92GRE/FUG], from experimental values obtained at high ionic strength, refer to:

$UO_2^{2+} + HPO_4^{2-} \rightleftharpoons UO_2HPO_4$ (aq) $\qquad \log_{10} K° = (7.24 \pm 0.26),$

$UO_2^{2+} + H_3PO_4(aq) \rightleftharpoons UO_2H_2PO_4^+ + H^+$ $\qquad \log_{10}{}^* K° = (1.12 \pm 0.06),$

$UO_2^{2+} + 2\ H_3PO_4(aq) \rightleftharpoons UO_2(H_2PO_4)_2(aq) + 2H^+$ $\qquad \log_{10}{}^* K° = (0.64 \pm 0.11).$

Using the data from the spectroscopic measurements of [96BRE/GEI] and the auxiliary data on phosphoric acid [92GRE/FUG] page 387, the following $\log_{10} K°$ values are obtained:

$$\log_{10} K°(UO_2HPO_4, aq) = (7.18 \pm 0.14) \text{ and } \log_{10} K°(UO_2H_2PO_4^+) = (0.61 \pm 0.16).$$

The potentiometric value for the formation of $UO_2(H_2PO_4)_2(aq)$ is $\log_{10}{}^* K° = 3.50$.

The equilibrium constant for the formation of $UO_2HPO_4(aq)$ in [96BRE/GEI] is in excellent agreement with the value of [92GRE/FUG]; the latter value has therefore been retained. The agreement between the equilibrium constants for the formation of $UO_2H_2PO_4^+$ is poorer and this review has chosen to retain the value in [92GRE/FUG]. As indicated in Appendix A, this review is not confident about the potentiometric data obtained by Brendler et al. [96BRE/GEI]. This review also retains the value of $\log_{10}{}^* K°(UO_2(H_2PO_4)_2, aq)$ selected by [92GRE/FUG].

Sandino and Bruno [92SAN/BRU] studied the solubility of $(UO_2)_3(PO_4)_2 \cdot 4H_2O$ in the pH range 3 to 9 in 0.5 M NaClO$_4$ at 298.15 K. They interpreted the variation of the solubility curve between pH 4 to 9 as being due to the formation of $UO_2HPO_4(aq)$ and $UO_2PO_4^-$ for which they derived the equilibrium constants: $\log_{10} \beta = (6.03 \pm 0.09)$ and $\log_{10} \beta = (11.29 \pm 0.08)$, respectively, for the reactions:

$$UO_2^{2+} + HPO_4^{2-} \rightleftharpoons UO_2HPO_4(aq),$$
$$UO_2^{2+} + PO_4^{3-} \rightleftharpoons UO_2PO_4^-.$$

SIT–extrapolation to zero ionic strength using the interaction coefficients of [92GRE/FUG] gives $\log_{10} \beta^\circ = (7.28 \pm 0.10)$ and $\log_{10} \beta^\circ = (13.25 \pm 0.09)$, respectively.

The solubility product of the dioxouranium(VI) orthophosphate is $\log_{10} K_{s,0} = -(48.48 \pm 0.16)$, at $I = 0.5$ M and $\log_{10} K_{s,0}^\circ = -(53.32 \pm 0.17)$ at zero ionic strength. These values are derived using the equilibrium value for the formation of $UO_2H_2PO_4^+$ from phosphoric acid as given in [92GRE/FUG].

All these data have been already included in [92GRE/FUG] but under the reference [91SAN].

9.6.3.1.1.2 Complex formation in the U(IV)–H₃PO₄ system

There is no quantitative information to add to that given in [92GRE/FUG]. However, there are two studies reported by Baes [56BAE] and Louis and Bessière [87LOU/BES] that were not discussed in [92GRE/FUG]. Both are based on the observation that U(VI) is reduced to U(IV) by Fe(II) in phosphoric acid solutions according to:

$$U(VI) + 2 Fe(II) \rightleftharpoons U(IV) + 2 Fe(III), \tag{9.31}$$

where the composition of the phosphate complexes on the reactant and product sides are not known. [56BAE] and [87LOU/BES] report conditional equilibrium constants for the equation (9.31), however it is impossible to deduce the stoichiometry of the species and therefore no equilibrium constants can be selected. From the experimental results, the authors conclude that the phosphate complexes of U(IV) are stronger than those of U(VI). Addition of fluoride to the test solutions containing phosphoric acid increases the ease of reduction of U(VI) even more. The experiments referred to above have been performed at low pH; at higher pH sparingly soluble phosphate phases are formed.

9.6.3.2 Solid uranium phosphorus compounds (V.6.2.2)

9.6.3.2.1 Uranium orthophosphates (V.6.2.2.5)

9.6.3.2.1.1 U(VI) orthophosphates

[92GRE/FUG] give data for two solid normal dioxouranium(VI) phosphate hydrates, $(UO_2)_3(PO_4)_2 \cdot 4H_2O(cr)$ and $(UO_2)_3(PO_4)_2 \cdot 6H_2O(cr)$. The Gibbs energy of formation of the tetrahydrate is calculated from the solubility data discussed in section V.6.2.1.1.b of [92GRE/FUG], while the data for the hexahydrate are estimated from two observations of its decomposition to the tetrahydrate and water. However, there seems to have been an error in this estimation, since the Gibbs energy at 298.15 K of the reaction:

$$(UO_2)_3(PO_4)_2 \cdot 4H_2O(cr) + 2 H_2O(l) \rightleftharpoons (UO_2)_3(PO_4)_2 \cdot 6H_2O(cr),$$

from the values selected by [92GRE/FUG] is -4.753 kJ $\cdot$ mol^{-1}, implying that the hexahydrate is the stable hydrate in equilibrium with the saturated solution at 298.15 K.

This review has reworked the calculation, using the two items of information from Appendix A in [92GRE/FUG]:

- [54SCH/BAE]. The hexahydrate decomposes to the tetrahydrate below 373 K.

- [78KOB/KOL]. Both the tetra- and hexa-hydrates are "stable" at 298.15 K; we have replaced their suggested $\log_{10}K = ca.-3.54$ for the reaction:

$$(UO_2)_3(PO_4)_2 \cdot 6H_2O(cr) \rightleftharpoons (UO_2)_3(PO_4)_2 \cdot 4H_2O(cr) + 2\,H_2O(g)$$

by the more accurate $\log_{10}K > -3.54$ (if the tetrahydrate is the phase in equilibrium with the saturated solution at 298.15 K).

For these calculations, we have estimated the standard entropy contribution of H$_2$O in hydrates to be (40 ± 8) J $\cdot$ K^{-1} $\cdot$ mol (H$_2$O)$^{-1}$ by comparison with other hydrates with well-known entropies. Thus,

$$S^\circ_m((UO_2)_3(PO_4)_2 \cdot 6H_2O, cr, 298.15\,K) = S^\circ_m((UO_2)_3(PO_4)_2 \cdot 4H_2O, cr, 298.15K)$$
$$+ (80 \pm 16)\,J \cdot K^{-1} \cdot mol^{-1}$$

$S^\circ_m((UO_2)_3(PO_4)_2 \cdot 6H_2O, cr, 298.15\,K) = (669 \pm 27)$ J $\cdot$ K^{-1} $\cdot$ mol^{-1}.

A value of $\Delta_f H^\circ_m((UO_2)_3(PO_4)_2 \cdot 6H_2O, cr, 298.15\,K) = -7328.4$ kJ $\cdot$ mol^{-1}, then gives consistency with the above observations and keeps the hexahydrate unstable with respect to the tetrahydrate and water at 298.15 K.

The following are thus the revised selected values for the hexahydrate:

$\Delta_f H^\circ_m((UO_2)_3(PO_4)_2 \cdot 6H_2O, cr, 298.15\,K) = -(7328.4 \pm 10.7)$ kJ $\cdot$ mol^{-1},

$S^\circ_m((UO_2)_3(PO_4)_2 \cdot 6H_2O, cr, 298.15\,K)\ \ = (669 \pm 27)$ J $\cdot$ K^{-1} $\cdot$ mol^{-1},

$\Delta_f G^\circ_m((UO_2)_3(PO_4)_2 \cdot 6H_2O, cr, 298.15\,K) = -(6613 \pm 13)$ kJ $\cdot$ mol^{-1}.

9.6.3.2.2 U(VI) phosphates and vanadates (V.6.2.2.10)

Langmuir [97LAN] proposed Gibbs energies of formation for some crystalline U(VI) minerals like autunite, $H_2(UO_2)_2(PO_4)_2$, carnotite, $K_2(UO_2)(VO_4)_2$ and tyuyamunite, $Ca(UO_2)_2(VO_4)_2$ which can be used as examples of calculations (see Appendix A). For U(IV) he gave the value for ningyoite, $CaU(PO_4)_2 \cdot 2H_2O(cr)$.

9.6.3.2.3 U(VI) fluorophosphates

Numerous reactions leading to uranium fluorophosphates are described by [89PET/SEL] and their enthalpy effects listed. However, as no details were given, the paper is cited for information only.

9.6.4 U(VI) arsenates (V.6.3.2.1)

9.6.4.1 Aqueous U(VI) arsenates

Rutsch *et al.* [97RUT/GEI], [99RUT/GEI] have used fluorescence spectroscopy to identify the species $UO_2(H_2AsO_4)^+$, $UO_2(HAsO_4)(aq)$ and $UO_2(H_2AsO_4)_2(aq)$ in 0.1 M $NaClO_4$ between pH 1.5 and 5. The values of $\log_{10} \beta_{q,r}$ and $\log_{10} \beta^o_{q,r}$ according to:

$$UO_2^{2+} + r\, H^+ + q\, AsO_4^{3-} \rightleftharpoons UO_2H_r(AsO_4)_q^{2+r-3q} \qquad (9.32)$$

are given in Table 9-32. The values at $I = 0$ are calculated by the authors using Davies equation; these values are not significantly different when the SIT model is used.

Table 9-32: Equilibrium constants related to the reaction (9.32) [99RUT/GEI].

Species	$\log_{10} \beta_{q,r}$ (I = 0.1 M)	$\log_{10} \beta^o_{q,r}$
$UO_2(H_2AsO_4)^+$	(20.39 ± 0.24)	(21.96 ± 0.24)
$UO_2(HAsO_4)$ (aq)	(17.19 ± 0.31)	(18.76 ± 0.31)
$UO_2(H_2AsO_4)_2$ (aq)	(38.61 ± 0.20)	(41.53 ± 0.20)

All equilibrium formation constants agree within the uncertainty with those in the corresponding phosphate system (see 9.6.3.1).

These values are new and are selected by this review. This review has calculated the values of the constants with UO_2^{2+}, $HAsO_4^{2-}$ and H_3AsO_4 as components to compare the complexation of U(VI) with H_3AsO_4 and H_3PO_4 as described in [92GRE/FUG]. We have used the protonation constants of H_3AsO_4 derived from [92GRE/FUG] (see Appendix A).

$H_2AsO_4^- + H^+ \rightleftharpoons H_3AsO_4(aq)$ $\qquad \log_{10} {}^*K_1^o = (2.26 \pm 0.20)$,

$HAsO_4^{2-} + H^+ \rightleftharpoons H_2AsO_4^-$ $\qquad \log_{10} {}^*K_2^o = (6.76 \pm 0.20)$,

$AsO_4^{3-} + H^+ \rightleftharpoons HAsO_4^{2-}$ $\qquad \log_{10} {}^*K_3^o = (11.60 \pm 0.20)$.

The results are the following:

$UO_2^{2+} + HAsO_4^{2-} \rightleftharpoons UO_2(HAsO_4)(aq)$ $\qquad \log_{10} K^o = (7.16 \pm 0.37)$,

$UO_2^{2+} + H_3AsO_4(aq) \rightleftharpoons UO_2(H_2AsO_4)^+ + H^+$ $\qquad \log_{10} {}^*K^o = (1.34 \pm 0.42)$,

$UO_2^{2+} + 2\,H_3AsO_4(aq) \rightleftharpoons UO_2(H_2AsO_4)_2(aq) + 2\,H^+$ $\log_{10} {}^*K^o = (0.29 \pm 0.53)$.

9.6.5 USb(cr) (V.6.4.1)

Ochiai *et al.* [94OCH/SUZ] have measured the heat capacity of an uncharacterised sample of USb(cr) from 2 to 300 K. The material showed a pronounced ferromagnetic peak at 218 K, slightly different from earlier values. The data are presented only in the form of a small graph, from which the selected value:

$$C_{p,m}^{\circ}(\text{USb, cr, 298.15 K}) = (75 \pm 7) \text{ J} \cdot \text{K}^{-1} \cdot \text{mol}^{-1}$$

is obtained, the relatively high uncertainty reflecting the lack of characterisation of the sample.

9.7 Uranium group 14 compounds and complexes (V.7)

9.7.1 Uranium carbides (V.7.1.1)

9.7.1.1 UC(cr) (V.7.1.1.1)

As part of a study of the vaporisation behaviour in the U–Ce–C system, Naik *et al.* [93NAI/VEN] have measured, by Knudsen effusion mass-spectrometry, the uranium pressures over monophasic UC(cr) (1922 to 2247 K) and the diphasic U(l) + UC(cr) region (1571 to 2317 K). These studies, which within the combined uncertainties, agree well with earlier data, do not provide any new data on the Gibbs energy of formation of UC(cr).

9.7.2 The aqueous uranium carbonate system (V.7.1.2.1)

9.7.2.1 Uranium(VI) carbonates

A large number of additional papers dealing with various aspects of the chemistry of the aqueous uranium(VI)–carbonate system have appeared since the previous review. These papers contain structural information, equilibrium data, and discussions of the rates and mechanisms of ligand exchange reactions. There is also new information on the chemical properties of the ternary U(VI)–carbonate–fluoride, and U(VI)–carbonate–hydroxide systems. The new experimental data refer mainly to experiments made in 0.1 M NaClO$_4$ and the corresponding equilibrium constants, including the ones recalculated to zero ionic strength, are given in Table 9-33.

Table 9-33: Equilibrium constants in the U(VI) carbonate systems.

Reaction	Ionic medium NaClO$_4$	$t°C$	$\log_{10} K$	$\log_{10} K°$	Reference
$UO_2CO_3 (s) \rightleftharpoons$ $UO_2^{2+} + CO_3^{2-}$	0.10 M	25	−(13.50 ± 0.22)	−(14.34 ± 0.22)	[96MEI/KLE]
	0.10 M	(22±2)	−(13.55 ± 0.14)	−(14.39 ± 0.14)	[97PAS/CZE]*
	0.10 M	25	−(14.18 ± 0.03)	−(15.02 ± 0.06)	[93MEI/KIM]
	0.10 M	24	−(13.89 ± 0.11)	−(14.73 ± 0.11)	[93MEI/KIM2]
	0.10 M	25	−(14.10 ± 0.14)	−(14.94 ± 0.14)	[96KAT/KIM]
	0.10 M	22	−(13.55 ± 0.14)	−(14.39 ± 0.14)	[93PAS/RUN]**
				−(14.49 ± 0.04)	[92GRE/FUG]#
$UO_2^{2+} + CO_3^{2-} \rightleftharpoons$ $UO_2CO_3 (aq)$	0.10 M	25	(8.81 ± 0.08)	(9.65 ± 0.08)	[96MEI/KLE]
	0.10 M	(22±2)	(9.13 ± 0.05)	(9.97 ± 0.05)	[97PAS/CZE]*
	0.10 M	25	(9.23 ± 0.04)	(10.07 ± 0.08)	[93MEI/KIM]
	3.0 M	25	(8.60 ± 0.18)	(9.14 ± 0.23)	[79CIA/FER]
				(9.67 ± 0.05)	[92GRE/FUG]#
$UO_2^{2+} + 2 CO_3^{2-} \rightleftharpoons$ $UO_2(CO_3)_2^{2-}$	0.10 M	25	(15.5 ± 0.8)	(16.3 ± 0.8)	[96MEI/KLE]
	0.10 M	(22±2)	(15.7 ± 0.2)	(16.5 ± 0.2)	[97PAS/CZE]*
	0.10 M	25	(15.38 ± 0.17)	(16.22 ± 0.34)	[93MEI/KIM]
				(16.94 ± 0.12)	[92GRE/FUG]
$UO_2^{2+} + 3 CO_3^{2-} \rightleftharpoons$ $UO_2(CO_3)_3^{4-}$	0.10 M	25	(21.74 ± 0.44)	(21.74 ± 0.44)	[96MEI/KLE]
	0.10 M	(22±2)	(21.6 ± 0.3)	(21.6 ± 0.3)	[97PAS/CZE]*
	0.10 M	25	(21.86 ± 0.05)	(21.86 ± 0.10)	[93MEI/KIM]
				(21.60 ± 0.05)	[92GRE/FUG]
$UO_2(CO_3)_2^{2-} + CO_3^{2-} \rightleftharpoons$ $UO_2(CO_3)_3^{4-}$	0.50 M	25	(6.35 ± 0.05)	(4.98 ± 0.09)	[91BID/CAV]
	0.1–2m	25-80		4.49	[89SER/SAV]
$3UO_2(CO_3)_3^{4-} + 3 H^+ \rightleftharpoons$ $(UO_2)_3(CO_3)_6^{6-} + 3 HCO_3^-$	2.5 m	25	(18.1 ± 0.5)	(12.4 ± 0.7)	[95ALL/BUC]
$3 UO_2^{2+} + 6 CO_3^{2-} \rightleftharpoons$ $(UO_2)_3(CO_3)_6^{6-}$	2.5 m	25	(55.6 ± 0.5)	(55.6 ± 0.5)	[95ALL/BUC]
$UO_2(OH)_3^- + 3 HCO_3^- \rightleftharpoons$ $3 H_2O + UO_2(CO_3)_3^{4-}$	0.1 M	25(?)	(8.5 ± 0.5)	(7.2 ± 0.5)*	[98GEI/BER]
$2UO_2^{2+} + CO_2(g) + 4H_2O(l)$ $\rightleftharpoons (UO_2)_2CO_3(OH)_3^- + 5H^+$	< 10^{-3} M	25(?)	−(18.9 ± 1.0)		[98GEI/BER3]
$UO_2^{2+} + CO_3^{2-} + 2OH^- \rightleftharpoons$ $(UO_2)CO_3(OH)_2^{2-}$	0.5–5 M	25	<22.6		[98YAM/KIT]
$UO_2^{2+} + 2CO_3^{2-} + 2OH^- \rightleftharpoons$ $UO_2(CO_3)_2(OH)_2^{4-}$	0.5–5 M	25	<23.5		[98YAM/KIT]

* recalculated by this review.
\# these data are those revised in Appendix D of [95SIL/BID].
** this is the same value as in [97PAS/CZE].

9.7.2.1.1 Binary U(VI) carbonate complexes

The weighted average of the constants are given in Table 9-33 with previous data from [92GRE/FUG] and the estimated uncertainties have been calculated using the SIT method, as described in [92GRE/FUG]. The following values at 298.15 K were obtained and selected:

$UO_2CO_3(cr) \rightleftharpoons UO_2^{2+} + CO_3^{2-}$ $\quad \log_{10} K_{s,0}^\circ = -(14.76 \pm 0.02)$,
$\Delta\varepsilon = -(0.330 \pm 0.011)$ mol·kg^{-1},
$\Delta\varepsilon_{cal} = -(0.38 \pm 0.06)$ mol·kg^{-1}.

$UO_2^{2+} + CO_3^{2-} \rightleftharpoons UO_2CO_3(aq)$ $\quad \log_{10} \beta_1^\circ = (9.94 \pm 0.03)$,
$\Delta\varepsilon = (0.232 \pm 0.027)$ mol·kg^{-1},
$\Delta\varepsilon_{cal} = (0.38 \pm 0.06)$ mol·kg^{-1}.

$UO_2^{2+} + 2\,CO_3^{2-} \rightleftharpoons UO_2(CO_3)_2^{2-}$ $\quad \log_{10} \beta_2^\circ = (16.61 \pm 0.09)$,
$\Delta\varepsilon = (0.454 \pm 0.052)$ mol·kg^{-1},
$\Delta\varepsilon_{cal} = (0.32 \pm 0.15)$ mol·kg^{-1}.

$UO_2^{2+} + 3\,CO_3^{2-} \rightleftharpoons UO_2(CO_3)_3^{4-}$ $\quad \log_{10} \beta_3^\circ = (21.84 \pm 0.04)$,
$\Delta\varepsilon = (0.233 \pm 0.046)$ mol·kg^{-1},
$\Delta\varepsilon_{cal} = (0.24 \pm 0.18)$ mol·kg^{-1}.

The equilibrium constant for the reaction,

$$3\,UO_2^{2+} + 6\,CO_3^{2-} \rightleftharpoons (UO_2)_3(CO_3)_6^{6-}$$

determined in [95ALL/BUC], $\log_{10} \beta_{6,3}^\circ = (55.6 \pm 0.5)$, is in good agreement with the value selected in [92GRE/FUG] and so the latter is therefore retained.

The equilibrium constants given above are weighted average values with their corresponding uncertainties estimated as described in Appendix C. The addition of new experimental data at $I = 0.1$ M results in a change in most of the values given in [92GRE/FUG]. As judged by the estimated uncertainty in the average values these deviations are significant; however, considering the largest uncertainty in the individual experimental determinations, the difference in the two averages is acceptable. It is clear that the uncertainty estimates must be looked upon with caution as discussed in the introduction in Appendix C. The uncertainty reported is a measure of the precision of an experiment, not its accuracy. The values of $\Delta\varepsilon$ for the various reactions are in fair agreement with the tabulated values for the individual reactants/products as seen above. In view of the uncertainty in these parameters this review has not considered a revision of the individual ε values.

The solubility product $\log_{10} K_{s,0}^\circ (UO_2CO_3(cr)) = -(14.76 \pm 0.02)$ deserves a special comment; all new values of the solubility product have been obtained in 0.1 M NaClO$_4$ and most of them are systematically somewhat lower than those obtained by previous investigators. This review has no explanation for this discrepancy; it may possibly be due to differences in the degree of crystallinity of the solid phases. This is supported by the fact that most of the stability constants for the complexes deduced from the solubility data are in good agreement with data from other sources.

The only experimental study of a solution where $UO_2(CO_3)_2^{2-}$ is a predominant complex has been made by Bidoglio et al. [91BID/CAV], who studied the equilibrium:

$$UO_2(CO_3)_2^{2-} + CO_3^{2-} \rightleftharpoons UO_2(CO_3)_3^{4-} \quad (9.33).$$

Their data give $\log_{10} K_3^\circ = (4.98 \pm 0.09)$.

By combining the value of Bidoglio et al. [91BID/CAV] with the value $\log_{10} \beta_3^\circ = (21.84 \pm 0.04)$ we obtain $\log_{10} \beta_2^\circ = (16.86 \pm 0.10)$, in agreement both with the values obtained in direct determinations of $\log_{10} \beta_2^\circ$ (cf. Table 9-33) and with the value selected in [92GRE/FUG].

The enthalpy of reaction for the equilibrium (9.33) at 298.15 K has also been studied by Sergeyeva et al. [89SER/SAV], and the value $\Delta_r H_m^\circ (9.33) = -61$ kJ·mol^{-1} is in good agreement with the value -59.1 kJ·mol^{-1} selected in [92GRE/FUG]. However, it is not clear from the abstract of [89SER/SAV], if these data refer to experiments that are based on previous experimental information, or not.

The study of Sergeyeva et al. [89SER/SAV] has not been taken into account because of lack of experimental information.

Additional data on U(VI) carbonate complexes are given in [91BRU/GLA], [91CAR/BRU], [92KIM/SER], [92LIE/HIL] and [99MEI/KAT].

9.7.2.1.2 Ternary U(VI) hydroxide carbonate complexes

The experimental solubility studies of Yamamura et al. [98YAM/KIT] in the ternary U(VI)–OH$^-$–CO$_3^{2-}$ system, vide infra, confirm the value for the equilibrium constant for formation of $UO_2(CO_3)_3^{4-}$ given in [92GRE/FUG]. The structure and dynamics in the ternary complex $(UO_2)_2(CO_3)(OH)_3^-$ have been studied using EXAFS and NMR [2000SZA/MOL], which provide an independent confirmation of the stoichiometry of this species. The following structure shows the bridge arrangement in the complex, where U denotes the UO$_2$–unit with the axial "yl" oxygen atoms perpendicular to the plane of the paper:

Figure 9-5: Structure of the ternary complex $(UO_2)_2(CO_3)(OH)_3^-$.

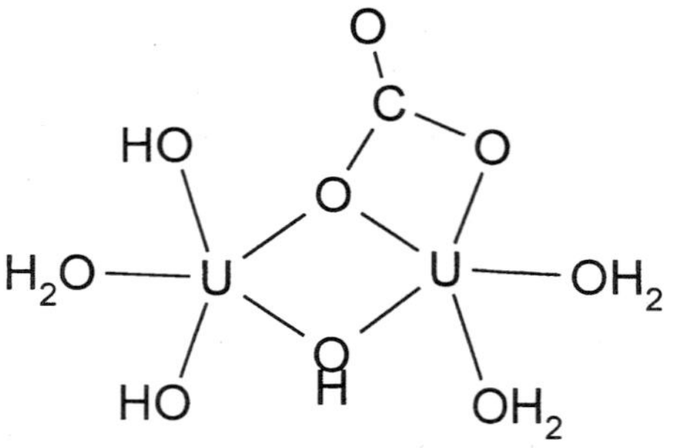

9.7.2.1.3 Ternary U(VI) fluoride carbonate complexes

Aas et al. [98AAS/MOU] investigated the formation of ternary carbonate complexes $UO_2(CO_3)_p F_q^{2-2p-q}$ in 1.00 M NaClO$_4$ at 298.15 K according to:

$$UO_2^{2+} + p\, CO_3^{2-} + qF^- \rightleftharpoons UO_2(CO_3)_p F_q^{2-2p-q}$$

and give $\log_{10} \beta_{1,1,1} = (12.56 \pm 0.05)$, $\log_{10} \beta_{1,1,2} = (14.86 \pm 0.08)$ and $\log_{10} \beta_{1,1,3} = (16.77 \pm 0.06)$ where the uncertainty is given at the three sigma level (see Appendix A). This is the only experimental determination and the following equilibrium constants, recalculated to zero ionic strength, are selected by this review (see Appendix A):

$$\log_{10} \beta_{1,1,1}^\circ = (13.75 \pm 0.09),$$
$$\log_{10} \beta_{1,1,2}^\circ = (15.57 \pm 0.14),$$
$$\log_{10} \beta_{1,1,3}^\circ = (16.38 \pm 0.11).$$

9.7.2.1.4 Calcium uranium carbonate complex

Complex formation in the system Ca(II) – U(VI) – CO_3^{2-} – H_2O has been studied by Bernhard et al. [96BER/GEI], [97BER/GEI], Geipel et al. [98GEI/BER3], [98GEI/BER4] and Amayri et al. [97AMA/GEI]. In a more recent paper, [2001BER/GEI], Bernhard et al. have used additional spectroscopic studies and an EXAFS investigation to corroborate their earlier findings. The authors present convincing evidence for the formation of an uncharged complex $Ca_2UO_2(CO_3)_3$(aq), and propose the equilibrium constant at zero ionic strength for the reaction:

$$2\, Ca^{2+} + UO_2^{2+} + 3\, CO_3^{2-} \rightleftharpoons Ca_2UO_2(CO_3)_3(aq) \qquad (9.34).$$

$\log_{10} K°(9.34) = (26.5 \pm 0.3)$. In the most recent paper [2001BER/GEI] this value has been changed to $\log_{10} K°(9.34) = (30.55 \pm 0.25)$. The methods used to analyse the experimental data are discussed in Appendix A and for reasons given there this review has not selected the equilibrium constant proposed by Bernhard et al. Kalmykov and Choppin [2000KAL/CHO] have studied the same system using a similar experimental method. These data in 0.1 M NaClO$_4$ are in excellent agreement with the more recent value of Bernhard *et al.* Kalmykov and Choppin have also studied the ionic strength dependence of the reaction and used the SIT model to calculate the value, $\log_{10} K^\circ (9.34) = (29.22 \pm 0.25)$, which is based on information over a larger ionic strength range than that proposed by Bernhard *et al.* [97BER/GEI]. The agreement between the equilibrium constants at zero ionic strength proposed in [2000KAL/CHO] and [2001BER/GEI] is good. However, for reasons given in the discussion of [2000KAL/CHO] and [2001BER/GEI] in Appendix A, this review does not accept these data. There is no doubt that extensive ionic interactions occur between $UO_2(CO_3)_3^{4-}$ and counter-ions; the constant proposed in [2000KAL/CHO] and [2001BER/GEI] may be used as guidance.

9.7.2.2 Solid U(VI) carbonates compounds

The solubilities of schröckingerite, $NaCa_3UO_2(CO_3)_3 SO_4 F \cdot 10H_2O$, and grimselite, $NaK_3UO_2(CO_3)_3 \cdot H_2O$, were measured by O'Brien and Williams [83OBR/WIL] at temperatures of 293.15 and 298.15 K in the former case, and at 278.75 to 298.15 K for the latter solid. The compositions of the test solution were speciated into as many as 35 species resulting in the equilibria:

$$NaCa_3UO_2(CO_3)_3 SO_4 F \cdot 10H_2O(cr) \rightleftharpoons Na^+ + 3Ca^{2+} + UO_2^{2+} + 3CO_3^{2-}$$
$$+ SO_4^{2-} + F^- + 10H_2O(l),$$

$$NaK_3UO_2(CO_3)_3 \cdot H_2O(cr) \rightleftharpoons Na^+ + 3K^+ + UO_2^{2+} + 3CO_3^{2-} + H_2O(l).$$

For the case of schröckingerite the ionic strength for the three solutions at 298.15 K was *ca.* 0.1 m, with $\log_{10} K_{s,0} = -(35.53 \pm 0.06)$. The infinite dilution value based solely on the Debye–Hückel slope gave $\log_{10} K_{s,0}^\circ = -(38.53 \pm 0.06)$, which in turn yielded:

$$\Delta_f G_m^\circ (NaCa_3UO_2(CO_3)_3SO_4F \cdot 10H_2O, 298.15 \text{ K}) = -(8073.5 \pm 2.8) \text{ kJ} \cdot \text{mol}^{-1},$$

essentially the same as that reported by O'Brien and Williams [83OBR/WIL], $-(8077.3 \pm 8.7) \text{ kJ mol}^{-1}$.

The corresponding equilibrium constant for grimselite was determined at ionic strengths to 0.439 m, which presents a significant problem in extrapolation to infinite dilution compounded by a lack of knowledge of the major anions present. At 298.15 K and 0.428 m ionic strength, $\log_{10} K_{s,0} = -(26.47 \pm 0.06)$. O'Brien and Williams also report:

$$\Delta_f G_m^\circ (NaK_3UO_2(CO_3)_3 \cdot H_2O, 298.15 \text{ K}) = -(4051.3 \pm 1.8) \text{ kJ} \cdot \text{mol}^{-1},$$

$$\Delta_f H_m^\circ (NaK_3UO_2(CO_3)_3 \cdot H_2O, 298.15 \text{ K}) = -(4359.0 \pm 1.8) \text{ kJ} \cdot \text{mol}^{-1},$$

but these values appear to be without substantiation.

9.7.2.3 Uranium(V) carbonates

9.7.2.3.1 Aqueous U(V) carbonates complexes

Mizuguchi *et al.* [93MIZ/PAR] and Capdevila and Vitorge [99CAP/VIT] have studied the redox potential U(VI)/U(V) in carbonate media using cyclic voltammetry. The measured potential from Mizuguchi *et al.* has been recalculated to the NHE scale by this review and the value in a 1 M Na_2CO_3 ionic medium is -0.50 V, in good agreement with that reported by Caja and Pradvic [69CAJ/PRA], $E° = -0.492$ V. The paper of Capdevila and Vitorge contains a reinterpretation of their previous data [90CAP/VIT] and [92CAP], and the authors propose $\log_{10} \beta_3° = (6.95 \pm 0.36)$ for the reaction:

$$UO_2^+ + 3\,CO_3^{2-} \rightleftharpoons UO_2(CO_3)_3^{5-},$$

where the uncertainty is given at the 1.96σ level. Capdevila and Vitorge have made a more extensive study of the ionic strength dependence of the equilibrium constant than in the previous study. As a result the $\Delta\varepsilon$ value for the half–cell reaction:

$$UO_2(CO_3)_3^{4-} + e^- \rightleftharpoons UO_2(CO_3)_3^{5-}$$

is changed from $-(0.62 \pm 0.15)$ kg · mol^{-1} selected in the [95SIL/BID] revision of [92GRE/FUG] to $-(0.97 \pm 0.2)$ kg · mol^{-1}. The difference between the two sets of values is within the estimated uncertainty ranges. In addition, the SIT extrapolation is uncertain for ions with the high charges encountered here. Nevertheless, the present review selects the revised value,

$$\log_{10} \beta_3° = (6.95 \pm 0.36).$$

9.7.2.4 Uranium(IV) carbonates

9.7.2.4.1 Binary U(IV) carbonate complexes

The new experimental data concern the structure and stoichiometry of the limiting complex $U(CO_3)_5^{6-}$, which has been confirmed both in the solid state using single crystal X–ray diffraction on the corresponding Pu(IV) complex [96CLA/CON] and in solution using EXAFS [98RAI/FEL]. The uranium(IV)–hydroxide–carbonate complex has been studied by Rai *et al.* [98RAI/FEL]. This group also did similar investigations on the corresponding Th(IV), [2000RAI/MOO], and Np(IV), [99RAI/HES], systems. These data are discussed in Appendix A, and the conclusion of this review is that the stoichiometry of the limiting complex, $An(CO_3)_5^{6-}$, An = Th, U, and Np, and its structure, are well established.

The equilibrium constants for the reactions:

$$UO_2(am) + 4\,H^+ + 5\,CO_3^{2-} \rightleftharpoons U(CO_3)_5^{6-} + 2\,H_2O(l), \tag{9.35}$$

$$U^{4+} + 5\,CO_3^{2-} \rightleftharpoons U(CO_3)_5^{6-} \tag{9.36}$$

are reported to be $\log_{10} {}^*K_{s,5}^\circ$ (9.35) = 33.8 and $\log_{10} \beta_5^\circ$ (9.36) = 31.3 [98RAI/FEL]. The latter value depends on the solubility constant of UO_2(am, hydr.). Using $\log_{10} K_{s,0}^\circ$ (UO_2, am, hydr., (9.11)) = $-(54.5 \pm 1.0)$ selected in the present review (*cf.* section 9.3.2.2.1) and an estimated uncertainty for the $\log_{10} {}^*K_{s,5}^\circ$ value of Rai *et al.* [98RAI/FEL] (*cf.* Appendix A), the resulting equilibrium constant, $\log_{10} \beta_5^\circ$ (9.36) = (32.3 ± 1.4), is in fair agreement with the value, $\log_{10} \beta_5^\circ (U(CO_3)_5^{6-}, 298.15\ K) = (34.0 \pm 0.9)$ selected in [92GRE/FUG].

The chemistry of the An(IV)–carbonate–hydroxide system is complicated. In addition, the use of solubility data to deduce the chemical speciation complicates matters further. The solubility data of Rai *et al.* are of high quality and the chemical model proposed gives a reasonable representation of them. However, this does not mean unequivocally that this chemical model is correct as discussed in Appendix A. For this reason and the lack of an analysis of the uncertainty of the constants proposed, this review has not accepted the equilibrium constants proposed in [98RAI/FEL].

9.7.2.4.2 Ternary U(IV) hydroxide carbonate complexes

Rai *et al.* [95RAI/FEL] determined the solubility of UO_2(am, hydr.) and ThO_2(am, hydr.) in alkaline carbonate and bicarbonate solutions of varying composition which clearly showed the formation of ternary hydroxide-carbonate complexes, but [95RAI/FEL] do not propose a thermodynamic interpretation. In a later paper containing some additional experimental data, Rai *et al.* [98RAI/FEL] suggest the formation of one ternary complex $U(OH)_2(CO_3)_2^{2-}$, with equilibrium constants, $\log_{10} K^\circ$ (9.37) = -4.8 and $\log_{10} \beta_{1,2,2}^\circ$ (9.38) = 41.3, for the reactions:

$$UO_2(am) + 2\ HCO_3^- \rightleftharpoons U(OH)_2(CO_3)_2^{2-}, \quad (9.37)$$

$$U^{4+} + 2\ CO_3^{2-} + 2\ OH^- \rightleftharpoons U(OH)_2(CO_3)_2^{2-}. \quad (9.38)$$

In combination with the solubility constant selected in the present review for UO_2(am, hydr.), the latter value becomes $\log_{10} \beta_{1,2,2}^\circ$ (9.38) = 42.4.

The stoichiometries of ternary Np(IV) hydroxide–carbonate complexes have been determined by Eriksen *et al.* [93ERI/NDA]. The experimental results obtained for the Np(IV) and the U(IV) systems are not concordant as one would expect for these chemically very similar systems. There is additional experimental evidence for the formation of a ternary hydroxide-carbonate complex with the composition $M(OH)_3(CO_3)^-$ from studies of the corresponding Th(IV) complex [94OST/BRU]. This review accepts the evidence for the formation of ternary complexes containing hydroxide. However, the experimental method used by Rai *et al.* does not allow a determination of the number of coordinated carbonate ions, *cf.* Appendix A. The proposed equilibrium constants in [98RAI/FEL] can be used as phenomenological parameters to describe the solubility at high carbonate and hydroxide concentrations.

The transformation from the limiting complex $An(CO_3)_5^{6-}$, via ternary hydroxide-carbonate complexes to $An(OH)_4(s)$, takes place over a rather narrow pH, pCO_3^{2-} region. The stoichiometry of the complexes formed and their equilibrium constants are not known in sufficient detail to allow chemical modelling. This is an important area for future research, which will require the use of methods other than the solubility technique used so far.

9.7.3 Silicon compounds and complexes (V.7.2)

9.7.3.1 Aqueous uranium silicates (V.7.2.1)

The only additional information on aqueous uranium silicates since the previous review of [92GRE/FUG] refers to the reaction:

$$UO_2^{2+} + Si(OH)_4(aq) \rightleftharpoons UO_2SiO(OH)_3^+ + H^+ \tag{9.39}$$

Two factors make it difficult to identify the stoichiometry and the equilibrium constants of the complexes formed:

- $Si(OH)_4(aq)$ is a very weak acid, and the formation of hydroxide complexes must be considered when interpreting the experimental data.

- The formation of silicate polymers in solution, *cf.* section (VI.4.2) in [92GRE/FUG].

The dissolved silica is often prepared by hydrolysis of tetramethyl-orthosilicate [98MOL/GEI], [98JEN/CHO] in order to avoid polymerisation.

Fluorescence spectroscopy is the most sensitive tool to detect differences in ligand bonding, and the two studies made by Moll et al. [97MOL/GEI], [98MOL/GEI] seem to make it possible to differentiate between monomers and polymers through the fluorescence lifetime of the U(VI) complexes. The experimental data are collected in Table 9-34:

Table 9-34: Equilibrium constants of the reaction (9.39).

Reaction	ionic medium NaClO$_4$	$t(°C)$	$\log_{10} {}^*K$	$\log_{10} {}^*K°$	Reference
	0.3 M	20	$-(1.74 \pm 0.20)$	$-(1.44 \pm 0.20)$	[98MOL/GEI]
$UO_2^{2+} + Si(OH)_4(aq) \rightleftharpoons$	0.2 M	25	$-(2.01 \pm 0.09)$	$-(1.74 \pm 0.09)$	[92SAT/CHO]
$UO_2SiO(OH)_3^+ + H^+$	0.1 M	25	$-(2.92 \pm 0.06)$	$-(2.65 \pm 0.06)$	[98JEN/CHO]
	0.2 M	25	$-(2.21 \pm 0.06)$	$-(1.94 \pm 0.06)$	[99HRN/IRL][a]
	0.2 M	25	$-(1.98 \pm 0.13)$	$-(1.71 \pm 0.13)$	[71POR/WEB]

(a): The uncertainty has been increased by a factor of three by the reviewers.

From Table 9-34 it is obvious that there is a large scatter in the experimental data that should be resolved from new experiments.

Moll et al. [97MOL/GEI], [98MOL/GEI], studied equilibrium (9.39) at 293.15 K in a 0.3 M NaClO$_4$ ionic medium, using Time Resolved Laser-induced Fluorescence Spectroscopy (TRLFS). A detailed discussion of the two investigations is given in Appendix A. The first study provides qualitative evidence for the formation of complexes between U(VI) and polysilicates, while the second study provides the most complete experimental data and also quantitative information of the equilibrium constant for reaction (9.39). For reasons given in Appendix A, it was necessary to re-evaluate the experimental data. This results in the following value of the equilibrium constant for reaction (9.39) in 0.3 M NaClO$_4$ and at zero ionic strength, $\log_{10}^{*} K$ (9.39) = $-(1.74 \pm 0.20)$ and $\log_{10}^{*} K^{\circ}$ (9.39) = $-(1.44 \pm 0.20)$, respectively.

There are three additional experimental determinations of the equilibrium constants for reaction (9.39) that are discussed in Appendix A, Satoh and Choppin [92SAT/CHO], Jensen and Choppin [98JEN/CHO], and Hrnecek and Irlweck [99HRN/IRL]. In the first study solvent–solvent extraction methodology was used to evaluate the equilibrium constant for reaction (9.39). These data, later re-interpreted by Jensen and Choppin, are given in the Table 9-34. The resulting equilibrium constant is in excellent agreement with the previous determination by [71POR/WEB], while the agreement is just within the estimated uncertainty range with that of [98MOL/GEI]. The equilibrium constant for reaction (9.39) from the second study [98JEN/CHO] is about one order of magnitude smaller than the previous values; the discrepancy between the two sets of data was ascribed to the presence of polysilicates in [92SAT/CHO]. For reasons given in Appendix A, this review has chosen not to use this value when selecting the equilibrium constant for reaction (9.39). Hrnecek and Irlweck [99HRN/IRL] have also used a solvent extraction technique to determine the equilibrium constants in the U(VI) silicate system. They have taken care to identify both equilibria between the silicate monomer OSi(OH)$_3^-$ and polymers ($\equiv$ Si(OH))$_j$, and presented evidence that the latter reaction resulted in the formation of an uncharged U(VI) silicate complex. Because the stoichiometry of the polymer is not known, this review does not select the proposed equilibrium constant, $\log_{10}^{*} K = -(5.85 \pm 0.06)$, for the reaction:

$$UO_2^{2+} + (\equiv SiOH)_j \rightleftharpoons \left[(\equiv SiOH)_{j-2}(\equiv SiO)_2 UO_2\right] + 2 H^+ \ .$$

However, it can be used as a guideline to evaluate the effect of silicate polymers on speciation. Hrnecek and Irlweck point out that the dioxouranium(VI) ion shows similar binding strength to polymeric and low oligomeric silicic acids.

The extrapolations to zero ionic strength have been made using an estimated interaction coefficient, $\varepsilon(UO_2SiO(OH)_3^+, ClO_4^-) = 0.3$ kg·mol^{-1}.

Moll et al. [96MOL/GEI] have determined the solubility and speciation of $(UO_2)_2SiO_4 \cdot 2H_2O$ in a 0.1 M NaClO$_4$ ionic medium at 298.15 K. These data are discussed in Appendix A, and in section 9.7.3.2.3. The experimental results of this study are consistent with the selected equilibrium constant for the formation of $UO_2SiO(OH)_3^+$. No value for the equilibrium constant for the formation of

$UO_2SiO(OH)_3^+$ was selected in [92GRE/FUG], because only one experimental determination, [71POR/WEB], was available. Using the additional information from [71POR/WEB], [92SAT/CHO], [98MOL/GEI] and [99HRN/IRL], listed in Table 9-34, this review has selected for the reaction:

$$UO_2^{2+} + Si(OH)_4(aq) \rightleftharpoons UO_2SiO(OH)_3^+ + H^+$$

$\log_{10}{}^* K° = - (1.84 \pm 0.10)$, where the uncertainty is at the 1.96σ level.

9.7.3.2 Solid uranium silicates (V.7.2.2)

9.7.3.2.1 Solid U(IV) silicates

On the basis of solubility measurements Langmuir [97LAN] proposed a value of $\Delta_f G_m^°(USiO_4, am)$ which is not selected as explained in Appendix A.

9.7.3.2.2 Solid U(VI) silicates

Clark et al. [98CLA/EWI], Chen et al. [99CHE/EWI], [99CHE/EWI2] describe a method for estimation of standard Gibbs energies of formation of U(VI) layered oxide hydrates and silicates. Because of the very few experimental thermodynamic data available for minerals, we found it relevant to include this study, despite the fact that it is based on estimations. Clark et al. have extended a method originally suggested by Tardy and Garrels [74TAR/GAR] for layer silicates, by using data on the structural hierarchy for a large number of uranium(VI) solid phases from Burns et al. [96BUR/MIL]. The estimation method uses the standard energies of formation of crystalline schoepite, silica and oxides of Na, K, Mg and Ca from the NEA TDB and CODATA databases together with an estimated value for the standard Gibbs energy contribution of structural water equal to -247 kJ·mol^{-1}. The authors have made an error propagation analysis of the estimated energies of formation and a comparison with experimental data. The uncertainty in the estimates is fairly large, ranging from 8 to 90 kJ·mol^{-1}. The difference between experimental and estimated data is larger, ranging from 34 to 124 kJ·mol^{-1}. The discrepancy for becquerelite is unexpectedly large, 440 kJ·mol^{-1}. The authors do not offer any explanation for this discrepancy. Clark et al. [98CLA/EWI] also report the Gibbs energy contribution of structural water that is useful for other uranium phases.

9.7.3.2.3 $(UO_2)_2SiO_4·2H_2O$ (soddyite) (V.7.2.2.1)

Moll et al. [96MOL/GEI] determined the solubility constant of soddyite in 0.1 M NaClO$_4$ at 298.15 K, using a well characterised crystalline solid. Because the test solutions at pH = 3 used by Moll et al. are over-saturated with respect to silica, this review has made a reinterpretation of their data and propose a slightly different value of the solubility product for the reaction:

$$(UO_2)_2SiO_4 \cdot 2H_2O(sodd) + 4 H^+ \rightleftharpoons 2 UO_2^{2+} + Si(OH)_4(aq) + 2 H_2O(l) \quad (9.40).$$

$\log_{10} {}^*K_s (9.40) = (7.1 \pm 0.5)$ in 0.1 M NaClO$_4$ and $\log_{10} {}^*K_s^\circ (9.40) = (6.7 \pm 0.5)$ at zero ionic strength, cf. Appendix A.

Moroni and Glasser [95MOR/GLA] have studied reactions between cement components and U(VI) oxide at 85°C, in aqueous solutions at high pH, cf. Appendix A. Soddyite is one of the phases formed and this review has used the data where this is the single solid present to evaluate the equilibrium constant for the reaction:

$$(UO_2)_2SiO_4(cr) + SiO_2 + 2H_2O(l) + 2H^+ \rightleftharpoons 2[UO_2(H_3SiO_4)^+],$$

from the experimental values of the total concentrations of dissolved uranium and silica. By combining this equilibrium constant with the equilibrium constant for the reaction:

$$UO_2^{2+} + Si(OH)_4(aq) \rightleftharpoons UO_2(H_3SiO_4)^+ + H^+,$$

selected in this review and assuming that the solubility reaction is independent of temperature, we obtain $\log_{10} {}^*K_s (9.40) = 6.4$, in fair agreement with the value proposed by Moll et al. The equilibrium constant obtained from the data [95MOR/GLA] is not selected by this review, although it supports the value reported by Moll et al.

Pérez et al. [97PER/CAS] studied the solubility of soddyite in bicarbonate solutions and their proposed solubility constant is $\log_{10} {}^*K_s^\circ (9.40) = (3.9 \pm 0.7)$. However, there is a systematic variation in the solubility product, cf. Appendix A, and the proposed equilibrium constant has therefore not been accepted by this review.

Nguyen et al. [92NGU/SIL] measured the solubilities of soddyite $(UO_2)_2SiO_4 \cdot 2H_2O$, in water under an inert atmosphere at pH = (3.00 ± 0.05) and $T = 303$ K. Calculations of solubility products are performed according to the following equilibrium:

$$(UO_2)_2SiO_4 \cdot 2H_2O(cr) + 4H^+ \rightleftharpoons 2UO_2^{2+} + Si(OH)_4(aq) + 2H_2O(l) \quad (9.41)$$

The authors assume that, in the pH range 3 to 4.5, Si is only present as H$_4$SiO$_4$(aq). Using the total measured concentrations of U and Si in the equilibrated solutions, free concentrations of all the species are calculated using thermodynamic data on U(VI) hydrolysis, silicon and other cation species taken from [92GRE/FUG]. The $\log_{10} {}^*K_s (9.41)$ value, $T = 303$ K, is derived and then that for $\log_{10} {}^*K_s^\circ$, according to the TDB Guidelines. The value for soddyite is: $\log_{10} {}^*K_s^\circ (9.41) = (5.74 \pm 0.21)$.

In view of non-concordant solubility constants, this review does not recommend a value, but suggests that the average value from Nguyen et al. [92NGU/SIL] and Moll et al. [96MOL/GEI], with increased uncertainty,

$$\log_{10} {}^*K_s^\circ (9.41) = (6.2 \pm 1.0)$$

may be used as a guideline until it has been confirmed. The estimated uncertainty covers the uncertainty ranges of the two studies.

9.7.3.2.4 Ca(UO₂)₂(SiO₃OH)₂·5 H₂O (uranophane), Na(UO₂)(SiO₃OH) ·2 H₂O (sodium boltwoodite) and Na₂(UO₂)₂(Si₂O₅)₃ · 4H₂O (sodium weeksite)

Nguyen et al. [92NGU/SIL] measured the solubilities of synthetic uranophane $Ca(UO_2)_2(SiO_3OH)_2 \cdot 5\,H_2O$ sodium boltwoodite, $Na(UO_2)(SiO_3OH) \cdot 2\,H_2O$ and sodium weeksite $Na_2(UO_2)_2(Si_2O_5)_3 \cdot 4H_2O$ in water under an inert atmosphere at pH = (3.50 ± 0.05) for uranophane and pH = (4.50 ± 0.05) for the other salts ($T = 303$ K). Calculations of the solubility constants are made according to the following equilibria[1]:

$$Ca(H_3O)_2(UO_2)_2(SiO_4)_2 \cdot 3H_2O(cr) + 6H^+ \rightleftharpoons Ca^{2+} + 2UO_2^{2+}(aq) + 2Si(OH)_4(aq) + 5H_2O(l) \quad (9.42)$$

$$Na(H_3O)UO_2SiO_4 \cdot H_2O(cr) + 3H^+ \rightleftharpoons Na^+ + UO_2^{2+} + Si(OH)_4(aq) + 2H_2O(l) \quad (9.43)$$

$$Na_2(UO_2)_2(Si_2O_5)_3 \cdot 4H_2O(cr) + 6H^+ + 5H_2O(l) \rightleftharpoons 2Na^+ + 2UO_2^{2+} + 6Si(OH)_4(aq) \quad (9.44).$$

The solubility data were analysed as described above for soddyite to derive $\log_{10} {}^*K_s$ values at $T = 303$ K and then those for $\log_{10} {}^*K_s^\circ$, according to the TDB Guidelines. The values are:

$\log_{10} {}^*K_s^\circ$ (9.42) = (9.42 ± 0.48)   for uranophane,

$\log_{10} {}^*K_s^\circ$ (9.43) > (5.82 ± 0.16)   for Na–boltwoodite,

$\log_{10} {}^*K_s^\circ$ (9.44) = (1.50 ± 0.08)   for Na–weeksite.

For Na–boltwoodite the equilibrium phase is assumed to be pure.

For reasons discussed in Appendix A concerning the purity of the phases and the calculations, and the fact that solutions are probably supersaturated with respect to silica, these values are not selected, but can be used in scoping calculations.

Pérez et al. [2000PER/CAS] have also investigated the solubility of a synthetic uranophane in bicarbonate solutions at 298.15 K of low ionic strength ($2 \cdot 10^{-2}$ M or less, pH 8 to 9, 10^{-3} M < [HCO$_3^-$] < $2 \cdot 10^{-2}$ M). The solid phase was characterised by both X-ray diffraction and partial chemical analysis. However, this review finds it unlikely that the solid used has the composition suggested by the authors (for a discussion see Appendix A). The average value of $\log_{10} {}^*K_s^\circ$ (9.42) = (11.7 ± 0.6) differs from the value of [92NGU/SIL]. This large discrepancy of $\log_{10} {}^*K_s^\circ$ for uranophane leads this review to not select any value for this mineral.

---

[1] Nguyen et al. [92NGU/SIL] used the chemical formula $Ca(H_3O)_2(UO_2)_2(SiO_4)_2 \cdot 3H_2O$ for uranophane and $Na(H_3O)(UO_2)(SiO_4) \cdot H_2O$ for sodium boltwoodite, which are equivalent to $Ca(UO_2)_2(SiO_3OH)_2 \cdot 5H_2O$ and $Na(UO_2)(SiO_3OH) \cdot 2H_2O$, respectively.

Uranophane is one of the solid phases found in the study by Moroni and Glasser [95MOR/GLA], but these data do not allow a determination of thermodynamic constants.

Murphy [97MUR] attempts to solve the problem of UO_2 alteration by water under environmental conditions according to the well-known sequence: schoepite (in fact metaschoepite, $UO_2(OH)_2 \cdot H_2O$), soddyite $(UO_2)_2 SiO_4 \cdot 2H_2O$ and uranophane $Ca(UO_2)_2(SiO_3OH)_2 \cdot 5H_2O$. He used [92GRE/FUG] data to predict the solubilities as a function of temperature, but this approach was not successful.

9.8 Uranium actinide complexes (V.8)

9.8.1 Actinides–actinides interactions (V. 8.1)

Stout et al. [93STO/CHO] give a new value of the formation constant of the cation-cation $UO_2^{2+} \cdot NpO_2^+$ in 6 M $NaClO_4$, pH = 1 to 2:

$$\beta = (2.25 \pm 0.03) \text{ L} \cdot \text{mol}^{-1},$$

which is in agreement with previous values reported in [92GRE/FUG], and the following values:

$$\Delta_f H_m^\circ = -(12.0 \pm 1.7) \text{ kJ} \cdot \text{mol}^{-1},$$
$$\Delta_r S_m^\circ = -(34 \pm 6) \text{ J} \cdot \text{K}^{-1} \cdot \text{mol}^{-1},$$

at (298.15 ± 0.10) K, for this association reaction. Additional data, $\beta = (2.4 \pm 0.2)$ $\text{L} \cdot \text{mol}^{-1}$ ($I = 6$ M, $NaClO_4$) of Stout et al. [2000STO/HOF] do not change this result.

The value of $\beta = (2.2 \pm 1.5) \text{ L} \cdot \text{mol}^{-1}$ has been reported by Stout et al [2000STO/HOF] for the $UO_2^{2+} \cdot PuO_2^+$ association reaction.

9.9 Uranium group 2 (alkaline-earth) compounds (V.9)

9.9.1 Magnesium compounds (V.9.2)

Two papers ([92DUE/FLE], [93PAT/DUE]) describe the chemical and electrochemical insertion of Mg in $\alpha-U_3O_8$ to form $Mg_xU_3O_8$ ($0 < x < 0.6$) and in $\alpha-UO_{2.95}$ to form $Mg_xUO_{2.95}$ ($0 < x < 0.26$). Similar studies were also carried out on the insertion of Li, Na and Zn [95DUE/PAT].

Such species are single phases over a wide range of composition; for instance, the phase $Mg_xU_3O_8$ retains the orthorhombic U_3O_8 structure with x up to 0.5. However, the experimental phase boundaries depend greatly on the preparation method (chemical or electrochemical) and conditions (temperature, applied potential).

In the case of $\alpha-UO_{2.95}$ insertion compounds, $Mg_{0.17}UO_{2.95}$ was described [93PAT/DUE] as a pure phase (hexagonal, isomorphous with $\alpha-UO_{2.95}$), based on

X–ray diffraction data. Its standard enthalpy of formation (using well characterised samples) was obtained from dissolution data of MgO(cr), γ–UO$_3$, and α–U$_3$O$_8$ in: 0.274 mol·dm^{-3} Ce(SO$_4$)$_2$ + 0.484 mol·dm^{-3} H$_3$BO$_3$ + 0.93 mol·dm^{-3} H$_2$SO$_4$, which is the same medium as that used by Cordfunke and Ouweltjes [81COR/OUW2].

The authors [93PAT/DUE] measured only the dissolution enthalpies of:

$Mg_{0.17}UO_{2.95}$: $\Delta_{sol}H_m(Mg_{0.17}UO_{2.95}) = -(143.99 \pm 0.60)$ kJ·mol^{-1},

MgO(cr): $\Delta_{sol}H_m(MgO) = -(149.50 \pm 0.55)$ kJ·mol^{-1}.

The values for the dissolution enthalpies of:

γ–UO$_3$: $\Delta_{sol}H_m(\gamma-UO_3) = -(79.94 \pm 0.48)$ kJ·mol^{-1},

α–U$_3$O$_8$: $\Delta_{sol}H_m(\alpha-U_3O_8) = -(354.61 \pm 1.70)$ kJ·mol^{-1},

were taken from [81COR/OUW2].

The enthalpy change corresponding to the formation of the insertion compound from the stoichiometric oxides:

$$0.17 \text{ MgO(cr)} + 0.34 \ \gamma\text{–UO}_3 + 0.22 \ \alpha\text{–U}_3\text{O}_8 \rightleftharpoons Mg_{0.17}UO_{2.95}(\text{cr}) \quad (9.45)$$

is calculated to be $\Delta_r H_m^\circ ((9.45), 298.15 \text{ K}) = (13.38 \pm 0.73)$ kJ·mol^{-1}, suggesting it is metastable with respect to the component oxides. With NEA adopted values, this gives:

$$\Delta_f H_m^\circ (Mg_{0.17}UO_{2.95}, \alpha, 298.15 \text{ K}) = -(1291.44 \pm 0.99) \text{ kJ·mol}^{-1}.$$

Using the same experimental scheme, the authors also measured: $\Delta_f H_m^\circ (UO_{2.95}, \alpha, 298.15 \text{ K}) = -(1211.28 \pm 1.28)$ kJ·mol^{-1}, (recalculated in section 9.3.3.3.3).

From this value and the above, one calculates the enthalpy effect associated with the insertion:

$$0.17 \text{ Mg} + UO_{2.95}(\alpha) \rightleftharpoons Mg_{0.17}UO_{2.95}(\text{cr}) \quad (9.46)$$

as $\Delta_r H_m^\circ ((9.46), 298.15 \text{ K}) = -(80.16 \pm 1.62)$ kJ · mol^{-1}.

Electrochemical measurements (charging and discharging curves, open circuit voltages) using the cell:

$$\text{Mg(amalgam)} \mid 0.5 \text{ mol·dm}^{-3} \text{ Mg(ClO}_4)_2 \text{ (DMF)} \mid Mg_xUO_{2.95}$$

(DMF = dimethylformamide) allowed the authors [93PAT/DUE] to calculate the Gibbs energy change for reaction:

$$x \text{ Mg} + \alpha-UO_{2.95} \rightleftharpoons Mg_xUO_{2.95}(\text{cr}) \quad (9.47)$$

as $\Delta_r G_m$ ((9.47), 294 K) = $-473 \cdot x$ kJ·mol^{-1}. For x = 0.17, this relation yields -80.4 kJ·mol^{-1}, very close to the value of the enthalpy change deduced by calorimetry, indicating a negligible entropy change.

Oxygen potentials in the (Mg, Ln, U) – O systems have been measured by Fujino and co-workers ([2001FUJ/PAR] for Ln = Ce and [2001FUJ/SAT], Ln = Eu); such studies are beyond the scope of the current review.

9.9.2 Calcium compounds (V.9.3)

9.9.2.1 Calcium uranates (V.9.3.1.)

Although no thermodynamic data are reported, Pialoux and Touzelin [98PIA/TOU] give interesting information on the complex phase relationships in the system U – Ca – O, based on X–ray diffraction studies up to about 1800 K and at oxygen pressures over the range $10^{-19} \leq p_{O_2}$ /bar ≤ 1. Fuller details are given in Appendix A. The authors present a schematic pseudo-binary phase diagram of the UO_2 – CaO system; they, like others, were unable to prepare pure stoichiometric $CaUO_3$.

9.9.2.1.1 CaUO$_4$(cr) (V.9.3.1.1)

Moroni and Glasser [95MOR/GLA] have studied the solubility of U(VI) oxides at 358 K at high pH, in the presence of Ca^{2+}. As noted in Appendix A, these data provide an approximate value of $\Delta_f G_m^\circ$ (CaUO$_4$, cr, 298.15 K) of -1848 kJ·mol^{-1}, which, with all the assumptions involved, gives a qualitative confirmation of the well-founded value of $-(1888.7 \pm 2.4)$ kJ·mol^{-1} selected in [92GRE/FUG].

9.9.2.1.2 CaU$_6$O$_{19}$·11H$_2$O(cr)

The solubility of this compound is discussed in section 9.3.2.1.5.2.1.

9.9.2.1.3 Other calcium uranates

Microprobe analysis and X–ray diffraction [97VAL/RAG] showed that the well-crystallised form of Ca$_3$U$_{4.5}$O$_{16.5}$(cr) was obtained by treatment, at 573 K and 0.5 kbar, of a poorly crystallised material, CaU$_{1.6}$O$_{5.8}$·2.5H$_2$O in pure water or a 0.01 mol·dm^{-3} solution of CaCl$_2$. The solubility of the final product of the treatment ranged, in all experiments, between $10^{-5.9}$ and $10^{-6.3}$ mol·dm^{-3}. These data cannot be used to deduce thermodynamic constants and are given for information only.

As noted in Appendix A, Takahashi *et al.* [93TAK/FUJ] have made Gibbs energy measurements on non–stoichiometric calcium uranates which are beyond the scope of the present review.

9.9.3 Strontium compounds (V.9.4)

Although no thermodynamic data are reported, Pialoux and Touzelin [99PIA/TOU] give interesting information on the complex phase relationships in the system U – Sr – O, based on X–ray diffraction studies up to about 1800 K and at oxygen pressures over the range $10^{-19} \leq p_{O_2} /$ bar ≤ 1. Fuller details are given in Appendix A. The authors present a schematic pseudo-binary phase diagram between "SrUO$_3$" and SrUO$_4$, although they, like others, were unable to prepare pure stoichiometric SrUO$_3$.

9.9.3.1 SrUO$_3$(cr)

Fuger et al. [93FUG/HAI] have used the Goldschmidt tolerance factor [70GOO/LON], page 132, to correlate experimental values of the enthalpy of formation of MM'O$_3$ oxides from the binary oxides (M = Ba, and M' = Ti, Hf, Zr, Ce, Tb, U, Pu, Am, Cm, and M = Sr, and M' = Ti, Mo, Zr, Ce, Tb, Am). They thus estimated the enthalpy of formation of SrUO$_3$ to be 3 kJ·mol^{-1} less negative than the sum of the enthalpies of formation of the binary oxides. Thus, using [92GRE/FUG] accepted data, we find: $\Delta_f H_m^\circ$ (SrUO$_3$, cr, 298.15 K) = $-$ (1672.6 $\pm$ 8.6) kJ·mol^{-1}, (including the uncertainty arising from the correlation). This value is in accord with that, $-$ (1675 $\pm$ 10) kJ·mol^{-1}, estimated by Cordfunke and Ijdo [94COR/IJD], using experimentally available data for SrM'O$_3$ compounds only.

Huang et al. [97HUA/YAM] carried out vaporisation studies over the temperature range 1534 – 1917 K using a mixture of "SrUO$_3$" and UO$_2$, due to the reported difficulties in preparing single phase SrUO$_3$. From the mass spectrometric measurements, the composition of the ternary oxide was estimated to be SrUO$_{3.1}$, based on X–ray diffraction data. Using an estimated enthalpy increment and values of $H_m(T) - H_m^\circ(298.15 \text{ K})$ estimated from those of SrZrO$_3$, they obtained the standard enthalpy of formation of "SrUO$_3$". We have assumed this applies to the final oxide (see full discussion in Appendix A):

$$\Delta_f H_m^\circ (\text{SrUO}_{3.1}, \text{cr}, 298.15 \text{ K}) = - (1785 \pm 60) \text{ and} - (1698.1 \pm 5.0) \text{ kJ·mol}^{-1},$$

from the second- and third-law treatments, respectively, where the latter is recalculated by this review. The latter value is (6 $\pm$ 10) kJ·mol^{-1} more negative than the sum of $\Delta_f H_m^\circ$ (UO$_{2.1}$, cr, 298.15 K) = $-$ (1102 $\pm$ 5) kJ·mol^{-1} (our estimate) and $\Delta_f H_m^\circ$ (SrO, cr, 298.15 K) = $-$ (590.6 $\pm$ 0.9) kJ·mol^{-1} [92GRE/FUG]. The suggestion that stoichiometric SrUO$_3$ is of very marginal stability compared to the binary oxides is supported by the difficulty in obtaining a pure phase, as acknowledged by [97HUA/YAM] and by the observations of [94COR/IJD]. We therefore select for the stoichiometric oxide:

$$\Delta_f H_m^\circ (\text{SrUO}_3, \text{cr}, 298.15 \text{ K}) = - (1672.6 \pm 8.6) \text{ kJ·mol}^{-1},$$

as proposed by [93FUG/HAI]. This value is not incompatible with the conclusions of [97HUA/YAM]. Yamashita and Fujino [89YAM/FUJ] have measured the oxygen potential of Sr$_y$Y$_y$U$_{1-2y}$O$_{2+x}$ (y = 0.05 and 0.025) solid solutions by thermogravimetric methods between 1123 and 1673 K, but such studies are not part of the current NEA review effort.

9.9.3.2 Strontium uranates (VI)

9.9.3.2.1 α–SrUO$_4$ (V.9.4.1.1)

Takahashi et al. [93TAK/FUJ] have determined the molar enthalpy of formation of α-SrUO$_{4-x}$ at 298.15 K for $0 < x < 0.478$, based on the dissolution of well-characterised compounds in 5.94 mol·dm^{-3} HCl. The authors also made use of the enthalpy of dissolution of SrCl$_2$(cr) in 1 mol·dm^{-3} HCl [83MOR/WIL] and of the enthalpy of transfer of SrCl$_2$(cr) from 1 to 5.94 mol·dm^{-3} HCl [78PER/THO]. Their results, recalculated for a 95% confidence interval with the NEA accepted auxiliary values, are summarised below.

Table 9-35: Enthalpies $\Delta_{sol}H_m$ and $\Delta_f H_m$ of SrUO$_{4-x}$.

x in SrUO$_{4-x}$	$\Delta_{sol}H_m$ (kJ·mol^{-1})	$\Delta_f H_m$ (kJ·mol^{-1})
0.478	$-(170.55 \pm 1.89)$	$-(1894.31 \pm 2.83)$
0.380	$-(152.71 \pm 0.60)$	$-(1926.14 \pm 2.09)$
0.297	$-(149.24 \pm 0.99)$	$-(1941.47 \pm 2.19)$
0.127	$-(143.40 \pm 7.90)$	$-(1971.61 \pm 8.11)$
0.0	$-(142.86 \pm 2.45)$	$-(1990.28 \pm 3.22)$

The value for the enthalpy of formation of the stoichiometric α–SrUO$_4$ supports that selected in [92GRE/FUG]:

$$\Delta_f H_m^\circ (\text{SrUO}_4, \alpha, 298.15 \text{ K}) = -(1989.6 \pm 2.8) \text{ kJ·mol}^{-1},$$

which was based on the results of [67COR/LOO]. We thus maintain this selection.

Following the authors, we have fitted the five data points in Table 9-35 to the quadratic function:

$$\Delta_f H_m (\text{SrUO}_{4-x}, \alpha, 298.15 \text{ K}) = -1989.3 + 98.1 \, x + 202.8 \, x^2,$$

from which the partial molar enthalpy of solution of oxygen can be approximated as a function of x.

9.9.3.2.2 β–SrUO$_4$ (V.9.4.1.2)

In a rather comprehensive paper, Cordfunke et al. [99COR/BOO] recently reported the enthalpies of formation of β–SrUO$_4$ together with those of a series of strontium uranates, based on dissolution of well-characterised compounds in appropriate media at 298.15 K. For β–SrUO$_4$ and for Sr$_3$UO$_6$, Sr$_3$U$_{11}$O$_{36}$ and Sr$_5$U$_3$O$_{14}$, the medium was 5.075 mol·dm^{-3} HCl (sln. A). In the same paper, the authors also reported the enthalpies of formation of Sr$_2$UO$_{4.5}$ and Sr$_2$UO$_5$ from dissolution in 1.00 mol·dm^{-3} HCl + 0.0470 mol·dm^{-3} FeCl$_3$ (sln. B). The authors also give a recalculation of an earlier study [67COR/LOO] of the dissolution of β–SrUO$_4$ in 6.00 mol·dm^{-3} HNO$_3$, discussed below. The auxiliary data on the enthalpies of formation of the relevant solids and the partial molar enthalpies of formation of HCl and HNO$_3$ in the solutions are listed in the discussion of [99COR/BOO] in Appendix A. The dissolution in the appropriate media of mixtures of $SrCl_2$ and $\gamma\text{–}UO_3$ with variable stoichiometric ratios was also carried out in order to close the relevant thermodynamic cycles.

The authors also report a new determination of the enthalpy of solution of $\gamma\text{–}UO_3$ in 6.00 mol·dm^{-3} HNO$_3$ as:

$$\Delta_{sol}H_m(UO_3, \gamma, 6.00 \text{ mol·dm}^{-3} \text{ HNO}_3) = -(72.10 \pm 0.33) \text{ kJ·mol}^{-1}.$$

Previous determinations by the same group yielded $-(71.30 \pm 0.13)$ kJ·mol^{-1} [64COR] and $-(72.05 \pm 0.25)$ kJ·mol^{-1} [75COR], cited by [78COR/OHA], for this quantity. In subsequent calculations involving $\gamma\text{–}UO_3$ in this medium, we will use the weighted average $-(71.53 \pm 0.50)$ kJ·mol^{-1}, keeping conservative uncertainty limits.

As discussed in the entry for [99COR/BOO] in Appendix A, the two recalculated values for $\Delta_f H_m^\circ(SrUO_4, \beta, 298.15 \text{ K})$ are $-(1985.75 \pm 1.57)$ kJ·mol^{-1} from dissolution in (sln. A) and $-(1991.13 \pm 2.72)$ kJ·mol^{-1} from dissolution in 6.00 mol·dm^{-3} HNO$_3$.

As these two values do not overlap within their uncertainties, and since we have no reason to prefer one to the other, we shall accept, with increased uncertainty limits, their average,

$$\Delta_f H_m^\circ(SrUO_4, \beta, 298.15 \text{ K}) = -(1988.4 \pm 5.4) \text{ kJ·mol}^{-1},$$

as the selected value. This value is slightly different from that adopted by [92GRE/FUG], $-(1990.8 \pm 2.8)$ kJ·mol^{-1}.

9.9.3.2.3 SrU_4O_{13} (V.9.4.2)

In their recent publication on strontium uranates, containing new structural results in the range $0.25 < Sr/U < 0.33$, [99COR/BOO] note that a compound previously believed [67COR/LOO] to be SrU_4O_{13} (but always containing some U_3O_8) could not correspond to a pure phase. Consequently, we shall no longer give data for this species.

9.9.3.2.4 Sr_2UO_5 (V.9.4.3)

As for β–SrUO$_4$, discussed in section 9.9.3.2.2, we have recalculated $\Delta_f H_m^\circ(Sr_2UO_5, \text{cr}, 298.15 \text{ K})$ to be $-(2631.50 \pm 2.31)$ kJ·mol^{-1} from dissolution in (sln. B) and $-(2635.88 \pm 3.38)$ kJ·mol^{-1} from dissolution in 6.00 mol·dm^{-3} HNO$_3$. The two values overlap within their uncertainties and we adopt the weighted mean:

$$\Delta_f H_m^\circ(Sr_2UO_5, \text{cr}, 298.15 \text{ K}) = -(2632.9 \pm 1.9) \text{ kJ·mol}^{-1}$$

which overlaps with the value in [92GRE/FUG], $-(2635.6 \pm 3.4)$ kJ·mol^{-1} within the uncertainties.

9.9.3.2.5 Sr$_2$U$_3$O$_{11}$ (V.9.4.4)

The value of the enthalpy of formation of this compound selected by [92GRE/FUG] on the basis of the experimental results of [67COR/LOO] has been recalculated to account for the small change in the adopted value for the enthalpy of solution of γ–UO$_3$ in 6.00 mol·dm^{-3} HNO$_3$ (see section 9.9.3.2.2). We obtain:

$$\Delta_f H_m^\circ (\text{Sr}_2\text{U}_3\text{O}_{11}, \text{cr}, 298.15 \text{ K}) = -(5243.7 \pm 5.0) \text{ kJ} \cdot \text{mol}^{-1},$$

with a slightly larger uncertainty (5.0 instead of 4.1 kJ·mol^{-1}).

9.9.3.2.6 Sr$_3$UO$_6$ (V.9.4.5)

Cordfunke et al. [99COR/BOO] recently re-measured the enthalpy of formation of this compound using 5.075 mol·dm^{-3} HCl (sln. A) as solvent. The cycle used is analogous to that for β–SrUO$_4$ (see section 9.9.3.2.2 and Appendix A). Using NEA adopted auxiliary values, we recalculate $\Delta_f H_m^\circ (\text{Sr}_3\text{UO}_6, \text{cr}, 298.15 \text{ K}) = -(3255.39 \pm 2.92)$ kJ·mol^{-1}. This value is in marginal disagreement (just beyond the combined uncertainties) with the value $-(3263.08 \pm 4.24)$ kJ·mol^{-1}, recalculated from the data of [67COR/LOO] in 6.00 mol·dm^{-3} HNO$_3$, and with the value of $-(3263.95 \pm 4.39)$ kJ·mol^{-1}, based on the results of [83MOR/WIL2] using 1.00 mol·dm^{-3} HCl as solvent. Since the latest value is barely consistent with the two earlier results, we have preferred to adopt the mean of these two values, namely,

$$\Delta_f H_m^\circ (\text{Sr}_3\text{UO}_6, \text{cr}, 298.15 \text{ K}) = -(3263.5 \pm 3.0) \text{ kJ} \cdot \text{mol}^{-1}.$$

The value selected by [92GRE/FUG] was $\Delta_f H_m^\circ (\text{Sr}_3\text{UO}_6, \text{cr}, 298.15 \text{ K}) = -(3263.4 \pm 3.0)$ kJ·mol^{-1}.

9.9.3.2.7 Sr$_3$U$_{11}$O$_{36}$

Cordfunke et al. [99COR/BOO] recently measured the enthalpy of formation of this compound using 5.075 mol^{-1}·dm^{-3} HCl (sln. A) as solvent. The cycle used is analogous to that for β–SrUO$_4$ (see section 9.9.3.2.2). The experimentally reported enthalpy of solution is $\Delta_{sol} H_m (\text{Sr}_3\text{U}_{11}\text{O}_{36}, \text{sln A}) = -(795.11 \pm 1.17)$ kJ·mol^{-1}.

We recalculate the enthalpy of formation to be:

$$\Delta_f H_m^\circ (\text{Sr}_3\text{U}_{11}\text{O}_{36}, \text{cr}, 298.15 \text{ K}) = -(15903.8 \pm 16.5) \text{ kJ} \cdot \text{mol}^{-1},$$

which is the selected value.

It should be noted that in a previous study by the same group [91COR/VLA] the dissolution of apparently the same sample (judging from the analytical results) of Sr$_3$U$_{11}$O$_{36}$ in 5.0 mol·dm^{-3} HCl was given without details as $-(794.9 \pm 1.2)$ kJ·mol^{-1}; we shall not make use here of this poorly documented result.

Dash et al. [2000DAS/SIN] have measured enthalpy increments of this compound from 300 to 1000 K, from which they derived the heat capacity expression:

$$C_{p,m}(Sr_3U_{11}O_{36}, cr, T) = 962.72 + 0.35526\,T - 3.954\times10^5\,T^{-2}\ \text{J}\cdot\text{K}^{-1}\cdot\text{mol}^{-1}$$
(298.15 to 1000 K).

and thus

$$C_{p,m}^\circ(Sr_3U_{11}O_{36}, cr, 298.15\ \text{K}) = (1064.2 \pm 10.6)\ \text{J}\cdot\text{K}^{-1}\cdot\text{mol}^{-1}.$$

9.9.3.2.8   Sr$_5$U$_3$O$_{14}$

Cordfunke et al. [99COR/BOO] investigated the enthalpy of formation of this compound. They report that it could not be obtained as a pure phase: on the basis of the analyses and of the X–ray diffraction pattern, they concluded that their sample contained 4.4 % of β–SrUO$_4$. They corrected their experimental enthalpy of solution for the known enthalpy of solution of β–SrUO$_4$ in the same medium (sln. A, see section 9.9.3.2.2 above and Appendix A). Given the uncertainties in the assumptions made for those corrections, we will adopt here conservative uncertainty limits, viz. one half of the correction for the impurities and take:

$$\Delta_{sol}H_m(Sr_5U_3O_{14}, \text{pure, sln. A}) = -(829.63 \pm 5.30)\ \text{kJ}\cdot\text{mol}^{-1}.$$

With the adopted auxiliary data, we recalculate and select:

$$\Delta_f H_m^\circ(Sr_5U_3O_{14}, cr, 298.15\ \text{K}) = -(7248.6 \pm 7.5)\ \text{kJ}\cdot\text{mol}^{-1}.$$

This value is noticeably different from that reported, $-(7265.8 \pm 7.5)$ kJ·mol^{-1}, even taking into account the different auxiliary data, for an unknown reason.

9.9.3.2.9   Sr$_3$U$_2$O$_9$

Dash et al. [2000DAS/SIN] have estimated the enthalpy of formation of Sr$_3$U$_2$O$_9$(cr) from electronegativity considerations. However, as described in the Appendix A entry for that reference, we have preferred to estimate this value from the enthalpies of other strontium uranates(VI) considered above. If the formation of Sr$_3$U$_2$O$_9$(cr) from its neighbours is exothermic, $\Delta_f H_m^\circ(Sr_3U_2O_9, cr, 298.15\ \text{K})$ is $< -(4619.9 \pm 4.0)$ kJ·mol^{-1}, whereas a similar condition for the decomposition of Sr$_5$U$_3$O$_{11}$(cr) implies $\Delta_f H_m^\circ(Sr_3U_2O_9, cr, 298.15\ \text{K}) > -(4616.0 \pm 8.0)$ kJ·mol^{-1}. Although these values are formally incompatible, the selected value:

$$\Delta_f H_m^\circ(Sr_3U_2O_9, cr, 298.15\ \text{K}) = -(4620 \pm 8)\ \text{kJ}\cdot\text{mol}^{-1}$$

is consistent with these conditions within their uncertainties. This value is somewhat more negative than that proposed by [2000DAS/SIN] (whose paper however was written before the extensive new data of [99COR/BOO] had appeared).

Dash et al. [2000DAS/SIN] have measured enthalpy increments of this compound from 300 to 1000 K, from which they derived the heat capacity expression

$$C_{p,m}(Sr_3U_2O_9, cr, T) = 319.18 + 0.11602\,T - 4.6201\times10^6\,T^{-2}\ \text{J}\cdot\text{K}^{-1}\cdot\text{mol}^{-1}$$
(298.15 to 1000 K),

$$C_{p,m}^\circ(Sr_3U_2O_9, cr, 298.15\ \text{K}) = (301.8 \pm 3.0)\ \text{J}\cdot\text{K}^{-1}\cdot\text{mol}^{-1}.$$

9.9.3.3 Strontium uranates (V)

9.9.3.3.1 $Sr_2UO_{4.5}$

Cordfunke *et al.* [99COR/BOO] dissolved this U(V) compound in 1.00 mol·dm^{-3} HCl + 0.0470 mol·dm^{-3} FeCl$_3$ (sln. B), obtaining $\Delta_{sol}H_m$ (Sr$_2$UO$_{4.5}$, sln. B) = $-(389.97 \pm 0.91)$ kJ·mol^{-1}. The enthalpy of dissolution of apparently the same sample (as judged by the analytical results) in (HCl + 0.0470 FeCl$_3$ + 82.16 H$_2$O), was given as $-(390.66 \pm 0.51)$ kJ·mol^{-1}, without any details, by the same group of authors [94COR/IJD]. We shall not make use here of this less well-documented result.

The enthalpy of solution of the appropriate mixture of SrCl$_2$ and γ–UO$_3$ was also reported as $\Delta_{sol}H_m$ (4 SrCl$_2$ + γ–UO$_3$, sln. B) = $-(259.25 \pm 0.50)$ kJ·(mol UO$_3$)$^{-1}$. The enthalpy of solution of UCl$_4$ in (HCl, 0.0419 FeCl$_3$ + 70.66 H$_2$O), $\Delta_{sol}H_m$ (UCl$_4$, sln. B) = $-(186.67 \pm 0.60)$ kJ·mol^{-1} (taken from [88COR/OUW]) is also used in this cycle. Using the same cycle as the authors, we recalculate:

$$\Delta_f H_m^\circ (Sr_2UO_{4.5}, cr, 298.15\ K) = -(2494.0 \pm 2.8)\ kJ \cdot mol^{-1}.$$

The uncertainty limits on this value have been increased slightly because of the minor differences in the media involved in the cycle used by [99COR/BOO].

9.9.3.4 Strontium uranium tellurites

Sali *et al.* [97SAL/KRI] have prepared and characterised, by X-Ray diffraction, the new U(VI) compound SrUTe$_2$O$_8$(cr), from a 1:2 molar ratio mixture of SrUO$_4$(cr) and TeO$_2$(cr) in an argon atmosphere at 1073 K. The same compound could be obtained by heating a 1:1 molar mixture of SrTeO$_3$(cr) and UTeO$_5$(cr). No thermodynamic data have been reported for this phase.

9.9.4 Barium compounds (V.9.5)

Although no thermodynamic data are reported, Touzelin and Pialoux [94TOU/PIA] give interesting information on the complex phase relationships in the system U – Ba – O, based on X-ray diffraction studies up to about 1800 K and at oxygen pressures over the range $10^{-19} \leq p_{O_2}/\text{bar} \leq 1$. Fuller details are given in Appendix A. The authors present a schematic pseudo-binary phase diagram between "BaUO$_3$" and BaUO$_4$, although they, like others, were unable to prepare pure stoichiometric BaUO$_3$.

9.9.4.1 BaUO$_3$(cr) (V.9.5.1.1)

In a paper predominantly dealing with the structural chemistry of barium uranates, Cordfunke *et al.* [97COR/BOO] reported integral enthalpies of formation of five compounds Ba$_y$UO$_{3+x}$(cr) with $1.033 \leq y \leq 1.553$ and $0.134 \leq x \leq 0.866$, determined by solution calorimetry, the calorimetric solvent being (HCl + 0.0400 FeCl$_3$ + 70.68 H$_2$O).

As noted in Appendix A, our calculated values of the enthalpies of formation are appreciably less negative than those reported by the authors, whose reaction scheme contained an error. We obtain:

$$\Delta_f H_m^\circ (Ba_{1.033}UO_{3.134}, cr, 298.15\ K) = -(1721.68 \pm 3.57)\ kJ \cdot mol^{-1},$$

$$\Delta_f H_m^\circ (Ba_{1.065}UO_{3.172}, cr, 298.15\ K) = -(1733.43 \pm 3.65)\ kJ \cdot mol^{-1},$$

$$\Delta_f H_m^\circ (Ba_{1.238}UO_{3.407}, cr, 298.15\ K) = -(1868.05 \pm 4.16)\ kJ \cdot mol^{-1},$$

$$\Delta_f H_m^\circ (Ba_{1.400}UO_{3.604}, cr, 298.15\ K) = -(1982.10 \pm 4.38)\ kJ \cdot mol^{-1},$$

$$\Delta_f H_m^\circ (Ba_{1.553}UO_{3.866}, cr, 298.15\ K) = -(2116.90 \pm 4.38)\ kJ \cdot mol^{-1}.$$

The authors note a linear change of the enthalpy of formation as a function of the Ba/U from Ba_3UO_6 to $Ba_{1.033}UO_{3.134}$ in these compounds, including the values of $\Delta_f H_m^\circ (BaUO_{3.05}, cr, 298.15\ K) = -(1700.4 \pm 3.1)\ kJ \cdot mol^{-1}$ and $\Delta_f H_m^\circ (BaUO_{3.08}, cr, 298.15\ K) = -(1710.0 \pm 3.0)\ kJ \cdot mol^{-1}$ of unspecified origin. Since these two results may suffer from the same error as the detailed measurements, they have been discounted.

However, since both the Ba/U ratio and the uranium valence are changing, this approximately linear relation for the five samples measured must be fortuitous (or perhaps related to the oxygen pressure during preparation). Thus the authors' derived value for the enthalpy of formation of $BaUO_3(cr)$ as $-(1680 \pm 10)\ kJ \cdot mol^{-1}$ (which becomes $-(1688 \pm 10)\ kJ \cdot mol^{-1}$ with the corrected values) by extrapolation of this linear relationship must be treated with some circumspection.

At this stage, we see no argument to modify the previously adopted value for stoichiometric $BaUO_3(cr)$ [92GRE/FUG],

$$\Delta_f H_m^\circ (BaUO_3, cr, 298.15\ K) = -(1690 \pm 10)\ kJ \cdot mol^{-1},$$

based on the solution calorimetry by [84WIL/MOR] who used two samples of $BaUO_{3+x}(cr)$ with x = 0.20 and 0.06.

Cordfunke et al. [97COR/BOO] also measured the oxygen potential as a function of the composition of $Ba_{1.033}UO_{3+x}$ samples using a reversible emf cell of the type:

$$(Pt)\ Ba_{1.033}UO_{3+x}(cr)\ |\ CaO-stabilized\ ZrO_2\ |\ air\ (Pt),\ p_{O_2} = 0.202\ bar$$

The authors used a linear extrapolation of the potentials to x = 0 to approximate the value of the equilibrium oxygen potential of stoichiometric $BaUO_3(cr)$ at 1060 K, which becomes $-629\ kJ \cdot mol^{-1}$ with our refitted equation (see Appendix A). This is, in fact, likely to be an upper limit for the oxygen potential, which probably drops sharply near x = 0 as all the uranium becomes quadrivalent (cf. UO_{2+x}). Such a value is low enough to make the preparation of unoxidised stoichiometric $BaUO_3$ rather difficult, thus supporting the long reported failure ([75BRA/KEM], [82BAR/JAC],

[84WIL/MOR]) to obtain this phase. Hinatsu [93HIN] also reported a composition of $BaUO_{3.023}$ as the lowest oxygen content that he could attain when reducing $BaUO_4$ in a flow of hydrogen. The thermophysical properties of the "$BaUO_3$" phase reported by Yamanaka et al. [2001YAM/KUR] also concern a compound with an oxygen content greater than $BaUO_3$, similar to those reported by other authors, as judged by the lattice parameter.

Yamawaki et al. [96YAM/HUA] have studied the vaporisation of a sample of "$BaUO_3$" by Knudsen effusion mass spectrometry, finding $BaO(g)$, $Ba(g)$ and $UO_2(g)$ in the vapour. They processed their results (with estimated thermal functions for $BaUO_3(cr)$) to derive second- and third-law enthalpies of formation $\Delta_f H_m^\circ$ ($BaUO_3$, cr, 298.15 K) of $-(1742.5 \pm 10.0)$ kJ·mol^{-1} (uncertainty increased) and $-(1681.8 \pm 20.8)$ kJ·mol^{-1}, respectively. However, as noted in Appendix A, it is very probable that the vaporisation was non-congruent, with likely changes in the Ba/U as well as the O/U ratio.

The calculated values are in general agreement with calorimetric data discussed above, but clearly cannot be used to establish the properties of a barium uranate of any defined composition.

Fujino et al. [91FUJ/YAM] have measured the oxygen potential of solid solutions of composition $Ba_{0.05}Y_{0.05}U_{0.9}O_{2+x}$ between 1173 and 1573 K and made an extensive comparison with UO_{2+x} and lanthanide ternary and quaternary oxide systems. These non–stoichiometric quaternary phases are beyond the scope of the present review.

9.10 Uranium group 1 (alkali) compounds (V.10)

9.10.1 General

Iyer, Jayanthi and Venugopal [92VEN/IYE], [97IYE/JAY], [97JAY/IYE], [98JAY/IYE], have made a number of studies on the thermodynamic properties (enthalpy increments and Gibbs energies) of numerous uranates of the alkali metals. Experimental details are given in Appendix A, and the Gibbs energy studies are discussed both in Appendix A and under the headings for the various compounds. For convenience, the published data on the enthalpy increments of these uranates are summarised and compared here.

The authors have fitted their experimental increments $H_m(T) - H_m^\circ(298.15 \text{ K})$ to simple polynomials in T; unfortunately, in most of the studies, they have not imposed the necessary constraint of $H_m(T) - H_m^\circ(298.15 \text{ K}) = 0$ at $T = 298.15$ K, so their calculated heat capacities near 298.15 K will be in error. All their enthalpy increment measurements have therefore been recalculated with the proper constraint. The data points are somewhat scattered (deviations of 5% from the fitted curve are not uncommon), and the heat capacities at the ends of the (rather narrow) temperature ranges depend appreciably on the equation used in the fit. Thus the heat capacities at 298.15 K have appreciable uncertainties.

The best fits are shown in Table 9-36 and Figure 9-6, which also include data for related compounds from [92GRE/FUG].

Figure 9-6: Heat capacity *versus* temperature for some alkali uranates.

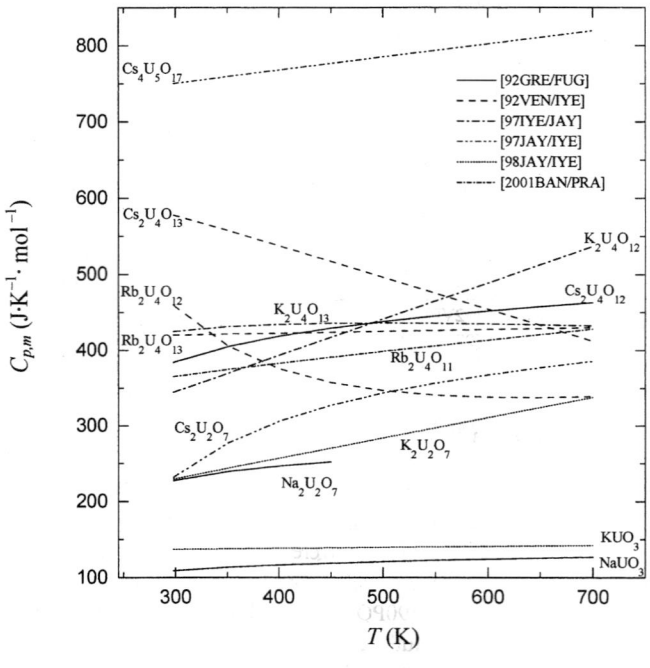

It is clear that while much of the data forms a consistent pattern, some of the variations of $C_{p,m}$ with temperature are not realistic, perhaps due to decomposition during the measurements. In particular, the data for $K_2U_4O_{12}(cr)$, $Rb_2U_4O_{12}(cr)$, $Cs_2U_4O_{13}(cr)$, appear anomalous and have not been accepted.

Heat capacity data are selected only for those compounds in bold and are included in Table 9-36.

Table 9-36: Derived heat capacity equations.

Phase		T range (K)	\multicolumn{3}{c}{Heat capacity coefficients $C_{p,m} = a + b \cdot T + e \cdot T^{-2}$ J·K^{-1}·mol^{-1}}			$C^°_{p,m}$, 298.15 K J·K^{-1}·mol^{-1}
			a	b	e	
KUO$_3$(cr)	(a)	369 – 714	1.332577 10^2	1.2558 10^{-2}	0	(137 ± 10)
K$_2$U$_2$O$_7$(cr)	(a)	391 – 683	1.490840 10^2	2.6950 10^{-1}	0	(229 ± 20)
K$_2$U$_4$O$_{12}$(cr)	(b)	426 – 770	2.022090 10^2	4.7780 10^{-1}	0	
K$_2$U$_4$O$_{13}$(cr)	(b)	411 – 888	4.710680 10^2	– 4.6890 10^{-2}	– 2.8754 10^6	(425 ± 50)
Rb$_2$U$_4$O$_{11}$(cr)	(c)	396 – 735	3.304 10^2	1.4134 10^{-2}	– 6.198 10^5	(365.6 ± 25.0)
Rb$_2$U$_4$O$_{11}$(cr)	(c)	365 – 735	3.321400 10^2	7.1252 10^{-2}	0	(353 ± 25)
Rb$_2$U$_4$O$_{12}$(cr)	(d)	375 – 755	1.897590 10^2	1.5487 10^{-1}	1.9903 10^7	
Rb$_2$U$_4$O$_{13}$(cr)	(d)	325 – 805	4.125600 10^2	2.5000 10^{-2}	0	(420 ± 50)
Cs$_2$U$_2$O$_7$(cr)	(e, f)	298.15 – 852	3.553248 10^2	8.15759 10^{-2}	– 1.31959 10^7	(231.2 ± 0.5)
Cs$_2$U$_4$O$_{12}$(cr)	(d, g)	361 – 719	4.237260 10^2	7.19405 10^{-2}	– 5.43750 10^6	(384.0 ± 1.0)
Cs$_2$U$_4$O$_{13}$(cr)	(d)	347 – 753	7.128220 10^2	– 4.2745 10^{-1}	– 6.8805 10^5	
Cs$_4$U$_5$O$_{17}$(cr)	(e)	368 – 906	6.992110 10^2	1.7199 10^{-1}	0	(750 ± 50)

(a) [98JAY/IYE] (d) [92VEN/IYE] (g) [80COR/WES]
(b) [97IYE/JAY] (e) [97JAY/IYE]
(c) [2001BAN/PRA] (f) [81OHA/FLO]

9.10.2 Lithium compounds (V.10.1)

9.10.2.1 Other uranates

The preparation, characterisation, thermochemical and electrochemical properties of δ–Li$_{0.69}$UO$_3$, γ–Li$_{0.55}$UO$_3$ and Li$_{0.88}$U$_3$O$_8$(cr) were reported by Dickens et al. [89DIC/LAW]. More experimental details on this work are given in Appendix A in the reviews of [86DIC/PEN, 89DIC/LAW], [90POW] as well as in the original references. The calorimetric reagent for the dissolution, 0.0350 mol·dm^{-3} Ce(SO$_4$)$_2$ + 1.505 mol·dm^{-3} H$_2$SO$_4$, was the same as that used in the studies by [77COR/OUW2], [81COR/OUW], [88DIC/POW] and [90POW].

The enthalpies of solution were given as:

$\Delta_{sol}H_m$ (Li$_{0.69}$UO$_3$, δ) = – (165.24 ± 1.94) kJ·mol^{-1},

$\Delta_{sol}H_m$ (Li$_{0.55}$UO$_3$, γ) = – (146.12 ± 2.52) kJ·mol^{-1},

$\Delta_{sol}H_m$ (Li$_{0.88}$U$_3$O$_8$, cr) = – (460.52 ± 1.60) kJ·mol^{-1}.

With the auxiliary enthalpies of solution of the binary uranium oxides used also in [90POW], and that for LiUO$_3$(cr), – (208.35 ± 0.53) kJ·mol^{-1} (all original data are from [77COR/OUW2], [81COR/OUW]), the selected enthalpies of the reactions from the stoichiometric phases are calculated as in Table 9-37.

Table 9-37: Derived enthalpies of reactions from stoichiometric phases.

Phase	Reaction	$\Delta_r H_m$ (kJ · mol^{-1})
$Li_{0.69}UO_3(\delta)$	$0.69\,LiUO_3(cr) + 0.31\,UO_3(\gamma) \rightleftharpoons Li_{0.69}UO_3(\delta)$	$-(4.76 \pm 1.98)$ [a]
$Li_{0.55}UO_3(\gamma)$	$0.55\,LiUO_3(cr) + 0.45\,UO_3(\gamma) \rightleftharpoons Li_{0.55}UO_3(\gamma)$	$-(6.56 \pm 2.54)$ [a]
$Li_{0.88}U_3O_8(cr)$	$0.88\,LiUO_3(cr) + U_3O_8(cr) \rightleftharpoons Li_{0.88}U_3O_8(cr) + 0.88\,UO_3(\gamma)$	(0.39 ± 2.16) [a]
$Li_{0.12}UO_{2.95}(cr)$	$0.12\,LiUO_3(cr) + 0.05\,U_3O_8(\alpha) + 0.73\,UO_3(\gamma) \rightleftharpoons Li_{0.12}UO_{2.95}(cr)$	$-(0.18 \pm 0.82)$ [b]
$Li_{0.19}U_3O_8(cr)$	$0.19\,LiUO_3(cr) + U_3O_8(\alpha) \rightleftharpoons 0.19\,UO_3(\gamma) + Li_{0.19}U_3O_8(cr)$	$-(1.44 \pm 1.94)$ [b]

(a) [89DIC/LAW]   (b) [95DUE/PAT]

Using NEA accepted values, we obtain the selected enthalpies of formation:

$$\Delta_f H_m^\circ (Li_{0.69}UO_3, \delta, 298.15K) = -(1434.5 \pm 2.4)\,kJ \cdot mol^{-1},$$
$$\Delta_f H_m^\circ (Li_{0.55}UO_3, \gamma, 298.15K) = -(1394.5 \pm 2.8)\,kJ \cdot mol^{-1},$$
$$\Delta_f H_m^\circ (Li_{0.88}U_3O_8, cr, 298.15K) = -(3837.1 \pm 3.8)\,kJ \cdot mol^{-1}.$$

For the insertion reactions one then obtains, using NEA accepted values, the values in Table 9-38. The electrochemical results are commented upon in Appendix A and will not be further used here.

Dueber et al. [95DUE/PAT] report similar calorimetric and electrochemical studies on two single phased compounds $Li_{0.12}UO_{2.95}$ and $Li_{0.19}U_3O_8$. The enthalpies of solution were given as:

$$\Delta_{sol}H_m (Li_{0.12}UO_{2.95}, cr) = -(100.91 \pm 0.73)\,kJ \cdot mol^{-1},$$
$$\Delta_{sol}H_m (Li_{0.19}U_3O_8, cr) = -(377.57 \pm 0.91)\,kJ \cdot mol^{-1}.$$

With the above auxiliary data, the enthalpies of the reactions from uranium oxides are calculated to be those given in Table 9-38.

Table 9-38: Derived enthalpies of insertion reactions.

Phase	Reaction	$\Delta_r H_m$ kJ · mol^{-1}	$\Delta_r H_m$ kJ · (mol Li)$^{-1}$
$Li_{0.69}UO_3(\delta)$	$0.69\,Li(cr) + UO_3(\gamma) \rightleftharpoons Li_{0.69}UO_3(\delta)$	$-(220.79 \pm 2.77)$	$-(319.99 \pm 4.01)$
$Li_{0.55}UO_3(\alpha)$	$0.55\,Li(cr) + UO_3(\gamma) \rightleftharpoons Li_{0.55}UO_3(\gamma)$	$-(170.73 \pm 3.03)$	$-(310.42 \pm 5.51)$
$Li_{0.88}U_3O_8(cr)$	$0.88\,Li(cr) + U_3O_8(\alpha) \rightleftharpoons Li_{0.88}U_3O_8(cr)$	$-(262.29 \pm 4.45)$	$-(298.06 \pm 5.06)$
$Li_{0.12}UO_{2.95}(cr)$	$0.12\,Li(cr) + UO_{2.95}(\alpha) \rightleftharpoons Li_{0.12}UO_{2.95}(cr)$	$-(43.69 \pm 1.78)$	$-(364.08 \pm 14.83)$
$Li_{0.19}U_3O_8(cr)$	$0.19\,Li(cr) + U_3O_8(\alpha) \rightleftharpoons Li_{0.19}U_3O_8(cr)$	$-(58.16 \pm 3.11)$	$-(306.10 \pm 16.37)$

Using NEA accepted values, we recalculate from the values in Table 9-37 the selected enthalpies of formation:

$$\Delta_f H_m^\circ (Li_{0.12}UO_{2.95}, cr, 298.15\,K) = -(1255.0 \pm 1.2)\,kJ \cdot mol^{-1},$$

$\Delta_f H_m^\circ$ (Li$_{0.19}$U$_3$O$_8$, cr, 298.15 K) = $-$ (3633.0 ± 3.1) kJ · mol^{-1}.

The enthalpy changes corresponding to the insertion of lithium into the binary oxides are calculated to be those in Table 9-38, where the selected value of $\Delta_f H_m^\circ$ (UO$_{2.95}$, α, 298.15 K) = $-$ (1211.28 ± 1.28) kJ · mol^{-1} (see 9.3.3.3.3) from the determination of [93PAT/DUE] has been used. The values per mole of lithium are in general agreement with the more precise values of [89DIC/LAW].

From electrochemical measurements analogous to those reported by [93PAT/DUE] for MgUO$_{2.95}$ (see section 9.9.1), Dueber *et al.* [95DUE/PAT] also report $\Delta_f G_m^\circ$ (Li$_{0.12}$UO$_{2.95}$, 298.15 K) = $-$ 42.8 kJ · mol^{-1} and $\Delta_f G_m^\circ$ (Li$_{0.19}$U$_3$O$_8$, 298.15 K) = $-$ 55.2 kJ · mol^{-1}, in agreement with the calorimetric values, indicating a negligible entropy effect.

9.10.3 Sodium compounds (V.10.2)

9.10.3.1 Sodium monouranates (VI) (V.10.2.1.1)

The data for four monouranates of U(VI) selected by [92GRE/FUG] were mainly based on the study by Cordfunke and Loopstra [71COR/LOO]. However, small changes to the auxiliary data used in their reaction schemes have necessitated the recalculation of the enthalpies of formation; full details are given in Appendix A. For the final selections of the data for the monouranates, for convenience given below,

$\Delta_f H_m^\circ$ (Na$_4$UO$_5$, cr, 298.15 K) = $-$ (2457.0 ± 2.2) kJ · mol^{-1},
$\Delta_f H_m^\circ$ (Na$_2$UO$_4$, α, 298.15 K) = $-$ (1897.7 ± 3.5) kJ · mol^{-1},
$\Delta_f H_m^\circ$ (Na$_2$UO$_4$, β, 298.15 K) = $-$ (1884.6 ± 3.6) kJ · mol^{-1},

only those for Na$_4$UO$_5$(cr) have been changed from the previously selected values [92GRE/FUG],

Note however that a substantial correction has been made to the data for Na$_6$U$_7$O$_{24}$(cr) (see section 9.10.3.4.2).

9.10.3.2 Sodium monouranates (V) (V.10.2.1.2)

9.10.3.2.1 NaUO$_3$(cr)

Jayanthi *et al.* [94JAY/IYE] have reported measurements of the pressure of Na(g) (by both mass-loss and mass-spectroscopic Knudsen effusion) and the oxygen activity (emf with CaO–ZrO$_2$ electrolyte) in the three–phase field NaUO$_3$(cr) + Na$_2$UO$_4$(cr) + Na$_2$U$_2$O$_7$(cr). They use these measurements to derive values for the Gibbs energy of formation of NaUO$_3$(cr), more positive than those selected by [92GRE/FUG] by about 30 kJ · mol^{-1}. As noted in Appendix A, their oxygen potential values seem to be much too negative, and we have therefore preferred to maintain the values selected by [92GRE/FUG] for the enthalpy of formation of NaUO$_3$(cr), based on an enthalpy of solution. However, in view of the above discrepancy, and the existence of another discrepant enthalpy of solution measurement by O'Hare and Hoekstra [74OHA/HOE3] (which gives an enthalpy of formation of $NaUO_3(cr)$ which is more negative than the selected value by ca. 25 kJ · mol^{-1}, see [92GRE/FUG]), we have increased the uncertainty of the selected value substantially:

$$\Delta_f H_m^\circ (NaUO_3, cr, 298.15 \text{ K}) = -(1494.9 \pm 10.0) \text{ kJ} \cdot \text{mol}^{-1}.$$

9.10.3.3 Other sodium uranates

The preparation, characterisation, thermochemical and electrochemical properties of α–$Na_{0.14}UO_3$ and δ–$Na_{0.54}UO_3$ were reported by Powell [90POW] and Dickens et al. [88DIC/POW]. Full experimental details in these references are given in Appendix A. The calorimetric reagent for the dissolution, 0.0350 mol · dm^{-3} $Ce(SO_4)_2$ + 1.505 mol · dm^{-3} H_2SO_4, was the same as that used in the studies by [77COR/OUW2], [81COR/OUW].

The enthalpies of solution were given as:

$$\Delta_{sol} H_m (Na_{0.14}UO_3, \alpha) = -(95.32 \pm 0.35) \text{ kJ} \cdot \text{mol}^{-1},$$

$$\Delta_{sol} H_m (Na_{0.54}UO_3, \delta) = -(140.2 \pm 0.7) \text{ kJ} \cdot \text{mol}^{-1}.$$

With the auxiliary enthalpies of solution of the binary uranium oxides used in [90POW], and that for $NaUO_3(cr)$, – (200.04 ± 0.51) kJ · mol^{-1} (all original data from [77COR/OUW2], [81COR/OUW]), the enthalpies of the reactions from the stoichiometric phases are calculated as in Table 9-39.

Using NEA adopted values, we obtain the selected enthalpies of formation

$$\Delta_f H_m^\circ (Na_{0.54}UO_3, \delta, 298.15 \text{ K}) = -(1376.9 \pm 5.5) \text{ kJ} \cdot \text{mol}^{-1},$$

$$\Delta_f H_m^\circ (Na_{0.14}UO_3, \alpha, 298.15 \text{ K}) = -(1267.2 \pm 1.8) \text{ kJ} \cdot \text{mol}^{-1}.$$

Table 9-39: Derived enthalpies of reactions from stoichiometric phases.

Phase	Reaction	$\Delta_r H_m^\circ$ (kJ · mol^{-1})
$Na_{0.54}UO_3(\delta)$	$0.54 \, NaUO_3(cr) + 0.46 \, UO_3(\gamma) \rightleftharpoons Na_{0.54}UO_3(\delta)$	– (6.76 ± 0.77) [a]
$Na_{0.14}UO_3(\alpha)$	$0.14 \, NaUO_3(cr) + 0.86 \, UO_3(\gamma) \rightleftharpoons Na_{0.14}UO_3(\alpha)$	– (5.48 ± 0.48) [a]
$Na_{0.12}UO_{2.95}(cr)$	$0.12 \, NaUO_3(cr) + 0.05 \, U_3O_8(\alpha) + 0.73 \, UO_3(\gamma) \rightleftharpoons Na_{0.12}UO_{2.95}(cr)$	(4.13 ± 1.14) [b]
$Na_{0.20}U_3O_8(cr)$	$0.20 \, NaUO_3(cr) + U_3O_8(\alpha) \rightleftharpoons 0.20 \, UO_3(\gamma) + Na_{0.20}U_3O_8(cr)$	(2.57 ± 0.87) [b]

(a) [88DIC/POW]
(b) [95DUE/PAT]

For the insertion reactions one then obtains, using NEA data, the values in Table 9-40. The electrochemical results are commented upon in Appendix A and will not be further used here.

Dueber et al. [95DUE/PAT] report similar calorimetric and electrochemical studies on two single phased compounds $Na_{0.12}UO_{2.95}$ and $Na_{0.20}U_3O_8$. The enthalpies of solution, using 0.274 mol · dm^{-3} $Ce(SO_4)_2$ + 0.484 mol · dm^{-3} H_3BO_3 + 0.93 mol · dm^{-3} H_2SO_4 as solvent, were given as:

$$\Delta_{sol}H_m (Na_{0.12}UO_{2.95}, cr) = -(103.00 \pm 1.08) \text{ kJ} \cdot \text{mol}^{-1},$$

$$\Delta_{sol}H_m (Na_{0.20}U_3O_8, cr) = -(379.17 \pm 0.78) \text{ kJ} \cdot \text{mol}^{-1}.$$

With the auxiliary data as above, the enthalpies of the reactions from the stoichiometric phases are calculated to be those given in Table 9-39.

Table 9-40: Derived enthalpies of insertion reactions.

Phase	Reaction	$\Delta_r H_m$ kJ · mol^{-1}	$\Delta_r H_m$ kJ · (mol Na)$^{-1}$
$Na_{0.54}UO_3(\delta)$	$0.54\, Na(cr) + UO_3(\delta) \rightleftharpoons Na_{0.54}UO_3(\delta)$	$-(163.2 \pm 5.7)$	$-(302.2 \pm 10.5)$
$Na_{0.14}UO_3(\alpha)$	$0.14\, Na(cr) + UO_3(\alpha) \rightleftharpoons Na_{0.14}UO_3(\alpha)$	$-(54.8 \pm 2.3)$	$-(391.4 \pm 16.4)$
$Na_{0.12}UO_{2.95}$	$0.12\, Na(cr) + UO_{2.95}(\alpha) \rightleftharpoons Na_{0.12}UO_{2.95}(cr)$	$-(36.1 \pm 2.3)$	$-(300.8 \pm 19.2)$
$Na_{0.20}U_3O_8$	$0.20\, Na(cr) + U_3O_8(\alpha) \rightleftharpoons Na_{0.14}U_3O_8(cr)$	$-(51.6 \pm 3.2)$	$-(258.0 \pm 16.0)$

Using NEA adopted values, we recalculate from the values in Table 9-39 the selected enthalpies of formation:

$$\Delta_f H_m^\circ (Na_{0.12}UO_{2.95}, cr, 298.15 \text{ K}) = -(1247.4 \pm 1.9) \text{ kJ} \cdot \text{mol}^{-1},$$

$$\Delta_f H_m^\circ (Na_{0.20}U_3O_8, cr, 298.15 \text{ K}) = -(3626.4 \pm 3.2) \text{ kJ} \cdot \text{mol}^{-1}.$$

The enthalpy changes corresponding to the insertion of sodium into the binary oxides are calculated to be those in Table 9-40, where the recalculated value of $\Delta_f H_m^\circ (UO_{2.95}, \alpha, 298.15 \text{ K}) = -(1211.28 \pm 1.28) \text{ kJ} \cdot \text{mol}^{-1}$ from the determination of [93PAT/DUE] (see 9.3.3.3.3) has been used.

[95DUE/PAT] deduced from electrochemical measurements (cf. section 9.9.1) the Gibbs energies at 298.15 K of the insertion reactions in the last two rows of Table 9-40 to be -37.7 kJ · mol^{-1} and -50.7 kJ · mol^{-1}. These values are close to the corresponding enthalpies of reaction, showing that the entropies of reaction are small, as expected for a condensed phase reaction.

From electrochemical measurements analogous to those reported by [93PAT/DUE] for $MgUO_{2.95}$ (see section 9.9.1), Dueber et al. [95DUE/PAT] also report:

$$\Delta_r G_m^\circ (Na_{0.12}UO_{2.95}, 298.15 \text{ K}) = -42.8 \text{ kJ} \cdot \text{mol}^{-1},$$

$$\Delta_r G_m^\circ (Na_{0.20}U_3O_8, 298.15 \text{ K}) = -55.2 \text{ kJ} \cdot \text{mol}^{-1},$$

9.10.3.4 Sodium polyuranates (V.10.2.2)

9.10.3.4.1 Na$_2$U$_2$O$_7$(cr) (V.10.2.2.1)

The solubility of this compound is also discussed in 9.3.2.1.5.2.2.

For reasons indicated in Appendix A, the data obtained by Díaz Arocas and Grambow [98DIA/GRA], and by Valsami-Jones and Ragnarsdottir [97VAL/RAG], who measured the solubilities of polyuranates, are not selected by this review. [97ALL/SHU] gives spectroscopic information on uranates precipitated at pH above 8.

9.10.3.4.2 Na$_6$U$_7$O$_{24}$(cr) (V.10.2.2.3)

A substantial error has been found in the value of the enthalpy of formation of Na$_6$U$_7$O$_{24}$(cr) selected by [92GRE/FUG], due to a mis-reporting of a datum in the original paper by Cordfunke and Loopstra [71COR/LOO]. As noted in Appendix A, their reported value for the enthalpy of solution of this compound is clearly that for 1/7 of a mole of Na$_6$U$_7$O$_{24}$(cr), rather than 1 mole. The recalculated enthalpy of formation becomes:

$$\Delta_f H_m^\circ (\text{Na}_6\text{U}_7\text{O}_{24}, \text{cr}, 298.15 \text{ K}) = - (10841.7 \pm 10.0) \text{ kJ} \cdot \text{mol}^{-1}.$$

With this value, the phase relationships in the Na–U–O system at around 800 K are in good accord with experimental observations, as summarised by Lindemer et al. [81LIN/BES].

9.10.4 Potassium compounds (V.10.3)

9.10.4.1 KUO$_3$ (cr) (V.10.3.1)

Jayanthi et al. [99JAY/IYE] have reported measurements of the pressure of K(g) (by mass-loss Knudsen effusion) and the oxygen activity (emf with CaO–ZrO$_2$ electrolyte) in the three–phase field KUO$_3$(cr) + K$_2$UO$_4$(cr) + K$_2$U$_2$O$_7$(cr), similar to the studies on the corresponding sodium compounds by [94JAY/IYE]. The authors use these measurements to derive values for the Gibbs energy of formation of KUO$_3$(cr), using estimated values [81LIN/BES], [92GRE/FUG] for the entropies of the hexavalent uranates. We have reversed this procedure, as described in Appendix A, to compare the experimental oxygen and potassium activities in the three phase field with those calculated with the enthalpies of formation selected by [92GRE/FUG], the estimated entropies (as above) and the heat capacities also measured by [99JAY/IYE]. The agreement with the oxygen potentials is quite good, whereas that of the potassium activities is rather poor. In view of the need to use estimated entropies in the calculation and possible sources of error, such as lack of complete equilibrium in the experimentation, we shall retain the enthalpies of formation selected by [92GRE/FUG].

In addition to the Gibbs energy measurements, [99JAY/IYE] measured enthalpy increments of KUO_3(cr) (and $K_2U_2O_7$(cr)) up to ca. 700 K using a high temperature Calvet calorimeter. As noted in section 9.10.1 these data have been refitted to a polynomial, which gives on differentiation:

$$C_{p,m}(KUO_3, cr, T) = 133.258 + 1.2558 \times 10^{-2} T \quad (J \cdot K^{-1} \cdot mol^{-1})$$
(298.15 to 714 K).

9.10.4.2 $K_2U_2O_7$(cr) (V.10.3.3)

Jayanthi et al. [98JAY/IYE], [99JAY/IYE], have reported enthalpy increment measurements for this phase. As described in Appendix A, their data have been refitted to obtain heat capacities given by the equation summarised in section 9.10.1.

The heat capacity increases with temperature appreciably more rapidly than would be expected for a compound containing hexavalent uranium.

9.10.4.3 $K_2U_4O_{12}$(cr), $K_2U_4O_{13}$(cr)

Iyer et al. [97IYE/JAY] have reported enthalpy increment measurements for these phases. As described in Appendix A, their data have been refitted to obtain heat capacities given by the equations summarised in section 9.10.1. However, as shown in the plot of heat capacities in section 9.10.1, the expressions for $K_2U_4O_{12}$(cr) are not consistent with the general behaviour of the heat capacities of the alkali–metal uranates, and the data for this compound have not been selected in this review.

Iyer et al. [97IYE/JAY] also give details of measurements of the oxygen activity (emf with CaO–ZrO_2 electrolyte) in the mixtures $K_2U_4O_{12}$(cr) + $K_2U_4O_{13}$(cr) from 1053 to 1222 K with air (p_{O_2} = 0.21 bar) as the reference electrode. The authors have used these Gibbs energies of reaction to derive data at 298.15 K for $K_2U_4O_{12}$(cr), using their heat capacity data (extrapolated from ca. 750 K) and literature data at 298.15 K for the hexavalent compounds. However, we have not pursued this approach since the data for the enthalpy of formation and entropy of $K_2U_4O_{13}$(cr) at 298.15 K are estimated, so any derived data for $K_2U_4O_{12}$(cr) would have appreciable uncertainties.

9.10.4.4 $K_2U_6O_{19} \cdot 11H_2O$(cr)

The solubility of this compound is discussed in section 9.3.2.1.5.2.4.

9.10.5 Rubidium compounds (V.10.4)

9.10.5.1 $RbUO_3$(cr) (V.10.4.1)

Jayanthi et al. [96JAY/IYE] have reported measurements of the pressure of Rb(g) (by mass-loss Knudsen effusion) and the oxygen activity (emf with CaO–ZrO_2 electrolyte) in the three–phase field $RbUO_3$(cr) + Rb_2UO_4(cr) + $Rb_2U_2O_7$(cr), similar to the studies from the same laboratory on the corresponding sodium [94JAY/IYE] and potassium compounds [99JAY/IYE]. The authors use these measurements to derive values for the

Gibbs energy of formation of RbUO$_3$(cr), using estimated values [81LIN/BES], [92GRE/FUG] for the entropies of the hexavalent uranates.

However, as noted in Appendix A, the entropy term in the expression for the vaporisation of Rb(g) according to the reaction:

$$RbUO_3(cr) + Rb_2UO_4(cr) \rightleftharpoons Rb_2U_2O_7(cr) + Rb(g)$$

is far too small (18.2 J · K^{-1} · mol^{-1}) to relate to a vaporisation reaction. In view of this, and the need to use estimated data for many of the properties of the solids in further calculations, we have not processed the data further, and will retain the calorimetric enthalpy of formation selected by [92GRE/FUG] namely,

$$\Delta_f H_m^\circ (RbUO_3, cr, 298.15\ K) = -(1520.9 \pm 1.8)\ kJ \cdot mol^{-1}.$$

9.10.5.2 Rb$_2$U$_3$O$_{8.5}$(cr), Rb$_2$U$_4$O$_{11}$(cr), Rb$_2$U$_4$O$_{12}$(cr) and Rb$_2$U$_4$O$_{13}$(cr)

Sali et al. [94SAL/KUL] reported the preparation and characterisation, by X–ray diffraction, of the first two phases, but no thermodynamic data were presented. Banerjee et al. [2001BAN/PRA], (which is assumed to supersede the conference paper by [98BAN/SAL]), Iyer et al. [90IYE/VEN], and Venugopal et al. [92VEN/IYE] have reported enthalpy increment measurements for the last three phases. As described in Appendix A, their data have mostly been refitted, to obtain heat capacities given by the equations summarised in section 9.10.1. However, as shown in the plot of heat capacities in that section, the heat capacity data for Rb$_2$U$_4$O$_{12}$(cr) are inconsistent with the general behaviour of the heat capacities of the alkali-metal uranates (indeed $C_{p,m}$ (Rb$_2$U$_4$O$_{12}$, cr) decreases as the temperature increases), and no data for this compound have been selected in this review.

Venugopal et al. [92VEN/IYE] also give details of measurements of the oxygen activity (emf with CaO–ZrO$_2$ electrolyte) in the mixtures Rb$_2$U$_4$O$_{12}$(cr) + Rb$_2$U$_4$O$_{13}$(cr) from 1075 to 1203 K with air (p_{O_2} = 0.21 bar) as the reference electrode. The authors have used these Gibbs energies of reaction to derive data at 298.15 K for Rb$_2$U$_4$O$_{12}$(cr), using their heat capacity data (extrapolated from ca. 750 K) and literature data at 298.15 K for the hexavalent compound. However, we have not pursued this approach since the data for the enthalpy of formation and entropy of Rb$_2$U$_4$O$_{13}$(cr) at 298.15 K are estimated, so any derived data for Rb$_2$U$_4$O$_{12}$(cr) would have appreciable uncertainties.

For the same reason the data on the oxygen potentials in the diphasic Rb$_2$U$_4$O$_{11}$(cr) + Rb$_2$U$_4$O$_{12}$(cr) region from 985 to 1186 K reported by Sali et al. [95SAL/JAY] and Iyer et al. [96IYE/JAY], reviewed in Appendix A, have not be used to derive any room temperature data for Rb$_2$U$_4$O$_{11}$(cr). Sali et al. [95SAL/JAY] have also reported the enthalpy of oxidation of Rb$_2$U$_4$O$_{11}$(cr) to Rb$_2$U$_4$O$_{13}$(cr) at 673 K:

$$Rb_2U_4O_{11}(cr) + O_2(g) \rightleftharpoons Rb_2U_4O_{13}(cr),$$

$$\Delta_r H_m (673 \text{ K}) = -(279 \pm 15) \text{ kJ} \cdot \text{mol}^{-1}.$$

Again, in the absence any experimental data on the enthalpy of formation of $Rb_2U_4O_{13}$(cr), this value cannot be processed further.

9.10.5.3 $Rb_2U(SO_4)_3$(cr)

Jayanthi *et al.* [93JAY/IYE] and Saxena *et al.* [99SAX/RAM] have studied rubidium uranium sulphate using DSC, DTA and TGC techniques, and measured enthalpy increments from 373 to 803 K using a H.T. Calvet calorimeter.

Two solid state transitions were observed at 616 and 773 K (DSC) and 628 and 780 K by DTA/TG. The compound starts to decompose at 893 K.

The enthalpy data for the low–temperature polymorph [93JAY/IYE] are quite consistent with the subsequent heat capacities from 273 to 623 K measured by DSC, reported briefly by Saxena *et al.* [99SAX/RAM], and these two sets of data have been combined to define the enthalpy increments and heat capacity of this phase:

$$C_{p,m}(Rb_2U(SO_4)_3, \text{cr}, T) = 386.672 + 6.883 \cdot 10^{-2} T - 5.25917 \cdot 10^6 T^{-2} \quad (J \cdot K^{-1} \cdot mol^{-1})$$
(298.15 to 628 K),

and thus

$$C^\circ_{p,m}(Rb_2U(SO_4)_3, \text{cr}, 298.15 \text{ K}) = (348.0 \pm 30.0) \text{ J} \cdot K^{-1} \cdot mol^{-1}.$$

As noted in Appendix A, the measurements of the enthalpy increments for the two polymorphs stable above 628 K are not precise enough to define reliable heat capacities, and our calculated enthalpy of the transition at *ca.* 628 K, 3.7 kJ · mol^{-1}, is smaller than that given by the authors.

9.10.6 Caesium compounds (V.10.5)

9.10.6.1 Caesium monouranate (V.10.5.1)

Huang *et al.* [98HUA/YAM] have measured the vapour phase composition in the decomposition of Cs_2UO_4(s) from 873 to 1373 K, in Pt cells both *in vacuo* and in a D_2/D_2O environment. As discussed in detail in Appendix A, no data relevant to the review can be obtained from this study.

9.10.6.2 Caesium polyuranates (V.10.5.2)

9.10.6.2.1 $Cs_2U_2O_7$(cr) (V.10.5.2.1)

Jayanthi *et al.* [97JAY/IYE] have reported enthalpy increment measurements for this phase. As described in Appendix A, the data have been refitted with a constraint of $C^\circ_{p,m}(Cs_2U_2O_7, \text{cr}, 298.15 \text{ K}) = 231.2 \text{ J} \cdot K^{-1} \cdot mol^{-1}$, selected in [92GRE/FUG] from the low–temperature measurements of [81OHA/FLO]. The derived heat capacity expressions are given in the Table 9-36 in section 9.10.1; the heat capacity increases surprisingly rapidly with temperature.

The enthalpy of formation and the entropy are unchanged from those selected in [92GRE/FUG]:

$$\Delta_f H_m^\circ (Cs_2U_2O_7, cr, 298.15\ K) = -(3220 \pm 10)\ kJ \cdot mol^{-1},$$

$$S_m^\circ (Cs_2U_2O_7, cr, 298.15\ K) = (327.75 \pm 0.66)\ J \cdot K^{-1} \cdot mol^{-1}.$$

9.10.6.2.2 $Cs_4U_5O_{17}(cr)$

Jayanthi et al. [97JAY/IYE] have reported enthalpy increment measurements for this phase. As described in Appendix A, the data have been refitted with a constraint of $H_m(T) - H_m^\circ(298.15\ K) = 0$ at $T = 298.15$ K. The derived heat capacity expressions are given in the Table 9-36 in section 9.10.1.

These authors also studied the oxygen activity (emf with a CaO–ZrO$_2$ electrolyte) in the three–phase region $Cs_4U_5O_{17}(cr) + Cs_2U_2O_7(cr) + Cs_2U_4O_{12}(cr)$ with air (p_{O_2} = 0.21 bar) as the reference electrode. Their recalculated results (see Appendix A) for the Gibbs energy of the relevant reaction:

$$Cs_2U_4O_{12}(cr) + 3\ Cs_2U_2O_7 + 0.5\ O_2(g) \rightleftharpoons 2\ Cs_4U_5O_{17}(cr) \quad (9.48)$$

are:

$$\Delta_r G_m\ (9.48) = -136044 + 96.985\ T \quad J \cdot mol^{-1} \quad (1048\ to\ 1206\ K).$$

The authors have used these Gibbs energies of reaction to derive data at 298.15 K for $Cs_4U_5O_{17}(cr)$, using their heat capacity data (extrapolated from ca. 800 K) and literature data at 298.15 K for the hexavalent compounds. However, we have not pursued this approach since many of the relevant data have not been selected for this review (see for example comments in [92VEN/IYE]).

9.10.6.2.3 $Cs_2U_4O_{12}(cr)$ and $Cs_2U_4O_{13}(cr)$

Venugopal et al. [92VEN/IYE], have reported enthalpy increment measurements for these phases. As described in Appendix A, the data for $Cs_2U_4O_{12}(cr)$, which extend to 898 K, agree excellently with those reported by Cordfunke and Westrum [80COR/WES]. Since the latter authors gave a fitted equation for $C_{p,m}(T)$ which is a smooth continuation of their low–temperature data, this equation, adopted by [92GRE/FUG], has been retained. For $Cs_2U_4O_{13}(cr)$, however, the heat capacity data for $Cs_2U_4O_{13}(cr)$ derived from the enthalpy increments of [92VEN/IYE] are inconsistent with the general behaviour of the heat capacities of the alkali–metal uranates as shown in the plot of heat capacities in section 9.10.1, (indeed the heat capacity of $Cs_2U_4O_{13}(cr)$ decreases sharply as the temperature increases), and no data for this compound have been selected in this review.

Une [85UNE] and Venugopal et al. [92VEN/IYE] have both studied the oxygen activity (emf with solid electrolytes) in the mixtures $Cs_2U_4O_{12}(cr) + Cs_2U_4O_{13}(cr)$ with air (p_{O_2} = 0.21 bar) as the reference electrode. Their recalculated results (see Appendix A) for the Gibbs energy of the relevant reaction:

$$\text{CsU}_4\text{O}_{12} + 0.5\text{O}_2(\text{g}) \rightleftharpoons \text{Cs}_2\text{U}_4\text{O}_{13}(\text{cr}) \tag{9.49}$$

are in good agreement, although the corresponding individual enthalpy and entropy terms differ somewhat:

[85UNE]: $\Delta_r G_m$ (9.49) = $-$ 190108 + 151.758 T J · mol^{-1} (1048 to 1173 K),
[92VEN/IYE]: $\Delta_r G_m$ (9.49) = $-$ 174638 + 136.100 T J · mol^{-1} (1019 to 1283 K).

At 1100 K, these equations give $\Delta_r G_m$ ((9.49), 1100 K) = $-$ 23.17 and $-$ 24.93 kJ · mol^{-1} respectively.

The authors [92VEN/IYE] have used their Gibbs energies of reaction to derive a $\Delta_f G_m^\circ$ value at 298.15 K for Cs$_2$U$_4$O$_{13}$(cr), using their heat capacity data (extrapolated from ca. 750 K) and literature data at 298.15 K for Cs$_2$U$_4$O$_{12}$(cr). However, for reasons given in detail in Appendix A, we have not pursued this approach since the uncertainties in the additional data required means that any derived data for Cs$_2$U$_4$O$_{13}$(cr) would have appreciable uncertainties.

In summary, the following data for Cs$_2$U$_4$O$_{12}$(cr) are unchanged from those selected in [92GRE/FUG]:

$\Delta_f H_m$ (Cs$_2$U$_4$O$_{12}$, cr, 298.15 K) = $-$ (5571.8 ± 3.6) kJ · mol^{-1},

S_m° (Cs$_2$U$_4$O$_{12}$, cr, 298.15 K) = (526.4 ± 1.0) J · K^{-1} · mol^{-1},

$C_{p,m}^\circ$ (Cs$_2$U$_4$O$_{12}$, cr, 298.15 K) = (384.0 ± 1.0) J · K^{-1} · mol^{-1},

$C_{p,m}$ (Cs$_2$U$_4$O$_{12}$, cr, T) = 423.7260 + 7.19405×10^{-2}T $-$ 5.43750×10$^6 T^{-2}$ J · K^{-1} · mol^{-1}
(298.15 – 898 K).

No data for Cs$_2$U$_4$O$_{13}$(cr) are selected in this review.

9.11 Uranium compounds with elements from other groups

9.11.1 Thallium compounds

9.11.1.1 Thallium uranates

Sali et al. [93SAL/KUL], [96SAL/IYE] have reported the preparation of two new thallium uranates Tl$_2$U$_3$O$_9$ and Tl$_2$U$_4$O$_{11}$, and give their structural data. The latter paper also gives details of measurements of the oxygen potential in the diphasic fields, Tl$_2$U$_4$O$_{11}$ + Tl$_2$U$_4$O$_{12}$ and Tl$_2$U$_4$O$_{12}$ + Tl$_2$U$_4$O$_{13}$, obtained from solid-state emf cells. As described in Appendix A, the recalculated values of the Gibbs energies of the two relevant reactions:

$$\text{Tl}_2\text{U}_4\text{O}_{11}(\text{cr}) + \frac{1}{2}\text{O}_2(\text{g}) \rightleftharpoons \text{Tl}_2\text{U}_4\text{O}_{12}(\text{cr}), \tag{9.50}$$

$$\text{Tl}_2\text{U}_4\text{O}_{12}(\text{cr}) + \frac{1}{2}\text{O}_2(\text{g}) \rightleftharpoons \text{Tl}_2\text{U}_4\text{O}_{13}(\text{cr}), \tag{9.51}$$

are finally:

$$\Delta_r G_m \text{ (9.50)} = -390.53 + 0.2032 \cdot T \text{ (kJ} \cdot \text{mol}^{-1}),$$

$$\Delta_r G_m \text{ (9.51)} = -136.85 + 0.0841 \cdot T \text{ (kJ} \cdot \text{mol}^{-1}).$$

However, in the absence of any experimental data for any of the three individual thallium uranates involved, no reliable data for any of these can be derived from this study.

More recently, Banerjee et al. [2001BAN/PRA] measured enthalpy increments of pelleted $Tl_2U_4O_{11}$(cr) in a Calvet calorimeter from 301 to 735 K. Values of $H_m(T) - H_m^\circ(298.15\,\text{K})$ were tabulated for thirteen temperatures from 301 to 673 K and fitted by a Shomate analysis, using an estimated value of $C_{p,m}^\circ(Tl_2U_4O_{11}, \text{cr}, 298.15\,\text{K}) = (360.1 \pm 20.0)\,\text{J} \cdot \text{K}^{-1} \cdot \text{mol}^{-1}$ from the sum of the heat capacities of the component oxides, Tl_2O(cr) $+2\,UO_2$(cr) $+ 2\,UO_3$(cr). The resulting (corrected) equation for $H_m(T) - H_m^\circ(298.15\,\text{K})$ gives on differentiation:

$$C_{p,m} = (Tl_2U_4O_{11}, \text{cr}, T) = 368.2 + 2.4886 \cdot 10^{-2} \cdot T - 1.375 \cdot 10^6 \cdot T^{-2}\,\text{J} \cdot \text{K}^{-1} \cdot \text{mol}^{-1}$$

$$(298.15 \text{ to } 673\,\text{K}).$$

9.11.2 Zinc compounds.

Dueber et al. [95DUE/PAT] report calorimetric and electrochemical measurements on the formation of the insertion compound $Zn_{0.12}UO_{2.95}$(cr), similar to the corresponding compounds of magnesium (section 9.9.1), lithium (section 9.10.2.1) and sodium (section 9.10.3.3). From the enthalpies of solution of the constituents of the equation:

$$0.12\,ZnO(\text{cr}) + 0.49\,UO_3(\gamma) + 0.17\,U_3O_8(\alpha) \rightleftharpoons Zn_{0.12}UO_{2.95}(\text{cr}) \quad (9.52)$$

these authors report $\Delta_r H_m^\circ((9.52), 298.15\,\text{K}) = +(11.36 \pm 0.78)\,\text{kJ} \cdot \text{mol}^{-1}$. Using NEA adopted values, we recalculate:

$$\Delta_f H_m^\circ (Zn_{0.12}UO_{2.95}, \text{cr}, 298.15\,\text{K}) = -(1238.07 \pm 1.06)\,\text{kJ} \cdot \text{mol}^{-1}.$$

Using the recalculated value (see section 9.3.3.3) $\Delta_f H_m^\circ (UO_{2.95}, \alpha, 298.15\,\text{K}) = -(1211.28 \pm 1.28)\,\text{kJ} \cdot \text{mol}^{-1}$ from the determination of [93PAT/DUE], the enthalpy effect corresponding to the insertion of zinc according to reaction:

$$0.12\,Zn(\text{cr}) + UO_{2.95}(\alpha) \rightleftharpoons Zn_{0.12}UO_{2.95}(\text{cr}) \quad (9.53)$$

is obtained as $\Delta_r H_m^\circ ((9.53), 298.15\,\text{K}) = -(26.79 \pm 1.50)\,\text{kJ} \cdot \text{mol}^{-1}$.

From electrochemical measurements, [95DUE/PAT] also reported $\Delta_r G_m^\circ ((9.53), 298.15\,\text{K}) = -25.2\,\text{kJ} \cdot \text{mol}^{-1}$. This value is in agreement with the calorimetric result if $\Delta_r S_m^\circ ((9.53), 298.15\,\text{K})$ is assumed to be negligible.

9.11.3 Polyoxometallates

The actinides form complexes with different heteropolyanions and some of these are very strong. There is an extensive literature on the chemistry of these systems with a good review article by Yusov and Shilov [99YUS/SHI] that can be consulted for details. The non-protonated anions have very high negative charges and are also fairly strong bases, the basicity varying with the degree of protonation. There are no experimental studies that have identified the proton content of the complexes studied experimentally and this review has therefore not selected any equilibrium constants. Some conditional equilibrium constants are listed below. Shilov [91SHI] estimated the stability of the U(IV) complex with the polyoxo anion $P_2W_{17}O_{61}^{10-}$ (PWO). This anion does not form complexes with U(VI) in acidic media, but it is a strong complexing anion for U(IV). As explained in Appendix A, Bion et al. [95BIO/MOI] determined the equilibrium constants in HNO_3 for the reactions:

$$U(IV) + PWO \rightleftharpoons U(PWO) \qquad K_1,$$
$$U(IV) + 2\,PWO \rightleftharpoons U(PWO)_2 \qquad K_2.$$

The U(IV) is the sum of U^{4+} and nitrato complexes. Their results are reported in Table 9-41.

Table 9-41: Conditional equilibrium constants K_1 and K_2 at different concentrations of H^+

$\log_{10}[H^+]$	− 0.3	0	0.3
$\log_{10}K_1$	(7 ± 0.25)	(8 ± 0.25)	(9 ± 0.25)
$\log_{10}K_2$	(11.5 ± 0.25)	(13.5 ± 0.25)	(15.5 ± 0.25)

Saito and Choppin [98SAI/CHO] investigated the complex formation between U(VI) and different polyoxometallate anions, L^{x-}, which form stable complexes around pH = 4 (0.1 M $NaClO_4$). They used a liquid-liquid extraction technique and these data are discussed in Appendix A. These authors have assumed simple association reactions between UO_2^{2+} and L^{x-}, an assumption that we find too simplistic in view of the fairly strong basicity of the polyoxometallate ions as mentioned above.

The equilibrium constants given in Table 9-41 and Table 9-42 are conditional equilibrium constants and are therefore not selected by this review.

Table 9-42: Conditional equilibrium constants of U(VI) with some polyoxometalate anions. These are ternary systems where the species formed have the general composition $(UO_2)_pH_qL_r$, where in this case $p = 1$, q is not known and $r = 1$ and 2. The values of $\log_{10}K$ are only valid at pH ≈ 4 in this study.

	$V_{10}O_{28}^{6-}$	$Mo_7O_{24}^{6-}$	$CrMo_6O_{24}H_6^{3-}$	$IMo_6O_{24}^{5-}$	$TeMo_6O_{24}^{6-}$	$MnMo_9O_{32}^{6-}$
$\log_{10} K_1$	(2.40 ± 0.1)	(3.88 ± 0.3)	(2.05 ± 0.06)	(2.57 ± 0.07)	(3.16 ± 0.04)	(3.53 ± 0.2)
$\log_{10} K_2$					(5.25 ± 0.3)	

9.12 Uranium mineral phases

9.12.1 Introduction

In view of the importance of the mineral systems, the NEA TDB review has extended the mineral section in [92GRE/FUG] and [95SIL/BID] with a discussion of thermodynamic data obtained by using estimation methods. However, no data of this type have been selected; they have been included to provide estimates when no experimental information is available.

Minerals are characterised by their specific structure type and chemical composition; the latter is rarely stoichiometric. The variable stoichiometry is a complicating factor for experimental studies of their thermodynamics. Thermodynamic data for minerals are therefore often estimates from the precisely known structures and experimental thermodynamic information for a few pure phases containing structure elements of different types present in the more complex minerals. The thermodynamic properties of the minerals are considered to be a sum of the contributions from these different constituents (see [97FIN] and references within in Appendix A).

9.12.2 Review of papers

Only part of the literature on mineral phases was reviewed by Grenthe *et al.* [92GRE/FUG] and the following studies [81OBR/WIL], [81VOC/PIR], [83OBR/WIL], [83VOC/PEL], [84VOC/GRA], [84VOC/GOE], [86VOC/GRA], [88ATK/BEC], [90VOC/HAV] were not included. For reasons given below this omission does not affect the set of selected thermodynamic uranium data given in the [92GRE/FUG].

Two studies presented by O'Brien and Williams [81OBR/WIL], [83OBR/WIL] deal with the stability of different secondary uranyl minerals. These references are Parts 3 and 4 of a series of papers; Parts 1 and 2 [79HAA/WIL], [80ALW/WIL] were reviewed and discussed by Grenthe *et al.* [92GRE/FUG] (Sections V.5.1.3.2.c and V.7.1.2.2.b, *pp.* 254–255 and 328, and in Appendix A, *pp.* 646–648).

O'Brien and Williams [81OBR/WIL] reported the Gibbs energies of formation of sodium, potassium and ammonium zippeites, (basic dioxouranium(VI) sulphates).

There are almost no experimental details on the technique and data analysis used in this work and no information about the auxiliary data used. In [83OBR/WIL], O'Brien and Williams reported the Gibbs energy of formation, of schröckingerite, $Ca_3NaUO_2(CO_3)_3FSO_4·10H_2O(cr)$, and grimselite, $K_3NaUO_2(CO_3)_3·H_2O(cr)$, calculated from solubility experiments. The method used is essentially the same as previously described by Haacke et al., [79HAA/WIL] in Part 1 of the series. The authors performed corrections for ionic strength effects on the equilibrium constants for the aqueous complexes, but the data suffer from the same flaw as previous studies, i.e. neglect of the formation of hydrolysis species, which in the investigated pH range (7.69 to 10.00) are expected to dominate the uranium speciation. As a result the calculated solubility products and thermodynamic data deduced from them are not reliable.

The solubility products of cobalt, nickel, and copper uranyl phosphate, (meta-torbenite) were measured by Vochten et al., [81VOC/PIR]. The solid compounds were synthesised and adequately characterised. However, the solubility experiments and the corresponding data analysis have serious limitations, e.g., lack of uranium and phosphorus analyses, standardisation of the pH electrode with buffer solutions which were outside the pH range under investigation, lack of correction for ionic strength effects on the auxiliary thermodynamic data (which were taken from Sillén and Martell [64SIL/MAR], and only included uranyl phosphate complexes), etc. Furthermore, the authors did not specify in which ionic medium the solubility experiments were made. In view of these limitations the data of Vochten et al., [81VOC/PIR] could not be used for an evaluation of reliable thermodynamic data.

Sabugalite is a hydrated acid aluminium uranyl phosphate, with an ideal composition corresponding to $HAl(UO_2)_4(PO_4)_4·16H_2O(cr)$. Vochten and Pelsmaekers, [83VOC/PEL], have measured the solubility of the mineral in phosphoric and hydrochloric acid, but with no experimental details and no primary experimental data. This study can therefore not be used to determine thermodynamic data for the solid.

Bassetite is a secondary uranyl phosphate and has the ideal composition $Fe(UO_2)_2(PO_4)_2·8H_2O(cr)$. Vochten et al. [84VOC/GRA] synthesised this mineral and its fully oxidised form and made extensive studies of them using X–ray powder diffraction, Mössbauer and infrared spectroscopy, zeta-potential measurements and thermal analysis. They also measured the solubility as a function of the acidity in the range $2.4 < pH < 6.0$, but the ionic strength of the experiments is not reported. The solubility value was based on measuring the total iron concentration in solution. No information is given about the speciation of the solution phase under the conditions used. Hence it is impossible to calculate the thermodynamic data for the dissolution reaction or for the solid phases studied.

Vochten and Goeminne [84VOC/GOE] and Vochten et al. [86VOC/GRA] have reported the synthesis and properties of several uranyl arsenates. In [84VOC/GOE] they describe solubility measurements of copper (meta-zeunerite), cobalt (meta-kirchheimerite) and nickel uranyl arsenates at room temperature, but without information on ionic strength and speciation. The experimental methodology and data analysis were basically the same as described in other publications by the same authors (see the comments on [81VOC/PIR]). These studies do not contain sufficient experimental detail to allow a calculation of precise thermodynamic data. A similar study of manganese and iron (meta–kahlerite and its fully oxidised form) uranyl arsenates [86VOC/GRA] suffers from the same shortcomings and cannot be used to provide thermodynamic data for the mineral phases.

Atkins et al. [88ATK/BEC] studied the variation of the stability field of becquerelite, $CaU_6O_{19} \cdot 11H_2O$(cr), as a function of temperature and calcium concentration; they also measured the solubility of becquerelite and $Ca_2UO_5 \cdot (1.3–1.7) H_2O$(cr) at 293.15 K, in two ionic media, pure water and a 0.5 M NaOH. These data do not provide sufficient information to extract thermodynamic data because of lack of experimental details.

Vochten and van Haverbeke, [90VOC/HAV], investigated the reaction of schoepite, chemically equivalent to $UO_3 \cdot 2H_2O$(cr), to form becquerelite, billietite and wölsendorfite, which correspond to $CaU_6O_{19} \cdot 11H_2O$(cr), $BaU_6O_{19} \cdot 11H_2O$(cr), and $PbU_2O_7 \cdot 2H_2O$(cr), respectively. The reactions were studied at 333.15 K. In addition, the solubility of the different solids in water was measured at 298.15 K as a function of the pH. Care was taken to avoid the presence of CO_2(g). However, the stability of the solids in the absence of calcium, barium, and lead, in the aqueous solution at the pH values investigated was not addressed. Based on the solubility measured under the given experimental conditions, the authors calculated the solubility products for becquerelite and billietite. They used auxiliary data from various sources of published thermodynamic data for the uranyl hydroxo species, but do not report the actual values used. They present two distribution diagrams for the aqueous uranyl species involved in these two systems, which are clearly incorrect - for example, the complex UO_2OH^+ is shown to be predominant at pH 9. This indicates that the equilibrium model used by the authors is incomplete and that the thermodynamic data derived are erroneous. Rai et al. [2002RAI/FEL] have made a very careful study of the solubility product of a synthetic becquerelite. The value they give is close to that reported by Vochten et al. [90VOC/HAV] and has been retained by this review (see section 9.3.2.1.5.2.1).

9.12.3 Summary of mineral information

Table D.6 on pp. 361 – 367 in [95SIL/BID] has been updated, Table 9-43, with new information. This table has been included to facilitate the task of geochemists searching for thermodynamic data to model uranium migration in aquatic environments. It contains the tables in Chapters IX and X of [92GRE/FUG], which have been combined, reorganised, updated with new references and purged of aqueous complexes. Only solid phases, which this review judges to be important for modelling radionuclide migration, were retained in Table 9-43. Thus materials like inter-metallic compounds, alloys, halides and nitrates have been excluded, as have compounds which have been discussed in [92GRE/FUG] and in earlier sections of this review. The solid phases have been arranged into families (phosphates, sulphates, *etc.*). Note that Table 9-43 does not bring new information to the reader and is provided only as a convenience to geochemical modellers; in combination with Chapter IX in [92GRE/FUG] it provides a survey of the bibliographic information used by the NEA TDB review.

The users have to consult the main text to obtain information on the thermodynamic data in Table 9-43 for which selected values are available.

As noted above, geochemists have often been forced to use estimated thermodynamic data for mineral phases. There are several methods used and Saxena has given a survey of some of them in Chapter VIII in [97ALL/BAN]. A recent paper by Chen, Ewing and Clark [99CHE/EWI], followed by [99CHE/EWI2], provides additional information of the potential of structure-based estimation methods. The paper describes the estimation of Gibbs energies and enthalpies of formation of a large number of U(VI) phases, also those formed as a result of weathering of uraninite, UO_2(cr), and is therefore highly relevant both for uranium geology and for the modelling of transformation of spent nuclear fuel. The agreement between estimated and measured values of the enthalpy of formation and the Gibbs energy of formation is very good, often less than 0.2 %. With these rather high molecular weight species, however, even such small percentage differences are not insignificant. For instance, in the case of andersonite, $Na_2Ca(UO_2)(CO_3)_2 \cdot 6H_2O$, the difference between estimated and measured Gibbs energy of formation is only 0.08%, corresponding to 4.7 kJ · mol^{-1}. The method relies on the use of proper "end-member" structures for the calibration of the model, but data for them are not always available. The agreement between model and experiment is then less satisfactory, *cf.* Table 7 in [99CHE/EWI].

Thermodynamic data for uranium silicate minerals are discussed in the main text, section 9.7.3.2.

Table 9-43: Minerals of uranium and related solid phases of interest for geochemical modelling. The inclusion of these formulae in this table is to be understood as information on the existence of published material. It does not imply that the authors of this table give any credit to either the thermodynamic data or the chemical composition or existence of these species. The compounds for which selected data are presented in Chapter III of [92GRE/FUG], as well as those which are discussed, but for which no data are recommended in Ref. [92GRE/FUG], are marked correspondingly.

Ternary and quaternary oxides and hydroxides[a]		Reference
$PbU_2O_7 \cdot 2H_2O(cr)$	wölsendorfite	[90VOC/HAV]
$Ni(UO_2)_3O_3(OH)_2^{4-} \cdot 6H_2O(cr)$		[91VOC/HAV]
$Mn(UO_2)_3O_3(OH)_2^{4-} \cdot 6H_2O(cr)$		[91VOC/HAV]
$MgUO_4(cr)$		data selected in [92GRE/FUG]
$MgU_2O_6(cr)$		[81GOL/TRE2]
$MgU_2O_7(cr)$		[81GOL/TRE2]
$MgU_3O_{10}(cr)$		data selected in [92GRE/FUG]
$Mg(UO_2)_6O_4(OH)_6^{10-} \cdot 13H_2O(cr)$		[91VOC/HAV]
$Mg_3U_3O_{10}(cr)$		[83FUG]
$CaUO_4(cr)$		data selected in [92GRE/FUG]
$\alpha\text{-}CaUO_4$		data selected in [71NAU/RYZ]
$\beta\text{-}CaUO_4$		data selected in [92GRE/FUG]
$Ca_3UO_6(cr)$		data selected in [92GRE/FUG]
$CaU_2O_6(cr)$		[81GOL/TRE], [82MOR], [83FUG], [85PHI/PHI], [86MOR], [88PHI/HAL]
$CaU_2O_7(cr)$	calciouranoite, anhydr.	[81GOL/TRE], [82MOR], [83FUG], [85PHI/PHI], [86MOR], [88PHI/HAL]
$CaU_6O_{19} \cdot 11H_2O(cr)$	becquerelite	data selected in this review [94CAS/BRU], [94SAN/GRA]
$\alpha\text{-}SrUO_4(cr)$		data selected in [92GRE/FUG]
$\beta\text{-}SrUO_4(cr)$		data selected in [92GRE/FUG] [79TAG/FUJ], [86MOR]
$SrUO_4(cr)$		[83KOH]
$Sr_2UO_5(cr)$		data selected in [92GRE/FUG]
$Sr_3UO_6(cr)$		data selected in [92GRE/FUG]
$Sr_2U_3O_{11}(cr)$		data selected in [92GRE/FUG]
$SrU_4O_{13}(cr)$		data selected in [92GRE/FUG]
$BaUO_3(cr)$		data selected in [92GRE/FUG]
$BaUO_4(cr)$		data selected in [92GRE/FUG]
$Ba_3UO_6(cr)$		data selected in [92GRE/FUG]
$BaU_2O_7(cr)$	bauranoite, anhydr.	data selected in [92GRE/FUG]
$Ba_2U_2O_7(cr)$		data selected in [92GRE/FUG]
$Ba_6Dy_2(UO_6)_3(cr)$		[82MOR]

(Continued on next page)

Table 9-43 (continued)

Ternary and quaternary oxides and hydroxides[a]		Reference
Ba_2MgUO_6(cr)		data selected in [92GRE/FUG]
Ba_2CaUO_6(cr)		data selected in [92GRE/FUG]
Ba_2SrUO_6(cr)		data selected in [92GRE/FUG]
$BaU_6O_{19} \cdot 11H_2O$(cr)	billietite	[90VOC/HAV]
$NaUO_3$(cr)		data selected in [92GRE/FUG]
α-Na_2UO_4		data selected in [92GRE/FUG]
ß-Na_2UO_4		data selected in [92GRE/FUG]
Na_3UO_4(cr)		data selected in [92GRE/FUG]
Na_4UO_5(cr)		data selected in [92GRE/FUG]
		[71COR/LOO], [78COR/OHA], [81LIN/BES], [82HEM], [82WAG/EVA], [83FUG], [85PHI/PHI], [85TSO/BRO],[86MOR], [88PHI/HAL]
$Na_2U_2O_7$(cr)		data selected in [92GRE/FUG]
$Na_6U_7O_{21}$(cr)		[78COR/OHA]
$Na_6U_7O_{24}$(cr)		data selected in [92GRE/FUG]
		[71COR/LOO], [81LIN/BES, [82HEM], [82MOR], [82WAG/EVA], [83FUG], [86MOR]
$Na_4O_4UO_4 \cdot 9H_2O$(cr)		[82WAG/EVA]
$Na_2U_2O_7 \cdot 1.5H_2O$(cr)		[65MUT], [82WAG/EVA]
KUO_3(cr)		data selected in [92GRE/FUG]
		[81COR/OUW], [81LIN/BES], [82HEM], [82MOR], [83FUG], [83KAG/KYS], [85PHI/PHI],[86MOR], [88PHI/HAL]
K_2UO_4(cr)		data selected in [92GRE/FUG]
K_4UO_5(cr)		[81LIN/BES]
$K_2U_2O_7$(cr)		data selected in [92GRE/FUG]
		[75OHA/HOE2], [81LIN/BES, [85FUG], [86MOR]
$K_2U_4O_{13}$(cr)		[81LIN/BES]
$K_2U_6O_{19} \cdot 11H_2O$(cr)	compreignacite	data selected in this review
$K_2U_7O_{22}$(cr)		[81LIN/BES]
Sulphates[b,c]:		*References*
$Ca_3NaUO_2(CO_3)_3FSO_4 \cdot 10H_2O$(cr)	schröckingerite	[82HEM], [83OBR/WIL]
$Zn_2(UO_2)_6(SO_4)_3(OH)_{10} \cdot 8H_2O$(cr)	Zn-zippeite	[79HAA/WIL], [81OBR/WIL], [82HEM][d]
$Co_2(UO_2)_6(SO_4)_3(OH)_{10} \cdot 8H_2O$(cr)	Co-zippeite	[79HAA/WIL], [81OBR/WIL], [82HEM][d]
$Ni_2(UO_2)_6(SO_4)_3(OH)_{10} \cdot 8H_2O$(cr)	Ni-zippeite	[79HAA/WIL], [81OBR/WIL], [82HEM][d]
$Mg_2(UO_2)_6(SO_4)_3(OH)_{10} \cdot 8H_2O$(cr)	Mg-zippeite	[79HAA/WIL],[81OBR/WIL], [82HEM][d]
$Na_4(UO_2)_6(SO_4)_3(OH)_{10} \cdot 4H_2O$(cr)	Na-zippeite	[81OBR/WIL], [82HEM]
$K_4(UO_2)_6(SO_4)_3(OH)_{10} \cdot 4H_2O$(cr)	zippeite	[81OBR/WIL], [82HEM]

(Continued on next page)

Table 9-43 (continued)

Phosphates		References
$U_3(PO_4)_4(cr)$		[73MOS], [78ALL/BEA][e]
$U(HPO_4)_2 \cdot 4H_2O(cr)$		**data selected** in [92GRE/FUG]
$U(HPO_4)_2(cr)$		[71MOS], [84VIE/TAR], [86WAN][f]
$U(HPO_4)_2H_3PO_4 \cdot H_2O(cr)$		[55SCH][g]
$UO_2HPO_4 \cdot 4H_2O(cr)$	H-autunite	**data selected** in [92GRE/FUG]
$UO_2HPO_4(cr)$		See footnote (h)
$(UO_2)_2(HPO_4)_2(cr)$		See footnote (h)
$H_2(UO_2)_2(PO_4)_2 \cdot 10H_2O(cr)$		See footnotes (h) and (i)
$UO_2(H_2PO_4)_2 \cdot 3H_2O(cr)$		
$(UO_2)_3(PO_4)_2(cr)$		**data selected** in [92GRE/FUG]
$(UO_2)_3(PO_4)_2 \cdot 4H_2O(cr)$		**data selected** in [92GRE/FUG]
		[92SAN/BRU]
$(UO_2)_3(PO_4)_2 \cdot 6H_2O(cr)$		**data selected** in [92GRE/FUG]
$NH_4UO_2PO_4 \cdot 3H_2O(cr)$	uraniphite	[61KAR], [71NAU/RYZ],
		[84VIE/TAR], [88PHI/HAL]
$NH_4UO_2PO_4(cr)$	NH_4-autunite, anhydr.	[56CHU/STE], [65VES/PEK],
		[84VIE/TAR], [85PHI/PHI], [86WAN][k]
$(NH_4)_2(UO_2)_2(PO_4)_2(cr)$		[78LAN], [88PHI/HAL]
$HAl(UO_2)_4(PO_4)_4 \cdot 16H_2O(cr)$	sabugalite	[83VOC/PEL]
$HAl(UO_2)_4(PO_4)_4(cr)$	sabugalite, anhydr.	[84GEN/WEI]
$Pb_2UO_2(PO_4)_2 \cdot 2H_2O(cr)$	parsonsite	[84NRI2]
$Pb_2UO_2(PO_4)_2(cr)$	parsonsite, anhydr.	[84GEN/WEI]
$Pb(UO_2)_2(PO_4)_2 \cdot 10H_2O(cr)$		[65MUT/HIR]
$Pb(UO_2)_2(PO_4)_2 \cdot 2H_2O(cr)$	przhevalskite	[84NRI2][l]
$Pb(UO_2)_2(PO_4)_2(cr)$	przhevalskite, anhydr.	[78LAN], [85PHI/PHI], [88PHI/HAL]
$Pb(UO_2)_3(PO_4)_2(OH)_2(cr)$		[84GEN/WEI]
$Pb_2(UO_2)_3(PO_4)_2(OH)_4 \cdot 3H_2O(cr)$	dumontite	[84NRI2]
$Pb(UO_2)_4(PO_4)_2(OH)_4 \cdot 7H_2O(cr)$	renardite[m]	[84NRI2]
$Pb(UO_2)_4(PO_4)_2(OH)_4 \cdot 8H_2O(cr)$		[84NRI2]
$Cu(UO_2)_2(PO_4)_2 \cdot 12H_2O(cr)$	torbernite	[65MUT/HIR], [84NRI2]
$Cu(UO_2)_2(PO_4)_2 \cdot 8H_2O(cr)$	meta-torbernite	[81VOC/PIR], [84NRI2], [84VIE/TAR]
$Cu(UO_2)_2(PO_4)_2(cr)$	torbernite, anhydr.	[78LAN], [85PHI/PHI], [88PHI/HAL]
$Ni(UO_2)_2(PO_4)_2 \cdot 7H_2O(cr)$		[81VOC/PIR], [84VIE/TAR]
$Co(UO_2)_2(PO_4)_2 \cdot 7H_2O(cr)$		[81VOC/PIR], [84VIE/TAR]
$Fe(UO_2)_2(PO_4)_2OH \cdot 6H_2O(cr)$		[84VOC/GRA]
$Fe(UO_2)_2(PO_4)_2OH \cdot 8H_2O(cr)$	bassetite	[65MUT/HIR], [84VOC/GRA][n]

(Continued on next page)

Table 9-43 (continued)

Phosphates		References
$Fe(UO_2)_2(PO_4)_2(cr)$	bassetite, anhydr.	[78LAN], [85PHI/PHI], [86WAN], [88PHI/HAL]
$Mg(UO_2)_2(PO_4)_2 \cdot 10H_2O(cr)$	saleeite	[65MUT/HIR]
$Mg(UO_2)_2(PO_4)_2(cr)$	saleeite, anhydr.	[78LAN], [85PHI/PHI, [86WAN], [88PHI/HAL]
$CaU(PO_4)_2 \cdot 2H_2O(cr)$	ningyoite	[65MUT], [65MUT/HIR], [78LAN], [84VIE/TAR], [88PHI/HAL][i]
$CaU(PO_4)_2(cr)$		[86WAN]
$Ca(UO_2)_2(PO_4)_2 \cdot 10H_2O(cr)$	autunite	[65MUT], [65MUT/HIR], [84VIE/TAR][i]
$Ca(UO_2)_2(PO_4)_2(cr)$	autunite, anhydr.	[78LAN], [85PHI/PHI, [86WAN], [88PHI/HAL]
$Ca(UO_2)_4(PO_4)_2(OH)_4(cr)$		[84GEN/WEI]
$Sr(UO_2)_2(PO_4)_2(cr)$	Sr-autunite, anhydr.	[78LAN], [85PHI/PHI], [86WAN], [88PHI/HAL]
$Sr(UO_2)_2(PO_4)_2 \cdot 10H_2O(cr)\cdot$		[65MUT/HIR]
$Ba(UO_2)_2(PO_4)_2(cr)$	uranocircite, anhydr.	[78LAN], [84GEN/WEI], [85PHI/PHI], [86WAN], [88PHI/HAL]
$Ba(UO_2)_2(PO_4)_2 \cdot 10H_2O(cr)$	uranocircite II	[65MUT/HIR]
$NaUO_2PO_4(cr)$	Na-autunite, anhydr.	[65VES/PEK], [85PHI/PHI], [86WAN][k]
$Na_2(UO_2)_2(PO_4)_2(cr)$		[78LAN], [88PHI/HAL][k]
$Na_2(UO_2)_2(PO_4)_2 \cdot 10H_2O(cr)$		[65MUT/HIR]
$KUO_2PO_4(cr)$	K-autunite, anhydr.	[56CHU/STE], [65VES/PEK], [82WAG/EVA], [84VIE/TAR], [85PHI/PHI], [86WAN][k]
$K_2(UO_2)_2(PO_4)_2(cr)$		[78LAN], [88PHI/HAL][k]
$KUO_2PO_4 \cdot 3H_2O(cr)$	See footnote [o]	[61KAR], [71NAU/RYZ], [84VIE/TAR], [85PHI/PHI], [88PHI/HAL]
$K_2(UO_2)_2(PO_4)_2 \cdot 10H_2O(cr)$		[65MUT/HIR]

Pyrophosphates	References
$UP_2O_7 \cdot xH_2O(cr)$	[67MER/SKO][p]
$UP_2O_7(cr)$	data selected in [92GRE/FUG]
$(UO_2)_2P_2O_7(cr)$	data selected in [92GRE/FUG]

(Continued on next page)

Table 9-43 (continued)

Arsenates		References
$(UO_2)_3(AsO_4)_2(cr)$	troegerite, anhydr.	**data selected** in [92GRE/FUG]
$UO_2HAsO_4(cr)$	hydrogen spinite	[56CHU/SHA], [84GEN/WEI]
$NH_4UO_2AsO_4(cr)$		[56CHU/SHA]
$Pb_2UO_2(AsO_4)_2(cr)$	hallimondite	[84GEN/WEI]
$Zn(UO_2)_2(AsO_4)_2(cr)$	meta-lodevite, anhydr.	[84GEN/WEI]
$Cu(UO_2)_2(AsO_4)_2(cr)$	meta-zeunerite, anhydr.	[84GEN/WEI]
$Cu(UO_2)_2(AsO_4)_2 \cdot 8H_2O(cr)$	meta-zeunerite	[84VOC/GOE]
$Ni(UO_2)_2(AsO_4)_2 \cdot 7H_2O(cr)$		[84VOC/GOE]
$Co(UO_2)_2(AsO_4)_2(cr)$	meta-kirchheimerite, anhydr.	[84GEN/WEI]
$Co(UO_2)_2(AsO_4)_2 \cdot 7H_2O(cr)$	meta-kirchheimerite, heptahydrate	[84VOC/GOE]
$Fe(UO_2)_2(AsO_4)_2(cr)$	kahlerite, anhydr.	[84GEN/WEI]
$Fe(UO_2)_2(AsO_4)_2 \cdot 8H_2O(cr)$	meta-kahlerite	[86VOC/GRA]
$Mn(UO_2)_2(AsO_4)_2 \cdot 8H_2O(cr)$		[86VOC/GRA]
$Mn(UO_2)_2(AsO_4)_2(cr)$	novacekite, anhydr.	[84GEN/WEI]
$Ca(UO_2)_2(AsO_4)_2(cr)$	uranospinite, anhydr.	[84GEN/WEI], [91FAL/HOO]
$Ca(UO_2)_2(AsO_4)_2(OH)_4(cr)$	arsenuranylite, anhydr.	[84GEN/WEI]
$Ba(UO_2)_2(AsO_4)_2(cr)$	meta-heinrichite, anhydr.	[84GEN/WEI]
$NaUO_2AsO_4 \cdot 4H_2O(cr)$	Na-uranospinite	[71NAU/RYZ]
$NaUO_2AsO_4(cr)$		[56CHU/SHA], [91FAL/HOO]
$KUO_2AsO_4(cr)$	abernathyite, anhydr.	[56CHU/SHA], [71NAU/RYZ], [82WAG/EVA]

Carbonates		References
$UO_2CO_3(cr)$	rutherfordine	**data selected** in [92GRE/FUG]
$UO_2CO_3 \cdot H_2O(cr)$		[82HEM][(q)]
$UO_2(HCO_3)_2 \cdot H_2O(cr)$		[76BOU/BON], [78COR/OHA], [82WAG/EVA], [83FUG] [86MOR]
$Ca_3NaUO_2(CO_3)_3FSO_4 \cdot 10H_2O(cr)$	schröckingerite	[82HEM], [83OBR/WIL]
$Ca_2UO_2(CO_3)_3 \cdot 10\text{-}11H_2O(cr)$	liebigite	[80BEN/TEA], [80ALW/WIL], [82HEM], [83OBR/WIL][(r)]
$CaMgUO_2(CO_3)_3 \cdot 12H_2O(cr)$	swartzite	[80ALW/WIL], [80BEN/TEA], [82HEM][(r)]
$Mg_2UO_2(CO_3)_3 \cdot 18H_2O(cr)$	bayleyite	[80ALW/WIL], [80BEN/TEA], [82HEM][(r)]

(Continued on next page)

Table 9-43 (continued)

Carbonates		References
$CaNa_2UO_2(CO_3)_3 \cdot 6H_2O(cr)$	andersonite	[80ALW/WIL], [80BEN/TEA], [83OBR/WIL][r]
$Ca_2CuUO_2(CO_3)_4 \cdot 6H_2O(cr)$	voglite	[82HEM]
$Ca_3Mg_3(UO_2)_2(CO_3)_6(OH)_4 \cdot 18H_2O(cr)$	rabbittite	[82HEM]
$Na_4UO_2(CO_3)_3(cr)$		data selected in [92GRE/FUG]
$K_3NaUO_2(CO_3)_3 \cdot H_2O(cr)$	grimselite	[83OBR/WIL]

Silicates		References
$USiO_4(cr)$	coffinite	data selected in [92GRE/FUG]
$(UO_2)_2SiO_4 \cdot 2H_2O(cr)$	soddyite	[82HEM], [92NGU/SIL], [94CAS/BRU]
$PbUO_2SiO_4 \cdot H_2O(cr)$	kasolite	[82HEM]
$K(H_3O)UO_2SiO_4(cr)$	boltwoodite[s]	[82HEM]
$Na_{0.7}K_{0.3}(H_3O)UO_2SiO_4 \cdot H_2O(cr)$	Na-boltwoodite[s]	[82HEM], [92NGU/SIL]
$Cu(UO_2)_2(SiO_3OH)_2 \cdot 6H_2O(cr)$	cupro sklodowskite	[82HEM]
$Mg(UO_2)_2(SiO_3OH)_2 \cdot 5H_2O(cr)$	sklodowskite	[82HEM][t]
$Ca(UO_2)_2(SiO_3OH)_2(cr)$	uranophane, anhydr.	[78LAN], [80BEN/TEA], [86WAN], [88LEM], [88PHI/HAL]
$Ca(UO_2)_2(SiO_3OH)_2 \, 5H_2O(cr)$	uranophane	[82HEM], [92NGU/SIL], [94CAS/BRU]
$Ca(UO_2)_2Si_6O_{15} \cdot 5H_2O(cr)$	haiweeite	[82HEM]
$Na_2(UO_2)_2Si_6O_{15} \cdot 4H_2O(cr)$	Na-weeksite	[92NGU/SIL]
$K_2(UO_2)_2Si_6O_{15} \cdot 4H_2O(cr)$	weeksite	[82HEM]

Vanadates		References
$Pb(UO_2)_2(VO_4)_2(cr)$	curienite, anhydr.	[84GEN/WEI]
$Al(UO_2)_2(VO_4)_2OH(cr)$	vanuralite, anhydr.	[84GEN/WEI]
$CuUO_2VO_4(cr)$	sengierite, anhydr.	[84GEN/WEI]
$Ca(UO_2)_2(VO_4)_2(cr)$	tyuyamunite, anhydr.	[78LAN], [80BEN/TEA], [85PHI/PHI], [88PHI/HAL]
$Ba(UO_2)_2(VO_4)_2(cr)$	francevillite, anhydr.	[84GEN/WEI]
$NaUO_2VO_4(cr)$	strelkinite, anhydr.	[84GEN/WEI]
$K_2(UO_2)_2(VO_4)_2 \cdot H_2O(cr)$	carnotite	[62HOS/GAR]
$K_2(UO_2)_2(VO_4)_2(cr)$	carnotite, anhydr.	[78LAN], [80BEN/TEA], [85PHI/PHI], [88PHI/HAL]

(a) Simple oxides and hydroxides are discussed in Section V.3.3 (pp.131–148) of [92GRE/FUG].
(b) Only binary and ternary sulphates are listed. Simple uranium sulphates are discussed in Section V.5.1.3.2. (pp.249–254) of [92GRE/FUG].
(c) It should be noted that there is some disagreement in the literature on the number of water molecules in the formulae of the zippeite family.
(d) See Section V.5.1.3.2.c (pp.254–255) and the discussion of [79HAA/WIL] in Appendix A (p.646) of [92GRE/FUG].
(e) See the discussion in Section V.6.2.2.5.b (p.294) of [92GRE/FUG].
(f) See the discussion in Section V.6.2.2.7.b (p.297) of [92GRE/FUG].
(g) See the discussion in Section V.6.2.2.9 (p.298) of [92GRE/FUG].
(h) Compounds with formula $H_2(UO_2)_2(PO_4)_2 \cdot xH_2O(cr)$ ($x = 0$ to 10) are discussed in Sections V.6.2.1.1.b and V.6.2.2.10.c (pp.284–286 and 299–300 respectively) of [92GRE/FUG]. Grenthe *et al.* selected thermodynamic data for $UO_2HPO_4 \cdot 4H_2O(cr)$, *cf.* Table III.1 in [92GRE/FUG].
(i) See the discussion of [65MUT] in Appendix A (p.599) of [92GRE/FUG].
(j) See the discussions in Sections V.6.2.1.1.b and V.6.2.2.8 (pp.284–286 and 298) and the comments of [54SCH/BAE] in Appendix A (p.564) of [92GRE/FUG].
(k) Veselý *et al.* ([65VES/PEK]) reported solubility products for *hydrated* alkali phosphates. Grenthe *et al.* ([92GRE/FUG]) reinterpreted the results of [65VES/PEK], *cf.* Table V.40 (p.283), Section V.6.2.1.1.b (p.286), and Appendix A (pp.600–601) of [92GRE/FUG]. It should be noted that Langmuir ([78LAN]) incorrectly referred to the anhydrous compounds as sodium and potassium autunite.
(l) Nriagu ([84NRI2]) gives four waters of hydration for this mineral (Table 1).
(m) The stoichiometry of the mineral renardite is not clear, although it is related to dewindtite: $Pb_3(UO_2)_6(PO_4)_4O_2(OH)_2 \cdot 12H_2O(s)$ ([91FIN/EWI]).
(n) Muto *et al.* ([65MUT/HIR]) give ten molecules of water of hydration for this mineral.
(o) This formula is related to meta–ankoleite: $K_2(UO_2)_2(PO_4)_2 \cdot 6H_2O(cr)$.
(p) See the discussion of [67MER/SKO] in Appendix A (p.603) of [92GRE/FUG].
(q) Hemingway ([82HEM]) incorrectly assigned the name of sharpite to this solid.
(r) Discussed in section V.7.1.2.2.b (pp.327–328) of [92GRE/FUG].
(s) There is some uncertainty in the composition of boltwoodite and Na–boltwoodite, *cf.* [81STO/SMI], [91FIN/EWI], [92NGU/SIL].
(t) Hemingway ([82HEM]) reports estimated data for a solid with six water molecules.

Chapter 10

Discussion of new data selection for Neptunium

10.1 Neptunium aqua ions (7)

There are few experimental data that have appeared after those discussed in [2001LEM/FUG].

EXAFS spectra of NpO_2^+ and Np^{4+} obtained by Allen *et al.* [97ALL/BUC] and those of Np(IV), NpO_2^+ and NpO_2^{2+} obtained by Reich *et al.* [2000REI/BER] do not give consistent results on the hydration number of NpO_2^+, respectively, five and (3.6 ± 0.6). It should be noted that coordination numbers determined by EXAFS are not particularly precise. Hence, the coordination number for Np^{4+} given as (10 ± 1), [97ALL/BUC], is uncertain and this review suggests a coordination number of (9 ± 1).

Soderholm *et al.* [99SOD/ANT] have also given the full EXAFS spectra of the aqueous ions of Np(III) to Np(VI) (see Appendix A). Soderholm *et al.* [99SOD/ANT] have combined EXAFS and electrochemical measurements and propose values of the formal redox potential of the NpO_2^{2+}/NpO_2^+ and Np^{4+}/Np^{3+} couples in 1 M HClO$_4$. This is an interesting approach, but in the opinion of this review, it does not result in more precise data than the traditional electrochemical methods. The review also notices that the values obtained are very far from those obtained in other studies discussed in [2001LEM/FUG].

The values selected by [2001LEM/FUG] are therefore retained in the present review (see Appendix A).

Bolvin *et al.* [2001BOL/WAH] have given the structure of the Np(VII) species predominant in strong alkaline solution: $NpO_4(OH)_2^{3-}$, thus closing the debate that has gone on since the discovery of the Np(VII) species (see [2001LEM/FUG] page 92 and in Appendix A [2001WIL/BLA] and [97CLA/CON]). Consequently, the equation 7.2 of [2001LEM/FUG] must be written:

$$NpO_4(OH)_2^{3-} + 2\,H_2O(l) + e^- \rightleftharpoons NpO_2(OH)_4^{2-} + 2\,OH^-.$$

The close similarity in the coordination of Np in the species involved in the redox couple Np(VII)/Np(VI) explains why it is stable and reproducible in alkaline solutions.

Evaluations of $\Delta_f G_m^\circ(NpO_2^+, 298.15\text{ K})$ and $\Delta_f G_m^\circ(Np^{4+}, 298.15\text{ K})$ from literature data have been conducted by Kaszuba and Runde [99KAS/RUN]. These data are the same as in [2001LEM/FUG], but the method used to calculate the Gibbs energy is different. The values obtained in the two evaluations are very similar and, as explained in Appendix A, there is no reason to revise the selected values of [2001LEM/FUG], in favour of these new evaluations.

10.2 Neptunium oxygen and hydrogen compounds and complexes (8)

10.2.1 Aqueous neptunium hydroxide complexes (8.1)

10.2.1.1 Neptunium (VI) hydroxide complex (8.1.2)

An error of sign for the values of $\varepsilon((NpO_2)_3(OH)_5^+, ClO_4^-)$ and $\varepsilon((NpO_2)_2(OH)_2^{2+}, ClO_4^-)$ exists in [2001LEM/FUG] on pages 105 and 814. The correct values are:

$$\varepsilon((NpO_2)_3(OH)_5^+, ClO_4^-) = (0.45 \pm 0.20)\text{ kg} \cdot \text{mol}^{-1},$$

$$\varepsilon((NpO_2)_2(OH)_2^{2+}, ClO_4^-) = (0.57 \pm 0.10)\text{ kg} \cdot \text{mol}^{-1}.$$

These errors do not affect the calculated equilibrium constants or Gibbs energy values given on pages 105, 106, and 639 of [2001LEM/FUG] which have been calculated with the correct $\Delta\varepsilon$ values.

10.2.1.2 Neptunium(V) hydroxide complexes (8.1.3)

An analysis of the literature data concerning Np(V) behaviour in aqueous solutions was carried out by Kaszuba and Runde [99KAS/RUN]. These authors give values for $\log_{10} \beta_{1,1}^\circ$ and $\log_{10} \beta_{2,1}^\circ$ in very close agreement with those of [2001LEM/FUG], which are retained by this review as selected values (see Appendix A).

Experimental results on the solubility of Np(V) hydroxide compounds have been reported by Peretrukhin et al. [96PER/KRY]. As discussed in Appendix A no thermodynamic data can be extracted from this study.

10.2.1.3 Neptunium(IV) hydroxide complexes (8.1.4)

The main advance in knowledge since the previous review concerns the hydrolysis of Np(IV). As discussed in [2001LEM/FUG] the situation is the same as for the other tetravalent actinides [92GRE/FUG]. Due to phenomena difficult to control, like formation of colloids and the evolution of the limiting solubility of solid phases between crystalline and amorphous forms of the dioxide or tetrahydroxide, the description of actinide (IV) hydrolysis in terms of thermodynamic constants is rather uncertain.

Only the value $\log_{10} {}^*\beta^\circ_{1,1} = -(0.29 \pm 1.00)$ was selected by [2001LEM/FUG]. The large uncertainty comes from the unweighted average of three scattered experimental values. Due to large uncertainties in the literature data, [2001LEM/FUG] did not select values for the hydrolysis constants $\beta_{2,1}$ and $\beta_{3,1}$. A value of $\log_{10} \beta^\circ_{4,1} = (46.17 \pm 1.12)$ is also reported by [2001LEM/FUG] from solubility data, or can be calculated as an alternative route from the selected values of the Gibbs energy of formation of NpO$_2$(am) and Np(OH)$_4$(aq) (pages 115 and 128 of [2001LEM/FUG], see also the review of [99NEC/KIM] in Appendix A). Indeed $\Delta_f G^\circ_m$ (NpO$_2$, am, 298.15 K) and $\Delta_f G^\circ_m$ (Np(OH)$_4$(aq), 298.15 K) given by [2001LEM/FUG] are derived from the same solubility data (see 10.2.2.2.). This review has clearly taken the position not to select Gibbs energies of amorphous phases (see section 9.3.2.2).

None of the spectroscopic studies considered in [2001LEM/FUG] for the selection of $\log_{10} {}^*\beta^\circ_{1,1}$ could identify two distinct absorption bands, *i.e.*, one for Np^{4+} and one for Np(OH)$^{3+}$. There is not even a shift of the characteristic Np(IV) absorption bands. (The same is true for the spectroscopic studies on Pu(IV) hydrolysis, *cf.* section 11.2.1.2). The only effect observed was the decrease of the absorption band when pH was increased. Neck *et al.* [2001NEC/KIM2] demonstrated by spectroscopic and LIBD measurements that (1) the decrease of the Np(IV) absorption bands is a function of both pH and total Np(IV) concentration, *i.e.*, it cannot be due to mononuclear hydrolysis, and (2) the onset of the decrease of absorption is consistent with the onset of colloid formation.

Neck and Kim [99NEC/KIM], [2001NEC/KIM] have made critical surveys of the literature data of actinide(IV) hydrolysis as done partially by [2001LEM/FUG], but from the standpoint of correlations and semi-theoretical considerations to predict the values of successive hydrolysis constants of M^{4+} aqueous ions. The parameters from these theoretical methods are fitted from sets of experimental thermodynamic constants of trivalent (Am, Cm) and tetravalent (Pu) actinides with the first hydrolysis constant, $\log_{10} \beta^\circ_{1,1}$, as the input parameter. For Np^{4+}, starting with $\log_{10} \beta^\circ_{1,1} = (14.55 \pm 0.2)$ ($\log_{10} {}^*\beta^\circ_{1,1} = 0.55 \pm 0.2$) obtained at very low neptunium concentrations [77DUP/GUI], they predicted, depending on the method used (correlation or semi-empirical model), $\log_{10} \beta^\circ_{2,1} = 28.0$ or 28.3 (in close agreement with the experimental value (28.35 ± 0.3) from [77DUP/GUI]), $\log_{10} \beta^\circ_{3,1} = 39.0$ or 39.4 and $\log_{10} \beta^\circ_{4,1} = 46.6$ or 47.5. The choice of the value of $\beta^\circ_{1,1}$ reflects the necessity of using data from experiments where the presence of colloids has been avoided as in [77DUP/GUI] where the Np concentration is at tracer levels. The $\log_{10} \beta^\circ_{4,1}$ value is consistent with the value obtained from reinterpretation of the solubility data of amorphous neptunium dioxide, as shown in the following section 10.2.2.2 ($\log_{10} \beta^\circ_{4,1} = (47.7 \pm 1.1)$).

The reinterpretation of the available literature data on Np(IV) hydrolysis made by Neck and Kim [2001NEC/KIM] changes the $\log_{10} \beta^\circ_{1,1}$ value selected by [2001LEM/FUG] and confirms the value of $\log_{10} \beta^\circ_{4,1}$, which can be calculated, for instance, from the appropriate Gibbs energies of formation selected by [2001LEM/FUG].

A revision of the hydrolysis constants of Np^{4+} is required.

This review considers the hydrolysis constants $\log_{10} \beta_{1,1}^\circ$ and $\log_{10} \beta_{2,1}^\circ$ determined in the solvent extraction study of Duplessis and Guillaumont [77DUP/GUI], and the $\log_{10} \beta_{3,1}^\circ$ value estimated by [2001NEC/KIM] as the best available. The $\log_{10} \beta_{4,1}^\circ$ value estimated by [2001NEC/KIM] is close to the value calculated from solubility data (see section 10.2.2.2.2 and Table 11-1, footnote (b)).

The following equilibrium constants are selected at 298.15 K for the reactions:

$$Np^{4+} + n\,OH^- \rightleftharpoons Np(OH)_n^{4-n} \qquad (10.1)$$

$$Np^{4+} + n\,H_2O(l) \rightleftharpoons Np(OH)_n^{4-n} + n\,H^+, \qquad (10.2)$$

$\log_{10} \beta_1^\circ \,(10.1) = (14.55 \pm 0.20)$ $\qquad \log_{10} {}^*\beta_1^\circ \,(10.2) = (0.55 \pm 0.20)$,

$\log_{10} \beta_2^\circ \,(10.1) = (28.35 \pm 0.30)$ $\qquad \log_{10} {}^*\beta_2^\circ \,(10.2) = (0.35 \pm 0.30)$,

$\log_{10} \beta_4^\circ \,(10.1) = (47.7 \pm 1.1)$ $\qquad \log_{10} {}^*\beta_4^\circ \,(10.2) = -(8.3 \pm 1.1)$.

According to NEA TDB guidelines the estimated values, $\log_{10} \beta_3^\circ \,(10.1) = (39.2 \pm 1.0)$, $\log_{10} {}^*\beta_3^\circ \,(10.2) = -(2.8 \pm 1.0)$, are not selected by this review.

Solubility data of Neck et al. [2001NEC/KIM2] obtained by using spectroscopic methods for determining the Np(IV) concentration show that the species $Np(OH)_2^{2+}$ is dominant in the range $\log_{10}[H^+] = -1.6$ to -2.7 at very low concentrations of Np(IV). This confirms the hydrolysis results obtained at tracer scale by [77DUP/GUI] (see Appendix A).

Corresponding $\Delta_f G_m^\circ (Np(OH)_n^{4-n}, 298.15\,K)$ values can be calculated (see on section 11.2.3 of this review which gives a comparison of thermodynamic data on tetravalent actinides).

Peretrukhin et al. [96PER/KRY] proposed the formation of $Np(OH)_5^-$ from the increase in the solubility of Np(IV) hydroxide in 0.5 to 14.1 M NaOH. However, for the reasons given in Appendix A the present review does not extract thermodynamic quantities from these data.

10.2.2 Crystalline and amorphous neptunium oxides and hydroxides (8.2)

10.2.2.1 Solubility of neptunium(V) oxides and hydroxides (8.2.4)

Kaszuka and Runde [99KAS/RUN] re-analysed the solubility data on NpO_2OH reported by [92NEC/KIM], [95NEC/FAN] and [96RUN/NEU], which agree within 0.02 log-units for fresh, and 0.07 log-units for aged NpO_2OH, and give, respectively, the averaged solubility product $\log_{10} K_{s,0}^\circ = -(8.77 \pm 0.09)$ and $\log_{10} K_{s,0}^\circ = -(9.48 \pm 0.16)$. These values are very close to the values selected by [2001LEM/FUG] from the same experimental data, (Table 8.5, p. 126 in [2001LEM/FUG]), $\log_{10}{}^{*}K_{s,0}^{\circ}$ (NpO$_2$OH, am, fresh) = (5.3 ± 0.2), ($\log_{10} K_{s,0}^{\circ}$ = – (8.7 ± 0.2)), $\log_{10}{}^{*}K_{s,0}^{\circ}$ (NpO$_2$OH, am, aged) = (4.7 ± 0.5), ($\log_{10} K_{s,0}^{\circ}$ = – (9.3 ± 0.5)).

Therefore, there is no reason to change the values selected by [2001LEM/FUG].

Based on enthalpies of solution measurements with well-crystallised anhydrous Np$_2$O$_5$, [2001LEM/FUG] selected $\Delta_f H_m^{\circ}$ (Np$_2$O$_5$, cr, 298.15 K) = – (2162.7 ± 9.5) kJ·mol^{-1} and, using the estimate S_m° (Np$_2$O$_5$, cr, 298.15 K) = (174 ± 20) J·K^{-1}·mol^{-1}, obtained $\Delta_f G_m^{\circ}$ (Np$_2$O$_5$, cr, 298.15 K) = – (2031.6 ± 11.2) kJ·mol^{-1}, corresponding to $\log_{10}{}^{*}K_{s,0}^{\circ}$ = (1.8 ± 1.0) for the reaction:

$$0.5\,\text{Np}_2\text{O}_5(\text{cr}) + \text{H}^+ \rightleftharpoons \text{NpO}_2^+ + 0.5\,\text{H}_2\text{O}(\text{l}).$$

Recently Efurd et al. [98EFU/RUN], not cited in [2001LEM/FUG], reported a value of $\log_{10}{}^{*}K_{s,0}^{\circ}$ = (2.6 ± 0.4) for the solubility of a solid described as a poorly crystalline hydrated Np$_2$O$_5$·xH$_2$O (see Appendix A). Although not corresponding to compounds of the same crystallinity and water content, the results of Efurd et al. are not incompatible with the adopted values for crystalline anhydrous Np$_2$O$_5$(cr). However, this lack of knowledge of the exact composition of the compounds in equilibrium with the aqueous phase, makes it impossible to deduce Gibbs energies from the data of Efurd et al. [98EFU/RUN].

This review retains the value of $\Delta_f G_m^{\circ}$ (Np$_2$O$_5$, cr, 298.15 K) selected by [2001LEM/FUG].

10.2.2.2 Neptunium(IV) oxides and hydroxides (8.2.5)

10.2.2.2.1 Solubility of crystalline oxide NpO$_2$(cr) (8.2.5.1)

No new experimental data have appeared since the last review.

According to the values of Gibbs energy of formation of NpO$_2$(cr) selected by [2001LEM/FUG], $\Delta_f G_m^{\circ}$ (NpO$_2$, cr, 298.15 K) = – (1021.731 ± 2.514) kJ·mol^{-1} and $\Delta_f G_m^{\circ}$ (Np^{4+}, 298.15 K) = – (491.8 ± 5.6) kJ·mol^{-1} and the auxiliary values from [92GRE/FUG], this review calculates $\log_{10} K_{s,0}^{\circ}$ (NpO$_2$, cr) = – (65.75 ± 1.07). A somewhat different value has been proposed in the recent literature [99NEC/KIM], [2001NEC/KIM] with reference to [87RAI/SWA], as $\log_{10} K_{s,0}^{\circ}$ (NpO$_2$, cr) = – (63.7 ± 1.8). The difference comes mainly from the value of $\Delta_f G_m^{\circ}$ (Np^{4+}, 298.15 K) = – (502.9 ± 7.5) kJ·mol^{-1} [76FUG/OET] used to calculate $\log_{10} K_{s,0}^{\circ}$ (see section 11.2.3).

This review does not change the value of Gibbs energy of formation of NpO$_2$(cr) selected by [2001LEM/FUG].

10.2.2.2.2 Solubility of amorphous oxide NpO$_2$(am, hydr.) (8.2.5.2)

The constant for the reaction,

$$NpO_2(am, hydr.) + 2H_2O(l) \rightleftharpoons Np^{4+} + 4OH^- \qquad (10.3)$$

consistent with the selections in [2001LEM/FUG] is $\log_{10} K_{s,0}^\circ$ (NpO$_2$, am, hydr., (10.3)) $= -(54.5 \pm 1.0)$ derived from $\log_{10}{}^*K_{s,0}^\circ$ (NpO$_2$, am, hydr.) $= (1.53 \pm 1.0)$. This latter value is calculated from the concentration of Np(IV) in equilibrium with the solid and $\log_{10}{}^*\beta_{1,1}^\circ = -(0.29 \pm 1.00)$. The large uncertainty in $\log_{10}{}^*K_{s,0}^\circ$ is estimated. The experimental solubility data come from [87RAI/SWA]. Lemire et al. [2001LEM/FUG] calculated and selected $\Delta_f G_m^\circ$ (NpO$_2$, am, 298.15 K) $= -(953.3 \pm 8.0)$ kJ·mol^{-1}. This method of calculating the Gibbs energy of formation of an amorphous phase does not appear valid to this review for the reason put forward in the section 9.3.2.2.

Neck and Kim [99NEC/KIM], [2001NEC/KIM], (see Appendix A) reinterpreted the solubility data of [87RAI/SWA] at pH below 3 with the hydrolysis constants $\log_{10} \beta_{1,1}^\circ$, $\log_{10} \beta_{2,1}^\circ$ from [77DUP/GUI] and an estimated value for $\log_{10} \beta_{3,1}^\circ$ (see section 10.2.1.3), and give $\log_{10} K_{s,0}^\circ$ (NpO$_2$, am, hydr., (10.3)) $= -(56.7 \pm 0.5)$. With this value and all the solubility data of [85RAI/RYA], [93ERI/NDA] and [96NAK/YAM] corresponding to pH = 5 to 13, they calculate $\log_{10} \beta_{4,1}^\circ = (47.7 \pm 1.1)$ according to the relationship:

$$\log_{10}[Np(OH)_4(aq)] = \log_{10} K_{s,0}^\circ (NpO_2, am, hydr., (10.3)) + \log_{10} \beta_{4,1}^\circ = -(9 \pm 1).$$

The large uncertainty covers the rather large scatter in the data. This value is close to the one obtained by semi-empirical estimation methods, (see section 10.2.1.3), but, as suggested, higher than that selected by [2001LEM/FUG]. The recent experimental data obtained by Neck et al. [2001NEC/KIM2], using spectroscopic techniques to measure the Np concentration, after removing colloids, in solutions equilibrated with Np(OH)$_4$(am) is $\log_{10} K_{s,0}^\circ$ (NpO$_2$, am, hydr., (10.3)) $= -(56.5 \pm 0.4)$.

The authors of [2001LEM/FUG] suggested that the solubility of NpO$_2$(am) retained to derive the thermodynamic values $\log_{10} K_{s,0}^\circ$ (NpO$_2$, am, hydr., (10.3)) and $\log_{10} \beta_{4,1}^\circ$ "might require revision" (p. 114). The present review selects the following equilibrium constants reported by [2001NEC/KIM]:

$\log_{10} K_{s,0}^\circ$ (NpO$_2$, am, hydr., (10.3), 298.15 K) $= -(56.7 \pm 0.5)$,

$\log_{10} \beta_{4,1}^\circ$ (Np(OH)$_4$, aq, 298.15 K) $= (47.7 \pm 1.1)$.

However, for the reasons already given in sections 9.3.2.2, the present review does not select a value of $\Delta_f G_m^\circ$ (NpO$_2$, am, hydr.) for this Np(IV) compound, since it is not chemically well-defined - it could be the hydroxide, amorphous hydrous oxide or oxyhydroxide.

An interpretation of the solubility data of NpO$_2$(am) by Kaszuba and Runde [99KAS/RUN] does not add additional information.

The new value selected for the solubility constant of NpO$_2$(am, hydr.), $\log_{10} K_{s,0}^\circ$ (NpO$_2$, am, hydr., (10.3)) = – (56.7 ± 0.5) or $\log_{10}{}^* K_{s,0}^\circ$ (NpO$_2$, am, hydr.,) = – (0.7 ± 0.5) would correspond to a standard Gibbs energy of $\Delta_f G_m^\circ$ (NpO$_2$, am, 298.15 K) = – (970.0 ± 6.3) which is 12.7 kJ · mol^{-1} lower than the value selected in [2001LEM/FUG].

10.3 Neptunium halide compounds (9.1)

10.3.1 Neptunium fluoride compounds (9.1.2)

10.3.1.1 NpF(g), NpF$_2$(g), NpF$_3$(g) (9.1.2.1, 9.1.2.3)

There are no experimental data for these species and their enthalpies of formation were estimated in [2001LEM/FUG] to be close to the mean of the enthalpies of formation of the corresponding UF$_n$(g) and PuF$_n$(g) compounds from [92GRE/FUG] and [2001LEM/FUG], respectively. Since, in this review, the enthalpies of formation of the lower uranium fluoride gaseous species have been revised (see section 9.4.2.1.1), for the sake of consistency, we have made the corresponding changes to the estimated enthalpies of formation of the neptunium fluoride gases, although the changes are well within the assigned uncertainties.

The revised values are:

$\Delta_f H_m^\circ$ (NpF, g, 298.15 K) = – (80 ± 25) kJ · mol^{-1},

$\Delta_f H_m^\circ$ (NpF$_2$, g, 298.15 K) = – (575 ± 30) kJ · mol^{-1},

$\Delta_f H_m^\circ$ (NpF$_3$, g, 298.15 K) = – (1115 ± 25) kJ · mol^{-1}.

The standard entropies and heat capacities of these species are unchanged.

10.4 Neptunium group 15 compounds and complexes (11)

10.4.1 Neptunium nitrogen compounds (11.1)

10.4.1.1 NpN(cr) (11.1.1.1)

The data for NpN(cr) in [2001LEM/FUG] were essentially estimated to be close to the mean of UN and PuN at 298.15 K, where the order of stability is thus UN < NpN < PuN. However, the thermal data for the elements and the nitrides are such that with the current data for NpN(cr), this does not hold for $\Delta_f G_m$ at higher temperatures. For example, at 2100 K, the order of stability is NpN < PuN < UN. As a matter of interest, the main reasons for this are the different thermal properties of the elements, rather than the nitrides. This weakens the argument for taking 298.15 K as the base temperature for the "interpolation".

As noted in [2001LEM/FUG], Nakajima et al. [97NAK/ARA] confirmed the earlier findings of Olson and Mulford [66OLS/MUL2] that pure NpN vaporises to Np(liq) and N_2(g). Their paper actually gives little further data on $\Delta_f G_m$ (NpN), since calculation of this requires a knowledge of the nitrogen pressure in the Np(l) + NpN(cr) system, rather than p_{Np}, which is close to that of pure Np. Because the measurements of Olson and Mulford [66OLS/MUL2] were at appreciably higher temperatures (2480 – 3100 K), where the solubility of nitrogen in Np(l) becomes appreciable, the data of [97NAK/ARA] were not used in [2001LEM/FUG] to define $\Delta_f G_m$ (NpN).

However, the same authors [99NAK/ARA] have now measured p_{Np} in a system of NpN "co-loaded" with PuN. Under these conditions, PuN dissociates into Pu(g) + 0.5 N_2(g), which provides a known p_{N_2}, viz. half that of p_{Pu} (also determined), after allowing for the faster effusion of N_2(g). This is high enough to suppress the formation of Np, so $\Delta_f G_m$ (NpN) can be calculated from the known pressures in the vaporisation reaction NpN(cr) $\rightleftharpoons$ Np(g) + 0.5 N_2(g). [99NAK/ARA] checked that no solid solutions of NpN and PuN were formed.

The results of the two studies [97NAK/ARA], [99NAK/ARA] are in reasonable agreement and suggest that around 2000 K, $\Delta_f G_m$ (NpN, cr) does indeed lie between those of UN(cr) and PuN(cr) (with UN the most stable), and is ca. 10 kJ · mol^{-1} more negative than the values derived from the data in [2001LEM/FUG].

We therefore select:

$$\Delta_f H_m^\circ \text{ (NpN, cr, 298.15 K)} = -(305 \pm 10) \text{ kJ} \cdot \text{mol}^{-1},$$

with unchanged values for S_m° (298.15 K) and $C_{p,m}^\circ$ (298.15 K).

10.4.2 Neptunium antimony compounds

10.4.2.1 Neptunium antimonides

10.4.2.1.1 Neptunium monoantimonide

Hall *et al.* [91HAL/HAR] measured the heat capacity of a sample of high purity NpSb from 2 to 300 K. Small single crystals of NpSb were pressed into a cylinder which was sealed into a silver calorimeter under helium. A sharp λ-peak in the $C_{p,m}$ curve at 198 K is attributed to the known anti-ferromagnetic transition; an additional broad anomaly, of unknown origin, was observed in the $C_{p,m}$ curve at *ca.* 150 K. The values obtained for the heat capacity and entropy at 298.15 K were:

$$C_{p,m}^\circ (\text{NpSb, cr, 298.15 K}) = (48.9 \pm 2.9) \text{ J} \cdot \text{K}^{-1} \cdot \text{mol}^{-1},$$

$$S_m^\circ (\text{NpSb, cr, 298.15 K}) = (101.4 \pm 6.1) \text{ J} \cdot \text{K}^{-1} \cdot \text{mol}^{-1},$$

and these are the selected values. The authors note that the value of $C_{p,m}^\circ$ (298.15 K) is lower than that calculated from the low-temperature electronic and lattice heat capacities, which may suggest a change in the lattice structure below room temperature.

10.5 Neptunium chloride (9.2.2), nitrate (11.1.4.1) complexes and complexes with other actinides

The recent papers on the complexation of Np (and Pu) in various oxidation states deal mainly with chloride complexation [97RUN/NEU], [97ALL/BUC], but also with nitrate complexation [98SPA/PUI]. However, these papers were not included in the discussion of the previous review, but do not contain information that would justify changing the values selected by [2001LEM/FUG] (see Appendix A). Moreover, this review has pointed out that for trivalent actinides, the association constants for these weak 1:1 and 2:1 complexes have been overestimated. Therefore, this review considers that the selections of [2001LEM/FUG] of chloride and nitrate complexes for trivalent Np (and Pu) must be used with caution.

New data exist for the cation-cation complexes $NpO_2^+ \cdot UO_2^{2+}$ (see 9.8.1) and $NpO_2^+ \cdot Th^{4+}$, for which Stoyer *et al.* [2000STO/HOF] reported equilibrium constants of $\beta = (2.4 \pm 0.2) \text{ L} \cdot \text{mol}^{-1}$ and $(1.8 \pm 0.9) \text{ L} \cdot \text{mol}^{-1}$, respectively in 6 M NaClO$_4$.

10.6 Neptunium carbon compounds and complexes (12.1)

10.6.1 Aqueous neptunium carbonate complexes (12.1.2.1)

All currently available studies of the binary carbonate complexes of Np(III), Np(IV), Np(V) and Np(VI) have already been discussed in [2001LEM/FUG] and the selected values are retained in the present review.

10.6.1.1 Ternary Np(IV) hydroxide-carbonate complexes

For the reasons discussed in Appendix A in the reviews of [93ERI/NDA], [99RAI/HES], and [2001KIT/KOH], the equilibrium constants determined for the ternary Np(IV) hydroxide-carbonate complexes $Np(OH)_3(CO_3)^-$ [93ERI/NDA], $Np(OH)_4(CO_3)^{2-}$ [93ERI/NDA], $Np(OH)_4(CO_3)_2^{4-}$ [2001KIT/KOH], and $Np(OH)_2(CO_3)_2^{2-}$ [99RAI/HES], [2001KIT/KOH] are not selected.

Figure 10-1 shows experimental solubility data for NpO_2(am, hydr.) determined by Eriksen et al. [93ERI/NDA] and by Kitamura and Kohara [2001KIT/KOH] at total carbonate concentrations of $([HCO_3^-] + [CO_3^{2-}]) = 0.1$ and 0.01 mol·L^{-1} in 0.5 M $NaClO_4$ solution. For comparison the corresponding solubility curves calculated with the equilibrium constants proposed by Rai et al. [99RAI/HES] are included as well. The comparison in Figure 10-1 shows that not only are the proposed stoichiometries of the ternary Np(IV) complexes contradictory, but so are the underlying experimental solubility data.

The increased solubility at $-\log_{10}[H^+] = 8 - 11$ (compared to a carbonate-free solution) may be explained by the formation of either $Np(OH)_3(CO_3)^-$ [93ERI/NDA] or $Np(OH)_2(CO_3)_2^{2-}$ [99RAI/HES], [2001KIT/KOH]. According to the equilibrium constant of $\log_{10} K_{s,5}^\circ = -21.15$ [99RAI/HES], the complex $Np(CO_3)_5^{6-}$ has no significant contribution to the total Np(IV) concentration. At $-\log_{10}[H^+] = 8 - 9.5$ and 0.1 M total carbonate (Figure 10-1-a), the solubilities calculated with the equilibrium constants of [93ERI/NDA] and [99RAI/HES] are consistent, whereas those of [2001KIT/KOH] are about one order of magnitude higher. In solutions containing 0.01 M total carbonate (Figure 10-1-b), the results of [93ERI/NDA] agree with those of [2001KIT/KOH], whereas the model of [99RAI/HES] predicts no significant effect of Np(IV) hydroxide-carbonate complexes. It should be noted that in a former solubility study of Rai and Ryan [85RAI/RYA], the Np(IV) concentration in 0.01 M carbonate solutions of pH 7 – 14 was found to be at the detection limit for ^{237}Np ($10^{-8.3} - 10^{-8.4}$ mol·L^{-1}).

The pH-independent solubility data at $-\log_{10}[H^+] > 11$ can be ascribed either to the formation of $Np(OH)_4(CO_3)^{2-}$ [93ERI/NDA] or $Np(OH)_4(CO_3)_2^{4-}$ [2001KIT/KOH]. However, the experimental data of Eriksen et al. [93ERI/NDA] are one order of magnitude higher than those of Kitamura and Kohara [2001KIT/KOH], Rai and Ryan [85RAI/RYA], and Rai et al. [99RAI/HES] found no evidence for the formation of $Np(OH)_4(CO_3)^{2-}$ or $Np(OH)_4(CO_3)_2^{4-}$.

The solubility behaviour of NpO_2(am, hydr.) in carbonate solution observed by Eriksen et al. [93ERI/NDA] is clearly different from that reported by Rai et al. [99RAI/HES] and Kitamura and Kohara [2001KIT/KOH]. The results of the two latter studies show a comparable dependence on pH. However, the solubility curves are shifted by about 1.3 units in $\log_{10}[Np(IV)]_{tot}$. This indicates a systematic difference in the solubility product of the two solid phases, but having the same stoichiometry of the complexes and reflects the difference between $\log_{10} K_{s,1,2,2}^\circ = -(10.4 \pm 0.4)$ calculated in this review from experimental data of [2001KIT/KOH] (*cf.* Appendix A) and $\log_{10} K^\circ_{s,1,2,2} = -11.75$ [99RAI/HES]. Kitamura and Kohara [2001KIT/KOH] determined their solubility data from the direction of over-saturation after precipitation of possibly very small solid particles at a total Np concentration of 10^{-5} mol · L^{-1} in 0.5 M NaClO$_4$. Rai *et al.* [99RAI/HES] determined the solubility of NpO$_2$(am, hydr.) from under-saturation and the particle size of the solid used was probably larger. However, there is no information available to confirm this possible explanation in terms of particle size effect. Eriksen *et al.* [93ERI/NDA] used electrodeposited NpO$_2$(am, hydr.) and their solubility data was determined from under-saturation. As indicated above, the data of Eriksen *et al.* are inconsistent with those of [99RAI/HES] and [2001KIT/KOH]. This review has not selected any equilibrium constants for the ternary Np(IV) hydroxide carbonate complexes because of the uncertainty of the solubility product of the solid phase used in the solubility experiments. However, the experimental data in [99RAI/HES] and [2001KIT/KOH] strongly indicate that complexes with the stoichiometry Np(OH)$_2$(CO$_3$)$_2^{2-}$ and Np(OH)$_4$(CO$_3$)$^{2-}$ or Np(OH)$_4$(CO$_3$)$_2^{4-}$ are formed.

Figure 10-1: Solubility of NpO_2(am, hydr.) at $I = 0.5$ mol·L^{-1} ($NaClO_4$ solution) and total carbonate concentrations of a) 0.1 mol·L^{-1} and b) 0.01 mol·L^{-1}. Experimental data from Eriksen *et al.* [93ERI/NDA] at 20°C, and Kitamura and Kohara [2001KIT/KOH] at (22 ± 3)°C in comparison with the solubility calculated according to equilibrium constants given by Rai *et al.* [99RAI/HES].

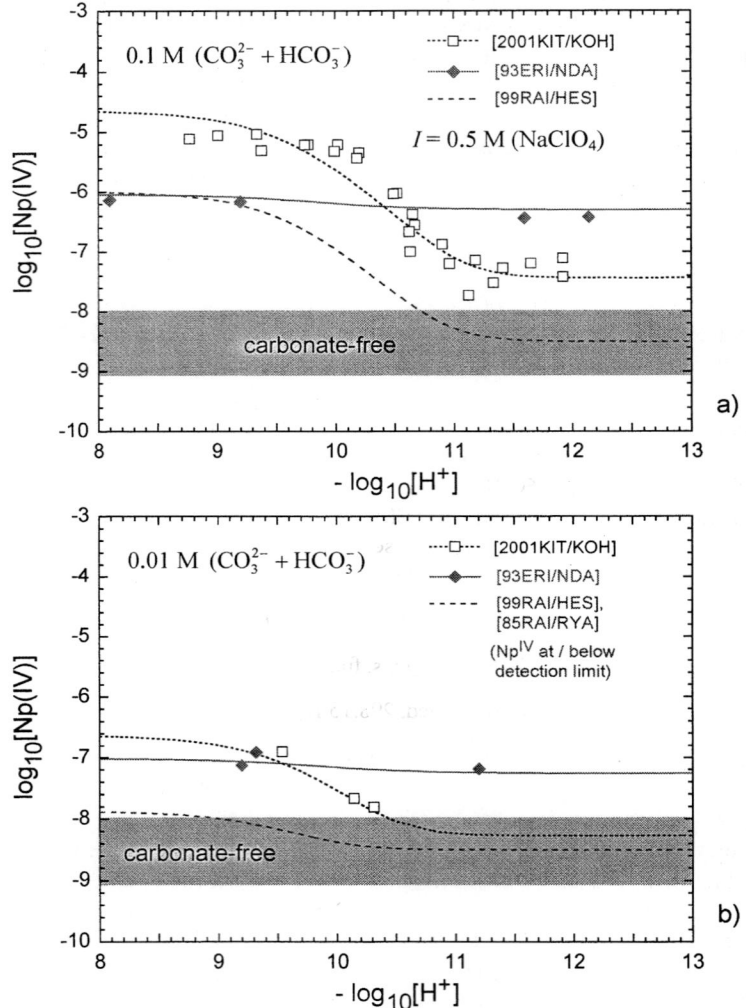

10.6.2 Solid neptunium carbonates (12.1.2.2)

10.6.2.1 Solid alkali metal neptunium(V) carbonates hydrates (12.1.2.2.2)

The published literature on solid alkali metal and ammonium neptunium(V) carbonates was reviewed in [2001LEM/FUG], but thermodynamic data were selected only for the sodium dioxoneptunium(V) carbonates, $NaNpO_2CO_3 \cdot xH_2O(s)$ and $Na_3NpO_2(CO_3)_2 \cdot xH_2O(s)$, (pages 275 and 279). However, there is further information to be gleaned from the extensive study by [94RUN/KIM] which was only briefly discussed in [2001LEM/FUG]. Also, as discussed in Appendix D (Section D.2.2), there are difficulties in assigning the most appropriate SIT interaction coefficients in assessing values for the solubility products of the sodium neptunium(V) carbonates. Therefore, the solubility constants for the sodium dioxoneptunium(V) carbonates are reinvestigated in the present review.

10.6.2.1.1 Sodium neptunium(V) carbonates

In the previous review [2001LEM/FUG], the data selection for $NaNpO_2CO_3 \cdot xH_2O(s)$ is mainly based on the solubility studies of [83MAY] in 1 M $NaClO_4$ and of Grenthe et al. [86GRE/ROB] in 3 M $NaClO_4$. The latter results were also published in an IAEA seminar report [84VIT] and shown in a figure of [90RIG]. As a consequence of the different $\log_{10} K_{s,0}^\circ$ values obtained by using the selected SIT coefficients for conversion to $I = 0$, [2001LEM/FUG] concluded that the study of [83MAY] refers to a hydrated solid phase, whereas the data of [86GRE/ROB] and the lowest of the widely scattered solubility data measured by Lemire et al. [93LEM/BOY] at 30°C in 1 M $NaClO_4$, were ascribed to an aged, less hydrated solid phase. The following solubility constants were selected in [2001LEM/FUG] for the reaction:

$$NaNpO_2CO_3(s) \rightleftharpoons Na^+ + NpO_2^+ + CO_3^{2-} \qquad (10.4)$$

$\log_{10} K_{s,0}^\circ$ ($NaNpO_2CO_3 \cdot 3.5H_2O$, s, fresh, 298.15 K) = $-(11.16 \pm 0.35)$

$\log_{10} K_{s,0}^\circ$ ($NaNpO_2CO_3$, s, aged, 298.15 K) = $-(11.66 \pm 0.50)$.

The solubility constants determined by Neck et al. [94NEC/RUN], [94NEC/KIM], [95NEC/RUN] in 0.1 – 5 M $NaClO_4$ (partly published in an earlier report [91KIM/KLE]), were suspected by [2001LEM/FUG] to include systematic errors, because they yield a value of $\Delta\varepsilon$ which is not consistent with the SIT coefficients selected in the NEA TDB and because the carbonic acid dissociation constants determined in $\geq$ 1 M $NaClO_4$ deviate from those calculated with NEA TDB auxiliary data. From these studies, Lemire et al. [2001LEM/FUG] accepted only the value extrapolated to $I = 0$. The results of Meinrath [94MEI] in 0.1 M $NaClO_4$ were found to be consistent with those of [94NEC/RUN], but were not considered further [2001LEM/FUG]. The solubility constants determined by [94NEC/KIM] in 5 M NaCl and by Runde et al. [94RUN/KIM], [96RUN/NEU] in 0.1 – 5 M NaCl were also not included in the discussion of data selection, despite the fact that the corresponding $\Delta\varepsilon(10.4)$ value (see Figure

10-2) and the carbonic acid dissociation constants determined in these studies are in agreement with NEA TDB auxiliary data and other widely accepted databases [84HAR/MOL], [91PIT].

The carbonic acid dissociation constants in $NaClO_4$ solutions from [94NEC/RUN] were later used to derive ternary Pitzer parameters for CO_3^{2-} in $NaClO_4$ solutions [96FAN/NEC] and the $\varepsilon(Na^+, CO_3^{2-})_{NaClO_4}$ value discussed in Appendix D. Table 10-1 lists the reported experimental solubility constants. The values grouped under (A) and (B) in Table 10-1 were converted to $I = 0$ with the SIT (the uncertainties are omitted for easier comparison):

(A) exclusively based on the interaction coefficients selected in [2001LEM/FUG]

(B) based on the same interaction coefficients as calculation (A), with the exception that $\varepsilon(Na^+, CO_3^{2-})_{NaCl} = -(0.08 \pm 0.03)$ and $\varepsilon(Na^+, CO_3^{2-})_{NaClO_4} = (0.04 \pm 0.05)$ kg·mol^{-1} (the latter value is derived from the carbonic acid dissociation constants reported by [96FAN/NEC]) are used instead of the single unique value of $\varepsilon(Na^+, CO_3^{2-}) = -(0.08 \pm 0.03)$ kg·mol^{-1}.

Calculation (A), with a unique value of $\varepsilon(Na^+, CO_3^{2-}) = -(0.08 \pm 0.03)$ kg·mol^{-1} leads to consistent $\log_{10} K_{s,0}^\circ$ values from the data in dilute to concentrated NaCl media and in 0.1 M $NaClO_4$, while the $\log_{10} K_{s,0}^\circ$ values calculated from data in $NaClO_4$ media decrease systematically with increasing $NaClO_4$ concentration. Calculation (B), with the individual values of $\varepsilon(Na^+, CO_3^{2-})_{NaCl} = -(0.08 \pm 0.03)$ kg·mol^{-1} and $\varepsilon(Na^+, CO_3^{2-})_{NaClO_4} = (0.04 \pm 0.05)$ kg·mol^{-1}, yields consistent values at $I = 0$, (for all data in both media) with a mean value of $\log_{10} K_{s,0}^\circ (NaNpO_2CO_3 \cdot 3.5H_2O, cr) = -(11.0 \pm 0.2)$.

An analogous observation is made if the conditional solubility constants reported for $Na_3NpO_2(CO_3)_2(s)$ in 1.0, 3.0 and 5 M $NaClO_4$ [86GRE/ROB], [95NEC/RUN] and in 5 M NaCl [94RUN/KIM], [95NEC/RUN] for the reaction:

$$Na_3NpO_2(CO_3)_2(s) \rightleftharpoons 3Na^+ + NpO_2^+ + 2CO_3^{2-} \quad (10.5)$$

are converted to $I = 0$ by calculations (A) and (B) (Table 10-1). Calculation (B) with $\varepsilon(Na^+, CO_3^{2-})_{NaCl} = -(0.08 \pm 0.03)$ kg·mol^{-1} and $\varepsilon(Na^+, CO_3^{2-})_{NaClO_4} = (0.04 \pm 0.05)$ kg·mol^{-1} yields fairly consistent $\log_{10} K_{s,0}^\circ$ values with a mean value of $\log_{10} K_{s,0}^\circ (Na_3NpO_2(CO_3)_2, s) = -(14.0 \pm 0.4)$. The solubility constant selected by [2001LEM/FUG], $-(14.70 \pm 0.66)$, is lower and the assigned uncertainty is larger.

However, as noted in Appendix D the present review does not accept the value of $\varepsilon(Na^+, CO_3^{2-})_{NaClO_4} = (0.04 \pm 0.05)$ kg·mol^{-1} derived from the carbonic acid dissociation constants of Fanghänel et al. [96FAN/NEC]. Therefore the equilibrium constants calculated with this interaction coefficient are not selected.

10.6 Neptunium carbon compounds and complexes (12.1)

Table 10-1: Solubility constants of NaNpO$_2$CO$_3$·3.5H$_2$O(s) and Na$_3$NpO$_2$(CO$_3$)$_2$(s) at 20 – 25°C and conversion to $I = 0$ with the SIT

Reference	Medium	t (°C)	$\log_{10} K_{s,0}$	$\log_{10} K^\circ_{s,0}$ (A)	$\log_{10} K^\circ_{s,0}$ (B)
NaNpO$_2$CO$_3$·3.5H$_2$O(s)					
[96RUN/NEU]	0.1 M NaCl	23	– (10.40 ± 0.2)	– 11.05	– 11.05
[94RUN/KIM]	1.0 M NaCl	22	– (9.77 ± 0.16)	– 10.98	– 10.98
[96RUN/NEU]	3.0 M NaCl	23	– (9.40 ± 0.2)	– 10.86	– 10.86
[94RUN/KIM]	5.0 M NaCl	22	– (9.61 ± 0.11)	– 11.21 *	– 11.21 *
[94NEC/KIM]	5.0 M NaCl	25	– (9.52 ± 0.04)	– 11.12 *	– 11.12 *
[94NEC/RUN]	0.1 M NaClO$_4$	25	– (10.28 ± 0.04)	– 10.92	– 10.91
[94MEI]	0.1 M NaClO$_4$	25	– (10.22 ± 0.02)	– 10.86	– 10.85
[94NEC/RUN]	1.0 M NaClO$_4$	25	– (10.10 ± 0.03)	– 11.13	– 10.99
[83MAY]	1.0 M NaClO$_4$	25	– (10.14 ± 0.04)	– 11.17	– 11.04
[94NEC/RUN]	3.0 M NaClO$_4$	25	– (10.45 ± 0.04)	– 11.31	– 10.89
[86GRE/ROB]	3.0 M NaClO$_4$	20	– 10.56	– 11.42	– 11.00
[94NEC/RUN]	5.0 M NaClO$_4$	25	– (11.06 ± 0.06)	– 11.54 *	– 10.76 *
Na$_3$NpO$_2$(CO$_3$)$_2$(s)					
[94RUN/KIM]	5.0 M NaCl	22	– (11.46 ± 0.23)	– 14.22 *	– 14.22 *
[95NEC/RUN]	1.0 M NaClO$_4$	25	– (12.23 ± 0.15)	– 14.44	– 14.19
[86GRE/ROB]	3.0 M NaClO$_4$	20	– 12.44	– 14.62	– 13.78
[95NEC/RUN]	3.0 M NaClO$_4$	25	– (12.59 ± 0.10)	– 14.77	– 13.93
[95NEC/RUN]	5.0 M NaClO$_4$	22	– (13.57 ± 0.11)	– 15.34 *	– 13.78 *

(A) calculated with the SIT coefficients of the NEA TDB:
$\varepsilon(Na^+, CO_3^{2-}) = -0.08$ kg · mol^{-1}, $\varepsilon(Na^+, ClO_4^-) = 0.01$ kg · mol^{-1}, $\varepsilon(Na^+, Cl^-) = 0.03$ kg · mol^{-1}, $\varepsilon(NpO_2^+, ClO_4^-) = 0.25$ kg · mol^{-1} and $\varepsilon(NpO_2^+, Cl^-) = 0.09$ kg · mol^{-1}.

(B) calculated with the same SIT coefficients as in (A), except for CO_3^{2-} in NaClO$_4$ solution: $\varepsilon(Na^+, CO_3^{2-})_{NaClO_4} = 0.04$ kg · mol^{-1} and $\varepsilon(Na^+, CO_3^{2-})_{NaCl} = -0.08$ kg · mol^{-1}.

* At this high ionic strength the SIT calculation gives rise to relatively large uncertainties.

Figure 10-2: SIT extrapolation to $I = 0$ for the reaction:

$$NaNpO_2CO_3 \cdot 3.5H_2O(s) \rightleftharpoons Na^+ + NpO_2^+ + CO_3^{2-} + 3.5\,H_2O(l).$$

The SIT regression lines are not fitted to the data but calculated with fixed $\Delta\varepsilon$ values: $\Delta\varepsilon = 0.04$ kg·mol^{-1} in NaCl solution is calculated using exclusively data from [2001LEM/FUG], while $\Delta\varepsilon = 0.30$ kg·mol^{-1} in NaClO$_4$ solution is calculated with $\varepsilon(Na^+, CO_3^{2-})_{NaClO_4} = 0.04$ kg·mol^{-1} derived from [96FAN/NEC], corresponding to calculation (B) in Table 10-1. Calculation (A), with $\Delta\varepsilon = 0.18$ kg·mol^{-1} in NaClO$_4$ solution [2001LEM/FUG] is shown as dotted lines.

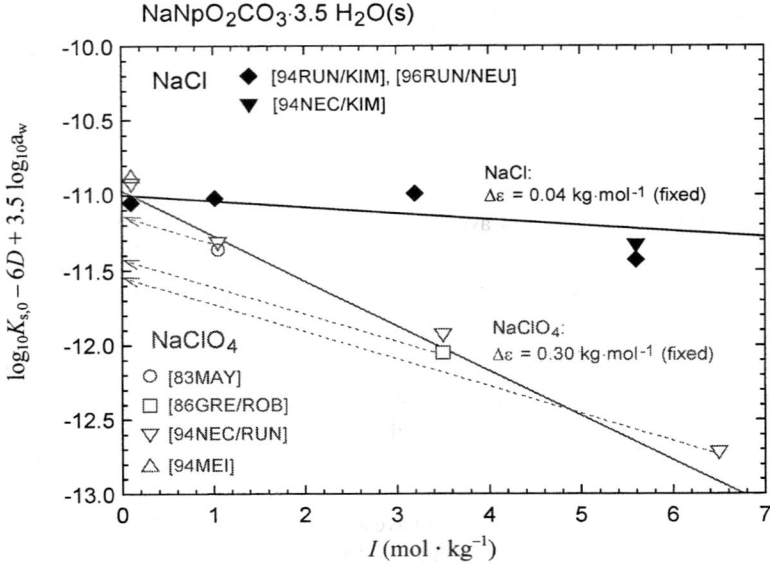

Concerning the chemical nature of NaNpO$_2$CO$_3$ · xH$_2$O(s), it has been shown by X–ray powder diffraction that the solubility data in [94NEC/RUN], [94NEC/KIM], [94RUN/KIM], [95NEC/RUN], [96RUN/NEU] refer to the same solid characterised earlier by Volkov et al. [77VOL/VIS] and Maya [83MAY]: hydrated NaNpO$_2$CO$_3$ · 3.5H$_2$O(cr). Only Meinrath [94MEI] reported another (hexagonal) modification with a different X–ray pattern. Furthermore, the solids used in [94NEC/RUN], [94NEC/KIM], [94RUN/KIM], [95NEC/RUN], [96RUN/NEU] were left to age at least several weeks up to more than six months. The solid used in these studies and that of [83MAY], which was also left to age to a crystalline solid prior to the solubility measurement, are not fresh precipitates as assumed in [2001LEM/FUG], but are well-crystallised compounds.

For this reason and because of the possible ambiguities in $\log_{10} K_{s,0}^\circ$ values derived from data at higher $NaClO_4$ concentrations, the present review does not agree with the two values selected by [2001LEM/FUG] for $NaNpO_2CO_3 \cdot 3.5H_2O$(s, fresh) and $NaNpO_2CO_3$(s, aged). The revised selection of $\log_{10} K^\circ_{s,0}$ is based exclusively on experimental studies which fulfil the following prerequisites:

- The solubility study must refer to $NaNpO_2CO_3 \cdot 3.5H_2O$(cr) described by [77VOL/VIS] and [83MAY], and include sufficient data at low carbonate concentrations, with free NpO_2^+ as the predominant species.

- The carbonic acid dissociation constants used by the authors to calculate $\log_{10}[CO_3^{2-}]$ from the measured $-\log_{10}[H^+]$ values must be consistent with auxiliary data from the NEA TDB.

- The recalculation to $I = 0$ must not include data at high ionic strengths where large uncertainties arise because the validity range of the SIT is exceeded or ambiguity exists concerning the interaction coefficient of the carbonate ion.

All these prerequisites are fulfilled for the five studies in 0.1 M $NaClO_4$ [94NEC/RUN], 1 M $NaClO_4$ [83MAY], and 0.1, 1.0 and 3.0 M NaCl [94RUN/KIM], [96RUN/NEU]. The unweighted average of the $\log_{10} K^\circ_{s,0}$ values calculated in Table 10-1 with the SIT coefficients from [2001LEM/FUG] is selected:

$$\log_{10} K^\circ_{s,0} (NaNpO_2CO_3 \cdot 3.5H_2O, \text{cr}, 298.15 \text{ K}) = -(11.00 \pm 0.24)$$

and

$$\Delta_f G^\circ_m (NaNpO_2CO_3 \cdot 3.5H_2O, \text{cr}, 298.15 \text{ K}) = -(2590.4 \pm 5.8) \text{ kJ} \cdot \text{mol}^{-1}.$$

With regard to the changed selection compared to the previous review [2001LEM/FUG], it has to be pointed out that all experimental studies used for the re-evaluation of thermodynamic data for $NaNpO_2CO_3 \cdot 3.5H_2O$(cr) were performed with a crystalline compound aged at least for several weeks and not with a fresh, gelatinous precipitate. Such fresh precipitates might have a different water content or even a different chemical composition, possibly with a stoichiometry of $Na_{0.6}NpO_2(CO_3)_{0.8} \cdot xH_2O$(cr) as reported by [79VOL/VIS2]. They are assumed to be metastable and to convert gradually into $NaNpO_2CO_3 \cdot xH_2O$(cr) (cf. discussion in [2001LEM/FUG], p. 274). Thermal treatment of $NaNpO_2CO_3 \cdot xH_2O$(cr) can lead to a decrease of the water content, i.e., $3.5 \geq x \geq 0$ at $20 - 130°C$, and to changing XRD patterns as shown by Volkov et al. [77VOL/VIS].

The solubility constant selected in [2001LEM/FUG] for $Na_3NpO_2(CO_3)_2$(s) is based on the solubility data reported by Simakin [77SIM] for the reaction:

$$Na_3NpO_2(CO_3)_2(s) + CO_3^{2-} \rightleftharpoons NpO_2(CO_3)_3^{5-} + 3 Na^+ \qquad (10.6)$$

with $\log_{10} K_{s,3}(10.6) = -(1.46 \pm 0.09)$ at $I = 3$ M $(NaNO_3 - Na_2CO_3)$. There is a problem with this study that was not mentioned in the discussion in [2001LEM/FUG]. The CO_3^{2-} concentration was varied in the range of $0.25 - 1$ mol$\cdot L^{-1}$, which does not allow both the Na^+ concentration and ionic strength to be kept constant. If the ionic strength were kept constant at $I = 3$ M, then [Na$^+$], which is involved directly in the equilibrium constant of reaction (10.6), decreases from 2.75 to 2.0 mol · L^{-1}. If [Na$^+$] were kept constant at 3.0 mol · L^{-1} then I increases from 3.25 to 4.0 mol · L^{-1}.

Moreover, as shown in Appendix D.2.1, the SIT coefficient for the complex $NpO_2(CO_3)_3^{5-}$ depends significantly on the anion of the electrolyte medium. [2001LEM/FUG] used the value of $\varepsilon(Na^+, NpO_2(CO_3)_3^{5-}) = -(0.53 \pm 0.19)$ kg · mol^{-1} determined from data in NaClO$_4$ solutions and calculated from the data of Simakin [77SIM]:

$$\log_{10} K_{s,3}^\circ (10.6) = -(9.20 \pm 0.65)$$

$$\log_{10} K_{s,0}^\circ (Na_3NpO_2(CO_3)_2, s, 298.15 \text{ K}) = -(14.70 \pm 0.66).$$

With regard to the NaNO$_3$-Na$_2$CO$_3$ solutions used in [77SIM], the value of $\varepsilon(Na^+, NpO_2(CO_3)_3^{5-}) = -(0.29 \pm 0.11)$ kg · mol^{-1} derived from data of [94RUN/KIM] in NaCl solution and close to $\varepsilon(K^+, NpO_2(CO_3)_3^{5-}) = -(0.23 \pm 0.02)$ kg · mol^{-1} derived from data of [97NOV/ALM] in pure carbonate solution (cf. Appendix A), might be a better approximation. It leads to considerably different equilibrium constants:

$$\log_{10} K_{s,3} (10.6) = -(8.39 \pm 0.50)$$

$$\log_{10} K_{s,0}^\circ (Na_3NpO_2(CO_3)_2, s, 298.15 \text{ K}) = -(13.89 \pm 0.52).$$

The values of $\varepsilon(Na^+, CO_3^{2-}) = -(0.08 \pm 0.03)$ kg · mol^{-1}, $\varepsilon(Na^+, NO_3^-) = -(0.04 \pm 0.03)$ kg · mol^{-1} and $\log_{10} \beta_3^\circ (NpO_2(CO_3)_3^{5-}, 298.15 \text{ K}) = (5.50 \pm 0.15)$ used in these calculations are taken from [2001LEM/FUG].

The solubility constant derived in Table 10-1 from data determined by [94RUN/KIM] in 5 M NaCl at CO$_3^{2-}$ concentrations in the range of $10^{-3} - 10^{-1}$ mol · L^{-1} does not involve the ambiguities discussed above and, hence, appears more reliable. The solid used by [94RUN/KIM] was also well characterised by X–ray diffraction. Therefore, the present review selects:

$$\log_{10} K_{s,0}^\circ (Na_3NpO_2(CO_3)_2, cr, 298.15 \text{ K}) = -(14.22 \pm 0.50)$$

$$\Delta_f G_m^\circ (Na_3NpO_2(CO_3)_2, cr, 298.15 \text{ K}) = -(2830.6 \pm 6.4) \text{ kJ · mol}^{-1}.$$

The large uncertainty for $\log_{10} K_{s,0}^\circ$ arises mainly from the term $\Delta\varepsilon(10.5) = (0.02 \pm 0.08)$ kg · mol^{-1} applied to the data in 5.6 m NaCl. It is noteworthy that this solubility constant agrees with the values of $\log_{10} K_{s,0}^\circ (10.5) = -(14.1 \pm 0.3)$ and $-(14.3 \pm 0.2)$, calculated from the value in 5.6 m NaCl by using the Pitzer models of [95NEC/FAN] and [96RUN/NEU], respectively. The auxiliary data used in these papers for Np(V) and CO$_3^{2-}$ in NaCl solution are consistent with the NEA TDB selections.

10.6.2.1.2 Potassium neptunium(V) carbonates

In two recent studies not discussed in [2001LEM/FUG], Novak et al. [97NOV/ALM] and Al Mahamid et al. [98ALM/NOV] determined accurate solubility data for

10.6 Neptunium carbon compounds and complexes (12.1)

$KNpO_2CO_3 \cdot xH_2O(s)$ and $K_3NpO_2(CO_3)_2 \cdot xH_2O(s)$ in K_2CO_3 solutions and $KCl-K_2CO_3$ mixtures.

The solids formed in K_2CO_3 solutions below 0.4 molal gave X–ray powder diffraction patterns comparable with those reported by Keenan and Kruse [64KEE/KRU] for $KNpO_2CO_3 \cdot xH_2O(s)$. The same solid was stable in $KCl-K_2CO_3$ mixtures with m_{KCl} = 0.003 – 3.2 mol·kg^{-1} and $m_{K_2CO_3}$ = 0.001, 0.01, and 0.1 mol·kg^{-1}. At higher K_2CO_3 concentrations, two modifications of $K_3NpO_2(CO_3)_2 \cdot xH_2O(s)$ were identified as the phases A and B described by Volkov et al. [74VOL/KAP2] and Visyashcheva et al. [74VIS/VOL]. Phase B is stable over a broad range of K_2CO_3 concentrations and phase A is preferentially formed at K_2CO_3 concentrations above 2 molal [97NOV/ALM], [74VOL/KAP2]. Their solubilities do not differ markedly.

The water content in $KNpO_2CO_3 \cdot xH_2O(s)$ and $K_3NpO_2(CO_3)_2 \cdot xH_2O(s)$, phase B, was estimated in [74VIS/VOL] to be $x \leq 2$ and $x \approx 1.6$, respectively, but it is not exactly known. Therefore, this study disregarded in the evaluation of thermodynamic data as was done by [2001LEM/FUG] in the case of sodium neptunium(V) carbonates.

Novak et al. [97NOV/ALM] extended the set of equilibrium constants and ion interaction Pitzer parameters reported in [95FAN/NEC] for the system Np(V)-Na-OH-CO_3-Cl-H_2O (298.15 K) and calculated $\log_{10} K^\circ_{s,0} = -(13.6 \pm 0.1)$ for $KNpO_2CO_3(s)$ and $\log_{10} K^\circ_{s,0} = -(15.9 \pm 0.1)$ for $K_3NpO_2(CO_3)_2(s)$.

In the present review, the experimental data of [97NOV/ALM] and [98ALM/NOV] are re-evaluated using the SIT model (cf. Appendix A). Neglecting the small difference in temperature (experimental data at (295.15 ± 1.00) K compared to the standard state of 298.15 K) the following results are obtained for the reactions, (where the hydration water molecules are omitted):

$$KNpO_2CO_3(s) + 2CO_3^{2-} \rightleftharpoons NpO_2(CO_3)_3^{5-} + K^+, \qquad (10.7)$$

$$\log_{10} K^\circ_{s,3}((10.7), 298.15\text{ K}) = -(7.65 \pm 0.11),$$

$$\Delta\varepsilon(10.7) = -(0.25 \pm 0.06) \text{ kg} \cdot \text{mol}^{-1}$$

and

$$K_3NpO_2(CO_3)_2(s) + CO_3^{2-} \rightleftharpoons NpO_2(CO_3)_3^{5-} + 3K^+, \qquad (10.8)$$

$$\log_{10} K^\circ_{s,3}((10.8), 298.15\text{ K}) = -(9.96 \pm 0.06),$$

$$\Delta\varepsilon(10.8) = -(0.22 \pm 0.02) \text{ kg} \cdot \text{mol}^{-1}.$$

Combined with auxiliary data from [2001LEM/FUG], the values of $\Delta\varepsilon(10.7)$ and $\Delta\varepsilon(10.8)$ lead to a consistent interaction coefficient which is selected in this review:

$$\varepsilon(NpO_2(CO_3)_3^{5-}, K^+) = -(0.22 \pm 0.03) \text{ kg} \cdot \text{mol}^{-1}.$$

By combining $\log_{10} K^\circ_{s,3}((10.7), 298.15\text{ K})$ and $\log_{10} K^\circ_{s,3}((10.8), 298.15\text{ K})$

with $\log_{10} \beta_3^\circ$ ($NpO_2(CO_3)_3^{5-}$, 298.15 K) = (5.50 ± 0.15) [2001LEM/FUG], we obtain the following solubility constants which are selected in the present review:

$$KNpO_2CO_3(s) \rightleftharpoons K^+ + NpO_2^+ + CO_3^{2-} \qquad (10.9)$$

$$\log_{10} K_{s,0}^\circ (KNpO_2CO_3, s, 298.15 \text{ K}) = -(13.15 \pm 0.19)$$

and

$$K_3NpO_2(CO_3)_2(s) \rightleftharpoons 3K^+ + NpO_2^+ + 2CO_3^{2-} \qquad (10.10)$$

$$\log_{10} K_{s,0}^\circ (K_3NpO_2(CO_3)_2, s, 298.15 \text{ K}) = -(15.46 \pm 0.16).$$

The solubility of the potassium dioxoneptunium(V) carbonates is about two orders of magnitude lower than that of the corresponding sodium dioxoneptunium(V) carbonates. Therefore, the solubility of Np(V) in quaternary Na-K-Cl-CO$_3$ solutions will be controlled by KNpO$_2$CO$_3$(s) even if the concentration of Na$^+$ is much higher than that of K$^+$ [98ALM/NOV].

Chapter 11

Discussion of new data selection for Plutonium

11.1 Plutonium aqua ions (16.1)

EXAFS fine structure spectra of aqueous ions Pu(III to VI): Pu^{3+}, Pu^{4+}, PuO_2^{2+} in 1 M $HClO_4$ and PuO_2^+ at pH = 6 are given by Conradson et al. [98CON/ALM].

11.2 Plutonium oxygen and hydrogen compounds and complexes (17)

11.2.1 Aqueous hydroxide complexes (17.1)

11.2.1.1 Plutonium(V) hydroxide complexes (17.1.2)

The disproportionation of Pu(V) in concentrated NaOH solutions has been investigated by Budantseva et al. [98BUD/TAN] but, as explained in Appendix A, no thermodynamic values can be derived from the published data.

11.2.1.2 Plutonium(IV) hydroxides complexes (17.1.3)

New experimental data on the solubility of $Pu(OH)_4$ leading to thermodynamic values have appeared since the previous review in [98CAP/VIT], [98CHA/TRI], [98EFU/RUN], [99KNO/NEC], [99RAI/HES2], and re-interpretations of most of the data on Pu^{4+} hydrolysis, already considered in [2001LEM/FUG], have been proposed in [99KNO/NEC], [99NEC/KIM] and [2001NEC/KIM].

With Pu(IV), the phenomena already mentioned for tetravalent actinides concerning hydrolysis are exacerbated, among others with colloid formation; in addition, it is difficult to prevent Pu(IV) from oxidising with subsequent disproportionation of Pu(V) in acidic media. With regard to colloid formation, Knopp et al. [99KNO/NEC] showed by LIBD (Laser-Induced Breakdown Detection) and ultrafiltration that when the Pu(IV) concentration exceeds the solubility of amorphous PuO_2(am, hydr.), colloids are the predominant species in solution (see Appendix A).

Considering this particular tendency of Pu(IV), this review starts with the problem of the selection of hydrolysis constants. The values of $\log_{10} K^\circ_{s,0}$ are given in section 11.2.2.1.

Knopp et al. [99KNO/NEC], and Neck and Kim [99NEC/KIM], [2001NEC/KIM], pointed out that to make theoretical calculations according to the methods already mentioned (see Chapters on U and Np), the initial choice of $\beta_{1,1}^\circ$ must involve an experimental value corresponding to experimental conditions where colloids are not likely to be formed. These conditions seem not to be fulfilled for solutions in which spectroscopic measurements have been made, because the total concentrations of Pu are higher than the solubility of PuO_2(am, hydr.). Therefore, the value retained by Neck and Kim is $\log_{10} \beta_{1,1}^\circ = (14.6 \pm 0.2)$ derived from experiments at tracer levels [72MET/GUI], [73MET]. All the other values reported in the literature mentioned in [2001LEM/FUG] are smaller. The model calculation in [2001NEC/KIM] led to $\log_{10} \beta_{1,1}^\circ = 14.6$, $\log_{10} \beta_{2,1}^\circ = 28.3$, $\log_{10} \beta_{3,1}^\circ = 39.4$ and $\log_{10} \beta_{4,1}^\circ = 47.5$, which are in close agreement with the experimental values of [72MET/GUI], [73MET]: $\log_{10} \beta_{1,1}^\circ = (14.6 \pm 0.2)$, $\log_{10} \beta_{2,1}^\circ = (28.6 \pm 0.3)$, $\log_{10} \beta_{3,1}^\circ = (39.7 \pm 0.4)$ and $\log_{10} \beta_{4,1}^\circ = (47.5 \pm 0.5)$. There are no experimental data other than those of [72MET/GUI], [73MET] to check the values of $\beta_{2,1}^\circ$ and $\beta_{3,1}^\circ$, but the value of $\beta_{4,1}^\circ$ was confirmed from solubility experiments (see section 11.2.2.1). [2001LEM/FUG] selected for Pu(IV), $\log_{10} {}^*\beta_{1,1}^\circ = -(0.78 \pm 0.60)$, ($\log_{10} \beta_{1,1}^\circ = (13.22 \pm 0.60)$) as the average value of the most reliable results of spectroscopic measurements. No other values are selected for the hydrolysis constants. None of the spectroscopic studies considered in [2001LEM/FUG] could identify two distinct absorption bands for Pu^{4+} and $Pu(OH)^{3+}$ or a shift of the absorption maximum. The only effect observed was the decrease of the absorption band when pH was increased and the solubility limit of PuO_2(am, hydr.) was exceeded [99KNO/NEC]. For these reasons, and despite the fact that [2001LEM/FUG] noticed some ambiguities in the interpretation of the solvent extraction data of Metivier and Guillaumont [72MET/GUI] and analogous data for Np(IV) [77DUP/GUI] (cf. section 10.2.1.3), we rely on these hydrolysis constants.

As mentioned for U(IV) and Np(IV), the hydrolysis constants reported by Neck and Kim [2001NEC/KIM] are derived in a systematic and consistent manner (and also for Th) taking into account colloid formation, and appear to this review to be the "best values" that can be derived from the available data.

This review retains the equilibrium constants from the solvent extraction study of Metivier and Guillaumont [72MET/GUI] converted to $I = 0$ using the SIT model and selects the values proposed by [99KNO/NEC], [99NEC/KIM] and [2001NEC/KIM] at 298.15 K for the reactions:

$$Pu^{4+} + n\,OH^- \rightleftharpoons Pu(OH)_n^{4-n}, \qquad (11.1)$$

$$Pu^{4+} + n\,H_2O(l) \rightleftharpoons Pu(OH)_n^{4-n} + n\,H^+. \qquad (11.2)$$

$\log_{10} \beta_1^\circ\,(11.1) = (14.6 \pm 0.2)$ $\log_{10} {}^*\beta_1^\circ\,(11.2) = (0.6 \pm 0.2)$,

$\log_{10} \beta_2^\circ\,(11.1) = (28.6 \pm 0.3)$ $\log_{10} {}^*\beta_2^\circ\,(11.2) = (0.6 \pm 0.3)$,

$\log_{10} \beta_3^\circ\,(11.1) = (39.7 \pm 0.4)$ $\log_{10} {}^*\beta_3^\circ\,(11.2) = -(2.3 \pm 0.4)$,

$\log_{10} \beta_4^\circ (11.1) = (47.5 \pm 0.5)$ $\log_{10} {}^*\beta_4^\circ (11.2) = -(8.5 \pm 0.5)$.

Peretrukhin et al. [96PER/KRY] determined the formation of negatively charged hydroxide complexes $Pu(OH)_6^{2-}$, from the increase in the solubility of Pu(IV) hydroxide in 0.5 to 14.1 M NaOH. However, for the reasons given in Appendix A, the present review does not derive thermodynamic data from the results reported by [96PER/KRY].

The standard Gibbs energy values $\Delta_f G_m^\circ (Pu(OH)_n^{4-n}$, 298.15 K) can be calculated from $\Delta_r G_m^\circ (11.2)$ and $\Delta_f G_m^\circ (Pu^{4+}$, 298.15 K) $= -(478.0 \pm 2.7)$ kJ·mol^{-1} (see 11.2.3).

The previous NEA TDB review [2001LEM/FUG] used the temperature dependence of the first hydrolysis constant determined in [57RAB], [60RAB/KLI] to calculate $\Delta_r H_m^\circ ((11.2), n=1, 298.15 K) = (36 \pm 10)$ kJ·mol^{-1} and $S_m^\circ (PuOH^{3+}$, 298.15 K) $= -(239 \pm 37)$ J·K^{-1}·mol^{-1}. However, as shown by [99KNO/NEC], these are two typical studies where the total Pu concentration exceeds the solubility of PuO$_2$(am, hydr.). Therefore, these values are rejected by the present review.

11.2.2 Solid plutonium oxides and hydroxides (17.2)

11.2.2.1 Plutonium(IV) oxides and hydroxides

Haschke et al. [2000HAS/ALL], [2001HAS/ALL], [2002HAS/ALL] have suggested that a solid solution, PuO$_{2+x}$ (0 < x < 0.25), can be formed by exposing an oxide with a composition close to PuO$_{1.97}$ to water vapour at temperatures from 300 to 623 K for some days. They cite as evidence the formation of H$_2$(g) and a small increase in the lattice parameter of the oxide; there was no independent analysis of the composition of the oxide. However, PuO$_2$ is known to adsorb oxygen and/or water to give an apparent bulk composition of at least PuO$_{2.09}$ [61WAT/DOU], [63JAC/RAN] and of course, hydrogen is one of the major products of water radiolysis. Moreover, the increase in lattice parameters (from an ill-defined starting value) is within the range found in PuO$_{2.00}$ due to radiation damage [62RAN/FOX]. Thus, the evidence for the formation of a thermodynamically stable bulk phase with O/Pu > 2 is far from conclusive. Such non-stoichiometric oxides (like UO$_{2+x}$) fall outside the scope of this review, but further work is clearly required on the interesting phenomena observed by Haschke et al.

11.2.2.1.1 Solubility of crystalline oxide PuO$_2$(cr) (17.2.1.2)

No new experimental data have appeared since the last review. According to the values of Gibbs energy of formation of PuO$_2$(cr) selected by [2001LEM/FUG]:

$\Delta_f G_m^\circ (PuO_2, cr, 298.15 K) = -(998.1 \pm 1.0)$ kJ·mol^{-1},

$\Delta_f G_m^\circ (Pu^{4+}, 298.15 K) = -(478.0 \pm 2.7)$ kJ·mol^{-1}

and the auxiliary values discussed in [92GRE/FUG], this review calculates:

$$\log_{10} K_{s,0}^{\circ} (PuO_2, cr) = -(64.04 \pm 0.51).$$

The solubility product values calculated by other authors who used somewhat different thermodynamic data for Pu^{4+} and $PuO_2(cr)$ are $-(63.8 \pm 1.0)$ [89KIM/KAN] and $-(64.1 \pm 0.7)$ [87RAI/SWA], respectively.

This review retains the thermodynamic data selected by [2001LEM/FUG].

11.2.2.1.2 Solubility of amorphous oxide and hydroxide $PuO_2(am, hydr.)$

The value selected by [2001LEM/FUG] for the solubility equilibrium for aged $PuO_2(am, hydr.)$,

$$PuO_2(am, hydr.) + 2H_2O(l) \rightleftharpoons 4OH^- + Pu^{4+}, \qquad (11.3)$$

is $\log_{10} K_{s,0}^{\circ} (PuO_2, am, hydr., (11.3)) = -(58 \pm 1)$. All the values recently reported agree with this value, but with reduced uncertainty.

The literature data on the solubility of $PuO_2(am, hydr.)$ for acidic and basic media (pH = 0 to 13) have been reviewed by Neck and Kim [99NEC/KIM], [2001NEC/KIM], and Knopp et al. [99KNO/NEC], and re-interpreted on the basis of the hydrolysis constants (see above) from [72MET/GUI] resulting in $\log_{10} K_{s,0}^{\circ} (PuO_2, am, hydr., (11.3)) = -(58.7 \pm 0.9)$ (see Appendix A). The analysis in [99KNO/NEC], [99NEC/KIM] and [2001NEC/KIM] incorporated experimental solubility data of [49KAS], [65PER], [84RAI], [86LIE/KIM] and [89KIM/KAN], where the aqueous Pu(IV) concentration was confirmed experimentally.

To calculate $\log_{10} K_{s,0}^{\circ} (PuO_2, am, hydr., (11.3))$, Capdevila and Vitorge [98CAP/VIT] used a non-conventional method explained in Appendix A. They calculated the $[Pu^{4+}]$ in equilibrium with Pu(IV) hydroxide from the measured equilibrium concentrations of the aqueous ions PuO_2^+, PuO_2^{2+} and Pu^{3+}, at constant $-\log_{10}[H^+]$. The solutions studied were 0.1 M $HClO_4$ + (x − 0.1) M $NaClO_4$ media, x = 0.1, 0.5, 1, 2 and 3. $[Pu^{4+}]$ remains always lower than 10^{-4} M, but the electrochemical method used to derive $[Pu^{4+}]$ is independent of the hydrolysis of that ion. As explained in Appendix A, the values obtained are $\log_{10} K_{s,0}^{\circ} (PuO_2, am, hydr., (11.3)) = -(58.3 \pm 0.5)$ and $\varepsilon(Pu^{4+}, ClO_4^-) = (0.85 \pm 0.20)$ kg·mol^{-1}.

In three very recent studies by Fujiwara et al. [2001FUJ/YAM], Rai et al. [2001RAI/MOO] and [2002RAI/GOR], the solubility of $PuO_2(am, hydr.)$ in the range $4 < -\log_{10}[H^+] < 9$ was determined under reducing and oxidizing conditions, respectively.

Fujiwara et al. [2001FUJ/YAM] measured the solubility of Pu(IV) hydrous oxide under reducing conditions in 1.0 M $NaClO_4$ solutions. The data refer to the reaction:

$$PuO_2(am) + 2H_2O + e^- \rightleftharpoons Pu^{3+} + 4OH^-. \qquad (11.4)$$

Extrapolation to $I = 0$ with the SIT model and combining with the redox potentials selected in [2001LEM/FUG] for the couple Pu^{4+}/Pu^{3+} leads to a solubility product of $\log_{10} K^\circ_{s,0}$ (PuO$_2$, am, hydr., 298.15 K, (11.3)) = $-$ (58.1 $\pm$ 0.4).

Rai et al. [2001RAI/MOO] determined the solubility of amorphous Pu(IV) hydrous oxide in different air-equilibrated NaClO$_4$ and NaCl solutions at $I = 0.4$ and 4.0 mol·L^{-1}. A similar previous study in 0.0015 M CaCl$_2$ [84RAI] is also included to evaluate the equilibrium constant for the reaction:

$$PuO_2(am, hydr.) \rightleftharpoons PuO_2^+ + e^-. \qquad (11.5)$$

By combining the values of $\log_{10} K^\circ$ (11.5) from [84RAI] and [2001RAI/MOO] with the PuO_2^+/Pu^{4+} redox potential and other auxiliary data selected in [2001LEM/FUG], the solubility product is calculated to be $\log_{10} K^\circ_{s,0}$ (PuO$_2$, am, hydr., (11.3)) = $-$ 58.0 and $-$ 57.4, respectively. As the equilibrium constants are given without uncertainty limits and include unknown uncertainties arising from the measured pe values [2001RAI/MOO] (cf. Appendix A), they are not used in this review for the selection of $\log_{10} K^\circ_{s,0}$ (PuO$_2$, am, hydr., 298.15 K, (11.3)).

Rai et al. [2002RAI/GOR] have made a precise and well described experimental study of the solubility of PuO$_2$(am) under controlled reducing conditions in NaCl solutions at low ionic strength, less than 0.025 M. They measured the equilibrium constant of reaction (11.4). Using a Pitzer approach and combining appropriate data from [2001LEM/FUG] they calculated $\log_{10} K^\circ_{s,0}$ (PuO$_2$, am, hydr., (11.3)) = $-$ (58.2 $\pm$ 0.97) (see Appendix A) nearly the same as that given by Fujiwara et al. [2001FUJ/YAM].

Based on the solubility constants determined for PuO$_2$(am, hydr.) by [99KNO/NEC], [98CAP/VIT], [2001FUJ/YAM] and [2002RAI/GOR] who used three different, independent methods, the present review selects the following unweighted mean value:

$$\log_{10} K^\circ_{s,0} (PuO_2, am, hydr., 298.15 K, (11.3)) = -(58.33 \pm 0.52)$$

with the uncertainty given as 1.96σ. This equilibrium constant describes best the solubility of tetravalent plutonium hydrous oxides.

However, for the reasons already given in section 9.3.2.2, the present review does not select a value of $\Delta_f G^\circ_m$ (PuO$_2$, am, hydr.) for this chemically ill-defined Pu(IV) hydroxide, amorphous hydrous oxide or oxyhydroxide.

For the solubility of PuO$_2$(am, hydr.), more or less well-characterised in a set of solutions of various compositions, the values of $\log_{10}$[Pu(OH)$_4$(aq)] = $-$ (7.7 $\pm$ 0.4), $-$ (10.4 $\pm$ 0.5) and $-$ (10.2 $\pm$ 0.8) have been reported, respectively, by Efurd et al. [98EFU/RUN], Rai et al. [99RAI/HES2] and Chandratillake et al. [98CHA/TRI] in the range of pH = 8 to 13. As discussed in Appendix A, the concentration of Pu(IV) of ca. $10^{-7.7}$ M in basic solutions indicates the probable presence of colloids. With $\log_{10} K^\circ_{s,0}$ (PuO$_2$, am, hydr., 298.15 K, (11.3)) = $-$ (58.33 $\pm$ 0.52) as selected above, these new solubility data suggest a value of $\log_{10} \beta_{4,1}^\circ = (47.9 \pm 0.7)$, somewhat larger than the value obtained from the solvent extraction study of Metivier and Guillaumont [72MET/GUI], $\log_{10} \beta_{4,1}^\circ = (47.5 \pm 0.5)$.

This review considers $\log_{10} \beta_{4,1}^\circ = (47.5 \pm 0.5)$ as the best experimental value.

11.2.3 Comparison of thermodynamic data for oxides and hydroxide complexes of tetravalent actinides

11.2.3.1 Solid oxides and aqueous hydroxide complexes of tetravalent actinides

Table 11-1 and Table 11-2 give a summary of the thermodynamic constants retained in this review governing the behaviour of tetravalent actinides (except Pa(IV)). Data are given in relation to the studies of [92GRE/FUG], [95SIL/BID] and [2001LEM/FUG].

Table 11-1: Summary of solubility and equilibrium constants for tetravalent actinides retained for discussion or selected by this review.

	Th	U	Np	Pu	Am
			$\log_{10} K_{s,0}^\circ$ (a)		
$AnO_2(cr)$	$-(54.3 \pm 1.1)^{(e)}$	$-(60.86 \pm 0.36)^{(c)}$	$-(65.75 \pm 1.07)^{(d)}$ $\{-(63.7 \pm 1.8)\}^{(f)}$	$-(64.04 \pm 0.51)^{(d)}$	$-(65.4 \pm 1.7)^{(h)}$
$AnO_2(am,hydr.)$	$-(47.0 \pm 0.8)^{(d)}$	$-(54.5 \pm 1.0)^*$	$-(56.7 \pm 0.5)^*$	$-(58.33 \pm 0.52)^*$	
$\log_{10} K_{s,4}^\circ$ (b)	$-(8.5 \pm 0.6)^{(g)}$	$-(8.5 \pm 1.0)$	$-(9.0 \pm 1.0)$	$-(10.4 \pm 0.5)$	
$\log_{10} \beta_{1,1}^\circ$	$(11.8 \pm 0.2)^{(g)}$	$(13.46 \pm 0.06)^{(c)}$	$(14.55 \pm 0.20)^{(g)*}$	$(14.6 \pm 0.2)^*$	
$\log_{10} \beta_{2,1}^\circ$	$(22.0 \pm 0.6)^{(g)}$	$(26.9 \pm 1.0)^{(g)}$	$(28.35 \pm 0.30)^{(g)*}$	$(28.6 \pm 0.3)^*$	
$\log_{10} \beta_{3,1}^\circ$	$(31.0 \pm 1.0)^{(g)}$	$(37.3 \pm 1.0)^{(g)}$	$(39.2 \pm 1.0)^{(g)}$	$(39.7 \pm 0.4)^*$	
$\log_{10} \beta_{4,1}^\circ$	$(39.0 \pm 0.5)^{(g)}$	$(46.0 \pm 1.4)^*$	$(47.7 \pm 1.1)^*$	$(47.5 \pm 0.5)^*$	

* selected by this review as described in the corresponding section.
(a) solubility product at 298.15 K for the equilibrium: $AnO_2 \text{(am or cr)} + 2H_2O(l) \rightleftharpoons An^{4+} + 4OH^-$
$- RT \ln K_{s,0}^\circ = \Delta_f G_m^\circ(An^{4+}) + 4\Delta_f G_m^\circ(OH^-) - \Delta_f G_m^\circ(AnO_2, \text{am or cr}) - 2\Delta_f G_m^\circ(H_2O, l)$.
(b) for equilibrium $AnO_2(\text{am}) + 2H_2O(l) \rightleftharpoons An(OH)_4(aq)$
$\log_{10} K_{s,4}^\circ = \log_{10}[An(OH)_4(aq)] = \log_{10} K_{s,0}^\circ(AnO_2(\text{am, hydr.})) + \log_{10} \beta_{4,1}^\circ$.
The $\log_{10} K_{s,4}^\circ$ values and hence the $\log_{10} \beta_{4,1}^\circ$ values derived for Th(IV), U(IV) and Np(IV) from experimental solubilities of $AnO_2(\text{am, hydr.})$ in neutral and alkaline solutions might represent upper limits because it is difficult to ascertain that the experimental data are not affected by contributions from colloids or oxidation state conversions.
(c) selected by [92GRE/FUG].
(d) selected by [2001LEM/FUG].
(e) calculated by this review using $\Delta_f G_m^\circ(ThO_2, cr, 298.15 K) = -(1169.238 \pm 3.504)$ kJ·mol^{-1} from CODATA [89COX/WAG] selected by [92GRE/FUG] and $\Delta_f G_m^\circ(Th^{4+}, 298.15 K) = -(704.6 \pm 5.4)$ kJ·mol^{-1} from [76FUG/OET].
(f) selected by [76FUG/OET].
(g) equilibrium constants from [2001NEC/KIM].
(h) selected by [95SIL/BID].

11.2 Plutonium oxygen and hydrogen compounds and complexes (17)

Table 11-2: Standard Gibbs energy of formation of tetravalent actinides ions and oxides

	Th	U	Np	Pu	Am
	$\Delta_f G_m^\circ (An(OH)_n^{4-n}, 298.15\text{ K})$ kJ·mol^{-1} (a)				
n = 0	−(704.6 ± 5.4)$^{(d)}$	−(529.9 ± 1.8)$^{(b)}$	−(491.8 ± 5.6)$^{(c)*}$	−(478.0 ± 2.7)$^{(c)*}$	−(346.4 ± 8.7)$^{(f)}$
n = 1	−(929.2 ± 5.6)$^{(e)}$	−(763.9 ± 1.8)$^{(b)}$	−(732.1 ± 5.7)*	−(718.6 ± 2.9)*	
n = 2	−(1144.6 ± 6.4)$^{(e)}$	−(997.8 ± 6.0)$^{(e)}$	−(968.1 ± 5.8)*	−(955.7 ± 3.2)*	
n = 3	−(1353.2 ± 7.9)$^{(e)}$	−(1214.4 ± 6.0)$^{(e)}$	−(1187.2 ± 8.0)$^{(e)}$	−(1176.3 ± 3.5)*	
n = 4	−(1553.2 ± 7.9)$^{(e)}$	−(1421.3 ± 8.2)*	−(1392.9 ± 8.4)*	−(1378.0 ± 3.9)*	
	$\Delta_f G_m^\circ (AnO_2, \text{cr}, 298.15\text{ K})$ kJ·mol^{-1}				
	−(1169.2 ± 3.5)$^{(b)}$	−(1031.8 ± 1.0)$^{(b)}$	−(1021.7 ± 2.5)$^{(c)*}$	−(998.1 ± 1.0)$^{(c)*}$	−(877.7 ± 4.3)$^{(f)}$

* selected by this review as described in the corresponding section (note that the values for n=2 and n=3 in the U-column are not selected).

(a) $\Delta_f G_m^\circ (An(OH)_n^{4-n}) = \Delta_f G_m^\circ (An^{4+}) + n \Delta_f G_m^\circ (OH^-) - RT \ln \beta_{n,1}^\circ$, the values are given with only one significant digit (the values given have an uncertainty of, at most, 0.5 kJ·mol^{-1}).
(b) selected by [92GRE/FUG].
(c) selected by [2001LEM/FUG].
(d) selected by [76FUG/OET].
(e) standard Gibbs energy calculated from equilibrium constants from [2001NEC/KIM] and auxiliary data from [92GRE/FUG].
(f) selected by [95SIL/BID].

The variation of the solubility products for the amorphous An(IV) oxides, $\log_{10} K_{s,0}^\circ$ (AnO$_2$, am, hydr.), reported in Table 11-1 follows a smoothly decreasing trend from thorium to the heaver actinides. The same smooth variation is observed for the solubility products, $\log_{10} K_{s,0}^\circ$ (AnO$_2$, cr), calculated for the crystalline An(IV) oxides from thermochemical data for Th, U, Pu and Am. However, a noticeable deviation is observed for $\log_{10} K_{s,0}^\circ$ (NpO$_2$, cr) = − (65.75 ± 1.07), which is based on data selected by [2001LEM/FUG]. This deviation disappears for the value, $\log_{10} K_{s,0}^\circ$ (NpO$_2$, cr) = − (63.7 ± 1.8), given in brackets in Table 11-1. The latter value corresponds to $\Delta_f G_m^\circ$ (Np^{4+}, 298.15 K) = − (502.9 ± 7.5) kJ·mol^{-1} [76FUG/OET], which is different from the value of − (491.8 ± 5.6) kJ·mol^{-1} selected by [2001LEM/FUG]. However, there is no further evidence from other experimental data that would justify changing the standard Gibbs energies selected by [2001LEM/FUG], $\Delta_f G_m^\circ$ (Np^{4+}, 298.15 K) = − (491.8 ± 5.6) kJ·mol^{-1}, which is retained by this review.

According to Neck and Kim [99NEC/KIM], [2001NEC/KIM], the following comments can be made about An(IV) hydrolysis:

- truly low solubility of U, Np and Pu (excluding colloids) can be explained by the presence of monomeric species in contrast to the case of Th which forms polymers,
- with regard to crystalline oxide, the An(IV) concentrations in acidic media depends on the crystallinity of the solid and are considerably lower than those corresponding to amorphous AnO_2 even after several months or years of equilibration,
- in near-neutral and alkaline solutions where $An(OH)_4$(aq) predominates, the limiting solubility phase is always an amorphous oxide giving higher concentrations than those predicted according to $\log_{10} K_{s,0}^\circ (AnO_2, cr)$, which is 6 - 7 orders of magnitude lower than $\log_{10} K_{s,0}^\circ (AnO_2, am, hydr.)$. Such affirmation about concentration is valuable provided that colloids have been removed and the oxidation state of the actinide checked.

Among the points to be clarified to have a better understanding of An(IV) solubility, the identification of the solid phases and particularly their surface properties are of great importance because solubility depends strongly on surface effects.

Surface composition is dependent on the redox properties of the investigated system and can be non-stoichiometric. The size of the solid particles also plays a role as Bundschuh *et al.* [2000BUN/KNO] have recently proven (see Appendix A), as does radioactivity, which can lead to alteration of the surface.

More precise knowledge is required of the speciation of the aqueous species in the equilibrium solution, as well as of any colloids that may be present.

11.2.3.2 Crystalline and amorphous oxides

As already mentioned, the thermodynamics of solubility reactions are determined by the properties of the surface phase, which can differ significantly from that of the bulk, in particular for crystalline oxides. The information on the stoichiometry, particle size and the crystallinity of the surface phases is often incomplete, and it is therefore not surprising that the equilibrium constants for solubility reactions obtained directly may differ substantially from those calculated from calorimetric data. In general the former are larger than the latter. Some examples are given in the Table 11-3, which includes equilibrium constants, $\log_{10} K_{s,0}^\circ$, for the dissolution of actinide oxides and hydroxides, defined as amorphous, or crystalline in the original papers. The data for "crystalline" and "amorphous" selected in this review may indicate the magnitude of the effect of an incomplete knowledge of the surface state of the solid. For this reason, the selected values of $\log_{10} K_{s,0}^\circ$ are not used to calculate the standard Gibbs energies of formation, $\Delta_f G_m^\circ$, but the recommended equilibrium constants, $\log_{10} K_{s,0}^\circ$, are useful for the geochemical modelling of actinide solubilities.

Table 11-3: Comparison of equilibrium constants $\log_{10} K^\circ_{s,0}$ (298.15 K) derived from thermochemical data and solubility experiments.

Solid	from thermochemical data		from solubility data [a]	
$UO_3 \cdot 2H_2O$(cr)	$-(23.19 \pm 0.43)$	[92GRE/FUG]	$-(22.03 \pm 0.14)$	[92SAN/BRU]
$UO_2(OH)_2$(am)			$-(21.67 \pm 0.14)$	[92SAN/BRU]]
$UO_3 \cdot 2H_2O$(s) or			$-(22.81 \pm 0.01)$	[92KRA/BIS]
$UO_2(OH)_2$(s, hydr)			$-(22.88 \pm 0.19)$	[93MEI/KIM]
			$-(22.90 \pm 0.07)$	[96MEI/KAT]
			$-(22.44 \pm 0.02)$	[96MEI/KAT]
			$-(22.75 \pm 0.06)$	[96KAT/KIM]
			$-(22.41 \pm 0.03)$	[98MEI2]
			$-(22.62 \pm 0.20)$	[94TOR/CAS]
			$-(23.14)$	[98DIA/GRA]
$Na_2U_2O_7$(cr)	$-(30.7 \pm 0.5)$	[92GRE/FUG]		
$Na_2U_2O_7$(s, hydr)			$-(29.45 \pm 1.04)$	[98YAM/KIT]
Np_2O_5(cr)	$-(12.2 \pm 0.8)$	[2001LEM/FUG]		
Np_2O_5(cr)			$-(11.4 \pm 0.4)$	[98EFU/RUN]
			$-(10.10 \pm 0.02)$	[98PAN/CAM]
NpO_2OH("aged")			$-(9.3 \pm 0.5)$	[2001LEM/FUG]
NpO_2OH(am)			$-(8.7 \pm 0.2)$	[2001LEM/FUG]
$Am(OH)_3$(cr)	$-(25.2 \pm 1.0)$	[97MER/LAM]	$-(26.4 \pm 0.6)$	present review
$Am(OH)_3$(am)			$-(25.1 \pm 0.8)$	present review
ThO_2(cr)	$-(54.3 \pm 1.1)$	(b)	(c)	
ThO_2(cr, hydr) (20 nm particles)			$-(52.8 \pm 0.3)$	[2000BUN/KNO]
ThO_2(am, hydr)			$-(47.0 \pm 0.8)$	[2001NEC/KIM]
UO_2(cr)	$-(60.86 \pm 0.36)$	[92GRE/FUG]	(c)	
UO_2(am, hydr)			$-(54.5 \pm 1.0)$	present review [c]
NpO_2(cr)	$-(65.75 \pm 1.07)$	[2001LEM/FUG]	(c)	
NpO_2(am, hydr)			$-(56.7 \pm 0.5)$	present review [c]
PuO_2(cr)	$-(64.04 \pm 0.51)$	[2001LEM/FUG]	(c)	
PuO_2(am, hydr)			$-(58.33 \pm 0.52)$	present review [c]

(a) The conditional constants given in the original papers are converted to $I = 0$ with the SIT model.
(b) Recalculated in this review using for S°_m(Th, cr, 298.15 K) = (51.8 ± 0.5) J·K^{-1}·mol^{-1}, S°_m(ThO$_2$, cr, 298.15 K) = (65.23 ± 0.20) J·K^{-1}·mol^{-1} and $\Delta_f H^\circ_m$(ThO$_2$, cr, 298.15 K) = $-(1226.4 \pm 3.5)$ kJ·mol^{-1} from CODATA [89COX/WAG] and $\Delta_f G^\circ_m$(Th^{4+}, cr, 298.15 K) = $-(704.6 \pm 5.4)$ kJ·mol^{-1} from [76FUG/OET].
(c) Not relevant for the solubility in neutral to alkaline solutions.

The solubility studies in neutral to alkaline solutions generally lead to solubilities which are, for tetravalent actinides compounds, six and more orders of magnitude higher and consistent with those of the amorphous solids, AnO_2(am, hydr). This is surprising as the least soluble phase should have the highest thermodynamic stability. The experimental information indicates that this is not the case, which suggests that there are chemical processes that are not understood. Therefore, they are not included in the thermodynamic description of these systems (see selected values in Chapter 5).

11.3 Plutonium group 17 complexes

11.3.1 Aqueous plutonium chloride complexes (18.2.2.1)

11.3.1.1 Pu(III) chloride complexes

An EXAFS study of Allen et al. [97ALL/BUC] shows no evidence for the formation of Pu(III) chloride complexes up to LiCl concentrations of 12 mol · L^{-1}, while a later EXAFS study [2000ALL/BUC] indicates the formation of Am(III) and Cm(III) chloride complexes according to the reaction:

$$An^{3+} + nCl^- \rightleftharpoons AnCl_n^{3-n} \tag{11.6}$$

with $n = 1 - 2$ in 8 – 12.5 M LiCl. The EXAFS studies of Allen et al. [97ALL/BUC], [2000ALL/BUC] and the new spectroscopic data for Am(III) and Cm(III) chloride complexes lead to the selection of $\log_{10} \beta_1^\circ$ (11.6) = 0.24 and $\log_{10} \beta_2^\circ$ (11.6) = – 0.74 (cf. section 12.4.1.3). The equilibrium constant selected by [2001LEM/FUG], $\log_{10} \beta_1^\circ$ (PuCl^{2+}, 298.15 K) = (1.2 ± 0.2) based on the study of Ward et al. [56WAR/WEL], appears to be overestimated. It is therefore rejected by the present review, which prefers to base its selection on EXAFS data rather on ionic exchange data. As the value of the formation constant for PuCl^{2+} seems smaller than was assessed in [2001LEM/FUG], the value for the formation constant of PuCl^{3+} based, in part on the study of Rabideau and Cowan [55RAB/COW], may also need to be reassessed.

Additional data are given in [95NOV/CRA] and [99CHO].

11.3.1.2 Pu(VI) chloride complexes

Runde et al. [97RUN/NEU], [99RUN/REI] investigated the formation of Pu(VI) chloride complexes by absorption spectroscopy and EXAFS. In 1 M HCl or HClO$_4$ solutions containing 1.5 NaCl – 16 M LiCl they observed evidence for the reactions:

$$PuO_2^{2+} + nCl^- \rightleftharpoons PuO_2Cl_n^{2-n} \quad (11.7)$$

up to $n = 4$ [97RUN/NEU]. Runde et al. [99RUN/REI] have obtained, by absorption spectroscopy, the formation constants for the reactions of PuO$_2$Cl$^+$ and PuO$_2$Cl$_2$(aq) as $\log_{10} \beta_1°$ (11.7) = (0.23 ± 0.03) and $\log_{10} \beta_2°$ (11.7) = – (1.7 ± 0.2). As discussed in detail in Appendix A, [99RUN/REI], this review selects:

$$\log_{10} \beta_1° (PuO_2Cl^+, 298.15\ K) = (0.23 \pm 0.03),$$

with $\Delta\varepsilon((11.7), n = 1) = -(0.13 \pm 0.03)$ kg · mol^{-1} in NaCl solution and

$$\log_{10} \beta_2° (PuO_2Cl_2, aq, 298.15\ K) = -(1.15 \pm 0.30).$$

These values are significantly different from $\log_{10} \beta_1°$ (11.7) = (0.70 ± 0.13) and $\log_{10} \beta_2°$ (11.7)= – (0.6 ± 0.2) selected by [2001LEM/FUG] from the spectroscopic study of Giffaut [94GIF], but are close to the values selected by [92GRE/FUG] from numerous and well-ascertained data for the analogous U(VI) chloride complexes. The reasons for the revised data selection and a discussion of the data in [94GIF] are also included in Appendix A, discussion of [99RUN/REI].

11.3.1.3 Cation-cation complexes of Pu(V)

New papers on plutonium complexation have dealt also with neptunium complexation (see 10.5). New data are available concerning the cation-cation complex PuO$_2^+$ · UO$_2^{2+}$. Stoyer et al. [2000STO/HOF] determined a value of β = (2.2 ± 1.5) L · mol^{-1} for the formation of PuO$_2^+$ · UO$_2^{2+}$ in 6 M NaClO$_4$. This result was already discussed in section 9.8.1.

11.4 Plutonium group 16 compounds and complexes (19)

11.4.1 Plutonium selenium compounds

11.4.1.1 Plutonium selenides

11.4.1.1.1 Plutonium monoselenide

Hall et al. [91HAL/MOR] have measured the heat capacity of a sample of high purity PuSe from 7 to 300 K. The sample was annealed at 1273 K for 2 hours in high purity argon immediately prior to the measurements to remove any stored damage to the crystalline structure due to radioactive decay. In keeping with the lack of magnetism in

PuSe, no anomalies were found in the $C_{p,m}$ curve. The values obtained for the heat capacity and entropy at 298.15 K were:

$C_{p,m}^{\circ}$ (PuSe, cr, 298.15 K) = (59.7 ± 1.2) J · K^{-1} · mol^{-1},
S_{m}° (PuSe, cr, 298.15 K) = (92.1 ± 1.8) J · K^{-1} · mol^{-1},

and these are the selected values.

11.4.2 Plutonium tellurium compounds

11.4.2.1 Plutonium tellurides

11.4.2.1.1 Plutonium monotelluride

Hall et al. [90HAL/JEF2] have measured the heat capacity of a sample of high purity PuTe from 10 to 300 K. Small single crystals of PuTe were pressed into a cylinder which was sealed into a silver calorimeter under helium. A very small anomaly, of unknown origin, was observed in the $C_{p,m}$ curve at 269 K. The values obtained for the heat capacity and entropy at 298.15 K were:

$C_{p,m}^{\circ}$ (PuTe, cr, 298.15 K) = (73.1 ± 2.9) J · K^{-1} · mol^{-1},
S_{m}° (PuTe, cr, 298.15 K) = (107.9 ± 4.3) J · K^{-1} · mol^{-1},

and these are the selected values.

11.5 Plutonium nitrogen compounds and complexes

11.5.1 Plutonium nitrides

11.5.1.1 PuN(cr)

Ogawa et al. [98OGA/KOB] have given an equation for $\Delta_f G_m$ (PuN, T), based on an assessment of the existing data. However, as for their similar suggestion for UN(cr), their equation does not reproduce the reasonably well-defined entropy of formation of PuN(cr) (differing by ca. 6.2 J·K^{-1}·mol^{-1}), and we prefer to retain the data selected in [2001LEM/FUG].

11.6 Plutonium carbon compounds and complexes (21.1.2)

11.6.1 Aqueous plutonium carbonates (21.1.2.1)

The aqueous plutonium carbonate systems were discussed in [2001LEM/FUG]; however, two important experimental studies published in 1993 [93PAS/RUN] and 1997 [97PAS/CZE] were overlooked. This review does not share some of the judgments made by the prior NEA TDB review as discussed in the following section. These facts result in important changes in some of the selected values.

11.6.1.1 Pu(VI) carbonate complexes (21.1.2.1.1)

Pashalidis et al. [93PAS/RUN], and Pashalidis et al. [97PAS/CZE] studied the solubility of UO_2CO_3 and PuO_2CO_3 in aqueous carbonate solutions as discussed in Appendix A. These are two precise experimental studies where the uranium(VI) data are in excellent agreement with other studies, cf. section 9.7.2.1. This indicates that the procedures used are satisfactory and that the experiments are not affected by systematic errors. However, the precision in the plutonium(VI) data is much lower than that of the corresponding uranium(VI) data. The equilibrium constants proposed in [93PAS/RUN] and [97PAS/CZE] have been recalculated to zero ionic strength using the SIT model by this review and are:

$$PuO_2^{2+} + CO_3^{2-} \rightleftharpoons PuO_2CO_3(aq) \qquad \log_{10} \beta_1^\circ = (9.7 \pm 0.3),$$

$$PuO_2^{2+} + 2\, CO_3^{2-} \rightleftharpoons PuO_2(CO_3)_2^{2-} \qquad \log_{10} \beta_2^\circ = (15.3 \pm 0.5),$$

$$PuO_2^{2+} + 3\, CO_3^{2-} \rightleftharpoons PuO_2(CO_3)_3^{4-} \qquad \log_{10} \beta_3^\circ = (18.4 \pm 0.2).$$

This review has scrutinised the discussions of previous experimental data given in [2001LEM/FUG]; the data reported in [62GEL/MOS] and [67GEL/MOS] are not reliable as confirmed by this review. The data of Sullivan and Woods [82SUL/WOO] were accepted in [2001LEM/FUG], but the value of $\log_{10} \beta_1^\circ = 13.2$ for Pu is very different from the values of the corresponding uranium(VI) and neptunium(VI) complexes. There is no chemical reason to expect this and according to the present review, this is sufficient reason to disregard the results in [82SUL/WOO]. The only two experimental studies that are reliable are those of Robouch and Vitorge [87ROB/VIT], and Ullman and Schreiner [88ULL/SCH]. Robouch and Vitorge report $\log_{10} \beta_1^\circ = 9.3$. The values of $\log_{10} \beta_2^\circ$ reported in [87ROB/VIT] and [88ULL/SCH] are 13.6 and 15.1, respectively, whereas the values of $\log_{10} \beta_3^\circ$ are 17.0 and 18.5. The equilibrium constants in [88ULL/SCH] are in good agreement with those of [97PAS/CZE], while the constants from [87ROB/VIT] differ slightly. These data were determined in 3 M $NaClO_4$ and the error in the extrapolation to zero ionic strength, ± 0.5, is therefore larger than that in the other two studies.

This review selects the following average values of [87ROB/VIT], [88ULL/SCH], [93PAS/RUN] and [97PAS/CZE]:

$$\log_{10} \beta_1^\circ = (9.5 \pm 0.5),$$

$$\log_{10} \beta_2^\circ = (14.7 \pm 0.5),$$

$$\log_{10} \beta_3^\circ = (18.0 \pm 0.5).$$

The formation constant for the polynuclear Pu(VI) hydroxide complex $(PuO_2)_3(CO_3)_6^{6-}$ is discussed in [2001LEM/FUG] where it is pointed out that, by comparison to similar U and Np complexes, no good data exist for the interaction coefficient. As extrapolation to zero ionic strength cannot done properly, this review does not retain this data in the table of selected values.

11.6.1.2 Pu(IV) carbonate complexes (21.1.2.1.3)

Capdevila et al. [96CAP/VIT] investigated the formation of the limiting carbonate complex:

$$Pu(CO_3)_4^{4-} + CO_3^{2-} \rightleftharpoons Pu(CO_3)_5^{6-}. \tag{11.8}$$

by absorption spectroscopy in $NaClO_4/NaHCO_3/Na_2CO_3$ solutions of varying ionic strength. They estimated an equilibrium constant of $\log_{10} K_5^\circ (11.8) = -(1.36 \pm 0.09)$ which was selected in [2001LEM/FUG]. [96CAP/VIT] also reported the formation constant of the pentacarbonate complex, $\log_{10} \beta_5$ ($Pu(CO_3)_5^{6-}$, 298.15 K) = (35.8 ± 1.3) in 3 M $NaClO_4/NaHCO_3/Na_2CO_3$. The latter value is based on redox equilibria including the disproportionation of Pu(V) and was not used by [2001LEM/FUG] for the evaluation of thermodynamic data.

Clark et al. [98CLA/CON] studied the structure of the $Pu(CO_3)_5^{6-}$ ion in the solid state and in aqueous solution; this study confirms the stoichiometry of the limiting carbonate complex deduced from potentiometric and spectroscopic studies [92CAP], [96CAP/VIT].

Rai et al. [99RAI/HES2] have studied the solubility of $PuO_2(am)$ in the $K^+ - HCO_3^- - CO_3^{2-} - OH^- - H_2O$ system. The methodology is the same as used in prior studies on Th(IV) [97FEL/RAI], U(IV) [98RAI/FEL] and Np(IV) [99RAI/HES]. The authors used the ion interaction approach of Pitzer to determine the following equilibrium constants (given without uncertainty limits) for the reaction:

$$AnO_2(am) + 2 H_2O + 5 CO_3^{2-} \rightleftharpoons An(CO_3)_5^{6-} + 4 OH^- \tag{11.9}$$

$\log_{10} K_{s,5}^\circ ((11.9), An = Th) = -18.4$ [97FEL/RAI],

$\log_{10} K_{s,5}^\circ ((11.9), An = U) = -22.2$ [98RAI/FEL],

$\log_{10} K_{s,5}^\circ ((11.9), An = Np) = -21.15$ [99RAI/HES],

$\log_{10} K_{s,5}^\circ ((11.9), An = Pu) = -22.68$ [99RAI/HES2].

The reported Pitzer parameters for the pairs Na^+ or K^+ and $An(CO_3)_5^{6-}$ are reasonable for ions of this charge. However, the solubility constants at $I = 0$ are highly correlated with the simultaneously calculated ion interaction coefficients. Therefore, this review ascribes rather large uncertainty estimates of ± 1 $\log_{10}$-units to the reported equilibrium constants. By combining the $\log_{10} K_{s,5}^\circ$ values with the solubility constants of $AnO_2(am, hydr.)$ selected in the present review:

$\log_{10} K_{s,0}^\circ$ (ThO_2, am, hydr., 298.15 K) = $-(47.0 \pm 0.8)$,

$\log_{10} K_{s,0}^\circ$ (UO_2, am, hydr., 298.15 K) = $-(54.5 \pm 1.0)$,

$\log_{10} K_{s,0}^\circ$ (NpO_2, am, hydr., 298.15 K) = $-(56.7 \pm 0.5)$,

$\log_{10} K_{s,0}^\circ$ (PuO_2, am, hydr., 298.15 K) = $-(58.33 \pm 0.52)$, we obtain the following formation constants of the limiting An(IV) carbonate complexes:

$$\log_{10} \beta_5^\circ \, (\text{Th(CO}_3)_5^{6-}, 298.15 \text{ K}) = (28.6 \pm 1.3),$$
$$\log_{10} \beta_5^\circ \, (\text{U(CO}_3)_5^{6-}, 298.15 \text{ K}) = (32.3 \pm 1.4),$$
$$\log_{10} \beta_5^\circ \, (\text{Np(CO}_3)_5^{6-}, 298.15 \text{ K}) = (35.6 \pm 1.1),$$
$$\log_{10} \beta_5^\circ \, (\text{Pu(CO}_3)_5^{6-}, 298.15 \text{ K}) = (35.65 \pm 1.13).$$

The formation constant of $\text{U(CO}_3)_5^{6-}$ derived from the data in [98RAI/FEL] and the value selected in [92GRE/FUG], $\log_{10} \beta_5^\circ \, (\text{U(CO}_3)_5^{6-}, 298.15 \text{ K}) = (34.0 \pm 0.9)$, overlap within the uncertainty limits. The formation constant of $\text{Np(CO}_3)_5^{6-}$ selected in [2001LEM/FUG], $\log_{10} \beta_5^\circ \, (\text{Np(CO}_3)_5^{6-}, 298.15 \text{ K}) = (35.61 \pm 1.09)$, is identical with the value calculated above, although the former value results from a re-interpretation of the experimental data in [99RAI/HES] combined with other equilibrium constants selected in [2001LEM/FUG]. For these reasons and the systematic trend in the series of the tetravalent actinides, the present review relies also on the value for $\text{Pu(CO}_3)_5^{6-}$ and selects:

$$\log_{10} \beta_5^\circ \, (\text{Pu(CO}_3)_5^{6-}, 298.15 \text{ K}) = (36.65 \pm 1.13).$$

Combining this value with $\log_{10} K_5^\circ \, (11.8) = - (1.36 \pm 0.09)$ [2001LEM/FUG] leads to:

$$\log_{10} \beta_4^\circ \, (\text{Pu(CO}_3)_4^{4-}, 298.15 \text{ K}) = (37.0 \pm 1.1).$$

Felmy et al. [97FEL/RAI] and Rai et al. [98RAI/FEL], [99RAI/HES], [99RAI/HES2] also studied the formation of the mixed An(IV) hydroxide-carbonate complexes based on their solubility measurements. In the case of Th(IV), [97FEL/RAI] adopted the equilibrium constant determined by Östhols et al. [94OST/BRU] for the complex $\text{Th(OH)}_3(\text{CO}_3)^-$, whereas the solubility data for U(IV), Np(IV) and Pu(IV) were modelled by assuming the reaction:

$$\text{AnO}_2(\text{am}) + 2\text{H}_2\text{O} + 2\text{CO}_3^{2-} \rightleftharpoons \text{An(OH)}_2(\text{CO}_3)_2^{2-} + 2\text{OH}^- \qquad (11.10)$$

with $\log_{10} K_{s,1,2,2}^\circ \, (11.10) = -12.11$ for U(IV) [98RAI/FEL], -11.75 for Np(IV) [99RAI/HES] and -12.09 for Pu(IV) [99RAI/HES2]. Uncertainty limits and Pitzer parameters for these complexes were not reported.

The solubilities of ternary Np(IV) hydroxide-carbonate complexes have been discussed in section 10.6.1.1. A comparison of these solubility data with those from the corresponding Pu(IV) system is given in Figure 11-1 that shows the solubility, $\log_{10}[\text{An(IV)}]$, versus $-\log_{10}[\text{H}^+]$ for solubility data for Pu(IV) from [94YAM/SAK] and [99RAI/HES], and corresponding data for Np(IV) from [2001KIT/KOH]. It is obvious that the solubility is the same in these systems, in particular for the data determined from the direction of over-saturation, i.e., for the Pu(IV) data of [94YAM/SAK] and the Np(IV) data of [2001KIT/KOH]. This indicates that the composition of the ternary complexes is the same.

The calculated curves in Figure 11-1 are based on the equilibrium constants proposed by Rai et al. [99RAI/HES2] ($\log_{10} K^\circ_{s,1,2,2} = -12.1$ for $Pu(OH)_2(CO_3)_2^{2-}$) and Yamaguchi et al. [94YAM/SAK] ($\log_{10} K^\circ_{s,1,2,2} = -(10.2 \pm 0.5)$ for $Pu(OH)_2(CO_3)_2^{2-}$ and $\log_{10} K^\circ_{s,1,4,2} = -(5.9 \pm 0.3)$ for $Pu(OH)_4(CO_3)_2^{4-}$, converted to $I = 0$ in this review by using the SIT). The equilibrium constant $\log_{10} K^\circ_{s,4} = -(10.4 \pm 0.5)$ for $Pu(OH)_4$(aq) is taken from the study of [99RAI/HES2]. It should be noted that [99RAI/HES2] reported experimental data in K_2CO_3 solutions containing 0.01 M KOH which are concordant with those of [94YAM/SAK], but they did not use these results to evaluate an equilibrium constant for the complex $Pu(OH)_4(CO_3)_2^{4-}$.

As the solubility product, $\log_{10} K^\circ_{s,0}$ (PuO_2, am, hydr.) $= -(58.33 \pm 0.52)$, is somewhat lower than $\log_{10} K^\circ_{s,0}$ (NpO_2, am, hydr.) $= -(56.7 \pm 0.5)$, the formation constants of the ternary Pu(IV) hydroxide-carbonate complexes are accordingly higher than those of the analogous Np(IV) complexes. However, as the solubility product of the amorphous oxide phases used in these experiments is uncertain, this review has not selected equilibrium constants for the ternary complexes. Additional work is necessary to quantify their formation constants.

Figure 11-1. Solubility data for PuO$_2$(am, hydr.) from Yamaguchi et al. [94YAM/SAK] in 0.1 M NaHCO$_3$-NaClO$_4$ and Na$_2$CO$_3$-NaClO$_4$-NaOH, and from Rai et al. [99RAI/HES2] in KHCO$_3$ and K$_2$CO$_3$ (+ 0.01 M KOH) at total carbonate concentrations of a) 0.1 mol·L^{-1} and b) 0.01 mol·L^{-1}. Experimental Np(IV) solubility data from Kitamura and Kohara [2001KIT/KOH] are shown for comparison. The solubility curves are calculated for $I = 0.1$ mol·L^{-1} according to the equilibrium constants reported by [94YAM/SAK] and [99RAI/HES2].

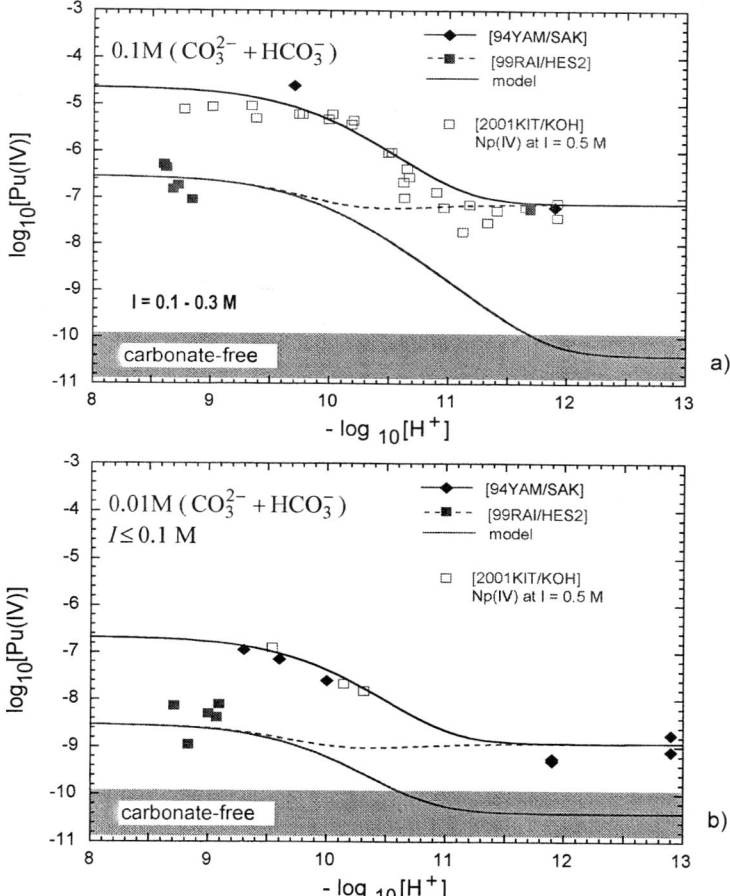

11.6.2 Solid plutonium carbonate (21.1.2.2)

11.6.2.1 Solid Pu(VI) carbonates (21.1.2.2.1)

In the previous NEA TDB review [2001LEM/FUG] the solubility constant for the reaction:

$$PuO_2CO_3(s) \rightleftharpoons PuO_2^{2+} + CO_3^{2-} \tag{11.11}$$

was selected from a study of Robouch and Vitorge [87ROB/VIT] in 3 M NaClO$_4$ at 293.15 K:

$$\log_{10} K_{s,0}^\circ ((11.11), 298.15 \text{ K}) = -(14.2 \pm 0.3).$$

Pashalidis et al. [93PAS/RUN] reported a solubility study in 0.1 M NaClO$_4$ at 295.15 K, which was not discussed in [2001LEM/FUG]. A second paper [97PAS/CZE], which arrived too late for a critical evaluation in the previous review [2001LEM/FUG], includes the experimental data of [93PAS/RUN] and extends the former solubility study to higher carbonate concentrations. The solubility constant recalculated in the present review from these data and extrapolated to $I = 0$ with the SIT (cf. Appendix A), $\log_{10} K_{s,0}^\circ (11.11) = -(15.0 \pm 0.2)$, is lower than that from [87ROB/VIT] indicating a less crystalline solid in the former study. However, as the formation constants of the Pu(VI) carbonate complexes in [87ROB/VIT] and [97PAS/CZE] are in reasonable agreement, the crystallinity does not seem to have changed during the experiments.

In a later study, Reilly et al. [2000REI/NEU] determined the solubility constant of PuO$_2$CO$_3$(s) in 0.1, 0.2, 0.5, 1.0, 2.1, 3.8 and 5.6 m NaCl, and for comparison in 5.6 m NaClO$_4$. The experimental $\log_{10} K_{s,0}$ values are given without correction for chloride complexation, which explains the difference between the values in 5.6 m NaCl ($\log_{10} K_{s,0} = -(14.0 \pm 0.1)$) and 5.6 M NaClO$_4$ ($\log_{10} K_{s,0} = -(14.5 \pm 0.1)$). The ionic strength dependence of the experimental $\log_{10} K_{s,0}$ values at $I < 3$ mol·kg^{-1} is in excellent agreement with $\Delta\varepsilon = (0.13 \pm 0.04)$ kg·mol^{-1}, as predicted with the interaction coefficients $\varepsilon(\text{Na+, CO}_3^{2-}) = -(0.08 \pm 0.03)$ kg·mol^{-1} and $\varepsilon(\text{PuO}_2^{2+}, \text{Cl}^-) = \varepsilon(\text{UO}_2^{2+}, \text{Cl}^-) = (0.21 \pm 0.02)$ kg·mol^{-1} [2001LEM/FUG]. In the present review the equilibrium constant at $I = 0$ is calculated to be $\log_{10} K_{s,0}^\circ (11.11) = -(14.67 \pm 0.10)$, cf. Appendix A.

Although the uncertainty in the extrapolation to $I = 0$ is larger for the experimental data of [87ROB/VIT] in 3.5 m NaClO$_4$ than for the later studies in 0.1 m NaClO$_4$ [93PAS/RUN], [97PAS/CZE] and 0.1 – 2.1 m NaCl solutions [2000REI/NEU], the present review selects the unweighted mean value from these studies:

$$\log_{10} K_{s,0}^\circ ((11.11), 298.15 \text{ K}) = -(14.65 \pm 0.47).$$

The uncertainty (95% confidence interval) may also reflect possible differences in the crystallinity of the solid. Similar differences have been observed in the solubility data for the isostructural solid UO$_2$CO$_3$(s) (cf. discussion in 9.7.2.1.1).

11.7 Plutonium group 2 (alkaline-earth) compounds

11.7.1 Plutonium-strontium compounds (22.1)

The compound $SrPu_2Ti_4O_{12}(cr)$, along with $Pu_2Ti_3O_{8.79}(cr)$ and $Pu_2Ti_2O_7(cr)$, in which Pu is in the trivalent state, has been prepared and identified by X-ray diffraction [96SHO/BAM]. No thermodynamic data exist for these compounds.

11.7.2 Plutonium-barium compounds (22.1)

11.7.2.1 BaPuO₃(cr) (22.2.2)

Nakajima *et al.* [99NAK/ARA2] have measured the partial pressures of BaO(g) and Ba(g) over a diphasic sample containing a mixture of $BaPuO_3$ and PuO_2 by Knudsen-cell mass-spectrometry from 1673 to 1873 K. As noted in Appendix A, they analysed their results in terms of the reaction:

$$BaPuO_3(cr) \rightleftharpoons BaO(g) + PuO_2(cr)$$

although in practice, the vaporisation is likely to be more complex than this.

Using estimated thermal functions for $BaPuO_3(cr)$, [99NAK/ARA2] derived second- and third-law enthalpies of formation, $\Delta_f H_m^\circ$ ($BaPuO_3$, cr, 298.15 K) of -1661 and -1673 kJ·mol⁻¹, respectively, with no quoted uncertainties.

Considering the use of estimated thermal functions for $BaPuO_3(cr)$ and the likely complexity of the actual vaporisation process, these values are in good accord with the relatively precise calorimetric value from the study by Morss and Eller [89MOR/ELL], adopted by [2001LEM/FUG]:

$$\Delta_f H_m^\circ (BaPuO_3, cr, 298.15 \text{ K}) = -(1654.2 \pm 8.3) \text{ kJ} \cdot \text{mol}^{-1}$$

which is retained here.

Chapter 12

Discussion of new data selection for Americium

12.1 Introductory remarks

12.1.1 Estimation of standard entropies (V.4.2.1.2)

The standard entropies of many americium ionic solid compounds have been revised from those suggested in [95SIL/BID] to take account of the improved procedure suggested by Konings [2001KON]. As noted in Appendix A, in this procedure the entropy is assumed to be the sum of a lattice contribution, estimated from known values of similar actinide species, and an excess contribution, representing mainly electronic effects. Such a scheme gives good agreement for many lanthanide and actinide compounds, and indeed is the basis of the existing NEA estimates for the aqueous americium ions. However, since the $^7F_0$ ground state of Am^{3+} is non-degenerate and any excited states do not contribute to the entropy at 298.15 K, the excess entropy is zero for most Am(III) compounds. This leads to appreciably smaller standard entropies than estimated in [95SIL/BID] for these species.

To avoid repetition, the values used for the lattice and excess contributions are summarised in Table 12-1, with references to fuller text discussions.

Table 12-1: Summary of entropy contributions for some Am solid compounds (J·K^{-1}·mol^{-1}).

Species	S_m (lattice)	S_m (excess)	S_m (total)	[95SIL/BID]	Comment
AmO$_2$(cr)	(65.2 ± 8.0)	(12.5 ± 6.0)	(77.7 ± 10.0)	(67.0 ± 10.0)	[2001KON2]
Am$_2$O$_3$(cr)	(133.6 ± 6.0)	0	(133.6 ± 6.0)	(160.0 ± 15.0)	[2001KON]
Am(OH)$_3$(cr)	(116 ± 8)	0	(116 ± 8)	none	See 12.3.2.2
AmF$_3$(cr)	(110.6 ± 6.0)	0	(110.6 ± 6.0)	(127.6 ± 5.0)	[2001KON]
AmF$_4$(cr)	(141.6 ± 6.0)	(12.5 ± 5.0)	(154.1 ± 8.0)	(148.5 ± 5.0)	See 12.4.2.2.2
AmCl$_3$(cr)	(146.2 ± 6.0)	0	(146.2 ± 6.0)	(164.8 ± 6.0)	[2001KON]
AmOCl(cr)	---	---	(92.6 ± 10.0)	(111.0 ± 10.0)	See 12.4.2.3.2
Cs$_2$NaAmCl$_6$(cr)	---	---	(421 ± 15)	(440 ± 15)	See 12.4.2.3.3
AmBr$_3$(cr)	(182.0 ± 10.0)	0	(182.0 ± 10.0)	(205 ± 17)	See 12.4.2.4.1.1
AmOBr(cr)	---	---	(104.9 ± 12.8)	(128.0 ± 20.0)	See 12.4.2.4.2
AmI$_3$(cr)	(211 ± 15)	0	(211 ± 15)	(234 ± 20)	See 12.4.2.5.1.1
Am(OH)CO$_3$·0.5H$_2$O(cr)	(141 ± 21)	0	(141 ± 21)	none	See 12.6.1.1.3.1

12.1.2 Introductory remarks on the inclusion of thermodynamic data of aqueous Cm(III) complexes in this review

It is well known that chemical compounds and aqueous species of the trivalent actinides and lanthanides have similar chemical properties [94CHO/RIZ]. This is true both for their coordination chemistry and their complex formation reactions in solution; a noticeable and very important exception is their redox properties. Certain systematic trends in the thermodynamic data and equilibrium constants often correlate with the variation of the ionic radius, *e.g.*, with the decrease of the ionic radius from La^{3+} to Lu^{3+} [94CHO/RIZ], [96BYR/SHO]. In particular Cm(III), but also the lanthanides Nd(III) and Eu(III), are often used as analogs for Am(III), because of the similarity in ionic size: *e.g.*, for coordination number CN = 8, the crystal radii are 1.10 Å (Am^{3+}), 1.09 Å (Cm^{3+}), 1.11 Å (Nd^{3+}) and 1.07 Å (Eu^{3+}) [76SHA], [94CHO/RIZ].

In the same way there are pronounced similarities and trends in most of the chemical properties of the actinide elements in the oxidation states An(IV, V or VI), with the exception of their redox properties [80COT/WIL], [84GRE/EAR], [83CHO2] and [94CHO/RIZ]. This chemically well-established principle has been used by [2001LEM/FUG] for the comparison and partly for the estimation of thermodynamic data for aqueous species of pentavalent and hexavalent U, Np and Pu, and recently for the tetravalent actinides [99NEC/KIM], [2001NEC/KIM].

Making use of this oxidation state analogy principle, the present review includes experimental data for Cm(III) when evaluating thermodynamic data for aqueous Am(III) complexes, because the former are often more accurate than the latter. Particularly in the case of complex formation reactions of Cm(III) there is a large amount of accurate information available from spectroscopic data obtained with the very sensitive time-resolved laser fluorescence spectroscopy (TRLFS). In some cases time-resolved laser fluorescence spectroscopy (TRLFS). In some cases the discussion of data evaluation is supported by comparing analogous chemical reactions and data for the trivalent lanthanides. We have used these oxidation state analogies for the following purposes:

- To evaluate activity coefficients and ion interaction coefficients of aqueous Am(III) species by using experimental data for Nd^{3+} or aqueous Cm(III) complexes.

- To select equilibrium constants for aqueous Am(III) complexes by using spectroscopic data for analogous complex formation reactions of Cm(III).

- To compare the chemical behaviour and thermodynamic data of some isostructural solids of Am(III) and the lanthanides, Nd(III) and Eu(III), and to estimate standard entropies of crystalline Am(III) hydroxide and hydroxycarbonate.

The differences in the activity coefficients and ion interaction coefficients of Am(III), Cm(III), Nd(III) and Eu(III) species are considered to be negligible [90FEL/RAI], [97KON/FAN], [98FAN/KIM] and the differences in the formation constants of analogous aqueous complexes are in most cases smaller than the experimental uncertainties [98NEC/FAN], whereas the solubility constants of isostructural solids can differ considerably.

12.2 Elemental americium (V.1)

12.2.1 Americium ideal monatomic gas (V.1.2)

12.2.1.1 Heat capacity and entropy (V.1.2.1)

The thermal functions of Am(g) in [95SIL/BID] were calculated using 33 spectroscopic levels up to 18000 cm^{-1} given by Brewer [84BRE], plus 18 levels between 18000 and 30000 cm^{-1} previously communicated by Fred [75FRE] to Oetting *et al.* [76OET/RAN]. Since these data were assembled, Blaise and Wyart [92BLA/WYA] have published a comprehensive listing of the known energy levels of all the actinide gases containing 204 levels for Am(g) up to 40600 cm^{-1}. However, these 204 levels do not include three relatively low-lying levels estimated by Brewer [84BRE] at 14539, 16639 and 18294 cm^{-1}. These have been added to the energy level list, together with seven further missing levels estimated by this review to lie between 17800 and 25000 cm^{-1}.

The revised values of the thermal functions have been calculated from these 214 levels with a total statistical weight of 1625; the Gibbs energy functions begin to differ from those used by [95SIL/BID] by more than 0.005 $J \cdot K^{-1} \cdot mol^{-1}$ at about 2500 K.

The values at 298.15 K for the ^{241}Am isotope (molar mass = 241.0568 g · mol^{-1} [99FIR]) are:

$C_{p,m}^{\circ}$ (Am, g, 298.15 K) = (20.786 ± 0.010) J · K^{-1} · mol^{-1},

S_{m}° (Am, g, 298.15 K) = (194.55 ± 0.05) J · K^{-1} · mol^{-1}.

These are unchanged from the values calculated from the energy levels used in [95SIL/BID]. However, owing to a typographical error, the entropy was misquoted (as (195.6 ± 2.0) J · mol^{-1} · K^{-1}) in both the text and tables. Note that the uncertainty has also been greatly reduced, due to the much larger number of electronic levels now available.

The calculated heat capacity of Am(g) from 298.15 to 1100 K is essentially constant as given in Table 6.3.

12.2.1.2   Enthalpy of formation (V.1.2.2)

The enthalpy of formation selected by [95SIL/BID] is based on three reasonably consistent measurements of the vapour pressure from ca 1000 to 1600 K. Over this temperature range, the revised values of ($G_{m}(T) - H_{m}^{\circ}$(298.15 K))/T(Am, g, 298.15 K) differ from those used in the earlier assessment by only 0.001 J · K^{-1} · mol^{-1}, so the derived value of the enthalpy of sublimation using the revised thermal functions will differ negligibly from the value selected by [95SIL/BID] (which utilised the correct value of S_{m}° (Am, g, 298.15 K) – see above). This value is therefore retained.

$\Delta_{f} H_{m}^{\circ}$ (Am, g, 298.15 K) = (283.8 ± 1.5) kJ · mol^{-1}.

The calculated Gibbs energy of formation thus becomes:

$\Delta_{f} G_{m}^{\circ}$ (Am, g, 298.15 K) = (242.3 ± 1.6) kJ · mol^{-1}.

12.3 Americium oxygen and hydrogen compounds and complexes (V.3)

12.3.1 Aqueous americium hydroxide complexes (V.3.1)

12.3.1.1   Aqueous Am(III) and Cm(III) hydroxide complexes (V.3.1.1)

The hydrolysis of americium(III) has been critically discussed in the previous NEA review [95SIL/BID]. As the published data vary by orders of magnitude, as a result of experimental shortcomings, the values from most publications were rejected by [95SIL/BID]. A number of new experimental studies have appeared since the previous review. They will be discussed below together with the studies used in the previous review for the selection of thermodynamic data.

12.3 Americium oxygen and hydrogen compounds and complexes (V.3)

The hydrolysis reactions of Am(III) can be described by:

$$Am^{3+} + n\,H_2O(l) \rightleftharpoons Am(OH)_n^{(3-n)} + n\,H^+, \qquad (12.1)$$

with n = 1, 2 and 3. As discussed in the previous review [95SIL/BID] there is neither evidence for the formation of polynuclear complexes nor for anionic hydroxide complexes up to pH = 13 – 14. The lack of evidence for polynuclear species is to some extent a result of the low total concentrations of Am(III) used in the experiments.

The selection of thermodynamic data in [95SIL/BID] was based on the following studies in NaClO$_4$ solution: the solubility experiments of Silva [82SIL], and Stadler and Kim [88STA/KIM] in 0.1 M NaClO$_4$, the solvent extraction study of Lundqvist [82LUN] in 1 M NaClO$_4$ and the potentiometric study of Nair et al. [82NAI/CHA] in 1 M NaClO$_4$. After reinterpretation of these studies the following equilibrium constants at $I = 0$ were selected by Silva et al. [95SIL/BID]:

$$\log_{10} {}^*\!\beta_1^\circ\,((12.1),\,298.15\ \text{K}) = -\,(6.4 \pm 0.7),$$
$$\log_{10} {}^*\!\beta_2^\circ\,((12.1),\,298.15\ \text{K}) = -\,(14.1 \pm 0.6),$$
$$\log_{10} {}^*\!\beta_3^\circ\,((12.1),\,298.15\ \text{K}) = -\,(25.7 \pm 0.5).$$

In an earlier review, Fuger [92FUG] selected lower formation constants from the same studies, although the uncertainty limits of the two sets overlap:

$$\log_{10} {}^*\!\beta_1^\circ\,((12.1),\,298.15\ \text{K}) = -\,(7.1 \pm 0.5),$$
$$\log_{10} {}^*\!\beta_2^\circ\,((12.1),\,298.15\ \text{K}) = -\,(14.8 \pm 0.5).$$

Experimental data in NaCl solutions, from the solubility of Am(OH)$_3$(s) in 0.1 and 0.6 M NaCl solutions [88STA/KIM] and a solvent extraction study in 0.7 M NaCl [83CAC/CHO], were not included or disregarded in the discussion and the data selection of Silva et al. [95SIL/BID]. Additional data are available from a solubility study of Runde and Kim [94RUN/KIM] in 5 M NaCl.

As an extension of the previous review [95SIL/BID], the present review also discusses experimental studies on curium (III). Systematic investigations by means of TRLFS clearly demonstrated that chloride complexation of the Cm^{3+} ion is negligible even in concentrated NaCl [95FAN/KIM] and hence the formation of ternary hydroxide-chloride complexes can be ruled out as well. Moreover, Fanghänel et al. [94FAN/KIM] reported a TRLFS study on the hydrolysis of curium (III) ranging from very dilute to concentrated NaCl. The new data selection in the present review is based on both the studies already discussed in [95SIL/BID] and the large amount of new experimental data for Am(III) and Cm(III) hydroxide complexes. A significant change between the evaluation in [95SIL/BID] and the present review concerns the inclusion of additional data in NaCl solutions from the study of Stadler and Kim [88STA/KIM]. The literature data used in [95SIL/BID] and in the present review for the evaluation of hydrolysis constants at $I = 0$ are summarised in Table 12-2.

Table 12-2: Literature data for Am(III) and Cm(III) hydroxide complexes used in the previous [95SIL/BID] and present reviews for the evaluation of equilibrium constants at $I = 0$.

An	Medium	$\log_{10} {}^*\beta_1$	$\log_{10} {}^*\beta_2$	$\log_{10} {}^*\beta_3$	Method	Reference
Am	0.1 M NaClO$_4$	−(7.7 ± 0.3) −(6.9 ± 0.6) [a]	−(16.7 ± 0.7) −(15.1 ± 0.6) [a]	−(25.0 ± 0.3)	sol	[82SIL]
Am	0.1 M NaClO$_4$	−(7.5 ± 0.3) −(7.0 ± 0.4) [a]	−(15.4 ± 0.4) −(15.1 ± 0.4) [a]	−(27.0 ± 0.5) −(26.4 ± 0.5) [a]	sol	[88STA/KIM]
Cm	0.1 M NaClO$_4$	−(7.13 ± 0.18)	−(15.54 ± 0.28)		TRLFS	[92WIM/KLE]
Am	1.0 M NaClO$_4$	−(7.5 ± 0.3) −(7.3 ± 0.4) [a]			dis	[82LUN]
Am	1.0 M NaClO$_4$	−(7.03 ± 0.04) −(7.2 ± 0.2) [a]			pot	[82NAI/CHA]
Cm	0.1 M KCl	−(7.7 ± 0.3)			pot	[83EDE/BUC]
Am	0.1 M NaCl	−(7.8 ± 0.4)	−(15.4 ± 0.5)	−(26.6 ± 0.5)	sol	[88STA/KIM]
Am	0.6 M NaCl	−(8.1 ± 0.3) −(8.3 ± 0.3) [b]	−(15.8 ± 0.4) −(16.2 ± 0.4) [b]	−(27.1 ± 0.5) −(27.6 ± 0.5) [b]	sol	[88STA/KIM]
Am	0.7 M NaCl *	−(7.54 ± 0.2) [c]			dis	[83CAC/CHO]
Am	5 M NaCl **	−(7.6 ± 0.6)	−(16.3 ± 0.7)	−(27.1 ± 0.5)	sol	[94RUN/KIM]
Cm	0–6.2 m NaCl				TRLFS	[94FAN/KIM]
	$I = 0$	−(7.6 ± 0.1)	−(15.7 ± 0.2)			
Am	$I = 0$	−(7.1 ± 0.5)	−(14.8 ± 0.5)		review	[92FUG]
Am	$I = 0$	−(6.4 ± 0.7)	−(14.1 ± 0.6)	−(25.7 ± 0.5)	review	[95SIL/BID]
Am+Cm	$I = 0$	−(7.2 ± 0.5)	−(15.1 ± 0.7)	−(26.2 ± 0.5)	present review	

(a) Recalculated in [95SIL/BID].
(b) Recalculated in the present review.
(c) Not used for data selection because of shortcomings discussed in [95SIL/BID].
*, ** All the data are given for $t = 25°C$ except *: $t = 21°C$ and **: $t = 22°C$.

Figure 12-1 and Figure 12-2 show the application of the SIT model to the formation constants of Am(III) and Cm(III) hydroxide complexes in chloride solution. Experimental data in 0.1 M NaClO$_4$ are also included because the differences between the activity coefficients in NaCl and NaClO$_4$ solutions at this low ionic strength are negligible compared to other uncertainties.

Since the experimental uncertainties of the different experimental methods often are considerably larger than differences due to the physico-chemical properties of americium(III) and curium(III), the data for these two actinide elements are treated and weighted together. In order not to give too large weight to the many data points from the study of Fanghänel et al. [94FAN/KIM] as compared to those of the other authors, only the values extrapolated to $I = 0$ from Fanghänel et al. [94FAN/KIM] and the other literature data are used for the evaluation of the average values of $\log_{10} {}^*\beta_1^\circ$ and $\log_{10} {}^*\beta_2^\circ$.

The study of Fanghänel et al. [94FAN/KIM], which covers a wide range of NaCl concentrations, was used to determine $\Delta\varepsilon$ for the reactions ((12.1), $n = 1, 2$) and the ion interaction coefficients of the different An(III) hydroxide complexes.

The experimental equilibrium constants for the first hydroxide complex of Am(III) or Cm(III) are based on the following experimental methods: solubility studies with Am(OH)$_3$(s) [88STA/KIM], [94RUN/KIM], potentiometric titration [83EDE/BUC], solvent extraction [83CAC/CHO], and laser fluorescence spectroscopy [92WIM/KLE], [94FAN/KIM]. For the reaction:

$$An^{3+} + H_2O(l) \rightleftharpoons AnOH^{2+} + H^+, \tag{12.2}$$

where An^{3+} stands for Am^{3+} and Cm^{3+}, the values of $\log_{10} {}^*\beta_1^\circ$ and $\Delta\varepsilon_1$ (Figure 12-1) are:

$$\log_{10} {}^*\beta_1^\circ ((12.2), 298.15 \text{ K}) = -(7.2 \pm 0.5),$$

$$\Delta\varepsilon_1 (12.2) = -(0.15 \pm 0.06) \text{ kg} \cdot \text{mol}^{-1}.$$

The uncertainty within and between the different studies is covered. With $\varepsilon(\text{Am}^{3+}, \text{Cl}^-) = (0.23 \pm 0.02) \text{ kg} \cdot \text{mol}^{-1}$ and $\varepsilon(\text{H}^+, \text{Cl}^-) = (0.12 \pm 0.01) \text{ kg} \cdot \text{mol}^{-1}$, the ion interaction coefficient between AmOH^{2+} and Cl$^-$ is calculated to be:

$$\varepsilon(\text{AmOH}^{2+}, \text{Cl}^-) = -(0.04 \pm 0.07) \text{ kg} \cdot \text{mol}^{-1}.$$

If the $\Delta\varepsilon_1$ value in NaCl solution or $\Delta\varepsilon_1 = (0.04 \pm 0.05) \text{ kg} \cdot \text{mol}^{-1}$ as estimated in [95SIL/BID] for NaClO$_4$ solution, is used to calculate $\log_{10} {}^*\beta_1^\circ$ from the two studies in 1 M NaClO$_4$ (the solvent extraction study of Lundqvist [82LUN] and the potentiometric study of Nair et al. [82NAI/CHA]), the calculated $\log_{10} {}^*\beta_1^\circ$ values are in the range -6.2 to -6.5. Certain effects like sorption or precipitation (cf. discussion in [95SIL/BID]) might have led to overestimated hydrolysis constants in these studies. For these reasons the data in [82LUN], [82NAI/CHA] are disregarded in the present review.

Figure 12-1: Application of the SIT model to literature data in dilute to concentrated NaCl and 0.1 m NaClO$_4$, for the equilibrium: An^{3+} + H$_2$O(l) ⇌ An(OH)$^{2+}$ + H$^+$. The filled and open symbols denote experimental data for Am(III) and Cm(III), respectively. The solid line is calculated with $\log_{10} {}^*\beta_1^\circ = -(7.2 \pm 0.5)$ and $\Delta\varepsilon_1 = -(0.15 \pm 0.06)$ kg · mol^{-1}. The dotted lines show the associated uncertainties (95 % confidence interval).

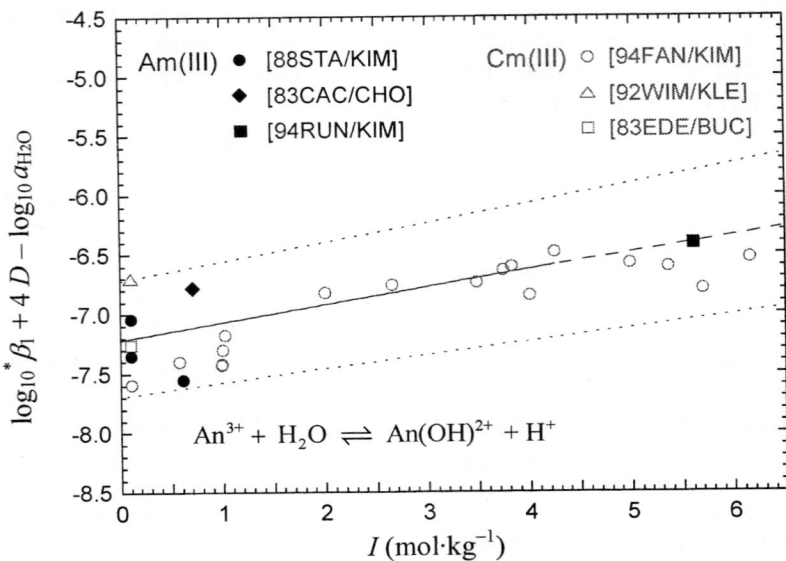

The experimental data for the formation of the second hydroxide complex have large uncertainties. This holds for both solubility studies with Am(OH)$_3$(s), [88STA/KIM], [94RUN/KIM] and fluorescence spectroscopy with Cm(III), [92WIM/KLE], [94FAN/KIM], in the latter case, because of the low solubility at increased pH values and sorption effects.

For the reaction:

$$An^{3+} + 2\,H_2O(l) \rightleftharpoons An(OH)_2^+ + 2\,H^+, \qquad (12.3)$$

the following values of $\log_{10} {}^*\beta_2^\circ$ and $\Delta\varepsilon_2$ are selected (Figure 12-2):

$$\log_{10} {}^*\beta_2^\circ\,((12.3),\,298.15\ \text{K}) = -(15.1 \pm 0.7),$$

$$\Delta\varepsilon_2\,(12.3) = -(0.26 \pm 0.20)\ \text{kg} \cdot \text{mol}^{-1}.$$

The assigned uncertainties cover the whole range of expectancy and the evaluated value of $\Delta\varepsilon_2$ corresponds to $\varepsilon(Am(OH)_2^+, Cl^-) = -(0.27 \pm 0.20)$ kg·mol^{-1}.

Figure 12-2: Application of the SIT model to literature data in dilute to concentrated NaCl and 0.1 m NaClO$_4$, for the equilibrium: $An^{3+} + 2 H_2O(l) \rightleftharpoons An(OH)_2^+ + 2 H^+$. The filled and open symbols denote experimental data for Am(III) and Cm(III), respectively. The solid line is calculated with $\log_{10} {}^*\beta_2^\circ = -(15.1 \pm 0.7)$ and $\Delta\varepsilon_2 = -(0.26 \pm 0.20)$ kg·mol^{-1}. The dotted lines show the associated uncertainties (95 % confidence interval).

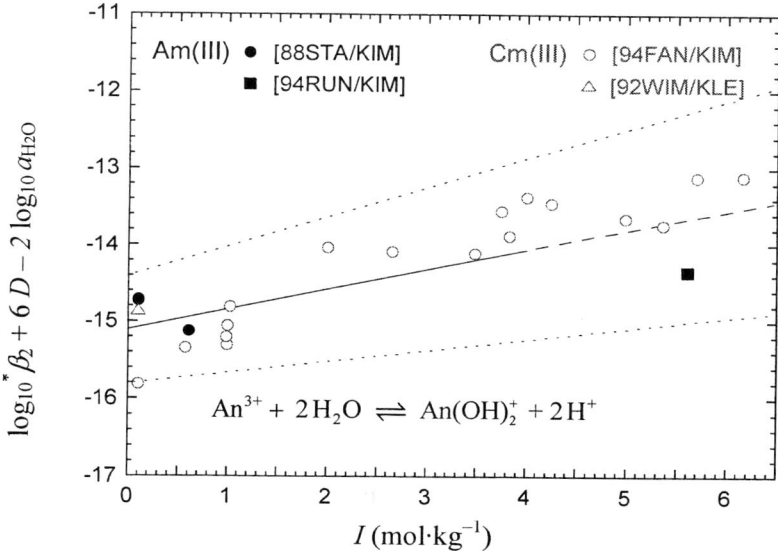

Because of the low solubility of Am(III) and Cm(III) in the alkaline range, the formation of the neutral hydroxide complex, An(OH)$_3$(aq), by reaction (12.4):

$$An^{3+} + 3 H_2O(l) \rightleftharpoons An(OH)_3(aq) + 3 H^+ \quad (12.4)$$

cannot be investigated by spectroscopic or potentiometric methods.

In the present review the equilibrium constant is evaluated from the different sets of solubility data reported by Stadler and Kim [88STA/KIM] (cf. Appendix A). The mean value of the $\log_{10} {}^*\beta_3$ values in 0.1 M NaClO$_4$, 0.1 and 0.6 M NaCl, converted to $I = 0$ with the SIT model, is given by:

$$\log_{10} {}^*\beta_3^\circ ((12.4), 298.15 \text{ K}) = -(26.2 \pm 0.5).$$

The upper limit value of $\log_{10} \beta_3^\circ < 13.4$ ($\log_{10} {}^*\beta_3^\circ < -28.6$) given in [90FEL/RAI], already mentioned in [95SIL/BID], is not accepted in the present review. The value of [90FEL/RAI] was derived from solubility data determined with $AmOHCO_3(cr)$ at pH $12.5 - 13.1$ and carbonate concentrations of 0.07 mol $\cdot$ L^{-1}. With the equilibrium constants selected for $AmOHCO_3(cr)$ and $Am(OH)_3(aq)$ one would expect a solubility of about 10^{-7} mol $\cdot$ L^{-1} at $[OH^-] = 0.1$ mol $\cdot$ L^{-1} and $[CO_3^{2-}] = 0.07$ mol $\cdot$ L^{-1}. However, according to the data selected in [95SIL/BID] and the present review, the initial solid $AmOHCO_3(cr)$ must be expected to convert into $Am(OH)_3(cr)$, which is more stable and has hence a lower solubility under these conditions. This would explain the lower solubilities in the range of $10^{-7.5} - 10^{-9.1}$ mol $\cdot$ L^{-1} measured by [90FEL/RAI].

The equilibrium constants, $\log_{10} {}^*\beta_n^\circ$ ((12.1), 298.15 K), selected in the present review, correspond to the following standard Gibbs energies of formation:

$$\Delta_f G_m^\circ (AmOH^{2+}, 298.15 \text{ K}) = -(794.7 \pm 5.5) \text{ kJ} \cdot \text{mol}^{-1},$$

$$\Delta_f G_m^\circ (Am(OH)_2^+, 298.15 \text{ K}) = -(986.8 \pm 6.2) \text{ kJ} \cdot \text{mol}^{-1},$$

$$\Delta_f G_m^\circ (Am(OH)_3, aq, 298.15 \text{ K}) = -(1160.6 \pm 5.5) \text{ kJ} \cdot \text{mol}^{-1}.$$

As discussed in the previous review [95SIL/BID] the solubility data of Stadler and Kim [88STA/KIM] show that anionic hydroxide complexes, e.g., the reaction:

$$Am(OH)_3(aq) + OH^- \rightleftharpoons Am(OH)_4^- \tag{12.5}$$

are negligible at OH$^-$ concentrations below 0.1 mol $\cdot$ L^{-1}.

12.3.1.2 Aqueous Am(V) hydroxide complexes (V.3.1.2)

In the previous NEA review, [95SIL/BID], no thermodynamic data were selected for Am(V) hydroxide complexes. The only experimental data available, formation constants for the complex $AmO_2OH(aq)$, which were derived from solubility studies in 5 M NaCl ($\log_{10} \beta_1 = (1.5 \pm 0.5)$, [85MAG/CAR]) and 3 M NaCl ($\log_{10} \beta_1 = (1.7 \pm 0.6)$, [88STA/KIM], [88STA/KIM2]), were not accepted because of experimental shortcomings [95SIL/BID].

More reliable data were reported by Runde and Kim [94RUN/KIM], who measured the solubility of amorphous $AmO_2OH(am)$ over a wide range of H$^+$ concentrations in 5 M NaCl ($8.0 < -\log_{10}[H^+] < 13.5$) at 295.15 K:

$$\log_{10} \beta_1 (AmO_2OH, aq, 5 \text{ M NaCl}) = (3.62 \pm 0.27),$$

$$\log_{10} \beta_2 (AmO_2(OH)_2^-, 5 \text{ M NaCl}) = (5.89 \pm 0.22).$$

An analogous solubility study with an aged precipitate of Np(V) hydroxide in the same medium led to almost the same formation constants for the analogous Np(V) hydroxide complexes:

$$\log_{10} \beta_1 (NpO_2OH, aq, 5 \text{ M NaCl}) = (3.66 \pm 0.22) \qquad [94RUN/KIM],$$

$$\log_{10} \beta_2 (\text{NpO}_2(\text{OH})_2^-, 5 \text{ M NaCl}) = (5.98 \pm 0.19) \qquad \text{[94RUN/KIM]}.$$

There are no experimental data for Am(V) hydroxide complexes at low ionic strength and the application of the SIT model to convert the formation constants obtained by [94RUN/KIM] in 5.6 m NaCl to $I = 0$ has a large uncertainty (*cf.* Appendix A). However, as the formation constants of Am(V) and Np(V) hydroxide complexes are the same within uncertainty limits [94RUN/KIM], [96RUN/NEU], the present review recommends the formation constants of the aqueous Np(V) hydroxide complexes selected in [2001LEM/FUG]:

$$\log_{10} {}^*\beta_1^\circ ((12.6), 298.15 \text{ K}) = -(11.3 \pm 0.7), \qquad \text{[2001LEM/FUG]}$$

$$\log_{10} {}^*\beta_2^\circ ((12.6), 298.15 \text{ K}) = -(23.6 \pm 0.5), \qquad \text{[2001LEM/FUG]}$$

for the reaction:

$$\text{AnO}_2^+ + n\,\text{H}_2\text{O(l)} \rightleftharpoons \text{AnO}_2(\text{OH})_n^{1-n} + n\,\text{H}^+, \qquad (12.6)$$

as reasonable estimates for the analogous Am(V) hydroxide complexes. The constants for the Np(V) hydroxide complexes are based on numerous experimental data at different ionic strengths.

12.3.2 Solid americium oxides and hydroxides (V.3.2)

12.3.2.1 Americium oxides

As discussed in section 12.1.1, the selected values of the standard entropies of $\text{Am}_2\text{O}_3(\text{cr})$ and $\text{AmO}_2(\text{cr})$ estimated by Konings [2001KON], [2001KON2], but with increased uncertainties,

$$S_m^\circ (\text{Am}_2\text{O}_3, \text{cr}, 298.15 \text{ K}) = (133.6 \pm 6.0) \text{ J} \cdot \text{K}^{-1} \cdot \text{mol}^{-1},$$

$$S_m^\circ (\text{AmO}_2, \text{cr}, 298.15 \text{ K}) = (77.7 \pm 10.0) \text{ J} \cdot \text{K}^{-1} \cdot \text{mol}^{-1},$$

have been preferred to earlier values. These lead to revised Gibbs energies of formation:

$$\Delta_f G_m^\circ (\text{Am}_2\text{O}_3, \text{cr}, 298.15 \text{ K}) = -(1605.4 \pm 8.3) \text{ kJ} \cdot \text{mol}^{-1},$$

$$\Delta_f G_m^\circ (\text{AmO}_2, \text{cr}, 298.15 \text{ K}) = -(877.7 \pm 4.3) \text{ kJ} \cdot \text{mol}^{-1}.$$

12.3.2.2 Solid Am(III) hydroxides (V.3.2.4)

As outlined in the previous NEA TDB review [95SIL/BID], the thermodynamic properties and hence the solubility of $\text{Am}(\text{OH})_3(\text{s})$ depends on the degree of crystallinity, which can vary with the time of aging. Another important factor is the particle size, which is also affected by aging processes and by self–irradiation from the α–activity of americium, particularly in studies with $^{241}\text{Am}(\text{OH})_3(\text{s})$. The available literature data on the solubility constant of Am(III) hydroxides are summarised in Table 12-3. For the reasons already pointed out in the review of Silva *et al.* [95SIL/BID], the data in

[84BER/KIM], [89PAZ/KOC], [90PER/SAP] are disregarded and not included in the present discussion.

Table 12-3: Solubility constants for Am(III) hydroxides and conversion to $I = 0$, with the SIT coefficients in Appendix B.

Medium	t (°C)	Solid	$\log_{10}{}^*K_{s,0}$	$\log_{10}{}^*K^\circ_{s,0}$	Reference
dilute solutions (I = 0.005 M)	22	^{243}Am(OH)$_3$(am) ^{241}Am(OH)$_3$(am)	(17.6 ± 0.5)[b]	(17.5 ± 0.3) (17.0 ± 0.5)[a] (17.4 ± 0.5)[b]	[83RAI/STR]
0.1 M NaClO$_4$	25	^{243}Am(OH)$_3$(am)	(17.5 ± 0.3) (17.5 ± 0.5)[b]	(16.8 ± 0.5)[b]	[83EDE/BUC]
0.1 M NaClO$_4$	25	^{243}Am(OH)$_3$(am)	(17.3 ± 0.5)[b]	(16.6 ± 0.5)[b]	[85NIT/EDE]
0.1 M NaClO$_4$	25	^{243}Am(OH)$_3$(cr)	(16.6 ± 0.4) (15.9 ± 0.6)[a] (16.3 ± 0.6)[b]	(15.9 ± 0.4) (15.2 ± 0.6)[a] (15.6 ± 0.6)[b]	[82SIL]
0.1 M NaClO$_4$ (≤ 3.7 GBq·L^{-1}) 0.1 M NaClO$_4$ (44 – 185 GBq·L^{-1})	25 25	^{241}Am(OH)$_3$(s) (aged)	(15.7 ± 0.3)[c] (15.5 ± 0.6)[a] (16.4 ± 0.3)[c]	(15.0 ± 0.3) (14.8 ± 0.6)[a] (15.7 ± 0.3)	[88STA/KIM] [88STA/KIM2]
0.1 M NaCl (74 – 185 GBq·L^{-1})	25		(16.3 ± 0.5)[c]	(15.6 ± 0.5)	
0.6 M NaCl (74 – 185 GBq·L^{-1})	25		(16.7 ± 0.2)[c]	(15.5 ± 0.2)	
5 M NaCl	22	^{241}Am(OH)$_3$(am)	(17.9 ± 0.4)	(15.2 ± 0.7)[d]	[94RUN/KIM]

(a) Recalculated in [95SIL/BID].
(b) Recalculated on the basis of the hydrolysis constants selected in the present review.
(c) Original data ($\log_{10} K_{s,0}$) converted to $\log_{10}{}^*K_{s,0}$ with NEA TDB auxiliary data for the ion product of water and corrected for the conversion of pH$_{exp}$ to $-\log_{10}[H^+]$ (cf. Appendix A, discussion of [88STA/KIM]).
(d) At this high NaCl concentration, the SIT extrapolation can become inaccurate. Conversion to $I = 0$ with the Pitzer parameters in [94FAN/KIM], [97KON/FAN] leads to $\log_{10}{}^*K^\circ_{s,0} = (14.6 ± 0.5)$.

The solubility study of Silva [82SIL] is the only one performed with a crystalline Am(III) hydroxide characterised by X–ray diffraction. The damage by α–radiation was diminished by the use of ^{243}Am. Concordant solubilities were measured by Stadler and Kim [88STA/KIM] in 0.1 M NaClO$_4$, 0.1 M NaCl and 0.6 M NaCl at rather high specific α–activities of ^{241}Am (cf. Table 12-3 and Figure 12-3). The solubility constant derived from a series of experiments at lower specific α–activity is about 0.6 log$_{10}$ units lower. However, the solids used by [88STA/KIM] were not characterised by X–ray diffraction. In a later study, Runde and Kim [94RUN/KIM] investigated the solubility of ^{241}Am(OH)$_3$(s) in 5 M NaCl. The precipitate was amorphous to X–rays, after an aging period of four months. Despite this, the solubility constant at zero ionic strength is within the range of uncertainty of the values from [82SIL] and [88STA/KIM].

In the previous review [95SIL/BID], the experimental data of Silva [82SIL] on Am(OH)$_3$(cr) characterised by X–ray diffraction were used to calculate,

$$\log_{10}{}^*K_{s,0}^\circ \,(\text{Am(OH)}_3, \text{cr}, 298.15 \text{ K}) = (15.2 \pm 0.6),$$

for the reaction:

$$\text{Am(OH)}_3(\text{s}) + 3\,\text{H}^+ \rightleftharpoons \text{Am}^{3+} + 3\,\text{H}_2\text{O(l)} \tag{12.7}$$

and

$$\Delta_f G_m^\circ (\text{Am(OH)}_3, \text{cr}, 298.15 \text{ K}) = -(1223.4 \pm 5.9) \text{ kJ} \cdot \text{mol}^{-1}.$$

However, as all data in [82SIL] were measured at $-\log_{10}[\text{H}^+] > 7$, the calculated solubility constant depends on the selected formation constants of the Am(III) hydroxide complexes, in particular on the value of $\log_{10}{}^*\beta_1$. Applying the hydrolysis constants selected in the present review, the solubility data of Silva [82SIL] in the range $-\log_{10}[\text{H}^+] = 7$ to 7.5, gives $\log_{10}{}^*K_{s,0} = (16.3 \pm 0.6)$ in 0.1 M NaClO$_4$. Extrapolation to zero ionic strength with the SIT equation leads to:

$$\log_{10}{}^*K_{s,0}^\circ \,(\text{Am(OH)}_3, \text{cr}, 298.15 \text{ K}) = (15.6 \pm 0.6)$$

and

$$\Delta_f G_m^\circ (\text{Am(OH)}_3, \text{cr}, 298.15 \text{ K}) = -(1221.1 \pm 5.9) \text{ kJ} \cdot \text{mol}^{-1}.$$

The experimental data reported in [82SIL], [88STA/KIM] and the solubility calculated with the selected equilibrium constants are shown in Figure 12-3.

Rai et al. [83RAI/STR] determined the solubility of X–ray amorphous Am(OH)$_3$(am) precipitates in dilute solutions containing $1.5 \cdot 10^{-3}$ M CaCl$_2$, but in the absence of additional background electrolyte. A comparable solubility value was determined by Nitsche and Edelstein [85NIT/EDE] at pH = (7.0 ± 0.1) in 0.1 M NaClO$_4$. Their precipitate was also X–ray amorphous. Edelstein et al. [83EDE/BUC] reported similar results for Am(OH)$_3$(am) in 0.1 M NaClO$_4$. The experimental data given in these papers are shown in Figure 12-4.

Based on the high solubility measured in [83RAI/STR] at pH < 7.5, Silva et al. [95SIL/BID] selected a solubility constant of $\log_{10}{}^*K_{s,0}^\circ (\text{Am(OH)}_3, \text{am}, 298.15 \text{ K}) = (17.0 \pm 0.6)$ for amorphous Am(III) hydroxide. In the present review, the $\log_{10}{}^*K_{s,0}^\circ$ values from the data at pH < 7.5 in [83RAI/STR] and the data at pH 7 – 8 in [83EDE/BUC], [85NIT/EDE] are re-calculated using the re-evaluated hydrolysis constants values (cf. Table 12-3). The mean value is selected:

$$\log_{10}{}^*K_{s,0}^\circ \,((12.7), \text{Am(OH)}_3, \text{am}, 298.15 \text{ K}) = (16.9 \pm 0.8).$$

Figure 12-3: Solubility measurements of crystalline Am(III) hydroxide [82SIL] and aged Am(OH)$_3$(s) [88STA/KIM] in 0.1 M NaClO$_4$ and NaCl at 298.15 K. The solid curve is calculated with $\log_{10}{}^*K^\circ_{s,0}$ (Am(OH)$_3$, cr) = (15.6 ± 0.6) and the hydrolysis constants selected in the present review, corrected to I = 0.1 mol · L^{-1}. The dotted lines show the associated uncertainty range (95% confidence interval). The dashed line is calculated with the constants selected in the previous review [95SIL/BID].

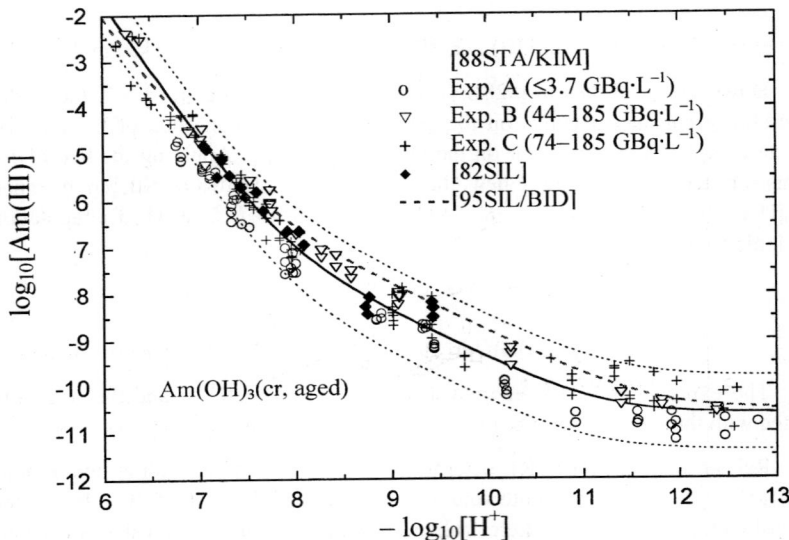

Figure 12-4: Solubility measurements of amorphous Am(III) hydroxide in pH–adjusted dilute solutions [83RAI/STR] and 0.1 M NaClO$_4$ [83EDE/BUC], [85NIT/EDE]. The solid curve is calculated for Am(OH)$_3$(am) in 0.1 M NaClO$_4$, with the constants selected in the present review and the dotted lines show the associated uncertainty. The solubility of crystalline (aged) Am(III) hydroxide is shown for comparison as the lower solid line. The dashed line is based on the constants selected in the previous review [95SIL/BID] and refers to the low ionic strength in the study of Rai *et al.* [83RAI/STR].

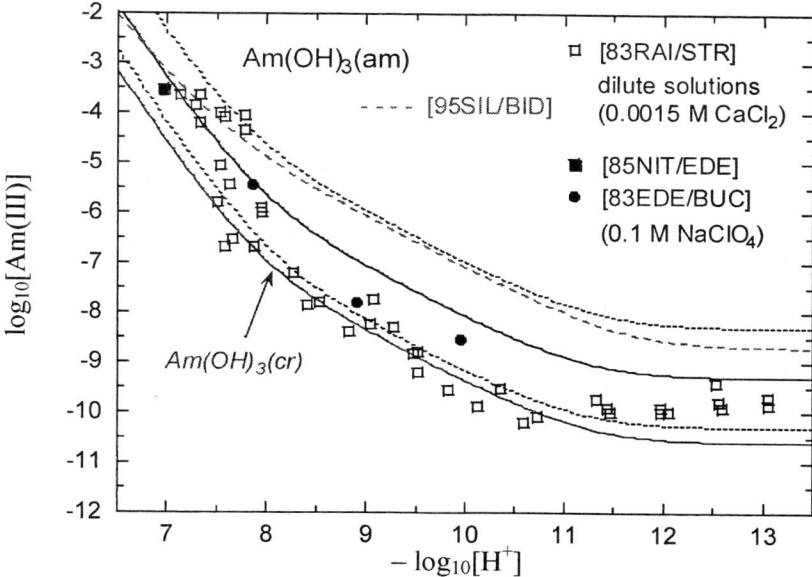

The use of the notation "crystalline" and "amorphous" to describe a solid phase might indicate an oversimplified model. The X–ray data give information on the bulk structure, while the solubility is determined by the surface characteristics. These are not necessarily identical. However, the data for "crystalline" and "amorphous" Am(OH)$_3$(s) selected in this review may indicate the magnitude of the effect of an incomplete knowledge of the surface state of the solid.

This is confirmed by the solubility data of Rai et al. [83RAI/STR] (cf. Figure 12-4) that do not correlate with the X–rays investigations carried out in parallel. All the investigated fresh or aged precipitates at pH < 9.6 were found amorphous to X–rays. However, the solubility in the pH range 8 – 11 is comparable with that calculated for Am(OH)$_3$(cr) (lower solid line in Figure 12-4). Only some data at pH 7 – 8 show a markedly increased solubility. On the other hand, Rai et al. [83RAI/STR] observed the appearance of X–ray diffraction peaks for aged precipitates at pH = 13, while the measured solubility remained in the range of the value for Am(OH)$_3$(am) (upper solid line in Figure 12-4).

In numerous studies Am(III) solubilities at pH = 11 – 13 were found in the range [Am(III)] = $10^{-10} - 10^{-11}$ mol · L^{-1} (Table 12-4), independent of the α–activity and the crystallinity of the bulk solid. Combining $\log_{10} {}^*\beta_3^\circ$ ((12.4), 298.15 K) = – (26.2 ± 0.5) with the solubility constants for amorphous and crystalline or aged Am(OH)$_3$(s), respectively, the equilibrium constant for the reaction:

$$Am(OH)_3(s) \rightleftharpoons Am(OH)_3(aq) \tag{12.8}$$

is given by:

$\log_{10} K_{s,3}^\circ$ ((12.8), 298.15 K) = – (9.3 ± 1.0) for Am(OH)$_3$(am),

$\log_{10} K_{s,3}^\circ$ ((12.8), 298.15 K) = – (10.6 ± 0.8) for Am(OH)$_3$(cr).

The solubilities reported in [86EWA/HOW], [84BER/KIM], [88STA/KIM], [91VIT/TRA] for alkaline solutions of low or moderate ionic strength (Table 12-4), are concordant with the value calculated for crystalline (or aged) Am(OH)$_3$(s). The data of Rai et al. ($\log_{10}$[Am(III)] = – (9.8 ± 0.3), [83RAI/STR]) are closer to the value calculated for Am(OH)$_3$(am). The increased solubility measured by Runde and Kim [94RUN/KIM] in alkaline 5 M NaCl solutions with an X–ray amorphous hydroxide, $\log_{10}$[Am(OH)$_3$(aq)] = – (9.3 ± 0.3), is in accord with the value calculated for Am(OH)$_3$(am). Alternatively, it could be interpreted by assuming the presence of aged Am(OH)$_3$(s) and an ionic strength dependence of:

$\Delta\varepsilon(12.8) = \varepsilon(Am(OH)_3(aq), Na^+) + \varepsilon(Am(OH)_3(aq), Cl^-) = -(0.23 \pm 0.15)$ kg · mol^{-1}.

However, as the nature of the Am(OH)$_3$(s) in [94RUN/KIM] is not definitely known, the ionic strength dependence of reaction (12.8) requires further experimental investigation.

12.3 Americium oxygen and hydrogen compounds and complexes (V.3)

Table 12-4: Literature data for the solubility of Am(III) in alkaline solution (pH = 11 to 13).

Medium	$t(°C)$	Solid	$\log_{10}[\text{Am(III)}]$	Reference
dilute NaOH	22	$^{243}\text{Am(OH)}_3\text{(am)}/^{241}\text{Am(OH)}_3\text{(am)}$	$-(9.8 \pm 0.3)$	[83RAI/STR]
0.1 M Na(ClO$_4$/OH)	25	$^{241}\text{Am(OH)}_3\text{(s)}$	$-(10.9 \pm 0.2)$	[84BER/KIM]
0.1 M Na(ClO$_4$/OH) (≤ 3.7 GBq·L^{-1})	25	$^{241}\text{Am(OH)}_3\text{(s)}$	$-(10.9 \pm 0.4)$	[88STA/KIM] [88STA/KIM2]
0.1 M Na(ClO$_4$/OH) (44 – 185 GBq·L^{-1})	25		$-(10.6 \pm 0.3)$	
0.1 M Na(Cl/OH) (74 – 185 GBq·L^{-1})	25		$-(10.3 \pm 0.5)$	
0.6 M Na(Cl/OH) (74 – 185 GBq·L^{-1})	25		$-(10.9 \pm 0.6)$	
water equilibrated with concrete + NaOH + $3 \cdot 10^{-5}$ M CO$_3^{2-}$	r.t.?	?	$-(10.5 \pm 0.6)$	[86EWA/HOW]
KOH + Ca(OH)$_2$(s) cement leachates	r.t.?	$^{241}\text{Am(OH)}_3\text{(s)}$?	-11.1	[91VIT/TRA]
3 M Na(ClO$_4$/OH)	25	$^{241}\text{Am(OH)}_3\text{(s)}$	$-(5.7 \pm 0.2)^{(a)}$	[89PAZ/KOC]
5 M Na(Cl/OH)	22	$^{241}\text{Am(OH)}_3\text{(s)}$	$-(9.3 \pm 0.3)$	[94RUN/KIM]

(a) Solutions were centrifuged but not filtered. As discussed in [95SIL/BID], the high americium concentration is probably caused by the presence of colloidal particles.

r.t. room temperature.

Based on calorimetric measurements of the enthalpy of solution for crystalline ^{243}Am(OH)$_3$(cr) in 6 M HCl, and an entropy value of $S_m^°$(Am(OH)$_3$, cr, 298.15 K) = (129 ± 10) J·K^{-1}·mol^{-1}, estimated by analogy with rare earth hydroxides, Morss and Williams [94MOR/WIL] calculated $\log_{10}{}^* K_{s,0}^°$(Am(OH)$_3$, cr, 298.15 K) = (12.5 ± 1.6). This solubility constant is three orders of magnitude lower than that derived from the solubility data of Silva [82SIL] for ^{243}Am(OH)$_3$(cr). However, as discussed in Appendix A, the results from different samples of solids are not consistent within the uncertainty limits:

[94MOR/WIL] Batch A:

$\Delta_{sol}H_m$ (Am(OH)$_3$, cr, 298.15 K, 6 M HCl) = $-(112.6 \pm 2.7)$ kJ·mol^{-1},

$\Delta_f H_m^°$ (Am(OH)$_3$, cr, 298.15 K) = $-(1360.4 \pm 3.0)$ kJ·mol^{-1}.

[94MOR/WIL] Batch B:

$\Delta_{sol}H_m$ (Am(OH)$_3$, cr, 298.15 K, 6 M HCl) = $-$ (101.9 ± 5.1) kJ · mol^{-1},

$\Delta_f H_m^\circ$ (Am(OH)$_3$, cr, 298.15 K) = $-$ (1371.1 ± 5.3) kJ · mol^{-1}.

These discrepancies indicate that there are chemical differences in the solid samples, which possibly do not represent homogeneous Am(OH)$_3$(cr).

In an analogous study Merli et al. [97MER/LAM] determined the enthalpy of solution for ^{241}Am(OH)$_3$(cr) in 6 M HCl. Their experimental data differ considerably from those in [94MOR/WIL] as they obtain:

$\Delta_{sol}H_m$ (Am(OH)$_3$, cr, 298.15 K, 6 M HCl) = $-$ (129.4 ± 1.1) kJ · mol^{-1},

$\Delta_f H_m^\circ$ (Am(OH)$_3$, cr, 298.15 K) = $-$ (1343.6 ± 1.8) kJ · mol^{-1}.

Merli et al. [97MER/LAM] used a similar entropy estimate of S_m° (Am(OH)$_3$, cr, 298.15 K) = (126 ± 8) J · K^{-1} · mol^{-1} and calculated a solubility constant of $\log_{10}{}^*K_{s,0}^\circ$ (Am(OH)$_3$, cr, 298.15 K) = (16.8 ± 1.0) for crystalline americium(III) hydroxide. The entropy estimate used by [97MER/LAM] was obtained from S_m° (Eu(OH)$_3$, cr, 298.15 K) = (119.9 ± 0.2) J · K^{-1} · mol^{-1}, experimentally determined by Chirico and Westrum [80CHI/WES], by adding a correction of 5.7 J · K^{-1} · mol^{-1} for the greater mass of americium. However, in trivalent europium, but not in americium, the $^7F_1$ and $^7F_2$ electronic levels contribute to the entropy. For americium compounds, therefore, this contribution of 9.5 J · K^{-1} · mol^{-1} should be deduced from the Eu-based values (cf. Konings [2001KON]). Inclusion of this additional correction leads to:

S_m° (Am(OH)$_3$, cr, 298.15 K) = (116 ± 8) J · K^{-1} · mol^{-1}.

With this value the calorimetric results of [97MER/LAM] lead to the solubility constant:

$\log_{10}{}^*K_{s,0}^\circ$ (Am(OH)$_3$, cr, 298.15 K) = (17.3 ± 1.0).

This latter value is 1.7 $\log_{10}$ units larger than $\log_{10}{}^*K_{s,0}^\circ$ = (15.6 ± 0.6) calculated from the solubility data reported in [82SIL] for ^{243}Am(OH)$_3$(cr), but closer to $\log_{10}{}^*K_{s,0}^\circ$ = (16.9 ± 0.8) selected for Am(OH)$_3$(am). This is a rather unexpected result, because calorimetric data determined with well-crystallised solids usually lead to lower solubility constants than experimental solubility data. The reason for this discrepancy is not obvious, because the solid phase characterisation in [97MER/LAM] indicates that the solid phase is crystalline, not amorphous as a result of α-radiation damage.

Similar observations are reported for analogous lanthanide hydroxides Ln(OH)$_3$(cr) with Ln = Pr, Nd, Sm, Eu [98DIA/RAG]. The solubility constants calculated from thermochemical data for the crystalline hydroxides of these lanthanides exceed the values from solubility measurements by 1 - 3 orders of magnitude. As pointed out in [97MER/LAM] and [98DIA/TAG], particularly for Nd(OH)$_3$(cr), the entropy and enthalpy of solution are well determined and there is good agreement between experimental data from different laboratories. Diakonov et al. [98DIA/RAG] concluded that most likely the discrepancies between $\log_{10}{}^{*} K_{s,0}^{\circ}$ values from solubility and thermochemical data arise from uncertainties or erroneous data for the standard enthalpies and entropies of the aqueous ions. However, the present review is convinced that there is no reason to cast doubt on:

$$\Delta_f H_m^{\circ} (\text{Am}^{3+}, 298.15 \text{ K}) = -(616.7 \pm 1.5) \text{ kJ} \cdot \text{mol}^{-1}$$

and

$$S_m^{\circ} (\text{Am}^{3+}, 298.15 \text{ K}) = -(201 \pm 15) \text{ J} \cdot \text{K}^{-1} \cdot \text{mol}^{-1},$$

selected in [95SIL/BID]. The standard molar entropy of Am^{3+} selected in [95SIL/BID] is based on the experimental value of Pu^{3+} taking into account the difference in electronic configuration and a parallel correlation for the lanthanide aqueous ions.

Combining $\log_{10}{}^{*} K_{s,0}^{\circ} (\text{Am(OH)}_3, \text{cr}, 298.15 \text{ K}) = (15.6 \pm 0.6)$ and $\Delta_r G_m^{\circ} ((12.7)\ 298.15 \text{ K}) = -(89.045 \pm 3.424) \text{ kJ} \cdot \text{mol}^{-1}$ selected from the solubility measurements in [82SIL] with $\Delta_f H_m^{\circ} (\text{Am(OH)}_3, \text{cr}, 298.15 \text{ K}) = -(1343.6 \pm 1.8)$ kJ·mol^{-1} determined in [97MER/LAM] and $S_m^{\circ} (\text{Am}^{3+}, 298.15 \text{ K}) = -(201 \pm 15)$ J·K^{-1}·mol^{-1} as adopted in [95SIL/BID], the standard molar entropy of crystalline Am(III) hydroxide would be: $S_m^{\circ} (\text{Am(OH)}_3, \text{cr}, 298.15 \text{ K}) = (149 \pm 20)$ J·K^{-1}·mol^{-1}. The uncertainty limits of this value and the estimate used above, $S_m^{\circ} (\text{Am(OH)}_3, \text{cr}, 298.15 \text{ K}) = (116 \pm 8)$ J·K^{-1}·mol^{-1}, do not overlap. Moreover, an entropy of $S_m^{\circ} (\text{Am(OH)}_3, \text{cr}, 298.15 \text{ K}) = (149 \pm 20)$ J·K^{-1}·mol^{-1} is not compatible with the standard molar entropy of crystalline Am(III) hydroxycarbonate, $S_m^{\circ} (\text{AmOHCO}_3 \cdot 0.5\text{H}_2\text{O}, \text{cr}, 298.15 \text{ K}) = (144 \pm 10)$ J·K^{-1}·mol^{-1} or (141 ± 21) J·K^{-1}·mol^{-1} as re-calculated for reasons of internal consistency from $\Delta_f G_m^{\circ} (\text{AmOHCO}_3 \cdot 0.5\text{H}_2\text{O}, \text{cr}, 298.15 \text{ K}) = -(1530.2 \pm 5.6)$ kJ·mol^{-1} selected in this review (cf. section 12.6.1.1.3.1). In contrast to the good agreement between thermochemical and solubility data for $\text{AmOHCO}_3 \cdot 0.5\text{H}_2\text{O}(\text{cr})$, the data for the crystalline Am(III) hydroxide are conflicting. Accepting the selected value:

$$\Delta_f G_m^{\circ} (\text{Am(OH)}_3, \text{cr}, 298.15 \text{ K}) = -(1221.1 \pm 5.9) \text{ kJ} \cdot \text{mol}^{-1},$$

derived from the solubility study in [82SIL] and the entropy estimate selected in the present review:

$$S_m^{\circ} (\text{Am(OH)}_3, \text{cr}, 298.15 \text{ K}) = (116 \pm 8) \text{ J} \cdot \text{K}^{-1} \cdot \text{mol}^{-1},$$

the standard molar enthalpy of crystalline Am(III) hydroxide is calculated to be:

$$\Delta_f H_m^{\circ} (\text{Am(OH)}_3, \text{cr}, 298.15 \text{ K}) = -(1353.2 \pm 6.4) \text{ kJ} \cdot \text{mol}^{-1}.$$

This value is selected, but it is in disagreement with $\Delta_f H_m^{\circ} (\text{Am(OH)}_3, \text{cr}, 298.15 \text{ K}) = -(1343.6 \pm 1.8)$ kJ·mol^{-1} determined in [97MER/LAM]. However, the available solubility and thermochemical data do not allow a better selection of $\Delta_f H_m^{\circ} (\text{Am(OH)}_3, \text{cr}, 298.15 \text{ K})$.

12.3.2.3 Solid Am(V) hydroxides (V.3.2.5)

There are only a few studies of the solubility of Am(V) precipitates in near neutral and alkaline solutions. The solubility constant for the reaction:

$$AmO_2OH(am) \rightleftharpoons AmO_2^+ + OH^- \qquad (12.9)$$

was reported to be $\log_{10} K_{s,0}$ ((12.9), 298.15 K) $= -(9.3 \pm 0.5)$ in both 5 M NaCl [85MAG/CAR] and 3 M NaCl [88STA/KIM], [88STA/KIM2]. The solid phase was not characterised in these studies. It was already pointed out in the previous review [95SIL/BID] that since it is not clear whether the pH measurements in these papers were corrected for the liquid junction potential, the data reported in [85MAG/CAR], [88STA/KIM], [88STA/KIM2] cannot be used for the selection of thermodynamic data. In a less ambiguous later study, Runde and Kim [94RUN/KIM] investigated the solubility of an X–ray amorphous ^{241}Am(V) hydroxide precipitate at 295.15 K in 5 M NaCl. From the solubility data in the range $8.0 < -\log_{10}[H^+] < 9.5$, the solubility constant was calculated to be: $\log_{10} K_{s,0}$ (12.9) $= -(8.94 \pm 0.42)$. This value is consistent with an analogous value reported in [96ROB/SIL] for NpO$_2$OH(am) and with the thermodynamic data selected in [2001LEM/FUG] for fresh precipitates of amorphous Np(V) hydroxide (cf. Figure 12-5). Accordingly, the solubility constants for NpO$_2$OH(am) and AmO$_2$OH(am) are the same within the range of uncertainty.

Neglecting the slight difference in temperature between the standard state and 22°C in the study of [94RUN/KIM] and increasing the uncertainty to ± 0.5 $\log_{10}$ units, because there are no Am(V) data at lower ionic strength for comparison, the present review selects:

$$\log_{10}{}^* K_{s,0}^\circ (AmO_2OH, am, 298.15 \text{ K}) = (5.3 \pm 0.5),$$

for the reaction:

$$AmO_2OH(am) + H^+ \rightleftharpoons AmO_2^+ + H_2O(l). \qquad (12.10)$$

For the ternary Am(V) hydroxides $MAmO_2(OH)_2 \cdot x H_2O(cr)$, M = Li, Na, K, and $M_2AmO_2(OH)_3 \cdot x H_2O(cr)$, M = Na, K prepared by Tananaev [90TAN] and characterised by X–ray diffraction, no thermodynamic data are available. Peretrukhin et al. [96PER/KRY] measured the solubility of $Na_2AmO_2(OH)_3 \cdot x H_2O(s)$ in 1.6 – 11.7 M NaOH, but because of the shortcomings and ambiguities discussed in Appendix A, the present review does not derive thermodynamic data from this source. Additional data on the behaviour of Am(V) hydroxides are given in [96KUL/MAL].

Figure 12-5: Solubility constants of amorphous Np(V) and Am(V) hydroxides in 0.3 – 5.6 m NaCl solution (for the reaction: $AnO_2OH(am) \rightleftharpoons AnO_2^+ + OH^-$). The data in [96ROB/SIL] for Np(V) and in [94RUN/KIM] for Am(V) are in good agreement with those calculated (solid line) for fresh amorphous Np(V) hydroxide (dotted lines show the associated uncertainty) with $\log_{10} K_{s,0}^\circ$ (NpO$_2$OH, am, 298.15 K) = – (8.7 ± 0.2) and $\epsilon(NpO_2^+, Cl^-) = (0.09 \pm 0.05)$ kg · mol^{-1} [2001LEM/FUG] and $\epsilon(Na^+, OH^-) = (0.04 \pm 0.01)$ kg · mol^{-1} [95SIL/BID]. For the reasons given in the previous review [95SIL/BID], the solubility constants from [85MAG/CAR], [88STA/KIM] are not used for the selection of thermodynamic data.

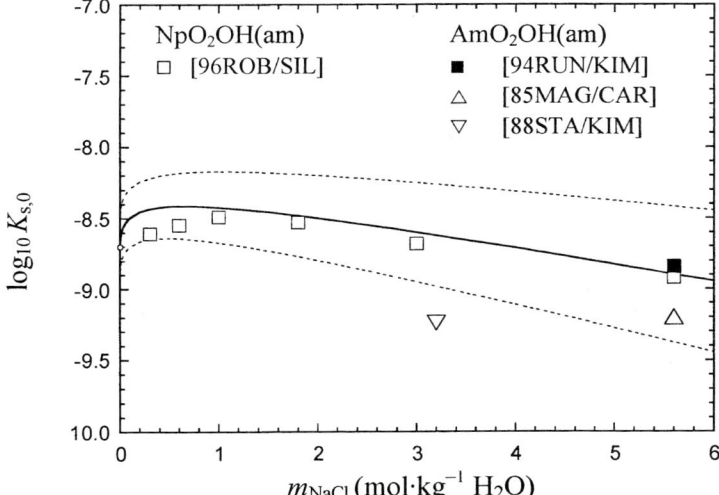

12.4 Americium group 17 (halogen) compounds and complexes (V.4)

12.4.1 Aqueous group 17 (halogen) complexes (V.4.1)

12.4.1.1 Aqueous Am(III) fluorides (V.4.1.1)

Few additional studies on Am(III) fluoride complexation have been published since the previous review [95SIL/BID]. For reasons given earlier this review has also considered experimental studies on the complexation of Cm(III) with fluoride. The experimental data for the reactions:

$$An^{3+} + nF^- \rightleftharpoons AnF_n^{(3-n)}, \qquad (12.11)$$

with An = Am(III), Cm(III) are summarised in Table 12-5.

Suganuma et al. [97SUG/SAT], [97SUG/SAT2] determined formation constants of AmF^{2+} in water and mixed water/solvent systems using a solvent extraction technique. This review considers only the complexation constant in pure aqueous solutions. In both papers the same value is presented for the formation constant of AmF^{2+}: $\log_{10} \beta_1 = (2.51 \pm 0.06)$ at a constant ionic strength of 0.1 mol·L^{-1}, corresponding to $\log_{10} \beta_1^\circ = (3.15 \pm 0.1)$ when converted to $I = 0$ with the SIT. This value agrees fairly well with the selected data of the previous review [95SIL/BID], $\log_{10} \beta_1^\circ = (3.4 \pm 0.4)$.

A compilation of literature data is reported in [97CHA/SAW], but no new experimental data on the fluoride complexation of Am(III) and Cm(III) are presented.

Experimental data for the formation of Cm(III) fluoride complexes have been published by Feay [54FEA], Aziz and Lyle [69AZI/LYL], Degischer and Choppin [75DEG/CHO], and Choppin and Unrein [76CHO/UNR]; these papers were discussed in the previous review. For the reasons given in [95SIL/BID], the results of [54FEA], [69AZI/LYL], [75DEG/CHO] are not used for the evaluation of thermodynamic data.

Recently the fluoride complexation of Cm(III) was studied by TRLFS in 0 – 5 mol·kg^{-1} NaCl solutions [99AAS/STE]. The data were extrapolated to zero ionic strength using the Pitzer model to give:

$$\log_{10} \beta_1^\circ (CmF^{2+}, 298.15\ K) = (3.44 \pm 0.05).$$

This value is in excellent agreement with the formation constant at $I = 0$ selected by [95SIL/BID] for AmF^{2+}:

$$\log_{10} \beta_1^\circ (AmF^{2+}, 298.15\ K) = (3.4 \pm 0.4).$$

In addition to the experimental data for AmF^{2+} of Choppin and Unrein [76CHO/UNR], and Nash and Cleveland [84NAS/CLE], which were selected by the previous review, the data determined by Suganuma et al. [97SUG/SAT], [97SUG/SAT2], Choppin and Unrein, and Aas et al. [99AAS/STE] for CmF^{2+} are used to calculate the thermodynamic formation constant of the monofluoro complex of trivalent actinides. The estimated value of the ion interaction coefficient given in the previous review, $\Delta\varepsilon = -(0.12 \pm 0.1)$ kg·mol^{-1}, is used for the extrapolation to infinite dilution. The unweighted average of the selected experimental data extrapolated to $I = 0$ with the SIT and the $\log_{10}\beta_1°$ value given by Aas et al. [99AAS/STE] lead to the following equilibrium constant:

$$\log_{10}\beta_1°(AnF^{2+}, 298.15\ K) = (3.4 \pm 0.3).$$

No new data are available for other Am(III) and Cm(III) fluoride complexes. The present review retains the equilibrium constant for AmF_2^+ selected by [95SIL/BID]:

$$\log_{10}\beta_2°(AmF_2^+, 298.15\ K) = (5.8 \pm 0.2).$$

Table 12-5: Literature values of the formation constants for $AnF_n^{(3-n)}$ and $CmF_n^{(3-n)}$ complexes.

Method	Medium	t (°C)	An	$\log_{10}\beta_1$	$\log_{10}\beta_2$	$\log_{10}\beta_3$	Reference
sol	0.1 M HClO$_4$	23	Cm			(a)	[54FEA]
			Am			(a)	
dis	0.5 M NaClO$_4$	25	Cm	3.34	6.18	9.08	[69AZI/LYL]
			Am	3.39	6.11	9.0	
dis	1.0 M NaClO$_4$	25	Cm	(2.93 ± 0.10)			[75DEG/CHO]
			Am	(2.93 ± 0.10)			
dis	1.0 M NaClO$_4$	10	Cm	(2.50 ± 0.02)			[76CHO/UNR]
			Am	(2.39 ± 0.01)			
		25	Cm	(2.61 ± 0.02)			
			Am	(2.49 ± 0.02)			
		40	Cm	(2.68 ± 0.05)			
			Am	(2.57 ± 0.02)			
		55	Cm	(2.81 ± 0.02)			
			Am	(2.71 ± 0.03)			
dis	0.1 M NaClO$_4$	25	Am	(2.51 ± 0.06)			[97SUG/SAT] [97SUG/SAT2]
ix	0.1 M NaClO$_4$	25	Am	(2.59 ± 0.01)	(4.75 ± 0.04)		[84NAS/CLE]
TRLFS	0–5 m NaCl $I = 0$	25	Cm	(3.44 ± 0.05)			[99AAS/STE]

(a) For the stepwise reaction $AnF_2^+ + F^- \rightleftharpoons AnF_3(aq)$, Feay reports $\log_{10} K_3 = 3.90$ for Cm and 3.11 for Am.

12.4.1.2 Aqueous Am(III) chloride complexes (V.4.1.2)

12.4.1.3 Aqueous Am(III) and Cm(III) chlorides (V.4.1.2.1)

The published data on the chloride complexation of trivalent actinides can be divided into two groups. Most of the experimental data have been determined by methods based on two–phase equilibria such as ion exchange or liquid–liquid extraction. The other group of data is based on spectroscopic methods. The evaluated complexation constants are similar within each group. However, the spectroscopically determined complexation constants are about two orders of magnitude smaller than those deduced from solvent extraction and ion exchange equilibria. The latter methods are unable to distinguish between ion–ion interaction and inner–sphere complexation. Changes in activity coefficients caused by the replacement of most of the background electrolyte by a weak ligand like chloride are often misinterpreted as complex formation. The effects observed in phase equilibrium studies, *i.e.*, when the $NaClO_4$ background medium is successively replaced by NaCl, can be described by using the correct ion interaction term of the SIT (instead of constant activity coefficients) to calculate the activity coefficients of the Am^{3+} ion:

$$\log_{10} \gamma_{Am^{3+}} = -9D + \varepsilon(Am^{3+}, ClO_4^-)\cdot m_{ClO_4^-} + \varepsilon(Am^{3+}, Cl^-)\cdot m_{Cl^-}$$

with $\varepsilon(Am^{3+}, ClO_4^-) = (0.49 \pm 0.03)$ kg · mol^{-1} and $\varepsilon(Am^{3+}, Cl^-) = (0.23 \pm 0.02)$ kg · mol^{-1} (Table B-4, Appendix B). An interpretation in terms of chloride complexation would mean that the observed effect on phase equilibria is accounted for twice. To avoid such inconsistencies the present review prefers to rely on equilibrium constants derived from spectroscopic data.

The spectroscopic studies of Barbanel and Mikhailova [69BAR/MIK] on Am(III), of Shiloh and Marcus [64SHI/MAR], and Marcus and Shiloh [69MAR/SHI] on Am(III), Np(III) and Pu(III) in concentrated HCl and LiCl solutions, as well as the recent detailed TRLFS studies of Fanghänel *et al.* [95FAN/KIM] and Könnecke *et al.* [97KON/FAN] on Cm(III) in concentrated $CaCl_2$ solutions, have demonstrated that inner–sphere chloride complexes are formed only at very high chloride concentrations (above 4–5 molal), as indicated by changes in the absorption and fluorescence spectra. This is in contradiction to the complexation constant for the monochloro complex, $AmCl^{2+}$ recommended by the previous review [95SIL/BID], $\log_{10} \beta_1^\circ = (1.05 \pm 0.06)$. An equilibrium constant of this magnitude would result in the presence of about 50% of the monochloro complex in a 1 molal chloride solution. In the TRLFS study of Fanghänel *et al.* [95FAN/KIM], no detectable complex formation was observed even in 4 m chloride concentrations. The fluorescence emission spectra of Cm(III) in 4 molal chloride (2 m $CaCl_2$) and in dilute $HClO_4$ solutions show no significant differences in the peak location and the peak shape. Similar observations have been made in the absorption spectra of Am(III) by Giffaut [94GIF] in 3 M NaCl and $NaClO_4$, and Runde *et al.* [97RUN/NEU] in 5 M NaCl and $HClO_4$ solutions. In both cases the absorption spectra are identical.

The spectroscopically determined species' distribution of Cm(III) in 0–6 m $CaCl_2$ is shown in Figure 12-6. Significant amounts of the monochloro complex are detected above 4 m chloride and at higher chloride concentrations the dichloro complex is stabilised.

The spectroscopic data obtained by TRLFS have been confirmed by an EXAFS study of Allen et al. [2000ALL/BUC]. The chloride complexation of lanthanides and trivalent actinides (Am(III), Cm(III) and Pu(III)) was studied in concentrated LiCl solutions of up to 12 m. No significant chloride complexation was found for Pu(III) even at this extremely high chloride molality. The data for Am(III) and Cm(III) are in agreement with the results of Fanghänel et al. [95FAN/KIM]. The two studies are not directly comparable as they were performed in different ionic media, $CaCl_2$ [95FAN/KIM] and LiCl [2000ALL/BUC]. However, as shown in [2000ALL/BUC], the results overlap if the chloride coordination numbers determined in the two studies are compared as a function of the water activity.

Figure 12-6: Distribution of Cm species at 298.15 K as a function of the $CaCl_2$ concentration.

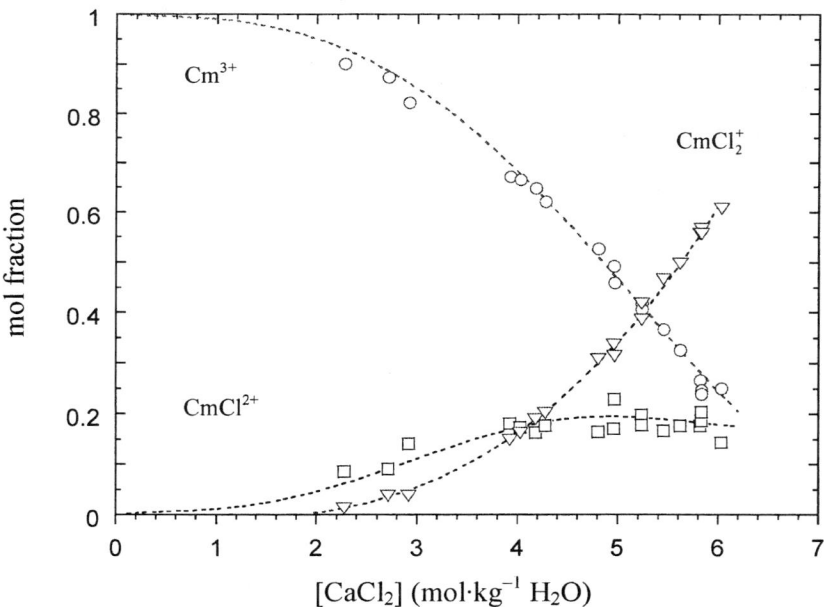

Fanghänel et al. [95FAN/KIM] and Könnecke et al. [97KON/FAN] have used their data to derive a quantitative model of chloride complexation of trivalent actinides in chloride solutions over a wide concentration range applying the ion–interaction approach of Pitzer. The Pitzer approach was applied as the ionic strength range covered by this study is far beyond the range of applicability of the SIT approach. The lowest ionic strength at which chloride complexation was detected (about I = 12 mol·kg^{-1}) considerably exceeds the validity range of the SIT approach, which is in general considered to be limited to $I \leq 3 - 4$ mol · kg^{-1}. The equilibrium constants at zero ionic strength for the two complexes formed were determined to be:

$$\log_{10} \beta_1^\circ (\text{AnCl}^{2+}, 298.15 \text{ K}) = (0.24 \pm 0.03)$$

and

$$\log_{10} \beta_2^\circ (\text{AnCl}_2^+, 298.15 \text{ K}) = -(0.74 \pm 0.05).$$

The equilibrium constants and the uncertainty estimates given by [97KON/FAN] represent the best estimates for the complexes AnCl^{2+} and AnCl$_2^+$, both for Cm and Am, and are therefore selected by this review.

The data from Yeh et al. [2000YEH/MAD] are not considered for reasons given in Appendix A.

12.4.2 Americium halide compounds (V.4.2)

12.4.2.1 Enthalpies of formation

Silva et al. [95SIL/BID] estimated the enthalpies of formation of the solid americium trihalides except AmCl$_3$(cr), by plotting the enthalpy difference ($\Delta_f H_m^\circ$ (MX$_3$, cr, 298.15 K) – $\Delta_f H_m^\circ$ (M^{3+}, 298.15 K)) (where M = U, Np, Pu and X = F, Br, I) against the ionic radius of the M^{3+} ion. The values for the relevant data for all the species were taken from [76FUG/OET], [83FUG/PAR], since the NEA review for Np and Pu was not then available. We have therefore repeated these estimations, using the current NEA data in this review and including, in the estimation of $\Delta_f H_m^\circ$ (AmBr$_3$, cr, 298.15 K), experimental results on the enthalpy of formation of CfBr$_3$(cr) [90FUG/HAI] and Cf^{3+} [84FUG/HAI].

This has resulted in the small differences in the estimated enthalpies of formation given in Table 12-6.

The change in $\Delta_f H_m^\circ$ (AmBr$_3$, cr, 298.15 K) results in the same change in $\Delta_f H_m^\circ$ (AmOBr, cr, 298.15 K), which thus becomes – (887.0 ± 9.0) kJ · mol^{-1}.

A similar procedure was used to estimate the enthalpy of formation of AmF$_4$(cr), using ThF$_4$ and UF$_4$. The difference from the earlier estimate is somewhat bigger in this case, since (unlike [95SIL/BID]), we have excluded the estimated value for $\Delta_f H_m^\circ$ (PuF$_4$, cr, 298.15 K) from the analysis, (although the estimated data for this phase are consistent with the extrapolation procedure within its appreciable uncertainty).

Table 12-6: Estimated enthalpies of formation of americium halides at 298.15 K.

Species	$\Delta_f H_m^\circ$ (kJ · mol^{-1})	
	[95SIL/BID]	This review
AmF$_3$(cr)	− (1588.0 ± 13.0)	− (1594.0 ± 14.0)
AmBr$_3$(cr)	− (810.0 ± 10.0)	− (804.0 ± 6.0)
AmI$_3$(cr)	− (612.0 ± 7.0)	− (615.0 ± 9.0)
AmF$_4$(cr)	− (1710.0 ± 21.0)	− (1724.0 ± 17.0)

12.4.2.2 Americium fluoride compounds (V.4.2.2)

12.4.2.2.1 Americium trifluoride (V.4.2.2.1)

12.4.2.2.1.1 AmF$_3$(cr)

The revised standard entropy from the estimate of [2001KON], (see Table 12-1) is:

$$S_m^\circ (AmF_3, cr, 298.15\ K) = (110.6 \pm 6.0)\ J \cdot K^{-1} \cdot mol^{-1}.$$

As noted in section 12.4.2.1, the enthalpy of formation has been revised to:

$$\Delta_f H_m^\circ (AmF_3, cr, 298.15\ K) = - (1594.0 \pm 14.0)\ kJ \cdot mol^{-1},$$

leading to the Gibbs energy of formation:

$$\Delta_f G_m^\circ (AmF_3, cr, 298.15\ K) = - (1519.8 \pm 14.1)\ kJ \cdot mol^{-1}.$$

The heat capacities of AmF$_3$(cr), required for the analysis of the vapour pressure data, have been estimated to be the same as those of PuF$_3$(cr):

$$C_{p,m} (AmF_3, cr, T) = 104.078 + 0.707 \cdot 10^{-3}\ T - 10.355 \cdot 10^5\ T^{-2}\ J \cdot K^{-1} \cdot mol^{-1},$$

from 298.15 to 1500 K.

As already mentioned in the previous review [95SIL/BID], there is a considerably discrepancy between the thermochemical data for AmF$_3$(cr) and results from solubility measurements. Combining the standard Gibbs energies selected for AmF$_3$(cr) and Am^{3+} with the auxiliary data for F$^-$ yields a solubility constant of $\log_{10} K_{s,0}^\circ$ (AmF$_3$, cr, 298.15 K) = − (13.3 ± 2.5). Nash and Cleveland [84NAS/CLE2] observed a decrease of the Am concentration (initially 10^{-8} M in 0.1 M NaClO$_4$ solutions of pH 3.5) at fluoride concentrations above 0.01 M. This was ascribed to the formation of solid AmF$_3$ particles or "merely to the adsorption of Am^{3+} by negatively charged colloidal particles which stick to the walls of the polyethylene container". There was no positive observation of a solid phase. However, judging from the behaviour of the PuF$_3$–H$_2$O [53JON] and LnF$_3$–H$_2$O [75STO/KHA] systems, any precipitated solid was very probably hydrated.

Silva et al. [95SIL/BID] calculated a solubility constant of $\log_{10} K_{s,0}^\circ = -(16.5 \pm 0.3)$ from the data of [84NAS/CLE2], which therefore probably applies to an amorphous AmF_3 hydrate of undefined composition. The discrepancy with the solubility of $AmF_3(cr)$ calculated from the thermochemical data, although not directly comparable, is unsatisfactory, but must be accepted at present, given the uncertainties in the interpretation of the study by [84NAS/CLE2] and the thermochemical data for $AmF_3(cr)$. A definitive study of the AmF_3–H_2O system, including well-defined solubility measurements, is clearly required.

12.4.2.2.1.2 $AmF_3(g)$

In order to analyse more completely the vapour pressure data (see below), we have computed the thermal functions of $AmF_3(g)$ by statistical-mechanical calculations using estimated molecular parameters. The geometry and vibration frequencies were assumed to be the same as those estimated for $PuF_3(g)$, [85HIL/GUR], [2000RAN/FUG], with no electronic contributions (in keeping with the philosophy of Konings [2001KON] for Am(III) compounds). The ground-state degeneracy was taken to be unity. These calculations give:

$$S_m^\circ (AmF_3, g, 298.15 \text{ K}) = (330.4 \pm 8.0) \text{ J} \cdot \text{K}^{-1} \cdot \text{mol}^{-1},$$

$$C_{p,m}^\circ (AmF_3, g, 298.15 \text{ K}) = (72.2 \pm 5.0) \text{ J} \cdot \text{K}^{-1} \cdot \text{mol}^{-1}.$$

As noted in [95SIL/BID], Carniglia and Cunningham [55CAR/CUN] made precise measurements by Knudsen effusion of the vapour pressure of $AmF_3(cr)$. After correction for the half-life of ^{241}Am, the Gibbs energy of the sublimation process:

$$AmF_3(cr) \rightleftharpoons AmF_3(g) \tag{12.12}$$

is

$$\Delta_r G_m (12.12) = (398462 - 173.663 \, T) \text{ J} \cdot \text{mol}^{-1} \quad (1140 \text{ to } 1469 \text{ K}).$$

With the thermal functions for $AmF_3(cr)$ and $AmF_3(g)$ estimated above, these data can be analysed by second-law and third-law methods. The derived enthalpies of sublimation at 298.15 K are respectively (420.3 ± 7.0) and (437.5 ± 8.9) kJ·mol^{-1}. This difference reflects the fact that the calculated entropy of sublimation at the mean temperature, $\Delta_r S_m ((12.12), 1305 \text{ K})$ is (188.0 ± 4.1) J·K^{-1}·mol^{-1}, compared to the experimental value of (173.7 ± 4.1) J·K^{-1}·mol^{-1}. However, the second-law analysis would imply $S_m^\circ (AmF_3, g, 298.15 \text{ K}) = (317.2 \pm 11.7)$ J·K^{-1}·mol^{-1}, which in turn would require appreciably larger vibration frequencies than assumed for other actinide trifluoride gaseous species. The third-law analysis is thus preferred, giving finally:

$$\Delta_f H_m^\circ (AmF_3, g, 298.15 \text{ K}) = -(1156.5 \pm 16.6) \text{ kJ} \cdot \text{mol}^{-1},$$

and hence

$$\Delta_f G_m^\circ (AmF_3, g, 298.15 \text{ K}) = -(1147.8 \pm 16.8) \text{ kJ} \cdot \text{mol}^{-1}.$$

The experimental vapour pressures and the fitted line are shown in Figure 12-7.

Figure 12-7: Vapour Pressure of AmF$_3$(cr).

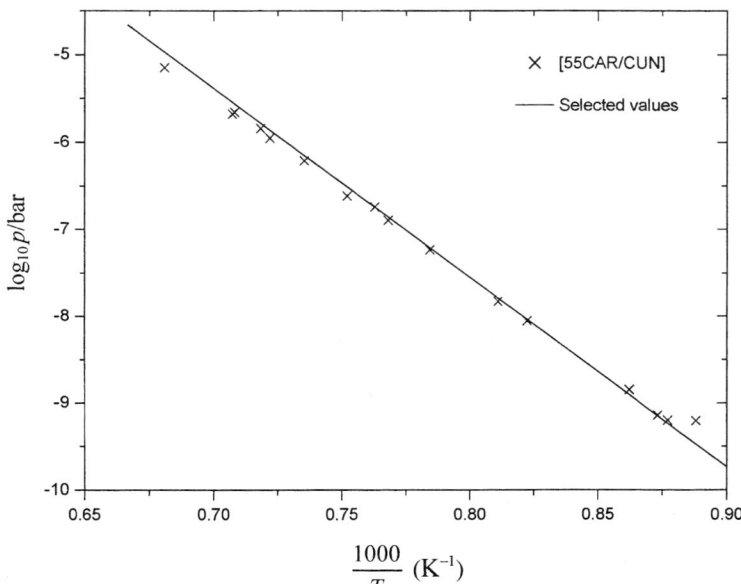

12.4.2.2.2 Americium tetrafluoride (V.4.2.2.2)

The standard entropy has been estimated by the same procedure as that used by Konings [2001KON] for Am(III) compounds. The lattice contribution is estimated to be (141.59 ± 6.00) J·K^{-1}·mol^{-1} from the average of the lattice contributions of ThF$_4$(cr), UF$_4$(cr), PuF$_4$(cr) by subtracting the excess entropy contribution from the measured entropies, as in Table 12-7. The excess contributions are calculated from the splitting of the energy levels of the ground state, as computed by Carnall *et al.* [91CAR/LIU]. S_m° (AmF$_4$, cr, 298.15 K) is then obtained from the sum of the lattice and excess contributions, the latter being calculated from the energy levels of the ground state, computed by Liu *et al.* [94LIU/CAR].

Table 12-7: Calculations of S_m^o (AnF$_4$, cr, 298.15 K).

Solid	S_m^o (298.15 K) total J·K^{-1}·mol^{-1}	S_m^o (298.15 K) excess J·K^{-1}·mol^{-1}	S_m^o (298.15 K) lattice J·K^{-1}·mol^{-1}
ThF$_4$	142.05	0.0	142.05
UF$_4$	151.70	7.75	143.95
PuF$_4$	147.25	8.48	138.77
AmF$_4$	154.10	12.51	141.59

The selected value is thus:

S_m^o (AmF$_4$, cr, 298.15 K) = (154.1 ± 8.0) J·K^{-1}·mol^{-1}.

As noted in section 12.4.2.1, the enthalpy of formation has been revised to:

$\Delta_f H_m^o$ (AmF$_4$, cr, 298.15 K) = − (1724.0 ± 17.0) kJ·mol^{-1},

leading to the Gibbs energy of formation:

$\Delta_f G_m^o$ (AmF$_4$, cr, 298.15 K) = − (1632.5 ± 17.2) kJ·mol^{-1}.

12.4.2.3 Americium chlorides (V.4.2.3)

12.4.2.3.1 Americium trichloride (V.4.2.3.2)

12.4.2.3.1.1 AmCl$_3$(cr)

The revised standard entropy from the estimate of [2001KON], (see section 12.1.1) is:

S_m^o (AmCl$_3$, cr, 298.15 K) = (146.2 ± 6.0) J·K^{-1}·mol^{-1}.

The enthalpy of formation remains unchanged:

$\Delta_f H_m^o$ (AmCl$_3$, cr, 298.15 K) = − (977.8 ± 1.3) kJ·mol^{-1},

and thus

$\Delta_f G_m^o$ (AmCl$_3$, cr, 298.15 K) = − (905.1 ± 2.3) kJ·mol^{-1}.

12.4.2.3.2 Americium oxychloride (V.4.2.3.3)

Two studies [54KOC/CUN], [76WEI/WIS] of the equilibrium constants of the reaction:

$$AmCl_3(cr) + H_2O(g) \rightleftharpoons AmOCl(cr) + 2\ HCl(g) \qquad (12.13)$$

were analysed in [95SIL/BID] to give:

$\Delta_r S_m^o$ ((12.13), 298.15 K) = 127.81 + 3.57 = (131.38 ± 8.00) J·K^{-1}·mol^{-1}.

With the revised value of S_m^o (AmCl$_3$, cr, 298.15 K) = (146.2 ± 6.0) J·K^{-1}·mol^{-1} (see section 12.4.2.3.1), the modified value of the standard entropy of AmOCl(cr) becomes:

S_m° (AmOCl, cr, 298.15 K) = (92.6 ± 10.0) J · K^{-1} · mol^{-1}.

This is close to the weighted sum of the entropies of Am$_2$O$_3$(cr) and AmCl$_3$(cr), (93.3 ± 2.8) J · K^{-1} · mol^{-1}.

The enthalpy of formation remains unchanged:

$\Delta_f H_m^\circ$ (AmOCl, cr, 298.15 K) = – (949.8 ± 6.0) kJ · mol^{-1},

and thus

$\Delta_f G_m^\circ$ (AmOCl, cr, 298.15 K) = – (897.1 ± 6.7) kJ · mol^{-1}.

12.4.2.3.3 Quaternary chloride Cs$_2$NaAmCl$_6$(cr) (V.4.2.3.4.2)

As in [95SIL/BID], the entropy of Cs$_2$NaAmCl$_6$(cr) was estimated to be close to the sum of the constituent chlorides; the revised standard entropy of Cs$_2$NaAmCl$_6$(cr) becomes (420.7 ± 15.0) J · K^{-1} · mol^{-1}, which we round to:

S_m° (Cs$_2$NaAmCl$_6$, cr, 298.15 K) = (421 ± 15) J · K^{-1} · mol^{-1}.

12.4.2.4 Americium bromides (V.4.2.4)

12.4.2.4.1 Americium tribromide (V.4.2.4.2)

12.4.2.4.1.1 AmBr$_3$(cr)

The standard entropy has been revised to be consistent with those of the other Am(III) compounds (see section 12.1.1). The lattice entropy has been estimated from that of S_m° (UBr$_3$, cr, 298.15 K) [92GRE/FUG] by subtracting the excess entropy from the experimental value (in this context, increased by R·ln2, see [2001KON]). In the absence of any data on the electronic levels for UBr$_3$(cr), its excess entropy is estimated to be the mean of those for UF$_3$(cr) and UCl$_3$(cr), (16.74 ± 6.0) J · K^{-1} · mol^{-1} [2001KON].

Thus,

S_m° (AmBr$_3$, cr, 298.15 K) = (182.0 ± 10.0) J · K^{-1} · mol^{-1},

where the uncertainty has been increased to allow for the uncertainty of the estimation procedure.

As noted in section 12.4.2.1, the enthalpy of formation has been revised to:

$\Delta_f H_m^\circ$ (AmBr$_3$, cr, 298.15 K) = – (804.0 ± 6.0) kJ · mol^{-1},

the Gibbs energy of formation is:

$\Delta_f G_m^\circ$ (AmBr$_3$, cr, 298.15 K) = – (773.7 ± 6.7) kJ · mol^{-1}.

12.4.2.4.2 Americium oxybromide (V.4.2.4.3)

The study [82WEI/WIS] of the equilibrium constants of the reaction:

$$AmBr_3(cr) + H_2O(g) \rightleftharpoons AmOBr(cr) + 2\,HBr(g) \qquad (12.14)$$

was analysed in [95SIL/BID] to give:

$\Delta_r S_m^\circ$ ((12.14), 298.15 K) = 127.72 + 3.71 = (131.43 ± 8.00) J · K^{-1} · mol^{-1}.

With the revised value of S_m° (AmBr$_3$, cr, 298.15 K) = (182.0 ± 10.0) J · K^{-1} · mol^{-1} (see section 12.4.2.4.1), the modified value of the standard entropy of AmOBr(cr) becomes:

S_m° (AmOBr, cr, 298.15 K) = (104.9 ± 12.8) J · K^{-1} · mol^{-1}.

This is close to the weighted sum of the entropies of Am$_2$O$_3$(cr) and AmBr$_3$(cr), (105.2 ± 3.9) J · K^{-1} · mol^{-1}.

As a consequence of the change in the enthalpy of formation of AmBr$_3$(cr), the enthalpy of formation has been revised to:

$\Delta_f H_m^\circ$ (AmOBr, cr, 298.15 K) = – (887.0 ± 9.0) kJ · mol^{-1}

leading to the Gibbs energy of formation:

$\Delta_f G_m^\circ$ (AmOBr, cr, 298.15 K) = – (848.5 ± 9.8) kJ · mol^{-1}.

12.4.2.5 Americium iodides (V.4.2.5)

12.4.2.5.1 Americium triiodide (V.4.2.5.2)

12.4.2.5.1.1 AmI$_3$(cr)

The standard entropy has been revised to be consistent with those of the other Am(III) compounds (see section 12.1.1). The lattice entropy has been estimated from that of S_m° (UI$_3$, cr, 298.15 K) by subtracting its excess entropy from the selected value [92GRE/FUG] (in this context, increased by R·ln2, see [2001KON]). In the absence of any data on the electronic levels for UI$_3$(cr), the excess entropy is estimated to be the mean of those for UF$_3$(cr) and UCl$_3$(cr), (16.74 ± 6.0) J · K^{-1} · mol^{-1} [2001KON].

Thus,

S_m° (AmI$_3$, cr, 298.15 K) = (211 ± 15) J · K^{-1} · mol^{-1},

and as noted in section 12.4.2.1, the enthalpy of formation has been revised to:

$\Delta_f H_m^\circ$ (AmI$_3$, cr, 298.15 K) = – (615.0 ± 9.0) kJ · mol^{-1},

leading to the Gibbs energy of formation:

$\Delta_f G_m^\circ$ (AmI$_3$, cr, 298.15 K) = – (609.5 ± 10.1) kJ · mol^{-1}.

12.5 Americium group 16 (chalcogen) compounds and complexes (V.5)

12.5.1 Americium sulphates (V.5.1.2)

12.5.1.1 Aqueous Am(III) and Cm(III) sulphate complexes(V.5.1.2.1)

There are no new experimental data on the sulphate complexation of Am(III) since the previous review [95SIL/BID] was published. The equilibrium constants selected by the previous review [95SIL/BID] for the formation of the monosulphate and disulphate complexes of Am(III) for the following reactions:

$$Am^{3+} + n\, SO_4^{2-} \rightleftharpoons Am(SO_4)_n^{(3-2n)} \qquad (12.15)$$

are
$$\log_{10} \beta_1^\circ\, ((12.15),\, 298.15\text{ K}) = (3.85 \pm 0.03),$$

$$\log_{10} \beta_2^\circ\, ((12.15),\, 298.15\text{ K}) = (5.4 \pm 0.7).$$

They are based on experimental data determined at pH = 3 – 4 by using solvent extraction techniques [64SEK], [65SEK2], [67CAR/CHO], [68AZI/LYL], [72MCD/COL], [80KHO/MAT], ion exchange [68AZI/LYL], [68NAI], and electromigration [90ROS/REI].

More recently, [96PAV/FAN] and [98NEC/FAN] investigated sulphate complexes of Cm(III) by using TRLFS to identify the complexes, *viz.* $CmSO_4^+$, $Cm(SO_4)_2^-$ and $Cm(SO_4)_3^{3-}$. The highest sulphate concentration studied by [96PAV/FAN] was a 4.7 m Cs_2SO_4 solution. The formation of the trisulphate complex was observed only at sulphate concentrations above 1 mol · kg^{-1}. The formation constants of the complexes $CmSO_4^+$ and $Cm(SO_4)_2^-$ were determined in 3 m NaCl/Na$_2$SO$_4$ at pH 2, with the sulphate concentration varying in the range [SO_4^{2-}] = 0.03 – 0.37 mol · kg^{-1}. In another study reported in [98NEC/FAN], Cm(III) sulphate complexation was investigated as a function of the ionic medium composition (0 – 5.8 m NaCl) at two values of constant sulphate concentration (0.15 and 0.55 m Na$_2$SO$_4$).

The spectroscopically determined $\log_{10} \beta_1$ values in [96PAV/FAN] and [98NEC/FAN] are about 0.6 logarithmic units lower than the literature data for Am(III) and Cm(III) in NaClO$_4$–Na$_2$SO$_4$ or NH$_4$ClO$_4$–(NH$_4$)$_2$SO$_4$ solutions of comparable ionic strength and the $\log_{10} \beta_2$ values are 1.5 to 2 orders of magnitude smaller (*cf.* Figure 12-8). These differences can only partly be attributed to the different activity coefficients in NaCl compared to NaClO$_4$ or NH$_4$ClO$_4$ solutions. The large discrepancies in the data from the two groups of methods are far beyond the experimental uncertainties of two types of methods, and are typical for systems where weak complexes are formed (see also section 12.4.1.3.).

The experimental techniques based on two–phase equilibria such as ion exchange and solvent extraction are in these cases unable to distinguish between ion–ion interaction and inner–sphere complexation. The replacement of large parts of the background electrolyte by the ligand causes changes in the activity coefficients, which are misinterpreted as complex formation. The same can hold for activity changes caused by the formation of outer–sphere complexes or ion pairs. In some of the ion exchange and solvent extraction studies [67CAR/CHO], [68AZI/LYL], [72MCD/COL] analogous experiments with both Am(III) and Cm(III) gave very similar equilibrium constants for the two actinides (cf. Figure 12-8) indicating that discrepancies between the spectroscopic results of [96PAV/FAN] and [98NEC/FAN], and previous non–spectroscopic methods are primarily not a result of chemical differences, but rather that the two types of experiments measure different phenomena. This review has based the selection of equilibrium constants for aqueous sulphate complexes on the spectroscopically determined data rather than on those from other sources.

These considerations are supported by a comparison of thermodynamic data for aqueous complexes of Am(III) and U(VI). The formation constants selected in the previous review for the Am(III) sulphate complexes ($\log_{10} \beta_1^\circ$ ($AmSO_4^+$, 298.15 K) = (3.85 ± 0.03), $\log_{10} \beta_2^\circ$ ($Am(SO_4)_2^-$, 298.15 K) = (5.4 ± 0.7) [95SIL/BID]) are significantly higher than the selected data for U(VI) sulphate complexes ($\log_{10} \beta_1^\circ$ (UO_2SO_4, aq, 298.15 K) = (3.15 ± 0.02), $\log_{10} \beta_2^\circ$ ($UO_2(SO_4)_2^{2-}$, 298.15 K) = (4.14 ± 0.07) [92GRE/FUG], and this review). This finding is inconsistent with the known chemical systematics and correlations among formation constants of actinide complexes (cf. [97ALL/BAN]). The formation constants of Am(III) complexes with other inorganic ligands (e.g., carbonate, hydroxide, fluoride, phosphate) are generally close to, or smaller than, those of the corresponding U(VI) complexes. We therefore conclude that the equilibrium constants of [95SIL/BID] for the aqueous Am(III) sulphate complexes are systematically too large, cf. Figure 12-8.

In Figure 12-8, the formation constants reported in [98NEC/FAN] for the monosulphate and disulphate complexes of Cm(III) are extrapolated to zero ionic strength. Application of the SIT to the data in Na_2SO_4–NaCl solutions of $I \leq 4$ mol · kg^{-1} leads to:

$$\log_{10} \beta_1^\circ = (3.28 \pm 0.03), \quad \Delta\varepsilon_1 = -(0.14 \pm 0.02) \text{ kg} \cdot \text{mol}^{-1},$$

$$\log_{10} \beta_2^\circ = (3.59 \pm 0.03), \quad \Delta\varepsilon_2 = -(0.24 \pm 0.01) \text{ kg} \cdot \text{mol}^{-1}.$$

If the SIT is applied only to the two $\log_{10} \beta_1^\circ$ values in 0.15 and 0.55 m Na_2SO_4 without additions of NaCl, a somewhat lower formation constant of $\log_{10} \beta_1^\circ = (3.15 \pm 0.03)$ is calculated, but the value of $\Delta\varepsilon_1 = -0.14$ kg · mol^{-1} remains the same.

The formation constants determined in 3 m NaCl – Na_2SO_4 solutions, $\log_{10} \beta_1 = (0.93 \pm 0.08)$ and $\log_{10} \beta_2 = (0.61 \pm 0.08)$ [96PAV/FAN] are somewhat higher (cf. Figure 12-8). With the values of $\Delta\varepsilon_1$ and $\Delta\varepsilon_2$ derived from the data in Na_2SO_4–NaCl solutions of varying ionic strength, the formation constants at $I = 0$ are calculated to be $\log_{10} \beta_1^\circ = (3.45 \pm 0.10)$ and $\log_{10} \beta_2^\circ = (3.81 \pm 0.09)$.

Figure 12-8: Extrapolation to $I = 0$ of experimental data for the formation of Am(III) and Cm(III) sulphate complexes $AnSO_4^+$ and $An(SO_4)_2^-$ at pH = 2–4. Circles refer to literature data in $NaClO_4$–Na_2SO_4 [64BAN/PAT], [64SEK], [65SEK2], [67CAR/CHO], [68NAI], [68AZI/LYL], [72MCD/COL], [80KHO/MAT], [90ROS/REI], which are based on ion exchange or solvent extraction methods. These results were used in the previous review for the determination of thermodynamic data. Open triangles are derived from spectroscopic data in 0.15 and 0.55 m Na_2SO_4 solutions containing additions of NaCl (from [98NEC/FAN]), and the squares refer to a series of spectra in 3 m NaCl–Na_2SO_4 [96PAV/FAN]. The solid lines are calculated with the data selected in the previous and present reviews and the dotted lines show the associated uncertainties.

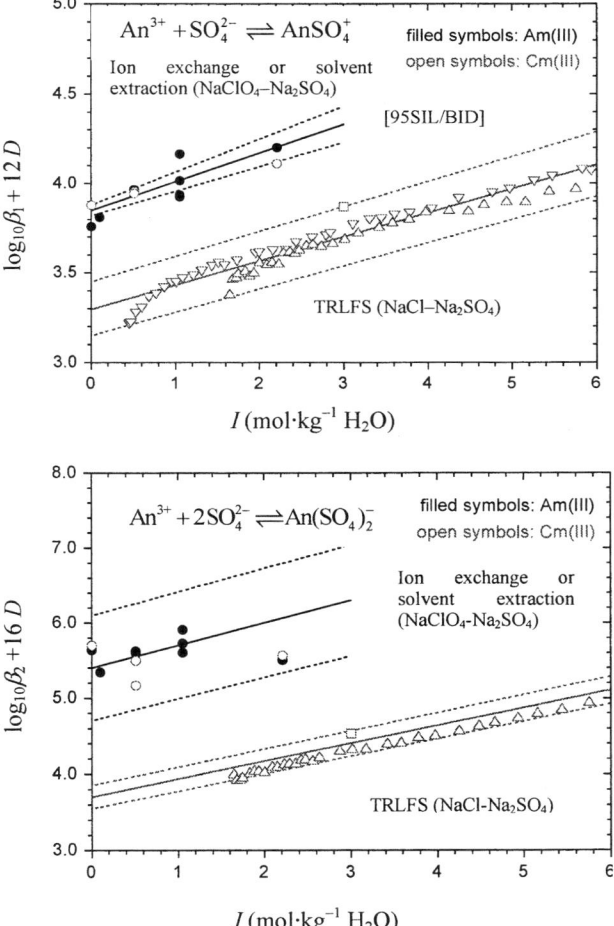

The conditional constants determined in [96PAV/FAN] at $I = 3$ mol · kg^{-1} are based on a series of spectra at varying sulphate concentration ($[SO_4^{2-}] = 0.03 - 0.37$ m), whereas each of the data points reported in [98NEC/FAN] is derived from a single spectrum at the corresponding ionic strength.

Therefore, the present review selects the mean values:

$$\log_{10} \beta_1^\circ (AmSO_4^+, 298.15 \text{ K}) = (3.30 \pm 0.15),$$

$$\log_{10} \beta_2^\circ (Am(SO_4)_2^-, 298.15 \text{ K}) = (3.70 \pm 0.15),$$

with the estimated uncertainties covering the whole range of the spectroscopic data determined by [96PAV/FAN] and [98NEC/FAN] (*cf.* Figure 12-8). However, no specific ion interaction coefficients are selected. The investigation of An(III) sulphate complexation requires electrolyte mixtures with fairly high sulphate concentrations and it is not possible to estimate the unknown interaction coefficients $\varepsilon(An^{3+}, SO_4^{2-})$ and $\varepsilon(AnSO_4^+, SO_4^{2-})$ from these data in NaCl-Na$_2$SO$_4$ mixtures where $\varepsilon(An^{3+}, Cl^-)$ and $\varepsilon(AnSO_4^+, Cl^-)$ have a simultaneous effect on the activity coefficients.

12.6 Americium group 14 compounds and complexes (V.7)

12.6.1 Carbon compounds and complexes (V.7.1)

12.6.1.1 Americium carbonate compounds and complexes (V.7.1.2)

12.6.1.1.1 Aqueous Am(III) and Cm(III) carbonate complexes

In the previous review [95SIL/BID] it was shown that the available experimental studies of aqueous Am(III) complexes in carbonate solutions can be described by the reactions:

$$Am^{3+} + n\,CO_3^{2-} \rightleftharpoons Am(CO_3)_n^{(3-2n)}, \qquad (12.16)$$

with $n = 1, 2$ and 3. Based on the spectroscopic studies of [89NIT/STA], [91MEI/KIM] in 0.1 M NaClO$_4$, the solvent extraction study of Lundqvist [82LUN] in 1 M NaClO$_4$ and three solubility studies, a) in dilute solution [90FEL/RAI], b) in 0.1 M NaClO$_4$ [91MEI/KIM] and c) in 3.0 M NaClO$_4$ [89ROB], the following formation constants were selected at zero ionic strength:

$\log_{10} \beta_1^\circ (AmCO_3^+, 298.15 \text{ K}) = (7.8 \pm 0.3)$ (based on 6 studies),
$\log_{10} \beta_2^\circ (Am(CO_3)_2^-, 298.15 \text{ K}) = (12.4 \pm 0.4)$ (based on 3 studies),
$\log_{10} \beta_3^\circ (Am(CO_3)_3^{3-}, 298.15 \text{ K}) = (15.2 \pm 0.6)$ (based on 1 study).

Additional new data for Am(III) carbonate complexes are available from the solubility studies by Giffaut [94GIF] in 0.1 and 4 M NaCl, Runde and Kim [94RUN/KIM] in 5 M NaCl, and from a spectroscopic study by Wruck *et al.* [97WRU/PAL] in 0.1 M NaClO$_4$. As already noted in the previous review [95SIL/BID], Am(III) solubility studies in carbonate solutions are associated with ambiguities and uncertainties arising from possible solid phase transformations and/or alterations. This can lead to considerable errors in the calculated formation constants. In order to minimise such errors, the present review uses the stepwise formation constants for the evaluation of thermodynamic data.

The present review also considers the available experimental data for the analogous Cm(III) carbonate complexes [94KIM/KLE], [98FAN/WEG2], [99FAN/KON]. These studies are based on TRLFS and provide a systematic investigation of the stepwise carbonate complexation of Cm(III) in dilute to concentrated NaCl [99FAN/KON]. In another TRLFS study [95FAN/KIM] demonstrated that chloride complexation of Cm^{3+} is negligible even in concentrated NaCl and hence the formation of ternary carbonate–chloride complexes can be ruled out. Because of the considerably increased experimental information for Am(III) and Cm(III) carbonate complexes, particularly in NaCl solutions, the thermodynamic data and ion–interaction coefficients of the carbonate complexes are re-evaluated in the present review. The results of the new studies and the former studies used in [95SIL/BID] for the selection of thermodynamic data are summarised in Table 12-8.

Table 12-8: Data for Am(III) and Cm(III) carbonate complexes discussed in the present and previous [95SIL/BID] reviews for the evaluation of stepwise formation constants at $I = 0$.

reaction: $An^{3+} + CO_3^{2-} \rightleftharpoons AnCO_3^+$

An	Medium	t(°C)	Method	$\log_{10} \beta_1$	$\log_{10} \beta_1^\circ$	Reference
Am	1.0 M NaClO$_4$	25	dis	(5.81 ± 0.04)	(8.00 ± 0.10)[a]	[82LUN]
Am	0.1 M NaClO$_4$	22.5		(6.69 ± 0.16)	(7.98 ± 0.16)[b]	[89NIT/STA]
Am	0.1 M NaClO$_4$	25		(6.48 ± 0.03)	(7.77 ± 0.1)[b]	[91MEI/KIM]
Am	0.1 M NaClO$_4$	25	LIPAS	(6.26 ± 0.12)	(7.55 ± 0.24)[b]	[97WRU/PAL]
		50		(6.68 ± 0.12)		
		75		(7.54 ± 0.43)		
Cm	0.1 M NaClO$_4$	r.t	TRLFS	(6.65 ± 0.07)	(7.95 ± 0.14)[b]	[94KIM/KLE]
Cm	1.0 M NaCl	25	TRLFS	(5.90 ± 0.1)	(8.21 ± 0.1)[b]	[98FAN/WEG2]
Cm	0.1 – 6 m NaCl	25	TRLFS			[99FAN/KON]
	$I = 0$				8.30	
Am	$I = 0$	r.t.	sol		7.6	[90FEL/RAI]
Am	0.1 M NaClO$_4$	25	sol	(5.97 ± 0.15)		[91MEI/KIM]
				(6.1 ± 0.3)[a]	(7.3 ± 0.3)[a]	
Am	3.0 M NaClO$_4$	20	sol	(5.45 ± 0.12)		[89ROB]
				(5.73 ± 0.24)[a]	(7.80 ± 0.32)[a]	
Am	0.1 M NaCl	21	sol	(7.7 ± 0.4)	(9.0 ± 0.4)[b]	[94GIF]
Am	4 M NaCl	21	sol	(7.6 ± 0.4)[c]	(10.0 ± 0.4)[b][c]	[94GIF]
Am	5 M NaCl	22	sol	(5.7 ± 0.4)	(8.04 ± 0.43)[b]	[94RUN/KIM]
Am	$I = 0$	25	review		(7.8 ± 0.3)	[95SIL/BID]
Am+Cm	$I = 0$	25	review		(8.1 ± 0.3)	[98NEC/FAN], [99FAN/KON]
Am+Cm	$I = 0$	25	review		(8.0 ± 0.4)	present review

(continued next page)

Table 12-8 (continued)

reaction: $AnCO_3^+ + CO_3^{2-} \rightleftharpoons An(CO_3)_2^-$

An	Medium	t(°C)	Method	$\log_{10} K_2$	$\log_{10} K_2^\circ$	Reference
Am	1.0 M NaClO$_4$	25	dis	(3.91 ± 0.11)	(4.57 ± 0.23)[a]	[82LUN]
Cm	1.0 M NaCl	25	TRLFS	(4.37 ± 0.2)	(5.11 ± 0.2)[b]	[98FAN/WEG2]
Cm	0.1 – 6 m NaCl	25	TRLFS			
	I = 0				5.22	[99FAN/KON]
Am	I = 0	r.t.	sol		4.7	[90FEL/RAI]
Am	0.1 M NaClO$_4$	25	sol	(3.61 ± 0.28)[c]	(4.0 ± 0.3)[a][c]	[91MEI/KIM]
Am	0.1 M NaCl	21	sol	(4.30 ± 0.26)	(4.7 ± 0.3)[b]	[94GIF]
Am	4 M NaCl	21	sol	(4.24 ± 0.26)[c]	(4.9 ± 0.3)[b][c]	[94GIF]
Am	5 M NaCl	22	sol	(4.0 ± 0.6)	(4.6 ± 0.6)[b]	[94RUN/KIM]
Am	I = 0	25	review		(4.5 ± 0.2)	[95SIL/BID]
Am+Cm	I = 0	25	review		(5.1 ± 0.7)	[98NEC/FAN], [99FAN/KON]
Am+Cm	I = 0	25	review		(4.9 ± 0.5)	present review

reaction: $An(CO_3)_2^- + CO_3^{2-} \rightleftharpoons An(CO_3)_3^{3-}$

An	Medium	t(°C)	Method	$\log_{10} K_3$	$\log_{10} K_3^\circ$	Reference
Cm	1.0 M NaCl	25	TRLFS	(2.91 ± 0.15)	(2.09 ± 0.15)[b]	[98FAN/WEG2]
Cm	0.01– 6 m NaCl	25	TRLFS			
	I = 0				2.00	[99FAN/KON]
Am	I = 0	r.t.	sol		2.9	[90FEL/RAI]
Am	0.1 M NaCl	21	sol	(1.80 ± 0.26)	(1.4 ± 0.3)[b]	[94GIF]
Am	4 M NaCl	21	sol	(2.24 ± 0.25)[c]	(0.9 ± 0.3)[b][c]	[94GIF]
Am	5 M NaCl	22	sol	(3.2 ± 0.5)	(2.0 ± 0.5)[b]	[94RUN/KIM]
Am	I = 0	25	review		(2.9 ± 0.5)	[95SIL/BID]
Am+Cm	I = 0	25	review		(2.2 ± 0.7)	[98NEC/FAN], [99FAN/KON]
Am+Cm	I = 0	25	review		(2.1 ± 0.8)	present review

reaction: $An(CO_3)_3^{3-} + CO_3^{2-} \rightleftharpoons An(CO_3)_4^{5-}$

An	Medium	t(°C)	Method	$\log_{10} K_4$	$\log_{10} K_4^\circ$	Reference
Cm	1.0 M NaCl	25	TRLFS	(1.0 ± 0.2)		[98FAN/WEG2]
Cm	1 – 6 m NaCl	25	TRLFS			
	I = 0				– 2.16	[99FAN/KON]

(a) Recalculated in [95SIL/BID].
(b) Recalculated with the SIT coefficients selected in the present review.
(c) Not used for the selection of thermodynamic data.
r.t. room temperature.

Figure 12-9 shows the application of the SIT regression to the stepwise formation constants of Am(III) and Cm(III) carbonate complexes in chloride solutions. Experimental data in NaClO$_4$ solutions are also included for comparison. At low ionic strength, the differences between the activity coefficients in NaCl and NaClO$_4$ solutions are negligible compared to other uncertainties. Since the experimental uncertainties of the different experimental methods are considerably larger than the chemical differences between Am(III) and Cm(III), the data for these two actinide elements are treated and weighted together.

The spectroscopic data for the Cm(III) carbonate complexes in 0.1 – 6 m NaCl [99FAN/KON] follow the linear SIT regression up to high ionic strength and allow the evaluation of accurate $\Delta\varepsilon$ values. In Table 12-8, the conditional equilibrium constants reported in the different experimental studies are converted to $I = 0$ and the mean values are selected. In order not to give too large weight to the many data points from the study of Fanghänel et al. [99FAN/KON], only their values extrapolated to $I = 0$ are included in the calculation of mean values for $\log_{10} \beta_1^\circ$, $\log_{10} K_2^\circ$ and $\log_{10} K_3^\circ$. The experimental data on the formation of the first carbonate complex:

$$An^{3+} + CO_3^{2-} \rightleftharpoons AnCO_3^+, \qquad (12.17)$$

yield $\Delta\varepsilon(12.17) = -(0.14 \pm 0.03)$ kg·mol^{-1} and $\log_{10} \beta_1^\circ$ ((12.17), 298.15 K) = (8.0 ± 0.4), if only the data based on spectroscopic and solvent extraction methods are considered. If the data derived in solubility studies are considered as well, the mean value remains the same, but the uncertainty increases to ± 0.8. As mentioned above, this may be related to problems associated with changes of the solid phases involved.

For the stepwise complex formation equilibria:

$$AnCO_3^+ + CO_3^{2-} \rightleftharpoons An(CO_3)_2^- \qquad (12.18)$$

$$An(CO_3)_2^- + CO_3^{2-} \rightleftharpoons An(CO_3)_3^{3-} \qquad (12.19)$$

the present review calculates:

$$\Delta\varepsilon(12.18) = -(0.07 \pm 0.02) \text{ kg·mol}^{-1},$$
$$\log_{10} K_2^\circ ((12.18), 298.15 \text{ K}) = (4.9 \pm 0.5),$$

which is the mean value from spectroscopic and solubility studies, and

$$\Delta\varepsilon(12.19) = -(0.01 \pm 0.02) \text{ kg·mol}^{-1},$$
$$\log_{10} K_3^\circ ((12.19), 298.15 \text{ K}) = (2.1 \pm 0.8)$$

with the uncertainty covering the individual uncertainties within and between the different studies.

Figure 12-9: Application of the SIT method to the stepwise formation constants of Am(III) (filled symbols) and Cm(III) (open symbols) carbonate complexes in dilute to concentrated NaCl solutions [94GIF], [94RUN/KIM], [98FAN/WEG2], [99FAN/KON]. The data in dilute solutions [90FEL/RAI], 0.1 m NaClO$_4$ [89NIT/STA], [91MEI/KIM], [94KIM/KLE], [97WRU/PAL], 1.05 m NaClO$_4$ [82LUN] and 3.5 m NaClO$_4$ [89ROB] are included for comparison. The solid and dotted lines are calculated with the selected values and uncertainty limits of $\log_{10} K_n°$ ((12.17)-(12.19)) and $\Delta\varepsilon$((12.17)-(12.19)).

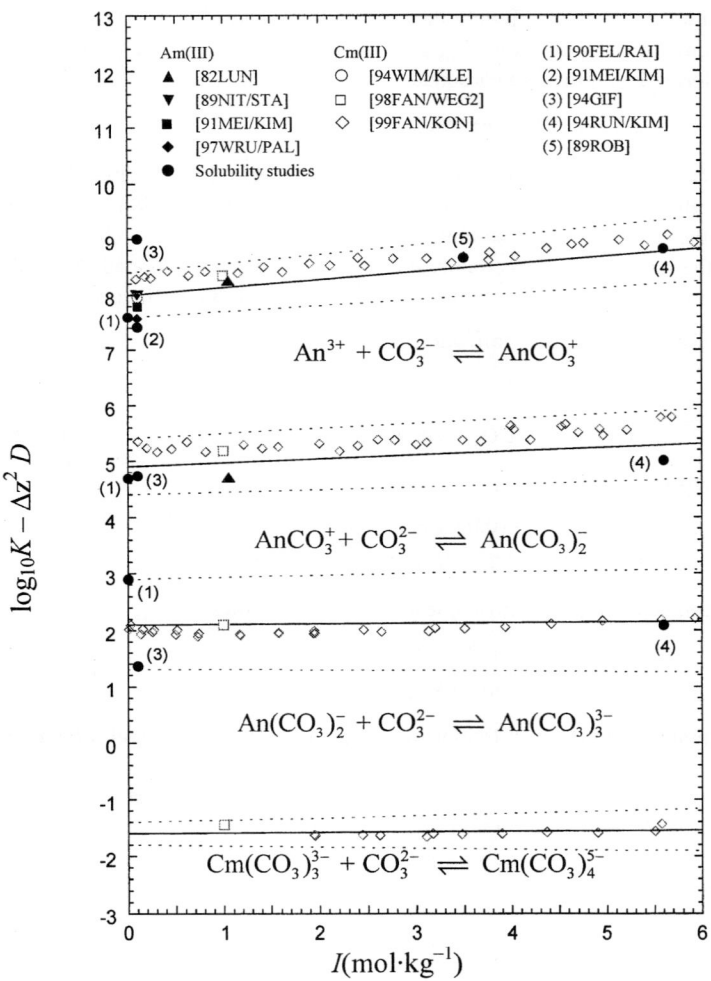

Because of the lack of sufficient experimental data, the previous review [95SIL/BID] used estimated ion interaction coefficients of:

$\varepsilon(\text{AmCO}_3^+, \text{ClO}_4^-) = (0.17 \pm 0.04) \text{ kg} \cdot \text{mol}^{-1}$,
$\varepsilon(\text{Am(CO}_3)_2^-, \text{Na}^+) = - (0.05 \pm 0.05) \text{ kg} \cdot \text{mol}^{-1}$
$\varepsilon(\text{Am(CO}_3)_3^{3-}, \text{Na}^+) = - (0.15 \pm 0.05) \text{ kg} \cdot \text{mol}^{-1}$.

Based on $\Delta\varepsilon((12.16)-(12.18))$ in NaCl solutions, $\varepsilon(\text{Am}^{3+}, \text{Cl}^-) = (0.23 \pm 0.02) \text{ kg} \cdot \text{mol}^{-1}$ (selected in this review, Appendix B) and $\varepsilon(\text{Na}^+, \text{CO}_3^{2-}) = - (0.08 \pm 0.03) \text{ kg} \cdot \text{mol}^{-1}$, the present review calculates the following ion interaction coefficients for the carbonate complexes of Cm(III) and Am(III):

$\varepsilon(\text{AmCO}_3^+, \text{Cl}^-) = (0.01 \pm 0.05) \text{ kg} \cdot \text{mol}^{-1}$,
$\varepsilon(\text{Am(CO}_3)_2^-, \text{Na}^+) = - (0.14 \pm 0.06) \text{ kg} \cdot \text{mol}^{-1}$,
$\varepsilon(\text{Am(CO}_3)_3^{3-}, \text{Na}^+) = - (0.23 \pm 0.07) \text{ kg} \cdot \text{mol}^{-1}$.

They are consistent with values of corresponding charge types or estimates given in [95SIL/BID].

The selected values of $\log_{10} \beta_1^\circ$ and the stepwise equilibrium constants $\log_{10} K_2^\circ$ and $\log_{10} K_3^\circ$ lead to the following overall formation constants:

$\log_{10} \beta_1^\circ \ (\text{AnCO}_3^+, 298.15 \text{ K}) = (8.0 \pm 0.4)$,
$\log_{10} \beta_2^\circ \ (\text{An(CO}_3)_2^-, 298.15 \text{ K}) = (12.9 \pm 0.6)$,
$\log_{10} \beta_3^\circ \ (\text{An(CO}_3)_3^{3-}, 298.15 \text{ K}) = (15.0 \pm 1.0)$.

The standard molar Gibbs energies of formation for the carbonate complexes of Am(III) derived from the selected formation constants are:

$\Delta_f G_m^\circ (\text{AmCO}_3^+, 298.15 \text{ K}) = - (1172.3 \pm 5.3) \text{ kJ} \cdot \text{mol}^{-1}$,
$\Delta_f G_m^\circ (\text{Am(CO}_3)_2^-, 298.15 \text{ K}) = - (1728.1 \pm 5.9) \text{ kJ} \cdot \text{mol}^{-1}$,
$\Delta_f G_m^\circ (\text{Am(CO}_3)_3^{3-}, 298.15 \text{ K}) = - (2268.0 \pm 7.5) \text{ kJ} \cdot \text{mol}^{-1}$.

12.6.1.1.1.1 Tetracarbonato complex

The formation of the tetracarbonato complex, $\text{M(CO}_3)_4^{5-}$, has been observed for Cm(III) by TRLFS [98FAN/WEG2], [99FAN/KON] and also for lanthanide ions, from the limiting slopes in a solubility study with Ce(III) [83FER/GRE] and solvent extraction experiments with Eu(III) [89CHA/RAO], [91RAO/CHA]. For chemical reasons one would expect the same behaviour for Am(III). However, the available spectroscopic and solubility data for Am(III) do not provide evidence for the reaction:

$$\text{Am}^{3+} + 4 \text{ CO}_3^{2-} \rightleftharpoons \text{Am(CO}_3)_4^{5-} \qquad (12.20)$$

up to carbonate concentrations of 1 mol · L^{-1}. We cannot exclude that, in the case of the complex, $M(CO_3)_4^{5-}$, where the central ion is highly coordinated by bidentate carbonate ligands, the very small differences in the ionic radii of Cm^{3+} and Am^{3+} may cause somewhat larger differences in the formation constant. (Such effects are known for chelate ligands used for the separation of trivalent actinides and lanthanides). Therefore, the present review selects no thermodynamic data for the complex, $Am(CO_3)_4^{5-}$.

Moreover, the conditional equilibrium constants for $Cm(CO_3)_4^{5-}$ determined in the TRLFS studies of Fanghänel et al. yield $\log_{10} K_4^\circ = -2.16$ as calculated in [99FAN/KON] with the Pitzer approach, whereas the linear SIT extrapolation to $I = 0$ would lead to a considerably different value, $\log_{10} K_4^\circ = -(1.6 \pm 0.1)$, as indicated by the dashed line in Figure 12-9. This reflects a general problem concerning the evaluation of equilibrium constants at zero ionic strength when highly charged aqueous species are involved in the reaction.

12.6.1.1.1.2 Am(III) and Cm(III) bicarbonate complexes

After a critical discussion of the published literature, the previous review [95SIL/BID] concluded that there is no experimental evidence for the existence of strong americium bicarbonate complexes. However, evidence for rather weak bicarbonate complexes was found for a number of trivalent lanthanides with $\log_{10} \beta_1$ ($LnHCO_3^{2+}$, 0.7 M $NaClO_4$, 298.15 K) in the range of 1.6 to 1.9 [93LEE/BYR], and for yttrium(III) with $\log_{10} \beta_1^\circ$ ($YHCO_3^{2+}$, 298.15 K) = (2.4 ± 0.1) [85SPA].

A spectroscopic study of the first Cm(III) bicarbonate complex was reported later by Fanghänel et al. [98FAN/WEG]. Using TRLFS in 1 M NaCl solutions of pH 3 – 6 under carbon dioxide partial pressures of 0.5 to 11 bar, Fanghänel et al. [98FAN/WEG] demonstrated the formation of a Cm(III) bicarbonate complex, $CmHCO_3^{2+}$:

$$Cm^{3+} + HCO_3^- \rightleftharpoons CmHCO_3^{2+}. \tag{12.21}$$

The reported formation constant ($\log_{10} \beta_1$ ($CmHCO_3^{2+}$, 1 m NaCl, 298.15 K) = (1.9 ± 0.2) [98FAN/WEG]) is converted using the SIT to $I = 0$ (cf. Appendix A) and accepted in the present review for the analogous Am(III) complex:

$$\log_{10} \beta_1^\circ ((12.21), 298.15 \text{ K}) = (3.1 \pm 0.3)$$

and

$$\Delta_f G_m^\circ (AmHCO_3^{2+}, 298.15 \text{ K}) = -(1203.2 \pm 5.1) \text{ kJ} \cdot \text{mol}^{-1}.$$

As the formation constant of $AmHCO_3^{2+}$ is about five orders of magnitude lower than that of the carbonate complex, $AmCO_3^+$, the weak bicarbonate complex becomes predominant only under high carbon dioxide partial pressures above 1 bar.

12.6.1.1.1.3 Mixed Am(III) and Cm(III) hydroxide–carbonate complexes

In the previous review [95SIL/BID] it was shown that all experimental studies claiming the formation of mixed Am(III) hydroxide-carbonate complexes [69SHI/GIV], [82BID], [84BER/KIM], [86EWA/HOW], [87CRO/EWA] could be reinterpreted with chemical models that included only binary carbonate and/or hydroxide complexes. This conclusion is supported by TRLFS studies of Cm(III) in carbonate-bicarbonate solutions [98FAN/WEG], [98FAN/WEG2], [99FAN/KON]. In these studies, pH and carbonate concentration were varied over a wide range, but there was no evidence for the formation of ternary hydroxide-carbonate complexes.

The formation of Cm(III) hydroxide-carbonate complexes could also be ruled out under conditions corresponding to carbon dioxide partial pressures $\geq 10^{-3.5}$ bar in 1 m NaCl. Neck et al. [98NEC/FAN] calculated upper limits in 1 m NaCl of $\log_{10} \beta_{1,1}$ (12.22) < 11, $\log_{10} \beta_{1,2}$ (12.22) < 15 and $\log_{10} \beta_{2,1}$ (12.22) < 16.5 for the reactions:

$$Cm^{3+} + p\,OH^- + q\,CO_3^{2-} \rightleftharpoons Cm(OH)_p(CO_3)_q^{3-p-2q} \qquad (12.22).$$

12.6.1.1.2 Am(V) carbonate complexes

The two reports [94GIF] and [94RUN/KIM] were published after the previous review [95SIL/BID] was finalised. They include solubility data for $NaAmO_2CO_3(s)$ over a wide range of carbonate concentrations in 3, 4 and 5 M NaCl where the pentavalent americium, $^{241}Am(V)$, is formed by radiolytic oxidation.

Giffaut [94GIF] measured batch solubilities at 294.15 K in 4 M NaCl containing carbonate concentrations in the range $-5.5 < \log_{10}[CO_3^{2-}] < -1.5$. Preliminary results of this study [93GIF/VIT] were not credited in [95SIL/BID] because of the lack of experimental details.

Runde and Kim [94RUN/KIM] measured the solubility in 3 and 5 M NaCl at 295.15 K as a function of $\log_{10}[H^+]$ under an atmosphere of 0.01 bar CO_2 in argon. The same data were also reported in a later paper [96RUN/NEU].

Although the solid phases were not characterised by X–ray diffraction, there is sufficient evidence from analogous studies with Np(V) that the measured solubilities actually refer to $NaAmO_2CO_3(s)$ as assumed by the authors (cf. Appendix A, discussion of [94GIF] and [94RUN/KIM]).

The present review uses the stepwise formation constants determined in these studies to evaluate thermodynamic data for the aqueous carbonate complexes of Am(V). In Table 12-9, they are converted to zero ionic strength with the ion interaction coefficients selected in [2001LEM/FUG] for the analogous complexes of Np(V):

$\varepsilon(\text{NpO}_2^+, \text{Cl}^-) = (0.09 \pm 0.05) \text{ kg} \cdot \text{mol}^{-1}$,
$\varepsilon(\text{NpO}_2\text{CO}_3^-, \text{Na}^+) = -(0.18 \pm 0.15) \text{ kg} \cdot \text{mol}^{-1}$,
$\varepsilon(\text{NpO}_2(\text{CO}_3)_2^{3-}, \text{Na}^+) = -(0.33 \pm 0.17) \text{ kg} \cdot \text{mol}^{-1}$,
$\varepsilon(\text{NpO}_2(\text{CO}_3)_3^{5-}, \text{Na}^+) = -(0.53 \pm 0.19) \text{ kg} \cdot \text{mol}^{-1}$.

Because of the high ionic strength employed, application of the SIT equation for the extrapolation to zero ionic strength leads to large uncertainties which increase with the NaCl concentration. Therefore, the present review selects the weighted averages of the values converted to $I = 0$. As can be seen from the comparison in Table 12-9 they are close to the equilibrium constants selected in [2001LEM/FUG] for the analogous Np(V) complexes.

The selected values of the stepwise equilibrium constants correspond to the following overall formation constants for the reactions:

$$\text{AmO}_2^+ + n\,\text{CO}_3^{2-} \rightleftharpoons \text{AmO}_2(\text{CO}_3)_n^{1-2n}, \qquad (12.23)$$

$\log_{10} \beta_1^\circ\,(\text{AmO}_2\text{CO}_3^-, 298.15 \text{ K}) = (5.1 \pm 0.5)$,
$\log_{10} \beta_2^\circ\,(\text{AmO}_2(\text{CO}_3)_2^{3-}, 298.15 \text{ K}) = (6.7 \pm 0.8)$,
$\log_{10} \beta_3^\circ\,(\text{AmO}_2(\text{CO}_3)_3^{5-}, 298.15 \text{ K}) = (5.1 \pm 1.0)$.

The possible error due to the somewhat lower temperature in [94GIF], [94RUN/KIM] compared to the standard state of 298.15 K is negligible compared to other uncertainties. The standard molar Gibbs energies of formation for the carbonate complexes of Am(V) derived from the selected formation constants are:

$\Delta_f G_m^\circ\,(\text{AmO}_2\text{CO}_3^-, 298.15 \text{ K}) = -(1296.8 \pm 6.8) \text{ kJ} \cdot \text{mol}^{-1}$,
$\Delta_f G_m^\circ\,(\text{AmO}_2(\text{CO}_3)_2^{3-}, 298.15 \text{ K}) = -(1833.8 \pm 7.7) \text{ kJ} \cdot \text{mol}^{-1}$,
$\Delta_f G_m^\circ\,(\text{AmO}_2(\text{CO}_3)_3^{5-}, 298.15 \text{ K}) = -(2352.6 \pm 8.5) \text{ kJ} \cdot \text{mol}^{-1}$.

Table 12-9: Stepwise formation constants of aqueous Am(V) carbonate complexes from the solubility studies in [94GIF], [94RUN/KIM]

reaction:	$AmO_2^+ + CO_3^{2-} \rightleftharpoons AmO_2CO_3^-$			
Medium,	$t(°C)$	$\log_{10} \beta_1$	$\log_{10} \beta_1°$	Reference
3.0 M NaCl	22	(4.74 ± 0.09)	(5.09 ± 0.74) [a]	[94RUN/KIM]
4.0 M NaCl	21	(5.4 ± 0.3)	(5.56 ± 0.91) [a]	[94GIF]
5.0 M NaCl	22	(4.69 ± 0.04)	(4.63 ± 0.95) [a]	[94RUN/KIM]
$I = 0$, weighted average			(5.1 ± 0.5)	present review
$I = 0$, Np(V)	25		(4.96 ± 0.06)	[2001LEM/FUG]

reaction:	$AmO_2CO_3^- + CO_3^{2-} \rightleftharpoons AmO_2(CO_3)_2^{3-}$			
Medium,	$t(°C)$	$\log_{10} K_2$	$\log_{10} K_2°$	Reference
3.0 M NaCl	22	(2.68 ± 0.10)	(1.44 ± 0.74) [a]	[94RUN/KIM]
4.0 M NaCl	21	(3.6 ± 0.2)	(2.22 ± 1.04) [a]	[94GIF]
5.0 M NaCl	22	(2.85 ± 0.08)	(1.35 ± 1.29) [a]	[94RUN/KIM]
$I = 0$, weighted average			(1.6 ± 0.6)	present review
$I = 0$, Np(V)	25		(1.57 ± 0.08)	[2001LEM/FUG]

reaction:	$AmO_2(CO_3)_2^{3-} + CO_3^{2-} \rightleftharpoons AmO_2(CO_3)_3^{5-}$			
Medium,	$t(°C)$	$\log_{10} K_3$	$\log_{10} K_3°$	Reference
3.0 M NaCl	22	(2.12 ± 0.13)	$-(1.16 \pm 0.83)$ [a]	[94RUN/KIM]
4.0 M NaCl	21	(1.4 ± 0.3)	$-(2.25 \pm 1.16)$ [a]	[94GIF]
5.0 M NaCl	22	(2.09 ± 0.09)	$-(1.81 \pm 1.45)$ [a]	[94RUN/KIM]
$I = 0$, weighted average			$-(1.6 \pm 0.6)$	present review
$I = 0$, Np(V)	25		$-(1.03 \pm 0.11)$	[2001LEM/FUG]

(a) Converted to the molal units and extrapolated to $I = 0$ with the ion interaction coefficients selected in [2001LEM/FUG] for the analogous Np(V) complexes.

12.6.1.1.3 Solid americium carbonates (V.7.1.2.2)

12.6.1.1.3.1 Americium(III) hydroxycarbonate AmOHCO₃(s)

The solubility constant selected in the previous NEA review [95SIL/BID] for the reaction:

$$AmOHCO_3(cr) \rightleftharpoons Am^{3+} + OH^- + CO_3^{2-} \qquad (12.24)$$

is based on two largely discrepant values derived in the solubility studies of Silva and Nitsche [84SIL/NIT] and Runde et al. [92RUN/MEI], and thus has a large uncertainty:

$$\log_{10} K_{s,0}° (AmOHCO_3, cr, 298.15 K) = -(21.2 \pm 1.4).$$

In both studies [84SIL/NIT], [92RUN/MEI] the solid was characterised by X-ray diffraction and identified as orthorhombic Am(III) hydroxycarbonate. No solubility data are available for the hexagonal form observed by Standifer and Nitsche [88STA/NIT] at 333.15 K. The solubility study in [84BER/KIM] was disregarded, because the solid was not characterised. The authors' assumption that, under the constant CO_2 partial pressure of $p_{CO_2} = 10^{-3.5}$ bar, the initial solid $Am(OH)_3(s)$ converted rapidly enough into the more stable $AmOHCO_3(s)$, was not considered as sufficiently proven.

Table 12-10: Solubility constants reported for Am(III) hydroxycarbonate. All data for crystalline solids refer to the orthorhombic modification.

Medium	$t(°C)$	solid	$\log_{10} K_{s,0}$	$\log_{10} K^°_{s,0}$	Reference
0.1 M $NaClO_4$	25	$^{241}AmOHCO_3(s)$	$-(18.70 \pm 0.12)$	$-(20.2 \pm 0.2)^{(b)}$	[92RUN/MEI]
0.1 M $NaClO_4$	25	$^{243}AmOHCO_3(cr)$	$-(20.8 \pm 0.3)^{(a)}$	$-(22.3 \pm 0.3)^{(a)}$	[84SIL/NIT]
dilute solutions (corr. to $I = 0$)	r.t.	$^{243}AmOHCO_3(cr)$		$-22.5^{(c)}$	[90FEL/RAI]
0.1 M $NaClO_4$	25	$^{241}AmOHCO_3(s)$? not characterised	$-(21.03 \pm 0.11)$	$-(22.5 \pm 0.2)^{(b)}$	[84BER/KIM]
0.1 M NaCl	21	$^{241}AmOHCO_3(s)$? not characterised	$-(21.0 \pm 0.4)$	$-(22.5 \pm 0.4)^{(b)}$	[94GIF]
4 M NaCl	21	$^{241}AmOHCO_3(s)$? not characterised	$-(20.7 \pm 0.4)$	$-(23.3 \pm 0.5)^{(b)}$	[94GIF]
$I = 0$		$^{241}AmOHCO_3 \cdot 0.5\, H_2O(cr)$		$-(23.1 \pm 1.0)^{(d)}$ $-(22.6 \pm 1.0)^{(d,e)}$	[96MER/FUG]

(a) Re-calculated in [95SIL/BID].
(b) Re-calculated with the SIT coefficients in Appendix B. The uncertainty corresponds to the 95% confidence interval.
(c) Felmy et al. [90FEL/RAI] combined their solubility measurements at pH > 6.5 with those of [84SIL/NIT], from which the solubility constant at $I = 0$ is re-calculated with the ion–interaction (Pitzer) approach.
(d) Calculated from thermochemical data
(e) Re-calculated in the present review.
r.t. Room temperature.

In a later study, Giffaut [94GIF] performed batch solubility experiments at $I = 0.1$ and 4.0 mol · L^{-1} ($NaHCO_3/Na_2CO_3/NaCl$). The precipitates were not characterised. $AmOHCO_3(s)$ was assumed to be the solubility limiting solid as concluded from the observed dependence of the americium concentration on $\log_{10}[H^+]$, $\log_{10}[CO_3^{2-}]$, and the CO_2 partial pressure.

Merli and Fuger [96MER/FUG] determined the standard enthalpy of formation of $^{241}AmOHCO_3 \cdot 0.5H_2O(cr)$ and of crystalline and amorphous lanthanide hydroxycarbonates ($LnOHCO_3 \cdot 0.5H_2O(cr)$ with Ln = Nd, Sm; $LnOHCO_3 \cdot 0.5H_2O(am)$ with Ln = Dy, Yb) from their enthalpies of solution in 1 M HCl (cf. Appendix A). The reported standard enthalpy of formation, deduced by a thermodynamic cycle and NEA auxiliary data selected in [95SIL/BID]:

$$\Delta_f H_m^\circ (\text{AmOHCO}_3 \cdot 0.5\text{H}_2\text{O, cr, 298.15 K}) = -(1682.9 \pm 2.6) \text{ kJ} \cdot \text{mol}^{-1}$$

is retained in the present review. Merli and Fuger [96MER/FUG] estimated a value of $S_m^\circ(\text{AmOHCO}_3 \cdot 0.5\text{H}_2\text{O, cr, 298.15 K}) = (154 \pm 10) \text{ J} \cdot \text{K}^{-1} \cdot \text{mol}^{-1}$ which is based on their selection of $S_m^\circ(\text{Am(OH)}_3\text{, cr, 298.15 K})$, and ultimately on the experimental value of $S_m^\circ(\text{Eu(OH)}_3\text{, cr, 298.15 K})$. The present review includes, as in the case of Am(OH)$_3$(cr) (cf. section 12.3.2.2), an entropy correction of $-9.5 \text{ J} \cdot \text{K}^{-1} \cdot \text{mol}^{-1}$ arising from the contribution of thermally populated higher electronic states in europium compounds lacking in Am(III) species, leading to:

$$S_m^\circ (\text{AmOHCO}_3 \cdot 0.5\text{H}_2\text{O, cr, 298.15 K}) = (144 \pm 10) \text{ J} \cdot \text{K}^{-1} \cdot \text{mol}^{-1}.$$

With this value, the Gibbs energy for the reaction:

$$\text{AmOHCO}_3 \cdot 0.5\text{H}_2\text{O(cr)} \rightleftharpoons \text{Am}^{3+} + \text{OH}^- + \text{CO}_3^{2-} + 0.5\text{H}_2\text{O(l)} \qquad (12.25)$$

and a solubility constant of:

$$\log_{10} K_{s,0}^\circ (\text{AmOHCO}_3 \cdot 0.5\text{H}_2\text{O, cr, 298.15 K}) = -(22.6 \pm 1.0)$$

are calculated. This latter value is in excellent agreement with the results from solubility measurements in [84SIL/NIT], [90FEL/RAI], [84BER/KIM], [94GIF] (cf. Table 12-10). The good agreement between thermochemical and solubility data for crystalline Am(III) hydroxycarbonate is further corroborated by analogous results for the orthorhombic modification of crystalline Nd(III) hydroxycarbonate. The solubility constant determined in [96MER/FUG] from thermochemical data,

$$\log_{10} K_{s,0}^\circ (\text{NdOHCO}_3 \cdot 0.5\text{H}_2\text{O(cr)}, 298.15 \text{ K}) = -(21.3 \pm 0.7),$$

is in excellent agreement with those derived from solubility experiments with Nd hydroxycarbonate (re-calculated to $I = 0$ as described in Appendix A, discussion of [96MER/FUG]):

$$\log_{10} K_{s,0}^\circ = -(20.7 \pm 0.2) \quad \text{[91MEI/KIM2]},$$

$$\log_{10} K_{s,0}^\circ = -(21.4 \pm 0.3) \quad \text{[92RUN/MEI]},$$

$$\log_{10} K_{s,0}^\circ = -(21.6 \pm 0.2) \quad \text{[93MEI/TAK]},$$

$$\log_{10} K_{s,0}^\circ = -(21.75 \pm 0.3) \quad \text{[93CAR]}.$$

For crystalline Am(III) hydroxycarbonate, AmOHCO$_3 \cdot$ 0.5H$_2$O(cr), we select the mean value of the solubility constants determined by [84SIL/NIT], who did not analyse their solid for the water content, and [96MER/FUG]:

and
$$\log_{10} K_{s,0}^\circ \text{ (AmOHCO}_3 \cdot 0.5\text{H}_2\text{O, cr, 298.15 K)} = -(22.4 \pm 0.5)$$

$$\Delta_f G_m^\circ \text{ (AmOHCO}_3 \cdot 0.5\text{H}_2\text{O, cr, 298.15 K)} = -(1530.2 \pm 5.6) \text{ kJ} \cdot \text{mol}^{-1}.$$

By combining the standard molar Gibbs energy with the selected value of:

$$\Delta_f H_m^\circ \text{ (AmOHCO}_3 \cdot 0.5\text{H}_2\text{O, cr, 298.15 K)} = -(1682.9 \pm 2.6) \text{ kJ} \cdot \text{mol}^{-1}$$

from [96MER/FUG] and auxiliary data from [95SIL/BID], the internally consistent and thus selected value of the standard molar entropy is calculated to be:

$$S_m^\circ \text{ (AmOHCO}_3 \cdot 0.5\text{H}_2\text{O, cr, 298.15 K)} = (141 \pm 21) \text{ J} \cdot \text{K}^{-1} \cdot \text{mol}^{-1}.$$

The enthalpy of solution determined in [96MER/FUG] with a sample of ^{241}AmOHCO$_3$(cr) aged for 40 days was about 40 kJ · mol^{-1} more negative than for samples of freshly prepared solids. As noted in Appendix A, [96MER/FUG] indicated that the difference with a fresh solid could have been caused by partial loss of crystallinity and very likely chemical alteration as a result of α–irradiation. In any case, differences of the same magnitude were observed between crystalline and amorphous lanthanide hydroxycarbonates. The standard entropies of the amorphous hydroxycarbonates can hardly be estimated, but their solubility constants are expected to be at least two orders of magnitude higher [96MER/FUG]. Such an effect would explain the difference between the solubility data measured in [92RUN/MEI] with ^{241}AmOHCO$_3$(s) and in [84SIL/NIT] with ^{243}AmOHCO$_3$(cr). Therefore, the solubility constant determined by Runde et al. [92RUN/MEI] is ascribed to an aged solid, rendered amorphous by α–irradiation. As there are no other data for comparison and because solubility data for amorphous solids are often widely scattered, the solubility constant given in [92RUN/MEI] is selected with the uncertainty increased to ± 1 log$_{10}$ unit:

$$\log_{10} K_{s,0}^\circ \text{ (AmOHCO}_3 \cdot x\text{H}_2\text{O, am, hydr., 298.15 K)} = -(20.2 \pm 1.0).$$

12.6.1.1.3.2 Americium(III) carbonate Am$_2$(CO$_3$)$_3$(s)

Solubility data for hydrated Am$_2$(CO$_3$)$_3$ · xH$_2$O(s) have been determined by Meinrath and Kim [91MEI/KIM], [91MEI/KIM2], and Runde et al. [92RUN/MEI] in 0.1 M NaClO$_4$ and by Robouch [89ROB] in 3.0 M NaClO$_4$ (Table 12-11). The reported content of crystal water, which can vary over a wide range (from x = 2 to 8), was disregarded in the previous review [95SIL/BID]. The reported conditional constants for the reaction:

$$\frac{1}{2}\text{Am}_2(\text{CO}_3)_3(s) \rightleftharpoons \text{Am}^{3+} + \frac{3}{2}\text{CO}_3^{2-} \quad (12.26)$$

were extrapolated to $I = 0$ with the SIT coefficients in Appendix B. The $\log_{10} K_{s,0}^\circ$ value calculated from the data in [89ROB] was found to be strongly discrepant, but there was no reason to discard it. The Am$_2$(CO$_3$)$_3$·xH$_2$O(s) prepared in [92RUN/MEI] was X–ray amorphous, whereas Robouch [89ROB] reported an X–ray diffraction pattern analogous to those of lanthanide carbonates. Silva *et al.* [95SIL/BID] recommended a mean value of:

$$\log_{10} K^\circ_{s,0} ((12.26), 298.15 \text{ K}) = -(16.7 \pm 1.1),$$

for crystalline $Am_2(CO_3)_3(cr)$. No new experimental investigations appeared since the publication of the previous review [95SIL/BID]. However, the solid in [92RUN/MEI] was explicitly described as X-ray amorphous and the solids in [91MEI/KIM], [91MEI/KIM2] were only identified by their chemical behaviour, which was analogous to that of lanthanide carbonates. In addition it is not clear whether the lower solubility constant, $\log_{10} K^\circ_{s,0}$, which was calculated by extrapolating the data of [89ROB] in 3 M $NaClO_4$ to zero ionic strength is actually due to a higher degree of crystallinity[1]. Therefore, contrary to [95SIL/BID], the present review ascribes the mean value of the reported solubility constants not to a well-defined crystalline $Am_2(CO_3)_3(cr)$, but to an amorphous solid phase, $Am_2(CO_3)_3 \cdot xH_2O(am)$, and selects:

$$\log_{10} K^\circ_{s,0} (Am_2(CO_3)_3 \cdot xH_2O, am, 298.15 \text{ K}) = -(16.7 \pm 1.1).$$

Table 12-11: Solubility constants reported for the reaction: $1/2 \, Am_2(CO_3)_3(s) \rightleftharpoons Am^{3+} + 3/2 \, CO_3^{2-}$

Medium	t (°C)	$\log_{10} K_{s,0}$	$\log_{10} K^\circ_{s,0}$	Reference
0.1 M $NaClO_4$	25	$-(14.90 \pm 0.13)$	$-(16.54 \pm 0.18)$ [a]	[91MEI/KIM]
0.1 M $NaClO_4$	22	$-(14.785 \pm 0.05)$	$-(16.38 \pm 0.10)$ [a]	[91MEI/KIM2]
0.1 M $NaClO_4$	25	$-(14.725 \pm 0.09)$	$-(16.32 \pm 0.18)$ [a]	[92RUN/MEI]
3.0 M $NaClO_4$	20	$-(15.08 \pm 0.15)$		[89ROB]
		$-(15.27 \pm 0.15)$ [b]	$-(17.54 \pm 0.24)$ [a,b]	

(a) Converted to $I = 0$ in [95SIL/BID] with:
$\varepsilon(Am^{3+}, ClO_4^-) = (0.49 \pm 0.03)$ kg·mol^{-1} and $\varepsilon(Na^+, CO_3^{2-}) = -(0.08 \pm 0.03)$ kg·mol^{-1}.
(b) Recalculated in [95SIL/BID].

12.6.1.1.3.3 Sodium americium(III) carbonates

Keller and Fang [69KEL/FAN] synthesised $NaAm(CO_3)_2 \cdot xH_2O(s)$ and $Na_3Am(CO_3)_3 \cdot xH_2O(s)$ and investigated their thermal decomposition. For the latter solid no thermodynamic data are available. The only thermodynamic data for $NaAm(CO_3)_2 \cdot xH_2O(s)$ cited in the previous review [95SIL/BID] are based on solubil-

---

[1] If the conditional solubility constant determined by [89ROB] in 3 M $NaClO_4$ is converted to $I = 0$ using $\varepsilon(Am^{3+}, ClO_4^-) = (0.49 \pm 0.03)$ kg·mol^{-1} [95SIL/BID] and $\varepsilon(Na^+, CO_3^{2-})_{NaClO_4} = (0.04 \pm 0.05)$ kg·mol^{-1} derived from [96FAN/NEC] instead of $\varepsilon(Na^+, CO_3^{2-}) = -(0.08 \pm 0.03)$ kg·mol^{-1} [95SIL/BID] (*cf.* Appendix D.2.2), the resulting value of $\log_{10} K^\circ_{s,0} (12.26) = -(16.74 \pm 0.32)$ is consistent with the $\log_{10} K^\circ_{s,0}$ values derived from the studies of [91MEI/KIM], [91MEI/KIM2], [92RUN/MEI] in 0.1 M $NaClO_4$ (Table 12-11).

ity experiments performed by Vitorge [84VIT] within the MIRAGE project, but these data are not published in the open literature. Due to the lack of experimental details, the solubility constants calculated from these data, $\log_{10} K_{s,0}$ (NaAm(CO$_3$)$_2$(s)) = $-$ 18.32 at I = 0.1 mol · L^{-1} [85KIM], [85KIM2] corresponding to $\log_{10} K^\circ_{s,0}$ (12.28) = $-$ 20.3 if converted to I = 0 with the SIT coefficients, $\log_{10} K^\circ_{s,0}$ (12.28) = $-$ 17.56 at I = 0 [85KIM2] and $\log_{10} K^\circ_{s,0}$ (12.28) = $-$ 17.38 at I = 0 [86AVO/BIL] are accepted neither in the previous nor in the present review.

Meinrath [91MEI], and Runde and Kim, [94RUN/KIM] investigated the solubility of NaAm(CO$_3$)$_2$·xH$_2$O(s) in carbonate solutions containing 5 M NaCl, under an atmosphere of 1 % CO$_2$ in argon (p_{CO_2} = 0.01 bar). The solids were characterised by thermogravimetry, IR spectroscopy and X–ray powder diffraction patterns. Well known analogous lanthanide compounds were prepared for comparison, NaNd(CO$_3$)$_2$·xH$_2$O(s) in [91MEI], [94RUN/KIM], NaEu(CO$_3$)$_2$·xH$_2$O(s) in [94RUN/KIM], and used as reference compounds. At low carbonate concentrations, Meinrath [91MEI], and Runde and Kim [94RUN/KIM] reported comparable solubilities, but at $\log_{10}[CO_3^{2-}] > -4$, Meinrath obtained two sets of higher and lower solubility data. He assumed the formation of two solids with different crystallinity and evaluated only the constant $\log_{10} K_{s,3}$ ((12.27), 5 M NaCl) = $-$ (3.85 ± 0.20) for the reaction:

$$\text{NaAm(CO}_3\text{)}_2 \cdot x\text{H}_2\text{O(cr)} + \text{CO}_3^{2-} \rightleftharpoons \text{Na}^+ + \text{Am(CO}_3\text{)}_3^{3-} + x\text{H}_2\text{O(l)} \quad (12.27)$$

Because of these ambiguities and shortcomings in the pH measurement for the determination of carbonate concentrations (*cf.* Appendix A), the experimental data of Meinrath [91MEI] are not used for the selection of thermodynamic data. Runde and Kim [94RUN/KIM] evaluated a solubility constant of $\log_{10} K_{s,0}$ ((12.28), 5 M NaCl) = $-$ (16.5 ± 0.3) for the reaction:

$$\text{NaAm(CO}_3\text{)}_2 \cdot x\text{H}_2\text{O(cr)} \rightleftharpoons \text{Na}^+ + \text{Am}^{3+} + 2\,\text{CO}_3^{2-} + x\,\text{H}_2\text{O(l)} \quad (12.28)$$

In order to calculate the solubility constant at I = 0, the number of water molecules of crystallisation is set equal to x = (5 ± 1). (The reported values for NaAm(CO$_3$)$_2$ · xH$_2$O(s) and the analogous Eu and Nd compounds are x = 4 [69KEL/FAN], x = 5 [91MEI], [94RUN/KIM] or x = 6 [91MEI], [74MOC/NAG]). The water activity in 5.6 m NaCl, a_{H_2O} = 0.7786, is estimated using the Pitzer model [91PIT]. Applying the SIT equation and a value of $\Delta\varepsilon(12.28)$ = (0.10 ± 0.06) kg · mol^{-1} according to the selected interaction coefficients in Appendix B, the solubility constant at I = 0 is calculated to be $\log_{10} K^\circ_{s,0}$ ((12.28), 298.15 K) = $-$ (21.0 ± 0.5). In regard to the high ionic strength it is noteworthy that the same value, $\log_{10} K^\circ_{s,0}$ ((12.28), 298.15 K) = $-$ (21.0 ± 0.4), has been calculated in [98NEC/FAN] with the Pitzer model. Accordingly, the present review recommends the following thermodynamic data:

$$\log_{10} K^\circ_{s,0} \text{ (NaAm(CO}_3\text{)}_2 \cdot 5\text{H}_2\text{O, cr, 298.15 K)} = -(21.0 \pm 0.5)$$

and

$$\Delta_f G^\circ_m \text{ (NaAm(CO}_3\text{)}_2 \cdot 5\text{H}_2\text{O, cr, 298.15 K)} = -(3222.0 \pm 5.6) \text{ kJ} \cdot \text{mol}^{-1}.$$

The selected solubility constant is supported by a similar value reported for the analogous Nd(III) compound. Rao et al. [96RAO/RAI], [99RAO/RAI] performed extensive solubility studies with $NaNd(CO_3)_2 \cdot 6H_2O$(cr), as a function of pH in $NaHCO_3$ and Na_2CO_3 solutions and mixtures with NaCl. Using the Pitzer model they evaluated a solubility constant of $\log_{10} K^°_{s,0}$ ($NaNd(CO_3)_2 \cdot 6H_2O$, cr, 298.15 K) = -21.39 (given without uncertainty).

12.6.1.1.3.4 Sodium dioxoamericium(V) carbonate $NaAmO_2CO_3$(s)

Solubility data for $NaAmO_2CO_3$(s) in 3, 4 and 5 M NaCl have been reported in [93GIF/VIT], [94GIF], [94RUN/KIM], [96RUN/NEU]. The preliminary results reported by Giffaut and Vitorge [93GIF/VIT] for the solubility of ^{241}Am(V), formed by radiolytic oxidation of ^{241}Am(III) in 4 M NaCl at 294.15 K, were not credited by the previous review [95SIL/BID], because no details were given in that paper.

Slightly different data were presented in Giffaut's doctoral thesis [94GIF], in which the batch experiments are described in detail. The solid phase formed in these solutions was not characterised, but the assumption of $NaAmO_2CO_3$(s) as the solubility controlling solid is justified, because the analogous Np(V) carbonate compound is known to be the stable solid under the conditions used in [94GIF] (cf. [2001LEM/FUG]). This assumption is further corroborated by Runde and Kim [94RUN/KIM], and Runde et al. [96RUN/NEU], who measured the solubility of both $NaAmO_2CO_3$(s) and $NaNpO_2CO_3$(s) in 3 and 5 M NaCl at 295.15 K as a function of $\log_{10}[H^+]$ under an atmosphere of 0.01 bar CO_2 in argon. The solubility data for ^{241}Am(V) are similar to those obtained with $NaNpO_2CO_3$(s) characterised by X–ray diffraction. The dependence of the solubility on $\log_{10}[CO_3^{2-}]$ is the same for both Am(V) and Np(V), which indicates that the solids have the same composition.

The experimental solubilities measured by Giffaut [94GIF] in the range $-5.5 < \log_{10}[CO_3^{2-}] < -1.5$ in 4 M NaCl compare well with those of Runde and Kim [94RUN/KIM] at the same carbonate concentrations in 3 and 5 M NaCl. However, the solubility constant given in [94GIF], $\log_{10} K_{s,0} = -(10.4 \pm 0.25)$, in 4 M NaCl is highly speculative because of the lack of data in the range $\log_{10}[CO_3^{2-}] < -5.5$, where the uncomplexed AmO_2^+(aq) ion contributes significantly to the total Am(V) concentration. The solubility constants given in [94RUN/KIM], [96RUN/NEU], $\log_{10} K_{s,0}$ (12.29) = $-(9.65 \pm 0.19)$ in 3 M NaCl and $-(9.56 \pm 0.13)$ in 5 M NaCl, are based on a sufficiently large number of solubility data in the range $-7 < \log_{10}[CO_3^{2-}] < -5.5$, where the solid is in equilibrium with AmO_2^+. Conversion to $I = 0$ with: $\varepsilon(Na^+, Cl^-) = (0.03 \pm 0.01)$ kg·mol^{-1}, $\varepsilon(Na^+, CO_3^{2-}) = -(0.08 \pm 0.03)$ kg·mol^{-1}, $\varepsilon(AmO_2^+, Cl^-)$ set equal to $\varepsilon(NpO_2^+, Cl^-) = (0.09 \pm 0.05)$ kg·mol^{-1} [2001LEM/FUG] leads to $\log_{10} K^°_{s,0}$ (12.29) = $-(10.93 \pm 0.27)$ and $-(10.77 \pm 0.36)$, respectively, for the reaction,

$$NaAmO_2CO_3(s) \rightleftharpoons Na^+ + AmO_2^+ + CO_3^{2-}. \qquad (12.29)$$

Accordingly the following thermodynamic data are selected:

$$\log_{10} K_{s,0}^\circ \text{ (NaAmO}_2\text{CO}_3, \text{ s, 298.15 K)} = - (10.9 \pm 0.4),$$

$$\Delta_f G_m^\circ \text{ (NaAmO}_2\text{CO}_3, \text{ s, 298.15 K)} = - (1591.9 \pm 6.6) \text{ kJ} \cdot \text{mol}^{-1}.$$

12.6.2 Aqueous americium silicates (V.7.2.2)

Up to the release of the previous review [95SIL/BID] no data were available for aqueous actinide(III) silicate complexes. Recently, Wadsak et al. [2000WAD/HRN] applied a solvent extraction method to study the interaction between silicate and Am(III) in 0.2 M NaClO$_4$ at 298.15 K. Keeping the total silica concentration constant at 0.03 mol $\cdot$ L^{-1} and varying pH in the range 3.0 – 3.8, they determined an equilibrium constant of $\log_{10}{}^* K$ ((12.30), 0.2 M NaClO$_4$) = – (2.16 ± 0.04) for the reaction:

$$\text{Am}^{3+} + \text{Si(OH)}_4\text{(aq)} \rightleftharpoons \text{AmSiO(OH)}_3^{2+} + \text{H}^+. \quad (12.30)$$

This review increases the uncertainty to the 1.96σ level (cf. Appendix A) and recalculates the values at $I = 0$ with the SIT using interaction coefficients of:

$$\varepsilon(\text{Am}^{3+}, \text{ClO}_4^-) = (0.49 \pm 0.03) \text{ kg} \cdot \text{mol}^{-1} \quad \text{[95SIL/BID]},$$
$$\varepsilon(\text{Na}^+, \text{SiO(OH)}_3^-) = - (0.08 \pm 0.03) \text{ kg} \cdot \text{mol}^{-1} \quad \text{[95SIL/BID]},$$
$$\varepsilon(\text{AmSiO(OH)}_3^{2+}, \text{ClO}_4^-) = (0.39 \pm 0.04) \text{ kg} \cdot \text{mol}^{-1} \text{ [95SIL/BID]},$$

for other Am(III) complexes with monovalent anions. This leads to:

$$\log_{10}{}^* K^\circ ((12.30), 298.15 \text{ K}) = - (1.61 \pm 0.08).$$

Steinle et al. [97STE/FAN] investigated the complexation of Cm(III) with monosilicic acid at room temperature by TRLFS. From the spectra at pH = 5.0 – 5.5 in 0.1 M NaClO$_4$ the authors calculated $\log_{10} K$ ((12.31), 0.1 M NaClO$_4$) = 7.4 (with an uncertainty of ± 0.1 as estimated in this review, Appendix A) for the reaction:

$$\text{Cm}^{3+} + \text{SiO(OH)}_3^- \rightleftharpoons \text{CmSiO(OH)}_3^{2+}. \quad (12.31)$$

Applying the known first dissociation constant of Si(OH)$_4$(aq) (Table 8-2), the equilibrium constant according to reaction (12.30) is calculated to be:

$$\log_{10}{}^* K \text{ ((12.30), 0.1 M NaClO}_4) = - (2.2 \pm 0.1),$$
$$\log_{10}{}^* K^\circ ((12.30), 298.15 \text{ K}) = - (1.76 \pm 0.10).$$

The formation constants reported in [2000WAD/HRN] for AmSiO(OH)$_3^{2+}$ and in [97STE/FAN] for the analogous Cm(III) complex, overlap within the uncertainty limits. The present review selects the unweighted average of:

$$\log_{10}{}^* K^\circ ((12.30), 298.15 \text{ K}) = - (1.68 \pm 0.18)$$

and

$$\Delta_f G_m^\circ (\text{AmSiO(OH)}_3^{2+}, 298.15 \text{ K}) = - (1896.8 \pm 5.0) \text{ kJ} \cdot \text{mol}^{-1}.$$

The results in [97STE/FAN], [2000WAD/HRN] show that the formation of aqueous Am(III) silicate complexes may be relevant in natural groundwater systems. However, in neutral and alkaline solutions with total silica concentrations above 10^{-3} mol · L^{-1}, the thermodynamic modelling is complicated by the formation of unknown complexes with polynuclear silicate anions.

12.7 Americium group 6 compounds and complexes (V.10)

12.7.1 Americium(III) molybdate compounds and complexes

There are no thermodynamic data available for americium(III) molybdate compounds and complexes. Using Nd(III) as an analog, Felmy et al. [95FEL/RAI] performed a solubility study over a wide range of pH, Na$_2$MoO$_4$ and NdCl$_3$ concentrations. The speciation is complicated by the large variety of possible solid phases and aqueous complexes with both monomeric molybdate and polymolybdate species. Therefore, the evaluated data and the stoichiometries of the proposed complexes are not definitely ascertained. However, the thermodynamic model of [95FEL/RAI], including ion interaction Pitzer parameters and the equilibrium constants listed below, may be used as guidance for modelling aqueous Am(III) molybdate systems.

$\text{Nd}^{3+} + 2\,\text{MoO}_4^{2-} \rightleftharpoons \text{Nd}(\text{MoO}_4)_2^{-}$ $\qquad \log_{10} K° = 11.2$

$2\,\text{Nd}^{3+} + \text{Mo}_7\text{O}_{20}(\text{OH})_4^{2-} \rightleftharpoons \text{Nd}_2\text{Mo}_7\text{O}_{24}(\text{aq}) + 4\,\text{H}^+$ $\qquad \log_{10} K° = 3.85$

$\text{NaNd}(\text{MoO}_4)_2(\text{cr}) \rightleftharpoons \text{Na}^+ + \text{Nd}^{3+} + 2\,\text{MoO}_4^{2-}$ $\qquad \log_{10} K° = -20.5$

$\text{Nd}_2(\text{MoO}_4)_3(\text{s, hydr.}) \rightleftharpoons 2\,\text{Nd}^{3+} + 3\,\text{MoO}_4^{2-}$ $\qquad \log_{10} K° = -26.1$.

12.7.2 Aqueous complexes with tungstophosphate and tungstosilicate heteropolyanions (V.10.2)

The stabilisation of Am(IV), Am(V) and Am(VI) in aqueous solutions of polyphosphatotungstate has been described in a number of publications already mentioned in the previous review [95SIL/BID]. The complexation of Am(III) and Am(IV) with the anions $W_{10}O_{36}^{12-}$, $P_2W_{17}O_{61}^{10-}$, $PW_{11}O_{39}^{7-}$ and $SiW_{11}O_{39}^{8-}$ is already reported in the earlier literature and studied in more detail in some recent papers [98CHA/DON], [98ERI/BAR], [99CHA/DON], [99YUS/SHI]. Erine et al. [98ERI/BAR] investigated the An(IV) – An(III) potentials, and the kinetics and activation parameters for redox reactions of Am, Cm, Bk and Cf in aqueous solutions of $K_{10}P_2W_{17}O_{61}$. The review article of Yusov and Shilov [99YUS/SHI] describes the Keggin or Dawson structures of lanthanide and actinide complexes with heteropolyanions. These authors also discuss the redox properties and present a number of formation constants.

Chartier, Donnet and Adnet [98CHA/DON], [99CHA/DON] investigated the kinetics and mechanism of redox processes involving Am(III), Am(IV), Am(V) and Am(VI) in 1 M HNO$_3$ solutions containing $\alpha_2 - P_2W_{17}O_{61}^{10-}$ and $\alpha - SiW_{11}O_{39}^{8-}$. The Am(V) complexes were found to be intermediate species of minor importance. The conditional formation constants of the Am(III) and Am(IV) complexes in 1 M HNO_3 determined in [98CHA/DON], [99CHA/DON] by spectroscopy and redox potential measurements at 298.15 K are summarised in Table 12-12, together with the constants for Am(III) complexes with $\alpha_2\text{-}P_2W_{17}O_{61}^{10-}$ and $\alpha\text{-}SiW_{11}O_{39}^{8-}$ in 2 M HNO_3 reported in conference abstracts by the same group of authors as cited in [99YUS/SHI].

Applying Laser Induced Fluorescence Spectroscopy at varying acid concentrations (0.1 – 2 M HNO_3), Ioussov and Krupa [97IOU/KRU] obtained comparable results for the 1:1 complex of Cm(III) with $SiW_{11}O_{39}^{8-}$ and, in addition, the corresponding equilibrium constants for the complex, $Cm(PW_{11}O_{39})^{4-}$.

For all of these ligands and complexes the degree of protonation is unknown. The constants in Table 12-12 represent conditional equilibrium constants calculated with the total concentration, without taking the degree of protonation into account. As a consequence the conditional constants depend strongly on the acid concentration. Moreover, the high charge of the heteropolyanions does not allow a reliable extrapolation to zero ionic strength. Therefore, no thermodynamic data are selected in this review, but the conditional constants given in Table 12-12 may be considered as guidelines.

Table 12-12: Conditional formation constants of Am(III), Cm(III) and Am(IV) complexes with tungstophosphate and tungstosilicate heteropolyanions. (The constants refer to the total ligand and complex concentrations, independent of the degree of protonation, which is not known).

$Am^{3+} + n\, P_2W_{17}O_{61}^{10-} \rightleftharpoons Am(P_2W_{17}O_{61})_n^{3-10n}$			
Medium	$\log_{10} \beta_1$	$\log_{10} \beta_2$	Reference
1 M HNO_3, $t = 25°C$	(2.7 ± 0.1)		[98CHA/DON],[99CHA/DON]
2 M HNO_3, $t = ?$	1.9	3	[92ADN/MAD]
$Cm^{3+} + PW_{11}O_{39}^{7-} \rightleftharpoons Cm(PW_{11}O_{39})^{4-}$			
Medium	$\log_{10} \beta_1$		Reference
HNO_3–$NaNO_3$, room temp.			[97IOU/KRU]
0.1 M HNO_3	(6.7 ± 0.2)		
0.1 M HNO_3 + 1.0 M $NaNO_3$	(6.6 ± 0.2)		
0.2 M HNO_3	(6.3 ± 0.2)		
0.2 M HNO_3 + 1.0 M $NaNO_3$	(6.3 ± 0.2)		
0.5 M HNO_3	(6.0 ± 0.2)		
0.5 M HNO_3 + 1.0 M $NaNO_3$	(6.1 ± 0.2)		
1.0 M HNO_3	(5.6 ± 0.3)		
1.0 M HNO_3 + 1.0 M $NaNO_3$	(5.5 ± 0.3)		
2.0 M HNO_3	(4.9 ± 0.3)		

(Continued on next page)

Table 12-12 (continued)

$Am^{3+} + n\, SiW_{11}O_{39}^{8-} \rightleftharpoons Am(SiW_{11}O_{39})_n^{3-8n}$			
Medium	$\log_{10} \beta_1$	$\log_{10} \beta_2$	Reference
1 M HNO$_3$, 25°C	(4.4 ± 0.4)	(6.7 ± 0.4)	[98CHA/DON],[99CHA/DON]
2 M HNO$_3$, t (?)	3.3		[99YUS/SHI]

$Cm^{3+} + n\, SiW_{11}O_{39}^{8-} \rightleftharpoons Cm(SiW_{11}O_{39})_n^{3-8n}$		
Medium	$\log_{10} \beta_1$	Reference.
HNO$_3$–NaNO$_3$, room temp.		[97IOU/KRU]
0.1 M HNO$_3$	(6.5 ± 0.2)	
0.1 M HNO$_3$ + 1.0 M NaNO$_3$	(6.4 ± 0.2)	
0.2 M HNO$_3$	(6.0 ± 0.2)	
0.2 M HNO$_3$ + 1.0 M NaNO$_3$	(6.0 ± 0.2)	
0.5 M HNO$_3$	(5.2 ± 0.2)	
0.5 M HNO$_3$ + 1.0 M NaNO$_3$	(5.1 ± 0.2)	
1.0 M HNO$_3$	(4.4 ± 0.3)	
1.0 M HNO$_3$ + 1.0 M NaNO$_3$	(4.3 ± 0.3)	
2.0 M HNO$_3$	(3.7 ± 0.3)	

$Am^{4+} + n\, P_2W_{17}O_{61}^{10-} \rightleftharpoons Am(P_2W_{17}O_{61})_n^{4-10n}$			
Medium	$\log_{10} \beta_1$	$\log_{10} \beta_2$	Reference
1 M HNO$_3$, 25°C	(19.3 ± 0.2) [a]	(22.9 ± 0.2) [a]	[98CHA/DON]
	(19.2 ± 0.2)	(22.8 ± 0.2)	[99CHA/DON]

$Am^{4+} + n\, SiW_{11}O_{39}^{8-} \rightleftharpoons Am(SiW_{11}O_{39})_n^{4-8n}$			
Medium	$\log_{10} \beta_1$	$\log_{10} \beta_2$	Reference
1 M HNO$_3$, 25°C	(21.3 ± 0.3) [a]	(26.1 ± 0.4) [a]	[98CHA/DON]
	(21.3 ± 0.3)	(26.2 ± 0.2)	[99CHA/DON]

[a] Preliminary results

Chapter 13

Discussion of new data selection for Technetium

13.1 Elemental technetium (V.1)

13.1.1 Heat capacity and entropy (V.1.1.2)

There are three recent papers concerning the heat capacity of Tc(cr). In her thesis, Boucharat [97BOU] reports DSC measurements from 673 to 1583 K, as described in Appendix A. However, her results show a maximum in $C_{p,m}$ around 1080 K and a minimum around 1540 K, and are *ca.* 10 - 20% lower than the experimental data by Spitsyn *et al.* [75SPI/ZIN]. There is no structural reason why Tc(cr) should have extrema in its heat capacity in this temperature range and these results have been discounted.

Van der Laan and Konings [2000LAA/KON] have studied the heat capacity of an alloy of approximate composition $Tc_{0.85}Ru_{0.15}$ from *ca.* 300 to 973 K, using a differential scanning calorimeter. The sample was formed by neutron irradiation of a disc of pure ^{99}Tc metal, (from the same source as the metal used by Boucharat [97BOU]), which transmutes some of the ^{99}Tc to ^{100}Ru. Isotope Dilution Mass Spectrometry (IDMS) indicated that the average ruthenium content was (15.0 ± 0.4) atom-%, although EPMA measurements showed a higher concentration of Ru at the rim of the sample, owing to shielding effects.

Heat capacities of pure Tc were derived by assuming a zero change of heat capacity upon alloying. As noted in Appendix A, these derived heat capacities are 3 − 7 % greater than the values estimated by [99RAR/RAN]. Such differences are well within the combined uncertainties.

Very recently, Shirasu and Minato [2002SHI/MIN] have determined the heat capacity of ^{99}Tc(cr) by DSC measurements from room temperature up to about 1100 K. Their measurements were made on a disk of Tc(cr) metal, prepared from highly-pure technetium metal powder (15 ppm total metallic impurities) by arc melting of the powder under an atmosphere of purified argon. Their data, which are presented only in the form of a figure and a fitted equation, agree excellently with those estimated in

[99RAR/RAN]. Since the data of [99RAR/RAN] form a completely consistent set for $C_{p,m}^\circ$, S_m°, and enthalpy increments, these values are retained.

[2002SHI/MIN] also reported $C_{p,m}^\circ$ data for an alloy of composition $Tc_{0.51}Ru_{0.49}(cr)$, which are close to the corresponding weighted sums of the heat capacities of Tc(cr) and Ru(cr).

Puigdomènech and Bruno [95PUI/BRU] provides no new thermodynamic data and their calculated thermodynamic quantities for aqueous and solid compounds were not used in the assessment of [99RAR/RAN].

13.2 Simple technetium aqua ions of each oxidation state (V.2)

13.2.1 TcO_4^- (V.2.1)

Ben Said et al. [98BEN/FAT] measured E for the $TcO_4^- / Tc(IV)$ couple in HCl/NaCl media in the presence of variable ratios of Fe(III)/Fe(II) using solvent extraction to determine the TcO_4^- concentration in the aqueous phase. The potential readings were as expected, highly dependent on the nature of the Tc(IV) species in solution (hydrolysis and chloride complexation). Therefore, $E^\circ > 0.844 \pm 0.006$ V/NHE is suggested for the standard potential of the $TcO_4^- / Tc(IV)$ couple in 1 M H^+ /1.3 M Cl$^-$ solutions.

During preparation of this review, abstracts from "The Third Russian-Japanese Seminar on Technetium" [2002GER/KOD] were published that contained a number of interesting summaries of structural, [2002GER/GRI], [2002KIR/GER] and [2002MAS/PER], spectroscopic and electrochemical studies of Tc(VII). However, as no detailed experimental data were provided, these results have not be considered further in this review.

13.3 Oxide and hydrogen compounds and complexes (V.3)

13.3.1 The acid/base chemistry of Tc(IV) (V.3.1.1)

A partial charge model, which is based on the electronegativity equalisation principle, was used by Henry and Merceron [94HEN/MER] to predict the speciation of Tc(IV) solutions as a function of pH at ambient conditions. The speciation is in qualitative agreement with [99RAR/RAN], but no quantitative equilibrium data are provided.

13.3.2 The protonation of TcO_4^- (V.3.1.2.1)

Solvent extraction experiments [94OMO/MUR] at 298.15 K of TcO_4^- from 1 M HCl/NaCl solutions produced a $\log_{10} K_C$ value of (1.02 ± 0.18), which is higher than any listed in [99RAR/RAN], but also appears to be dubious in view of methodology employed and the relatively high acidity at which this acid dissociates. Suzuki et al. [99SUZ/TAM] described solvent partitioning experiments involving $HTcO_4$, but these

13.3.3 General properties, hydration number (V.3.2.5.1)

Lefort [63LEF] had reported that pulse radiolysis of TcO_4^- in H_2SO_4 solutions led to reduction to Tc(IV) [99RAR/RAN]. However, Ben Said et al. [2001BEN/SEI] observed that no reduction occurred when 10^{-4} M TcO_4^- in 0.01 M CO_3^{2-} (pH = 11) was irradiated from a gamma source. The $CO_3^{\bullet -}$ radical is believed to oxidise Tc(IV) to Tc(VII), whereas irradiation in the presence of 0.01 M $HCOO^-$ produces colloidal TcO_2.

Maslennikov et al. [97MAS/COU] describe electrochemical measurements for the irreversible reduction of Tc(VII) to Tc(III) which subsequently catalysed the reduction of the nitrate medium at more negative potentials. Maslennikov et al. [98MAS/MAS] report electrolysis experiments performed on Tc(VII) solutions containing nitrate and formate at a graphite electrode. No thermodynamic data are contained in this paper although the rate of reduction is tentatively described in terms of two dimeric Tc-formate complexes.

Kremer et al. [97KRE/GAN] contains single crystal X-ray diffraction data for a Tc(V) coordination compound, $[TcO_2(tn)_2]I \cdot H_2O$, where tn is trimethylenediamine.

13.4 Group 17 (halogen) compounds and complexes (V.4)

13.4.1 Aqueous Tc(IV) halides (V.4.2.1.1)

The speciation of Tc(IV) was studied in 1 – 6 M HCl solutions by UV-visible and Raman spectrophotometry, EXAFS and electrochemistry [2000BEN/FAT]. These results confirm that in 6.0 M HCl, $TcCl_6^{2-}$ is the predominant species [99RAR/RAN], but that aging for 10 days in 1.0 M HCl produced mainly $TcCl_5(OH_2)^-$ with minor contributions form $TcCl_4(OH_2)_2(aq)$ and $TcCl_6^{2-}$. No thermodynamic data are provided. Vichot et al. [2000VIC/FAT] attempted to synthesize a sulphate complex by substitution in $TcCl_6^{2-}$, but no direct evidence for its existence was provided.

13.4.2 Other aqueous halides (V.4.2.1.3)

Gorshkov et al. [2000GOR/MIR] give formation constants (see Table A-49) for the Tc(I) complexes, $Tc(CO)_5(H_2O)_{3-n}X_n^{(1-n)+}$, where X = Cl, Br, I, SCN, from $Tc(CO)_3(H_2O)_3^+$ based on NMR and potentiometric titrations. However, no supporting results are presented and the constants for the halides are too small to be meaningful, whereas those for the thiocyanide complexes appear to be unsubstantiated.

13.5 Technetium nitrido compounds (V.6.1.3.1)

Solvent extraction experiments at 298.15 K were used to investigate HCl/NaCl solutions ($I = 1$ M) believed to contain the dimer $\left[\{(H_2O)Cl_3NTc)\}_2 (\mu-O)\right]^{2-}$, which is in equilibrium with the tetrachloronitrido monomer at [H$^+$] > 0.2 M according to the following reaction [97ASA/SUG]:

$$2\,TcNCl_4^- + 3\,H_2O(l) \rightleftharpoons \left[\{(H_2O)Cl_3NTc)\}_2 (\mu-O)\right]^{2-} + 2\,H^+ + 2\,Cl^-$$

These authors reported a $\log_{10} {}^*K$ value in molar units of (5.455 ± 0.01). At [H$^+$] < 0.2 M, further hydrolysis is believed to result in the formation of $\left[\{Cl_2NTc)\}_2 (\mu-O)_2\right]^{2-}$, but no equilibrium constant was determined. Baldas *et al.* [98BAL/HEA] used a sophisticated combination of cyclic and alternating current voltametry in conjunction with *in situ* spectrophotometric observations of the Tc(V/IV) couple of technetium oxy and nitrido halides. However, these measurements were conducted in a non-aqueous medium and are therefore not relevant to this review.

Chapter 14

Discussion of new auxiliary data selection

14.1 Group 16 (chalcogen) auxiliary species

14.1.1 Tellurium auxiliary species

14.1.1.1 TeO$_2$(cr)

This review accepts the values assessed by Cordfunke *et al.* [90COR/KON], but with increased uncertainties for the standard entropy and heat capacity. Note that there is a typographical error in Table 65, p. 389 of [90COR/KON]; the value of the standard enthalpy of formation of TeO$_2$(cr) derived by Mallika and Sreedharan [86MAL/SRE] is $-(321.1 \pm 1.3)$ kJ·mol^{-1}, not $-(327.1 \pm 1.3)$ kJ·mol^{-1}. The selected values are:

$$\Delta_f H_m^\circ (\text{TeO}_2, \text{cr}, 298.15 \text{ K}) = -(321.0 \pm 2.5) \text{ kJ} \cdot \text{mol}^{-1},$$

$$S_m^\circ (\text{TeO}_2, \text{cr}, 298.15 \text{ K}) = (69.89 \pm 0.15) \text{ J} \cdot \text{K}^{-1} \cdot \text{mol}^{-1},$$

$$C_{p,m}^\circ (\text{TeO}_2, \text{cr}, 298.15 \text{ K}) = (60.67 \pm 0.15) \text{ J} \cdot \text{K}^{-1} \cdot \text{mol}^{-1},$$

$$C_{p,m} (\text{TeO}_2, \text{cr}, T) = 63.271 + 2.1893 \cdot 10^{-2} \cdot T - 8.1142 \cdot 10^5 \cdot T^{-2} \text{ J} \cdot \text{K}^{-1} \cdot \text{mol}^{-1}$$
$$(298.15 \text{ to } 1000 \text{ K}).$$

14.2 Other auxiliary species

14.2.1 Copper auxiliary species

14.2.1.1 CuCl(g)

CuCl(g) is used in one study [84LAU/HIL] to determine the stability of UCl$_3$(g). The enthalpy of formation of CuCl(g) has been calculated from the two measurements of the dissociation energy at 0 K, $D^0(0 \text{ K}) = (374.9 \pm 8.4)$ kJ·mol^{-1} by [96HIL/LAU] from the reaction, Ag(g) + CuCl(g) $\rightleftharpoons$ Cu(g) + AgCl(g), and $D^0(0 \text{ K}) = (382.4 \pm 8.4)$ kJ·mol^{-1} by [72GUI/GIG] from the polymerisation reaction, 3CuCl(g) $\rightleftharpoons$ Cu$_3$Cl$_3$(g); the uncertainty in the latter study has been increased from the purely statistical value of 3.3 kJ·mol^{-1} given by the authors.

For the enthalpy of the relevant reaction,

$$CuCl(g) \rightleftharpoons Cu(g) + Cl(g), \tag{14.1}$$

the mean value of $\Delta_r H_m$ ((14.1), g, 0 K) = (378.7 ± 10.0) kJ · mol^{-1} (where the uncertainty has been increased to allow for uncertainties in the ion cross-sections) leads finally, with enthalpies of formation of the gaseous atoms from [89COX/WAG], to the selected value:

$$\Delta_f H_m^\circ (CuCl, g, 298.15 \text{ K}) = (77.0 \pm 10.0) \text{ kJ} \cdot \text{mol}^{-1}$$

14.3 Group 2 (alkaline earth) auxiliary data

14.3.1 Calcium Auxiliary data

14.3.1.1 CaF(g)

All data for CaF(g) are taken from the assessment by Glushko *et al.* [81GLU/GUR], which are based on early CODATA compatible data.

14.3.1.2 CaCl(g)

All data for CaCl(g) are taken from the assessment by Glushko *et al.* [81GLU/GUR], which are based on early CODATA compatible data.

14.3.2 Barium Auxiliary data

14.3.2.1 Ba(g)

All data for Ba(g) are taken from the assessment by Glushko *et al.* [81GLU/GUR], which are based on early CODATA compatible data.

14.3.2.2 BaF(g)

All data for BaF(g) are taken from the assessment by Glushko *et al.* [81GLU/GUR], which are based on early CODATA compatible data.

14.4 Sodium auxiliary data

14.4.1 NaNO$_3$(cr) and NaNO$_3$(aq).

The following data were assessed and employed in [92GRE/FUG], page 620 and are collected here for reference. The enthalpy of formation of NaNO$_3$(cr)

$$\Delta_f H_m^\circ (\text{NaNO}_3, \text{cr}, 298.15\text{K}) = -(467.58 \pm 0.41) \text{ kJ mol}^{-1}$$

was based on the CODATA values for Na$^+$ and NO$_3^-$ and the weighted average of the enthalpies of solution from Table II-1 of the same reference [89COX/WAG], while the enthalpy of solution of NaNO$_3$ in 6 M HNO$_3$

$$\Delta_{sol} H_m (\text{NaNO}_3, 6 \text{ M HNO}_3, 298.15\text{K}) = (15.44 \pm 0.21) \text{ kJ} \cdot \text{mol}^{-1}$$

was taken from the work of Cordfunke, as cited in [75COR/OUW].

Appendices

Appendix

Appendix A

Discussion of selected references

This appendix comprises discussions relating to a number of key publications which contain experimental information cited in this review. These discussions are fundamental in explaining the accuracy of the data concerned and the interpretation of the experiments, but they are too lengthy or are related to too many different sections to be included in the main text. The notation used in this appendix is consistent with that used throughout the present book, and not necessarily consistent with that used in the publication under discussion.

[00PIS]

The relevant part of this calorimetric study deals with $UO_4 \cdot 2H_2O(s)$. Analysed samples of this compound were dissolved, at an unspecified temperature, in 1 M H_2SO_4 (two measurements) and 2 M H_2SO_4 (one measurement), according to reaction:

$$UO_4 \cdot 2H_2O(s) + H_2SO_4(sln) \rightleftharpoons (UO_2SO_4 + H_2O_2 + 2H_2O)(sln). \quad (A.1)$$

The author also dissolved (two measurements) $UO_3 \cdot H_2O(s)$ in 1 M H_2SO_4,

$$UO_3 \cdot H_2O(s) + H_2SO_4(sln) \rightleftharpoons (UO_2SO_4 + 2H_2O)(sln). \quad (A.2)$$

For the enthalpy of reaction (A.1), the author accepted as valid for 1 M H_2SO_4 the average of the values obtained from the two media, which we also take with increased uncertainty limits, from which $\Delta_r H_m^\circ((A.1), 298.15 \text{ K}) = (0.427 \pm 0.500)$ kJ · mol^{-1}. The uncertainty in the reported value of $\Delta_r H_m^\circ (A.2)$, has also been increased, $\Delta_r H_m^\circ (A.2) = -(64.218 \pm 0.700)$ kJ · mol^{-1}. To obtain $\Delta_f H_m^\circ (UO_4 \cdot 2H_2O, \text{cr}, 298.15 \text{ K})$ from these results, we make the assumption that the peroxide used by the author was crystalline and that his results are valid for 298.15 K.

[92GRE/FUG] give $\Delta_f H_m^\circ (UO_3 \cdot H_2O, \beta, 298.15 \text{ K}) = -(1533.8 \pm 1.3)$ kJ · mol^{-1}, but since it is not known which polymorph of this phase was used by the author, we increase the uncertainty to ± 4.0 kJ · mol^{-1}. We also assume the enthalpy of formation of H_2O_2 in 1 M H_2SO_4 to be the same, within the uncertainty limits, as the infinite dilution value, $\Delta_f H_m^\circ (H_2O_2, \text{aq}, 298.15 \text{ K}) = -(191.17 \pm 0.10)$ kJ · mol^{-1} and thus obtain $\Delta_f H_m^\circ$ ($UO_4 \cdot 2H_2O$, cr, 298.15 K) = $-$ (1789.6 ± 4.1) kJ · mol^{-1}. This value gives very good support to the value selected by [92GRE/FUG] on the basis of the results of [63COR/ALI] and [66COR], $\Delta_f H_m^\circ$ ($UO_4 \cdot 2H_2O$, cr, 298.15 K) = $-$ (1784.0 ± 4.2) kJ · mol^{-1}.

The author gives results on the dissolution of the mixed peroxy compound $(Na_2O_2)_2UO_4 \cdot 9H_2O$(s), also previously described in the literature. This review will not consider this species further.

[71COR/LOO]

It recently became clear that the value of the enthalpy of formation of the compound $Na_6U_7O_{24}$(cr) accepted in [92GRE/FUG], based on its enthalpy of solution in 6.00 mol·dm^{-3} HNO$_3$ as reported by [71COR/LOO], was quite incompatible with the stabilities of the other Na–U(VI) uranates. In fact, the calculated value given by [71COR/LOO] for the enthalpy of formation of the compound was itself not in agreement with their enthalpy of solution. The situation was clarified when it became apparent that the enthalpy of solution of the compound was mistakenly reported by the authors for a formula unit containing only one uranium atom, namely, $Na_{6/7}UO_{24/7}$.

A recalculation of the enthalpy of formation using $\Delta_{sol}H_m$ ($Na_6U_7O_{24}$, cr) = $-$ (595.72 ± 2.05) kJ · mol^{-1}, which is seven times the value listed by [71COR/LOO], was made using the following auxiliary data: $\Delta_f H_m^\circ$ (NaNO$_3$, cr) = $-$ (467.58 ± 0.41) kJ · mol^{-1} from [89COX/WAG]; $\Delta_{sol}H_m$ (NaNO$_3$, 6.00 mol·dm^{-3} HNO$_3$) = (15.44 ± 0.21) kJ · mol^{-1} from [75COR/OUW], as in [92GRE/FUG]; $\Delta_{sol}H_m$ (γ–UO$_3$, 6.00 mol·dm^{-3} HNO$_3$) = $-$ (71.53 ± 0.50) kJ · mol^{-1}, slightly different from the value adopted in [92GRE/FUG] as discussed in the comments on [99COR/BOO]; $\Delta_f H_m$ (HNO$_3$, partial, 6.00 mol·dm^{-3} HNO$_3$) = $-$ (200.315 ± 0.402) kJ · mol^{-1} and $\Delta_f H_m$ (H$_2$O, partial, 6.00 mol·dm^{-3} HNO$_3$) = $-$ (286.372 ± 0.040) kJ · mol^{-1}, both interpolated from the enthalpy of dilution data given by Parker [65PAR], using densities from Table 2.5. The value obtained,

$$\Delta_f H_m^\circ (Na_6U_7O_{24}, cr, 298.15 \text{ K}) = -(10841.7 \pm 10.0) \text{ kJ} \cdot \text{mol}^{-1}$$

adopted here, is appreciably different from that accepted in [92GRE/FUG], $-$ (11351 ± 14) kJ · mol^{-1}. It does, however, remove the discrepancy with the other sodium uranates reported by [71COR/LOO], which are discussed below.

As seen above, we use a slightly different value, in this assessment for the enthalpy of solution of γ–UO$_3$ in 6.00 mol·dm^{-3} HNO$_3$. We have thus recalculated the values for the enthalpies of formation of the uranates originating from the HNO$_3$ cycle used in [71COR/LOO].

For α–Na$_2$UO$_4$, we obtain $\Delta_f H_m^\circ$ (Na$_2$UO$_4$, α, 298.15 K) = $-$ (1892.55 ± 2.28) kJ · mol^{-1}, instead of $-$ (1892.4 ± 2.3) kJ · mol^{-1} in [92GRE/FUG]. In the latter assessment the final value was obtained by making a weighted average of the HNO_3 cycle with the results from four other cycles (in HCl and H_2SO_4 media) unaffected by the present discussion. The same procedure yields here,

$$\Delta_f H_m^\circ (Na_2UO_4, \alpha, 298.15 \text{ K}) = -(1897.7 \pm 3.5) \text{ kJ} \cdot \text{mol}^{-1}$$

which is the adopted value, identical to that accepted in [92GRE/FUG].

For β–Na_2UO_4, we recalculate, from the [71COR/LOO] cycle, $\Delta_f H_m^\circ (Na_2UO_4, \beta, 298.15 \text{ K}) = -(1886.35 \pm 2.05)$ kJ · mol^{-1}. However, the value adopted in [92GRE/FUG] was based on the best value for the difference of the enthalpies of solution of the α– and β–phases, (13.1 ± 0.8) kJ · mol^{-1}. Consequently, the value adopted here remains the same as in [92GRE/FUG]:

$$\Delta_f H_m^\circ (Na_2UO_4, \beta, 298.15 \text{ K}) = -(1884.6 \pm 3.6) \text{ kJ} \cdot \text{mol}^{-1}.$$

For β–Na_4UO_5, we recalculate $\Delta_f H_m^\circ (Na_4UO_5, \text{cr}, 298.15 \text{ K}) = -(2456.47 \pm 3.47)$ kJ · mol^{-1}, only slightly different (as expected) from the value of $-(2456.2 \pm 3.0)$ kJ · mol^{-1} derived in Appendix A of [92GRE/FUG]. In the main text however, uncertainty limits of only ± 2.1 kJ · mol^{-1} were given that are obviously too small. The assessment of the final value for the enthalpy of formation of the compound is made, as in [92GRE/FUG], by taking the weighted average of the result originating from the measurements of [71COR/LOO] and of those from [85TSO/BRO] who reported $-(2457.3 \pm 2.8)$ kJ · mol^{-1}. This yields:

$$\Delta_f H_m^\circ (Na_4UO_5, \text{cr}, 298.15 \text{ K}) = -(2457.0 \pm 2.2) \text{ kJ} \cdot \text{mol}^{-1},$$

which is the value adopted here, as compared to $-(2456.6 \pm 1.7)$ kJ · mol^{-1} in [92GRE/FUG]. For $Na_2U_2O_7$(cr), we recalculate $\Delta_f H_m^\circ (Na_2U_2O_7, \text{cr}, 298.15 \text{ K}) = -(3196.46 \pm 3.09)$ kJ · mol^{-1} while [92GRE/FUG] give $-(3196.1 \pm 3.9)$ kJ · mol^{-1}, from the [71COR/LOO] data. However, for their final selection, [92GRE/FUG] preferred a value based on the results of [85TSO/BRO], namely:

$$\Delta_f H_m^\circ (Na_2U_2O_7, \text{cr}, 298.15 \text{ K}) = -(3203.8 \pm 4.0) \text{ kJ} \cdot \text{mol}^{-1}.$$

We see no reason to alter this choice.

With these revised data, and reasonable estimates for the missing entropies and heat capacities of the Na–U(VI) uranates, the phase relationships in the whole Na–U–O system around 800 K are in good agreement with those summarised by [81LIN/BES].

[71NIK/PIR]

This paper has been cited but was not reviewed in [92GRE/FUG]. It gives first a survey of the literature data on the solubility of dehydrated schoepite $UO_3 \cdot H_2O$, $UO_2(OH)_2$, up to 1971. Some references [55GAY/LEI], [58BRU], [60BAB/KOD] have been taken into account and discussed in [92GRE/FUG]. The data contained in all the quoted papers are reported in Table A-1. As shown, up to 1971, the data are controversial.

Table A-1: Literature data for the solubility of dehydrated schoepite.

T (K)	$\log_{10} K^\circ_{s,0}$	Reference
298.15	− 22.950	[54MIL]
298.15	− 23.176	[60OOS]
298.15	− 19.698	[62PER/BER]
298.15	− 22.130	[67GRY/KOR]
293 – 323	− 21.12 to − 20.6	[67GRY/KOR]
298.15	− 21.04	[55GAY/LEI]
293.15	− 17.778	[58BRU]
298.15	− 21.255	[60BAB/KOD]

Nikolaeva and Pirozhkov give also in the paper considered here values of the constant for the equilibrium:

$$UO_2(OH)_2 \rightleftharpoons UO_2^{2+} + 2\, OH^-,$$

in the range 295 to 423 K.

Experiments were conducted starting with $UO_3 \cdot H_2O$ (precipitated UO_3 hydrate heated to 393 K). Excess amounts of this compound with regard to complete neutralisation of the acids were equilibrated with $HClO_4$ (HNO_3 and H_2SO_4) for which initial concentrations were known, and were always less than 10^{-2} M. Equilibrium pH measurements (ranging from 3 to 5) as a function of the temperature T ($\pm$ 0.5 K), were conducted by two complementary methods. Repeated measurements on a given solution gave a reproducibility of ΔpH = 0.02 to 0.08. U concentrations were obtained by U_3O_8 gravimetry for solutions up to 363 K. Uncertainties in the acid and U concentrations, and in pH, were not given.

For $HClO_4$ and HNO_3 solutions, free concentrations of UO_2^{2+} were calculated from the initial acid concentration, $[H^+]$, taken as proton activity at each T and the concentrations of the species $(UO_2)_2(OH)_2^{2+}$, $UO_2(OH)^+$ and $UO_2(OH)_2(aq)$ were calculated from their formation constants at each T according to the data of [71NIK]. For sulphuric acid solutions, the complex $UO_2SO_4(aq)$ was also taken into account in [71NIK]. Solubility products were calculated using $[OH^-]$ which were derived from $[H^+]$ values on the basis of K_w values determined at the test temperature, according to the data of [67PER/KRY].

Using activity instead of $[H^+]$ gives mixed solubility products. For the conditions of this work γ_{H^+} is not less than 0.911 (0.01 M $HClO_4$) and does not change by more than 3 to 4 % within the range of temperature. Another point is that the calculated values of the U concentrations, when measured, are higher than the experimental values (7 to 10 %). It is difficult to check if this discrepancy is due to the values of the constants used or to an incomplete separation of the solid phase from the solution (as quoted by the authors in some cases). The review of [71NIK] on page 621 of

[92GRE/FUG] considers the hydrolysis constants of UO_2^{2+} as well as the sulphate complexation constant reported by Nikolaeva and Pirozhkov as not reliable and only gives these values as indicative. Ionic strengths are sufficiently low to consider that the obtained solubility products correspond to $I = 0$.

Taking all the data for $HClO_4$ (295 to 423 K) and HNO_3 (295 to 363 K) media, the authors give the following variation of $\log_{10} K_{s,0}^{\circ}$ with T:

$$\log_{10} K_{s,0}^{\circ} = -26.050 + 5349.583\, T^{-1} - 1253404\, T^{-2}.$$

The experimentally reported values of the solubility in these two media agree to within 0.1 $\log_{10}$ units. At 295.15 K these values are ($\log_{10}$ units) -22.284 and -22.320, respectively (taken as the average reported experimental data). At the same temperature the value for the H_2SO_4 medium is reported as -22.093. The temperature dependence of the values in this medium is also slightly different. Because the corrections made for sulphate complexation and sulphuric acid dissociation result in additional uncertainty in the corrections of the experimental data, the values for this medium will not be considered further here.

With the accepted NEA values for the ionic product of water [75OLO/HEP], we calculate for reaction:

$$UO_3 \cdot H_2O \text{ (orth, } T) + 2\,H^+ \rightleftharpoons UO_2^{2+} + 2\,H_2O(l) \tag{A.3}$$

at 298.15 K for the perchloric medium, $\log_{10} {}^*K_{s,0}$ (A.3) $= -(5.80 \pm 0.10)$ and obtain:

$$\Delta_f G_m^{\circ}(UO_3 \cdot H_2O, \alpha, \text{orth}, 298.15 \text{ K}) = -(1393.72 \pm 1.84)\text{ kJ} \cdot \text{mol}^{-1}.$$

Using the experimental results in perchloric media as a function of temperature, we have also calculated the $\log_{10} K$ (A.3) value as (5.05 ± 0.1) at 323.15 K; (4.47 ± 0.1) at 343.15 K; (4.05 ± 0.1) at 363.15 K; (3.75 ± 0.2) at 373.15 K; (3.39 ± 0.2) at 398.15 K and (2.94 ± 0.2) at 423.15 K.

The uncertainty limits affecting all of these recalculated values represent only an estimate of the internal consistency of the experimental results of the authors, as ample caution was given above concerning the complexing and hydrolysis constants they used. In fact, the values at 298.15 K are only used to help assess, in combination with data from another source (see the main text), the Gibbs energy of formation of the compound, and the values above room temperature are given here for information only.

[74DHA/TRI]

The authors report transpiration measurements of the total pressures of uranium bearing species over $U_3O_8(s)$ in the presence of both dry and moist oxygen which show that the mass loss in the presence of water vapour is much larger than the loss of $UO_3(g)$ in dry oxygen, due to the formation of a $UO_3 \cdot (H_2O)_n$ vapour species.

By variation of the water vapour pressure in the transpiring oxygen, the H_2O/UO_3 ratio in the hydroxide gas was shown to be (1.08 ± 0.10), indicating the formation of $UO_2(OH)_2(g)$.

The authors' pressures of $UO_3(g)$ from the reaction:

$$\frac{1}{3}U_3O_8(s) + \frac{1}{6}O_2(g) \rightleftharpoons UO_3(g) \quad (A.4)$$

from 1525 to 1675 K are in good agreement with the literature data which form the basis of the choice of $\Delta_f G_m^\circ (UO_3, g)$ in [82GLU/GUR] and [92GRE/FUG].

The authors tabulate the results of 14 experiments (with replicated mass losses) using oxygen with known amounts of water vapour from 1323 to 1623 K in the relevant reaction:

$$\frac{1}{3}U_3O_8(s) + \frac{1}{6}O_2(g) + H_2O(g) \rightleftharpoons UO_2(OH)_2(g) \quad (A.5)$$

In calculating the pressure of $UO_2(OH)_2(g)$, allowance was made for the (much smaller) mass loss due to the simultaneous vaporisation of $UO_3(g)$.

The authors' equation for the Gibbs energy of reaction (A.5) seems to be based solely on the series of experiments with a partial pressure of water vapour of 23 Torr. It is clear from the abstract, though not from the text, that their equation (3) for the pressure of $UO_2(OH)_2(g)$, equivalent to:

$$\log_{10}(p_{UO_2(OH)_2}/\text{bar}) = -9612\, T^{-1} + 1.13$$

refers to this series only.

Since the paper was published, two somewhat different estimates of the thermal functions of $UO_2(OH)_2(g)$ have been published: [95EBB], [98GOR/SID].

We have therefore preferred to use all the fourteen tabulated values to calculate the equilibrium constant and thus the Gibbs energy of reaction (A.5), and further calculate the second– and third–law values for $\Delta_r H_m^\circ ((A.5), 298.15\ \text{K})$ and hence $\Delta_f H_m^\circ (UO_2(OH)_2, g, 298.15\ \text{K})$.

Our calculations for the Gibbs energies of reaction (A.5) can be fitted to:

$$\Delta_r G_m (A.5) = 200050 - 60.422\, T \quad \text{J} \cdot \text{mol}^{-1} \quad (T = 1323 \text{ to } 1623\ \text{K}) \quad (A.6)$$

in reasonable agreement with the authors' equation, recalculated from their pressure equation to be

$$\Delta_r G_m (A.5) = 184020 - 50.638\, T \quad \text{J} \cdot \text{mol}^{-1} \quad (T = 1273 \text{ to } 1623\ \text{K}). \quad (A.7)$$

The results of the calculations of $\Delta_f H_m^\circ (UO_2(OH)_2, g, 298.15\ \text{K})$ are shown in Table A-2. As noted in section 9.3.1.2.1 the big differences in this table are compounded by the very much smaller equilibrium constants of reaction (A.5) obtained in the more recent measurements by [93KRI/EBB]. In view of these large and unexplained differences, no data for this compound can be selected in this review.

Table A-2: Derived values of $\Delta_f H_m^\circ$ (UO$_2$(OH)$_2$, g, 298.15 K), kJ · mol^{-1}.

Method	Thermal functions	
	[95EBB]	[98GOR/SID]
Second–law, Eq.(A.6)	– (1212.9 ± 21.5)	– (1218.2 ± 21.5)
Second–law, Eq.(A.7)	– (1229.2 ± 15.4)	– (1234.5 ± 15.4)
Third–law	– (1262.6 ± 7.8)	– (1291.8 ± 10.4)

[76MOR/MCC]

This paper reports as experimental data i) the values of the integral heat of dissolution of Th(NO$_3$)$_4$·5H$_2$O(cr) in HClO$_4$ 10^{-2} M at 15, 25 and 35°C (± 0.02°C), the final solutions being 0.303 to 2.270·10^{-3} m (15°C), 0.516 to 6.350·10^{-3} m (25°C) and 0.547 to 2.052·10^{-3} m (35°C) and ii) the molality of the saturated solution of Th(NO$_3$)$_4$·5H$_2$O(cr) in HClO$_4$ 10^{-2} M, m(Th(NO$_3$)$_4$) = (3.66 ± 0.02) m. Thorium nitrate pentahydrate is the stable hydrate at 25°C, its properties including $C_{p,m}^\circ$ values are well known and it was chemically characterised before being used in the calorimetric experiments. All molar integral heats of dissolution were corrected to give the $\Delta_r H_m^\circ$(A.8) values for the reaction:

$$\text{Th(NO}_3\text{)}_4 \cdot 5 \text{ H}_2\text{O(cr)} \rightleftharpoons \{\text{Th}^{4+} + 4 \text{ NO}_3^-, \text{ infinite dilution}\} \quad (A.8)$$

Morss and McCue calculated $\Delta_r H_m^\circ$(A.8) using a second-order Debye-Hückel treatment. By means of the Van't Hoff equation, these values were corrected, again at each temperature, for the first hydrolysis of Th^{4+}, using $^*\beta_{1,1}^\circ$ = 1.4·10^{-4} (log$_{10}$ $^*\beta_{1,1}^\circ$ = – 3.85) at 25°C and the enthalpy corresponding to this first hydrolysis, as given by Baes et al. [65BAE/MEY]. Formation of ThNO$_3^{3+}$ was included in the calculation of $\Delta_r H_m^\circ$(A.8). The corrected values of $\Delta_r H_m^\circ$(A.8) are the following: – (14066 ± 154) J · mol^{-1} for 15°C, – (19807 ± 71) J · mol^{-1} for 25°C and – (23094 ± 38) J · mol^{-1} for 35°C, giving [$\Delta C_{p,m}$]$_{15}^{25}$ = – 574 J · K^{-1} · mol^{-1} and [$\Delta C_{p,m}$]$_{25}^{35}$ = – 329 J · K^{-1} · mol^{-1}. The average value of $\Delta C_{p,m}^\circ$ for 25°C is taken as $\Delta C_{p,m}^\circ$ = – (450 ± 10) J · K^{-1} · mol^{-1}. Using $C_{p,m}^\circ$(NO$_3^-$, 298.15 K) = – 86.61 J · K^{-1} · mol^{-1} and auxiliary data for heat capacities, leads to $C_{p,m}^\circ$(Th^{4+}, 298.15 K) = – (1 ± 11) J · K^{-1} · mol^{-1}.

To obtain the entropy of Th^{4+}, Morss and McCue calculated first $\Delta_r G_m^\circ$((A.8), 298.15 K) = – 10970 J · mol^{-1} for the equilibrium mentioned above. They used (3.7 ± 0.1) mol · kg^{-1} for the molal concentration of the saturated solution of thorium nitrate in 10^{-2} M HClO$_4$ and the data of Robinson and Levien [47ROB/LEV] for the mean activity coefficient of thorium nitrate and water derived from vapour pressure measurements. Combining $\Delta_r G_m^\circ$ (298.15 K) and $\Delta_r H_m^\circ$ (298.15 K) gives S_m° (Th^{4+}, 298.15 K) = $-$ (424.0 ± 3.6) J · mol^{-1} · K^{-1}.

The semi-empirical Morss and Cobble relationship (correlation of the entropy of monoatomic aqueous ions at 25°C with their charges and radii [70MOR/COB]) briefly discussed in [76MOR/MCC] predicts S_m° (Th^{4+}, 298.15 K) = $-$ (390 ± 30) J · K^{-1} · mol^{-1}, essentially in agreement with the experimental value reported by [76MOR/MCC]. Using this last value, the Criss and Cobble relation, which gives $C_{p,m}^\circ$ as a function of S_m°, yields $C_{p,m}^\circ$ (Th^{4+}, 298.15 K) = $-$ 28 J · K^{-1} · mol^{-1}, a value outside the uncertainty limits given for the experimental value.

This paper gives the first valuable data on $C_{p,m}^\circ$ of the aqueous thorium ion. The later study of Hovey [97HOV] (see this review) gave different values, of which $C_{p,m}^\circ$ (Th^{4+}, 298.15 K) = $-$ (224 ± 5) J · K^{-1} · mol^{-1} is used to question the value of $C_{p,m}^\circ$ (U^{4+}, 298.15 K) selected by [92GRE/FUG].

[83OBR/WIL]

Schröckingerite, $NaCa_3UO_2(CO_3)_3SO_4F \cdot 10H_2O$, was synthesised and recrystallised from water, but still contained gypsum as an impurity. The solubility was measured at 25°C whereby the dioxouranium(VI) and sulphate concentrations were measured and the excess sulphate over the stoichiometric ratio of 1:1 was assumed to be the level of gypsum impurity. X–ray diffraction of the solid phase following the dissolution experiments was used as the criterion that this solid dissolved congruently. Grimselite, $NaK_3UO_2(CO_3)_3 \cdot H_2O$, was synthesised and characterised based on its water content and XRD pattern.

The solubilities were measured at temperatures of 20 and 25°C in the former case (I = 0.105 – 0.108 M), and from 5.6 to 25.0°C for the latter solid (I = 0.428 – 0.439 M). The solutions were speciated based on data taken from a variety of sources and activity coefficients were derived from the methods of Reardon and Langmuir [76REA/LAN], Kielland [37KIE], and Alwan [80ALW]. The average $\log_{10} K_{s,0}$ value at 25°C for the dissolution of schröckingerite:

$$NaCa_3UO_2(CO_3)_3SO_4F \cdot 10\ H_2O(cr) \rightleftharpoons Na^+ + 3Ca^{2+} + UO_2^{2+} + 3\ CO_3^{2-} + SO_4^{2-} + F^- + 10\ H_2O(l)$$

was reported as $-$ (35.16 ± 0.04), which led to a $\log_{10} K_{s,0}^\circ$ value of $-$ 38.76 using their activity coefficient model, whereas the simple Debye-Hückel approach [92GRE/FUG] gives $-$ 38.97. A value for $\Delta_f G_m^\circ$ of $-$ (8077.3 ± 8.7) kJ · mol^{-1} (schröckingerite) was reported, which is well within the uncertainty range of that obtained from the latter approach. In the case of grimselite, the ionic strengths are too high to use only the Debye-Hückel term when calculating activity coefficients. For the dissolution of grimselite:

$$NaK_3UO_2(CO_3)_3 \cdot H_2O(cr) \rightleftharpoons Na^+ + 3K^+ + UO_2^{2+} + 3\ CO_3^{2-} + H_2O(l)$$

the $\log_{10} K_{s,0}$ value at 25°C is reported as -26.45, from which they derive a $\log_{10} K_{s,0}^{\circ}$ value of -38.76 and subsequently the following thermodynamic parameters, $\Delta_f G_m^{\circ} = -(4051.3 \pm 1.8)$ kJ·mol^{-1} and $\Delta_f H_m^{\circ} = -(4359.0 \pm 1.8)$ kJ·mol^{-1} (grimselite). However, considering the uncertainties in the speciation and activity coefficients, particularly at these higher ionic strengths, and the relative narrow range of temperature investigated, these estimates must be considered as only being provisional.

[85UNE]

The paper reports measurements of the oxygen activity (emf with Y_2O_3–ZrO_2 electrolyte) in the mixtures of $Cs_2U_4O_{12}$(cr) and $Cs_2U_4O_{13}$(cr).

$Cs_2U_4O_{13}$(cr) was prepared by heating appropriate amounts of β–UO_3 with Cs_2CO_3 in air at 873 K for 12 hours; the 3.10 % weight loss was slightly greater than that calculated (3.00%) The lower uranate was obtained by decomposing the U(VI) compound in high–purity Ar at 1223 K for six hours; the 1.08 % weight loss was in good agreement with the theoretical prediction (1.10%). The products were characterised by X–ray diffraction.

The oxygen potentials were studied using a Y_2O_3–ZrO_2 electrolyte tube in flowing argon, from 1048 to 1198 K with air ($p_{O_2(g)} = 0.203$ bar) as the reference electrode. Correction of their emf values (omitting the somewhat deviant point at 1198 K) to the standard pressure gives, after recalculation:

$$\Delta_r G_m \text{ (A.9)} = -190108 + 151.758\, T \quad \text{J·mol}^{-1} \qquad (T = 1048 \text{ to } 1173 \text{ K})$$

for the reaction:

$$Cs_2U_4O_{12}(cr) + 0.5\, O_2(g) \rightleftharpoons Cs_2U_4O_{13}(cr) \qquad (A.9).$$

The author's expression is slightly different. The calculated decomposition temperature of $Cs_2U_4O_{13}$(cr) in air is 1201 K, noticeably lower than the value of 1310 K indicated by Cordfunke et al. [75COR/EGM].

The Gibbs energy values agree well with the later work by Venugopal et al. [92VEN/IYE] although the individual enthalpy and entropy terms differ by ca. 10%.

These authors have used these Gibbs energies of reaction to derive data for 298.15 K for $Cs_2U_4O_{12}$(cr), using the literature data [80COR/WES] for $Cs_2U_4O_{13}$(cr). However, we have not pursued this approach for reasons given in detail in the discussion of [92VEN/IYE].

[86DIC/PEN]

This paper reports the synthesis of novel compounds involving lithium insertion into several uranium oxides: U_3O_8(s) and α–, γ– and δ–UO_3. The lithiating agent was LiI dissolved in organic solvents. The progress of the reaction was determined by titration of the liberated iodine with sodium thiosulphate, the products being characterised by powder X-ray diffraction. Galvanostatic and open circuit discharge curves were used to define the various (sometimes narrow) monophasic regions in the oxides, within the overall range of $0 < \text{Li}/\text{U} < 1.13$. Diffusion coefficients of Li in the various phases were obtained by electrochemical pulse methods. From Gibbs–Duhem integrations of the open circuit data, the authors deduce a value of about -300 kJ per mol of Li for the Gibbs energy of insertion into UO_n, in line with later results by the same group [89DIC/LAW], [95DUE/PAT]. No further quantitative thermodynamic information can be extracted from this paper.

[87ALE/OGD]

This is a paper describing project work on fission product release from irradiated fuel, using a modulated beam operable up to 2400 K, coupled to a mass spectrometer. The only section of the paper relevant to the current review is that on the formation of $UO_2(OH)_2(g)$, in which the authors give the Gibbs energy of the gaseous reaction:

$$UO_3(g) + H_2O(g) \rightleftharpoons UO_2(OH)_2(g) \quad (A.10)$$

to be $\Delta_r G_m (A.10) = -333500 + 156.9 \; T$ J·mol^{-1} with no indication of the relevant temperature range or experimental description of any kind.

We are thus quite unable to ascertain the experimental validity this expression, but because of the considerable discrepancy in the other thermodynamic data for $UO_2(OH)_2(g)$, have nevertheless derived the enthalpies of formation $\Delta_f H_m^\circ (UO_2(OH)_2, g, 298.15 K)$ by second- and third- law analyses, assuming a temperature range of $1900 - 2100$ K. The second- and third-law values vary between -1301 and -1398 kJ·mol^{-1} with unknown (but large) uncertainties and are discussed further in section 9.3.1.2.1.

[88DIC/POW]

Except for the compound δ–$Na_{0.54}UO_3$, details on the preparation and characterisation of the compounds studied are given in [90POW]. This latter thesis, however, does not contain work on the thermochemistry of α–UO_3 and δ–$Na_{0.54}UO_3$.

δ–$Na_{0.54}UO_3$ was obtained by ambient temperature insertion of Na by reaction of sodium benzophenone in tetrahydrofurane with δ–UO_3. It was characterised as described in [90POW]. The dissolution medium was the same as for α–$Na_{0.14}UO_3$ [90POW] and the enthalpy of solution was reported as $-(140.2 \pm 0.7)$ kJ·mol^{-1}. From this value, one obtains $\Delta_f H_m^\circ (Na_{0.54}UO_3, \alpha, 298.15 K) = -(1377.0 \pm 5.5)$ kJ·mol^{-1}. For the insertion reaction:

$$0.54 \, Na(cr) + UO_3(\delta) \rightleftharpoons Na_{0.54}UO_3(\delta) \quad (A.11)$$

we obtain using $\Delta_f H_m^\circ (UO_3, \delta, 298.15 K)$ recommended in this review (section 9.3.3.1), $\Delta_r H_m^\circ ((A.11), 298.15 K) = -(163.2 \pm 5.7)$ kJ·mol^{-1}.

The enthalpy of solution of α–UO, prepared as described previously [85DIC/LAW] and characterised by X–ray diffraction, is reported as $-(96.03 \pm 0.72)$ kJ · mol^{-1}, to be compared with $\Delta_{sol}H_m$ (UO$_3$, γ) $= -(84.64 \pm 0.38)$ kJ · mol^{-1}, obtained in the same medium by [77COR/OUW2]. With the NEA adopted value for the enthalpy of formation of γ–UO$_3$, we obtain $\Delta_f H_m^\circ$ (UO$_3$, α, 298.15 K) = $-(1212.41 \pm 1.45)$ kJ · mol^{-1}, compared with the value accepted by [92GRE/FUG], $-(1217.5 \pm 3.0)$ kJ · mol^{-1} for a compound which, as noted by [66LOO/COR] could not have had the exact UO$_{3.00}$ composition. The former value is accepted here, because, given the enthalpies of solution, the α and the δ phases should have enthalpies of formation close to each other and about 10 kJ · mol^{-1} less negative than the γ phase.

[88KEN/MIK]

This paper presents electrochemical and chemical preparations of technetium formate complexes for use as radiological image enhancers. The former technique is based on controlled potential electrolysis (-600 mV *versus* SCE) at a graphite foil working electrode of millimolar KTcO$_4$ in 4 M formate at pH = 3.21 under aerobic and anaerobic conditions. The reduced species formed at the working electrode were separated by anion exchange chromatography. Chemical production involved reduction with stannous chloride and sodium borohydride. Although three bands were observed by chromatography that were assigned to a yellow reduced complex, a colourless intermediate complex and a red oxidised technetium formate complex, no further detailed characterisation was forthcoming and no thermodynamic information was presented.

[88OHA/LEW]

This paper is cited on pages 139 and 140 of [92GRE/FUG] but is not reviewed in Appendix A of that publication. It deals with:

- the determination of $\Delta_f H_m^\circ$(UO$_3$ · 0.9H$_2$O, α, 298.15 K) $= -(1506.3 \pm 2.1)$ kJ · mol^{-1} and $\Delta_f G_m^\circ$(UO$_3$ · 0.9H$_2$O, α, 298.15 K) $= -(1374.4 \pm 2.6)$ kJ · mol^{-1} on the basis of an estimated entropy value of S_m°(UO$_3$ · 0.9H$_2$O, α, 298.15 K) = (125 ± 5) J · K^{-1} · mol^{-1}. [92GRE/FUG] keep the enthalpy value but take a slightly higher value of entropy, (126 ± 7) J · K^{-1} · mol^{-1} and consequently arrive at the selected value $\Delta_f G_m^\circ$(UO$_3$ · 0.9H$_2$O, α, 298.15 K) $= -(1374.4 \pm 2.6)$ kJ · mol^{-1}, that differs slightly from that of O'Hare *et al.*,

- the derivation of thermodynamic values for the dehydration steps of hydrates of uranium trioxide, and

- the derivation of thermodynamic values for aqueous dissolution of UO$_3$·xH$_2$O compounds.

The synthesised compound used in this work is the alpha form of $UO_3 \cdot 0.9H_2O$, well characterised by X–ray diffraction and chemical analysis. The enthalpy change for the reaction:

$$UO_3 \cdot 0.9H_2O \text{ (cr)} + 2 \text{ HF(aq)} \rightleftharpoons UO_2F_2\text{(aq)} + 1.9 \text{ H}_2O\text{(l)}$$

is measured as $\Delta_r H_m = -(80.91 \pm 0.34)$ kJ · mol^{-1}.

The value $\Delta_f H_m^\circ (UO_3 \cdot 0.9H_2O, \alpha, 298.15 \text{ K}) = -(1506.3 \pm 2.1)$ kJ · mol^{-1} is calculated through a thermochemical cycle including the measured $\Delta_r H_m$ value, previous data on the dissolution of $UO_3(\gamma)$ in HF(aq) and auxiliary CODATA values, as done in [88TAS/OHA]. So data of the two papers are consistent and the value of $\Delta_r H_m^\circ$ for:

$$UO_3 \cdot 2H_2O\text{(cr)} \rightleftharpoons UO_3 \cdot 0.9H_2O\text{(cr)} + 1.1H_2O\text{(g)} \quad (A.12)$$

is calculated as $\Delta_r H_m^\circ$ (A.12) = (54.08 ± 0.49) kJ · mol^{-1}.

On the basis of previous literature data the authors give a set of values of $\Delta_f H_m^\circ$ at 298.15 K for the compounds $UO_3 \cdot xH_2O$, x = 2, 1, 0.9, 0.85, 0.64 and 0.39 (not systematically reported in [92GRE/FUG]).

The question of $S_m^\circ (UO_3 \cdot 0.9H_2O, \alpha, 298.15 \text{ K})$ is addressed considering $\Delta_r S_m^\circ$ at 298.15 K for reaction (A.12) for which the vapour pressure equation in equilibrium is used:

$$\log_{10} \frac{p}{p_0} = -\frac{2568.9}{T} + 6.843$$

to deduce $\Delta_r H_m^\circ = (54.08 \pm 0.49)$ kJ · mol^{-1} and the values of entropy of dehydration $\Delta_r S_m^\circ$ of many oxides hydrates (reflected by the term 6.843, but this point is rather vague in the paper). From that equation it follows $\Delta_r S_m^\circ$ = 144 J · K^{-1} · mol^{-1} from which $S_m^\circ (UO_3 \cdot 0.9H_2O, \alpha, 298.15 \text{ K})$ = (125 ± 5) J · K^{-1} · mol^{-1} is derived using known values for $S_m^\circ (UO_3 \cdot 2H_2O, \text{cr}, 298.15 \text{ K})$ and $S_m^\circ (H_2O, \text{g}, 298.15 \text{ K})$.

The authors handle all the literature data concerning the dehydration of $UO_3 \cdot xH_2O$ in the same way as for x = 0.9 and give S_m° and $\Delta_f G_m^\circ$ at 298.15 K for x = 2, 1, 0.9, 0.85, 0.64 and 0.39. These data are not mentioned in [92GRE/FUG].

The third part of the paper deals with estimation of $\Delta_r G_m^\circ$ of schoepite for the reactions:

$$UO_3 \cdot 2H_2O \text{ (cr)} + 2 \text{ H}^+ \rightleftharpoons UO_2^{2+} + 3 \text{ H}_2O\text{(l)} \quad (A.13)$$

$$UO_3 \cdot 0.9H_2O \text{ (cr)} + 2 \text{ H}^+ \rightleftharpoons UO_2^{2+} + 1.9 \text{ H}_2O\text{(l)} \quad (A.14)$$

at 298.15 K (and other higher temperatures) using values of Gibbs energy of formation obtained in this paper, in [88TAS/OHA] and auxiliary data: $\Delta_f G_m^\circ (UO_2^{2+}, 298.15 \text{K}) = -(952.7 \pm 2.1)$ kJ · mol^{-1} (the value selected by [92GRE/FUG] is $-(952.551 \pm 1.747)$ kJ · mol^{-1}) and the CODATA value $\Delta_f G_m^\circ (H_2O, \text{l}, 298.15 \text{ K}) = -(237.14 \pm 0.04)$ kJ · mol^{-1}. The values for $\log_{10} {}^*K_{s,0}^\circ$ for the reactions derived from

$\Delta_r G_m^\circ = -27.3$ and -29.0 kJ·mol^{-1} (for reactions (A.13) and (A.14), respectively) are 4.78 and 5.08. The value given for reaction (A.13) by [80LEM/TRE] is (5.6 ± 0.5). The value of 4.78 [88OHA/LEW] is consistent with the value of [92GRE/FUG], $\log_{10} K^\circ = -5.5$, for reaction (V.9) on page 113 of [92GRE/FUG]:

$$UO_3 \cdot 2H_2O \text{ (cr)} \rightleftharpoons UO_2(OH)_2 \text{(aq)} + H_2O\text{(l)} \qquad (A.15)$$

and $\log_{10} {}^*\beta_{2,1}^\circ < -10.3$ selected for:

$$UO_2^{2+} + 2\,H_2O\text{(l)} \rightleftharpoons UO_2(OH)_2\text{(aq)} + 2\,H^+$$

which give together, $\log_{10} {}^*K_{s,0}^\circ = 4.8$. The value of $\log_{10} {}^*K_{s,0}^\circ$ (A.13) calculated from the data of [92GRE/FUG] is $-(4.81 \pm 0.43)$. This review has selected $\log_{10} {}^*\beta_{2,1}^\circ = -(12.15 \pm 0.07)$, which gives $\log_{10} K^\circ$ (A.15) $= -(7.35 \pm 0.43)$ and $\Delta_f G_m^\circ$ (UO$_2$(OH)$_2$, aq, 298.15 K) $= -1357.5$ kJ·mol^{-1}.

[88STA/KIM]

This paper is reviewed together with [88STA/KIM2].

[88STA/KIM2]

This review also includes the review of [88STA/KIM].

The authors investigated the pH dependence of the solubility of ^{241}Am(OH)$_3$(s) at (25 ± 0.5)°C. The interference of CO$_2$ was carefully excluded. According to the experimental pH titration procedure the precipitates had several weeks for ageing, but no attempts were made to characterise the solid phases. The effect of α–radiation on the solubility was studied using different total concentrations of ^{241}Am. One set of solubility data (experiment A) was performed in 0.1 M NaClO$_4$ (pH = 6.9 – 12.9) at low specific α–radiation ($\leq 1.2 \cdot 10^{-4}$ M). In three additional series in 0.1 M NaClO$_4$ (B), 0.1 M NaCl (C), and 0.6 M NaCl (D), larger amounts of ^{241}Am ($1.42 \cdot 10^{-3}$ M to $6 \cdot 10^{-3}$ M) were used in order to extend the solubility measurements to lower pH values in the range 6.2 – 6.6. A few additional data measured in 5 M NaClO$_4$ do not provide sufficient information to derive reliable equilibrium constants, and an additional series in 3 M NaCl resulted in radiolytic oxidation of Am(III) to Am(V).

Series B, C and D include data at pH = 6.2 – 6.6, where the effect of mononuclear hydrolysis is small or negligible. In addition, the Am^{3+} ion concentration was determined by spectroscopy to exclude a possible interference of polynuclear species in the calculation of the solubility constant. The hydrolysis constants were then calculated from the data in the whole pH range by least squares fitting. Experiment A (in 0.1 M NaClO$_4$ and lower specific α–activity) contains only data at pH ≥ 7. As there are no data with a significant amount of free Am^{3+}, the fitted solubility and hydrolysis constants are highly correlated (cf. comparison in Figure A.1). Hence the hydrolysis constants were adopted from series B (also in 0.1 M NaClO$_4$) and only $\log_{10} K_{s,0}$ was fitted. In the review of Silva et al. [95SIL/BID], the results from series B, C, and D were not discussed within the data selection. Only the data from experiment A (at lower specific α–activity) were used and recalculated by fitting simultaneously $\log_{10} K_{s,0}$ and the hydrolysis constants. The fitted constants differ considerably from those proposed in [88STA/KIM]. In the present review the constants given in the original paper as $\log_{10} K_{s,0}$ and $\log_{10} \beta_n$ (reactions formulated with OH⁻ ions, cf. Table A-3) are converted to $\log_{10}{}^* K_{s,0}$ and $\log_{10}{}^* \beta_n$ values with the NEA–TDB auxiliary data for the ion product of water and corrected with respect the relations between the experimental pH values (pH$_{exp}$) and $-\log_{10}[H^+]$ or $\log_{10}[OH^-]$. For the 0.1 M NaClO$_4$ and NaCl solutions, the corrections are very small (0.02 units in $\log_{10}[H^+]$), but in the case of 0.6 M NaCl they become significant (0.17 units in $\log_{10}[H^+]$). Using the same equipment for pH measurement in 0.6 M NaCl as in [88STA/KIM], Felmy et al. [91FEL/RAI] reported the relation $-\log_{10}[H^+] = $ pH$_{exp}$ + 0.04, which has been confirmed in the laboratories of the present reviewer.

Table A-3: Equilibrium constants for Am(OH)$_3$(s) and Am(III) hydroxide complexes derived from the solubility experiments of Stadler and Kim [88STA/KIM] (all data at 25°C).

		Original data given in [88STA/KIM] [a]			
Exp	Medium	$\log_{10} K_{s,0}$	$\log_{10} \beta_1$	$\log_{10} \beta_2$	$\log_{10} \beta_3$
A	0.1 M NaClO$_4$ (≤ 3.7 GBq·L⁻¹)	$-(25.7 \pm 0.3)$	(6.3 ± 0.3) [b]	(12.2 ± 0.4) [b]	(14.4 ± 0.5) [b]
B	0.1 M NaClO$_4$ (44–185 GBq·L⁻¹)	$-(25.0 \pm 0.2)$	(6.3 ± 0.3)	(12.2 ± 0.3)	(14.4 ± 0.2)
C	0.1 M NaCl (74–185 GBq·L⁻¹)	$-(25.1 \pm 0.5)$	(6.0 ± 0.4)	(12.2 ± 0.5)	(14.8 ± 0.5)
D	0.6 M NaCl (74–185 GBq·L⁻¹)	$-(25.0 \pm 0.1)$	(5.6 ± 0.3)	(11.6 ± 0.4)	(14.1 ± 0.5)
		present review			
Exp	Medium	$\log_{10}{}^* K_{s,0}$	$\log_{10}{}^* \beta_1$	$\log_{10}{}^* \beta_2$	$\log_{10}{}^* \beta_3$
A	0.1 M NaClO$_4$ (≤ 3.7 GBq·L⁻¹)	(15.7 ± 0.3)	$-(7.5 \pm 0.3)$ [b]	$-(15.4 \pm 0.4)$ [b]	$-(27.0 \pm 0.5)$ [b]
B	0.1 M NaClO$_4$ (44–185 GBq·L⁻¹)	(16.4 ± 0.3)	$-(7.5 \pm 0.3)$	$-(15.4 \pm 0.3)$	$-(27.0 \pm 0.2)$
C	0.1 M NaCl (74–185 GBq·L⁻¹)	(16.3 ± 0.5)	$-(7.8 \pm 0.4)$	$-(15.4 \pm 0.5)$	$-(26.6 \pm 0.5)$
D	0.6 M NaCl (74 – 185 GBq·L⁻¹)	(16.7 ± 0.2)	$-(8.3 \pm 0.3)$	$-(16.2 \pm 0.4)$	$-(27.6 \pm 0.5)$

(a) $\log_{10} K_{s,0}$ refers to the reaction Am(OH)$_3$(s) $\rightleftharpoons$ Am^{3+} + 3 OH⁻
$\log_{10} \beta_n$ refers to the reaction Am^{3+} + n OH⁻ $\rightleftharpoons$ Am(OH)$_n^{3-n}$.

(b) adopted from experiment B.

Figure A-1 shows the good agreement between solubility data in experiment B of [88STA/KIM] and those determined by Silva [82SIL] for ^{243}Am(OH)$_3$(cr) in the range $7 \leq -\log_{10}[H^+] \leq 9.5$. However, the solubility curve calculated with the equilibrium constants re-evaluated by Silva et al. [95SIL/BID] from the data in [82SIL] (solid curve) deviates 0.5 $\log_{10}$ units from the data of [88STA/KIM] at $-\log_{10}[H^+] < 6.5$, where the Am^{3+} ion is the predominant aqueous species. Although Silva et al. [95SIL/BID] applied a correct fitting procedure, the redundancy between the fitted constants led to a solubility constant, which is somewhat too low, and consequently to overestimated hydrolysis constants $\log_{10} {}^*\beta_1$ and $\log_{10} {}^*\beta_2$ compared, for example, to those determined by spectroscopic methods [92WIM/KLE], [94FAN/KIM]. On the other hand it is noteworthy that the discrepancies do not exceed the uncertainty limits of 0.6 $\log_{10}$ units given in [95SIL/BID].

Figure A-1: Comparison of solubility measurements for Am(OH)$_3$(cr) [82SIL] and aged Am(OH)$_3$(s) [88STA/KIM] in 0.1 M NaClO$_4$ at 25°C. The solid curve is calculated with the equilibrium constants evaluated in [95SIL/BID] from the solubility data of Silva [82SIL]: $\log_{10} {}^*K_{s,0} = (15.9 \pm 0.6)$; $\log_{10} {}^*\beta_1 = -(6.9 \pm 0.6)$; $\log_{10} {}^*\beta_2 = -(15.1 \pm 0.6)$. The dashed curve is calculated with the constants derived by Stadler and Kim [88STA/KIM] from the data in experiment B: $\log_{10} {}^*K_{s,0} = (16.4 \pm 0.2)$, $\log_{10} {}^*\beta_1 = -(7.5 \pm 0.3)$, $\log_{10} {}^*\beta_2 = -(15.4 \pm 0.3)$, $\log_{10} {}^*\beta_3 = -(27.0 \pm 0.2)$.

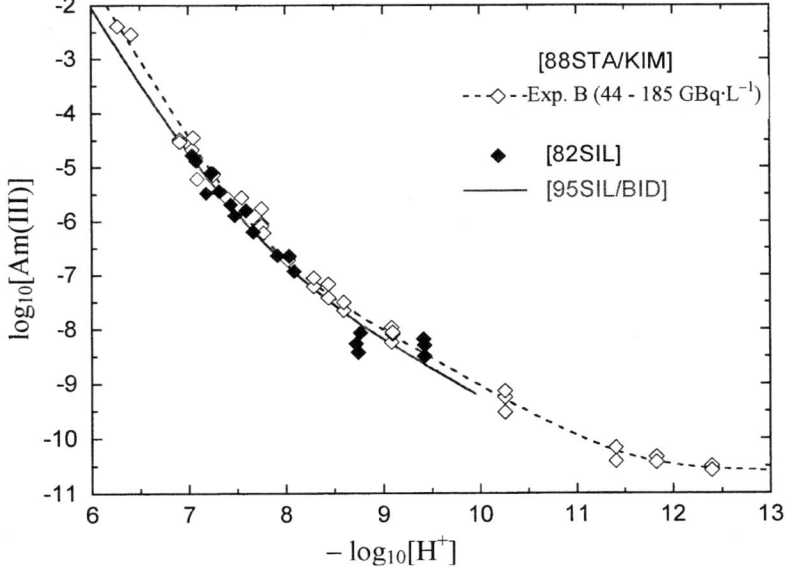

[88TAS/OHA]

The results presented in this paper are considered on pages 136–137 of [92GRE/FUG]. They concern: $C_{p,m}(T)$, (T ranging from 7.59 to 343.88 K and from 350 to 400 K) and $\Delta_f H_m^\circ$ (UO$_3 \cdot$2H$_2$O, cr, 298.15 K). The derived values of:

$$S_m^\circ \text{ (UO}_3 \cdot \text{2H}_2\text{O, cr, 298.15 K)} = (188.54 \pm 0.38) \text{ J} \cdot \text{K}^{-1} \cdot \text{mol}^{-1}$$

$$C_{p,m} \text{ (UO}_3 \cdot \text{2H}_2\text{O, cr, 298.15 K} < T < 400 \text{ K)} = 84.238 + 0.294592 \, T \text{ J} \cdot \text{K}^{-1} \cdot \text{mol}^{-1}$$

and the proposed values by the authors of:

$$\Delta_f H_m^\circ \text{ (UO}_3 \cdot \text{2H}_2\text{O, cr, 298.15 K)} = -(1826.1 \pm 1.7) \text{ kJ} \cdot \text{mol}^{-1} \text{ and}$$

$$\Delta_f G_m^\circ \text{ (UO}_3 \cdot \text{2H}_2\text{O, cr, 298.15 K)} = -(1636.5 \pm 1.7) \text{ kJ} \cdot \text{mol}^{-1}$$

were selected. But no discussion of this key paper is given in Appendix A of [92GRE/FUG], except for a short comment on page 675 concerning the values of standard entropies of O$_2$(g) and H$_2$(g) used to calculate the Gibbs energy of formation, $\Delta_f G_m^\circ$.

In this work, schoepite, UO$_2$(OH)$_2 \cdot$H$_2$O $\equiv$ (UO$_3 \cdot$2H$_2$O) is well characterised. Direct data involving low temperature measurements of $C_{p,m}(T)$ are corrected to take into account the small degree of dehydration of schoepite above 270 K. No anomaly in the sigmoid curve for $C_{p,m}(T)$ appears. Incremental values of $S_m^\circ(T)$, $H_m(T) - H_m(0)$ and $-(G_m(T) - H_m(0))/T$ are given (10 to 360 K). Enthalpy measurements of $H_m(T) - H_m^\circ(298.15\text{K})$ up to 400 K are also corrected for the vapor pressure over schoepite. A smoothed polynomial variation is given for $H_m(T) - H_m^\circ(298.15\text{K})$ and $C_{p,m}(T)$ is derived (see above) with the conditional limit $C_{p,m}^\circ(298.15\text{K}) = 172.07$ J $\cdot$ K^{-1} $\cdot$ mol^{-1} given by the low temperature data.

The standard molar enthalpy for:

$$\text{UO}_3 \cdot \text{2H}_2\text{O (cr)} + 2 \text{ HF(aq)} \rightleftharpoons \text{UO}_2\text{F}_2\text{(aq)} + 3 \text{ H}_2\text{O(sln)}$$

is measured as $\Delta_r H_m^\circ = -(76.25 \pm 0.35)$ kJ $\cdot$ mol^{-1}. The value $\Delta_f H_m^\circ$ (UO$_3 \cdot$2H$_2$O, cr, 298.15 K) $= -(1825.4 \pm 2.1)$ kJ $\cdot$ mol^{-1} is calculated through a thermochemical cycle including that $\Delta_r H_m^\circ$ value, previous data on dissolution of UO$_3$(γ) in HF(aq) and auxiliary CODATA values.

The authors also discuss three previous sets of results from [64COR], [66DRO/KOL] and [71SAN/VID] leading to $\Delta_f H_m^\circ$ (UO$_3 \cdot$2H$_2$O, cr, 298.15 K) values. Taking the weighed average of these values and the value obtained by them, they give $-(1826.1 \pm 1.7)$ kJ$\cdot$mol^{-1} as a recommended value to be used. [92GRE/FUG] followed the authors' recommendation.

However, [92GRE/FUG] state (page 136 in V.3.3.1.5): "The recent work of Tasker et al. [88TAS/OHA] report a similarly determined experimental value (enthalpy of hydration of γ–UO$_3$ to give schoepite) of $-(29.93 \pm 0.52)$ kJ $\cdot$ mol^{-1}". Such an explicit value is not reported in this paper. But from the value $\Delta_f H_m^\circ (UO_3 \cdot 2H_2O, cr, 298.15 K) = -(1825.4 \pm 2.1)$ kJ·mol^{-1} (see above) and $\Delta_f H_m^\circ (UO_3, \gamma, 298.15 K) = -(1223.8 \pm 1.2)$ kJ·mol^{-1} and $\Delta_f H_m^\circ (H_2O, l, 298.15 K) = (285.83 \pm 0.04)$ kJ·mol^{-1}, it is possible to obtain -29.4 kJ·mol^{-1} for enthalpy of hydration of γ–UO$_3$ to give schoepite.

[89DIC/LAW]

These authors report the preparation, characterisation, and thermochemical studies on δ–UO$_3$, δ–Li$_{0.69}$UO$_3$, γ–Li$_{0.55}$UO$_3$ and Li$_{0.88}$U$_3$O$_8$(cr). The three latter compounds were also studied using electrochemical techniques. Insertion of lithium was achieved [86DIC/PEN] through reaction with anhydrous LiI or, for the latter compound, with n-butyl lithium. Characterisation techniques were the same as those described in [90POW], with, in addition, the determination of Li by atomic absorption spectrometry. The calorimetric reagent for the dissolution, 0.0350 mol·dm^{-3} Ce(SO$_4$)$_2$ +1.505 mol·dm^{-3} H$_2$SO$_4$ [77COR/OUW2], [81COR/OUW], was the same as in [88DIC/POW] and in [90POW]. Uncertainty limits are reported as twice the standard deviation of the mean. The calorimetric results are discussed in greater detail in [90POW] and will not be repeated here. The other values for the enthalpies of solution were reported as:

$\Delta_{sol} H_m (Li_{0.69}UO_3, \delta) = -(165.24 \pm 1.94)$ kJ·mol^{-1}

$\Delta_{sol} H_m (Li_{0.55}UO_3, \gamma) = -(146.12 \pm 2.52)$ kJ·mol^{-1}

$\Delta_{sol} H_m (Li_{0.88}U_3O_8, cr) = -(460.52 \pm 1.60)$ kJ·mol^{-1}.

Using the auxiliary enthalpies of solution of the binary uranium oxides used also in [90POW], and that for LiUO$_3$(cr), $-(208.35 \pm 0.53)$ kJ·mol^{-1} (all original data from [77COR/OUW2], [81COR/OUW]), the enthalpies of the following reactions:

$$0.69 \, LiUO_3(cr) + 0.31 \, \gamma\text{–}UO_3 \rightleftharpoons \delta\text{–}Li_{0.69}UO_3 \quad (A.16)$$

$$0.55 \, LiUO_3(cr) + 0.45 \, \gamma\text{–}UO_3 \rightleftharpoons \gamma\text{–}Li_{0.55}UO_3 \quad (A.17)$$

$$0.88 \, LiUO_3(cr) + U_3O_8(cr) \rightleftharpoons Li_{0.88}U_3O_8(cr) + 0.88 \, \gamma\text{–}UO_3 \quad (A.18)$$

were obtained as $\Delta_r H_m^\circ ((A.16), 298.15 K) = -(4.76 \pm 1.98)$ kJ·mol^{-1}, $\Delta_r H_m^\circ ((A.17), 298.15 K) = -(6.56 \pm 2.54)$ kJ·mol^{-1} and $\Delta_r H_m^\circ ((A.18), 298.15 K) = (0.39 \pm 2.16)$ kJ·mol^{-1}. Using NEA adopted values, we obtain:

$\Delta_f H_m^\circ (Li_{0.69}UO_3, \delta, 298.15 K) = -(1434.52 \pm 2.37)$ kJ·mol^{-1}

$\Delta_f H_m^\circ (Li_{0.55}UO_3, \gamma, 298.15 K) = -(1394.53 \pm 2.78)$ kJ·mol^{-1}

$\Delta_f H_m^\circ (Li_{0.88}U_3O_8, cr, 298.15 K) = -(3837.09 \pm 3.75)$ kJ·mol^{-1}.

For the insertion reactions:

$$0.69 \, Li(cr) + \gamma\text{–}UO_3 \rightleftharpoons \delta\text{–}Li_{0.69}UO_3 \quad (A.19)$$

$$0.55 \text{ Li(cr)} + \gamma\text{–UO}_3 \rightleftharpoons \gamma\text{–Li}_{0.55}\text{UO}_3 \quad (A.20)$$

$$0.88 \text{ Li(cr)} + \text{U}_3\text{O}_8\text{(cr)} \rightleftharpoons \text{Li}_{0.88}\text{U}_3\text{O}_8\text{(cr)} \quad (A.21)$$

one then obtains, using NEA accepted values:

$$\Delta_r H_m^\circ ((A.19), 298.15 \text{ K}) = - (220.79 \pm 2.77) \text{ kJ} \cdot \text{mol}^{-1}$$

$$\Delta_r H_m^\circ ((A.20), 298.15 \text{ K}) = - (170.73 \pm 3.03) \text{ kJ} \cdot \text{mol}^{-1}$$

$$\Delta_r H_m^\circ ((A.21), 298.15 \text{ K}) = - (262.29 \pm 4.45) \text{ kJ} \cdot \text{mol}^{-1}.$$

Per mole of inserted lithium, these values correspond to $- (319.99 \pm 4.01)$ kJ · mol^{-1}, $- (310.42 \pm 5.51)$ kJ · mol^{-1}, and $- (298.06 \pm 5.06)$ kJ · mol^{-1}, respectively.

The authors also deduce integral Gibbs energy changes for the insertion reactions from discharge and open circuit voltage curves using cells of the type:

$$\text{Li(cr)} \mid 1 \text{ M LiClO}_4 \text{ (1:1 P.C./DME)} \mid \text{Li}_x\text{UO}_n$$

where P.C./DME is presumably a propylene carbonate / dimethyletherethylene glycol mixture.

The Gibbs energies of the insertion reactions:

$$x \text{ Li(cr)} + \text{UO}_n\text{(cr)} \rightleftharpoons \text{Li}_x\text{UO}_n\text{(cr)} \quad (A.22)$$

obtained by graphical integration, (presented only in a graph) for δ–Li$_x$UO$_3$ appear to be an approximately linear function of x, but the calorimetric result indicated on their plot, for comparison, at $x = 0.55$, actually corresponds to γ–Li$_{0.55}$UO$_3$, (for which, no equilibrium emf value could be obtained) and the value for δ–Li$_{0.69}$UO$_3$ is omitted (it falls close to the curve for the Gibbs energies). The curve for $\Delta_r G_m$ (A.22) for Li$_x$U$_3$O$_8$, as a function of x flattens off to have a slope close to zero as x approaches 1.0. These electrochemical results will not be considered further in this review.

[89GUR/DEV]

This abstract gives results of calorimetric measurements on pure samples of UO$_2$(OH)$_2$(cr) and UO$_2$CO$_3$(cr) (10 to 340 K). The standard enthalpy of formation of UO$_2$CO$_3$(cr) was calculated from solution calorimetry data. By combining the results with literature data, recommended values of the standard molar thermodynamic properties of the compounds and the reactions:

$$\text{UO}_2(\text{OH})_2 \cdot \text{H}_2\text{O(cr)} \rightleftharpoons \text{UO}_2(\text{OH})_2\text{(cr)} + \text{H}_2\text{O(l)}$$

$$\text{UO}_2\text{CO}_3\text{(cr)} + \text{H}_2\text{O(l)} \rightleftharpoons \text{UO}_2(\text{OH})_2\text{(cr)} + \text{CO}_2\text{(g)}$$

were obtained. These results are reported and discussed in [97GUR/SER] (see the review of this paper).

[89PET/SEL]

Numerous reactions leading to uranium fluorophosphates are described. IR spectra and X–ray diffraction were used to characterise the compounds, but no details were given.

The enthalpy effect associated with the thermal decomposition (in an argon atmosphere) of some compounds is given without any details, as follows:

$$UO_2(PO_2F_2)_2(s) \rightleftharpoons UO_2PO_3F(s) + POF_3(g) \qquad (423-773 \text{ K})$$
$$\Delta_r H_m^\circ(298 \text{ K}) = (89.6 \pm 1.2) \text{ kJ} \cdot \text{mol}^{-1}$$

$$U(PO_2F_2)_4(s) \rightleftharpoons U(PO_3F)_2(s) + 2 \, POF_3(g) \qquad (423-773 \text{ K})$$
$$\Delta_r H_m^\circ(298 \text{ K}) = (218.8 \pm 10.0) \text{ kJ} \cdot \text{mol}^{-1}$$

$$UO_2(PO_2F_2)F(s) \rightleftharpoons \tfrac{1}{2}UO_2PO_3F(s) + \tfrac{1}{2}UO_2F_2(s) + \tfrac{1}{2}POF_3(g) \qquad (T > 403 \text{ K})$$
$$\Delta_r H_m^\circ(298 \text{ K}) = (39.1 \pm 0.95) \text{ kJ} \cdot \text{mol}^{-1}$$

$$UO_2PO_3F(s) \rightleftharpoons \tfrac{1}{3}(UO_2)_3(PO_4)_2(s) + \tfrac{1}{3}POF_3(g) \qquad (953-993 \text{ K})$$
$$\Delta_r H_m^\circ(298 \text{ K}) = (64.6 \pm 1.0) \text{ kJ} \cdot \text{mol}^{-1}$$

[89SER/SAV]

The solubility of uraninite was measured in "pure" water and aqueous HCl (0.0001 – 1 molal), KCl + HCl (1 – 1.1 molal Cl$^-$) at 300, 400, 500 and 600°C at 1 kbar, whereby the oxygen fugacity was controlled with either magnetite/haematite or nickel/nickel oxide buffers. The solubility in "pure" water was independent of temperature and the nature of the buffer with an average value of $10^{-(9.0 \pm 0.5)}$. The formation of $U(OH)_4(aq)$ and $U(OH)_3^+$ is proposed over the entire range of temperature and pH with the apparent formation of undisclosed chloride complexes.

The solubility of $UO_2(OH)_2(cr)$ in water was measured at 100°C under argon. In addition, the solubility of $UO_2CO_3(cr)$ was measured in water and $NaHCO_3$ solutions (0.001 – 0.1 molar) under a CO_2 atmosphere (p_{CO_2} = 7.7–20.93 atm) at 100 and 200°C.

The $UO_3 - CO_2 - H_2O$ system was studied spectrophotometrically from 25 to 80°C, under a p_{CO_2} ranging from 0.52 to 0.97 atm at ionic strengths of 0.1 to 2.0 molal ($NaClO_4$). The concentrations measured ranged as follows: m(U) = 0.002 molal, m($NaHCO_3$) = 0.015 – 0.15 molar, and pH = 5.15 – 7.35. From these data it was suggested that the following formation reaction dominated:

$$UO_2(CO_3)_2^{2-} + CO_3^{2-} \rightleftharpoons UO_2(CO_3)_3^{4-} \qquad (A.23)$$

and the stepwise formation constant $\log_{10} K = 4.49$, at infinite dilution was given by the expression: $3193\ T^{-1} - 6.22$. Access to the experimental data is needed before the results summarised here can be evaluated.

At 25°C this gives $\Delta_r H_m^\circ(A.23) = -61$ kJ · mol^{-1}, which is in good agreement with the value -59.1 kJ · mol^{-1} selected in [92GRE/FUG]. However, it is not clear from the abstract if the new data are based on the older ones obtained by some of the authors. The old data were used in the selection of enthalpies of reaction in [92GRE/FUG].

[89STA/MAK]

Salts of sulphamic acid, $HSO_3NH_2 = HX$, designated as MX_z, have an exceptionally high solubility in water while HX solubility itself is low. The standard Gibbs energy for the dissolution of crystalline MX_z is given at 298.15 K by:

$$\Delta_{sol}G_m^\circ(kJ \cdot mol^{-1}) = 5.71 \qquad \log_{10} K_{s,0}^\circ = \Delta_{sol}H_m^\circ - 298.15 \cdot \Delta_{sol}S_m^\circ$$

$K_{s,0}^\circ$ being the solubility product of sulphamates, which also gives the solubility in water if M^{z+} is not hydrolysed to allow aqueous ion M^{z+} to exist.

In this paper a good fit is shown between calculated $\Delta_{sol}G_m^\circ$ and $\Delta_{sol}H_m^\circ$ (calculated through $\Delta_{sol}S_m^\circ$) for M, M^{2+} and M^{3+} single ions and some experimental data according to an electrostatic model. The ionic radius of M^{z+} is the driving parameter. Giving the ionic radius of 146 pm to UO_2^{2+}, the authors calculate $\Delta_{sol}G_m^\circ = -169.1$ kJ · mol^{-1} and $\Delta_{sol}H_m^\circ = -40.5$ kJ · mol^{-1}, $S_m^\circ = 431.3$ J · K^{-1} · mol^{-1} (which means a solubility of (63 ± 10) g/(100g water), but do not take into account U(VI) hydrolysis). For U(III) with a radius of 103 pm for M^{z+} they obtain $\Delta_{sol}G_m^\circ = -603.6$ kJ · mol^{-1} and $\Delta_{sol}H_m^\circ = -447.7$ kJ · mol^{-1}, $S_m^\circ = 525.8$ J · K^{-1} · mol^{-1} (which means a solubility of (144 ± 20) g/(100 g water)).

The estimated data cannot be included in the TDB II selected data because of the lack of several experimental values for the oxidation of heavy elements which could have given some confidence in the application of the model used for U(VI). For UX$_3$ the situation is better. Indeed experimental and calculated values for lanthanides are very close ($\pm$ 5%). The estimated solubility of U(VI) sulphamate: UO_2X_2 is less than for sulphate (157.7 g/(100 g water)), chloride (328 g/(100 g water)) and nitrate (121.8 g/(100 g water)).

[89TAT/SER]

The solid phases, $Li_2XO_4 \cdot H_2O$ and $UO_2XO_4 \cdot 2.5H_2O$ (X = S, Se) were prepared by interaction of the appropriate acid with either Li_2CO_3 or $\beta - UO_3 \cdot H_2O$, and were recrystallised from water. Chemical analyses, XRD (powder and single crystal), and TGA were used to characterise the solids. Equilibrium was reached at 25°C with respect to the solid phases in 8–10 hours for the sulphate system. The phase diagram shows the formation of $UO_2SO_4 \cdot 3.5H_2O(s)$. Solutions saturated at 0 and 5°C give the same solid phases, $Li_2UO_2(SO_4)_2 \cdot H_2O(s)$, and tend to refute earlier claims of Oechsner and Chauvenet [05OEC/CHA] for the formation of $Li_2UO_2(SO_4)_2 \cdot 4H_2O(s)$. The selenium system is more viscous and consequently attained equilibrium only after 70 – 80 to 150 hours. The congruently soluble compounds, $Li_2UO_2(SeO_4)_2 \cdot 5H_2O(cr)$ and $Li_4(UO_2)_3(SeO_4)_5 \cdot 16H_2O(cr)$, are formed with solubilities of 65.7 and 67.0 %, respectively. The former is triclinic, whereas the latter is monoclinic and isomorphic with $Mg_2(UO_2)_3(SeO_4)_5 \cdot 16H_2O(cr)$.

[89YAM/FUJ]

This abstract concerns the oxygen potential of $Sr_yY_yU_{1-2y}O_{2+x}$ (y = 0.05 and 0.025) solid solutions measured by thermogravimetric methods between 1123 and 1673 K.

This paper (and related papers by the same group) will not be considered further in this review as the work reported deals only with non–stoichiometric compounds, which are not part of the current NEA review effort.

[90COS/LAK]

The authors have prepared high density essentially stoichiometric UH_3 and UD_3 by heating U in high pressure H_2 or D_2 (125 MPa) above the melting point of U (1408 K) and cooling slowly still under high pressures (decreasing with T). No thermodynamic data are given, except to report that UH_{3-x} with H/U $ca.$ 2.5 melts at $ca.$ 1338 K, but this is probably taken from the earlier work by [78LAK].

[90HAY/THO]

This paper, which is a comprehensive review/assessment of the thermodynamics and vaporisation of UN(cr), is mentioned only briefly in [92GRE/FUG], presumably because it appeared only just before finalisation of [92GRE/FUG]. Table A-4 is a comparison of the two sets of selected values $(J \cdot K^{-1} \cdot mol^{-1} \cdot)$.

Table A-4: Comparison of selected values from [90HAY/THO] and [92GRE/FUG].

Property	[90HAY/THO]	[92GRE/FUG]
$C_{p,m}^\circ$ (298.15 K)	47.96	(47.57 ± 0.40)
S_m° (298.15 K)	62.68	(62.43 ± 0.22)

The two assessments overlap within the combined uncertainties.

[90IYE/VEN]

This conference paper reports measurements of the oxygen activity (emf with CaO–ZrO$_2$ electrolyte) in the mixtures Rb$_2$U$_4$O$_{12}$(cr) + Rb$_2$U$_4$O$_{13}$(cr) and preliminary data on the enthalpy increments of the compounds. The oxygen potential data are the same as those presented in [92VEN/IYE], where the more complete data for the enthalpy measurements are also given. There are thus no additional data to be derived from this paper.

[90KUM/BAT]

Isopiestic measurements were carried out at 25°C by Kumok and Batyreva [90KUM/BAT] on only five concentrated K$_2$SeO$_4$ solutions (2.0097–6.2602 mol·kg^{-1}) using NaCl(aq) as the standard. Mean stoichiometric activity coefficients were calculated from an equation for the osmotic coefficient by [62LIE/STO] as a function of molality and the resulting Gibbs–Duhem equation. Given the limited number of concentrations studied and the high concentrations, the usefulness of the reported recalculated solubility product (from an untraceable reference) and Gibbs energy of formation of UO$_2$SeO$_4$·4H$_2$O (cr) appear suspect, *i.e.*, $\log K_{s,0}^\circ = -(2.25 \pm 0.04)$ and $\Delta_f G_m^\circ$ (298.15 K) = $-(2357.4 \pm 3.0)$ kJ · mol^{-1}.

[90PHI]

This report is not explicitly mentioned in [92GRE/FUG], although all the data are considered in it. The values selected by the authors are similar to those selected in [92GRE/FUG]. They concern mainly $\Delta_f G_m^\circ$, $\Delta_f H_m^\circ$, S_m° of UOH^{3+}, U(OH)$_2^{2+}$, U(OH)$_3^+$, U(OH)$_4$(aq), U(OH)$_5^-$ and U(OH)$_4$(s).

Experimental data are processed in the same way as in [92GRE/FUG] using the same CODATA auxiliary values. Nevertheless slight differences exist (about 1 kJ · mol^{-1} or less for Gibbs energy and enthalpy) between the two sets of values (except for U(OH)$_2^{2+}$, U(OH)$_3^+$ which are not selected in [92GRE/FUG]). For instance, the analysis by Philips of a part of the experimental data considered by [92GRE/FUG] led him to propose $\log_{10} {}^*\beta_1^\circ = -(0.61 \pm 0.05)$ while [92GRE/FUG] selected $\log_{10} {}^*\beta_1^\circ = -(0.54 \pm 0.06)$. In other cases, the differences seem to be due to small differences in the relevant thermodynamic values concerning the U^{4+} aqua ion or UO$_2$(cr). In the opinion of the reviewer, this report does not provide information that makes it necessary to revise the analysis by [92GRE/FUG]. The following values, mostly estimates, are given by the author: $\log_{10} {}^*\beta_2^\circ = -2.76$ (without uncertainty), $\log_{10} {}^*\beta_3^\circ = -3.79$ (without uncertainty), $\log_{10} {}^*\beta_4^\circ = -(4.83 \pm 0.50)$, $\log_{10} {}^*\beta_5^\circ = -(15.09 \pm 0.50)$ and $\log_{10} K_{s,4}^\circ = -(9.47 \pm 0.30)$.

[90POW]

Various hydrogen, lithium, and sodium insertion compounds with uranium oxides were prepared and studied using X-ray diffraction, neutron scattering, and thermogravimetric techniques. Related work on other oxides is also reported.

Enthalpies of formation derived from enthalpies of solution have been reported for δ–UO_3, α–$Na_{0.14}UO_3$ and δ–$H_{0.83}UO_3$. The medium used for the dissolution of the sample was identical to that used by [77COR/OUW2], [81COR/OUW], namely 0.0350 mol·dm^{-3} $Ce(SO_4)_2$ + 1.505 mol·dm^{-3} H_2SO_4. In the study by [90POW], use was made of these authors' data for $\Delta_{sol}H_m$ (UO_3, γ) = – (84.64 ± 0.38) kJ·mol^{-1}, $\Delta_{sol}H_m$ (U_3O_8, α) = – (351.27 ± 1.34) kJ·mol^{-1} and $\Delta_{sol}H_m$ ($NaUO_3$, cr) = – (200.04 ± 0.51) kJ·mol^{-1} (uncertainties are quoted at the 1.96σ level).

The sample of δ–UO_3 was prepared by thermal decomposition, at 648 K, of β–$UO_2(OH)_2$. The X-ray powder diffraction of the sample agreed with that in the literature [55WAI] and showed no phase other than δ–UO_3. The authors reported $\Delta_{sol}H_m$ (UO_3, δ) = – (94.71 ± 0.72) kJ·mol^{-1} (our 1.96σ uncertainty limits). The difference between the enthalpies of solution of δ– and γ–UO_3, together with the NEA accepted value for the standard enthalpy of formation of γ–UO_3 leads to $\Delta_f H_m^\circ$ (UO_3, δ, 298.15 K) = – (1213.73 ± 1.44) kJ·mol^{-1}.

The α–$Na_{0.14}UO_3$ sample was prepared by heating a mixture of α–UO_3 and $NaUO_3$ in an evacuated silica tube [73GRE/CHE]. The compound was characterised by X-ray diffraction and the uranium mean oxidation state was ascertained by redox titration (dissolution in a potassium dichromate solution and back titration with Fe^{2+}). The author deduced from open circuit voltage curves that the sample was monophasic.

The determination is based on the enthalpy of solution of the constituents of the reaction:

$$0.14\ NaUO_3(cr) + 0.86\ \gamma\text{–}UO_3 \rightleftharpoons \alpha\text{–}Na_{0.14}UO_3 \qquad (A.24).$$

The enthalpy of solution of α–$Na_{0.14}UO_3$ was determined as – (95.32 ± 0.35) kJ·mol^{-1} (1.96σ). Thus, $\Delta_r H_m^\circ$ ((A.24), 298.15 K) = – (5.48 ± 0.48) kJ·mol^{-1}, (not – (6.76 ± 0.77) kJ·mol^{-1} as given in the thesis). Using NEA adopted values, we obtain:

$$\Delta_f H_m^\circ (Na_{0.14}UO_3, \alpha, 298.15\ K) = -(1267.23 \pm 1.80)\ kJ\cdot mol^{-1};$$

the author's value of – (1276.19 ± 0.87) kJ·mol^{-1} seems to be in error.

For the insertion reaction:

$$0.14\ Na\ (cr) + \alpha\text{–}UO_3 \rightleftharpoons \alpha\text{–}Na_{0.14}UO_3 \qquad (A.25)$$

we obtain, $\Delta_r H_m^\circ$ ((A.25), 298.15 K) = – (54.82 ± 2.31) kJ·mol^{-1}, in general agreement with other insertion compounds of Na in uranium oxides.

[90POW] also cites earlier data [88DIC/POW] on the enthalpy of formation of cubic δ–$Na_{0.54}UO_3$.

The δ–$H_{0.83}UO_3$ sample was obtained by a process described as hydrogen spill–over and characterised by X–ray diffraction and redox titration (mean oxidation state of uranium). The determination is based on the enthalpy of solution of the constituents of the equation:

$$0.415 \text{ α–}U_3O_8 + 0.415 \text{ }H_2O(l) \rightleftharpoons \text{δ–}H_{0.83}UO_3 + 0.245 \text{ γ–}UO_3 \quad (A.26).$$

The enthalpy of transfer of H_2O, -0.05 kJ·mol^{-1}, was taken from [81COR/OUW]. The enthalpy of solution of the compound is reported as $-(142.25 \pm 1.64)$ kJ·mol^{-1}, which leads to $\Delta_r H_m^\circ ((A.26), 298.15 \text{ K}) = -(17.19 \pm 1.73)$ kJ·mol^{-1}. With NEA adopted values, we obtain:

$$\Delta_f H_m^\circ (H_{0.83}UO_3, \delta, 298.15 \text{ K}) = -(1285.14 \pm 2.02) \text{ kJ·mol}^{-1}.$$

For the insertion reaction:

$$0.415 \text{ }H_2(g) + \text{δ–}UO_3 \rightleftharpoons \text{δ–}H_{0.83}UO_3 \quad (A.27)$$

using the accepted value for the standard enthalpy of formation of δ–UO_3 (see section 9.3.3.1), we obtain $\Delta_r H_m^\circ ((A.27), 298.15 \text{ K}) = -(71.41 \pm 2.48)$ kJ·mol^{-1}.

Powell also cites without any reference earlier results on α–$H_{1.08}UO_3$. which we shall not consider further.

[90SAW/CHA2]

This paper was not reviewed in [92GRE/FUG]. It deals with the complexation of U(IV) with fluoride ion. A commercial combination (fluoride sensitive (FSE)/ silver–silver chloride) electrode was employed with NaCl/ $NaClO_4$ filling solution at (23 ± 1) °C. The ionic strength was maintained at 1.0 M ($HClO_4$, NaF) with the calculated total hydrogen ion molarity as high as 0.9273 M. The fluoride concentration generally reached was less than 0.5 M. Liquid junction potentials were measured by a method described previously [85SAW/RIZ] and the absolute values were generally large (-9.0 to -29.8 mV). The correction for HF (aq) formation was considered. It is not clear from the paper if the authors used weighted $\bar{n}$ values in the least squares refinement of the data. Two fitting algorithms used by the authors were an in-house program and MINIQUAD. The final tabulated values converted to the molal scale using a density of 1.0568 g·mL^{-1}, corresponding to an ionic strength of 1.046 m ($HClO_4$) are reported in Table A-5 for the formation constants of the MF_n^{4-n} species. The correction factor used to convert from the molar to the molal scale is $\log_{10}\rho = 0.02$.

Table A-5. $\log_{10} \beta_n$ values corresponding to an ionic strength of 1.046 m $HClO_4$. The values in parentheses indicate the number of titrations from which these values were derived.

Metal	$\log_{10} \beta_1$	$\log_{10} \beta_2$	$\log_{10} \beta_3$	$\log_{10} \beta_4$
Th(IV)	(7.59 ± 0.01) (6)	(13.38 ± 0.05) (6)	(17.29 ± 0.20) (4)	(23.55 ± 0.11) (4)
U(IV)	(8.46 ± 0.01) (5)	(14.62 ± 0.01) (5)	(19.45 ± 0.03) (5)	(23.84 ± 0.06) (5)
Np(IV)	(8.15 ± 0.04) (3)	(14.48 ± 0.13) (3)	(19.99 ± 0.13) (2)	(25.87 ± 0.18) (2)
Pu(IV)	(7.59 ± 0.12) (3)	(14.73 ± 0.10) (3)	(20.05 ± 0.37) (3)	(25.99 ± 0.16) (3)

After conversion to $I = 0$ using the SIT and appropriate auxiliary data of [92GRE/FUG], that review finds $\log_{10} \beta_n^\circ$ values, which agree with those selected by [92GRE/FUG] (cf. pages 166 and 167). They are, respectively: $\log_{10} \beta_1^\circ = (9.78 \pm 0.12)$ vs (9.28 ± 0.09), $\log_{10} \beta_2^\circ = (16.97 \pm 0.13)$ vs (16.23 ± 0.15), $\log_{10} \beta_3^\circ = (22.38 \pm 0.14)$ vs (21.6 ± 1.0) and $\log_{10} \beta_4^\circ = (27.06 \pm 0.12)$ vs (25.6 ± 1.0).

For Np(IV) and Pu(IV), the data are discussed in [2001LEM/FUG]. The values calculated for $\log_{10} \beta_n^\circ$ differ slightly from those reported in that review possibly due to the use of a different correction factor and different SIT interaction coefficients.

[90SEV/ALI]

Sevast'yanov et al. prepared UCl_6(cr) by reacting "uranium oxide" mixed with activated charcoal with a gas mixture of CCl_4(g) and Cl_2(g) at 753 K. The volatile products UCl_5(g) and UCl_6(g) were condensed in separate zones in a glass apparatus so that the synthesis, separation, purification and collection were carried out in a closed system. The final purification was by two sublimations at 363 – 373 K. The product was shown to be UCl_6 by "analytical data (error ± 0.3 mass %)", with no further details.

The vapour pressure of UCl_6(cr) was measured by Knudsen effusion from a molybdenum cell, combined with mass–spectrometric analysis of the vapour. In three measurements at 285, 320 and 353 K the vapour was shown to be UCl_6(g) without perceptible thermal dissociation, the vapour pressures being $1.66 \cdot 10^{-7}$, $6.21 \cdot 10^{-6}$ and $5.71 \cdot 10^{-5}$ bar, respectively. It is not entirely clear how these values were obtained. From the slope of the plot of $\log_{10}(p/\text{bar})$ vs. $1/T$, a second–law enthalpy of sublimation of (75.7 ± 3.3) kJ · mol^{-1} was obtained.

The authors have also calculated the thermal functions of UCl_6(g) using a U–Cl distance of 2.54 Å, and vibration frequencies of 355(1), 326(2), 330(3), 110(3), 130(3) and 120(3) cm^{-1} (where the numbers in parentheses are the vibrational degeneracies). These values are slightly different from those selected here (see Table 9-1), but the derived values of $S_m(T)$ differ only by ca 2.4 J · K^{-1} · mol^{-1} (assuming the multiplicity of the ground state is unity, rather than their quoted value of two). From these thermal functions the authors report a third–law enthalpy of (88.8 ± 4.0) kJ · mol^{-1}, though this is probably a misprint for (80.8 ± 4.0) kJ · mol^{-1}, since their average of their second- and third-law values is quoted to be (78.2 ± 5.0) kJ · mol^{-1} and the third-law enthalpy of sublimation using our slightly different thermal functions (see section 9.4.3.1.1.4) is (81.2 ± 4.0) kJ · mol^{-1}.

[90VOC/HAV]

This paper was not reviewed in [92GRE/FUG] nor in [95SIL/BID] but it is cited in [95GRE/PUI] and in the Appendix on "Thermodynamics of Uranium" in [95SIL/BID]. Synthetic becquerelite is obtained by reacting schoepite with a solution of CaCl$_2$ at 60°C for one week and characterised by chemical analyses and X-ray diffraction. The solubility of Ca[(UO$_2$)$_6$O$_4$(OH)$_6$]·8 H$_2$O in water is measured in the absence of CO$_2$ for five pH values in the range 4.50 to 6.28 after one week of equilibration at 25°C. Total concentrations of Ca and U are measured and the values of solubility of becquerelite are calculated to change from $1.45 \cdot 10^{-3}$ to $0.25 \cdot 10^{-5}$ M with the increase in pH. The chemical model used to process the data takes into account the U(VI) species, $(UO_2)_m(OH)_n^{2m-n}$, with $(n:m) = (1,1), (2:1), (2:2), (4:3)$ and $(5:3)$. The values of $\log_{10} \beta_{n,m}$ are those given by Sillen and Martell [71SIL/MAR], and Högfeldt [82HOG]. The value of the equilibrium constant for:

$$Ca[(UO_2)_6O_4(OH)_6] \cdot 8 \, H_2O \rightleftharpoons Ca^{2+} + 6 \, UO_2^{2+} + 14 \, OH^- + 4 \, H_2O$$

is given as $\log_{10} K_{s,0} = -(152.4 \pm 0.3)$.

In this study the ionic strength is less than $2 \cdot 10^{-2}$ M and variable. The values of $\log_{10} \beta_{m,n}$ are not given, but Vochten and Van Haverbeke report the distribution of the U(VI) species as a function of pH which clearly show that they are incorrect. For instance, the species UO$_2$OH$^+$ is predominant at pH 9. The solid phase in the presence of water after one week was not characterised.

From $\log_{10} K_{s,0} = -(152.4 \pm 0.3)$ one can calculate $\log_{10} {}^*K_{s,0} = (43.6 \pm 0.3)$ which is close to the values reported by [2002RAI/FEL] (see this review). Casas et al. [97CAS/BRU] calculated from the data of Vochten and Van Haverbeke a value of $\log_{10} {}^*K_{s,0}^\circ = 41.4$ using the aqueous thermodynamic model of Grenthe et al. [92GRE/FUG]. Due to the lack of information in this study and the use of erroneous auxiliary data, this review does not consider the value reported by Vochten and Van Haverbeke for the selection of a solubility product for becquerelite.

This paper also gives additional information on the solubility of billietite, Ba[(UO$_2$)$_6$O$_4$(OH)$_6$]·8 H$_2$O, but as the data are processed in the same manner as were those of becquerelite, the value $\log_{10} K_{s,0} = -(158.7 \pm 0.6)$ is not retained by this review.

[91AGU/CAS]

This paper gives some information on the solubility of nanocrystalline UO_2 particles (30 to 60 Å) obtained by reducing U(VI) in acidic solution by H_2 in presence of Pd then raising its pH to 7. In the presence of NaCl the equilibrium concentration of U decreases and stabilises to $(4.65 \pm 0.06) \cdot 10^{-6}$ M in the range $5 \cdot 10^{-2}$ to 4 M NaCl (pH = 7.5). This decrease (from $4 \cdot 10^{-5}$ M in $NaClO_4$) cannot be explained by the formation of chlorocomplexes. The authors propose the presence of UO_2Cl_2 (or at least the presence of a Cl containing phase) from X–ray diffraction of the solid phase. This is very unlikely as this compound is very soluble.

[91BID/CAV]

The carbonate complexation of U(VI) has been investigated at 25°C by thermal lensing spectroscopy (TLS) at a wavelength of 448 nm. This very sensitive laser spectroscopic method was used to determine the stepwise equilibrium:

$$UO_2(CO_3)_2^{2-} + CO_3^{2-} \rightleftharpoons UO_2(CO_3)_3^{4-} \qquad (A.28)$$

at a total U(VI) concentration of $4 \cdot 10^{-6}$ mol·L^{-1}, which is low enough to exclude the interference of the trimer, $(UO_2)_3(CO_3)_6^{6-}$. Experimental details and auxiliary data used for the carbonic acid dissociation constants and fitting procedures are well described. Solutions of the composition 0.02 M NaHCO$_3$/0.48 M NaClO$_4$ were titrated with 0.2 M HClO$_4$/0.3 M NaClO$_4$ and equilibrated at 25°C with a gas stream of 100% CO_2 or a CO_2/N_2 (10.33 % CO_2) mixture to vary the CO_3^{2-} concentration in the range 10^{-5} to $10^{-8.5}$ mol·L^{-1}. All solutions contained a constant U(VI) concentration of $4 \cdot 10^{-6}$ mol·L^{-1}.

The evaluation of the spectroscopic results requires knowledge of the normalised TLS signals, S_n, at 448 nm for the different species, $UO_2(CO_3)_n^{2-2n}$. The value of S_3 was determined from a solution at higher carbonate concentration and the ratio of S_2/S_3 was fitted simultaneously with the equilibrium constant for reaction (A.28). The variation of estimations for the ratios S_1/S_3 and S_0/S_3 had no significant effect on the results. The best fitted value for the stepwise carbonate complexation constant was reported to be $\log_{10} K_3$ ((A.28), 0.5 M NaClO$_4$, 298.15 K) = (6.35 ± 0.05). Conversion to $I = 0$ with $\Delta\varepsilon$(A.28) = (0.09 ± 0.15) kg·mol^{-1} according to the SIT coefficients of the NEA TDB [95SIL/BID] leads to $\log_{10} K_3^\circ$ ((A.28), 298.15 K) = (4.98 ± 0.09), which is somewhat higher than the value of $\log_{10} K_3^\circ$ ((A.28), 298.15 K) = (4.66 ± 0.13), resulting from the constants $\log_{10} \beta_2^\circ$ (298.15 K) = (16.94 ± 0.12) and $\log_{10} \beta_3^\circ$ (298.15 K) = (21.60 ± 0.05) selected in [92GRE/FUG].

[91BRU/GLA]

This is a kinetic study of the rate of exchange of carbonate between $UO_2(CO_3)_3^{4-}$ and free carbonate. The study confirms the stoichiometry of the limiting complex and indicates according to the reviewers that the exchange mechanism is dissociative. This conclusion is drawn from the fact that the entropy of activation has a fairly large positive value.

[91CAR/BRU]

Two sets of experiments were performed. The first involved measurements of U(VI) sorption isotherms on calcite suspensions in 0.005 M $NaHCO_3$ in a flow reactor in the presence of $0-10^{-2}$ M U(VI). Initially a relatively high p_{CO_2} was maintained, followed by degassing to 10^{-2} to 10^{-3} bar to encourage co-precipitation of a uranium–calcium carbonate phase. Apparently the equilibrations were done at room temperature.

The second type of experiment involved measurements in a flow reactor of the sorption kinetics of U(VI) onto calcite at $p_{CO_2} = 0.97$ atm as a function of [U(VI)], [Ca] and pH in a solution of 0.005 M $NaHCO_3$. Less than 2% of the uranium in solution was sorbed by the calcite and no co-precipitation was observed. Sorption was rapid and apparently reversible.

The sorption experiments could be described by either the formation of a solid solution, or by an exchange reaction, which is limited by the surface adsorption sites on the calcite crystals according to the equilibrium:

$$UO_2^{2+} + CaCO_3(s) \rightleftharpoons UO_2CO_3(\text{surface}) + Ca^{2+}.$$

This equilibrium constant is independent of pH (5.4 to 7.7) with a value of: $\log K = (5.12 \pm 0.53)$. The kinetic study shows that both Ca^{2+} and H^+ approach steady-state values with time at lower [U(VI)]. Rates of U(VI) adsorption are given at lower [U(VI)] where the percentage adsorption is significant, but no time units are indicated.

[91CHO/MAT]

In [92GRE/FUG], this paper was incorrectly listed as belonging to vol. 53/54 of the same Journal. As the paper appeared when [92GRE/FUG] was at the proof stage, the relevant values were inserted as footnotes on pages 111 and 113, with a brief indication of their consistency with those selected.

The first and second hydrolysis constants for the mononuclear hydroxo species were studied at 25°C by solvent extraction techniques from aqueous solutions of 0.1 and 1.0 molar $NaClO_4$. This method involved the use of diglycolate as a competitive ligand in the aqueous phase and dibenzoylmethane (DBM) as the extractant. The pH ranged from 5.34 to 7.29 at 0.1 M ionic strength and from 6.22 to 7.22 at 1.0 M. Uranium concentrations were monitored by liquid scintillation counting. The pH was calibrated versus dilute $HClO_4$ at the appropriate ionic strengths. The extracted species was confirmed to be $UO_2(DBM)_2$, whereas in the aqueous phase, only the 1:1 $UO_2DGA(aq)$ species was formed as confirmed by variations in the DGA concentration at different pH values. The mean values of $\log_{10} \beta_1$ were given as (5.31 ± 0.02), $I = 0.1$ M and (4.94 ± 0.02), $I = 1.0$ M, where the uncertainties represent 1σ. The concentration of free dioxouranium(VI) was estimated to be on the order of 10^{-8} M. However, the concentration of total dioxouranium(VI) in either phase was not given. The resulting hydrolysis constants for the formation of $UO_2(OH)^+$ are $\log_{10} {}^*\beta_1 = -(5.91\pm0.08)$, $I = 0.1$ M and $\log_{10} {}^*\beta_1 = -(5.75\pm0.07)$, $I = 1.0$ M, whereas for $UO_2(OH)_2(aq)$ the corresponding values $\log_{10} {}^*\beta_2$ are $-(12.43\pm0.09)$ and $-(12.29\pm0.09)$.

Based on the SIT model and corresponding parameters for the ion interaction parameters in [92GRE/FUG], the infinite dilution molal hydrolysis constants are:

$\log_{10} {}^*\beta_1^\circ = -(5.74\pm0.88)$, from $I = 1.05$ m and $-(5.73 \pm 0.38)$, from $I = 0.10$ m

$\log_{10} {}^*\beta_2^\circ = -(12.07\pm0.10)$, from $I = 1.05$ m and $-(12.23 \pm 0.09)$, from $I = 0.10$ m

The large uncertainty in $\log_{10} {}^*\beta_1^\circ$ stems from the uncertainty in $\Delta\varepsilon(UO_2(OH)^+, ClO_4^-)$. The selected values from Table V.7 of [92GRE/FUG] are $-(5.2 \pm 0.3)$ and ≤ -10.3. Those selected by this current review are $-(5.25 \pm 0.24)$ and $\leq -(12.15 \pm 0.07)$, (see Table 9-6).

[91COR/KON]

This paper, which was included in the references quoted in [92GRE/FUG], reports data on the vapour pressure of $UCl_4(cr)$ by the transpiration technique from 699 to 842 K (18 points). Their fitted vapour pressure equation, $\log_{10}(p/\text{bar}) = -10084\, T^{-1} + 9.878$, is in good agreement with the many earlier studies. The data were processed to give $\Delta_{sub}H_m^\circ(298.15\text{ K}) = (200.7\pm2.0)$ kJ·mol^{-1}. However, this value was based on a distorted tetrahedral structure for $UCl_4(g)$; since it is now known that the molecule is almost certainly tetrahedral (see [95HAA/MAR]), this value has changed in the new assessment (see section 9.4.3.1.1.2).

[91COR/VLA]

This is a renewed study of $Sr_3U_{11}O_{36}$, a compound previously described by the same group ([67COR/LOO]) as $SrU_4O_{13}(cr)$, within a general study of strontium uranates. The present (predominantly structural) study shows the composition of the compound to be $Sr_3U_{11}O_{36}(cr)$.

The compound was prepared by heating a stoichiometric mixture of SrO(s) and $U_3O_8(cr)$ in a gold boat for 20 hours at 1273 K with repeated cycles of heating and grinding. The authors report that in previous studies [67COR/LOO] with Sr/U close to 0.25, U_3O_8 (cr) was always present.

The compound was fully analysed and showed no impurities. The structure obtained from X–ray, electron and neutron diffraction techniques, is related to that of U_3O_8, orthorhombic, space group *Pmmm*, in which U atoms are found with both octahedral and pentagonal-bipyramidal coordination. Heated in air above 1400 K, the compound decomposes to $Sr_2U_3O_{11}$ (cr) and U_3O_8 (cr) [67COR/LOO].

The enthalpy of formation of the compound was obtained from its dissolution in 5.0 mol·dm^{-3} HCl and reported, without any experimental details, as $-(794.9 \pm 1.2)$ kJ·mol^{-1}. Combination of this value with the enthalpy of solution of $SrCl_2$ (cr) in the same medium, $-(34.19 \pm 0.49)$ kJ·mol^{-1}, and the enthalpy of formation of $SrCl_2$ (cr) as determined by [90COR/KON2] (see [92GRE/FUG]), leads to a reported $\Delta_f H_m^\circ (Sr_3U_{11}O_{36}, cr, 298 K) = -(15905.6 \pm 22.5)$ kJ·mol^{-1}. This calculation must also involve the dissolution of a polymorph of UO_3 in the same medium, but no details of which polymorph, nor its enthalpies of formation and dissolution, are given by the authors.

However, the value of $\Delta_f H_m^\circ (SrCl_2, cr)$ used by this group, $-(832.43 \pm 0.85)$ kJ·mol^{-1} [90COR/KON2], is slightly different from the value accepted in [92GRE/FUG], $-(833.850 \pm 0.70)$ kJ·mol^{-1}; with the latter, $\Delta_f H_m^\circ (Sr_3U_{11}O_{36}, cr, 298 K)$ would become *ca.* 4.2 kJ·mol^{-1} more negative.

More recently [99COR/BOO], the same group of authors reported, together with results on other strontium uranates, details on the dissolution of $Sr_3U_{11}O_{36}$ in 5.075 mol·dm^{-3} HCl and on a cycle leading to the enthalpy of formation of this compound. As the sample used was apparently the same (as judged by the analytical results) as that of [91COR/VLA], we will not consider further the thermochemical results reported in this paper.

The results given in [99COR/BOO] lead to a recalculated value of $-(15903.81 \pm 16.45)$ kJ·mol^{-1} for the standard enthalpy of $Sr_3U_{11}O_{36}$(cr), as discussed in this Appendix.

[91FUJ/YAM]

This paper deals with the solubility of Ba in UO_{2+x} in the presence of Y and of the oxygen potential of the solid solutions of composition, $Ba_{0.05}Y_{0.05}U_{0.9}O_{2+x}$. Solid solutions were formed between 1273 and 1673 K if x is smaller than or equal to 0.1. Phase characterisation and the lattice parameter (for the fcc phase) are given. Oxygen non-stoichiometry was determined by redox titration of uranium. Oxygen potentials were determined between 1173 and 1573 K.

Extensive comparison with UO_{2+x} and lanthanide ternary and quaternary systems is made.

Such interesting studies on non–stoichiometric polynary oxides fall outside the scope of this review.

[91GIR/LAN]

Giridhar and Langmuir [91GIR/LAN] repeated the study [85BRU/GRE] of the reaction: $UO_2^{2+} + Cu(s) + 4 H^+ \rightleftharpoons U^{4+} + Cu^{2+} + 2 H_2O(l)$ at 25°C (0.3 to 2.6 molal $HClO_4 - NaClO_4$). However, they relied on accurate analyses of the molalities of the component ions, rather than measuring the cell potential. They also ran one equilibration from supersaturation and allowed 3 to 17 days to achieve equilibrium. This approach removed the need to estimate liquid junction potentials and ensured complete reaction, which were issues in the previous study [92GRE/FUG]. Activity coefficients were computed using the SIT model with any required data taken from [92GRE/FUG] with $\varepsilon(UOH^{3+}, ClO_4^-)$ set equal to $\varepsilon(La^{3+}, ClO_4^-)$. Their analysis led to a value of $\varepsilon(UOH^{3+}, ClO_4^-)$ of 1.42 kg · mol^{-1} as giving the minimum standard deviation in $E°$ for the couple, UO_2^{2+}/U^{4+}, and a related $E°$ value that is independent of the perchlorate molality. The authors claim without giving any details that a non-linearity exists in the [92GRE/FUG] derivation of $\varepsilon(UOH^{3+}, ClO_4^-)$ giving rise to the lower recommended value of (0.76 ± 0.06) kg · mol^{-1}. This latter value is significantly lower than the one proposed by Giridhar and Langmuir. The "non–linearity" is due to a systematic deviation of the $\log_{10} {}^*\beta_1°$ value at the lowest ionic strengths were the activity factor variations are the largest (cf. the discussion in [92GRE/FUG]). They also chose a value for $\log_{10} {}^*\beta_1°$ of -0.65 from [76BAE/MES], cf. $-(0.54 \pm 0.06)$ [92GRE/FUG]. In keeping with [92GRE/FUG], the authors adopted the [92GRE/FUG] hydrolysis constant value for the formation of the $U(OH)_4(aq)$ species, as the only other hydrolysis product in this system (actually representing < 0.02% of the total reduced uranium). The final computed value of $E°$ derived in an iterative fashion for the couple, UO_2^{2+}/U^{4+}, is (0.263 ± 0.004) V, which is within the combined uncertainties of the [92GRE/FUG] recommended value of (0.2673 ± 0.0012) V.

[91HIL/LAU]

This paper was included in the references quoted in [92GRE/FUG]. It reports new torsion–effusion measurements for the sublimation of $UCl_4(cr)$ (26 points from 588 to 674 K) and $UBr_4(cr)$ (43 points from 579 to 693 K). Data are also given for the sublimation of $ThI_4(cr)$. Two orifice sizes were used, and a small correction made for orifice size in the data for $UCl_4(cr)$. Detailed data points as well as fitted equations are given.

Comparison is made with the assessed values given by [83FUG/PAR] and two other studies. The authors' interpretation of the significance of the entropies of sublimation has been overtaken by new data on the symmetry of $UX_4(g)$ species, but their basic data have been included in the reassessment of the vapour pressures of the tetrahalides.

[91HIL/LAU2]

This paper was included in the references (but not the reviews) in [92GRE/FUG]. It reports mass–spectrometric measurements of the stability of $UF_5(g)$ via the gaseous reaction: $Ag(g) + UF_5(g) \rightleftharpoons AgF(g) + UF_4(g)$ (1091 to 1384 K), and of the gaseous reaction: $U(g) + UF_2(g) \rightleftharpoons 2\ UF(g)$ (2103 to 2405 K). Additional data on the gaseous reaction: $AgF(g) + Cu(g) \rightleftharpoons Ag(g) + CuF(g)$ (1217 to 1336 K) are used to define the stability of $AgF(g)$. Detailed data points as well as fitted equations are given.

All these results are included in the assessment of the gaseous uranium fluorides in section 9.4.2.1.

[91MEI]

Most of the investigations reported in this Ph.D. thesis are published in a series of papers on the solubility and complexation of Am(III) in carbonate solution [91MEI/KIM], [91MEI/KIM2], [92RUN/MEI]. These papers were already discussed in the previous review [95SIL/BID]. In addition, Meinrath [91MEI] investigated the solubility of $NaAm(CO_3)_2 \cdot xH_2O(s)$ at 25°C in 5 M NaCl carbonate solution, under an atmosphere of 1% CO_2 in argon (p_{CO_2} = 0.01 bar). The carbonate concentration was varied by pH titration followed by equilibration with the CO_2/Ar mixture. The ^{231}Am solid and an analogous Nd compound were characterised by chemical analysis, thermogravimetry, IR spectroscopy and X–ray powder diffraction. The prepared solids of $NaAm(CO_3)_2 \cdot xH_2O(s)$ and $NaNd(CO_3)_2 \cdot xH_2O(s)$ gave comparable X–ray patterns. Meinrath used Ross glass electrodes calibrated against pH buffers containing 5 M NaCl. As the addition of NaCl does not have an equivalent effect on the different pH buffers, the measured pH values are meaningless and cannot be converted into H^+ or OH^- concentrations. On the other hand, the relations between the measured pH values and the analytical bicarbonate and carbonate concentrations were determined experimentally by potentiometric titration, so that the determination of $\log_{10}[CO_3^{2-}]$ from measured pH is internally consistent. The present reviewer assumes that the error in $\log_{10}[CO_3^{2-}]$ is less than 0.2 $\log_{10}$ units.

In a later study of the same laboratory, Runde and Kim [94RUN/KIM] repeated the solubility experiment of Meinrath [91MEI] at 22°C with both, $NaAm(CO_3)_2 \cdot xH_2O(s)$ and $NaEu(CO_3)_2 \cdot xH_2O(s)$. The sodium americium carbonate was characterised by comparing the X–ray powder diffraction pattern with those of the analogous Eu and Nd compounds. In contrast to the procedure applied in [91MEI], Runde and Kim [94RUN/KIM] calibrated the pH electrode against H^+ and OH^- concentration in 5 M NaCl solutions. The carbonic acid dissociation constants determined in

[94RUN/KIM] are in reasonable agreement with widely accepted literature data (cf. discussion of [94RUN/KIM]).

The experimental solubility data reported in the two studies [91MEI], [94RUN/KIM] are shown in Figure A-2. At low carbonate concentrations, the authors measured comparable solubilities. At $\log_{10}[CO_3^{2-}] > -4$, Meinrath [91MEI] measured an upper curve, when the carbonate concentration was decreased by adding HCl and a lower curve by back titration with NaOH. Particularly in the case of the latter data, the present review doubts whether the equilibrium state was reached by bubbling the CO_2/Ar mixture through the solutions titrated with NaOH. Meinrath ascribed the discrepant data to the presence of two solids with different crystallinity or the alteration of amorphous components of the initial solid.

Figure A-2. Solubility of $NaAm(CO_3)_2 \cdot xH_2O(s)$ determined in [91MEI], [94RUN/KIM] at 25°C and 22°C, respectively, in 5.6 m NaCl solution under an atmosphere of $p_{CO_2} = 10^{-2}$ bar in argon. The solid line is calculated with the thermodynamic constants selected in the present review, converted to $I = 5.6$ mol·kg^{-1} with the SIT coefficients in Appendix B. The dotted lines show the associated uncertainty (95 % confidence level).

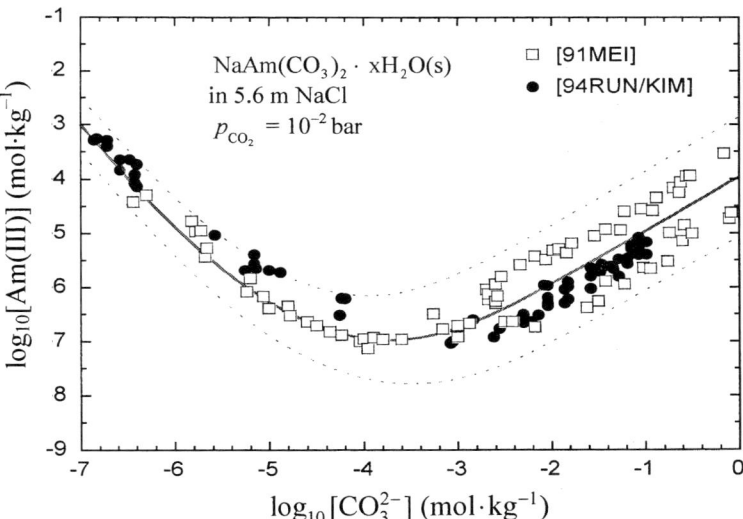

Runde and Kim [94RUN/KIM] calculated a solubility constant of $\log_{10} K_{s,0}$ (A.29), 5 M NaCl) = $-(16.5 \pm 0.3)$ for the reaction:

$$NaAm(CO_3)_2 \cdot xH_2O(s) \rightleftharpoons Na^+ + Am^{3+} + 2\,CO_3^{2-} + xH_2O(l) \quad (A.29)$$

In order to calculate the solubility constant at $I = 0$, the present review sets the number of crystal water molecules equal to $x = (5 \pm 1)$. (The reported values for $NaAm(CO_3)_2 \cdot xH_2O(s)$ and the analogous Eu and Nd compounds are $x = 4$ [69KEL/FAN], $x = 5$ [91MEI], [94RUN/KIM] or $x = 6$ [91MEI], [74MOC/NAG]). The water activity in 5.6 m NaCl, $a_w = 0.7786$, is calculated with the ion–interaction approach of Pitzer [91PIT]. Applying the SIT, with $\Delta\varepsilon(A.29) = (0.10 \pm 0.06)$ kg · mol^{-1} according to the selected interaction coefficients in Appendix B, the solubility constant at $I = 0$ is calculated to be $\log_{10} K^\circ_{s,0}$ ((A.29), 298.15 K) $= - (21.0 \pm 0.5)$.

At $\log_{10}[CO_3^{2-}] > -2$, the solubility data of both authors, $\log_{10}[Am]_{tot}$ vs. $\log_{10}[CO_3^{2-}]$, increase with a slope of $(+1)$, indicating the formation of $Am(CO_3)_3^{3-}$ as the limiting carbonate complex:

$$NaAm(CO_3)_2 \cdot 5H_2O(s) + CO_3^{2-} \rightleftharpoons Na^+ + Am(CO_3)_3^{3-} + 5 H_2O(l) \qquad (A.30).$$

From the solubility data converted to the molal scale, $\log_{10} K_{s,3}$ ((A.30), 5.6 m NaCl) is calculated to be $-(2.75 \pm 0.2)$ and $-(3.8 \pm 0.2)$ for the upper and lower curves in [91MEI], respectively, and $-(3.55 \pm 0.25)$ [94RUN/KIM]. Combining $\log_{10} K^\circ_{s,0}$ ((A.29), 298.15 K) $= -(21.0 \pm 0.5)$ and $\log_{10} \beta^\circ_3$ (298.15 K) $= (15.0 \pm 1.0)$ as selected in the present review with the SIT coefficients in Appendix B and $a_w = 0.7786$, the solubility constant is calculated to be $\log_{10} K_{s,3}$ ((A.30), 5.6 m NaCl, 298.15 K) $= -(3.2 \pm 1.1)$ and covers the experimental data of both, Meinrath [91MEI] and Runde and Kim [94RUN/KIM].

[91SHE/MUL]

The authors report a correction to an earlier paper [88MUL/SHE] in which properties, including the $C_{p,m}$ of U(l), were measured up to *ca.* 5000 K using rapid heat pulsing. The later paper reports (appreciable) corrections to the temperatures involved due to improved measurements of emissivity used to convert the brightness temperatures to thermodynamic temperatures.

The derived constant, $C_p(U, l, 1408 \text{ to } 5000 \text{ K}) = 50.2$ J·K^{-1}·mol^{-1}, is to be contrasted with the values given by [82GLU/GUR] used by the CODATA Key Values team [89COX/WAG] to derive $\Delta_f H^\circ_m$ (U, g, 298.15 K) from the vapour pressure. These increase steadily from 47.9 J·K^{-1}·mol^{-1} at 1408 K to 58.4 J·K^{-1}·mol^{-1} at 5000 K.

Moreover, this emissivity correction removes the previous good agreement with the enthalpy data of Stephens [74STE] from levitation calorimetry, and corresponds to a higher emissivity at the melting point. However, since Stephens' data at the lower temperatures, of relevance to the vapour pressure measurements, are in good accord with the conventional enthalpy drop measurements by Levinson [64LEV], the currently selected values of the heat capacity of U(l) have been retained until the discrepancy in the emissivity is resolved.

[92ADN/MAD]

In this conference abstract no experimental details are given. The authors report the following apparent formation constants for 1:1 and 2:1 complexes of Am(III), Pu(IV) and U(VI) with $P_2W_{17}O_{61}^{10-}$ in 2 M HNO_3:

$$\begin{array}{lll} \text{Am(III)} & \log_{10}\beta_1 = 1.9, & \log_{10}\beta_2 = 3, \\ \text{Pu(IV)} & \log_{10}\beta_1 = 8, & \log_{10}\beta_2 = 13, \\ \text{U(VI)} & \log_{10}\beta_1 = 0.99. & \end{array}$$

Due to the lack of experimental data, these values are not considered by this review.

[92BLA/WYA]

This is a comprehensive (480 pp.) listing of the known energy levels of all the gaseous actinide elements (Ac to Es) and their gaseous ions. 2252 levels are given for U(g) and 204 for Am(g), substantially more than the number of levels that were available in the calculations on which the values in [92GRE/FUG] and [95SIL/BID] were based (ca. 1596 for U and only 51 for Am). The values of $C_{p,m}^\circ$ and S_m° at 298.15 K for U(g) and Am(g) calculated by this review from the larger number of levels are essentially unchanged from earlier calculations (although S_m° (Am, g, 298.15 K) is misquoted in [95SIL/BID]) and the values in the temperature range of the measurements of vapour pressures change so little that no changes in $\Delta_f H_m^\circ$ (U, g, 298.15 K) and $\Delta_f H_m^\circ$ (Am, g, 298.15 K) are required.

[92CHO/DU]

Solvent extraction was used to determine the stability constants for UO_2^{2+} and Eu^{3+} with Cl^- and NO_3^- at ionic strengths of 3.5, 6.5, 10.0, and 14.1 m ($NaClO_4$) at 25°C. ^{152}Eu and ^{233}U tracers were used in conjunction with dinonylnaphthalenesulphonic acid (HDNNS) in heptane at concentrations optimised as a function of ionic strength to give distribution ratios (D = total metal in the organic phase/total metal in the aqueous phase) in the range 0.1 to 10. For the first two solutions, [H^+] = 0.1 M and 0.5 M for the latter. Equilibration times were four hours with constant shaking. Values of the stability constants, β_1 and β_2, were obtained from the dependence of D on ligand concentration according to the relationship:

$$\frac{1}{D} = \frac{1}{D^\circ} + \frac{\beta_1}{D^\circ} \times [L] + \frac{\beta_2}{D^\circ} \times [L]^2$$

where D° is the distribution ratio in the absence of ligand, L (Cl^- or NO_3^-). The experimental results are shown in figure form (1/D versus [Cl^-]) at 3.5 and 10.0 m chloride, and the resulting $\log_{10} \beta_n$ values are given in tabular form (Table A-6), summarised here for the dioxouranium(VI) case only.

Table A-6: Equilibrium constant $\log_{10} \beta_n$ for UO_2^{2+} with Cl^- and NO_3^-.

I	$\log_{10} \beta_1$		$\log_{10} \beta_2$	
(molal)	(molar)	(molal)	(molar)	(molal)
Chloride				
3.5 (3.0 M)	(0.090 ± 0.032)	(0.054 ± 0.049)	$-(0.36 \pm 0.05)$	$-(0.42 \pm 0.09)$
6.5 (5.0 M)	$-(0.092 \pm 0.050)$	$-(0.152 \pm 0.068)$	$-(0.01 \pm 0.006)$	$-(0.018 \pm 0.073)$
10.0 (7.0 M)	$-(0.707 \pm 0.104)$	(0.570 ± 0.190)	$-(0.560 \pm 0.038)$	(0.407 ± 0.053)
14.1 (9.0 M)	$-(0.930 \pm 0.190)$	(0.749 ± 0.140)		
Nitrate				
3.5 (3.0 M)	$-(0.180 \pm 0.033)$	$-(0.208 \pm 0.033)$		
6.5 (5.0 M)	$-(0.092 \pm 0.050)$	$-(0.152 \pm 0.068)$	$-(0.728 \pm 0.079)$	$-(0.790 \pm 0.054)$
10.0 (7.0 M)	$-(0.01 \pm 0.04)$	$-(0.082 \pm 0.089)$		

The authors used the SIT analysis and some (only nine for the chloride complex and three for nitrate) values from [92GRE/FUG] that are shown in two plots following the OECD guidelines. For the former, the values of $\log_{10} \beta_1 + 4 D$ above $I = 3.5$ molal deviate substantially from linearity based on the recommended slope of $\Delta\varepsilon = -0.25$ kg·mol^{-1}. On the other hand, the three values for nitrate are highly scattered around the recommended slope corresponding to $\Delta\varepsilon = -0.09$ kg·mol^{-1}. In view of the deviation and scatter in these new values and the fact that undisclosed values are taken from [92GRE/FUG] in these plots, this study does not appear to warrant addition to the current database. There is an error in the ordinates of Figures 5 and 6, which read $\log_{10} \beta_1 + 6 D$ rather than $\log_{10} \beta_1 + 4 D$, but this must be a carry-over from the corresponding Eu^{3+} data in the previous two figures and the correct values are plotted in these figures.

[92DUC/SAN]

A modified version of the Wilcox's equation [62WIL] is used to estimate the values of $\Delta_f H_m^\circ$ (4800 compounds tested) and $\Delta_f G_m^\circ$ (2700 compounds tested) of binary inorganic compounds in the solid state. The maximum deviation between the calculated and the experimentally determined values is less than 15.4 kcal·mol^{-1}. The values used to test the equations are those assessed by [82WAG/EVA], and those published later in Journal of Chemical Thermodynamics up to 1988. The method is based on the use of structural increments which are the numbers of anions, n_a, cations, n_b, and the apparent number of bonds, n_L. The model depends on these three parameters and three constants, X_A, Y_A and W_A, for the anion and three constants, X_B, Y_B and W_B, for the cation. The equations are:

$$-\Delta_f H_m^\circ (\text{kcal} \cdot \text{mol}^{-1}) = (n_a + n_b) \cdot (X_A - X_B)^2 + n_a Y_A + n_b Y_B + n_L \frac{W_B}{W_A}$$

$$-\Delta_f G_m^o (\text{kcal} \cdot \text{mol}^{-1}) = (n_a + n_b) \cdot (X_A' - X_B')^2 + n_a Y_A' + n_b Y_B' + n_L \frac{W_B'}{W_A'}$$

For the different oxidation states of the cation B = U, the parameters are given in the Table A-7:

Table A-7: Values of X_A, Y_A and W_A for the calculation of enthalpy and Gibbs energy from the Wilcox equation.

	U(III)	U(IV)	U(V)	U(VI)
Enthalpy				
X	10.054	8.988	12.792	7.938
Y	132.635	157.650	138.170	68.553
W	2.945	2.796	3.151	3.748
Gibbs Energy				
X'	14.625	13.621	12.452	12.055
Y'	130.224	160.967	145.193	121.567
W'	29.361	25.037	13.496	9.971

With regard to the anion parameters, the situation is more complicated owing to the number of anions selected (54 for enthalpy and 45 for Gibbs energy). Table A-8 gives a selection for A = X$^-$ (halides) and A = O^{2-}.

Table A-8: Values of X_B, Y_B and W_B for the calculation of enthalpy and Gibbs energy from the Wilcox equation.

	F	Cl	Br	I	O
Enthalpy					
X	8.803	8.459	8.462	8.631	11.028
Y	32.862	10.996	6.122	− 1.325	10.397
W	121.546	27.532	3.666	− 22.115	47.658
Gibbs energy					
X'	13.700	12.604	12.386	12.273	15.377
Y'	75.996	10.971	− 0.469	− 19.395	37.892
W'	− 172.251	72.721	110.962	193.493	− 7.626

The authors point out that some estimations including $\Delta_f H_m^o$ (UF$_6$, cr), are unsatisfactory. For this compound, one has: experimental $\Delta_f H_m^o$ (UF$_6$, cr) = − 525.1 kcal · mol^{-1}, (− (525.26 ± 0.43) kcal · mol^{-1} in [92GRE/FUG]), compared with the calculated value, − 535.2 kcal · mol^{-1}. However, the agreement for all the binary alkaline metal compounds was better than 10 kcal · mol^{-1}.

Values of the X, Y and W parameters are also given for the anion UO_4^{2-} and many cations.

Better estimation methods are available for the compounds of the actinide series, based on trends within the actinide and lanthanide series alone, and no use has been made of the reported correlations.

[92DUE/FLE]

The chemical and electrochemical insertion of Mg into $\alpha - U_3O_8$ to form $Mg_xU_3O_8$ ($0 < x < 0.6$) is described. These phases have an orthorhombic structure, with parameters similar to $\alpha - U_3O_8$; the IR absorption spectra are also very similar. Compounds with $x = 0.20 - 0.27$ and $x > 0.40$ were single phase, although the observed phase boundaries depend greatly on the method of preparation (chemical or electrochemical) and conditions (temperature, applied potential).

The Gibbs energies of formation of these phases from Mg and $\alpha - U_3O_8$ were determined at "ambient temperature" (294 ± 2) K from an electrochemical cell using Mg amalgam as a reference electrode, and were described by a linear function of x.

[92FIN/MIL]

Structural and thermodynamic data on synthetic or natural uranium trioxide hydrates are important as products of leaching or weathering of UO_2 or uraninite (UO_{2+x}). However, the situation is quite complicated as discussed in [92GRE/FUG]; it is generally difficult to identify in a given situation which compound is really present. This paper focuses on polytypes of schoepite studied by X-ray diffraction and scanning electron microscopy, but does not give new thermodynamic data with respect to those [88OHA/LEW], [89BRU/SAN] already reported in [92GRE/FUG].

According to the review by the authors to clarify the situation, synthetic uranium trioxide dihydrate $UO_3 \cdot 2H_2O$ presents two polytypes, referred to as schoepite and metaschoepite ($UO_3 \cdot 2H_2O$), both orthorhombic, and naturally occurring "uranyl dehydrate" presents three polytypes, also orthorhombic, corresponding to slight losses of water, whose characteristics are established in this paper. Four synthetic polymorphs of uranium trioxide monohydrate, $UO_3 \cdot H_2O$, are well identified, $UO_3 \cdot 0.8H_2O$, $\alpha-$, $\beta-$ and $\gamma-UO_2(OH)_2$. The authors point out that the problem of inter-relationships between all these phases is not yet clear.

Two of the three natural polytypes of uranium trioxide dihydrate might be identical with schoepite and metaschoepite, the third one being a mixture of schoepite, the monohydrate and fluid inclusions.

$UO_3 \cdot 2H_2O$ (cr) is stable in water below 40°C. Its structure is not known. The ideal structure of $\beta - UO_3 \cdot 2H_2O$ (cr) is described as UO_2 coordinated to six equatorial hydroxyl groups, forming a layer structure and having the composition $UO_2(OH)_2 \cdot H_2O$.

[92FUG]

This descriptive critical assessment of available experimental results is divided into sections for each valence state and those referring to uranium species are mentioned here.

For the U(IV) species, values of the hydrolysis constants are presented from [92GRE/FUG] and Fuger et al. [92FUG/KHO], both in press at the time of [92FUG], whereby the latter give, for the reaction,

$$U^{4+} + nH_2O \rightleftharpoons U(OH)_n^{4-n} + nH^+,$$

at 25°C, $\log_{10} {}^*\beta_1^\circ = -(0.34 \pm 0.02)$, $\log_{10} {}^*\beta_3^\circ = -(1.1 \pm 0.3)$, and $\log_{10} {}^*\beta_4^\circ = -(5.4 \pm 0.2)$. Corresponding values cited from [92GRE/FUG] were $\log_{10} {}^*\beta_{1,1}^\circ = -(0.54 \pm 0.06)$ and $\log_{10} {}^*\beta_4^\circ = -4.7$. All these results are based on data prior to those of [90RAI/FEL]. The author concurred with the opinion of [92GRE/FUG] that the results of [90RAI/FEL] would lead to a much lower value for $\log_{10} {}^*\beta_4^\circ$ and that further experimental work was needed to resolve such a discrepancy. The author also concluded that "the existence of $U(OH)_3^+$ was not fully demonstrated".

For the hexavalent state, the author states that this system was analysed in greater depth in [92GRE/FUG] compared to the [92FUG/KHO] study. The only contentious issues appear to be the formation constants of two mononuclear species and the dimer, $(UO_2)_2OH^{3+}$. The IAEA compilation [92FUG/KHO] prefers values:

$$\log_{10} {}^*\beta_{1,1}^\circ = -(5.76 \pm 0.10)$$
$$\log_{10} {}^*\beta_{2,1}^\circ = -(13.0 \pm 0.25)$$
$$\log_{10} {}^*\beta_{1,2}^\circ = -(4.06 \pm 0.15)$$

for the reaction, $mUO_2^{2+} + nH_2O(l) \rightleftharpoons (UO_2)_m(OH)_n^{2m-n} + nH^+$, while [92GRE/FUG] selected the values:

$$\log_{10} {}^*\beta_{1,1}^\circ = -(5.2 \pm 0.3),$$
$$\log_{10} {}^*\beta_{2,1}^\circ \leq -10.3 \text{ (estimated)},$$
$$\log_{10} {}^*\beta_{1,2}^\circ \leq -(2.7 \pm 1.0) \text{ (estimated)}.$$

The author also notes, without discussion, that the results on the temperature dependence of the hydrolysis reactions of UO_2^{2+} are rather controversial.

For the quadrivalent uranium carbonate complexes there is unanimity for the formation constants of $UO_2(CO_3)_3^{4-}$ and $U(CO_3)_5^{6-}$ between the author and the recommended values in [92GRE/FUG]. For the pentavalent complex, $UO_2(CO_3)_3^{5-}$, the $\log_{10} \beta_3$ value selected by [92FUG/KHO] was (13.3 ± 1.0) for the reaction, $UO_2^+ + 3CO_3^{2-} \rightleftharpoons UO_2(CO_3)_3^{5-}$, at an ionic strength of 3.0 while [92GRE/FUG] recommended $\log_{10} \beta_3^\circ = (7.41 \pm 0.27)$. For the hexavalent species (1:1, 2:1, 3:1 and ommended $\log_{10} \beta_3^\circ = (7.41 \pm 0.27)$. For the hexavalent species (1:1, 2:1, 3:1 and 6:3), the [92GRE/FUG] and [92FUG/KHO] reviews recommended formation constants and enthalpies of reaction at zero ionic strength that are in agreement within the assigned uncertainties.

[92HAS2]

The paper deals with the mechanism of oxidation of tetravalent aqueous uranium ion by oxidizing species of Pt(IV), Ir(IV), Np(IV, V, VI), Fe(III), Pu(IV,VI), Cr(VII), Ce(IV) and V(V). Through a compilation of literature data it is shown that there is a linear correlation between the activation Gibbs energy and values of the standard redox potential for the oxidation reactions.

[92HIS/BEN]

This paper deals with correlations in $\Delta_f H_m^\circ (MX_n)$. The function $\Delta(MX_n)$ is defined as $\Delta_f H_m^\circ (MX_n, cr, 298.15\ K) - \Delta_f H_m^\circ (X^-, g, 298.15\ K)$. For two different metals M and M', and given n, and any halogen X, the correlation $\Delta(MX_n) = a\ \Delta(M'X_n) + b$ is found to hold within about 10 kJ·mol^{-1}, but with some more substantial deviations.

NaX(cr), CaX$_2$ (cr), AlX$_3$ (cr), UX$_4$ (cr), and NbX$_5$ (cr) were used for the correlations according to the valency.

Values for 28 unknown enthalpies of formation are predicted, including $\Delta_f H_m^\circ (UI_5, cr) = -523$ kJ·mol^{-1} and $\Delta_f H_m^\circ (UO_2I_2, cr) = -992$ kJ·mol^{-1}, compared with values of -112 and -252 kJ·mol^{-1} estimated from a different correlation [89LIE/GRE].

However, neither of these phases is likely to be stable under normal conditions, so these enthalpy values have not been considered in the current review.

[92KIM/SER]

This paper discusses the application of three different Photoacoustic Spectroscopy (PAS) methods (FT–LIPAS for solution, and UV–Vis–NIR PAS, FT–IR PAS for solid phase) for speciation of U(VI) in NaClO$_4$ / NaHCO$_3$ solutions ($I = 0.1$ M) and for characterisation of U(VI) precipitates obtained from these solutions. Additional UV–Vis absorption spectroscopic measurements are made to identify some species using their known spectra and the composition of the solutions and solid phases according to the thermodynamic data selected by [92GRE/FUG] using the MINTEQ geochemical code.

Five solutions, all 10^{-2} M in U, are studied by classical spectroscopy:

1. 0.1 M $NaHCO_3$, pH = 6.8 to 9.1,
2. 0.05 M $NaHCO_3/NaClO_4$, pH = 5.9 to 8.2,
3. 0.02 M $NaHCO_3/NaClO_4$, pH = 4.4 held constant,
4. 0.01 M $NaHCO_3/NaClO_4$, pH = 4.1 held constant,
5. 0.1 M $NaClO_4$, pH = 3.0 held constant.

Species identified in the above test solutions are:

1. $UO_2(CO_3)_3^{4-}$,
2. $UO_2(CO_3)_2^{2-}$, which transforms with time to $UO_2(CO_3)_3^{4-}$ and then a precipitation occurs at pH 8.2 (solid 1, see discussion below),
3. $(UO_2)_3(OH)_5^+$, which transforms with time to $(UO_2)_2(OH)_2^{2+}$ and then a precipitation occurs at pH 4.4 (solid 2, see discussion below),
4. $(UO_2)_2(OH)_2^{2+}$,
5. UO_2^{2+}

Two solutions both 10^{-4} M in U were studied by PAS:

6. 0.01 M $NaHCO_3/NaClO_4$, pH = 9.1,
7. 0.1 M $NaClO_4$, pH = 4.8.

For the lowest U concentration the species identified are:

6. $UO_2(CO_3)_3^{4-}$,
7. UO_2^{2+}.

The speciation calculation agrees with these experimental observations except for solutions 2 and 3. For solution 2, precipitation is not predicted at pH = 8.2 and for solution 3, the species $(UO_2)_3(OH)_5^+$ is not predicted. A possible reason for these discrepancies might be ionic strength effects, as it is not indicated if the thermodynamic data used have been corrected for ionic strength.

UV-Vis-NIR PAS spectrum of solid 1 is the same as observed for the precipitate produced from 1 M $NaHCO_3$ (probably amorphous $UO_2CO_3(s)$) and the spectrum of solid 2 is the same as that obtained from the precipitate produced from 1 M $NaClO_4$ (poorly crystalline $UO_3(s)$ identified by X-ray diffraction). According to the thermodynamic data, precipitation of solid 1 is expected at pH less than 6.6, while solid 2 is predicted to appear at a pH more than 7. At pH = 4.4, rutherfordine is expected to be the stable phase. These discrepancies are probably due to lack of information on the composition of the solids and may indicate that the phases are not pure.

This paper gives only some confirmation on the values of selected formation constants in [92GRE/FUG].

[92KRA/BIS]

The solubility of $UO_2(OH)_2 \cdot H_2O(cr)$ and rutherfordine UO_2CO_3(ruth.) were studied [92KRA/BIS] at 25°C in 0.1M $NaClO_4$ in the absence and presence of carbonate (equilibration times 3 –14 days). The relevant molar concentrations are provided. The pH range was generally 4.5 – 5.5 for the study with no added carbonate and of all combinations of species considered in fitting the data, only those which represented more than 5% of the total uranium in solution were considered. Using additional information from a spectrophotometric characterisation of the solutions, which showed peaks at 420 and 428 nm, it was concluded that the 2:2 and 5:3 species dominated with $\log_{10} \beta_{m,n}$ values of (22.16 ± 0.03) and (53.05 ± 0.04), respectively, for the reaction:

$$n UO_2^{2+} + m OH^- \rightleftharpoons (UO_2)_n(OH)_m^{(2n-m)}.$$

These stability constants are given in Table 3 of [92KRA/BIS] and refer to 0.1 M ionic strength. Using $pK_w = -(13.78 \pm 0.01)$ in 0.1 M $NaClO_4$ and application of the SIT to make the conversion to infinite dilution in the hydrogen ion form gives values of $\log_{10} {}^*\beta_{2,2} = -(5.40 \pm 0.04)$ and $\log_{10} {}^*\beta_{5,3} - (15.85 \pm 0.06)$ in 0.1 M $NaClO_4$ and $\log_{10} {}^*\beta^°_{2,2} = -(5.19 \pm 0.04)$ and $\log_{10} {}^*\beta^°_{5,3} - (15.21 \pm 0.06)$.

The authors also reported for crystalline schoepite, $\log_{10} K_{s,0} = -(22.21 \pm 0.01)$ at $I = 0$ M which gives, using the SIT, $\log_{10} K^°_{s,0} = -(22.81 \pm 0.01)$.

[92LIE/HIL]

In this paper, there are considerations of the variation of the ratios R:

$$R = \frac{\left[UO_2(CO_3)_2^{2-}\right]}{\left[UO_2OH^+\right]} \text{ and } R = \frac{\left[UO_2(CO_3)_3^{4-}\right]}{\left[UO_2OH^+\right]}$$

for U(VI) aqueous solutions, at different pH (pH = 6, 7, 8) as a function of the total carbonate concentration, $[C] = [CO_{2(aq)}] + [HCO_3^-] + [CO_3^{2-}]$. The thermodynamic constants are those selected in [71SIL/MAR]. All $\log_{10}R$ vs. $\log_{10}C$ plots are straight lines which intersect at significant R values, at C_i. It is concluded that:

- above pH = 6 and above $C = 10^{-2}$ M the carbonato complexes prevail over the monohydroxocomplexes;
- at pH = 8, this holds for $C < 10^{-4}$ M (sea water);
- at a given pH the tricarbonato complex prevails for $C > C_i$.

None of these observations provides new data. Other observations developed in this paper concern thorium species.

[92MOR/WIL]

This paper contains an estimation of thermochemical properties for Am(III) and Cm(III) hydroxides. The authors applied an empirical acid–base correlation to $\Delta_r H_m^\circ$ for the reaction:

$$\tfrac{1}{3}M_2O_3(s) + H_2O(l) \rightleftharpoons \tfrac{2}{3}M(OH)_3(s).$$

The required empirical acidity parameter was calculated from known thermochemical data for Pu(III) and the solubility constant for Pu(III) hydroxide and adopted for Am(III) and Cm(III). For Am(OH)$_3$(s) the following estimates are given without uncertainties:

$$S_m^\circ (Am(OH)_3, s, 298.15\ K) = 129\ J \cdot K^{-1} \cdot mol^{-1},$$

$$\Delta_f H_m^\circ (Am(OH)_3, s, 298.15\ K) = -1346\ kJ \cdot mol^{-1},$$

$$\Delta_f G_m^\circ (Am(OH)_3, s, 298.15\ K) = -1218\ kJ \cdot mol^{-1},$$

$$\log_{10} K_{s,0}^\circ (Am(OH)_3, s, 298.15\ K) = -25.7,$$

$$(i.e.,\ \log_{10}{}^* K_{s,0}^\circ (Am(OH)_3, s, 298.15\ K) = 16.3).$$

These estimates agree well with the data obtained later by Merli *et al.* [97MER/LAM] from their calorimetric measurements with Am(OH)$_3$(cr).

[92NGU/BEG]

Raman spectroscopy records the symmetric stretching vibration of the linear O=U=O group. The corresponding wave number has been shown to decrease with a change in the inner coordination sphere of complexes of U. For instance, with increasing value of n, the number of OH groups per U, it decreases according to $v_1(cm^{-1}) = -A \cdot n + 870$, with $A = (21.5 \pm 1.0)$ and n up to 4. This is due to the weakening of the O=U=O bonds. This paper represents an extension of such a linear relationship for the following ligands of interest in the context of the present review: fluoro, chloro, bromo, sulphato, hydrogenosulphato, carbonato, nitrato and perchlorato.

Solutions of different ionic strengths, pH and compositions, depending on the ligand, are examined, $5 \cdot 10^{-3}$ to 0.5 m in U. pH is adjusted with trifluoromethanesulfonic acid, HCF_3SO_3, and tetramethyl-ammonium hydroxide. $(CH_3)_4NOH$ is used to avoid the precipitation of ternary U(VI) oxides. This allows investigations at higher U(VI) concentrations in alkaline solutions. The calculated speciation is based on data selected by [92GRE/FUG]. v_1 frequencies are related to the nature of the species, including $UO_2ClO_4^+$ which appears as an exception because v_1 increases with respect to UO_2^{2+} ($A = -15$). That probably means a decrease of the hydration shell of the UO_2^{2+} core and a coordination of ClO_4^- with an energy similar to that of water.

Br^-, HSO_4^- and NO_3^- ligands give outer–sphere complexes. The A values are (12 ± 1) for F^- $(4 \geq n)$, (4 ± 1) for Cl^- $(5 \geq n)$, (9 ± 1) for SO_4^{2-} $(3 \geq n)$ and (19 ± 1) for CO_3^{2-} $(4 \geq n)$.

Furthermore, the following general relationships are established for mononuclear complexes on the basis of data including OH^-, F^-, Cl^-, SO_4^{2-} and CO_3^{2-}:

$$\log_{10} \beta(UO_2L) = -0.52(\Delta v_1) - 1.61,$$
$$\log_{10} \beta(UO_2L_2) = -0.50(\Delta v_1) - 4.10,$$
$$\log_{10} \beta(UO_2L_3) = -0.46(\Delta v_1) - 5.86.$$

These data do not bring new thermodynamic information on U(VI) complexation.

[92NGU/SIL]

Soddyite, $(UO_2)_2SiO_4 \cdot 2H_2O$, and uranophane, $Ca(H_3O)_2(UO_2)_2(SiO_4)_2 \cdot 3H_2O$, are natural uranyl silicate minerals whose structures are known. Pure natural sodium uranyl boltwoodite, $Na(H_3O)UO_2SiO_4 \cdot H_2O$, is often mixed with the potassium form, but a synthetic mineral, as well as a synthetic sodium weeksite mineral, $Na_2(UO_2)_2(Si_2O_5)_3 \cdot 4H_2O$, are known. All have been identified as secondary products of UO_2 spent fuel or the leaching of nuclear glasses. Procedures for the synthesis of these compounds and XRD data for checking their structure are available. Some different formula of these minerals are reported in the literature as pointed out by [99CHE/EWI], which differ by the number of H_2O or the inclusion of a "H_3O versus SiO_3OH" unit. This can influence the Gibbs energy of formation, but has probably little influence on the solubility product.

The authors have prepared synthetic well-crystallised minerals and checked their composition by X–ray diffraction, FTIR (comparison with natural minerals except for weeksite) and chemical analysis. There are some discrepancies regarding the theoretical and experimental mass fractions of U (6% for uranophane and 1% for the other), Si (up to 10%), Ca (20%) and Na (up to 10%), depending on the mineral, but X–ray and IR data fit very well with reference compounds and give confidence that the minerals are the ones expected with slightly distorted stoichiometry. Nevertheless, these deviations could be a sign of a perturbation in the system caused by silica precipitation (see below). Chemical analyses are made after dissolution in acidic media (0.1 M $HClO_4$ + traces of HF).

Solubility measurements from under-saturation of several washed samples under Ar (mass fraction less than $5 \cdot 10^{-5}$ in O_2 and 10^{-7} CO_2) at a temperature of (30 ± 0.5) °C in a pH controlled device are carried out over a period of 150 days. Ionic strength is not kept constant to avoid uncontrolled precipitation. pH is continuously checked (calibrated at pH = 7 and 4) and adjusted. The chosen values for solubility measurements are pH = (3.00±0.05) for soddyite, pH = (3.50±0.05) for uranophane and pH = (4.50±0.05) for the other salts. Solutions are filtered down to a 4.1 nm cut off and analysed for U, Si, Na and Ca. The U concentration is determined by atomic absorption spectrometry, absorption spectroscopy for the uranyl cation and alpha liquid scintillation counting. Other elements are measured by atomic absorption spectrometry. After 100 days the increase in U concentration stops, after which the systems are considered to be in equilibrium. The increase in steady-state U concentration is as follows, for Na–weeksite: $(3.61\pm 0.09)\cdot 10^{-4}$ M, for Na–boltwoodite: $(4.64\pm 0.12)\cdot 10^{-4}$ M, for uranophane: $(1.64\pm 0.03)\cdot 10^{-2}$ M and for soddyite: $(1.93\pm 0.03)\cdot 10^{-2}$ M. Other total concentrations of Si, Ca and Na are also given which sometimes do not match exactly the expected values for congruent dissolution. This could be a result of the composition of the compounds (see above).

X–ray diffraction data recorded on the solid phases in equilibrium with the solutions show that no phase change and no secondary phases occur with soddyite, uranophane and Na–weeskite, but that secondary phase(s) occur with Na–boltwoodite (probably soddyite).

All experimental solubility data seem correct to the reviewer.

Calculations of the solubility products are made according to equilibria involving solid compounds and, in aqueous solution, the species, UO_2^{2+}, $SiO_2(aq)$, H^+, Na^+ and Ca^{2+}:

$(UO_2)_2SiO_4\cdot 2H_2O(sodd) + 4H^+ \rightleftharpoons 2UO_2^{2+} + SiO_2(aq) + 4H_2O(l)$

$Ca(H_3O)_2(UO_2)_2(SiO_4)_2\cdot 3H_2O(uran) + 6H^+ \rightleftharpoons Ca^{2+} + 2UO_2^{2+} + 2SiO_2(aq) + 9H_2O(l)$

$Na(H_3O)UO_2SiO_4\cdot H_2O(bolt) + 3H^+ \rightleftharpoons Na^+ + UO_2^{2+} + SiO_2(aq) + 4H_2O(l)$

$Na_2(UO_2)_2(Si_2O_5)_3\cdot 4H_2O(week) + 6H^+ \rightleftharpoons 2Na^+ + 2UO_2^{2+} + 6SiO_2(aq) + 7H_2O(l)$

Free concentrations of these species are calculated using a code (HALTAFALL) and thermodynamic data on U(VI) hydrolysis, silicate anions and other cationic species taken from a draft version of [92GRE/FUG]. This review has verified that the hydrolysis constants used are those finally selected in [92GRE/FUG] (all the m and n values are taken into account) and that the other constants are those selected as auxiliary data in [92GRE/FUG] for $I = 0$. No more details are given by the authors of the calculations. They accepted (or the code shows) that, in the pH range 3 to 4.5, Si is only present as $SiO_2(aq)$. So the $[UO_2^{2+}]$ seems to be the only calculated free concentration in each case, but it is not clear how the values of $\log_{10}\beta$ and I ($mol\cdot kg^{-1}$) are obtained. From the $[UO_2^{2+}]$ values and the other experimental concentration values, $\log_{10}{}^*K_s(I)$ ($t = 30°C$) are derived and then the $\log_{10}{}^*K_s^\circ$ values, according to [2000GRE/WAN]:

$$\log_{10}{}^*K_s^\circ = \log_{10}{}^*K_s(I) - D\Delta Z^2$$

The values $\Delta_r G_m^\circ = -RT \ln {}^*K_s^\circ$ are further used to calculate $\Delta_f G_m^\circ$ for the compounds with auxiliary data for the following standard molar Gibbs energies:

- $-(237.53 \pm 0.04)$ kJ·mol^{-1} for H$_2$O(l)
- $-(552.5 \pm 1.05)$ kJ·mol^{-1} for Ca^{2+}
- $-(262.17 \pm 0.08)$ kJ·mol^{-1} for Na$^+$
- $-(833.79 \pm 3.01)$ kJ·mol^{-1} for SiO$_2$(aq)
- $-(953.70 \pm 1.76)$ kJ·mol^{-1} for UO$_2^{2+}$

which are slightly different from the values selected in [92GRE/FUG]. The reported equilibrium constants are listed in Table A-9.

Table A-9: Solubility product and Gibbs energy of formation of soddyite, uranophane, Na–boltwoodite and Na–weeksite.

	$\log_{10} {}^*K_s^\circ$	$\Delta_f G_m^\circ$ (kJ·mol^{-1})
Soddyite	(5.74 ± 0.21)	$-(3658.0 \pm 4.8)$
Uranophane	(9.42 ± 0.48)	$-(6210.6 \pm 7.6)$
Na–boltwoodite	$\geq (5.82 \pm 0.16)$	$\geq -(2966.0 \pm 3.6)$
Na–weeksite	(1.50 ± 0.08)	$-(9088.5 \pm 18.4)$

Uncertainties are propagated standard deviations of the analytical results. For Na–boltwoodite the equilibrium phase is assumed to be pure.

The authors have not considered complex formation between silicate and UO$_2^{2+}$, but this has only a minor influence on speciation at pH = 3 and 3.5. A problematic point in the experiment and the analysis is the possible formation of a precipitate of SiO$_2$(am), which would be difficult to detect by X–ray diffraction. The mole ratio U/Si in solutions of soddyite, uranophane and weeksite are 3.26, 2.30 and 0.04, respectively. These values differ considerably from the theoretical values for congruent dissolution of the solids used, which are 2, 1 and 0.33, respectively. This may be indicative of precipitation of amorphous silica in the test solutions. Precipitation will not influence the determination of the solubility product, unless there is sorption of dissolved uranium on the silica. According to the data in [92GRE/FUG], the solubility of fine–grained crystalline quartz is $1.8 \cdot 10^{-4}$ M. The solubility of amorphous silica is expected to be higher and the analytical value of the total concentrations of silica reported in [92NGU/SIL] seems reasonable. The low uranium concentration in the weeksite sample may be due to strong sorption of uranium(VI) at the higher pH used in this experiment. To conclude, this review considers the reported values of the Gibbs energy of formation of soddyite and uranophane as the best available estimates, but in the opinion of the reviewer the thermodynamic data must be re-evaluated. This has tentatively been done by [99CHE/EWI] taking into account the data of [96MOL/GEI] and [97PER/CAS], and selecting $\log_{10} {}^*K_s^\circ = (5.96 \pm 0.5)$ for soddyite, $\log_{10} {}^*K_s^\circ = (11.7 \pm 0.6)$ for uranophane and giving, finally:

$$\Delta_f G_m^\circ = - (3653.0 \pm 2.9) \text{ kJ} \cdot \text{mol}^{-1} \text{ for soddyite,}$$

$$\Delta_f G_m^\circ = - (6192.3 \pm 3.94) \text{ kJ} \cdot \text{mol}^{-1} \text{ for uranophane,}$$

$$\Delta_f G_m^\circ = - (2844.8 \pm 3.9) \text{ kJ} \cdot \text{mol}^{-1} \text{ for sodium boltwoodite,}$$

$$\Delta_f G_m^\circ = - (7993.9 \pm 9.8) \text{ kJ} \cdot \text{mol}^{-1} \text{ for sodium weeksite.}$$

[92SAN/BRU]

This is a publication based on Sandino's thesis [91SAN] that was reviewed in [92GRE/FUG]. The experimental data are the same, but the publication contains a more extensive literature review of solid phosphate phases. The title of the paper is somewhat misleading because the paper also contains a solubility study of schoepite (in fact metaschoepite) that is used to deduce information on hydroxide complexes of U(VI). The authors report data on the solubility of $(UO_2)_3(PO_4)_2 \cdot 4H_2O(cr)$ as a function of $-\log_{10}[H^+]$ in 0.5 M NaClO$_4$ under a N$_2$(g) atmosphere at 25°C. The concentration of phosphate was maintained at 10^{-2} M and $-\log_{10}[H^+]$ varied from 2.5 to 9.5. From these data, the following equilibrium constants were determined at $I = 0.5$ M and $I = 0$ using the SIT:

$UO_2^{2+} + HPO_4^{2-} \rightleftharpoons UO_2HPO_4(aq)$ $\log_{10} \beta = (6.03 \pm 0.09)$
                                                              $\log_{10} \beta^\circ = (7.28 \pm 0.10)$

$UO_2^{2+} + PO_4^{3-} \rightleftharpoons UO_2PO_4^-$ $\log_{10} \beta = (11.29 \pm 0.08)$
                                                         $\log_{10} \beta^\circ = (13.25 \pm 0.09)$

$UO_2^{2+} + 3 H_2O(l) \rightleftharpoons UO_2(OH)_3^- + 3 H^+$ $\log_{10} {}^*\beta_{3,1} = - (19.67 \pm 0.17)$
                                                                    $\log_{10} {}^*\beta_{3,1}^\circ = - (19.74 \pm 0.18)$

$(UO_2)_3(PO_4)_2 \cdot 4H_2O(cr) \rightleftharpoons 3 UO_2^{2+} + 2 PO_4^{3-} + 4 H_2O(l)$ $\log_{10} K_{s,0} = - (48.48 \pm 0.16)$
                                                                                                 $\log_{10} K_{s,0}^\circ = - (53.32 \pm 0.17)$.

The uncertainties are at the 2σ level. The equilibrium constants for the formation of UO_2HPO_4 and $UO_2PO_4^-$, and the solubility product of the phosphates have been considered in [92GRE/FUG] under the reference [91SAN]. The values reported by [92GRE/FUG] differ very slightly from those quoted here.

In order to model the solubility of $(UO_2)_3(PO_4)_2 \cdot 4H_2O(cr)$, the authors had to use equilibrium constants for the formation of the hydrolysed U(VI) species. They reanalysed the data of [89BRU/SAN] using the value selected by [92GRE/FUG], except for ${}^*\beta_{7,3}^\circ$ and ${}^*\beta_{3,1}^\circ$. The data of [89BRU/SAN] concern the solubility of crystalline and amorphous schoepite in neutral and alkaline media ($-\log_{10}[H^+] = 6.9$ to 8.2) obtained by potentiometric titration at 25°C in 0.5 M NaClO$_4$. The best fit gives the following values, respectively, for amorphous and crystalline schoepite as the solubility limiting phase:

$$\log_{10} {}^*\beta_{7,3} = - (32.00 \pm 0.17) \text{ and } \log_{10} {}^*\beta_{7,3} = - (33.32 \pm 0.22),$$
$$\log_{10} {}^*\beta_{3,1} = - (19.83 \pm 0.34) \text{ and } \log_{10} {}^*\beta_{3,1} = - (20.18 \pm 0.19),$$
$$\log_{10} {}^*K_{s,0} = (6.59 \pm 0.14) \quad \text{and } \log_{10} {}^*K_{s,0} = (6.23 \pm 0.14),$$

where the uncertainties are at the 2σ level. The table below summarises all the data of the authors at $I = 0$.

Table A-10: Equilibrium constants at zero ionic strength for the U(VI) hydroxide system [92SAN/BRU].

	$(UO_2)_3(PO_4)_2 \cdot 4H_2O$	$UO_3 \cdot 2H_2O(am)$	$UO_3 \cdot 2H_2O(cr)$
$\log_{10} {}^*\beta_{7,3}^\circ$		$-(31.55 \pm 0.17)$	$-(32.87 \pm 0.22)$
$\log_{10} {}^*\beta_{3,1}^\circ$	$-(19.74 \pm 0.18)$	$-(19.90 \pm 0.34)$	$-(20.25 \pm 0.19)$
$\log_{10} {}^*K_{s,0}$	$-(53.32 \pm 0.17)$	(6.33 ± 0.14)	(5.97 ± 0.14)

The weighed average of the values of $\log_{10} {}^*\beta_{3,1}^\circ$ (three values) and $\log_{10} {}^*\beta_{7,3}^\circ$ (two values) calculated by this review are $\log_{10} {}^*\beta_{3,1}^\circ = -(20.14 \pm 0.16)$ and $\log_{10} {}^*\beta_{7,3}^\circ = -(32.04 \pm 0.13)$.

Sandino and Bruno gave unweighted average values of $\log_{10} {}^*\beta_{3,1}^\circ$ corresponding to amorphous and crystalline schoepite as $\log_{10} {}^*\beta_{3,1}^\circ = -(20.1 \pm 0.5)$ and $\log_{10} {}^*\beta_{7,3}^\circ = -(32.2 \pm 0.8)$ taking uncertainties to cover the whole range of individual uncertainties. These average values are in good agreement. So, this review keeps the values from Sandino and Bruno to select equilibrium values of $\log_{10} {}^*\beta_{3,1}^\circ$ and $\log_{10} {}^*\beta_{7,3}^\circ$.

The previous values given by [89BRU/SAN] were $\log_{10} {}^*\beta_{3,1}^\circ = -(19.69 \pm 0.01)$ and $\log_{10} {}^*\beta_{7,3}^\circ = -(31.9 \pm 0.1)$, whereas [92GRE/FUG] selected $\log_{10} {}^*\beta_{3,1}^\circ = -(19.2 \pm 0.4)$ and $\log_{10} {}^*\beta_{7,3}^\circ = -(31 \pm 2)$.

[92SAT/CHO]

This paper is discussed together with [98JEN/CHO].

[92VEN/IYE]

The paper reports measurements of the oxygen activity (emf with CaO–ZrO$_2$ electrolyte) in the mixtures, A$_2$U$_4$O$_{12}$(cr) + A$_2$U$_4$O$_{13}$(cr) (A = Cs or Rb), and complementary data on the enthalpy increments of the four uranates.

The U(VI) compounds were prepared by heating appropriate amounts of U_3O_8 with the alkali metal carbonates in air at 1173 K for 20 hours in gold boats; the lower uranates were prepared by decomposing the U(VI) compounds in purified Ar at 1273 K for 200 hours. The products were characterised by X–ray diffraction. The uranium content in $Rb_2U_4O_{12}(cr)$ was determined potentiometrically to be (72.69 ± 0.31) mass %, in good agreement with the theoretical value of 72.40 %.

The oxygen potentials were studied using a CaO–ZrO$_2$ electrolyte tube in flowing argon, from ca. 1020 to 1283 K with air ($p_{O_2} = 0.2121$ bar) as the reference electrode.

Correction of their emf values to the standard pressure gives, after recalculation:

$$\Delta_r G_m \text{ (A.31)} = -174638 + 136.100\, T \quad \text{J·mol}^{-1} \quad (1019 \text{ to } 1283 \text{ K})$$

$$\Delta_r G_m \text{ (A.32)} = -102082 + 54.378\, T \quad \text{J·mol}^{-1} \quad (1075 \text{ to } 1203 \text{ K})$$

for the reactions:

$$Cs_2U_4O_{12}(cr) + 0.5\, O_2(g) \rightleftharpoons Cs_2U_4O_{13}(cr) \tag{A.31}$$

$$Rb_2U_4O_{12}(cr) + 0.5\, O_2(g) \rightleftharpoons Rb_2U_4O_{13}(cr) \tag{A.32}$$

The data below 1075 K for reaction (A.32) were not used, because a non-equilibrium state exists. The considerable differences in the entropies of these reactions and in the calculated decomposition temperatures of the U(VI) compounds in air (1225 K for $Cs_2U_4O_{13}(cr)$ and 1680 K for $Rb_2U_4O_{13}(cr)$) are somewhat surprising.

Drop calorimetric measurements are reported for the four compounds, using a Calvet high temperature calorimeter. The data were fitted by a simple polynomial, but the authors' expressions for $H_m(T) - H_m^\circ(298.15\,\text{K})$ are not equal to zero at $T = 298.15$ K, so the enthalpy data have been refitted with this constraint. The derived heat capacity expressions are given in Table A-11.

Table A-11: Heat capacity coefficients and standard heat capacity for $Rb_2U_4O_{12}(cr)$ $Rb_2U_4O_{13}(cr)$, $Cs_2U_4O_{12}(cr)$ and $Cs_2U_4O_{13}(cr)$.

Phase	T range (K)	Heat capacity coefficients $C_{p,m} = a + b \cdot T + e \cdot T^{-2}$ J·K^{-1}·mol^{-1}			$C^\circ_{p,m}(298.15\,\text{K})$ J·K^{-1}·mol^{-1}
		a	b	e	
$Rb_2U_4O_{12}(cr)$	375 – 755	$1.897590 \cdot 10^2$	$1.5487 \cdot 10^{-1}$	$1.9903 \cdot 10^7$	–
$Rb_2U_4O_{13}(cr)$	325 – 805	$4.125600 \cdot 10^2$	$2.5000 \cdot 10^{-2}$	0.0000	(420 ± 50)
$Cs_2U_4O_{12}(cr)$	361 – 719	$3.926650 \cdot 10^2$	$1.0180 \cdot 10^{-1}$	0.0000	(423 ± 40)
$Cs_2U_4O_{13}(cr)$	347 – 753	$7.128220 \cdot 10^2$	$-4.2745 \cdot 10^{-1}$	$-6.8805 \cdot 10^5$	–

However, as shown in the plot of heat capacities in section 9.10.1, the expressions for $Rb_2U_4O_{12}(cr)$ and $Cs_2U_4O_{13}(cr)$ are not consistent with the general behaviour of the heat capacities of the alkali–metal uranates, and these data have not been selected in this review.

The authors have used these Gibbs energies of reaction to derive data at 298.15 K for $Cs_2U_4O_{12}(cr)$ and $Rb_2U_4O_{12}(cr)$, using their heat capacity data (extrapolated up from ca. 750 K) and literature data at 298.15 K for the hexavalent compounds. However, we have not pursued this approach since:

- there are two transformations in $Cs_2U_4O_{13}(cr)$ at 898 and 968 K [92GRE/FUG], which the authors have neglected;
- all of the data for $Rb_2U_4O_{13}(cr)$ at 298.15 K are estimated;
- there are unexpected differences in the Gibbs energies of the similar reactions (A.31), for Cs and (A.32), for Rb.

Thus any derived data would have quite large uncertainties.

[92VEN/KUL]

The authors investigated the vaporisation of UN(cr) from 1757 to 2400 K by Knudsen effusion mass spectrometry. The principal reaction is the loss of $N_2(g)$ to form a nitrogen-saturated U(l), but U(g) and UN(g) are also present in the vapour. The pressures of U(g) were measured from 1757 to 2396 K and the lower pressures of UN(g) from 2190 to 2400 K.

The UN(cr) was in the form of sintered microspheres prepared by heating microspheres of UO_2 + C in flowing nitrogen in unspecified conditions. The uranium content was (94.42 ± 0.40) mass %, in good agreement with the theoretical value of 94.44%; the residual oxygen level was (0.08 ± 0.01) mass %.

The mass–spectrometric measurements were carried out in a tantalum effusion cell inside a tantalum cell; silver was added to the sample as a calibrant. N_2, U, UN and UO were detected in the vapour. The pressure of UO(g) was larger than for either U(g) or UN(g) in fresh samples, but decreased as oxygen was lost from the sample. The background nitrogen pressure was too large for reliable nitrogen pressures to be determined. The ion currents due to U(g) and UN(g) were converted to pressures by calibration with Ag and atomic cross–sections given by Mann [70MAN]; that for UN was assumed to be a factor of 0.75 smaller than the sum of the cross–sections of U and N.

The uranium ion currents were steady after an initial period (perhaps due to the time needed to establish the UN(cr) + U(l) phase field and loss of the initial oxygen impurity), and corresponded to the equation:

$$\log_{10}(p(U)/\text{bar}) = 5.59 - 26857 \cdot T^{-1} \qquad (1757 \text{ to } 2396 \text{ K})$$

with very good consistency between six different runs. The reported pressure at 2000 K is at the lower end of the range of pressures given by five previous studies.

The authors used the thermal functions from Hultgren et al. [73HUL/DES] to calculate the third–law enthalpy of sublimation of uranium at 298.15 K, $\Delta_{sub}H_m^o$ (U, 298.15 K) = (564.4 ± 7.4) kJ · mol^{-1}. The second–law value is somewhat smaller, $\Delta_{sub}H_m^o$ (U, 298.15 K) = (554.6 ± 10.3) kJ · mol^{-1}. These values relate to vaporisation from a uranium liquid saturated with nitrogen; it will also contain some tantalum, since this metal is appreciably soluble in U(l) [81CHI/AKH]. It is therefore not surprising that the authors' values are appreciably more positive than the CODATA value of (533 ± 8) kJ · mol^{-1} for pure uranium adopted by [92GRE/FUG].

The pressures for UN(g), corresponding to the sublimation of UN,

$$UN(cr) \rightleftharpoons UN(g) \tag{A.33}$$

(although the solid will in fact be slightly hypostoichiometric) were fitted to the equation:

$$\log_{10}(p(UN)/bar) = 7.19 - 37347 \cdot T^{-1} \quad (2190 \text{ to } 2400 \text{ K})$$

The experimental pressures were rather more scattered than the (appreciably larger) pressures of U(g). Thus $\Delta_r G_m$ (A.33) = 715003 – 137.65·T J·mol^{-1}.

The authors combine this equation with values of $\Delta_f G_m$ (UN, cr) derived from the assessment by Matsui and Ohse [87MAT/OHS], $\Delta_f G_m$ (UN, cr, T) = – 304890 + 88.2·T J·mol^{-1} to define $\Delta_f G_m$ (UN, g, T) (but seem to have made a numerical error in their first term). We have changed the equation for $\Delta_f G_m$ (UN, cr, T) to that which is consistent with the value selected in [92GRE/FUG] for $\Delta_f H_m$ (UN, cr, 298.15 K) and to relate to the mid-temperature of the measurements involving UN(g), 2300 K. The NEA-TDB–compatible equation from 2200 to 2400 K is then $\Delta_f G_m$ (UN, cr, T) = −297596 + 87.53·T J·mol^{-1}, from which we derive

$$\Delta_f G_m (UN, g, T) = 417407 - 50.12 \cdot T \ (J \cdot mol^{-1}) \quad (2200 - 2400 \text{ K}).$$

This is the first significant experimental determination of the stability of UN(g). However, since no thermal functions are available for UN(g), these Gibbs energy values cannot be reliably converted to provide standard data at 298.15 K.

[92WIM/KLE]

The equilibrium constants for the formation of CmOH^{2+} and Cm(OH)$_2^+$ have been determined by TRLFS at 25°C in 0.1 M NaClO$_4$. The method allows a direct determination of the concentrations of Cm^{3+} and the two complexes. The accuracy of the equilibrium constant thus depends mainly on the experimental determination of the free hydrogen ion concentration. There are few details on the electrode calibration and the reviewers have therefore discussed the procedures used with the authors. The electrode was calibrated against commercial standard buffers at pH = 4, 5, 6, 7, 8, 9 and 10 and the measured potentials were fitted by linear regression. The concentration of OH⁻ was calculated from the measured pH by using the relation: $-\log_{10}[\text{OH}^-] = -13.78 + \text{pH} + \log_{10}\gamma_{\pm}$, where -13.78 is the ionic product of water in 0.1 M NaClO$_4$, and $\gamma_{\pm}$ is the mean activity coefficient of NaClO$_4$. This means that the single ion activity coefficient of H⁺ is assumed to be equal to the mean activity coefficient of NaClO$_4$. This assumption has not been justified and this review has therefore calculated the single ion activity coefficient for H⁺ using the SIT and finds $\gamma_{\text{H}^+} = 0.80$. The recalculated values are $\log_{10} \beta_1 = (6.67 \pm 0.18)$ and $\log_{10} \beta_2 = (12.06 \pm 0.28)$. The authors point out that these values are very different from those obtained by solvent extraction and electro–migration, but in good agreement with the values obtained by Stadler and Kim [88STA/KIM] for americium using solubility measurements. The study of Wimmer et al. gives more direct information on the species formed than the earlier studies and the constants reported are therefore retained by this review and considered in the discussion of selected data. Recalculation of the constants to zero ionic strength, using the interaction coefficients for the americium system gives $\log_{10} \beta_1° = (7.31 \pm 0.18)$ and $\log_{10} \beta_2° = (13.11 \pm 0.28)$. These values are consistent with those selected in [95SIL/BID], $\log_{10} \beta_1° = (7.6 \pm 0.7)$ and $\log_{10} \beta_2° = (13.9 \pm 0.6)$, but are more precise.

[93BOI/ARL]

Results are reported of rapid (ca. 100 μsec) heating of a uranium wire under a constant pressure of 120 MPa, to determine a number of thermophysical properties. However, no temperature measurements are reported, the temperatures being derived from the enthalpies given by [76OET/RAN] and [56STU/SIN]. Thus, no information of interest to the present review is given. It may be noted that their enthalpy increments as a function of volume are appreciably smaller than those given by [88MUL/SHE], [91SHE/MUL] and others.

[93ERI/NDA]

The authors have studied the solubility of Tc(IV) and Np(IV) hydrous oxides as a function of pH and carbonate concentration and from these data deduced the stoichiometry and equilibrium constants of the following complex formation reactions:

$\text{TcO}_2 \cdot n\text{H}_2\text{O(s)} \rightleftharpoons \text{TcO(OH)}_2\text{(aq)} + (n-1)\,\text{H}_2\text{O(l)}$ $\log_{10} K = -(8.17 \pm 0.05)$

$\text{TcO}_2 \cdot n\text{H}_2\text{O(s)} \rightleftharpoons \text{TcO(OH)}_3^- + (n-2)\,\text{H}_2\text{O(l)} + \text{H}^+$ $\log_{10} K = -(19.06 \pm 0.24)$

$\text{Np(OH)}_4\text{(s)} \rightleftharpoons \text{Np(OH)}_4\text{(aq)}$ $\log_{10} K = -(8.28 \pm 0.23)$

$\text{Np(OH)}_4\text{(s)} \rightleftharpoons \text{NpO}_2^+ + e^- + 2\text{H}_2\text{O(l)}$ $\log_{10} K = -(9.40 \pm 0.50)$

$\text{TcO}_2 \cdot n\text{H}_2\text{O(s)} + \text{CO}_2\text{(g)} \rightleftharpoons \text{Tc(OH)}_2\text{CO}_3\text{(aq)} + (n-1)\,\text{H}_2\text{O(l)}$ $\log_{10} K = -(7.09 \pm 0.08)$

$\text{TcO}_2 \cdot n\text{H}_2\text{O(s)} + \text{CO}_2\text{(g)} \rightleftharpoons \text{Tc(OH)}_3\,\text{CO}_3^- + \text{H}^+ + (n-2)\,\text{H}_2\text{O(l)}$ $\log_{10} K = -(15.35 \pm 0.07)$

$Np(OH)_4(aq) + CO_3^{2-} \rightleftharpoons Np(OH)_4 CO_3^{2-}$ $\log_{10} K = (3.00 \pm 0.12)$

$Np(OH)_4(aq) + HCO_3^- \rightleftharpoons Np(OH)_3 CO_3^- + H_2O(l)$ $\log_{10} K = (3.23 \pm 0.12)$.

The experiments have been carried out at room temperature and in a medium of low ionic strength. The solid phases were prepared by electroreduction of TcO_4^- and Np(V) on a Pt–net that was used in the following solubility measurements. This seems to be a practical method to prepare the solid phases and the scatter of the solubility data is small. The formula of the hydroxo carbonato complexes proposed by Eriksen *et al.* are surprising as discussed in [2001LEM/FUG].

A test of the methodology can be obtained by comparing the value of the equilibrium constant for the reaction:

$$Np(OH)_4(s, am) \rightleftharpoons NpO_2^+ + e^- + 2H_2O(l),$$

obtained in [93ERI/NDA], with that given in [2001LEM/FUG], $\log_{10} K = -(8.68 \pm 1.71)$. The latter value is deduced from the standard potential of the couple:

$$Np(OH)_4(s, am) \rightleftharpoons NpO_2^+ + e^- + 2H_2O(l), \quad \log_{10} K = -(17.68 \pm 0.17)$$

and the solubility product of amorphous $Np(OH)_4(s)$. The very large uncertainty ranges of the two values, $\log_{10} K = -(8.68 \pm 1.71)$ and $\log_{10} K = -(9.4 \pm 0.5)$, respectively, overlap, but the uncertainties remain large.

The following speciation diagram (Figure A-3 and Figure A-4) show the influence of the proposed carbonate complexes on the speciation of Np(IV) in solutions studied by Rai *et al.* [98RAI/FEL]. The equilibrium constants for the complexes $U(OH)_4(CO_3)^{2-}$ and $U(OH)_3(CO_3)^-$ are those for the corresponding Np(IV) species given by Eriksen *et al.*, (substitution of neptunium to uranium) while the equilibrium constant for $U(OH)_2(CO_3)_2^{2-}$ corresponds to the reaction:

$$U^{4+} + 2 CO_3^{2-} + 2 H_2O(l) \rightleftharpoons U(OH)_2(CO_3)_2^{2-} + 2 H^+$$

with $\log_{10} K = 20.0$, calculated from the equilibrium constants determined by Rai *et al.* [98RAI/FEL], but changing the solubility product for $UO_2(am)$ so that the value of the equilibrium constant for the formation of $U(CO_3)_5^{6-}$ conforms with the value in [92GRE/FUG]. This change is 2.7 $\log_{10}$ units and the corresponding change has also been made for the equilibrium above. The two figures below have been calculated at a total carbonate concentration of 0.2 and 0.02 M.

The modelling results indicate that the complex, $U(OH)_2(CO_3)_2^{2-}$, is the predominant species under the conditions used by Eriksen *et al.*, but this is inconsistent with their experimental results. These two experimental determinations are obviously not concordant and the review has therefore not used them.

The conclusions above are valid both for the Tc(IV) and Np(IV) systems.

Figure A-3: Speciation diagram of $[U^{4+}] = 10^{-7}$ M and $[CO_3^{2-}] = 1.5 \cdot 10^{-3}$ M, using equilibrium constants of neptunium given by Eriksen et al.

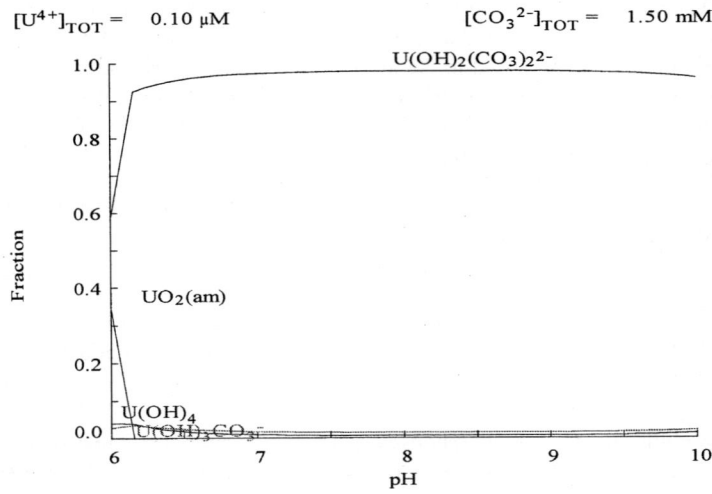

Figure A-4: Speciation diagram of $[U^{4+}] = 10^{-7}$ M and $[CO_3^{2-}] = 2.1 \cdot 10^{-1}$ M.

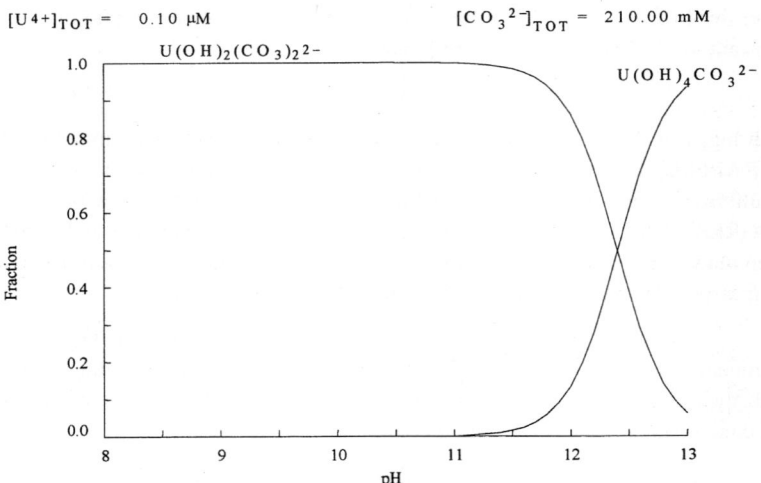

[93FER/SAL]

This paper reports mainly the determination of the formation constant of U(VI) fluoride complexes. It also gives new values of the hydrolysis constants for the species, $(UO_2)_2(OH)_2^+$ and $(UO_2)_3(OH)_5^+$.

Potentiometric titrations utilised an Ag | AgCl and a glass electrode cell without liquid junction at 3.0 M ionic strength ($[NaClO_4] \equiv 3.496$ mol·kg^{-1}) at 25°C. One series of titrations was carried out at constant U(VI) and Ag$^+$ concentrations with varying $[H^+]$ in the absence of fluoride, while another series varied fluoride at constant U(VI) and $[H^+]$ concentrations. Seven titrations in the latter mode were performed at four H$^+$ concentrations ($-\log_{10}[H^+] = 3.5-4.5$, $[U(VI)] = 0.001-0.075$ M). In addition, ^{19}F NMR spectra were recorded for dioxouranium(VI) solutions of 3 M NaClO$_4$ containing 0.1 M NaF and showed chemical shifts to higher frequency relative to the solution in the absence of uranium.

The potentiometric titrations in the absence of fluoride were interpreted in terms of three species, yielding the hydrolysis constants: $\log_{10} \beta_{2,2} = -(5.98 \pm 0.02)$ and $\log_{10} \beta_{5,3} = -(16.23 \pm 0.05)$. From the values of $\varepsilon(nm, ClO_4^-, Na^+)$ in [92GRE/FUG], the values of $\log_{10} {}^*\beta_{2,2}^\circ = -(5.54 \pm 0.04)$ and $\log_{10} {}^*\beta_{5,3}^\circ = -(15.08 \pm 0.70)$ were calculated. They are in good agreement with the values selected by [92GRE/FUG].

From the titrations with $[U(VI)] = 10^{-3}$ M and $-\log_{10}[H^+] = 3.50$, the following $\log_{10} \beta_n$ values were obtained for the formation of the binary complexes, $UO_2F_n^{2-n}$, $n = 1-4$: $\log_{10} \beta_1 = (4.86 \pm 0.02)$; $\log_{10} \beta_2 = (8.62 \pm 0.04)$; $\log_{10} \beta_3 = (11.71 \pm 0.06)$ and $\log_{10} \beta_4 = (13.78 \pm 0.08)$. The fit derived from these constants in terms of z (the average number of coordinated fluoride ligands per U(VI)) as a function of the free fluoride concentration is given. However, the individual values only reach a maximum of ca. 2.35, so that it is perhaps surprising that β_4 for $UO_2F_4^{2-}$ could be determined with such a small estimated uncertainty. The reason is that small uncertainties are often an artefact introduced by the choice of error variables used in the least squares refinement and the weights of the different experimental data. Using ^{19}F NMR, evidence is presented for the formation of the fifth complex, $UO_2F_5^{3-}$, for which a stepwise formation constant of 3 M^{-1} was reported. This value is in error as indicated in a later study by Vallet et al. [2001VAL/WAH]. The error is due to the spectrometer and software system. By using a more advanced spectrometer and better software, Vallet et al. find K = 0.6 M^{-1} at – 4°C. This value has been accepted by this review. From [71AHR/KUL] the values of $\log_{10} \beta_n$ at 1 M ionic strength were used in conjunction with the present values and the ion interaction parameters for the reactants (from [92GRE/FUG]) to give the following (in kg · mol^{-1}):

$$\varepsilon(UO_2F^+, ClO_4^-) = (0.28 \pm 0.04),$$
$$\varepsilon(UO_2F_2 \text{ (aq)}, Na^+ \text{ or } ClO_4^-) = (0.13 \pm 0.05),$$

$$\varepsilon(UO_2F_3^-, Na^+) = -(0.14 \pm 0.05),$$
$$\varepsilon(UO_2F_4^{2-}, Na^+) = -(0.30 \pm 0.06).$$

In a simple specific ion interaction theory, $\varepsilon(UO_2F_2 \text{(aq)}, Na^+ \text{or } ClO_4^-)$ should be equal to zero.

For the ternary system, analysis of the titrations performed at the three higher H^+ concentrations revealed the formation of multinuclear species. However, at the higher fluoride concentrations the titration curves became independent of the dioxouranium(VI) concentration. At $-\log_{10}[H^+] = 3.6$ and 3.9, the data could be resolved simply in terms of the above–mentioned hydroxo and fluoro complexes. The results at the highest H^+ concentration and $[U(VI)] = 0.003$ M deviated from this speciation model, but could not be rationalised in terms of mixed hydroxofluoro complexes and were ascribed to the formation of a fine precipitate.

[93GRE/LAG]

Glass electrode potentiometric titrations of the $U(VI) - SO_4^{2-} - OH^-$ system were carried out at 25.0°C in solutions containing 0.500 M Na_2SO_4 and 2.00 M $NaClO_4$ with U(VI) concentrations in the range 0.003487 to 0.02000 M. The glass electrode was calibrated on the molar scale in the same ionic medium. A least–squares analysis of the conditional equilibrium constants for the reactions:

$$p"UO_2^{2+}" + q H_2O(l) + r SO_4^{2-} \rightleftharpoons "(UO_2)_p(OH)_q" + q H^+ \quad (A.34)$$

where $"UO_2^{2+}"$ represents $\{UO_2^{2+} + UO_2SO_4\text{(aq)} + UO_2(SO_4)_2^{2-}\}$ and $"(UO_2)_p(OH)_q"$ represents the complexes, $\{(UO_2)_p(OH)_q(SO_4)_r^{2p-q-2r}\}$, was carried out first to determine the values of the coefficients p and q based on the speciation model of Peterson [61PET]. The general equation (A.34), stands for:

$$p UO_2^{2+} + q H_2O(l) + r SO_4^{2-} \rightleftharpoons (UO_2)_p(OH)_q(SO_4)_r^{2p-q-2r} + q H^+$$
$$p UO_2SO_4\text{(aq)} + q H_2O(l) + (r-p) SO_4^{2-} \rightleftharpoons (UO_2)_p(OH)_q(SO_4)_r^{2p-q-2r} + q H^+$$
$$p UO_2(SO_4)_2^{2-}\text{(aq)} + q H_2O(l) + (r-2p) SO_4^{2-} \rightleftharpoons (UO_2)_p(OH)_q(SO_4)_r^{2p-q-2r} + q H^+$$

The data from Peterson [61PET], who studied this system in 1.5 M Na_2SO_4, were also used in a least-squares analysis (note that in the review of [92GRE/FUG] this work was not considered due to the high ionic strength and difficulty in estimating the uncertainties, although conditional formation constants were recalculated). The primary experimental data, on which this data analysis was based, are not given in the paper; only $\bar{n}$ versus $-\log_{10}[H^+]$ data. The original speciation was finally limited to three hydrolyzed U(VI) species (viz., $p:q$ values of: 2:–2; 3:–4; and 3:–8) giving an improved fitting factor such that the 3:–5 and 4:–6 species proposed by Peterson were considered to be redundant. The conditional constants were then further refined using β_1 and β_2 values taken from [92GRE/FUG] for the two unhydrolysed sulphate complexes and applying the SIT to adjust these values to the ionic strengths of both studies. The final regression to obtain the coefficient r and the corresponding $^*\beta_{p,q,r}$ values for the equilibrium:

$$p\,UO_2^{2+} + q\,H_2O + r\,SO_4^{2-} \rightleftharpoons (UO_2)_p(OH)_q(SO_4)_r^{2p-q-2r} + q\,H^+$$

was carried out by assuming that one ternary complex containing the maximum number of sulphate ligands was formed for each set of p:q values in the two studies, which when treated together provide two different sulphate concentrations for the regression. The SIT was used to reconcile the two different ionic strengths and $^*\beta_{p,q,r}$ values were estimated, viz. $\log_{10} {}^*\beta_{2,-2,2} = -(3.26 \pm 0.46)$, $\log_{10} {}^*\beta_{3,-4,3} = -(8.64 \pm 0.69)$ or $\log_{10} {}^*\beta_{3,-4,4} = -(8.81 \pm 0.69)$ and $\log_{10} {}^*\beta_{5,-8,6} = -(19.79 \pm 1.20)$ in 1.500 M Na$_2$SO$_4$; $\log_{10} {}^*\beta_{2,-2,2} = -(2.73 \pm 0.09)$, $\log_{10} {}^*\beta_{3,-4,3} = -(8.15 \pm 0.17)$ or $\log_{10} {}^*\beta_{3,-4,4} = -(7.84 \pm 0.17)$ and $\log_{10} {}^*\beta_{5,-8,6} = -(18.53 \pm 0.25)$ in 0.500 M Na$_2$SO$_4$, 2.00 M NaClO$_4$. Semi–quantitative and structural arguments are given in support of the proposed speciation.

The text of this paper shows that the two 3:–4 (p:q) complexes are given as alternative species, but in the main text of this review they are both listed as coexisting species. The constants given in the footnote on page 241 of [92GRE/FUG] are not the same as in [93GRE/LAG] and should therefore be disregarded. The constants in this latter publication agree well with those in two later studies ([2000COM/BRO] and [2000MOL/REI]).

[93JAY/IYE]

Enthalpy increments for Rb$_2$U(SO$_4$)$_3$(cr) were measured from 373 to 803 K using a high temperature Calvet calorimeter, supplemented by DSC and simultaneous DTA and TGA measurements. Rb$_2$U(SO$_4$)$_3$(s) was prepared by dissolution of UO$_3$(cr) in an H$_2$SO$_4$ solution, electrolytic reduction, addition of Rb$_2$CO$_3$(cr) dissolved in 0.5 M H$_2$SO$_4$, followed by crystallisation.

The product was analysed for total U and sulphate, and showed good agreement with theoretical values and by AES, which showed < 300 ppm impurities (mainly Al and Si). X–ray diffraction was used to check that the material had not been modified during the experiment. Pellets, dried at 400 K, were used for the measurements, NBS standard sapphire being used to calibrate the calorimeter.

Two solid state transitions were observed at 616 and 773 K (DSC), and 628 and 780 K by DTA and TGA. The compound starts to decompose at 893 K.

The results of the enthalpy increment measurements were tabulated and fitted into three separate ranges from 370 to 628 K, 630 to 760 K and 765 to 800 K.

The enthalpy data for the low–temperature polymorph are consistent with the values of the heat capacities from 273 to 623 K subsequently reported briefly by Saxena *et al.* [99SAX/RAM], and these two sets of data have been combined to define the enthalpy increments and heat capacity of this phase – see section 9.10.5.3.

The nine measurements of the enthalpy increments for the phase stable from *ca.* 628 to 763 K are not precise enough to define a reliable heat capacity for this phase. The authors fitted their measurements to the expression:

$$H_m (T) - H_m^\circ (298.15 \text{ K}) = 3279 + 13.4 \cdot T + 0.3063 \cdot T^2 \quad \text{J} \cdot \text{mol}^{-1} \text{ (630–760 K)}$$

which corresponds to a heat capacity rising sharply from 399 J · K^{-1} · mol^{-1} at 630 K to 479 J · K^{-1} · mol^{-1} at 760 K, but a constant heat capacity of 437.6 J · K^{-1} · mol^{-1} fits the data almost as well.

The authors's calculated enthalpy of transition at *ca.* 628 K seems to include some sensible heat also; our calculated values for H_m (628 K) – H_m° (298.15 K) for the two polymorphs are 128793 and 132494 J · mol^{-1}, giving a transition enthalpy of 3.7 kJ · mol^{-1}.

The reported values for the polymorph stable above 780 K correspond to an implausibly large heat capacity, presumably due to incipient decomposition by loss of SO$_2$(g).

The authors also estimated the entropy of Rb$_3$U(SO$_4$)$_3$ (cr) from the sum of values for Rb$_2$SO$_4$(cr) and U(SO$_4$)$_2$.

[93KRI/EBB]

This is a report of project work involving the vaporisation of actinide species from waste oxide processors. The only work relevant to the current review is a study of the vaporisation behaviour of U$_3$O$_8$(cr) in oxygen and water vapour, similar to the study by Dharwadkar *et al.* [74DHA/TRI]; like this, it comprises transpiration measurements of the total pressures of uranium–bearing species over U$_3$O$_8$ in the presence of both dry and moist oxygen. The mass loss in the presence of water vapour is larger than the loss of UO$_3$(g) in dry oxygen, due to the formation of a UO$_3$·(H$_2$O)$_n$ vapour species.

The variation of the water content in the transpiring oxygen suggested that the H$_2$O/UO$_3$ ratio in the hydroxide gas formed was close to 1, although the relevant data are quite scattered.

The authors' pressures of UO$_3$(g) from the reaction:

$$\tfrac{1}{3}\text{U}_3\text{O}_8(\text{cr}) + \tfrac{1}{6}\text{O}_2(\text{g}) \rightleftharpoons \text{UO}_3(\text{g}) \tag{A.35}$$

from 1273 to 1573 K are in good agreement with the similar data by [74DHA/TRI], and other literature data which form the basis of the choice of $\Delta_f G_m$ (UO$_3$, g) in [82GLU/GUR] and [92GRE/FUG].

The authors tabulate the results of 20 experiments (including replicates) using oxygen with known amounts of water vapour from 1173 to 1373 K (results at higher temperatures were discounted by the authors) involving the reaction:

$$\frac{1}{3}U_3O_8(cr) + \frac{1}{6}O_2(g) + H_2O(g) \rightleftharpoons UO_2(OH)_2(g) \qquad (A.36)$$

In calculating the pressure of $UO_2(OH)_2(g)$, allowance was made for the smaller mass loss due to the simultaneous vaporisation of $UO_3(g)$.

The authors' results are somewhat scattered (replicated pressures differ by as much as a factor of 7), but give values of K_p for reaction (A.36) which are lower by a factor of *ca.* 330 than those of the similar study by Dharwadkar *et al.* [74DHA/TRI]. However, recently, Krikorian *et al.* [97KRI/FON] studied the release of uranium-containing species from a $^{238}PuO_2$ sample containing 3.3 mol% of $^{234}UO_2$ under similar conditions. Despite some uncertainties in the calculation (particularly of the activity of UO_2 in the $(U,Pu)O_{2+x}$ solid solution presumably formed), the releases were in general agreement with those predicted from the data of [93KRI/EBB].

As noted in the review of [74DHA/TRI], there are two appreciably different estimates of the thermal functions of $UO_2(OH)_2(g)$ by Ebbinghaus [95EBB] and Gorokhov and Sidorova [98GOR/SID], with which these results can be combined.

We have used the twenty tabulated values of K_p to calculate the Gibbs energy of reaction (A.36), and further calculate the second– and third–law values for $\Delta_r H_m^\circ$ ((A.36), 298.15 K) and hence $\Delta_f H_m^\circ$ ($UO_2(OH)_2$, g, 298.15 K).

As can be seen, there is a considerable variation in the values for the derived enthalpy of formation of $UO_2(OH)_2(g)$. The big differences in the Table A-12 and the unexplained large differences between the results of this study and those of [74DHA/TRI], mean that no reliable data for this compound can be selected in this review.

Table A-12: Derived values of $\Delta_f H_m^\circ$ ($UO_2(OH)_2$, g, 298.15 K), kJ · mol^{-1}

Method	Thermal functions	
	[95EBB]	[98GOR/SID]
Second-law	− (1153.6 ± 58.1)	− (1157.6 ± 58.1)
Third-law	− (1197.2 ± 12.5)	− (1224.5 ± 13.3)

[93MCB/GOR]

Assessed data for U(α, β, γ, liq) are given as tables, graphs and $C_{p,m}$ coefficients. The values are essentially identical to those given by [82GLU/GUR], which form the basis of the CODATA Key Values selection used in [92GRE/FUG]. In fact the only values other than those at 298.15 K quoted in [92GRE/FUG] are for $C_{p,m}$(U, α) up to 942 K, which are the very similar to values given by [76OET/RAN].

[93MEI/KAT]

This paper supplements/complements [93MEI/KIM].

This paper gives: i) the proof of the existence of $(UO_2)_2(OH)_2^{2+}$ (abbreviated as 2:2) on the basis of solubility and spectroscopic measurements, and ii) the fine structure of both absorption and emission spectra of this species. Many of the additional studies from this group after 1993 will use these data. In this paper a less precise value of $\log_{10} {}^*\beta_{2,2}$ (with respect to the 1998 values for instance) is given from only spectroscopy data coming from both saturated and undersaturated U(VI) solutions. Finally this paper bridges hydrolysis data to those reported in [92GRE/FUG]. The new spectroscopic data also confirm previous characterisations of hydrolysed U(VI) species (2:2 and 5:3, $(UO_2)_3(OH)_5^+$) [63RUS/JOH], [63BAR/SOM] (Cited in [92GRE/FUG]).

- Identification of $(UO_2)_2(OH)_2^{2+}$ species:

The solubility measurements of rutherfordine, $UO_2CO_3(cr)$, in 0.1 M $NaClO_4$ at $(24 \pm 2)°C$ as a function of pH = 2.8 to 4.15, under 8% and 100% CO_2 and of schoepite, $UO_3 \cdot 2H_2O(cr)$ (in fact metaschoepite), pH = 3.4 to 4.8 under 1%, 0.3% and 0.03% CO_2 show that in the presence of the oxide phase, CO_2 has no influence on the solubility whereas the solubility does depend on CO_2 when $UO_2CO_3(cr)$ is the solubility-limiting solid phase. The slope of the straight line, $\log[U(VI)]_t$ versus pH, is -2 which is compatible with species bearing a charge 2+: UO_2^{2+} or $(UO_2)_2(OH)_2^{2+}$. Indeed, according to:

$$UO_3 \cdot 2H_2O(scho) \rightleftharpoons UO_2^{2+} + 2OH^- + H_2O \ (l)$$
$$2UO_3 \cdot 2H_2O(scho) \rightleftharpoons (UO_2)_2(OH)_2^{2+} + 2OH^- + 2H_2O(l)$$
$$UO_2CO_3(ruth) \rightleftharpoons UO_2^{2+} + CO_3^{2-}$$
$$2UO_2CO_3(ruth) + 2H_2O(l) \rightleftharpoons (UO_2)_2(OH)_2^{2+} + 2HCO_3^-$$

and the definitions of ${}^*\beta_{2,2}$ and $K_{s,0}$, it can be derived:

$$\log_{10}[UO_2^{2+}] = \log_{10} K_{s,0}(UO_3 \cdot 2H_2O) - 2\log_{10} K_W - 2\log_{10} \gamma_{H^+} - 2pH$$

$$\log_{10}[(UO_2)_2(OH)_2^{2+}] = 2\log_{10} K_{s,0}(UO_3 \cdot 2H_2O) - 4\log_{10} K_W$$
$$+ \log_{10} {}^*\beta_{2,2} - 2\log_{10} \gamma_{H^+} - 2pH$$

$$\log_{10}[UO_2^{2+}] = \log_{10} K_{s,0}(UO_2CO_3) - \sum\log_{10} K - \log p_{CO_2} - 2pH$$

$$\log_{10}[(UO_2)_2(OH)_2^{2+}] = 2\log_{10} K_{s,0}(UO_2CO_3) - 2\sum\log_{10} K - 2\log p_{CO_2}$$
$$+ 2\log_{10} \gamma_{H^+} + \log_{10} {}^*\beta_{2,2} - 2pH$$

with $\sum\log_{10} K = -17.62$ (sum of the $\log_{10}$ of the Henry constant for CO_2 and the $\log_{10}$ of the dissociation constants of aqueous CO_2), $\log \gamma_{H^+} = -0.11$ and $\log_{10} K_w$

$= -(13.78 \pm 0.01)$. According to these relationships the dependence of the solubility on pH is, in each case, the same. Hence from solubility experiments alone it is not possible to deduce a chemical model; that is to say to choose one species or another. But as pH of the saturated solutions increases, the UV–Vis absorption spectra of UO_2^{2+} (340 to 520 nm) disappears and a new spectrum appears. This could be identified as arising from the $(UO_2)_2(OH)_2^{2+}$ species in the solutions. The single component spectrum of the dimer is then derived with $\lambda_{max} = 420$ nm and a value of the extinction coefficient 20 times higher than that of UO_2^{2+}. From the peak areas of the spectra of the two species the ratio:

$$R = \frac{[(UO_2)_2(OH)_2^{2+}]}{[UO_2^{2+}]}$$

is calculated and shown to be independent of pH as expected from the relationships:

$$\log_{10} R = \log_{10} K_{s,0}(UO_3 \cdot 2H_2O) + \log_{10} {}^*\beta_{2,2} - 2\log_{10} K_W$$

or

$$\log_{10} R = \log K_{s,0}(UO_2CO_3) + \log_{10} {}^*\beta_{2,2} + 2\log \gamma_{H^+} - \sum \log_{10} K - \log_{10} p_{CO_2}$$

depending on the solid phase controlling the solubility. $R = (4.9 \pm 0.9)$ for pH = 3.7 to 4.2 under 1, 0.3 and 0.03% of CO_2, and $R = (1.9 \pm 0.4)$ for pH = 3.2 to 4.4 under 8% of CO_2 (For 100% CO_2 only the dioxouranium(VI) cation is present). All these data provide proof of the existence of $(UO_2)_2(OH)_2^{2+}$.

Furthermore, the value $\log_{10} {}^*\beta_{2,2} = -(5.89 \pm 0.12)$ is evaluated from the ratio of the values of the free concentration of the two species (to be compared with the value -5.94 given by [63RUS/JOH] in 1 M $NaClO_4$). No attempt to calculate the solubility products of the solids is made in this paper. Conversely, previous $\log_{10} K_{s,0}$ values [93MEI/KIM] of rutherfordine and schoepite are used to check the R values with a good accuracy.

In the same way, fluorescence characteristics of the saturated solutions change drastically with increasing pH and deconvolution of the spectra on the basis of the emission spectrum of the well known UO_2^{2+} ion at the same concentration as that used to calculate ${}^*\beta_{2,2}$ gives the single emission spectrum of $(UO_2)_2(OH)_2^{2+}$, as well as the fluorescence lifetime (UO_2^{2+} : 473, 488, 510, 534, 560 and 587 nm, $\tau = 1 \mu s$; $(UO_2)_2(OH)_2^{2+}$: 499, 519, 542, 566 nm, $\tau = (2.9 \pm 0.9) \mu s$). A short explanation of the origin of the emission spectra is given.

All these data obtained from more than 30 test solutions give confidence in the identification of the dimer.

- Determination of $\log_{10} {}^*\beta_{2,2}$ from undersaturated solutions.

According to the thermodynamic data known in 1993, and listed in [92GRE/FUG], the authors defined the conditions where U(VI) can be found in solutions containing the species UO_2^{2+} and 2:2 (less than 5% of 1:1, UO_2OH^+ and 5:3, $(UO_2)_3(OH)_5^+$). The conditions are: pH and uranium concentration, respectively, less than 4 and $5 \cdot 10^{-3}$ M. They investigated by absorption spectroscopy 34 test solutions with pH from 2.07 to 3.85 at different U concentrations ($5 \cdot 10^{-4}$ to 10^{-2} M) and reported $\log_{10} {}^*\beta_{2,2} = -(5.97 \pm 0.06)$. The reviewer accepts the $\log_{10} {}^*\beta_{2,2}$ values reported here as being new and independent.

[93MEI/KIM]

This is a solubility study using $UO_2CO_3(s)$ made in 0.1 M $NaClO_4$ at 25°C and p_{CO_2} = 1 bar. The solid is well characterised and the authors have assumed that equilibrium was attained in the test solutions. The chemical model is consistent with the one described in [92GRE/FUG]. The authors have measured pH instead of $-\log_{10}[H^+]$, but they do not describe how the conversion was made. They give a reference but this was not available to the reviewers. We have therefore compared the dissociation constant for the reaction: $CO_2(g) + H_2O(l) \rightleftharpoons CO_3^{2-} + 2H^+$, with the value calculated from the data at zero ionic strength (largely based on CODATA) in [92GRE/FUG]. This indicates that the authors have made a correction of pH to $-\log_{10}[H^+]$. There is, however, a small difference between the two values, $\log_{10}K = -(17.62 \pm 0.07)$, vs. $\log_{10}K = -(17.52 \pm 0.02)$. This indicates that the calculated values of $\log_{10}[CO_3^{2-}]$ are too low, which in turn results in a small systematic error in the equilibrium constants. If this suggestion is correct it will result in $\log_{10}\beta_n$ values that are 0.1, 0.2 and 0.3 units lower than those reported by the authors, explaining some of the (small) differences between the equilibrium constants found by Meinrath and Kimura in this paper and those reported in [92GRE/FUG]. We have recalculated the equilibrium constants in [93MEI/KIM] to zero ionic strength and find $\log_{10}\beta_1^\circ = (10.07 \pm 0.08)$, $\log_{10}\beta_2^\circ = (16.22 \pm 0.34)$, $\log_{10}\beta_3^\circ = (21.84 \pm 0.10)$, and $\log_{10}K_{s,0}^\circ = -(15.02 \pm 0.06)$, where the uncertainty estimates are given as 1.96σ. The value of $\log_{10}\beta_1^\circ$ is higher, and that of $\log_{10}\beta_2^\circ$ lower, than those selected in [92GRE/FUG]. The solubility product of $UO_2CO_3(s)$ reported in [93MEI/KIM] is significantly lower than that in [92GRE/FUG].

[93MEI/KIM2]

This paper supplements/complements [93MEI/KAT].

This paper provides: i) the identification of $UO_2CO_3(cr)$ and $UO_3 \cdot 2H_2O(s)$, and their solubilities in aqueous solutions, at pH = 2 to 5, under variable pressures of CO_2 (air atmosphere to 100% CO_2) and ii) values of their solubility products on the basis of solubility measurements, qualitative spectroscopic measurements and previous thermodynamic data on U(VI) hydrolysis. This paper couples hydrolysis and solubility product data with those reported in [92GRE/FUG].

Titration of $2 \cdot 10^{-3}$ M U(VI) solutions with a $5 \cdot 10^{-2}$ M Na_2CO_3 solution under air, 0.98% and 100% CO_2, ($I = 0.1$ M $NaClO_4$, $t = (24 \pm 2)°C$) leads to precipitation of solid phases in the pH range 2.8 to 4.6 and to solutions in which the uranium concentration is measured by classical absorption spectroscopy. pH values are carefully measured. However, throughout this paper, the authors do not distinguish between pH and $- \log_{10}[H^+]$. The saturated solutions were aged for 3 weeks. Additional solubility data are obtained both from undersaturation and supersaturation, the pH being adjusted with 10^{-1} M $HClO_4$, 10^{-1} M NaOH or $5 \cdot 10^{-2}$ M Na_2CO_3 solutions (equilibrium time three days for $UO_2CO_3(cr)$ and 15 days for $UO_3 \cdot 2H_2O(s)$). Phases separation is done by ultrafiltration, 0.45 or 0.2 μm and no colloids are detected down to 1.3 nm.

- Solid phases and solutions in equilibrium.

Well crystallised rutherfordine, $UO_2CO_3(cr)$, and less crystallised schoepite (in fact metaschoepite), $UO_3 \cdot 2H_2O(s)$, are shown by X–ray diffraction, DTA/DGA, FTIR and PAS to be the solid phases formed under 100, 0.98 and 0.03% CO_2 atmospheres, respectively. The transformation from schoepite to rutherfordine is estimated to take place at a partial pressure of p_{CO_2} around 0.03 atm in the pH range covered. The dioxouranium(VI) ion is shown by UV–Vis spectroscopy to be present in solutions of pH = 2 to 4.4, but is precipitated as rutherfordine in 100% CO_2. A change in the spectra occurs between pH = 3.5 to 4.6 in the presence of schoepite (0.98 and 0.03% CO_2): a red shift of the UO_2^{2+} absorption maxima up to 420 nm and a four fold increase in absorption compared to dioxouranium(VI) cation. These changes are due to the hydrolysis of U(VI). The absorption ratios of some acidified solutions, in which dioxouranium(VI) ions are formed, remain unchanged between pH = 3.5 to 4.6.

The data are consistent with the equilibria:

$$UO_3 \cdot 2H_2O(scho) \rightleftharpoons UO_2^{2+} + 2OH^- + H_2O(l) \quad (0.98 \text{ and } 0.03\% \ CO_2, \text{ pH} = 2 \text{ to } 4.4)$$

$$UO_2CO_3(ruth) \rightleftharpoons UO_2^{2+} + CO_3^{2-} \quad (100\% \ CO_2, \text{ pH} = 3.5 \text{ to } 4.6)$$

i.e.:

$$\log_{10}[UO_2^{2+}] = \log_{10} K_{s,0}(UO_3 \cdot 2H_2O) - 2\log_{10} K_w - 2\text{pH}$$

$$\log_{10}[UO_2^{2+}] = \log_{10} K_{s,0}(UO_2CO_3) - \sum \log_{10} K' - \log_{10} p_{CO_2} - 2\text{pH}$$

from which solubility products can be deduced using the following relationships:

$$\log_{10}[OH^-] = \log_{10} K_w + \text{pH}$$

$$\log_{10}[CO_3^{2-}] = \sum \log_{10} K' + \log_{10} p_{CO_2} + 2\text{pH}$$

with $\log_{10} K_w = -(13.78 \pm 0.01)$ and $\sum \log_{10} K' = -(17.62 \pm 0.07)$ is the sum of the values of $\log_{10} K_H$, $\log_{10}{}^* K$, $\log_{10} K_1$ and $\log_{10} K_2$, the constants being associated with the following equilibria:

$$CO_2(g) \rightleftharpoons CO_2(aq) \qquad K_H$$

$$CO_2(aq) + H_2O(l) \rightleftharpoons H_2CO_3(aq) \qquad {}^*K$$

$$H_2CO_3(aq) \rightleftharpoons HCO_3^- + H^+ \qquad K_1$$

$$HCO_3^- + H^+ \rightleftharpoons CO_3^{2-} + 2H^+ \qquad K_2$$

The dependence of $\log_{10}[UO_2^{2+}]$ versus pH is a straight line with slope -2, and vs. $\log_{10}[CO_3^{2-}]$ is a straight line with slope -1 in the appropriate pH range at constant p_{CO_2} if UO_2^{2+} (or a species bearing a 2+ charge) is predominant in the solutions. That is the case for solutions under 100% CO_2, which contain only the aqueous dioxouranium(VI) ion. In solutions under 0.98 and 0.03% CO_2, taking into account the spectral change, the pH dependence is a thermodynamic indication of the presence of $(UO_2)_2(OH)_2^{2+}$ (see below). This point is shown more clearly in [93MEI/KAT].

The value of $K_{s,0}(UO_2CO_3)$ can be determined in a unique way from data obtained under 100% CO_2 pressure as $\log_{10} K_{s,0}(UO_2CO_3) = -(13.89 \pm 0.11)$. This value is comparable with previous ones quoted in [92GRE/FUG].

To calculate the value of $K_{s,0}(UO_3 \cdot 2H_2O)$ one has to take into account the presence of the hydrolysed U(VI) species identified by their spectra. On the basis of U(VI) hydrolysis data reported in [92GRE/FUG], the authors select the presence of $(UO_2)_2(OH)_2^{2+}$, in the range of pH studied, where the concentration is given by:

$$\log_{10}[(UO_2)_2(OH)_2^{2+}] = \log_{10} \beta_{2,2} + 2\log_{10}[UO_2^{2+}] + 2\log_{10}[OH^-]$$
$$= \log_{10} \beta_{2,2} + 2\log_{10} K_{s,0}(UO_3 \cdot 2H_2O) - 2\log_{10} K_w - 2pH.$$

Furthermore the authors determine:

$$R = \log_{10} \frac{[(UO_2)_2(OH)_2^{2+}]}{[UO_2^{2+}]},$$

$$R = \log_{10} K_{s,0}(UO_3 \cdot 2H_2O) + \log_{10} \beta_{2,2}.$$

To calculate $K_{s,0}(UO_3 \cdot 2H_2O)$, one needs at least a $\beta_{2,2}$ value in addition to a value for R which is not pH dependent in the presence of the solid phase. The authors take for $\log_{10} \beta_{2,2}$ an average value of the published $\log_{10} \beta_{2,2}$ values up to 1993, $\log_{10} {}^*\beta_{2,2} = -(5.97 \pm 0.16)$ ($I = 0.1$ M, $t = 25°C$) which corresponds to $\log_{10} \beta_{2,2} = (21.59 \pm 0.18)$, according to the K_w value used. The value of R is determined from the spectroscopic measurements (see above) to be $-(0.69 \pm 0.37)$. The calculated solubility product is $\log_{10} K_{s,0}(UO_3 \cdot 2H_2O, s) = -(22.28 \pm 0.19)$.

The only unclear point in the study is the conversion of measured pH values to concentrations. The experimental information indicates that the electrodes are calibrated against standard buffers. The authors then calculate the hydroxide concentration from the ionic product of water in 0.1 M $NaClO_4$, using the expression:

$\log_{10}[OH^-] = -13.78 + pH$, indicating that they have mixed concentrations and activities. The carbonate concentration is then calculated from the measured dissociation constant of carbonic acid in 0.1 M NaClO$_4$, again using pH instead of $-\log_{10}[H^+]$. In this way, the two errors cancel, and the solubility product reported for UO$_2$CO$_3$(s), $\log_{10}K_{s,0} = -(13.89 \pm 0.11)$ is therefore accepted by the reviewers. Conversion to $I = 0$ with the SIT gives $\log_{10} K_{s,0}^\circ(UO_2CO_3, s) = -(14.73 \pm 0.11)$. The solubility data for schoepite has to be corrected for the presence of the dimer, $(UO_2)_2(OH)_2^{2+}$. This was done using averaged literature data for the concentration equilibrium constant that are in excellent agreement with the value in [92GRE/FUG]. In this way, cf. Equation (12) on p. 83 in [93MEI/KIM2], the error in using pH instead of $-\log_{10}[H^+]$ also cancels out in this case and the equilibrium constant reported by the authors is accepted by this review, $\log_{10} K_{s,0}(UO_2 \cdot 2H_2O, s) = -(22.28 \pm 0.19)$ in 0.1 M NaClO$_4$ and, using the SIT $\log_{10} K_{s,0}^\circ(UO_2 \cdot 2H_2O, s) = -(22.88 \pm 0.19)$.

- Thermodynamic evaluation.

From the obtained solubility products the authors calculated $\Delta_r G_m^\circ$ for the reactions:

$$UO_2^{2+} + CO_3^{2-} \rightleftharpoons UO_2CO_3(s) \tag{A.37}$$

$$UO_2^{2+} + 2OH^- + H_2O(l) \rightleftharpoons UO_3 \cdot 2H_2O(s), \tag{A.38}$$

$$\Delta_r G_m^\circ (A.37) = (79.28 \pm 0.62) \text{ kJ} \cdot \text{mol}^{-1}$$

$$\Delta_r G_m^\circ (A.38) = (127.17 \pm 1.10) \text{ kJ} \cdot \text{mol}^{-1}.$$

According to the following auxiliary data:

$$\Delta_f G_m^\circ (UO_2^{2+}) = -(952 \pm 2) \text{ kJ} \cdot \text{mol}^{-1},$$
$$\Delta_f G_m^\circ (CO_3^{2-}) = -(530.94 \pm 0.58) \text{ kJ} \cdot \text{mol}^{-1},$$
$$\Delta_f G_m^\circ (H_2O, l) = -(237.140 \pm 0.040) \text{ kJ} \cdot \text{mol}^{-1},$$
$$\Delta_f G_m^\circ (OH^-) = -(158.49 \pm 0.10) \text{ kJ} \cdot \text{mol}^{-1},$$

they calculate:

$$\Delta_f G_m^\circ (UO_2CO_3, s) = -(1562.2 \pm 3.3) \text{ kJ} \cdot \text{mol}^{-1},$$
$$\Delta_f G_m^\circ (UO_3 \cdot 2H_2O, s) = -(1633.3 \pm 3.5) \text{ kJ} \cdot \text{mol}^{-1}.$$

We have made a recalculation using the selected data in [92GRE/FUG] and find $\Delta_f G_m^\circ (UO_2CO_3, s) = -(1564.5 \pm 1.9) \text{ kJ} \cdot \text{mol}^{-1}$, consistent with the value $-(1563.05 \pm 1.8) \text{ kJ} \cdot \text{mol}^{-1}$ selected in [92GRE/FUG]. In the same way we have recalculated the data for schoepite and find $\Delta_f G_m^\circ (UO_3 \cdot 2H_2O, s) = -(1634.7 \pm 2.1) \text{ kJ} \cdot \text{mol}^{-1}$, consistent with the value, $-(1636.5 \pm 1.7) \text{ kJ} \cdot \text{mol}^{-1}$, selected in [92GRE/FUG].

[93MIZ/PAR]

The authors give a summary of what is known about the reduction of U(VI) in bicarbonate and carbonate solutions. In the former case the U(V) produced disproportionates to U(VI) and U(IV), and in the later case U(V) remains stable. Using classical cyclic voltammetry and spectrophotometry through a grid Pt cathode (path length (0.78 ± 0.02) mm, wavelength: 230 to 800 nm) they show that in solutions 10^{-2} M U(VI), 1 M Na_2CO_3 (where $UO_2(CO_3)_3^{4-}$ is present) the reaction:

$$UO_2(CO_3)_3^{4-} + e^- \rightleftharpoons U(V)$$

holds quasi-reversibly at $E° = -(0.85 \pm 0.01)$ V (vs. Ag/AgCl, 3 M NaCl) with the exchange of $n = (0.99 \pm 0.05)$ electrons. Absorption spectroscopic observations are consistent with the presence of only U(V) and allow calculation of $\log_{10}[U(VI)]/[U(V)]$ vs. the applied potential, E (five experimental points), which gives the n and $E°$ values.

In acidic solution 10^{-2} M U(VI), 10^{-1} M $HClO_4$ or $LiClO_4$ (pH < 1.5, probably) they confirm produced by cyclic voltammetry about the quasi–reversible reduction of UO_2^{2+} to UO_2^+ with $E° = -0.138$ V (vs. Ag/AgCl, 3 M NaCl).

The quasi–reversibility is due, as reported previously, to further reduction of U(V) to U(IV) through a mechanism which cannot be validated by the spectroelectrochemistry data.

The reviewer retains the data in carbonate solutions.

[93NAI/VEN]

As part of a study of the vaporisation behaviour in the U–Ce–C system, Naik et al. have measured, by Knudsen effusion mass–spectrometry, the uranium pressures over monophasic UC(cr) (1922 to 2247 K) and the diphasic U(l) + UC(cr) region (1571 to 2317 K). The samples were contained in a carburised tantalum cup inside a molybdenum effusion cell, and silver was added as a pressure calibrant. The carbide samples, prepared from UO_2 + C microspheres, clearly initially contained some residual oxygen, since the intensity of the UO^+ signal was initially larger than that of all other species, but "became less later".

The pressure of U over $UC_{0.981 \pm 0.033}$ at 2100 K agreed with those from earlier studies, within the combined uncertainties of composition and pressure, as did that over the U(l) + UC region, which is, of course, somewhat lower than that over pure U(l) due to carbon dissolution.

No new information on the thermodynamics of the U–C system can be derived from this study.

[93NIK/TSI]

Mass–spectrometric data are presented from which (with a number of corrections/assumptions for fragmentation, dissolution, *etc.*) the equilibrium constants of the reaction:

$$\tfrac{1}{2} \text{NiF}_2(\text{cr}) + \text{UF}_4(\text{g}) \rightleftharpoons \tfrac{1}{2} \text{Ni}(\text{cr}) + \text{UF}_5(\text{g})$$

are calculated over the temperature range 1017 – 1109 K. Their derived value of $\Delta_f H_m$ (UF$_5$, g, 0 K) $= - (1914.8 \pm 16.0)$ kJ · mol^{-1} is compared comprehensively with ten other values in the literature, which range from $-$ 1806 to $-$ 1949 kJ · mol^{-1}. Their recommended value is $\Delta_f H_m$ (UF$_5$, g, 0 K) $= - (1913.9 \pm 15.2)$ kJ · mol^{-1}. (The corresponding value from [92GRE/FUG] would be $\Delta_f H_m$ (UF$_5$, g, 0 K) $= - (1905.2 \pm 15.0)$ kJ · mol^{-1}). These data have been incorporated into the reassessment of the properties of UF$_5$(g) – see section 9.4.2.1.3.

[93ODA]

This report contains calculations of the populations of the rotational and vibrational states and band contours of UF$_6$(g) around 300 K, based on the molecular parameters given by [85ALD/BRO], except for a slightly larger interatomic distance, from an earlier interpretation of the spectra. The data selected by [92GRE/FUG] are based on the values of [85ALD/BRO], (although the actual values used are not stated there) and there is no reason to make any changes.

[93OGA]

This is an interesting paper on modelling of substoichiometric (U, Pu) nitrides. However, since the present review does not deal with such ternary or non–stoichiometric phases, this paper is of no immediate relevance.

[93OGO/ROG]

The authors attempt to establish regularities in the relation between six thermodynamic parameters at 298.15 K (volume coefficient of thermal expansion, the bulk modulus, molar volume, atomisation energy, $C_{p,m}$ and Gruneisen parameter) for the mononitrides of 5d transition elements plus Th, U and Pu. From six universal relations between these parameters, they predict missing values, including C_v, the specific heat capacity. The predicted values of C_v are: C_v(UN, cr, 298.15 K) = 48.97 J ·K^{-1} · mol^{-1} and C_v(PuN, cr, 298.15 K) = 50.05 J· K^{-1} · mol^{-1}, compared to the currently assessed values for $C_{p,m}$ of (47.57 ± 0.40) and (49.6 ± 1.00) J · K^{-1} · mol^{-1}, respectively. Their "regularised" value of C_v for UN seems reasonable, but that for PuN is probably too small (since $C_{p,m} > C_v$). No new thermodynamic data of relevance to the review are presented.

[93PAS/RUN]

This paper is reviewed together with [97PAS/CZE].

[93PAT/DUE]

This paper yields thermochemical information on $\alpha-UO_{2.95}$ and on its insertion with Mg. $\alpha-UO_{2.95}$ was prepared by heating $UO_4 \cdot 2H_2O$ to 792 K for eight hours, following Cordfunke and Aling [63COR/ALI]. Reaction with dibutylmagnesium at ambient temperatures for 48 hours achieved the insertion to form $Mg_xUO_{2.95} (0 < x < 0.26)$. The hexagonal $\alpha-UO_{2.95}$ was characterised by powder X–ray diffraction and the O/U ratio was accurately determined by potentiometric titration (dissolution in potassium dichromate solution and back titration using a Fe^{2+} standard solution).

$Mg_{0.17}UO_{2.95}$ was shown to be a pure phase by X–ray diffraction, with a hexagonal lattice, a = 3.964 Å, c = 4.172 Å, similar to the values for $\alpha-UO_{2.95}$, (a = 3.985 Å, c = 4.163 Å).

The determination of the enthalpy of formation of $\alpha-UO_{2.95}$ was based on the enthalpy of solution of the constituents of the reaction

$$0.05 \ \alpha-U_3O_8 + 0.85 \ \gamma-UO_3 \rightleftharpoons \alpha-UO_{2.95} \qquad (A.39)$$

The calorimetric reagent was 0.274 mol·dm^{-3} $Ce(SO_4)_2$ + 0.484 mol·dm^{-3} H_3BO_3 + 0.93 mol·dm^{-3} H_2SO_4, the same as that used by [81COR/OUW2] for the dissolution of $\gamma-UO_3$ and $\alpha-U_3O_8$. In fact, the values for the enthalpy of dissolution of these two latter reagents were taken from these authors as $-(79.94 \pm 0.48)$ kJ·mol^{-1} and $-(354.61 \pm 1.70)$ kJ·mol^{-1}, respectively. $\Delta_{sol}H_m(UO_{2.95}, \alpha)$ was determined as $-(93.37 \pm 0.65)$ kJ·mol^{-1}, the reported uncertainty limits being for the 1.96σ interval. These values lead to $\Delta_r H_m^\circ ((A.39), 298.15 \text{ K}) = (7.69 \pm 0.77)$ kJ·mol^{-1}.

Using NEA accepted values for the standard enthalpy of formation of $\gamma-UO_3$ and $\alpha-U_3O_8$, we recalculate $\Delta_f H_m^\circ (UO_{2.95}, \alpha, 298.15 \text{ K}) = -(1211.28 \pm 1.28)$ kJ·mol^{-1}, identical, except for the uncertainty limits, to that reported by [93PAT/DUE]. This value, accepted here, is consistent with $\Delta_f H_m^\circ (UO_3, \alpha, 298.15 \text{ K}) = -(1212.41 \pm 1.45)$ kJ·mol^{-1} (see section 9.3.3.1).

The same reagent was used for the determination of the enthalpy of formation of $Mg_{0.17}UO_{2.95}$. The values for the dissolution of $Mg_{0.17}UO_{2.95}$ and MgO were determined as $-(143.99 \pm 0.60)$ and $-(149.50 \pm 0.55)$ kJ·mol^{-1}, respectively. The enthalpy change corresponding to the reaction:

$$0.17 \ MgO(cr) + 0.34 \ \gamma-UO_3 + 0.22 \ \alpha-U_3O_8 \rightleftharpoons Mg_{0.17}UO_{2.95}(cr) \qquad (A.40)$$

was obtained as $\Delta_r H_m^\circ ((A.40), 298.15 \text{ K}) = (13.38 \pm 0.73)$ kJ·mol^{-1}. This value corresponds to the formation of the insertion compound from the stoichiometric oxides, with respect to which it is therefore presumably unstable.

Using NEA adopted values, we recalculate, from the equation above:

$\Delta_f H_m^\circ$ (Mg$_{0.17}$UO$_{2.95}$, cr, 298.15 K) = $-$ (1291.44 $\pm$ 0.99) kJ $\cdot$ mol^{-1}.

With $\Delta_f H_m^\circ$ (UO$_{2.95}$, α, 298.15 K) = $-$ (1211.28 $\pm$ 1.28) kJ $\cdot$ mol^{-1} discussed above, the enthalpic effect associated with the insertion,

$$0.17\,\text{Mg} + \alpha\text{–UO}_{2.95} \rightleftharpoons \text{Mg}_{0.17}\text{UO}_{2.95}\,(\text{cr}) \qquad (A.41)$$

is obtained as

$\Delta_r H_m$ ((A.41), 298.15 K) = $-$ (80.16 $\pm$ 1.62) kJ $\cdot$ mol^{-1}.

Electrochemical measurements (discharge and charge curves, open circuit voltages) using the cell:

$$\text{Mg(amalgam)}\,\big|\,0.5\,\text{mol}\cdot\text{dm}^{-3}\,\text{Mg(ClO}_4)_2\,,\,\text{DMF}\,\big|\,\text{Mg}_x\text{UO}_{2.95}$$

allowed the calculation of the Gibbs energy change for reaction:

$$x\,\text{Mg} + \alpha\text{–UO}_{2.95} \rightleftharpoons \text{Mg}_x\text{UO}_{2.95}(\text{cr}) \qquad (A.42)$$

as $\Delta_r G_m$ ((A.42), 294 K) = $-$ 473$\cdot x$ kJ $\cdot$ mol^{-1}, with x up to 0.27. For x = 0.17, this expression yields $-$ 80.4 kJ $\cdot$ mol^{-1}, very close to the calorimetric enthalpy of reaction, indicating a negligible entropy change.

[93STO/CHO]

UO_2^{2+} and NpO_2^+ are known to give cation–cation species with β values increasing with ionic strength: (0.69 $\pm$ 0.01) L $\cdot$ mol^{-1} at I = 3.0 M, (2.5 $\pm$ 0.5) L $\cdot$ mol^{-1} at I = 6.26 M and (3.7 $\pm$ 0.1) L $\cdot$ mol^{-1} at I = 7 M. This paper gives β = (2.25 $\pm$ 0.03) L $\cdot$ mol^{-1} at I = 6.0 M as well as $\Delta_r H_m$ = $-$ (12.0 $\pm$ 1.7) kJ $\cdot$ mol^{-1} and $\Delta_r S_m$ = $-$ (34 $\pm$ 6) J $\cdot$ K^{-1} $\cdot$ mol^{-1} at (25 $\pm$ 0.1) °C for the association reaction.

All these data are obtained by spectrophotometric determinations in the system: NaClO$_4$, pH = 1 to 2, U concentration: 6.1$\cdot$10^{-2} to 0.489 M, Np concentration held constant at 10^{-3} M, t = (25 $\pm$ 0.1) to (65 $\pm$ 0.1) °C. Free NpO_2^+ concentration is measured at 978.4 nm, ε_{max} = 368 L $\cdot$ mol^{-1} $\cdot$ cm^{-1}. Purity of the Np and U solutions before mixing are checked by spectrophotometry, U at 416 nm, ε = 7.82 L $\cdot$ mol^{-1} $\cdot$ cm^{-1} (HClO$_4$ > 10^{-2} M), Np(V) at 980.4 nm, ε = 395 L $\cdot$ mol^{-1} $\cdot$ cm^{-1}, Np(IV) at 960 nm and Np(VI) at 1222 nm (HClO$_4$ 0.2 M).

[93TAK/FUJ]

The first part of this paper is concerned with the change of the crystallographic properties (lattice parameters, atomic positions, oxygen vacancies) of substoichiometric calcium and strontium uranates (rhombohedral, space group $R\overline{3}m$) as a function of the ionic radii of the alkaline earth cation and the oxygen non-stoichiometry.

The second part is devoted to the determination of the molar enthalpy of formation of $SrUO_{4-x}$ (α, 298.15 K) for $x = 0$ to 0.478.

The procedures for the preparation and characterisation of the samples are detailed. Orthorhombic β–$SrUO_4$ was the starting material for rhombohedral α–$SrUO_4$ and α–$SrUO_{3.5}$. It was obtained by heating a stoichiometric mixture of high purity $SrCO_3$(cr) and UO_2(cr) in air at 1273 K for 48 hours. Its lattice parameters were in good agreement with literature values. α–$SrUO_{3.5}$ was obtained by reducing β–$SrUO_4$ in a stream of H_2 at 1073 K for six hours. α–$SrUO_4$ was obtained by heating α–$SrUO_{3.5}$ in air for six hours, at 773 K; this temperature being low enough to avoid the transition to β–$SrUO_4$. The intermediate oxides α–$SrUO_{4-x}$, ($0 < x < 0.5$) were obtained by heating the calculated amounts of α–$SrUO_4$ and α–$SrUO_{3.5}$ in vacuum sealed quartz ampoules at 1273 K for 12 hours. The oxygen non-stoichiometry in the compounds was determined by Ce(IV) back titration against Fe^{2+}; the error in x is estimated to be ± 0.003. X–ray diffraction analysis provided information on the oxygen positional parameters as a function of x.

The enthalpies of solution of the compounds in 5.94 mol·dm^{-3} HCl, recalculated for the 95 % confidence interval, are given in Table A-13. The enthalpies of formation, also given in Table A-13, were recalculated using the cycle given by the authors, with auxiliary enthalpies of formation either adopted by NEA:

$\Delta_f H_m^\circ$ ($SrCl_2$, cr, 298.15 K) = $- (833.85 \pm 0.70)$ kJ·mol^{-1},

$\Delta_f H_m^\circ$ (UO_2Cl_2, cr, 298.15 K) = $- (1243.6 \pm 1.3)$ kJ·mol^{-1},

$\Delta_f H_m^\circ$ (UCl_4, cr, 298.15 K) = $- (1018.8 \pm 2.5)$ kJ·mol^{-1},

or compatible with the NEA values:

$\Delta_{sol} H_m$ ($SrCl_2$, cr, in 5.94 mol·dm^{-3} HCl) = $- (38.1 \pm 1.1)$ kJ·mol^{-1}

(calculated from the enthalpy of solution of $SrCl_2$ in 1 mol·dm^{-3} HCl for the 95% confidence interval by [83MOR/WIL] and the enthalpy of transfer from 1 to 5.94 mol·dm^{-3} HCl by [78PER/THO][1]),

$\Delta_{sol} H_m$ (UO_2Cl_2, cr, 5.94 mol·dm^{-3} HCl) = $- (64.2 \pm 0.5)$ kJ · mol^{-1}
(interpolated from [83FUG/PAR], the uncertainty limits being ours),

$\Delta_{sol} H_m$ (UCl_4, cr, 5.94 mol·dm^{-3} HCl) = $- (64.2 \pm 0.5)$ kJ · mol^{-1}
(*idem*),

$\Delta_f H_m$ (H_2O, 5.94 mol·dm^{-3} HCl, partial) = $- (286.646 \pm 0.150)$ kJ · mol^{-1}
(extrapolated from the value for 6.00 mol·dm^{-3} HCl [2000RAN/FUG])

$\Delta_f H_m$ (HCl, 5.94 mol·dm^{-3} HCl, partial) = $- (153.400 \pm 0.150)$ kJ · mol^{-1}
(*idem*).

---

[1] The confidence limits of the alkaline earth chlorides dilution values by [78PER/THO] could not be ascertained, as these authors do not report individual results.

Table A-13: Enthalpies of solution and formation of $SrUO_{4-x}$.

x in $SrUO_{4-x}$	$\Delta_{sol}H_m$, kJ·mol^{-1}	$\Delta_f H_m$ kJ·mol^{-1}
0.478	$-(170.55 \pm 1.89)$	$-(1894.31 \pm 2.83)$
0.380	$-(152.71 \pm 0.60)$	$-(1926.14 \pm 2.09)$
0.297	$-(149.24 \pm 0.99)$	$-(1941.47 \pm 2.19)$
0.127	$-(143.40 \pm 7.90)$	$-(1971.61 \pm 8.11)$
0	$-(142.86 \pm 2.45)$	$-(1990.28 \pm 3.22)$

The value for the enthalpy of formation of the stoichiometric α–$SrUO_4$ is in excellent agreement with that selected in [92GRE/FUG], $-(1989.6 \pm 2.8)$ kJ · mol^{-1}, based on the results of [67COR/LOO]. If the five data points above are fitted (as did the authors) to a quadratic function of x, the best fit is given by:

$$\Delta_f H_m (SrUO_{4-x}, α, 298.15 \text{ K}) (\text{kJ} \cdot \text{mol}^{-1}) = -1989.3 + 98.1 \cdot x + 202.8 \cdot x^2,$$

a relation from which the partial molar enthalpy of solution of oxygen can be approximated as a function of x.

[94AHO/ERV]

Dissolved U(VI) and U(IV) concentrations, E_H and pH values, as well as chemical compositions are available for filtered undergroundwater (< 0.45 μm) taken from several boreholes drilled in the Palmottu site. Seven samples are well characterised. Modelling on the basis of thermodynamical data selected by [92GRE/FUG] and from the CHEMVAL database (1992) is compared to these experimental data. Species involved in the pH range 6.87 to 9.05 are mono-, di-, and tri-dioxouranium(VI)carbonates and $U(OH)_4$(aq) for U(IV). In this pH range both databases give the same predominant species. The equilibria, which account for experimental data, are those between $U(OH)_4$(aq) and the di- and tri-carbonate U(VI) complexes. Indeed experimental and calculated U(VI)/U(IV) ratios using [92GRE/FUG] data are in good agreement, as well as the slope, E_H /pH, deduced from these experimental data and the one calculated from the equilibria.

[94COR/IJD]

$Sr_2UO_{4.5}$(s) was obtained from Sr_2UO_5(s) by reduction in an Ar/H_2 gas mixture at temperatures from 1200 to 1300 K. The sample characterisation is described in detail; these are given also in the paper by [99COR/BOO], in both cases with similar information for other strontium uranates. The compound is indexed as pseudo–orthorhombic, isostructural with the high temperature form of $Ca_2UO_{4.5}$(s) and lattice parameters nearly equal to those of Sr_3UO_6(s). The enthalpy of dissolution of the compound in (HCl + 0.0470

FeCl$_3$ + 82.16 H$_2$O) is given as – (390.66 ± 0.51) kJ · mol^{-1}, without any details. As a better documented value is given in [99COR/BOO], using the same sample, we shall not make use of the value mentioned here. In a discussion of the relative stability of the various strontium uranates, the authors state that, unlike BaUO$_3$(s), SrUO$_3$(s) is not stable as a pure phase: stabilisation can only be achieved through oxidation of U^{4+}.

[94FAN/KIM]

This is an important study from a methodological point of view. The authors have used TRFLS, with peak deconvolution to isolate the spectra of Cm^{3+}, CmOH^{2+} and Cm(OH)$_2^+$, and thereby obtaining a direct experimental determination of the concentration ratio between these species. The experiments have been made at 25°C and in test solutions with varying NaCl concentrations. The electrode has been calibrated to measure the pH, and these data were then used with the Pitzer model to calculate the activity coefficients of all reactants and products. The result is a precise speciation model and a quantitative description of the equilibria in the system in the pH range 3 – 10 over a NaCl concentration range from 0.01 to 6.15 mol·kg^{-1}. An important result is that the equilibrium constants at zero ionic strength are very near the same both when using the SIT and Pitzer models. However, the data indicate very clearly that the SIT model is not satisfactory at concentrations over 4.2 m. The results of this study provide support for the use of the SIT model at ionic strengths below 4 m, as stated in Appendix B. An interesting observation is that the second hydrolysis complex, Cm(OH)$_2^+$, is strongly stabilised in comparison with Cm(OH)$^{2+}$ at high chloride concentrations. Both the values of $\log_{10} \beta_1^\circ$ = (6.38 ± 0.09) and $\log_{10} \beta_2^\circ$ = (13.9 ± 0.6) reported by [94FAN/KIM] are smaller than the values given in [95SIL/BID] for the corresponding americium species. In addition, the estimated uncertainty is much smaller. There is no chemical reason for such large differences between these two actinide elements and this review suggests that the values given in [94FAN/KIM] should be accepted both for the americium and curium systems, and considered in the discussion of selected data.

[94GIF]

The author measured the solubility of ^{241}Am, initially added as Am$_2$(CO$_3$)$_3$ · xH$_2$O(s), in 0.1 and 4 M NaCl batch solutions containing known amounts of NaHCO$_3$ and Na$_2$CO$_3$. Most of the experiments were performed at 21°C and the americium concentrations were measured over periods up to 115 days. The procedures and constants used to calculate $\log_{10}$[CO$_3^{2-}$] from $\log_{10}$[H$^+$] measured with a glass electrode are in accord with the NEA TDB [92GRE/FUG]. The results in 4 M NaCl provided evidence for the radiolytic oxidation of Am(III) to Am(V), which resulted in considerably increased americium concentrations. The addition of metallic iron led to the reduction of Am(V) to Am(III) and solubilities similar to those in 0.1 M NaCl.

In addition to the small difference between the Am(III) solubilities measured in

0.1 and 4 M NaCl, UV–Vis–NIR absorption spectra of Am(III) solutions in 0.2 M $HClO_4$ containing 3 M NaCl or 3 M $NaClO_4$ did not show any differences. The author concluded that the formation of Am(III) chloride complexes requires NaCl concentrations above 4 mol · L^{-1}. This is in accord with a spectroscopic (TRLFS) investigation of the chloride complexation of Cm(III) [95FAN/KIM] and the thermodynamic data selected in the present review. (Giffaut's studies on the chloride complexation of neptunium and plutonium in the oxidation states III to VI were already considered and discussed in the corresponding sections of [2001LEM/FUG]. In the present volume, a more complete analysis for the chloride complexation of Pu(VI), encompassing results of this study, the work of [99RUN/REI] and corresponding U(VI) studies [92GRE/FUG], is provided in the Appendix A entry for [99RUN/REI].)

- Am(III) carbonates

The solids formed in the solubility studies were not characterised by X–ray diffraction or other methods. In a preliminary paper [93GIF/VIT], the initially added $Am_2(CO_3)_3 \cdot xH_2O$(s) was assumed to be the equilibrium solid phase and the solubility data were interpreted accordingly. The suggested results were not credited in the previous review [95SIL/BID]. In his thesis, Giffaut re-analysed the Am(III) solubilities and concluded that the formation of $AmOHCO_3$(s) was the solubility-limiting solid. The present review accepts this interpretation. The evaluated solubility constants for the reaction:

$$AmOHCO_3(s) \rightleftharpoons Am^{3+} + OH^- + CO_3^{2-} \quad (A.43)$$

are $\log_{10}K_{s,0} = - (21.0 \pm 0.4)$ in 0.1 M NaCl and $- (20.7 \pm 0.4)$ in 4 M NaCl corresponding to $\log_{10} K^{\circ}_{s,0} = - (22.5 \pm 0.4)$ and $- (23.3 \pm 0.5)$, respectively, if converted to $I = 0$ with the SIT coefficients selected in Appendix B. They are in reasonable agreement with other data for $AmOHCO_3$(cr), [84SIL/NIT], [84BER/KIM], [90FEL/RAI], [96MER/FUG] (*cf.* discussion of data selection, section 12.6.1.1.3.1). The formation constants for the complexes, $AmCO_3^+$, $Am(CO_3)_2^-$ and $Am(CO_3)_3^{3-}$, evaluated from the solubility in 0.1 M NaCl are also included in the selection of thermodynamic data (*cf.* section 12.6.1.1.1). The formation constants calculated from the solubility in 4 M NaCl are outside the range of those from other studies. The deviations are possibly caused by the formation of $NaAm(CO_3)_2$(s), particularly at carbonate concentrations above 10^{-2} mol·L^{-1}.

Giffaut also measured Am(III) solubilities at different temperatures in the range 20–70°C. The carbonate concentration in 4 M NaCl–$NaHCO_3$–Na_2CO_3 batch solutions (partly controlled by $p_{CO_2} = 1$ bar) was varied in the range, $\log_{10}[CO_3^{2-}] = -5$ to 0, and the solubility data were interpreted assuming the formation of either $AmOHCO_3$(s) or $NaAm(CO_3)_2$(s) as equilibrium solid phases. As the solids were not unambiguously characterised or identified, these measurements do not allow the determination of thermodynamic data. However, it is noteworthy, that the observed effect of temperature over the range studied is rather small. The measured americium concentrations vary less than 0.5 $\log_{10}$ units.

- Am(V) carbonates

The solid Am(V) carbonate formed in the 4 M NaCl solutions was not characterised, but the assumption of NaAmO$_2$CO$_3$(s) as the solubility-controlling solid is justified, because the analogous Np(V) carbonate compound is well known to be the stable solid under the conditions in [94GIF] (cf. [2001LEM/FUG]). The experimental solubilities measured by Giffaut in the range, $-5.5 < \log_{10}[\text{CO}_3^{2-}] < -1.5$, in 4 M NaCl compare well with those of Runde and Kim [94RUN/KIM], and Runde et al. [96RUN/NEU] at the same carbonate concentrations in 3 and 5 M NaCl. The reported equilibrium constants $\log_{10}K_{s,n}$ ($n = 1, 2, 3$) for the reaction:

$$\text{NaAmO}_2\text{CO}_3(\text{s}) + (n-1)\,\text{CO}_3^{2-} \rightleftharpoons \text{Na}^+ + \text{AmO}_2(\text{CO}_3)_n^{1-2n} \quad \text{(A.44)}$$

are shown in Table A-14 in comparison with the corresponding data of [94RUN/KIM]. The values of $\log_{10}K_{s,1}$ and $\log_{10}K_{s,3}$ are in good agreement, whereas the larger deviations in the $\log_{10}K_{s,2}$ values indicate that the uncertainties are underestimated in [94GIF], [94RUN/KIM].

The stepwise formation constants of Am(V) carbonate complexes given by:

$$\log_{10}K_n = \log_{10}K_{s,n} - \log_{10}K_{s,(n-1)}$$

are used in section 12.6.1.1.2, together with the corresponding data of [94RUN/KIM] for the selection of thermodynamic data for the Am(V) carbonate complexes.

Because of the lack of data in the range, $\log_{10}[\text{CO}_3^{2-}] < -5.5$, i.e., with the uncomplexed AmO$_2^+$ ion as the predominant solution species, the solubility constant of $\log_{10}K_{s,0}$(4 M NaCl) $= -(10.4 \pm 0.25)$ [94GIF] proposed by Giffaut for the reaction:

$$\text{NaAmO}_2\text{CO}_3(\text{s}) \rightleftharpoons \text{Na}^+ + \text{AmO}_2^+ + \text{CO}_3^{2-} \quad \text{(A.45)}$$

is highly speculative. This becomes evident in Figure IV.4 (p. 144) in [94GIF], where the solubility curve is extrapolated to lower carbonate concentrations with a value of $\log_{10}K_{s,0} = -9.6$ in 4 M NaCl, which is consistent with the solubility constants given in [94RUN/KIM], [96RUN/NEU], ($\log_{10}K_{s,0} = -(9.65 \pm 0.19)$ in 3 M NaCl and $-(9.56 \pm 0.13)$ in 5 M NaCl. There is no reason to prefer the calculations in Figure IV.5 (p. 145) in [94GIF], where the solubility curve is extrapolated to lower carbonate concentrations with the proposed value of $\log_{10}K_{s,0} = -10.4$.

Table A-14: Equilibrium constants for the dissolution of $NaAmO_2CO_3(s)$.

	$NaAmO_2CO_3(s) \rightleftharpoons Na^+ + AmO_2^+ + CO_3^{2-}$	
Medium, $t(°C)$	$\log_{10}K_{s,0}$	Ref.
3.0 M NaCl, 22°C	$-(9.65 \pm 0.19)$	[94RUN/KIM]
4.0 M NaCl, 21°C	$-(10.4 \pm 0.25)$ [a]	[94GIF]
5.0 M NaCl, 22°C	$-(9.56 \pm 0.13)$	[94RUN/KIM]
	$NaAmO_2CO_3(s) \rightleftharpoons Na^+ + AmO_2CO_3^-$	
Medium, $t(°C)$	$\log_{10}K_{s,1}$	Ref.
3.0 M NaCl, 22°C	$-(4.91 \pm 0.21)$	[94RUN/KIM]
4.0 M NaCl, 21°C	$-(5.00 \pm 0.17)$	[94GIF]
5.0 M NaCl, 22°C	$-(4.87 \pm 0.14)$	[94RUN/KIM]
	$NaAmO_2CO_3(s) + CO_3^{2-} \rightleftharpoons Na^+ + AmO_2(CO_3)_2^{3-}$	
Medium, $t(°C)$	$\log_{10}K_{s,2}$	Ref.
3.0 M NaCl, 22°C	$-(2.23 \pm 0.19)$	[94RUN/KIM]
4.0 M NaCl, 21°C	$-(1.4 \pm 0.12)$	[94GIF]
5.0 M NaCl, 22°C	$-(2.02 \pm 0.15)$	[94RUN/KIM]
	$NaAmO_2CO_3(s) + 2 CO_3^{2-} \rightleftharpoons Na^+ + AmO_2(CO_3)_3^{5-}$	
Medium, $t(°C)$	$\log_{10}K_{s,3}$	Ref.
3.0 M NaCl, 22°C	$-(0.11 \pm 0.23)$	[94RUN/KIM]
4.0 M NaCl, 21°C	(0.0 ± 0.25)	[94GIF]
5.0 M NaCl, 22°C	(0.09 ± 0.13)	[94RUN/KIM]

[a] Not accepted for the reasons described in the text.

[94HEN/MER]

This paper describes the partial charge model, which is based on the electronegativity equalisation principle (EEP) to predict the speciation of Al, U, and Tc solutions as a function of pH. For the uranium (IV and VI) systems the models deal only with mononuclear species, but do predict the existence of $U(OH)_5^-$ and $UO_2(OH)_3^-$ (presumably in extremely dilute solutions), respectively, in dilute to strong base. Although there is no mention of further hydrolysis at high pH, the figures do indicate the formation of doubly–charged anions by a pH of 14. The speciation of Tc(IV) is predicted to vary from being dominated by TcO^{2+} at pH = -1 to $TcO(OH)_3^-$ at pH = 14. No thermodynamic information is forthcoming, but the speciation models may be compared to those suggested by Raman and EXAFS results.

[94JAY/IYE]

This paper reports measurements of the pressure of Na(g) (by both mass–loss and mass–spectroscopic Knudsen effusion) and the oxygen activity (emf with CaO–ZrO_2 electrolyte) in the three–phase field $NaUO_3(cr) + Na_2UO_4(cr) + Na_2U_2O_7(cr)$.

$Na_2UO_4(cr)$ and $Na_2U_2O_7(cr)$ were prepared by heating appropriate amounts of $U_3O_8(cr)$ or $UO_3(cr)$ with $Na_2CO_3(s)$ in air at 1400 and 1000 K, respectively, and $NaUO_3(cr)$ by heating the relevant proportions of $Na_2UO_4(cr)$ and $UO_2(cr)$ in purified argon at 1050 K. The products were characterised by X–ray diffraction.

The mass–spectrometric Knudsen effusion measurements were carried out in a graphite cell, further contained in a molybdenum cell, from 1117 to 1290 K, with Ag as a standard. Graphite could possibly have reduced at least the surface U(VI) uranates, although the X–ray diffraction pattern of the sample was unchanged after the measurements. The mass–loss Knudsen effusion measurements (in an unspecified cell, but probably again graphite) extended from 1283 to 1330 K, using a Cahn microbalance.

The oxygen potentials were studied using a CaO–ZrO_2 electrolyte tube in flowing argon, from 923 to 1118 K. Air (p_{O_2} = 0.21 bar) and Ni/NiO were used as the reference electrodes.

The sodium pressures from the mass–loss and mass–spectrometric studies agreed well; the latter showed a distinct change of slope (of the $\log_{10} p$ vs. $1/T$ plot) at around 1195 K, close to the temperature of the α to β transition in $Na_2UO_4(cr)$ (1193 K, [82COR/MUI]). However, the change in slope corresponds to a bigger enthalpy than that expected for a simple phase transformation, and like the authors, we have preferred to use only the data below 1193 K, with an increased uncertainty since these are not absolute pressure measurements. These give:

$$\log_{10}(p_{Na}/\text{bar}) = 12.92 - 22701 \cdot T^{-1} \qquad (1117 \text{ to } 1193 \text{ K})$$

over the three–phase field, $NaUO_3(cr) + Na_2UO_4(cr) + Na_2U_2O_7(cr)$.

The authors do not state the Gibbs energies of formation of NiO(cr) used to convert the study with a Ni–NiO reference electrode to oxygen potentials. We have used the values assessed by Taylor and Dinsdale [90TAY/DIN], which include consideration of the magnetic transitions in Ni and NiO. Their data in the relevant temperature range can be expressed as $\Delta_f G_m$(NiO, cr, T) = $-234220 + 85.042 \cdot T$ (J·mol^{-1}) from 900 to 1200 K. Our derived values for the sodium and oxygen potentials in the relevant three–phase field are then:

$RT \ln(p_{Na} / \text{bar}) = -434608 + 247.352\, T \quad \text{J} \cdot \text{mol}^{-1}$ (1117 to 1197 K)

$RT \ln(p_{O_2} / \text{bar}) = -694307 + 314.770\, T \quad \text{J} \cdot \text{mol}^{-1}$ (968 to 1118 K) (air reference)

$RT \ln(p_{O_2} / \text{bar}) = -744232 + 358.577\, T \quad \text{J} \cdot \text{mol}^{-1}$ (923 to 1023 K) (Ni/NiO reference)

The coefficients of the two expressions for the oxygen potential differ appreciably, but the actual values are reasonably consistent, so the mean of these expressions will be used in the subsequent analysis:

$RT \ln(p_{O_2} / \text{bar}) = -719270 + 336.674\, T \quad \text{J} \cdot \text{mol}^{-1}$ (923 to 1118 K) (mean).

The authors chose to combine the experimental oxygen and sodium potentials to define the Gibbs energy of the reaction:

$$3\, NaUO_3(cr) + Na_2UO_4(cr) + 0.5\, O_2(g) \rightleftharpoons 2\, Na_2U_2O_7(cr) + Na(g) \qquad \text{(A.46)}.$$

In fact, the individual measurements can be used to derive (or check their consistency with existing) thermodynamic data for the three uranates. The oxygen potential will be defined by the reaction:

$$2\, NaUO_3(cr) + 0.5\, O_2(g) \rightleftharpoons Na_2U_2O_7(cr) \qquad \text{(A.47)}$$

independently of the presence or absence of $Na_2UO_4(cr)$, while the sodium potential in the three-phase field will be defined by the reaction:

$$NaUO_3(cr) + Na_2UO_4(cr) \rightleftharpoons Na_2U_2O_7(cr) + Na(g) \qquad \text{(A.48)}.$$

Reaction (A.46) is just reaction (A.47) plus reaction (A.48). (There is an infinity of similar reactions which could be written, involving linear combinations of reactions (A.47) and (A.48). This redundancy arises because reactions such as (A.46) contain more than ($n+1$) species, where is n is the number of different elements, three in this case).

We have therefore compared the experimental values of reaction (A.46), (A.47) and (A.48) with the values calculated from the data in [92GRE/FUG]. For this, the heat capacity data for $Na_2U_2O_7$ have been extended above the upper limit of 540 K in [92GRE/FUG] to include those for the β–polymorph from the data by Cordfunke *et al.* [82COR/MUI]. The unknown entropy of transformation was taken to be zero.

The results are shown in Table A-15, where the experimental uncertainties are twice the authors' estimates and those for the calculated values are estimated here.

Table A-15: Experimental and calculated Gibbs energies of sodium uranate reactions.

T/K	$\Delta_r G_m$ (A.47) (kJ·mol^{-1})		$\Delta_r G_m$ (A.48) (kJ·mol^{-1})		$\Delta_r G_m$ (A.46) (kJ·mol^{-1})	
	Exp.	Cal.	Exp.	Cal.	Exp.	Cal.
900	−(208.1 ± 0.2)	−(131.0 ± 0.3)	(212.0 ± 13.4)	(181.8 ± 50.7)	(3.9 ± 13.4)	(50.8 ± 10.2)
1000	−(191.3 ± 0.2)	−(122.0 ± 0.3)	(187.3 ± 13.4)	(169.7 ± 50.7)	−(4.0 ± 13.4)	(47.7 ± 10.3)
1100	−(174.5 ± 0.2)	−(112.9 ± 0.3)	(162.5 ± 13.4)	(157.7 ± 50.7)	−(11.9 ± 13.4)	(44. ± 10.3)
1200	−(157.6 ± 0.2)	−(103.1 ± 0.3)	(137.8 ± 13.4)	(146.0 ± 50.8)	−(19.8 ± 13.4)	(42.9 ± 10.4)

The measured sodium pressures (reaction (A.48)) agree reasonably well with the calculated values, especially in the range of temperature where they were measured (1100 – 1200 K). However, the oxygen potentials (reaction (A.47)) seem to be much too negative, perhaps due to a lack of equilibrium between the three solids. The authors' measurements for reaction (A.47) would suggest an enthalpy of formation of NaUO$_3$ which is ca. 30 kJ·mol^{-1} more positive than the selected value. However, this would mean that NaUO$_3$ would not be a stable phase on the Na–U–O isothermal section at 800 K, cf. Figure 1 in the paper.

It may be noted that there is a discrepant calorimetric value for $\Delta_f H_m$ (NaUO$_3$) (by reaction with XeO$_3$(g)) from the study by O'Hare and Hoekstra [74OHA/HOE3], but this is 25 kJ·mol^{-1} more negative than the selected value.

We do not see any reason to change the currently selected enthalpy of formation of NaUO$_3$, but in view of these two discrepant values, we have increased its uncertainty substantially.

[94KAL/MCC]

This paper incorporates the data on the electronic levels of UO(g), derived by laser spectroscopy, from a number of earlier papers by the same authors (then at different Institutions); further levels between 15000 and 21000 cm^{-1} have been identified in a later paper [97KAL/HEA]. All these data have been used to refine the thermal functions of UO(g) - see section 9.3.1.1.1.

[94KAT/MEI]

Kato et al. describe the de-convolution of the spectra from six U(VI) solutions, $I = 0.1$ M NaClO$_4$, pH = 3.87, 4.06, 4.98, 5.77, 6.02 and 7, and $5 \cdot 10^{-6}$ to $2 \cdot 10^{-2}$ M in U, obtained by equilibration with schoepite in an open atmosphere, on the basis of single component emission spectra of the species $n{:}m$ (UO$_2$)$_m$(OH)$_n^{2m-n}$, 0:1, 2:2, 5:3 and the monocarbonato dioxouranium(VI), 1:0:1, 1:0:q $UO_2(CO_3)_q^{2-2q}$. The single component spectra are given in [93MEI/KIM].

The results seem compatible (however, the uranium concentrations of the solutions are not given) with the prediction of the composition of the solutions on the basis of formation constant of species: 1:1 [83CAC/CHO2], 2:2 [93MEI/KAT], 5:3 [93MEI/KIM], 1:0:1, 1:0:2 and 1:0:3 [93MEI/KIM]. The species 1:0:2 and 1:0:3, which predominate above pH = 7, do not fluoresce.

This paper does not give any new information. It confirms previous results on fluorescence lifetime values.

[94KIM/CHO]

Titrations were performed at 25°C and 0.1 M ionic strength (NaCl) in two identical cells housed in a CO_2 free glove box. Equilibrium was approached from supersaturation by addition of 0.1 M NaOH to $9.789 \cdot 10^{-4}$ M U(VI) to give a pH of 9.5. After overnight equilibration, $9.824 \cdot 10^{-4}$ M NaClO was added to one cell. A steady state was achieved after four days whereupon a sample was taken and the pH of the remaining solution was readjusted. Batch experiments with schoepite were also performed with different ratios of hypochlorite. X-ray diffraction was used to characterise the precipitates. The presence of hypochlorite increased the solubility of schoepite by a factor of 100. However, although the effect of ClO$^-$ is clearly an oxidation process and spectroscopic evidence is supplied to confirm the presence of UO_2^{2+}, $(UO_2)_2(OH)_2^{2+}$ and $(UO_2)_3(OH)_5^+$, this is by the authors' admission, not an equilibrium reaction. Moreover, the pH dependence of the schoepite solubility profile for those experiments without hypochlorite present shows an increase in solubility with pH that is inconsistent with the established thermodynamics.

[94KIM/KLE]

The formation constants for the complexes, $Cm(CO_3)_n^{3-2n}$, $n = 1, 2, 3$, have been determined at room temperature (no information is given in the paper and we assume that the temperature is close to 25°C) in a 0.1 M NaClO$_4$ ionic medium.

There is ample information on the experimental details and the chemical model proposed by the authors is in good agreement with previous information on similar systems. It is not clear whether the authors measure pH or $-\log_{10}[H^+]$. As all concentrations are given in molar units, the reviewers are confident that the equilibrium constants reported are concentration constants, valid in 0.1 M sodium perchlorate. The free carbonate concentration is calculated from the expression, $\log_{10}[CO_3^{2-}] = \log_{10}K + \log_{10} p_{CO_2} + 2\,pH$, where $\log_{10}K = -(17.62 \pm 0.07)$ refers to the reaction:

$$CO_2(g) + H_2O(l) \rightleftharpoons CO_3^{2-} + 2H^+.$$

This value is in fair agreement with the one calculated from the selected values in [92GRE/FUG], $\log_{10} K = -(17.52 \pm 0.04)$. The formation constant for $CmCO_3^+$ in 0.1 M $NaClO_4$ is reported as $\log_{10} \beta_1 = (6.65 \pm 0.07)$ at room temperature (25°C?), practically the same as for Eu(III), $\log_{10} \beta_1 = (6.57 \pm 0.08)$. These values are in excellent agreement with most of the previous literature information, as discussed by the authors, cf. their Table 5. The reviewers have recalculated these data to zero ionic strength using the SIT, with the following interaction constants: $\varepsilon(Cm^{3+}, ClO_4^-) = 0.49$, $\varepsilon(CmCO_3^+, ClO_4^-) = 0.2$ and $\varepsilon(CO_3^{2-}, Na^+) = -0.08$ kg·mol^{-1}, and find $\log_{10} \beta_1^\circ = (7.94 \pm 0.10)$, in good agreement with the selected value for Am(III), $\log_{10} \beta_1^\circ = (7.8 \pm 0.3)$, selected in [95SIL/BID].

[94MOR/WIL]

The authors determined the enthalpy of solution of ^{243}Am(OH)$_3$(cr) in 6 M HCl and calculated the standard molar enthalpy, standard molar Gibbs energy, and solubility constant for crystalline Am(III) hydroxide. Am(III) oxalate was precipitated and calcined to AmO_2(s). The AmO_2(s) was reduced to Am_2O_3(s) under H_2(g) at 800°C and finally hydrated in the presence of steam at 230°C. The reported X–ray powder diffraction and IR investigations confirmed that the prepared solids refer to hexagonal Am(OH)$_3$(cr). However, the X–ray powder patterns showed in addition two weak unassigned lines at d = 0.2678 and 0.1904 nm.

The experimental measurements with solid samples from the same batch of preparation (batch A: $\Delta_{sol}H_m$ (Am(OH)$_3$, cr, 298.15 K, 6 M HCl) = $-$ 113.5 and $-$ 111.6 kJ·mol^{-1}; batch B: $-$ 103.8, $-$ 103.0, and $-$ 99.0 kJ·mol^{-1}) are in reasonable agreement, while the mean values from A ($-$(112.6 ± 2.7)) and B ($-$(101.9 ± 5.1)) are not. This indicates that the solids in batches A and B are not chemically identical. The authors proposed an overall mean value of $\Delta_{sol}H_m$ (Am(OH)$_3$, cr, 298.15 K, 6 M HCl) = $-$ (106.2 ± 7.6) kJ·mol^{-1}. Based on a value of $\Delta_{sol}H_m$ (Am, cr, 6 M HCl) = $-$ (617.5 ± 2.1) kJ·mol^{-1} for the enthalpy of solution of americium metal, the standard molar enthalpy of formation of well–crystallised Am(OH)$_3$(cr) was calculated to be $\Delta_f H_m^\circ$ (Am(OH)$_3$, cr, 298.15 K) = $-$ (1371.2 ± 7.9) kJ·mol^{-1}. An entropy value of S_m° (Am(OH)$_3$, cr, 298.15 K) = (129 ± 10) J·K^{-1}·mol^{-1} was estimated by analogy with Nd and Eu hydroxides and used together with S_m° (Am^{3+}, 298.15 K), S_m° (OH$^-$, 298.15 K) and $\Delta_f H_m^\circ$ (OH$^-$, 298.15 K) consistent with the values selected in [95SIL/BID], to calculate $\Delta_r G_m^\circ = (168.2 \pm 9.2)$ kJ·mol^{-1} for the reaction:

$$Am(OH)_3(cr) \rightleftharpoons Am^{3+} + 3\,OH^-.$$

Accordingly, the solubility constant was calculated to be $\log_{10} K_{s,0}^\circ$ (Am(OH)$_3$, cr, 298.15 K) = $-$ (29.5 ± 1.6), which corresponds to $\log_{10} {}^*K_{s,0}^\circ$ (Am(OH)$_3$, cr, 298.15 K) = (12.5 ± 1.6). In this calculation, however, Morss and Williams used for $\Delta_f H_m^\circ$ (Am^{3+}) a value 4.5 kJ·mol^{-1} more negative than that adopted by [95SIL/BID]. For reasons of consistency, the present review combines the experimental data of Morss and Willams, $\Delta_{sol}H_m$ (Am, cr, 6 M HCl) = $-(613.1 \pm 1.4)$ kJ · mol^{-1} and $\Delta_f H_m^o$ (Am^{3+}) = $-(616.7 \pm 1.5)$ kJ · mol^{-1} as selected in [95SIL/BID]. The standard molar enthalpy of formation is then calculated to be $\Delta_f H_m^o$ (Am(OH)$_3$, cr, 298.15 K) = $-(1366.8 \pm 7.7)$ kJ · mol^{-1}. With S_m^o (Am(OH)$_3$, cr, 298.15 K) = (116 ± 8) J · K^{-1} · mol^{-1} as estimated in the present review (cf. section 12.3.2.2) the solubility constant becomes $\log_{10} K_{s,0}^o$ (Am(OH)$_3$, cr, 298.15 K) = $-(28.7 \pm 1.6)$ or $\log_{10}{}^* K_{s,0}^o$ (Am(OH)$_3$, cr, 298.15 K) = (13.3 ± 1.6).

As their calculated solubility constant was considerably lower than those derived from solubility measurements, e.g., in [82SIL], Morss and Williams [94MOR/WIL] proposed a "working value" of $\log_{10}{}^* K_{s,0}^o$ (Am(OH)$_3$, s, 298.15 K) = (14.5 ± 2.0) for a less crystalline Am(OH)$_3$(s), typical for a hydroxide in contact with an aqueous solution. Applying a correlation of the relative basicity of actinide and lanthanide hydroxides as a function of ionic size, the authors proposed further "working values" of $\log_{10}{}^* K_{s,0}^o$ (Pu(OH)$_3$, s, 298.15 K) = $(15.5 \pm .2.0)$ and $\log_{10}{}^* K_{s,0}^o$ (Cm(OH)$_3$, s, 298.15 K) = (14.0 ± 2.0). The experimental value for $\Delta_f H_m^o$ (Am(OH)$_3$, cr, 298.15K) reported by [94MOR/WIL] differs significantly from an earlier estimate by the same authors [92MOR/WIL].

[94OCH/SUZ]

Single crystals of UTe and USb were grown by the Bridgman technique from high purity elements. 10% excess of Te and Sb were added to allow for their evaporation, but no further details of the preparation, or elemental analyses of the products are given. The lattice constants of 6.160 and 6.210 Å were a little larger than previously reported values.

Electrical resistivity and heat capacity were measured from 2 to 300 K, with very few details of the apparatus used. Preliminary results of the heat capacity are presented in small graphs. Both materials show pronounced ferromagnetic peaks at 102 and 218 K, respectively, several K different from earlier values. The values of $C_{p,m}$ (298.15 K) read from the graphs are about 75 J · K^{-1} · mol^{-1} for USb(cr) and 84 J · K^{-1} · mol^{-1} for UTe(cr). Although the authors state that the results were sample dependent and that higher quality materials are desirable for more precise measurements, these data have been included in the review with fairly high uncertainties.

[94OMO/MUR]

Omori et al. determined the distribution coefficient for pertechnetate between chloroform and water (1 molar ionic strength, HCl and NaCl) at 298.15 K. Tetraphenylarsonium chloride (TPAC) was the extraction agent employed and it was detected spectrophotometrically, whereas the ^{99}Tc tracer was measured by liquid scintillation counting. TcO$_4^-$ was identified by paper chromatography. The distribution coefficient for

TPAC ($[TPAC]_{org}/[TPAC]_{aq}$) was determined to be 29.5 from the [H$^+$] dependence at constant chloride molality (viz., 1 M). The protonation constant for TPAC in the aqueous phase was also calculated in this manner to be 1.36 M^{-1}, using a value of 0.079 M for the chloride dissociation of TPAC. Addition of TcO$_4^-$ yielded an extraction constant (K) for the equilibrium, TcO$_4^-$ + TPAC(org) $\rightleftharpoons$ TPATcO$_4$ (org) + Cl$^-$, of $(3.0 \pm 0.2) \cdot 10^4$ leading to a value for the acid (molar) dissociation constant of HTcO$_4$(aq) of $\log_{10} K_C = (1.02 \pm 0.18)$. This value is even higher than those in Table V.10 of [99RAR/RAN] who considered all of these values to be unreliable and possible artefacts of misrepresented medium effects. Indeed, although no data are tabulated, the plot provided of 1/D' versus [H$^+$] is quite flat such that the slope $(= 1/{K_C \cdot K})$ is virtually zero within the experimental uncertainty. The slight deviation from zero slope could be accounted for by activity coefficient variations as Na$^+$ is substituted systematically by H$^+$ over the range 0.1 to about 0.95 M.

[94RIZ/RAO]

This study describes a direct calorimetric determination of the enthalpy changes for the formation of $(UO_2)_2(OH)_2^{2+}$, PuO$_2$OH$^+$ and PuO$_2$(OH)$_2$(aq), the uranium system having been studied in 1.0 M tetramethylammonium chloride and the plutonium system in 1.0 M sodium perchlorate, both at 25°C. The experimental procedures are satisfactory and the experimental results for the uranium system are in good agreement with previous determinations in other ionic media. The U(VI) concentration has been varied from 1.6 mM to 149.9 mM in the $-\log_{10}[H^+]$ range 3 to 5. Corrections for heats of dilution were made using experimental data and corrections for the enthalpy of formation of water from H$^+$ and OH$^-$ were made using pK_w = 13.79 and $\Delta_r H_m$ = 56.6 kJ·mol^{-1} (according to the authors and probably corresponding to 1 M tetramethylammonium chloride). The ratio of added hydroxide to the concentration of U(VI) was at most 0.86. The formation constant for $(UO_2)_2(OH)_2^{2+}$ in 1 M tetramethylammonium chloride is assumed to be the same as in 1 M sodium perchlorate. The value used, $\log_{10} {}^*\beta_{2,2} = -5.89$ from [62BAE/MEY], is in fair agreement with the value selected in [92GRE/FUG].

The average value for the enthalpy of formation of the (2:2) complex $(UO_2)_2(OH)_2^{2+}$, $\Delta_r H_m$ (2:2) is (44.4 ± 1.9) kJ·mol^{-1} and the value of $\Delta_r S_m$ (2:2) is (36 ± 6) J·K^{-1}·mol^{-1}. The enthalpy and entropy data for $(UO_2)_2(OH)_2^{2+}$ are in good agreement with the values given in Table V.8 of [92GRE/FUG], despite the fact that the ionic strength is different. The ionic strength dependence of enthalpies of reaction are discussed in Chapter IX.10 of [97ALL/BAN]. Using the selected value $\log_{10} {}^*\beta_{2,2}^\circ = -(5.62 \pm 0.04)$, in [92GRE/FUG] and auxiliary data (D, I_m, $\Delta\varepsilon$) corresponding to a 1 M NaClO$_4$ medium, we obtain $\log_{10} {}^*\beta_{2,2} = -(5.93 \pm 0.06)$ and $\Delta_r G_m (2:2) = 33.8$ kJ·mol^{-1}, which in turn results in $\Delta_r H_m (2:2) = 42.7$ kJ·mol^{-1} and $\Delta_r S_m (2:2) = 30$ J·K^{-1}·mol^{-1}. These values are the same as those of [62BAE/MEY].

The plutonium data have been discussed in [2001LEM/FUG].

[94RUN/KIM]

This report (thesis of W. Runde) contains experimental data on the solubility, hydrolysis and carbonate complexation of trivalent and pentavalent americium and pentavalent neptunium in sodium chloride solution. All experiments were performed at room temperature $(22 \pm 1)°C$ with the nuclides ^{241}Am and ^{237}Np, mainly in 5 M NaCl. The solubility and carbonate complexation of Am(V) and Np(V) were also investigated in 3 M NaCl and 1 M NaCl, respectively. The formation constant of the monocarbonate complex of Np(V) was additionally investigated by absorption spectroscopy in 0.1, 3 and 5 M NaCl. The solubility experiments were performed in titration cells, either under an inert argon atmosphere or under a defined carbon dioxide partial pressure of 0.01 bar in argon. The oxidation states of Am(III) and radiolytically oxidised Am(V) were confirmed by absorption spectroscopy. A combined pH glass electrode was calibrated against H^+ and OH^- concentrations in the NaCl solutions. The equilibrium constants between H^+, OH^-, HCO_3^- and CO_3^{2-} determined in [94RUN/KIM] are in reasonable agreement with widely accepted literature data in NaCl solution (*e.g.*, with those calculated with SIT coefficients of the NEA TDB or with those calculated for higher NaCl molalities with the Pitzer coefficients in [84HAR/MOL], [91PIT]).

An attempt was made to correct the apparent solubility constants of solid Am(III) and Am(V) hydroxides and carbonates in 5 M NaCl for the formation of chloride complexes. As the applied chloride complexation constants were too large, these considerations must be rejected. (According to the corresponding data selected in the present review for Am(III) and in [2001LEM/FUG] for Np(V), chloride complexation is negligible up to 5 M NaCl). In a later paper, modelling Np(V) and Am(V) in alkaline and carbonate containing solutions with the Pitzer approach, Runde *et al.* [96RUN/NEU] also neglected the formation of An(V) chloride complexes and interpreted the different equilibrium constants in chloride and perchlorate media in terms of ion interaction and activity coefficient effects.

- Am(III) hydroxides

Runde and Kim measured the solubility of an X-ray amorphous precipitate Am(OH)$_3$(am), in the range $7 < -\log_{10}[H^+] < 14$ and evaluated a solubility constant of $\log_{10}K_{s,0}((A.49), 5\ M\ NaCl) = -(25.8 \pm 0.4)$ for the reaction:

$$Am(OH)_3(am) \rightleftharpoons Am^{3+} + 3\ OH^- \tag{A.49}$$

corresponding to $\log_{10}{}^*K_{s,0}^\circ = (15.2 \pm 0.7)$, if converted to $I = 0$ with the SIT coefficients in Appendix B, or (14.6 ± 0.5) as calculated in [94FAN/KIM], [97KON/FAN] with the Pitzer approach. These values correspond rather to the solubility constant selected in this review for the crystalline hydroxide, $\log_{10}{}^*K_{s,0}^\circ$ (Am(OH)$_3$, cr, 298.15 K) = (15.6 ± 0.6). On the other hand, the solubility at $-\log_{10}[H^+] > 11$ corresponds to $\log_{10} K_{s,3}$(A.50) = $-(9.3 \pm 0.3)$,

$$Am(OH)_3(s) \rightleftharpoons Am(OH)_3(aq) \tag{A.50}$$

which is consistent with the data selected for Am(OH)$_3$(am) (cf. section 12.3.2.2, Fig. 12-4). Because of these ambiguities and possible inaccuracies arising from the ionic strength corrections, the constants $\log_{10} K_{s,0}$ ((A.49), 5 M NaCl) and $\log_{10} K_{s,3}$ ((A.50), 5 M NaCl) derived in [94RUN/KIM] are not used for the selection of thermodynamic data. The formation constants reported for the first and second hydroxide complex, $\log_{10} \beta_1$ (AmOH^{2+}, 5 M NaCl) = (6.9 ± 0.6) and $\log_{10} \beta_2$ (Am(OH)$_2^+$, 5 M NaCl) = (12.8 ± 0.7), are in good or at least fair agreement with data for the analogous Cm(III) complexes [94FAN/KIM], determined by time resolved laser fluorescence spectroscopy (cf. discussion of data selection, section 12.3.1.1, Fig.12-2 and 12-2).

- Am(III) carbonates

Measuring the solubility of NaAm(CO$_3$)$_2$·xH$_2$O(s) as a function of $\log_{10}$[H$^+$] under an atmosphere of p_{CO_2} = 0.01 bar in argon, Runde and Kim repeated the solubility study of Meinrath [91MEI]. In addition, they performed an analogous experiment with NaEu(CO$_3$)$_2$·xH$_2$O(s). The sodium americium carbonate was characterised by comparing its X–ray powder diffraction pattern with those of the analogous Eu and Nd compounds. The evaluated solubility constant was $\log_{10} K_{s,0}$ ((A.51), 5 M NaCl) = – (16.5 ± 0.3) for the reaction:

$$\text{NaAm(CO}_3\text{)}_2 \cdot x\text{H}_2\text{O(s)} \rightleftharpoons \text{Na}^+ + \text{Am}^{3+} + 2\,\text{CO}_3^{2-} + x\text{H}_2\text{O(l)} \qquad (A.51).$$

With the number of crystal water molecules, $x = (5 \pm 1)$ (cf. discussion of [91MEI]), a_w = 0.7786 in 5.6 m NaCl [91PIT], and $\Delta\varepsilon$(A.51) = (0.10 ± 0.06) kg · mol^{-1} according to the SIT coefficients in Appendix B, the solubility constant at I = 0 is calculated in this review to be $\log_{10} K_{s,0}^\circ$ (A.51) = – (21.0 ± 0.5). The uncertainty reflects those of the ε values, but the use of the SIT equation in 5.6 m NaCl may also include other uncertainties. However, this value at I = 0 is consistent with $\log_{10} K_{s,0}^\circ$ (A.51) = – (21.0 ± 0.4) as calculated in [98NEC/FAN], [99FAN/KON] with the Pitzer approach and is therefore selected in the present review. The carbonate complexation constants evaluated from this solubility study, $\log_{10} \beta_1$ (AmCO$_3^+$, 5 M NaCl) = (5.7 ± 0.4), $\log_{10} \beta_2$ (Am(CO$_3$)$_2^-$, 5 M NaCl) = (9.7 ± 0.5), and $\log_{10} \beta_3$ (Am(CO$_3$)$_3^{3-}$, 5 M NaCl) = (12.9 ± 0.2) are comparable with those determined for aqueous Cm(III) carbonate complexes by time resolved laser fluorescence spectroscopy [99FAN/KON].

- Am(V) and Np(V) hydroxides

Runde and Kim measured the solubility of amorphous Am(V) and Np(V) hydroxides in 5 M NaCl as a function of the H$^+$ concentration in the range 7.8 < – $\log_{10}$[H$^+$] < 13.5. Both data sets indicate the formation of a neutral and an anionic hydroxide complex, AnO$_2$OH(aq) and AnO$_2$(OH)$_2^-$. The solubility data for the Np(V) hydroxide are shifted to lower concentrations compared to those of the Am(V) hydroxide. A difference of 0.6 $\log_{10}$ units is constant over the whole pH range investigated. This observation was ascribed to the aging of NpO$_2$OH(am) as already claimed in [92NEC/KIM] for analogous data in 1 and 3 M NaClO$_4$. The evaluated solubility constants and formation constants of the hydroxide complexes, given for the reactions:

$$AnO_2OH(am) \rightleftharpoons AnO_2^+ + OH^- \quad (A.52)$$

and

$$AnO_2^+ + nOH^- \rightleftharpoons AnO_2(OH)_n^{1-n} \quad (A.53)$$

are summarised in Table A-16. For comparison the corresponding constants selected in [2001LEM/FUG], $\log_{10}{}^*K_{s,0}^\circ$ (NpO$_2$OH, am, 298.15 K) = (5.3 ± 0.2) and (4.7 ± 0.5) for fresh and aged precipitates, respectively, $\log_{10}{}^*\beta_1^\circ$ (NpO$_2$OH, aq, 298.15 K) = – (11.3 ± 0.7), and $\log_{10}{}^*\beta_2^\circ$ (NpO$_2$(OH)$_2^-$, 298.15 K) = – (23.6 ± 0.5) are extrapolated to 5 M NaCl with the SIT coefficients selected in [2001LEM/FUG] (ε(Na$^+$,OH$^-$) = (0.04 ± 0.01) kg · mol^{-1}, ε(NpO$_2^+$,Cl$^-$) = (0.09 ± 0.05) kg · mol^{-1}, ε(Na$^+$,NpO$_2$(OH)$_2^-$) = – (0.01 ± 0.07) kg · mol^{-1}). Runde and Kim's solubility constant for NpO$_2$OH(aged) is consistent with the SIT calculation based on the data in [2001LEM/FUG] and the solubility constant for AmO$_2$OH(am) is consistent with the SIT calculation for NpO$_2$OH(am, fresh) (cf. Table A-16). However, the calculations for the hydroxide complexes, NpO$_2$OH(aq) and NpO$_2$(OH)$_2^-$, are considerably less in accord with Runde and Kim's experimental data in 5 M NaCl. An accurate fit would require additional interaction coefficients, ε(NpO$_2$OH(aq),Cl$^-$) = – (0.12 ± 0.13) kg · mol^{-1} and ε(NpO$_2$(OH)$_2^-$,Cl$^-$) = – (0.17 ± 0.13) kg · mol^{-1}. However, the interactions of neutral complexes or anion–anion interactions are usually neglected in the SIT, whereas they are included in the Pitzer parameterisations proposed in [96RUN/NEU] and [95FAN/NEC]. It is noteworthy that the equilibrium constants at $I = 0$ calculated in [96RUN/NEU] with the Pitzer approach are in excellent agreement with the thermodynamic data selected in [2001LEM/FUG] from experimental data in NaClO$_4$ solution (cf. Table A-16).

Besides data from Runde and Kim in 5 M NaCl there are no other reliable data for Am(V) hydroxide complexes, but according to the results in [94RUN/KIM] it seems justifiable to use the formation constants of the aqueous Np(V) hydroxide complexes selected in [2001LEM/FUG] as reasonable estimates for the analogous Am(V) hydroxide complexes.

Table A-16: Equilibrium constants derived from the solubility of AmO$_2$OH(am) and NpO$_2$OH(am, aged) in 5 M NaCl at 22°C [94RUN/KIM], and comparison of the equilibrium constants at $I = 0$ recommended in [96RUN/NEU] and [2001LEM/FUG].

In 5 M NaCl	$\log_{10} K_{s,0}$	$\log_{10} \beta_1$	$\log_{10} \beta_2$	
AmO$_2$OH(am)	−(8.94 ± 0.42)	(3.62 ± 0.27)	(5.89 ± 0.22)	[94RUN/KIM]
NpO$_2$OH(am, aged)	− (9.0 ± 0.4)			a)
NpO$_2$OH(aged)	− (9.56 ± 0.18)	(3.66 ± 0.22)	(5.98 ± 0.19)	[94RUN/KIM]
NpO$_2$OH(am, aged)	− (9.6 ± 0.6)	(3.0 ± 0.7)	(5.0 ± 0.7)	a)

$I = 0$	[96RUN/NEU] (Pitzer approach)	[2001LEM/FUG] (SIT approach)
$\log_{10} {}^*K^\circ_{s,0}$ (NpO$_2$OH, am, fresh)	5.24	(5.3 ± 0.2)
$\log_{10} {}^*K^\circ_{s,0}$ (NpO$_2$OH, am, aged)	4.49	(4.7 ± 0.5)
$\log_{10} {}^*\beta^\circ_1$ (NpO$_2$OH, aq)	− 11.41	− (11.3 ± 0.7)
$\log_{10} {}^*\beta^\circ_2$ (NpO$_2$(OH)$_2^-$, aq)	− 23.59	− (23.6 ± 0.5)

a) Calculated for 5 M NaCl with the thermodynamic data and SIT coefficients selected in [2001LEM/FUG].

- Am(V) and Np(V) carbonates

The numerous equilibrium constants reported in [94RUN/KIM] for solid and aqueous carbonates of Np(V) and Am(V) in 0.1–5 M NaCl solutions are summarised in Table A-2. They are based on solubility studies with NaNpO$_2$CO$_3$(s) and NaAmO$_2$CO$_3$(s) under an atmosphere of p_{CO_2} = 0.01 bar in argon and carbonate concentrations in the range − 7.0 < log$_{10}$[CO$_3^{2-}$] < − 1.4. A few results reported later in [96RUN/NEU] are included as well. Some of the formation constants for Np(V) carbonate complexes were additionally confirmed by absorption spectroscopy. The solid sodium dioxoneptunium(V) carbonate was characterised by X–ray diffraction and identified as the hydrated NaNpO$_2$CO$_3$·3.5H$_2$O(s) previously characterised in [77VOL/VIS], [83MAY]. For the corresponding americium(V) compound no X–ray diffraction pattern is reported, but the very similar solubilities for Am(V) and Np(V), and the analogous dependence on log$_{10}$[CO$_3^{2-}$] justifies the assumption that the analogous NaAmO$_2$CO$_3$ · xH$_2$O(s) is the solubility limiting solid.

Applying the Pitzer approach to solid and aqueous Np(V) carbonates, Runde *et al.* [96RUN/NEU] also evaluated the equilibrium constants at $I = 0$. The calculated solubility constant of NaNpO$_2$CO$_3$ · 3.5H$_2$O(s) and the formation constants of the Np(V) carbonate complexes are in excellent agreement with the thermodynamic data selected in [2001LEM/FUG] (*cf.* Table A-17). Runde and Kim's experimental data for Am(V) in 3 and 5 M NaCl are consistent with those of Giffaut [94GIF] in 4 M NaCl. The results reported in these two theses are used in the present review to evaluate thermodynamic data for Am(V) carbonate complexes from experimental data for Am(V), and so are not based solely by analogy with Np(V) (cf. discussion of selected data, sections 12.6.1.1.2. and 12.6.1.1.3.4). The conditional constants in 3 – 5 M NaCl are converted to $I = 0$ with the SIT coefficients selected in [2001LEM/FUG] for the analogous Np(V) species, and the weighted average values are selected for the Am(V) carbonates:

$$\log_{10} K^\circ_{s,0} (\text{NaAmO}_2\text{CO}_3, \text{s}, 298.15 \text{ K}) = -(10.9 \pm 0.4)$$

$$\log_{10} \beta^\circ_1 (\text{AmO}_2\text{CO}_3^-, 298.15 \text{ K}) = \quad (5.1 \pm 0.5)$$

$$\log_{10} \beta^\circ_2 (\text{AmO}_2(\text{CO}_3)_2^{3-}, 298.15 \text{ K}) = \quad (6.6 \pm 0.8)$$

$$\log_{10} \beta^\circ_3 (\text{AmO}_2(\text{CO}_3)_3^{5-}, 298.15 \text{ K}) = \quad (5.1 \pm 1.0).$$

They are similar to the thermodynamic data selected in [2001LEM/FUG] for the analogous Np(V) compounds (cf. Table A-17), but the uncertainty limits are considerably larger.

Table A-17: Equilibrium constants derived in [94RUN/KIM], [96RUN/NEU] from the solubility studies with $\text{NaNpO}_2\text{CO}_3(\text{s})$ and $\text{NaAmO}_2\text{CO}_3(\text{s})$ in NaCl solutions at 22°C.

Medium	An	$\log_{10} K_{s,0}$	$\log_{10} \beta_1$	$\log_{10} \beta_2$	$\log_{10} \beta_3$
0.1 M NaCl	Np	$-(10.4 \pm 0.2)$ [a]	(4.8 ± 0.1) [a]		
	Np		(4.68 ± 0.03) [b]		
1.0 M NaCl	Np	$-(9.77 \pm 0.16)$	(4.32 ± 0.07)	(6.49 ± 0.09)	(8.43 ± 0.06)
3.0 M NaCl	Np	$-(9.4 \pm 0.2)$ [a]	(4.3 ± 0.1) [a]	(7.1 ± 0.2) [a]	(9.2 ± 0.2) [a]
	Np		(4.67 ± 0.07) [b]		
	Am	$-(9.65 \pm 0.19)$	(4.74 ± 0.09)	(7.42 ± 0.03)	(9.54 ± 0.13)
5.0 M NaCl	Np	$-(9.61 \pm 0.11)$	(4.71 ± 0.04)	(7.54 ± 0.05)	(9.63 ± 0.05)
	Np		(4.72 ± 0.13) [b]	(7.63 ± 0.19) [b]	
	Am	$-(9.56 \pm 0.13)$	(4.69 ± 0.04)	(7.54 ± 0.07)	(9.65 ± 0.05)
$I = 0$					
[96RUN/NEU]	Np	-11.14 [c]	5.06	6.49	5.43
[2001LEM/FUG]	Np	$-(11.16 \pm 0.35)$ [c]	(4.96 ± 0.06)	(6.53 ± 0.10)	(5.50 ± 0.15)
This review	Am	$-(10.9 \pm 0.4)$ [d]	(5.1 ± 0.5) [d]	(6.6 ± 0.8) [d]	(5.1 ± 1.0) [d]

(a) Determined in [96RUN/NEU], all other data were determined in [94RUN/KIM].

(b) Determined by absorption spectroscopy

(c) $\text{NaNpO}_2\text{CO}_3 \cdot 3.5\text{H}_2\text{O}(\text{s})$

(d) Calculated from the data reported in [94GIF], [94RUN/KIM] for Am(V) in 3 – 5 M NaCl, (cf. section 12.6.1.1.2.).

[94SAL/KUL]

During a study of the reaction of $UO_2(cr)$ with alkali metal chromates up to 1473 K in an inert atmosphere, new uranate phases, $Rb_2U_4O_{11}$, $Rb_2U_3O_{8.5}$ and $Na_2U_3O_9$, were observed, based on their characterisation by X–ray diffraction, thermal and chemical methods. $Na_yU_{1-y}O_{2-x}$ (y = 0 – 0.2) was reported as a solid solution of Na_2O in UO_2.

No thermodynamic information on ternary uranium oxides can be deduced directly from this investigation.

[94SAN/GRA]

Solid phases (schoepite, $UO_3 \cdot 2H_2O$ (in fact metaschoepite); becquerelite, $CaU_6O_{19} \cdot 11H_2O$ and compreignacite, $K_2U_6O_{19} \cdot 11H_2O$) were characterised by XRD, SEM and SEM–EDX. Synthetic methods for the latter two phases involving hydrothermal precipitation are described. The solubility experiments with these solids were performed at 25°C in either 1 molal $CaCl_2$ or KCl at the following conditions: schoepite in $CaCl_2$ at pH = 3.97, 7.47 and in KCl at pH = 4.25, 6.00 with samples taken after three, five and nine months; becquerelite in $CaCl_2$ at pH 4.16, 4.46, 5.85 and compreignacite in KCl at pH = 3.12, 4.46, 5.83 with samples taken after three months. The pH measurements were corrected for liquid junction potentials. Conversion of the schoepite was complete within three months and these solids did not change further in more than one year. The general equilibrium can be written:

$$M_xU_6O_{19} \cdot 11H_2O(cr) + (12 - xn)H^+ \rightleftharpoons xM^{n+} + 6UO_2^{2+} + 18H_2O(l).$$

The authors followed the SIT extrapolation model to derive the infinite dilution solubility constants. Consequently, the authors ignored the presence of $UO_2(OH)^+$ and $(UO_2)(OH)_2(aq)$, and after the undisclosed results of a Pitzer treatment, they also neglected specific chloride complexation. The concentrations of dioxouranium(VI) and pH are displayed in the form of a figure, but the values are not tabulated. The results of duplicate runs with the Ca– and K–containing solids yielded the following solubility constants:

M = Ca:

$\log_{10}{}^*K_{s,0} = (42.9 \pm 0.2)$ and (44.7 ± 0.1)

$\log_{10}{}^*K^\circ_{s,0} = (41.9 \pm 0.5)$ and (43.7 ± 0.5)

M = K:

$\log_{10}{}^*K_{s,0} = (38.2 \pm 0.2)$ and (40.5 ± 0.2)

$\log_{10}{}^*K^\circ_{s,0} = (36.8 \pm 0.3)$ and (39.2 ± 0.3).

The values for becquerelite compare favourably with that of [90VOC/HAV], 43.2 in "pure" water at 25°C. The authors state that using the NEA recommended values for the solubility constant of schoepite and the solubility data from their study, equilibrium constants for the transformation reaction:

$$M_xU_6O_{19} \cdot 11H_2O(cr) + x\,nH^+ \rightleftharpoons xM^{n+} + 6UO_2(OH)_2 \cdot H_2O(cr)$$

are $\log_{10} {}^*K^\circ_{s,0}$ = 13.03 and 7.96 for becquerelite and compreignacite, respectively.

[94SER/DEV]

The first part of the paper is a history on the work done on actinide chemistry in the former USSR, as well as on the international attempts to select databases on these elements. [92GRE/FUG] is included as the latest attempt. Then, the paper presents the characteristics of the DiaNIK database aimed at selecting recommended values for 145 actinide species (Ac to Am) on the basis of the examination of 500 references dealing with aqueous inorganic complexes, updated from 1989 to 1993 and including all the references of [92FUG/KHO]. [92GRE/FUG] also includes references and material from [92FUG/KHO].

Analysis of the same data by [94SER/DEV] and [92GRE/FUG] leads generally to slight differences for values at zero ionic strength (298.15 K, 1 bar), due probably to different methods for extrapolation to I = 0. For, $UO_2^{2+} + Cl^- \rightleftharpoons UO_2Cl^+$, for instance $\log_{10} K^\circ$ is (0.27 ± 0.15) versus (0.17 ± 0.02) [92GRE/FUG], using the SIT theory.

The set of recommended $\log_{10} K^\circ$ values (as compared to those selected in [92GRE/FUG]) are:

$U^{4+} + H_2O(l) \rightleftharpoons UOH^{3+} + H^+$ $\log_{10} {}^*\beta_1^\circ = -0.4 \pm 0.2\ (-0.54 \pm 0.06)$

$UO_2^{2+} + H_2O(l) \rightleftharpoons UO_2OH^+ + H^+$ $\log_{10} {}^*\beta_1^\circ = -5.8 \pm 0.1\ (-5.2 \pm 0.3)$

$UO_2^{2+} + 3\,H_2O(l) \rightleftharpoons UO_2(OH)_3^- + 3\,H^+$ $\log_{10} {}^*\beta_{3,1}^\circ = -19.7 \pm 0.5\ (-19.2 \pm 0.4)$

$3UO_2^{2+} + 5\,H_2O(l) \rightleftharpoons (UO_2)_3(OH)_5^+ + 5\,H^+$ $\log_{10} {}^*\beta_{5,3}^\circ = -15.4 \pm 0.21\ (-15.55 \pm 0.12)$

$3\,UO_2^{2+} + 7\,H_2O(l) \rightleftharpoons (UO_2)_3(OH)_7^- + 7\,H^+$ $\log_{10} {}^*\beta_{7,3}^\circ = -31.9 \pm 0.8\ (-31 \pm 2)$

$UO_2^{2+} + HPO_4^{2-} \rightleftharpoons UO_2HPO_4(aq)$ $\log_{10} \beta_1^\circ = 7.1 \pm 0.2\ (7.24 \pm 0.26)$

$UO_2^{2+} + PO_4^{3-} \rightleftharpoons UO_2PO_4^-$ $\log_{10} \beta_1^\circ = 12.9 \pm 0.3\ (13.23 \pm 0.15)$.

But more importantly, the authors question the validity of the CODATA selection ([89COX/WAG]), $S_m^\circ(UO_2^{2+}) = -(98.2 \pm 3.0)$ J·K^{-1}·mol^{-1}. According to them, the uncertainty in this value will affect the calculated equilibrium constant involving the aqueous dioxouranium(VI) ion by one order of magnitude and might have some bearings also on other uranium and actinyl ions.

The CODATA value (adopted in NEA TDB) is based on the following experimental data: the heat capacity of $UO_2(NO_3)_2 \cdot 6H_2O(cr)$ [40COU/PIT] down to temperatures approaching 0 K; its enthalpy of solution at infinite dilution according to equation:

$$UO_2(NO_3)_2 \cdot 6H_2O(cr) \rightleftharpoons UO_2^{2+} + 2NO_3^- + 6H_2O(l)$$

and its solubility as discussed in [92GRE/FUG].

The thermodynamic routes suggested by Sergeyeva et al. [94SER/DEV] to calculate $S_m^\circ(UO_2^{2+})$ lead to higher values. For instance, the route:

$$UO_2(OH)_2 \cdot H_2O(cr) + 2 H^+ \rightleftharpoons UO_2^{2+} + 3 H_2O(l)$$

gives $-(77.3 \pm 3.0)$ J · K^{-1} · mol^{-1}. This important difference was raised previously in [92KHO]. At that time, Pitzer gave the following comment: "... for the system $UO_2(OH)_2 \cdot H_2O(cr) + 2 H^+ \rightleftharpoons UO_2^{2+} + 3 H_2O(l)$, the entropy of the solid appears to be accurate, but the ΔS_m of the reaction is quite uncertain, in my view. There are various OH$^-$ complexes of UO_2^{2+} which must be involved in the saturated solution of $UO_2(OH)_2 \cdot H_2O(cr)$. [92GRE/FUG] in Table III.1 pp. 32–33 give values of $S_m^\circ = (17 \pm 50)$ for $UO_2(OH)^+$, $-(38 \pm 15)$ for $(UO_2)_2(OH)_2^{2+}$ and (83 ± 30) J · K^{-1} · mol^{-1} for $(UO_2)_3(OH)_5^+$. These values suggest a rather large uncertainty for $S_m^\circ(UO_2^{2+})$ via the $UO_2(OH)_2 \cdot H_2O(cr)$ route. In summary, $UO_2(NO_3)_2$(aq) is a far simpler system which is dominated by the uranyl ion."

In the paper under review, no mention is made of these comments which are available in a written form.

As no other data (apparently) have appeared in the literature since [92GRE/FUG] to support a re-evaluation of the value of $S_m^\circ(UO_2^{2+})$ obtained via the nitrate route, the value accepted by CODATA, $-(98.2 \pm 3.0)$ J · K^{-1} · mol^{-1}, is retained in this update.

The thermodynamic values of $UO_2(OH)_2(cr)$, $UO_2CO_3(cr)$ and $UO_2SO_4 \cdot 2.5H_2O(cr)$ used by [94SER/DEV] in their discussion are discussed in [92GRE/FUG].

In addition, Sergeyeva et al. recommended the following $\log_{10} K^\circ$ values for Np(V) complexes where the values in [2001LEM/FUG] are given in parenthesis:

$NpO_2^+ + H_2O(l) \rightleftharpoons NpO_2OH + H^+$	$\log_{10} {}^*\beta_1^\circ = -9.0 \pm 0.3$	(-11.3 ± 0.7)
$NpO_2^+ + CO_3^{2-} \rightleftharpoons NpO_2CO_3^-$	$\log_{10} \beta_1^\circ = 4.7 \pm 0.2$	(4.96 ± 0.06)
$NpO_2^+ + 2 CO_3^{2-} \rightleftharpoons NpO_2(CO_3)_2^{3-}$	$\log_{10} \beta_2^\circ = 6.7 \pm 0.4$	(6.53 ± 0.10)

The agreement is good except for the first hydrolysis constant.

For Pu, they propose the following equilibrium constants with the values from [2001LEM/FUG] within parenthesis:

$Pu^{3+} + SO_4^{2-} \rightleftharpoons PuSO_4^+$	$\log_{10} \beta_1^\circ = 4.5 \pm 0.5$	(3.91 ± 0.61)
$Pu^{3+} + 2 SO_4^{2-} \rightleftharpoons Pu(SO_4)_2^-$	$\log_{10} \beta_2^\circ = 6.7 \pm 0.6$	(5.70 ± 0.77)
$Pu^{4+} + HSO_4^- \rightleftharpoons PuSO_4^{2+} + H^+$	$\log_{10} {}^*\beta_1^\circ = 5.5 \pm 0.5$	(4.91 ± 0.22)

$$Pu^{4+} + 2\,HSO_4^- \rightleftharpoons Pu(SO_4)_2 + 2\,H^+ \qquad \log_{10}{^*\beta_2^\circ} = 7.7 \pm 0.7 \qquad (7.18 \pm 0.32)$$

For Pu(III) the values have been calculated from $\log_{10}{^*\beta_1^\circ} = (1.93 \pm 0.61)$ and $\log_{10}{^*\beta_2^\circ} = (1.74 \pm 0.76)$ related to the formation of complexes from HSO_4^- [2001LEM/FUG] with $\log_{10} K^\circ(HSO_4^-) = (1.980 \pm 0.050)$.

For Am, Sergeyeva *et al.* recommend:

$$Am^{3+} + SO_4^{2-} \rightleftharpoons AmSO_4^+ \qquad \log_{10}\beta_1^\circ = (4.2 \pm 0.4)$$

$$Am^{3+} + 2\,SO_4^{2-} \rightleftharpoons Am(SO_4)_2^- \qquad \log_{10}\beta_2^\circ = (6.1 \pm 0.5).$$

[95SIL/BID] selected $\log_{10}\beta_1^\circ = (3.85 \pm 0.03)$ and $\log_{10}\beta_2^\circ = (5.4 \pm 0.7)$, respectively.

All the data for Np and Pu are not considered further in this review because they have been discussed in [2001LEM/FUG].

[94SER/SER]

From this study, it can be concluded that an aqueous saturated solution of dioxouranium(VI) selenate obtained by dissolution of $UO_2SeO_4 \cdot 2H_2O$ (cr) is in equilibrium with $UO_2SeO_4 \cdot 4.5H_2O$ (cr) and at 25°C the composition of the solution is 68.7 wt% of UO_2SeO_4 (cr). Other data are not relevant for this review.

[94TOR/CAS]

Three sets of solubility experiments carried out at (298 ± 0.5) K in 1 M NaCl (pH = 4 to 9, oxic condition) are discussed. The first two sets of experiments involved unirradiated, crystalline UO_2 (powder of 10–50 μm size) under an oxidizing atmosphere of 5% O_2 in N_2. In the first case, the solid was equilibrated at pH = 5.0, whereupon HCl was added to achieve a pH = 3.5. After one month the pH was raised to 8.2 leading to the precipitation of a "yellow solid phase", which was characterised as being amorphous. This secondary phase was metaschoepite ($UO_3 \cdot 2H_2O$). The pH of the solution in contact with these two phases was then adjusted with acid or base and sampled after about three weeks such that "equilibrium" was approached from under- and super-saturation. The second set of experiments were similar, but did not involve the precipitation of the secondary yellow phase.

For comparison, a "poorly crystalline dioxouranium(VI) hydroxide hydrated phase" was prepared with an X–ray diffraction pattern corresponding to metaschoepite. The third set of experiments involved contacting this phase with 1 M NaCl for up to nine months prior to adding NaOH in a stepwise manner with each equilibration taking about three weeks. Equilibrium was apparently established within the initial three months with some enhancement in the crystallinity of the solid phase. The experimental data are only shown graphically as U(VI) concentrations (10^{-3}–10^{-6} M) as a function of pH with two distinct curves corresponding to data set 1 and 3. They were modelled using the equilibrium constants from [92GRE/FUG] and the SIT approach for all the potentially important hydrolysed U(VI) species, UO_2Cl^+ and $UO_2Cl_2^+$, except for tentially important hydrolysed U(VI) species, UO_2Cl^+ and $UO_2Cl_2^+$, except for $\log_{10} \beta_{2,1}^\circ$ ($UO_2(OH)_2$) taken as $\log_{10}{}^*\beta_{2,1}^\circ = -11.5$ from [92SIL].

The values of $\log_{10}{}^* K_{s,0}$ for the dissolution of metaschoepite:

$$UO_3 \cdot 2H_2O(s) + 2 H^+ \rightleftharpoons UO_2^{2+} + 3 H_2O (l)$$

were extracted from the data sets 1 and 3 as described above, namely, $\log_{10}{}^* K_{s,0} = (5.92 \pm 0.08)$ for metaschoepite and (5.57 ± 0.08) for prepared metaschoepite, which give, respectively, $\log_{10}{}^* K_{s,0}^\circ = (5.73 \pm 0.28)$ and (5.38 ± 0.20), accordingly $\log_{10} K_{s,0}^\circ = -(22.27 \pm 0.28)$ and $\log_{10} K_{s,0}^\circ = -(22.62 \pm 0.20)$.

The fits to these two data sets indicate that the upper limit for the stability constant of $UO_2(OH)_2(aq)$ in [92GRE/FUG] may be overestimated, but this finding is really only based on three data points.

The solubilities of $UO_2(s)$ in oxidizing, basic solutions are in reasonable agreement with those for crystalline metaschoepite corresponding to the [92GRE/FUG] model for crystalline schoepite (in fact, metaschoepite). However, in acidic to neutral solutions the solubilities are almost independent of pH within the large scatter of the data, perhaps indicating the formation of metastable surface phases or a high degree of supersaturation.

[94TOU/PIA]

Although no thermodynamic data are reported, this paper gives interesting information on the complexity of the phase relationships in the system U–Ba–O, based on X–ray diffraction studies up to about 1800 K. The two instruments used (either a graphite resistance furnace or a Pt/Rh resistance furnace) allowed measurements over the range $10^{-19} \leq p_{O_2} / \text{bar} \leq 1$.

For the studies with the Pt/Rh resistance furnace, compounds of nominal composition, Ba_3UO_6, $BaUO_4$ and $BaUO_3$, were obtained by reacting stoichiometric quantities of $BaCO_3$ and U_3O_8 under the appropriate oxygen pressure. The samples were pelleted and sintered prior to use. The lattice parameters of the "as prepared" compounds are in general agreement with literature results. For the studies in the graphite furnace, mixtures of $BaCO_3$ and UO_2 were pelleted and reacted in the furnace itself.

The authors indicate that the exact composition of the phases (in particular, as noted by this review, their oxygen content) was not known, but was inferred from the change of lattice parameters as a function of temperature and oxygen pressure.

For "Ba_3UO_6", a tetragonal to cubic transformation is observed above 1063 K. For the samples with a Ba/U ratio near unity, the orthorhombic phase "$BaUO_4$" can be reduced above 1423 K to cubic $BaUO_{3+x}$ in oxygen pressures below 10^{-12} bar. At this temperature, an order-disorder transformation is observed when $x \leq 0.10$. At lower temperatures, the homogeneity range is smaller, with the lowest oxygen content at 293 K (obtained by extrapolation) corresponding to x = 0.17.

The authors present a schematic pseudo-binary phase diagram between "BaUO$_3$" and "BaUO$_4$".

[95ALL/BUC]

This study consists of two parts, one structure determination of the complex $(UO_2)_3(CO_3)_6^{6-}$ in the solid state and in solution, the other a determination of the equilibrium constant for the reaction:

$$3\ UO_2(CO_3)_3^{4-} + 3\ H^+ \rightleftharpoons (UO_2)_3(CO_3)_6^{6-} + 3\ HCO_3^-$$

at 25°C and I = 2.5 m. The authors report $\log_{10} \beta_{6,3}$ = (18.1 ± 0.5). From this value and the protonation constants of CO_3^{2-} the authors calculate $\log_{10} \beta_{6,3}^\circ$ = (55.6 ± 0.5) for the reaction:

$$3\ UO_2^{2+} + 6\ CO_3^{2-} \rightleftharpoons (UO_2)_3(CO_3)_6^{6-}$$

which is in fair agreement with the value $\log_{10} \beta_{6,3}^\circ$ = (54 ± 1) given in [92GRE/FUG]. In addition, there is information on the U–O stretching frequency in the dioxouranium(VI) group. The structure determinations are excellent and provide independent non-thermodynamic proof of the existence and structure of the tri-nuclear complex.

[95BAN/GLA]

This paper deals with the rates of exchange of the carbonato group between free CO_3^{2-} and $(UO_2)_3(CO_3)_6^{6-}$ or $UO_2(CO_3)_3^{4-}$. The authors have studied the exchange reactions using ^{17}O enriched U(VI) and ^{13}C enriched carbonate in solutions I = 1 M (HClO$_4$, NaOH, Na$_2$CO$_3$) under a CO$_2$ atmosphere (p_{CO_2} = 1 atm). There are no new determinations of thermodynamic data, but the findings confirm the stoichiometry and structure of the carbonate complexes and also the previously determined equilibrium constants discussed in [92GRE/FUG].

[95BIO/MOI]

The unsaturated polyanions, PWO = $X_2W_{17}O_{61}^{10-}$ series (X = P, As), are strong aqueous complexing agents for actinides(IV), forming the limiting complex, M(PWO)$_2$. The structure of $U(P_2W_{17}O_{61})_2^{16-}$ written as $U(PWO)_2^{16-}$ and $K_{16}[U(P_2W_{17}O_{61})_2]\cdot 38H_2O$ written as $K_{16}[U(PWO)_2]\cdot 38H_2O$ are well established [97BIO/MOI]. The aqueous complex displays a charge transfer band (U^{4+} to W^{6+}), with $\varepsilon_{510\ nm}$ = 1009 M$^{-1}\cdot$cm^{-1}.

Complexometric titrations of U(IV), 5.9·10^{-4} M, with PWO, 3.1·10^{-3} M (as K$_{10}$P$_2$W$_{17}$O$_{61}$) are conducted in 2 M H/KNO$_3$ media, [H$^+$] = 0.5, 1 and 2 M at (21 ± 2)°C. The U(IV) solutions contain trace amounts of sulphuric acid (1 to 5 10^{-3} M) and zinc ($3 \cdot 10^{-5}$ M) and are 0.05 M in hydrazinium nitrate. Due to low sulphate concentration and weak complexation with Zn^{2+} their presence is inconsequential. But the presence of 2 M nitrate ions can not be dismissed. The constants for:

$$U(IV) + PWO \rightleftharpoons U(PWO) \qquad K_1$$

$$U(IV) + 2\,PWO \rightleftharpoons U(PWO)_2 \qquad K_2.$$

where U(IV) represents the sum of the concentrations of U^{4+} and nitrato complexes. The plots of $\log_{10} K_1$ and $\log_{10} K_2$ vs. $-\log_{10}[H^+]$ are linear with slopes 3.3 and 5.8. Ignoring nitrato complexes, the complexation of U^{4+} is as follows, considering the $pK_{a1} = 4.49$, $pK_{a2} = 3.75$ and $pK_{a3} = 2.0$ values of the PWO anion:

$$U^{4+} + H_3PWO^{7-} \rightleftharpoons U(PWO)^{6-} + 3\,H^+$$

$$U^{4+} + 2\,H_3PWO^{7-} \rightleftharpoons U(PWO)_2^{16-} + 6\,H^+$$

or

$$U^{4+} + PWO^{10-} \rightleftharpoons U(PWO)^{6-} \qquad \beta_1$$

$$U^{4+} + 2\,PWO^{10-} \rightleftharpoons U(PWO)_2^{16-} \qquad \beta_2$$

with

$$\log_{10} \beta_1 = \log_{10} K_1 + (pK_{a1} + pK_{a2} + pK_{a3}) - 3\,pH$$

$$\log_{10} \beta_2 = \log_{10} K_2 + 2\,(pK_{a1} + pK_{a2} + pK_{a3}) - 6\,pH.$$

All the results appear in Table A-18:

Table A-18: Equilibrium constants obtained in various experimental conditions.

$-\log_{10}[H^+]$	-0.3	0	0.3
$\log_{10} K_2$	(7 ± 0.25)	(8 ± 0.25)	(9 ± 0.25)
$\log_{10} K_1$	(11.5 ± 0.25)	(13.5 ± 0.25)	(15.5 ± 0.25)
$\log_{10} \beta_1$		(18.3 ± 0.3)	
$\log_{10} \beta_2$		(34.0 ± 1.1)	

PWO does not form complexes with U(VI) as shown by spectrophotometry, cyclic voltammetry (0.1 M $HClO_4$ in the presence or absence of PWO, $E_{1/2} = 0.065$ V for U(VI)/U(V) reduction), polarography ($E_{1/2}$ value remains constant) and ^{31}P–NMR.

Complexometric titrations of Pu(IV), $6.3 \cdot 10^{-3}$ M, at the same acidity and ionic strength as for U(IV) yielded only $\log_{10} \beta_1 > 8$ and $\log_{10} \beta_2 > 14$. In the presence of PWO, a nitric acid solution of Pu(III) becomes intense blue (an unstable mixed valency anion Pu(III)–W(V)–W(IV) is formed). When the colour has disappeared, the absorption spectrum shows the presence of the Pu(IV)PWO complex. The change in the redox potential value of the Pu(IV)/Pu(III) couple in 1 M HNO_3 in the presence of PWO is estimated on the basis of literature data on Ce, Am, Bk, Np, Cm and Cf. The redox potential shifts from 0.92 V to 0 V and this is why PWO can oxidise Pu(III) to Pu(IV).

[95CLA/HOB]

This paper gives a complete review of structural data of actinide carbonates and on thermodynamic data, including both equilibria between solids and species in aqueous solution, and equilibria between species in aqueous solution, the latter being taken from [92GRE/FUG]. Classification is done according to oxidation states. A short part is devoted to the hydrolysis of Th, U, Np, Pu and Am.

For U hydrolysis there are no new references that were not included in [92GRE/FUG]. A short discussion of the behaviour of U(IV) and U(VI) is given.

For U carbonate compounds and complexes there are no new references that have not been examined in [92GRE/FUG] or discussed in the present review.

[95COH/LOR]

This paper gives two examples of how the solubility data of pure compounds in water as a function of temperature are compiled, selected and tabulated for IUPAC publications ([94SIE/PHI]) concerning actinides, particularly nitrates; many ternary systems are also considered.

From semi-empirical considerations it is suggested that the solubility of a salt $MX \cdot r H_2O$ can be modelled by the equation:

$$\ln(\frac{m}{m_0}) - (\frac{m}{m_0} - 1) = \frac{A}{T} + B \ln(\frac{T}{T_{fus}}) + CT + D$$

where A, B, C and D are adjustable coefficients from the *liquidus* curve of the binary system, T is the temperature (K), T_{fus} is the melting temperature of the salt, m is the molality of the saturated solution (mol · kg^{-1}), and m_0 is equal to $1/r M_{H_2O}$, and M_{H_2O} is the molar mass of water in kg · mol^{-1} and r is the hydration number. The reference state for the activity coefficient is the infinitely dilute solution.

It is not trivial to calculate m and then, the solubility. For the binary system $UO_2(NO_3)_2 \cdot 6H_2O - H_2O$ ($r = 6$ is the stable hydrate at 298.15 K) in the temperature range 253 – 334 K, the experimental data (from 1900 to 1984) are fitted to give:

$A = (964.618 \pm 0.0149)$ mol · kg^{-1} · K

$B = -(23711.09 \pm 4.57)$ mol · kg^{-1}

$C = -(172.094 \pm 0.00261)$ mol · kg^{-1} · K^{-1}

$D = (0.3187 \pm 0.0000489)$ mol · kg^{-1}

$m_0 = (9.25 \pm 0.0149)$ mol · kg^{-1}

from which the solubility of $UO_2(NO_3)_2 \cdot 6H_2O$ (cr) can be derived as a function of temperature with $T_{fus} = 333.4$ K. The uncertainties are one standard deviation; the appreciable uncertainty in m_o arises from the fact that r may exceed six below 298.15 K.

At $T = 298.15$ K the calculated solubility is 3.21 mol·kg^{-1}. In a later paper Apelblat and Korin, [98APE/KOR] have used a value of $m_{sat} = 3.323$ mol·kg^{-1}.

However, both these values are at (or beyond) the extremes of the twelve experimental values of m_{sat} at 298.15 K quoted by [95COH/LOR]. These comprise ten very consistent values ranging from 3.213 to 3.26 mol·kg^{-1}, with two outliers at 3.04 and 3.3 mol·kg^{-1}. We therefore prefer to retain the solubility selected for the CODATA Key Values [89COX/WAG], namely m_{sat}(298.15 K) = 3.24 mol·kg^{-1}. This value is essential in the derivation of $S_m^\circ(UO_2^{2+}, aq)$.

This paper [95COH/LOR] contains helpful phase diagrams of the solubility as a function of temperature up to 450 K, where the lower hydrates with $r = 3$ and 2 are in equilibrium with the saturated solution.

[95DUE/PAT]

This paper covers similar ground to the earlier papers by the same group on insertion of uranium oxides with hydrogen, alkali and alkaline earth metals. Here, new enthalpy and Gibbs energy values are given for $Li_{0.12}UO_{2.95}$(cr), $Na_{0.12}UO_{2.95}$(cr), $Zn_{0.12}UO_{2.95}$(cr), $Li_{0.19}U_3O_8$(cr) and $Na_{0.20}U_3O_8$(cr).

As in [93PAT/DUE], the calorimetric reagent was 0.274 mol·dm^{-3} Ce(SO$_4$)$_2$ + 0.484 mol·dm^{-3} H$_3$BO$_3$ + 0.93 mol·dm^{-3} H$_2$SO$_4$, identical to that used by [81COR/OUW2] for the dissolution of UO$_3$(γ) and U$_3$O$_8$(α). The enthalpy change corresponding to reactions:

$$0.12\ LiUO_3(cr) + 0.05\ U_3O_8(\alpha) + 0.73\ UO_3(\gamma) \rightleftharpoons Li_{0.12}UO_{2.95}(cr) \quad (A.54)$$

and

$$0.19\ LiUO_3(cr) + U_3O_8(\alpha) \rightleftharpoons 0.19\ UO_3(\gamma) + Li_{0.19}U_3O_8(cr) \quad (A.55)$$

were obtained as $\Delta_r H_m^\circ((A.54), 298.15\ K) = -(0.18 \pm 0.82)$ kJ·mol^{-1} and $\Delta_r H_m^\circ((A.55), 298.15\ K) = -(1.44 \pm 1.94)$ kJ·mol^{-1}, respectively, from the enthalpies of dissolution of the various components into the same medium as that used by the same group for $Mg_{0.19}UO_{2.95}$ (see section 9.9.1). It should be noted that, in [95DUE/PAT], (Table 6), the value for the enthalpy of dissolution of U$_3$O$_8$ was misprinted as $-(345.61 \pm 1.94)$ kJ·mol^{-1} instead of $-(354.61 \pm 1.94)$ kJ·mol^{-1}.

Using values selected in [92GRE/FUG], we recalculate from equations (A.54) and (A.55):

$$\Delta_f H_m^\circ(Li_{0.12}UO_{2.95}, cr, 298.15\ K) = -(1254.97 \pm 1.23)\ kJ \cdot mol^{-1}$$

and

$\Delta_f H_m^\circ$ (Li$_{0.19}$U$_3$O$_8$, cr, 298.15 K) = $-$ (3632.95 $\pm$ 3.11) kJ $\cdot$ mol^{-1}.

Using the recalculated value (see section 9.3.3.3.3.), $\Delta_f H_m^\circ$ (UO$_{2.95}$, α, 298.15 K) = $-$ (1211.28 $\pm$ 1.28) kJ $\cdot$ mol^{-1}, from the determination of [93PAT/DUE], the enthalpy change corresponding to the insertion of the lithium according to reactions:

$$0.12 \text{ Li(cr)} + \text{UO}_{2.95}(\alpha) \rightleftharpoons \text{Li}_{0.12}\text{UO}_{2.95} \text{ (cr)} \quad (A.56)$$

$$0.19 \text{ Li(cr)} + \text{U}_3\text{O}_8(\alpha) \rightleftharpoons \text{Li}_{0.19}\text{U}_3\text{O}_8(\text{cr}) \quad (A.57)$$

are obtained as $\Delta_r H_m^\circ$ ((A.56), 298.15 K) = $-$ (43.69 $\pm$ 1.78) kJ $\cdot$ mol^{-1} and $\Delta_r H_m^\circ$ ((A.57), 298.15 K) = $-$ (58.16 $\pm$ 3.11) kJ $\cdot$ mol^{-1}, respectively. Calculated per mole of inserted lithium, these values become $-$ (364.08 $\pm$ 14.83) and $-$ (306.10 $\pm$ 16.37) kJ $\cdot$ mol^{-1}, in general agreement with the more precise values of [89DIC/LAW].

From electrochemical measurements analogous to those reported by [93PAT/DUE] for MgUO$_{2.95}$ (see section 9.9.1.), the authors also report $\Delta_r G_m^\circ$ ((A.56), 298.15 K) = $-$ 42.8 kJ $\cdot$ mol^{-1} and $\Delta_r G_m^\circ$ ((A.57), 298.15 K) = $-$ 55.2 kJ $\cdot$ mol^{-1}, close to the calorimetric values, indicating a small entropy effect.

[95DUE/PAT] also report calorimetric and electrochemical measurements on the species, Na$_{0.12}$UO$_{2.95}$(cr) and Na$_{0.20}$U$_3$O$_8$(cr). The enthalpies of solution of the constituents of the equations:

$$0.12 \text{ NaUO}_3(\text{cr}) + 0.05 \text{ U}_3\text{O}_8(\alpha) + 0.73 \text{ UO}_3(\gamma) \rightleftharpoons \text{Na}_{0.12}\text{UO}_{2.95} \text{ (cr)} \quad (A.58)$$

and

$$0.20 \text{ NaUO}_3(\text{cr}) + \text{U}_3\text{O}_8(\alpha) \rightleftharpoons \text{Na}_{0.20}\text{U}_3\text{O}_8 \text{ (cr)} + 0.20 \text{ UO}_3(\gamma) \quad (A.59)$$

led to the values, $\Delta_r H_m^\circ$ ((A.58), 298.15 K) = (4.13 $\pm$ 1.14) kJ $\cdot$ mol^{-1} and $\Delta_r H_m^\circ$ ((A.59)298.15 K) = (2.57 $\pm$ 0.87) kJ $\cdot$ mol^{-1}, respectively. Using NEA adopted enthalpies of formation, we recalculate:

$$\Delta_f H_m^\circ \text{ (Na}_{0.12}\text{UO}_{2.95}, \text{cr, 298.15 K)} = - (1247.37 \pm 1.88 \text{) kJ} \cdot \text{mol}^{-1}$$

and

$$\Delta_f H_m^\circ \text{ (Na}_{0.20}\text{U}_3\text{O}_8, \text{cr, 298.15 K)} = - (3626.45 \pm 3.25) \text{ kJ} \cdot \text{mol}^{-1}.$$

The latter value is *ca.* 3 kJ $\cdot$ mol^{-1} more positive than that listed by the authors.

The enthalpies of insertion according to the reactions:

$$0.12 \text{ Na(cr)} + \text{UO}_{2.95}(\alpha) \rightleftharpoons \text{Na}_{0.12}\text{UO}_{2.95}(\text{cr}) \quad (A.60)$$

and

$$0.20 \text{ Na(cr)} + \text{U}_3\text{O}_8(\alpha) \rightleftharpoons \text{Na}_{0.20}\text{U}_3\text{O}_8(\alpha) \quad (A.61)$$

are obtained as $\Delta_r H_m^\circ$ ((A.60), 298.15 K) = $-$ (36.09 $\pm$ 1.94) kJ $\cdot$ mol^{-1} and $\Delta_r H_m^\circ$ ((A.61), 298.15 K) = $-$ (51.65 $\pm$ 3.14) kJ $\cdot$ mol^{-1}, respectively.

The authors deduce from electrochemical measurements (*cf.* section 9.10.3.3) the values $\Delta_r G_m^\circ$ ((A.60), 298.15 K) = $-$ 37.7 kJ $\cdot$ mol^{-1} and $\Delta_r G_m^\circ$ ((A.61), 298.15 K) =

−50.7 kJ · mol^{-1}. These values are close to the corresponding enthalpies of reaction, showing that the entropies of reaction are small, as expected for a condensed phase reaction.

From the enthalpies of solution of the constituents of the equation:

$$0.12\ ZnO(cr) + 0.49\ UO_3(\gamma) + 0.17\ U_3O_8(\alpha) \rightleftharpoons Zn_{0.12}UO_{2.95}(cr) \quad (A.62)$$

the authors report $\Delta_r H_m^\circ$ ((A.62), 298.15 K) = (11.36 ± 0.78) kJ · mol^{-1}. This value corresponds to the enthalpy of formation of the insertion compound from the binary stoichiometric oxides. Using NEA adopted values, we recalculate:

$$\Delta_f H_m^\circ (Zn_{0.12}UO_{2.95},\ cr,\ 298.15\ K) = -(1238.07 \pm 1.06)\ kJ \cdot mol^{-1}.$$

Using the recalculated value (see section 9.3.3.3.3), $\Delta_f H_m^\circ$ (UO$_{2.95}$, α, 298.15 K) = − (1211.28 ± 1.28) kJ · mol^{-1}, from the determination of [93PAT/DUE], the enthalpy effect corresponding to the insertion of zinc according to the reaction,

$$0.12\ Zn\ (cr) + UO_{2.95}(\alpha) \rightleftharpoons Zn_{0.12}UO_{2.95}(cr) \quad (A.63)$$

is obtained as $\Delta_r H_m^\circ$ ((A.63), 298.15 K) = − (26.79 ± 1.50) kJ · mol^{-1}.

From electrochemical measurements, [95DUE/PAT] also report $\Delta_r G_m^\circ$ ((A.63), 298.15 K) = − 25.2 kJ · mol^{-1}. This value is compatible with the calorimetric result if $\Delta_r S_m^\circ$ ((A.63), 298.15 K) is small, as is very likely.

[95EBB]

Thermal functions for the species, AnO$_3$(g), AnO$_2$(OH)$_2$(g), AnO$_2$Cl$_2$(g) and AnO$_2$F$_2$(g), (with An = U, Np, Pu and Am), have been calculated using known and estimated molecular constants. The trioxide gases were taken to be T–shaped (as previously shown for UO$_3$(g)) and the MO$_2$X$_2$(g) species were assigned C$_{2v}$ symmetry, based on the known structures and bond lengths of the corresponding Cr, Mo and W analogues. The vibration frequencies were also based on the same analogies, and the earlier data for UCl$_4$(g) (not the revised values reported in [95HAA/MAR]). No higher electronic levels were included for the U compounds; those for the Np and Pu compounds were estimated from the levels of the corresponding ions, MO$_2^{2+}$, in solids or in aqueous solution.

The hydroxyl groups were assumed to undergo hindered rotation, with estimated internal rotation barriers.

No uncertainty values are quoted, nor is there any discussion of enthalpies of formation, even where these data are available.

The data for UO$_3$(g), being based on the same experimental data, are very similar to those adopted in [82GLU/GUR], [92GRE/FUG]. For UO$_2$F$_2$(g) and UO$_2$Cl$_2$(g), the calculated entropies are smaller than those adopted in [92GRE/FUG] (and the current update), due predominantly to the assumption of higher vibration frequencies. For these molecules, however, where reasonably well–defined vaporisation data are available, the entropies of the vaporisation reactions are consistent with the existing entropies of the UO_2F_2 species, which are therefore retained (but with increased uncertainties) until experimental vibration frequencies become available.

The author's calculated entropies for $UO_2(OH)_2(g)$ are, however, greater than those calculated by Gorokhov and Sidorova [98GOR/SID] (by as much as 22.6 J · K^{-1} · mol^{-1} at 298.15 K), principally because Ebbinghaus has assumed internal rotation of the OH groups (conceptually equivalent to a very small bending frequency), whereas Gorokhov and Sidorova [98GOR/SID] have assumed a higher bending frequency.

As noted in section 9.3.1.2.1, no data for $UO_2(OH)_2(g)$ are selected for this review.

[95ELI/BID]

This paper is a careful analysis of the emission spectra of aerated U(VI) solutions (mainly at $I = 0.5$ M $HClO_4/NaClO_4$, room temperature) which are know to contain the hydrolysis species, 1:1, 2:1, 3:1, 2:2, 5:3, 7:3, in addition to the dioxouranium(VI) cation. Test solutions studied cover large ranges of pH and uranium concentration. Some data refer to 1 M $HClO_4$ or to basic solutions at variable ionic strength. The assignment of the species is made on the basis of the $\log_{10} {}^*\beta^\circ_{m,n}$ values selected by [92GRE/FUG] (except for the 2:1 species) and corrected for $I = 0.5$ M. pH measurements are given with a precision of 0.04 units (glass electrode filled with 0.5 M $NaClO_4$). Data obtained for pH less than five give clear results on the assignment of lifetimes and spectra of the 1:1, 3:1, 2:2 and 5:3 complexes (where the reviewer has some doubts about the species 2:1). Species 3:1 is predominant in pH range 10–12.

Characteristics of the emission spectrum of UO_2^{2+} are confirmed (emission lines, fluorescence lifetime, activation energy).

By combining data obtained from solutions at constant pH (3, 4, 5, 9.8 and 11) and variable U concentration (3: 10^{-5} to 10^{-3} M; 4: 10^{-4} to 10^{-3} M; 5: 10^{-6} to 10^{-3} M, 9.8: 10^{-6} to 10^{-4} and 11: 5 10^{-5} M) and the reverse (pH: 2.3 to 4.05 at U: 10^{-4} M), fluorescence lifetimes are identified and measured for all species, but the individual spectra of some of the complexes are poorly defined. Some indications are given on the fluorescence intensities.

The given selected lifetime values (in μs) are the following: (1.7 ± 0.2) for 0:1 species, (32.8 ± 2) for 1:1 species, (3.2 ± 0.2) for 1:2 species, (0.4 ± 0.1) for 1:3 species, (9.5 ± 0.3) for 2:2 species, (6.6 ± 0.3) for 5:3 species and (10 ± 2) for 7:3 species.

Comparisons are made with previous literature on the same topic. Some data in the presence of phosphoric acid are also given.

[95FAN/KIM]

The formation of chloride complexes of Cm(III) has been studied at 25°C using time resolved laser fluorescence spectroscopy over the chloride concentration range 0 to 20 mol·kg^{-1}. The authors present a table of concentration equilibrium constants and note that chloride complexes are only important at chloride concentrations larger than 3 m. A detailed thermodynamic analysis of these data is given in [97KON/FAN].

[95HAA/MAR]

The results of a study of the geometry of $UCl_4(g)$ by electron diffraction at a (nozzle) temperature of 900 K are in agreement with a strictly tetrahedral structure with a U–Cl bond distance of 2.503 Å. The gas–phase IR adsorption spectra were recorded from 25 to 3400 cm^{-1} at temperatures from 700 to 900 K, and the stretching and deformation vibrations have been determined to be 337.4 and 71.7 cm^{-1}, respectively.

These findings are confirmed by density function calculations, which suggest a tetrahedral structure with r(U–Cl) = 2.51 Å, and frequencies of 68 and 341 cm^{-1}. Refinement of these calculations, taking into account the experimental IR data allow the two IR inactive modes to be predicted (326.6 and 61.5 cm^{-1}).

There is a short discussion of the comparison of the calculated and experimental entropies of vaporisation, which are in reasonable agreement. These are important new data on the properties of $UCl_4(g)$ which have required that the data for this species (and all the other gaseous U–halides) have been re-evaluated.

[95HOB/KAR]

Data are reported on the solubilities of U and Pu in strong basic media, but highly concentrated in salts (nine different salts are present in the studied solutions). The identified solid phases are $Na_2U_2O_7$ and $PuO_2·xH_2O$. The variations of the concentrations of U and Pu with sodium hydroxide concentration do not allow derivation of thermodynamic values.

[95MOL/MAT]

This paper is reviewed together with [96MOL/GEI].

[95MOR/GLA]

The authors describe the solubility of uranium(VI) oxide at 85°C at high pH, under conditions that may be encountered in cement systems. A careful phase analysis has been made and a number of new uranium silicate phases have been identified. The phase characterisation has been made using electron microscopy, X–ray diffraction and energy dispersive X–ray analysis. The analytical methods have been calibrated against phases of known composition. The composition of the aqueous phase has also been determined. The analytical methods are described in detail. This review has used some of the data in attempt to extract thermodynamic information.

Samples 1–5, in the range 11.0 < pH < 13.1, contain, in addition to the aqueous phase, two solid phases, soddyite $(UO_2)_2SiO_4 \cdot 2 H_2O(s)$ calcium uranate $CaUO_4(s)$. However, the data do not permit a calculation of the equilibrium constant for the two-phase equilibrium,

$$(UO_2)_2SiO_4 \cdot 2H_2O(s) + 2\,Ca^{2+}(aq) + 2\,H_2O(l) \rightleftharpoons 2\,CaUO_4(s) + H_2SiO_4^{2-}(aq) + 6\,H^+(aq) \quad (A.64)$$

because there is no information on the total concentration of dissolved silica. A rough value of the solubility product of $CaUO_4(s)$ can be estimated from the solubility equilibrium,

$$CaUO_4(s) + 2\,H_2O(l) \rightleftharpoons Ca^{2+} + UO_2(OH)_4^{2-} \quad (A.65)$$

using $\log_{10} K_{s,4}^\circ = \log_{10}[U(VI)]_{tot} + \log_{10}[Ca]_{tot}$, where, in the pH range studied, $[U(VI)]_{tot} = [UO_2(OH)_4^{2-}]$. The average value of $\log_{10} K_{s,4}^\circ$ for the first four experimental data is $-(9.3 \pm 0.7)$. By combination of this value with the equilibrium constant for the reaction:

$$UO_2^{2+} + 4\,H_2O(l) \rightleftharpoons UO_2(OH)_4^{2-} + 4\,H^+$$

$\log_{10} {}^*\beta_4^\circ = -(32.4 \pm 0.7)$, from the present review, we obtain for the reaction:

$$CaUO_4(s) + 4\,H^+ \rightleftharpoons Ca^{2+} + UO_2^{2+} + 2\,H_2O(l), \quad (A.66)$$

$\log_{10} {}^*K_{s,0}^\circ = (23.1 \pm 0.9)$.

If this value is taken to be the same at 298.15 K, the Gibbs energy of formation of $CaUO_4(s)$ is calculated to be -1848 kJ $\cdot$ mol^{-1}. This value is considerably different from the value $-(1888.7 \pm 2.0)$ kJ $\cdot$ mol^{-1} selected in [92GRE/FUG], based on sound thermochemical measurements.

The experimental points 16, 34, 38 and 41 [95MOR/GLA] also refer to solutions in equilibrium with two solid phases, soddyite and uranophane. From the determination of the pH and the concentration of Ca it is possible to decide if the system has attained equilibrium, or not. We have:

$$(UO_2)_2SiO_4(s) + H_4SiO_4 + Ca^{2+}(aq) \rightleftharpoons Ca(UO_2)_2(HSiO_4)_2(s) + 2\,H^+(aq) \quad (A.67)$$

for which the ratio $[H^+]^2 / [H_4SiO_4][Ca^{2+}]$ should be constant at equilibrium. This is not the case, as the ratio varies by five orders of magnitude. We have therefore not attempted to extract any thermodynamic data. Finally, the experimental points 44, 45 and 46 of [95MOR/GLA] refer to a system with only one solid phase, soddyite. These data have been used to estimate the equilibrium constant, $\log_{10} {}^*K_s$ (A.68), for the reaction:

$$(UO_2)_2SiO_4(s) + 4\,H^+ \rightleftharpoons 2\,UO_2^{2+} + H_4SiO_4 \tag{A.68}$$

by combining equilibrium constants for the following reactions:

$$(UO_2)_2SiO_4(s) + SiO_2 + 2H_2O(l) + 2\,H^+ \rightleftharpoons 2[UO_2(H_3SiO_4)^+] \tag{A.69}$$

$\log_{10} {}^*K_s$ (A.69) = 3.1 and the equilibrium constant $\log_{10} {}^*K_s$ (A.70) = – 1.84 is selected by this review for the reaction :

$$UO_2^{2+} + H_4SiO_4 \rightleftharpoons UO_2(H_3SiO_4)^+ + H^+ \tag{A.70}$$

This calculation has been made assuming that the equilibrium constant for equation (A.69) is independent of temperature. The value $\log_{10} K_s$ (A.68) = 6.8 is in good agreement with the value $\log_{10} K_s$ (A.68) = (6.3 ± 0.5), which is the mean value calculated in this review from [92NGU/SIL] and [96MOL/GEI] (cf. discussion in section 9.7.3.2.3).

[95NOV/CRA]

A thermodynamic database is presented for predicting the behaviour of trivalent actinides (Am(III) and Pu(III)) in concentrated brines of the Waste Isolation Pilot Plant (WIPP). The ion-interaction approach (Pitzer equations) is applied for activity coefficient calculations. The database provides a set of Pitzer parameters and standard chemical potentials of the relevant species. For the background electrolyte solutions (sea water system) the models of Harvie et al. [84HAR/MOL] and Felmy and Weare [86FEL/WEA] are applied. Pitzer parameters and chemical potentials for the actinide species are taken from various literature publications on Nd(III), Am(III) and Pu(III). The database is used to predict the solubility of trivalent actinides under various conditions (NaCl concentration, pH, CO_2 – fugacity). There are no new experimental data presented in the paper.

[95NOV/ROB]

The Pitzer equations were applied to the thermodynamic modelling of Np(V) solubilities in carbonate-free and carbonate-containing $NaClO_4$ and NaCl solutions. A set of Pitzer parameters and formation constants at $I = 0$ was proposed for the aqueous Np(V) hydroxide and carbonate complexes, and solubility constants at $I = 0$ for $NpO_2OH(s)$ and $NaNpO_2CO_3(s)$. In later papers [97NOV/ALM], [98ALM/NOV], which are based on more experimental data, the proposed parameters and constants were refined and extended. Experimental data are reported for the solubility of amorphous Np(V) hydroxide in 0.3–5.6 m NaCl. The details of the solubility study are published in [96ROB/SIL].

[95PAL/NGU]

This paper gives formation constants for f the species, $(UO_2)_m(OH)_n^{2m-n}$, from UO_2^{2+} and H_2O as reactants. The following $(n;m)$ values are (2;2), (5;3), (7;3), (8;3) and (10;3). The precipitation of sparingly soluble sodium uranate makes it impossible to study the hydrolysis of uranium(VI) in sodium perchlorate media at higher pH. To circumvent this problem Palmer and Nguyen-Trung have used tetramethylammonium trifluoromethanesulphonate $((CH_3)_4NCF_3SO_3)$ as the ionic medium and tetramethylammonium hydroxide $((CH_3)_4NOH)$ to change the hydrogen ion concentration. The ionic strength was 0.1 mol · kg^{-1} and the range of the variations of the U(VI) and $-\log_{10}[H^+]$ concentrations were, respectively, $0.475 \cdot 10^{-3}$ to $4.94 \cdot 10^{-3}$ mol · kg^{-1} and 2.429 to 11.898, at $t = (25 \pm 0.1)°C$. The values of $\log_{10} {}^*\beta_{n,m}$ and $\log_{10} {}^*\beta_{n,m}^\circ$, obtained by the authors using an extended Debye-Hückel approach to calculate the values at zero ionic strength are given in Table A-19 on a molal scale.

Table A-19: Equilibrium constants at $I = 0.1$ m and $I = 0$ for $(UO_2)_m(OH)_n^{2m-n}$ species.

n:m	$\log_{10} {}^*\beta_{n,m}$	$\log_{10} {}^*\beta_{n,m}^\circ$
2;2	$-(5.77 \pm 0.01)$	$-(5.51 \pm 0.04)$
5;3	$-(16.10 \pm 0.01)$	$-(15.33 \pm 0.12)$
7;3	$-(28.80 \pm 0.04)$	$-(27.77 \pm 0.09)$
8;3	$-(37.64 \pm 0.07)$	$-(37.65 \pm 0.14)$
10;3	$-(60.56 \pm 0.08)$	$-(62.4 \pm 0.3)$

Another set of $\log_{10} {}^*\beta_{n,m}$ constants is given fixing $\log_{10} {}^*\beta_{1,1}$ and $\log_{10} {}^*\beta_{7,4}$ (see below).

The experiments have been made in a laboratory with extensive experience of potentiometric measurements and their interpretation, as also detailed in the paper.

The reviewers obtained the primary experimental data from the authors and these are shown as Z (the average number of coordinated hydroxides per uranium) vs. $-\log_{10}[H^+]$ in the following Figure A-5 and Figure A-6. The Figure A-5 gives the experimental data Z vs $(-\log_{10}[H^+])$ from [95PAL/NGU]. The corresponding calculated values given by the curves using the equilibrium constants of Palmer and Nguyen-Trung are given in Figure A-6 for $-\log_{10}[H^+]$ in the range 3 to 7.5.

Figure A-5. Experimental data of [95PAL/NGU] as Z versus − $\log_{10}[H^+]$

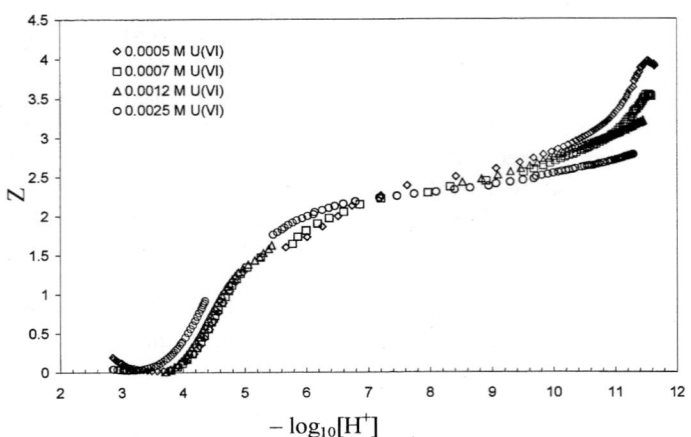

− $\log_{10}[H^+]$

Figure A-5 and Figure A-6 immediately indicate that the chemical model used by the authors is not satisfactory at pH > 7. We have therefore reinterpreted the experimental data using the LETAGROP program, with the same error carrying variables as in [95PAL/NGU], but giving all data equal weight. In Table A-20 and Table A-21, we describe the re-interpretation in some detail.

In the first attempts, the set of equilibrium constants given by Palmer and Nguyen-Trung, with fixed values of two constants: $\log_{10}{}^*\beta_{1,1} = -5.50$ (fixed), $\log_{10}{}^*\beta_{2,2} = -(5.7 \pm 0.01)$, $\log_{10}{}^*\beta_{5,3} = -(16.18 \pm 0.01)$, $\log_{10}{}^*\beta_{7,3} = -(28.25 \pm 0.04)$, $\log_{10}{}^*\beta_{8,3} = -(37.62 \pm 0.07)$, $\log_{10}{}^*\beta_{10,3} = -(60.53 \pm 0.08)$, $\log_{10}{}^*\beta_{7,4} = -22.76$ (fixed) (Model 1). We have also re-interpreted the data without any fixed values of the constants (Model 2). σ(Z) is the standard deviation in Z in the LETAGROP refinement. Titrations 5, 6, 7, 8 and 10 up to a value of $Z_{max} = -3.33$, were used in the refinement. The result are given in Table A-22.

Table A-20: − $\log_{10}[H^+]$ ranges used in the various titrations.

Titration	1	2	3	4	5	6	7	8	9	10	11
− $\log_{10}[H^+]$	3.1– 6.06	3.0– 6.06	3.4– 6.09	3.9– 5.7	3.7– 11.45	3.7– 11.30	2.8– 11.3	2.8– 4.15	2.85– 11.02	2.82– 11.86	2.49– 5.3

Table A-21: Tests of various chemical models and refinement of the corresponding equilibrium constants.

	$\log_{10} {}^*\beta^\circ_{n,m}$	
Species (n,m)	Model 1	Model 2
1,1	− 5.50 (fixed)	
2,2	− (5.61 ± 0.25)	− (5.50 ± 0.25)
5,3	− (16.15 ± 0.29)	− (15.94 ± 0.25)
7,3	− (28.35 ± 0.16)	− (28.35 ± 0.14)
8,3	− 41 (max–value)	− 41.5 (max–value)
10,3	− 64.6 (max–value)	− 64.1 (max–value)
7,4	− 22.76 (fixed)	
	$\sigma(Z) = 0.40$	$\sigma(Z) = 0.40$

In both models, the magnitude of the equilibrium constants for (2:2), (5:3) and (7:3) are in fair agreement with the values in [95PAL/NGU]. For (8:3) and (10:3) only an upper limit was obtained. The uncertainties in the equilibrium constants and the value of $\sigma(Z)$ are large. We also tried to replace the tri-nuclear complexes by $UO_2(OH)_3^-$ and $UO_2(OH)_4^{2-}$, but the fitting did not improve.

In a second attempt we made a refinement using all titrations, but including only the experimental data up to pH = 7.2. The following result was obtained: $\log_{10} {}^*\beta_{1,1} = -5.50$ (fixed), $\log_{10} {}^*\beta_{2,2} = -(5.79 \pm 0.10)$, $\log_{10} {}^*\beta_{5,3} = -(16.38 \pm 0.15)$, $\log_{10} {}^*\beta_{7,3} = -(28.68 \pm 0.23)$, $\log_{10} {}^*\beta_{7,4} = -22.76$ (fixed), $\sigma(Z) = 0.073$. This result is in good agreement with the results of Palmer and Nguyen-Trung, but the uncertainty in the constants is approximately ten times larger than they report. A comparison between experimental data and calculated Z-curves using these constants is shown in Figure A-6.

Figure A-6. Experimental data Z vs ($-\log[H^+]$) from [95PAL/NGU] and the corresponding calculated values (the curves) from the equilibrium constants obtained by this review using the LETAGROP least-squares program. The calculated curves using the equilibrium constants from [95PAL/NGU] practically coincide with the experimental point.

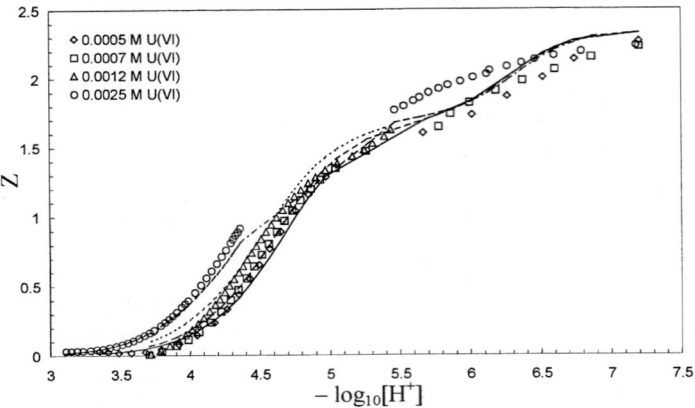

The stoichiometry of the uranium(VI) complexes at high pH has also been studied by direct structure methods using EXAFS technique, [98CLA/CON], [99WAH/MOL] and [2000MOL/REI2]. These studies, *vide infra*, show conclusively that no polynuclear species are formed even at ten times higher total concentrations than used by Palmer and Nguyen-Trung. However, the EXAFS studies were made at much higher hydroxide concentrations than used by Palmer and Nguyen-Trung. The conclusion of this review is that the data in [95PAL/NGU] are affected by large errors at higher pH, while the experimental data at pH < 6 seem to be more precise. The reviewers do not find that the proposed stoichiometry and equilibrium constants for the (8:3) and (10:3) complexes have been established. The data at lower pH can be used to evaluate equilibrium constants that are in good agreement with previously published data.

In conclusion, this review includes the value of $\log_{10} {}^*\beta_{7,3} = -(28.80 \pm 0.04)$ in the data for the selection of the equilibrium constants. Palmer and Nguyen-Trung give $\log_{10} {}^*\beta_{7,3}^\circ = -(27.77 \pm 0.09)$. This review calculates, using the SIT and auxiliary data for NaClO$_4$, with $\varepsilon((UO_2)_3(OH)_7^-, Na^+) = (0.00 \pm 0.05)$ kg · mol^{-1}, $\log_{10} {}^*\beta_{7,3}^\circ = -(28.40 \pm 0.04)$.

[95PUI/BRU]

This paper contains no new data, but presents a review of existing studies on Tc aqueous and solid-state thermodynamic properties with estimations of heat capacities and entropies not available in the literature. The final results are presented in terms of tabulated data for $\Delta_f G_m^\circ$, S_m° and $C_{p,m}^\circ$ with input modules that can be directly imported into the EQ3/6 database. It should be noted that [99RAR/RAN] considered this and earlier reviews, but derived their own thermodynamic parameters based on more recent data. Some species are considered in this review that were not assigned values by [99RAR/RAN], such as those for $HTcO_4$(cr), $HTcO_4$(aq), $HTcO_4^-$ and H_2TcO_4(aq), which must be considered suspect due to the difficulty in obtaining significant quantities of these species under unambiguous experimental conditions.

[95RAI/FEL]

The authors have measured the solubility of amorphous MO_2(am) for M = Th and U, as a function of the concentration of carbonate and NaOH. The experiments are described in some detail, which is important because there are a number of experimental difficulties in studies of this type. The authors have taken adequate precautions to avoid oxidation of U(IV) and they have tried to eliminate colloidal particles when sampling. A remaining source of uncertainty is the rate of change of the crystallinity of the solid used. However, the procedures used are in the opinion of this review adequate to draw qualitative or semi-quantitative conclusions from the data. Most of the experiments have been made at very high pH and it seems questionable to use pH measurements calibrated against buffers under these conditions. The hydroxide concentrations can be well estimated from the analytical composition of the test solutions. The authors confirm that the limiting carbonate complex formed by the two M(IV) ions is $M(CO_3)_5^{6-}$; they also give strong evidence for the formation of mixed hydroxide/carbonate complexes at high hydroxide concentrations. Rai et al. propose compositions such as $M(OH)_4CO_3^{2-}$ and $M(OH)_3CO_3^-$. The solubility of ThO_2(am) at constant carbonate concentration and varying total concentrations of NaOH has a slope of -2 (on a logarithmic scale), indicating a stoichiometry, $Th(CO_3)_n(OH)_2$, where the stoichiometric coefficient n, cannot be determined as the carbonate concentration is constant. Other investigators have proposed the formation of ternary hydroxide/carbonate complexes in the Th(IV) system [94OST/BRU], but at a much lower pH. In this case the number of coordinated carbonate ligands could be determined by varying the partial pressure of CO_2. It should be pointed out that this complex should be present in very small amounts at the high pH used by Rai et al. It is not straight-forward to suggest the stoichiometric composition of these complexes using information on the co-ordination chemistry of these elements. The limiting carbonato complex, $M(CO_3)_5^{6-}$, is ten-coordinate, while the coordination number of the limiting hydroxide complex, $M(OH)_4$(aq), is unknown. There might be substantial changes both in coordination number and coordination geometry when ternary complexes are formed. The reviewers find it unlikely that a moderate increase in the concentration of hydroxide would result in the replacement of four carbonate ligands from $M(CO_3)_5^{6-}$; a species such as $M(CO_3)_4(OH)_2^{6-}$ seems more plausible. This is certainly an area where more experimental work is required. Some additional studies on this system have been made in a later study by Rai et al. [98RAI/FEL].

[95RAI/FEL2]

The paper describes a study on the sulphate complexation of trivalent actinides. For this purpose existing literature data for the Am(III) sulphate system have been re-examined and new experimental solubility data have been measured for $NdPO_4(cr)$ in the presence of sulphate. $NdPO_4(cr)$ is considered as analog for $AmPO_4(cr)$. The experimental and literature data are evaluated by applying the ion–interaction (Pitzer) approach. The relatively weak sulphate complexation of Am(III) is treated by introducing a $\beta^{(2)}$ Pitzer parameter for the interaction $Am^{3+}-SO_4^{2-}$. Am(III)–sulphate complexation constants are not incorporated into the thermodynamic model explicitly. The presented model describes the experimental solubility data fairly well.

[95SAL/JAY]

This conference paper reports measurements of the oxygen potential over mixtures of $Rb_2U_4O_{11}(cr) + Rb_2U_4O_{12}(cr)$, most of which have been reported subsequently in more detail in the journal paper [96IYE/JAY]. However, one result reported in this paper, but not in [96IYE/JAY], is the enthalpy of oxidation to $Rb_2U_4O_{13}(cr)$ at 673 K:

$$Rb_2U_4O_{11}(cr) + O_2(g) \rightleftharpoons Rb_2U_4O_{13}(cr)$$

$$\Delta_r H_m (673 \text{ K}) = -(279 \pm 15) \text{ kJ} \cdot \text{mol}^{-1}.$$

In the absence any experimental data on the enthalpy of formation of $Rb_2U_4O_{13}(cr)$, this value cannot be processed further.

[95YAJ/KAW]

The solubility of $UO_2(cr)$, which was prepared by reduction of ammonium diuranate at 650°C for two hours (50:50 mixture of N_2 and H_2), was measured in 0.1 M $NaClO_4$ (not as stated in the abstract, NaCl) at 25°C. The crystallinity was confirmed by XRD and the observed pattern is shown.

The approach to equilibrium was varied from under- and over-saturation. In the former experiments, a uranium(IV) stock solution was diluted with 0.01 M $Na_2S_2O_4$ to a concentration of ca. 0.003 M (apparently no seeds of $UO_2(cr)$ were added). In the case of under-saturation, 30 mg of $UO_2(cr)$ were added to 30 mL of 0.01 M $Na_2S_2O_4$. The ionic strength was adjusted, then either $HClO_4$ or NaOH was added to give pH values of 2 – 12, which were measured after equilibration lasting 7, 14 and 28 days. No details were given as to how the pH was buffered in the intermediate range when uranium concentrations of ca. 10^{-9} M were measured by ICPMS. Experimental results are only given as figures. In general these authors obtained lower solubilities than Rai et al.

[90RAI/FEL]. Below pH = 2 the slope of $\log_{10}[U]$ versus pH was approximately -4, whereas above pH = 4 the uranium concentrations were independent of pH, corresponding to the reactions:

$$UO_2(cr) + 4H^+ \rightleftharpoons U^{4+} + 2 H_2O(l)$$

$$UO_2(cr) + 2 H_2O(l) \rightleftharpoons U(OH)_4(aq).$$

Activity coefficients were estimated with the Davies equation resulting in reported values of $\log_{10} {}^*K_{s,0}^\circ = (0.34 \pm 0.4)$ and $\log_{10} K_{s,4}^\circ = -(8.7 \pm 0.4)$. Given the form of the Davies equation in the paper, the $\log_{10} {}^*K_{s,0}$ value at 0.1 M ionic strength would have been (1.63 ± 0.4). Applying the SIT to this value gives: $\log_{10} {}^*K_{s,0}^\circ = (0.52 \pm 0.41)$. This indicates that the UO_2 used in this study was only partially crystallised, as the recommended value for $UO_2(cr)$ is $\log_{10} {}^*K_{s,0}^\circ = -(4.86 \pm 0.36)$, [92GRE/FUG].

The value for $\log_{10} K_{s,4}^\circ$ is higher than the solubility of $-(9.47 \pm 0.56)$ averaged over all temperatures in [92GRE/FUG], but is within the combined uncertainties. This apparent agreement for the neutral species in equilibrium with a crystalline solid is at odds with the result for $\log_{10} {}^*K_{s,0}$, being more consistent with an amorphous like solid phase.

[96ALL/SHU]

These papers deal with the identification of precipitates at pH = 7.2, 8.2 and 11.6 (9 weeks ageing) obtained by addition of 1 M NaOH to three U(VI) solutions 0.6 mM in U, HCl 6 mM and Na_2SO_4 1 mM under a CO_2-free Ar atmosphere. U(VI) oxide/hydroxide precipitates are often more or less amorphous so EXAFS is an appropriated technique to identify it by the U–O and U–U bond lengths (R), and O and U coordination number (N). The reference used in this work to identify solid phases is metaschoepite (given as $UO_2(OH)_2 \cdot H_2O$ by the authors) characterised by XRD and obtained at pH 7 by precipitation of uranium with NaOH from a supersaturated 0.1 mM U(VI), 0.1 M HCl solution. Schoepite and its polymorph are described by [92FIN/MIL].

U–O and U–U distances in the solid compound prepared at pH = 7.2, as well as O and U coordination numbers, confirm that it is metaschoepite while U(VI) oxides precipitated at higher pH values probably contain some uranates or polyuranates (bond lengths close to those found in the structural determination of $Na_2UO_4(cr)$, $Na_2U_2O_7(cr)$ or $K_2U_7O_{22}(cr)$, precipitated at pH = 8.2, and $Na_2U_3O_{10} \cdot xH_2O$, precipitated at pH = 11.2). Table 1 gives R and N values and 95 % confidence limits.

Analysis of the EXAFS data is made with the code FEFF6 based on the structure of α–$UO_2(OH)_2$ published in [71TAY/HUR].

[96BER/GEI]

This paper gives spectroscopic evidence (TRLFS) of the existence of the complex, $Ca_2[UO_2(CO_3)_3]$(aq) (di-calcium tri-carbonato uranium(VI)), under such conditions as: 10^{-5} M U, $2 \cdot 10^{-2}$ M Ca^{2+}, 10^{-3} M CO_3^{2-}, pH = 8, which can arise in some seepage waters from U mine tailing. The evidence presented include:

- a specific TRLFS spectrum for the solution which is supposed to contain the complex. It differs from those of UO_2^{2+}, hydrolysed U(VI) species, and sulphato or phosphato U(VI) complexes (recall the carbonato complex, $UO_2(CO_3)_3^{4-}$, does not show any emission spectrum). The characteristic peaks of the spectrum (natural water) are located at 463.5, 483.6, 502,8, 524.03 and 555.4 nm and the lifetime is $\tau = (64 \pm 17)$ ns (much shorter than for the previously cited U(VI) species, which is in the μs range). These values depend on the origin of the solution, natural or synthetic. In the latter case, τ is shorter, (43 ± 12) ns and the position of the emission maxima differ from one to five nm depending on the wavelength. Furthermore, the substitution of Mg for Ca drastically modifies the spectrum,

- similarities in the main emission maxima of the postulated species with those of the spectra of the solid compounds: liebigite $Ca_2[UO_2(CO_3)_3] \cdot 10\ H_2O$ and zellerite $Ca[UO_2(CO_3)_2] \cdot 5\ H_2O$. Corresponding τ values are however different,

- the postulated species does not sorb on an anionic exchanger, in contrast to the complex, $UO_2(CO_3)_3^{4-}$, that sorbs strongly.

The equilibrium constant for :

$$2\ Ca^{2+} + UO_2(CO_3)_3^{4-} \rightleftharpoons Ca_2[UO_2(CO_3)_3]$$

at $I = 0.1$ M, $t = 25°C$, derived from the spectroscopic measurements, is $\log_{10} K = (5.0 \pm 0.7)$. Using $\log_{10} \beta(UO_2(CO_3)_3^{4-}) = (21.8 \pm 0.1)$ [82MAY], (the selected value by [92GRE/FUG] is (21.6 ± 0.05) at zero ionic strength), $\log_{10} \beta = (26.8 \pm 0.8)$ is calculated for:

$$2\ Ca^{2+} + UO_2^{2+} + 3\ CO_3^{2-} \rightleftharpoons Ca_2[UO_2(CO_3)_3].$$

The identification of this complex comes from the impossibility of fitting the TRLFS measurements of a natural sample of water with regard to the known TRLFS characteristics of the expected species on the basis of [92GRE/FUG] selected data.

The paper gives spectroscopic parameters for the U species.

The paper [97AMA/GEI] reports a synthesis of liebigite (orthorhombic) with fluorescence signals the same as for the natural mineral. In a further study, Amayri *et al.* [98AMA/BER] report the solubility of liebigite to be (9.9 ± 0.5) g · L^{-1} (0.1 M NaClO$_4$, 25°C , pH = 8), U being present in the solution as $Ca_2[UO_2(CO_3)_3]$(aq).

[96BRE/GEI]

In this work, precipitation of uranium in the presence of phosphate anions is avoided by using low concentrations of the element, less than 10^{-4} M (potentiometry) and 10^{-5} M (TRLFS). For potentiometric measurements a stock solution is used, 10^{-4} M $UO_2(NO_3)_2 \cdot 6H_2O$. Aliquots with variable contents of U(VI), 1.641 to $8.071 \cdot 10^{-5}$ M and 10^{-3} M H_3PO_4 are titrated with NaOH at 21.8 to 22.4°C, pH ranging from 2.65 to 4.97 (variable ionic strength 7.2 to 9.9 10^{-3} M). pH is carefully measured with a mean deviation of 0.037 pH units. The selected glass electrode is calibrated with six different NIST buffers (pH = 6.86 to 1.00). More than 300 experimental data from 20 titrations are used to calculate $\log_{10} \beta°$ values for the species identified as $UO_2(HPO_4)(aq)$, $UO_2(H_2PO_4)^+$, $UO_2(H_2PO_4)_2(aq)$ (at 22°C). Correction for ionic strength is made according to the Davies equation. The value are :

$$\log_{10} \beta° (UO_2(HPO_4), aq) = (19.87 \pm 0.29),$$

$$\log_{10} \beta° (UO_2(H_2PO_4)^+) = (22.58 \pm 0.17),$$

$$\log_{10} \beta° (UO_2(H_2PO_4)_2, aq) = (46.9 \pm 0.22),$$

where uncertainties are quoted at the 3σ level.

The composition and equilibrium constants in the potentiometric study were deduced by measuring the free ligand concentration by means of the free hydrogen ion concentration. The accuracy of the data is determined by the difference between the total phosphate concentration and the sum of concentrations of $H_3PO_4(aq)$, $H_2PO_4^-$, HPO_4^{2-} and PO_4^{3-}. As the total concentration of U(VI) is only 10% of the total phosphate concentration and the accuracy is low. The error is largest in the solutions below pH = 4. The experiments have been made at low ionic strength; this does not mean that the extrapolation to zero ionic strength is facilitated. On the contrary, the variation in the activity coefficients of reactants and products are larger at low ionic strength than at higher when the total concentration of the reactants are varied. For this reason, the review has not used the potentiometric equilibrium constants when evaluating the selected values. The situation is very different in the spectroscopic measurements where the concentrations of the different complexes are measured directly without any thermodynamic assumptions. Therefore, only these data have been considered by the review.

TRLFS measurements are made on two sets of solutions: 10^{-5} M U(VI), PO_4^{3-} concentrations up to 10^{-3} M, pH = 3.02 – 3.06 and $5 \cdot 10^{-6}$ M U(VI), 10^{-3} M PO_4^{3-}, pH ranging from 3.06 to 3.46, both at (20 ± 0.5)°C. It is expected that three species are present: UO_2^{2+}, $UO_2(H_2PO_4)^+$ and $UO_2(HPO_4)(aq)$. Characteristics of the fluorescence spectra of the dioxouranium(VI) cation agree with what is known ($\tau = (1.1 \pm 0.1)$ µs, from previously reported values are 0.83 µs [93MEI/KAT] and 1.0 µs [94KAT/MEI]). Fluorescence spectra of the complexes cannot be resolved with regard to the maxima wavelengths and lifetime (494.39, 515.92, 536.84 and 564.37 nm, $\tau = (14.0 \pm 1.3)$ µs) when the conditions of recording the spectra are the same. But varying the delay time and gate width when recording the spectra gives the possibility of differentiating between the two complexes and consequently calculating the $\log_{10} \beta°$ values. They are:

$$\log_{10} \beta°(UO_2(HPO_4), aq) = (19.53 \pm 0.14),$$
$$\log_{10} \beta°(UO_2(H_2PO_4)^+) = (22.31 \pm 0.16),$$

where the ionic strength correction has been made using the Davies equations for I from 0.9 to $1.1 \cdot 10^{-4}$ M. This review has recalculated them using the SIT and finds the same values for I ranging from 10^{-4} to 0.9 M.

No valuable comparisons of fluorescence spectra in phosphoric solutions can be made with previous results where the species are not identified.

[96DIA/GAR]

The aim of this work is to measure the solubilities of elements coprecipitated with uranium from U solutions as the pH is increased. Solubilities of these elements from such solids are less than the solubilities measured from the pure phase obtained from single component solutions. The paper also gives data on the solubilities of Na-polyuranates. The solids are prepared from titration under an inert atmosphere (Ar or N_2) of solutions 2.5 or $4.5 \cdot 10^{-3}$ m in uranium / 5 m NaCl with carbonate-free 0.1 m NaOH / 5 m NaCl (25 ± 0.2)°C. Solids are selected for pH = 5.5, and pH between 5.7 and 10.2, and they are considered ready for starting solubilities experiments when the U solubility concentration in the mother liquor remains constant for two weeks. All precipitates have a Na/U ratio of 0.3. pH is measured and corrected to take into account the high ionic strength.

Solubilities of U(VI) at pH = 10.25, in 5 m NaCl are $2 \cdot 10^{-5}$ m (filtration through 1.2 nm pore size) up to 400 hours and $5 \cdot 10^{-5}$ m at pH = 5.5 (filtration through 0.2 μm pore size) after 500 hours. The U(VI) concentration decreases in between, but the data are scattered.

No thermodynamic data can be derived from the solubility values reported in this paper.

[96GEI/BRA]

Two sets of uranium(VI) sulphate solutions and one set without sulphate are studied by TRLFS. In each case the integrated fluorescence signal (over 450 to 600 nm to include the four emission maxima) is fitted with a sum of experimental decay functions to calculate lifetimes and fractions of the different species.

Solutions are prepared from a stock solution made with $Na_2U_2O_7(s)$ dissolved in 1 M $HClO_4$, NaOH, $NaClO_4$ and 1 M H_2SO_4. pH is measured within ± 0.05 pH units at 20°C.

Solutions 10^{-4} M U(VI) at pH 1, 10^{-5} M U(VI) at pH = 2.17, both I = 0.1 M (HClO$_4$ + NaClO$_4$) give lifetimes for UO_2^{2+}, respectively, equal to (1.57 ± 0.15) μs and (1.14 ± 0.10) μs (maxima are at 488.3, 509.8, 533.5 and 488, 510, and 534 nm). These data are in agreement with previous work under similar conditions [94KAT/MEI], [95MOU/DEC], [95ELI/BID], [94CZE/BUC].

Data from solutions 10^{-5} M U(VI), 0.05 M sulphate, pH = 2 to 6, I = 0.2 M show clearly the presence of five species. Their relative amounts agree with the speciation diagram established with the selected data from [92GRE/FUG] (p. 241 and 683), although seven are predicted following [92GRE/FUG]. Lifetimes are given in Table A-22. Thermodynamic data are derived for the first and second complexes at 25°C.

Table A-22: Equilibrium constants and lifetime measurements of UO_2^{2+}, $UO_2SO_4(aq)$, $UO_2(SO_4)_2^{2-}$, UO_2OH^+ and $UO_2(OH)_2(aq)$.

Species	$\log_{10} \beta_n$, (I = 0.2 M)	$\log_{10} \beta_n^o$	lifetime (μs)
UO_2^{2+}			< 3
$UO_2SO_4(aq)$	(2.42 ± 0.14)	(3.35 ± 0.15)	(4.7 ± 0.3)
$UO_2(SO_4)_2^{2-}$	(3.30 ± 0.17)	(4.21 ± 0.17)	(11.5 ± 0.3)
UO_2OH^+			(8.3 ± 0.3)
$UO_2(OH)_2(aq)$			(18.1 ± 0.3)

Data from solutions: 10^{-5} M U(VI), 10^{-3} to 0.25 M sulphate, pH = 2, I = 1 M and 0.5 M sulphate, I = 1.5 M show clearly the presence of four species: UO_2^{2+}, $UO_2SO_4(aq)$, $UO_2(SO_4)_2^{2-}$ and $UO_2(SO_4)_3^{4-}$ whose domains of predominance agree with the selected data from [92GRE/FUG] (pp. 241 and 683), but do not fit exactly for the first and second complexes. Lifetimes and $\log_{10} \beta_n$ (n =1 – 3) values (25°C) derived from the spectra are given in Table A-23:

Table A-23: Equilibrium constants and lifetime measurements of UO_2^{2+}, $UO_2SO_4(aq)$, $UO_2(SO_4)_2^{2-}$ and $UO_2(SO_4)_3^{4-}$.

Species	$\log_{10} \beta_n$, (I = 1M)	$\log_{10} \beta_n^o$	lifetime (μs)
UO_2^{2+}			(2.7 ± 0.3)
$UO_2SO_4(aq)$	(1.88 ± 0.27)	(3.33 ± 0.29)	(4.3 ± 0.5)
$UO_2(SO_4)_2^{2-}$	(2.9 ± 0.4)	(4.29 ± 0.45)	(11.0 ± 1.0)
$UO_2(SO_4)_3^{4-}$	(3.2 ± 0.25)	(2.62 ± 0.45)	(18.3 ± 1.0)

The values of $\log_{10} \beta_n^o$ at zero ionic strength are calculated by the authors according to the SIT theory following :

$$\log_{10} \beta_n^o = \log_{10} \beta_n - \Delta z^2 D + \Delta \varepsilon I_m$$

the symbols having their usual meaning. The ε values are: $\varepsilon(UO_2^{2+}, ClO_4^-) = 0.46$, $\varepsilon(SO_4^{2-}, Na^+) = -0.12$, $\varepsilon(UO_2(SO_4)_2^{2-}, Na^+) = -0.12$ and $\varepsilon(UO_2(SO_4)_3^{4-}, Na^+) = -0.24$ kg · mol^{-1} estimated from $P_2O_7^{4-}$. The calculated values are given in Table A-22 and Table A-23. They are somewhat higher than those selected in [92GRE/FUG] for $n = 1$ and 2, respectively, $\log_{10} \beta_1^\circ = (3.15 \pm 0.02)$ and $\log_{10} \beta_2^\circ = (4.14 \pm 0.07)$. This review suggests using $\varepsilon(UO_2(CO_3)_3^{4-}, Na^+) = -(0.01 \pm 0.11)$ kg · mol^{-1} as an estimate for $\varepsilon(UO_2(SO_4)_3^{4-}, Na^+)$ and increasing the uncertainty to ± 0.2 kg · mol^{-1} which will give $\log_{10} \beta_3^\circ = (3.02 \pm 0.38)$.

[96IYE/JAY]

This paper reports the preparation and identification of $Rb_2U_4O_{11}$(cr) and measurements of oxygen potential over its mixtures with $Rb_2U_4O_{12}$(cr).

The reported new compound, $Rb_2U_4O_{11}$(cr), was obtained by reacting $Rb_2U_2O_7$(cr) and UO_2(cr) under argon at 1473 K in an alumina boat. The X–ray diffraction pattern was indexed and indicates a tetragonal cell, with $a = 9.378$ Å and $c = 19.922$ Å. Unspecified analyses for Rb and U were said to confirm the suggested composition.

As the cell used was:

$$Pt \mid Rb_2U_4O_{11} + Rb_2U_4O_{12} \mid CSZ \mid Ni + NiO \mid Pt$$

where CSZ is calcia (15 mol %) stabilised zirconia, the cell reaction corresponds to:

$$Rb_2U_4O_{11} (cr) + NiO (cr) \rightleftharpoons Rb_2U_4O_{12} (cr) + Ni (cr) \qquad (A.71).$$

A least square fit of consistent individual results on two samples was given as:

$$E \text{ (mV)} = 802 - 0.645 \cdot T \qquad (985 - 1186 \text{ K}).$$

The recalculated Gibbs energy for reaction (A.71) is:

$$\Delta_r G_m (A.71) = -154.76 + 0.1245 \cdot T \text{ kJ} \cdot \text{mol}^{-1}.$$

The authors use the molar Gibbs energy of NiO from [87VEN/IYE], $\Delta_f G_m (NiO, cr, T) = -231.76 + 0.0839 \cdot T$ kJ · mol^{-1}, but as for similar papers from the same group, we prefer to use that derived from the assessment of [90TAY/DIN], based on a larger number of Gibbs energy studies involving NiO. This, in the relevant temperature range, gives $\Delta_f G_m (NiO, cr, T) = -234.22 + 0.08504 \cdot T$ kJ · mol^{-1}.

Hence the derived Gibbs energy of the oxidation reaction (A.72):

$$Rb_2U_4O_{11} (cr) + 0.5 O_2 (g) \rightleftharpoons Rb_2U_4O_{12} (cr) \qquad (A.72)$$

becomes $\qquad \Delta_r G_m (A.72) = -388.98 + 0.2095\, T \text{ kJ} \cdot \text{mol}^{-1}.$

The authors combine their results with their values for $\Delta_f G_m(Rb_2U_4O_{12}, cr)$ suggested in [92VEN/IYE], but for reasons quoted in the review of that paper, the data for this compound have not been selected for the current assessment, so the oxygen potentials in the current paper cannot be processed further.

[96JAY/IYE]

The authors report measurements of the pressure of Rb(g) by mass–loss Knudsen effusion and the oxygen activity (emf with CaO–ZrO$_2$ electrolyte) in the three–phase field RbUO$_3$(cr) + Rb$_2$UO$_4$(cr) + Rb$_2$U$_2$O$_7$(cr).

Rb$_2$UO$_4$(cr) and Rb$_2$U$_2$O$_7$(cr) were prepared by heating appropriate amounts of U$_3$O$_8$(cr) with Rb$_2$CO$_3$(cr) in air at 1000 K, respectively, for 16 hours in alumina boats, and RbUO$_3$(cr) by reducing Rb$_2$U$_2$O$_7$(cr) with hydrogen at 1000 K. The products were characterised only by X–ray diffraction.

The effusion measurements were carried out from 1305 to 1459 K in a boron nitride cell, attached to a Cahn microbalance.

The oxygen potentials were studied using a CaO–ZrO$_2$ electrolyte tube in flowing argon, from 920 to 1100K. Air ($p_{O_2(g)}$ = 0.21 bar) and Ni/NiO were used as the reference electrodes.

The rubidium pressures were fitted to the equation:

$$\log_{10}(p_{Rb}/\text{bar}) = 0.95 - 7977 \cdot T^{-1} \qquad (1117 \text{ to } 1193 \text{ K})$$

over the three–phase field, RbUO$_3$(cr) + Rb$_2$UO$_4$(cr) + Rb$_2$U$_2$O$_7$(cr).

As for the earlier similar study on the Na compounds, we have used the values assessed by Taylor and Dinsdale [90TAY/DIN], for the Gibbs energies of formation of NiO(cr), $\Delta_f G_m$(NiO, cr, T) = $-$ 234220 + 85.042 T J mol^{-1} from 900 to 1200 K, to convert the study with a Ni–NiO reference electrode to oxygen potentials. Our derived values for the rubidium and oxygen potentials in the relevant three–phase field are then:

$$RT \ln(p_{Rb}/\text{bar}) = -152719 + 18.188\, T \quad \text{J} \cdot \text{mol}^{-1} \qquad (1305 \text{ to } 1495 \text{ K})$$

$$RT \ln(p_{O_2}/\text{bar}) = -531440 + 189.726\, T \quad \text{J} \cdot \text{mol}^{-1} \qquad (981 \text{ to } 1090 \text{ K})$$
(air ref.)

$$RT \ln(p_{O_2}/\text{bar}) = -558749 + 214.081\, T \quad \text{J} \cdot \text{mol}^{-1} \qquad (920 \text{ to } 1100 \text{ K})$$
(Ni/NiO ref.)

The first expression for the oxygen potential is different from that given by the authors, who seem to have neglected the correction from air to pure oxygen. The mean of the last two expressions for the oxygen potentials is the preferred relation:

$$RT \ln(p_{O_2}/\text{bar}) = -545095 + 201.903\, T \quad \text{J} \cdot \text{mol}^{-1} \qquad (920 \text{ to } 1100 \text{ K})$$
(mean)

However, the coefficient of T in the expression for $RT \ln(p_{Rb}/\text{bar})$, which should correspond to the entropy change for the relevant reaction:

$$RbUO_3(cr) + Rb_2UO_4(cr) \rightleftharpoons Rb_2U_2O_7(cr) + Rb(g)$$

is far too small to refer to a vaporisation process – the corresponding terms in the similar studies on $NaUO_3$ [94JAY/IYE] and KUO_3 [99JAY/IYE] are 247.4 and 243.0 $J \cdot K^{-1} \cdot mol^{-1}$, respectively. This suggests that either equilibrium was not reached or maintained in the Knudsen cell, or some adventitious high temperature reaction was occurring.

No further processing of these data has therefore been carried out.

[96KAT/KIM]

This is a carefully made solubility study with U(VI) and Np(VI) at 25°C in a 0.1 M $NaClO_4$ medium. The authors have taken precautions to assure the composition of the solid phases and that the oxidation state of Np(VI) does not change during the experiments. The conversion of pH to $-\log_{10}[H^+]$ has been properly made and both the activity coefficient for H^+, and the value for $\log_{10} K_W$ are close to the values calculated from the SIT and the interaction coefficients in [92GRE/FUG]. Some of the experimental solubility data deviate from the expected linear relations shown in Figure 3 and Figure 4 of this publication, indicating experimental shortcomings. However, the main part of the experiments conform well to the proposed solubility model. The equilibrium constants of Kato et al. have been recalculated to zero ionic strength by this review, using the uncertainty estimate of the authors. We find $\log_{10} K^\circ_{s,0}$ (NpO_2CO_3, cr) = $-(15.46 \pm 0.12)$, $\log_{10} K^\circ_{s,0}$ (UO_2CO_3, cr) = $-(14.94 \pm 0.14)$, and $\log_{10} K^\circ_{s,0}$ ($NpO_3 \cdot H_2O$, cr) = $-(22.32 \pm 0.14)$, and $-(22.34 \pm 0.22)$, $\log_{10} K^\circ_{s,0}$ ($UO_3 \cdot 2H_2O$, cr) = $-(22.75 \pm 0.06)$. The experimental value in 0.1 M $NaClO_4$ converted to $I = 0$ with the SIT, taking into account hydrolysis of U(VI) is $\log_{10} K^\circ_{s,0} = -(22.90 \pm 0.05)$, which is more accurate than that given in [93MEI/KIM2], $\log_{10} K^\circ_{s,0} = -(22.88 \pm 0.19)$. The latter quantity can be compared with the corresponding quantity, $\log_{10} K^\circ_{s,0}$ ($UO_3 \cdot 2H_2O$, cr) = $-(23.19 \pm 0.43)$ calculated from the Gibbs energy of formation of $UO_3 \cdot 2H_2O(cr)$, obtained from calorimetric data selected by [92GRE/FUG]. Sandino [91SAN] has reported $\log_{10} K^\circ_{s,0}$ ($UO_3 \cdot 2H_2O$, cr) = $-(23.06 \pm 0.18)$ and after re-interpretation of the solubility data of [89BRU/SAN] for amorphous and crystalline schoepite (in fact metaschoepite) [92SAN/BRU], $\log_{10} K^\circ_{s,0}$ ($UO_3 \cdot 2H_2O$, am) = $-(21.67 \pm 0.14)$ and $\log_{10} K^\circ_{s,0}$ ($UO_3 \cdot 2H_2O$, cr) = $-(22.03 \pm 0.14)$. These values differ considerably indicating differences in crystallinity between the samples used.

[96KON/BOO]

The IR spectrum of $UF_4(g)$ has been measured from $75-700$ cm^{-1} between 1300 and 1370 K. Absorptions were found at $v_3 = 537$ and $v_4 = 114$ cm^{-1} in good agreement with the (less reliable) matrix isolation data. Earlier electron–diffraction data utilised appreciably higher estimated values of v_4 and these were re–interpreted with the lower experimentally determined data. The electron diffraction data were consistent with a tetrahedral structure with $r(U-F) = 2.056$ Å, and vibration frequencies of 625, 123, 539 and 114 cm^{-1}. With the same electronic contributions as for UCl$_4$, the calculated entropy of UF$_4$(g, 1050 K) is (498.8 ± 2.3) J·K^{-1}·mol^{-1}, in tolerable agreement with that determined from the vapour pressure data and S_m(UF$_4$, cr, 1050 K), (492.8 ± 3.0) J·K^{-1}·mol^{-1}.

The data in this paper have been used in the re–assessment of the vapour pressure of UF$_4$ and the stability of UF$_4$(g) and other uranium fluoride gaseous species.

[96KOV/BOO]

The melting point and heat of fusion of UCl$_3$ were determined by DTA, in a molybdenum crucible.

The UCl$_3$ sample was prepared from UH$_3$ and HCl at 623 to 673 K and purified by transport reaction with I$_2$(g); it was stored in an Ar glove box to minimise reaction with water and air. Six determinations of the melting point gave a mean of (1115 ± 2) K in good agreement with earlier, somewhat less precise data. The integration of the sharp melting peaks gave $\Delta_{fus}H_m^\circ = (49.0 \pm 2.0)$ kJ·mol^{-1}.

Two runs were also made of the IR spectrum of UCl$_3$ at an unspecified temperature. The band at 338 cm^{-1} was attributed to UCl$_4$(g) (formed by the known disproportionation); in the second run where the Mo crucible was outgassed to a pressure of 10^{-8} bar at 1273 K prior to the measurements, an additional band at 275 cm^{-1} was observed. This was attributed to the stretching band of UCl$_3$(g).

[96KUL/MAL]

The behaviour of transplutonium (Am, Cm, Bk, Cf) and rare earth elements was studied in acidic and alkaline ferricyanide solutions. In $0.01 - 1$ M mineral acid solutions containing $0.1 - 0.5$ M K$_3$Fe(CN)$_6$, the precipitation of An(III) ferricyanides, AnFe(CN)$_6 \cdot$xH$_2$O(s), was observed for An = Am, Cm, Bk and Cf. The ratio of metal: ferricyanide in the isolated precipitates was analysed to be 1:1. Gradual reduction of Fe(CN)$_6^{3-}$ to Fe(CN)$_6^{4-}$ in the precipitated transplutonium ferricyanides was ascribed to radiolysis effects in the solids. In alkaline solutions, the different redox behaviour of Am enables its separation from Cm, Bk and Cf. A precipitate of americium(III) hydroxide suspended in $2 - 8$ M NaOH / $0.1 - 0.5$ M K$_3$Fe(CN)$_6$ was dissolved as the AmO$_2^+$ ferricyanide complex. The oxidation of Am(III) to Am(V) was confirmed by spectroscopy. After one day, Am(V) precipitated as a hydroxide assumed to be Na$_2$AmO$_2$(OH)$_3 \cdot$xH$_2$O(s). Neither solubilities nor thermodynamic data are reported in this paper.

[96MEI]

This is a survey article where the author discusses the structure and coordination number of uranium carbonate complexes, both in the solid state and in solution. The review of the solid state structures is useful, but the models used to deduce coordination numbers of complexes in solution are too simplistic in the opinion of the reviewers.

[96MEI/KAT]

The paper includes two sets of data. All experiments are performed in 0.1 M $NaClO_4$ solution.

Stock solutions 10^{-2} M in U are prepared from depleted U_3O_8. Solutions corresponding to the second set of data are spiked with ^{233}U (less than 0.5% daughters) to improve measurements at very low U concentrations by liquid scintillation counting. The glass electrode is calibrated with five standard (pH = 1.7 to 10) solutions as discussed in [97MEI]. p_{CO_2} is imposed by flushing continuously certified N_2/CO_2 gas mixtures into the test solutions (except for 0.03% when air is used). Phase separation is made by ultrafiltration (0.45 or 0.2 μm) and no colloids with size above 1.3 nm are detected.

The first set of data deals with solubilities of UO_2CO_3(s) under 8% CO_2 and $UO_3 \cdot 2H_2O$(s) under 0.3% CO_2 as a function of pH between 2.8 and 4.6 (I = 0.1 M $NaClO_4$, t = 25°C). It supplements the data presented and discussed in [93MEI/KIM2], [93MEI/KAT] on the solubility of $UO_3 \cdot 2H_2O$(s) under 0.03% air and UO_2CO_3(s) under 1% and 100% CO_2. Reinterpretation of all the data is given on the basis of the solubility product of the solid phases, characterised by XRD as schoepite and rutherfordine, and the presence in the solution of two species, UO_2^{2+} and $(UO_2)_2(OH)_2^{2+}$. The dependencies of solubility on pH (3 to 5) and on $\log_{10}[CO_3^{2-}]$ (-14 to -10) are checked and found to be in agreement with the well-established existence of the two species, as discussed in [93MEI/KAT] and [93MEI/KIM].

For calculation of the solubility products the authors take $\log_{10} {}^*\beta_{2,2}$ = $-(6.00 \pm 0.06)$ which is derived from their previous works as the average values of $-(5.89 \pm 0.12)$, $-(5.97 \pm 0.06)$ [93MEI/KAT] and $-(6.14 \pm 0.08)$ [96MEI/SCH]. They used the following relationships:

$$R = \frac{[(UO_2)_2(OH)_2^{2+}]}{[UO_2^{2+}]} = \frac{[2:2]}{[0:1]}$$

$$\log_{10} R = \log_{10} {}^*K_{s,0}(UO_3 \cdot 2H_2O) + \log_{10} {}^*\beta_{2,2} \text{ or,}$$

$$\log_{10} R = \log_{10} K_{s,0}(UO_2CO_3) + \log_{10} {}^*\beta_{2,2} - 2\log_{10}\gamma_{H^+} - \sum \log_{10} {}^*K - \log_{10} p_{CO_2}$$

depending on the solid phase controlling solubility at a given pH and p_{CO_2}, with $\Sigma \log_{10} {}^*K° = -18.145^1$ (sum of the log of Henry constant for CO_2 and the logs of the first and second dissociation constants of H_2CO_3 and $\log_{10} \gamma_{H^+} = -0.09^1$). R values, independent of pH, could be obtained by spectroscopy [93MEI/KAT] or by solving a mass balance, according to:

$$[U]_t = [0:1] + 2 \cdot [2:2]$$

$${}^*\beta_{2,2} = \frac{[2:2][H^+]^2}{[0:1]^2} = R \cdot \frac{[H^+]^2}{[0:1]}$$

The equation:

$$2R^2 + R - \frac{[U]_t \, {}^*\beta_{2,2}}{[H^+]^2} = 0$$

gives:

$$R = \frac{1}{4}\left[\left(1 + 8 \cdot \frac{[U]_t \, {}^*\beta_{2,2}}{[H^+]^2}\right)^{1/2} - 1\right].$$

R will be independent of pH if $[U]_t \cdot {}^*\beta_{2,2} = \alpha \cdot [H^+]^2$, where α is a constant available from the experimental data.

The authors give as selected values, $\log_{10} {}^*K_{s,0}(UO_3 \cdot 2H_2O, cr) = (5.26 \pm 0.07)$ and (5.72 ± 0.02) determined under 0.3 and 0.03 % CO_2, and $\log_{10} K_{s,0}(UO_2CO_3, cr) = -(14.05 \pm 0.08)$ (8% CO_2) according to the equilibria:

$$UO_3 \cdot 2H_2O(cr) + 2H^+ \rightleftharpoons UO_2^{2+} + 3H_2O(l)$$

$$UO_2CO_3(cr) \rightleftharpoons UO_2^{2+} + CO_3^{2-}$$

The two values of $\log_{10} {}^*K_{s,0}(UO_3 \cdot 2H_2O, cr)$ determined under $p_{CO_2} = 3 \cdot 10^{-3}$ and $3 \cdot 10^{-4}$ bar, extrapolated to $I = 0$ with the SIT, correspond to $\log_{10} K°_{s,0}(UO_3 \cdot 2H_2O, cr) = -(22.90 \pm 0.07)$ and $-(22.44 \pm 0.02)$, respectively, and differ somewhat from the values determined previously by [93MEI/KIM], [93MEI/KIM2].

Because of the ambiguities in the calculation procedure, they are not retained by the reviewers.

The second set of data is related to the solubility of $UO_3 \cdot 2H_2O(cr)$ between pH = 3.8 and 8.5 in an open atmosphere. For pH > 7, schoepite transforms to sodium uranate so these data are discarded. Here the U(VI) solubility is interpreted on the basis of the formation of the species: 0:1, UO_2^{2+}; 2:2, $(UO_2)_2(OH)_2^{2+}$; 5:3, $(UO_2)_3(OH)_5^+$; and UO_2CO_3(aq) identified by TRLFS according to the data of [94KAT/MEI], mainly by fluorescence lifetime measurements. The presence of UO_2OH^+, which has a low inten-

---

[1] these values differ from that given in [93MEI/KAT].

sity emission spectrum [94KAT/MEI], [95MOU/DEC] and [96BER/GEI], is questionable. So, a tentative fit of the solubility is made as follows.

- As a first step, the authors neglect UO_2OH^+ and take $\log_{10} {}^*\beta_{2,2} = -(6.00 \pm 0.06)$ and $\log_{10} \beta(UO_2CO_3, aq) = -(9.23 \pm 0.04)$ [93MEI/KIM]. Then, they determine two parameters, $\log_{10} {}^*K_{s,0}(UO_3 \cdot 2H_2O, cr)$ and $\log_{10} {}^*\beta_{5,3}$. The result is $\log_{10} {}^*K_{s,0}(UO_3 \cdot 2H_2O, cr) = (5.72 \pm 0.03)$ and $\log_{10} {}^*\beta_{5,3} = -(17.06 \pm 0.12)$.

- In the second approach the authors include the UO_2OH^+ species and take $\log_{10} {}^*\beta_{2,2} = -(6.00 \pm 0.06)$ and $\log_{10} \beta(UO_2CO_3, aq) = -(9.23 \pm 0.04)$ and neglect the species 5:3. The result is $\log_{10} {}^*K_{s,0}(UO_3 \cdot 2H_2O, cr) = (5.77 \pm 0.03)$ and $\log_{10} {}^*\beta_{1,1} = -(5.29 \pm 0.14)$. But as the species 5:3 is identified in the solutions by TRLFS, these results give only an upper limit of $\log_{10} {}^*\beta_{1,1}$. [92GRE/FUG] gives $\log_{10} {}^*\beta_{1,1} = -(5.4 \pm 0.6)$ (note the large uncertainty).

- The third approach of the authors consists of decreasing the $\log_{10} {}^*\beta_{1,1}$ value and taking into account all the species, while $\log_{10} {}^*\beta_{2,2}$ and $\log_{10} \beta(UO_2CO_3, aq)$ are fixed. For $\log_{10} {}^*\beta_{1,1}$ less than -5.6 down to -6.08 both the value of $\log_{10} {}^*K_{s,0}(UO_3 \cdot 2H_2O, cr)$ and $\log_{10} {}^*\beta_{5,3}$ are insensitive to the chosen value of $\log_{10} {}^*\beta_{1,1}$. They have the following values: $5.72 < \log_{10} {}^*K_{s,0}(UO_3 \cdot 2H_2O, cr) < 5.74$, and $-17.13 < \log_{10} {}^*\beta_{5,3} < -17.74$.

This review accepts these values of $\log_{10} {}^*K_{s,0}$ with an uncertainty of 0.03.

The present solubility data obtained in an open atmosphere agree very well with [92KRA/BIS], but shows a slight but distinct discrepancy (0.4 $\log_{10}$ units) both between the data of [96KAT/KIM] and [93MEI/KIM] (pH = 3.5 to 4.8). This may possibly be an effect of the crystal size.

The available literature values of $\log_{10} {}^*K^\circ_{s,0}$ and $\log_{10} K^\circ_{s,0}$ are collected in a table of this paper.

[96MEI/KLE]

This is a combined spectrophotometric and solubility study made at 25°C in a 0.1 M $NaClO_4$ medium. The authors have used an appropriate method to recalculate measured pH values to $-\log_{10}[H^+]$. This paper gives detailed information on the deconvolution of the measured absorption spectra, and thereby the contributions of the different species to the measured absorptivity. This is useful information for other spectroscopic studies, particularly fluorescence spectroscopy.

The authors also report values for a number of equilibrium constants: $\log_{10} K_{s,0}(UO_2CO_3(s)) = -(13.50 \pm 0.22)$, $\log_{10} \beta_1 (UO_2CO_3(aq)) = (8.81 \pm 0.08)$, $\log_{10} \beta_2 (UO_2(CO_3)_2^{2-}) = (15.5 \pm 0.8)$, and $\log_{10} \beta_3 (UO_2(CO_3)_3^{4-}) = (21.74 \pm 0.44)$. The corresponding values recalculated to zero ionic strength are -14.33, 9.68, 16.4 and 21.7, respectively. All constants are within the estimated uncertainty and in excellent agreement with the values selected in [92GRE/FUG]. The equilibrium constants have been deduced from the spectrophotometric data and the fairly large uncertainty is a result of the fact that both the equilibrium constants and molar absorptivities (through the deconvolution) have to be determined from the experimental data for each absorbing species. The large uncertainty for the second complex is probably due to the small amounts present, cf. Figure 1 on p. 82 in the publication. The small uncertainty for the first complex is due to the fact that the absorption spectrum of UO_2^{2+} can be determined separately. The authors have used the equilibrium constant for the formation of $(UO_2)_2(OH)_2^{2+}$ from [96MEI/SCH] discussed in the present review: $\log_{10}{}^*\beta_{2,2}$ (0.1 M NaClO$_4$, 298.15 K) = $-(6.14 \pm 0.08)$ corresponding to $\log_{10}{}^*\beta_{2,2}^\circ$ (298.15 K) = $-(5.93 \pm 0.08)$, which is slightly smaller than the value proposed in [92GRE/FUG], $-(5.62 \pm 0.04)$. However, this value will not affect the magnitude of the equilibrium constants for the carbonate species under the experimental conditions used.

[96MEI/SCH]

This paper is discussed together with [97MEI/SCH].

In these papers the authors analyse the UV–Vis (340–520 nm) spectra of 23 U(VI) aerated solutions (0.03% partial pressure of CO_2, p_{CO_2} = 0.3 kPa), I = 0.1 M NaClO$_4$, pH = 2.9 to 4.8, 10^{-4} to 10^{-2} in uranium. Solutions are allowed to stand for three weeks at 25°C to reach equilibrium and are then filtered through 220 nm pore filters. They are considered as under-saturated. The focus of this review is on the determination of $\log_{10}{}^*\beta_{2,2}$.

The analysis is conducted with a factor analysis method which is well explained. It shows the existence of two species, UO_2^{2+} in the more acidic solutions and a second one at higher pH, to be identified. Each one has a characteristic spectrum with absorption maxima for, UO_2^{2+}, $\varepsilon_{413.88nm}$ = (9.7 ± 0.2) cm$^2 \cdot$ mol^{-1}, and for the second, $\varepsilon_{421.8nm}$ = (101 ± 2) cm$^2 \cdot$ mol^{-1}. The last one is similar to the spectra of the hydrolysed 2:2 species identified in [93MEI/KAT]. Furthermore a test assuming $\log_{10}{}^*\beta_{1,1}$ = $-(4.67 \pm 0.17)$ for the unknown species, UO_2OH^+, shows that this value is too large as compared with the previous data, for instance [92GRE/FUG], [93MEI/KAT] and [96MEI/KAT]. The data from ten solutions (pH = 3.5 to 4.04) containing the two species are used to calculate $\log_{10}{}^*\beta_{2,2}$ = $-(6.14 \pm 0.08)$. This value agrees with [93MEI/KAT] and is discussed by the authors with respect to the previous value in [98MEI].

The reviewer accepts this improved $\log_{10}{}^*\beta_{2,2}$ value, which when recalculated to I = 0 with the SIT, $\log_{10}{}^*\beta_{2,2}^\circ$ = $-(5.93 \pm 0.08)$, is in reasonable agreement with the value of $-(5.62 \pm 0.04)$ selected by [92GRE/FUG].

[96MER/FUG]

The standard enthalpy of formation of crystalline (orthorhombic) and amorphous lanthanide hydroxycarbonates, LnOHCO$_3$ · 0.5H$_2$O(cr) with Ln = Nd, Sm, LnOHCO$_3$ · 0.5H$_2$O(am), with Ln = Dy, Yb and, ^{241}AmOHCO$_3$ · 0.5H$_2$O(cr) have been determined by calorimetry at 298.15 K, $p°$ = 101.325 kPa, in 1 M HCl. In addition, the enthalpy of formation of Nd$_2$(CO$_3$)$_3$(am) has been determined under the same conditions. All experiments are described in detail. The solid phases have been well characterised by gravimetry, thermogravimetry, IR spectroscopy and X–ray powder diffraction. The enthalpy of solution of AmOHCO$_3$ · 0.5H$_2$O(cr) was determined from three samples. The average value from samples A and B, with the measurements performed in triplicate, was $\Delta_{sol}H_m$ (AmOHCO$_3$ · 0.5H$_2$O, cr, 298.15 K, 1 M HCl) = – (61.3 ± 2.5) kJ · mol^{-1}. The results from preparation C (– 104.9 and – 102.4 kJ · mol^{-1}) refer to samples aged for 40 days. The more negative enthalpy of solution determined for the aged samples was considered to be caused by partial loss of crystallinity (and very likely chemical alteration) as a result of α-irradiation. The aged solid was not further characterised.

The enthalpy of solution from samples A and B is combined with auxiliary data consistent with the NEA TDB, to deduce the standard molar enthalpy of formation of crystalline americium hydroxycarbonate by a well-defined thermodynamic cycle. The calculated value of:

$$\Delta_f H_m° (\text{AmOHCO}_3 · 0.5\text{H}_2\text{O, cr, 298.15 K}) = - (1682.9 \pm 2.6) \text{ kJ} \cdot \text{mol}^{-1}$$

is accepted in the present review. The authors have further used an estimated standard entropy of $S_m°$ (AmOHCO$_3$ · 0.5H$_2$O, cr, 298.15 K) = (154 ± 10) J · K^{-1} · mol^{-1} and $S_m°$ (Am^{3+}, aq, 298.15 K) = – (201 ± 15) J · K^{-1} · mol^{-1} as estimated in [95SIL/BID] to calculate the Gibbs energy for the reaction:

$$\text{AmOHCO}_3 · 0.5\text{H}_2\text{O(cr)} \rightleftharpoons \text{Am}^{3+} + \text{OH}^- + \text{CO}_3^{2-} + 0.5\text{H}_2\text{O(l)} \quad (A.73)$$

and the solubility constant of $\log_{10} K_{s,0}°$ (AmOHCO$_3$ · 0.5H$_2$O, cr, 298.15 K) = – (23.1 ± 1.0).

For the reasons discussed in sections 12.1.1 and 12.6.1.1.3 the present review includes an entropy correction of – 9.5 J · K^{-1} · mol^{-1} arising from the thermally populated higher electronic states in europium compounds, leading to:

$$S_m° (\text{AmOHCO}_3 · 0.5\text{H}_2\text{O, cr, 298.15 K}) = (144 \pm 10) \text{ J} \cdot \text{K}^{-1} \cdot \text{mol}^{-1}$$

and

$$\log_{10} K_{s,0}° (\text{AmOHCO}_3 · 0.5\text{H}_2\text{O, cr, 298.15 K}) = - (22.6 \pm 1.0).$$

This latter value is in excellent agreement with the results from solubility measurements in [84SIL/NIT], [90FEL/RAI], [84BER/KIM], [94GIF] (cf. Table 12-9 in section 12.6.1.1.3.1).

It should be noted that the analogous results and calculations for the orthorhombic Nd hydroxycarbonate lead to $\log_{10} K^°_{s,0}$ (NdOHCO$_3$ · 0.5H$_2$O, cr, 298.15 K) = $-(21.3 \pm 0.7)$, which is in excellent agreement with the solubility constants derived from various solubility measurements (cf. Table A-24).

It is difficult to estimate accurate standard entropies of the amorphous hydroxycarbonates, but their distinctly more negative enthalpies of solution compared to those of the crystalline compounds point to solubility constants at least two orders of magnitude higher [96MER/FUG]. This gives a reasonable explanation for the different solubility constants determined for crystalline ^{243}AmOHCO$_3$(cr), $\log_{10} K^°_{s,0}$ (A.73) = $-(22.3 \pm 0.3)$ as re-calculated in [95SIL/BID] from the data in [84SIL/NIT], and a solid of ^{241}AmOHCO$_3$(s) possibly rendered amorphous with $\log_{10} K^°_{s,0}$ (A.73) = $-(20.2 \pm 0.2)$ during the solubility study of Runde et al. [92RUN/MEI].

Table A-24: Comparison of the solubility constants reported for orthorhombic NdOHCO$_3$ · 0.5H$_2$O(cr).

Reference	Medium	t (°C)	$\log_{10} K_{s,0}$	$\log_{10} K^°_{s,0}$
[96MER/FUG]	$I = 0$	25		$-(21.3 \pm 0.7)$ [a]
[91MEI/KIM2]	0.1 M NaClO$_4$,	(22 ± 1)	$-(19.19 \pm 0.08)$ [b]	$-(20.7 \pm 0.2)$ [c]
[92RUN/MEI]	0.1 M NaClO$_4$,	25	$-(19.94 \pm 0.16)$ [b]	$-(21.4 \pm 0.3)$ [c]
[93MEI/TAK]	0.1 M NaClO$_4$,	(24 ± 2)	$-(20.12 \pm 0.09)$ [b]	$-(21.6 \pm 0.2)$ [c]
[93CAR]	dilute solutions, corrected to $I = 0$	(25)		$-(21.75 \pm 0.29)$ [b]

[a] Calculated from thermochemical data
[b] Solubility measurements
[c] Corrected to $I = 0$ with the SIT coefficients in Appendix B. The uncertainty is increased to the 95 % confidence interval.

[96MOL/GEI]

Moll et al. [96MOL/GEI] have determined the solubility and speciation of (UO$_2$)$_2$SiO$_4$ · 2H$_2$O in a 0.1 M NaClO$_4$ ionic medium at 25°C. The solid phase has been well characterised, as described in [95MOL/MAT]. The authors have studied the solubility over the pH range from 3 to 9. The experiments are described in detail and the solid phase has been characterised both before and after equilibrium using X–ray diffraction. No phase transformation could be observed. The data at pH = 3, 4 and 5 indicate that the system is in, or close to, equilibrium. The data at higher pH are much more scattered and we have not found it meaningful to use them. Moll et al. only used the data at pH 3 to determine the solubility constant. This is not satisfactory and we have therefore reinterpreted their data. The reaction studied is:

$$(UO_2)_2SiO_4 \cdot 2H_2O(s) + 4\,H^+ \rightleftharpoons 2\,UO_2^{2+} + Si(OH)_4(aq) + 2\,H_2O(l) \quad (A.74).$$

At pH 3 the concentrations of dioxouranium(VI) hydroxide and silicate complexes are so small that they can be neglected in comparison with that of UO_2^{2+}. Hence, the solubility constant is equal to:

$$^*K_s(A.74) = \frac{[c]^3}{2[H^+]^4}$$

where c is the measured total concentration of uranium. At higher pH the possible formation of hydroxide and silicate complexes has to be taken into account using known equilibrium constants. The experimental data in [96MOL/GEI], plotted as $\log_{10}[c]$ versus pH fall quite well on a straight line between pH 3 and 5. From Figure 3 in [96MOL/GEI] the reviewers have estimated a slope equal to 1.3, in good agreement with the value 1.33 expected from reaction (A.74) if the amounts of hydroxide and silicate complexes are small in comparison with that of [UO_2^{2+}]. Using the known equilibrium constants for the silicate and hydroxide complexes we can estimate the ratio [UO_2^{2+}]/[UO_2OH^+] and [UO_2^{2+}]/[$UO_2SiO(OH)_3^+$]. The first is 263, 26 and 2.6 at pH = 3, 4 and 5, respectively, *i.e.*, the hydroxide complex is a minor species in this pH range. The second ratio was calculated using the selected equilibrium constant proposed in this review, $\log_{10}K = -1.95$. We find that the ratio at pH = 3, 4 and 5 is equal to 10, 5.6 and 1.8, respectively. These numbers indicate that the concentrations of UO_2OH^+ and $UO_2SiO(OH)_3^+$ are so small in comparison with the experimental errors, that they do not affect the slope of the solubility curve. The experimental data of [96MOL/GEI] are consistent with the known equilibrium constant of the silicate and hydroxide complexes and the solid phase seems to be stable at least up to pH = 6. The equilibrium constant reported by the authors has been recalculated as described below. The authors have measured hydrogen activities and not hydrogen concentrations and the calculated solubility constant must be corrected accordingly, as described in the comments to [98MOL/GEI]. By fitting the experimental data up to pH = 6 with a straight line of slope 1.3, we obtain a solubility constant of $\log_{10} {^*K_s}(A.74) = (7.1 \pm 0.5)$ at $I = 0.1$ M. Recalculation to zero ionic strength gives $\log_{10} {^*K_s^\circ}(A.74) = (6.7 \pm 0.5)$. This value differs slightly from those reported by Moll *et al.*, $\log_{10} {^*K_s}((A.74), 0.1$ M $NaClO_4) = (6.46 \pm 0.45)$ (under a N_2 atmosphere) and (6.60 ± 0.53) (in air) and $\log_{10} {^*K_s^\circ}(A.74) = (6.03 \pm 0.45)$ and (6.15 ± 0.53), respectively.

[96PAR/PYO]

This paper is written in Korean, with abstract, Figures and Tables in English. The authors prepare U(VI) hydroxide precipitates at pH = 6.4 denoted (P1) and 9.7 denoted (P2) and then, they measure the concentration of U in groundwater, groundwater colloid-free, synthetic groundwater and 0.1 M NaCl as a function of pH (5 to 11) and as a function of the concentrations of SO_4^{2-} and HCO_3^- ($2.7 \cdot 10^{-4}$ to $2.2 \cdot 10^{-3}$ M, pH = 8.35 – 8.7), which are the main anions present in the groundwater. Precipitates are not well defined. The only data which could provide some information for this review are related to solubilities in 0.1 M NaCl and in HCO_3^- solutions (ionic strength not controlled). In NaCl the solubility of P2 lies around 10^{-6} M from pH = 7 to 11. In bicarbonate solutions, the P1 solubility increases from $7 \cdot 10^{-5}$ to $2 \cdot 10^{-4}$ M, but the P2 solubility remains constant around $2 \cdot 10^{-3}$ M.

No modelling is done. Furthermore, no characterisation of the solid phases has been made. The groundwater was filtered before use to remove colloids, but no attempts seem to have been made to remove colloids before analysis of the uranium solubility. The solubility of the phases prepared at pH = 9.7 differ by three orders of magnitude between the groundwaters and the 0.1 M NaCl. The latter has the lowest solubility. In addition, the solubility of both phases is independent of pH in the range 7 < pH < 11. These results are not compatible with the known equilibrium constants, indicating that experimental artefacts affect the results. From the abstract it is impossible to obtain sufficient information to suggest possible flaws in the procedures used. The data in this paper are not credited by the present review.

[96PAV/FAN]

The authors have used TRLFS to evaluate the stoichiometry for the complexes, $Cm(SO_4)_n^{3-2n}$, n = 1, 2, 3 in a 3 m Na_2SO_4/NaCl ionic medium. The individual spectra of the different complexes are well resolved, allowing a precise determination of the stoichiometry and equilibrium constants for the first two complexes. There are pronounced variations in the lifetime of the various species, ranging from 65.2 μs for Cm^{3+} to 171 μs in 4.12 m Na_2SO_4. This is a clear indication that the complexes formed are of inner-sphere type with bidentate coordination of sulphate. The equilibrium constants reported, $\log_{10} \beta_1$ = (0.93 ± 0.08) and $\log_{10} \beta_2$ = (0.61 ± 0.08), refer to the ionic strength, 3 m. They were recalculated to zero ionic strength, using $\Delta\varepsilon_1$ = − (0.14 ± 0.02) and $\Delta\varepsilon_2$ = − (0.24 ± 0.01) kg · mol^{-1} derived from data in Na_2SO_4-NaCl mixtures of varying ionic strength [98NEC/FAN] (*cf.* Figure 12-8, section 12.5.1.1) to give $\log_{10} \beta_1^\circ$ = (3.45 ± 0.10) and $\log_{10} \beta_2^\circ$ = (3.81 ± 0.09). The very large change in the equilibrium constants is a result of the large Debye–Hückel term for these reactions, which involve species with high charges.

[96PER/KRY]

This report was not analysed in [2001LEM/FUG]. It gives an overview of the literature data on the solubility of Np, Pu, Am, and Tc hydroxo compounds in alkaline solutions, and was partially covered in [95SIL/BID], [99RAR/RAN] and [2001LEM/FUG], and also provides experimental results. With regard to actinides, they concern mainly the solubility of Np(IV), Pu(IV), Np(V) and Pu(V) hydroxides. No thermodynamic constants can be extracted from the data, which were obtained for operational purposes.

The authors report experimental solubility data for Am(III), Np(IV), Pu(IV), Tc(IV), Np(V), Pu(V), Am(V) and Tc(V) hydroxides in 0.5 – 14 M NaOH solutions at (25 ± 2)°C. In addition, the solubilities of Np(IV), Pu(IV) and Tc(IV) hydroxides were measured in 1.0 and 4.0 M NaOH containing additions of fluoride, phosphate, carbonate, oxalate and some organic ligands. The experimental procedures involve an essential shortcoming: the authors prepared NaOH stock solutions without special precautions to prevent contact with atmospheric carbon dioxide. They assumed that a maximum carbonate concentration of about 0.01 M in concentrated NaOH should not have a significant effect and interpreted their results exclusively with the formation of solid hydroxides and aqueous hydroxide complexes. However, such a carbonate contamination can also lead to the formation of carbonate or ternary hydroxide-carbonate complexes. Another possible source of error, *i.e.*, too high apparent solubilities, could arise from insufficient separation of small colloidal species from true aqueous actinide species. The authors mentioned the application of centrifugation without giving details, but do not mention the use of ultrafiltration techniques.

- Americium(III) and americium(V)

The solubility of freshly precipitated $Am(OH)_3$(am) was determined in 1 and 5 M NaOH after equilibration for three days and found to be $6.0 \cdot 10^{-6}$ and $6.9 \cdot 10^{-6}$ mol·L^{-1}, respectively. These concentrations are similar to those reported in [89PAZ/KOC] at pH > 11 in 3 M $NaClO_4$, but are several orders of magnitude higher than those expected according to the thermodynamic data selected in the present review. Analogous experiments with a precipitate aged by boiling for three hours led to americium concentrations of $5.9 \cdot 10^{-7}$ and $3.6 \cdot 10^{-7}$ mol · L^{-1}. Solubilities of Am(V) were measured in 1.6 – 11.7 M NaOH with $Na_2AmO_2(OH)_3 \cdot xH_2O$(s) as the starting material. They increased only slightly with the NaOH concentration, from $2.0 \cdot 10^{-4}$ to $2.9 \cdot 10^{-4}$ mol · L^{-1}.

- Neptunium(IV) and plutonium(IV)

NaOH solutions of Np(IV) were prepared from Cs_2NpCl_6 and those of Pu(IV) from a stock solution of $Pu(NO_3)_4$ using $2 \cdot 10^{-4}$ mole of the element. Np(IV) is stabilised with 0.1 M $Na_2S_2O_4$ and Pu(IV) with 0.1 M N_2H_5OH. The solubilities of fresh precipitates of $Np(OH)_4$(am) and $Pu(OH)_4$(am) were measured in the presence of 0.1 M N_2H_5OH as the reducing agent. The $Np(OH)_4$(am) suspensions were additionally purged with argon to avoid oxidation by traces of oxygen, whereas Pu(IV) was considered to be stable against oxidation by atmospheric oxygen. The concentrations measured after contact times of 3 – 4 hours increased from $\log_{10}[Np] = -6.5$ to -4.7 and from $\log_{10}[Pu] = -7.8$ to -5.0 in 0.5 to 14.1 M NaOH, $t = (26 \pm 0.2)$°C. When plotted *versus* $\log_{10} a_{NaOH}$, slopes of +1 and +2 were observed and ascribed to the formation of anionic hydroxide complexes according to:

$$Np(OH)_4(aq) + OH^- \rightleftharpoons Np(OH)_5^- \quad (A.75)$$

and

$$Pu(OH)_4(aq) + 2\, OH^- \rightleftharpoons Pu(OH)_6^{2-} \quad (A.76)$$

respectively. It is to be noted that the Np(IV) solubilities measured by Peretrukhin *et al.* in 0.5 – 2 M NaOH are about two orders of magnitude higher than those reported by Rai and Ryan [85RAI/RYA] for corresponding NaOH solutions. The reverse is true for Pu(IV) when comparing to literature data. This could be due to traces of carbonate or oxygen in the first case and only oxygen in the second case. If we assume a carbonate content of about $0.001 - 0.01$ mol $\cdot$ L^{-1} in the study of Peretrukhin et al., approximately proportional to the NaOH concentration, the solubility increase could be caused by the formation of ternary hydroxide-carbonate complexes, e.g., by the reactions:

$$Np(OH)_4(aq) + CO_3^{2-} \rightleftharpoons Np(OH)_4(CO_3)^{2-} \quad (A.77)$$

or

$$Pu(OH)_4(aq) + 2\ CO_3^{2-} \rightleftharpoons Pu(OH)_4(CO_3)_2^{4-} \quad (A.78).$$

The reactions (A.77) and (A.78) were proposed by other authors to interpret the increasing solubility of Np(OH)$_4$(am) [93ERI/NDA] and Pu(OH)$_4$(am) [94YAM/SAK] in alkaline solutions (pH = 12 – 13) containing well-defined carbonate concentrations in the range $0.001 - 0.1$ mol $\cdot$ L^{-1}. Because of the ambiguities connected with a very probable but unknown carbonate contamination, the data of Peretrukhin et al. are not considered as sufficiently reliable to estimate thermodynamic data for anionic hydroxide complexes of Np(IV) or Pu(IV).

In 1 and 4 M NaOH containing additions of 0.05 and 0.5 M Na$_2$CO$_3$, Peretrukhin et al. measured Np and Pu concentrations which were enhanced by factors between 2 and 26. In similar experiments, the solubility enhancing effect was studied for additions of 0.01 and 0.05 M phosphate, 0.02 and 0.1 M fluoride, 0.1 and 0.5 M glycolate, 0.03 and 0.3 M citrate, 0.05 and 0.2 M EDTA, and 0.03 and 0.075 M oxalate. However, the experimental data do not allow the identification of the aqueous complexes formed.

- Neptunium(V) and plutonium(V)

The solubilities of Np(V) and Pu(V) hydroxide compounds were determined after an equilibration time of three days. The initial solid phases were Na$_2$NpO$_2$(OH)$_3 \cdot$ xH$_2$O(s), which was characterised by X-ray powder diffraction, and Na$_2$PuO$_2$(OH)$_3 \cdot$ xH$_2$O(s). The measured concentrations increased from log$_{10}$[Np] = -3.9 to -3.25 in 0.5 to 17.5 M NaOH and from log$_{10}$[Pu] = -4.3 to -3.4 in 0.6 to 14.0 M NaOH. The authors explained the somewhat lower solubility of Pu(V) in $0.6 - 6$ M NaOH with a possible disproportionation of Pu(V) or the instability of Na$_2$PuO$_2$(OH)$_3 \cdot$ xH$_2$O(s). The stability of the solid Np(V) hydroxides is discussed in more detail. In 5 M NaOH, NpO$_2$OH $\cdot$ xH$_2$O(s) was transformed into Na$_2$NpO$_2$(OH)$_3 \cdot$ xH$_2$O(s), whereas NaNpO$_2$(OH)$_2 \cdot$ xH$_2$O(s) was the more stable solid at lower NaOH concentrations. At the highest hydroxide concentration of 17.5 mol $\cdot$ L^{-1} the authors assumed the formation of Na$_3$NpO$_2$(OH)$_4 \cdot$ xH$_2$O(s). At (25 ± 2)°C the variation of log$_{10}$[Np(V)] versus log$_{10}$[OH] (0.5 to 14 M NaOH) show more or less two slopes which were assigned to a change of Na$_2$NpO$_2$(OH)$_3 \cdot$ xH$_2$O to NaNpO$_2$(OH)$_2 \cdot$ xH$_2$O (checked by X-ray diffraction) and further interpreted as the presence in solution of NpO$_2$(OH)$_3^{2-}$ and NpO$_2$(OH)$_4^{3-}$. The behaviour of Pu under the same conditions is not so clear as that of Np. Some data concern the solubility of $Na_2AmO_2(OH)_3 \cdot xH_2O$ and $Am(OH)_3$ in the range 1.6–12 M and 1–5 M NaOH, respectively.

The present review does not rely on the proposed aqueous speciation involving hydroxide complexes, $NpO_2(OH)_n^{1-n}$, up to $n = 4$. This speciation scheme was adopted from a previous spectroscopic study of Tananaev [90TAN], which is not in accord with the thermodynamic data selected in [2001LEM/FUG]. Moreover, it was shown in [97NEC/FAN] that two of the Np(V) absorption bands observed in [90TAN] were actually caused by carbonate and hydroxide–carbonate complexes.

Because of the shortcomings and ambiguities discussed above, the present review does not derive thermodynamic data from the reported solubility data.

- Technetium(IV) and (V)

Solubility measurements of Tc(IV) and Tc(V) oxides/hydroxides were made in 0.5 to 15 M NaOH solutions after three days exposure using β emission by ^{99}Tc as the analytical tool. As cited above, although the NaOH solutions were pre-treated to remove carbonate impurities, sample preparations were conducted in air, which would result in an indeterminant amount of CO_2 uptake. Immediately before use the Tc compounds were washed to remove any soluble Tc(VII) that may have formed. The solid samples were characterised by XRD, IR, X-ray photoelectron spectroscopy and magnetic susceptibility, although apparently no particle size measurements were made, other than reference to collecting the fraction that settled after one day from water and then repeated, re-suspension and centrifugation. In fact, later in the paper the authors demonstrate that four-month aged $TcO_2(s)$ and $TcO_2 \cdot nH_2O(s)$ exhibit not only significantly lower solubilities than solids aged for one day at low hydroxide concentrations, but also the trend with increasing NaOH concentration is more consistent with $Tc(OH)_5^-$, and perhaps higher-order hydroxide species, forming in solution. The distinction was less pronounced for the Tc(V) analogues.

The equilibrations were conducted in the absence and presence of hydrazine, hydroxylamine and sodium sulphite with $TcO_2 \cdot nH_2O(s)$ and $Tc_2O_5(s)$ (no mention of agitation is given), whereby the presence of the weak reducing agents, NH_2OH and SO_3^{2-}, apparently promoted colloid formation following slow oxidation to Tc(VII). On the other hand, only the presence of N_2H_4 was believed to result mainly in stabilisation of the Tc oxidation state corresponding to that of the solid present. In light of the prevailing conditions outlined above, the conclusions are only qualitative, the most significant of which is that the solubilities of the Tc(IV) and Tc(V) oxides are 1 to 2.5 $\log_{10}$ units lower in the presence of hydrazine than when oxidizing conditions prevail. Hydrazine was shown independently to reduce pertechnetate ion, whereas sulphite and NH_2OH led to the formation of intermediates that slowly produced colloids. The kinetics of dissolution of ($TcO_2(s)$ and $Tc_2O_5(s)$) and ($TcO_2 \cdot nH_2O(s)$ and $Tc_2O_5 \cdot nH_2O(s)$)

were also followed for 30 days. In the latter case, the kinetics were also monitored in the presence of 0.041 M NH_2OH and 0.022 M N_2H_4 (noting that the Tc(V) species are apparently mislabelled in the Figures). Significantly, in the presence of N_2H_4, a plateau in Tc concentration is reached within *ca.* 10 – 20 days indicating that an equilibrium was attained, whereas at all other conditions the concentrations increased linearly, consistent with continuous oxidation. However, the former finding would indicate that a three-day equilibration was insufficient to reach a stable concentration.

The references cited in this review are discussed in [89PAZ/KOC] for [95SIL/BID] and for all the others in [2001LEM/FUG].

[96RAK/TSY]

The interaction of uranium dioxide (grain size (12 ± 2) μm) of varying stoichiometries (O:U = 2.005 – 1.985) with Artesian aqueous solutions containing up to $1.5 \cdot 10^{-3}$ M, Ca^{2+}; $1.4 \cdot 10^{-3}$ M, Cl^- and other ions and nominally pH = 7, was measured at 300 K. The duration of the experiments in stirred and static reactors was 100, 240, 350, 600, 1440, and 7560 hours with no difference observed either in the method used, or the uranium isotope used (^{235}U or ^{238}U). Higher uranium concentrations were found at higher O:U ratios, *e.g.*, after 7560 hours, 0.13–0.35 mg · L^{-1} for a ratio of 1.985 compared with 15 mg · L^{-1} for 2.005. This trend was thought to be due to the higher content of U(VI) in the latter solid and is consistent with the observed decrease in the O:U ratio after the dissolution experiment. The diffusion of oxygen ions to the surface during dissolution is discussed. The higher uranium content in the aqueous phase, compared to solubility values ($2 \cdot 10^{-4}$ to $2 \cdot 10^{-1}$ mg · L^{-1}) in pure water reported in the literature, is ascribed to the presence of various cations and anions in the aqueous solution samples. No allusion is made to the effect of oxygen: in fact, no mention is made of any attempt to control or verify the oxygen content. However, as neither the exact solution composition and pH are given, nor is it established that equilibrium was definitely reached after 7560 hours, no reliable solubility data can be derived from this study.

[96ROB/SIL]

This paper was briefly reviewed in [2001LEM/FUG]. The solubility of amorphous dioxoneptunium(V) hydroxide was measured in 0.30, 0.60, 1.0, 1.8, 3.0, and 5.6 molal NaCl at room temperature (21 ± 2)°C over a period of 39 days. The solid was prepared by precipitation from NpO_2^+ solutions with carbonate-free NaOH. The H^+ and OH^- concentrations were determined by an appropriate method. The absence of carbonate and hydroxide complexes was ascertained by spectroscopy and by keeping the solutions at pH < 10. The SIT was applied to evaluate the solubility constant $\log_{10} {}^*K_{s,0}^\circ = (5.21 \pm 0.12)$, which is in good agreement with the value selected in [2001LEM/FUG], $\log_{10} {}^*K_{s,0}^\circ = (5.3 \pm 0.2)$. The same holds for the interaction coefficient of $\varepsilon(NpO_2^+, Cl^-) = (0.08 \pm 0.05)$ kg · mol^{-1} derived from the ionic strength dependence of the solubility product. It is almost equal to $\varepsilon(NpO_2^+, Cl^-) = (0.09 \pm 0.05)$ kg · mol^{-1} as calculated in [2001LEM/FUG] from the NpO_2^+ trace activity coefficients determined in [95NEC/FAN] with a solvent extraction method.

[96SAL/IYE]

This paper reports the preparation and identification of $Tl_2U_3O_9(cr)$ and $Tl_2U_4O_{11}(cr)$, together with measurements of the oxygen potential over the diphasic fields, $Tl_2U_4O_{11} + Tl_2U_4O_{12}$ and $Tl_2U_4O_{12} + Tl_2U_4O_{13}$.

The U(VI) uranates, $Tl_2U_2O_7(cr)$ and $Tl_2U_4O_{13}(cr)$, were first prepared by reacting appropriate amounts of $Tl_2CO_3(cr)$ and $U_3O_8(cr)$ in dry air at 973 K; $Tl_2U_4O_{12}(cr)$ was then obtained by reduction of $Tl_2U_4O_{13}(cr)$ in flowing Ar at 1000 K, as for the analogous rubidium compound. The products were characterised by X–ray diffraction.

The new compounds $Tl_2U_3O_9(cr)$ and $Tl_2U_4O_{11}(cr)$ were prepared by annealing mixtures of $Tl_2U_2O_7(cr)$ and $UO_2(cr)$ in a platinum boat under Ar at 1000 K for eight hours. Redox and X–ray fluorescence analyses confirmed the composition of the samples. Their X–ray diffraction patterns were indexed as hexagonal, with $a = 3.972$ Å and $c = 9.960$ Å for $Tl_2U_3O_9(cr)$ and tetragonal, with $a = 9.741$ Å and $c = 19.922$ Å for $Tl_2U_4O_{11}(cr)$, isostructural with $Rb_2U_4O_{11}(cr)$.

The cells used for the emf measurements were:

$$Pt\,|\,Tl_2U_4O_{11} + Tl_2U_4O_{12}\,|\,CSZ\,|\,Ni + NiO\,|\,Pt \qquad (I)$$

and $\quad Pt\,|\,Tl_2U_4O_{12} + Tl_2U_4O_{13}\,|\,CSZ\,|\,\text{air}\,(p_{O_2}=0.2121\text{ bar})\,|\,Pt \qquad (II)$

where CSZ is calcia (15 mol %) stabilised zirconia.

Least square fits of consistent individual results on the two cells were given as

E (mV) = 709.2 – 0.4694·T (805 – 1179 K)   for cell (I)

E (mV) = 810.0 – 0.6124·T (996 – 1141 K)   for cell (II).

The reaction in cell (I) is:

$$Tl_2U_4O_{11}(cr) + NiO(cr) \rightleftharpoons Tl_2U_4O_{12}(cr) + Ni(cr) \qquad (A.79)$$

and the recalculated Gibbs energy for this reaction is:

$$\Delta_r G_m\,(A.79) = -156.31 + 0.1182\,T \text{ kJ} \cdot \text{mol}^{-1}.$$

As for similar papers from the same group, we have used the Gibbs energies of formation of NiO(cr) assessed by [90TAY/DIN], to convert the emf of cell (I) to oxygen potentials.

The derived Gibbs energies of the oxidation reactions (A.80) and (A.81):

$$Tl_2U_4O_{11}(cr) + 0.5\ O_2(g) \rightleftharpoons Tl_2U_4O_{12}(cr) \quad (A.80)$$

$$Tl_2U_4O_{12}(cr) + 0.5\ O_2(g) \rightleftharpoons Tl_2U_4O_{13}(cr) \quad (A.81)$$

are finally:

$$\Delta_r G_m\ (A.80) = -390.53 + 0.2032\ T\ kJ \cdot mol^{-1}$$

$$\Delta_r G_m\ (A.81) = -136.85 + 0.0841\ T\ kJ \cdot mol^{-1}.$$

The authors' expression for $2 \cdot \Delta_r G_m$ (A.81) seems to have an incorrect entropy term.

The authors combine their oxygen potential results with values for $\Delta_f G_m$ ($Tl_2U_4O_{13}$, cr) estimated by [81LIN/BES], (by calculations which need correction for the above error) to derive Gibbs energy expressions for the lower uranates. However, in the absence of any experimental data for a thallium uranate, we do not select these for the review.

[96SHO/BAM]

The compounds $SrPu_2Ti_4O_{12}(cr)$, $Pu_2Ti_3O_{8.79}(cr)$, and $Pu_2Ti_2O_7(cr)$, in which plutonium is in the (III) oxidation state, have been prepared and identified by X–ray diffraction. The solid solubility limits of $Pu_2Ti_2O_7(cr)$ in $Ln_2Ti_2O_7(cr)$ were shown to increase as the size of the lanthanide ion decreases. Attempts to synthesise analogous compounds with Pu(IV) or solid solutions containing Pu(IV) and Ce(IV) were unsuccessful.

This paper contains no thermodynamic data.

[96YAM/HUA]

A Knudsen effusion mass–spectrometric study showed that the major gaseous species in the vaporisation of a sample of '$BaUO_3$' are, in decreasing order, $BaO(g)$, $Ba(g)$ and $UO_2(g)$. The starting '$BaUO_3$' sample was prepared by hydrogen reduction of $BaUO_4$ for eight hours at 1673 K. Analyses of both $BaUO_4$ and $BaUO_3$ were by X–ray diffraction only, so the precise Ba/U and O/U ratios in the solid were not established unambiguously; the lattice parameter of 4.416 Å corresponds to $BaUO_{3.02}$. The following reactions were assumed to occur in the Knudsen cell:

$$BaUO_3(cr) \rightleftharpoons BaO(g) + UO_2(cr) \quad (A.82)$$

$$BaO(g) \rightleftharpoons Ba(g) + 1/2\ O_2(g) \quad (A.83)$$

$$UO_2(cr) \rightleftharpoons UO_2(g) \quad (A.84)$$

but in view of the complexities of the chemistry of the Ba–U–O system, other reactions could easily have occurred – see below. As usual in this type of study, estimated cross sections for BaO(g) and UO_2(g) were used to convert ion currents to pressures, with an attendant uncertainty.

The authors processed their reported pressure equations:

$$\log_{10}(p_{BaO} / \text{bar}) = 18.37 - 53246 \cdot T^{-1}$$
$$\log_{10}(p_{Ba} / \text{bar}) = 19.14 - 58202 \cdot T^{-1}$$
$$\log_{10}(p_{UO_2} / \text{bar}) = 21.22 - 63078 \, T^{-1}$$

using estimated thermal functions for $BaUO_3$(cr) and values for BaO(g) and UO_2(g) from [90COR/KON] to derive third- and second-law enthalpies of formation of $BaUO_3$(cr), $\Delta_f H$ ($BaUO_3$, cr, 298,15 K) of $-(1742.5 \pm 10.0)$ kJ·mol^{-1} (uncertainty increased) and $-(1681.8 \pm 20.8)$ kJ·mol^{-1}, respectively. However, the lattice parameter of the sample after the experiments had increased to 4.405 Å, corresponding to a composition of $BaUO_{3.12}$. This, and the lack of agreement between the second- and third-law enthalpies of formation, indicates that the vaporisation was non-congruent, with probable changes in the Ba/U as well as the O/U ratio.

The calculated values of the enthalpy of formation are in general agreement with more recent calorimetric data of barium uranates given by [97COR/BOO], but clearly cannot be used to define the properties of a barium uranate of any defined composition.

[97ALL/BUC]

Formation of weak complexes between UO_2^{2+}, NpO_2^+, Np^{4+}, Pu^{3+} and Cl^- is investigated at (25 ± 1)°C by EXAFS in appropriate HCl or LiCl solutions to avoid hydrolysis and under conditions to avoid redox reactions. The chloride concentration was varied up to 14 M. Systematic changes of the EXAFS spectra obtained up to 13 Å^{-1} in the k-space in transmission mode (U) or fluorescence mode (Np and Pu) with an increase in [Cl$^-$] show the build up of inner sphere complexes except for Pu^{3+}. The limiting complexes, determined from the EXAFS spectra, are $UO_2Cl_3^-$ (Cl coordination number, N_{Cl}, equal to 2.6), NpO_2Cl, (N_{Cl} = 1.0) and $NpCl_2^{2+}$, (N_{Cl} = 2.0).

From this study only $\log_{10} \beta$ values are estimated by the authors from data corresponding to N_{Cl} = 0.5: $\log_{10} \beta = -0.48$ for UO_2Cl^+, $\log_{10} \beta = -0.78$ for $NpCl^{3+}$ and $\log_{10} \beta = -0.85$ for NpO_2Cl (95% limits are ± 12%). The chloride concentration at which they are formed is given by [Cl$^-$] = $1/\beta$ in acidic media.

These values are lower than those previously reported. Although no interactions M–Cl are detected at a distance from M above 4 Å, the presence of outer sphere complexes in the solution cannot be excluded. Furthermore, the EXAFS show that on Dowex 50 equilibrated with the more concentrated chloride solutions, the anionic species, $UO_2Cl_4^{2-}$ or $NpCl_6^{2-}$, are captured.

This paper gives the hydration numbers of aqueous ions, $N = 5.3$ for UO_2^{2+}, $N = 5$ for NpO_2^+, $N = 11.2$ for Np^{4+} and $N = 10.2$ for Pu^{3+}. They are derived from spectra corresponding to solutions with the lowest chloride concentration. For Np^{4+} the authors suggest that the number of water molecules around this ion is between 9 and 11. These values are in agreement with previous results. This paper was not quoted in [2001LEM/FUG].

[97ASA/SUG]

The extraction experiments were initiated by equilibrating tetraphenylarsonium tetrachloronitridotechnetate(VI) anion, $[\{(H_2O)Cl_3NTc\}_2(\mu-O)]^{2-}$, dissolved in chloroform for 210 minutes at 298.15 K with an aqueous HCl/NaCl solution at 1 M ionic strength. ^{99}Tc was used as a tracer. The distribution ratio (concentration in the organic to aqueous phase) increases with acid and TPAC concentration. The overall extraction constant (K_{ex}), which can be defined by the equilibrium:

$$TcNCl_4^- + RCl(org) \rightleftharpoons RTcNCl_4(org) + Cl^-, \text{ with } R = \text{tetraphenylarsonium},$$

which in the aqueous phase can be broken down into a dimerisation step, followed by two hydrolysis steps each involving the loss of two chloride ions. The $[\{(H_2O)Cl_3NTc\}_2(\mu-O)]^{2-}$ anion has a peak at 540 nm, which is observed at $[H^+] = 0.2–1.0$ M, whereas no peak was seen at 557 nm, which would correspond to the formation of the tetrachloro analogue, i.e., $[\{Cl_4NTc\}_2(\mu-O)]^{4-}$. Hence, in this range of acidity the dominant reaction in the aqueous phase is believed to be:

$$2\ TcNCl_4^- + 3\ H_2O(l) \rightleftharpoons [\{(H_2O)Cl_3NTc\}_2(\mu-O)]^{2-} + 2\ H^+ + 2\ Cl^-.$$

Assuming that these are the dominant aqueous species and in excess TPAC(org), the data were regressed to yield a value for this equilibrium constant β_2 of $(2.85 \pm 0.06)\cdot 10^5$ M^3 with $K_{ex} = (8.8 \pm 0.2)\cdot 10^3$. For comparison, $K_{ex} = (2.99 \pm 0.19)\cdot 10^4$ for TcO_4^- extraction of water by TPAC in chloroform. In 1 M HCl, 95% of the aqueous Tc is in the form of the monomer. At $[H^+] < 0.2$ M, further hydrolysis was suggested due to the appearance of a peak at 342 nm, which is believed to result from the formation of $[\{Cl_2NTc\}_2(\mu-O)_2]^{2-}$.

[97BER/GEI]

This paper validates the stoichiometry and the formation constant of the complex $Ca_2[UO_2(CO_3)_3]$ identified in [96BER/GEI]. TRLFS and LIPAS are used to study solutions of the following composition: U, $2\cdot 10^{-5}$ M; HCO_3^-/CO_3^{2-}, $8\cdot 10^{-3}$ M; Ca^{2+} 10^{-4} to $5\cdot 10^{-3}$ M; pH = 8.0 (I = 0.1 M, NaClO$_4$) where two equilibria are supposed to exist:

$$b\ Ca^{2+} + UO_2(CO_3)_3^{4-} \rightleftharpoons Ca_b UO_2(CO_3)_3^{(4-2b)-}(aq) \qquad K$$

$$Ca^{2+} + CO_3^{2-} \rightleftharpoons CaCO_3(aq) \qquad \log_{10}\beta° = 3.1$$

From the function:

$$\log_{10}\frac{[Ca_bUO_2(CO_3)_3^{(4-2b)-}(aq)]}{[U_{tot}]-[Ca_bUO_2(CO_3)_3^{(4-2b)-}]} = \log_{10}R = b\log_{10}[Ca^{2+}] + \log_{10}K$$

where R is measured through TRLFS, the authors obtain $b = (1.80 \pm 0.2)$ and $\log_{10}K = (6.8 \pm 0.7)$. Using the Davies equation, they obtain $\log_{10}K° = 4.1$ and with $\log_{10}\beta°(UO_2(CO)_3^{4-}) = (21.6 \pm 0.05)$ [92GRE/FUG] they obtain $\log_{10}\beta°_{2,3} = (25.7 \pm 0.7)$ for the equilibrium:

$$2\,Ca^{2+} + UO_2^{2+} + 3\,CO_3^{2-} \rightleftharpoons Ca_2[UO_2(CO_3)_3](aq).$$

This value is considered by the authors to be better than the previous one (26.8 ± 0.8) given in [96BER/GEI].

Additional proof of the existence of the di-calcium tri-carbonato uranium(VI) complex is given by a comparison with the LIPAS spectrum which is different from that of $UO_2(CO)_3^{4-}$: a slight red shift of 0.35 at 462 nm and 458 nm, outside of the measuring errors.

[97BOU]

This thesis gives data on the thermophysical properties, including heat capacity and thermal diffusivity, of Tc and some Nb–Tc alloys. Tc was prepared by heating $NH_4TcO_4(cr)$ in Ar–H$_2$ mixtures at 873 K, followed by arc–melting in Cu under an argon atmosphere. Chemical analyses showed the arc–melted samples were 99.94 % pure, the major impurities being oxygen and rhenium; however, no analysis for nitrogen was reported.

Heat capacities were measured in a DSC apparatus from 673 to 1583 K; the results are presented only in the form of a graph. For Tc(cr), the heat capacity has a maximum around 1080 K and a minimum around 1540 K and are ca. 10–20% lower than the only other experimental values derived from the thermal diffusivity measurements of Spitsyn et al. [75SPI/ZIN] (which have an uncertainty of at least 5%). Moreover, Boucharat's thermal diffusivity measurements give values which are also 10-30% lower than those from two earlier (and consistent) studies. Since there is no physical reason why there should be extrema in $C_{p,m}$ (Tc, cr) in the relevant temperature range, these results have not been accepted for this review.

It may be noted that Van der Laan and Konings [2000LAA/KON] have studied the heat capacity up to 973 K of a Tc–Ru alloy made by neutron irradiation of a similar batch of Tc(cr) to that used by Boucharat.

[97BRU/CAS]

Eh/pH data collected from Cigare Lake ore deposit boreholes along the path of undergroundwater, together with analysis data of uraninite, are consistent with equilibrium of the water with UO_2 (reducing water in contact with the ore, pH = 7) and then U_3O_7 (less reducing water, pH < 7 due to U(VI) hydrolysis). The database used is included in the HARPHRQ code, which is a geochemical code speciation based on PHREEQE, 1991.

[97CAS/BRU]

The solubility of a natural becquerelite, $Ca(UO_2)_6O_4(OH)_6 \cdot 8H_2O$ or $CaU_6O_{19} \cdot 11H_2O$ (see [94SAN/GRA]), was measured at 25°C under a nitrogen atmosphere. XRD, EMPA, SEM and EDS were used to characterise treated (pre-washed for 3 – 4 hours with doubly-distilled water) and untreated samples; no schoepite was detected (< 5% vol.) after the experiments. A sequence of collecting four data points over a period of 622 days was reported including an initial period 81 days exposure of 0.0215 g of solid to 450 mL of water with constant stirring. The pH ($\pm$ 0.2) was adjusted with NaOH and $HClO_4$ after each equilibration. The speciation of U(VI) and formation constants were determined according to [92GRE/FUG] with the species considered being: UO_2^{2+}, $UO_2(OH)^+$, $UO_2(OH)_2(aq)$, $UO_2(OH)_3^-$, $UO_2(OH)_4^{2-}$, $(UO_2)_2(OH)^{3+}$, $(UO_2)_2(OH)_2^{2+}$, $(UO_2)_3(OH)_4^{2+}$, $(UO_2)_3(OH)_5^+$, $(UO_2)_3(OH)_7^-$ and $(UO_2)_4(OH)_7^+$ (except for $\log_{10} {}^*\beta_{2,1}$ which was assigned a value of – 12.05 (*cf.* ≤ – 10.3 [92GRE/FUG] and – (12.15 $\pm$ 0.07) in the current review). The SIT was used to represent the activity coefficients at these low ionic strengths. The observed total U concentration was regressed in terms of the measured hydrogen ion and calcium concentrations including the formation constants and activity coefficients mentioned above to yield a $\log_{10} {}^*K_{s,0}^\circ$ value of (29 $\pm$ 1) for the dissolution of becquerelite, *viz.*,

$$Ca(UO_2)_6O_4(OH)_6 \cdot 8H_2O)(cr) + 14\, H^+ \rightleftharpoons Ca^{2+} + 6\, UO_2^{2+} + 18\, H_2O(l).$$

The calcium levels exceeded the 1:6 stoichiometry (Ca:U) indicating either incongruent dissolution or the presence of a calcium containing solid phase impurity. The uranium concentrations are significantly less than would be expected from schoepite, uranophane, or soddyite. One contribution for the much lower solubility observed in this study compared with those reported previously, $\log_{10} {}^*K_{s,0}^\circ$ = 41.9 – 43.7 [94SAN/GRA] and $\log_{10} {}^*K_{s,0}^\circ$ = 43.2, [90VOC/HAV], was suggested to be the larger crystal size used in their experiments, but the difference would appear much too large to be accounted for solely by this effect. No explanation can be offered at this time for the disparity between these results. The solubility constant, $\log_{10} {}^*K_{s,0}^\circ$ = (29 $\pm$ 1), is not accepted in this review (see [90VOC/HAV]).

[97CLA/CON]

This is the first attempt to identify by X ray Absorption Spectroscopy, XAS, a species of Np(VII) in strong alkaline solution where Np(VII) is stable. EXAFS fluorescence spectra of a test solution $3 \cdot 10^{-2}$ M in Np in 2.5 M NaOH under an ozone atmosphere were recorded in the k space up to 10 Å^{-1}. The Fourier Transform (FT) of these spectra is interpreted as resulting from a coordination of Np by two oxo oxygen atoms, four hydroxo oxygen (N = (4 ± 1)) atoms and by one molecule of water. This FT is close to that of $NpO_2(CO_3)_3^{5-}$, and this similarity confirms, according to the authors, the presence of the NpO_2 core in the species which is proposed to be $NpO_2(OH)_4H_2O^{2-}$. These results were refuted by [2001WIL/BLA] and [2001BOL/WAH].

[97COR/BOO]

The structure of the perovskite-related phases, $Ba_{1+y}UO_{3+x}$ and $(Ba,Sr))_{1+y}UO_{3+x}$, was investigated over a wide range of x and y by X–ray and neutron diffraction. The structural properties within the series, $BaUO_3(cr)$ to $Ba_3UO_6(cr)$, are discussed in detail. Integral enthalpies of formation of the compounds, $Ba_{1+y}UO_{3+x}$, with $1.033 \leq y \leq 1.553$ and $3.134 \leq x \leq 3.866$ were determined by solution calorimetry; in addition, the stability of $BaUO_{3+x}$ was investigated by oxygen potential determinations.

The alkaline earth uranates were prepared in various Ba/U ratios by heating the purified binary oxides (BaO, UO_2) in a Ar + H_2 atmosphere at 950°C followed by heating at 1300°C in stabilised zirconia crucibles. Chemical analyses were carried after ion exchange separation of the alkaline earth ions. Total uranium was determined titrimetrically with dichromate and Ba by complexometry. The oxygen content was obtained from the analysed Ba/U ratio and the weight increase to $Ba_3UO_6(cr) + BaUO_4(cr)$ upon heating the samples in O_2 to 1000°C.

The calorimetric dissolution medium was (HCl + 0.0400 $FeCl_3$ + 70.68 H_2O) referred to below as (sln). In addition to the samples, mixtures of $BaCl_2$ + UCl_4 with Ba/U ratios varying from 0.83 to 2.49 were also dissolved and the values corresponding to the $Ba_{1+y}UO_{3+x}$, needed to close the thermodynamic cycles, were interpolated. Also used in the cycles were the values of $\Delta_{sol}H_m$ (UO_2Cl_2, cr) = – (102.18 ± 0.63) kJ · mol^{-1} from [88COR/OUW], the standard enthalpies of formation of UCl_4, UO_2Cl_2, $BaCl_2$ and the partial molar enthalpies of formation of HCl and H_2O in the medium.

In our recalculations, we have used the values $\Delta_f H_m$ (HCl, sln, partial) = – (164.71 ± 0.11) kJ · mol^{-1} and $\Delta_f H_m$ (H_2O, sln, partial) = – (285.84 ± 0.04) kJ · mol^{-1}. These values are based on a free chloride concentration of (0.81 ± 0.015) mol · dm^{-3} which takes into account the complexing effect of the ferric ion. For the interpolation of the $BaCl_2/UCl_4$ enthalpies of solution, we have fitted the experimental results to a quadratic function. We obtain:

$\Delta_f H_m^\circ$ (Ba$_{1.033}$UO$_{3.134}$, cr, 298.15 K) = $-$ (1721.68 $\pm$ 3.57) kJ $\cdot$ mol^{-1},

$\Delta_f H_m^\circ$ (Ba$_{1.065}$UO$_{3.172}$, cr, 298.15 K) = $-$ (1733.43 $\pm$ 3.65) kJ $\cdot$ mol^{-1},

$\Delta_f H_m^\circ$ (Ba$_{1.238}$UO$_{3.407}$, cr, 298.15 K) = $-$ (1868.05 $\pm$ 4.16) kJ $\cdot$ mol^{-1},

$\Delta_f H_m^\circ$ (Ba$_{1.400}$UO$_{3.604}$, cr, 298.15 K) = $-$ (1982.10 $\pm$ 4.38) kJ $\cdot$ mol^{-1},

$\Delta_f H_m^\circ$ (Ba$_{1.553}$UO$_{3.866}$, cr, 298.15 K) = $-$ (2116.90 $\pm$ 4.38) kJ $\cdot$ mol^{-1}.

These enthalpies of formation are appreciably less negative than those reported by the authors, whose reaction scheme (Table 7 of their paper) contains a numerical error - the value used in reaction 2 in that table corresponds to the oxidation of one mole of UCl$_4$(cr), not the 0.899 mole of the relevant equation.

The authors report a linear change of the enthalpy of formation from Ba$_3$UO$_6$ to Ba$_{1.033}$UO$_{3.134}$ as a function of the Ba/U(total) ratio in the compounds. In this relationship, they include two further values, $\Delta_f H_m^\circ$ (BaUO$_{3.05}$, cr, 298.15 K) = $-$ (1700.4 $\pm$ 3.1) kJ $\cdot$ mol^{-1} and $\Delta_f H_m^\circ$ (BaUO$_{3.08}$, cr, 298.15 K) = $-$ (1710.0 $\pm$ 3.0) kJ $\cdot$ mol^{-1}, of unspecified origin. Since these two results may suffer from the same error as the detailed measurements, they have been discounted.

However, since both the Ba/U ratio and the uranium valency are changing in these compounds, this approximately linear relation for the five samples measured must be by chance (or perhaps related to the oxygen pressure during preparation). Thus the authors' derived value for the enthalpy of formation of BaUO$_3$(cr) is $-$ (1680 $\pm$ 10) kJ $\cdot$ mol^{-1} (which becomes $-$ (1688 $\pm$ 10) kJ $\cdot$ mol^{-1} from the corrected values) by extrapolation of this linear relationship and must be treated with some circumspection.

At this stage, we see no argument to modify the previously selected value

$\Delta_f H_m^\circ$ (BaUO$_3$, cr, 298.15 K) = $-$ (1690 $\pm$ 10) kJ $\cdot$ mol^{-1}.

The authors also measured the oxygen potential as a function of the oxygen content of Ba$_{1.033}$UO$_{3+x}$(cr), using a reversible emf cell of the type:

(Pt)Ba$_{1.033}$UO$_{3+x}$ | Calcia $-$ Stabilized ZrO$_2$ | O$_2$ (Pt), p_{O_2} = 0.202 bar

After each equilibration, the actual composition of the sample was determined on the rapidly cooled sample, assuming that during cooling the composition did not change. The fitted equations were:

E(mV) = 1595.9 $-$ 1138.2$\cdot$x   at 1060 K

E(mV) = 1494.4 $-$ 874.8$\cdot$x   at 1090 K

for 0.15 < x < 0.335

The authors used a linear extrapolation of the potentials to x = 0 to give an approximate value of the equilibrium oxygen potential of stoichiometric BaUO$_3$(cr) at 1060 K, which becomes $-$ 629 kJ $\cdot$ mol^{-1} with our refitted equation. This is, in fact, likely to be an upper limit for the oxygen potential, which probably drops sharply near x = 0 as all the uranium becomes quadrivalent (cf. UO_{2+x}). Such a value is low enough to support the long reported failure [75BRA/KEM], [82BAR/JAC], [84WIL/MOR] to obtain stoichiometric $BaUO_3(cr)$.

[97DEM/SER]

Aqueous saturated solutions of dioxouranium(VI) sulphate, $UO_2SO_4 \cdot 2.5\,H_2O$, are in equilibrium with $UO_2SO_4 \cdot 3.5\,H_2O(cr)$ and at 25°C the composition of the solution is 60.8 wt% of UO_2SO_4. These are the only relevant data which can be derived from this paper.

[97ELL/ARM]

This paper deals with the solubility of uranium from well-characterised (according to particle size) U-contaminated soils in synthetic acid rain water ($[H_2SO_4] = 5 \cdot 10^{-5}$ M and $[HNO_3] = 10^{-4}$ M in a 2/1 ratio, pH = 4), simulated groundwater ($[NaHCO_4] = [MgSO_4] = 2 \cdot 10^{-3}$ M, $[CaCl_2] = [CaSO_4] = 10^{-3}$ M, pH = 7.9 and $I = 1.7 \cdot 10^{-2}$) and other solutions (lung serum and stomach-like solutions) which are of no interest for this review. The equilibrium chemistry of the experiments were modelled using Geochemist's Workbench version 2 code[1] to predict speciation and precipitation of saturated phases. Solubility data (^{235}U and ^{238}U in a natural isotopic ratio) are rather good, but could not be used to deduce thermodynamic data. The species and minerals put forward to explain solubilities are: $UO_2(CO_3)_2^{2-}$, $UO_2(CO_3)_3^{4-}$, $UO_2(CO_3)(OH)_3^-$ and uranophane $Ca(UO_2)_2Si_2O_7 \cdot 5H_2O$, haiweeite $Ca(UO_2)_2Si_6O_{15} \cdot 5H_2O$, soddyite $(UO_2)_2SiO_4 \cdot 2H_2O$. This review suggests that the species, $UO_2(CO_3)(OH)_3^-$, should be $(UO_2)_2(CO_3)(OH)_3^-$. The formula of uranophane is $Ca(UO_2)_2(SiO_3OH)_2 \cdot 5H_2O$.

[97FIN]

Accurate thermodynamic data for minerals are rare and such data therefore often have to be estimated. This paper describes one estimation method, originally proposed by Tardy and Garrels [76TAR/GAR], where the $\Delta_f G_m^\circ$ value for each mineral is estimated from the sum of the constituent oxide contributions, and developed by Saxena [97SAX]. The $\Delta_f G_m^\circ$ values given in Table 1 in [97FIN] refer to the molar contributions of these constituents in various mineral phases. The values are listed in Table A-25, where the selected values for the pure oxides are also given:

---

[1] Bethke, C. M., Urbana–Champaign, University of Illinois, USA

Table A-25: Comparison between $\Delta_f G_m^\circ$ calculated from the sum of constituent oxide contributions and $\Delta_f G_m^\circ$ selected in [92GRE/FUG].

Oxide	Calculated $\Delta_f G_m^\circ$ (kJ·mol⁻¹) at 298.15 K for the oxide constituents in mineral phases	Selected $\Delta_f G_m^\circ$ (kJ·mol⁻¹) at 298.15 K for the pure phases from [92GRE/FUG]
UO_3(s)	−1153.5	−1145.74 (UO_3–γ)
H_2O(l)	−242.6	−237.14
CaO(s)	−716.4	−603.30
SiO_2(s)	−865.9	−856.29 (quartz)
CO_2(g)	−404.6	−394.37

The paper by Finch contains a list of $\Delta_f G_m^\circ$ values for oxide, carbonate and silicate minerals, most of which have a layer structure. Some of these values refer to experimental data discussed in this review under [92NGU/SIL] or in [92GRE/FUG] under [88OHA/LEW]. The Gibbs energy of formation for other minerals listed has been calculated using the method outlined above, but without estimated uncertainty ranges. The stability ranges of the different minerals can be calculated using the Gibbs energy of formation and presented graphically in activity – activity diagrams; Finch gives several examples. It is important to notice that the phase boundaries are strongly dependent on the accuracy of the estimated Gibbs energy of formation. To take one example: the phase boundary between metaschoepite and becquerelite is at $\log_{10}[Ca^{2+}]/[H^+]^2 = 11.58$ using the values given by Finch. A change in the estimated Gibbs energy of formation of the two phases by 10 kJ·mol⁻¹ results in a shift of the phase boundary of 3.5 units along the $\log_{10}[Ca^{2+}]/[H^+]^2$ axis. Diagrams of this type are useful for putting constraints on thermodynamic data, estimated or experimental, but also when discussing mineral formation. It is essential that the method is used only for structurally related phases, e.g., the layer structures discussed by Finch. The $\Delta_f G_m^\circ$ values given in Table II of [97FIN] can be used as estimates of the Gibbs energy of formation for the listed minerals, but they are not sufficiently precise to be included as selected values. Table A-26 lists the Gibbs energy of formation proposed by Finch, however the user should consult the original publication for more details. The values given in parentheses are not considered reliable by Finch. For the experimental value of liebigite this is due to the large surface area of the mineral samples. The estimated values that are considered unreliable have different structures from those of the minerals used for calibration.

Table A-26: Estimated and experimental Gibbs energy of formation for some uranium(VI) minerals. All data refer to 298.15 K and are given in kJ · mol^{-1}.

Mineral	Formula	$\Delta_f G_m^\circ$ (kJ·mol^{-1})	Estimated (Est) or Experimental (Exp)
Schoepite	$[(UO_2)_8O_2(OH)_{12}](H_2O)_{12}$	− 13334.7	Est.
Metaschoepite	$[(UO_2)_8O_2(OH)_{12}](H_2O)_{10}$	− 13092.1	Exp.
Dehydrated schoepite [a]	$[(UO_2)O_{0.25-x}(OH)_{1.5+2x}]$, 0<x<0.15	− 1362.3	Exp.
Becquerelite	$Ca[(UO_2)_6O_4(OH)_6](H_2O)_8$	− 10305.8	Est.
Ca–protasite	$Ca_2[(UO_2)_6O_6(OH)_4](H_2O)_8$	− 10779.6	Est.
Metacalciouranoite	$Ca_3[(UO_2)_6O_8(OH)_2](H_2O)_6$	− 10768.3	Est.
Calciouranoite	$Ca_3[(UO_2)_6O_8(OH)_2](H_2O)_9$	− 11496.1	Est.
Soddyite	$(UO_2)_2SiO_4(H_2O)_2$	− 3658.0	Exp.
Swamboite	$(UO_2)[(UO_2)(SiO_3OH)]_6(H_2O)_{30}$	− 21092.4	Est.
Uranosilite	$(UO_2)(Si_7O_{15})$	(− 7214.5)	Est.
Uranophane	$Ca[(UO_2)(SiO_3OH)]_2(H_2O)_5$	− 6210.6	Exp.
Haiweeite	$Ca[(UO_2)(Si_2O_5)_3](H_2O)_5$	(− 9431.4)	Est.
Ursilite	$Ca_4[(UO_2)_4(Si_2O_5)_5(OH)_6](H_2O)_{15}$	(− 20504.6)	Est.
Rutherfordine	UO_2CO_3	− 1563.1	Exp.
Joliotite	$UO_2CO_3(H_2O)_{1-2}$	− 2043.3	Est.
Sharpite	$Ca(UO_2)_6(CO_3)_5(OH)_4(H_2O)_3$	− 11601.1	Est.
Fontanite	$Ca(UO_2)_3(CO_3)_4(H_2O)_3$	− 6523.1	Est.
Urancalcarite	$Ca(UO_2)_2(CO_3)(OH)_6(H_2O)_3$	− 6037.0	Est.
Zellerite	$Ca(UO_2)(CO_3)_2(H_2O)_3$	(− 3892.1)	Est.
Liebigite	$Ca_2(UO_2)(CO_3)_3(H_2O)_{10}$	(− 6226.0)	Exp.

a): This formulation corresponds to UO$_3$·yH$_2$O, y = 0.75 to 0.9. In [98FIN/HAW] the possible range of variation of y for the dehydrated schoepite is 0.75 to 1.00.

[97GEI/BER]

This paper gives fluorescence data on species which are assumed to exist in natural solutions according to the data selected by [92GRE/FUG] for U concentration between 9·10^{-3} and 7.3·10^{-2} M.

Table A-27: Main characteristics of fluorescence spectra of some aqueous complexes.

pH	Species	Main fluorescence lines, nm	fluorescence lifetime, µs
7.13–7.82	$Ca_2UO_2(CO_3)_3(aq)$	465, 484, 504, 524	(0.043 ± 0.012)
9.76	$UO_2(CO_3)_3^{4-}$	no fluorescence	
2.6	$UO_2SO_4(aq)$ + traces of $UO_2(SO)_2^{2-}$	477, 493, 514, 538	(4.3 ± 0.4)
	UO_2^{2+}	472, 488, 510, 533	(1.6 ± 0.2)

These data can be used for identification of species.

[97GEI/RUT]

The issue of this paper is to decide if the dependence of the fluorescence lifetime, τ, of U(VI) in perchloric acid or perchloric acid/perchlorate solutions can be a result of a U(VI) perchlorate complex. This question has also been raised in [99BOU/BIL]. Geipel et al. show that there is no evidence of the existence of such a complex. On the contrary, it shows that the lifetime, τ, decreases linearly with increasing concentration of H_2O (from 3 M in concentrated perchloric and sulphuric acid solutions to 55.5 M in pure water) as a result of the quenching effect of water (Stern-Vollmer mechanism). Furthermore the emission wavelength of the fluorescence spectra in $HClO_4$ (up to 11.5 M) around 510.5 nm remains constant ($\pm$ 1 nm) and the fluorescence decay is always described by a single exponential.

[97GUR/SER]

It should be noted that throughout the English version of this paper, the term "rhomb" is erroneously used instead of "orth". In this discussion we have adopted the correct terminology. Partial results were given in [89GUR/DEV].

The paper has four parts.

• The first part deals with the synthesis, characterisation and measurement of $C_{p,m}$ of $UO_2(OH)_2(\alpha, orth)$ from 14 to 316 K, from which $S_m^\circ(T)$ and $H_m(T) - H_m(0)$ are derived by calculation.

• The second part is a survey of published experimental results (structure of compounds but mainly thermochemical data) on the systems O–U(VI), O–H–U(VI), O–C–U(VI), O–H–S–U(VI), O–H–N–U(VI) and O–H–C–U(VI) leading to a selection of thermodynamic values. Of the 70 references cited, none are more recent than 1992, (which is surprising for a paper submitted at the end of 1995 and published in 1997), except for a personal communication of Estigneev from 1996 (see below). Consequently, this survey does not add much to the discussions of [92GRE/FUG] about the compounds and complexes belonging to these systems.

- The third part concerns derived thermodynamic data obtained according to the new experimental results on $UO_2(OH)_2(\alpha, orth)$.

- The last part is a discussion about a possible error in $S_m^\circ(UO_2^{2+}, 298.15\ K)$ which again questions the validity of all the data based on the entropy of that ion. This topic was already discussed in the review of [94SER/DEV] (see this Appendix). Some of the results in the first and third parts were given briefly in an earlier paper [89GUR/DEV].

There are several misprints in the text.

- Part 1: The system O–H–U(VI), to which belongs $UO_3 \cdot xH_2O$, $0 < x < 1$, $UO_2(OH)_2(\alpha, orth)$, $UO_2(OH)_2(\beta, orth)$, schoepite: $UO_2(OH)_2 \cdot H_2O(cr, ortho)$, is complicated because many polymorph compounds can appear and transform under chemical or physical conditions. Two samples of $UO_2(OH)_2(\alpha, orth)$ are synthesised and characterised as the member $x = 1$ of the series: $UO_3 \cdot xH_2O(\alpha, orth)$ by XRD (lattice parameters are given) and chemical analysis (amount of U: 78.24 %). Adiabatic calorimetric measurements on $m = 3.29$ g lead to $C_{p,m}^\circ = f(T, 14.67\ to\ 316.82\ K)$ from which Gurevich et al. calculate:

$$C_{p,m}^\circ (298.15\ K) = (113.96 \pm 0.12)\ J \cdot K^{-1} \cdot mol^{-1}$$

$$S_m^\circ (298.15\ K) = (128.10 \pm 0.20)\ J \cdot K^{-1} \cdot mol^{-1}$$

$$H_m^\circ (298.15\ K) - H_m(0) = (19.703 \pm 0.015)\ kJ \cdot mol^{-1}.$$

Expansion of $C_{p,m}^\circ = f(T)$ is given as a sum of $A_x[1 - \exp(-0.001\ T)^x]$, x ranging from 0 to 5. From this function $C_{p,m}^\circ$ is calculated.

- Part 2: According to literature data, $\Delta_r H_m^\circ$ and/or $\Delta_r G_m^\circ$ are selected at 298.15 K for 16 heterogeneous reactions belonging to the systems mentioned above. All the used data are known and given in [92GRE/FUG]. This review notices, however, some minor discrepancies between the given data and those obtained using the selected data in [92GRE/FUG].

The authors claim that $UO_2(OH)_2(\beta, orth)$ is clearly metastable with respect to $UO_2(OH)_2(\alpha, orth)$ and that confusion existed between the β and α forms in [72NIK/SER]. This mistake is said to have been replicated in several papers and particularly in [92GRE/FUG]. Nevertheless, this has no influence on the estimated value of $S_m^\circ(UO_2(OH)_2(\beta, orth))$ given on p.138 [92GRE/FUG], because the final selected value comes from [66ROB].

[97GUR/SER] also cite their extrapolation to zero ionic strength of the solubility measurements (by potentiometry) on $UO_3 \cdot H_2O(\alpha, orth)$ in the temperature range 295 – 523 K by [71NIK/PIR], not cited in [92GRE/FUG]. For the reaction:

$$UO_3 \cdot H_2O\ (\alpha, orth, T) + 2H^+ \rightleftharpoons UO_2^{2+} + 2H_2O(l)$$

at infinite dilution the following $\log_{10} {}^*K°$ values are obtained: (5.5 ± 0.1), at 295 K, (5.0 ± 0.1), at 323 K, (4.4 ± 0.1), at 352 K, (3.9 ± 0.1), at 363 K, (3.6 ± 0.2), at 373 K, (3.1 ± 0.2), at 398 K and (2.7 ± 0.2) at 423 K.

By assuming that $\log_{10} {}^*K°$ has the same value at 295 and 298.15 K, we obtain:

$$\Delta_f G_m^°(UO_3 \cdot H_2O, \alpha, \text{orth}, 298.15 \text{ K}) = -(1395.44 \pm 1.75) \text{ kJ} \cdot \text{mol}^{-1}.$$

- Part 3: This part describes calculation of the following new thermodynamic values:

$$\Delta_f H_m^°(UO_2(OH)_2, \alpha, \text{orth}, 298.15 \text{ K}) = -(1536.87 \pm 1.30) \text{ kJ} \cdot \text{mol}^{-1}$$

and also of:

$$\Delta_f H_m^°(UO_2(OH)_2, \beta, \text{orth}, 298.15 \text{ K}) = -(1533.87 \pm 1.30) \text{ kJ} \cdot \text{mol}^{-1}$$

$$\Delta_f H_m^°(UO_2(OH)_2 \cdot H_2O, \text{scho}, 298.15 \text{ K}) = -(1825.86 \pm 1.20) \text{ kJ} \cdot \text{mol}^{-1}$$

which both agree with selected values in [92GRE/FUG], $-(1533.8 \pm 1.3)$ kJ · mol^{-1} and $-(1826.1 \pm 1.7)$ kJ · mol^{-1}, respectively.

For the calculation of $\Delta_f H_m^°(UO_2(OH)_2, \alpha, \text{orth}, 298.15 \text{ K})$, [97GUR/SER] extrapolate to $x = 1$, the literature values for $\Delta_f H_m^°$ for the reaction:

$$UO_3(\gamma) + x H_2O(l) \rightleftharpoons UO_3 \cdot xH_2O(\alpha, \text{orth})$$

being based on the differences in the enthalpies of solution of the two compounds and calculate a value of $-(27.4 \pm 0.2)$ kJ · mol^{-1}. Literature values used for these differences were, for $x = 0.393$ and $x = 0.648$, $-(11.71 \pm 0.52)$ and $-(15.61 \pm 0.57)$ kJ · mol^{-1}, respectively, from [72SAN/VID]; for $x = 0.9$, $-(25.27 \pm 0.51)$ kJ · mol^{-1} from [88OHA/LEW]; for $x = 0.85$, [97GUR/SER] used a value of $-(23.60 \pm 0.18)$ kJ · mol^{-1} incorrectly cited from [64COR], instead of $-(24.65 \pm 0.25)$ kJ · mol^{-1}.

The difference in the enthalpies of solution between UO$_3$·xH$_2$O(α, orth) and UO$_3$(γ) is an approximation of the difference in their enthalpies of formation when the enthalpy of transfer of water from infinite dilution to the dissolution medium is negligible. This was the case for the experiments of [72SAN/VID] and [88OHA/LEW] who used a dilute HF medium. In the case of [64COR], the difference in enthalpy of transfer of 0.85 mole of H$_2$O from an infinitely dilute medium to 6.0 M HNO$_3$ amounts to $-(0.54 \pm 0.04)$ kJ · mol^{-1}. Thus, from the results of [64COR], the difference in enthalpy of formation between UO$_3$ · 0.85H$_2$O(α, orth) and UO$_3$(γ) becomes $-(25.17 \pm 0.25)$ kJ · mol^{-1}. A new extrapolation to $x = 1$ of the difference between the enthalpies of formation of UO$_3$·xH$_2$O(α, orth) and UO$_3$(γ), giving equal weight to each experimental point, gives $-(26.8 \pm 1.5)$ kJ · mol^{-1}, which is very similar to that selected by [92GRE/FUG], but with substantially increased uncertainty limits. Keeping in mind the precarious character of this extrapolation, we calculate $\Delta_f H_m^°$ (UO$_2$(OH)$_2$, α, orth, 298.15 K) = $-(1536.4 \pm 1.9)$ kJ · mol^{-1}.

The authors also use the new entropy value, $S_m^°(UO_2(OH)_2, \alpha, \text{orth}, 298.15 \text{ K})$ = (128.10 ± 0.20) J·K^{-1}·mol^{-1}, in order to calculate $S_m^°(UO_3 \cdot 0.5H_2O, \text{cr})$ by taking:

$$S_m^°(UO_3 \cdot 0.5H_2O, \text{cr}, 298.15 \text{ K}) = \frac{1}{2}[S_m^°(UO_3(\gamma)) + S_m^°(UO_2(OH)_2(\alpha, \text{orth})]$$
$$= (112 \pm 2) \text{ J·K}^{-1}\text{·mol}^{-1}.$$

Interpolation of $\Delta_r H_m^°$ (298.15 K) for:

$$UO_3(\gamma) + x H_2O(l) \rightleftharpoons UO_3 \cdot x H_2O(\alpha, \text{orth})$$

from experimental values for x equal to 0.648 and 0.393 gives:

$$\Delta_r H_m^° (x = 0.5, 298.15 \text{ K}) = -(13.35 \pm 0.61) \text{ kJ·mol}^{-1}$$

and leads to:

$$\Delta_r H_m^° (UO_3 \cdot 0.5H_2O, \text{cr}, 298.15 \text{ K}) = -(1380.07 \pm 1.30) \text{ kJ·mol}^{-1}.$$

The reported value for $UO_3 \cdot 0.5H_2O$ is not selected by this review.

• Part 4: The value of $S_m^°(UO_2^{2+}, 298.15 \text{ K}) = -(98.2 \pm 3.0)$ J·K^{-1}·mol^{-1} selected in [92GRE/FUG] is derived from data related to:

$$UO_2(NO_3)_2 \cdot 6H_2O(\text{cr}) \rightleftharpoons UO_2^{2+} + 2 NO_3^- + 6 H_2O(l)$$
$$\Delta_{sol} H_m^° (298.15 \text{ K}) = (19.35 \pm 0.20) \text{ kJ·mol}^{-1}$$
$$\Delta_{sol} G_m^° (298.15 \text{ K}) = -(13.24 \pm 0.21) \text{ kJ·mol}^{-1}.$$

Following the authors, three other independent alternative routes allow calculations of $S_m^°(UO_2^{2+})$, which all lead to higher values. These are:

$$UO_2(OH)_2 \cdot H_2O + 2 H^+ \rightleftharpoons UO_2^{2+} + 3 H_2O(l)$$

leading to $S_m^°(UO_2^{2+}) = -(77.3 \pm 3.0)$ J·K^{-1}·mol^{-1},

$$UO_2CO_3(\text{ruth}) \rightleftharpoons UO_2^{2+} + CO_3^{2-}$$

leading to $S_m^°(UO_2^{2+}) = -(71.8 \pm 10.0)$ J·K^{-1}·mol^{-1} and, finally:

$$UO_2SO_4 \cdot 2.5H_2O(\text{cr}) \rightleftharpoons UO_2^{2+} + SO_4^{2-} + 2.5 H_2O(l)$$

leading to $S_m^°(UO_2^{2+}) = -(77.2 \pm 15.0)$ J·K^{-1}·mol^{-1}.

These routes take into account solubility products derived from solubility measurements of the appropriate sparingly soluble compounds as the key primary data. These data are less reliable than the calorimetric heat of dissolution of dioxouranium(VI) nitrate hexahydrate measured by numerous authors and given by Gurevich *et al.*

In this publication, the authors reproduce the argument already developed by [94SER/DEV] and in a more condensed manner by [92KHO]. These references have already been considered by this review with the conclusion that the value of the entropy of the dioxouranium(VI) ion selected in [92GRE/FUG] should be maintained.

[97HOV]

This is an important study as it is a precise determination of the partial molar heat capacity and volume of a M^{4+} ion. By selecting Th(IV), the author has been able to control the experimental difficulties due to hydrolysis. The procedure, the data and the data treatment are detailed in the paper. The constancy of the results for the system $Th(ClO_4)_4 + HClO_4$ has been tested using Young's rule and the Pitzer ion-interaction model with a concordant result over the entire concentration range 1.0 to 2.9 m investigated. Hovey reports $C^o_{p,m}(Th^{4+}) = -(224 \pm 5)$ J·K^{-1}·mol^{-1}. The author has also reviewed the previous attempts [76MOR/MCC], [75APE/SAH] to estimate the partial molar heat capacity of Th^{4+} and finds that the results deviate strongly from his own experimental data. The data of Morss and McCue [76MOR/MCC] based on the integral heat of dilution of $Th(NO_3)_4$ at 15, 25 and 35°C, recalculated using a new value of the partial molar heat capacity of NO_3^-, $C^o_{p,m} = -72$ J·K^{-1}·mol^{-1} [88HOV/HEP3], gives $C^o_{p,m}(Th^{4+}, 298.15\,K) = -(60 \pm 11)$ J·K^{-1}·mol^{-1} close to the value selected by [92GRE/FUG] for U^{4+}, $C_{p,m}(U^{4+}, T = 298$ to $473\,K) = -(48 \pm 15)$ J·K^{-1}·mol^{-1}. The data of Apelbat and Sahar [75APE/SAH] at 30°C based on measurement of bulk heat capacity of $Th(NO_3)_4$ solutions given as a linear function of $m^{1/2}$ for the range 0 to 2.9 m lead to $C^o_{p,m}(Th^{4+}) = 111$ J·K^{-1}·mol^{-1}, as calculated by Hovey, using $C^o_{p,m}(NO_3^-, 298.15\,K) = -63.7$ J·K^{-1}·mol^{-1} [89HOV/HEP], who plausibly argue that the discrepancy is not surprising, given the fact that specific molalities and heat capacities are not reported and also the improbable ability of a simple linear equation to represent heat capacities over such a wide range of temperature.

The experimental conditions chosen by Hovey are preferable to those of [76MOR/MCC], *i.e.*, little hydrolysis and no complexation.

A parent paper [86HOV/TRE] discusses $C^o_{p,m}(Al^{3+})$, where the $C^o_{p,m}(Al^{3+})$ value is found to be -119 J·K^{-1}·mol^{-1}, while the Criss-Cobble equation [64CRI/COB2] leads to 16 J·K^{-1}·mol^{-1}.

In addition, the correlation methods of [97SHO/SAS2] (see this Appendix) give $C^o_{p,m}(U^{4+}$ aq, $298.15\,K) = 0.8$ J·K^{-1}·mol^{-1} (based on the results of [76MOR/MCC] for the thorium ion) in disagreement with the selected value in [92GRE/FUG].

It seems that the Criss–Cobble equation for calculation of heat capacity from the entropy breaks down for multicharged ions. As Hovey noticed, the correspondence between the result of Morss and McCue for Th^{4+} and the Criss-Cobble prediction for U^{4+} could be fortuitous.

Hovey's studies cast serious doubt on the value selected on p.95 of [92GRE/FUG] based on the selected partial molar entropy, $S_m^o(U^{4+})$, and the Criss-Cobble estimate by Lemire and Tremaine [80LEM/TRE], as well as on $C_{p,m}^o(U^{3+}, T = 298$ to 473 K$) = -(64 \pm 22)$ J·K^{-1}·mol^{-1} although a comparison between values at 298.15 K and average values for the temperature range 298.15 to 473 K may not be strictly correct. Indeed, it should be pointed out that the value selected by [92GRE/FUG] is an average value for this temperature range. This review prefers the value of Hovey in the temperature range 273 to 473 K.

[97HUA/YAM]

The enthalpy effect associated with the formation of SrUO$_3$(cr) has been deduced from vaporisation studies of this compound over the temperature range 1534 – 1917 K, using Knudsen effusion mass spectrometry. The sample was obtained by reduction of SrUO$_4$(cr) with hydrogen. As stated in the text, "A mixture of SrUO$_3$(cr) and UO$_2$(cr) was employed for the mass spectrometry study, since it was not easy to prepare single phase SrUO$_3$". X–ray diffraction analysis yielded a content of 23wt.% UO$_2$ in the mixture. After the mass spectrometric measurements, the formula of the compound was estimated to have changed to SrUO$_{3.1}$(cr), based on X–ray diffraction data.

On the basis of the main vapour species detected, the following reactions were postulated to occur in the Knudsen cell:

$$SrUO_3(cr) \rightleftharpoons Sr(g) + 0.5\ O_2(g) + UO_2(cr)$$

$$Sr(g) + 0.5\ O_2(g) \rightleftharpoons SrO(g)$$

$$UO_2(cr) \rightleftharpoons UO_2(g)$$

but the change in lattice parameters of the sample suggests that this is a considerable simplification, and the vaporisation is in fact non–congruent. The corresponding partial vapour pressures were fitted to the following expressions (recalculated in bar):

$$\log_{10} p_{Sr}\ (bar) = 1.115 - 13405.7 \cdot T^{-1}$$

$$\log_{10} p_{SrO}\ (bar) = 8.984 - 33827.0 \cdot T^{-1}$$

$$\log_{10} p_{UO_2}\ (bar) = 7.334 - 29918.8 \cdot T^{-1}.$$

To obtain the standard enthalpy of formation of "SrUO$_3$(cr)", its enthalpy increment and Gibbs energy function, $[(G_m(T) - H_m(298.15\ K)]/T$, were estimated by assuming the values of $H_m(T) - H_m^o(298.15\ K)$ for (SrMO$_3$(cr)–SrO(cr)–MO$_2$(cr)) were the same for M = Zr and U. This assumption appears reasonable in view of the closeness of crystal structures of SrUO$_3$ and SrZrO$_3$. All the necessary thermodynamic data were taken from [90COR/KON]. The values for the standard enthalpy of formation of "SrUO$_3$"(cr) were $-(1785 \pm 60)$ and $-(1698.1 \pm 5.0)$ kJ·mol^{-1} from the second- and third-law treatments, respectively, where the latter is recalculated by this review. Given the change in composition of the sample during the experiment, it is not clear to which composition these values refer; we have assumed they apply to the final composition, so $\Delta_f H_m^\circ$ (SrUO$_{3.1}$, cr, 298.15 K) = $-$ (1785 ± 60) and $-$ (1698.1 ± 5.0) kJ·mol^{-1}.

[97HUA/YAM] also estimated values for the enthalpies of formation of a series of compounds with the formula, SrMO$_3$(cr), based on experimental values for the Ti, Hf, Zr [93KUB/ALC] and U compounds, by plotting their enthalpies of formation from the corresponding binary oxides as a function of M^{4+} ionic radii. Unfortunately, these authors were not aware of the fact that experimental values for the standard enthalpies of formation of other SrMO$_3$ compounds had been published, namely with M = Ce, Tb, Am [90GOU/HAI] and that relationships based on experimental values for the enthalpies of formation of SrMO$_3$ and BaMO$_3$, and the packing of the ions in the lattice (Goldschmidt tolerance factor t [70GOO/LON], page 132) had been also published [93FUG/HAI].

From their relationship [97HUA/YAM] obtain for the standard enthalpy of formation of "SrUO$_3$"(cr), an estimated value of $-$ (1684 ± 20) kJ · mol^{-1}, compared with the third-law treatment of $\Delta_f H_m^\circ$ (SrUO$_{3.1}$, cr, 298.15 K) = $-$ (1698.6 ± 2.2) kJ·mol^{-1} from the vaporisation study. Using $\Delta_f H_m^\circ$ (SrO, cr, 298.15 K) = $-$ (590.6 ± 0.9) kJ · mol^{-1} from [92GRE/FUG] and $\Delta_f H_m^\circ$ (UO$_{2.1}$, cr, 298.15K) = $-$ (1102.0 ± 5) kJ · mol^{-1} (our estimate), the above third-law value is (6 ± 10) kJ · mol^{-1} more negative than the sum of the enthalpies of formation of SrO(cr) and UO$_{2.1}$(cr).

The relationship used by [93FUG/HAI] leads to a value 3 kJ · mol^{-1} more positive than the sum of the enthalpies of formation of the binary oxides, with an average standard deviation of ± 8.5 kJ · mol^{-1} (calculated for all experimental points). The opinion that SrUO$_3$(cr) is of very marginal stability compared to the binary oxides is supported by the difficulty in obtaining a pure phase as acknowledged (see above) by [97HUA/YAM]. As a consequence, we find it preferable to adopt the less negative value:

$$\Delta_f H_m^\circ (\text{SrUO}_3, \text{cr}, 298.15 \text{ K}) = - (1672.6 \pm 8.6) \text{ kJ} \cdot \text{mol}^{-1}$$

obtained by using the relationship proposed by [93FUG/HAI] and the NEA adopted values for the binary oxides. This value is not incompatible with the conclusions of [97HUA/YAM].

[97ION/MAD]

This paper shows that, taking into account covalency effects, the values for the enthalpies of formation of the lanthanide and actinide trihalides can be estimated with a better accuracy than when a purely ionic model ([85BRA/LAG], [86BRA/LAG]) is used. For instance, for the tribromides of U, Np and Pu, the predicted values are within 0 – 4 kJ · mol^{-1} of the experimental ones when using the covalent model, while the ionic model leads to differences ranging between 16 and 26 kJ · mol^{-1}. For the corresponding iodides, the covalent model yields predictions within 2 to 14 kJ · mol^{-1}, while the ionic model gives estimates within 2 to 29 kJ · mol^{-1} of the experimental values.

In the case of the americium trihalides, the predictions of the covalent model are $-$ 1609, $-$ 978, $-$ 809 and $-$ 585 kJ · mol^{-1} for the enthalpies of formation of the trifluoride, trichloride, tribromide and triiodide, respectively, while the values adopted by NEA [95SIL/BID] (and not changed by this review) are $-$ (1588 ± 13), $-$ (977.8 ± 1.3) (experimental), $-$ (810 ± 10) and $-$ (612 ± 7) kJ · mol^{-1}.

The procedure adopted by the NEA TDB for the estimation of a number of experimentally missing enthalpies of formation of halides is described in [83FUG/PAR] and is based on an extrapolation from neighbouring actinides of the difference $[\Delta_f H_m^\circ(MX_n, cr, 298.15 K) - \Delta_f H_m^\circ(M^{n+}, aq, 298.15 K)]$ versus the actinide ionic radius. It has also been adopted [2001LEM/FUG] for the estimation of the enthalpies of formation of NpF$_3$ (cr) and NpCl$_3$ (cr).

[97IOU/KRU]

The authors applied laser–induced fluorescence spectroscopy to investigate the complex formation of ^{244}Cm(III) with the heteropolyanions $PW_{11}O_{39}^{7-}$ and $SiW_{11}O_{39}^{8-}$ in 0.1 – 2 M nitric acid solutions with and without addition of 1 M NaNO$_3$. The study was performed at room temperature with Cm(III) concentrations of 10^{-3} mol · L^{-1} and ligand concentrations varying from 10^{-6} to 10^{-3} mol · L^{-1}. The experimental techniques, excitation and luminescence spectra are well described. In the solutions investigated, only the formation of 1:1 complexes could be ascertained unambiguously, but a notable decrease of the luminescence intensity at high $PW_{11}O_{39}^{7-}$ concentrations was interpreted as indication for the formation of the 2:1 complex $Cm(PW_{11}O_{39})_2^{11-}$. Since the degree of protonation of the ligands and complexes is unknown, the authors calculated apparent equilibrium constants (based on the total concentrations of the complex and uncomplexed ligand) which decrease by 2 – 3 orders of magnitude when the HNO$_3$ concentration is increased from 0.1 to 2.0 mol · L^{-1} (cf. Table 12-12., section 12.7.2).

[97IYE/JAY]

The paper reports measurements of the oxygen activity (emf with CaO–ZrO$_2$ electrolyte) in the mixtures, K$_2$U$_4$O$_{12}$(cr) + K$_2$U$_4$O$_{13}$(cr), and complementary data on the enthalpy increments of the two uranates.

K$_2$U$_4$O$_{13}$(cr) was prepared by heating appropriate amounts of UO$_3$ with potassium carbonate in air at 900 K for several hours in platinum boats; the lower uranate by decomposing the U(VI) compound in high-purity Ar at an unspecified temperature. The products were characterised by X–ray diffraction, and oxidation of the lower uranate in dry air corresponded to the uptake of 0.5 mole of O$_2$(g) per mole of K$_2$U$_4$O$_{12}$.

The oxygen potentials were studied using a CaO–ZrO$_2$ electrolyte tube in flowing argon, from 1053 to 1222 K with air (p_{O_2} = 0.2121 bar) as the reference electrode.

Correction of their emf values to the standard pressure gives, after recalculation

$$\Delta_r G_m \text{ (A.85)} = -30640 + 78.46 \cdot T \quad \text{J} \cdot \text{mol}^{-1} \quad (1053 \text{ to } 1222 \text{ K})$$

$$K_2U_4O_{12}(cr) + 0.5\, O_2(g) \rightleftharpoons K_2U_4O_{13}(cr). \tag{A.85}$$

This expression is different from that given by the authors, who seem to have neglected the correction from air to pure oxygen. The calculated decomposition temperature of K$_2$U$_4$O$_{13}$(cr) in air is 1538 K.

Drop calorimetric measurements are reported for the two compounds, using a Calvet high temperature calorimeter. The data were fitted to a simple polynomial, but the authors' expressions for $H_m(T) - H_m^\circ(298.15 \text{ K})$ are not equal to zero at T = 298.15 K, so the enthalpy data have been refitted with this constraint. The derived heat capacity expressions are given in the Table A-28.

Table A-28: Temperature coefficients for K$_2$U$_4$O$_{12}$(cr) and K$_2$U$_4$O$_{13}$(cr) compounds.

Phase	T range (K)	Heat capacity coefficients $C_{p,m} = a + b \cdot T + d \cdot T^{-2}$ J·K^{-1}·mol^{-1}			$C_{p,m}^\circ$ (298.15 K) J·K^{-1}·mol^{-1}
		a	b	d	
K$_2$U$_4$O$_{12}$(cr)	426 – 770	2.022090·10^2	4.7780·10^{-1}	0	–
K$_2$U$_4$O$_{13}$(cr)	411 – 888	4.710680·10^2	– 4.6890·10^{-2}	– 2.8754·10^6	(425 ± 50)

However, as shown in the plot of heat capacities in section 9.10.1, the expression for K$_2$U$_4$O$_{12}$(cr) is not consistent with the general behaviour of the heat capacities of the alkali-metal uranates, and the data for this compound have not been selected in this review.

The authors have used these Gibbs energies of reaction to derive data at 298.15 K for K$_2$U$_4$O$_{12}$(cr), using their heat capacity data (extrapolated from *ca.* 750 K) and literature data at 298.15 K for the hexavalent compounds. However, we have not pursued this approach since the data for the enthalpy of formation and entropy of K$_2$U$_4$O$_{13}$(cr) at 298.15 K are estimated, so any derived data for K$_2$U$_4$O$_{12}$(cr) would have appreciable uncertainties.

[97JAY/IYE]

The paper reports measurements of the oxygen activity (emf with CaO–ZrO$_2$ electrolyte) in the mixtures, Cs$_4$U$_5$O$_{17}$(cr) + Cs$_2$U$_2$O$_7$(cr) + Cs$_2$U$_4$O$_{12}$(cr), and complementary data on the enthalpy increments of the first two uranates.

The U(VI) uranates, Cs$_2$U$_4$O$_{13}$(cr), Cs$_4$U$_5$O$_{17}$(cr) and Cs$_2$U$_2$O$_7$(cr), were prepared by heating appropriate amounts of U$_3$O$_8$ with caesium carbonate in air at 1000 K in alumina boats; the lower uranate by decomposing the U(VI) compound in high–purity Ar at 1200 K. The products were characterised by X–ray diffraction.

The oxygen potentials were studied using a CaO–ZrO$_2$ electrolyte tube in flowing argon, from 1048 to 1206 K with air (p_{O_2} = 0.2121 bar) as the reference electrode.

Correction of their emf values to the standard pressure gives, after recalculation:

$$\Delta_r G_m = -136044 + 96.985 \cdot T \quad \text{J} \cdot \text{mol}^{-1} \qquad (1048 \text{ to } 1206 \text{ K})$$

for the reaction:

$$Cs_2U_4O_{12}(cr) + 3Cs_2U_2O_7(cr) + 0.5\ O_2(g) \rightleftharpoons 2\ Cs_4U_5O_{17}(cr) \qquad (A.86)$$

This expression is different from that given by the authors, who seem to have neglected the correction from air to pure oxygen.

Drop calorimetric measurements are reported for $Cs_2U_2O_7(cr)$ and $Cs_4U_5O_{17}(cr)$, using a Calvet high temperature calorimeter. The authors fitted the data to simple polynomials, but these expressions for $H_m(T) - H_m(298.15\ \text{K})$ are not equal to zero at $T = 298.15$ K, so the enthalpy data have been refitted with this constraint. In addition, there are precise measurements of the heat capacity of $Cs_2U_2O_7(cr)$ up to 350 K [81OHA/FLO], so for this phase the additional constraint of $C_{p,m}(Cs_2U_2O_7, \text{cr}, 298.15\ \text{K}) = 231.2\ \text{J} \cdot \text{K}^{-1} \cdot \text{mol}^{-1}$ from this work, was also imposed. The derived heat capacity expressions are given in the Table A-29.

Like $K_2U_2O_7(cr)$ (see [99JAY/IYE]), but not $Na_2U_2O_7(cr)$, the heat capacity increases more rapidly with temperature than expected for a U(VI) compound.

Table A-29: Temperature coefficient for $Cs_2U_2O_7(cr)$ and $Cs_4U_5O_{17}(cr)$ compounds.

Phase	T range (K)	Heat capacity coefficients $C_{p,m} = a + b \cdot T + d \cdot T^{-2}$ J·K^{-1}·mol^{-1}			$C^\circ_{p,m}$ (298.15 K) J·K^{-1}·mol^{-1}
		a	b	d	
$Cs_2U_2O_7(cr)$	298.15 – 852	3.553248·10^2	8.15759·10^{-2}	–1.31959·10^7	(231.2 ± 0.5)
$Cs_4U_5O_{17}(cr)$	368 – 906	6.992110·10^2	1.7199·10^{-1}	0	(750 ± 50)

The authors have used these Gibbs energies of reaction to derive data for 298.15 K for $Cs_4U_5O_{17}(cr)$, using their heat capacity data (extrapolated from ca. 800 K) and literature data at 298.15 K for the hexavalent compounds. However, we have not pursued this approach since many of the relevant data have not been selected for this review – see for example comments in [92VEN/IYE].

[97KON/FAN]

The two studies [95FAN/KIM] and [97KON/FAN] address a classical problem in solution coordination chemistry to distinguish between complex formation and other types of ionic interactions. In these two studies the problem is resolved by using fluorescence spectroscopy to identify the very weak complexes $CmCl^{2+}$ and $CmCl_2^+$, formed in the Cm(III)–chloride system and to measure their equilibrium constants. The data have been used to give a quantitative model of chloride complexation in $CaCl_2$ solutions over a wide chloride concentration range using the Pitzer formalism. For this purpose they have used estimated values of $\beta^{(0)}$ and $\beta^{(1)}$ for Cm^{3+} using data from the $NdCl_3$–H_2O system, and an estimated value of $\beta^{(1)}$ for $CmCl^{2+}$ and $CmCl_2^+$ using a method described in [97ALL/BAN], Ch.IX.8, p. 378, extended by the authors to 1:2 electrolytes. The equilibrium constants at zero ionic strength and the additional Pitzer parameters were determined by regression. In this way a unified description of the thermodynamics and speciation is given for concentrations of $CaCl_2$ ranging from 0 to 6 mol $\cdot$ kg^{-1}. The equilibrium constants at zero ionic strength for the two complexes formed are $\log_{10} \beta_1^\circ = (0.24 \pm 0.03)$ and $\log_{10} \beta_2^\circ = -(0.74 \pm 0.05)$. The value of $\log_{10} \beta_1^\circ$ differs considerably from the one given in [95SIL/BID], who do not report a value for $\log_{10} \beta_2^\circ$. This discrepancy illustrates the large error that can arise when using the SIT method to treat the ionic strength dependence of weak complexes. The numerical values of equilibrium constants and interaction parameters from these two studies are far superior to data obtained by other methods such as solvent extraction and ion exchange. The equilibrium constants and the uncertainty estimate given by [97KON/FAN] represent the best estimates for the complexes, $AnCl^{2+}$ and $AnCl_2^+$, both for Cm and Am.

[97KON/NEC]

Isopiestic measurements of $NaTcO_4$ solutions and mixtures with NaCl were made *versus* NaCl reference solutions at 298.15 K. By combining their results with those of Boyd [78BOY], recalculated by Rard and Miller [91RAR/MIL], for concentrations to ≤ 8.0 m $NaTcO_4$ the following ion interaction Pitzer parameters were determined: $\beta^{(0)}_{Na^+/TcO_4^-} = 0.01111$, $\beta^{(1)}_{Na^+/TcO_4^-} = 0.1595$, $C^\Phi_{Na^+/TcO_4^-} = 0.00236$, $\theta_{TcO_4^-/Cl^-} = 0.067$, and $\Psi_{TcO_4^-/Cl^-/Na^+} = -0.0085$. The solubility of $CsTcO_4(cr)$ was measured in CsCl solutions (0 – 7.4 m) at 298.15 K. In pure water the solubility was found to be (0.0184 ± 0.0004) m. The parameters resulting from fitting these data are: $\log_{10} K^\circ_{s,0} (CsTcO_4) = -(3.607 \pm 0.023)$, $\beta^{(0)}_{Cs^+/TcO_4^-} = -0.1884$, $\beta^{(1)}_{Cs^+/TcO_4^-} = -0.1588$, $C^\phi_{Cs^+/TcO_4^-} = 0$, and $\Psi_{TcO_4^-/Cl^-/Cs^+} = -0.0011$, whereby $C^\phi_{Cs^+/TcO_4^-}$ could be determined independently from $\Psi_{TcO_4^-/Cl^-/Cs^+}$. Further combination of the two sets of experimental data yielded Pitzer parameters in good agreement with the independently obtained values and also gave rise to estimates of the solubility of $CsTcO_4(cr)$ in NaCl solution to the saturation limit of the latter. The paper concludes with a discussion of the "like anion" interactions in the mixing terms being different for TcO_4^- as compared to ClO_4^-, a difference that tends to zero in their trace activity coefficients.

[97KRE/GAN]

X–ray diffraction measurements of single crystals of $[TcO_2(tn)_2]I \cdot H_2O$, tn = trimethylenediamine, were made. Similar to other Tc(V) and Re(V) analogues, the *trans*–TcO_2^+ core has two N–bonded tn ligands in the equatorial plane. The calculated differences in conformational energies for the different geometries of the tn ligands are too small (*ca.* 0.4 kJ · mol^{-1}) to infer which structure could be the most stable in solution.

[97KRI/RAM]

The mass loss Knudsen effusion measurements of the reaction:

$$3 \, UTeO_5(cr) \rightleftharpoons U_3O_8(cr) + 3 \, TeO_2(g) + \tfrac{1}{2} O_2(g) \qquad (A.87)$$

from 1063 to 1155 K in this conference report have been reported more fully in the journal paper by the same authors [98KRI/RAM].

[97LAN]

This book presents selected thermodynamic constants for aqueous species of U(IV) and U(VI), and common natural compounds (minerals) on the basis of the literature data available up to 1995. Most of them come from [92GRE/FUG], but Langmuir questions the validity of some constants selected by [92GRE/FUG] and suggests other selected data, which are given without uncertainty. He discusses the geochemistry of uranium.

- The first question is for $\Delta_f G_m^\circ (U^{4+})$. He gives the value $- 529.066$ kJ · mol^{-1}, 0.794 kJ · mol^{-1} higher than that selected by [92GRE/FUG] as $- (529.860 \pm 1.765)$ kJ · mol^{-1}. The value of Langmuir is consistent with the value of the $E^\circ (UO_2^{2+}/U^{4+}) = (0.263 \pm 0.004)$ V couple and $\log_{10} {^*\beta_1^\circ} = - 0.65$ for the reaction $U^{4+} + H_2O(l) \rightleftharpoons U(OH)^{3+} + H^+$ ($t = 25°C$) discussed in [91GIR/LAN] (see this review). The corresponding values selected by [92GRE/FUG] are (0.2673 ± 0.0012) V and $- (0.54 \pm 0.06)$ [90BRU/GRE]. Langmuir states that data leading to $E^\circ (UO_2^{2+}/U^{4+})$ were not corrected for the complexation of U^{4+} by hydroxide and sulphate anions. This statement is erroneous. Bruno *et al.* [90BRU/GRE] made the appropriate corrections and also corrected the required auxiliary data. The E° value of Langmuir is within the uncertainty given by Grenthe *et al.* [92GRE/FUG]. The present review retains $\log_{10} {^*\beta_1^\circ} = - (0.54 \pm 0.06)$ from [92GRE/FUG]. So this review does not change the selected value of $\Delta_f G_m^\circ (U^{4+})$ given by [92GRE/FUG].

- Langmuir also questions the values given by [92GRE/FUG] for $\Delta_f G_m^\circ (U(OH)_4, aq)$ and $\Delta_f G_m^\circ (UO_2, am)$, on the basis of new literature data that appeared since 1992. The present review has an extensive discussion of new data up to 2001, and selects new values different from both [92GRE/FUG] and those proposed by Langmuir. This review does not agree that one should remove of $U(OH)_n^{4-n}$, $n = 2$ and 3, species from speciation codes as proposed by Langmuir.

- For USiO$_4$, Langmuir proposes to take $\Delta_f G_m^\circ$(USiO$_4$, cr) = − 1885.98 kJ · mol^{-1}, a value estimated by Hemingway [82HEM]. [92GRE/FUG] selected − (1883.6 ± 4.0) kJ · mol^{-1}. These values are compatible. For the amorphous coffinite a value of $\Delta_f G_m^\circ$(USiO$_4$, am) = − 1835.227 kJ · mol^{-1} is given by Langmuir on the basis of an ion activity product in three groundwaters from coffinite-bearing formations. This value is new but not selected by this review due to the lack of identification of the solid phase.

- For U(IV) the value of the Gibbs energy of formation of ningyoite, CaU(PO$_4$)$_2$·2H$_2$O(cr), from [65MUT] is proposed.

- For $\Delta_f G_m^\circ$(UO$_2$(OH)$_2$, aq), Langmuir suggests a value of − 1358.126 kJ · mol^{-1}, 10 kJ · mol^{-1} higher that the limit given by [92GRE/FUG]. The value selected by Langmuir is derived from the data of [91CHO/MAT] (see Appendix A) for the equilibrium:

$$UO_2^{2+} + 2\,H_2O(l) \rightleftharpoons UO_2(OH)_2 + 2\,H^+$$

which is calculated to be $\log_{10} {}^*\beta_{2,1}^\circ$ = − 12.0. The present review selects for this equilibrium, $\log_{10} {}^*\beta_{2,1}^\circ$ = − (12.15 ± 0.07) leading to $\Delta_f G_m^\circ$(UO$_2$(OH)$_2$, aq) = − (1357.5 ± 1.8) kJ · mol^{-1}.

- Langmuir gives the Gibbs energy of formation of several dioxouranium(VI) phosphates starting with the value for H−autunite, H$_2$(UO$_2$)$_2$(PO$_4$)$_2$(cr), and using equilibrium constants of the exchange reaction of this mineral with several cations. $\Delta_f G_m^\circ$(H$_2$(UO$_2$)$_2$(PO$_4$)$_2$, cr) is calculated assuming $\Delta_r G_m^\circ$ = 0 for:

$$2\,U(HPO_4)_2 \cdot 4H_2O \rightleftharpoons H_2(UO_2)_2(PO_4)_2(cr) + 8\,H_2O(l)$$

and with $\Delta_f G_m^\circ$(U(HPO$_4$)$_2$·4H$_2$O, cr) = − (3844.453 ± 3.717) kJ · mol^{-1}, the value selected by [92GRE/FUG]. Values of the Gibbs energy of formation for crystalline dioxouranium(VI) vanadates, carnotite, K$_2$(UO$_2$)$_2$(VO$_4$)$_2$ and tyuyamunite, Ca(UO$_2$)$_2$(VO$_4$)$_2$ are calculated from the solubility data of [62HOS/GAR]. As the derivation of these values is rather speculative this review does not select any of them, although they can be used as guidance. Finally, values for several dioxouranium(VI) silicates are taken from [92NGU/SIL] (see Appendix A).

[97LUB/HAV]

Selenate solutions of U(VI), I = 3 M (NaClO$_4$ + Na$_2$SeO$_4$) are investigated by spectrophotometry at (298.2 ± 0.5) K after stabilisation (more than 24 hours). Log$_{10}$[H$^+$] is calculated from potentiometric measurements using a glass electrode filled with an inner 2.99 M NaClO$_4$ + 0.01 NaCl solution saturated by AgCl. The electrode was calibrated according to the Gran method ($E = E^\circ + \frac{2.30259\,RT}{F} \cdot (-\log_{10}[H^+] + E_j)$). UO$_2$(ClO$_4$)$_2$ · xH$_2$O(cr) is the starting material used to prepare the stock solution with a dioxouranium(VI) content determined by gravimetry as U$_3$O$_8$(cr) by the 8–hydroxy-

8–hydroxyquinoline method. The selenate concentration varied between 0 and 0.6 M while $-\log_{10}[H^+]$ varied from 2.26 to 2.86. The U concentration was kept constant at $3.31 \cdot 10^{-2}$ M.

Absorbances of 21 solutions at 50 wavelengths in the range 360 to 500 nm are analysed by factor analysis. Three species, UO_2^{2+}, $UO_2(SeO_4)(aq)$ and $UO_2(SeO_4)_2^{2-}$ are identified by their absorption spectra which show very close absorption maxima, but have different molar absorptivities. The authors found no evidence for the formation of hydroxy species in the range $-\log_{10}[H^+] = 1.4$ to 3.0.

The data for the complexes, $UO_2(SeO_4)$ and $UO_2(SeO_4)_2^{2-}$, are $\log_{10}\beta_1 = (1.576 \pm 0.016)$ and $\log_{10}\beta_2 = (2.423 \pm 0.013)$, $(I = 3M)$.

The authors also report some attempts to measure $\log_{10}\beta_1$ and $\log_{10}\beta_2$ from potentiometric measurements using a uranium electrode, the response mechanism of which is not completely clear, but which can apparently be used successfully at low ionic strength. Experiments are conducted without adjusting the ionic strength, at $-\log_{10}[H^+] = 3.5$, U and selenate concentrations ranging from 0.1 to $0.5 \cdot 10^{-3}$ M and up to $3 \cdot 10^{-2}$ M, respectively. Only $\log_{10}\beta_1 = (2.64 \pm 0.01)$ can be calculated. If it is assumed, following the authors, that this value is equal to $\log_{10}\beta_1^\circ$ it is possible to calculate $\varepsilon(SeO_4^{2-}, Na^+)$ according to, $\log_{10}\beta_1 + 8D = \log_{10}\beta_1^\circ - \Delta\varepsilon\, I_m$, with $\Delta\varepsilon = -\varepsilon(SeO_4^{2-}, Na^+) - \varepsilon(UO_2^{2+}, ClO_4^-)$, but this is highly debatable.

Lubal et al. used an erroneous value, $\varepsilon(UO_2^{2+}, ClO_4^-) = 0.26$ kg·mol^{-1}, in this calculation and arrived at $\varepsilon(SeO_4^{2-}, Na^+) = -0.009$ kg·mol^{-1}. With the correct interaction coefficient, $\varepsilon(UO_2^{2+}, ClO_4^-) = 0.46$ kg·mol^{-1}, this review obtains $\varepsilon(SeO_4^{2-}, Na^+) = -0.21$ kg·mol^{-1}, very different from the value $\varepsilon(SO_4^{2-}, Na^+) = -0.12$ kg·mol^{-1}. Consequently, the mean activity coefficient of sodium selenate can be calculated from:

$$\log_{10}\gamma_\pm = \frac{-0.51052 \cdot (3 \cdot m)^{1/2}}{1 + 1.5 \cdot (3 \cdot m)^{1/2}} + 2\,\varepsilon(SeO_4^{2-}, Na^+) \cdot m,$$

where m is the molal concentration. The values of $\log_{10}\beta_1$ and $\log_{10}\beta_2$ obtained by spectrophotometry are comparable with those of the sulphate complexes, $UO_2SO_4(aq)$ and $UO_2(SO_4)_2^{2-}$. This provides a certain confidence. However, because of the ambiguities discussed above, the present review does not select the data reported by [97LUB/HAV].

[97LUB/HAV2]

This paper concerns the determination of equilibrium constants of species, $(UO_2)_m(OH)_n^{2m-n}$, with the following (n,m) values (2,2), (5,3) and (4,3) using several computer programs to process spectrophotometric and potentiometric data obtained at $(25 \pm 0.2)°C$ in 3 M NaClO$_4$ solutions with U(VI) concentrations in the range $8.3 \cdot 10^{-3}$ to $3.3 \cdot 10^{-3}$ M and $-\log_{10}[H^+]$ in the ranges 2 to 4 (spectrophotometric studies), or 3 to 5 (potentiometric studies). The proton concentration is measured by calibrating a modified glass electrode using the Gran method (see [97LUB/HAV]). Chemometric analysis of the absorbance of ten test solutions *versus* wavelength shows that three hydrolysis species are responsible for the changes in the absorption spectra. This result agrees well with the fit of the potentiometric data. The best fit is obtained with the chemical model of [92GRE/FUG] based on the existence of the species (1,2), (2,2), (5,3) and (4,3) and by fixing $\log_{10} {}^*\beta_{1,2} = -3.70$ (the species (1,2) is known to be a minor species in the conditions of this study). The following values are obtained: $\log_{10} {}^*\beta_{2,2} = -(6.24 \pm 0.02)$, $\log_{10} {}^*\beta_{5,3} = -(16.80 \pm 0.04)$ and $\log_{10} {}^*\beta_{4,3} = -(12.8 \pm 0.1)$. Spectroscopic data gives $\log_{10} {}^*\beta_{2,2} = -(6.13 \pm 0.02)$, $\log_{10} {}^*\beta_{5,3} = -(16.81 \pm 0.02)$ and $\log_{10} {}^*\beta_{4,3} = -(12.57 \pm 0.02)$. Each of the individual absorption spectra can be described with one broad and two narrow bands, each with Gaussian profiles. The λ_{max} and ε_{max} values are closer to those found by [2000NGU/BUR] than those obtained by [97MEI2]. The $\log_{10} {}^*\beta_{2,2}$ and $\log_{10} {}^*\beta_{5,3}$ values differ from those of [93FER/SAL] (determined in the same medium) by 0.25 and 0.55 units, respectively. Lubal and Havel have not considered all the studies on U(VI) hydrolysis published between 1992 and 1996 (see Table 9-3 of this review).

This paper gives experimental data only in Figures, and without uncertainties. It seems more direct to test different ways to process data, particularly spectrophotometric ones, than to provide a careful study of U(VI) hydrolysis. This review does not retain the equilibrium constants given by [97LUB/HAV2] for the selection of new values despite the fact that they are close to those obtained by other authors.

This review has calculated using auxiliary data from [92GRE/FUG] the following values at $I = 0$: $\log_{10} {}^*\beta_{2,2}^\circ = -(5.79 \pm 0.35)$, $\log_{10} {}^*\beta_{5,3}^\circ = -(15.64 \pm 0.70)$ and $\log_{10} {}^*\beta_{4,3}^\circ = -(11.21 \pm 0.92)$ (potentiometric data) and $\log_{10} {}^*\beta_{2,2}^\circ = -(5.70 \pm 0.35)$, $\log_{10} {}^*\beta_{5,3}^\circ = -(15.65 \pm 0.70)$ and $\log_{10} {}^*\beta_{4,3}^\circ = -(10.98 \pm 0.91)$ (spectrophotometric data). Within the large uncertainties coming from the extrapolation from 3.5 m NaClO$_4$ to $I = 0$, these values of $\log_{10} {}^*\beta_{2,2}^\circ$, $\log_{10} {}^*\beta_{4,3}^\circ$ and $\log_{10} {}^*\beta_{5,3}^\circ$ are consistent with those selected by [92GRE/FUG] and retained in the present review: $\log_{10} {}^*\beta_{2,2}^\circ = -(5.62 \pm 0.04)$, $\log_{10} {}^*\beta_{5,3}^\circ = -(15.55 \pm 0.12)$ and $\log_{10} {}^*\beta_{4,3}^\circ = -(11.9 \pm 0.3)$.

[97MAS/COU]

Polarographic measurements with a hanging mercury drop electrode (HMDE) *vs.* a ACE were made at 298.15 K in TcO$_4^-$, HNO$_3$/KNO$_3$ solutions (0.1 M ionic strength). Two electrochemical reduction steps were observed, with the $E_{1/2}$ for the first wave being linear in pH such that $E_{1/2} = (0.070 \pm 0.003)$ V/SCE at 1 M HNO$_3$. This first diffusion-controlled step corresponds to the reduction of Tc(VII) to Tc(III), occurring in the pH range 0 – 4. Polarographic and cyclic voltametry in the range 0 to – 0.6 V/SCE showed the reduction to be completely irreversible with no participation of NO$_3^-$. The second reduction process in 0.1 to 1.0 M HNO$_3$ solutions took place at HMDE potentials of -0.6 to -0.8 V/SCE corresponding to the Tc(III) catalysed reduction of nitrate ions. Using a graphite cathode at -0.5 V/SCE in 4 M HNO_3 ($5 \cdot 10^{-4}$ M TcO_4^-) caused 9% of the Tc to electrodeposit on the cathode with the formation of a brown solution consistent with Tc(V) formation characterised by an absorption band at 480 nm.

[97MEI]

As discussed in detail in the analysis of [98MEI] this paper deals with the formation of the two hydrolysis species: $(UO_2)_2(OH)_2^{2+}$ (2:2) and $(UO_2)_3(OH)_5^+$ (5:3) from UO_2^{2+} studied in 26 test solutions undersaturated with U(VI), 0.1 M $NaClO_4$, $6 \cdot 10^{-3}$ to $2 \cdot 10^{-4}$ M in uranium, pH = 2.4 to 4.8. Meinrath has used UV–Vis spectral data processed by factor analysis, a model-free multivariate technique for analysing multiple observations simultaneously. What is specific and new in this paper, is the discussion of pH measurements and the analysis of the contribution of pH errors and then effects on the free concentration of the different species at the 95% confidence level attributed to $\log_{10} {}^*\beta_{2,2}$ and $\log_{10} {}^*\beta_{5,3}$. From 21 solutions (pH from (3.504 ± 0.032) to (4.718 ± 0.019)) and 14 solutions (pH from (3.939 ± 0.029) to (4.776 ± 0.014)), the weighted average values and standard deviations given by Meinrath are, respectively, $\log_{10} {}^*\beta_{2,2} = -(6.145 \pm 0.088)$ and $\log_{10} {}^*\beta_{5,3} = -(17.142 \pm 0.138)$ (the last digit in all these values to be taken with caution). Some of these data are used in [98MEI].

Under the conditions used in the present work, the species 1:1, 2:1, 7:4, or 4:3 were not observed probably as a result of the low U concentrations employed. Their existence seemed improbable to Meinrath or their concentrations were too low to be detected by spectroscopy. Based on the values of $\log_{10} {}^*\beta_{n,m}$ in [92GRE/FUG] one expects the species 1:1, 2:1, 7:4 and 4:3 to be present in small amounts. Whether they can be detected or not depends on the molar absorptivity, $\log_{10}\varepsilon$, of the complexes. If $\log_{10}\varepsilon$ = 2 to 3, as for the other species identified by Meinrath, it should have been possible to detect them.

Due to the high sensitivity of $\log_{10} {}^*\beta_{n,m}$ to [H^+], the pH measurement is of major importance. In all experiments a multiple point calibration of the glass electrode (mV vs pH of reference solutions) is done which allows use of a calibration line set up by linear regression. This gives in turn the statistical ΔpH (random error) for a given potential. For instance from five different pH calibration points (1.7 to 10),

pH = (7.078 ± 0.047) – (0.017029 ± 0.000339) emf (mV).

For the $\log_{10} {}^*\beta_{n,m}$ values, uncertainties in pH and free concentrations have similar contributions. This review notes that the equation obtained by Meinrath bears an uncertainty as large as ± 0.047 log units and that calibration with standard buffers is not the best that can be recommended. A better approach is the use of potentiometric titration. It is unfortunate that Meinrath has used pH buffers for the calibration of the glass electrode, but when determining concentration equilibrium constants it is obviously better to work with concentrations throughout *i.e.*, $-\log_{10}[H^+]$ instead of pH.

A comparison of the $\log_{10} {}^*\beta_{n,m}$ values with previous ones reported in [92GRE/FUG] is given in the analysis of [98MEI].

[97MEI2]

Meinrath reviews here the spectral characteristics of species: 0:1, UO_2^{2+}, 2:2, $(UO_2)_2(OH)_2^{2+}$, 5:3, $(UO_2)_3(OH)_5^+$ and $UO_2CO_3(aq)$ and discusses the advantages/disadvantages of UV–Vis and TRLFS to determine the U(VI) speciation in environmental samples. The possible origin of the absorption and emission spectra of U(VI) in aqueous solution is discussed. Some examples are shown of deconvolution of the spectra of U(VI) into single components taken from previous results. There are no new experimental data in this paper.

The UV–Vis spectra of hydrolysed species are dominated by the spectroscopic properties of the O=U=O group (electronic and vibrational). The red shifts of the maximum of absorption of UO_2^{2+} increase very little with increasing number of coordinated OH groups, but the molar extinction coefficient increases by more than one order of magnitude. The emission spectra clearly show the vibrational structure and that the lifetime of the excited state is rather long. The U(VI) monocarbonato complex spectrum presents only shoulders on the absorption edge toward the UV region, but a well-defined emission spectrum. Free carbonate ion concentrations of more than 10^{-7} M quench the fluorescence very efficiently and the di- and tri-carbonato dioxouranium(VI) complexes can therefore not be detected by TRLFS. Table A-30 gives of the main characteristics of species under consideration.

Table A-30: Spectroscopic data of some U(VI) complexes.

	UO_2^{2+}	$UO_2(OH)_2^{2+}$	$UO_3(OH)_5^+$	$UO_2CO_3(aq)$
Absorption max, nm	413.8	421.8	429.0	400(sh)
ε, L·mol^{-1}·cm^{-1}	(9.7 ± 0.2)	(101 ± 2)	(474 ± 1)	(36 ± 3)
Lifetime, µs	(0.9 ± 0.3)	(2.9 ± 0.4)	(7 ± 1)	(35 ± 5)
Emission max, nm	473, 488, 509, 534, 560, 588	499, 519, 542, 556	500, 516, 533, 554	450(sh), 464, 481, 504, 532(sh)

[97MEI/SCH]

This paper is reviewed together with [96MEI/SCH].

[97MER/LAM]

The authors carried out an experiment similar to that reported by Morss and Williams

[94MOR/WIL]. ^{241}Am(IV) dioxide was reduced by a H$_2$ flow at 1090 K. The resulting hexagonal Am$_2$O$_3$(cr) was then hydrated to Am(OH)$_3$(cr) in an autoclave at (403 ± 5) K and a water saturation pressure of 275 kPa. The prepared Am(OH)$_3$(cr) was investigated and characterised by gravimetry, X–ray diffraction and FTIR spectroscopy. One weak extraneous line at d = 0.289 nm was attributed to small impurities of Am$_2$O$_3$(cr). The molar enthalpy of solution of Am(OH)$_3$(cr) was determined in 6 M HCl to be $\Delta_{sol}H^\circ_m$ (Am(OH)$_3$, cr, 298.15 K, 6 M HCl) = – (129.4 ± 1.1) kJ · mol^{-1} (mean value of five measurements with samples from the same solid), which differs from the mean value reported in [94MOR/WIL] by (23 ± 8) kJ · mol^{-1}. Combining $\Delta_{sol}H^\circ_m$ (Am(OH)$_3$, cr, 298.15 K, 6 M HCl) with the molar enthalpy of solution for americium metal selected in [95SIL/BID], $\Delta_{sol}H^\circ_m$ (Am, cr, 298.15 K, 6 M HCl) = – (613.1 ± 1.4) kJ · mol^{-1}, the standard molar enthalpy of formation was calculated to be $\Delta_f H^\circ_m$ (Am(OH)$_3$, cr, 298.15 K) = – (1343.6 ± 1.8) kJ · mol^{-1}. Using an estimate of S°_m (Am(OH)$_3$, cr, 298.15 K) = (126 ± 8) J · K^{-1} · mol^{-1} and thermochemical data for Am^{3+} and OH$^-$ selected in [95SIL/BID], the following data were calculated: $\Delta_r G^\circ_m$ = (144.1 ± 5.6) kJ · mol^{-1} for the reaction:

$$Am(OH)_3(cr) \rightleftharpoons Am^{3+} + 3\ OH^-$$

corresponding to $\log_{10} K^\circ_{s,0}$ (Am(OH)$_3$, cr, 298.15 K) = – (25.2 ± 1.0) and $\log_{10}{}^* K^\circ_{s,0}$ (Am(OH)$_3$, cr, 298.15 K) = (16.8 ± 1.0). With S°_m (Am(OH)$_3$, cr, 298.15 K) = (116 ± 8) J · K^{-1} · mol^{-1} as estimated in the present review (cf. section 12.3.2.2) the solubility constant becomes:

$$\log_{10}{}^* K^\circ_{s,0}\ (Am(OH)_3, cr, 298.15\ K) = (17.3 \pm 1.0).$$

This value is 1.7 log$_{10}$ units larger than $\log_{10}{}^* K^\circ_{s,0}$ = (15.6 ± 0.6) as re-calculated in the present review from the solubility data reported in [82SIL] for ^{243}Am(OH)$_3$(cr) and close to $\log_{10}{}^* K^\circ_{s,0}$ = (16.9 ± 0.8) selected for Am(OH)$_3$(am). There is no clear-cut explanation for this contradiction, because the solid phase characterisation reported in [97MER/LAM] does not indicate a degradation of ^{241}Am(OH)$_3$(cr) to an amorphous phase by α–radiation damage.

[97MOL/GEI]

This paper is reviewed together with [98MOL/GEI].

[97MUR]

The paper deals with the problem of UO_2 alteration by water in environmental conditions according to the sequence schoepite (in fact metaschoepite $UO_3 \cdot 2H_2O$), soddyite $((UO_2)_2SiO_2 \cdot 2H_2O)$ and uranophane $(Ca(UO_2)_2[SiO_3(OH)]_2 \cdot 5H_2O)$.

The calculated solubility of schoepite decreases from $10^{-3.1}$ to $10^{-4.8}$ M when T decreases from 100 to 25°C. All data used to calculate $\log_{10} {}^*K_{s,0} = f(T)$ for

$$UO_3 \cdot 2H_2O + 2\,H^+ \rightleftharpoons UO_2^{2+} + 3\,H_2O(l)$$

are from [92GRE/FUG], $UO_3 \cdot 2H_2O$ being considered as crystalline dioxouranium(VI) hydroxide hydrate.

Uncertainties in the data are too large to calculate the dissolution of soddyite according to:

$$(UO_2)_2SiO_4 \cdot 2H_2O + 4\,H^+ \rightleftharpoons 2\,UO_2^{2+} + SiO_2(aq) + 4\,H_2O(l).$$

The values given by [92NGU/SIL] are $\log_{10} {}^*K_s = (5.74 \pm 0.21)$ at 30°C and $\log_{10} {}^*K_s = (6.03 \pm 0.45)$ at 25°C. Furthermore, the experimental conditions do not guarantee that equilibrium is obtained. For uranophane data are lacking. Considerations of possible retrograde solubilities of these minerals are developed.

[97NEU/REI]

This paper describes solubility and speciation studies in the Pu(VI) carbonate system. Precaution was taken by the authors to maintain the hexavalent oxidation state during the experiments. The bi- and tri-carbonato complexes are identified by spectrophotometric titration experiments and confirm previously published results. Equilibrium constants are reported for both complexes in 0.1 M $NaClO_4$ at 25°C. The values given are in agreement with previous measurements. However, the authors have not presented the experimental data in detail. Information on the accuracy of the measurements and the evaluated complexation constants are missing as well. Therefore, this review is unable to judge the published complexation constants.

In the second part of the paper, the solubility of solid dioxoplutonium(VI) carbonate, $PuO_2CO_3(cr)$, has been studied. The synthesised solid, which is characterised by XRD and EXAFS analysis, is isostructural with the homologous uranium compound. The solubility product of $PuO_2CO_3(cr)$ was determined as a function of the NaCl concentration. The experimental data are of a preliminary character as the authors themselves have pointed out. More detailed results are presented in [2000REI/NEU], (*cf.* discussion of that paper).

[97OMO/MIY]

This paper deals with the extraction of pertechnetate from aqueous solution by solutions of chloroform containing tetraphenylarsonium chloride and 3,5-dichlorophenol at 25°C. The effect of association of the latter two solutes on the extraction constant is demonstrated yielding values for these two constants of $(3.53 \pm 0.08) \cdot 10^4$ M and $(1.33 \pm 0.03) \cdot 10^3$ M, respectively. No experimental results are given and this paper may be considered to be outside the scope of this review.

[97PAS/CZE]

This review includes the review of [93PAS/RUN] and concerns U(VI) and Pu(VI).

These two studies have been made at 22°C using the same solubility technique in a 0.1 M $NaClO_4$ ionic medium. The correction of the measured pH values to $-\log_{10}[H^+]$ is not well described. [97PAS/CZE] used the expression:

$$\log_{10}[CO_3^{2-}] = -17.55 - \log_{10} p_{CO_2} + 2\ \text{pH},$$

to calculate the carbonate concentration from the measured pH, while [93PAS/RUN] used:

$$\log_{10}[CO_3^{2-}] = -17.65 - \log_{10} p_{CO_2} + 2\ \text{pH}.$$

The experimental pH values have to be recalculated to $-\log_{10}[H^+]$. The error made in [97PAS/CZE] seems to be due to an erroneous citation of [96FAN/NEC]. The correct expression for calculating the carbonate concentration given in [96FAN/NEC] is:

$$\log_{10}[CO_3^{2-}] = -17.56 - \log_{10} p_{CO_2} - 2\log_{10}[H^+].$$

In order to account for the systematic error made in [97PAS/CZE] their equilibrium constants have been recalculated by this review using the auxiliary data from [92GRE/FUG]. The original equilibrium constants for the U(VI) carbonate system, $\log_{10} K_{s,0}$ for $UO_2CO_3(s)$ and $\log_{10} \beta_n$ for the complexes $UO_2(CO_3)_n^{2-2n}$, and the recalculated values in 0.1 M $NaClO_4$ are:

$\log_{10} K_{s,0} = -(13.35 \pm 0.14)$, recalculated: $-(13.55 \pm 0.14)$,

$\log_{10} \beta_1 = (8.93 \pm 0.05)$, recalculated: (9.13 ± 0.05),

$\log_{10} \beta_2 = (15.3 \pm 0.2)$, recalculated: (15.7 ± 0.2),

$\log_{10} \beta_3 = (21.0 \pm 0.3)$, recalculated: (21.6 ± 0.3).

These recalculated values are converted to zero ionic strength using the SIT:

$\log_{10} K_{s,0}^\circ = -(14.39 \pm 0.14)$,

$\log_{10} \beta_1^\circ = (9.97 \pm 0.05)$,

$\log_{10} \beta_2^\circ = (16.5 \pm 0.2)$,

$\log_{10} \beta_3^\circ = (21.6 \pm 0.3)$.

The investigations have been made at different partial pressures of CO_2 (p_{CO_2} = $10^{-3.5}$, 10^{-2} and 1 bar) and with structure and spectroscopic characterisation of the solids. This together with the slope analysis ensures that the chemical model chosen is correct. The authors discuss the possibility of a phase transformation in the Pu(VI) system, but the evidence for this is unconvincing. The observed change of slope at p_{CO_2} = $10^{-3.5}$ bar in [93PAS/RUN] may also be due to the formation of hydroxide complexes in solution. The agreement between the recalculated value of $\log_{10} K^\circ_{s,0}$ (UO_2CO_3, s) = $-(14.39 \pm 0.14)$ and the value of $-(14.49 \pm 0.04)$ recalculated in [95SIL/BID] from [92GRE/FUG], is excellent.

The equilibrium constant given by Pashalidis et al. in 0.1 M $NaClO_4$ fits the experimental data very well (Figs. 1 and 2), because the fit is independent of the systematic error in the determination of the carbonate concentration. The authors have also estimated the Pitzer parameters for the uranium carbonate complexes using the equilibrium constants at different ionic strength given in [92GRE/FUG] and using the equilibrium constants for the complexes, UO_2CO_3(aq) and $UO_2(CO_3)_2^{2-}$, at zero ionic strength given there. For the third complex, $UO_2(CO_3)_3^{4-}$, the authors preferred to make a re-evaluation and suggest that $\log_{10} \beta_3^\circ = (21.3 \pm 0.3)$, instead of 21.6, given in [92GRE/FUG]. This choice together with the selected Pitzer parameters describe the experimental data with good accuracy, but not better than the SIT model used in [92GRE/FUG]. The selected parameter set was tested against solubility data determined by the authors, and the agreement was reasonably good, although there is a systematic difference of about 0.2 units in $\log_{10}[CO_3^{2-}]$ between the experimental and calculated solubility curves, as explained above. After recalculation, the experimental results in [97PAS/CZE] are in accord with the constants selected in the previous NEA review.

The equilibrium constants for the plutonium(VI) carbonate system are affected by the same systematic error as in the uranium(VI) system; the reported equilibrium constants and the corresponding recalculated values for the Pu(VI) system in 0.1 M $NaClO_4$ are given below:

$\log_{10} K_{s,0} = -(13.98 \pm 0.12)$, recalculated: $-(14.18 \pm 0.12)$,

$\log_{10} \beta_1 = (8.7 \pm 0.3)$, recalculated: (8.9 ± 0.3),

$\log_{10} \beta_2 = (14.1 \pm 0.5)$, recalculated: (14.5 ± 0.5),

$\log_{10} \beta_3 = (17.8 \pm 0.2)$, recalculated: (18.4 ± 0.2).

From the recalculated values the following constants at zero ionic strength are obtained by using the SIT:

$\log_{10} K^\circ_{s,0} = -(15.02 \pm 0.12)$,

$\log_{10} \beta_1^\circ = (9.7 \pm 0.3)$,

$\log_{10} \beta_2^\circ = (15.3 \pm 0.5)$,

$$\log_{10} \beta_3^\circ = (18.4 \pm 0.2).$$

The equilibrium constant for the first complex does not differ much from the corresponding uranium system and the differences for the second and third complexes can be rationalised in terms of differences in ionic radii between the two MO_2^{2+} ions. The equilibrium constants at zero ionic strength do not agree well with those proposed in [2001LEM/FUG]. The value of $\log_{10} \beta_1^\circ = (9.7 \pm 0.3)$ is very different from the value $\log_{10} \beta_1^\circ = 13.8$ obtained from the data of Sullivan and Woods [82SUL/WOO]. There are no chemical reasons to assume large differences in this stability constant between UO_2^{2+} and PuO_2^{2+}, indicating that the data in [82SUL/WOO] are affected by some systematic errors. The solubility product of $PuO_2CO_3(s)$ is larger in [87ROB/VIT] than in [97PAS/CZE], indicating a less crystalline solid in the former study. However, the crystallinity does not seem to have changed during the experiments as indicated by the good agreement between the stability constants for the carbonate complexes in the two studies.

[97PER/CAS]

Soddyite was synthesised [92NGU/SIL] and characterised as a pure single phase before and after the experiments by X–ray diffraction, SEM and IR methods, (BET surface area = (25.4 ± 0.2) $m^2 \cdot g^{-1}$). Dissolution was studied at room temperature in 0.001 M Na_2SiO_3 and 0.007 M $NaClO_4$, with added Na_2CO_3 in the range 0.001 to 0.02 M (0.25 g of soddyite in 100 mL of stirred solution). The effluent was analysed for uranium and presumably pH was measured on the activity scale (8.53 to 9.11). Data are analysed according to:

$$(UO_2)_2SiO_4 \cdot 2H_2O \text{ (cr)} + 6 HCO_3^- \rightleftharpoons 2 UO_2(CO_3)_3^{4-} + H_4SiO_4 \text{(aq)} + 2 H^+ + 2 H_2O(l) \quad \text{(A.88)}$$

Steady-state U(VI) concentrations were reached after *ca.* 400 hours at low carbonate concentrations and within 100 hours at the highest concentrations. The silicic acid concentration was determined from mass balance according to the above equation and the uranium speciation was determined by the HARPHRQ code and extrapolation of the resulting equilibrium constants was performed with the SIT [92GRE/FUG], from which $\log_{10}{}^*K_s$ values were estimated for the equilibrium:

$$(UO_2)_2SiO_4 \cdot 2H_2O \text{ (cr)} + 4H^+ \rightleftharpoons 2 UO_2^{2+} + H_4SiO_4 \text{(aq)} + 2 H_2O(l) \quad \text{(A.89)}$$

The authors state that complexes other than $UO_2(CO_3)_3^{4-}$ are important at $[HCO_3^-] = 0.002$ M, and therefore an average $\log_{10}{}^*K_s^\circ$ value, (3.9 ± 0.7), was calculated from the higher concentrations of carbonate. However, to this review there is a clear monotonic trend in the $\log_{10}{}^*K_s^\circ$ values even at $[HCO_3^-] > 0.002$ M, e.g., $[HCO_3^-] = 0.005$, 0.005, 0.008, 0.010, 0.015, 0.020; $\log_{10}{}^*K_s^\circ = 4.60, 4.42, 3.80, 4.24, 3.77, 2.58$, respectively. The authors refer to two previous studies conducted at only one pH with $\log_{10}{}^*K_s^\circ$ values of 5.74 [92NGU/SIL] and 6.15 [96MOL/GEI].

The kinetics of reaction (A.88) were treated by the detailed balancing equation, which expresses the rate of the net reaction as:

$$\frac{d[U]}{dt} = k_1[HCO_3^-]^6 - k_{-1}[UO_2(CO_3)_3^{4-}]^2[H_4SiO_4(aq)][H^+]^2 \quad (A.90).$$

This equation simply relies on the stoichiometry of the equilibrium (A.88) to fix the rate order for each component, such that $K_1 = k_1/k_{-1}$. Further assumptions were made that the bicarbonate, silicic acid and pH remained constant during the reaction. Based on the integrated rate equation (A.90), the ratio of the two rate constants gave a value of K_1 which was converted to a $\log_{10}{}^*K_s$ value of (4.3 ± 0.6). Although this kinetic approach appears to favour the solubility constant determined by direct measurement in this study, the concept of a rate law involving sixth- and fifth-order rate determining steps is very questionable and likely represents a misuse of the theory of microscopic reversibility to obtain rate constants.

The authors have taken only the formation of $UO_2(CO_3)_3^{4-}$ into account and claim that the observed variation in the solubility product is due to this assumption. The reviewers have repeated this calculation by taking the formation of all carbonate complexes into account. There is still a large systematic variation in the experimental $\log_{10}{}^*K_s$ values, indicating some systematic error in the measurements. In addition, one notices very large uncertainties in the pH determinations that result in a large uncertainty in the solubility product. In view of these shortcomings this review has not considered the solubility constant, $\log_{10}{}^*K_s^\circ = (3.9 \pm 0.7)$, suggested by Pérez et al. when selecting the recommended value.

[97RAI/FEL]

Solubility experiments were conducted on freshly-precipitated amorphous $UO_2(am)$ and $ThO_2(am)$ whereby equilibrium was approached from undersaturation. Detailed precautions were taken to prevent oxidation of the $UO_2(am)$ and dissolved uranium, including the addition of $EuCl_2$. Experiments with $UO_2(am)$ were performed in 0.03, 0.1, 0.2, 1.0, and 6.0 m NaCl and in 1.0, 2.0 and 3.0 m $MgCl_2$, whereas for $ThO_2(am)$ runs were made in 4.0 and 6.0 m NaCl, and in 1.0, 1.82 and 3.0 m $MgCl_2$. Molal hydrogen ion concentrations were determined from glass electrode measurements corrected according to an empirical Gran plot ($pm_{H^+} = pH_{obs} + A$). E_h of the U solutions was measured with platinum electrodes and confirmed by either UV–Vis–IR or solvent extraction. X-ray diffraction patterns of the products show predominantly one, low-intensity, broad peak at 3.157 Å consistent with previous findings for amorphous solids [87BRU/CAS], [90RAI/FEL]. All the solubility results presented could be fitted with a single line ($\log_{10}C_U$ versus pc_H) with a slope of ca. -3 and little variation for samples taken after 1 to 438 days. However, U(VI) appears to have been present in some solutions of low chloride concentrations (< 0.1 m) leading to higher scatter than observed at high chloride concentrations. The authors evaluated binary Pitzer parameters for U^{4+}–Cl^-: $\beta^{(0)} = 1.644$; $\beta^{(1)} = 15.5$; $C^\phi = 0.0995$ and $U(OH)^{3+}$–Cl^-: $\beta^{(0)} = 1.0$; $\beta^{(1)} = 7.856$; $C^\phi = 0.0$. No mixing terms are provided.

Values of $\Delta_f G_m^\circ / RT$ are given for U^{4+}, $U(OH)^{3+}$ and $UO_2(am)$. They correspond to $\Delta_f G_m^\circ (U^{4+}, 298.15 \text{ K}) = -530.97 \text{ kJ} \cdot \text{mol}^{-1}$ (cf., recommended value in [92GRE/FUG]:

$-(529.9 \pm 1.8)$ kJ · mol^{-1}), $\Delta_f G_m^\circ$ (U(OH)$^{3+}$, 298.15 K) = $-$ 765.26 kJ · mol^{-1} (cf. recommended value in [92GRE/FUG]: $-(763.9 \pm 1.8)$ kJ · mol^{-1}), and $\Delta_f G_m^\circ$ (UO$_2$(am), 298.15 K) = $-$ 990.77 kJ · mol^{-1}.

For the solubility product of the reaction:

$$UO_2(am) + 2 H_2O(l) \rightleftharpoons U^{4+} + 4 OH^-$$

the authors calculated $\log_{10} K_{s,0}^\circ = -53.45$. The Pitzer parameters yielded excellent agreement for the formation constant of U(OH)$^{3+}$ with the study of Kraus and Nelson [50KRA/NEL].

[97RED]

This paper, in which few experimental details are given, examines the solubility of UO$_2$(cr) in pure water, acid chloride mixtures, hydrofluoric acids, sodium carbonate and mixtures of sodium carbonate and oxalate solutions, at various redox potentials. The behaviour of UO$_3$·H$_2$O(cr) in pure water is also examined.

In the temperature range 300–600°C, with f(O$_2$) controlled by Ni–NiO and Fe$_2$O$_3$–Fe$_3$O$_4$ couples, the author confirmed earlier results ([87DUB/RAM], [89RED/SAV]) on the negligible temperature dependence of the solubility over the temperature range covered. The reported solubility was 2·10^{-9} mol U/kg H$_2$O. For the generally accepted dissolution reaction, UO$_2$(cr) + 2 H$_2$O(l) $\rightleftharpoons$ U(OH)$_4$(aq), this gives $\log_{10} K_{s,4} = -8.7$ (1 kbar, pH = 7) compared to an average value of $\log_{10} K_{s,4} = -(9.47 \pm 0.56)$ reported by [88PAR/POH] at pH > 4 over the temperature range 100 to 300°C (50 MPa H$_2$).

The solubility of UO$_2$(cr) at 500°C (Ni–NiO buffer) in HCl ranged between $\log_{10} m_U = -(7.2 \pm 0.1)$ in 10^{-4} m HCl and $\log_{10} m_U = -(3.0 \pm 0.3)$ for 1 m HCl and corresponded roughly to the relation $m_U = 10^{-3}$ m_{HCl}. From these results, the author suggested the presence of U(OH)$_3$Cl(aq) as the predominant species.

For chloride media of composition $m_{KCl} + m_{HCl} = 1$ mol · kg^{-1} H$_2$O, various equations were given over the temperature range 400 to 600°C (Ni–NiO buffer), but the acknowledged existence of higher uranium oxidation states prevents us, in the opinion of the reviewer, from drawing further conclusions.

In HF solutions (10^{-3} – 0.1 mol · kg^{-1} H$_2$O) at 500°C and 1 kbar, the following solubility equations were given:

$$m_U = 2.353 \cdot 10^{-2} (m_{HF})^2 + 8.293 \cdot 10^{-4} m_{HF} + 10^{-6} \text{ (Ni–NiO)}$$

$$m_U = 0.1158 (m_{HF})^2 + 3.207 \cdot 10^{-2} m_{HF} + 10^{-6} \text{ (Fe}_2\text{O}_3\text{–Fe}_3\text{O}_4\text{)}.$$

From this, the author suggested the presence of U(OH)$_2$F$_2$(aq) as the predominant species.

The solubility of $UO_2(cr)$ in 0.25 m NaOH + 0.25 m $Na_2C_2O_4$, and 0.5 m Na_2CO_3 (1 kbar, Ni–NiO, 14 days) is reported as $3 \cdot 10^{-6}$ mol U · kg^{-1} H$_2$O. The authors indicated that this is three orders of magnitude lower than those reported by [76NGU/POT].

Starting with trioxide U(VI) monohydrate, $UO_3 \cdot H_2O(cr)$, at room temperature, the progressive dehydration to $UO_3 \cdot 0.33\ H_2O$ in contact with water was reported between 150 and 300°C (in the E_h domain of existence of $Cu_2O(cr)$ or $CuO(cr)$). After an undefined period during which "U(VI) hydroxides" precipitated, the following solubility relationships were given for the range 200 to 500°C:

$\log_{10} m_U = 15375.3\ T^{-1} + 0.0647\ T - 67.05$ 473–573 K

$\log_{10} m_U = -221.7\ T^{-1} - 2.71$ 573–773 K.

The predominant presence of $(UO_2)_3(OH)_7^-$ was deduced by the author from these results. Except for the $U(OH)_4(aq)$ species, very little quantitative thermodynamic information can be extracted from this paper.

[97RUN/NEU]

Spectroscopic techniques (TRLFS, Raman, UV–Vis absorption, EXAFS) have been applied to investigate the speciation of actinides in concentrated chloride solutions. The interactions of actinides in various oxidation states, Am(III), Np(IV), Np(V), Pu(VI), U(VI) with chloride were studied. Spectra of trivalent Am, tetravalent and pentavalent Np in 5 M NaCl and 5 M NaClO$_4$ show no significant differences. The authors conclude that under this condition inner-sphere chloride complexes are not stable. On the contrary, the hexavalent actinides U(VI) and Pu(VI) show substantial changes in the spectroscopic data when the chloride concentration is changed, adding LiCl up to 16 M. Inner-sphere chloro complexes of the type, $AnO_2Cl_n^{2-n}$, with $n = 1, 2, 3$ and 4, are formed. The tetrachloro complex was identified only in solutions of very high chloride concentrations (> 15 M). There are no thermodynamic data given in the paper, but the spectroscopic results contain useful information for the interpretation of measured thermodynamic data.

[97RUT/GEI]

This paper is discussed together with [99RUT/GEI].

[97SAL/KRI]

This conference paper describes X–ray and thermal studies of the new compound $SrUTe_2O_8(cr)$, which is reported as resulting from the 1:2 molar ratio mixture of a $SrUO_4(cr)$ and $TeO_2(cr)$ mixture in an argon atmosphere at 1073 K. The same compound could be obtained by heating a 1:1 molar mixture of $SrTeO_3(cr)$ and $UTeO_5(cr)$.

The X–ray diffraction pattern was indexed on the basis of a monoclinic cell with $a_0 = 13.028$ Å, $b_0 = 5.434$ Å, $c_0 = 12.576$ Å and $\beta = 93.21°$. No details are given.

From thermal measurements in air, the melting point was given as 1213 K with an enthalpy of melting of (93.92 ± 10.00) kJ · mol^{-1}. No further details are given.

Above 1243 K, the compound decomposes by vaporisation of TeO$_2$(g), as indicated by weight loss curves and X–ray identification of the condensed material collected in the cooler zone of the reaction tube. The mass loss *vs.* temperature relationship is similar to that of a sample of TeO$_2$(l) under the same conditions. Decomposition was complete around 1473 K. No further thermodynamic results could be extracted from this paper.

[97SCA/ANS]

Using the selected data in [92GRE/FUG] the authors calculated the predominant U(VI) species for a 3.6·10^{-3} M U(VI) solution in 0.1 M NaClO$_4$ in air at pH 1.9 (UO_2^{2+}) and pH = 5 (species 2:2: $UO_2(OH)_2^{2+}$ and 5:3: $UO_3(OH)_5^+$). They separated the species at pH = 5 in a creatine buffered solution using the capillary electrophoresis method. Retention times of the 2:2 and 5:3 species are in agreement with expected the electrophoretic mobility, which depends on charge and size of each species. The fluorescence data reported in this paper are not directly connected to the solution fraction which contains the species.

This paper gives an original qualitative direct confirmation for the existence of two species (2:2 and 5:3) at pH = 5.

[97SHO/SAS]

This paper builds upon the database of [88SHO/HEL] containing the thermodynamic properties of inorganic ions using the revised Helgerson-Kirkham-Flowers (HKF) equation of state permitting estimation of standard partial molal volumes, heat capacities and entropies, as well as apparent partial molal enthalpies and Gibbs energies of formation of more than 300 aqueous ions to 1000°C and 5 kbar. These extrapolations are based on the thermodynamic properties at 25°C and 1 bar, and for those ions for which no experimental data exist, these properties are predicted from correlations for similarly charged ions with known properties. The $\Delta_f G_m^\circ$, $\Delta_f H_m^\circ$ and S_m° values for UO_2^{2+}, U^{3+} and U^{4+} were taken from [92GRE/FUG].

The experimental values of V_m° and $C_{p,m}$ for UO_2^{2+} from [89HOV/NGU] were regressed according to the revised-HKF equations for cations to yield values of 5.73 cm^3 · mol^{-1} and 10.2 cal · K^{-1} · mol^{-1} (42.7 J · K^{-1} · mol^{-1}), respectively, at 25°C and 1 bar. The recommended $C_{p,m}^\circ$ value at 25°C and 1 bar from [92GRE/FUG] is (42.4 ± 3.0) J · K^{-1} · mol^{-1}, which is virtually identical to the regressed value given here.

The V_m° and $C_{p,m}^\circ$ values (25°C and 1 bar) for U^{3+} were obtained from linear correlations for the trivalent ions, Al^{3+}, Ga^{3+} and Cr^{3+} (also Fe^{3+} and Rh^{3+} for V_m°) with S_m°, *i.e.*, – 39.3 cm^3 · mol^{-1} and – 152.3 J · K^{-1} · mol^{-1}, respectively. These linear correlations are only shown graphically and no statistics of fit are given. Moreover, the fit is particularly poor for $C_{p,m}^\circ$ versus S_m° such that there is no reason to prefer this value over the recommended value of $-(64 \pm 22)$ $J \cdot K^{-1} \cdot mol^{-1}$ [92GRE/FUG] or $-(150 \pm 50)$ $J \cdot K^{-1} \cdot mol^{-1}$ adopted in the current review.

The V_m^o and $C_{p,m}^o$ values (25°C and 1 bar) for U^{4+} were obtained by first using a linear correlation (of the slopes of V_m^o and $C_{p,m}^o$ versus S_m^o) with ionic charge for groups of alkali metal ions, divalent transition metal ions and trivalent cations (see previous paragraph) to obtain the corresponding slope for tetravalent ions. Then by taking this slope and the V_m^o, $C_{p,m}^o$, S_m^o data for Th^{4+}, two general linear equations were proposed:

$$V_m^o \text{(tetravalent cations in cm}^3 \cdot mol^{-1}) = 0.10 \, S_m^o + 43.3,$$

$$C_{p,m}^o \text{(tetravalent cations in cal} \cdot K^{-1} \cdot mol^{-1}) = 0.31 \, S_m^o + 31.1.$$

The resulting values are -53.3 $cm^3 \cdot mol^{-1}$ and 0.8 $J \cdot K^{-1} \cdot mol^{-1}$, respectively, (cf. $-(48 \pm 15)$ $J \cdot K^{-1} \cdot mol^{-1}$ [92GRE/FUG]). The current review adopts the value for $C_{p,m}^o$ of $-(220 \pm 50)$ $J \cdot K^{-1} \cdot mol^{-1}$, which is even more discrepant from the predicted value of Shock et al.

Although the correlations given in [97SHO/SAS] are of value and appear more consistent for predicting values of V_m^o compared to values of $C_{p,m}^o$, the thermodynamic data obtained from these empirical relationships have not been selected in this review.

[97SHO/SAS2]

This paper deals with application of the revised Helgeson-Flowers-Kirkham (HKF) equations of state for aqueous uranium species using standard state thermodynamic data at 25°C and 1 bar to obtain equilibrium constants for redox and hydrolysis reactions (U(III), U(IV), U(V), and U(VI)) to 1000°C and 0.5 GPa. Standard partial molal Gibbs energy of formation, enthalpy of formation and entropy of U^{3+}, U^{4+}, UO_2^+ and UO_2^{2+} are taken from [92GRE/FUG]. The correlation methods of [97SHO/SAS2] were used to derive the standard partial molal heat capacity and standard partial molal volume of U^{3+}, U^{4+}, and UO_2^+, whereas those for UO_2^{2+} were based on the experimental results of [89HOV/NGU].

Table A-31: Standard state thermodynamic data (in bold, [92GRE/FUG]).

Ion	$\Delta_f G_m^\circ$ kJ·mol^{-1}	$\Delta_f H_m^\circ$ kJ·mol^{-1}	S_m° J·K^{-1}·mol^{-1}	$C_{p,m}^\circ$ J·K^{-1}·mol^{-1}	V_m° cm^3·mol^{-1}
U^{3+}	− 475.093	− 489.110	− 192.9	− 152.3	− 39.3
	− (476.473 ± 1.810)	− (489.100 ± 3.712)	− (188.170 ± 13.853)	− (64.000 ± 22.000)	
U^{4+}	− 529.904	− 591.199	− 416.7	0.8	− 53.3
	− (529.860 ± 1.765)	− (591.200 ± 3.300)	− (416.896 ± 12.553)	− (48.000 ± 15.000)	
UO$_2^+$	− 961.023	− 1025.080	− 25.1	− 92.5	10.2
	− (961.021 ± 1.752)	− (1025.127 ± 2.960)	− (25.00 ± 8.000)		
UO$_2^{2+}$	− 952.613	− 1019.013	− 98.3	42.7	5.73
	− (952.551 ± 1.747)	− (1019.000 ± 1.500)	− (98.200 ± 3.000)	(42.4 ± 3.000)	

The values for $\Delta_f G_m^\circ$, $\Delta_f H_m^\circ$ and S_m° are well within the suggested uncertainties given by [92GRE/FUG], and in fact stem largely from the same source. Note that the current values of $\Delta_f G_m^\circ$, $\Delta_f H_m^\circ$ and S_m° for these ions are the same as proposed by [92GRE/FUG], but the $C_{p,m}^\circ$ values have been changed markedly. Shock et al. claim to have better estimates of $C_{p,m}^\circ$ (with the exception of that for UO$_2^{2+}$, where both groups use the calorimetric data of [89HOV/NGU]) and point out the limitation of the Criss–Cobble relationship used in [92GRE/FUG] when applied to tri– and tetravalent ions. In this case Shock et al. used their correlation of heat capacity with entropy, based on measured values for the different sets of charged cations ([97SHO/SAS]). The standard partial molar volumes were estimated from similar linear correlations against the corresponding entropies for each charge type.

The $\Delta_f G_m^\circ$ values of U(VI) hydrolysed mononuclear species are taken from [92GRE/FUG]. Similarly, for those of U^{4+} [92GRE/FUG] provided the estimates for U(OH)$^{3+}$, U(OH)$_4$(aq), and U(OH)$_5^-$. The $\Delta_f G_m^\circ$ values for the two remaining hydrolysed cations were taken from [80LEM/TRE], i.e., − 755.630 kJ·mol^{-1} (U(OH)$_2^{2+}$) and − 975.709 kJ·mol^{-1} (U(OH)$_3^+$). On the other hand, for the hydrolysis products of U^{3+} and UO$_2^+$, empirical linear correlations were employed. For the 1:1 hydroxo complexes, the correlations were based on the $\Delta_f G_m^\circ$ values for the parent ion: $\Delta_r G_m^\circ = -0.1732 \cdot \Delta_f G_m^\circ - 64183$ for the equilibrium (U^{3+} + OH$^-$ ⇌ U(OH)$^{2+}$) and $\Delta_r G_m^\circ = -0.1368 \cdot \Delta_f G_m^\circ - 7.531$ for the equilibrium (UO$_2^{2+}$ + OH$^-$ ⇌ UO$_2$(OH)$^+$). The corresponding equilibrium constants were calculated from the relationships:

$$\log_{10} \beta_2^\circ = 1.98 \log_{10} \beta_1 - 0.16$$
$$\log_{10} \beta_3^\circ = 2.89 \log_{10} \beta_1 - 1.74$$
$$\log_{10} \beta_4^\circ = 3.58 \log_{10} \beta_1 - 3.58,$$

where only the estimation of $\log_{10} \beta_2$ is considered valid for UO$_2^+$. The heat capacities and volumes were then estimated from the entropies as mentioned above for the unhydrolysed cations. The values of these estimated thermodynamic properties are given in Table A-32.

Table A-32: Standard state thermodynamic data (in bold, [92GRE/FUG]).

Ion	$\Delta_f G_m^\circ$ kJ·mol^{-1}	$\Delta_f H_m^\circ$ kJ·mol^{-1}	S_m° J·K^{-1}·mol^{-1}	$C_{p,m}^\circ$ J·K^{-1}·mol^{-1}	V_m° cm^3·mol^{-1}
U(OH)$^{2+}$	− 676.97	− 701.66	5.0	− 161.5	5.1
U(OH)$_2^+$	− 914.10	− 6929.75	143.1	− 263	26
U(OH)$_3$ (aq)	− 1065.6	− 1144.0	291.2	− 530	43
U(OH)$_4^-$	− 1242.9	− 1381.7	321.0	669	59
U(OH)$^{3+}$	− 763.99	− 830.11	− 200.0	72.4	− 2.8
	− (763.918 ± 1.798)	− (830.120 ± 9.540)	−(199.946 ± 32.521)		
U(OH)$_2^{2+}$	− 1012.8	− 1089.6	− 69.8	− 170	18
U(OH)$_3^+$	− 1212.8	− 1351.9	22.2	84	27
U(OH)$_4$ (aq)	− 1452.5	− 1658.3	30.8	231	41
	− (1452.500 ± 8.000)	− (1655.798 ± 10.934)	(40.000 ± 25.000)	(205.000 ± 80.000)	
U(OH)$_5^-$	− 1621.1	− 1915.6	− 33.0	444	41
	≥− 1621.144				
UO$_2$(OH)(aq)	− 1094.5	− 1238.0	− 58.2	103.8	14.5
UO$_2$(OH)$_2^-$	− 1227.1	− 1430.6	− 11.7	103	30.6
UO$_2$(OH)$^+$	− 1160.01$^{(a)}$	− 1261.5$^{(a)}$	17.2$^{(a)}$	18.0	
	− (1160.009 ± 2.447)	− (1261.657 ± 15.107)	(17.000 ± 50.000)		
UO$_2$(OH)$_2$ (aq)	− 1368.0	− 1539.4	− 15.9	72	32.8
	≥− 1368.038				
UO$_2$(OH)$_3^-$	− 1554.2	− 1804.2	− 14.2	168	36.4
	− (1554.377 ± 2.878)				
UO$_2$(OH)$_4^{2-}$	− 1712.8	− 2020.2	26.4	− 35.2	51.4
	− (1712.746 ± 11.550)				

(a) attributed to [92GRE/FUG] in the original

In the absence of experimental results, these correlations are at least indicative of the relative stabilities of these species and were also used by Shock *et al.* to construct predictive Pourbaix diagrams involving these hydrolysed species and also incorporating uranium complexes.

[97STE/FAN]

This paper, included in a report, describes the complex formation between Cm(III) and monosilicic acid as studied by fluorescence spectroscopy. The experimental method has been detailed in previous papers from the laboratory. The key uncertainty in this study is the speciation of silica in the system (monosilicic acid has been prepared by rapid acidification of sodium silicate, which may result in the presence of polymers). The concentration of monosilicic acid in the experiments is $3.6 \cdot 10^{-4}$ M, which is only slightly larger than the solubility of silica in water. The stoichiometry of the complex formed is well established and the equilibrium constant is reported as $\log_{10} \beta_1 = (7.4 \pm 0.1)$, at room temperature in 0.1 M NaClO$_4$, where the uncertainty is estimated by the reviewers.

[97TAK]

In this description of a new device to measure heat capacities at high temperatures, Takahashi reports (in the form of a small graph) new measurements of the heat capacity of UO$_2$(cr) from ca. 300 to 1500 K. The results, deemed to have an uncertainty of $\pm 3\%$, are about 2% lower than earlier values from enthalpy drop measurements.

[97TOR/BAR]

No thermodynamic data could be inferred from these experiments on the mechanism of UO$_2$ dissolution. Nevertheless it is shown that the dissolution rate of UO$_2$(cr) in NaClO$_4$ and O$_2$ concentration is dependent (pH = 3 to 6.7, p_{O_2} = 5, 21 and 100 % in N$_2$), but in a rather complex way. In this pH range, adsorbed oxygen acts as the oxidant. Above pH = 6.7 a layer of UO$_{2.25}$ appears due to oxygen diffusion, as shown by deconvolution of the XPS U$_{4f_{7/2}}$ peak and the data do not convey evidence for UO$_2$(cr) dissolution. This point is interesting to note with regard to solubility data for uranium dioxide at high pH in oxic conditions. These results confirm numerous previous observations.

[97VAL/RAG]

The solubility measurements were carried out at 300°C and 0.5 kbar, in pure water and a 0.01 mol · dm^{-3} Ca(OH)$_2$ solution. Starting materials were microcrystalline γ–UO$_3$ and poorly crystallised CaU$_{1.6}$O$_{5.8}$·2.5H$_2$O. As demonstrated by the authors, the starting materials changed with time during contact with the aqueous phase. The UO$_3$ samples recrystallised markedly, although microprobe analysis indicated a composition of pure UO$_3$, the X–ray diffraction pattern was different from that of the initial material. CaU$_{1.6}$O$_{5.8}$·2.5H$_2$O evolved into Ca$_3$U$_{4.5}$O$_{16.5}$ of good crystallinity. For all the experiments, equilibrium, post recrystallisation solubilities span a narrow range between $10^{-5.9}$ and $10^{-6.3}$ mol · dm^{-3}. Solutions were not filtered prior to the analysis. There was essentially no difference between experiments in water (equilibrium pH = 5.7–5.8) and 0.01 mol · dm^{-3} Ca(OH)$_2$ (equilibrium pH = 9.4). Because of the lack of thermodynamic information on the uranate phases involved in this study, the authors did not attempt to model their results. In the case of UO$_3$, the lack of pH dependence of the results contrasted with the changes that could be calculated using the geochemical model EQ3NR with the NEA database. The authors obtained a better fit of their results by including in their calculations the stability constants for UO$_2$(OH)$_2$(aq) from [72NIK/SER]. However, in [92GRE/FUG], the results of these latter authors were extensively analysed and were not credited.

[97WRU/PAL]

The formation constant of the Am(III) carbonate complex, AmCO$_3^+$, from Am^{3+} and

CO_3^{2-} has been investigated in 0.1 m $NaClO_4/NaHCO_3$ by LIPAS at 25, 50 and 75°C. At a constant Am concentration of $1.0 \cdot 10^{-6}$ mol · kg^{-1}, the pH was varied in the range 4.0 to 6.5 and $\log_{10}[CO_3^{2-}]$ in the range -8.2 to -4.8. In this paper, an extended conference abstract, the spectroscopic measurements and LIPAS apparatus are well described and a series of spectra are shown in a Figure, but no details are reported on the pH measurements at elevated temperature and the calculation of $\log_{10}[CO_3^{2-}]$. From the evaluated formation constants at $I = 0.1$ mol · kg^{-1} and different temperatures, $\log_{10} \beta_1 = (6.26 \pm 0.12)$ at (25 ± 0.5)°C, (6.68 ± 0.12) at (50 ± 2)°C, and (7.54 ± 0.43) at (75 ± 3)°C, the authors estimated an enthalpy of reaction ($Am^{3+} + CO_3^{2-} \rightleftharpoons AmCO_3^+$) of $\Delta_r H_m$ (0.1 m $NaClO_4$) = (37 ± 11) kJ · mol^{-1} by applying the Van't Hoff relation.

The formation constant at 25°C is considered together with other literature values for the selection of $\log_{10} \beta_1^o$ ($AmCO_3^+$, 298.15 K). In contrast to the series of spectra recorded at 25°C, those at 50 and 75°C do not show an isosbestic point. As this is neither discussed nor explained, the reported $\log_{10} \beta_1$ values at 50 and 75°C, and enthalpy of reaction are not selected in the present review.

Laser–Induced Photoacoustic Spectroscopy (LIPAS), is used to study the hydrolysis of UO_2^{2+} in 0.1 m $NaClO_4$ at (25 ± 0.5)°C between pH = 3.0 and 5.6. Two uranium concentrations are used, 10^{-6} and 10^{-4} M. In the first case species 0:1 and 1:1 (UO_2OH^+) are expected to be present, but the LIPAS spectra do not show any change with increasing pH. At higher U(VI) concentrations a clear increase in absorption is measured. A fit with three species 0:1, 1:1 (with fixed $\log_{10} {}^*\beta_{1,1} = -5.8$) and 2:2 ($(UO_2)_2(OH)_2^{2+}$) gives $\log_{10} {}^*\beta_{2,2} = -(5.45 \pm 0.05)$. This value is not very sensitive to the choice of $\log_{10} {}^*\beta_{1,1}$. Extrapolation to zero ionic strength according to the Davies equation written as:

$$\log_{10} {}^*\beta_{2,2}^o = \log_{10} {}^*\beta_{2,2} - A \cdot \Delta Z^2 \left(\frac{I^{1/2}}{1+I^{1/2}} - 0.3 I \right)$$

with $A = 0.511$, $I = 0.1$, $\Delta Z^2 = -2$, gives, according to the authors, $\log_{10} {}^*\beta_{2,2}^o = -(5.56 \pm 0.05)$, which agrees with the selected value of $\log_{10} {}^*\beta_{2,2}^o = -(5.62 \pm 0.04)$ in [92GRE/FUG]. However, the authors seem to have made a calculation error. Using the Davies equation given by [97WRU/PAL] this review calculates $\log_{10} {}^*\beta_{2,2}^o = -(5.23 \pm 0.05)$ and according to the SIT, we obtain practically the same value, $\log_{10} {}^*\beta_{2,2}^o = -(5.24 \pm 0.05)$, which is considerably different from the value selected by [92GRE/FUG].

[98AAS/MOU]

Aqueous U(VI) ternary complexes including F and CO_3 ligands, $UO_2(CO_3)_p F_q^{2-2p-q}$, $p = 1$ and $q = 1, 2$ and 3, have been the subject of two studies, one dealing with the elucidation of structure for $q = 2$ and 3, as well as for $UO_2(CO_3)_3^{4-}$, [97SZA/AAS], [95BAN/GLA] and the present work dealing with thermodynamic data. Potentiometric titrations of U(VI) solutions in 1 M NaClO$_4$ were made using a quinhydrone and a specific fluoride electrode against a Ag/AgCl electrode, to measure [F$^-$] and log$_{10}$[H$^+$] ($t = (25.00 \pm 0.05)$°C). The p_{CO_2} was maintained constant with a purified N$_2$–CO$_2$ gas mixture. U and F total concentrations were fixed ([U] = (2.3 to 5.7)·10^{-3} M and [F$^-$] = (8 to 40)·10^{-3} M) and the carbonate concentration was varied by adding HClO$_4$ (pH varied from 3.7 to 5.8) to the test solutions having fixed amounts of NaHCO$_3$.

The data were processed to give the average number of carbonato ligands, $\bar{n}$, bound to the UO$_2$ core *versus* log$_{10}$[HCO$_3^-$] taking into account the equilibria:

$$UO_2^{2+} + p\,H_2O(l) + p\,CO_2(g) + q\,F^- \rightleftharpoons UO_2(CO_3)_p F_q + 2p\,H^+$$

$$CO_2(g) + H_2O(l) \rightleftharpoons HCO_3^- + H^+ \qquad K_{a_1}$$

$$CO_2(g) + H_2O(l) \rightleftharpoons CO_3^{2-} + 2\,H^+ \qquad K_{a_1} K_{a_2}$$

with log$_{10} K_{a_1}$ (1M NaClO$_4$) = $-(7.625 \pm 0.01)$ (measured in this work) and K_{a_2} (1 M NaClO$_4$) = -9.57 from [58FRY/NIL]. The value of $\bar{n}$ can reach 1.5 at log$_{10}$[HCO$_3^-$] = -1.5 when both U and F concentrations are the highest. q values are refined with LETAGROP program.

The formation of these complexes was described by:

$$UO_2^{2+} + p\,CO_3^{2-} + q\,F^- \rightleftharpoons UO_2(CO_3)_p F_q^{2-2p-q}$$

because no polynuclear complexes ($\bar{n}$ greater than 1) were indicated by NMR (^{19}F) or the LETAGROP program. The contribution of the species, UO$_2$(CO$_3$)$_2$F^{3-}, is estimated to be less than 4%.

The values of the formation constants are log$_{10}\beta_{1,1,1} = (12.56 \pm 0.05)$, log$_{10}\beta_{1,1,2} = (14.86 \pm 0.08)$, log$_{10}\beta_{1,1,3} = (16.77 \pm 0.06)$ (uncertainty 3σ). These results indicate that these complexes have a pentagonal bipyramidal geometry.

This review has calculated the value of the equilibrium constant at zero ionic strength using the SIT and assuming ε(UO$_2$CO$_3$F$^-$, Na$^+$) = (0.00 ± 0.05), ε(UO$_2$CO$_3$F$_2^{2-}$, Na$^+$) = $-(0.02 \pm 0.09)$ and ε(UO$_2$CO$_3$F$_3^{3-}$, Na$^+$) = $-(0.25 \pm 0.05)$ kg · mol^{-1}. The values are: log$_{10}\beta^\circ_{1,1,1} = (13.75 \pm 0.09)$, log$_{10}\beta^\circ_{1,1,2} = (15.57 \pm 0.14)$ and log$_{10}\beta^\circ_{1,1,3} = (16.38 \pm 0.11)$.

[98ALM/NOV]

This paper is discussed together with [95NOV/ROB] and [97NOV/ALM].

The two papers report reliable and accurate solubility data for the potassium dioxoneptunium(V) carbonates, KNpO$_2$CO$_3$(s) and K$_3$NpO$_2$(CO$_3$)$_2$(s), which are used in the present review to evaluate the solubility constants at $I = 0$ and 25°C. The authors also proposed a comprehensive set of ion interaction Pitzer parameters for Np(V) carbonate complexes. All solubility experiments were performed batch-wise at room temperature,

(22 ± 1)°C. The measurements were conducted over periods of 150 – 200 days. The time was sufficient to ensure a steady-state equilibrium, when the initial solid was transformed into a more stable, less soluble one. The solids formed under the experimental conditions (*cf.* Table A-33) were identified and characterised by X-ray powder diffraction.

In the first paper [97NOV/ALM], the solubility of Np(V) was determined in 0.01 – 4.8 M K_2CO_3 solutions, with either freshly precipitated NpO_2OH(am) or $KNpO_2CO_3 \cdot xH_2O$(s) as initial solids. For K_2CO_3 solutions up to 0.4 M, a $KNpO_2CO_3$(s) solid was formed, which gave X-ray powder diffraction patterns comparable with those of the $KNpO_2CO_3 \cdot xH_2O$(s) solid reported by Keenan and Kruse [64KEE/KRU]. At higher K_2CO_3 concentrations, two modifications of $K_3NpO_2(CO_3)_2 \cdot xH_2O$(s) were formed, the phases A and B described in the earlier papers of Volkov *et al.* [74VOL/KAP2] and Visyashcheva *et al.* [74VIS/VOL]. Phase B was formed over a broad range of K_2CO_3 concentrations, whereas phase A only formed at the highest concentration (4.8 M K_2CO_3) and unexpectedly in one solution of relatively low concentration (0.25 M K_2CO_3). According to [74VOL/KAP2] phase A is preferentially formed at K_2CO_3 concentrations above 2 molal. In contrast to phase A, where the crystal lattice parameters change with hydration in a moist atmosphere or when wetted and with dehydration by heating up to 400°C, those of form B remain unaffected [74VOL/KAP2]. The water content in $KNpO_2CO_3 \cdot xH_2O$(s) and $K_3NpO_2(CO_3)_2 \cdot xH_2O$(s), phase B, was estimated in [74VIS/VOL] to be $x \leq 2$ and $x \approx 1.6$, respectively. As the number of hydration water molecules is not exactly known, they are disregarded in the calculations of Novak *et al.* and also in the present review.

In the second paper of this group, [98ALM/NOV], the solubility of $KNpO_2CO_3$(s) was determined in KCl–K_2CO_3 mixtures with $m_{KCl} = 0.003 - 3.2$ mol·kg^{-1} and $m_{K_2CO_3}$ = 0.001, 0.01, and 0.1 mol·kg^{-1}. In these experiments the initial $KNpO_2CO_3$(s) remained the stable solid over the whole period of 160 days. In two additional experiments at higher carbonate concentrations (1 m KCl + 1 m K_2CO_3), the initial solids used, $KNpO_2CO_3$(s) and NpO_2OH(am), were both transformed into $K_3NpO_2(CO_3)_2$(s) and the same steady state Np(V) concentration was obtained. The solubility was also investigated in quaternary Na–K–Cl–CO$_3$ solutions of high Na$^+$ and low K$^+$ concentrations (5 m NaCl + 0.1 m KCl + 3.5·10^{-4} to 1.0 m Na$_2CO_3$). In these solutions, the solubility was either controlled by $KNpO_2CO_3$(s) (at $m_{CO_3^{2-}} < 0.01$ mol·kg^{-1}) or by $Na_3NpO_2(CO_3)_2$(s) (at $m_{CO_3^{2-}} > 0.01$ mol·kg^{-1}). The X-ray pattern of a solid formed at the highest ionic strength could not be identified by comparison with literature data.

Table A-33: Np(V) carbonate solids observed in [97NOV/ALM], [98ALM/NOV].

Solution	initial solid	final solid
0.01 – 0.1 M K_2CO_3	$NpO_2OH(am)$	$KNpO_2CO_3(s)$
0.03 – 0.4 M K_2CO_3	$KNpO_2CO_3(s)$	$KNpO_2CO_3(s)$
0.5 – 2.0 M K_2CO_3	$NpO_2OH(am)$	$K_3NpO_2(CO_3)_2(s)$ B
0.25 M K_2CO_3	$NpO_2OH(am)$	$K_3NpO_2(CO_3)_2(s)$ A
4.8 M K_2CO_3	$NpO_2OH(am)$	$K_3NpO_2(CO_3)_2(s)$ A
4.8 M K_2CO_3	$KNpO_2CO_3(s)$	$K_3NpO_2(CO_3)_2(s)$ A
(0.003 – 3.2) m KCl + (0.001 – 0.1) m K_2CO_3	$KNpO_2CO_3(s)$	$KNpO_2CO_3(s)$
1 m KCl + 1 m K_2CO_3	$NpO_2OH(am)$	$K_3NpO_2(CO_3)_2(s)$
1 m KCl + 1 m K_2CO_3	$KNpO_2CO_3(s)$	$K_3NpO_2(CO_3)_2(s)$
5 m NaCl + 0.1 m KCl + ($3.5 \cdot 10^{-4}$ – 0.01) m Na_2CO_3	$KNpO_2CO_3(s)$	$KNpO_2CO_3(s)$
5 m NaCl + 0.1 m KCl + (0.01 – 0.22) m Na_2CO_3	$Na_3NpO_2(CO_3)_2(s)$	$Na_3NpO_2(CO_3)_2(s)$
5 m NaCl + 0.1 m KCl + 1.0 m Na_2CO_3	$Na_3NpO_2(CO_3)_2(s)$	unidentified

- Thermodynamic modelling.

Adopting the parameter set of Harvie et al. [84HAR/MOL] for the seawater salt system, Novak et al. [97NOV/ALM] extended the Pitzer parameters reported in [95FAN/NEC] for the system Np(V)–Na–OH–CO_3–Cl–H_2O (25°C) to K^+ containing solutions and calculated binary Pitzer parameters $\beta^{(0)}$, $\beta^{(1)}$, $\beta^{(2)}$, and C^ϕ for the interactions between K^+ and $NpO_2(CO_3)_3^{5-}$. Since the parameter set of Fanghänel et al. [95FAN/NEC] was designed for actinide and carbonate trace concentrations, it can be inaccurate at $m_{CO_3^{2-}} > 0.1$ mol·kg^{-1}. Modelling the solubility data of Ueno and Saito [75UEN/SAI] in 0.05 – 1.6 M Na_2CO_3 and their own experimental data at high K_2CO_3 concentrations, Novak et al. [97NOV/ALM] evaluated in addition a ternary parameter, θ, accounting for anion–anion interactions between $NpO_2(CO_3)_3^{5-}$ and CO_3^{2-}. Based on the Np(V) carbonate complexation constants at $I = 0$ from [95FAN/NEC], the following solubility constants of the potassium dioxoneptunium(V) carbonates were fitted simultaneously from the solubility data in K_2CO_3 solution: $\log_{10} K_{s,0}^\circ = -(13.6 \pm 0.1)$ for $KNpO_2CO_3(s)$ and $\log_{10} K_{s,0}^\circ = -(15.9 \pm 0.1)$ for $K_3NpO_2(CO_3)_2(s)$. The combined set of Pitzer parameters and equilibrium constants at $I = 0$ (solubility constants of solid Np(V) hydroxide and carbonates, and formation constants of Np(V) hydroxide and carbonate complexes) from [95FAN/NEC], [97NOV/ALM], was then applied in [98ALM/NOV] to calculate the Np(V) solubilities in more complex systems. For all K–Cl–CO_3 and Na–K–Cl–CO_3 solutions, and even for three synthetic Na–Mg–K–Cl brines of various compositions and $I = 0.86$, 3.0, and 7.8, the model calculations were, within the uncertainty range, consistent with the experimental solubilities. Both the stable carbonate solid and the total Np(V) concentration were well predicted.

For reasons of consistency with the NEA–TDB system, the experimental data reported in the two papers are re-evaluated in the present review on the basis of the SIT. The small difference in temperature, experimental data at $(22 \pm 1)°C$ compared to the standard state of 25°C, is considered to be within the range of other uncertainties. As there are no solubility data at low carbonate concentrations, with the uncomplexed NpO_2^+ ion as the predominant aqueous species, the solubility constants, $\log_{10} K_{s,0}^°$, cannot be evaluated directly. In K_2CO_3 solutions above 0.17 mol·kg^{-1}, the limiting carbonate complex, $NpO_2(CO_3)_3^{5-}$, represents more than 95% of the total Np(V) concentration, and the dissolution reactions can be written as:

$$KNpO_2CO_3(s) + 2\ CO_3^{2-} \rightleftharpoons NpO_2(CO_3)_3^{5-} + K^+ \quad (A.91)$$

and

$$K_3NpO_2(CO_3)_2(s) + CO_3^{2-} \rightleftharpoons NpO_2(CO_3)_3^{5-} + 3\ K^+ \quad (A.92)$$

with the solubility constants, $\log_{10} K_{s,3}$, given by the sums of ($\log_{10} K_{s,0} + \log_{10} \beta_3$). In Figure A-7 and Figure A-8, the SIT extrapolation is applied to the experimental data in K_2CO_3 solutions ≥ 0.17 mol·kg^{-1}. In addition, Figure A-7 includes the solubility in 0.1 m K_2CO_3 + 3.2 m KCl, where the total Np(V) concentration is also practically equal to that of $NpO_2(CO_3)_3^{5-}$. (Data at $m_{CO_3^{2-}} = 0.1$ mol·kg^{-1} and lower ionic strengths are not included in the SIT extrapolation, because other Np(V) species may contribute considerably to the total Np(V) concentration). The following results are obtained by extrapolation to $I = 0$:

- Figure A-7:
  $\log_{10} K_{s,3}^°((A.91), 298.15\ K) = -(7.65 \pm 0.11)$, $\Delta\varepsilon = -(0.25 \pm 0.06)$ kg·mol^{-1} and hence $\varepsilon(NpO_2(CO_3)_3^{5-}, K^+) = -(0.21 \pm 0.07)$ kg·mol^{-1}

- Figure A-8:
  $\log_{10} K_{s,3}^°((A.92), 298.15\ K) = -(9.96 \pm 0.06)$, $\Delta\varepsilon = -(0.22 \pm 0.02)$ kg·mol^{-1} and hence $\varepsilon(NpO_2(CO_3)_3^{5-}, K^+) = -(0.23 \pm 0.02)$ kg·mol^{-1}.

The two values of $\varepsilon(NpO_2(CO_3)_3^{5-}, K^+)$ calculated with auxiliary SIT coefficients from [2001LEM/FUG] agree within the uncertainty range. This is noteworthy, because they are derived from data in different media (predominantly a KCl medium and pure K_2CO_3 solutions, respectively).

Combining $\log_{10} K_{s,3}^°((A.91), 298.15\ K)$ and $\log_{10} K_{s,3}^°((A.92), 298.15\ K)$ with $\log_{10} \beta_3^° = (5.50 \pm 0.15)$ selected in the previous NEA review [2001LEM/FUG], we obtain:

$$\log_{10} K_{s,0}^°((A.93), 298.15\ K) = -(13.15 \pm 0.19)$$

for the reaction: $KNpO_2CO_3(s) \rightleftharpoons K^+ + NpO_2^+ + CO_3^{2-} \quad (A.93)$

and $\log_{10} K_{s,0}^°((A.94), 298.15\ K) = -(15.46 \pm 0.16)$

for the reaction: $\quad K_3NpO_2(CO_3)_2(s) \rightleftharpoons 3K^+ + NpO_2^+ + 2CO_3^{2-}$ (A.94)

The derived constants, $\log_{10} K_{s,0}^\circ$, differ by 0.4 to 0.5 logarithmic units from the values given in [97NOV/ALM]. These discrepancies are mainly due to the different ion interaction approaches (SIT or Pitzer equations) used to calculate activity coefficients for the highly charged complex, $NpO_2(CO_3)_3^{5-}$. In particular, the Pitzer parameters $\beta^{(2)}$ and θ included in the model of [97NOV/ALM], [98ALM/NOV], which are of course correlated with the simultaneously fitted constants at $I = 0$, cause deviations from the SIT extrapolation to $I = 0$.

Figure A-7: Application of the SIT to Np(V) solubility data in K_2CO_3 solutions ≥ 0.17 mol · kg^{-1}, reaction: $KNpO_2(CO_3)(s) + 2\ CO_3^{2-} \rightleftharpoons NpO_2(CO_3)_3^{5-} + K^+$.

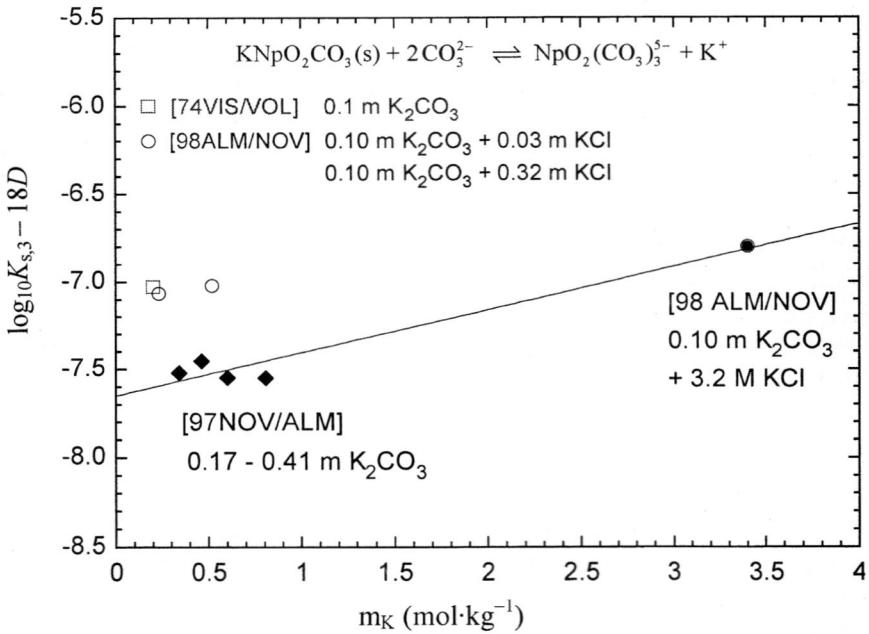

Figure A-8: Application of the SIT to Np(V) solubility data in K_2CO_3 solutions ≥ 0.17 mol·kg^{-1}, reaction: $K_3NpO_2(CO_3)_2(s) + CO_3^{2-} \rightleftharpoons NpO_2(CO_3)_3^{5-} + 3K^+$.

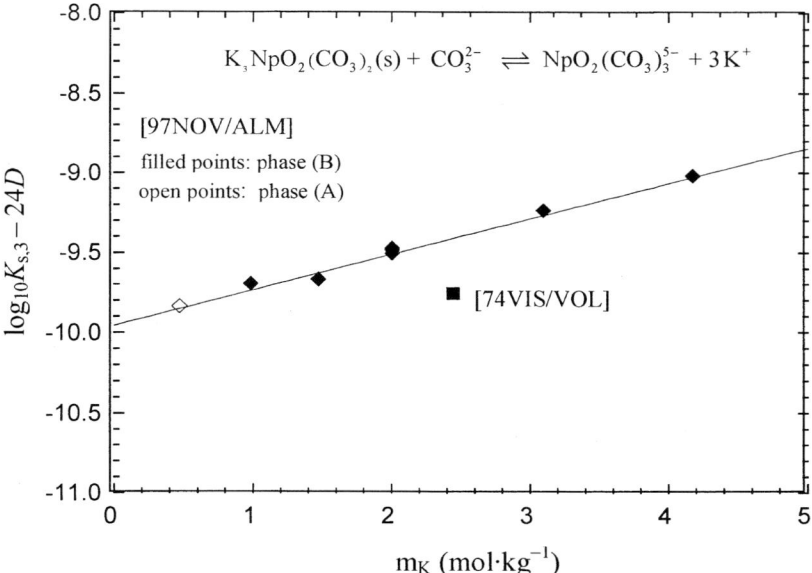

[98APE/KOR]

The vapour pressure of water over saturated aqueous solutions of dioxouranium(VI) nitrate hexahydrate (purity > 99 %) as a function of temperature (K) follows the relationship:

$$\ln(p/\text{kPa}) = (224.504 \pm 35.298) - (15151.1 \pm 1573.0) \cdot T^{-1} - (30.345 \pm 5.269) \cdot \ln T$$

in the range 278 to 323 K. The derived water activity and molar enthalpy of vaporisation of water at 298.15 K are (0.698 ± 0.020) and (50.75 ± 0.70) kJ · mol^{-1}, where the uncertainties are selected by this review. The water activity is lower than the value of 0.7325 selected for the CODATA Key Values [89COX/WAG] from the assessment by Goldberg [79GOL] of the literature data (principally from Robinson and Lim [51ROB/LIM] in this concentration range).

With the temperature variation of the solubility of dioxouranium(VI) nitrate hexahydrate from Broul et al. [81BRO/NYV] and the variation of osmotic coefficient with concentration from Robinson and Stokes [59ROB/STO], the authors finally derived the enthalpy of dissolution of dioxouranium(VI) nitrate hexahydrate in saturated solutions. Their calculated value at 298.15 K is: $\Delta_{sol}H_m^\circ$ (298.15 K, m_{sat} = 3.323 mol · kg^{-1}) = 43.4 kJ · mol^{-1}. The enthalpy of dilution to infinite dilution from m_{sat}, using data from [92GRE/FUG], is then -23.6 kJ · mol^{-1}, with unknown uncertainty.

The authors suggest there is a large variation in the reported solubilities of dioxouranium(VI) nitrate hexahydrate in water at 298.15 K, but this results from a misreading of the paper of Robinson and Lim [51ROB/LIM]. These authors did indeed make measurements on solutions up to 5.511 molal, but they make no statement that these are below the saturation limit, as implied by [98APE/KOR]. In fact, very considerable supersaturation of nitrate solutions can occur in isopiestic experiments of the type undertaken by Robinson and Lim [51ROB/LIM] (*cf.* for example the studies on Eu(NO$_3$)$_3$ solutions up to 6.3858 molal by Rard and Spedding [82RAR/SPE], although the saturation limit at 298.15 K is 4.2732 molal [84RAR]).

The saturation molality at 298.15 K used by the authors, 3.323 mol · kg^{-1}, is noticeably higher than that selected by [89COX/WAG], 3.24 mol · kg^{-1}. However, the latter is retained as the selected value, as detailed in the entry for [95COH/LOR] in Appendix A. The values of m_{sat} and the activity of water are of some importance since they form the basis of the derivation of S_m° (UO_2^{2+}) and by comparison, the entropies of other actinide ions.

[98BAL/HEA]

Cyclic and alternating current voltametry were employed with *in situ* spectrophotometric measurements made through an optically transparent electrode in the study of Bu$_4$N$^+$ salts of [Tc^{VI}NX$_4$] and [Tc^{V}OX$_4$] complexes, where X = Cl and Br, in anhydrous dichloromethane/Bu$_4$NBF$_6$ solutions. Half–wave potentials are listed (0.5 M Bu$_4$NBF$_6$ at 298.15 K) *versus* SCE (referenced to [Fe(C$_5$H$_5$)$_2$]$^{+/0}$): [TcOCl$_4$] +1.84 V (VI/V), -0.52 V (V/IV);

[TcOBr$_4$] + 1.73 V (VI/V), − 0.39 V (V/IV); [TcNCl$_4$] + 0.21 V (VI/V); [TcNBr$_4$] + 0.32 V (VI/V). Excess halide affected neither the $E_{\frac{1}{2}}$ values nor the visible reflectance signal indicating that these species do not change their five coordinate symmetry. The reversibility of the VI/V couple at 298.15 K allowed the spectral characteristics of the nitrido complexes to be tabulated. Reflectance spectra of the equivalent solid state complexes were also measured and tabulated. Geometries of the Tc(V) and Tc(VI) species were calculated assuming C$_{4v}$ symmetry and compared with experimentally derived values where available for the Tc(V) ions. Good correspondence was found. Reduction is manifested by a shortening of the Tc–X bond and a slight expansion of the N–Tc–X and O–Tc–X angles, whereas the Tc–N and Tc–O bonds are unaffected.

[98BAN/SAL]

This conference paper gives preliminary data on the thermal properties of Rb$_2$U$_4$O$_{11}$(cr) from 366 to 735 K. The final results and processing of the data to derive the heat capacity of this phase are described in the later full publication [2001BAN/PRA].

[98BAR/RUB]

U(VI) aqueous/acetone/HClO$_4$, CF$_3$SO$_3$H, HBF$_4$ solutions are prepared from UO$_3$·xH$_2$O(s) and investigated by proton NMR in the temperature range t = − 95 to 25°C. The use of a mixed acetone/water solvent is necessary in order to obtain separate peaks for coordinated and free solvent water (the rate of exchange is fast at room temperature). The peak integrals for bound and free water are used to determine the coordination number for water and the line broadening to determine the rate of exchange and the activation energy for the exchange reaction. NMR spectra obtained from solutions where the ratios M(VI):H$_2$O:acetone were varied all gave the coordination number five, within the experimental uncertainty of ± 5%. The authors used ^{35}Cl NMR to ascertain that no inner-sphere complexes with perchlorate were formed. Other counter ions such as CF$_3$SO$_3^-$ and BF$_4^-$ result in ion association and a lower number of coordinated water molecules. A similar study using Np(VI) gave concordant results, while the Pu(VI) system could not be studied due to extensive line broadening from the paramagnetic Pu(VI). However, there is no chemical reason to expect a different coordination number for Pu(VI). The experimental data have been obtained in a mixed solvent, which is unlikely to affect the stoichiometry and geometry of the MO$_2$(OH$_2$)$_5^{2+}$ complexes. However, the kinetic parameters deduced cannot directly be transferred to pure water. For Am, the number of coordinated water molecules is still unknown.

[98BEN/FAT]

The aim of this work is to establish the potential of the TcO$_4^-$/Tc(IV) couple in HCl/NaCl media (I = 1.00 to 1.66 M) under a N$_2$ atmosphere in a cell in which high Fe^{3+}/Fe^{2+} concentrations at variable ratios are maintained (*i.e.*, 0.001 − 0.12, *cf.* 10^{-11} to 10^{-4} M Tc). The measured potentials are reported relative to a saturated calomel electrode with a half-cell potential of 0.242 V (cf. 0.2412 V given in [99RAR/RAN]). $E°$ for the Fe(III)/Fe(II) couple is set at 0.770V ($E'° = 0.700$V in 1.0 M HCl). TcO_4^- is determined by extraction (TPACl in chloroform) under the assumptions that the extraction process (two minutes) is much faster than the oxidation of Tc(IV) (several hours) during extraction and only the Tc(VII) and Tc(IV) states exist in solution. The general redox reaction, in which several Tc(IV) species are considered, is defined as:

$$TcO_4^- + mH^+ + nCl^- + 3Fe^{2+} \rightleftharpoons TcH_{m+2p-8}O_pCl_n^{(m-n-4)+} + (4-p)H_2O(l) + 3Fe^{3+}$$

from which

$$E = E_i'^° + 0.0197 \cdot \log_{10}\frac{[TcO_4^-]}{[Tc(IV)aq]} + 0.0197\, m\, \log_{10}[H^+].$$

The authors report values of $E_i'^°$ in 1 M HCl (ionic strengths 1.31 – 1.66) that range from 0.602 to 0.844 V/NHE compared to calculated values for the individual groups of Tc(IV) containing species, viz., 0.584 V/NHE for Tc(IV) as $TcO(OH)_2(aq)$; 0.645 V/NHE for Tc(IV) in the form of mixed oxide/chloride complexes, $TcOCl_n^{2-n}$; and 0.676 V/NHE for Tc(IV) as chloro complexes. Hence the potential of the $TcO_4^-/Tc(IV)$ couple is highly dependent on the speciation of Tc(IV) in acidic chloride solutions, e.g., in 1 M HCl with an additional 0.3 M Cl$^-$, the formal potential of the $TcO_4^-/Tc(IV)$ couple is higher than (0.844 ± 0.006) V/NHE.

[98BUD/TAN]

Pu(V) behaviour in basic media is controversial. This paper gives data on the stability of Pu(V) in NaOH solutions (mainly at 40°C) prepared with $NH_4PuCO_3(cr)$. At 8 M NaOH and above, Pu(V) is stable for at least one week. Below 8 M NaOH, disproportionation of Pu(V) into Pu(IV) (of which a large amount precipitates as hydroxide) and Pu(VI) is achieved within five hours. Concentrations of the Pu species in each oxidation state can be measured, or estimated, to calculate the equilibrium constant, K_{dis}, for the reaction:

$$2\,PuO_2(OH)_4^{3-} + 2H_2O(l) \rightleftharpoons PuO_2(OH)_4^{2-} + Pu(OH)_5^- + 3\,OH^-$$

or

$$2\,PuO_2(OH)_4^{3-} + 2H_2O(l) \rightleftharpoons PuO_2(OH)_4^{2-} + Pu(OH)_4(aq) + 4\,OH^-$$

Values of K_{dis} are given in the paper. Pu(VI) and Pu(V) concentrations are measured by spectrophotometry, while [Pu(IV)] is estimated from the solubility of Pu(IV) as a function of the NaOH concentration [96PER/KRY] (see Appendix A). Variation of $\log_{10}K_{dis}$ is reported to be a linear function of sodium hydroxide concentration. Recombination of $Pu(OH)_4$ and Pu(VI) in 1 to 10 M NaOH gives an equilibrium constant K_{rep} that is very different from the one calculated from the reverse reaction. The system is apparently not in equilibrium and therefore, no reliable thermodynamic constant can be extracted.

[98CAP/VIT]

This paper gives a determination of $\log_{10} K^\circ_{s,0}$ for Pu(IV) following a path which is independent of the hydrolysis of Pu^{4+}. The authors use spectrophotometric and potentiometric measurements to determine indirectly the [Pu^{4+}] in equilibrium with Pu(IV) hydroxide. The solutions studied have the composition, 0.1 M HClO$_4$ + (x – 0.1) M NaClO$_4$, x = 0.1, 0.5, 1, 2 and 3, and contain PuO$_2^+$, PuO$_2^{2+}$, Pu^{3+} and Pu^{4+} with Pu(OH)$_4$(am) assumed as the solid phase. All free ion concentrations except for [Pu^{4+}] are directly measured by spectroscopy and [Pu^{4+}] is calculated by:

$$[\text{Pu}^{4+}] = \frac{[\text{Pu}^{3+}][\text{PuO}_2^{2+}]}{[\text{PuO}_2^+]} \cdot \exp\frac{F}{RT}(E'^o_{6/5}(I) - E'^o_{4/3}(I))$$

where $E'^o_{6/5}(I)$ and $E'^o_{4/3}(I)$ are the formal potentials of the couples at the ionic strength $I = x$. These potentials are known for the conditions of the experiments from earlier measurements [95CAP/VIT]. The solubility product is then calculated from the measured concentrations of Pu^{3+}, PuO$_2^+$, PuO$_2^{2+}$ and log$_{10}$[H$^+$]. The [Pu^{4+}] always remains lower than 10^{-4} M. The SIT is used in all calculations. The $\log_{10} K^\circ_{s,0}$ value obtained is $-(58.3 \pm 0.5)$ which agrees well with values obtained by solubility measurements. The authors have also determined a value for the interaction coefficient, $\varepsilon(\text{Pu}^{4+}, \text{ClO}_4^-) = (0.85 \pm 0.20)$ kg·mol^{-1} in good agreement with that given in a previous paper, $\varepsilon(\text{Pu}^{4+}, \text{ClO}_4^-) = (0.82 \pm 0.07)$ kg·mol^{-1} [95CAP/VIT]. To deduce these values the authors used an analogy of $\varepsilon(\text{Pu}^{3+}, \text{ClO}_4^-) = \varepsilon(\text{Nd}^{3+}, \text{ClO}_4^-) = (0.49 \pm 0.03)$ kg·mol^{-1}.

[98CAS/DIX]

(CH$_3$)$_4$NTcO$_2$F$_4$ and TcO$_2$F$_3$·CH$_3$CN were characterised by ^{19}F, ^{17}O, and ^{99}Tc NMR. The study of the first compound dissolved in CH$_3$CN established that the oxygen atoms are *cis* to one another. In HF the caesium salt showed rapid exchange of the *trans* fluoride ligands even at $-80°$C. From the ^{19}F NMR spectrum of TcO$_2$F$_3$·CH$_3$CN, it is assumed that the nitrogen bound the acetonotrile molecule lies *trans* to one oxygen, with the oxygen atoms again being in a *cis* configuration. The crystal structure of LiTcO$_2$F$_4$ was recorded and shown to exhibit a distorted octahedral symmetry consistent with the NMR assignment of ligand geometry. Raman spectra were also recorded for Li$^+$, Cs$^+$ and (CH$_3$)$_4$N$^+$·TcO$_2$F$_4^-$ and TcO$_2$F$_3$·CH$_3$CN in CH$_3$CN. Density functional theory calculations were used to predict that the *cis* dioxo isomers of TcO$_2$F$_4^-$ and TcO$_2$F$_3$·CH$_3$CN, as described above, are the minimum energy structures.

[98CAS/PAB]

The solubility of a 20–50 μm powder of UO$_2$(cr) (UO$_{2.01}$ in the bulk) is measured in 8·10^{-3} M NaClO$_4$ and 1 M NaCl solutions under more or less reducing conditions and over a large pH range from 1.0 to 9.5 with an uncertainty of ± 0.1. Reducing conditions are maintained using highly purified H$_2$ bubbling in the presence of Pd black, and monitored by pe values in the range from 3 to -3, ($t = 25°$C). The uranium concentrations, C$_U$, are measured from filtered solutions using a filter with a cut-off at 0.22 µm. In order to model the results of the $\log_{10}C_U$ versus pH data, all the U(IV) and U(VI) oxo-hydroxo and chloride complexes and the U(IV) and U(VI) oxides, UO_2, U_3O_7, U_4O_9, UO_2OH and UO_3 should be used. The selected formation constants and solubility products are those of [92GRE/FUG] using the SIT to get the values at a given ionic strength. The equilibrium constant, $\log_{10}K_{s,4}$, of the solubility reaction:

$$UO_2(cr) + 2H_2O(l) \rightleftharpoons U(OH)_4\,(aq)$$

is used as a fitting parameter. In perchlorate solutions the best fit is for $\log_{10}K_{s,4} = -7.3$. In sodium chloride medium this value also gives a good fit. The value selected by [92GRE/FUG] is $\log_{10}K°_{s,4} = -(9.5 \pm 1.0)$. To explain the difference the authors took into account the variation of the solubility with the molar surface area of the dioxide which is supposed to control the solubility. Taking the value of $\log_{10}\beta°_4$ selected by [92GRE/FUG] gives for:

$$UO_2(cr) + 4H^+ \rightleftharpoons U^{4+} + 2H_2O(l)$$

the solubility constant, $\log_{10}{}^*K°_{s,0} = -(2.3 \pm 0.2)$. If the value, $\log_{10}{}^*K°_{s,0}(UO_2, cr) = -(4.86 \pm 0.36)$, given by [92GRE/FUG] is corrected for a molar surface area of 9000 $m^2 \cdot mol^{-1}$, measured for a microcrystalline UO_2, a value of -2.7 is obtained, close to -2.3. It thus seems that the data selected by [92GRE/FUG] allow modelling of solubility data for amorphous phases if the molar surface area is known. This is interesting, however, such a modelling would still require information on the time dependence of the recrystallisation of the amorphous phases. This is outside the scope of the present review.

This review has revised the values selected by [92GRE/FUG] and has selected $\log_{10}K°_{s,4}(UO_2, am) = -(8.5 \pm 1.0)$ and $\log_{10}{}^*K°_{s,0}(UO_2, am) = (1.5 \pm 1.0)$.

Casas et al. also give data on the solubility of different uraninite minerals, $(U^{4+}, U^{6+}, M)O_{2+x}$, in a granitic groundwater for pH between 5.5 and 8.5, and under a H_2–CO_2 (1%) atmosphere, which are not considered in this review because no thermodynamic constants were derived.

[98CHA/DON]
This paper is reviewed together with [99CHA/DON].

[98CHA/TRI]
In order to select the best database to account for the measured concentrations of the actinides, U and Pu, in Drigg water, the authors compare three databases, CHEMVAL Version 6 Database [96FAL/REA], BNFL internal database and HATCHES NEA Version 9[1], the

---

[1] HATCHES NEA Version 9, Harwell/Nirex Thermodynamic Database for Chemical Equilibrium Studies, published to Nuclear Energy Agency, OECD as NEA Version 9, November 1996.

latter including the thermodynamic data selected in [92GRE/FUG] for U and in [95SIL/BID] for Am. They retain this database as the more appropriate because it includes all the ingredients of Drigg water.

Tests of U(VI) solubility (over-saturation and under-saturation) in Drigg water, pH = 8.2, are in agreement with modelling using either $\alpha - UO_2(OH)_2$ ($\log_{10} K_{s,0} = 4.94$) or $UO_3 \cdot 2H_2O(cr)$ ($\log_{10} K_{s,0} = 4.82$) as solubility limiting phases and all other selected equilibrium constants from [92GRE/FUG]. The experiments consisted of stepwise addition of a dioxouranium(VI) solution to Drigg water followed by addition of NaOH to bring the solution back to the initial pH value. The uranium concentration was measured after removal of colloids.

The authors mention that the NEA data do not fit the U(IV) solubility data given by [95YAJ/KAW] in contrast to the CHEMVAL database that gives a better fit.

It is difficult to draw any conclusions from these experiments. However, the results do not contradict the data selected in [92GRE/FUG].

The solubility of a $Pu(OH)_4(s)$ precipitate in cementitous water (pH = 9 to 12.5) is in agreement with modelling using $\log_{10} {}^*\beta_4^\circ = -10.54$ ($\log_{10} \beta_4^\circ = 45.46$, HATCHES NEA Version 9) and $\log_{10} K_{s,0}^\circ = -55.66$ giving $\log_{10}[Pu(OH)_4(aq)] = -(10.2 \pm 0.8)$ estimated from a figure). The $K_{s,0}^\circ$ value differs from the value selected by [2001LEM/FUG], $\log_{10} K_{s,0}^\circ = -(58 \pm 1)$, and other values given by [99NEC/KIM] and [2001NEC/KIM].

As there are few experimental data in the study of Chandratillake *et al.* and a lack of experimental details, this review does not take these Pu data into account.

[98CLA/CON]

The structure of $[Pu(CO_3)_5]^{6-}$ has been determined both in the solid state and in solution. The first study was performed by using single crystal X–ray diffraction of the solid $(Na_6Pu(CO_3)_5)_2Na_2CO_3 \cdot 33H_2O$. This is a precise structure determination from which accurate bond distances and angles could be deduced. The experimental EXAFS data extend to 12 Å^{-1} which allows the identification of three shells around Pu, corresponding to 10.2 Pu–O distances in the first coordination sphere, 5.0 Pu–C distances from the coordinated carbonate and 5.0 Pu–O distances from the non-coordinated carbonate oxygen, in excellent agreement with the X–ray structure. The electronic absorption spectrum of the solid agreed very well with the absorption spectrum in 2.5 M carbonate solution; a strong indication that the complexes are identical in both phases. The authors point out an interesting structural similarity between $[Pu(CO_3)_5]^{6-}$ and $[PuO_2(CO_3)_3]^{4-}$ where the former is obtained by replacing the "yl"– oxygen in the latter by a chelating carbonate groups. From this study both the stoichiometry and structure of the limiting carbonate complex of Pu(IV) are established definitively.

[98CON/ALM]

EXAFS fine structure spectra of aqueous ions Pu(III to VI) (Pu^{3+}, Pu^{4+}, PuO_2^{2+} in 1 M $HClO_4$ and PuO_2^+ at pH = 6) are given. The edge energy (L_{III}) variation is linear with the formal oxidation state of Pu. This paper gives literature references for U and Np EXAFS spectra.

[98DAI/BUR]

This paper reports the determination of the enthalpy of dimerisation of the dioxouranium(VI) cation according to the reaction:

$$2\,UO_2^{2+} + 2\,H_2O(l) \rightleftharpoons (UO_2)_2(OH)_2^{2+} + 2\,H^+$$

in a solution $8 \cdot 10^{-2}$ M in U(VI), 1 M CF_3SO_3H, the pH being adjusted with $(Bu)_4NOH$. The UV–Vis spectroscopic data (15 spectra in the range 390–470 nm) are processed using FA–SM methodology[1]. The pH–meter is equipped with a temperature compensation capability. As temperature increases (8 to 75°C) pH decreases slowly (3.20 to 2.95). Calibration of pH-meter is made at room temperature at pH = 3.16 (t around 25 °C). Two species are detected by FA–SM and confirmed by an isosbestic point (420 nm). The spectra of UO_2^{2+} and $(UO_2)_2(OH)_2^{2+}$ are identified according to known UV–Vis characteristics, according to [93MEI/KAT], [95ELI/BID], [96MEI/SCH]. K values are computed as a function of temperature. The given value is $\Delta_r H_m^\circ = 45.5$ kJ · mol^{-1}. All the activity coefficients are supposed to remain constant. The authors do not calculate or estimate the confidence level of the proposed enthalpy value which agrees with the [94RIZ/RAO] value and compares with older data [57HEA/WHI], [62BAE/MEY].

[98DAV/FOU]

EXAFS spectra of 1 M HCl solutions of M^{3+} = U, Np, Pu, Am, Cf (U(III) obtained by Zn reduction of U(IV)) show no inner-sphere complexes. The M–O bond distances are calculated within an uncertainty of (± 0.01) Å and the coordination numbers are given between 9 and 10. These data are used to discuss the covalent bonding in aqueous M^{3+} ions using a formula which gives ionic Gibbs hydration energies of gaseous M^{3+} cations, with $\Delta_{hyd} G_m$ on the absolute scale. This point is particularly relevant for updating the thermodynamic data for U^{3+}, which is difficult to stabilise in aqueous solutions. From a general point of view, thermodynamics of hydration of gaseous species are important to check the theoretical calculations on actinides, which are growing in the literature (see [99VAL/MAR] in this Appendix).

---

[1] FA–SM: Factor analysis and self modelling give the number of components contributing to a spectral response and permit the deconvolution of overlapped spectra to obtain a resolved spectrum of each individual spectrum.

Discussion of selected references

For monatomic ions with ionic character $\Delta_{hyd}G_m$ is supposed to depend on five quantities which are: the number of water molecules in the primary shell, N, and in the second hydration shell, H, of the aqueous ion; the M^{n+}, ion crystallographic radius, R_c; the effective charge, q, of the ion and the size of the water molecule, R_w. The experimental distance, d, between the ion and the closest oxygen atom, is $R_c(N) + R_w$. The N values could be non-integers and the distance, d, can be measured by EXAFS. These parameters are not all independent, for instance, R_c depends on N, R_w depends on R_c and so on. These relations are analysed by the authors in later papers (see below). An expression for $\Delta_{hyd}G_m$ (on the absolute scale) gives this review an indication that the hydration theory is:

$$\Delta_{hyd}G_m = aq^2 (R_c + 2R_w)^{-1} + bq N(R_c + R_w)^{-2} + cq N(R_c + R_w)^{-3}$$
$$+ dq^2 N(R_c + R_w)^{-4} + eH + w(R_c + R_w)^3 + \Delta G_{disp}$$

This expression shows that for a given temperature $\Delta_{hyd}G_m$ is a function of q, N, R_c, R_w, and H, as well as coefficients, a, b, c, *etc.*, and depends on seven different terms, successively, the Born term, the influences of dipole, quadrupole and induced dipole interactions, the influence of water molecules in the second hydration sphere, the cavity formation and the dispersion effect, $\Delta G(sub)disp.$, arising from dispersion forces. Most of these terms are discussed in many papers and text books.

The parameters a, b, c, *etc.*, have been evaluated using 27 experimental values of $\Delta_{hyd}G_m$ for halides, alkalis, alkaline-earths and trivalent lanthanides. The calculated values deviate from experimental values by less than 0.3%. Application to lanthanides and actinides needs to take into account the covalent effect, which is quantified by the decrease of the distance, d, compared to a pure ionic situation, and consequently the decrease of R_c and R_w as well. Change in the ground state energy of M^{3+} ion in the transition from gaseous to aqueous state (the so called nephelauxetic effect) must also be considered.

Discussion of the structure and thermodynamics of trivalent lanthanide and actinide aqueous ions is given in [85DAV/FOU], [86DAV] and [97DAV/FOU]. Another paper [2001DAV/VOK] deals with the last developments in calculating Gibbs energy of hydration for these elements. Data on actinides have been given in an abstract [2000DAV/FOU].

Calculated values, $\Delta_{hyd}G_m^\circ (M^{n+}, 298.15 \text{ K})$, given by David *et al.* for this review for the reaction :

$$M^{n+}(g) + H_2O(l) \rightleftharpoons M^{n+}(aq), \quad n = 3 \text{ and } 4$$

are given in Table A-34. They are derived from calculated values on the absolute scale on the basis of $\Delta_{hyd}G_m^\circ (H^+, 298.15 \text{ K}) = -1056 \text{ kJ} \cdot \text{mol}^{-1}$, according to:

$$\Delta_{hyd}G_m^\circ (\text{abs}) = -1056 \, q + \Delta_{hyd}G_m^\circ.$$

Table A-34: Enthalpy of the hydration reaction for trivalent and tetravalent ions, M^{n+}.

$M^{3+}(q)$	$\Delta_{hyd}G^\circ_m$ (kJ · mol^{-1})	$M^{4+}(q)$	$\Delta_{hyd}G^\circ_m$ (kJ · mol^{-1})
U (2.750)	236	Th (3.539)	−1124
Np (2.755)	191	U (3.869)	−1793
Pu (2.761)	147	Np (3.904)	−1940
Am (2.766)	102	Pu (3.985)	−2162
Cm (2.773)	72		

Derivation of $\Delta_{hyd}G_m$ versus T permits a definition of an entropy model which gives the absolute entropy of hydration, S_m (hyd), for a monatomic ion. Then adding the calculated entropy of the gaseous ion, the absolute entropy of the ion is derived. A proposed expression by the authors is:

$$S_m(\text{hyd}) = \sum_{i=1}^{i=6} M_i \cdot m_i$$

depending on six terms, M_i, each depending on parameters and coefficients included in $\Delta_{hyd}G_m$ (d, q, N, H, R_c, R_w, a, b, c, *etc.*) and five coefficients m_i to be fitted to experimental entropy data. These coefficients are the derivative of d, N, H, quadrupole moment and polarisability versus T. They have been fitted to the experimental data of S_m (hyd) for halides, alkalis, alkaline-earths and trivalent lanthanides.

The deviation between the 38 experimental values and those calculated is about some tens of $J \cdot K^{-1} \cdot mol^{-1}$.

The values obtained by David et al. for S°_m (M^{n+}, 298.15 K) are compared in Table A-35 with the values selected by TDB reviews and, for Th^{4+}, by [85BAR/PAR]. These are derived from the appropriate expression of absolute entropy, on the basis of S°_m (H$^+$, abs., 298.15 K) = − 22.2 J · K^{-1} · mol^{-1}.

Table A-35: Comparison between the calculated molar standard entropy (J · K^{-1} · mol^{-1}) and values selected by TDB review.

M^{n+}	S°_m calculated	S°_m selected	Reference
U^{3+}	−178	−(188.170 ± 13.853)	[92GRE/FUG]
Np^{3+}	−183	−(193.584 ± 20.253)	[2001LEM/FUG]
Pu^{3+}	−193	−(184.510 ± 6.154)	[2001LEM/FUG]
Am^{3+}	−201	−(201.00 ± 15.00)	[95SIL/BID]
Th^{4+}	−438	−423	[85BAR/PAR]
U^{4+}	−433	−(416.896 ± 12.553)	[92GRE/FUG]
Np^{4+}	−420	−(426.390 ± 12.386)	[2001LEM/FUG]
Pu^{4+}	−409	−(414.535 ± 10.192)	[2001LEM/FUG]
Am^{4+}		−(406.00 ± 21.00)	[95SIL/BID]

[98DIA/GRA]

In this paper, solubility experiments are described in which precipitation after NaOH addition is monitored under pH-stat conditions from supersaturated U(VI) solutions, where pH is calibrated on the molality scale. In regard to this review, it shows clearly that starting with a 10^{-2} m solution of U(VI) (UO_2Cl_2 in 0.5 m $NaClO_4$), crystalline metaschoepite plates about 1 mm in diameter, $UO_3 \cdot 2H_2O$, precipitate from $NaClO_4$ solutions ($I = 0.5$ m) when the pH is increased and maintained between 4.7 to 6.3. The equilibrium concentration of U(VI) and hydrolysis constants taken from [92GRE/FUG] using the SIT procedure give $\log_{10}{}^*K_{s,0} = 5.14$ at 25°C for the equilibrium:

$$UO_3 \cdot 2H_2O(\text{scho}) + 2H^+ \rightleftharpoons UO_2^{2+} + 3H_2O(l).$$

To derive the value at zero ionic strength according to :

$$\log_{10}{}^*K_{s,0}^\circ = \log_{10}{}^*K_{s,0} + 3\log_{10}\gamma_{\pm UO_2(ClO_4)_2} - 4\log_{10}\gamma_{\pm HClO_4} + 3\log_{10}a_w$$

the following data are used : $\gamma_{\pm UO_2(ClO_4)_2} = 0.522$, $\gamma_{\pm HClO_4} = 0.726$ and $a_w = 0.996$ to get $\log_{10}{}^*K_{s,0}^\circ = 4.7$. As no numerical experimental results are given in this paper, uncertainties in this value could not be determined. Using the SIT this review calculates $\log_{10}{}^*K_{s,0}^\circ = 4.86$ which gives $\log_{10}K_{s,0}^\circ = -23.14$.

Contrary to the results in 0.5 m $NaClO_4$ solution, polyuranates precipitate, in NaCl solutions ($I = 3$ and 5 m), according to:

$$\frac{1}{m}(UO_2)_m(OH)_n^{(2m-n)} + (y+1-\frac{n}{m}+\frac{x}{2})H_2O(l) + x\,Na^+ \rightleftharpoons$$

$$Na_xUO_{(3+x/2)} \cdot yH_2O + (2+x-\frac{n}{m})H^+$$

the Na/U ratio depending on the Na^+/H^+ ratio in solution.

From 3 m NaCl solutions at pH 5.2 to 7.2 precipitates with the average composition of $Na_{0.3}UO_{3.15} \cdot yH_2O$ ($y = 1.56$ to 5.5) were formed. The water content increased with pH and the particles were finer at pH 7.2. Precipitates formed from 5 m NaCl in the pH range 4.8 to 6.5 were similar to those obtained from 3 m NaCl, but were finer (0.2 μm in diameter). At higher pH, the same phase was formed initially, but lost crystallinity with time and had a higher sodium content and a slightly elevated uranium level. Three orders of magnitude variations in final uranium concentration were measured at a given pH and can be accounted for by variations in the size of the precipitates; stepwise addition of base led to the largest particle sizes and the lowest solubilities. An orange precipitate was formed at pH higher than 7.6 and exhibited a regular trend of decreasing solubility with increasing pH.

The phase, $Na_{0.33}UO_{3.16} \cdot 2\,H_2O$ ($Na_2U_6O_{19} \cdot 12\,H_2O$), was formed at $10^4 < Na^+/H^+ < 10^6$. To calculate the solubility product using only data in the pH range, 5.2 to 6.0, the hydrolysis constant of UO_2^{2+} in 3 m NaCl from [63DUN/SIL] was taken, and gave $\log_{10} {}^*K_{s,0} = (7.95 \pm 0.15)$ at 25°C. Extrapolation to zero ionic strength of the constant referring to:

$$Na_{0.33}UO_{3.165} \cdot 2H_2O + 2.33H^+ \rightleftharpoons UO_2^{2+} + 3.165H_2O(l) + 0.33Na^+$$

is done according to (Pitzer approach):

$$\log_{10} {}^*K_{s,0}^\circ = \log_{10} {}^*K_{s,0} + 3\log_{10}\gamma_{\pm UO_2Cl_2} + 0.66\log_{10}\gamma_{\pm NaCl} - 4.66\log_{10}\gamma_{\pm HCl} + 3.165\log_{10}a_W$$

with $\gamma_{\pm UO_2Cl_2} = 0.619$, $\gamma_{\pm HCl} = 0.974$, $\gamma_{\pm NaCl} = 0.714$ and $a_W = 0.893$. The result is $\log_{10} {}^*K_{s,0}^\circ = (7.13 \pm 0.15)$ (presumably misprinted in the paper as (13 ± 0.15)). Application of the SIT by the review yields (6.52 ± 0.15). The uncertainty in $\log_{10} {}^*K_{s,0}^\circ$ of 0.15 incorporates variations in the solid phases with pH by taking the mean value for each pH and the corresponding standard deviation in the mean.

Precipitation from NaCl solution results in the formation of schoepite after one year when the ratio, Na^+/H^+, is less than 10^3. The equilibrium value of the U concentration in equilibrium over schoepite in 3 m NaCl led the authors to give $\log_{10} {}^*K_{s,0} = 5.43$ at 25°C and $\log_{10} {}^*K_{s,0}^\circ = 4.7$. When the Na^+/H^+ ratio reaches 10^{12}, the Na/U ratio in the precipitate reaches 1 which could correspond to the stoichiometric phase, $Na_2U_2O_7$. The value of $\log_{10} {}^*K_{s,0}^\circ$ is estimated by the authors to be 26, which is higher than the value given by [92GRE/FUG].

The authors made a review of the solubility of schoepite including [91SAN], [93MEI/KIM], [92KRA/BIS], [94TOR/CAS] and other references before 1992. They did a comparison of all the $\log_{10} K_{s,0}^\circ$ values (without taking uncertainties into account) and concluded that an average value of the solubility product of schoepite (excluding their data) is $\log_{10} K_{s,0}^\circ = (5.37 \pm 0.25)$ (uncertainty estimated). They pointed out that with amorphous or microcrystalline material, higher solubilities of schoepite can be achieved than that corresponding to this solubility product. On the other hand, lower U concentrations can be encountered, approaching that given by the solubility constant of $\log_{10} K_{s,0}^\circ = 4.83$, derived from calorimetric measurements [92GRE/FUG]. A review of the solubility of Na-polyuranates is presented in this paper

This review considers only the data obtained for schoepite in perchlorate solutions converted to $I = 0$ using the SIT theory.

[98EFU/RUN]

This paper is not cited in [2001LEM/FUG].

In this study the solubilities of Np and Pu were investigated at 25, 60 and 90°C in a Yucca Mountain groundwater of low ionic strength ($I \approx 3.7 \cdot 10^{-3}$ mol·L^{-1}) and a total carbonate concentration of $2.8 \cdot 10^{-3}$ mol·L^{-1}. The solutions were spiked with small amounts of Np und Pu stock solutions and the pH was adjusted to 6, 7, and 8.5.

The solubility of Np was determined from both oversaturation (after 50–450 days) and undersaturation (after 25–160 days with the precipitates formed in the oversaturation experiments). The diffuse and somewhat ambiguous X-ray powder diffraction patterns of the dark greenish brown precipitates were interpreted with the formation of a poorly crystalline hydrated $Np_2O_5 \cdot xH_2O(s)$. The solubility data at pH = 6 were almost independent of temperature and those at pH = 7 and 8.5 decreased about half an order of magnitude when the temperature was increased from 25 to 90°C. Using the solubility data at pH = 6 and 7 and $t = 25$°C, where the formation of hydroxide and carbonate complexes could be excluded, the solubility constant for the reaction:

$$\frac{1}{2}Np_2O_5(s) + H^+ \rightleftharpoons NpO_2^+ + \frac{1}{2}H_2O(l)$$

was calculated to be $\log_{10} {}^*K^\circ_{s,0} = (2.6 \pm 0.4)$ for poorly crystalline, possibly hydrated $Np_2O_5(s)$. This solubility constant is 1.3 orders of magnitude lower than the value of $\log_{10} {}^*K^\circ_{s,0} = (3.90 \pm 0.02)$ from the solubility study of Pan and Campbell [95PAN/CAM], [98PAN/CAM], who claimed crystalline $Np_2O_5(cr)$ to be the equilibrium solid phase. The solubility constant of Efurd et al. is compatible with the data selected in the NEA review [2001LEM/FUG] from calorimetric data and the estimated entropy of crystalline $Np_2O_5(cr)$:

$$\Delta_f G^\circ_m (Np_2O_5, cr, 298.15 \text{ K}) = -(2031.6 \pm 11.2) \text{ kJ·mol}^{-1}$$

$$\log_{10} {}^*K^\circ_{s,0} = (1.85 \pm 1.0).$$

The solubility of plutonium was studied only from oversaturation. The diffuse and broad X-ray diffraction peaks of the precipitate at 25°C and the somewhat sharper peaks for the precipitate at 95°C were ascribed to an amorphous $PuO_2(s)$, which possibly includes an aged Pu(IV) polymer or $Pu(OH)_4(am)$. The Pu concentrations measured after 50 – 400 days decreased somewhat with increasing temperature: $\log_{10}[Pu] = -(7.7 \pm 0.4)$ at 25°C, $-(8.1 \pm 0.2)$ at 60°C and $-(8.4 \pm 0.2)$ at 90°C. Small Pu(IV) colloids (< 4 nm) were supposed to be the predominant species in solution. This might explain the higher Pu concentrations compared to those of $\log_{10}[Pu] = -(10.4 \pm 0.5)$ measured at room temperature by Rai et al. [99RAI/HES2] with $PuO_2(am)$ in dilute KOH solutions of pH = 8 – 13. In addition, when dissolving parts of the aged Pu(IV) solid in 3 M HCl, the authors detected impurities of Pu(VI) that were assumed to have originated from the solid.

[98FAN/KIM]

This is a review of earlier work and preliminary results published later by Fanghänel et al. [94FAN/KIM], [95FAN/KIM], [98FAN/WEG2], [98FAN/WEG], and [99FAN/KON], Könnecke et al. [97KON/FAN], and Paviet et al. [96PAV/FAN]. It contains a detailed description of experimental methodology and a useful table of Pitzer $\beta^{(0)}$ and $\beta^{(1)}$ parameters for Cm^{3+} and the various Cm(III) complexes studied. A discussion of the number of coordinated water molecules in the sulphate complexes gives strong evidence for bidentate sulphate coordination and coordination number of nine around the central ion. A comparison between the number of coordinated water molecules in the corresponding carbonate system ([98FAN/WEG], page 52) reveals fairly large differences, 1.6 coordinated water in $Cm(CO_3)_3^{3-}$, versus 2.4 in $Cm(SO_4)_3^{3-}$. These variations may indicate the accuracy of the semi-empirical method used to estimate the water coordination from the lifetimes of the various complexes.

[98FAN/WEG]

The equilibrium constant and stoichiometry of $CmHCO_3^{2+}$ have been determined using time resolved laser fluorescence spectroscopy (at room temperature, close to 25°C) in a 1 m NaCl ionic medium. The experimental methods are satisfactory and the authors identify the formation of one weak bicarbonate complex with $\log_{10} \beta = 1.9$. This assignment is supported by an independent potentiometric identification of this species for La(III), [81CIA/FER] and Y(III) by Spahiu, [85SPA]. The study by Bidoglio [82BID], claims the formation of strong bicarbonate complexes with Am^{3+}, however, this is the result of an erroneous interpretation of the experimental data, cf. [95SIL/BID]. The carbonate, bicarbonate concentrations and pH are calculated using the EQ3/6 code and Pitzer parameters and data from Harvie et al. [84HAR/MOL]; no details are given, but the equilibrium constant is accepted as a good estimate. It should be noted that this species is only formed at very high partial pressures of CO_2, several bars, or more.

The lifetime of the excited level of $CmHCO_3^{2+}$ is somewhat shorter than that for Cm^{3+}, indicating the formation of an inner-sphere complex with bicarbonate, presumably involving coordination of both carboxylate oxygen atoms. The equilibrium constant has been recalculated to zero ionic strength by the reviewers using $\epsilon(Cm^{3+}, Cl^-) = 0.23$, $\epsilon(CmHCO_3^{2+}, Cl^-) = 0.16$ and $\epsilon(HCO_3^-, Na^+) = 0.0$ kg·mol^{-1} to give $\log_{10} \beta° = (3.1 \pm 0.3)$.

[98FAN/WEG2]

The formation constants for $Cm(CO_3)_n^{3-2n}$, $n = 1-4$ have been determined at 25°C in 1 m NaCl, using time resolved laser fluorescence spectroscopy. Electrode calibration and measurements of "pH" $= -\log_{10} m_{H^+}$ are described in detail. The slope of the pH vs. emf line from the glass electrode is consistent with the theoretical value. The possible formation of ternary hydroxide carbonate complexes for trivalent actinides has been discussed by several investigators, [82BID], [84BER/KIM] and [95SIL/BID]. However, the spectroscopic evidence in the present study does not indicate the formation of such complexes. Fanghänel et al. also discuss the formation of bicarbonate complexes, but more details are given in [98FAN/WEG]. There is a discussion of the mode of coordination of the carbonate ion and the present review agrees with the conclusion that bidendate complexes are formed. However, the discussion on the number of coordinated water molecules, based on the empirical relationship (1) on page 50 in [98FAN/WEG2] is inconsistent with a coordination number of nine, indicating that equation (1) is only an approximation. The stepwise equilibrium constants reported are: $\log_{10} \beta_1 = (5.90 \pm 0.1)$, $\log_{10} K_2 = (4.37 \pm 0.2)$, $\log_{10} K_3 = (2.91 \pm 0.15)$, and $\log_{10} K_4 = (1.0 \pm 0.2)$. The stoichiometry of the limiting complex is in agreement with experimental data on trivalent lanthanides, as referenced in [98FAN/WEG2].

A thermodynamic analysis of the Cm(III) carbonate system in the concentration range $0 < m_{NaCl} < 6$ is given in a later paper [99FAN/KON].

[98FAZ/YAM]

This paper gives some spectroscopic data on UO_2^{2+} and fluoro complexes in solution: 1 M $NaClO_4$, $[U] = 10^{-2}$ M, $[NaF] = 0$ to $6 \cdot 10^{-2}$ M, initial pH = 2, temperature range $t = 17$ to 38°C. These results add to what is already known and do not contribute to the thermodynamic database. The lifetime of the excited level of UO_2^{2+} is found to be 1.83 μs at 25°C. Speciation is made according to the data of [93FER/SAL].

[98FIN/HAW]

The name schoepite is commonly applied to any mineral or synthetic preparation with a formula close to $UO_3 \cdot 2H_2O$, but a distinction should be made between schoepite $(UO_2)_8O_2(OH)_{12} \cdot 12 H_2O$ (formally $UO_3 \cdot 2.25 H_2O$), metaschoepite $(UO_2)_8O_2(OH)_{12} \cdot 10 H_2O$ (formally $UO_3 \cdot 2 H_2O$) and dehydrated schoepite, $(UO_2)_8O_{0.25}(OH)_{1.5}$ (formally $UO_3 \cdot 0.75 H_2O$), which are structurally and chemically distinct from the four dioxouranium(VI) hydroxides: α-$UO_2(OH)_2$, β-$UO_2(OH)_2$, γ-$UO_2(OH)_2$ (formally $UO_3 \cdot H_2O$) and $U_3O_8(OH)_2$. In fact dehydrated schoepite is a solid solution represented by $UO_2O_{0.25-x}(OH)_{1.5+2x}$ ($0 < x < 0.25$) (formally $UO_3 \cdot 0.75 H_2O$ to $UO_3 \cdot H_2O$). Schoepite and metaschoepite are very difficult to distinguish between on the basis of X-ray powder diffraction data alone: their diffraction patterns are very similar. Relationships between hydrated oxides and hydroxides of U(VI) are addressed in [92FIN/MIL] and [97GUR/SER] (see this review) in addition to the papers reviewed in [92GRE/FUG]. Many papers deal with the this issue and are well summarised in [98FIN/HAW]. This paper is a detailed study of the chemical and structural transformations of schoepite to metaschoepite and finally to dehydrated schoepite with connection to the structure of α-$UO_2(OH)_2$. It complements [92FIN/MIL] and confirms previous data.

The structure of schoepite was firmly established by Finch *et al.* [96FIN/COO] as sheets of pentagonal bipyramidal polyhedra of the form $[(UO_2)_xO_y(OH)_z]^{(2x-2y-z)}$ (general structure of dioxouranium(VI) oxide hydrates) with x = 8, y = 2 and z = 12, bonded together by 12 water molecules through hydrogen bonds. Using single-crystal X-ray diffraction techniques, Finch *et al.* show that schoepite transforms slowly in air at ambient temperature to metaschoepite losing two of the interlayer H_2O groups which are more weakly bonded than the remaining ten. The orthorhombic cell structure change only slightly (about 2% for the a parameter), which is why only a precise determination of unit-cell parameters can distinguish between schoepite and metaschoepite. Above 120°C, schoepite transforms to dehydrated schoepite by losing all the interlayer H_2O molecules $((UO_2)_8O_2(OH)_{12})$ resulting in the complete collapse of the layer structure giving the structure of α-$UO_2(OH)_2 \cdot (UO_2O_{0.25}(OH)_{1.5})$ according to:

$$(UO_2)_8O_2(OH)_{12} = 8\ UO_2O_{0.25}(OH)_{1.5}$$

but with anion vacancies. Finally, dehydrated schoepite, isostructural with α-$UO_2(OH)_2$, is formulated as $UO_2O_{0.25-x}(OH)_{1.5+2x}$ ($0 < x < 0.25$). It is known to be stable in air and does not react with water except under hydrothermal conditions. This paper gives a summary of the thermal stability in presence of water of all the phases identified here (less than 100°C) and of the three polymorphs of dioxouranium(VI) hydroxide, which do not contain H_2O bonded molecules (above 100°C). Table A-36 gives the parameters of the unit cells of some of these phases. According to Sowder et al. [99SOW/CLA] the dehydration of metaschoepite to dehydrated schoepite has an intermediate step involving $(UO_2)_8O_2(OH)_{12} \cdot 1.5\ H_2O$. The paper [98FIN/HAW] refutes the existence of a paraschoepite phase which has been postulated previously. The synthetic $UO_3 \cdot 2H_2O$ commonly prepared by heating in water above 50°C, is metaschoepite.

Table A-36: Unit cell parameters (Å) for schoepite, metaschoepite and dehydrated schoepite crystals according to [98FIN/HAW] and for α-$UO_2(OH)_2$ crystals according to [97GUR/SER].

	a(Å)	b(Å)	c(Å)	Space group
schoepite	14.337	16.813	14.731	$P2_1ca$
metaschoepite	13.99	16.72	14.73	Pbna
Dehydrated schoepite	6.86	4.26	10.20	Abcm (?)
α-$UO_2(OH)_2$, average of 2 values	6.898	4.225	10.204	

[98GEI/BER]

The authors have determined the formation constant for $UO_2(CO_3)_3^{4-}$ using photoacoustic spectroscopy. In this way the authors have been able to work at low total concentrations of U(VI), $5 \cdot 10^{-4}$ to 10^{-5} M. The experiments were conducted in the pH range 5 to 10 in 0.1 M $NaClO_4$. The main emphasis of this paper is a description of the experimental equipment and methodology. The authors have determined the equilibrium constant for the reaction:

$$UO_2(OH)_3^- + 3\ HCO_3^- \rightleftharpoons UO_2(CO_3)_3^{4-} + 3\ H_2O(l)$$

and report a value of $\log_{10} K = (8.5 \pm 0.7)$. Extrapolation to zero ionic strength using the Davies equation gives $\log_{10} K° = (9.8 \pm 0.7)$. The authors seem to have made a sign error, since extrapolation to zero ionic strength using the SIT and the interactions coefficients from [92GRE/FUG] gives $\log_{10} K° = (7.2 \pm 0.7)$. Using the equilibrium constant for the formation of $UO_2(OH)_3^-$ selected in the present review, $\log_{10} {^*\beta_{3,1}^°} = -(20.25 \pm 0.42)$, we obtain $\log_{10} \beta_3^° = (17.9 \pm 0.8)$ for the formation of $UO_2(CO_3)_3^{4-}$ from the components, which differs considerably from the selected value of (21.6 ± 0.05) in [92GRE/FUG] and (21.84 ± 0.04) in the present review. However, the uncertainty in the proposed constant is large and there are some experimental information missing, e.g. whether the measured pH refers to activities or concentrations. The most important information is that the pH range where $UO_2(OH)_3^-$ is predominant extends to much lower pH than indicated in [92GRE/FUG]. This in turn implies a larger range of existence of $UO_2(OH)_2(aq)$. Since this first review was published, the equilibrium constant for the formation of $UO_2(OH)_2(aq)$ has been reassessed [91CHO/MAT], [2002BRO] and the present review has selected $\log_{10} {^*\beta_2^°} = -(12.15 \pm 0.07)$, i.e. a value much lower than that used to calculate the speciation diagram given on page 116 of [92GRE/FUG]. This equilibrium constant is in much better agreement with the results of Geipel et al. because it indicates that $UO_2(OH)_3^-$ is the predominant hydroxide complex at pH > 6.7. In conclusion, the study of Geipel et al. gives strong support to the speciation model proposed already in [92GRE/FUG], and also to the re-evaluation of the equilibrium constant of $UO_2(OH)_2(aq)$. However, the numerical value is too uncertain to be included when selecting the "best" value of $\log_{10} \beta_3^°$ for $UO_2(CO_3)_3^{4-}$. Preliminary results are given in [98GEI/BER2].

[98GEI/BER2]

This is a short version of the [98GEI/BER] paper and does not contain any new information. LIPAS spectroscopic data are used to study the formation of $UO_2(CO_3)_3^{4-}$ in solutions at low total concentrations of uranium ($[U] = 5 \cdot 10^{-5}$ to 10^{-4} M) and at the partial pressure of CO_2 in air. The pH region explored begins at a pH greater than 7. Under these conditions it is expected that the LIPAS spectrum belongs to the species, $UO_2(CO_3)_x^{(2x-2)-}$. This spectrum shows six maxima in the range 380 to 480 nm. The concentration of the unknown species was measured from the absorptivity and the free concentration of UO_2^{2+} was then obtained from the uranium mass balance. A plot of,

$$\log \frac{[UO_2(CO_3)_x^{(2x-2)-}]}{[Utot]-[UO_2(CO_3)_x^{(2x-2)-}]} \text{ vs. } \log_{10}[HCO_3^-],$$

gives a straight line with a slope of $x = 2.9 \pm 0.3$, $(8 \cdot 10^{-4} < [HCO_3^-] < 5 \cdot 10^{-3}$ M).

The calculation of $\log_{10} \beta_3^°$ ($UO_2(CO_3)_3^{4-}$) gives (21.57 ± 0.70), which is very close to (21.60 ± 0.05) selected by [92GRE/FUG]. This value, $\log_{10} \beta_3^°$ ($UO_2(CO_3)_3^{4-}$) = (21.57 ± 0.70), is also reported in [98GEI/BER], but seems to be uncertain, as explained in this review.

[98GEI/BER3]

This is an extended abstract describing studies of complex formation in the U(VI)–carbonate-hydroxide system using laser spectroscopic methods. The authors report a value of $\log_{10} K = -(18.9 \pm 1.0)$ for the reaction:

$$2\,UO_2^{2+} + CO_2(g) + 4\,H_2O(l) \rightleftharpoons (UO_2)_2CO_3(OH)_3^- + 5\,H^+.$$

As this work is only published as an abstract the proposed equilibrium constant is not used in this update.

[98GEI/BER4]

The conditions of formation of the complex, $Ca_2[UO_2(CO_3)_3]$, [96BER/GEI], [97BER/GEI] are extended to solutions $I = 0.1$ M (NaClO$_4$), pH = 8 and 9, [U] = $5 \cdot 10^{-5}$ M, $HCO_3^-/CO_3^{2-} = 10^{-2}$ M (to ensure that the complex, $UO_2(CO_3)_3^{4-}$, is formed) and Ca^{2+}, 10^{-2} and 10^{-3} M (to ensure that the complex $Ca_2[UO_2(CO_3)_3]$ is formed). Na$_2$EDTA is added to these solutions to control the concentration of free Ca^{2+} through the reaction:

$$Ca^{2+} + Na_2EDTA \rightleftharpoons CaEDTA + 2\,Na^+, \qquad \log_{10}\beta° = 10.59.$$

The other equilibria considered are:

$$Ca^{2+} + CO_3^{2-} \rightleftharpoons CaCO_3(aq), \qquad \log_{10}\beta° = 3.1$$

$$UO_2^{2+} + 3\,CO_3^{2-} \rightleftharpoons UO_2(CO_3)_3^{4-}, \qquad \log_{10}\beta_3° = 21.6$$

$$b\,Ca^{2+} + UO_2(CO_3)_3^{4-} \rightleftharpoons Ca_b UO_2(CO_3)_3^{(4-2b)-}(aq) \qquad K$$

Increasing the concentration of EDTA reduces $[Ca^{2+}]$ and the intensity of TRLFS spectrum of $Ca_b UO_2(CO_3)_3^{(4-2b)-}$ decreases. As in [97BER/GEI], the variation of

$$\log_{10} R = \frac{[Ca_b UO_2(CO_3)_3^{(4-2b)-}(aq)]}{[UO_2(CO_3)_3^{4-}]} \quad vs.\ [Ca^{2+}], \text{ according to :}$$

$$\log_{10} R = b \log_{10}[Ca^{2+}] + \log_{10} K$$

gives

$$b = (1.66 \pm 0.35) \text{ and } \log_{10} K = (7.55 \pm 0.23).$$

Recalculation of $\log_{10}\beta_{2,1,3}$ for

$$2\,Ca^{2+} + UO_2^{2+} + 3\,CO_3^{2-} \rightleftharpoons Ca_2[UO_2(CO_3)_3](aq)$$

gives

$\log_{10}\beta_{2,1,3} = (29.15 \pm 0.30)$ and $\log_{10}\beta°_{2,1,3} = (25.7 \pm 0.7)$ already given in [97BER/GEI].

This paper confirms the result of [97BER/GEI].

[98GOR/SID]

Earlier experimental work [74DHA/TRI], [87ALE/OGD] has shown that the vaporisation of uranium-bearing species from uranium oxides increases in the presence of a water vapour, owing to the formation of $UO_2(OH)_2(g)$ in addition to $UO_3(g)$, with other possible molecules in the gas phase being $UO_2OH(g)$ and $UOOH(g)$.

Gorokhov and Sidorova have calculated the thermal functions of $UO_2(OH)_2(g)$ from estimated molecular parameters. $UO_2(OH)_2(g)$ is assumed to have a C_{2v} symmetry with 15 vibrational frequencies, v_i. Bond lengths, angles and v_i values are derived from the isoelectronic molecule $UO_2F_2(g)$ and other molecules with C_{2v} symmetry, MO_2F_2 (M = Cr, Mo, W). Since the molecule contains hexavalent uranium, there were assumed to be no low-lying excited electronic states. The classical thermodynamic functions are calculated in the approximation of the rigid rotator harmonic oscillator.

These data are used to compare the two experimental studies [74DHA/TRI], [87ALE/OGD] of the reaction: $UO_3(g) + H_2O(g) \rightleftharpoons UO_2(OH)_2(g)$, which give rather similar Gibbs energies, but very different entropies of reaction. The calculated entropy of this reaction at 1500 K is $- 132.1 \: J \cdot K^{-1} \cdot mol^{-1}$, compared with the experimental value of $- 93.0 \: J \cdot K^{-1} \cdot mol^{-1}$ from [74DHA/TRI], whereas at 1800 K, the calculated and experimental [87ALE/OGD] values are $- 129.4$ and $- 156.9 \: J \cdot K^{-1} \cdot mol^{-1}$. The (marginally) better agreement in the second case leads the authors to prefer these experimental data from Alexander and Ogden [87ALE/OGD] and they derive finally $\Delta_f H_m^\circ$ ($UO_2(OH)_2$, g, 298.15 K) = $- (1345.5 \pm 30.0) \: kJ \cdot mol^{-1}$.

It should be noted that there are no experimental details of any kind given in the paper by [87ALE/OGD] – see review in this Appendix A.

Ebbinghaus [95EBB] has estimated appreciably different thermal functions for $UO_2(OH)_2(g)$, and as noted in the review of this paper in Appendix A and in the discussion in section 9.3.1.2, this review does not select any data for this species.

[98HUA/YAM]

The authors report measurements on the vapour phase composition in the decomposition of $Cs_2UO_4(s)$ from 873 to 1373 K, in Pt cells both *in vacuo* and in a D_2/D_2O environment. In agreement with earlier work, $Cs(g)$ is the major decomposition product in the vapour, to give $Cs_2U_4O_{12}(cr)$ by the reaction:

$$4 \: Cs_2UO_4(cr) \rightleftharpoons Cs_2U_4O_{12}(cr) + 6 \: Cs(g) + 2 \: O_2(g).$$

Unlike earlier workers, they also found X–ray diffraction evidence of $UO_2(cr)$ in the residue, which was ascribed to the reaction:

$$Cs_2U_4O_{12}(cr) \rightleftharpoons 4 \: UO_2(cr) + 2 \: Cs(g) + 2 \: O_2(g).$$

No mention is made of the many other mixed valency uranates which could be formed.

Because the oxygen pressure in the Knudsen cell is not well defined, no reliable thermodynamic data can be derived from the measurements of the Cs(g) pressures.

In the presence of D_2/D_2O, the pressure of Cs(g) increases, presumably due to a decrease in the oxygen pressure. CsOD(g) is also observed, and the residue contains $Cs_2U_2O_7$, attributable to the reaction:

$$2\ Cs_2UO_4(cr) + D_2O(g) \rightleftharpoons Cs_2U_2O_7(cr) + 2\ CsOD(g).$$

Enthalpy changes for the above reaction at 1300 K from the mass spectrometric data for three runs with differing D_2/D_2O ratios were (364.5 ± 13.5), (444.2 ± 90.5) and (308.5 ± 14.2) kJ·mol^{-1}. The mean of these quite discrepant values, (372.4 ± 47.9) kJ·mol^{-1}, is close to that calculated by the authors from literature data, i.e., 366.2 kJ·mol^{-1}.

No data relevant to this review can be abstracted from this paper.

[98ITO/YAM]

This paper gives pressure composition isotherms for the hydrogenation of $U_6Mn(s)$ and $U_6Ni(s)$ at 573 and 673 K and compares them to those for U–H. Ternary hydrides of approximate composition, $U_6MnH_{18}(s)$ and $U_6NiH_{14}(s)$, are probably formed and the amount of hydrogen in the alloys in equilibrium with these is much higher than for unalloyed uranium.

The data for the U–H system are indistinguishable from the existing data, which are therefore retained.

[98JAY/IYE]

Drop calorimetric measurements are reported for $KUO_3(cr)$ (370 – 714 K) and $K_2U_2O_7(cr)$ (391 – 683 K). Preparative details (and the same data) are reported in [99JAY/IYE].

There are some inconsistencies in both papers, and the enthalpy data have been refitted with an exact constraint of $H_m(T) - H_m^\circ(298.15\ K) = 0$ at 298.15 K. The derived heat capacity expressions:

$$C_{p,m}(KUO_3,\ cr,\ T) = 133.2577 + 1.2558 \cdot 10^{-2} T \quad J \cdot K^{-1} \cdot mol^{-1}$$

$$C_{p,m}(K_2U_2O_7,\ cr,\ T) = 149.0840 + 2.6950 \cdot 10^{-1} T \quad J \cdot K^{-1} \cdot mol^{-1}$$

are assumed to be valid from 298.15 to 750 K, although the temperature coefficient of the heat capacity of $K_2U_2O_7(cr)$ is much greater than would be anticipated for a simple U(VI) compound.

[98JEN/CHO]

The investigations in [92SAT/CHO] and [98JEN/CHO] were carried out at constant silica concentration and varying H$^+$ concentrations. The first study was made with a silicic acid concentration of 6.7·10^{-2} M in a 0.2 M NaClO$_4$ ionic medium at 25°C. The experimental method was liquid–liquid extraction with thenoyl-trifluoroacetone, TTA, and dibenzoyl-methane as extractants and benzene as the organic solvent. There are a number of experimental problems in [92SAT/CHO] that were not considered by these authors. These are discussed in [98JEN/CHO] and the original equilibrium constant $\log_{10}{}^*K(A.95) = -(2.44 \pm 0.06)$, for the reaction (A.95):

$$UO_2^{2+} + Si(OH)_4(aq) \rightleftharpoons UO_2(OSi(OH)_3)^+ + H^+, \qquad (A.95)$$

was therefore recalculated to be $\log_{10}{}^*K(A.95) = -(2.01 \pm 0.09)$. Extrapolation to zero ionic strength gives $\log_{10}{}^*K°(A.95) = -(1.74 \pm 0.09)$. In the second study, Jensen and Choppin used a spectrophotometric competition method to determine the equilibrium constant for Equation (A.95). The authors have prepared the silica solutions used by hydrolysis of tetramethyl-orthosilicate to avoid polymer formation. The experiments were made at a silicic acid concentration of 1.7·10^{-3} M, in 0.1 M NaClO$_4$ at 25°C. The authors report a value, $\log_{10}{}^*K(A.95) = -(2.92 \pm 0.06)$, that is significantly smaller than other determinations. Jensen and Choppin suggest that the higher values reported by other investigators might be due to the presence of polysilicates. This review is not convinced by this argument, as Moll et al. [98MOL/GEI] used the same method as Jensen and Choppin to prepare their silicate solutions, but found an equilibrium constant that was about one order of magnitude larger. They also showed that the fluorescence lifetime was different from that in test solutions where polymer formation was suspected. [98JEN/CHO] discuss possible sources of error in their data. They found no experimental evidence for interactions between dissolved silica and pyrocathecols (the ligand used in [98JEN/CHO]) reported in the literature. Three different direct experimental methods give concordant results, when using silicic acid solutions prepared in two different ways, suggesting to this review that there might be undetected experimental errors in studies of this type. It has therefore not been considered when selecting the equilibrium constant for reaction (A.95), which is based on the average of the determinations of [71POR/WEB], [92SAT/CHO] (recalculated), and [98MOL/GEI].

[98KAP/GER]

This paper deals with U(VI) sorption on natural carbonate minerals at high pH values, in the 8.3 to 12 range. The uranium concentration varies but is always lower than 3.36·10^{-7} M. The ionic strength varies but is always smaller than 0.014 M. Experimental K_d values are correlated with the concentration of the different species in solution and with the saturation index of aragonite, calcite, schoepite and some silicates. The authors have used thermodynamic data from [92GRE/FUG] and the speciation is calculated with a version of the MINTEQ A2 code. The predominant species at pH < 10 are $UO_2(CO_3)_2^{2-}$ and

$UO_2(CO_3)_3^{4-}$, and at pH > 11.8, $UO_2(OH)_3^-$, with a mixture of $UO_2(CO_3)_3^{4-}$ and $UO_2(OH)_3^-$ in between. At high pH, uranium seems to coprecipitate with $CaCO_3$ (rhombohedric aragonite or orthorhombic calcite), but it is not clear if U(VI) precipitates also as haiweeite $(Ca(UO_2)_2Si_6O_{15}\cdot 5H_2O)$.

The paper does not give any new thermodynamic information.

[98KIT/YAM]

Characteristics of fluorescence spectra (peak emission and fluorescence lifetime) of the following $(UO_2)_p(OH)_q$ complexes denoted: 0:1, 2:2 (present in acid solutions) and 3:1, 4:1 (present in basic solutions) are reported. The test solutions have been prepared using the equilibrium constants given by [92GRE/FUG] for the 2:2 species and by [98YAM/KIT] for the 3:1 and 4:1 species. The ionic medium of the solutions was 1 M $HClO_4$ or $NaClO_4$, which were undersaturated and always kept under an Ar atmosphere. The following species predominate at 25°C. In 1 M $HClO_4$, UO_2^{2+} is present at a total concentration of 0.010 M. At the same total concentration of uranium, the 2:2 complex is predominant at pH = 2.7, the 3:1 complex at pH = 12.3 (the total concentration is 10^{-5} M). The complex 4:1, finally, is predominant in 1 M NaOH with a total concentration of $3\cdot 10^{-4}$ M.

Each emission spectrum is analysed on the basis of five Gaussian functions corresponding to decay from five vibrational levels of the fluorescent excited state to the ground state. The $\lambda(v_o - v_o)$ corresponds to a pure electronic transition located at the lower wavelength range of the spectra.

At 25°C only the aqueous dioxouranium(VI) ion and the 2:2 species fluoresce with emission peaks and lifetimes, τ, consistent with previous data (except for $\tau_{2:2}$ measured by Meinrath [93MEI/KAT], [98MEI]) (see Table A-37). The ν values are found to be rather constant, 860 - 870 cm^{-1}.

All species give characteristic fluorescence spectra at 77 K, but only the room temperature characteristics are listed in Table A-37.

Table A-37: Spectroscopic characteristic of 1:1 and 2:2 species at 25°C.

Species	pH	I M	[U] M	$\lambda(v_o - v_o)^*$nm	τ µs	Reference
1:0	3.6	0.1		488.04	0.83	[93MEI/KAT]
	2.0	3	10^{-2}	487.80	1.8	[90PAR/SAK]
	1 $HClO_4$	1	10^{-4}	490.19	5.9	[95ELI/BID]
	1 $HClO_4$	1	10^{-2}	489.47	7.5	[98KIT/YAM]
2:2	3.6	0.1		490	2.9	[93MEI/KAT]
	3.0	3	10^{-2}	495.04	8.3	[90PAR/SAK]
	4.0	0.5	$5\cdot 10^{-4}$	497.51	9.5	[95ELI/BID]
	2.7	1	10^{-2}	499.25	10	[98KIT/YAM]

* Rounded values for λ

[98KON/HIL]

The recent data on the spectra and structure of $UCl_4(g)$ [95HAA/MAR] and $UF_4(g)$ [96KON/BOO] suggest strongly that these halides are strictly tetrahedral in shape. The authors have reexamined the spectroscopic and vapour pressure data of the actinide tetrahalides in the light of this new information.

New estimates of the vibration frequencies of thirteen actinide halides, $MX_4(g)$, species have been used to calculate their thermal functions, including entropies which are compared with experimental entropies of sublimation calculated from vapour pressure measurements. Although few details are given, the comparison suggests strongly that all of the $MX_4(g)$ species are tetrahedral. This has been assumed in the Np/Pu review [2001LEM/FUG], and the data for the stabilities of the $UX_4(g)$ species have been reviewed in detail for the present review, leading to changes in almost all the values for gaseous uranium halides, since many of these are dependent on the values for tetrahalide gaseous species.

[98KRI/RAM]

Knudsen effusion measurements using mass loss in a Cahn microbalance were used to determine the vapour pressures of $TeO_2(g)$, in the presence of the co–existing pressure of oxygen, in the reactions:

$$UTeO_5(cr) \rightleftharpoons \frac{1}{3} U_3O_8(cr) + \frac{1}{6} O_2(g) + TeO_2(g) \tag{A.96}$$

and

$$UTe_3O_9(cr) \rightleftharpoons UTeO_5(cr) + 2\, TeO_2(g) \tag{A.97}.$$

The compounds were prepared by heating in air mixtures of $UO_3(s)$ and $TeO_2(s)$ in the appropriate ratios 1:1 and 1:3, in a ceramic boat for five hours at 903 and 970 K, respectively. The products were identified by X–ray diffraction.

There are, however, a number of uncertainties in the interpretation of first reaction, since the oxygen pressure in the system was not controlled independently. The authors assumed the oxygen pressures were those given by the emf measurements of the oxygen potential in a study from the same laboratory [99SIN/DAS]. However, such oxygen pressures are very much less than 1/6 of the pressure of $TeO_2(g)$, so the relevant decomposition would not be that of reaction (A.96). We have therefore preferred to assume that $p_{O_2}/p_{TeO_2} = 1/6$, as given by reaction (A.96), although this is not entirely consistent, since even at these higher oxygen pressures, the uranium oxide phase in equilibrium would be oxygen deficient, U_3O_{8-x}, introducing another small uncertainty into the interpretation of the results.

Nevertheless, we have adopted this procedure, and thus in calculating p_{TeO_2} (and from this, p_{O_2}), have decreased the authors' measured mass losses by 3.34% (corresponding to a mole ratio of $TeO_2:O_2 = 6:1$).

From the tabulated data of total mass loss, with the above assumptions, the Gibbs energy of reaction (A.96) is calculated to be:

$$\Delta_r G_m \text{ (A.96)} = 243194 - 145.402 \cdot T \text{ J} \cdot \text{mol}^{-1} \qquad (1063 - 1155 \text{ K}).$$

The studies relating to reaction (A.97) are more straightforward, and the measured pressures have been refitted to the equation:

$$\log_{10}(p_{\text{TeO}_2} / \text{bar}) = -14486 \, T^{-1} + 12.682 \qquad (888 - 948 \text{ K})$$

for reaction (A.97), corresponding to the Gibbs energies of reaction:

$$\Delta_r G_m \text{ (A.97)} = 558379 - 409.01 \cdot T \quad \text{J} \cdot \text{mol}^{-1} \qquad (888 - 948 \text{ K}).$$

The authors' equation for $\log_{10}(p_{\text{TeO}_2} / \text{bar})$ seems to give pressures lower than the tabulated values by a factor of four.

For consistency with the similar study by Mishra et al. [98MIS/NAM], we have used the data for TeO$_2$(g) derived from the study by [94SAM] to obtain values at 298.15 K from these data.

Thus, the experimental data for reaction (A.96) were processed by a third–law method, using the estimated thermal functions for UTeO$_5$(cr), as in the emf study by [99SIN/DAS]. The resulting third–law enthalpy of reaction (A.96) is then:

$$\Delta_r H_m^\circ ((\text{A.96}), 298.15 \text{ K}) = (363.4 \pm 1.6) \text{ kJ} \cdot \text{mol}^{-1}.$$

With NEA data for U$_3$O$_8$(α) and $\Delta_f H_m^\circ$ (TeO$_2$, g, 298.15 K) = $-(52.8 \pm 2.7)$ kJ $\cdot$ mol^{-1} derived above from the data of Samant, cited by [98MIS/NAM] (see review of the paper by [98MIS/NAM]), we obtain $\Delta_f H_m^\circ$ (UTeO$_5$, cr, 298.15 K) = $-(1607.8 \pm 5.2)$ kJ $\cdot$ mol^{-1}. The second-law value, however, is $\Delta_f H_m^\circ$ (UTeO$_5$, cr, 298.15 K) = -1532.0 kJ $\cdot$ mol^{-1}, with an unknown uncertainty. The first value is in good agreement with other studies, section 9.5.3.2.1.

This study, like that of [98MIS/NAM] thus supports the stability derived from the more precise calorimetric data, and the estimated entropy of UTeO$_5$(cr).

A similar analysis of pressures of TeO$_2$(g) from the decomposition of UTe$_3$O$_9$(cr) by reaction (A.97), has been carried out, using thermal functions estimated here: S_m° (UTe$_3$O$_9$, cr, 298.15 K) = (305.8 ± 15) J $\cdot$ K^{-1} $\cdot$ mol^{-1} and $C_{p,m}$ (UTe$_3$O$_9$, cr, T) = 278.5 + 8.017·10^{-2} T – 3.44·10^6 T^{-2}, J $\cdot$ K^{-1} $\cdot$ mol^{-1}, to calculate the thermal functions of UTe$_3$O$_9$(cr). The derived third- and second-law enthalpies of reaction (A.97) at 298.15 K are (540.3 ± 3.1) and 585.3 kJ $\cdot$ mol^{-1}, (uncertainty unknown) The derived enthalpies of formation of UTe$_3$O$_9$(cr) then depend on that taken for UTeO$_5$(cr). Using the value of $-(1603.1 \pm 2.8)$ kJ $\cdot$ mol^{-1} selected in section 9.5.3.2.1 the respective third- and second-law enthalpies of formation are $\Delta_f H_m^\circ$ (UTe$_3$O$_9$, cr, 298.15 K) = $-(2249.0 \pm 6.2)$ and -2274.9 kJ $\cdot$ mol^{-1}, in reasonable agreement (considering the use of estimated thermal functions) with the calorimetric value, recalculated from [99BAS/MIS] as $-(2275.8 \pm 8.0)$ kJ · mol^{-1}.

[98MAS/MAS]

Electrolysis experiments were performed on Tc(VII) solutions (1.56 < pH < 8.50) containing nitrate and formate (1 M) using a graphite cathode at potentials of -0.6 to -1.6 V/SCE at 298.15 K. At -0.4 to -0.6 V/SCE, a violet solution was produced with UV–Visible absorption bands at 530 and 290 nm, which are indicative of a Tc(IV) formate complex containing the Tc_2^{8+} group with contribution from TcO_4^- to the second band. Deposition of a black amorphous Tc material increased with increased pH (maximum near pH = 7.37) and decreased applied potential to about 1.2 – 1.3 V/SCE (60% recovery at pH = 7.37 and -1.3 V/SCE after 30 min). The solution became pale yellow during electrolysis (λ_{max} = 652 nm) is believed to indicate the presence of a Tc_2^{7+}–formate species in solution and the deposit is suggested to be in the form: $TcO_{(2-x)} \cdot yH_2O$, where x = 0 – 0.25 and y = 1.6 – 2.0. Electrodeposition yields of up to 95.7 % were achieved at the following condition: -1.4 V/SCE after two hours at 298.15 K, $[Tc]_0 = 2 \cdot 10^{-3}$ M, pH = 7.37 in 1 M HCOONa. The percentage of Tc deposited decreased with increasing Tc concentration in the bulk solution and the rate of deposition was found to be first order with respect to the Tc concentration. The rate-determining step is therefore thought to involve diffusion of $Tc_2O_2(HCOO)_6^{2-}$ and $Tc_2O_2(HCOO)_6^{3-}$ to the electrode surface. The authors speculate that these species hydrolyse completely at pH > 5.5, allowing the electrodeposition to proceed to a higher degree. Lowering the concentration of NO_3^- and the build up of $TcO_{(2-x)} \cdot yH_2O$ deposits on the electrode surface tend to limit further deposition.

[98MEI]

This is a more recent paper of a series of papers written by G. Meinrath and co-workers that give spectroscopic and thermodynamic data on the following hydrolysis species: 0:1, UO_2^{2+}; 2:2, $(UO_2)_2(OH)_2^{2+}$; and 5:3, $(UO_2)_3(OH)_5^+$ or $UO_2O(OH)_3^+$. The papers in the series are: [93MEI/KAT], [96MEI/SCH], [97MEI] (UV–Vis), [93MEI/KAT], [94KAT/MEI] (fluorescence), [97MEI2], [98MEI/KAT] (UV–Vis and fluorescence), [96KAT/KIM], [96MEI/KAT] (solubilities). Structural data are known [69ABE], [70ABE]. Other papers deal also with spectroscopic and thermodynamic data of hydrolysed U(VI) species, [95ELI/BID], [98KIT/YAM] (fluorescence).

The most important result to be discussed is the determination of $\log_{10} {}^*\beta_{2,2}$ and $\log_{10} {}^*\beta_{5,3}$ made at I = 0.1 M (NaClO$_4$) and t = 25°C using spectroscopic analysis of U(VI) solutions under-saturated with respect to $UO_3 \cdot 2H_2O$(s). [97MEI2] deals with the same topic but gives more experimental details. Equilibrium concentrations of the species 1:1, 2:2 and 5:3 refer to the reactions:

$$UO_2^{2+} + H_2O(l) \rightleftharpoons UO_2(OH)^+ + H^+$$

$$2\,UO_2^{2+} + 2\,H_2O(l) \rightleftharpoons (UO_2)_2(OH)_2^{2+} + 2\,H^+$$

$$3\,UO_2^{2+} + 5\,H_2O(l) \rightleftharpoons (UO_2)_3(OH)_5^+ + 5\,H^+$$

and are derived by using the following characteristics of individual spectra: 1:1, $\varepsilon_{413.8} = (9.7 \pm 0.2)$ L·mol^{-1}·cm^{-1}; 2:2, $\varepsilon_{421.8} = (101 \pm 2)$ L·mol^{-1}·cm^{-1}; and 5:3, $\varepsilon_{429.0} = (474 \pm 7)$ L·mol^{-1}·cm^{-1}. The pH measurements are discussed in [97MEI] and $\log_{10}[H^+]$ is taken as:

$$\log_{10}[H^+] = pH - \log_{10} \gamma_{H^+} = pH - (-0.09).$$

The concentration of the species $UO_2(OH)^+$ is in every case negligible.

The weighted average value of the equilibrium constants from 13 solutions (studied in the pH range (3.939 ± 0.029) to (4.776 ± 0.014)) are $\log_{10} {}^*\beta_{2,2} = -(6.237 \pm 0.103)$ and $\log_{10} {}^*\beta_{5,3} = -(17.203 \pm 0.157)$ (95% confidence limit including precisions in concentration and pH).

In this paper Meinrath uses a statistical analysis of data from 25 solutions in the concentration ranges indicated above. Only three species are needed to reproduce the measured absorption spectra of the solutions in the range 340 – 520 nm. That eliminates the presence of other species such as 1:2, $(UO_2)_2OH^{3+}$ and 4:3, $(UO_2)_3(OH)_4^{2+}$.

Meinrath points out that the 5:3 species is formed in minute amounts in the pH region close to saturation with respect to $UO_3 \cdot 2H_2O(s)$. It can be detected spectroscopically (down to $(1 \pm 0.25) \cdot 10^{-6}$ M) because of its high molar absorption coefficient but only with difficulty from solubility measurements.

The experimental values of $\log_{10} {}^*\beta_{2,2}$ and $\log_{10} {}^*\beta_{5,3}$ at $I = 0.1$ M reported in [92GRE/FUG] range from -5.68 to -6.45 and from -15.6 to -17.7, respectively. The weighted average of $\log_{10} {}^*\beta_{2,2}$ from Meinrath falls in this range.

This review accepts these refined data.

This review has compared the speciation diagrams obtained by using the equilibrium constants proposed by Meinrath and those selected in [92GRE/FUG]. The total concentration of U(VI), 10 mM, is within the range for the concentrations used in potentiometric studies of U(VI) hydrolysis and should optimise the chances of finding the 5:3 complex.

Figure A-9 shows a speciation using constants at zero ionic strength calculated from the constants given by Meinrath.

Figure A-9: Speciation at zero ionic strength calculated using equilibrium constants given by Meinrath.

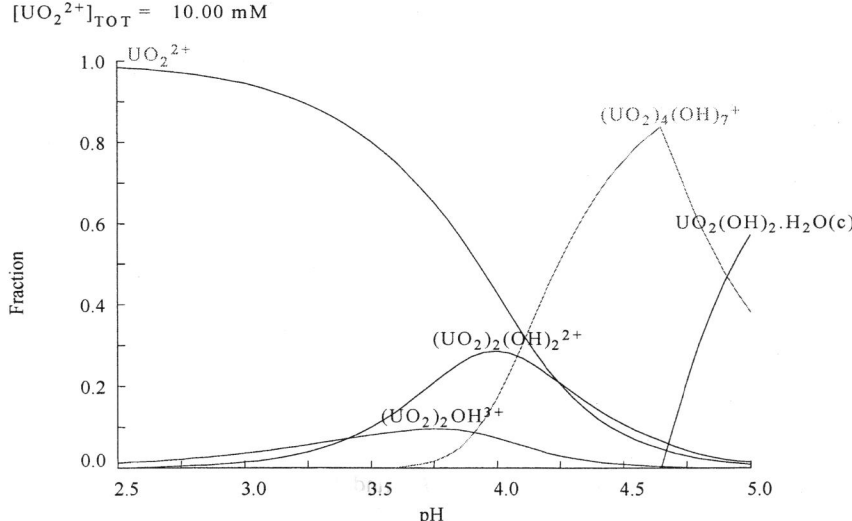

Figure A-10 has been calculated using the constants given in [92GRE/FUG] also at zero ionic strength. As it can be seen there is a pronounced difference in speciation.

Figure A-10: Speciation in the U(VI) hydroxide system at zero ionic strength using the equilibrium constants selected in [92GRE/FUG].

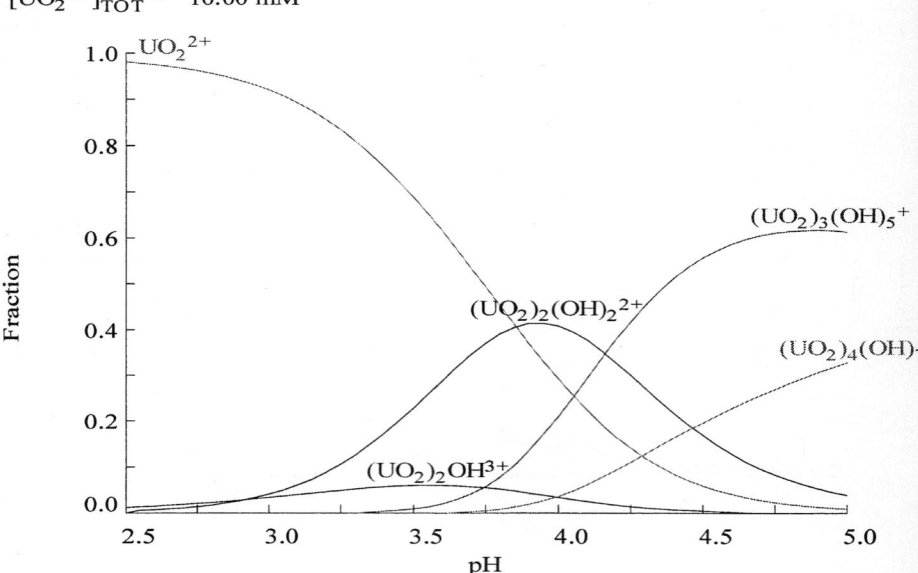

The following three figures (Figures A. 11, A. 12 and A.13) show speciation diagrams under the experimental conditions used by Meinrath (0.1 M NaClO$_4$) and by the Sillén school (3 M NaClO$_4$) with equilibrium constants given by Meinrath or [92GRE/FUG].

Figure A-11: Speciation in the U(VI) hydroxide system in 0.1 M NaClO$_4$ using the equilibrium constants of Meinrath.

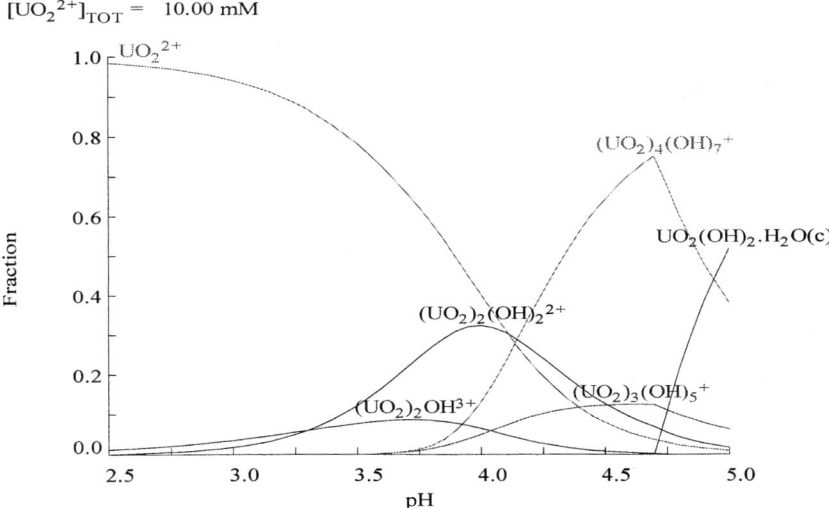

Figure A-12: Speciation in the U(VI) hydroxide system using the equilibrium constants from Meinrath recalculated to 3 M NaClO$_4$.

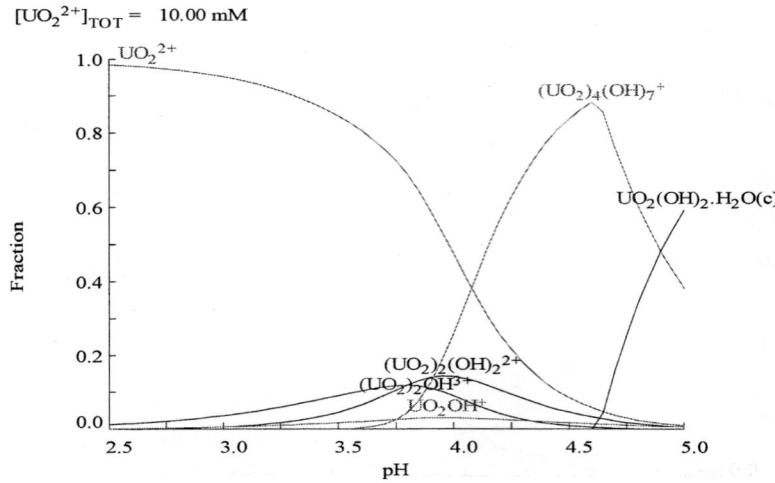

Figure A-13: Speciation in the U(VI) hydroxide system using the equilibrium constants from [92GRE/FUG] recalculated to 3 M NaClO$_4$.

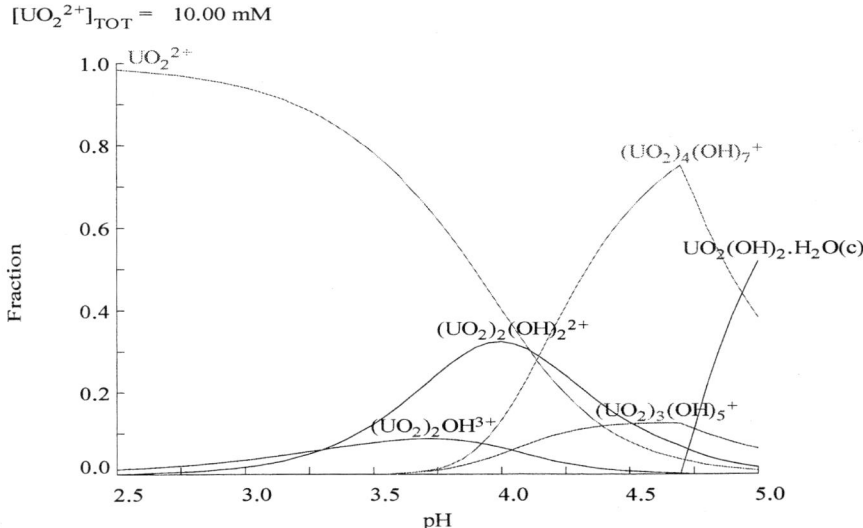

There is only a fairly small amount of the 5:3 species. However, this is related to the large amount of 7:4, a species that was not used by Sillén. The following figures (Figure A-14 and Figure A-15) show the effect when this complex is removed.

Figure A-14: Speciation in the U(VI) hydroxide system using the equilibrium constants from Meinrath recalculated to 3 M NaClO$_4$, but excluding the 7:4 species.

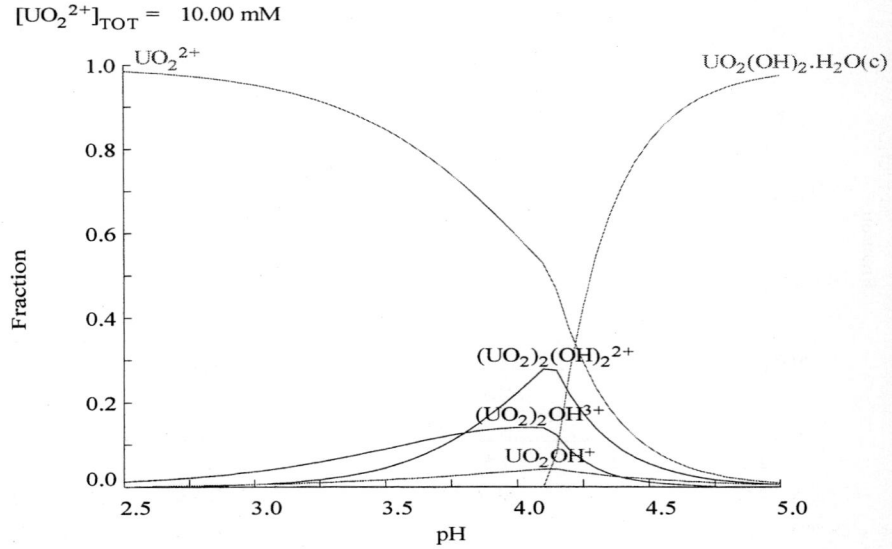

Figure A-15: Speciation in the U(VI) hydroxide system using the equilibrium constants from [92GRE/FUG] recalculated to 3 M NaClO$_4$, but excluding the 7:4 species.

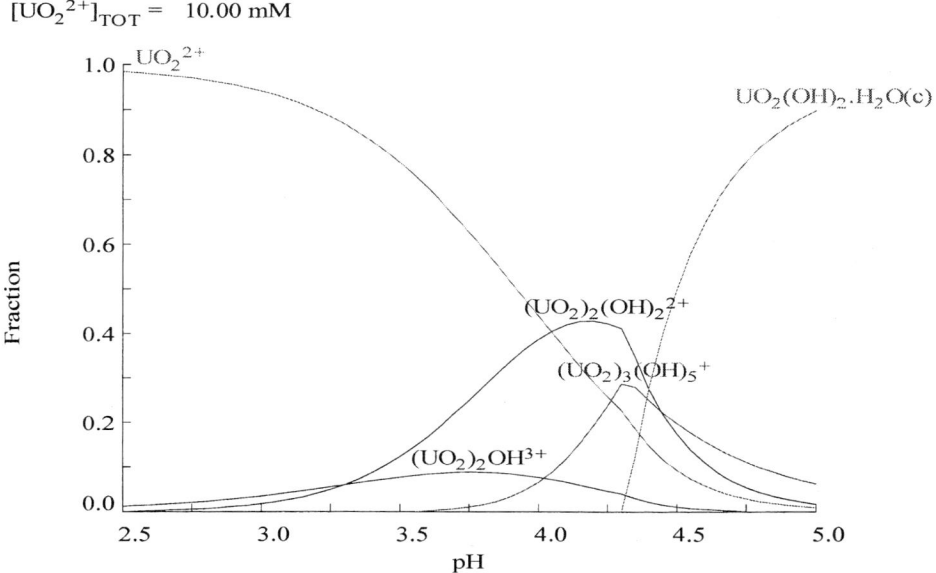

The last two diagrams, Figure A-14 and Figure A-15, indicate very clearly that the experimental data of Meinrath and Sillén are not consistent. The experimental method used by the Sillén group is a very sensitive indicator of the nuclearity of these systems.

With the equilibrium constants deduced by Meinrath it should not be possible to identify the complex $(UO_2)_3(OH)_5^+$ under the conditions used in the potentiometric studies.

However, the potentiometric studies have been made at much higher ionic strength than those of Meinrath. If the complex of higher charge, the (2:2), is stabilised by high concentrations of perchlorate, in the same way as complexes of negative charge, e.g., $UO_2(CO_3)_3^{2-}$ and $UO_2(CO_3)_3^{4-}$ are stabilised by high concentrations of Na$^+$, the observed differences are understandable.

[98MEI2]

This paper discusses previous papers on U(VI) hydrolysis and gives additional information to the study of [98MEI]. The experimental data considered here are solubility measurements of UO$_3$·2H$_2$O(cr), made at $t = (24 \pm 2)$°C in solutions in equilibrium with a gas phase with a partial pressure of CO$_2$ equal to 0.03 kPa (air). The ionic medium is 0.1 M

NaClO$_4$ and the pH range studied, 4 to 5.5. In addition, Meinrath uses data taken from [92KRA/BIS], [93MEI/KIM], [96MEI/KAT] and [96KAT/KIM], which relate to the same saturated system. In the present case, solutions are equilibrated for six months after which the solid phase is characterised as schoepite (yellow crystallite size > 0.1 μm). Solutions are filtered (220 nm pore size). The hydrogen ion concentrations are derived from pH using $\log_{10}\gamma_{H^+} = -0.092$ as explained previously.

The species UO_2^{2+}, $(UO_2)_2(OH)_2^{2+}$ and $(UO_2)_3(OH)_5^+$ are identified in the solutions from their spectra (340 to 540 nm) [96MEI/SCH], [97MEI2] (aqueous ion: λ_{max} = 413.8 nm, ε = (9.7 ± 0.2) L·mol^{-1}·cm^{-1}; (2:2) species: λ_{max} = 421.8 nm, ε = (101 ± 2) L·mol^{-1}·cm^{-1}; (5:3) species: λ_{max} = 429 nm, ε = (474 ± 7) L·mol^{-1}·cm^{-1}).

Analysis of the data is conducted using multivariate data analysis (chemometric methods) which shows that three species must be considered to model the data.

The main problem in deriving the equilibrium constants is the direct determination of the free concentration [UO_2^{2+}], due to its very low molar absorbance coefficient, compared to those of the other species. The modelling is done as follows. $\log_{10} {}^*\beta_{2,2} = -(6.145 \pm 0.088)$ is taken from [97MEI] to calculate $\log_{10} {}^*\beta_{5,3}$ from the measured free concentrations of the 2:2 and 5:3 species according to:

$$\log_{10} {}^*\beta_{5,3} = \log_{10}[5:3] - \frac{3}{2}\log_{10}[2:2] + \frac{3}{2}\log_{10} {}^*\beta_{2,2} - 2\,\text{pH} - 2\log_{10}\gamma_{H^+}.$$

The average value is $\log_{10} {}^*\beta_{5,3} = -(17.00 \pm 0.17)$. The concentration of free UO_2^{2+} is then calculated from ${}^*\beta_{2,2}$ and the measured concentration of the 2:2 species. A good fit of the data is obtained by summing up all the free concentrations of the different species and comparing this with the total analytical uranium concentration.

The solubility data and those of [92KRA/BIS] and [96MEI/KAT], which are very similar and depend on the equilibrium constant, ${}^*K_{s,0}$, ${}^*\beta_{2,2}$ and ${}^*\beta_{5,3}$, are modelled using $\log_{10} {}^*\beta_{2,2} = -(6.145 \pm 0.088)$ as a fixed parameter. It is impossible to determine both ${}^*K_{s,0}$ and ${}^*\beta_{2,2}$ from these data because of a very strong covariance between them. Meinrath obtains $\log_{10} {}^*K_{s,0} = (5.75 \pm 0.03)$ and $\log_{10} {}^*\beta_{5,3} = -(17.16 \pm 0.18)$ (and considers this value as "better" than that obtained by spectroscopy).

Reinterpretation of the [92KRA/BIS] data with fixed $\beta_{2,2}$ gives $\log_{10} {}^*K_{s,0} = (5.79 \pm 0.03)$ and $\log_{10} {}^*\beta_{5,3} = -(17.04 \pm 0.18)$.

The solubility constant, $\log_{10} {}^*K_{s,0}$, refers to the reaction [93MEI/KIM]:

$$UO_3 \cdot 2H_2O(cr) + 2\,H^+ \rightleftharpoons UO_2^{2+} + 3\,H_2O(l).$$

As the value for the solubility product $\log_{10}K_{s,0}$ refers to the reaction:

$$UO_3 \cdot 2H_2O\,(cr) \rightleftharpoons UO_2^{2+} + 2\,OH^- + H_2O(l),$$

we have, $\log_{10} {}^*K_{s,0} = \log_{10}K_{s,0} - 2\log_{10}K_w$, with $\log_{10}K_w = -(13.78 \pm 0.01)$.

Meinrath finds $\log_{10} K_{s,0}$ $(UO_3 \cdot 2H_2O$, cr) $= -(21.81 \pm 0.03)$ and $\log_{10} K_{s,0}$ $(UO_3 \cdot 2H_2O$, cr) $= -(21.77 \pm 0.03)$ from the re-interpretation of [92KRA/BIS] data.

[98MEI3]

The focus of this paper is on aspects of the environmental chemistry of U(VI). It has nine chapters. Two chapters are of a special interest for this review: hydrolytic behaviour of U(VI) and assessment of uncertainties in evaluation of pH.

Meinrath promotes the use of spectroscopy for the determination of the formation constants of $(UO_2)_n(OH)_m^{2n-m}$ species. He gives the example of statistical evaluation by factor analysis (FA) of 26 UV–Vis absorption spectra of U(VI) solutions (I = 0.1 M NaClO$_4$, t = 25°C, $-2.4 < \log_{10}[U] < -3.5$, and $2.4 < $ pH < 5), under-saturated with respect to schoepite. FA shows that three species must be selected to account for the absorptivity in the wavelength range 350 to 500 nm: 0:1, 2:2 and 5:3. The experimental data are not new. The application of FA is discussed at length. Values of the constants $^*\beta_{2,2}$ and $^*\beta_{5,3}$ defined as :

$$\log_{10} {}^*\beta_{2,2} = \log_{10}[(UO_2)_2(OH)_2^{2+}] - 2\log_{10}[UO_2^{2+}] - 2\text{pH} - 2\log_{10}\gamma_{H^+}$$

$$\log_{10} {}^*\beta_{5,3} = \log_{10}[(UO_2)_3(OH)_5^+] - 3\log_{10}[UO_2^{2+}] - 5\text{pH} - 5\log_{10}\gamma_{H^+}$$

are given for 26 pH values with uncertainties depending on whether pH errors ($\pm$ 0.025 at 95% confidence level, see below) are taken into account, or not. These values show that the uncertainties in pH determinations contribute in similar magnitude to the overall uncertainty in the constants, as do the uncertainties in the concentrations of the species. The final average values are, $\log_{10} {}^*\beta_{2,2} = -(6.168 \pm 0.0565)$ and $\log_{10} {}^*\beta_{5,3} = -(17.123 \pm 0.069)$, or alternatively $\log_{10} {}^*\beta_{2,2} = -(6.145 \pm 0.088)$ and $\log_{10} {}^*\beta_{5,3} = -(17.1424 \pm 0.1383)$ depending on the way followed to calculate both the mean values and the total uncertainties (within 95% confidence). The third digits must be taken with caution. They are quoted here only to avoid giving a rounded value after a statistical treatment of numerous data.

Schoepite (in fact metaschoepite) is the solubility limiting solid for U(VI) under ambient conditions and pH < 7, which readily transforms to a sodium uranate at pH > 7, even at low ionic strength. The species 5:3 is formed only in minor amounts in a pH region close to the saturation limit and is therefore difficult to detect by concentration dependent methods. In contrast, UV–Vis spectroscopy is more convenient due to the high molar absorption of this species.

This paper gives eight additional solubility data of schoepite (in fact metaschoepite), for an equilibration time of more than three months, to the data of [92KRA/BIS], [93MEI/KIM], [95KAT/KIM] and [96MEI]. Spectroscopy allows exploration only of the pH range 4 to 5.4. The value obtained from assessment of the spectra is $\log_{10} {}^*\beta_{5,3} = -(17.00 \pm 0.5)$, using $\log_{10} {}^*\beta_{2,2} = -(6.14 \pm 0.05)$ (value for pH = 5.34), where all the uncertainties are taken into account.

To measure the pH values a multiple calibration procedure (MCP) is used. A glass electrode is calibrated against several standard buffer solutions of NIST. The data pairs mV ($\pm$ 0.1 mV) *versus* pH ($\pm$ 0.02) are interpreted by linear least square regression (OLS) using mV = α + $\beta\cdot$pH. From data accumulated over years, uncertainties in α and β are derived. The β values never fit exactly with the theoretical Nernst value $-$ 59.16 mV pH^{-1}, but are higher.

To show the importance of neglecting pH uncertainty Meinrath gives the example of the CO$_2$–H$_2$O system where,

$$\log_{10}[CO_3^{2-}] = \log_{10}K + \log_{10}p_{CO_2} + 2\,\text{pH}.$$

Without taking into account the uncertainty in pH, $\log_{10}K = -(17.62 \pm 0.07)$ whereas by including it, $\log_{10}K = -(17.63 \pm 0.18)$ with an uncertainty that is almost three times larger.

A summary of the spectroscopic properties of U(VI) species can be found in this paper.

[98MEI/FIS]

The authors have studied the solubility of uranium solids as a function of pH in 0.1 M NaClO$_4$ and 1 M NaCl. Schoepite (in fact metaschoepite) is the solubility limiting phase at pH < 7.5 while at pH > 7.5 sodium uranates are formed. The latter solids are not well crystallised and it is therefore difficult to characterise them by X–ray diffraction; the authors propose both Na$_6$U$_7$O$_{24}$ and Na$_2$U$_2$O$_7\cdot$xH$_2$O. The authors have taken care in their attempts to identify the solid phase used in these solubility experiments and the problems encountered are clearly pointed out. The solubility has been measured in the pH region, 7.8 to 8.6, where schoepite is not stable and where the dominant uranium species in solution is UO$_2$(CO$_3$)$_3^{4-}$, *i.e.*, the reaction studied is then either of:

$$Na_2U_2O_7(s) + 6\,HCO_3^- \rightleftharpoons 2\,UO_2(CO_3)_3^{4-} + 3\,H_2O(l) + 2\,Na^+ \quad (A.98)$$

$$Na_6U_7O_{24}(s) + 21\,HCO_3^- \rightleftharpoons 7\,UO_2(CO_3)_3^{4-} + 6\,Na^+ + H^+ \quad (A.99).$$

The authors have not provided information of the total carbonate concentration and it is therefore impossible for the reviewers to analyse the data. A slope analysis of $\log_{10}s$ *versus* pH is not straightforward as the slope will depend on the solid phase selected. The authors have selected equation (A.98) and used their experimental solubility data in 0.1 M NaClO$_4$, the equilibrium constant, $\log_{10}\beta_3^\circ = 21.86$, for the formation of UO$_2$(CO$_3$)$_3^{4-}$ and the standard Gibbs energy of formation of UO$_2^{2+}$, Na$^+$ and H$_2$O(l), to obtain the solubility product for the reaction:

$$Na_2U_2O_7(s) + 6\,H^+ \rightleftharpoons 2\,Na^+ + 2\,UO_2^{2+} + 3\,H_2O(l) \quad (A.100).$$

These auxiliary data for UO$_2^{2+}$, Na$^+$ and H$_2$O(l) are not significantly different from the ones given in [92GRE/FUG]. Meinrath *et al.* obtains $\log_{10}{}^*K_{s,0}$ (A.100) =

(24.2 ± 0.2) (There is a mistake in the sign in this paper). This review has recalculated this value using the selected value of $\log_{10} \beta_3^\circ$ = 21.60 and obtains $\log_{10} {}^*K_{s,0}$ (A.100) = (23.7 ± 0.2), which recalculated to zero ionic strength gives $\log_{10} {}^*K_{s,0}^\circ$ (A.100) = (24.1 ± 0.2). This value is in fair agreement both with the value obtained by Yamamura et al. [98YAM/KIT], $\log_{10} {}^*K_{s,0}^\circ$ (A.100) = (25.1 ± 2.1), and the value obtained by using the standard Gibbs energy of formation of $Na_2U_2O_7$(cr) given in [92GRE/FUG]. In view of the lack of information of experimental details in [98MEI/FIS], these data have not been used for the selection of the solubility product of reaction (A.100).

[98MEI/KAT]

This work is the continuation of those reported in [93MEI/KAT], [94KAT/MEI], [96MEI/KAT] and gives additional spectroscopic data to [97MEI2].

Both absorption and emission spectra of U(VI) solutions containing the species: 0:1, UO_2^{2+}; 2:2, $UO_2(OH)_2^{2+}$; 5:3, $(UO_2)_3(OH)_5^+$; and UO_2CO_3(aq) have been collected. They permit measurement of the wavelength for the pure electronic transition between the ground state and excited levels of the species. Solutions containing $UO_2(CO_3)_2^{2-}$ and $UO_2(CO_3)_3^{4-}$ do not show emission spectra due to CO_3^{2-} quenching.

The $\lambda(v_o-v_o)$ transitions are the following in nm: 0:1, (488 ± 0.7); 2:2, (497 ± 1); 5:3, (499.5 ± 1.5); and UO_2CO_3(aq), (464 ± 0.9).

[98MIS/NAM]

Mass loss measurements (transpiration technique) were used to determine the vapour pressures of TeO_2(g), in the presence of one atmosphere of oxygen, in the reactions:

$$UTeO_5(cr) \rightleftharpoons \tfrac{1}{3} U_3O_8(cr) + \tfrac{1}{6} O_2(g) + TeO_2(g) \quad (A.101)$$

and

$$UTe_3O_9(cr) \rightleftharpoons UTeO_5(cr) + 2\,TeO_2(g) \quad (A.102)$$

The compounds were obtained by heating thoroughly ground mixtures of UO_2(cr) and TeO_2(cr) in the ratios 1:1 and 1:3, in air in a ceramic boat for eight hours at 1023 and 900 K, respectively, followed by isothermal heating (eight hours) with frequent grindings. The products were reported as showing only the characteristic X–ray diffraction lines of the expected species. In the study of reaction (A.101), the global mass loss was corrected for the loss of O_2 using the molar ratio, $TeO_2:O_2$ = 6:1.

From the tabulated data of TeO_2(g) mass loss, the following fitted equations were given (pressures converted to bar):

$$\log_{10}(p_{TeO_2}/\text{bar}) = -18512.1\,T^{-1} + 11.555 \qquad 1107\text{–}1207\text{ K}$$

for reaction (A.101), and:

$$\log_{10}(p_{TeO_2}/\text{bar}) = -14510.9\,T^{-1} + 10.422 \qquad 947\text{–}1011\text{ K}$$

for reaction (A.102), corresponding to the Gibbs energies of reaction:

$$\Delta_r G_m \text{ (A.101)} = 351500 - 218.62 \cdot T \quad \text{J} \cdot \text{mol}^{-1} \quad (1063 - 1155 \text{ K}).$$

$$\Delta_r G_m \text{ (A.102)} = 532602 - 376.02\, T \quad \text{J} \cdot \text{mol}^{-1} \quad (947 - 1011 \text{ K}).$$

For further processing of these data, the authors chose to combine their measured pressures of $TeO_2(g)$ with those over $TeO_2(cr)$ measured in the same apparatus in a thesis study by Samant, cited by [98MIS/NAM] as:

$$\log_{10}(p^\circ_{TeO_2} / \text{bar}) = -12632.9\, T^{-1} + 8.689 \quad \text{(A.103)}$$

for reaction (A.104):

$$TeO_2(cr) \rightleftharpoons TeO_2(g). \quad \text{(A.104)}$$

Since this involves extrapolation of the latter data (and subsequent processing involves data for metastable $TeO_2(cr)$ above its melting point), we have adopted a slightly different approach, which also allows a check on the consistency of the data in equation (A.103) with literature data.

The sublimation data in equation (A.103) have been combined with the thermal functions of $TeO_2(cr)$ and $TeO_2(g)$ from the assessment of [90COR/KON], using a 'pseudo' third–law processing, to derive the corresponding enthalpy of sublimation of $TeO_2(cr)$:

$$\Delta_{sub} H^\circ_m (TeO_2, cr, 298.15 \text{ K}) = (268.2 \pm 0.9) \text{ kJ} \cdot \text{mol}^{-1}$$

in good agreement with other vaporisation studies (cf. the table quoted by [90COR/KON]). Using $\Delta_f H^\circ_m (TeO_2, cr, 298.15 \text{ K}) = -(321.0 \pm 2.5)$ kJ·mol^{-1}, this corresponds to $\Delta_f H^\circ_m (TeO_2, g, 298.15 \text{ K}) = -(52.8 \pm 2.7)$ kJ·mol^{-1} (cf. $-(54.8 \pm 2.6)$ kJ·mol^{-1} selected by [90COR/KON]).

The experimental data for reaction (A.101) can now be processed by a third–law method, using the estimated thermal functions for $UTeO_5(cr)$, as in the emf study by [99SIN/DAS]. The resulting third–law enthalpy of reaction (A.101) is then:

$$\Delta_r H^\circ_m ((\text{A.101}), 298.15 \text{ K}) = (360.04 \pm 1.12) \text{ kJ} \cdot \text{mol}^{-1}.$$

With NEA data for $U_3O_8(\alpha)$ and $\Delta_f H_m (TeO_2, g, 298.15 \text{ K}) = -(52.8 \pm 2.7)$ kJ·mol^{-1}, we obtain $\Delta_f H^\circ_m (UTeO_5, cr, 298.15 \text{ K}) = -(1604.4 \pm 3.0)$ kJ·mol^{-1}. The second law value is $\Delta_f H^\circ_m (UTeO_5, cr, 298.15 \text{ K}) = -1611.8$ kJ·mol^{-1}, with an unknown uncertainty. These values, particularly the first, are in good agreement with other studies, see section 9.5.3.2.1.

This study thus supports the stability derived from the calorimetric data, and the estimated entropy of $UTeO_5(cr)$.

A similar analysis of pressures of $TeO_2(g)$ from the decomposition of $UTe_3O_9(cr)$ by reaction (A.102), has been carried out, using data estimated here: $S^\circ_m (UTe_3O_9, cr,$

298.15 K) = (305.8 ± 15) J · K^{-1} · mol^{-1} and $C_{p,m}$ (UTe$_3$O$_9$, cr, T) = 278.5 + 8.017 · 10^{-2} T − 3.44·10^6 T^{-2}, J · K^{-1} · mol^{-1}, to calculate the thermal functions of UTe$_3$O$_9$(cr). The derived third- and second-law enthalpies of reaction (A.102) at 298.15 K are (543.4 ± 2.1) and 562.1 kJ · mol^{-1} (uncertainty unknown). The derived enthalpies of formation of UTe$_3$O$_9$(cr) then depend on that taken for UTeO$_5$(cr). Using the value of − (1603.1 ± 2.8) kJ · mol^{-1} selected in section 9.5.3.2.1, the respective third- and second-law enthalpies of formation are $\Delta_f H_m$ (UTe$_3$O$_9$, cr, 298.15 K) = − (2252.1 ± 6.4) and − 2270.8 kJ · mol^{-1}, in tolerable agreement (considering the use of estimated thermal functions) with the calorimetric value, recalculated from [99BAS/MIS] as − (2275.8 ± 8.0) kJ · mol^{-1}.

[98MOL/GEI]

This review includes also the review of [97MOL/GEI].

[97MOL/GEI] is a preliminary version of [98MOL/GEI]. The following discussion will mainly refer to the second paper. The equilibrium constant for the reaction:

$$UO_2^{2+} + Si(OH)_4(aq) \rightleftharpoons UO_2SiO(OH)_3^+ + H^+$$

was studied using time resolved laser fluorescence spectroscopy (TRLFS) in solutions with constant pH equal to 3.9 and silicic acid concentrations varying from 10$^{-3.8}$ to 10$^{-2.2}$ M. The study was made at 20°C in a 0.3 M NaClO$_4$ ionic medium. In order to avoid the formation of polysilicates, the test solutions were prepared by hydrolysis of (CH$_3$O)$_4$Si, rather than by neutralizing sodium silicate. The TRLFS method gives information on both the fluorescence spectra and the fluorescence lifetime of the species present in the test solutions. Most of these showed bi–exponential fluorescence decay, indicating the presence of two emitting species. One of them is UO_2^{2+}, for which both the lifetime and fluorescence spectrum agree well with previous studies in the binary U(VI)–H$_2$O system *e.g.*, [95ELI/BID]. The second component had a lifetime of 19 μs, which is different from that of UO$_2$OH$^+$, 32.8 μs [95ELI/BID], the predominant hydrolysis species in the pH range studied. The silicic acid solutions in [97MOL/GEI] were prepared by acidifying solutions of sodium silicate (personal information from Dr. Moll). The test solutions contain polysilicates, which result in species with different fluorescence lifetimes. There are two components in solution, one with a lifetime of 36 μs, and the other with a lifetime longer than 95 μs. The experimental results of these two studies demonstrate the interaction between UO_2^{2+} and dissolved silica. They also show that one may neglect the formation of hydroxide complexes in most of the test solutions. UO$_2$OH$^+$ is the predominant uranium(VI) hydroxide complex under the experimental conditions used. Using the known equilibrium constant from [92GRE/FUG], we find that the ratio, [UO_2^{2+}]/[UO$_2$OH$^+$], is less than 0.05 at pH = 3.9 and hydrolysis will not affect the interpretation of the data significantly. The assumed complex formation reaction requires a slope close to one in the plot of ($\log_{10}$ [UO$_2$OSi(OH)$_3^+$]/[UO_2^{2+}]$_{free}$ − pH) *versus* $\log_{10}$[Si(OH)$_4$(aq)]; the authors report (0.80 ± 0.13), *cf.* Figure 4 in [98MOL/GEI]. This review prefers to put a smaller weight on the experimental point at the lowest concentration of silicic acid and thereby obtains a slope of 0.95, which gives better agreement between the experimental data and the proposed model. This will result in a larger equilibrium constant, $\log_{10} \beta(UO_2SiO(OH)_3^+) = -(1.85 \pm 0.20)$. The authors have made their experiments at a constant pH of 3.9. They have not recalculated the measured hydrogen ion activity to concentrations and their equilibrium constant must therefore be recalculated to obtain the concentration constant. This was done using the specific ion interaction theory with the interaction coefficient $\varepsilon(H^+, ClO_4^-) = 0.14$ kg · mol^{-1}. We find $\log_{10}\gamma_{H^+} = -0.111$ in the 0.3 M NaClO$_4$ ionic medium used. The recalculated value of $\log_{10} {}^*K(UO_2SiO(OH)_3^+)$ is then $-(1.74 \pm 0.20)$. The extrapolation to zero ionic strength was made using the SIT and the interaction coefficients given in [92GRE/FUG] and assuming $\varepsilon(UO_2SiO(OH)_3^+) = 0.3$ kg · mol^{-1}, and gave $\log_{10} {}^*K°(UO_2SiO(OH)_3^+) = -(1.44 \pm 0.20)$. The value of [71POR/WEB], recalculated to zero ionic strength is $\log_{10} {}^*K°(UO_2SiO(OH)_3^+) = -1.71$, and the two values agree within their estimated range of uncertainty. The term, $\log_{10} [UO_2OSi(OH)_3^+]/[UO_2^{2+}]$, in the various test solutions varies from -0.4 to 0.7, i.e., the silicate complex is a major species in most of the test solutions. With an equilibrium constant ten times smaller, as suggested in [98JEN/CHO], UO_2^{2+} would have been predominant, which is not consistent with the spectroscopic observations.

[98MOU/LAS]

This paper gives the characteristics of fluorescence spectra of the U(VI) species, 1:1, 2:1, 3:1, 2:2, 5:3 and 7:3, in carbonate-free solutions (I = 0.1 M NaClO$_4$, t = 20°C) obtained by dilutions and pH adjustments (NaOH) of a stock solution under a N$_2$ atmosphere. The stock solution is prepared by dissolution of U metal in HClO$_4$ and the initial uranium concentration is measured by mass spectrometry.

pH and [U(VI)]$_t$ concentrations of the test solutions submitted to laser excitation are chosen according to the thermodynamic data selected by [92GRE/FUG] for all the species, except for the species 2:1 [80LEM/TRE], to have the smallest number of species consistent with spectroscopic data (no more than three). Gate delays and duration time of the measurements are selected in such a way that only one complex, one fluorescence lifetime, could be registered. The authors did not correct the hydrolysis constant for I = 0.1 M and did not take into account the uncertainties of the constants.

The appropriate conditions to find the species in the solutions are the following: species 0:1, 4·10^{-7} M, pH = 4; species 1:1 + 0:1, 4·10^{-8} M, pH = 5; species 3:1, 4·10^{-7} M, pH = 8.5 and 4·10^{-8} M, pH = 10 and 4·10^{-6} M, pH = 9.5; species 2:1 + 1:1 + 3:1, 4·10^{-8} M, pH = 7; species 5:3, 4·10^{-6} M, pH = 6.8 and 4·10^{-5} M, pH = 6.5; species 7:3 + 3:1, 4·10^{-5} M, pH = 8.5.

Despite the fact that the thermodynamic data used are open to criticism, different well-resolved individual spectra and decay lifetimes (except for 2:1 species) of each species are given and compared to previous data. Good agreement is found for 0:1 and 2:2, but the other species' lifetimes are found to be longer. The longer the lifetime (single exponential), the smaller is the influence of artefacts (quenching, temperature effect, *etc.*). Table A-38 summarises the data.

Some spectroscopic features are also given. The vibrational quantum is rather constant for all spectra, which shows that hydrolysis has no (or little) effect on the ground state of the UO_2 group. Red shifts and broadening of peaks are observed as hydrolysis progresses, but no clear trend is observed on the lifetime of mononuclear species while for polymeric species, an increase is clearly seen. Peak positions do not depend on the excitation wavelength, but the intensities are dependent.

The methodology followed by the authors in setting up the data is the same as in [95ELI/BID] (but different from that of [94KAT/MEI] for species, 2:2 and 5:3).

Table A-38: Spectroscopic characterisation of 0:1, 1:1, 2:1, 3:1, 2:2, 5:3 and 7:3 species.

Species	Main fluorescence wavelength (nm)	Fluorescence lifetime (μs)	Reference
0:1	488, 509, 533, 559	(2 ± 0.1)	[98MOU/LAS]
	489, 510, 535, 560	(1.7 ± 0.2)	[95ELI/BID]
	488, 509, 534, 560	0.9	[94KAT/MEI]
1:1	497, 519, 544, 570	(80 ± 5)	[98MOU/LAS]
	496, 518, 542, 566	(32.8 ± 2)	[95ELI/BID]
2:1	488, 508, 534, 558	10–20	[98MOU/LAS]
		(3.2 ± 0.2)	[95ELI/BID]
3:1	499, 519, 543, 567	(0.8 ± 0.1)	[98MOU/LAS]
	506, 524, 555, 568	(0.4 ± 0.1)	[95ELI/BID]
2:2	497, 519, 542, 570	(9 ± 1)	[98MOU/LAS]
	498, 518, 544	(9.5 ± 0.3)	[95ELI/BID]
	499, 519, 542, 566	2.9	[94KAT/MEI]
5:3	496, 5514, 535, 556	(23 ± 3)	[98MOU/LAS]
	515, 536	(6.6 ± 0.3)	[95ELI/BID]
	500, 516, 533, 554	7	[94KAT/MEI]
7:3	503, 523, 547, 574	(230 ± 20)	[98MOU/LAS]
		(10 ± 2)	[95ELI/BID]

[98NAK/NIS]

Pressures of nitrogen were measured in the $\alpha - U_2N_{3+x}$ single-phase region for $0.26 < x < 0.52$. The starting material with $N/U > 1.80$ was prepared from UH_3 (s) and NH_3 (g) by heating at successive temperatures of 873 and 1273 K. Nitrogen pressures (0.001 to 0.2 bar) were measured from 673 to 1173 K; equilibrium was assumed to have been achieved when no pressure change was observed after 20 hours. The partial molal enthalpies and entropies of solution of N_2(g) were compared with earlier data, but not the pressures or Gibbs energies.

Such non-stoichiometric systems are not considered in any detail in the NEA-TDB reviews, and thus this paper contains no data relevant to this review.

[98NEC/FAN]

This report contains a review of the literature on the aqueous americium(III) and curium(III) complexes with numerous inorganic ligands. Based on selected literature data and (partly preliminary) experimental data of Fanghänel *et al.*, the authors applied the Pitzer approach for the thermodynamic modelling of trivalent actinides in NaCl solution. Considering Cm(III) as an analogue for Am(III) the authors proposed a comprehensive set of ion interaction Pitzer parameters and equilibrium constants at $I = 0$ and 25°C, without distinguishing between these two actinides. Literature data for aqueous Eu(III) and Nd(III) were also presented for comparison. The evaluated thermodynamic data and Pitzer parameters for aqueous An(III) ions and complexes were used to calculate the solubility of Am(III) hydroxide and carbonate solids as a function of the H^+ and CO_3^{2-} concentrations up to concentrated NaCl. The results were compared with solubility data available from the literature.

The cited experimental data of Fanghänel *et al.*, time resolved laser fluorescence studies of the aqueous Cm(III) complexes with chloride [95FAN/KIM], [97KON/FAN], hydroxide [94FAN/KIM], carbonate [98FAN/WEG], [98FAN/WEG2], [99FAN/KON], fluoride [99AAS/STE], silicate [97STE/FAN] and sulphate [96PAV/FAN], are also published in a series of papers. These papers are discussed in the corresponding sections and in Appendix A of the present review and the experimental data are re-analysed by applying the SIT instead of the Pitzer approach.

The only experimental data in this report, which were not published elsewhere, are cited as unpublished results of Könnecke, Paviet-Hartmann, Fanghänel and Kim. They are aimed at the study of Cm(III) sulphate complexes as a function of ionic strength in two series of 0.15 and 0.55 m Na_2SO_4 solutions containing 0 - 5.8 m NaCl at pH = 2. They are a continuation of the study at $I = 3$ mol $\cdot$ kg^{-1} (NaCl-Na_2SO_4) [96PAV/FAN], where the experimental procedures were described in detail.

Applying the Pitzer approach the authors calculated the formation constants, $\log_{10} \beta_1^\circ = 3.25$ and $\log_{10} \beta_2^\circ = 3.70$, for $AnSO_4^+$ and $An(SO_4)_2^-$, respectively, without giving uncertainty limits, for the reactions:

$$An^{3+} + n\ SO_4^{2-} \rightleftharpoons An(SO_4)_n^{3-2n} \quad \text{(A.105)}$$

The present review recalculates the formation constants at zero ionic strength by applying the SIT to the experimental data at $I \leq 4$ (*cf.* Figure 12-8, section 12.5.1.1) and obtains:

$$\log_{10} \beta_1^\circ \text{ (A.105)} = (3.28 \pm 0.03) \text{ and } \Delta\varepsilon_1 = -(0.14 \pm 0.02) \text{ kg} \cdot \text{mol}^{-1}$$

and

$\log_{10} \beta_2^{\circ}$ (A.105) = (3.59 ± 0.03) and $\Delta\varepsilon_2$ = − (0.24 ± 0.01) kg · mol^{-1}.

It is to be noted that the electrolyte mixtures include considerable sulphate concentrations. By applying the SIT only to the two $\log_{10} \beta_1^{\circ}$ values in 0.15 and 0.55 m Na$_2$SO$_4$ without additions of NaCl, a somewhat different constant of $\log_{10} \beta_1^{\circ}$ = (3.15 ± 0.03) is obtained, but the value of $\Delta\varepsilon_1$ = − 0.14 kg · mol^{-1} is the same.

[98NEC/KON]

These authors determined Pitzer ion–interaction parameters for the system: Cs$^+$/Na$^+$/K$^+$/Mg^{2+}/Cl$^-$/SO$_4^{2-}$/TcO$_4^-$/H$_2$O at 25°C and measured the solubility of CsTcO$_4$(cr) in pure water, and in MgCl$_2$, Mg$_2$SO$_4$, Na$_2$SO$_4$ and Cs$_2$SO$_4$ solutions at 298.15 K. The solubility of CsTcO$_4$(cr) is so low that its component ions were considered as trace components of the solution. The authors use a previously measured value of $\log_{10} K_{s,0}^{\circ}$ = − (3.607 ± 0.023) [97KON/NEC], which when combined with $\Psi_{Cs^+/TcO_4^-/Cl^-}$ = − 0.001 [97KON/NEC] and $\theta_{TcO_4^-/SO_4^{2-}}$ = 0.179 [98NEC/KON2], gives $\Psi_{Cs^+/TcO_4^-/SO_4^{2-}}$ = 0.0242. The remaining Pitzer parameters were derived mainly from [84HAR/MOL] and measurements of Rard and Miller [82RAR/MIL]. An analysis of the $\log_{10} K_{s,0}$ values for the four media using the SIT by this review gave a linear relationship with the ionic strength only for MgCl$_2$ solutions (truncated at an ionic strength of 7 m). This is not surprising. Activity coefficients in sulphate systems are notoriously difficult to model. Fixing $\log_{10} K_{s,0}^{\circ}$ at − 3.607 resulted in a $\Delta\varepsilon$ value of − (0.0335 ± 0.0027) kg · mol^{-1}. The mean activity coefficients of CsTcO$_4$(cr) derived from the ion-interaction treatment are shown graphically to reproduce closely the corresponding stoichiometric values, $\gamma_{\pm} = (K_{s,0}^{\circ} / m_{Cs} m_{TcO_4})^{1/2}$.

[98NEC/KON2]

Isopiestic measurements were performed at 298.15 K on pure Mg(TcO$_4$)$_2$ solutions and mixtures of NaTcO$_4$/NaCl, Mg(TcO$_4$)$_2$/MgCl$_2$, Mg(TcO$_4$)$_2$/NaTcO$_4$, Mg(TcO$_4$)$_2$/MgSO$_4$, and NaTcO$_4$/Na$_2$SO$_4$, using NaCl and CaCl$_2$ reference solutions, with the concentrations at equilibrium being determined gravimetrically. The authors list the binary Pitzer parameters for NaTcO$_4$(cr) from [97KON/NEC] as: $\beta_{Na^+/TcO_4^-}^{(0)}$ = 0.01111, $\beta_{Na^+/TcO_4^-}^{(1)}$ = 0.1595, and $C_{Na^+/TcO_4^-}^{\varphi}$ = 0.00236. From the present isopiestic study of Mg(TcO$_4$)$_2$ (0.16 – 3.8 m), the binary parameters are reported as: $\beta_{Mg^{2+}/TcO_4^-}^{(0)}$ = 0.3138, $\beta_{Mg^{2+}/TcO_4^-}^{(1)}$ = 1.840, and $C_{Mg^{2+}/TcO_4^-}^{\varphi}$ = 0.0114. The solubility of KTcO$_4$(cr) was measured at 25°C in KCl, K$_2$SO$_4$, MgCl$_2$ and CaCl$_2$ solutions. From the results in KCl (and taking $\theta_{TcO_4^-/Cl^-}$ = 0.067 [97KON/NEC]), it was determined that $\log_{10} K_{s,0}^{\circ}$ (KTcO$_4$) = − (2.239 ± 0.013), $\beta_{K^+/TcO_4^-}^{(0)}$ = − 0.0578, $\beta_{K^+/TcO_4^-}^{(1)}$ = 0.006, and $C_{K^+/TcO_4^-}^{\varphi}$ = 0, and $\Psi_{K^+/TcO_4^-/Cl^-}$ = − 0.0113 (where the covariance of the latter two parameters was so large that only one value could be justified). The solubility of KTcO$_4$(cr) in pure water was measured at (0.1040 ± 0.0015) mol · kg^{-1}. The solubility data for KTcO$_4$(cr) in CaCl$_2$ solutions (≤ 4.36 m) yielded the parameters: $\beta_{Ca^{2+}/TcO_4^-}^{(0)}$ = 0.2964, $\beta_{Ca^{2+}/TcO_4^-}^{(1)}$ = 1.661, $C_{Ca^{2+}/TcO_4^-}^{\varphi}$ = 0, and $\Psi_{Ca^{2+}/TcO_4^-/Cl^-}$ = − 0.033. The mixing parameter, $\Psi_{Mg^{2+}/TcO_4^-/Cl^-} = -0.0115$, was determined from the solubility data for KTcO$_4$(cr) in MgCl$_2$ solutions (≤ 4.65 m). From the osmotic coefficient measurements of the sulphate mixtures and the solubility of KTcO$_4$(cr) in K$_2$SO$_4$ (≤ 0.62 m), the following mixing parameters were reported: $\theta_{TcO_4^-/SO_4^{2-}} = 0.179$, $\Psi_{Na^+/TcO_4^-/SO_4^{2-}} = -0.003$, $\Psi_{K^+/TcO_4^-/SO_4^{2-}} = 0.002$, $\Psi_{Mg^{2+}/TcO_4^-/SO_4^{2-}} = -0.030$. Isopiestic measurements in Mg(TcO$_4$)$_2$/NaTcO$_4$ mixtures yielded the ternary parameter, $\Psi_{Na^+/Mg^{2+}/TcO_4^-} = -0.020$ given that $\theta_{Na^+/Mg^{2+}} = 0.07$ (cf. [84HAR/MOL]).

[98PAN/CAM]

The experimental data and results reported in this paper were already published in an earlier report of the same authors [95PAN/CAM], which has been extensively discussed in the NEA-TDB review of neptunium and plutonium [2001LEM/FUG]. The reported solubility data at pH < 8 refer to a solubility constant of $\log_{10} {}^*K^\circ_{s,0} = (3.90 \pm 0.02)$ for the reaction:

$$\frac{1}{2} Np_2O_5(cr) + H^+ \rightleftharpoons NpO_2^+ + \frac{1}{2} H_2O(l)$$

and $\Delta_f G^\circ_m$ (Np$_2$O$_5$, cr, 298.15 K) $= -(2008 \pm 11)$ kJ · mol^{-1}, which is considerably less negative than $\Delta_f G^\circ_m$ (Np$_2$O$_5$, cr, 298.15 K) $= -(2031.6 \pm 11.2)$ kJ · mol^{-1} calculated in [2001LEM/FUG] for ideal crystalline Np$_2$O$_5$(cr) with the selected values of $\Delta_f H^\circ_m$ (Np$_2$O$_5$, cr, 298.15 K) $= -(2162.7 \pm 9.5)$ kJ · mol^{-1} and S°_m (Np$_2$O$_5$, cr, 298.15 K) $= (174 \pm 20)$ J · K^{-1} · mol^{-1}.

The conclusion that the Np$_2$O$_5$(cr) samples used in the solubility experiments might have had large numbers of active sites on the surface [2001LEM/FUG] is supported by the lower solubilities measured by Efurd et al. [98EFU/RUN] with poorly crystalline Np$_2$O$_5$(s, hydr.) ($\log_{10} {}^*K^\circ_{s,0} = (2.6 \pm 0.4)$).

[98PIA/TOU]

Using the same high temperature X–ray diffraction techniques as in two other publications [94TOU/PIA], [99PIA/TOU], the authors studied the reaction between UO$_2$ and CaCO$_3$ (1/1), UO$_2$ and CaO (Ca/U = 1, 2 or 3), together with the thermal decomposition of Ca$_3$UO$_6$, CaUO$_4$ and Ca$_2$UO$_5$ at oxygen pressures in the range $10^{-19} \leq p_{O_2} /$bar ≤ 1. The rhombohedral phase, "α–CaUO$_4$", has a large homogeneity range, extending down to O/U = 3.15 according to the authors. The lattice parameters at 293 K given for the samples containing the highest and the lowest O/U ratios are in reasonable agreement with earlier literature results.

A monoclinic "CaUO$_3$" perovskite phase, not previously described in the literature, with an O/U ratio of ca. 3, is observed between 1573 and 1753 K, under the most reducing conditions. It is described as being isomorphous with Ca$_3$UO$_6$ (space group P2$_1$), which can be written Ca$_2$(Ca,U)O$_6$, but less distorted and with distinctly larger lattice parameters. The decrease in lattice dimensions from CaUO$_3$ to Ca$_2$(Ca,U)O$_6$ is attributed to the replacement of U(IV) by the smaller U(VI). The authors also attribute the limited stability of "CaUO$_3$", to the fact that the Goldschmidt tolerance factor ([70GOO/LON] page 132) is so close to the limit of existence of this structure. Given the uncertainty in the stoichiometry of the CaUO$_3$ described in this paper, the results of [98PIA/TOU] are not incompatible with the previously described trend (see, for instance, [90GOU/HAI], [93FUG/HAI]) of a decreasing stability, with reference to the binary component oxides, of the M(II)M'(IV)O$_3$ oxides, with the decreasing size of the M(II) ion and the increasing size of the M'(IV) ion. The authors also give arguments for a monoclinic – to – orthorhombic transformation of the "CaUO$_3$" phase around 1773 K.

Above *ca.* 1773 K, under reducing conditions, a fluorite solid solution U$_{1-\delta}$Ca$_\delta$O$_{2-\delta}$ is formed.

The authors also present a schematic pseudo-binary UO$_2$–CaO phase diagram, but provide no thermodynamic data.

[98RAI/FEL]

This is a continuation of [95RAI/FEL], where Rai *et al.* make an attempt to quantify previous experimental information and propose stoichiometric compositions and equilibrium constants for the ternary hydroxide/carbonate complexes of U(IV). One part of the paper describes an EXAFS study of the structure of U(CO$_3$)$_5^{6-}$, the limiting carbonate complex. This provides an excellent, non-thermodynamic, confirmation of the stoichiometry and mode of coordination in this complex. It was helpful to have more details on the EXAFS data, including an estimate of the uncertainty in the estimated parameters. The presence of U(CO$_3$)$_4^{4-}$ in the test solutions was confirmed by spectrophotometry. However, the quantitative interpretation of the solubility data is not satisfactory in the opinion of the reviewers. The experimental methods used are excellent and precautions have been taken to ensure that the uranium is present as U(IV). This turned out to be impossible at high concentrations of hydroxide (> 0.11 m) and at low carbonate concentrations, *cf.* Table 1 in [98RAI/FEL]. Despite the precautions taken one cannot rule out that the measured solubility might be affected by the presence of U(VI). The authors use the Pitzer model to interpret their experimental results. This is fine in principle. However, when using experimental data for a simultaneous determination of the equilibrium constant at zero ionic strength and a set of interaction parameters for the complexes, one is faced with a large correlation between the parameters. The problems are described in Chapter IX of [97ALL/BAN]. Hence it is essential to present an error estimate of the quantities deduced. Rai *et al.* [98RAI/FEL] have not done this. The agreement between the experimental solubility and that calculated from the proposed model is unsatisfactory as shown in the figures of this paper. The most important step in any interpretation of experimental data is to establish the chemical model, *i.e.*, the stoichiometry of the complexes. The experimental data in Na$_2$CO$_3$ solutions indicate the predominance of the well-established U(CO$_3$)$_5^{6-}$, but also a large deviation between the measured and calculated solubility at carbonate concentrations larger than 1 m. Figure 5 of [95RAI/FEL] indicates that this can be up to 0.5 logarithmic units. No explanation is given for this discrepancy. The deviations both in the slope and the measured solubility are even larger in the K_2CO_3 system. As the limiting complex is well established this review suggests that the observed deviations are a result of oxidation of U(IV) and/or the formation of additional complexes. The Pitzer parameters used are estimates but seem to be reasonable.

In regard to the data obtained in solutions of alkali carbonate and alkali hydroxide, the experimental data given in Figure 6 of [95RAI/FEL] show very clearly that the stoichiometry proposed by Rai *et al.* is incorrect. A slope of two is in much better agreement with the experimental data, indicating that the predominant equilibrium in these solutions may involve two OH$^-$ ions on the product side, *i.e.*,

$$UO_2(am) + xCO_3^{2-} + 2H_2O(l) \rightleftharpoons U(CO_3)_x(OH)_2^{(2-2x)} + 2OH^-$$

where the reviewers suggest the value of x is equal to three or four. This would retain a high coordination number for U(IV), eight or ten, and involve the replacement of only one or two coordinated carbonate ligands. Slope analysis can be used in this case because K_2CO_3 is present in much larger concentrations than NaOH, and thus acts as an ionic medium. The reviewers have used the data in their Table A2, which contains the two-day experiments to calculate the following solubility equilibrium:

$$UO_2(am) + 4CO_3^{2-} + 2H_2O(l) \rightleftharpoons U(CO_3)_4(OH)_2^{6-} + 2OH^-$$

for which

$$\log_{10} K_s = \log_{10} s + 2\log_{10}[OH^-] - 4\log_{10}[CO_3^{2-}] = -(5.47 \pm 0.25)$$

where s is the measured solubility. This is a concentration equilibrium constant valid in the ionic medium used, *i.e.*, 3 m K_2CO_3 + NaOH (0.01 – 0.1) m.

The data in bicarbonate solutions have also been reinterpreted. Slope analysis of the data in NaHCO$_3$ solutions indicates clearly the presence of $U(CO_3)_5^{6-}$ as the limiting complex. The decrease in slope observed at the lower bicarbonate concentrations is not due to the presence of $U(OH)_2(CO_3)_2^{2-}$ as suggested by Rai *et al.*, whereas the formation of $U(CO_3)_4^{4-}$ seems also probable. There may be problems with the experimental data, with evidence of extensive oxidation at low bicarbonate concentrations. The data in KHCO$_3$ solutions deviate strongly from the quantitative model used by Rai *et al.* The majority of the experimental data fall close on a line with the slope of five. To conclude, the quantitative equilibrium constants and the Pitzer parameters proposed by Rai *et al.* describe the measured solubility fairly well, but they are better regarded as phenomenological parameters until an independent verification of the speciation is available. Rai *et al.* proposed a solubility constant of $\log_{10} K_{s,5}^\circ$ for the reaction:

$$UO_2(am) + 2H_2O(l) + 5CO_3^{2-} \rightleftharpoons U(CO_3)_5^{6-} + 4OH^-$$

without giving uncertainty limits. If we combine the value $\log_{10} K_{s,5}^\circ = -22.15$ proposed by Rai *et al.* with the corresponding equilibrium constants for the reactions:

$$UO_2(am) + 4H^+ \rightleftharpoons U^{4+} + 2H_2O(l)$$

$$H^+ + OH^- \rightleftharpoons H_2O(l),$$

we obtain $\log_{10} \beta_5^\circ = (32.35 \pm 1.40)$, which is within the estimated uncertainties equal to the value, $\log_{10} \beta_5^\circ = (34.0 \pm 0.9)$ selected in [92GRE/FUG]. The uncertainty estimates for the data of Rai et al. have been made by assuming an uncertainty of $\pm 1.0 \log_{10}$ units in $\log_{10} K_{s,5}^\circ$ (the interaction coefficients for ions of such high charge have in general a large uncertainty). The discrepancy of 1.65 units in $\log_{10} \beta_5^\circ$ may partly be due to a systematic deviation between the activity coefficients calculated by Rai et al. and those obtained by using the SIT approach.

Conversely, combination of $\log_{10} K_{s,5}^\circ = -22.15$ with $\log_{10} \beta_5^\circ = (34.0 \pm 0.9)$ gives $\log_{10} K_{s,0}^\circ (UO_2, am) = -(56.15 \pm 1.3)$ or $\log_{10} {}^*K_{s,0}^\circ (UO_2, am) = -(0.15 \pm 1.3)$.

To conclude, the presence of $U(CO_3)_5^{6-}$ as the limiting complex has been confirmed and the suggested equilibrium constant is in fair agreement with the value proposed in other studies reviewed in [92GRE/FUG]. The experimental data show that ternary hydroxide/carbonate species are formed in strongly alkaline solutions and this review suggests that this species most likely contains two hydroxides per uranium.

[98SAI/CHO]

Some polyoxometalate anions, L^{X-}, are stable around pH = 4. This paper deals with the complexation of U(VI) for concentrations in the range $2 \cdot 10^{-6}$ to $2 \cdot 10^{-7}$ M (U labelled with ^{233}U) in solutions buffered at pH = 4 (10^{-2} M acetate) with sodium or ammonium salts of L^{X-} as counter ions. The ionic strength is kept constant at 0.1 M with $NaClO_4$. The polyoxoanions studied are two isopolyanions:

A: $V_{10}O_{28}^{6-}$, decavanadate and

B: $Mo_7O_{24}^{6-}$, heptamolybdate;

and four heteropolyanions:

C: $CrMo_6O_{24}H_6^{3-}$ hexahydrogenohexamolybdo(VI)chromate(III),

D: $IMo_6O_{24}^{5-}$, hexamolybdo(VI)iodate(VII),

E: $TeMo_6O_{24}^{6-}$, hexamolybdo(VI)tellurate(VI), and

F: $MnMo_9O_{32}^{6-}$, monomolybdo(VI)manganate(IV).

The complex formation was studied using competitive solvent extraction at room temperature $(23 \pm 1)°C$. The distribution coefficient, D, is measured as a function of the calculated free concentration of L in the aqueous phase (assuming complete dissociation of the salts which have pKs in the range 2 to 4). The data are analysed assuming the formation of outer-sphere mono-nuclear complexes between UO_2^{2+} and L^{X-}.

To derive $\log_{10} \beta_n$ values for the various ML_n complexes, corrections are made for competitive complexation by the acetate anion ($\log_{10} \beta_p$) = 2.6, 4.9 and 6.3 for $p = 1, 2$ and 3 and hydrolysis of UO_2^{2+}, $\log_{10} \beta_1 = 8.0$.

The range of variation of [L] and the equilibrium constants are given in the Table A-39, where the errors in $\log_{10} \beta_n$ are calculated using an uncertainty of $\pm 5\%$ in D values.

Table A-39: Equilibrium constants of U(VI) with polyanions L^{x-} at pH = 4.

	A	B	C	D	E	F
[L] (in mol · L^{-1})	< 8·10^{-3}	< 5·10^{-3}	< 1.6·10^{-2}	< 8·10^{-3}	< 10^{-5}	< 5·10^{-5}
$\log_{10} \beta_1$	(2.40 ± 0.1)	(3.88 ± 0.3)	(2.05 ± 0.06)	(2.57 ± 0.07)	(3.16 ± 0.04)	(3.53 ± 0.2)
$\log_{10} \beta_2$					(5.25 ± 0.3)	

[98SAV]

The author presents an empirical linear relation between the solubility constant, $\log_{10} K^\circ_{s,0}$ (M(OH)$_3$, cr), of trivalent transition metal and lanthanide ions, and the formation constant, $\log_{10} \beta^\circ_3$, of the complex, M(OH)$_3$(aq). They apply this relation to estimate $\log_{10} \beta^\circ_3$ for U(III), Pu(III) and Am(III). However, the solubility constants used in this paper ($\log_{10} K^\circ_{s,0} = -19.0$ for U(III), -19.7 for Pu(III), and -23.3 for Am(III), corresponding to $\log_{10} {}^*K^\circ_{s,0} = 23.0$, 22.3, and 18.7, respectively) are adopted without critical evaluation from a handbook on analytical chemistry. They deviate by several orders of magnitude from the solubility constants selected in the NEA reviews [92GRE/FUG], [95SIL/BID], [2001LEM/FUG] after critical discussion of the available experiment data. The same holds for the estimated values of $\log_{10} \beta^\circ_3$ for which the authors suggested an unreasonable increase of 4.3 orders of magnitude from Pu(III) to Am(III).

[98SCA/ANS]

The authors have performed TRLFS measurements on U(VI) solutions with [U] = 10^{-7} M and [H$_3$PO$_4$] = 10^{-3} and 10^{-4} M, at pH = 1.5, 5 and 7.5, (t = 25°C) open to the atmosphere. Under these conditions, according to the selected data of [92GRE/FUG] the following species are predicted to predominate in these solutions: UO_2^{2+} together with few percent of $UO_2H_2PO_4^+$ (for 10^{-4} M in uranium), UO_2HPO_4(aq) and $UO_2PO_4^-$ (both for 10^{-3} M in uranium). The known decay lifetimes of hydrolysed species give the fluorescence characteristics of each of the presumed phosphoric complexes (see Table A-40). The fluorescence intensities of the species, $UO_2H_2PO_4^+ / UO_2HPO_4$(aq) / $UO_2PO_4^-$ are in the ratio 6600/1200/800.

Table A-40: Emission wavelengths and lifetimes of emission of UO_2^{2+}, $UO_2H_2PO_4^+$, $UO_2HPO_4(aq)$ and $UO_2PO_4^-$.

Species	λ max, nm	lifetime, μs
UO_2^{2+}	488, 509, 533, 559	(2 ± 0.1)
$UO_2H_2PO_4^+$	494, 515, 539, 565	(11 ± 1)
UO_2HPO_4 (aq)	497, 519, 543, 570	(6 ± 0.5)
$UO_2PO_4^-$	499, 520, 544, 571	(24 ± 2)

Ionic strength is not given (I around 10^{-3} M) and the manner in which the thermodynamic data have been utilised is not provided. The lifetime of $UO_2HPO_4(aq)$ does not agree with data of [96BRE/GEI]. These results confirm indirectly the thermodynamic constant selected by [92GRE/FUG].

[98SER/RON]

This paper is a good review of the problems encountered with spent fuel leaching. It gives the cumulative fraction (%) and normalised cumulative fraction (g·cm^{-2}) of U released into deionised water from UOX, MOX, Simfuels and UO_2 as a function of time, over more than 1000 hours in contact with air. The normalised release of uranium is more or less the same, around 10^{-5} g·cm^{-2}, after 1000 hours. No other data are measured.

[98SHI]

The author reviews published data on the chemistry of actinides in oxidation states (III) to (VII) in alkaline solutions and suggests formal potentials for the couple An(V)/An(IV) in 1 M NaOH. The values suggested for U, Np, Pu, Am and Cm are, -0.13, 0.13, 0.53, 0.31 and 1.20 V, respectively, against the NHE. The values for U, Pu, Am and Cm are *estimated* from the formal potentials of the Np system in acidic and alkaline solutions by assuming that the species and their equilibrium constants are the same for all elements. The interpretation of the available experimental data given by Shilov [98SHI] is not straight forward, as activity variations and the formation of sparingly soluble phases have to be taken into account.

The first issue discussed is the formal potential of the Np(VII)/Np(VI) couple where the data (Figure 1 in [98SHI]) indicate a half cell reaction:

$$NpO_2(OH)_4^{2-} + 2OH^- \rightleftharpoons NpO_4(OH)_2^{3-} + 2H_2O(l) + e^- \qquad (A.106).$$

This speciation of Np(VI) is consistent both with the stoichiometry of the limiting hydroxide complex in the U(VI) system, discussed in section 9.3.2.1., and the X–ray structural data. Shilov discusses the stepwise formation of hydroxide complexes of Np(VII) based on slope analyses of the formal potentials as a function of the activity of OH$^-$. As the formal potential also includes the (unknown) activity coefficients of the Np(VI) and

Np(VII) species, only an approximate value of the formal potential for reaction (A.106) equal to 0.38 V, or $\log_{10}K(A.106) = 6.4$, can be estimated from the data in Figure 1 of [98SHI]. In view of these shortcomings, this review does not consider the proposed stoichiometry of the complexes, $NpO_4(H_2O)_2^-$ and $NpO_4OH(H_2O)^{2-}$ to be sufficiently well established to be included in the selected data. The same holds for the analogous considerations reported for the system Pu(VII)/Pu(VI).

Shilov [98SHI] also used the formal potential of the Np(VI)/Np(V) couple *versus* $\log_{10} a_{OH^-}$ to obtain information of the speciation and suggests that the following electrode reactions take place in dilute alkali:

$$NpO_2(H_2O)_5^+ + 3OH^- \rightleftharpoons NpO_2(OH)_3(H_2O)^- + e^- + 4H_2O(l) \qquad (A.107)$$

where "dilute alkali" corresponds to concentrations between 0.1 and 1 M. It is well known from data reviewed in [2001LEM/FUG] that Np(V) is extensively hydrolysed in this concentration region. Shilov [98SHI] suggests two other reactions at higher hydroxide concentrations:

$$NpO_2(OH)_2(H_2O)_n^- + 2OH^- \rightleftharpoons NpO_2(OH)_4^{2-} + nH_2O(l) + e^- \qquad (A.108)$$

and

$$NpO_2(OH)_3(H_2O)_m^{2-} + OH^- \rightleftharpoons NpO_2(OH)_4^{2-} + mH_2O(l) + e^-. \qquad (A.109)$$

As there is no evidence for the formation of Np(V) hydroxide complexes with the stoichiometry, $NpO_2(OH)_3(H_2O)_m^{2-}$, these proposals are at best speculative. Therefore, this review does not accept the conclusions drawn by Shilov on the speciation in the Np(VI)/Np(V) system. The selected value in [2001LEM/FUG] for the formal potential of the Np(VI)/Np(V) couple in 1 M $HClO_4$ is 0.743 V. The value in 1 M NaOH estimated from curves 2 and 3 in Figure 1 in [98SHI] is (0.15 ± 0.04) V. Using the selected value for the formation of $NpO_2(OH)_2^-$ and an estimated value for the formation of $NpO_2(OH)_4^{2-}$ from the corresponding U(VI) system, we obtain an estimated value of 0.26 V for the formal potential in 1 M NaOH. The deviation from the experimental value is large, indicating that no quantitative conclusions can be drawn from the analysis given by Shilov.

Analogous considerations are reported for the Pu(VI)/Pu(V) system in highly alkaline solution. The proposed redox equations are based on a Pu(V) speciation including hydroxide complexes with the stoichiometries, $PuO_2(OH)_3^{2-}$ and even $PuO_2(OH)_4^{3-}$. As mentioned above there is no proof for the existence of such aqueous An(V) hydroxides [2001LEM/FUG].

Shilov [98SHI] also discusses the formal potential of the Np(V)/Np(IV) couple, where the interpretation is complicated by precipitation of $Np(OH)_4(s)$. The formal potential was described using the following half-cell reactions:

$$(Np(OH)_4)_n + 2OH^- + H_2O(l) \rightleftharpoons (Np(OH)_4)_m + NpO_2(OH)_2(H_2O)_n^- + e^- \qquad (A.110)$$

and

$$\text{Np(OH)}_5(\text{H}_2\text{O})_x^- + 2\,\text{OH}^- \rightleftharpoons \text{NpO}_2(\text{OH})_3(\text{H}_2\text{O})_m^{2-} + (x-m)\,\text{H}_2\text{O(l)} + e^-. \quad (A.111)$$

The experimental slope is – 3 and deviates significantly from the values expected from equations (A.110) and (A.111). In addition there is no satisfactory experimental evidence for the formation of Np(OH)_5^-, cf [2001LEM/FUG].

The formal potential of the Pu(IV)/Pu(III) couple in 0.5 – 14 M NaOH, which is found to be – 0.96 V at $a_{\text{OH}^-} = 1$, was described with the reaction:

$$\text{Pu(OH)}_3(\text{H}_2\text{O})_m + 2\,\text{OH}^- \rightleftharpoons \text{Pu(OH)}_5(\text{H}_2\text{O})_x^- + (m-x)\text{H}_2\text{O} + e^-. \quad (A.112)$$

The formation of anionic Pu(IV) hydroxide complexes Pu(OH)_5^- is concluded from solubilities of Pu(OH)_4(am), which increase from $1.4 \cdot 10^{-8}$ to $9.8 \cdot 10^{-6}$ mol·L^{-1} in 0.5 – 14 M NaOH. The aqueous form of Pu(III) is assumed to be predominantly Pu(OH)_3(aq) in these solutions, in analogy with the speciation of Am(III). The solubility of Am(OH)_3(s) is claimed to remain at a constant value of about $1.6 \cdot 10^{-6}$ mol·L^{-1}, in dilute and 3–5 M NaOH. However, as discussed in the section 12.3.2.2 (cf. Table 12-3), these americium concentrations considerably exceed most of the experimental solubility data in alkaline solutions.

This paper was not quoted in [2001LEM/FUG], but it does not provide data that could lead changes in [2001LEM/FUG].

[98SPA/PUI]

The problem of analysing experimental data for weak complexes is discussed in the paper. A method is proposed which considers the change of activity coefficients when the ionic medium composition is changed. The method is applied for a reinterpretation of literature data on the formation of nitrate complexes of Pu and Np. It is shown that only the mononitrato complex of the tetravalent Np and Pu are relevant at $I < 6$ M and at $[\text{NO}_3^-] < 2$ M whereas the formation of nitrate complexes of actinides in the penta- and hexavalent state could not be confirmed. It is demonstrated that the complexation constants for An(V) and An(VI) nitrato complexes published in the literature are a result of variations in the activity coefficients caused by changes of the ionic medium when the ligand concentration is increased and approaches that of the ionic medium. This is again an example of the impossibility of distinguishing between complex formation and activity coefficient variations from experimental solution thermodynamics in systems where weak complexes are formed. The equilibrium constant for the formation of the first complex can be obtained from solutions where the ligand concentration is not too high and the numerical value is less affected by activity factor variations.

The equilibrium constants at infinite dilution are reported for the complexes NpNO_3^{3+} and PuNO_3^{3+}. Both values are consistent with the formation constants selected in [92GRE/FUG] for uranium.

[98WER/SPA]

Data on solubilities from two independent sources (The Swedish – SKB programme and USA–Pacific Northwest Lab.), of U, Np, Pu and Am (at 20 – 25°C, under oxic/anoxic conditions, using groundwater/deionised water, and with all samples filtered using the same technique) obtained from spent nuclear fuel are in rather good agreement, but cannot be explained on the basis of thermodynamic data. Both sets of experimental data come from spent nuclear fuel of comparable burn up that have been studied over a period of four to five years. The formation of secondary phases (schoepite ?) is demonstrated but these phases are not characterised.

The speciation of U in the equilibrium solutions was calculated using the data selected by [92GRE/FUG]. Under oxic conditions at pH = 7, where schoepite (in fact metaschoepite) is stable, one expects that the 5:3 species is predominant with a solubility of 10^{-4} M. However, the measured solubility in deionised water is three to four orders of magnitude lower. If the solubility is determined by the formation of U_3O_7 on the surface of UO_2, a better agreement is obtained (a solubility of 10^{-7} M as compared to an experimental value of 10^{-8} M). However, there is no evidence that this phase is formed.

Another factor that must be considered when trying to model chemical processes in this type of « real » system is that mixed phases are formed for which the thermodynamic data for pure phases are not applicable.

The authors have also made modelling calculations assuming $Pu(OH)_4$(am), $Np(OH)_4$(am) and $AmOHCO_3$(am) are the limiting solubility phases, but with little success.

This paper does not give any data useful for this review.

[98YAM/KIT]

This is a solubility study of the system, $Na^+ - OH^- - UO_2^{2+} - CO_3^{2-} - ClO_4^- - H_2O$ at 25°C, with [OH$^-$] ranging from $2 \cdot 10^{-3}$ to 1 M, and [CO_3^{2-}] from 10^{-3} to 0.5 M. Three different ionic strengths, 0.5, 1.0 and 2.0 M, were used. The authors have used the solid phase, $Na_2U_2O_7 \cdot xH_2O$(cr) (x = 3 – 5), which has been partly characterised by powder X-ray diffraction. In order to ensure that solubility equilibrium has been attained, the equilibrium was approached from both under- and over-saturated solutions, with concordant results. The hydrogen ion concentrations have been properly measured, with corrections for liquid junction effects. The experimental solubility equilibria refer to the following reactions:

$$\frac{1}{2}Na_2U_2O_7(s) + 3\,CO_3^{2-} + \frac{3}{2}H_2O(l) \rightleftharpoons UO_2(CO_3)_3^{4-} + 3\,OH^- + Na^+ \quad (A.113)$$

and

$$\frac{1}{2}Na_2U_2O_7(s) + (n-3)\,OH^- + \frac{3}{2}H_2O(l) \rightleftharpoons UO_2(OH)_n^{2-n} + Na^+, n = 3 \text{ and } 4 \tag{A.114}$$

i.e., the authors have only considered the formation of binary mononuclear complexes. It is unlikely that polynuclear complexes are formed at the low uranium concentrations in the test solutions. The absence of ternary complexes containing both carbonate and hydroxide is more unexpected, and the authors have only been able to make estimates of the maximum values of equilibrium constants for such hypothetical complexes. This review accepts the chemical model, but suggests that additional experimental efforts should be made to establish if ternary complexes are formed, and under what conditions they should be studied. The solid phase used in these experiments should also be better characterised.

The solubility product of the solid expressed as:

$$K_{s,0}^\circ = [UO_2^{2+}]\,[OH^-]^3\,[Na^+]$$

was determined indirectly using the experimental equilibrium constant for equation (A.113) and the value of the equilibrium constant for the formation of $UO_2(CO_3)_3^{4-}$ from [92GRE/FUG]. The experimental data were extrapolated to zero ionic strength using the SIT. The following values were obtained for the formation of the tri- and tetra-hydroxide complexes from the components UO_2^{2+} and OH^-: $\log_{10} \beta_3^\circ = (21.1 \pm 0.8)$ and $\log_{10} \beta_4^\circ = (23.6 \pm 0.7)$. The experimental value of $\Delta\varepsilon = 0.21\,\text{kg}\cdot\text{mol}^{-1}$ for reaction (A.113) is in good agreement with the value calculated from the revised interaction coefficients given in [95SIL/BID], $\Delta\varepsilon = 0.27\,\text{kg}\cdot\text{mol}^{-1}$. The interaction coefficients between $UO_2(OH)_3^-$ and $UO_2(OH)_4^{2-}$ and Na^+ are equal to $-(0.82 \pm 0.20)$ and $-0.16\,\text{kg}\cdot\text{mol}^{-1}$, respectively. The interaction coefficient for $UO_2(OH)_3^-$ has a surprisingly large negative value, which differs substantially from the estimate in [92GRE/FUG] $-(0.09 \pm 0.05)\,\text{kg}\cdot\text{mol}^{-1}$. The interaction coefficient for $UO_2(OH)_4^{2-}$, on the other hand, agrees well with the experimental values for both for U(VI) complexes and other anions with a charge of -2. Water is a reactant in the reactions studied, but the authors have not taken the variations in the activity of water into account when making the ionic strength corrections. The error is small, at most 0.02 $\log_{10}$ units in the reported equilibrium constants at zero ionic strength and less than 0.02 units in $\Delta\varepsilon$.

The value of $\log_{10} K_{s,0}^\circ$ is $-(29.45 \pm 1.04)$, as compared to the value $-(30.7 \pm 0.5)$, obtained from the standard Gibbs energies of formation given in [92GRE/FUG]. The latter value was erroneously calculated to be $-(28.09 \pm 0.47)$ in [98YAM/KIT]. The authors claim "a definite difference between both values may be … attributed to different solid phases". The difference, however, is small and within the estimated uncertainty of the two determinations.

The state of hydration of the solid phase after equilibrium does not seem to have been determined. If there is a variation in the water content it does not have a large effect on the solubility product. The thermodynamic data for $Na_2U_2O_7(cr)$ are discussed in [92GRE/FUG]. The solubility products obtained from calorimetry and solubility experiments are consistent with one another, however, the uncertainty is large in the solubility experiments. In view of the difficulties encountered in the characterisation of the solid phase the two sets of experiments may well refer to solids of different composition, at least of different water content. Yamamura et al. claim that there is a large discrepancy between the equilibrium constant for the formation of $UO_2(OH)_3^-$ between this study and a previous investigation by Sandino and Bruno [92SAN/BRU]. We do not agree with this statement. Sandino and Bruno used two different phases to determine the stoichiometry and equilibrium constants of anionic uranium(VI) complexes. By using schoepite (in fact metaschoepite) they obtain $\log_{10} {}^*\beta_3^\circ$ for the reaction:

$$UO_2^{2+} + 3\,H_2O(l) \rightleftharpoons UO_2(OH)_3^- + 3\,H^+ \qquad (A.115)$$

equal to $-(19.90 \pm 0.34)$ and $-(20.25 \pm 0.19)$, for an amorphous and crystalline schoepite phase, respectively. These values differ somewhat from those given in [92GRE/FUG], which were taken from Sandino's thesis [91SAN]. Using UO_2^{2+} and OH^- as components, $\log_{10} \beta_3^\circ$ is equal to (22.10 ± 0.34) and (21.75 ± 0.19), respectively. Sandino and Bruno [92SAN/BRU] also used solubility measurements of $(UO_2)_3(PO_4)_2 \cdot 4H_2O(s)$ to determine the equilibria in the hydroxide system. The equilibrium constant for reaction (A.115) is $\log_{10} {}^*\beta_3^\circ = -(19.74 \pm 0.18)$ and, $\log_{10} \beta_3^\circ = (22.26 \pm 0.18)$. The agreement between the different experimental determinations of $\log_{10} \beta_3^\circ$ in [92SAN/BRU] is satisfactory, taking both the experimental difficulties and the different chemical systems into account. The average value of $\log_{10} \beta_3^\circ$ is equal to (22.0 ± 0.4), where the uncertainty covers those in the different experiments. The solubility data for the three phases are thus consistent with the value (21.14 ± 0.79) in [98YAM/KIT]. Yamamura et al. also estimate upper values for the ternary complexes, $UO_2(CO_3)(OH)_2^{2-}$ ($\log_{10} \beta < 22.6$) and $UO_2(CO_3)_2(OH)_2^{4-}$ ($\log_{10} \beta < 23.5$) at $I = 0.5$ M, but give no justification for the proposed stoichiometry.

This is a study that clears up some of the contradictory information on the stoichiometry of uranium(VI) hydroxide complexes in alkaline solution. Yamamura et al. seem to be unaware of the study by Palmer and Nguyen–Trung [95PAL/NGU]. Because of the small total concentrations of uranium used in [98YAM/KIT], it is impossible to obtain information as to whether polynuclear species may form at high pH, or not, cf. [95PAL/NGU]. The formation of $UO_2(OH)_4^{2-}$ as the predominant complex, cf. [2000MOL/REI2], at high hydroxide concentrations is in agreement with a combined theoretical and experimental study by Wahlgren et al. [99WAH/MOL], but not with the work of Clark et al. [99CLA/CON].

[98YOO/CYN]

Uranium metal was studied by X–ray diffraction at pressures up to 100 GPa and temperatures up to 4300 K, in a diamond anvil cell, using laser heating. The tetragonal β–phase is stable only to ~3 GPa, (confirming previous work), but the orthorhombic α–phase is stable to ~70 GPa at about 2000 K; the phase in equilibrium with the liquid at all pressures is the γ–phase (bcc). Other findings are:

- $V_m(\gamma - \alpha)$ decreases markedly with pressure,
- the melting point of uranium is given as a function of pressure to 100 GPa, where $T_{fus} = \sim 4300$ K.

There are no thermodynamic data of relevance to the current review.

[99AAS/STE]

Fluoride complex formation of Cm(III) has been studied at 25°C using TRLFS. Two sets of experiments have been made, the first in 1 m NaCl to establish the chemical model, and the second by varying the NaCl concentration from 0 to 5 m to establish the parameters of the Pitzer model for the quantitative description of the system over a large NaCl concentration range. The first experiment indicates the formation of two fluoride complexes, CmF_n^{3-n} with n equal to 1 and 2. A precise equilibrium constant could only be determined for the first complex. No quantitative value could be obtained for the second complex due to precipitation of $CmF_3(s)$. The equilibrium constant at zero ionic strength, $\log_{10} \beta_1^\circ = (3.44 \pm 0.05)$ is in excellent agreement with the value selected in [95SIL/BID] for the corresponding Am(III) complex, $\log_{10} \beta_1^\circ = (3.4 \pm 0.4)$, but the uncertainty is much smaller. We select the value reported in [99AAS/STE] both for Cm and Am.

[99BAS/MIS]

From the molar enthalpies of solution of $UTeO_5(cr)$, $UTe_3O_9(cr)$, $\beta-UO_3$ and $TeO_2(cr)$ in 11 mol · dm^{-3} HCl, the authors calculated the standard molar enthalpies of formation of $UTeO_5(cr)$ and $UTe_3O_9(cr)$.

$UTeO_5(cr)$ and $UTe_3O_9(cr)$ were obtained by heating thoroughly ground $UO_2(cr)$ (mass fraction 0.9998) and $TeO_2(cr)$ (mass fraction 0.99995) in the molar ratios 1:1 and 1:3, respectively, for eight hours in air, in an alumina boat at 1023 and 900 K, respectively. $\beta-UO_3$ was prepared by thermal decomposition of ammonium diuranate in air at 773 K for 24 hours.

Each preparation involved progressive heating, isothermal treatments and multiple grindings. The products were reported as giving only the characteristic X-ray diffraction lines of the expected compounds. In addition, for $UTeO_5(cr)$ and $UTe_3O_9(cr)$, Te was determined by flame atomic absorption spectroscopy and U by absorption spectrophotometry of its peroxy complexes. Preparation and analytical procedures were also given with fewer details in [98MIS/NAM], which reported a study, by the same group, of the vaporisation behaviour and Gibbs energy of formation of the same two compounds.

The performances of the isoperibol solution calorimeter were duly verified with the dissolution of standard KCl and TRIS (tris-hydroxymethylaminomethane). The following molar enthalpies of solution were reported at 298.15 K:

$\Delta_{sol}H_m$ ($UTeO_5$, cr, in 11 mol · dm^{-3} HCl) = $-(69.6 \pm 0.7)$ kJ · mol^{-1},

$\Delta_{sol}H_m$ (UTe$_3$O$_9$, cr, in 11 mol $\cdot$ dm^{-3} HCl) = $-$ (142.5 $\pm$ 0.8) kJ $\cdot$ mol^{-1},

$\Delta_{sol}H_m$ (TeO$_2$, cr, in 11 mol $\cdot$ dm^{-3} HCl) = $-$ (52.1 $\pm$ 0.8) kJ $\cdot$ mol^{-1},

$\Delta_{sol}H_m$ (UO$_3$, β, in 11 mol $\cdot$ dm^{-3} HCl) = $-$ (78.7 $\pm$ 0.6) kJ $\cdot$ mol^{-1},

the uncertainties representing twice the standard deviation of the mean.

With $\Delta_f H_m^\circ$ (UO$_3$, β, 298.15 K) = $-$ (1220.3 $\pm$ 1.3) kJ $\cdot$ mol^{-1} (selected in [92GRE/FUG], and taken by the authors) and $\Delta_f H_m^\circ$ (TeO$_2$, cr, 298.15 K) = $-$ (321.0 $\pm$ 2.5) kJ $\cdot$ mol^{-1} (see section 14.1), the enthalpies of formation are recalculated to be:

$\Delta_f H_m^\circ$ (UTeO$_5$, cr, 298.15 K) = $-$ (1602.5 $\pm$ 2.8) kJ $\cdot$ mol^{-1},

$\Delta_f H_m^\circ$ (UTe$_3$O$_9$, cr, 298.15 K) = $-$ (2275.8 $\pm$ 8.0) kJ $\cdot$ mol^{-1}.

These correspond to enthalpies of formation from the binary oxides (γ–UO$_3$ and TeO$_2$(cr)) of $-$ 57.2 and $-$ 87.5 kJ $\cdot$ mol^{-1} for UTeO$_5$(cr) and UTe$_3$O$_9$(cr), respectively.

[99BOU/BIL]

The fluorescence lifetime of UO$_2^{2+}$ has been reported to vary from 1.4 to 59 μs and to increase with the increase of H$^+$ concentration. This paper gives data on τ(UO$_2^{2+}$) as a function of HClO$_4$ concentration (10^{-2} to 12.59 M) in perchloric acid solution and as a function of perchlorate anion concentration in HClO$_4$ + NaClO$_4$ mixtures, up to 8 M ([H$^+$] = 1, 3, 4 and 6 M which is the solubility limit of NaClO$_4$) and finally as the concentration of CF$_3$SO$_3$H (triflic acid, 10^{-2} to 11.3 M) at room temperature. Triflate forms weaker complexes than perchlorate. In perchlorate media the uranium concentration is kept constant at 5$\cdot$10^{-4} M (or 10^{-1} M for absorption spectra). The concentration is not indicated in the triflate study, but could be estimated to 10^{-4} – 10^{-3} M. All solutions are kept in polyethylene bottles under an inert atmosphere between each measurement.

In perchloric acid and perchlorate media the main fluorescence lines are at the same positions: 488, 509, 534, and 558 nm. τ increases from 1.6 μs (HClO$_4$, 10^{-2} M) to 91 μs (HClO$_4$, 10 M) and then decreases slightly up to 12 M. The absorption spectra remain unchanged. At constant ClO$_4^-$ concentration around 6 M, τ is more or less constant up to [H$^+$] = 3 to 4 M. At higher [H$^+$], the fluorescence decay is bi–exponential and can be resolved into two lifetimes: τ_1 = (16.6 $\pm$ 0.5) μs and τ_2 = (30 $\pm$ 2) μs.

Triflic acid and perchloric acid show the same behaviour at concentrations less than 1.5 M. For pure triflic acid, 11.3 M, there is a red shift in the fluorescence spectrum (494, 516, 544, 570 nm) which are those of dioxouranium(VI) triflate in powder form (wavelengths of 494, 516, 544, 570 nm), but the fluorescence lifetimes change in a complicated way as the acid concentration increases, being more or less constant up to 6 M (4.5 μs), followed by a drastic increase at 10 M and then a drastic decrease.

These data cannot be explained by the current mechanisms to account for the increase of fluorescence lifetimes of dioxouranium(VI) cation *versus* [H$^+$], which are discussed in some detail. The authors assume the formation of $UO_2ClO_4^+$ complex above 4.5 M ClO_4^-, the existence of which is mainly supported by the increase of both τ and the intensity of fluorescence. This paper raises the question of the coordination of the perchlorate anion to UO_2^{2+} in concentrated solutions of $HClO_4$ but the arguments presented are not clear and convincing. An answer to this question is given in additional studies [2001BIL/RUS] and [2001SEM/BOE]. In [2001BIL/RUS] it is shown that the dependencies of τ and emission spectra with $HClO_4$ and $NaClO_4$ concentrations (up to 10 M) cannot be attributed to the formation of an inner-sphere complex. That conclusion is based on the study of the quenching process of the fluorescence of UO_2^{2+} by chloride ion. The paper [2001SEM/BOE] deals with EXAFS measurements and shows definitively that ClO_4^- is not coordinated to UO_2^{2+}, even in 10 M $HClO_4$ (see Appendix A)

[99CAP/COL]

This paper is a comprehensive study of the ionisation and dissociation energies of the gaseous uranium and plutonium oxides by mass spectrometry of molecular beams produced by Knudsen effusion at high temperatures. The values obtained constitute a set of self–consistent quantities which agree with the existing thermodynamic data for these gaseous oxides within the combined uncertainties.

For this review, the only relevant data are the derived dissociation energies of UO(g), UO_2(g) and UO_3(g), which are, respectively, (7.81 ± 0.1), (15.7 ± 0.2) and (21.6 ± 0.3) eV, where the uncertainty for UO_2(g) is estimated here. The temperature for which such values apply is not very clear, but is conventionally taken to be 0 K. Correction to 298.15 K and SI units gives (757.2 ± 9.6), (1521.3 ± 19.3) and (2095.6 ± 28.9) kJ · mol^{-1} for the dissociation energies at 298.15 K and hence enthalpies of formation of:

$$\Delta_f H_m^\circ (UO, g, 298.15 K) = (25.3 \pm 12.5) \text{ kJ} \cdot \text{mol}^{-1},$$

$$\Delta_f H_m^\circ (UO_2, g, 298.15 K) = -(489.9 \pm 20.8) \text{ kJ} \cdot \text{mol}^{-1},$$

$$\Delta_f H_m^\circ (UO_3, g, 298.15 K) = -(815.1 \pm 30.0) \text{ kJ} \cdot \text{mol}^{-1}.$$

These values are consistent with the TDB-NEA selected values, (35.0 ± 10.0), $-(477.8 \pm 20.0)$ and $-(799.2 \pm 15.0)$ kJ · mol^{-1}.

[99CAP/VIT]

The authors have previously determined equilibrium constants at different temperatures and ionic strengths from measurements of the formal potential, E (the average of two half-wave potentials obtained in cyclic voltammetry). This paper gives results on the variation of $E°(T, I)$ (mV/SHE) (± 1 mV) for the M(VI)/M(V) redox potential for M = Pu in 0.3 (I = 0.9 M), 0.5, 1 and 1.5 M Na_2CO_3 (I = 4.5 M) solutions between 9 and 65°C, and gives a reinterpretation of data published in [90CAP/VIT] and [92CAP] for M = U. Under the conditions used the solutions are considered to contain the limiting tricarbonato complexes of M. The authors describe $E°(T,I)$ with a second degree polynomial and then deduce the temperature dependence of ε and $\Delta\varepsilon$ in the SIT model using these data. For the reaction: $MO_2(CO_3)_3^{4-} + e^- \rightleftharpoons MO_2(CO_3)_3^{5-}$, the standard redox potential for the $(MO_2(CO_3)_3^{4-} / MO_2(CO_3)_3^{5-})$ couple is obtained from the $E°(T,0)$ $(MO_2(CO_3)_3^{4-} / MO_2(CO_3)_3^{5-})$ and $\Delta\varepsilon(T,0)$ values, and then, $\Delta S°(T,I)$ and $\Delta C_p°(T,I)$ are obtained by plotting $E°(T,I)$ versus T at constant I, which in turn gives the ionic strength influence when these values are plotted versus I at constant T. Furthermore, considering the relation:

$$E°(T,0)(MO_2(CO_3)_3^{4-} / MO_2(CO_3)_3^{5-}) - E°(T,0)(MO_2^{2+} / MO_2^+) = \frac{RT \ln(10)}{F}\left(\log_{10}\beta_3°(V) - \log_{10}\beta_3°(VI)\right),$$

the standard redox potential involving aqueous cations can be derived from the values of the formation constants of tricarbonato complexes at $I = 0$, when known, or the reverse.

Data concerning U in 0.2 M Na_2CO_3 + 0.6 to 1.4 M $NaClO_4$ are reported as:

$$E°(SHE, 298.15 \text{ K}, 0)(UO_2(CO_3)_3^{4-} / UO_2(CO_3)_3^{5-}) = -(779 \pm 10) \text{ mV},$$

(the selected value in [92GRE/FUG] is $-(752 \pm 16)$),

$$\Delta S_m° (298.15 \text{ K}, 0) = -(174 \pm 5) \text{ J} \cdot \text{K}^{-1} \cdot \text{mol}^{-1},$$

$$\Delta C_{p,m}° (298.15 \text{ K}, 0) = -(414 \pm 176) \text{ J} \cdot \text{K}^{-1} \cdot \text{mol}^{-1}.$$

From:

$$E°(SHE, 298.15 \text{ K}, 0)(MO_2^{2+} / MO_2^+) = (87.9 \pm 1.3) \text{ mV},$$

selected in [92GRE/FUG], they derive:

$$\log_{10}\frac{\beta_3°(V)}{\beta_3°(VI)} = -(14.65 \pm 0.17),$$

from which using $\log_{10}\beta_3°(VI) = (21.60 \pm 0.05)$ [92GRE/FUG], the value $\log_{10}\beta_3°(V) = (6.95 \pm 0.18)$ is proposed.

The variation of $\Delta\varepsilon(T, 0) = \varepsilon(UO_2(CO_3)_3^{4-}, Na^+) - \varepsilon(UO_2(CO_3)_3^{5-}, Na^+)$ with T is also given. At 25°C the value is (0.97 ± 0.1) kg · mol^{-1}.

All uncertainties in these calculations correspond to 1 σ.

[99CHA/DON]

This review includes the review of [98CHA/DON].

[98CHA/DON] describe the mechanism of redox processes involving Am(III), Am(IV), Am(V) and Am(VI) in 1 M HNO_3. A table is presented including formation constants of Am(III) and Am(IV) complexes with $\alpha_2\text{-}P_2W_{17}O_{61}^{10-}$ and $\alpha\text{-}SiW_{11}O_{39}^{8-}$ ions. Slightly different recalculated values are given in [99CHA/DON].

[99CHA/DON] report apparent equilibrium constants for Am(III) and Am(IV), along with spectral data. The degree of protonation of the ligand is unknown because the experiments have been made only in 1 M HNO₃. These data are of little interest for processes in ground and surface water systems, but they may have laboratory applications.

[99CHO]

This paper discusses methods for estimating thermodynamic values of Pu(III to VI) from the corresponding values of the analogues, Eu(III), Am(III), Th(IV), Np(V) and U(VI), which are useful since the interconversion of the Pu species makes experiments difficult. A relationship which includes "effective" charge and dielectric constant of the aqueous ions and their ionic radius is proposed, but it is only useful for scoping calculations.

[99CLA/CON]

This is an important study where the authors have identified the structure of $UO_2(OH)_4^{2-}$ in the solid state and also made structural studies of strongly alkaline solutions of U(VI) using EXAFS. The authors show that no polynuclear species are present, and that the bond distances in $UO_2(OH)_4^{2-}$ are the same within the experimental uncertainty, in both solid and solution. Despite this fact they suggest that the complex in solution has the stoichiometry, $UO_2(OH)_5^{3-}$, no doubt guided by the prevalence of five co-ordinate uranium(VI) complexes. The authors have also discussed the possibility of equilibrium between these two species and use various spectroscopic data to support their case. Spectroscopic data from liquid nitrogen temperatures play an important part in these arguments. This review is not confident that one can draw conclusions valid in solution at 25°C from such data. Studies of the coordination geometry of uranium(VI) complexes have also been made by Wahlgren et al. [99WAH/MOL] and Vallet et al. [2001VAL/WAH] as discussed in Appendix A.

[99COR/BOO]

This paper deals with the enthalpy of formation at 298.15 K of $Sr_3U_{11}O_{36}$(cr), $Sr_2U_3O_{11}$(cr), β–SrUO₄, $Sr_5U_3O_{14}$(cr), Sr_2UO_5(cr), Sr_3UO_6(cr) and $Sr_2UO_{4.5}$(cr), obtained from their enthalpies of solution in 5.075 mol·dm⁻³ HCl (referred hereunder as sln. A) or in 1.00 mol·dm⁻³ HCl + 0.0470 mol·dm⁻³ FeCl₃ (referred hereunder as sln. B).

All the hexavalent compounds were obtained by reaction of the corresponding stoichiometric mixture of SrO(s) with U₃O₈(s) in oxygen. The mixtures were heated in a gold boat at 1300 K for 20 days, and then homogenised. Repetition of the procedure assured that the reaction was complete, as verified by X–ray diffraction. $Sr_5U_3O_{14}$(cr) required a reaction temperature of 1600 K and even then, the product was contaminated with

SrUO$_4$(cr) or Sr$_2$UO$_5$(cr). Pentavalent Sr$_2$UO$_{4.5}$(cr) was obtained by reduction in an Ar/H$_2$ gas mixture at 1200 < T < 1300 K. X–ray diffraction and chemical analyses were used to characterise the samples. Sr was determined spectrophotometrically with EDTA after ion exchange separation. U(IV) in the compounds arising from the disproportionation of U(V) was determined titrimetrically with K$_2$Cr$_2$O$_7$, back titrated with Fe^{2+}. U(VI) was also titrated with Fe^{2+} according to a described procedure [79LIN/KON]. All samples were handled in an O$_2$ - and H$_2$O - free Ar atmosphere.

The calorimeter and related procedure have been described in earlier publications. The results of the various dissolution experiments are given (including those of mixtures of SrCl$_2$ and UO$_3$ with variable stoichiometric ratios) with a confidence level of twice the standard deviation of the mean and are taken as such by this review. However, the enthalpies of formation are recalculated here using NEA adopted or NEA compatible auxiliary values, namely:

$\Delta_f H_m^\circ$ (SrCl$_2$, cr, 298.15 K) = – (833.85 ± 0.70) kJ · mol^{-1},

$\Delta_f H_m^\circ$ (Sr(NO$_3$)$_2$, cr, 298.15 K) = – (982.36 ± 0.80) kJ · mol^{-1},

$\Delta_f H_m^\circ$ (UO$_3$, γ, 298.15 K) = – (1223.8 ± 1.2) kJ · mol^{-1},

$\Delta_f H_m^\circ$ (UCl$_4$, cr, 298.15 K) = – (1018.8 ± 2.5) kJ · mol^{-1},

$\Delta_f H_m$ (HCl, partial, sln.A) = – (155.835 ± 0.102) kJ · mol^{-1},

$\Delta_f H_m$ (H$_2$O, partial, sln.A) = – (286.371 ± 0.040) kJ · mol^{-1},

$\Delta_f H_m$ (HCl, partial, sln.B) = – (164.260 ± 0.110) kJ · mol^{-1},

$\Delta_f H_m$ (H$_2$O, partial, sln.B) = – (285.850 ± 0.040) kJ · mol^{-1}.

Values in solution B are based on an effective chloride concentration of ca. 1.06 mol · dm^{-3}, calculated to be that arising from the chloride complexing by the ferric ion.

$\Delta_f H_m$ (HNO$_3$, partial, 6.00 mol·dm^{-3} HNO$_3$) = – (200.315 ± 0.402) kJ · mol^{-1},

$\Delta_f H_m$ (H$_2$O, partial, 6.00 mol·dm^{-3} HNO$_3$) = – (286.372 ± 0.040) kJ · mol^{-1}.

- β–SrUO$_4$.

The enthalpies of solution reported by the authors are:

$\Delta_{sol} H_m$ (SrUO$_4$, β, sln.A) = – (145.20 ± 0.32) kJ · mol^{-1},

$\Delta_{sol} H_m$ (1 SrCl$_2$ + 1 γ–UO$_3$, sln.A) = – (98.60 ± 0.62) kJ · (mol UO$_3$)$^{-1}$.

From these and the above relevant auxiliary values we recalculate:

$\Delta_f H_m^\circ$ (SrUO$_4$, β, 298.15 K) = – (1985.75 ± 1.57) kJ · mol^{-1}.

The authors also report a new determination of the enthalpy of solution of γ–UO$_3$ in 6.00 mol·dm^{-3} HNO$_3$ as $\Delta_{sol} H_m$ (UO$_3$, γ, 6.00 mol·dm^{-3} HNO$_3$) = – (72.10 ± 0.33)

kJ · mol^{-1}. Previous determinations by the same group yielded − (71.30 ± 0.13) kJ · mol^{-1} [64COR] and − (72.05 ± 0.25) kJ · mol^{-1} [75COR], cited by [78COR/OHA]. In subsequent calculations involving γ–UO$_3$ in this medium, we will use the weighted average − (71.53 ± 0.50) kJ · mol^{-1}, keeping conservative uncertainty limits. Our recalculation of earlier results on the dissolution of β–SrUO$_4$ in 6.00 mol·dm^{-3} HNO$_3$ by authors of the same laboratory [67COR/LOO] yield $\Delta_f H_m^\circ$ (β–SrUO$_4$, cr, 298.15 K) = − (1991.13 ± 2.72) kJ · mol^{-1}.

- Sr$_3$U$_{11}$O$_{36}$(cr).

The reported enthalpies of solution are:

$\Delta_{sol} H_m$ (Sr$_3$U$_{11}$O$_{36}$, sln.A) = − (795.11 ± 1.17) kJ · mol^{-1},

$\Delta_{sol} H_m$ (0.273 SrCl$_2$ + 1 UO$_3$, γ, sln.A) = − (73.77 ± 0.87) kJ · (mol UO$_3$)$^{-1}$.

It should be noted that in a previous study by the same group [91COR/VLA] the enthalpy of dissolution of apparently the same sample (as judged from the analytical results) of Sr$_3$U$_{11}$O$_{36}$(cr) in 5.0 mol·dm^{-3} HCl was given without details as − (794.9 ± 1.2) kJ · mol^{-1}; we shall not make use of this less documented result in the present review. We thus recalculate $\Delta_f H_m^\circ$ (Sr$_3$U$_{11}$O$_{36}$, cr, 298.15 K) = − (15903.81 ± 16.45) kJ · mol^{-1}.

The authors note that a compound previously believed [67COR/LOO] to be SrU$_4$O$_{13}$ (but always containing some U$_3$O$_8$) cannot correspond to a pure phase, in view of new structural results in the range 0.25 < Sr/U < 0.33.

- Sr$_2$U$_3$O$_{11}$(cr).

The authors only give a recalculation based on earlier results [67COR/LOO], which gave $\Delta_{sol} H_m$ (Sr$_2$U$_3$O$_{11}$, 6.00 mol·dm^{-3} HNO$_3$) = − (357.7 ± 2.1) kJ · mol^{-1}. Using auxiliary data accepted in this review, we obtain:

$\Delta_f H_m^\circ$ (Sr$_2$U$_3$O$_{11}$, cr, 298.15 K) = − (5243.73 ± 4.99) kJ · mol^{-1}.

The small difference from the value selected by [92GRE/FUG] is due to the fact that we have adopted here (see above) a slightly different value for the enthalpy of solution of γ–UO$_3$ in 6.00 mol · dm^{-3} HNO$_3$ and we calculate slightly larger uncertainty limits (4.99 instead of 4.1).

When using the above auxiliary data, the value for the enthalpy of formation assessed by [83FUG] is the same within 0.4 kJ · mol^{-1}.

- Sr$_5$U$_3$O$_{14}$(cr).

This compound could not be obtained as a pure phase; on the basis of the analyses and the X–ray diffraction pattern, the authors accepted the preparation as consisting of (0.9538 ± 0.0216) mass fraction Sr$_5$U$_3$O$_{14}$(cr), the remainder being β–SrUO$_4$. This assumed composition is necessarily approximate, especially as the Sr/U ratio is not the same in the two compounds. The enthalpy of solution of "as prepared" Sr$_5$U$_3$O$_{14}$ in solution A was reported as $-(592.21 \pm 1.08)$ J·g^{-1} and was corrected for the known enthalpy of solution of β–SrUO$_4$ to yield $-(602.85 \pm 1.08)$ J·g^{-1}. Given the uncertainties in the assumptions made for those corrections, we will adopt here conservative uncertainty limits, *i.e.*, half the correction for the impurities and take $\Delta_{sol}H_m$ (Sr$_5$U$_3$O$_{14}$, pure, sln.A) = $-(829.63 \pm 5.30)$ kJ·mol^{-1}. With $\Delta_{sol}H_m$ (1.667 SrCl$_2$ + 1 γ–UO$_3$, sln.A) = $-(121.36 \pm 0.45)$ kJ·(mol UO$_3$)$^{-1}$ and the adopted auxiliary data, we recalculate and select:

$$\Delta_f H_m^\circ (Sr_5U_3O_{14}, cr, 298.15 K) = -(7248.6 \pm 7.5) \text{ kJ·mol}^{-1}.$$

This value is noticeably different from that reported, $-(7265.8 \pm 7.5)$ kJ·mol^{-1}, even taking into account the different auxiliary data, for an unknown reason.

- Sr$_2$UO$_{4.5}$(cr).

This pentavalent compound was dissolved in solution B, yielding $\Delta_{sol}H_m$ (Sr$_2$UO$_{4.5}$, cr, sln.B) = $-(389.97 \pm 0.91)$ kJ·mol^{-1}. The enthalpy of dissolution of the same sample (as shown by the analytical results) in (HCl 0.0470 mol·dm^{-3} FeCl$_3$ + 82.16 H$_2$O) was given as $-(390.66 \pm 0.51)$ kJ·mol^{-1}, without any details, by the same group of authors [94COR/IJD]. We will not make use here of this less documented result.

The enthalpy of solution of the appropriate mixture of SrCl$_2$ and γ–UO$_3$ was also reported as $\Delta_{sol}H_m$ (4 SrCl$_2$ + 1 γ–UO$_3$, sln.B) = $-(259.25 \pm 0.50)$ kJ·(mol UO$_3$)$^{-1}$. Also used in the cycle is the enthalpy of solution of UCl$_4$(cr), $\Delta_{sol}H_m$ (UCl$_4$, cr, sln.B) = $-(186.67 \pm 0.60)$ kJ·mol^{-1} (taken from [88COR/OUW] for the dissolution in (HCl + 0.0419 mol·dm^{-3} FeCl$_3$ + 70.66 H$_2$O).

Using the same cycle as the authors, we recalculate:

$$\Delta_f H_m^\circ (Sr_2UO_{4.5}, cr, 298.15 K) = -(2493.99 \pm 2.75) \text{ kJ·mol}^{-1}.$$

The uncertainty limits on this value have been slightly increased to account for the small differences in the media involved in the cycle used by [99COR/BOO].

- Sr$_3$UO$_6$(cr).

The reported enthalpies of solution are $\Delta_{sol}H_m$ (Sr$_3$UO$_6$, cr, sln.A) = $-(560.94 \pm 1.43)$ kJ·mol^{-1} and $\Delta_{sol}H_m$ (3 SrCl$_2$ + 1 γ–UO$_3$, sln.A) = $-(166.88 \pm 0.49)$ kJ·(mol UO$_3$)$^{-1}$. Using a cycle analogous to those used for the other U(VI) compounds, we recalculate:

$$\Delta_f H_m^\circ (Sr_3UO_6, cr, 298.15 K) = -(3255.39 \pm 2.91) \text{ kJ·mol}^{-1}.$$

This value is in marginal agreement with the value $-(3263.08 \pm 4.24)$ kJ·mol^{-1}, recalculated from the data of [67COR/LOO] in 6.00 mol·dm^{-3} HNO$_3$, and with the value of $-(3263.95 \pm 4.39)$ kJ·mol^{-1}, based on the results of [83MOR/WIL2] working in 1.00 mol·dm^{-3} HCl. Note that, for the dissolution of γ–UO$_3$ in 6.00 mol·dm^{-3} HNO$_3$, a slightly different value $(-(71.53 \pm 0.50)$ kJ·mol^{-1}) was used here instead of $-(71.30 \pm 0.13)$

kJ·mol^{-1} taken by [92GRE/FUG]. The value selected by [92GRE/FUG] was $\Delta_f H_m^\circ$ (Sr$_3$UO$_6$, cr, 298.15 K) = – (3263.4 ± 3.0) kJ·mol^{-1}.

• Sr$_2$UO$_5$(cr).

The reported enthalpies of solution are $\Delta_{sol}H_m$ (Sr$_2$UO$_5$, cr, sln.B) = – (341.02 ± 1.10) kJ·mol^{-1} (erroneously listed in the text as – (341.96 ± 1.10)) and $\Delta_{sol}H_m$ (2SrCl$_2$ + γ–UO$_3$, sln.B) = – (166.36 ± 0.71) kJ·(mol UO$_3$)$^{-1}$.

From these, we recalculate $\Delta_f H_m^\circ$ (Sr$_2$UO$_5$, cr, 298.15 K) = – (2631.50 ± 2.31) kJ·mol^{-1}. The value selected in [92GRE/FUG] was – (2635.6 ± 3.4) kJ·mol^{-1} (which we recalculate as – (2635.88 ± 3.38) kJ·mol^{-1} with the value used here for the dissolution of γ–UO$_3$) on the basis of earlier results of the same group [67COR/LOO] working in 6.00 mol·dm^{-3} HNO$_3$. The two values are in marginal agreement.

[99DOC/MOS]

The authors report observations on the voltammetric behaviour of U in concentrated carbonate solutions that confirm the results of a previous study [93MIZ/PAR]. They report that the U(VI)/U(V) couple becomes irreversible compared to behaviour in less complexing media and report a formal potential at – 0.78 V *versus* SCE (the average of cathodic and anodic half-wave potentials $E_{1/2}$).

From a solution of 1 M Na$_2$CO$_3$ (pH = 11.95), and 10 mM U(VI), U(V) is generated by controlled potential coulometry at – 1.2 V *versus* SCE. This colourless solution is stable enough (two hours) to record EXAFS spectra under N$_2$ atmosphere The EXAFS spectra confirm the existence of UO$_2$(CO$_3$)$_3^{5-}$. Full cluster multiple scattering calculations were used to fit the data assuming D$_{3h}$ symmetry. These EXAFS structure parameters, bond distances, Debye–Waller factors and the frequencies of the different distances are reported.

[99FAN/KON]

This paper presents a careful thermodynamic analysis of the speciation in the Cm(III)–carbonate–NaCl system based on the Pitzer model. The authors report equilibrium constants at zero ionic strength and a set of interaction parameters. The Pitzer parameters for Cm^{3+} are the same as in [97KON/FAN], while the binary interaction parameters for the complexes were obtained in the fitting of model parameters to the experimental data. It is of interest to have uncertainty estimates of the various Pitzer parameters. In this way one could explore how sensitive the equilibrium constants at zero ionic strength are to variations in these parameters within the given uncertainty ranges. The authors calculated the following equilibrium constants at zero ionic strength, (A) using only their spectroscopic data for Cm(III) in 0 – 6 m NaCl and giving no uncertainty limits and (B) including, as well, literature data for Am(III) and Cm(III) carbonate complexes:

	(A)	(B)
$\log_{10} \beta_1^\circ$	8.30	(8.1 ± 0.3)
$\log_{10} \beta_2^\circ$	13.52	(13.0 ± 0.6)
$\log_{10} \beta_3^\circ$	15.52	(15.2 ± 0.4)
$\log_{10} \beta_4^\circ$	13.36	(13.0 ± 0.5)

In calculation (B) the ionic strength corrections are based on Pitzer parameters determined from the spectroscopic data in 0 – 6 m NaCl. A similar calculation is performed in the present review by using the SIT to select the best estimate of the various equilibrium constants. Using the derived parameter set the authors have made a reinterpretation of previous solubility experiments from [84BER/KIM], [90FEL/RAI], [91MEI], [94RUN/KIM] and [94GIF], which gives both a valuable comparison between the results of different investigators and also provides a validation of the parameters proposed by Fanghänel et al.

[99FEL/RAI]

In this paper Felmy and Rai review the application of Pitzer's equations for modelling the aqueous thermodynamics of actinide species in natural waters. The paper includes tables of ion interaction parameters and associated standard state equilibrium constants for tri-, tetra-, penta- and hexavalent actinides. Examples comparing experimental data and model calculations are presented for tri- and tetravalent actinides. The applicability of oxidation state analogues, e.g., the use of Nd(III) data for trivalent actinides and the use of Np(V) and U(VI) data for Pu(V) and Pu(VI), is also pointed out.

The solubilities of Am(III) and Nd(III) hydroxides (experimental data at I = 0.1 M from [82SIL], [96RAO/RAI2] and in dilute solutions from [83RAI/STR]) are shown to be the same within experimental uncertainties and similar to that of Pu(III) hydroxide [89FEL/RAI]. The authors present examples which show the applicability of the Pitzer approach for modelling the solubility of Pu(III) hydroxide in concentrated NaCl solutions and the solubility of $NdPO_4$(cr) in molybdate solutions (taken from [95FEL/RAI]). They also present a model for Am(III) in sulphate solutions, which is based solely on ion interaction parameters, without accounting for Am(III) sulphate complexes (cf. Appendix A, review of [95RAI/FEL2]).

The authors compare their own set of Pitzer parameters for trivalent actinide-carbonate systems with those of Fanghänel et al. [98NEC/FAN], [99FAN/KON] and point out the discrepancies. As a matter of fact, the parameters of these two groups are different because the underlying evaluation procedures are different. Felmy and Rai derived binary interaction parameters for the carbonate complexes of trivalent or tetravalent actinides from data in dilute to concentrated $NaHCO_3$ and Na_2CO_3 (or $KHCO_3$ and K_2CO_3) solutions

[90FEL/RAI], [96RAO/RAI], [97FEL/RAI], [98RAI/FEL], [99RAI/HES], [99RAI/HES2]. In a second step, additional ternary interaction parameters with chloride were calculated from data in carbonate-chloride mixtures [99FEL/RAI]. Conversely, Fanghänel et al. derived binary parameters for tri- and pentavalent actinide species from experimental data in dilute to concentrated NaCl solutions considering CO_3^{2-} and other ligands like OH^- and F^-, as trace components [95FAN/NEC] [94FAN/KIM], [98NEC/FAN], [99AAS/STE]. As a consequence the validity of their parameter sets for actinide carbonate systems are limited to carbonate concentrations < 0.1 mol·kg^{-1}. They are not appropriate for concentrated bicarbonate or carbonate solutions. Conversely, the parameter sets of Felmy and Rai are not appropriate for modelling actinides in chloride solutions of lower carbonate concentrations if they do not include ternary interaction parameters with chloride.

The importance of these ternary interaction parameters is demonstrated by new experimental solubility data for $ThO_2(am)$ in 0.1 – 2.3 m Na_2CO_3 solutions containing 2.33 m or 4.67 m NaCl. Using only the binary parameters for $Na^+ - Th(CO_3)_5^{6-}$ derived from solubility data in Na_2CO_3 solutions [97FEL/RAI], the solubility predicted for the $Na_2CO_3 - NaCl$ mixtures would be overestimated by orders of magnitude. Modelling of these data requires mixing parameters θ and ψ for the interactions $Th(CO_3)_5^{6-} - Cl^-$ and $Th(CO_3)_5^{6-} - Cl^- - Na^+$. Felmy and Rai [99FEL/RAI] also show that the corresponding ternary interaction parameters with perchlorate, derived in [97FEL/RAI] from $ThO_2(am)$ solubilities in $Na_2CO_3 - NaClO_4$ solutions [94OST/BRU], would underestimate the solubility in Na_2CO_3-NaCl solutions by orders of magnitude. This observation supports the discussion in Appendix D on the importance of ternary interaction coefficients and the different activity coefficients of negatively charged actinide complexes in NaCl and NaClO$_4$ solutions.

[99HAS/WAN]

The rates of volatilisation from $UO_2(s)$ in pure steam, steam/Ar, steam/He and steam/Ar/H$_2$ were measured thermogravimetrically at atmospheric pressure and temperatures from 1523 to 1873 K. The principal aim was to clarify the kinetics and mechanism of the volatilisation process and to assess the validity of selected thermodynamic data needed to interpret the experimental data. The volatilisation rates depended significantly on the flow rates and gas mixtures, and the thermodynamic data recommended by Olander [99OLA] were used to interpret the volatilisation data. Thus there are no new thermodynamic data to be derived from this paper.

[99HRN/IRL]

This is a detailed study of the interaction between silicate and uranium(VI) made by using a solvent extraction technique. The silicate concentrations varied between 0.01 and 0.067 M in the pH range 3.3 to 4.5. The experiments are made at 25°C in 0.2 M NaClO$_4$. The pH electrodes have been calibrated with buffer solutions and corrections made to convert measured pH to $-\log_{10}[H^+]$.

[99JAY/IYE]

This paper reports measurements of the pressure of K(g) (by mass–loss Knudsen effusion) and the oxygen activity (emf with $CaO - ZrO_2$ electrolyte) in the three-phase field, $KUO_3(cr) + K_2UO_4(cr) + K_2U_2O_7(cr)$, similar to the studies on the corresponding sodium compounds by [94JAY/IYE]. In addition, enthalpy increments of $KUO_3(cr)$ and $K_2U_2O_7(cr)$ up to ca. 700 K using a high temperature Calvet calorimeter are reported. The preparative and experimental details for the Gibbs energy measurements were similar to those described in [94JAY/IYE], except that a boron nitride effusion cell was used, rather than a graphite cell.

A number of errors in the paper have been corrected.

The enthalpy increment measurements (and fitted expressions) are those reported in the Conference Proceedings reviewed under [98JAY/IYE]. As noted there, and in section 9.10.4, these enthalpy data have been refitted.

The potassium pressures were fitted to the equation:

$$\log_{10}(p_K / \text{bar}) = 12.693 - 23198\, T^{-1}, \qquad (1265 \text{ to } 1328 \text{ K})$$

and hence,

$$RT \ln(p_K / \text{bar}) = -444123 + 243.006\, T \text{ kJ} \cdot \text{mol}^{-1}, \quad (1265 \text{ to } 1328 \text{ K})$$

over the three-phase field, $KUO_3(cr) + K_2UO_4(cr) + K_2U_2O_7(cr)$.

The authors used a reference electrode of Pt/air ($p_{O_2} = 0.21$ bar) in their emf cell. Correction of their emf values to the standard pressure gives, after recalculation:

$$RT \ln(p_{O_2} / \text{bar}) = -499021 + 276.176\, T \text{ kJ} \cdot \text{mol}^{-1} \quad (941 \text{ to } 1150 \text{ K}).$$

Since there are currently only estimated values [81LIN/BES], [92GRE/FUG], for the entropies of the potassium uranates involved, these data cannot immediately be used to derive further reliable thermodynamic quantities for any of the solids involved.

However, in view of the discrepancies between the experimental and calculated values for the reactions in the corresponding study of the sodium uranates [94JAY/IYE], (possibly due to lack of true equilibrium), we have made a similar comparison for the Gibbs energies, using estimated entropies [81LIN/BES], of the three reactions:

$$3\, KUO_3(cr) + K_2UO_4(cr) + 0.5\, O_2(g) \rightleftharpoons 2\, K_2U_2O_7(cr) + K(g), \qquad (A.116)$$

$$2\, KUO_3(cr) + 0.5\, O_2(g) \rightleftharpoons K_2U_2O_7(cr), \qquad (A.117)$$

$$KUO_3(cr) + K_2UO_4(cr) \rightleftharpoons K_2U_2O_7(cr) + K(g), \qquad (A.118)$$

(see the discussion in [94JAY/IYE] for the relation between these three equations). In addition to the entropy estimates, the heat capacity of $K_2UO_4(cr)$ was estimated from that of Na_2UO_4 (cr) (see review of [94JAY/IYE] in Appendix A).

The results are shown in the Table A-41, where the experimental uncertainties are twice the authors' estimates.

Table A-41: Experimental and calculated Gibbs energies of potassium uranate reactions.

$T(K)$	$\Delta_r G_m^\circ$ (A.117) (kJ · mol^{-1})		$\Delta_r G_m$ (A.118) (kJ · mol^{-1})		$\Delta_r G_m$ (A.116) (kJ · mol^{-1})	
	Exp.	Calc.	Exp.	Calc.	Exp.	Calc.
900	$-(125.1 \pm 1.2)$	$-(124.0 \pm 15.0)$	(225.4 ± 0.4)	(163.3 ± 15.0)	(100.2 ± 1.3)	(39.3 ± 22.0)
1000	$-(111.4 \pm 1.2)$	$-(116.4 \pm 15.0)$	(201.1 ± 0.4)	(148.4 ± 15.0)	(89.7 ± 1.3)	(32.0 ± 22.0)
1100	$-(97.6 \pm 1.2)$	$-(109.9 \pm 15.0)$	(176.8 ± 0.4)	(132.5 ± 15.0)	(79.2 ± 1.3)	(22.6 ± 22.0)
1200	$-(83.8 \pm 1.2)$	$-(104.6 \pm 15.0)$	(152.5 ± 0.4)	(115.3 ± 15.0)	(68.7 ± 1.3)	(10.6 ± 22.0)

[99KAS/RUN]

This paper is a critical review of the literature on thermodynamic data of Np complexes (hydroxo, carbonato, phosphato) and compounds (hydroxide, oxide, carbonate). Most of the scrutinised papers are considered in [2001LEM/FUG], except for the following: [84ALL/OLO], [85NIT/EDE], [97LAN], [97NOV/ALM], [98ALM/NOV] and [98EFU/RUN]. Zero ionic strength values are calculated for the standard thermodynamic constant values using the SIT, as described in [92GRE/FUG]. For reasons discussed in the paper the authors have selected thermodynamic data only for Np(V) and Np(IV).

- Aqueous ions.

The value, $\Delta_f G_m^\circ (NpO_2^+, 298.15 \text{ K}) = -(907.9 \pm 5.8)$ kJ · mol^{-1}, is calculated using $E^\circ(NpO_2^+/NpO_2^{2+}) = (1.161 \pm 0.014)$ V (an average value of [89RIG/ROB] and [70BRA/COB]) and $\Delta_f G_m^\circ (NpO_2^{2+}, 298.15 \text{ K})$ of [76FUG/OET]. The change in E° gives a value of $\Delta_f G_m^\circ (NpO_2^+, 298.15 \text{ K})$, 7.1 kJ · mol^{-1} greater than that of [76FUG/OET], but very close to that selected in [2001LEM/FUG], $\Delta_f G_m^\circ (NpO_2^+, 298.15 \text{ K}) = -(907.8 \pm 5.6)$ kJ · mol^{-1}.

The value, $\Delta_f G_m^\circ (Np^{4+}, 298.15 \text{ K}) = -(491.1 \pm 9.5)$ kJ · mol^{-1}, is calculated using $E^\circ(Np^{4+}/NpO_2^+) = (0.596 \pm 0.078)$ V and $\Delta_f G_m^\circ (NpO_2^+, 298.15 \text{ K}) = -(907.9 \pm 5.8)$ kJ · mol^{-1}. E° comes from a reinterpretation of the value of [52COH/HIN] that seems more precise than that in [76FUG/OET]. This gives a value, 11.8 kJ · mol^{-1}, greater than that of [76FUG/OET] and very close to that in [2001LEM/FUG], $\Delta_f G_m^\circ (Np^{4+}, 298.15 \text{ K}) = -(491.8 \pm 5.6)$ kJ · mol^{-1}.

The values proposed by the authors are essentially the same as those selected by [2001LEM/FUG] and no revision is required.

- Hydrolysis.

The values, $\log_{10} \beta_{1,1}^\circ = (2.7 \pm 0.2)$, $\log_{10} \beta_{2,1}^\circ = (4.35 \pm 0.15)$ ($\log_{10} {}^*\beta_{1,1}^\circ = -(11.3 \pm 0.2)$ and $\log_{10} {}^*\beta_{2,1}^\circ = -(23.65 \pm 0.15)$), for Np(V) are those of [92NEC/KIM], obtained from solubility experiments performed under a well-controlled CO_2-free argon atmosphere. [2001LEM/FUG] have selected $\log_{10} {}^*\beta_{1,1}^\circ = -(11.3 \pm 0.7)$ and $\log_{10} {}^*\beta_{2,1}^\circ = -(23.6 \pm 0.5)$. All these values are in close agreement.

Using $\log_{10} {}^*K_{s,0}^\circ = (1.5 \pm 0.3)$ for $NpO_2 \cdot xH_2O(cr)$ and a solubility of $10^{-8.3}$ M at pH greater than 7 from [87RAI/SWA], Kaszuba and Runde propose, $\log_{10} \beta_{4,1}^\circ = -(10 \pm 1)$. Neck and Kim [99NEC/KIM], [2001NEC/KIM], have proposed a value that the present review considers more accurate (see this Appendix). For the first hydrolysis constant an average value of $\log_{10} {}^*\beta_{1,1}^\circ = -(0.4 \pm 0.7)$ is chosen by the authors which is consistent with $\log_{10} {}^*\beta_{1,1}^\circ (UOH^{3+}) = -(0.54 \pm 0.06)$ selected by [92GRE/FUG]. The hydrolysis constants, $\beta_{2,1}^\circ$, $\beta_{3,1}^\circ$ and $\beta_{5,1}^\circ$, proposed in the literature are rejected. [2001LEM/FUG] have selected $\log_{10} {}^*\beta_{1,1}^\circ = -(0.29 \pm 1.00)$ and $\log_{10} {}^*K_{s,0}^\circ = (1.53 \pm 1.00)$, and a value for $\log_{10} \beta_{4,1}^\circ$ equal to $-(9.8 \pm 1.1)$.

- Carbonato complexes.

For the well-defined species $NpO_2(CO_3)_n^{(1-2n)}$, $n = 1, 2$ and 3, the proposed values of the constants are those of [94NEC/RUN] (and of other authors who give very similar values), $\log_{10} \beta_1^\circ = (4.81 \pm 0.15)$, $\log_{10} \beta_2^\circ = (6.55 \pm 0.23)$ and $\log_{10} \beta_3^\circ = (5.54 \pm 0.19)$, but increasing the uncertainty by 0.09 for the tricarbonato complex. [2001LEM/FUG] have selected (4.96 ± 0.06), (6.53 ± 0.10) and (5.50 ± 0.15), respectively.

The situation is less clear with carbonato complexes of Np(IV) with the existence of $Np(CO_3)_n^{(4-2n)}$, $n < 5$, having not been proven. The limiting complex $n = 5$ is isostructural with those of U(IV) and Pu(IV) [92GRE/FUG], [95CLA/HOB]. The authors of the present paper adjust $\log_{10} \beta_{1,0,5}^\circ (Np(IV))$ from $\log_{10} \beta_{1,0,5}^\circ (U(IV)) = (34.3 \pm 0.9)$ [92GRE/FUG] by the difference in the $\Delta_f G_m^\circ (298.15 \text{ K})/(298.15 \cdot R)$ between Np^{4+} and U^{4+} and give $\log_{10} \beta_{1,0,5}^\circ (Np(IV)) = (33.9 \pm 2.6)$. There is no rationale given for this and the present review does not accept this estimate. This value is in agreement with $\log_{10} \beta_{1,0,5}^\circ (Th) = (32.3 \pm 0.4)$ [94OST/BRU] and $\log_{10} \beta_{1,0,5}^\circ (Np(IV)) = 33.4$ obtained by [99RAI/HES] (the uncertainty is not given by these authors). By analogy with U(IV), $\log_{10} \beta_{1,0,4}^\circ (Np(IV)) = (35.1 \pm 2.6)$ is proposed.

[2001LEM/FUG] have not selected values for Np(IV) carbonate complexes, but give the following estimated values: $\log_{10} \beta_{1,0,5}^\circ = (38.98 \pm 1.97)$ (p. 261) and $\log_{10} \beta_{1,0,4}^\circ = (36.69 \pm 1.03)$ (p. 264). The value of $\log_{10} \beta_{1,0,5}^\circ$ comes from [99RAI/HES] who reported a value of (35.62 ± 1.15), which is not accepted in the present review.

- Hydroxocarbonato complexes.

The existence of mixed complexes has been discussed in several papers, but the arguments based on solubility experiments are not convincing. The recent spectroscopic data of [97NEC/FAN] obtained in 3 M NaOH/Na$_2$CO$_3$/NaClO$_4$ are corrected by the authors to zero ionic strength using the SIT parameters of NpO$_2$(CO$_3$)$_2^{3-}$ and NpO$_2$(CO$_3$)$_3^{5-}$ [94NEC/RUN] to get $\log_{10} \beta_{1,2,1}^\circ$ = (7.1 ± 0.8) and $\log_{10} \beta_{1,1,2}^\circ$ = (6.0 ± 0.6) for NpO$_2$(OH)$_2$(CO$_3$)$^{3-}$ and NpO$_2$(OH)(CO$_3$)$_2^{4-}$. These values are considered as a first approximation. [2001LEM/FUG] have not selected any values although the same literature data have been considered.

- Phosphato complexes.

Only one complex, NpO$_2$HPO$_4^-$, is well established. The value, $\log_{10} \beta^\circ$ = (2.9 ± 0.6), is an average from several literature values. [2001LEM/FUG] have selected $\log_{10} \beta^\circ$ = (2.95 ± 0.10) for this Np(V) species. For other complexes the information is conflicting.

No experimental data have been published for the Np(IV) phosphate system. There are only estimations.

- Solid phases.

It is well established that in Np(V) solutions Np$_2$O$_5$(s) and NpO$_2$OH(s) are the stable limiting solubility phases in the absence of carbonate, while in the presence of carbonate, MNpO$_2$CO$_3$·xH$_2$O(s) and M$_3$NpO$_2$(CO$_3$)$_2$(s) are formed, with the stability ranges depending on the concentrations of carbonate and M.

The thermodynamic solubility products reported by [92NEC/KIM], [95NEC/FAN] and [96RUN/NEU] agree within 0.02 log units for fresh, and 0.07 log units for aged, NpO$_2$OH(s). Average values are $\log_{10} K_{s,0}^\circ$ = − (8.77 ± 0.09) and $\log_{10} K_{s,0}^\circ$ = − (9.48 ± 0.16), respectively.

The recent value of $\log_{10} {}^*K_{s,0}^\circ$ = (2.6 ± 0.4) for Np$_2$O$_5$(s) [98EFU/RUN] obtained from solubility measurements agrees with that of [94MER/FUG], obtained from calorimetry, $\log_{10} {}^*K_{s,0}^\circ$ = (2.25 ± 0.95). It is retained by the authors of this paper.

[2001LEM/FUG] have selected (Table 8.5, p. 126) $\log_{10} {}^*K_{s,0}^\circ$ (NpO$_2$OH, fresh) = (5.3 ± 0.2), $\log_{10} {}^*K_{s,0}^\circ$ (NpO$_2$OH, aged) = (4.7 ± 0.5) for:

$$\text{NpO}_2\text{OH(s)} + \text{H}^+ \rightleftharpoons \text{NpO}_2^+ + \text{H}_2\text{O(l)},$$

and $\log_{10} {}^*K_{s,0}^\circ$ (Np$_2$O$_5$, cr) = (1.8 ± 1.0) for:

$$0.5\,\text{Np}_2\text{O}_5\text{(s)} + \text{H}^+ \rightleftharpoons \text{NpO}_2^+ + 0.5\,\text{H}_2\text{O(l)}.$$

The solubility products of NaNpO$_2$CO$_3$·3.5H$_2$O(s) and Na$_3$NpO$_2$(CO$_3$)$_2$(s) given by [95NEC/FAN] and [96RUN/NEU] agree within 0.02 log units. The recommended averaged values are $\log_{10} K_{s,0}^\circ$ = − (11.06 ± 0.17) and $\log_{10} K_{s,0}^\circ$ = − (14.28 ± 0.24), respectively. [2001LEM/FUG] have selected $\log_{10} K_{s,0}^\circ$ = − (11.16 ± 0.35) for fresh, and

$\log_{10} K_{s,0}^{\circ} = -(11.66 \pm 0.50)$ for aged (or less hydrated), sodium dioxoneptunium(V) monocarbonate (p. 276) and $\log_{10} K_{s,0}^{\circ} = -(14.70 \pm 0.66)$ for the trisodium dioxoneptunium(V) dicarbonate (p. 279). The two different values for the monocarbonato phase are an artefact of using erroneous activity coefficients for the carbonate ion at high NaClO$_4$ concentration. All the data from Neck et al. [95NEC/FAN] and Runde et al., [96RUN/NEU] were obtained with aged (more than half a year old) solid phases.

Solubility data for the corresponding potassium compounds have been obtained in conditions where $NpO_2(CO_3)_3^{5-}$ predominates, so the calculated solubility product includes uncertainties of aqueous Np(V) species ($\pm$ 0.2 log unit). The values, $\log_{10} K_{s,0}^{\circ} = -(13.6 \pm 0.1)$ and $\log_{10} K_{s,0}^{\circ} = -(15.9 \pm 0.1)$ [97NOV/ALM], are considered as a first estimate by Runde et al. [96RUN/NEU].

NpO$_2$(s) and Np(OH)$_4$(am) are the stable solid phases in Np(IV) systems at low carbonate concentrations. However, as formation constants are not known for carbonato and hydroxo carbonato complexes, their solubility products cannot be calculated, as is the case for Pu(IV). The solubility product of NpO$_2$·xH$_2$O(s) retained by Kaszuba and Runde is that of [87RAI/SWA].

[99KNO/NEC]

The first part of this paper deals with the critical survey of Pu(IV) hydrolysis behaviour in non-complexing media and reproduces the essentials of the discussion and literature data of [99NEC/KIM] and [2001NEC/KIM] as far as Pu is concerned (see the review in Appendix A). Additional comments bear on:

- the difficulty in preventing the Pu(IV) from oxidizing (and subsequently preventing Pu(V) from disproportionating in acidic media) and in identifying the nature of the limiting solubility phase, hydrated oxide or hydroxide,
- the close link between the determination of the solubility product of that phase and the values of hydrolysis constants of Pu(IV), as well as with the presence of colloids.

Extrapolation of literature data to zero ionic strength is done according to the SIT with $\varepsilon(Pu^{4+}, ClO_4^-) = (0.83 \pm 0.1)$ kg·mol^{-1}. This value comes from a recent work [98CAP/VIT] and is used in place of the value (1.03 ± 0.05) kg·mol^{-1} given in [92GRE/FUG]. Knopp et al. use the data of [49KAS], [65PER], [84RAI], [86LIE/KIM], [89KIM/KAN] and hydrolysis constants of [72MET/GUI] to calculate the corresponding $\log_{10} K_{s,0}^{\circ}$ values of which the average value is $\log_{10} K_{s,0}^{\circ} = -(58.7 \pm 0.9)$.

The second part of the paper is a study of colloid formation of Pu(IV) starting with a solution of Pu(VI), 3·10^{-3} M. The progressive reduction of Pu(VI) by H$_2$O$_2$ in 0.1 M HClO$_4$, which leads first to Pu(V), then to a colloid of Pu(IV), was monitored by UV–Vis spectroscopy and LIBD (Laser Induced Breakdown Detection) on filtered aliquots. When the Pu(IV) concentration is higher than the concentration calculated with $\log_{10} K_{s,0}^{\circ} =$

− (58.7 ± 0.9) combined with hydrolysis constants from [72MET/GUI], colloids are always present. This indicates that spectrophotometric data used to select $\log_{10} \beta_{1,1}^\circ$ in [2001LEM/FUG] are not the best ones.

[99MEI/KAT]

This paper compares literature data on the solubility products of U(VI) and Np(VI) carbonates as well as formation constants of the species, $AnO_2(CO_3)_n^{2-2n}$, $n = 1$, 2 and 3. The U data sets concerned are: set 1, [92KRA/BIS], set 2, [93MEI/KIM2], set 3, [93PAS/RUN], set 4, [93MEI/KIM], set 5, [97MEI2] and set 6, [97PAS/CZE], all referring to $I = 0.1$ M $NaClO_4$. It appears that the average values of $\log_{10} K_{s,0}$ is scattered between − 14.25 to − 13.25 (but data sets 1, 5 and 6 overlap). For AnO_2CO_3 (aq), $\log_{10} \beta_1$ values are in the range 8.7 to 9.3 (sets 1, 4, 5 and 6), for $AnO_2(CO_3)_2^{2-}$ all the $\log_{10} \beta_2$ values overlap at 15.3 (sets 4, 5 and 6) and for $AnO_2(CO_3)_3^{4-}$ the average value of $\log_{10} \beta_3$ from set 6 has no meaning while data sets 4 and 5 give similar values 21.8 and 22.0.

This is a useful paper with a good description of statistical methods for hypothesis testing and estimation of uncertainty ranges of published data. These are issues that have been a concern to experimental solution chemists since the nineteen forties. The early attempts to estimate the uncertainty involved the use of different experimental methods, a strategy used by solution chemists at the University of Lund (Lund school led by I. Leden, S. Fronaeus and S. Ahrland). These scientists also pointed out the importance of avoiding bias in the experiments by collecting about the same number of experimental data in the regions where the different complexes had their maximum concentrations. Species that were present in small amounts, say less than 5%, were looked upon with suspicion. The Stockholm school, with L. G. Sillén as the driving force, developed a very different strategy. Their work was dominated by one experimental technique, potentiometry, and a very large number of data points were measured for the system under scrutiny. These data were then treated by a least-squares program, LETAGROP. The uncertainty estimate was based on the assumption of normal distributed errors and that the error-square function could be approximated by a generalised second degree surface. It was not uncommon to find that the chemical models proposed using this methodology contained complexes that were present in such small amounts in the test solutions that they might be computational artefacts.

The estimation of the uncertainty of published data has been and still is a problem in the NEA-TDB reviews, because the primary experimental data are rarely available. The reviewers have therefore used both the authors' estimates and their own expert experience on the precision expected of a given experimental method when estimating the uncertainty of equilibrium constants. In systems where one can obtain independent experimental information on speciation by different methods (*e.g.*, potentiometry, ion-exchange, solubility measurements and spectroscopy), one often finds an excellent agreement between the methods, indicating that the uncertainty estimates are reasonable. It should also be pointed out that the uncertainty estimates rarely change the conclusions of predictive geochemical modelling.

The conclusion of this review is that the paper by Meinrath et al. provides very useful information to experimentalists both when planning an experiment and when interpreting experimental data. It seems particularly important to use the potential of these methods when only a fairly small number of data points are available. The following papers deal with the same topic, but they do not contain any new thermodynamic data: [2000MEI], [2000MEI2], [2000MEI/EKB] and [2000MEI/HUR].

[99MEI/VOL]

No new thermodynamic data other than those already published by Meinrath and reviewed in this Appendix, are reported in this paper. It is an attempt to explain the behaviour of U in aqueous field samples of old Saxonia (Germany) mines. The different samples have different concentrations of uranium and different compositions of the major components and pH. The authors compare these values with the predicted concentrations using E_h – pH predominance diagrams. The data (I = 0.1 and 1 M) used includes redox potential (U^{4+}/U^{3+}, UO_2^+/U^{4+}, UO_2^{2+}/UO_2^+, UO_2^{2+}/U^{4+}), formation constants for hydrolysis (U(IV) and U(VI)), carbonate (U(IV) and U(VI)) and sulphate (U(VI)) complexation, all selected by Meinrath from his early works, except those of U(IV) which are derived by analogy with Pu(IV). It is concluded that U solubilities under oxic conditions are mainly due to sulphato and carbonato U(VI) species.

[99MIK/RUM]

This paper deals with considerations on the electronic structures of M, M^{2+} and M^{3+} to explain why the divalent state of lanthanides and actinides can be observed in some particular redox conditions. The role of $\Delta_{hyd}G_m^\circ(M^{2+})$, $\Delta_{hyd}G_m^\circ(M^{3+})$ and the excitation energies from d to f shells are emphasised.

[99NAK/ARA]

The authors have measured the partial pressures of Np(g) and Pu(g) over a physical mixture of NpN(s) + PuN(s) by Knudsen–cell mass spectrometry from 1950 – 2070 K, supplementing their earlier study of pure NpN(s) [97NAK/ARA]. The partial pressures of Pu(g) agreed with the relatively consistent measurements of the congruent sublimation of PuN(s) summarised in [2001LEM/FUG]. Thus it was assumed that the partial pressure of $N_2(g)$ in the vaporisation of the mixture was the same as that involved in the congruent effusion of PuN(s):

$$p_{N_2} = 0.5\sqrt{\frac{M_{N_2}}{M_{Pu}}}\, p_{Pu},$$

where M_{N_2} and M_{Pu} are the relative molar masses.

This pressure of nitrogen is high enough to suppress the formation of Np(liq), so $\Delta_f G_m$ (NpN) can be calculated from the known pressures in the vaporisation reaction, NpN(cr) $\rightleftharpoons$ Np(g) + 0.5 N$_2$(g). [99NAK/ARA] checked that no solid solutions of NpN and PuN were formed.

The derived Gibbs energy of formation of NpN from this study is given by the equation:

$$\Delta_f G_m \text{ (NpN, cr, } T) = -269000 + 74.0\, T \quad \text{J} \cdot \text{mol}^{-1} \quad (1950 \text{ to } 2070 \text{ K}).$$

The Gibbs energies from the above equation are 3 – 6 kJ · mol^{-1} more negative than those derived from the earlier study [97NAK/ARA], with a smaller temperature dependence.

These two papers form the basis of a revision of the stability of NpN(cr), as described in section 13.4.1.

[99NAK/ARA2]

The authors have measured the partial pressures of BaO(g) and Ba(g) over a diphasic sample containing a 50:50 mixture of BaPuO$_3$(s) and PuO$_2$(s) by Knudsen–cell mass spectrometry from 1673 – 1873 K in Pt effusion cells inside a tantalum holder. The components of the mixture were identified by X–ray diffraction. The partial pressure of BaO(g) was about a factor of ten greater than that of Ba(g); so the predominant reaction during vaporisation was assumed to be:

$$\text{BaPuO}_3(\text{cr}) \rightleftharpoons \text{BaO(g)} + \text{PuO}_2(\text{cr}).$$

Ba(g) was assumed to be formed from BaO(g). In practice, the vaporisation is likely to be more complex than this, especially as PuO$_2$(cr) will certainly lose oxygen at these temperatures, particularly in an environment containing tantalum.

Using estimated thermal functions for BaPuO$_3$(cr), the authors have derived second– and third–law enthalpies of formation, $\Delta_f H_m^\circ$ (BaPuO$_3$, cr, 298.15 K) of – 1661 and – 1673 kJ · mol^{-1}, respectively, with no quoted uncertainties.

Considering the use of estimated thermal functions for BaPuO$_3$(cr) and the likely complexity of the actual vaporisation process, these values are in good accord with the relatively precise calorimetric value from the study by Morss and Eller [89MOR/ELL] adopted by [2001LEM/FUG]:

$$\Delta_f H_m^\circ \text{ (BaPuO}_3\text{, cr, 298.15 K)} = -(1654.2 \pm 8.3) \text{ kJ} \cdot \text{mol}^{-1},$$

which is thus retained here.

[99NEC/KIM]

This paper is very important in understanding the hydrolytic properties of tetravalent actinides, An(IV). Up until now, only separated and specific element descriptions of (Th, U, Np, Pu) in terms of thermodynamic constants have been presented in the literature. This paper gives a unified view.

Many problems complicate the experimental investigation and thermodynamic evaluation of the constants: polymerisation of monomeric species, colloid formation, stability of the tetravalent state in solution, lack of well-defined unique dioxide phase with a crystallinity that depends on many factors and the tendency of this dioxide to become amorphous with increasing pH of the equilibrium solutions. As aqueous An^{4+} ions hydrolyse readily at low pH, there are no solubility data for $AnO_2(cr)$ and $AnO_2(am)$, $An(OH)_4(am)$, $AnO_2 \cdot xH_2O(am)$ or $AnO_{2-n}(OH)_{2n}(am)$, $n = 0 - 2$ in the presence of An^{4+}. So a calculation of the solubility product of these materials requires a knowledge of the hydrolysis constants of the aqueous ions. Furthermore, solubility products of $AnO_2(cr)$ derived from acidic media depend on details of the method of preparation of the oxide, such as temperature, pre-treatment, *etc*.

Only data from low ionic strength ($I < 1$ M) and the most dilute available solutions are considered by the authors.

This review considers only the constants at zero ionic strength calculated using the SIT. Data related to Th are also reported.

In the following, data for monomeric U(IV) hydroxide complexes and solubility products of the oxide and amorphous oxide, are discussed first. Then, similar data for Np(IV) and Pu(IV) are considered.

Neck and Kim [99NEC/KIM] select $\log_{10} \beta_1^\circ = (13.6 \pm 0.2)$ for UOH^{3+} as an average value deduced from the selected values, $\log_{10} {}^*\beta_1^\circ = -(0.54 \pm 0.06)$ by [92GRE/FUG] and $-(0.34 \pm 0.20)$ by [92FUG/KHO]. The selection of [92FUG/KHO] is not explicitly discussed in [92GRE/FUG]. Taking the average value of the two selected data gives $\log_{10} {}^*\beta_1^\circ = -(0.4 \pm 0.2)$, which is the value recommended by [94SER/DEV] (see this Appendix). On the basis of an empirical correlation and a semi-empirical electrostatic model using only $\log_{10} \beta_1^\circ$ for UOH^{3+} as an input parameter, Neck and Kim [99NEC/KIM] predicted the value, $\log_{10} \beta_1^\circ = 13.8$, and derived $\log_{10} \beta_2^\circ = 27.5$, $\log_{10} \beta_3^\circ = 38.2$ and $\log_{10} \beta_4^\circ = 45.7$. The semi-theoretical considerations are based on experimental data for Pu(IV) and trivalent Am and Cm. The value, $\log_{10} \beta_1^\circ = 13.8$, is in close agreement with the experimental value. The value of $\log_{10} \beta_4^\circ$ is checked from solubility measurements of $UO_2(am)$ (see below).

The value selected by [92GRE/FUG] for the solubility product of crystalline UO_2 is $\log_{10} K_{s,0}^\circ (UO_2, cr) = -(60.86 \pm 0.36)$. To evaluate the corresponding value for amorphous dioxide, the data of [83RYA/RAI], [90RAI/FEL], [95YAJ/KAW], [97RAI/FEL] and [99GRA/MUL] (cited in [99NEC/KIM] as private communication) are analysed. They

Discussion of selected references

are all related to the solubility of U(IV) as a function of $-\log_{10}[H^+]$, from 1 to 13, at 25°C and different ionic strengths in perchlorate and chloride media.

From the data corresponding to pH < 5, and taking $\log_{10}\beta_1^\circ = 13.6$, $\log_{10}\beta_2^\circ = 27.5$ and $\log_{10}\beta_3^\circ = 38.2$, it is found that $\log_{10}K_{s,0}^\circ = -(55.2 \pm 1.0)$. This value differs from that of [90RAI/FEL] and [97RAI/FEL], $\log_{10}K_{s,0}^\circ = -53.45$, who included in the fitting only UOH^{3+} and U(OH)$_4$(aq), but is consistent with that of [95YAJ/KAW], who include only U(OH)$_4$(aq) in the fitting, $\log_{10}K_{s,0}^\circ = -(55.7 \pm 0.3)$.

From the data corresponding to pH > 5 and according to:

$$\log_{10}[\text{U(OH)}_4(\text{aq})] = \log_{10}K_{s,0}^\circ + \log_{10}\beta_4^\circ = -(8.5 \pm 1.0),$$

the authors derive $\log_{10}\beta_4^\circ = (46.7 \pm 1.0)$, in close agreement with the estimated value. This value differs from the value of 45.45 proposed by [90RAI/FEL] and [97RAI/FEL], but is consistent with that of [95YAJ/KAW], (47.0 ± 0.5).

The authors point out that most of the reported solubility data of UO$_2$(cr) ($t =$ 25°C and 100 to 300°C) correspond in fact to the solubility of a surface layer of amorphous oxide. This point is important because some data ($\Delta_f G_m^\circ$, S_m°, $C_{p,m}^\circ$) given in [92GRE/FUG] on U(OH)$_4$(aq) are based on the assumption that UO$_2$(cr) remains the equilibrium phase in the presence of aqueous solutions.

The value of the solubility product re-evaluated by Neck and Kim [99NEC/KIM], solves some inconsistencies in the thermodynamic data of U(IV) carbonate solutions of [98RAI/FEL]. For instance, the value for $\log_{10}\beta_5^\circ$ (U(CO$_3$)$_5^{6-}$) deduced from the experimental data using $\log_{10}K_{s,0}^\circ = -53.45$ [97RAI/FEL] for U(OH)$_4$(s), does not agree with the selected value of [92GRE/FUG], but does so using $\log_{10}K_{s,0}^\circ = -(55.2 \pm 1.0)$.

To discuss the hydrolysis of Np(IV), with regard to the nature of amorphous hydrated oxide, NpO$_2$(am), or NpO$_2$·xH$_2$O(am) or hydroxide, Np(OH)$_4$(am), of Np(IV), it is necessary to recall the meaning of some equilibrium constants:

$$\text{Np(OH)}_4(\text{am}) \rightleftharpoons \text{Np}^{4+} + 4\,\text{OH}^-$$
$$\text{NpO}_2(\text{am}) + 2\,\text{H}_2\text{O} \rightleftharpoons \text{Np}^{4+} + 4\,\text{OH}^- \quad \rightarrow K_{s,0}^\circ$$

$$\text{Np(OH)}_4(\text{am}) + 4\,\text{H}^+ \rightleftharpoons \text{Np}^{4+} + 4\,\text{H}_2\text{O(l)}$$
$$\text{NpO}_2(\text{am}) + 4\,\text{H}^+ \rightleftharpoons \text{Np}^{4+} + 2\,\text{H}_2\text{O(l)} \quad \rightarrow {}^*K_{s,0}^\circ$$

$$\text{Np}^{4+} + 4\,\text{OH}^- \rightleftharpoons \text{Np(OH)}_4(\text{aq}) \quad \rightarrow \beta_4^\circ$$

$$\text{Np}^{4+} + 4\,\text{H}_2\text{O(l)} \rightleftharpoons \text{Np(OH)}_4(\text{aq}) + 4\,\text{H}^+ \rightarrow {}^*\beta_4^\circ$$

$$\text{Np(OH)}_4(\text{am}) \rightleftharpoons \text{Np(OH)}_4(\text{aq}) \text{ or}$$
$$\text{NpO}_2(\text{am}) + 2\,\text{H}_2\text{O(l)} \rightleftharpoons \text{Np(OH)}_4(\text{aq}) \quad \rightarrow K_{s,4}^\circ$$

For Np(IV) the value of $\log_{10} {}^*\beta_1^\circ = -(0.29 \pm 1.00)$ is selected by [2001LEM/FUG]. The large uncertainty comes from the unweighted average of three experimental values. Due to large uncertainties in literature data [2001LEM/FUG] do not select a value for $\log_{10} \beta_2^\circ$. The value, $\log_{10} \beta_4^\circ = (46.2 \pm 1.1)$, is derived from solubility data of NpO$_2$(am) in acidic and basic aqueous solutions. The former data give $\log_{10} {}^*K_{s,0}^\circ = (1.53 \pm 1.0)$ using $\log_{10} {}^*\beta_1^\circ = -(0.29 \pm 1.00)$ which, when used with the value of $\log_{10}[\mathrm{Np(OH)_4(aq)}] = -(8.3 \pm 0.3)$, gives $\log_{10} {}^*\beta_4^\circ = -(9.8 \pm 1.1)$ ($\log_{10} \beta_4^\circ = (46.2 \pm 1.1)$). These data come from [87RAI/SWA].

[99NEC/KIM] selected the value, $\log_{10} \beta_1^\circ = (14.55 \pm 0.2)$ ($\log_{10} {}^*\beta_1^\circ = 0.55$), determined at very low neptunium concentrations [77DUP/GUI] to predict $\log_{10} \beta_2^\circ = 28.0$ or 28.2 depending on the evaluation method used (in close agreement with the experimental value of (28.35 ± 0.3) from [77DUP/GUI]), $\log_{10} \beta_3^\circ = 39.0$ or 39.2 and $\log_{10} \beta_4^\circ = 46.7$ or 47.2 (this last value agrees with the value selected by [2001LEM/FUG]). Neck and Kim pointed out that the validity of other literature data is suspect due to colloid formation.

The thermochemical value, $\log_{10} K_{s,0}^\circ$ (NpO$_2$, cr) $= -(63.7 \pm 1.8)$ [87RAI/SWA] cited by Neck and Kim [99NEC/KIM], comes from several critically evaluated standard data at 298.15 K: $\Delta_f G_m^\circ$(NpO$_2$, cr), S_m°(NpO$_2$, cr), [72FUG], S_m°(Np, cr) [76OET/RAN] and $\Delta_f G_m^\circ$(Np^{4+}) $= -(502.9 \pm 7.5)$ kJ·mol^{-1} [76FUG/OET]. According to the values of the Gibbs energy of formation of NpO$_2$(cr) selected by [2001LEM/FUG], $\Delta_f G_m^\circ$(NpO$_2$, cr, 298.15 K) $= -(1021.731 \pm 2.514)$ kJ·mol^{-1} and $\Delta_f G_m^\circ$(Np^{4+}, cr, 298.15 K) $= -(491.8 \pm 5.6)$ kJ·mol^{-1}, and with the auxiliary values from [92GRE/FUG], this review calculates $\log_{10} K_{s,0}^\circ$ (NpO$_2$, cr) $= -(65.75 \pm 1.07)$.

The value selected by [2001LEM/FUG] for $\log_{10} K_{s,0}^\circ$ (NpO$_2$, am) is $-(54.5 \pm 1.0)$ coming from $\log_{10} {}^*K_{s,0}^\circ$ (NpO$_2$, am) $= (1.53 \pm 1.0)$ (see above). The large uncertainty given is estimated.

Neck and Kim [99NEC/KIM] reinterpreted the solubility data of [87RAI/SWA] at pH below three with the new values of the β_1, β_2 and β_3 hydrolysis constants of Np^{4+} and give $\log_{10} K_{s,0}^\circ = -(56.7 \pm 0.4)$. With this value and all the solubility data of [85RAI/RYA], [93ERI/NDA] and [96NAK/YAM] corresponding to the range 5 to 13 they calculate $\log_{10} \beta_4^\circ = (47.7 \pm 1.1)$ according to $\log_{10}[\mathrm{Np(OH)_4(aq)}] = \log_{10} K_{s,0}^\circ + \log_{10} \beta_4^\circ = -(9.0 \pm 1.0)$. The large uncertainty covers the rather large spread of the data.

The values for the Np(IV) species revised by Neck and Kim [99NEC/KIM] are retained by this review, with the exception of the estimated value for $\log_{10} \beta_3^\circ$.

[2001LEM/FUG] selected for Pu(IV), $\log_{10} {}^*\beta_1^\circ = -(0.78 \pm 0.6)$, ($\log_{10} \beta_1^\circ = (13.22 \pm 0.6)$), as the average value of the most reliable results of spectroscopic measurements (see below). No other values are selected for the hydrolysis constants.

For Pu(IV) the situation is the same as for Np(IV). [99NEC/KIM] discuss the literature data and retain, to make predictions, those corresponding to experimental conditions where colloids are less likely to form [72MET/GUI]. These conditions seem not to have been fulfilled for solutions on which spectroscopic measurements have been made, because their total concentrations in Pu are higher than the solubility of PuO_2(am) allows. Taking $\log_{10} \beta_1^\circ = (14.6 \pm 0.2)$ (all the other values reported in the literature are smaller) they calculate $\log_{10} \beta_1^\circ = 14.4$, $\log_{10} \beta_2^\circ = 28.4$, $\log_{10} \beta_3^\circ = 39.4$ and $\log_{10} \beta_4^\circ = 47.5$, the experimental values being, respectively, (28.6 ± 0.3), (39.7 ± 0.4) and (47.5 ± 0.5). These results are a good test of the Neck and Kim model.

The thermochemical values of $\log_{10} K_{s,0}^\circ$ (PuO_2, cr) given by Neck and Kim [99NEC/KIM] are those of [89KIM/KAN], $-(63.8 \pm 1.0)$ or [87RAI/SWA], $-(64.1 \pm 0.7)$. From $\Delta_f G_m^\circ$(PuO_2, cr, 298.15 K) $= -(998.1 \pm 1.0)$ kJ · mol^{-1} and $\Delta_f G_m^\circ$(Pu^{4+}, cr, 298.15 K) $= -(478.0 \pm 2.7)$ kJ · mol^{-1}, and the auxiliary values from [92GRE/FUG] this review calculates $\log_{10} K_{s,0}^\circ$ (PuO_2, cr) $= -(64.04 \pm 0.51)$. The value selected by [2001LEM/FUG] for $\log_{10} K_{s,0}^\circ$ (PuO_2, am) is $-(58 \pm 1)$.

The literature data on the solubility of PuO_2(am) in acidic and basic media (pH = 0 to 13) are reviewed by [99NEC/KIM] and reinterpreted on the basis of the hydrolysis constants (see above) which gives $\log_{10} K_{s,0}^\circ = -(58.7 \pm 0.9)$. Neck and Kim [99NEC/KIM] select, as the best value, the average of that value and $\log_{10} K_{s,0}^\circ = -(58.3 \pm 0.5)$, the latter coming from a method independent of Pu(IV) hydrolysis [98CAP/VIT]. They recommend the value, $\log_{10} K_{s,0}^\circ = -(58.5 \pm 1.1)$. This review selects $\log_{10} K_{s,0}^\circ = -(58.33 \pm 0.52)$.

All the papers considered by Neck and Kim [99NEC/KIM] have been considered in [2001LEM/FUG].

[99OLA]

Assessments of the thermodynamic data for UO_{2+x}(cr), UO_3(g) and $UO_2(OH)_2$(g) are reviewed, with emphasis on those in computer codes used in the analysis of reactor accidents, in which the reaction of UO_2 with steam is important. No additional experimental data are presented.

For UO_3(g), Olander [99OLA] has considered the recent experimental work by Krikorian et al. [93KRI/EBB] as well as the earlier work on which the current selection of the stability by [92GRE/FUG] is based. For reasons that are not detailed, the author prefers the data of [93KRI/EBB], which are (presumably) the basis of the value of $\Delta_f H_m^\circ$ (UO_3, g, 298.15 K) $= -(796.7 \pm 3.5)$ kJ · mol^{-1}, given in an unpublished report by Ebbinghaus, as quoted by Olander (compared with the value of $-(799.2 \pm 15.0)$ kJ · mol^{-1} selected by [82GLU/GUR] and [92GRE/FUG]).

For $UO_2(OH)_2$(g), [99OLA] again selects, without comment, the value of $\Delta_f H_m^\circ$ ($UO_2(OH)_2$, g, 298.15 K) $= -(1200 \pm 10)$ kJ · mol^{-1}, from the same unpublished report by Ebbinghaus. As noted in the review of [93KRI/EBB] and in section 9.3.1.2., the data for the stability of this species are very discordant and no values are selected in this review. The uncertainties quoted by Olander [99OLA] for the enthalpies of formation of both gases seem to be far too small.

[99PIA/TOU]

Using the same high temperature X–ray diffraction techniques as in previous publications [94TOU/PIA], [98PIA/TOU], the authors studied the reaction between SrO and UO_2. No thermodynamic data are reported. With a 1:1 ratio, under an atmosphere of oxygen, orthorhombic "β-$SrUO_4$" (which the authors call α) is observed between 1123 and 1373 K. At 293 K, the reported lattice parameters are in agreement with the many literature values. The authors interpret the increase in lattice parameters between 1123 and 1373 K as being due to a change in the O:U ratio from 3.67 to 3.62.

Under more reducing conditions ($10^{-6} \leq p_{O_2} / \text{bar} \leq 10^{-5}$), a rhombohedral "α-$SrUO_4$" (called β by the authors) is stable above 1108 K, with an upper O:U value of 3.60, lower than usually reported in the literature for this phase. Under the most reducing conditions ($p_{O_2} / \text{bar} \leq 10^{-14}$), another rhombohedral phase is observed, with a lower O:U ratio of 3.11, also distinctly lower than reported in the literature.

The authors also report the progressive formation of an orthorhombic "$SrUO_3$" phase upon heating under $p_{O_2} / \text{bar} \leq 10^{-14}$ of $SrUO_4$ or a mixture UO_2/SrO at temperatures between 1173 and 1773 K. This phase could never be isolated in a pure state. Some evidence for a second-order monoclinic to orthorhombic transformation at 1073 K is presented. This monoclinic form does not appear to be isomorphous with Sr_3UO_6, in contrast with the situation for the corresponding Ca compounds [98PIA/TOU]. Arguments are presented for the isomorphism of both α- and β-$SrUO_3$ with Sr_3UO_5 (produced by reduction of Sr_3UO_6) with the same second-order transition temperature.

At temperatures above 1523 K, in a reducing atmosphere, the disordered fluorite phase, $U_{1-\delta}Sr_\delta O_{2-\delta}$ is formed, possibly after a transition to an intermediary tetragonal "$SrUO_3$" phase.

The authors also present a schematic pseudo-binary "$SrUO_3$"–$SrUO_4$ phase diagram.

[99RAI/HES]

This paper is a continuation of the studies of [95RAI/FEL], [97FEL/RAI] and [98RAI/FEL] of An(IV)–CO_3^{2-}–HCO_3^-–OH^- systems. The experiments are described in detail, but the interpretation of the data suffers from the same shortcomings as the previous papers, i.e., the lack of justification for the Pitzer parameters used, with the exception of those for $Np(CO_3)_5^{6-}$ for which the authors used the parameters for the analogous U(IV) complex; in addition, no uncertainty estimate has been made of the proposed equilibrium constants. This review finds sufficient evidence to support the assumption that $Np(OH)_2 (CO_3)_n^{2-2n}$ is formed; the authors' Figure 1c. refers to solubility data measured in an approximately constant ionic medium, 1.78 M K_2CO_3 where the concentration [OH⁻] has been varied from 0.01 to 1 M. The experimental data are close to a straight line with the slope − 2, indicating that the reaction studied is:

$$Np(OH)_4(am) + n\ CO_3^{2-} \rightleftharpoons Np(OH)_2(CO_3)_n^{2-2n} + 2\ OH^-$$

As the concentration of carbonate is high and constant in the experiment, it is impossible to determine the stoichiometric coefficient n, although the (conditional) equilibrium constant for the reaction above can be determined. In order to make some statements about the stoichiometry one must use known characteristics of the coordination chemistry of the M(IV) actinides. The limiting carbonate complex, $M(CO_3)_5^{6-}$, is ten-fold coordinated as shown by several different investigators and also by the authors of this paper. Some additional comments on the coordination geometry of these complexes are given in the comments to papers [95RAI/FEL] and [98RAI/FEL]. This paper has also been reviewed in [2001LEM/FUG]. The conclusions are essentially the same as drawn here, except for the comments on the composition of the ternary complexes. The paper contains useful data on the structure of the penta-carbonate complex, determined by EXAFS. In conclusion, the quantitative equilibrium data for this system are less reliable, with the exception of $\log_{10} \beta_5^\circ$ and $\log_{10} \beta_4^\circ$ given in [2001LEM/FUG] for the formation of $M(CO_3)_5^{6-}$ and $M(CO_3)_4^{4-}$. As the authors have used the same approach when determining the equilibrium constant for the $M(CO_3)_5^{6-}$ complexes of Th, U, Np and Pu [99RAI/HES2], the variation of the equilibrium constants is probably more precise.

[99RAI/HES2]

This is an excellent experimental study where the authors have analysed the solubility of amorphous $PuO_2(am)$ in carbonate/hydrogen carbonate solutions over a wide pH and concentration range. The difficult experiments are described in detail, as is the strategy for analysing the data. The analysis is anchored on the limiting complex, $Pu(CO_3)_5^{6-}$, which predominates over a wide concentration range where it has been characterised by EXAFS spectroscopy. By using the Pitzer model the authors then can determine both interaction parameters and the equilibrium constant for the complex; the latter requires in addition the solubility product for the amorphous hydroxide. This review accepts both the analysis and the conclusions drawn about the stoichiometry, structure and equilibrium constant of the limiting complex. The authors report for the reaction:

$$Pu^{4+} + 5\ CO_3^{2-} \rightleftharpoons Pu(CO_3)_5^{6-} \tag{A.119}$$

$\log_{10} \beta_5^\circ = 34.18$, but with no estimate of the uncertainty. This review has estimated the uncertainty to be at least ± 1.0 $\log_{10}$ units. In combination with $\log_{10} K_{s,0}^\circ$ (PuO_2, am) = − (58.33 ± 0.52) selected in the present review, the formation constant of the limiting carbonate complex is calculated to be $\log_{10} \beta_5^\circ = (35.65 ± 1.13)$.

The authors have shown that the limiting complex is also predominant at high bicarbonate concentrations and that the solubility can be described with the same equilibrium constant as deduced from the carbonate data. This review agrees with this general conclusion, although it can be observed from Figure 8 in [99RAI/HES2] that the calculated solubility at high [HCO_3^-] is systematically higher than the experimental data.

The authors have also analysed the solubility at lower concentrations of HCO_3^- and interpret these results with the complex, $Pu(OH)_2(CO_3)_2^{2-}$, with $\log_{10} K° = 44.76$, for the reaction:

$$Pu^{4+} + 2 CO_3^{2-} + 2 OH^- \rightleftharpoons Pu(OH)_2(CO_3)_2^{2-} \qquad (A.120)$$

By using the equilibrium constants for equations (A.119) and (A.120) above, together with the Pitzer model, the authors are able to describe the experimental solubility fairly well. However, according to this review, the authors have not demonstrated the stoichiometry of the mixed hydroxide-carbonate complex and this means that the result should be used with caution. The equilibrium can certainly be used to describe the solubility in the concentration range studied, but care must be exercised when estimating the solubility of amorphous PuO_2 outside the concentration range studied by Rai et al.

There is no doubt that ternary complexes are formed in these systems. In a previous study under comparable conditions, Yamaguchi et al. [94YAM/SAK] proposed the same complex stoichiometry. However, measuring solubility data from the direction of over-saturation, they obtained an equilibrium constant for $Pu(OH)_2(CO_3)_2^{2-}$, $\log_{10} K°_{s,(1,2,2)} = -(10.2 \pm 0.5)$ if converted to $I = 0$ with the SIT. This value is considerably higher than $- 12.1$ from [99RAI/HES2]. Further solubility data in 0.01 – 0.1 M carbonate solutions at pH 12 and 13 were ascribed to the formation of $Pu(OH)_4(CO_3)_2^{4-}$ [94YAM/SAK]. The latter experimental data are concordant with those determined by Rai et al. in 0.1 – 1 M K_2CO_3 solutions containing 0.01 M KOH. However, [99RAI/HES2] did not use these results to evaluate an equilibrium constant for the complex, $Pu(OH)_4(CO_3)_2^{4-}$.

In addition, Rai et al. report solubility data of PuO_2(am) in carbonate-free dilute KOH solutions at $(23 \pm 2)°C$. The total Pu concentrations determined by oxidation state analyses after equilibration for one week are illustrated in Figure 5 of [99RAI/HES2]. From the seven data points at pH 8.3 – 13.0, this review calculates a mean value of $\log_{10}[Pu(OH)_4(aq)] = -(10.4 \pm 0.5)$ (1.96 σ), which is in good agreement with experimental data of other authors, $\log_{10}[Pu(OH)_4(aq)] = -(10.4 \pm 0.2)$, [86LIE/KIM] and $-(10.2 \pm 0.8)$, [98CHA/TRI].

[99RUN/REI]

The paper [97RUN/NEU] (see this review) shows by spectroscopic measurements that in solutions, $[H^+] = 1$ M, $[Cl^-] = 2$ to 15 M, where complexes of Pu(VI) of the type, $PuO_2Cl_x^{2-x}$ with x = 1, 2, 3 and 4, are formed. The paper [99RUN/REI] is a study of the complexation of PuO_2^{2+} by Cl^- in $HClO_4$/NaCl solutions, $[H^+] = 0.1$ m, up to $I = 5.19$ m by conventional absorption spectrophotometry ($t = 23°C$).

The literature survey of the authors does not include the data of [94GIF], which have been used by [2001LEM/FUG] to select values of $\log_{10} \beta_1^\circ = (0.70 \pm 0.13)$ and $\log_{10} \beta_2^\circ = -(0.6 \pm 0.2)$, respectively, for the equilibria:

$$PuO_2^{2+} + Cl^- \rightleftharpoons PuO_2Cl^+ \quad (A.121)$$

and

$$PuO_2^{2+} + 2Cl^- \rightleftharpoons PuO_2Cl_2(aq) \quad (A.122)$$

with the following estimated values, $\Delta\varepsilon_{(1)}(A.121) = -(0.08 \pm 0.08)$ kg·mol^{-1} and $\Delta\varepsilon_{(2)}(A.122) = -(0.43 \pm 0.20)$ kg·mol^{-1}. The Giffaut [94GIF] $\log_{10} \beta$ values, from which the $\log_{10} \beta^\circ$ are derived, are in good agreement with the previously published values, at the same common ionic strengths, ≥ 2 M. Giffaut performed experiments up to $I = 4.5$ M (see review in Appendix A of [2001LEM/FUG]).

The study of [99RUN/REI] is a careful investigation which shows that to derive the free concentration of PuO_2^{2+}, the variation of the molar absorption coefficient of this species at (830.6 ± 0.1) nm with the ionic strength ([HClO$_4$] = 0.1 m + [NaClO$_4$] = 0.1 to 5.13 m) must be taken into account. This is the result of a decrease of the coordinated water molecules, as EXAFS spectra show also in NaCl solutions (1 to 5 M). The deconvolution of the spectra in solutions, [HClO$_4$] = 0.1 m + NaCl up to 5.19 m, is clear. The two chloride complexes have absorption maxima at (837.6 ± 0.2) and (843 ± 0.4) nm, respectively, with different molar absorptivities.

The values of $\log_{10} \beta^\circ$ given by Runde et al. using the SIT to extrapolate the values of $\log_{10} \beta$ at zero ionic strength are given in the Table A-42, as well as the corresponding values for U(VI) selected by [92GRE/FUG] for comparison. The $\Delta\varepsilon$ values used to calculate the SIT interaction coefficients are also given. The U and Pu values seem in reasonable agreement. There is no value for $\varepsilon(PuO_2^{2+}, Cl^-)$, but taking for this coefficient the corresponding value of uranium, $\varepsilon(UO_2^{2+}, Cl^-) = (0.21 \pm 0.02)$ kg·mol^{-1}, gives $\varepsilon(PuO_2Cl^+, Cl^-) = (0.11 \pm 0.04)$ kg·mol^{-1}.

Table A-42: $\Delta\varepsilon$ values used to calculate the SIT interaction coefficients and $\log_{10} \beta^\circ$ for PuO_2Cl^+, $PuO_2Cl_2(aq)$, and UO_2Cl^+, $UO_2Cl_2(aq)$.

	$\Delta\varepsilon$ (kg·mol^{-1})	$\log_{10} \beta^\circ$	Reference
PuO_2Cl^+	$-(0.13 \pm 0.03)$	(0.23 ± 0.03)	[99RUN/REI]
UO_2Cl^+	$-(0.25 \pm 0.02)$	(0.17 ± 0.02)	[92GRE/FUG]
$PuO_2Cl_2(aq)$	$-(0.4 \pm 0.1)$	$-(1.7 \pm 0.2)$	[99RUN/REI]
$UO_2Cl_2(aq)$	$-(0.62 \pm 0.17)$	$-(1.07 \pm 0.35)$	[92GRE/FUG]

All the thermodynamic data at $I = 0$ obtained by Runde et al. disagree significantly from those selected by [2001LEM/FUG], (see above). It is necessary to try to understand this disparity by comparing carefully the values of $\log_{10} \beta_n^\circ$ and $\varepsilon_{(n)}$ or $\Delta\varepsilon_{(n)}$ for the U(VI) and Pu(VI) chloride complexes, AnO_2Cl^+ and $AnO_2Cl_2(aq)$ in [92GRE/FUG], [94GIF] and [99RUN/REI] within the framework of the application of the SIT in the range of ionic strength where data are available.

The $\log_{10} \beta_n^\circ$ values and particularly the $\Delta\varepsilon_{(n)}$ values for U(VI) and Pu(VI) should be very similar. Monodentate complexes like AnO_2Cl^+ and $AnO_2Cl_2(aq)$ are typical examples where the oxidation state analogy principle should be valid. Fortunately, in the case of the uranium complexes there are many experimental data, down to low ionic strength, and application of the SIT to the accepted experimental data appears to be straightforward ([92GRE/FUG], p.192 – 196).

It is noteworthy that the data for U(VI) and those of Giffaut for Pu(VI) refer to mixed chloride-perchlorate media (at constant ionic strength). The evaluated SIT coefficients, $\varepsilon(AnO_2^{2+}, ClO_4^-)$ and $\varepsilon(AnO_2Cl^+, ClO_4^-)$, refer to pure perchlorate solutions. However, this is a simplification because firstly, the chloride concentration is considerably different from zero and not constant, and secondly the interaction coefficients, $\varepsilon(AnO_2^{2+}, ClO_4^-)$ and $\varepsilon(AnO_2Cl^+, ClO_4^-)$, are different, usually larger than $\varepsilon(AnO_2^{2+}, Cl^-)$ and $\varepsilon(AnO_2Cl^+, Cl^-)$. Therefore, $\Delta\varepsilon$ is not really constant, which makes a correct application of SIT complicated.

In the study of [99RUN/REI] this problem is avoided, because the spectroscopic data were determined in almost pure NaCl solutions (if we neglect the addition of 0.1 M $HClO_4$). However, we have to be aware that the SIT coefficients evaluated in [99RUN/REI] are $\varepsilon(PuO_2^{2+}, Cl^-)$ and $\varepsilon(PuO_2Cl^+, Cl^-)$, not those with perchlorate as given in [92GRE/FUG], [94GIF].

Table A-43 : Equilibrium constant, $\log_{10} \beta_1^\circ$, and interaction coefficient, $\Delta\varepsilon_{(1)}$, for the first chloride complexes of U(VI) and Pu(VI).

Reference	An	$\log_{10} \beta_1^\circ$	$\Delta\varepsilon_{(1)}$ (kg · mol^{-1})
[92GRE/FUG]	U	(0.17 ± 0.02)	$-(0.25 \pm 0.02)$[a]
[94GIF]	Pu	(0.70 ± 0.13)	$-(0.08 \pm 0.08)$[a]
[99RUN/REI]	Pu	(0.23 ± 0.03)	$-(0.13 \pm 0.03)$[b]

[a] for $NaClO_4$ medium   [b] for NaCl medium

As mentioned above, similar $\log_{10} \beta_1^\circ$ values must be expected for U(VI) and Pu(VI). The same should hold for $\Delta\varepsilon_{(1)}$ values in the case of U(VI) and Pu(VI) in $NaClO_4$ medium ([92GRE/FUG] and [94GIF]). This is, however, not the case (Table A-43). It is to be noted that Giffaut reported an experimental series only at relatively high ionic strength ($I = 2.2$ and 3.5 mol · kg^{-1}). If we do not apply the SIT extrapolation to $I = 0$ by linear regression to the Giffaut values, but use a fixed value of $\Delta\varepsilon_{(1)} = -(0.25 \pm 0.02)$ kg · mol^{-1} (the well-ascertained value for U(VI) from [92GRE/FUG]) for the conversion to $I = 0$, we obtain values similar to those in [92GRE/FUG] and [99RUN/REI]:

$\log_{10} \beta_1$ ($I = 2.2$ m) = $-(0.07 \pm 0.09)$ [94GIF] giving $\log_{10} \beta_1^\circ = (0.3 \pm 0.1)$

$\log_{10} \beta_1$ ($I = 3.5$ m) = $-(0.06 \pm 0.07)$ [94GIF] giving $\log_{10} \beta_1^\circ = (0.1 \pm 0.1)$.

The SIT regression plot in [99RUN/REI] is based on many data over a wide range of ionic strength ($I = 0.25 - 3.5$ mol · kg^{-1}) and hence is much more accurate. The fact that $\Delta\varepsilon_{(1)} = -(0.13 \pm 0.03)$ kg · mol^{-1} in NaCl [99RUN/REI] is different from $\Delta\varepsilon_{(1)} = -(0.25 \pm 0.02)$ kg · mol^{-1} in NaClO$_4$ medium [92GRE/FUG] is not surprising, because particularly $\varepsilon(AnO_2^{2+}, Cl^-) = 0.21$ kg · mol^{-1} is smaller than $\varepsilon(AnO_2^{2+}, ClO_4^-) = 0.46$ kg · mol^{-1}. The value of $\varepsilon(PuO_2Cl^+, Cl^-) = (0.11 \pm 0.04)$ kg · mol^{-1} [99RUN/REI] is reasonable for a 1:1 electrolyte and also reasonable in relation to $\varepsilon(UO_2Cl^+, ClO_4^-) = (0.33 \pm 0.04)$ kg · mol^{-1} [92GRE/FUG] (interaction coefficients of actinide cations with chloride are always smaller than those with perchlorate; this is a general trend).

For these reasons the values of $\log_{10} \beta_1^\circ$, $\Delta\varepsilon_{(1)}$ and $\varepsilon(PuO_2Cl^+, Cl^-)$ determined by [99RUN/REI] are the best available. They are selected by this review.

Table A-44: Equilibrium constant, $\log_{10} \beta_2^\circ$, and interaction coefficient, $\Delta\varepsilon_{(2)}$, for the second chloride complexes of U(VI) and Pu(VI).

Reference	An	$\log_{10} \beta_2^\circ$	$\Delta\varepsilon_{(2)}$ (kg · mol^{-1})
[92GRE/FUG]	U	$-(1.1 \pm 0.4)$	$-(0.62 \pm 0.17)$ [a]
[94GIF]	Pu	$-(0.6 \pm 0.2)$	$-(0.43 \pm 0.20)$ [a]
[99RUN/REI]	Pu	$-(1.7 \pm 0.2)$	$-(0.4 \pm 0.1)$ [b]

[a] for NaClO$_4$ medium   [b] for NaCl medium

Comparing Giffaut's $\log_{10} \beta_2^\circ$ and $\Delta\varepsilon_{(2)}$ (Table A-44) data with those for U(VI) discussed in [92GRE/FUG], similar comments could be made as in the case of the mono chloro-complex. It can be noted here that Giffaut [94GIF] did not measure absorption at the wavelengths 843 - 844 nm (PuO$_2$Cl$_2$(aq)) and deduced $\log_{10} \beta_2$ from the total concentration of Pu and that of PuO$_2$Cl$^+$. However, the situation is not so clear.

The $\log_{10} \beta_2^\circ$ value evaluated in [99RUN/REI] appears to be low. Unfortunately, in this case the ionic strength range of experimental data ($I = 1.8 - 3.5$ mol · kg^{-1}) is much narrower than used for the determination of $\log_{10} \beta_1^\circ$ and there are no data at low ionic strength. Therefore, the extrapolation to $I = 0$ depends appreciably on the $\Delta\varepsilon(2)$ value. In this respect the experimental data in Figure 6 of [99RUN/REI], which gives ($\log_{10} \beta_2 + \Delta z^2 \cdot D$) as a function of I, show large error bars. If we apply $\varepsilon(PuO_2Cl^+, Cl^-) = 0.21$ kg · mol^{-1} and set the SIT coefficients for the neutral complex, PuO$_2$Cl$_2$(aq), equal to zero as it is usually done in the NEA-TDB reviews, we obtain $\Delta\varepsilon_{(2)} = -(0.27 \pm 0.03)$ kg · mol^{-1} and $\log_{10} \beta_2^\circ = -(1.3 \pm 0.1)$.

For these reasons, this review selects as the best value for $\log_{10} \beta_2^\circ$ the average value of those of [94GIF] and [99RUN/REI], $\log_{10} \beta_2^\circ = -(1.15 \pm 0.30)$, taking a larger uncertainty than that given by these authors.

The study of Runde et al. appears more complete than the previous ones.

The reference [96RUN/NEU] in this paper should in fact be [97RUN/NEU].

[99RUT/GEI]

This paper reproduces mainly the results of [97RUT/GEI] and concerns the interaction of uranium(VI) with arsenate(V). Additional experimental information is given. The ranges of the variations of the parameters controlling the compositions of the solutions are slightly modified (arsenic acid, $5 \cdot 10^{-6}$ to $5 \cdot 10^{-2}$ M, pH = 1.5 to 3, [U] = 10^{-6} M).

Deconvolution of the TRLFS spectra shows the existence of three species in addition to that of UO_2^{2+} ($\tau = 1.70 \pm 0.50$ μs) for which characteristics are given in Table A-45. The species are taken to be the same as those in the U(VI)–phosphoric acid system, since the pK_a values of the acids are very similar. The pK_a of arsenic acid used by the authors are those derived from auxiliary data selected by [92GRE/FUG] but without considering uncertainties. For the neutral complexes the spectra are those of the mineral trogerite and the solids, $UO_2(HAsO_4) \cdot 4H_2O$ and $UO_2(H_2AsO_4)_2 \cdot H_2O$.

The stoichiometric U/As ratios of the different species are given by slope analyses of:

$$\frac{[\text{Complex}]}{[UO_2^{2+}]} = f([AsO_4^{3-}])$$

and their dependence on pH. The complex formation constants are reported for the following reaction:

$$UO_2^{2+} + r\,H^+ + q\,AsO_4^{3-} \rightleftharpoons UO_2H_r(AsO_4)_q^{2+r-3q}.$$

The values of $\log_{10} \beta$ and $\log_{10} \beta^\circ$ are given in Table A-45 (the values at $I = 0$ are calculated using the Davies equation).

Table A-45. Equilibrium constants derived from spectroscopic measurements.

Species	τ ($I = 0.1$) μs	Emission bands nm	$\log_{10} \beta$ ($I = 0.1$) 2σ	$\log_{10} \beta^\circ$ 2σ
$UO_2(H_2AsO_4)^+$	(12.25 ± 1.20)	478, 494, 514, 539, 563	(20.39 ± 0.24)	(21.96 ± 0.24)
$UO_2(HAsO_4)(aq)$	$0.1 < \tau < 1$	504, 525, 547	(17.19 ± 0.31)	(18.76 ± 0.31)
$UO_2(H_2AsO_4)_2(aq)$	(38.3 ± 3.50)	481, 497, 518, 541, 571	(38.61 ± 0.20)	(41.53 ± 0.20)

The concentration of the complex $UO_2(H_2AsO_4)_2(aq)$ is always very low (less than 1%) but it has a strong fluorescence intensity. Dioxouranium(VI) phosphate complexes are stronger than those for arsenic. This review has calculated the equilibrium constants for the formation of the complexes from UO_2^{2+} and $HAsO_4^{2-}$ or H_3AsO_4. The protonation constants of H_3AsO_4 have been derived from the auxiliary data selected by [92GRE/FUG], page 63. For each $\log_{10} K°$ value a large uncertainty: ± 0.99 is reported. This review has re-examined the experimental data of the protonation constants of H_3PO_4 (H_3L). The values collected in Table A-46, for $I = 0$, often without uncertainty, do not allow us to reduce the uncertainty.

Table A-46: Values of $\log_{10} K°$ for protonation of arsenic acid at 25°C.

Reaction	$\log_{10} K°$	Reference
$H_2L^- + H^+ \rightleftharpoons H_3L$	(2.26 ± 0.99)	[92GRE/FUG]
	2.15	[98MAR/SMI]
	2.223	[53AGA/AGA]
	2.30	[69SAL/HAK]
	2.26	[76TOS]
$HL^{2-} + H^+ \rightleftharpoons H_2L^-$	(6.76 ± 0.99)	[92GRE/FUG]
	6.65	[98MAR/SMI]
	6.980	[53AGA/AGA]
	5.76	[76TOS]
$L^{3-} + H^+ \rightleftharpoons HL^{2-}$	(11.60 ± 0.99)	[92GRE/FUG]
	11.80	[98MAR/SMI]
	11.29	[76TOS]

The $\log_{10} K°$ values from [92GRE/FUG] in Table A-46 have been calculated from the Gibbs energy of formation of the various species and not from the direct experimental determinations; this is the reason for the large uncertainty estimates which can be reduced ([92GRE/FUG], page 390). A more reasonable estimate is ± 0.20 in $\log_{10} K°$, which is used in the main text to select the equilibrium constants.

[99SAX/RAM]

This conference paper reports preliminary results of the heat capacity of rubidium uranium sulphate, $Rb_2U(SO_4)_3(s)$, measured by DSC. No details of the material under study are given. The DSC graph (373–673 K) shows a transition at *ca.* 630 K. The heat capacities are tabulated at rounded temperatures from 273 to 623 K. Except for the last point at 623 K, these are consistent with the enthalpy drop measurements from the same laboratory [93JAY/IYE] (which are not cited), and have been combined with them to provide heat capacities for this phase – see section 9.10.5.3.

[99SIN/DAS]

This paper reports measurements of the oxygen activity (emf with a CaO – ZrO$_2$ electrolyte) in the three-phase field, U$_3$O$_8$(cr) + TeO$_2$(cr) + UTeO$_5$(cr) from 821 to 994 K.

UTeO$_5$(cr) was prepared by repeatedly heating appropriate amounts of U$_3$O$_8$(cr) with TeO$_2$(cr) in dry air at 973 K for 400 hours, grinding and reheating until no phases other than UTeO$_5$(cr) were detected by X–ray diffraction. Equimolar amounts of the three phases were pelleted and annealed at 800 K before the emf measurements were made. These utilised a cell with a CaO – ZrO$_2$ electrolyte tube in flowing argon, with Ni/NiO as the reference electrode.

To convert the measurements to oxygen potentials, we have used the values assessed by Taylor and Dinsdale [90TAY/DIN], which include consideration of the magnetic transitions in Ni and NiO. Their data in the relevant temperature range can be expressed as $\Delta_f G_m$ (NiO, cr, T) = – 234220 + 85.042 T, J · mol^{-1}.

The recalculated values for the oxygen potentials in the relevant three-phase field are then:

0.5 $RT \ln (p_{O_2}$ / bar) = – 382363 + 206.420 T J · mol^{-1} (821 to 994 K)

corresponding to the Gibbs energy of the reaction:

$$U_3O_8(cr) + 3\ TeO_2(cr) + 0.5\ O_2(g) \rightleftharpoons 3\ UTeO_5(cr). \qquad (A.123)$$

It will be seen that the entropy change is appreciably larger than that expected for a reaction involving a change of 0.5 mole of a simple gas.

Any further processing of these data requires the values for S_m° (298.15 K) and $C_{p,m}(T)$ for UTeO$_5$(cr) up to 1000 K, which are not known. Following the authors, these can be estimated to be the sum of the values for UO$_3$(cr) [92GRE/FUG] and TeO$_2$(cr) [90COR/KON], giving:

S_m° (UTeO$_5$, cr, 298.15 K) = (166 ± 10) J · K^{-1} · mol^{-1}

$C_{p,m}$ (TeO$_2$, cr, T) = 151.845 + 3.6552·10^{-2} T – 1.8129·10^6 T^{-2} J · K^{-1} · mol^{-1}.

A third-law treatment of the data gives $\Delta_r H_m^\circ$ ((A.123), 298.15 K) = – (288.7 ± 10.7) kJ · mol^{-1} and thus $\Delta_f H_m^\circ$ (UTeO$_5$, cr, 298.15 K) = – (1608.8 ± 6.1) kJ · mol^{-1} where the standard enthalpy of formation of TeO$_2$(cr) is taken from the assessment by Cordfunke et al. [90COR/KON] – see section 14.1 This third-law value is close to two calorimetrically-determined values – see section 9.5.3.2.1. However, the second-law value of the enthalpy of reaction (A.123) is very different from the third-law value (in keeping with the unexpectedly large entropy change) $\Delta_r H_m^\circ$ ((A.123), 298.15 K) = – 372.8 kJ · mol^{-1}, with an unknown uncertainty, implying $\Delta_f H_m^\circ$ (UTeO$_5$, cr, 298.15 K) = – 1640.1 kJ · mol^{-1}.

This study thus supports the stabilities derived from the more precise (and thus preferred) calorimetric data, and the estimated entropy of UTeO$_5$(cr).

[99SOD/ANT]

XANES is used to identify and measure the concentrations of Np^{3+}, Np^{4+}, NpO$_2^+$ and NpO$_2^{2+}$ ($\pm$ 5%) in 1 M HClO$_4$ (22 $\pm$ 2°C). Spectra are recorded *versus* the applied potential, E, ($\pm$ 0.01 V) to an electrochemical cell working with a Pt wire electrode and a reference electrode Ag/AgCl, 3 M NaCl (0.196 V at 25°C) under a N$_2$/H$_2$O saturated atmosphere. The characteristics of the XANES spectra (edge position at $\pm$ 2 eV) are first established in solutions where aqueous ions are shown to exist by classical optical spectroscopy. All the XANES spectra are similar to those of Pu [98CON/ALM]. The data are analysed by diagonalisation of the absorption/energy matrix (17.595 to 17.665 eV) and as a result four components are identified in the aqueous ion spectra. A Nernst plot of E versus log$_{10}$[Np(x+1)]/[Np(x)] gives for x = 5 and 3 a transfer of 0.94 electrons and the following formal potentials in 1 M HClO$_4$:

$$E°_{(VI/V)} = (0.931 \pm 0.015) \text{ V}$$
$$E°_{(IV/III)} = -(0.053 \pm 0.001) \text{ V}.$$

The average values quoted by the authors from the literature are, respectively, (0.941 $\pm$ 0.001) V and $-$ (0.045 $\pm$ 0.005) V. [2001LEM/FUG] have selected the values $E°_{(VI/V)}$ = (1.137 $\pm$ 0.001) V (SHE) for the formal potential of NpO$_2^{2+}$/ NpO$_2^+$ (Table 7.2 p. 95) and (0.219 $\pm$ 0.010) V (SHE) for the standard potential of Np^{4+}/Np^{3+} (Table 7.3 p. 100). The calculated formal potential for the latter redox couple is (0.155 $\pm$ 0.001) V (SHE) (Figure 7.2 p. 103). The reported values in this paper ((1.127 $\pm$ 0.015) V (SHE) and (0.241 $\pm$ 0.001) V (SHE)) are far from those selected by [2001LEM/FUG].

The work of Soderholm *et al.* [99SOD/ANT] is more aimed at checking the applicability of the XANES method than to improve the method to measure redox potentials.

[99SUZ/TAM]

Distribution experiments on HTcO$_4$ were conducted at 298.2 K between aqueous solutions (0.01 – 7.0 M HNO$_3$ and 1.5·10^{-5} – 1.5·10^{-1} M) in contact with *n*–dodecane (ten minutes of agitation) containing cyclic amides and mixtures thereof (0.008 – 1.0 M). The amides investigated were:

- N–(2–ethyl)hexylbutyrolactam (EHBLA),
- N–(2–ethyl)hexylvalerolactam (EHVAL),
- N–(2–ethyl)hexyl–caprolactam (EHCLA),
- N–octylcaprolactam (OCLA),
- a mixture of 3–octyl–N–(2–ethyl)hexylvalerolactam and 4–octyl–N–(2–ethyl)hexylvalerolactam (3,4,OEHVLA),
- 2–octyl–N–(2–ethyl)hexylcaprolactam (2OEHCLA),

tion sphere for U(VI). The fact that Yamamura *et al.* [98YAM/KIT] find that the limiting complex is $UO_2(OH)_4^{2-}$ provides strong support for the interpretation of Wahlgren *et al.* [99WAH/MOL].

The paper gives much more structural information on U(VI) species, such as the probable non-existence of the species, $UO_2O(OH)_2^{2-}$. It shows that quantum chemical calculations are very useful in selecting between ambiguous formulations of the hydrolysed species.

[99YUS/SHI]

This is a review article describing the structure and properties of lanthanide and actinide complexes with unsaturated heteropolyanions with Keggin or Dawson structures. A number of equilibrium constants are reported but the degree of protonation of the complexes is not known. The charge of the heteropolyanions is very high, making it very difficult to make extrapolations to zero ionic strength. These results are of interest mainly for laboratory systems. However, it should be noted that these ligands bind most actinides strongly.

[2000ALL/BUC]

This EXAFS study discusses two important questions:

- the chemical similarity/difference between trivalent lanthanides and actinides, and the systematic variation in chloride complexation throughout these two groups of f–elements.
- the comparison of speciation as deduced from EXAFS, spectroscopic and other solution chemical data, *i.e.*, the possibility of distinguishing between complex formation and activity coefficient variations when weak complexes are formed.

The authors present convincing evidence for the formation of chloride complexes in both groups of elements, but with the important difference that the chloride complexation decreases with decreasing ionic radius for the lanthanide group, while it increases for the actinides. The authors have made a quantitative comparison between their data and spectroscopic information from the Cm(III) – Cl system [95FAN/KIM]. The agreement between the two sets of data, obtained by using very different experimental techniques, is excellent. The experimental data indicate that chloride complexes are not formed for Pu(III).

[2000BEN/FAT]

The speciation of Tc(IV) in HCl (1.0 – 6.0 M) was examined by UV–Visible and Raman spectroscopy, EXAFS and electrochemistry. In 6.0 M HCl the predominant species is $TcCl_6^{2-}$, whereas in 1.0 M HCl this anion slowly aquates (requiring approximately ten days at room temperature) to give the general composition of $TcO_n(H_2O)_qCl_p(OH)_{6-n-q-p}^{q-n-2}$ as manifested by a blue shift in the peak at 338 nm ($TcCl_6^{2-}$) to 320 nm. Peaks are also observed at ≈ 240 and ≈ 245 nm. The presence of dissolved oxygen leads to the formation of TcO_4^- (290 nm), which is a trace impurity in the starting solution.

An aged Tc(IV) solution in 1.0 M HCl showed bands at 332 – 344 cm^{-1}, indicative of Tc–Cl bonding and 911 cm^{-1} characteristic of Tc=O bonds (a band at 400 cm^{-1} may also be due to Tc–O stretch from H_2O). Pure $TcCl_6^{2-}$ exhibits bands at 332 – 344 cm^{-1} according to [69SCH/KRA], whereas TcO_4^- has bands at 323 and 912 cm^{-1}. EXAFS shows two types of Tc(IV) first neighbours: Cl at 2.32 – 2.34 Å and 2.48 – 2.51 Å; O at 1.63 – 1.64 Å and 1.90 Å. A better fit of the EXAFS data has five chlorine atoms and one oxygen bound to the Tc centre (the oxygen could be either Tc=O or Tc–O(H_2O)), rather than four chlorine and two oxygen atoms. The Raman and electrochemical information seem to reject the possibility of a Tc=O bond and therefore favour $TcCl_5(OH_2)^-$ as the dominant species in aged 1.0 M HCl solutions, with minor components being most likely $TcCl_4(OH_2)_2$ and $TcCl_6^{2-}$, although the general formula given above is not completely ruled out, *i.e.*, the presence of Tc=O bonds.

[2000BRU/CER]

This paper does not add new thermodynamic data to [92GRE/FUG], [95SIL/BID] and [2001LEM/FUG]. It aims to compare the solubilities of Th, U, Np, Pu, Am and Tc calculated from databases with values derived from spent fuel leaching and natural analogues. Sensitivity to variations in pH (7 to 9), E_h (− 180 to − 60 mV), carbonate concentration (from calcite equilibrium) and t (15 to 60°C) is checked for granite waters and granite water saturated with bentonite. Each radioelement has its own response to these parameters depending on the solubility limiting phase and the species in equilibrium.

[2000BUN/KNO]

Colloids of thorium are formed in non-complexing aqueous solutions when their concentrations reach the solubility limit. As the LIBD technique is a powerful tool to detect colloid generation, the solubility product can be determined by measuring the threshold of the breakdown probability for solutions of a given Th concentration, as a function of pH, in the range where Th^{4+} is the dominant species. Solutions with Th concentrations, $2.8·10^{-2}$ to $8.9·10^{-5}$ M ($Th(NO_3)_4$ in 0.5 M NaCl), have been investigated with increasing pH in the range 1.4 to 2.8. The mean from seven solutions gives $\log_{10} K_{s,0}$ ($ThO_2 · xH_2O$, coll) = − (49.54 ± 0.22). Extrapolation to $I = 0$ with the SIT gives $\log_{10} K_{s,0}^\circ$ = − (52.8 ± 0.3). This value is just consistent with $\log_{10} K_{s,0}^\circ$ (ThO_2, cr) = − (54.2 ± 1.3) calculated from thermodynamic data from [87RAI/SWA]. The difference is suggested to be a result of different particle sizes; the solubility of 20 nm colloids is calculated to be a factor of 10 to 25 larger than for a crystalline sample. This paper gives an overview of the solubility data of ThO_2(cr) and ThO_2(am).

[2000BUR/OLS]

This paper is highly relevant for the understanding of degradation of nuclear waste glass at elevated temperatures. The authors report the formation of a new uranium silicate and its structure, which has general interest; it has a layered structure consisting of Si_4O_{12} units which share four corners with adjacent tetramers. The uranium is located between the silica layers and is coordinated to four oxygen atoms, two from two adjacent tetramers in one plane and two from the plane below. This results in a less common octahedral coordination geometry around uranium, two oxygen and four silicate oxygen atoms in the plane perpendicular to the linear UO_2–unit; most uranium silicates have a pentagonal bipyramid coordination geometry.

[2000COM/BRO]

This paper deals with the hydrolysis of U(VI) in the presence of constant sulphate anion concentrations (0.1 and 1 M Na_2SO_4, $0.2 \cdot 10^{-3} < U < 2.2 \cdot 10^{-3}$ M, pH 3.8 to 6.8, $t = (25.0 \pm 0.1)$°C) and gives additional results to similar previous studies conducted at higher ionic strengths, 3 M (0.5 M Na_2SO_4 + 2 M $NaClO_4$) [93GRE/LAG] and 1.5 M Na_2SO_4 [61PET], [93GRE/LAG]. The aim of this paper is to determine formation constants of hydroxo–sulphato complexes.

In all the studies noted above, potentiometric titrations were used. In this work, computations were carried out using the MINIQUAD program and the selected data meet two criteria: standard deviation less than 10 % and normalised agreement factor less than 0.002.

The stoichiometric ratios UO_2/OH, which fit the data in 0.1 M Na_2SO_4, are 1/1, 2/2, 3/4, 3/5, 4/7 and 5/8. At a higher concentration, 1 M, the ratios are 2/2, 3/4, 4/7 and 5/8. The main difference with [93GRE/LAG] is the introduction to the model of the ratios 4/7 and 3/5, which also exist in non-complexing media. Stability constants, $\beta_{m,n}$, of possible complexes, $(UO_2)_m(OH)_n^{2m-n}$, in the presence of complexing sulphate anion are given, which cannot be compared to data selected in [92GRE/FUG] in non-complexing solution. These values differ, as expected, and recognised in [93GRE/LAG].

On the other hand, ternary dioxouranium(VI) hydroxo-sulphato and dioxouranium(VI) sulphato complexes can be identified and the corresponding formation constants calculated, according to:

$$p\,UO_2^{2+} + q\,H_2O(l) + r\,SO_4^{2-} \rightleftharpoons (UO_2)_p(OH)_q(SO_4)_r^{2p-q-2r} + q\,H^+$$

The p, q, r values selected in this work are (1, 0, 1), (1, 0, 2), (2, 2, 2), (3, 4, 3), (4, 7, 4) and (5, 8, 4). The $\beta_{p,q,r}$ values are given in Table A-47. The p, q, r set (3, 4, 4) is excluded on the basis of spectroscopic results of Moll et al. [2000MOL/REI2].

To determine the extrapolated values at zero ionic strength of the $\beta_{p,q,r}$ values of the hydroxo complexes according to the SIT, the authors have used the data of

[93GRE/LAG], [61PET] and [2000MOL/REI2]. From data at three different ionic strengths they calculate the $\log_{10} \beta^\circ_{p,q,r}$ and ε values shown in Table A-47.

Table A-47. $\log_{10} \beta^\circ_{p,q,r}$ and ε (p, q, r, Na$^+$) (kg · mol^{-1}).

p, q, r species	Sulphate concentration, mol · kg^{-1}			
	0	0.1004	1.027	1.566
2, 2, 2	– (0.64 ± 0.01)	– (2.17 ± 0.15)	– (3.02 ± 0.68)	– (3.20 ± 0.82)$^{(c)}$
ε	– (0.14 ± 0.22)			
3, 4, 3	– (5.9 ± 0.2)	– (6.60 ± 0.17)	– (7.18 ± 0.70)	– (9.01 ± 0.87)$^{(c)}$
ε	(0.6 ± 0.6)			
4, 7, 4	– (18.9 ± 0.2)	– (15.85 ± 0.28)	– (18.4 ± 1.3)	– (22.6 ± 1.2)$^{(c)}$
ε	(2.8 ± 0.7)			
5, 8, 4	– (18.7 ± 0.1)	– (17.69 ± 0.20)	– (19.61 ± 0.73)	– (20.14 ± 0.88)
ε	(1.1 ± 0.5)			
1, 0, 1	(3.15 ± 0.02)$^{(a)}$	(1.92 ± 0.03)	(1.06 ± 0.19)	(1.9 ± 0.19)$^{(b)}$
1, 0, 2	(4.14 ± 0.07)$^{(a)}$	(2.90 ± 0.08)	(2.06 ± 0.32)	(2.29 ± 0.24)$^{(b)}$

(a) From [92GRE/FUG].
(b) From [93GRE/LAG].
(c) Calculated from [2000MOL/REI2] and [93GRE/LAG].

Additional considerations are given in this paper on the structures of the complexes and the value, $\log_{10} \beta_{3,5,3} = -(11.7 \pm 0.2)$, is given for 0.1 m Na$_2SO_4$.

[2000DAS/SIN]

Enthalpy increments of two strontium uranates, Sr$_3$U$_2$O$_9$(cr) and Sr$_3$U$_{11}$O$_{36}$(cr), were measured from ca. 300 to 1000 K, using a Calvet microcalorimeter.

The compounds were prepared by heating the stoichiometric amounts of U$_3$O$_8$(s) with the SrCO$_3$(s) in air at 1100 K for ca. 20 hours in alumina boats and were characterised by X–ray diffraction.

The enthalpy increments were fitted to a simple polynomial, using the Shomate procedure. The heat capacities at 298.15 K were constrained to the values:

$C^\circ_{p,m}$ (Sr$_3$U$_2$O$_9$, cr, 298.15 K) = (301.8 ± 3.0) J · K^{-1} · mol^{-1}

$C^\circ_{p,m}$ (Sr$_3$U$_{11}$O$_{36}$, cr, 298.15 K) = (1064.2 ± 10.6) J · K^{-1} · mol^{-1}

calculated from their own detailed enthalpy increments from 299 to 339 K.

The derived enthalpy equations give on differentiation:

$C_{p,m}$ (Sr$_3$U$_2$O$_9$, cr, T) = 319.18 + 0.11602 T – 4.6201·10^6 T^{-2} J · K^{-1} · mol^{-1}
(298.15 to 1000 K)

$$C_{p,m}(Sr_3U_{11}O_{36}, cr, T) = 962.72 + 0.35526\,T - 3.954 \cdot 10^5\,T^{-2}\ \text{J} \cdot \text{K}^{-1} \cdot \text{mol}^{-1}$$
$$(298.15 \text{ to } 1000\text{ K}).$$

The enthalpy of formation of $Sr_3U_{11}O_{36}(cr)$ has been determined experimentally, but not that of $Sr_3U_2O_9(cr)$. The authors have estimated this from electronegativity considerations, to be -4594.9 kJ $\cdot$ mol^{-1}, with an unknown uncertainty. However, since the enthalpies of formation of several strontium uranates(VI) are now known fairly precisely, we have preferred to estimate this from the data for the other strontium uranates(VI). If the formation of $Sr_3U_2O_9(cr)$ from its neighbours is exothermic, $\Delta_f H_m^\circ(Sr_3U_2O_9, cr, 298.15\text{ K})$ is $< -(4619.9 \pm 4.0)$ kJ $\cdot$ mol^{-1}, whereas a similar condition for the decomposition of $Sr_5U_3O_{11}(cr)$ implies $\Delta_f H_m^\circ(Sr_3U_2O_9, cr, 298.15\text{ K}) > -(4616.0 \pm 8.0)$ kJ $\cdot$ mol^{-1}. Although these are formally incompatible, the selected value:

$$\Delta_f H_m^\circ(Sr_3U_2O_9, cr, 298.15\text{ K}) = -(4620 \pm 8)\ \text{kJ} \cdot \text{mol}^{-1},$$

is consistent with these conditions within their uncertainties. This value is somewhat more negative than that proposed by [99SIN/DAS] (whose paper was written before the extensive new data of [99COR/BOO] had appeared).

The authors also estimate the entropies of the two uranates to be 301.8 and 1277.9 J $\cdot$ K^{-1} $\cdot$ mol^{-1}, without uncertainties, but these estimates are not selected for the review at this time.

[2000FIN]

This paper is an update of the author's 1981 assessment [81FIN/CHA] of the thermodynamic and transport properties of UO_2(cr, l), to incorporate an appreciable amount of new data, principally on the liquid. For example, the heat capacity data for the liquid now extend up to 4500 K. The thermodynamic properties assessed include the enthalpy increment, $C_{p,m}$, $\Delta_{fus}H_m$, thermal expansion, density, surface tension and total vapour pressure. Only the thermal data are discussed here, with a minor comment on the vapour pressure.

Two fitting equations are given for $H_m(T) - H_m^\circ(298.15\text{K})$ as a function of T. The first has three terms, which give an approximate representation of the important contributions to the enthalpy and heat capacity, whereas the second is a simple polynomial fit.

The two equations for the enthalpy increment up to the melting point, 3120 K, are:

$$H_m(T) - H_m^\circ(298.15\text{K}) = c_1 \cdot \theta \cdot [(\exp(\theta/T) - 1)^{-1} - (\exp(\theta/298.15) - 1)^{-1}] + c_2(T^2 - 298.15^2) + c_3 \exp(-E_a/T) \quad \text{(A.127)}$$

$$H_m(T) - H_m^\circ(298.15\text{K}) = \sum_{n=0}^{n=5}(a_n(10^{-3}T)^n) + a_6(1000/T). \quad \text{(A.128)}$$

The coefficients are listed in Table A-48.

Table A-48: Values and units of the coefficients of equations (A.127) and (A.128).

Coefficient	Value	Units
c_1	81.613	$J \cdot K^{-1} \cdot mol^{-1}$
c_2	$2.285 \cdot 10^{-3}$	$J \cdot K^{-2} \cdot mol^{-1}$
c_3	$2.360 \cdot 10^7$	$J \cdot mol^{-1}$
θ	548.68	K
E_a	18531.7	K
a_0	-21.1762	$J \cdot mol^{-1}$
a_1	52.1743	$J \cdot K^{-1} \cdot mol^{-1}$
a_2	43.9735	$J \cdot K^{-2} \cdot mol^{-1}$
a_3	-28.0804	$J \cdot K^{-3} \cdot mol^{-1}$
a_4	7.88552	$J \cdot K^{-4} \cdot mol^{-1}$
a_5	-0.52668	$J \cdot K^{-5} \cdot mol^{-1}$
a_6	0.71391	$J \cdot K \cdot mol^{-1}$

These equations give very similar values for both $H_m(T) - H_m^\circ(298.15\,K)$ and $C_{p,m}(T)$, and either can be used to represent the data for $UO_2(cr)$ up to its melting point.

For $UO_2(l)$, all the experimental data on the enthalpy increment and heat capacity have been fitted to the expression:

$$H_m(T) - H_m^\circ(298.15\,K) = 80383 + 0.25136\,T - 1.3288 \cdot 10^9\,T^{-1} \quad J \cdot mol^{-1}$$
$$(3120 \text{ to } 4500\,K),$$

which gives a heat capacity decreasing from ca. 137 $J \cdot K^{-1} \cdot mol^{-1}$ at the melting point, to ca. 66 $J \cdot K^{-1} \cdot mol^{-1}$ at 4500 K. Combination of the two equations for the enthalpy increments gives for the enthalpy of fusion:

$$\Delta_{fus} H_m (UO_2, cr, 3120\,K) = (70 \pm 4)\,kJ \cdot mol^{-1}.$$

The equation for $C_{p,m}(UO_2, cr)$ in [92GRE/FUG] is valid only to 600 K and gives values which are close to those derived from the revised assessment by [2000FIN], so these data [92GRE/FUG] will not be adjusted.

The author also discusses the data for the total vapour over liquid UO_2, again to take account of additional data at very high temperatures, which in general are beyond the scope of the present review.

[2000GOR/MIR]

Tc(I) compounds are known to be good radiological imaging agents. Each test solution was prepared quantitatively according to the following reaction:

$$Tc(CO)_5X + 3 H_2O(l) \rightleftharpoons Tc(CO)_3(H_2O)_3^+ + X^- + 2 CO(g),$$

for X = Cl, Br, I, and SCN. ^{99}Tc NMR and potentiometric titrations were the methods of choice. At OH/Tc ratios < 1/1, three species, identified by NMR, were shown to form slowly and sequentially, namely: [Tc(CO)$_3$(OH)(H$_2$O)$_2$] (s), [Tc(CO)$_3$(μ–OH)(H$_2$O)$_2$]$_2$ (s), and [Tc(CO)$_3$(μ–OH)]$_4$ (s) characterised by chemical shifts at − 1055, − 763 and − 585 ppm, respectively. Formation constants (presumably stepwise) for Tc(CO)$_3$(H$_2$O)$_{3-n}$X$_n^{(1-n)+}$ complexes are listed (see Table A-49), but no description is given as to the method employed and no other supporting results are given.

Table A-49. Formation constants for Tc(CO)$_3$(H$_2$O)$_{3-n}$X$_n^{(1-n)+}$ complexes.

	Cl⁻	Br⁻	I⁻	SCN⁻
K_1	1.32	1.39	2.67	697
K_2	0.17	0.21	1.26	106
K_3	0.05	0.07	0.66	13

The values for the halides are too small to be meaningful and those for the thiocyanide complexes must be considered as unsubstantiated at this time. However, the IR spectrum of [Tc(CO)$_3$(H$_2$O)$_2$(NCS)] isolated from solution ([2000GOR/LUM]) apparently indicates that the thiocyanate ligand is bonded to Tc via the nitrogen. Weak complexes with acetate ions are claimed to have been observed, with a much stronger tridentate complex being formed with tartrate.

[2000HAS/ALL]

The authors have studied the reaction of an oxide of initial composition PuO$_{1.97}$(s) with adsorbed water at 298 to 623 K, for periods up to 30 days. Mass spectrometric analysis showed that H$_2$(g) is the only gaseous reaction product. Oxidation rates were measured at constant temperature and p_{H_2O} by microbalance and p–V–T techniques, but precise details of the experimentation are not given. A plot of the lattice parameter of the oxide as a function of the O/Pu ratio is given, but it is unclear how the ratios were determined – possibly from the weight gain (assuming only oxygen is absorbed) or from the volume of hydrogen produced (again from the assumed reaction: PuO$_2$(s) + x H$_2$O(ads) $\rightleftharpoons$ PuO$_{2+x}$(s) + x H$_2$(g)). The lattice parameter at first decreased sharply to 5.3975 Å at O/Pu = 2 and then increased slowly to a value of ca. 5.405 Å m at an indicated ratio, O/Pu = 2.25.

With such little experimental detail given, it is difficult to assess the results of this study. However, it may be noted that it is well known that PuO$_2$(cr) can adsorb oxygen or water to compositions up to at least PuO$_{2.09}$(s) [61WAT/DOU], [63JAC/RAN], and that the observed change in the lattice parameter is well within that known to occur in PuO$_{2.0}$(s) due to irradiation self–damage [62RAN/FOX].

Until more detailed analyses of the composition of the plutonium oxide in such studies are available, the evidence for the formation of a thermodynamically stable bulk phase with O/Pu > 2 cannot be regarded as conclusive. Further work is clearly required on the interesting phenomena observed by [2000HAS/ALL].

[2000HAY/MAR]

Theoretical studies on actinide species are increasing in the chemical literature. These calculations can be verified by comparing bond lengths with EXAFS and X-ray measurements, vibration frequencies with spectroscopic properties and energetics with thermodynamic constants in aqueous solutions, when solvation of the species can be estimated. This paper is an example of the state of the art in this area. It deals first with the structures and vibration frequencies of $AnO_2(H_2O)_5^{2+}$ and $AnO_2(H_2O)_5^+$, and then with the energy of these species together with $AnO_2(H_2O)_4(OH)^+$. Additional calculations bear on the energies of different species $AnO_2(H_2O)_n^{2+}$ with $n = 4$ and 6.

Updates of the literature both on calculation methods applied to actinides and experimental data on U, Np and Pu can be found in this paper.

This review focuses mainly on the predictive chemistry of the cited species in aqueous solution. Solvent effect are treated using a dielectric continuum model.

It is shown that $UO_2(H_2O)_6^{2+}$ must be the stable species in the gaseous phase, but that hydration stabilises the pentahydrate, as known experimentally. Hydrolysis of UO_2^{2+} in aqueous solution according to:

$$UO_2(H_2O)_5^{2+} + H_2O(l) \rightleftharpoons UO_2(H_2O)_4(OH)^+ + H_3O^+$$

is predicted to be endothermic by 55.2 kJ · mol^{-1}, but according to [92GRE/FUG] the experimental value is only 18.0 kJ · mol^{-1}. In the gas phase the reaction is predicted to be exothermic by $-$ 121 kJ · mol^{-1}. This difference emphasises the role of solvation but does not explain the discrepancy between the two $\Delta_r G_m^\circ$ (298.15 K) values.

Calculations on U(VI) (5f^0) do not need to take into account spin–orbit effects, which must be included in all other cases dealing with U(V) and Np or Pu in the V and VI oxidation states. Calculations of the redox potentials $E_{VI/V}^\circ$ show that the relativistic effect inhibits the reduction of AnO_2^{2+} in the gas phase (by 0.3 eV for U, 1.27 eV for Np and 0.2 eV for Pu). The predicted values of $E_{VI/V}^\circ$ are the following: 2.37 V for U, 4.00 V for Np and 3.28 V for Pu, far from the experimental values, but the overall trends of the values are the same. The systematic error of 2 – 3 eV is not explained.

[2000KAL/CHO]

This is a series of investigations where the complex formation between Ca^{2+} and $UO_2(CO_3)_3^{4-}$ has been investigated using spectroscopic techniques. All studies indicate a very strong interaction, resulting in a pronounced increase in the fluorescence intensity and fluorescence lifetime. Bernhard et al. [96BER/GEI], [97BER/GEI], Amayri et al. [97AMA/GEI] and Geipel et al. [98GEI/BER4] present convincing evidence for the formation of an uncharged complex, $Ca_2UO_2(CO_3)_3$(aq). These authors have estimated an equilibrium constant for the reaction:

$$2\,Ca^{2+} + UO_2^{2+} + 3\,CO_3^{2-} \rightleftharpoons Ca_2UO_2(CO_3)_3(aq) \qquad (A.129)$$

$\log_{10} K°$(A.129) = (26.5 ± 0.3). Kalmykov and Choppin [2000KAL/CHO] have studied the same system using a similar experimental method. The data in 0.1 M $NaClO_4$ are in excellent agreement with those of Geipel et al. [98GEI/BER4], [97BER/GEI]. Kalmykov and Choppin [2000KAL/CHO] have also studied the ionic strength dependence of the reaction and used the SIT to calculate the equilibrium constant at zero ionic strength. This value, $\log_{10} K°$(A.129) = (29.2 ± 0.25), is based on information over a larger ionic strength range than that proposed by Geipel et al. [98GEI/BER4], [97BER/GEI]. The present reviewers notice two features in the analysis of the data of Kalmykov and Choppin [2000KAL/CHO]. The first is that the equilibrium constant for the reaction:

$$2\,Ca^{2+} + UO_2(CO_3)_3^{4-} \rightleftharpoons Ca_2UO_2(CO_3)_3(aq)$$

is nearly independent of the ionic strength in the range 0.1 to 1 M; the second is that the value of $\Delta\varepsilon$ for this reaction has a large positive value, 2.67 kg·mol^{-1}, as estimated from Figure 3 in [2000KAL/CHO]. This results in an interaction coefficient for the uncharged complex, $\varepsilon Ca_2UO_2(CO_3)_3$(aq) ≈ 3.3 kg·mol^{-1}. This value is surprisingly high and suggests that the data treatment should be modified. We suggest the following alternative interpretation of the experimental data; the strong interaction between Ca^{2+} and $UO_2(CO_3)_3^{4-}$ indicates that there might be a similar, albeit weaker, specific interaction between Na^+ and the tris-carbonato complex. Hence the reaction studied is:

$$Na_xUO_2(CO_3)_3^{x-4} + 2\,Ca^{2+} \rightleftharpoons Ca_2UO_2(CO_3)_3(aq) + x\,Na^+,$$

where for simplicity, x is assumed to be equal to four such that both complexes are uncharged. With this assumption the reported equilibrium constants are conditional constants. The equilibrium constants given in the authors' Table 3 should be multiplied by $[Na^+]^4$ in order to obtain the equilibrium constant for the reaction given above. The resulting $\log_{10} K$ values are 0.90, 2.76, 4.41, 5.12 and 7.85 at the ionic strengths investigated. The resulting SIT plot is approximately linear over the entire ionic strength range investigated and gives $\log_{10} K° = (5.33 ± 0.10)$ and $\Delta\varepsilon = (0.56 ± 0.03)$ kg·mol^{-1} for this last reaction. This in turn means that the difference between the ε values for the uncharged complexes is at most 0.06, which seems quite reasonable. The conclusion is that great care should be used when applying the SIT, or any other specific ion interaction model, on systems where there are indications of a specific chemical interaction between anion and cation.

[2000KON/CLA]

This paper reports the identification by several spectroscopic techniques of a dimeric dioxouranium(VI) hydroxide species, $((UO_2)(OH)_3(m-OH))_2^{4-} = (UO_2)_2(OH)_8^{4-}$ in equilibrium with the already well-known 4:1 $UO_2(OH)_4^{2-}$ species (UV spectrum $\lambda_{max} = 412$ nm, $\varepsilon = 45$ $M^{-1} \cdot cm^{-1}$, Raman O=U=O frequency 816 cm^{-1}). This species has been identified at high U(VI) concentration ($5 \cdot 10^{-2}$ M) in tetramethylammonium hydroxide (TMAOH) solutions and in a gel corresponding to 0.5 M TMAOH. When the gel is washed with tetrahydrofuran, a yellow solid is obtained which also contains the dimer. No thermodynamic data are available. As these data have not been published, this review quotes them only as being indicative.

[2000LAA/KON]

This paper is a study of the heat capacity of an alloy of approximate composition, $Tc_{0.85}Ru_{0.15}$, from ca. 300 to 973 K, using a differential scanning calorimeter. The sample was unusual, in that it was formed by neutron irradiation of a disc of pure ^{99}Tc metal, which transmutes some of the ^{99}Tc to ^{100}Ru. Isotope dilution mass spectrometry (IDMS) indicated that the ruthenium content was (15.0 ± 0.4) atom%. However, EPMA showed that due to shielding effects there was substantially more Ru in the outer parts of the disc, with more than 30% Ru at the rim. The sample was thus not of uniform composition. Indeed, there is a minor discrepancy in the IDMS and EPMA results, since the integrated composition of the disc from the radial distribution plot (the authors' Figure 2) would be noticeably greater than 15 atom% of Ru, as given by EPMA.

Two DSC runs on the same sample gave results differing by *ca.* 4%, indicating an uncertainty of at least 2% in the derived heat capacities.

The authors calculated the heat capacity of pure ^{99}Tc assuming the change of $C_{p,m}$ upon mixing was zero. They used values for $C_{p,m}$(^{100}Ru, cr, T) obtained by combining their own DSC $C_{p,m}$ measurements with earlier enthalpy drop measurements by Cordfunke *et al.* [89COR/KON], both using natural ruthenium. The molar masses for ^{99}Tc and ^{100}Ru used by the authors in these calculations seem to be too low by *ca.* 1 g·mol^{-1}. We have therefore repeated these calculations, using the correct values of 98.9063 and 99.9042 g·mol^{-1}, respectively, together with the recent assessment of the heat capacity of Ru(cr) by Arblaster [95ARB]. The fitted equation for the heat capacity of ^{99}Tc, assuming $\Delta_f C_{p,m}$ is zero for the alloy is then:

$C_{p,m}$ (Tc, cr, T) = $24.383 + 6.6288 \cdot 10^{-3} T - 54995 T^{-2}$, $J \cdot K^{-1} \cdot mol^{-1}$ (300 to 973 K).

These values of $C_{p,m}$ are 3 to 7 % higher than those suggested in [99RAR/RAN] and were based on a slight reworking of the values proposed by Guillermet and Grimvall [89GUI/GRI]. Considering the combined uncertainties in the experimental results, the inhomogeneity of the $Tc_{0.85}Ru_{0.15}$ sample and the assumption of $\Delta_f C_{p,m} = 0$, this agreement is considered to be well within the uncertainty of the values selected in the technetium review [99RAR/RAN], and their values are retained.

As noted by [2000LAA/KON], Boucharat [97BOU] has made DSC measurements on samples of Tc(cr) from the same source as that used by [2000LAA/KON]. However, her derived heat capacities have a maximum around 1080 K and a minimum around 1540 K, for which there is no obvious explanation and have been discounted in this review.

[2000MOL/REI]

This paper complements [99WAH/MOL] and [99CLA/CON], which report contradictory results on the U(VI) species present in alkaline aqueous solutions: $UO_2(OH)_4^{2-}$ and/or $UO_2(OH)_5^{3-}$ (see this review). EXAFS spectra are recorded (in k space up to 17.5 $Å^{-1}$ and in the transmission mode) from three test solutions, 0.05 M (pH = 4.1), 0.5 M (pH = 13.7) and 3 M in $(CH_3)_4NOH$ and a total concentration of U(VI) equal to $5 \cdot 10^{-2}$ M. All testsolutions are kept under an argon atmosphere. Schoepite (in fact metaschoepite) precipitates from the first test solution when its pH is increased up to 7. Additional NMR (^{17}O-NMR) data are obtained on solutions, $5 \; 10^{-2}$ M U(VI) (^{17}O enriched "yl" oxygen), 5 % D_2O or 1 to 3.5 M $(CD_3)_4NOD$.

EXAFS data at pH 4.1 are consistent with previous structural data on $(UO_2)_2(OH)_2^{2+}$ and $(UO_2)_3(OH)_5^+$. The distances, UO_{ax}, UO_{eq} and U–U, in schoepite or metaschoepite are those of [96ALL/SHU]. At pH 13-14, EXAFS data show clearly an octahedral coordination for $UO_2(OH)_4^{2-}$ [99WAH/MOL] without coordinated water in the equatorial plane and that the concentration of $UO_2(OH)_5^{3-}$ in these solutions must be small. NMR measurements were used to probe the coexistence of these two species. ^{17}O-NMR spectra in 3 M $(CD_3)_4NOD$ show at 298 K a broadening of the single line obtained in 1 M $(CD_3)_4NOD$. Resolution of this signal in two peaks is obtained, at 258 K, by addition of methanol to the solutions. These peaks provide strong evidence of a rapid equilibrium between $UO_2(OH)_4^{2-}$ and $UO_2(OH)_5^{3-}$. A value of the stepwise equilibrium constant for :

$$UO_2(OH)_4^{2-} + OH^- \rightleftharpoons UO_2(OH)_5^{3-}$$

is tentatively given from the intensity of the deconvoluted peaks as 0.4. This value must be smaller in water in order to be consistent with EXAFS data and a value around 0.1 is proposed by Moll *et al.*

In conclusion these data confirm the formation and structure of $UO_2(OH)_4^{2-}$ in alkaline solutions, but also indicate the formation of small amounts of $UO_2(OH)_5^{3-}$ at $[OH^-] > 3$ M.

[2000MOL/REI2]

This paper gives numerous spectroscopic, EXAFS and ^{17}O NMR ($U\,^{17}O_2^{2+}$ enriched) data on sulphate solutions of U(VI), which are discussed with regard to the predominant species, UO_2SO_4(aq) and $UO_2(SO_4)_2^{2-}$, according to the data of [61PET], [93GRE/LAG] and [96GEI/BRA]. The aim of the work is to gain information on sulphato ligand bonding.

Of interest to this review are the potentiometric titrations (25.00 ± 0.05)°C performed in 0.5 M Na_2SO_4 + 2 M $NaClO_4$ and 1.5 M Na_2SO_4 (I = 3 M) at U concentrations, $2 \cdot 10^{-3}$ to $2 \cdot 10^{-2}$ M (conditions similar to the studies by [61PET] and [93GRE/LAG]), and in the pH range 4.3 to 5.8. These titrations were intended to characterise hydroxo sulphato dioxouranium(VI) complexes. The data are interpreted according to:

$$p\,UO_2^{2+} + q\,H_2O(l) + r\,SO_4^{2-} \rightleftharpoons (UO_2)_p(OH)_q(SO_4)_r^{(2p-q-2r)} + q\,H^+$$

using the LETAGROP least-squares program, which gives the number of OH groups bound to UO_2 groups. The UO_2/OH ratios that fit the data are 2/2, 3/4, 5/8 and possibly 4/7, while the p, q, r sets are (2, 2, 2), (3, 4, 3) and (5, 8, 6), and possibly (4, 7, 4), EXAFS data do not distinguish between polymeric complexes.

The $\log_{10} \beta_{q,p}$ values given by Moll et al. agree with those of [61PET] and [93GRE/LAG]. A fit of the potentiometric data is also obtained by incorporating the species with the ratios (OH/UO_2) of 7/4 and 8/5.

In conclusion, this work establishes the stoichiometry of the species, but does not give new thermodynamic data useful for this review.

[2000NEC/KIM]

This paper describes an empirical method for the prediction of stability constants using the relationship:

$$\log_{10} \beta_n^\circ = n \log_{10} \beta_1^\circ - \frac{{}^{rep}E_L(ML_n)}{RT\ln 10}$$

where ${}^{rep}E_L$ is an electrostatic repulsion term, calculated from Coulomb's law and requiring in addition to the formal charges, knowledge/estimates of the geometry of the complexes, and determination of an electrostatic shielding parameter (an "effective" dielectric constant). The latter is an empirical parameter that is obtained by calibration using known stability constants. The authors present a table of predicted and measured equilibrium constants for the carbonate complexes of Am(III), Cm(III), Np(V), U(VI), Pu(VI), U(IV) and Pu(IV). The agreement between the model and experimental data is surprisingly good, better than 0.5 $\log_{10}$ units. Using the model, the authors have proposed stability constants for the formation of MCO_3^{2+}, $M(CO_3)_2$(aq), and $M(CO_3)_3^{2-}$, where M is U(IV) and Pu(IV). The $\log_{10} \beta_n^\circ$ values, 13.7, 24.3 and 31.9 for U(IV) and, 13.6, 24.0 and 31.5 for Pu(IV), are reasonable. However, they are not included among the selected values as they are not based on experiments.

[2000NGU/BUR]

This extended abstract reports the individual UV–Visible spectra of $(UO_2)_m(OH)_n^{2m-n}$ species (in short n, m^{2m-n}) derived from an analysis of the spectra of $NaClO_4$ (pH < 4.5) or $(CH_3)_4NCF_3SO_3$ (pH > 11) solutions, 10^{-2} M in U (t = 25°C). Speciation is derived from

[2000NGU/PAL] and [95PAL/NGU] (see Appendix A). Seventy test solutions have been measured. These data, which are not yet published, are given here as being indicative only.

Table A-50: Values of the molar extinction coefficient, ε, for $(UO_2)_m(OH)_n^{2m-n}$ species.

Species (n, m)	λ max (nm)	ε (L·mol^{-1}·cm^{-1})
0,1^{2+}	414	8.5[a]
2,2^{2+}	421	58[b]
5,3^{+}	429	234[c]
7,3^{+}	429	223
8,3^{2-}	419	74
11,3^{5-}	420	82
4,2^{2-}	400	17

Meinrath [97MEI2] reported [a]: 9.7, [b]: 101, [c]: 474 l·mol^{-1}·cm^{-1}

[2000NGU/PAL]

This paper follows [92NGU/BEG] and [95PAL/NGU] and gives additional information on U(VI) hydrolysed species through careful examination of Raman spectra, the determination and analysis of which are described in detail. The authors used peak deconvolution of the overlapping Raman bands in their analysis of the data[1]. Characteristic v_1 values (symmetric stretching vibration of the linear UO_2 group) of the species, $(UO_2)_m(OH)_n^{2m-n}$ (in short n,m^{2m-n}), as well as their concentrations, are derived for the following solutions: U concentration $3.8 \cdot 10^{-3}$ to $6.74 \cdot 10^{-1}$ M, pH = 0.24 to 14.96, variable ionic strengths 0.02 to 2.3 M. pH is adjusted using HCF_3SO_3 and/or $(CH_3)_4NOH$, (t = 25°C). Corresponding values of v_1 and $n:m$ sets are selected by combining the spectroscopic data and the thermodynamic data of [95PAL/NGU], and taking also into account literature data for discussion.

This review first gives the main features of this work, according to the authors, and then additional comments.

For pH < 5.6 the three v_1 values of the most intense bands correspond to the well-known species, 0,1^{2+}, 2,2^{2+} and 5,3^{+}. In addition a weak band could be attributed to the species 1,2^{3+}. For pH > 12 only one v_1 value is found. According to [95PAL/NGU] the corresponding species is 4,1^{2-}, but [99CLA/CON] suggested that it could be 5,1^{3-}. Several reasons are advanced for rejecting the 5,1^{3-} species, the main one being that the v_1 values of the sets 0,1^{2+}, 2,2^{2+}, 5,3^{+}, and 4,1 fit well the relationship:

$$v_1 \text{ (cm}^{-1}) = -22\,r + 870 \tag{A.130}$$

where r is the ratio n/m = OH/UO_2, while the v_1 values for the species 0,1^{2+}, 2,2^{2+}, 5,3^{+} and 5,1^{3-} do not. On the basis of the relation (A.130) the species 1,2^{3+} is confirmed [92GRE/FUG].

Between pH = 5.6 and 12, four deconvoluted bands appear which are attributed to the sets: 7,3 (r = 2.33), 8,3 (2.67), 10,3 (3.33) [95PAL/NGU] and 11,3 (3.67), because the associated v_1 values fit exactly the equation (A.130), but the ratios $OH/UO_2 = 9/4$, $11/4$, $13/4$ and $15/4$ could be also selected to fit that relation. Reinterpretation of the data of [95PAL/NGU] by this review concludes (see Appendix A) that the existence of the species $8,3^{2-}$ and $10,3^{4-}$ are not proven by the potentiometric titrations of [95PAL/NGU]. So the question as to the presence of these species remains open.

The species $4,3^{2+}$ ($r = 1.33$) [92GRE/FUG], predominant at pH less than 3 and at high U concentration (> 1 m), should have v_1 at 841 cm^{-1}. Nguyen *et al.* [2000NGU/PAL] did not find any signal for the species $7,4^+$ ($r = 1.75$, $v_1 = 832$ cm^{-1}) which should exist in the solutions examined, because it is dominant between pH = 6 and 8 at U concentration higher than 10^{-3} M. The species $1,1^+$, $2,1$, $3,1^-$ are not present in the solutions that are not sufficiently dilute in U. The predicted v_1 values are respectively 848, 826 and 804 cm^{-1}.

Thermodynamic constants are deduced for all the species characterised by a v_1 band. They confirm the data of [95PAL/NGU] at $I = 0.1$, but are less accurate. The only new data concern $\log_{10} \beta (1,2^{3+})$, for which values range from (2.1 ± 0.2) to (2.7 ± 0.3) when I decreases from 2.3 to 0.1 M, and $\log_{10} \beta (11,3^{5-})$. For this constant, the value is (78 ± 8), $(0.02 < I < 0.56)$. But as no correction for ionic strength is possible, the value of the constant given in this paper is not considered for the revision of the values selected by [92GRE/FUG].

Finally the authors propose a structure for the species $8,3^{2-}$ and $11,3^{5-}$ (on the basis of those established for the species $5,3^+$ and $7,3^-$ and $10,3^{4-}$.

The comments are the following:

- This review considers that the success of the peak deconvolution depends on the ability to correct for possible changes in the base line, the number of species present and the spacing of their individual spectra.

- There is no problem with peak deconvolution for the major species present in the acid test solutions pH < 5.6. The matter is somewhat different for the minor species. $(UO_2)_2OH^{3+}$ seems well established from the spectrum in the authors' Figure 2-b, but the reviewers are less convinced by the minor peak at $v = 883$ cm^{-1} at pH = 0.24 which was assigned to UO_2^{2+} with less than five coordinated water molecules. The base line varies throughout the experiments, *cf.* the authors' Figures 3-a, b and 5-c, d. It would have been of value to know how the base line correction was made, *e.g.*, using spline functions, but no information is given.

- In the alkaline region there are many species present simultaneously according to the previous potentiometric study [95PAL/NGU], *cf.* Figures 4-a and b, 7-e and 9-b, c, d. It seems very difficult to make the deconvolution unless an assumption is made about the number of peaks (as done by the authors), hence it is doubtful if the Raman data provide independent proof of the speciation. However they do indicate very clearly that there are a number of different species formed between $(UO_2)_3(OH)_5^+$ and $UO_2(OH)_4^{2-}$.

- The equilibrium constants are consistent between potentiometry and the Raman data up to the complex 7,3⁻. For the higher complexes the difference between the two sets is gradually increasing, with a very large uncertainty reported for the Raman data. There is clear evidence for the formation of $UO_2(OH)_4^{2-}$ in very strongly alkaline solution, and that the predominant hydrolytic species in the acid range are $(UO_2)_2(OH)_2^{2+}$ and $(UO_2)_3(OH)_5^+$, with $(UO_2)_2(OH)^{3+}$ as a minor species. This finding and the proposed equilibrium constants are in good agreement with the conclusions in [92GRE/FUG]. However, the speciation in the alkaline region is still not conclusive. From a speciation diagram, cf. the Figure A-16, based on the proposed potentiometric equilibrium constants, this review finds that the polynuclear complexes should be predominant in the concentration range where EXAFS data [2000MOL/REI2] and [99CLA/CON] show that the system is mononuclear. To conclude, this paper gives an independent confirmation of the stoichiometry and equilibrium constants of known U(VI) species, and presents evidence for other species in the alkaline region.

- More precisely, this paper confirms the existence of the hydrolysed species $(UO_2)_m(OH)_n^{2m-n}$ with n,m sets: 1,2³⁺, 2,2²⁺, 5,3⁺, 7,3⁻ (respectively, $(UO_2)_2(OH)^{3+}$, $(UO_2)_2(OH)_2^{2+}$, $(UO_2)_3(OH)_5^+$, $(UO_2)_3(OH)_7^-$) and give strong evidence of the species, 4,1²⁻, $UO_2(OH)_4^{2-}$, in basic media, but do not resolve the ambiguities on the existence of 8,3²⁻ and 10,3⁴⁻ species, despite the fact that a nice correlation between the v_1 frequency of UO_2 with the ratio OH/UO_2 is established. All these species have been already considered to fit potentiometric data. Up until now the existence of the new species 11,3⁵⁻ is only supported by spectroscopic Raman measurements.

The stoichiometries and equilibrium constants of the species presented in this paper are not sufficiently well supported by experiments to be selected by this review.

¹Two internal standards, $(CH_3)_4N^+$ and $CF_3SO_3^-$, which were present in excess of the total dioxouranium(VI) concentration to fix the ionic strength, were used to determine the relative peak areas of the dioxouranium(VI) and dioxouranium(VI) hydroxide species. The relationship used in these calculations (equation 6 in [2000NGU/PAL]) involves the molar scattering coefficient, J_X (J_X = (6.3 ± 0.4) and (3.6 ± 0.3)), for the symmetrical stretching bands at 765 and 753 cm⁻¹, respectively,

[molarity of U– species] = J_X{(area of band)/(standard band area)} [molarity of standard].

This method yielded directly the molarities of the dominant dioxouranium(VI) species at pH ≤ 4.53, but at higher pH where the 0,1 species was minor, the UO_2^{2+} concentration was calculated from the hydrolysis quotients of the lower-order dioxouranium(VI) hydrolysed species determined previously at lower pH. The total molar stoichiometric uranium concentration was considered known to ± 1%. Chi squared values (χ = {Σ(obs² – calc²)}/calc) were generally < 0.05 for the total band areas compared to those calculated from the speciation at each pH value and total dioxouranium(VI) concentration. The hydrolysis quotients at each ionic strength were averaged with the mean value and the single standard deviation given in the paper.

Figure A-16. Speciation diagram of U(VI) in alkaline solutions according to constants given in [2000NGU/PAL] where precipitation of dioxouranium(VI) phases has been suppressed.

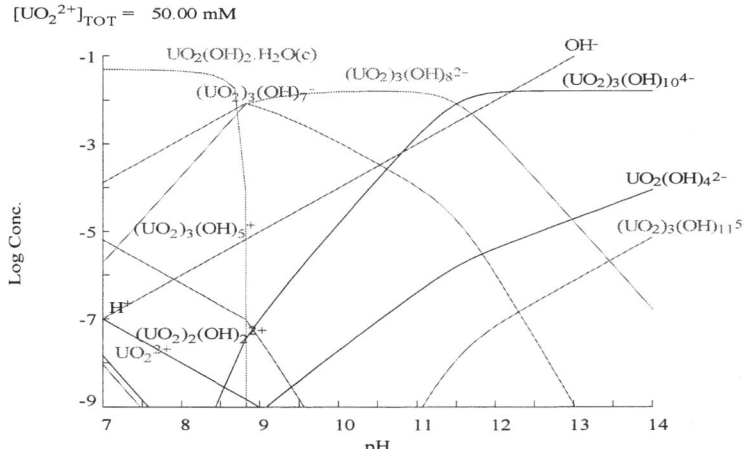

[2000PER/CAS]

The solubility and the rates of dissolution of uranophane ($Ca(UO_2)_2(SiO_3OH)_2 \cdot 5H_2O$), were studied as a function of bicarbonate concentration (0.001 to 0.02 M) near room temperature. The solid was synthesised as described by [92NGU/SIL] and characterised by XRD, SEM, FTIR and BET (surface area = 35.4 ± 0.5 m$^2 \cdot$ g^{-1}). ICP-MS analysis of the dissolved solid (preliminary experiments) showed that the U and Si levels were consistent with the stoichiometry of uranophane, but the Ca content corresponded only to (0.80 ± 0.05) mol Ca rather than the expected value of 1, and it was assumed that Na provided charge balance. Batch and flow-through reactors were used. The authors have also taken care to ascertain that no transformation of the solid phase had taken place during the solubility experiments. The pH on the activity scale was monitored continuously, whereas U, and occasionally Ca and Si, were analysed periodically. As in a previous study [97PER/CAS], at [HCO_3^-] = 0.005 M, the speciation calculations (HARPHRQ code) determined $UO_2(CO_3)_3^{4-}$ to be the dominant U(VI) species in solution and sporadic Ca and Si analyses confirmed congruent dissolution according to:

$$Ca(H_3O)_2(UO_2)_2(SiO_4)_2 \cdot 3H_2O\,(cr) + 6\,HCO_3^- \rightleftharpoons Ca^{2+} + 2\,UO_2(CO_3)_3^{4-} \\ + 2\,H_4SiO_4\,(aq) + 5\,H_2O(l). \quad (A.131)$$

The reviewer is surprised to find a solid phase containing H_3O^+ prepared from a solution at pH 8.8 and questions the stoichiometry proposed by the authors and suggests that an equivalent and more realistic chemical formula for uranophane would be $Ca(UO_2)_2(SiO_3OH)_2 \cdot 5H_2O$. The SIT model was used to account for activity coefficients and from $\log_{10} {}^*K_s$ (A.131), the equilibrium constant $\log_{10} {}^*K_s$ (A.132) was calculated for the reaction:

$$Ca(H_3O)_2(UO_2)_2(SiO_4)_2 \cdot 3H_2O(cr) + 6H_3O^+ \rightleftharpoons Ca^{2+} + 2UO_2^{2+} \\ + 2H_4SiO_4(aq) + 11H_2O(l). \quad (A.132)$$

The experimental values of $\log_{10} {}^*K_s$ (A.132) given in Table 3 of [2000PER/CAS] vary considerably, ranging from -10.75 to -12.74. The value of $\log_{10} {}^*K_s^\circ$ (A.132), averaged over nine measurements at $[HCO_3^-] = 0.005$ to 0.020 M, was reported at (11.7 ± 0.6) (note, taking the average of the actual values rather than their logarithms yields (12.0 ± 0.6)) and compares with a single value presented by Nguyen et al. [92NGU/SIL] of (9.4 ± 0.5). The authors comment on the large discrepancy between the existing values of the solubility product. Pérez et al. [2000PER/CAS] used a solid phase of the uncertain composition and their proposed value for the solubility constant of uranophane at zero ionic strength, $\log_{10} {}^*K_s^\circ$ (A.132)= (11.7 ± 0.6), is therefore not accepted by this review.

The rates of dissolution of uranophane, normalised to the mass ratio of solid to liquid, are expressed in units of $mol \cdot s^{-1} \cdot m^{-2}$. The batch rates are greater than those obtained from the flow-through technique by almost one order of magnitude, but this is slightly less than the combined experimental error, and were fitted according to the equation:

$$\log_{10} \text{rate (mol} \cdot s^{-1} \cdot m^{-2}) = a + b \cdot \log_{10}[HCO_3^-]$$

where $a = -(8.3 \pm 0.6)$ and $-(9.2 \pm 0.4)$; $b = (0.7 \pm 0.3)$ and (0.7 ± 0.2) for the batch and flow-through techniques, respectively. These are both reasoned to be "initial" rates such that the uranium concentration is much less than the equilibrium value. In view of the inherent difficulties in conducting kinetic experiments of this type, these results appear to be quite reasonable.

[2000QUI/BUR]

The free aqueous UO_2^{2+} ion displays symmetric (v_s) and antisymmetric (v_a) stretching modes in Raman scattering at 870 cm^{-1} and infrared absorption at 962 cm^{-1}. Replacement of water molecule in the equatorial plane weakens the oxo bond, resulting in an increase of the U–O distance and a decrease of the wave number of the v_s and v_a modes. For solutions containing a mixture of different species, this results in broad Raman and IR spectra. Equilibrium constants were calculated from the spectra using band fitting programs that includes assumptions on the number and profile of the individual species. IR-attenuated total reflectance (IR ATR) and Raman spectra have been obtained on 13 test solutions $HNO_3/NaNO_3$ (ionic strength, 0.23 to 0.35 M), 0.1 M in U(VI), in the pH range 1.55 to

4.20 at 25°C. All the data (IR and Raman) were processed together using a chemometric approach starting with the three $(n{:}m)^{(2m-n)}$ species $(UO_2)_m(OH)_n^{(2m-n)}$: $(0{:}1)^{2+}$, $(2{:}2)^{2+}$ and $(5{:}3)^+$. Mixtures of Lorentzian and Gaussian curves were used. Attempts to fit the spectra including other species in the chemical model failed. The set of values reported at zero ionic strength, but without explanation of the method used to correct for activity factors, are: $\log_{10} {}^*\beta_{2,2}^\circ = -(5.44 \pm 0.11)$ and $\log_{10} {}^*\beta_{2,2}^\circ = -(5.28 \pm 0.25)$; $\log_{10} {}^*\beta_{5,3}^\circ = -(15.16 \pm 0.09)$ and $\log_{10} {}^*\beta_{5,3}^\circ = -(14.62 \pm 0.33)$ from Raman and IR data, respectively. This study is more focused on the spectroscopy of U(VI) species than on the thermodynamics of U(VI) hydrolysis. Due to lack of information the equilibrium constants are not retained by this review.

[2000REI/BER]

This paper gives results on the coordination numbers of Tc(VII) and Np(IV, V and VI) derived from EXAFS spectra measured in transmission mode (except for diluted Tc solutions) in the k space up to 21 Å^{-1} for TcO_4^-, 12 Å^{-1} for Np(IV) and 18 Å^{-1} for NpO_2^+ and NpO_2^{2+}. The test solutions were 0.127 M or $1.3 \cdot 10^{-3}$ M in Tc (NaTcO$_4$ in water), $5 \cdot 10^{-2}$ M in Np in 0.1 M HNO$_3$ + 2 M H$_2$SO$_4$ for Np(IV) and in 0.1 M HNO$_3$ for NpO_2^+ and NpO_2^{2+}. The coordination of Tc by oxygen atoms is found to be N = (3.9 ± 0.6) and N = (4.1 ± 0.2) as expected. That of Np(IV) is N = (11 ± 1) which agrees with previous data in a 1 M HCl solution [97ALL/BUC]. The number of water molecules around NpO_2^+ and NpO_2^{2+} are respectively four (N= (3.6 ± 0.6) for the second shell of oxygen atoms), and five (N = (4.6 ± 0.6)). The calculated number of coordinated water molecules for Np(V) is one less than that reported by [97ALL/BUC]; the data of Reich et al. are more precise than those in [97ALL/BUC] due to the larger k-space range and better signal to noise ratio. This paper follows [97NEU/REI] (see Appendix A).

[2000REI/NEU]

Reilly, Neu and Runde [2000REI/NEU] determined the solubility of PuO$_2$CO$_3$(s) as a function of the H$^+$ concentration in 0.1, 0.2, 0.5, 1.0, 2.1, 3.8 and 5.6 m NaCl, and for comparison in 5.6 m NaClO$_4$. The solid phase is characterised by powder XRD, EXAFS and diffuse reflectance spectroscopy; however, the spectra and other experimental details are not given in this extended conference abstract. The reviewer obtained additional information by personal communication and considers the experimental and calculation procedures used in [2000REI/NEU] are reliable.

The experimental $\log_{10} K_{s,0}$ values are given in a Figure as a function of the molal ionic strength and compared to previous literature values [87ROB/VIT], [93PAS/RUN]. The solubility constants in NaCl solutions are not corrected for chloride complexation, although the presence of chloride complexes is assumed from the formation constants of the Pu(VI) mono- and dichloro complexes determined by this group of authors in another study (viz., $\log_{10} \beta_1^\circ = (0.23 \pm 0.03)$ and $\log_{10} \beta_2^\circ = -(1.7 \pm 0.2)$ [99RUN/REI]). Chloride complexation is given as an explanation for the higher solubility product in 5.6 m NaCl ($\log_{10} K_{s,0} = -(14.0 \pm 0.1)$) compared to that in 5.6 M NaClO$_4$ ($\log_{10} K_{s,0} = -(14.5 \pm 0.1)$). However, the difference might be, at least partly, also a result of activity coefficient differences in the two media.

The authors further report that Pu(VI) is stable in concentrated NaCl solutions, whereas in dilute NaCl or in NaClO$_4$ solutions Pu(V) and, after disproportionation, polymeric Pu(IV) is formed at a variable rate.

The present review has applied the SIT to convert the experimental data reported in [2000REI/NEU] and by other authors [87ROB/VIT], [93PAS/RUN], [97PAS/CZE] for the reaction:

$$PuO_2CO_3(s) \rightleftharpoons PuO_2^{2+} + CO_3^{2-} \tag{A.133}$$

to zero ionic strength (*cf.* Table A-51). As the reported experimental uncertainties are given only as error bars in the figure of [2000REI/NEU], and are generally in the range of 0.05 – 0.15 $\log_{10}$ units, they are omitted from Table A-53. The values of $\Delta\varepsilon = (0.13 \pm 0.04)$ kg·mol^{-1} and $\Delta\varepsilon = (0.38 \pm 0.06)$ kg·mol^{-1} in NaCl and NaClO$_4$ solutions, respectively, are calculated from the known interaction coefficient, $\varepsilon(Na^+, CO_3^{2-}) = -(0.08 \pm 0.03)$ kg·mol^{-1}, and by assuming $\varepsilon(PuO_2^{2+}, Cl^-) = \varepsilon(UO_2^{2+}, Cl^-) = (0.21 \pm 0.02)$ kg·mol^{-1} and $\varepsilon(PuO_2^{2+}, ClO_4^-) = \varepsilon(UO_2^{2+}, ClO_4^-) = (0.46 \pm 0.05)$ kg·mol^{-1} [2001LEM/FUG].

At NaCl molalities below 3 mol·kg^{-1}, the ionic strength dependence of the data reported by [2000REI/NEU] are in good agreement with the expected value of $\Delta\varepsilon = (0.13 \pm 0.04)$ kg·mol^{-1} and the SIT extrapolation to $I = 0$ leads to:

$$\log_{10} K_{s,0}^{\circ}((A.133), 298.15\ K) = -(14.67 \pm 0.10).$$

At higher NaCl and NaClO$_4$ molalities, the SIT equation may become inaccurate, in particular because of the increasing effect from Pu(VI) chloride complexation. Therefore, these data are not included in the linear SIT regression.

Table A-51: Solubility constants for $PuO_2CO_3(s)$ (molal scale) and conversion to $I = 0$ with the SIT coefficients in Appendix B.

Medium	$t(°C)$	$\log_{10} K_{s,0}$	$\log_{10} K°_{s,0}$	Reference
0.1 m NaCl	r.t.	− 13.8	− 14.7	[2000REI/NEU]
0.2 m NaCl		− 13.6	− 14.7	
0.5 m NaCl		− 13.4	− 14.7	
1.0 m NaCl		− 13.1	− 14.6	
2.1 m NaCl		− 13.1	− 14.7	
3.8 m NaCl		− 13.45	− 15.0 [a]	
5.6 m NaCl		− 14.0	− 15.4 [a]	
0.1 m NaClO$_4$	22	− 14.2 [b]	− 15.0	[93PAS/RUN], [97PAS/CZE]
3.5 m NaClO$_4$	20	− 13.5	− (14.2 ± 0.3) [a, c]	[87ROB/VIT]
5.6 m NaClO$_4$	r.t.	− 14.5	− 14.5 [a]	[2000REI/NEU]

[a] At this ionic strength the SIT extrapolation to $I = 0$ may include unknown large uncertainties.
[b] Recalculated in the present review.
[c] Recalculated and selected in [2001LEM/FUG].
r.t. room temperature.

[2000STO/HOF]

LIPAS is used to identify the complexes formed between PuO_2^+ or NpO_2^+ with Th^{4+} and UO_2^{2+}, and to quantify their stoichiometries and equilibrium constants. The technical advantage of LIPAS is its high sensitivity (which means that the study can be made at low concentrations of the reactant elements) and the possibility to perform experiments under an undisturbed inert atmosphere. In this way it is, for instance, possible to prevent disproportionation of PuO_2^+. In all experiments, concentrations of the pentavalent species are less than $0.25 \cdot 10^{-3}$ M for Np and 10^{-4} M for Pu, while those of Th and U are higher but could also reach millimolar level. The ionic strength ($I = 6$ M) is maintained with NaClO$_4$ and all experiments are conducted at pH < 3, $t = 25$ °C.

The results of the LIPAS technique are checked on the system, NpO_2^+ / UO_2^{2+}.

The equilibrium constant obtained for the formation of the cation-cation complex $NpO_2^+ \cdot UO_2^{2+}$ is $K = (2.4 \pm 0.2)$ L · mol^{-1}, very close to that previously reported by [93STO/CHO], using conventional absorption spectrophotometry ($K = (2.25 \pm 0.03)$ L · mol^{-1}, $I = 6$ M) and Raman spectroscopy ($K = (2.5 \pm 0.5)$ L · mol^{-1}, $I = 6.26$ M). There is previous evidence for the cation-cation complex, $NpO_2^+ \cdot Th^{4+}$, but the equilibrium conformation of chloride complexes (*cf.* section 12.4.1.2). In addition, the present review has doubts on the reliability of the experimental data presented, because the experimental uncertainty is at least an order of magnitude higher than the measured effects. Indeed the authors evaluate the complex formation constants from the increase of the quotient, D_0/D. This quotient typically increased in the range from 1.000 to 1.001 (Figure 2 in [2000YEH/MAD]). The measured effect is in the range of 0.1%. The error bars given in this figure refer to an uncertainty of about ± 0.02 %. Such extreme accuracy cannot be achieved by the experimental procedures described in the paper [2000YEH/MAD]. Distribution coefficients were measured from only a 1 mL of the aqueous and organic phases. However, volumetric sampling of 1 mL organic phase can hardly lead to an accuracy better than ± 1%, even if the authors were weighing their volumes (which is not mentioned in the paper); special precautions would have been necessary to achieve an accuracy of ± 0.1 %. Furthermore, the authors report that the analytical Am and Eu concentrations were determined by counting the gamma-activities until counting statistics at the 1σ level were achieved. This does certainly not allow the quantification of effects in the range of ± 0.1 %. Nevertheless, it should be noted that the reported results are in the expected range of magnitude if the change in distribution coefficient is due to a change in the activity coefficient of Am^{3+} from $HClO_4$ to HCl. Chloride complexes of Am(III) are only expected to be significant at high chloride concentrations.

[2001BAN/PRA]

The U(V) uranates $Rb_2U_4O_{11}$(cr) and $Tl_2U_4O_{11}$(cr), were prepared from the diuranates, $M_2U_2O_7$(cr) and UO_2(cr), at temperatures up to 1473 K in argon as described in references, [96IYE/JAY] for $Rb_2U_4O_{11}$(cr), and [94SAL/KUL] for $Tl_2U_4O_{11}$(cr). The principal analyses (before and after the experiments) were by X-ray diffraction, but undefined analyses for Rb and U (for $Rb_2U_4O_{11}$(cr)) and X-ray fluorescence analyses (for $Tl_2U_4O_{11}$(cr)) matched the indicated compositions. Enthalpy increments were determined in a Calvet calorimeter from 301 to 735 K on pelleted materials. The calorimeter was calibrated using synthetic sapphire (NIST standard reference material).

Values of $H_m(T) - H_m^\circ(298.15 K)$ are tabulated for ten temperatures from 396.3 to 735 K for $Rb_2U_4O_{11}$(cr) and thirteen temperatures from 301 to 673 K for $Tl_2U_4O_{11}$(cr). The resulting enthalpy differences were fitted by a Shomate analysis, using estimated values of $C_{p,m}^\circ$($Rb_2U_4O_{11}$, cr, 298.15 K) = 365.56 J · K^{-1} · mol^{-1} and $C_{p,m}^\circ$($Tl_2U_4O_{11}$, cr, 298.15 K) = 360.134 J · K^{-1} · mol^{-1} from the sum of the heat capacities of the component oxides, M_2O(cr) + $2UO_2$(cr) + $2UO_3$(cr).

The resulting equations for $H_m(T) - H_m^\circ(298.15 K)$ give on differentiation:

$C_{p,m}$($Rb_2U_4O_{11}$, cr, T) = 330.4 + 1.4134·10^{-2}·T − 6.198·10^5·T^{-2} (J · K^{-1} · mol^{-1})

(298.15 to 735 K)

and

$$C_{p,m}(\text{Tl}_2\text{U}_4\text{O}_{11}, \text{cr}, T) = 368.2 + 2.4886 \cdot 10^{-2} \cdot T - 1.375 \cdot 10^5 \cdot T^{-2} \; (\text{J} \cdot \text{K}^{-1} \cdot \text{mol}^{-1})$$
(298.15 to 673 K)

where in each case a wrong coefficient in the text has been corrected.

[2001BEN/SEI]

This paper deals with the effects of carbonate (with formate as a free radical scavenger) on the gamma radiation induced reduction of TcO_4^-. Lefort [63LEF] reported that γ-irradiation of TcO_2 in alkaline solution caused oxidation to TcO_4^-, whereas Tc(VII) is reduced to Tc(VI) by pulse radiolysis in alkaline and neutral solutions. The gamma source used in [2001BEN/SEI] was ^{60}Co. In $[\text{CO}_3^{2-}] = 10^{-2}$ M, with $[\text{TcO}_4^-] = 10^{-4}$ M at pH = 11 (argon or oxygen-saturated atmosphere) the absorption spectrum (244 nm, ε = 622 m$^2 \cdot$ mol^{-1}; 289 nm, ε = 236 m$^2 \cdot$ mol^{-1}) of TcO_4^- did not change, i.e., no reduction occurred. Conversely, in the presence of 10^{-2} M formate under an argon atmosphere, direct evidence was obtained for the formation of colloidal TcO_2. Addition of varying concentrations of carbonate to the starting solution did not inhibit the reduction reaction. The H• and OH• radical reactions with formate produce the reducing radical, $\text{CO}_2^{\bullet-}$, which in turn is responsible for the formation of Tc(VI), which undergoes a two-step disproportionation to give Tc(IV). On the other hand, it is proposed that the carbonate radical oxidises Tc(VI) back to Tc(VII), and under gamma radiolysis conditions, the concentrations of $\text{CO}_3^{\bullet-}$ radicals can build up while the concentration of Tc(VI) is limited to low steady-state concentrations. The conclusion from this study is that exposure of Tc to gamma radiation in the presence of carbonate may lead to enhanced mobilisation due to stabilisation of the Tc(VII) state.

[2001BER/GEI]

This is a continuation of previous investigations of the $\text{Ca}_2\text{UO}_2(\text{CO}_3)_3$(aq) complex as reported in [96BER/GEI], [97BER/GEI], [98GEI/BER3], [98GEI/BER4] and [97AMA/GEI]. The authors have used both TRLFS and EXAFS data to obtain information on the stoichiometry, structure and equilibrium constant of this species. The experiments are carefully described, but there are several unclear points, one being the ionic strength. From the description in the "Experimental" section one gets the impression that no ionic medium has been used; however in Table 4 they refer to the ionic strength, 0.1 M. The reviewer and the authors notice that the slope analysis results in a non-integral value. This clearly indicates large experimental errors, or that it has been impossible to resolve the spectra into different components. This means that the equilibrium constants proposed are not precise. It is not an acceptable procedure to use the experimental data and calculate an equilibrium constant assuming a slope of two, as the authors seem to have done. Additional evidence for problems with the calculation method used is provided by a comparison of the constants, $\log_{10} \beta_{2,1,3}^\circ$ = 30.55 (for $\text{Ca}_2\text{UO}_2(\text{CO}_3)_3$(aq)) and $\log_{10} \beta_{1,1,3}^\circ$ = 25.4 (for $\text{CaUO}_2(\text{CO}_3)_3^{2-}$). From these values we obtain the equilibrium constant for the reactions:

$$UO_2(CO_3)_3^{4-} + 2Ca^{2+} \rightleftharpoons Ca_2UO_2(CO_3)_3(aq) \qquad \log_{10} K_2 = 8.95$$
and
$$UO_2(CO_3)_3^{4-} + Ca^{2+} \rightleftharpoons CaUO_2(CO_3)_3^{2-} \qquad \log_{10} K_1 = 3.8$$

i.e., the binding constant of Ca^{2+} to $CaUO_2(CO_3)_3^{2-}$ is much larger than that to $UO_2(CO_3)_3^{4-}$; this is not likely in the opinion of the reviewer.

The EXAFS data are well analysed but unfortunately they do not provide strong support for the formation of Ca – bonding. A structural model could have foreseen this problem; it would have been better to study the corresponding Mg or Sr complexes. For reasons given above, this review does not accept the equilibrium constants proposed by Bernhard *et al.*; however their study provide additional evidence for the complex formation between $UO_2(CO_3)_3^{4-}$ and cations; this, itself, is an important observation.

[2001BOL/WAH]

This paper gives the structure of the Np(VII) species: $NpO_4(OH)_2^{3-}$ which is predominant in 2.5 M NaOH solution ($1.5 \cdot 10^{-2}$ M in Np). The structure is derived from XANES and EXAFS absorption spectra obtained in the transmission mode over an extended k space (up to 17 Å^{-1}) and from theoretical calculations using Hartree-Fock and Density Functional Theory methods. The test solution was sealed under an ozone atmosphere. This study; i) shows that the species postulated by Shilov [98SHI], $NpO_4(OH)_2^{3-}$, from the variation of the reversible formal potential of the Np(VII)/Np(VI) redox couple in strongly alkaline solution and by Peretrukhine *et al.* [95PER/SHI] (see [2001LEM/FUG] page 92), was correct, ii) confirms the suggestion concerning the existence of this species by Williams *et al.* [2001WIL/BLA] based also both on XAS data (fluorescence mode and k space data up to 12 Å^{-1}) and theoretical calculations, and, iii) refutes, as did the study of Williams *et al.* [2001WIL/BLA], the interpretation of Clark *et al.* [96CLA/CON] who gave $NpO_2(OH)_4^-$ as the major species present in alkaline solution of Np(VII), on the basis of XAS spectra (fluorescence in k space up to 10 Å^{-1}).

High quality EXAFS data obtained by [2001BOL/WAH] suggest four short oxo Np–O bonds (N = (3.6 ± 0.3)) and two long hydroxo Np–O (N = (3.3 ± 1.3)) bonds. The large uncertainty in the number of coordinated OH ligands does not allow an easy choice of N. The retained value, N = 2, comes from the excellent agreement of the bond distances calculated for a square bipyramidal O arrangement around Np (D_{2d} point group symmetry) with those measured on the compounds, $Co(NH_3)_6NpO_4(OH)_2 \cdot 2H_2O$ and $Na_3NpO_4(OH)_2 \cdot n\ H_2O$. But the main argument comes from theoretical calculations, both for the gaseous and solvated Np(VII)/Np(VI) species: $NpO_4(OH)_2^{3-} / NpO_4(OH)_2^{4-}$ and $NpO_2(OH)_4^- / NpO_2(OH)_4^{2-}$. Indeed, calculated bond lengths for the solvated species, $NpO_4(OH)_2^{3-}$, are in excellent agreement with single crystal X-ray and solution EXAFS data. Furthermore, the NpO_4^- entity has been shown to be square planar [2001BOL/WAH2].

The reversibility of the redox couple, $NpO_4(OH)_2^{3-} / NpO_2(OH)_4^{2-}$, could be due to fast proton exchanges between the two species, which have the same coordination geometry.

This paper closes the long story in existence since the eighties on the identification of the species of Np(VII) in strong alkaline solution.

[2001FUJ/YAM]

Fujiwara et al. determined the solubility of Pu(IV) hydrous oxide under reducing conditions in 1.0 M $NaClO_4$ solutions containing $Na_2S_2O_4$, under an Ar atmosphere at $(25 \pm 1)°C$. They used appropriate analytical methods for the determination of the H^+ and Pu concentrations. For solutions of high Pu concentration, the presence of Pu^{3+} was confirmed by absorption spectroscopy. Aqueous Pu species in other oxidation states could not be detected. The solubility data measured after 1 – 3 months in the range $4 < - \log_{10}[H^+] < 6$ refer to the reaction:

$$PuO_2(am, hydr.) + e^- + 2 H_2O(l) \rightleftharpoons Pu^{3+} + 4 OH^-. \qquad (A.134)$$

The authors calculated $\log_{10} K$ ((A.134), 298.15 K) $= -(38.39 \pm 0.19)$ in 1 M $NaClO_4$, with the uncertainty given as the standard deviation and extrapolated this value to zero ionic strength with the SIT coefficients of [92GRE/FUG]. The uncertainty given by [2001FUJ/YAM] is increased here to the 95% confidence interval (1.96 σ) and the result is $\log_{10} K°$ ((A.134), 298.15 K) $= -(40.44 \pm 0.39)$.

If this value is combined with $\log_{10} K°$ ((A.135), 298.15 K) $= -(17.69 \pm 0.04)$ selected in [2001LEM/FUG] or $-(17.64 \pm 0.17)$ [95CAP/VIT] as used by Fujiwara et al. for the reaction:

$$Pu^{4+} + e^- \rightleftharpoons Pu^{3+} \qquad (A.135)$$

the solubility constant for the reaction:

$$PuO_2(am, hydr.) + 2 H_2O(l) \rightleftharpoons Pu^{4+} + 4 OH^- \qquad (A.136)$$

is calculated to be:

$$\log_{10} K°_{s,0} (A.136) = -(58.1 \pm 0.4).$$

This value is comparable with other recently determined values of $\log_{10} K°_{s,0}$ ($Pu(OH)_4$, am, 298.15 K) $= -(58.3 \pm 0.5)$ [98CAP/VIT], and is based on a similar indirect experimental procedure, and $\log_{10} K°_{s,0}$ ($Pu(OH)_4$, am, 298.15 K) $= -(58.7 \pm 0.9)$ [99KNO/NEC], which has been calculated from the available literature data of experimental Pu(IV) concentrations corrected for hydrolysis with the hydrolysis constants of [72MET/GUI].

The Pu concentrations measured in the range $7.5 < - \log_{10}[H^+] < 9.0$, $\log_{10}[Pu] = -(8.9 \pm 0.2)$, were ascribed mainly to small Pu(IV) polymers, as supported by ultrafiltration. Fujiwara et al. used a simple approach to model the polymerisation equilibria and to estimate the concentration of mononuclear Pu(OH)$_4$(aq). The calculated result, $\log_{10}$[Pu(OH)$_4$(aq)] = − (10.3 ± 0.2), is in good agreement with experimental data determined by other authors, ($\log_{10}$[Pu(OH)$_4$(aq)] = − (10.4 ± 0.2) [86LIE/KIM], − (10.2 ± 0.8) [98CHA/TRI] and − (10.4 ± 0.5) [99RAI/HES2]), and the reported value of $\log_{10} \beta_4^\circ$ (Pu(OH)$_4$, aq, 298.15 K) ≥ 47.5 [2001FUJ/YAM] is consistent with the formation constant selected in the present review, $\log_{10} \beta_4^\circ = (47.5 \pm 0.5)$.

[2001KIT/KOH]

The authors present solubility measurements with Np(IV) in carbonate containing NaClO$_4$ solutions at (22 ± 3)°C. The batch experiments were performed from oversaturation. Reduction of $10^{-5} - 10^{-3}$ M Np(V) solutions led to the precipitation of Np(OH)$_4$(am). The reducing agent was Na$_2$S$_2$O$_4$. The ionic strength was adjusted to either 0.5 or 1.0 mol·L^{-1}. Appropriate analytical methods were used for the determination of the H$^+$ and Np(IV) concentrations. In batch samples containing an initial total carbonate concentration of C$_{tot}$ = [HCO$_3^-$] + [CO$_3^{2-}$] = 0.1 mol·L^{-1}, the H$^+$ concentration was varied in the range, $-\log_{10}$[H$^+$] = 8.5 - 12.5. In further samples at -log$_{10}$[H$^+$] = 8.5 - 10.5, C$_{tot}$ was varied from 0.005 to 0.1 mol·L^{-1}. The samples were shaken for two and four weeks and the Np(IV) concentration was determined after 10 000 Dalton ultrafiltration and subsequent solvent extraction with 0.5 M TTA in xylene.

The solubilities measured at varying total carbonate and H$^+$ concentrations in the range of - log$_{10}$[H$^+$] = 10 - 11 are highly scattered. Under these conditions, the solubility is very sensitive to a slight variation of C$_{tot}$ or log$_{10}$[H$^+$]. The solubility measured at C$_{tot}$ = 0.1 mol·L^{-1} in 0.5 and 1.0 M NaClO$_4$ was ascribed to the following reactions.

$- \log_{10}$[H$^+$] > 11:

$$\text{Np(OH)}_4\text{(am)} + 2\,\text{CO}_3^{2-} \rightleftharpoons \text{Np(OH)}_4(\text{CO}_3)_2^{4-} ; \quad (A.137)$$

$- \log_{10}$[H$^+$] < 11:

$$\text{Np(OH)}_4\text{(am)} + 2\,\text{HCO}_3^- \rightleftharpoons \text{Np(OH)}_2(\text{CO}_3)_2^{2-} + 2\,\text{H}_2\text{O(l)}. \quad (A.138)$$

The latter reaction can be rewritten as:

$$\text{Np(OH)}_4\text{(am)} + 2\,\text{CO}_3^{2-} \rightleftharpoons \text{Np(OH)}_2(\text{CO}_3)_2^{2-} + 2\,\text{OH}^-. \quad (A.139)$$

Using reasonable estimates for the SIT coefficients of the ternary complexes, the equilibrium constants at $I = 0$ were calculated to be $\log_{10} K^\circ_{s,(1,4,2)}$ (A.137) = − (6.82 ± 1.03) and $\log_{10} K^\circ_{s,(1,2,2)}$ (A.138) = − (3.81 ± 0.35) or $\log_{10} K^\circ_{s,(1,2,2)}$ (A.139) = − (11.15 ± 0.35).

The predominance of the complexes, Np(OH)$_2$(CO$_3$)$_2^{2-}$ and Np(OH)$_4$(CO$_3$)$_2^{4-}$, derived from the experimental data is consistent with the speciation proposed in [90PRA/MOR], [94YAM/SAK] and [99RAI/HES]. As also noted by Kitamura and Kohara, in particular the pH-independent solubility at $-\log_{10}[H^+] > 11$ could also be assumed by analogy to involve the same $Np(OH)_4(CO_3)^{2-}$ complex as proposed by [93ERI/NDA].

The calculation of the HCO_3^- and CO_3^{2-} concentrations from C_{tot} and $-\log_{10}[H^+]$ includes a substantial error. The authors obviously used the carbonic acid dissociation constants at $I = 0$ instead of the conditional constants in 0.5 and 1.0 M $NaClO_4$. This has a significant impact on $\log_{10} K_{s,(1,2,2)}$ derived from experimental data at $-\log_{10}[H^+] < 11$, while the calculation of $\log_{10} K_{s,(1,4,2)}$ is primarily based on experimental data at $-\log_{10}[H^+] > 11$ where $[CO_3^{2-}] \approx C_{tot}$, independent of whether $pK_2^\circ = 10.33$ or the pK_2 values in 0.5 and 1.0 M $NaClO_4$ are used for calculation. A re-evaluation of Kitamura and Kohara's most accurate set of data at $C_{tot} = 0.1$ mol·L^{-1} and $I = 0.5$ mol·L^{-1} yields:

for $Np(OH)_4(CO_3)_2^{4-}$: $\log_{10} K_{s,(1,4,2)} = \log_{10} K_{s,0} + \log_{10} \beta_{1,4,2}$
$= -(5.5 \pm 0.4)$ at $I = 0.5$ mol·L^{-1}

$\log_{10} K^\circ_{s,(1,4,2)} = -(6.9 \pm 0.4)$

or alternatively:

for $Np(OH)_4(CO_3)^{2-}$: $\log_{10} K_{s,(1,4,1)} = \log_{10} K_{s,0} + \log_{10} \beta_{1,4,1}$
$= -(6.5 \pm 0.4)$ at $I = 0.5$ mol·L^{-1}

$\log_{10} K^\circ_{s,(1,4,1)} = -(6.5 \pm 0.4)$

and

for $Np(OH)_2(CO_3)_2^{2-}$: $\log_{10} K_{s,(1,2,2)} = \log_{10} K_{s,0} + \log_{10} \beta_{1,2,2}$
$= -(10.8 \pm 0.4)$ at $I = 0.5$ mol·L^{-1}

$\log_{10} K^\circ_{s,(1,2,2)} = -(10.4 \pm 0.4)$.

These re-calculations include a fixed concentration of $\log_{10}[Np(OH)_4(aq)] = -8.3$ taken from [93ERI/NDA]. The values at $I = 0$ are obtained by using NEA-TDB auxiliary data and estimated SIT coefficients of $\varepsilon(Na^+, Np(IV)$ complex$) = -0.1$ kg·mol^{-1} for complexes with a charge of -2 and -0.2 kg·mol^{-1} for complexes with a charge of -4.

The difference between $\log_{10} K^\circ_{s,(1,2,2)} = -(10.4 \pm 0.4)$ calculated in this review from data in [2001KIT/KOH] and $\log_{10} K^\circ_{s,(1,2,2)} = -11.75$ determined by Rai et al. [99RAI/HES] might be due to a difference in $\log_{10} K^\circ_{s,0}$. The solubility data of Kitamura and Kohara refer to small solid particles formed from over-saturation at a total Np concentration of 10^{-5} mol·L^{-1} in 0.5 M $NaClO_4$, while those of Rai et al. [99RAI/HES] were determined from under-saturation with a solid of probably larger particle size. On the other hand, the pH-independent solubility at $-\log_{10}[H^+] > 11$ can be described with $\log_{10} K^\circ_{s,(1,4,1)} = -(6.5 \pm 0.4)$, which is clearly lower than the value of $\log_{10} K^\circ_{s,(1,4,1)} = -(5.3 \pm 0.3)$ determined by Eriksen et al. [93ERI/NDA].

Figure A-17. Experimental Np(IV) solubility data measured for freshly-formed solid particles of Np(OH)$_4$(am) as a function of the H$^+$ or D$^+$ concentration in 0.1 M perchlorate solution (above). The solid curve and the dashed speciation lines are calculated with $\log_{10} K_{s,0} = -54.4$ (at $I = 0.1$ M) and the hydrolysis constants from [77DUP/GUI] and [2001NEC/KIM].

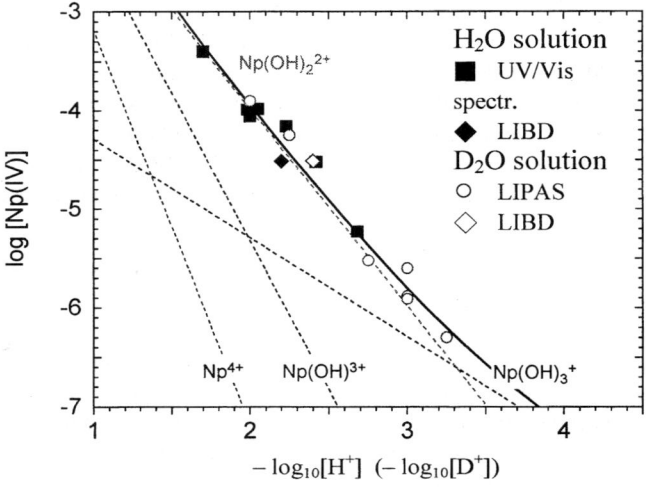

Figure A-18. Np(IV) species distribution in 0.1 M HClO$_4$–NaClO$_4$.

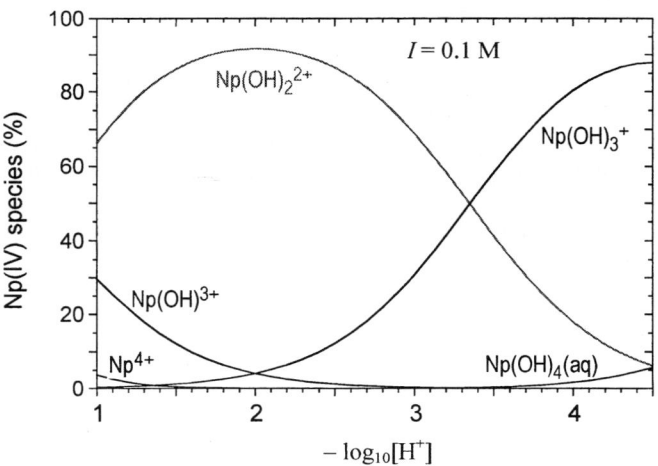

[2001RAI/MOO]

Rai et al. determined the solubility of amorphous Pu(IV) hydrous oxide in different air-equilibrated electrolyte media (0.4 and 4.0 M NaClO$_4$, and 0.4 and 4.0 M NaCl). The analytical methods, determination of the H$^+$ and Pu concentrations, and oxidation state analyses are well described. The solubility data in the range $4 < -\log_{10}[H^+] < 9$ are measured after equilibration periods of seven and 106 – 110 days. The high solubility data observed in the 4 M NaCl solutions equilibrated for a long time, where radiolytic oxidation leads to increased pe values, are caused by the formation of Pu(VI), while all other results are ascribed to the reaction:

$$PuO_2(am) \rightleftharpoons PuO_2^+ + e^- \qquad (A.140)$$

or

$$PuO_2(am) + H^+ + \frac{1}{4}O_2(g) \rightleftharpoons PuO_2^+ + \frac{1}{2}H_2O(l) \qquad (A.141)$$

This is consistent with the slope of -1 in the plots of $\log_{10}[Pu]$ versus $-\log_{10}[H^+]$. The variation of the solubility in the four different media is explained by the variation of the PuO$_2^+$ activity coefficient. This effect is well described by using the Pitzer ion interaction parameters reported in [95NEC/FAN] for the NpO$_2^+$ ion as analogues for the PuO$_2^+$ ion.

Similar solubility data from a previous study in 0.0015 M CaCl$_2$ solutions lead to $\log_{10} K°(A.140) = -19.45$ [84RAI] and have been used in [2001RAI/MOO] to estimate the effective oxygen fugacities. However, the results of the latter study yield a somewhat different equilibrium constant, $\log_{10} K°(A.140) = -18.85$ [2001RAI/MOO].

By combining these constants of [84RAI] and [2001RAI/MOO], which are given without uncertainty limits, with the PuO$_2^+$/Pu^{4+} redox potential and other auxiliary data selected in [2001LEM/FUG], the solubility constant for the reaction

$$PuO_2(am) + 2 H_2O(l) \rightleftharpoons Pu^{4+} + 4 OH^- \qquad (A.142)$$

is calculated to be $\log_{10} K°_{s,0}(A.142) = -58.0$ and -57.4, respectively. These values are slightly higher than those determined in other recent studies: $\log_{10} K°_{s,0}$ ((A.142), 298.15 K) $= -(58.3 \pm 0.5)$ [98CAP/VIT], $-(58.7 \pm 0.9)$ [99KNO/NEC] and $-(58.1 \pm 0.4)$ [2001FUJ/YAM]. As pointed out by [2001RAI/MOO], the measured pe values or the effective oxygen fugacities may include unknown uncertainties. Therefore, the data given in [2001RAI/MOO] are not used for the selection of thermodynamic equilibrium constants.

[2001SEM/BOE]

The possibility of the coordination of ClO$_4^-$ or CF$_3$SO$_3^-$ anions to UO$_2^{2+}$ in very acidic solutions of HClO$_4$ and CF$_3$SO$_3$H (up to 10 M) has been raised to account for the large variations of the fluorescence lifetime and of the intensity of fluorescence spectra of UO$_2^{2+}$ with the increase of the concentrations of these anions ([99BOU/BIL], [2001BIL/RUS]). No clear evidence of complexation of dioxouranium(VI) cation was found. To check the first coordination sphere of UO_2^{2+}, EXAFS spectra of 10^{-2} M U(VI) solutions in $HClO_4$ up to 11.5 M and in mixtures of $NaClO_4$ and $HClO_4$ (14 test solutions), as well as in CF_3SO_3H up to 10 M (5 test solutions) have been recorded in the transmission mode (k space up to 13 Å^{-1}). EXAFS spectra of U(VI) perchlorate and U(VI) triflate were taken as references. For all the aqueous perchlorate solutions, the experimental EXAFS oscillations are due to the contributions of the two axial oxygen atoms and the five equatorial oxygen atoms (N = (4.6 ± 0.5)) surrounding the U atom. No chlorine-scattering contribution is observed. The U–O distances ((1.75 ± 0.01) to (1.76 ± 0.01)) Å and ((2.40 ± 0.02) to (2.43 ± 0.02)) Å, respectively, agree with all the values measured for aqueous UO_2^{2+} ion in somewhat different conditions of acidity and ClO_4^-/U ratios and in U(VI) perchlorate heptahydrate. For solutions of U(VI) in 1 to 8 M triflic acid the two shells of oxygen atoms are found, as well as in 10 M, but for this concentration with a third shell of sulphur atoms (N = (1.3 ± 0.5)). The U–S distance is (3.62 ± 0.02) Å, near to the U-S distance of 3.67 Å in the dioxouranium(VI) triflato-hydrate benzo-[15]-crown-5, ($[UO_2(CF_3SO_3)_2]\cdot 2$ $C_{14}H_{20}O_5$). These data indicate clearly the absence of a perchlorate anion in the inner sphere of UO_2^{2+}, even in 11.5 M $HClO_4$. In contrast, one triflate anion is bound to UO_2^{2+} in 10 M CF_3SO_3H through an oxygen atom, giving the complex $UO_2CF_3SO_3(H_2O)_4^+$. Quantum mechanical calculations in the gas phase reported in this paper give some support to these conclusions.

[2001VAL/WAH]

The structures of the complexes, $UO_2F_n(H_2O)_{5-n}^{2-n}$ $n = 3 - 5$, have been studied by EXAFS. All have pentagonal bipyramidal geometry with U – F and U – O(H_2O) distances equal to 2.26 and 2.48 Å, respectively. On the other hand, the complex $UO_2(OH)_4^{2-}$ has a square bipyramidal geometry both in the solid state and in solution. The structures of hydroxide and fluoride complexes have also been investigated with wave function based and DFT methods in order to explore the possible reasons for the observed structural differences. These studies include models that describe the solvent by using a discrete second coordination sphere, a model with a spherical, or shape-adapted cavity in a conductor such as a polarisable continuum medium (CPCM), or a combination of the two. Solvent effects were shown to give the main contribution to the observed structure variations between the uranium(VI) tetrahydroxide and the tetrafluoride complexes. Without a solvent model both $UO_2(OH)_4(H_2O)^{2-}$ and $UO_2F_4(H_2O)^{2-}$ have the same square bipyramidal geometry, with the water molecule located at a distance of more than 4 Å from the uranium and with a charge distribution that is nearly identical in the two complexes. Of the models tested, only the CPCM models were able to describe the experimentally observed square- and pentagonal-bipyramidal geometry in the tetrahydroxide and tetrafluoride complexes. The geometries and the relative energies of the different isomers of $UO_2F_3(H_2O)_2^-$ are very similar, indicating that both isomers are present in comparable amounts in solution. All calculated bond distances are in good agreement with the experimental observations, provided that a proper model of the solvent is used. This study provides structural information but no thermodynamic data.

[2001WIL/BLA]

XANES and EXAFS spectra (fluorescence mode and k space up to 12 Å^{-1}) have been obtained in 1 M NaOH solutions, 6.5·10^{-3} M, during the electrochemical reduction of Np(VII) to Np(VI). They show clearly a change in the coordination number of Np. The Np(VI) coordination is consistent with, respectively, two and four oxygen atoms at short and long distances (as for U(VI) under similar conditions) while for Np(VII) the central ion is coordinated to four and two oxygen atoms, respectively, at short and long distances. This result refutes that of Clark et al. [97CLA/CON] who found the reverse for Np(VII) in similar conditions. Williams et al. [2001WIL/BLA] found a resolved peak in the Fourier transform of Np(VI) and Np(VII) spectra attributed to a strong Np–Na ion pairing. This is a very surprising finding and one does not expect such a short distance between two metal ions with formal oxidation states +VII and +I. In the more accurate EXAFS study of Bolvin et al. [2001BOL/WAH] this peak is absent. As tetraoxo coordination of Np(VII) was rather unusual, except in solid state, the authors calculated the energy of different configurations of the MO_4^- entity in different point group symmetries, T_d, D_{4h} and D_{2d} using Density Functional Theory. They showed that in the gaseous state, the D_{2d} structure of this ion has the lowest energy and suggested that $NpO_4(OH)_2^{3-}$ could be the species of Np(VII) present in alkaline solution. This suggestion was confirmed by Bolvin et al. [2001BOL/WAH] who obtained better XAS experimental data.

[2002BRO]

This study is a potentiometric determination of $\log_{10}\beta_{n,m}$ values for the equilibria:

$$m\,UO_2^{2+} + n\,H_2O(l) \rightleftharpoons (UO_2)_m(OH)_n^{(2m-n)} + n\,H^+$$

at 25°C in 0.1 M KCl, 0.1 M NaClO$_4$ and 1.0 M KNO$_3$, the concentration of U and the $-\log_{10}[H^+]$ values being, respectively, in the range (0.2 to 2 10^{-3} M, 3.5 to 5.9), (0.2 to 1.9 10^{-3} M, 3.7 to 6.1) and (0.2 to 4 10^{-3} M, 3.3 to 5.2). The potentiometric data were analysed with the computer program MINIQUAD. It gives the $\log_{10}{}^*\beta_{n,m}$ values and the associated uncertainties for the couples of (n,m) species, summarised in Table A-53.

Table A-53: Values of $\log_{10}{}^*\beta_{n,m}$ for (1,1), (2,2), (4,3), (5,3), (7,4) in 0.1 M KCl, 0.1 M NaClO$_4$ and 1.0 M KNO$_3$.

Species (n,m)	0.1 M KCl	0.1 M NaClO$_4$	1.0 M KNO$_3$
(1,1)	– (5.17 ± 0.03)	– (5.01 ±0.03)	
(2,2)	– (5.86 ± 0.04)	– (5.98 ± 0.04)	– (5.85 ± 0.03)
(4,3)	– (12.00 ± 0.06)	– (12.39 ± 0.05)	– (11.95 ± 0.05)
(5,3)	– (16.09 ± 0.06)	– (16.36 ± 0.05)	– (16.40 ± 0.06)
(7,4)			– (21.79 ± 0.06)

It is interesting to note that the species (7,4) is only found to be present in nitrate media. It exists also in sulphate media (see [2000COM/BRO] and [2000MOL/REI]). That could be the result of an oxoanion binding to form the complex.

This review has calculated, using the SIT model and the auxiliary data of [92GRE/FUG], the $\log_{10} {}^*\beta^\circ_{n,m}$ values from Brown's $\log_{10} {}^*\beta_{n,m}$ values in 0.1 M NaClO$_4$. These are $\log_{10} {}^*\beta^\circ_{1,1} = -(4.72 \pm 0.37)$, $\log_{10} {}^*\beta^\circ_{2,2} = -(5.76 \pm 0.04)$, $\log_{10} {}^*\beta^\circ_{4,3} = -(11.93 \pm 0.06)$ and $\log_{10} {}^*\beta^\circ_{5,3} = -(15.77 \pm 0.05)$, which can be directly compared to those of selected by [92GRE/FUG]. The large uncertainty in $\log_{10} {}^*\beta^\circ_{1,1}$ comes from that on $\varepsilon(UO_2OH^+, ClO_4^-)$. The value of $\log_{10} {}^*\beta^\circ_{1,1}$ is higher than that of [92GRE/FUG] and that of $\log_{10} {}^*\beta^\circ_{2,2}$ is lower, while those of $\log_{10} {}^*\beta^\circ_{4,3}$ and $\log_{10} {}^*\beta^\circ_{5,3}$ are in close agreement.

Brown used his data and the values of the appropriate equilibrium constants (with uncertainties) reported in [92GRE/FUG], and possibly from more recent literature, to calculate $\log_{10} \beta_{n,m}$ at zero ionic strength using the SIT model. He derived the interaction coefficients, but with some differences from the usual way of using the SIT extrapolation. For instance, from the experimental values of $\Delta\varepsilon$ for the different media he used the specific interaction coefficient $\varepsilon(UO_2^{2+}, Cl^-) = (0.27 \pm 0.12)$ kg·mol^{-1}, $\varepsilon(UO_2^{2+}, ClO_4^-) = (0.46 \pm 0.03)$ kg·mol^{-1}, and $\varepsilon(UO_2^{2+}, NO_3^-) = (0.50 \pm 0.03)$ kg·mol^{-1} instead of that for perchlorate media for all the media. He made corrections for the complexation of UO_2^{2+} by Cl$^-$ and NO$_3^-$ ions. As it is not clear which literature data have been used and as details of the calculations are lacking, the results of these extrapolations collected in Table A-54 are not considered by this review for the selection of the equilibrium constants.

Table A-54: $\log_{10} {}^*\beta^\circ_{n,m}$ and $\varepsilon(n,m)$ for UO_2^{2+} values, (1,1), (2,2), (4,3), (5,3) and (7,4) species.

Species	$\log_{10} {}^*\beta^\circ_{n,m}$	$\varepsilon (n,m)$ (kg·mol^{-1})		
		Cl$^-$	ClO$_4^-$	NO$_3^-$
UO$_2^{2+}$		(0.27 ± 0.12)	(0.46 ± 0.03)	(0.50 ± 0.03)
(1,1)	−(5.1 ± 0.1)	−(0.1 ± 0.2)	(0.1 ± 0.2)	(0.6 ± 0.4)
(2,2)	−(5.56 ± 0.06)	(0.20 ± 0.18)	(0.61 ± 0.07)	(0.33 ± 0.16)
(4,3)	−(11.7 ± 0.3)	−(0.2 ± 0.3)	(0.98 ± 0.14)	(0.51 ± 0.15)
(5,3)	−(15.46 ± 0.09)	(0.09 ± 0.23)	(0.53 ± 0.10)	(0.3 ± 0.3)
(7,4)	−(22.2 ± 0.2)			−(0.8 ± 0.3)

Nevertheless, the values of $\log_{10} {}^*\beta^\circ_{n,m}$ in Table A-54 are, within the uncertainties, in good agreement with those selected by this review (Table 9-6). However, some of these values do not agree with the values proposed previously. For instance, for the species (2,2) and (5,3), the Meinrath' values considered in this review are $\log_{10} {}^*\beta^\circ_{2,2} = -(5.93 \pm 0.09)$ and $\log_{10} {}^*\beta^\circ_{5,3} = -(16.51 \pm 0.18)$, and those of [92GRE/FUG] are $\log_{10} {}^*\beta^\circ_{2,2} =$

− (5.62 ± 0.04) and $\log_{10} {}^*\beta^\circ_{5,3} = -$ (15.55 ± 0.12). The interactions coefficients proposed by Brown (Table A-54) differ from those selected by [92GRE/FUG].

This review has calculated two speciation diagrams based on Brown's data for perchlorate and nitrate media (Figure A-19 and Figure A-20, respectively). They show some overlap between the (4,3) and (5,3) complexes. The equilibrium constant for the (5,3) complex would be larger if one does not include the (4,3) complex in the chemical model. Therefore, as the (5,3) complex predominates over a large pH region, the effect of including the (4,3) complex is not expected to be large. Presumably there is a large correlation between the equilibrium constants for the (5,3) and the (4,3) complexes.

Figure A-19: Speciation diagram obtained in perchlorate media assuming that no precipitation occurs.

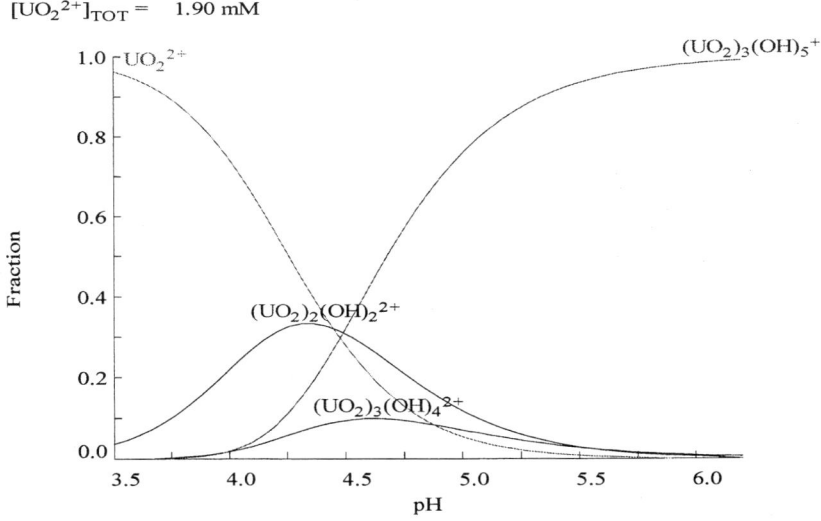

Figure A-20: Speciation diagram obtained in nitrate media (no precipitation of dioxouranium(VI) species is assumed).

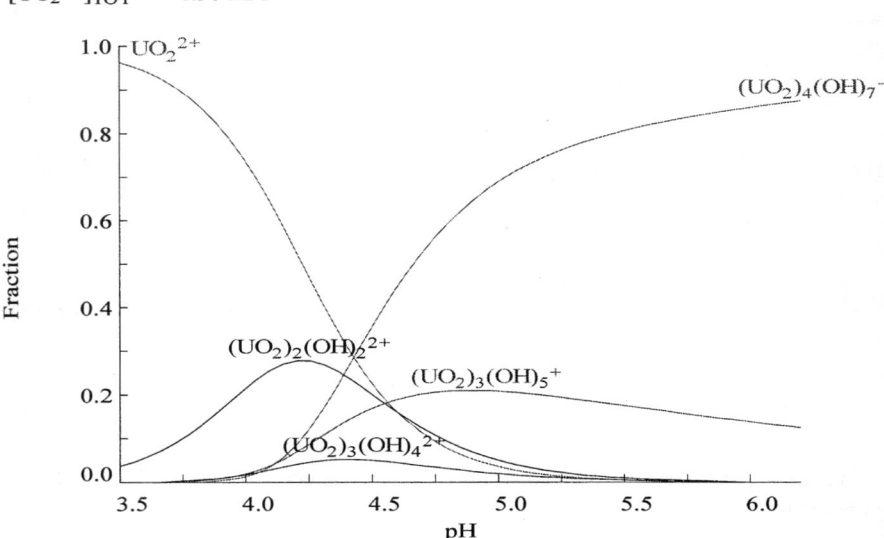

[2002GER/KOD]

During the review process, a volume of extended abstracts appeared summarizing the presentations at "The Third Russian-Japanese Seminar on Technetium" held in Dubna, Russia, June 23-July 1, 2002. The editors were K.G. German, G.E. Kodina and T. Sekine. The material presented in these abstracts is generally given in the form of figures and no thermodynamic information can be extracted. However, some of the abstracts contain interesting information that is relevant to the main to topic of technetium chemistry covered by this review and these are summarised below.

Interesting structural information were provided on $HTcO_4(cr)$, [2002KIR/GER], $(C_3H_7)_4NTcO_4$, [2002GRI/GER], $Zr(TcO_4)_4 \cdot 4H_2O$ [2002GER/GRI], $(UO_2)_2(TcO_4)_4 \cdot 3H_2O$, [2002GER/GRI], $(NpO_2)_2(TcO_4)_4 \cdot 3H_2O$ [2002GER/GRI].

NMR and spectroscopic data [2002KIR/GER] were also given in the various abstracts. Dark-red crystals of $HTcO_4(cr)$ were prepared and ^1H- and ^{99}Tc-NMR spectra indicated that the powder was not a hydrate, but consisted of a polymeric framework of four- and six-coordinated Tc.

Another work [2002GRI/GER] described solubility measurements on crystalline $(C_3H_7)_4NTcO_4$ and $(C_4H_9)_4NTcO_4$ in HNO_3 and the corresponding tetraalkylammonium hydroxide (the temperature was not given). The solubility product of the former salt was given as $1.93 \cdot 10^{-4}$ averaged over the ionic strength range of *ca.* 0.01 to unknown molarity. Clearly these results are informative, but are not quantitatively verifiable.

An interesting study was reported by Maslennikov *et al.* [2002MAS/PER], involving the electrochemical reduction of TcO_4^-. They observed a diffusion-controlled reduction of Tc(VII) to Tc(III) in $0.1 - 1.0$ M HNO_3 at 0.069 to -0.205 V/SCE (standard mercury-dropping electrode, SMDE) with a second reaction at -0.71 V/SCE associated with NO_3^- reduction catalysed by Tc(III). Time–resolved polarography and differential capacity measurements at pH $3.5 - 6.0$ (acetate buffer) led to the conclusion that the Tc(VII)/Tc(III) reduction wave reported in the literature is actually associated with Tc(VII)/Tc(V) reduction, followed by disproportionation to form insoluble Tc(IV), which is adsorbed by the electrode and there reduced to Tc(III). Fast-polarography, differential pulse polarography and cyclic voltammetry were used with a SMDE with 0.1 to 6.0 M NaOH solutions to observe the facile reduction of Tc(VII) to Tc(VI) at -0.775 to -0.705 V/SCE. Further reduction of TcO_4^{2-} was complicated by coupled disproportionation reactions of Tc(VI) and Tc(V), although below 2.0 M NaOH, Tc(IV) was the principal reduction product.

Topics of other papers presented at this conference appear to be beyond the scope of the present review.

[2002NGU]

This study is a published thesis, which includes all the results of [2000NGU/PAL], (see review in this Appendix A) and gives additional results on U(VI) hydrolysis. It deals mainly with the use of vibration spectroscopy (Raman and IR as IR ATR - Infra Red Attenuated Total Reflectance) to identify U(VI) species in aqueous non-complexing organic electrolytes ($2.70 < -\log_{10}m_{H+} < 14.42$) which allow U(VI) concentrations, C_U, up to 0.1 M to be attained in neutral and basic solutions.

Some drawbacks in the Raman spectra recorded to identify and quantify species, and consequently to calculate equilibrium constants were underlined in the review of [2000NGU/PAL]. In particular, the strong and broad Raman band centred at 753 cm^{-1} of one the two electrolytes used, $(CH_3)_4NOH$, (3 M) can overlap with the bands that can be assigned to the species $(UO_2)_m(OH)_n^{2m-n}$, in short $n:m^{2m-n}$, with $m = 3$ or $n > 4$ in basic solutions, which are located in the range 780 - 824 cm^{-1}. Furthermore, the half-width of some of the bands is variable, which is unexpected from a theoretical point of view. This review has considered that convincing data were not presented in [2000NGU/PAL] to retain the species $8:3^{2-}$, $10:3^{4-}$ and $11:3^{5-}$ in selecting equilibrium constants for U(VI) hydrolysis. These species have been detected at total concentrations of uranium above 0.3 M and at $-\log_{10}m_{H+}$ around 13.

To overcome the difficulties encountered in [2000NGU/PAL], Nguyen has made a new study, where the electrolyte $(C_2H_5)_4NOH$ ($2.92 < -\log_{10}m_{H+} < 14.50$) was used to prepare basic solutions of U(VI) ($C_U = 5 \cdot 10^{-3}$ to 10^{-1} M). The ionic strength varied between $6 \cdot 10^{-2}$ to 3.21 M. $(C_2H_5)_4NOH$ interferes much less with the Raman spectra of anionic U(VI) species than $(CH_3)_4NOH$. This electrolyte has a sharp Raman band at 718 cm^{-1}. It was possible to record spectra with a good signal to noise ratio. Two deconvolution methods (methods 1 and 2)* of the spectra have been employed after systematic subtraction of the background from the electrolyte in the spectra. In this way the spectra of U(VI) species in the range 700 to 950 cm^{-1} (domain of the symmetrical stretching mode of UO_2 core, v_1) were obtained. Additional data were obtained using IR ATR (anti-symmetrical stretching mode of UO_2 core, v_2) and UV-visible spectroscopy. They qualitatively confirm the Raman data, but they do not bring new information. IR ATR is used in an attempt to overcome the large IR absorption of H_2O below 900 cm^{-1} and permits spectra to be to recorded down to $C_U = 10^{-3}$ M.

The main results of this study in basic solutions are summarised in Table A-55 where the concentration of U(VI), C_U, in M, the ionic strength, I, in M, the wavelength of the v_1 mode in cm^{-1} and $\log_{10} {}^*\beta_{n,m}$ values are reported for the species identified in basic media. The results for the species 2:2^{2+} and 5:3$^+$ and 7:3$^-$ are the same as in [2000NGU/PAL] and are not given in the Table A-55.

These results do not change drastically the data of [2000NGU/PAL] about anionic species except that a new species, 6:1^{4-} is proposed at the highest concentration of $(C_2H_5)_4NOH$ used ($-\log_{10}m_{H+} > 14$); it also raises doubts about the species 10:3^{4-} for which the identification depends on the method used to process the data. The introduction of the complex, $UO_2(OH)_6^{4-}$, casts serious doubts about the method used. There is no structural evidence for U(VI) complexes with six coordinated unidentate ligands. As a matter of fact the preferred coordination geometry for mononuclear U(VI) hydroxide complexes seems to be a pentagonal bipyramid. There are indications of the formation of small amounts of $UO_2(OH)_5^{3-}$ from NMR studies [2000MOL/REI], but the NMR data indicate that the stepwise equilibrium constant from 4:1^{2-} to 5:1^{3-} could be between 0.1 and 1, far from the value calculated from Nguyen's data. The Raman and IR spectra of di- and tri-dioxouranium(VI) species are expected to have one elongated Raman and one IR single band provided that an equatorial pentagonal environment of U is respected. Nguyen obtained precise IR spectra in solutions of $(C_2H_5)_4NOD$ in D_2O. Deconvolution of IR spectra has not been done. So a unique assignment of the Raman and a IR spectrum to each species was impossible.

Table A-55: Species of U(VI) identified by Raman spectra in basic media according to [2002NGU].

		Species with equilibrium constant $\log_{10} {}^*\beta_{n,m}$					
C_U M	I, M	8:3	10:3	11:3	4:1	5:1	6:1
Method 1*							
ν/cm⁻¹		812	800	796	782	760	
0.005	0,06-0.08	(40±4)	(64 ± 6)				nd
0.010	0.16-0.47	(39 ± 4)	(66 ± 6)	(78 ± 8)	(31.5 ± 1.0)	(45.5 ± 1.0)	nd
0.010	0.93-2.02	(39 ± 4)	(64 ± 6)	(78 ± 8)	(31.6 ± 1.0)	(46.0 ± 1.0)	nd
0.023	0.07-0.08	(39 ± 4)	(67 ± 6)	(80 ± 8)	(32.0 ± 1.0)	(46.8 ± 1.0)	nd
0.023	0.04-3.02	(39 ± 4)	(65 ± 6)	(80 ± 8)	(31.8 ± 1.0)	(47.3 ± 1.0)	nd
0.050	0.61-3.10	(39 ± 4)	(68 ± 6)	(82 ± 8)	(32.6 ± 1.0)	(46.5 ± 1.0)	nd
0.100	1.27-3.21	(39 ± 4)	(69 ± 6)	(82 ± 8)	(32.5 ± 1.0)	(47.5 ± 1.0)	nd
Method 2*							
ν/cm⁻¹		815		796	784	767	748
0.010	0.93-2.02	35	nd	72	32	47	
0.050	3.10	41	nd	78	32	47	76
Data from [2000NGU/PAL]							
0.100		(39 ± 4)	(64 ± 6)	(78 ± 8)	(31.7 ± 1.6)		
nd : not detected							

In conclusion the large amount of data accumulated in this work does not lead us to change our conclusions about the species 8:3²⁻, 10:3⁴⁻ and 11:3⁵⁻ for data selection. The existence of the species 5:1³⁻ in measurable amount in very alkaline solutions is doubtful because data from different methods used by different authors are not concordant. This review does not accept the evidence presented for the formation of a species 6:1⁴⁻. As much of these data are not published they are given here to show that the behaviour of U(IV) in high basic media is still being studied.

* Method 1 - The Raman broad band of each given spectra is decomposed into individual bands based only at the inflexion points without any other constraint. A correct decomposition corresponds to a smallest residual peak.

Method 2 - The Raman broad band of all the spectra is decomposed giving to each species a Gaussian spectrum with a constant Full Width at Half Maximum (FWHM) centred at a given v_1 value (convergent self-consistent method). The procedure is iterative. It starts with the peak with the smallest FWHM and ends when the number of species with a given FWHM added by the procedure gives for the residual peak the background of the measurements.

[2002RAI/FEL]

The solubility of becquerelite, $Ca(UO_2)_6O_4(OH)_6 \cdot 8H_2O$ (or $CaU_6O_{19} \cdot 11H_2O$) is addressed in this review together with a discussion of previous studies [94SAN/GRA] and [97CAS/BRU], which covered a much more restricted pH and Ca-concentration range. The present study concerns the solubility of synthetic becquerelite in 0.02, 0.1 and 0.5 M $CaCl_2$ in the $-\log_{10}[H^+]$ range, 4.4 to 11.4 at $(22 \pm 2)°C$ under a N_2 atmosphere. The pH is adjusted with HNO_3 or $(C_2H_5)_4NOH$ and measured with a glass electrode calibrated by the Gran method. Tests solutions are filtered (0.0018 micrometer pore size), the U content is measured by ICPMS and the equilibrated solid phases are identified by X-ray diffraction, EXAFS (k space up to 13 Å^{-1}) and chemical analysis. The time for equilibration varied up to 70 days, but after four days a steady state is attained at all solution concentrations. The main problem noticed by the authors is the characterisation of the solid phases. For $-\log_{10}[H^+] < 8$, XRD and chemical analysis show that becquerelite is the dominant phase, essentially pure, but with a small amount of another phase. Above this value other phases with a lower U/Ca ratio appear. Variations of U concentration with both $-\log_{10}[H^+]$ and $\log_{10}[Ca^{2+}]$ confirm that equilibrium phase cannot be other uranates, such as $CaUO_4$, $Ca_2UO_5(H_2O)_{1.3-1.7}$, Ca_3UO_6, or CaU_2O_7. The solubility data are modelled in several steps using initially the data selected by [92GRE/FUG] for aqueous species and the Pitzer model for the ion-interaction parameters. The best fit up to $-\log_{10}[H^+] = 9$ is obtained with $\log_{10} {}^*K^°_{s,0} = (41.4 \pm 0.2)$ for the solubility product of becquerelite:

$$Ca(UO_2)_6O_4(OH)_6 \cdot 8H_2O(cr) + 14\,H^+ \rightleftharpoons Ca^{2+} + 6\,UO_2^{2+} + 18\,H_2O(l).$$

The large uncertainty in the solubility data in the concentration range where the species, $UO_2(OH)_2(aq)$, is expected to be dominant ($-\log_{10}[H^+]$ from six to around nine) explains the large uncertainty in the derived value of $\log_{10} {}^*\beta^°_{2,1} = -(11.3 \pm 1.0)$ for the formation of $UO_2(OH)_2(aq)$. This value is closer to those of [92SIL] and [91CHO/MAT] (-11.5 and -12.05, respectively). Note that no value was selected by [92GRE/FUG], only an upper limit $\log_{10} {}^*\beta^°_{2,1} < -10.3$ was given. The value of $\log_{10} {}^*\beta^°_{3,1} < -21.5$ is also derived for the formation of $UO_2(OH)_3^-$, but this species is not important when $-\log_{10}[H^+]$ is less than nine. For all the other aqueous U(VI) species ($n{:}m$ = 1:1, 4:1, 2:2, 4:3, 5:3 and 7:3) the values of the formation constants selected by [92GRE/FUG] are used in the modelling. The equilibrium constants obtained in this study are able to describe the solubility of U(VI) measured by Sandino et al. in 1 M $CaCl_2$ [94SAN/GRA] up to $-\log_{10}[H^+] = 6$. Sandino et al. report an average value of $\log_{10} {}^*K^°_{s,0} = (39.5 \pm 1.0)$ (see this review). Vochten et al. [90VOC/HAV] give, according to this paper, for becquerelite, $\log_{10} {}^*K_{s,0}^\circ = 43.2$. In fact the value given in [90VOC/HAV] is $\log_{10} {}^*K_{s,0} = (43.6 \pm 0.3)$, (see this review). Casas et al. [97CAS/BRU] calculated from the data of Vochten et al. a value of $\log_{10} {}^*K_{s,0}^\circ = 41.4$ using the aqueous thermodynamic model of Grenthe et al. [92GRE/FUG], but the value of $\log_{10} {}^*K_{s,0}^\circ$ they reported for natural becquerelite, (29 ± 1), is very far from those of [94SAN/GRA] [90VOC/HAV] and [2002RAI/FEL] for synthetic becquerelite. Clearly the study of Rai et al. is more detailed than the others. This is the only experimental study available where a careful phase analysis and a subsequent analysis of the experimental solubility data over a large concentration range have been carried out. However, the uncertainty of ± 0.2 on $\log_{10} {}^*K_{s,0}^\circ$ for becquerelite given in [2002RAI/FEL] appears to be too small, as it would correspond to an accuracy of ± 0.03 in $\log_{10}[U(VI)]_{tot}$ for the experimental data or calculated curves. In regard to Figures 1 to 4 in [2002RAI/FEL], an uncertainty of ± 0.2 in $\log_{10}[U(VI)]_{tot}$ is evaluated by this review and hence an uncertainty of ± 1.2 in $\log_{10} {}^*K_{s,0}^\circ$. This review therefore retains the value, $\log_{10} {}^*K_{s,0}^\circ$ (becquerelite) = (41.4 ± 1.2), for the selection of a solubility product for synthetic becquerelite, and notes that the value of $\log_{10} {}^*\beta_{2,1}^\circ = -(11.3 \pm 1.0)$ or $\log_{10} \beta_{2,1}^\circ = (16.7 \pm 1.0)$ is consistent with the value selected in this review from other data: $\log_{10} {}^*\beta_{2,1}^\circ = -(12.15 \pm 0.07)$, or $\log_{10} \beta_{2,1}^\circ = (15.85 \pm 0.07)$. These values just agree within their uncertainty limits. This is not the case for $\log_{10} {}^*\beta_{3,1}^\circ < -21.5$ or $\log_{10} \beta_{3,1}^\circ < 20.5$ compared to the selected values, $\log_{10} {}^*\beta_{3,1}^\circ = -(20.25 \pm 0.42)$, or $\log_{10} \beta_{3,1}^\circ = (21.75 \pm 0.42)$, which are inconsistent. This review agrees with the author's remark that above pH 8, the pH-independent solubility of U(VI) does not refer to becquerelite but to a calcium diuranate. The decrease of the U/Ca ratio from 5.5 to 2 could be due to the reaction:

$$1/2\ CaU_2O_7 \cdot x\ H_2O \rightleftharpoons 1/2\ Ca^{2+} + UO_2(OH)_3^- + (x-3)/2\ H_2O(l)$$

which could have affected the calculated $\log_{10} {}^*\beta_{2,1}^\circ$ value.

[2002RAI/GOR]

This paper reports data on the solubility of $PuO_2(am)$ in presence of $5.2 \cdot 10^{-4}$ M hydroquinone or 10^{-3} M $FeCl_2$ in the pH range 0.5 to 11 (ionic strength less than 0.025 M), at (23 ± 2) °C and under Ar atmosphere, with the identification of the oxidation state of Pu. The pH is adjusted with HCl or NaOH and measured with a glass electrode calibrated against standard buffers. The Eh is measured using a Pt electrode calibrated against standard redox buffers. Tests solutions are filtered (0.0036 µ pore size), the Pu content was measured by liquid scintillation (detection limit about $10^{-9.5}$ M) and the Pu oxidation state quantified by UV-Vis-IR spectroscopy and solvent extraction at the lowest Pu concentrations. All equilibria reach a steady state within four days, but measurements are performed up to 90 days. In both systems the measured potential, and pe values calculated from Eh values, follow the variations expected from thermodynamic data on Fe(II)/Fe(III) and hydroquinone systems with measured pH. It is concluded by the authors, from experimental data and thermodynamic arguments, that the dominant oxidation state of Pu in solution is

Pu(III) and that $PuO_2(am)$ is the solid equilibrium phase in the systems, at least up to pH 7 to 8. For higher pH, when the detection limit of Pu is reached, the data are not considered for modelling. On this basis the dominant reaction controlling the solubility is:

$$PuO_2(am) + 4 H^+ + e^- \rightleftharpoons Pu^{3+} + 2 H_2O(l) \qquad (A.143)$$

for which an equilibrium constant is calculated as $\log_{10} K° = (15.5 \pm 0.7)$ from pH, pe and Pu^{3+} concentrations using the Pitzer model and the values of $\log_{10} \beta_1°$ and $\log_{10} \beta_2°$ for Pu^{3+} taken from [99FEL/RAI]. In a similar work, Fujiwara et al. [2001FUJ/YAM] give $\log_{10} K° = (15.56 \pm 0.39)$ for reaction (A.143) (see this review). From the value $\log_{10} K° = (15.5 \pm 0.7)$, and using equilibrium constants, $\log_{10} \beta_1° = -0.78$ for Pu^{4+} and $\log_{10} K° = (17.70 \pm 0.67)$ for, $Pu^{4+} + e^- \rightleftharpoons Pu^{3+}$, selected by [2001LEM/FUG], Rai et al [2002RAI/GOR] report for the solubility product of $PuO_2(am)$, $\log_{10} K_{s,0}° = -(58.20 \pm 0.97)$ while the selected value by [2001LEM/FUG] is $-(58 \pm 1)$. These authors note that their values are close to those that they calculate from the data reported in [84RAI] and the values for the redox couple, Pu(V)/Pu(IV) and Pu(VI)/Pu(IV), selected by [2001LEM/FUG]. These values are $\log_{10} K_{s,0}° = -(57.99 \pm 0.77)$ and $-(58.33 \pm 0.85)$, respectively. The value given by Fujiwara et al. [2001FUJ/YAM] from similar measurements and calculations is $\log_{10} K_{s,0}° = -(58.1 \pm 0.4)$. Other values considered in this review for the solubility product of $PuO_2(am)$ are $-(58.7 \pm 0.9)$ [99KNO/NEC] and $-(58.3 \pm 0.5)$ [98CAP/VIT]. This work of Rai et al. is a precise and well-described experimental study and the equilibrium constants confirm previous results. The authors' value of $\log_{10} K_{s,0}° (PuO_2(am)) = -(58.2 \pm 0.97)$ is included by this review in the assessment of the solubility constant for $PuO_2(am, hydr.)$.

ten et al. [90VOC/HAV] give, according to this paper, for becquerelite, $\log_{10} {}^*K^\circ_{s,0} = 43.2$. In fact the value given in [90VOC/HAV] is $\log_{10} {}^*K_{s,0} = (43.6 \pm 0.3)$, (see this review). Casas et al. [97CAS/BRU] calculated from the data of Vochten et al. a value of $\log_{10} {}^*K^\circ_{s,0} = 41.4$ using the aqueous thermodynamic model of Grenthe et al. [92GRE/FUG], but the value of $\log_{10} {}^*K^\circ_{s,0}$ they reported for natural becquerelite, (29 ± 1), is very far from those of [94SAN/GRA] [90VOC/HAV] and [2002RAI/FEL] for synthetic becquerelite. Clearly the study of Rai et al. is more detailed than the others. This is the only experimental study available where a careful phase analysis and a subsequent analysis of the experimental solubility data over a large concentration range have been carried out. However, the uncertainty of ± 0.2 on $\log_{10} {}^*K^\circ_{s,0}$ for becquerelite given in [2002RAI/FEL] appears to be too small, as it would correspond to an accuracy of ± 0.03 in $\log_{10}[U(VI)]_{tot}$ for the experimental data or calculated curves. In regard to Figures 1 to 4 in [2002RAI/FEL], an uncertainty of ± 0.2 in $\log_{10}[U(VI)]_{tot}$ is evaluated by this review and hence an uncertainty of ± 1.2 in $\log_{10} {}^*K^\circ_{s,0}$. This review therefore retains the value, $\log_{10} {}^*K^\circ_{s,0}$ (becquerelite) $= (41.4 \pm 1.2)$, for the selection of a solubility product for synthetic becquerelite, and notes that the value of $\log_{10} {}^*\beta^\circ_{2,1} = -(11.3 \pm 1.0)$ or $\log_{10} \beta^\circ_{2,1} = (16.7 \pm 1.0)$ is consistent with the value selected in this review from other data: $\log_{10} {}^*\beta^\circ_{2,1} = -(12.15 \pm 0.07)$, or $\log_{10} \beta^\circ_{2,1} = (15.85 \pm 0.07)$. These values just agree within their uncertainty limits. This is not the case for $\log_{10} {}^*\beta^\circ_{3,1} < -21.5$ or $\log_{10} \beta^\circ_{3,1} < 20.5$ compared to the selected values, $\log_{10} {}^*\beta^\circ_{3,1} = -(20.25 \pm 0.42)$, or $\log_{10} \beta^\circ_{3,1} = (21.75 \pm 0.42)$, which are inconsistent. This review agrees with the author's remark that above pH 8, the pH-independent solubility of U(VI) does not refer to becquerelite but to a calcium diuranate. The decrease of the U/Ca ratio from 5.5 to 2 could be due to the reaction:

$$1/2\ CaU_2O_7 \cdot x\ H_2O \rightleftharpoons 1/2\ Ca^{2+} + UO_2(OH)_3^- + (x-3)/2\ H_2O(l)$$

which could have affected the calculated $\log_{10} {}^*\beta^\circ_{2,1}$ value.

[2002RAI/GOR]

This paper reports data on the solubility of $PuO_2(am)$ in presence of $5.2 \cdot 10^{-4}$ M hydroquinone or 10^{-3} M $FeCl_2$ in the pH range 0.5 to 11 (ionic strength less than 0.025 M), at (23 ± 2) °C and under Ar atmosphere, with the identification of the oxidation state of Pu. The pH is adjusted with HCl or NaOH and measured with a glass electrode calibrated against standard buffers. The Eh is measured using a Pt electrode calibrated against standard redox buffers. Tests solutions are filtered (0.0036 μ pore size), the Pu content was measured by liquid scintillation (detection limit about $10^{-9.5}$ M) and the Pu oxidation state quantified by UV-Vis-IR spectroscopy and solvent extraction at the lowest Pu concentrations. All equilibria reach a steady state within four days, but measurements are performed up to 90 days. In both systems the measured potential, and pe values calculated from Eh values, follow the variations expected from thermodynamic data on Fe(II)/Fe(III) and hydroquinone systems with measured pH. It is concluded by the authors, from experimental data and thermodynamic arguments, that the dominant oxidation state of Pu in solution is

Pu(III) and that PuO_2(am) is the solid equilibrium phase in the systems, at least up to pH 7 to 8. For higher pH, when the detection limit of Pu is reached, the data are not considered for modelling. On this basis the dominant reaction controlling the solubility is:

$$PuO_2(am) + 4 H^+ + e^- \rightleftharpoons Pu^{3+} + 2 H_2O(l) \qquad (A.143)$$

for which an equilibrium constant is calculated as $\log_{10} K° = (15.5 \pm 0.7)$ from pH, pe and Pu^{3+} concentrations using the Pitzer model and the values of $\log_{10} \beta_1°$ and $\log_{10} \beta_2°$ for Pu^{3+} taken from [99FEL/RAI]. In a similar work, Fujiwara et al. [2001FUJ/YAM] give $\log_{10} K° = (15.56 \pm 0.39)$ for reaction (A.143) (see this review). From the value $\log_{10} K° = (15.5 \pm 0.7)$, and using equilibrium constants, $\log_{10} \beta_1° = -0.78$ for Pu^{4+} and $\log_{10} K° = (17.70 \pm 0.67)$ for, $Pu^{4+} + e^- \rightleftharpoons Pu^{3+}$, selected by [2001LEM/FUG], Rai et al [2002RAI/GOR] report for the solubility product of PuO_2(am), $\log_{10} K_{s,0}° = -(58.20 \pm 0.97)$ while the selected value by [2001LEM/FUG] is $-(58 \pm 1)$. These authors note that their values are close to those that they calculate from the data reported in [84RAI] and the values for the redox couple, Pu(V)/Pu(IV) and Pu(VI)/Pu(IV), selected by [2001LEM/FUG]. These values are $\log_{10} K_{s,0}° = -(57.99 \pm 0.77)$ and $-(58.33 \pm 0.85)$, respectively. The value given by Fujiwara et al. [2001FUJ/YAM] from similar measurements and calculations is $\log_{10} K_{s,0}° = -(58.1 \pm 0.4)$. Other values considered in this review for the solubility product of PuO_2(am) are $-(58.7 \pm 0.9)$ [99KNO/NEC] and $-(58.3 \pm 0.5)$ [98CAP/VIT]. This work of Rai et al. is a precise and well-described experimental study and the equilibrium constants confirm previous results. The authors' value of $\log_{10} K_{s,0}° (PuO_2(am)) = -(58.2 \pm 0.97)$ is included by this review in the assessment of the solubility constant for PuO_2(am, hydr.).

Appendix B
Ionic strength corrections[1]

Thermodynamic data always refer to a selected standard state. The definition given by IUPAC [82LAF] is adopted in this review as outlined in Section 2.3.1. According to this definition, the standard state for a solute B in a solution is a hypothetical solution, at the standard state pressure, in which $m_B = m° = 1$ mol·kg^{-1}, and in which the activity coefficient γ_B is unity. However, for many reactions, measurements cannot be made accurately (or at all) in dilute solutions from which the necessary extrapolation to the standard state would be simple. This is invariably the case for reactions involving ions of high charge. Precise thermodynamic information for these systems can only be obtained in the presence of an inert electrolyte of sufficiently high concentration that ensures activity factors are reasonably constant throughout the measurements. This appendix describes and illustrates the method used in this review for the extrapolation of experimental equilibrium data to zero ionic strength.

The activity factors of all the species participating in reactions in high ionic strength media must be estimated in order to reduce the thermodynamic data obtained from the experiment to the state $I = 0$. Two alternative methods can be used to describe the ionic medium dependence of equilibrium constants:

- One method takes into account the individual characteristics of the ionic media by using a medium dependent expression for the activity coefficients of the species involved in the equilibrium reactions. The medium dependence is described by

---

[1] This Appendix essentially contains the text of the TDB-2 Guideline written by Grenthe and Wanner [00GRE/WAN], earlier versions of which have been printed in the previous NEA TDB reviews [92GRE/FUG], [95SIL/BID], [99RAR/RAN] and [2001LEM/FUG]. The equations presented here are an essential part of the review procedure and are required to use the selected thermodynamic values. The contents of Tables B.4 and B.5 have been revised.

virial or ion interaction coefficients as used in the Pitzer equations [73PIT] and the specific ion interaction theory.

- The other method uses an extended Debye–Hückel expression in which the activity coefficients of reactants and products depend only on the ionic charge and ionic strength, but it accounts for the medium specific properties by introducing ion pairing between the medium ions and the species involved in the equilibrium reactions. Earlier, this approach has been used extensively in marine chemistry, cf. Refs. [79JOH/PYT], [79MIL], [79PYT], [79WHI2].

The activity factor estimates are thus based on the use of Debye-Hückel type equations. The "extended" Debye-Hückel equations are either in the form of specific ion interaction methods or the Davies equation [62DAV]. However, the Davies equation should in general not be used at ionic strengths larger than 0.1 mol · kg^{-1}. The method preferred in the NEA Thermochemical Data Base review is a medium-dependent expression for the activity coefficients, which is the specific ion interaction theory in the form of the Brønsted-Guggenheim-Scatchard approach. Other forms of specific ion interaction methods (the Pitzer and Brewer "B-method" [61LEW/RAN] and the Pitzer virial coefficient method [79PIT]) are described in the NEA Guidelines for the extrapolation to zero ionic strength [2000GRE/WAN].

The specific ion interaction methods are reliable for intercomparison of experimental data in a given concentration range. In many cases this includes data at rather low ionic strengths, I = 0.01 to 0.1 M, cf. Figure B-1, while in other cases, notably for cations of high charge ($\geq$ + 4 and $\leq$ − 4), the lowest available ionic strength is often 0.2 M or higher, see for example Figures V.12 and V.13 in [92GRE/FUG]. It is reasonable to assume that the extrapolated equilibrium constants at I = 0 are more precise in the former than in the latter cases. The extrapolation error is composed of two parts, one due to experimental errors, and the other due to model errors. The model errors seem to be rather small for many systems, less than 0.1 units in $\log_{10} K°$. For reactions involving ions of high charge, which may be extensively hydrolyzed, one cannot perform experiments at low ionic strength. Hence, it is impossible to estimate the extrapolation error. This is true for all methods used to estimate activity corrections. Systematic model errors of this type are not included in the uncertainties assigned to the selected data in this review.

It should be emphasised that the specific ion interaction model is *approximate*. Modifying it, for example by introducing the equations suggested by Ciavatta ([90CIA], Eqs. (8–10), cf. Section B.1.4), would result in slightly different ion interaction coefficients and equilibrium constants. Both methods provide an internally consistent set of values. However, their absolute values may differ somewhat. Grenthe et al. [92GRE/FUG] e te that these differences in general are less than 0.2 units in $\log_{10} K°$, i.e., approximately J·mol^{-1} in derived $\Delta_f G_m°$ values.

1 The specific ion interaction equations

1.1 Background

e Debye-Hückel term, which is the dominant term in the expression for the activity coefients in dilute solution, accounts for electrostatic, non-specific long-range interactions. higher concentrations, short range, non-electrostatic interactions have to be taken into :ount. This is usually done by adding ionic strength dependent terms to the Debye-ckel expression. This method was first outlined by Brønsted [22BRO], [22BRO2] and borated by Scatchard [36SCA] and Guggenheim [66GUG]. Biedermann [75BIE] high-hted its practical value, especially for the estimation of ionic medium effects on equilibm constants. The two basic assumptions in the specific ion interaction theory are deibed below.

- **Assumption 1:** The activity coefficient γ_j of an ion j of charge z_j in the solution of ic strength I_m may be described by Eq. (B.1):

$$\log_{10} \gamma_j = -z_j^2 D + \sum_k \varepsilon(j,k,I_m) m_k \quad (B.1)$$

s the Debye-Hückel term:

$$D = \frac{A\sqrt{I_m}}{1+Ba_j\sqrt{I_m}} \quad (B.2)$$

ere I_m is the molal ionic strength:

$$I_m = \frac{1}{2}\sum_i m_i z_i^2$$

nd B are constants. which are temperature and pressure dependent, and a_j is an ion size ameter ("distance of closest approach") for the hydrated ion j. The Debye-Hückel limit-; slope, A, has a value of (0.509 ± 0.001) kg^{½}·mol$^{-½}$ at 25°C and 1 bar, (cf. Section .2). The term Ba_j in the denominator of the Debye-Hückel term has been assigned a ue of $Ba_j = 1.5$ kg^{½}·mol$^{-½}$ at 25°C and 1 bar, as proposed by Scatchard [76SCA] and :epted by Ciavatta [80CIA]. This value has been found to minimise, for several species, ionic strength dependence of $\varepsilon(j,k,I_m)$ between $I_m = 0.5$ m and $I_m = 3.5$ m. It should be ntioned that some authors have proposed different values for Ba_j ranging from $Ba_j = 1.0$ GUG] to $Ba_j = 1.6$ [62VAS]. However, the parameter Ba_j is empirical and as such is related to the value of $\varepsilon(j,k,I_m)$. Hence, this variety of values for Ba_j does not repret an uncertainty range, but rather indicates that several different sets of Ba_j and $j,k,I_m)$ may describe equally well the experimental mean activity coefficients of a given electrolyte. The ion interaction coefficients at 25°C listed in Table B-4, Table B-5 a Table B-6 have thus to be used with $Ba_j = 1.5$ kg$^{1/2}$·mol$^{-1/2}$.

The summation in Eq. (B.1) extends over all ions k present in solution. Their m lality is denoted by m_k, and the specific ion interaction parameters, $\varepsilon(j,k,I_m)$, in gene depend only slightly on the ionic strength. The concentrations of the ions of the ionic m dium are often very much larger than those of the reacting species. Hence, the ionic m dium ions will make the main contribution to the value of $\log_{10}\gamma_j$ for the reacting ions. Th fact often makes it possible to simplify the summation $\sum_k \varepsilon(j,k,I_m)m_k$, so that only i interaction coefficients between the participating ionic species and the ionic medium io are included, as shown in Eqs. (B.4) to (B.8).

- **Assumption 2:** The ion interaction coefficients, $\varepsilon(j,k,I_m)$ are zero for ions of t same charge sign and for uncharged species. The rationale behind this is that ε, which d scribes specific short-range interactions, must be small for ions of the same charge sir they are usually far from one another due to electrostatic repulsion. This holds to a less extent also for uncharged species.

Eq. (B.1) will allow fairly accurate estimates of the activity coefficients in m tures of electrolytes if the ion interaction coefficients are known. Ion interaction coef cients for simple ions can be obtained from tabulated data of mean activity coefficients strong electrolytes or from the corresponding osmotic coefficients. Ion interaction coef cients for complexes can either be estimated from the charge and size of the ion or det mined experimentally from the variation of the equilibrium constant with the ionic streng

Ion interaction coefficients are not strictly constant but may vary slightly with t ionic strength. The extent of this variation depends on the charge type and is small for 1 1:2 and 2:1 electrolytes for molalities less than 3.5 m. The concentration dependence of t ion interaction coefficients can thus often be neglected. This point was emphasised Guggenheim [66GUG], who has presented a considerable amount of experimental mater supporting this approach. The concentration dependence is larger for electrolytes of high charge. In order to reproduce accurately their activity coefficient data, concentration d pendent ion interaction coefficients have to be used, cf. Lewis et al. [61LEW/RAN], Ba and Mesmer [76BAE/MES], or Ciavatta [80CIA]. By using a more elaborate virial expa sion, Pitzer and co-workers [73PIT], [73PIT/MAY], [74PIT/KIM], [74PIT/MA [75PIT], [76PIT/SIL], [78PIT/PET], [79PIT] have managed to describe measured activ coefficients of a large number of electrolytes with high precision over a large concentrat range. Pitzer's model generally contains three parameters as compared to one in the speci ion interaction theory. The use of the theory requires knowledge of all these paramete The derivation of Pitzer coefficients for many complexes, such as those of the actini

B.1 The specific ion interaction equations

would require a very large amount of additional experimental work, since few data of this type are currently available.

The way in which the activity coefficient corrections are performed in this review according to the specific ion interaction theory is illustrated below for a general case of a complex formation reaction. Charges are omitted for brevity.

$$m\text{M} + q\text{L} + n\text{H}_2\text{O}(l) \rightleftharpoons \text{M}_m\text{L}_q(\text{OH})_n + n\text{H}^+$$

The formation constant of $\text{M}_m\text{L}_q(\text{OH})_n$, $^*\beta_{q,n,m}$, determined in an ionic medium of 1:1 salt NX of the ionic strength I_m, is related to the corresponding value at zero ionic strength, $^*\beta^\circ_{q,n,m}$ by Eq.(B.3).

$$\log_{10} {}^*\beta_{q,n,m} = \log_{10} {}^*\beta^\circ_{q,n,m} + m\log_{10}\gamma_\text{M} + q\log_{10}\gamma_\text{L} + n\log_{10}a_{\text{H}_2\text{O}} \quad (\text{B.3})$$
$$- \log_{10}\gamma_{q,n,m} - n\log_{10}\gamma_{\text{H}^+}$$

The subscript (q,n,m) denotes the complex ion, $\text{M}_m\text{L}_q(\text{OH})_n$. If the concentrations of N and X are much greater than the concentrations of M, L, $\text{M}_m\text{L}_q(\text{OH})_n$ and H^+, only the molalities m_N and m_X have to be taken into account for the calculation of the term, $\varepsilon(j,k,I_m)m_k$ in Eq. (B.1). For example, for the activity coefficient of the metal cation γ_M, Eq. (B.4) is obtained at 25°C and 1 bar.

$$\log_{10}\gamma_\text{M} = \frac{-z_\text{M}^2\, 0.509\sqrt{I_m}}{1 + 1.5\sqrt{I_m}} + \varepsilon(\text{M},\text{X},I_m)m_\text{X} \quad (\text{B.4})$$

Under these conditions, $I_m \approx m_\text{X} = m_\text{N}$. Substituting the $\log_{10}\gamma_j$ values in Eq. (B.3) with the corresponding forms of Eq. (B.4) and rearranging leads to:

$$\log_{10} {}^*\beta_{q,n,m} - \Delta z^2 D - n\log_{10}a_{\text{H}_2\text{O}} = \log_{10} {}^*\beta^\circ_{q,n,m} - \Delta\varepsilon\, I_m \quad (\text{B.5})$$

where, at 25°C and 1 bar:

$$\Delta z^2 = (mz_\text{M} - qz_\text{L} - n)^2 + n - mz_\text{M}^2 - qz_\text{L}^2 \quad (\text{B.6})$$

$$D = \frac{0.509\sqrt{I_m}}{1 + 1.5\sqrt{I_m}} \quad (\text{B.7})$$

$$\Delta\varepsilon = \varepsilon(q,n,m,\text{N or X}) + n\varepsilon(\text{H},\text{X}) - q\varepsilon(\text{N},\text{L}) - m\varepsilon(\text{M},\text{X}) \quad (\text{B.8})$$

Here $(mz_\text{M} - qz_\text{L} - n)$, z_M and z_L are the charges of the complex, $\text{M}_m\text{L}_q(\text{OH})_n$, the metal ion M and the ligand L, respectively.

Equilibria involving $\text{H}_2\text{O}(l)$ as a reactant or product require a correction for the activity of water, $a_{\text{H}_2\text{O}}$. The activity of water in an electrolyte mixture can be calculated as:

$$\log_{10} a_{H_2O} = \frac{-\phi \sum_k m_k}{\ln(10) \cdot 55.51} \quad \text{(B.9)}$$

where ϕ is the osmotic coefficient of the mixture and the summation extends over all solute species k with molality m_k present in the solution. In the presence of an ionic medium NX as the dominant species, Eq. (B.9) can be simplified by neglecting the contributions of all minor species, *i.e.*, the reacting ions. Hence, for a 1:1 electrolyte of ionic strength $I_m \approx m_{NX}$, Eq. (B.9) becomes:

$$\log_{10} a_{H_2O} = \frac{-2 m_{NX} \phi}{\ln(10) \times 55.51} \quad \text{(B.10)}$$

Alternatively, water activities can be taken from Table B-1. These have been calculated for the most common ionic media at various concentrations applying Pitzer's ion interaction model and the interaction parameters given in [91PIT]. Data in *italics* have been calculated for concentrations beyond the validity of the parameter set applied. These data are therefore extrapolations and should be used with care.

Table B-1: Water activities a_{H_2O} for the most common ionic media at various concentrations applying Pitzer's ion interaction approach and the interaction parameters given [91PIT]. Data in *italics* have been calculated for concentrations beyond the validity of the parameter set applied. These data are therefore extrapolations and should be used with care.

	Water activities a_{H_2O}							
c (M)	HClO$_4$	NaClO$_4$	LiClO$_4$	NH$_4$ClO$_4$	Ba(ClO$_4$)$_2$	HCl	NaCl	LiCl
0.10	0.9966	0.9966	0.9966	0.9967	0.9953	0.9966	0.9966	0.9966
0.25	0.9914	0.9917	0.9912	0.9920	0.9879	0.9914	0.9917	0.9915
0.50	0.9821	0.9833	0.9817	0.9844	0.9740	0.9823	0.9833	0.9826
0.75	0.9720	0.9747	0.9713	0.9769	0.9576	0.9726	0.9748	0.9731
1.00	0.9609	0.9660	0.9602	0.9694	0.9387	0.9620	0.9661	0.9631
1.50	0.9357	0.9476	0.9341	0.9542	0.8929	0.9386	0.9479	0.9412
2.00	0.9056	0.9279	0.9037		0.8383	0.9115	0.9284	0.9167
3.00	0.8285	0.8840	0.8280		*0.7226*	0.8459	0.8850	0.8589
4.00	0.7260	0.8331	*0.7309*			0.7643	0.8352	0.7991
5.00	*0.5982*	0.7744				0.6677	0.7782	0.7079
6.00	*0.4513*	0.7075				0.5592		*0.6169*

(continued on next page)

B.1 The specific ion interaction equations

Table B-1 (Continued)

c (M)	KCl	NH$_4$Cl	MgCl$_2$	CaCl$_2$	NaBr	HNO$_3$	NaNO$_3$	LiNO$_3$
0.10	0.9966	0.9966	0.9953	0.9954	0.9966	0.9966	0.9967	0.9966
0.25	0.9918	0.9918	0.9880	0.9882	0.9916	0.9915	0.9919	0.9915
0.50	0.9836	0.9836	0.9744	0.9753	0.9830	0.9827	0.9841	0.9827
0.75	0.9754	0.9753	0.9585	0.9605	0.9742	0.9736	0.9764	0.9733
1.00	0.9671	0.9669	0.9399	0.9436	0.9650	0.9641	0.9688	0.9635
1.50	0.9500	0.9494	0.8939	0.9024	0.9455	0.9439	0.9536	0.9422
2.00	0.9320	0.9311	0.8358	0.8507	0.9241	0.9221	0.9385	0.9188
3.00	0.8933	0.8918	0.6866	0.7168	0.8753	0.8737	0.9079	0.8657
4.00	0.8503	0.8491	0.5083	0.5511	*0.8174*	0.8196	0.8766	0.8052
5.00		*0.8037*		*0.3738*	*0.7499*	0.7612	0.8446	0.7390
6.00					*0.6728*	*0.7006*	*0.8120*	*0.6696*

c (M)	NH$_4$NO$_3$	Na$_2$SO$_4$	(NH$_4$)$_2$SO$_4$	Na$_2$CO$_3$	K$_2$CO$_3$	NaSCN
0.10	0.9967	0.9957	0.9958	0.9956	0.9955	0.9966
0.25	0.9920	0.9900	0.9902	0.9896	0.9892	0.9915
0.50	0.9843	0.9813	0.9814	0.9805	0.9789	0.9828
0.75	0.9768	0.9732	0.9728	0.9720	0.9683	0.9736
1.00	0.9694	0.9653	0.9640	0.9637	0.9570	0.9641
1.50	0.9548	0.9491	0.9455	0.9467	0.9316	0.9438
2.00	0.9403		0.9247	0.9283	0.9014	0.9215
3.00	0.9115		0.8735		0.8235	0.8708
4.00	0.8829		0.8050		0.7195	*0.8115*
5.00	*0.8545*				*0.5887*	*0.7436*
6.00	*0.8266*					*0.6685*

Values of osmotic coefficients for single electrolytes have been compiled by var ous authors, *e.g.*, Robinson and Stokes [59ROB/STO]. The activity of water can also calculated from the known activity coefficients of the dissolved species. In the presence an ionic medium, $N_{\nu_+} X_{\nu_-}$, of a concentration much larger than those of the reacting ior the osmotic coefficient can be calculated according to Eq. (B.11) (*cf.* Eqs. (23–39), (23–4 and (A4–2) in [61LEW/RAN]).

$$1 - \phi = \frac{A \ln(10) |z_+ z_-|}{I_m (B a_j)^3} \left[1 + B a_j \sqrt{I_m} - 2\ln(1 + B a_j \sqrt{I_m}) - \frac{1}{1 + B a_j \sqrt{I_m}} \right]$$
$$- \ln(10)\, \varepsilon(N,X)\, m_{NX} \left(\frac{\nu_+ \nu_-}{\nu_+ + \nu_-} \right)$$
(B.1

where ν_+ and ν_- are the number of cations and anions in the salt formula ($\nu_+ z_+ = \nu_- z$ and in this case:

$$I_m = \frac{1}{2} |z_+ z_-| m_{NX} (\nu_+ + \nu_-)$$

The activity of water is obtained by inserting Eq. (B.11) into Eq. (B.10). It shou be mentioned that in mixed electrolytes with several components at high concentrations, might be necessary to use Pitzer's equation to calculate the activity of water. On the oth hand, a_{H_2O} is nearly constant in most experimental studies of equilibria in dilute aqueo solutions, where an ionic medium is used in large excess with respect to the reactants. T medium electrolyte thus determines the osmotic coefficient of the solvent.

In natural waters the situation is similar; the ionic strength of most surface wate is so low that the activity of $H_2O(l)$ can be set equal to unity. A correction may be necessa in the case of seawater, where a sufficiently good approximation for the osmotic coefficie may be obtained by considering NaCl as the dominant electrolyte.

In more complex solutions of high ionic strengths with more than one electroly at significant concentrations, *e.g.*, (Na^+, Mg^{2+}, Ca^{2+}) (Cl^-, SO_4^{2-}), Pitzer's equation ([2000GRE/WAN]) may be used to estimate the osmotic coefficient; the necessary intera tion coefficients are known for most systems of geochemical interest.

Note that in all ion interaction approaches, the equation for the mean activity c effi-cients can be split up to give equations for conventional single ion activity coefficie in mixtures, *e.g.*, Eq. (B.1). The latter are strictly valid only when used in combinatic which yield electroneutrality. Thus, while estimating medium effects on standard potentia ombination of redox equilibria with, $H^+ + e^- \rightleftharpoons \frac{1}{2}H_2(g)$, is necessary (*cf.* Example).

.2 Ionic strength corrections at temperatures other than 298.15 K

ues of the Debye-Hückel parameters A and B in Eqs. (B.2) and (B.11) are listed in Table for a few temperatures at a pressure of 1 bar below 100°C and at the steam saturated ssure for $t \geq 100°C$. The values in Table B-2 may be calculated from the static dielectric constant and the density of water as a function of temperature and pressure, and are also nd for example in Refs. [74HEL/KIR], [79BRA/PIT], [81HEL/KIR], [84ANA/ATK], ARC/WAN].

The term, Ba_j, in the denominator of the Debye–Hückel term, D, *cf.* Eq. (B.2), has n assigned in this review a value of 1.5 $kg^{½} \cdot mol^{-½}$ at 25°C and 1 bar, *cf.* Section B.1.1 temperatures and pressures other than the reference and standard state, the following sibilities exist:

- The value of Ba_j is calculated at each temperature assuming that ion sizes are independent of temperature and using the values of B listed in Table B-2.

- The value Ba_j is kept constant at 1.5 $kg^{½} \cdot mol^{-½}$. Due the variation of B with temperature, *cf.* Table B-2, this implies a temperature dependence for ion size parameters. Assuming for the ion size is in reality constant, then it is seen that this simplification introduces an error in D, which increases with temperature and ionic strength (this error is less than ± 0.01 at $t \leq 100°C$ and $I < 6$ m, and less than ± 0.006 at $t \leq 50°C$ and $I \leq 4$ m).

- The value of Ba_j is calculated at each temperature assuming a given temperature variation for a_j and using the values of B listed in Table B-2. For example, in the aqueous ionic model of Helgeson and co–workers ([88TAN/HEL], [88SHO/HEL], [89SHO/HEL], [89SHO/HEL2]) ionic sizes follow the relation: $a_j(T) = a_j(298.15 \text{ K}, 1 \text{ bar}) + |z_j|g(T,p)$ [90OEL/HEL], where $g(T, p)$ is a temperature and pressure function which is tabulated in [88TAN/HEL], [92SHO/OEL], and is approximately zero at temperatures below 175°C.

The values of $\varepsilon(j,k,I_m)$, obtained with the methods described in Section B.1.3 at peratures other than 25°C, will depend on the value adopted for Ba_j. As long as a consent approach is followed, values of $\varepsilon(j,k,I_m)$ absorb the choice of Ba_j, and for moder- temperature intervals (between 0 and 200°C) the choice $Ba_j = 1.5 \text{ kg}^{½} \cdot mol^{-½}$ is the plest one and is recommended by this review.

The variation of $\varepsilon(j,k,I_m)$ with temperature is discussed by Lewis et [61LEW/RAN], Millero [79MIL], Helgeson et al. [81HEL/KIR], [90OEL/HEL], Giffaut al. [93GIF/VIT2] and Grenthe and Plyasunov [97GRE/PLY]. The absolute values for reported ion interaction parameters differ in these studies due to the fact that the D bye–Hückel term used by these authors is not exactly the same. Nevertheless, common all these studies is the fact that values of $(\partial \varepsilon / \partial T)_p$ are usually ≤ 0.005 kg·mol^{-1}·K^{-1} temperatures below 200°C. Therefore, if values of $\varepsilon(j,k,I_m)$ obtained at 25°C are used the temperature range 0 to 50°C to perform ionic strength corrections, the error $\log_{10} \gamma_j / I_m$ will be ≤ 0.13. It is clear that in order to reduce the uncertainties in solubil calculations at $t \neq 25°C$, studies on the variation of $\varepsilon(j,k,I_m)$ values with temperatu should be undertaken.

Table B-2: Debye-Hückel constants as a function of temperature at a pressure of 1 bar low 100°C and at the steam saturated pressure for $t \geq 100°C$. The uncertainty in the A rameter is estimated by this review to be ± 0.001 at 25°C, and ± 0.006 at 300°C, while the B parameter the estimated uncertainty ranges from ± 0.0003 at 25°C to ± 0.001 300°C.

$t(°C)$	p(bar)	$A\,((\text{kg} \cdot \text{mol}^{-1})^{\frac{1}{2}})$	$B \times 10^{-10}\,(\text{kg}^{\frac{1}{2}} \cdot \text{mol}^{-\frac{1}{2}} \cdot \text{m}^{-1})$
0	1.00	0.491	0.3246
5	1.00	0.494	0.3254
10	1.00	0.498	0.3261
15	1.00	0.501	0.3268
20	1.00	0.505	0.3277
25	1.00	0.509	0.3284
30	1.00	0.513	0.3292
35	1.00	0.518	0.3300
40	1.00	0.525	0.3312
50	1.00	0.534	0.3326
75	1.00	0.564	0.3371
100	1.013	0.600	0.3422
125	2.32	0.642	0.3476
150	4.76	0.690	0.3533
175	8.92	0.746	0.3593
200	15.5	0.810	0.365
250	29.7	0.980	0.379
300	85.8	1.252	0.396

B.3 Estimation of ion interaction coefficients

B.3.1 Estimation from mean activity coefficient data

Example B.1:

The ion interaction coefficient $\varepsilon(H^+, Cl^-)$ can be obtained from published values of $\gamma_{\pm, HCl}$ versus m_{HCl}:

$$2\log_{10} \gamma_{\pm, HCl} = \log_{10} \gamma_{H^+} + \log_{10} \gamma_{Cl^-}$$
$$= -D + \varepsilon(H^+, Cl^-) m_{Cl^-} - D + \varepsilon(Cl^-, H^+) m_{H^+}$$
$$\log_{10} \gamma_{\pm, HCl} = -D + \varepsilon(H^+, Cl^-) m_{HCl}$$

By plotting $(\log_{10} \gamma_{\pm, HCl} + D)$ versus m_{HCl} a straight line with the slope $\varepsilon(H^+, Cl^-)$ is obtained. The degree of linearity should in itself indicate the range of validity of the specific ion interaction approach. Osmotic coefficient data can be treated in an analogous way.

B.3.2 Estimations based on experimental values of equilibrium constants at different ionic strengths

Example B.2:

Equilibrium constants are given in Table B-3 for the reaction:

$$UO_2^{2+} + Cl^- \rightleftharpoons UO_2Cl^+ \tag{B.12}$$

The following formula is deduced from Eq. (B.5) for the extrapolation to $I = 0$:

$$\log_{10} \beta_1 + 4D = \log_{10} \beta_1^\circ - \Delta\varepsilon I_m \tag{B.13}$$

The linear regression is done as described in Appendix C. The following results are obtained:

$$\log_{10} \beta_1^\circ = (0.170 \pm 0.021)$$
$$\Delta\varepsilon(B.12) = -(0.248 \pm 0.022)\,kg \cdot mol^{-1}.$$

The experimental data are depicted in Figure B-1, where the dashed area represents the uncertainty range that is obtained by using the results in $\log_{10} \beta_1^\circ$ and $\Delta\varepsilon$ and correcting back to $I \neq 0$.

Table B-3: The preparation of the experimental equilibrium constants for the extrapolat to $I = 0$ with the specific ion interaction method at 25°C and 1 bar, according to react (B.12). The linear regression of this set of data is shown in Figure B-1.

I_m	$\log_{10} \beta_1$ (exp) [a]	$\log_{10} \beta_{1,m}$ [b]	$\log_{10} \beta_{1,m} + 4D$
0.1	$-(0.17 \pm 0.10)$	-0.174	(0.264 ± 0.100)
0.2	$-(0.25 \pm 0.10)$	-0.254	(0.292 ± 0.100)
0.26	$-(0.35 \pm 0.04)$	-0.357	(0.230 ± 0.040)
0.31	$-(0.39 \pm 0.04)$	-0.397	(0.220 ± 0.040)
0.41	$-(0.41 \pm 0.04)$	-0.420	(0.246 ± 0.040)
0.51	$-(0.32 \pm 0.10)$	-0.331	(0.371 ± 0.100)
0.57	$-(0.42 \pm 0.04)$	-0.432	(0.288 ± 0.040)
0.67	$-(0.34 \pm 0.04)$	-0.354	(0.395 ± 0.040)
0.89	$-(0.42 \pm 0.04)$	-0.438	(0.357 ± 0.040)
1.05	$-(0.31 \pm 0.10)$	-0.331	(0.491 ± 0.100)
1.05	$-(0.277 \pm 0.260)$	-0.298	(0.525 ± 0.260)
1.61	$-(0.24 \pm 0.10)$	-0.272	(0.618 ± 0.100)
2.21	$-(0.15 \pm 0.10)$	-0.193	(0.744 ± 0.100)
2.21	$-(0.12 \pm 0.10)$	-0.163	(0.774 ± 0.100)
2.82	$-(0.06 \pm 0.10)$	-0.021	(0.860 ± 0.100)
3.5	(0.04 ± 0.10)	-0.021	(0.974 ± 0.100)

[a] Equilibrium constants for reaction (B.12) with assigned uncertainties, corrected to 25°C where necessary.

[b] Equilibrium constants corrected from molarity to molality units, as described in Section 2.2

B.1 The specific ion interaction equations

Figure B-1: Plot of $\log_{10} \beta_1 + 4D$ versus I_m for reaction (B.12), at 25°C and 1 bar. The straight line shows the result of the weighted linear regression, and the dotted lines represent the uncertainty range obtained by propagating the resulting uncertainties at $I = 0$ back to $I = 4$ m.

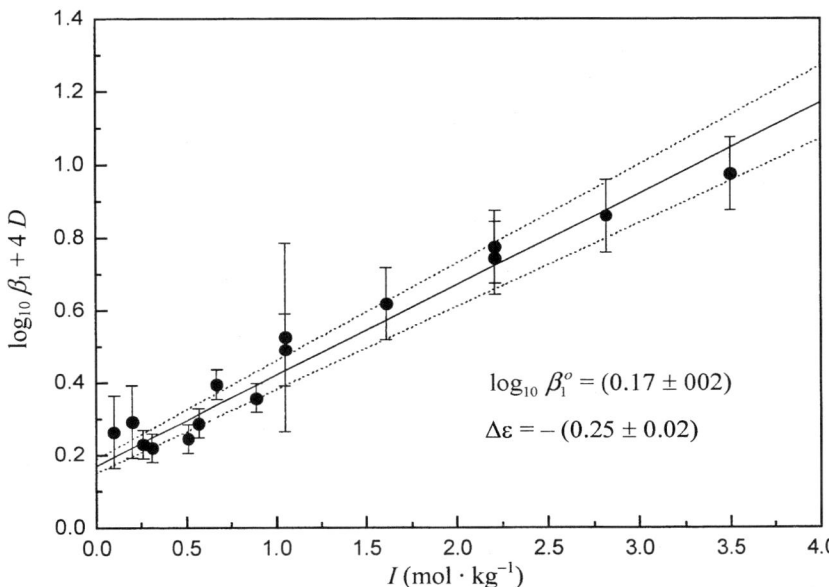

$$\log_{10} \beta_1^\circ = (0.17 \pm 002)$$
$$\Delta\varepsilon = -(0.25 \pm 0.02)$$

Example B.3:

When using the specific ion interaction theory, the relationship between the redox potential of the couple, PuO_2^{2+}/Pu^{4+}, in a medium of ionic strength, I_m, and the corresponding quantity at $I = 0$ should be calculated in the following way. The reaction in the galvanic cell:

$$Pt \mid H_2(g, r) \mid H^+(r) \parallel PuO_2^{2+}, Pu^{4+}, H^+, H_2O(l) \mid Pt$$

$$PuO_2^{2+} + H_2(g, r) + 4H^+ - 2H^+(r) \rightleftharpoons Pu^{4+} + 2H_2O(l) \tag{B.14}$$

where "r" is used to indicate that $H_2(g)$ and H^+ are at the chemical conditions in the reference electrode compartment, i.e., standard conditions when the reference electrode is the SHE. However, activities of H^+, $H_2O(l)$ and the ratio of activity of PuO_2^{2+} to Pu^{4+} depend on the conditions of the experimental measurements (*i.e.*, non–standard conditions, usua high ionic strength to improve the accuracy of the measurement).

For reaction (B.14):

$$\log_{10} K° = \log_{10}\left(\frac{a_{Pu^{4+}} \cdot a^2_{H_2O} \cdot a_{H^+(r)}}{a_{PuO_2^{2+}} \cdot a^4_{H^+} \cdot f_{H_2(r)}}\right).$$

Since by definition of the SHE, $f_{H_2(r)} = 1$ and $\gamma_{H^+(r)} = 1$,

$$\log_{10} K° = \log_{10} K + \log_{10}\gamma_{Pu^{4+}} - \log_{10}\gamma_{PuO_2^{2+}} - 4\log_{10}\gamma_{H^+} + 2\log_{10}a_{H_2O},$$

and

$$\log_{10}\gamma_{Pu^{4+}} = -16D + \varepsilon(Pu^{4+}, ClO_4^-)\, m_{ClO_4^-}$$
$$\log_{10}\gamma_{PuO_2^{2+}} = -4D + \varepsilon(PuO_2^{2+}, ClO_4^-)\, m_{ClO_4^-}$$
$$\log_{10}\gamma_{H^+} = -D + \varepsilon(H^+, ClO_4^-)\, m_{ClO_4^-}$$

Hence,

$$\log_{10} K° = \log_{10} K - 8D + (\varepsilon(Pu^{4+}, ClO_4^-) - \varepsilon(PuO_2^{2+}, ClO_4^-) - 4\varepsilon(H^+, ClO_4^-))m_{ClO_4^-}$$
$$+ 2\log_{10} a_{H_2O} \tag{B.1}$$

The relationship between the equilibrium constant and the redox potential is:

$$\ln K = \frac{nF}{RT} E^{o'} \tag{B.1}$$

$$\ln K° = \frac{nF}{RT} E°. \tag{B.1}$$

$E^{o'}$ is the redox potential in a medium of ionic strength I, $E°$ is the correspond standard potential at $I = 0$, and n is the number of transferred electrons in the reaction c sidered. Combining Eqs. (B.15), (B.16) and (B.17) and rearranging them leads to Eq.(B.1

$$E^{o'} - (8D - 2\log_{10} a_{H_2O})\left(\frac{RT\ln(10)}{nF}\right) = E° - \Delta\varepsilon\, m_{ClO_4^-}\left(\frac{RT\ln(10)}{nF}\right) \tag{B.1}$$

For $n = 2$ in the present example and $T = 298.15$ K, Eq.(B.18) becomes:

$$E^{o'}[mV] - 236.6\, D + 59.16\log_{10} a_{H_2O} = E°[mV] - 29.58\Delta\varepsilon\, m_{ClO_4^-}$$

where

$$\Delta\varepsilon = \varepsilon(Pu^{4+}, ClO_4^-) - \varepsilon(PuO_2^{2+}, ClO_4^-) - 4\varepsilon(H^+, ClO_4^-).$$

The value of a_{H_2O} can be taken from experimental data or calculated from equations (B.10) and (B.11).

In general, formal potentials are reported with reference to the standard hydrogen electrode, cf. Section 2.1.6.5, as exemplified in Tables V.2 and V.3 of the uranium NEA review [92GRE/FUG]. In that case, the H^+ appearing in the reduction reaction is already at standard conditions. For example, experimental data are available on the formal potentials of reactions:

$$PuO_2^{2+} + 4H^+ + 2e^- \rightleftharpoons Pu^{4+} + 2H_2O(l) \tag{B.19}$$

$$PuO_2^{2+} + e^- \rightleftharpoons PuO_2^+. \tag{B.20}$$

While reaction (B.19) corresponds to (B.14), reaction (B.20) is equivalent to:

$$PuO_2^{2+} + \frac{1}{2}H_2(g) \rightleftharpoons PuO_2^+ + H^+ \tag{B.21}$$

Here the designator "(r)" has been omitted, since in these equations only the H^+ in the reference compartment is relevant.

The cations in reaction (B.14) represent aqueous species in the ionic media used during the experiments. In reaction (B.21) H^+ represents the cation in the standard hydrogen electrode, and therefore it is already in standard conditions, and its activity coefficient must not be included in any extrapolation to $I = 0$ of experimental values for reaction (B.20). Reactions (B.20) and (B.21) are equivalent, as are reactions (B.14) and (B.19), as can be seen if any of these equations are combined with reaction (2.26). Hence Eq. (B.18) can be obtained more simply by using Eq. 2.33 for reaction (B.19).

B.1.4 On the magnitude of ion interaction coefficients

Ciavatta [80CIA] made a compilation of ion interaction coefficients for a large number of electrolytes. Similar data for complexations of various kinds were reported by Spahiu [83SPA] and Ferri et al. [83FER/GRE]. These and some other data for 25°C and 1 bar have been collected and are listed in Section B.3.

It is obvious from the data in these tables that the charge of an ion is of great importance for determining the magnitude of the ion interaction coefficient. Ions of the same charge type have similar ion interaction coefficients with a given counter-ion. Based on the tabulated data, Grenthe et al. [92GRE/FUG] proposed that it is possible to estimate, with an error of at most ± 0.1 kg · mol⁻¹ in ε, ion interaction coefficients for cases where there insufficient experimental data for an extrapolation to $I = 0$. The error that is made by t approximation is estimated to ± 0.1 kg · mol⁻¹ in Δε in most cases, based on comparis with Δε values of various reactions of the same charge type.

Ciavatta [90CIA] has proposed an alternative method to estimate values of ε fc first or second complex, ML or ML$_2$, in an ionic media NX, according to the following re tionships:

$$\varepsilon(ML, N \text{ or } X) \approx (\varepsilon(M,X) + \varepsilon(L,N))/2 \quad \text{(B.2}$$

$$\varepsilon(ML_2, N \text{ or } X) \approx (\varepsilon(M,X) + 2\varepsilon(L,N))/3 \quad \text{(B.2}$$

Ciavatta obtained [90CIA] an average deviation of ± 0.05 kg · mol⁻¹ betwee estimates according to Eqs. (B.22) and (B.23), and the ε values at 25°C obtained from ic strength dependency of equilibrium constants.

B.2 Ion interaction coefficients *versus* equilibrium constant: for ion pairs

It can be shown that the virial type of activity coefficient equations and the ionic pair model are equivalent provided that the ionic pairing is weak. In these cases the distinct between complex formation and activity coefficient variations is difficult or even arbitr unless independent experimental evidence for complex formation is available, *e.g.*, fr spectroscopic data, as is the case for the weak uranium(VI) chloride complexes. It shoulc noted that the ion interaction coefficients evaluated and tabulated by Ciavatta [80CIA] w obtained from experimental mean activity coefficient data without taking into account cc plex formation. However, it is known that many of the metal ions listed by Ciavatta fc weak complexes with chloride and nitrate ion. This fact is reflected by ion interaction cc ficients that are smaller than those for the non-complexing perchlorate ion, *cf.* Table F This review takes chloride and nitrate complex formation into account when these ions part of the ionic medium and uses the value of the ion interaction coefficient, $\varepsilon(M^{n+}, ClO$ as a substitute for $\varepsilon(M^{n+}, Cl^-)$ and $\varepsilon(M^{n+}, NO_3^-)$. In this way, the medium dependencc the activity coefficients is described with a combination of a specific ion interaction mc and an ion pairing model. It is evident that the use of NEA recommended data with ic strength correction models that differ from those used in the evaluation procedure can l to inconsistencies in the results of the speciation calculations.

It should be mentioned that complex formation may also occur between negativ charged complexes and the cation of the ionic medium. An example is the stabilisatior complex ion, $UO_2(CO_3)_3^{5-}$, at high ionic strength, see for example Section V.7.1.2.1.d 322) in the uranium review [92GRE/FUG].

3 Tables of ion interaction coefficients

ɔle B-4, Table B-5 and Table B-6 contain the selected specific ion interaction coefficients d in this review, according to the specific ion interaction theory described. Table B-4 itains cation interaction coefficients with Cl^-, ClO_4^- and NO_3^-. Table B-5 anion interac- coefficients with Li^+, Na^+ (or NH_4^+) and K^+. The coefficients have the units of kg·mol^{-1} are valid for 298.15 K and 1 bar. The species are ordered by charge and appear, within h charge class, in standard order of arrangement, cf. Section 2.1.9.

In some cases, the ionic interaction can be better described by assuming ion inter-ion coefficients as functions of the ionic strength rather than as constants. Ciavatta CIA] proposed the use of Eq. (B.24) for cases where the uncertainties in Table B-4 and ɔle B-5 are ± 0.03 kg·mol^{-1} or greater.

$$\varepsilon = \varepsilon_1 + \varepsilon_2 \log_{10} I_m \tag{B.24}$$

For these cases, and when the uncertainty can be improved with respect to the use ı constant value of ε, the values ε_1 and ε_2 given in Table B-6 should be used.

It should be noted that ion interaction coefficients tabulated in Table B-4, Table B-nd Table B-6 may also involve ion pairing effects, as described in Section B.3. In direct nparisons of ion interaction coefficients, or when estimates are made by analogy, this ect must be taken into account.

Table B-4: Ion interaction coefficients $\varepsilon(j,k)$ (kg·mol^{-1}) for cations j with k = Cl$^-$, ClO$_4^-$ and NO$_3^-$, taken from Ciavatta [80CIA], [88CIA] unless indicated otherwise. The uncertainties represent the 95% confidence level. The ion interaction coefficients marked with † can be described more accurately with an ionic strength dependent function, listed in Table B. The coefficients $\varepsilon(M^{n+}, Cl^-)$ and $\varepsilon(M^{n+}, NO_3^-)$ reported by Ciavatta [80CIA] were evaluated without taking chloride and nitrate complexation into account, as discussed in Section B. 2.

j $k \rightarrow$ $\downarrow$	Cl$^-$	ClO$_4^-$	NO$_3^-$
H$^+$	(0.12 ± 0.01)	(0.14 ± 0.02)	(0.07 ± 0.01)
NH$_4^+$	– (0.01 ± 0.01)	– (0.08 ± 0.04)†	– (0.06 ± 0.03)†
H$_2$gly$^+$	– (0.06 ± 0.02)		
Tl$^+$		– (0.21 ± 0.06)†	
ZnHCO$_3^+$	0.2$^{(a)}$		
CdCl$^+$		(0.25 ± 0.02)	
CdI$^+$		(0.27 ± 0.02)	
CdSCN$^+$		(0.31 ± 0.02)	
HgCl$^+$		(0.19 ± 0.02)	
Cu$^+$		(0.11 ± 0.01)	
Ag$^+$		(0.00 ± 0.01)	– (0.12 ± 0.05)†
YCO$_3^+$		(0.17 ± 0.04)$^{(b)}$	
Am(OH)$_2^+$	– (0.27 ± 0.20)$^{(q)}$	(0.17 ± 0.04)$^{(c)}$	
AmF$_2^+$		(0.17 ± 0.04)$^{(c)}$	
AmSO$_4^+$		(0.22 ± 0.08)$^{(d)}$	
AmCO$_3^+$	(0.01 ± 0.05)$^{(r)}$	(0.17 ± 0.04)$^{(c)}$	
PuO$_2^+$		(0.24 ± 0.05)$^{(e)}$	
PuO$_2$F$^+$		(0.29 ± 0.11)$^{(f)}$	
PuO$_2$Cl$^+$		(0.50 ± 0.09)$^{(g)}$	
NpO$_2^+$	(0.09 ± 0.05)	(0.25 ± 0.05)$^{(h)}$	
NpO$_2$OH$^+$		– (0.06 ± 0.40)$^{(i)}$	

(Continued on next page)

Table B-4 (Continued)

$j \quad k \rightarrow$ ↓	Cl^-	ClO_4^-	NO_3^-
$(NpO_2)_3(OH)_5^+$		(0.45 ± 0.20)	
NpO_2F^+		$(0.29 \pm 0.12)^{(j)}$	
NpO_2Cl^+		$(0.50 \pm 0.14)^{(k)}$	
$NpO_2IO_3^+$		$(0.33 \pm 0.04)^{(l)}$	
$Np(SCN)_3^+$		$(0.17 \pm 0.04)^{(m)}$	
UO_2^+		$(0.26 \pm 0.03)^{(n)}$	
UO_2OH^+		$-(0.06 \pm 0.40)^{(n)}$	$(0.51 \pm 1.4)^{(n)}$
$(UO_2)_3(OH)_5^+$	$(0.81 \pm 0.17)^{(n)}$	$(0.45 \pm 0.15)^{(n)}$	$(0.41 \pm 0.22)^{(n)}$
UF_3^+	$(0.1 \pm 0.1)^{(o)}$	$(0.1 \pm 0.1)^{(o)}$	
UO_2F^+	$(0.04 \pm 0.07)^{(p)}$	(0.28 ± 0.04)	
UO_2Cl^+		$(0.33 \pm 0.04)^{(n)}$	
$UO_2ClO_3^+$		$(0.33 \pm 0.04)^{(o)}$	
UO_2Br^+		$(0.24 \pm 0.04)^{(o)}$	
$UO_2BrO_3^+$		$(0.33 \pm 0.04)^{(o)}$	
$UO_2IO_3^+$		$(0.33 \pm 0.04)^{(o)}$	
$UO_2N_3^+$		$(0.3 \pm 0.1)^{(o)}$	
$UO_2NO_3^+$		$(0.33 \pm 0.04)^{(o)}$	
UO_2SCN^+		$(0.22 \pm 0.04)^{(o)}$	
Pb^{2+}		(0.15 ± 0.02)	$-(0.20 \pm 0.12)^\dagger$
$AlOH^{2+}$	$0.09^{(s)}$	$0.31^{(s)}$	
$Al_2CO_3(OH)_2^{2+}$	$0.26^{(s)}$		
Zn^{2+}		(0.33 ± 0.03)	(0.16 ± 0.02)
$ZnCO_3^{2+}$	$(0.35 \pm 0.05)^{(a)}$		
Cd^{2+}			(0.09 ± 0.02)
Hg^{2+}		(0.34 ± 0.03)	$-(0.1 \pm 0.1)^\dagger$
Hg_2^{2+}		(0.09 ± 0.02)	$-(0.2 \pm 0.1)^\dagger$
Cu^{2+}	(0.08 ± 0.01)	(0.32 ± 0.02)	(0.11 ± 0.01)
Ni^{2+}	(0.17 ± 0.02)		
Co^{2+}	(0.16 ± 0.02)	(0.34 ± 0.03)	(0.14 ± 0.01)

(Continued on next page)

Table B-4 (Continued)

$j \downarrow \quad k \rightarrow$	Cl$^-$	ClO$_4^-$	NO$_3^-$
FeOH^{2+}		0.38[b]	
FeSCN^{2+}		0.45[b]	
Mn^{2+}	(0.13 ± 0.01)		
YHCO$_3^{2+}$		(0.39 ± 0.04)[b]	
AmOH^{2+}	$-(0.04 \pm 0.07)$[q]	(0.39 ± 0.04)[c]	
AmF^{2+}		(0.39 ± 0.04)[c]	
AmCl^{2+}		(0.39 ± 0.04)[c]	
AmN$_3^{2+}$		(0.39 ± 0.04)[c]	
AmNO$_2^{2+}$		(0.39 ± 0.04)[c]	
AmNO$_3^{2+}$		(0.39 ± 0.04)[c]	
AmH$_2$PO$_4^{2+}$		(0.39 ± 0.04)[c]	
AmSCN^{2+}		(0.39 ± 0.04)[c]	
PuO$_2^{2+}$		(0.46 ± 0.05)[t]	
PuF$_2^{2+}$		(0.36 ± 0.17)[j]	
PuCl^{2+}		(0.39 ± 0.16)[u]	
PuI^{2+}		(0.39 ± 0.04)[v]	
NpO$_2^{2+}$		(0.46 ± 0.05)[w]	
(NpO$_2$)$_2$(OH)$_2^{2+}$		(0.57 ± 0.10)	
NpF$_2^{2+}$		(0.38 ± 0.17)[j]	
NpSO$_4^{2+}$		(0.48 ± 0.11)	
Np(SCN)$_2^{2+}$		(0.38 ± 0.20)[j]	
UO$_2^{2+}$	(0.21 ± 0.02)[x]	(0.46 ± 0.03)	(0.24 ± 0.03)[v]
(UO$_2$)$_2$(OH)$_2^{2+}$	(0.69 ± 0.07)[n]	(0.57 ± 0.07)[n]	(0.49 ± 0.09)[n]
(UO$_2$)$_3$(OH)$_4^{2+}$	(0.50 ± 0.18)[n]	(0.89 ± 0.23)[n]	(0.72 ± 1.0)[n]
UF$_2^{2+}$		(0.3 ± 0.1)[o]	
USO$_4^{2+}$		(0.3 ± 0.1)[o]	
U(NO$_3$)$_2^{2+}$		(0.49 ± 0.14)[y]	
Mg^{2+}	(0.19 ± 0.02)	(0.33 ± 0.03)	(0.17 ± 0.01)
Ca^{2+}	(0.14 ± 0.01)	(0.27 ± 0.03)	(0.02 ± 0.01)
Ba^{2+}	(0.07 ± 0.01)	(0.15 ± 0.02)	$-(0.28 \pm 0.03)$

(Continued on next page)

Table B-4 (Continued)

$j \downarrow \quad k \rightarrow$	Cl^-	ClO_4^-	NO_3^-
Al^{3+}	(0.33 ± 0.02)		
Fe^{3+}		(0.56 ± 0.03)	(0.42 ± 0.08)
Cr^{3+}	(0.30 ± 0.03)		(0.27 ± 0.02)
La^{3+}	(0.22 ± 0.02)	(0.47 ± 0.03)	
$La^{3+} \rightarrow Lu^{3+}$		$0.47 \rightarrow 0.52^{(b)}$	
Am^{3+}	$(0.23 \pm 0.02)^{(G)}$	$(0.49 \pm 0.03)^{(c)}$	
Pu^{3+}		$(0.49 \pm 0.05)^{(z)}$	
$PuOH^{3+}$		$(0.50 \pm 0.05)^{(i)}$	
PuF^{3+}		$(0.56 \pm 0.11)^{(i)}$	
$PuCl^{3+}$		$(0.85 \pm 0.09)^{(\#)}$	
$PuBr^{3+}$		$(0.58 \pm 0.16)^{(A)}$	
$PuSCN^{3+}$		$(0.39 \pm 0.04)^{(B)}$	
Np^{3+}		$(0.49 \pm 0.05)^{(z)}$	
$NpOH^{3+}$		$(0.50 \pm 0.05)^{(i)}$	
NpF^{3+}		$(0.58 \pm 0.07)^{(C)}$	
$NpCl^{3+}$		$(0.81 \pm 0.09)^{(D)}$	
NpI^{3+}		$(0.77 \pm 0.26)^{(E)}$	
$NpSCN^{3+}$		$(0.76 \pm 0.12)^{(j)}$	
U^{3+}		$(0.49 \pm 0.05)^{(y)}$	
UOH^{3+}		$(0.48 \pm 0.08)^{(y)}$	
UF^{3+}		$(0.48 \pm 0.08)^{(o)}$	
UCl^{3+}		$(0.50 \pm 0.10)^{(k)}$	
UBr^{3+}		$(0.52 \pm 0.10)^{(o)}$	
UI^{3+}		$(0.55 \pm 0.10)^{(o)}$	
UNO_3^{3+}		$(0.62 \pm 0.08)^{(y)}$	
Be_2OH^{3+}		$(0.50 \pm 0.05)^{(F)}$	
$Be_3(OH)_3^{3+}$	$(0.30 \pm 0.05)^{(F)}$	$(0.51 \pm 0.05)^{(y)}$	$(0.29 \pm 0.05)^{(F)}$
$Al_3CO_3(OH)_4^{4+}$	$0.41^{(S)}$		
$Fe_2(OH)_2^{4+}$		$0.82^{(b)}$	

(Continued on next page)

Table B-4: (Continued)

$j \downarrow \quad k \rightarrow$	Cl^-	ClO_4^-	NO_3^-
$Y_2CO_3^{4+}$		$(0.80 \pm 0.04)^{(b)}$	
Pu^{4+}		$(0.82 \pm 0.07)^{(H)}$	
Np^{4+}		$(0.84 \pm 0.06)^{(I)}$	
U^{4+}		$(0.76 \pm 0.06)^{(J)(o)}$	
Th^{4+}	(0.25 ± 0.03)		(0.11 ± 0.02)
$Al_3(OH)_4^{5+}$	$0.66^{(s)}$	$1.30^{(s)}$	

(a) Taken from Ferri et al. [85FER/GRE].

(b) Taken from Spahiu [83SPA].

(c) Estimated in [95SIL/BID].

(d) Evaluated in [95SIL/BID].

(e) Derived from $\Delta\varepsilon = \varepsilon(PuO_2^{2+},ClO_4^-) - \varepsilon(PuO_2^+,ClO_4^-) = (0.22 \pm 0.03)$ kg·mol^{-1} [95CAP/VIT]. [92GRE/FUG], $\varepsilon(PuO_2^+,ClO_4^-) = (0.17 \pm 0.05)$ kg·mol^{-1} was tabulated based on [89ROB], [89RIG/R and [90RIG]. Capdevila and Vitorge's data [92CAP], [94CAP/VIT] and [95CAP/VIT] were unavailab that time.

(f) Estimated in [2001LEM/FUG] by analogy with $\Delta\varepsilon$ of the corresponding Np(IV) reaction.

(g) From $\Delta\varepsilon$ evaluated by Giffaut [94GIF].

(h) As in [92GRE/FUG], derived from $\Delta\varepsilon = \varepsilon(NpO_2^{2+},ClO_4^-) - \varepsilon(NpO_2^+,ClO_4^-) = (0.21 \pm 0.03)$kg·r [87RIG/VIT], [89RIG/ROB] and [90RIG].

(i) Estimated in [2001LEM/FUG].

(j) Estimated in [2001LEM/FUG] by analogy with $\Delta\varepsilon$ of the corresponding U(IV) reaction.

(k) Estimated in [2001LEM/FUG] by analogy with $\Delta\varepsilon$ of the corresponding P(VI) reaction.

(l) Estimated in [2001LEM/FUG] by assuming $\varepsilon(NpO_2IO_3^+,ClO_4^-) \approx \varepsilon(UO_2IO_3^+,ClO_4^-)$.

(m) Estimated in [2001LEM/FUG] by assuming $\varepsilon(Np(SCN)_3^+,ClO_4^-) \approx \varepsilon(AmF_2^+,ClO_4^-)$.

(n) Evaluated in the uranium review [92GRE/FUG], using $\varepsilon(UO_2^{2+},X) = (0.46 \pm 0.03)$ kg·mol^{-1}, where Cl^-, ClO_4^- and NO_3^-.

(o) Estimated in the uranium review [92GRE/FUG].

(p) Taken from Riglet et al. [89RIG/ROB], where the following assumptions were n $\varepsilon(Np^{3+},ClO_4^-) \approx \varepsilon(Pu^{3+},ClO_4^-) = 0.49$ kg·mol^{-1} as for other (M^{3+}, ClO_4^-) interactions, $\varepsilon(NpO_2^{2+},ClO_4^-) \approx \varepsilon(PuO_2^{2+},ClO_4^-) \approx \varepsilon(UO_2^{2+},ClO_4^-) = 0.46$ kg·mol^{-1}.

Evaluated in the Section 12.3.1.1 from $\Delta\varepsilon$ (in NaCl solution) for the reactions $An^{3+} + nH_2O(l) \rightleftharpoons An(OH)_n^{(3-n)} + nH^+$.

Evaluated in the section (12.6.1.1.1) from $\Delta\varepsilon$ (in NaCl solution) for the reactions $An^{3+} + nCO_3^{2-} \rightleftharpoons An(CO_3)_n^{(3-n)}$ (based on $\varepsilon(Am^{3+}, Cl^-) = (0.23 \pm 0.02)$ kg·mol^{-1} and $\varepsilon(Na^+, CO_3^{2-}) = -(0.08 \pm 0.03)$ kg·mol^{-1}.

Taken from Hedlund [88HED].

By analogy with $\varepsilon(UO_2^{2+}, ClO_4^-)$ as derived from isopiestic measurements in [92GRE/FUG]. The uncertainty is increased because the value is estimated by analogy.

Estimated in [2001LEM/FUG] by analogy with $\Delta\varepsilon$ of the corresponding Am(III) reaction.

Estimated in [2001LEM/FUG] by assuming $\varepsilon(PuI^{2+}, ClO_4^-) \approx \varepsilon(AmSCN^{2+}, ClO_4^-)$ and $\varepsilon(I^-, NH_4^+) \approx \varepsilon(SCN^-, Na^+)$.

By analogy with $\varepsilon(UO_2^{2+}, ClO_4^-)$ as derived from isopiestic measurements noted in [92GRE/FUG]. The uncertainty is increased because the value is estimated by analogy.

These coefficients were not used in the NEA–TDB uranium review [92GRE/FUG] because they were evaluated by Ciavatta [80CIA] without taking chloride and nitrate complexation into account. Instead, Grenthe *et al.* used $\varepsilon(UO_2^{2+}, X) = (0.46 \pm 0.03)$ kg·mol^{-1}, for X = Cl$^-$, ClO$_4^-$ and NO$_3^-$.

Evaluated in the uranium review [92GRE/FUG] using $\varepsilon(U^{4+}, X) = (0.76 \pm 0.06)$ kg·mol^{-1}.

Estimated by analogy with $\varepsilon(Ho^{3+}, ClO_4^-)$ [83SPA] as in previous books in this series [92GRE/FUG], [95SIL/BID]. The uncertainty is increased because the value is estimated by analogy.

Derived from the $\Delta\varepsilon$ evaluated in [2001LEM/FUG].

Estimated in [2001LEM/FUG] by analogy with $\Delta\varepsilon$ of the corresponding U(IV) reaction, and by assuming $\varepsilon(Br^-, H^+) \approx \varepsilon(Br^-, Na^+)$.

Estimated in [2001LEM/FUG] by assuming $\varepsilon(PuSCN^{2+}, ClO_4^-) \approx \varepsilon(AmSCN^{2+}, ClO_4^-)$.

Evaluated in [2001LEM/FUG].

Derived from the $\Delta\varepsilon$ selected in [2001LEM/FUG].

Estimated in [2001LEM/FUG] by analogy with $\Delta\varepsilon$ of the corresponding Np(IV) chloride reaction, and by assuming $\varepsilon(I^-, H^+) \approx \varepsilon(I^-, Na^+)$.

Taken from Bruno [86BRU], where the following assumptions were made: $\varepsilon(Be^{2+}, ClO_4^-) = 0.30$ kg·mol^{-1} as for other $\varepsilon(M^{2+}, ClO_4^-)$; $\varepsilon(Be^{2+}, Cl^-) = 0.17$ kg·mol^{-1} as for other $\varepsilon(M^{2+}, Cl^-)$; and $\varepsilon(Be^{2+}, NO_3^-) = 0.17$ kg·mol^{-1} as for other $\varepsilon(M^{2+}, NO_3^-)$.

The ion interaction coefficient $\varepsilon(An^{3+}, Cl^-)$ for An = Am and Cm is assumed to equal to $\varepsilon(Nd^{3+}, Cl^-)$ which is calculated from trace activity coefficients of Nd^{3+} ion in 0–4 m NaCl. These trace activity coefficients are based on the ion interaction Pitzer parameters evaluated in [97KON/FAN] from osmotic coefficients in aqueous NdCl$_3$ – NaCl and NdCl$_3$ – CaCl$_2$.

(H) Derived from $\Delta\varepsilon = \varepsilon(Pu^{4+},ClO_4^-) - \varepsilon(Pu^{3+},ClO_4^-) = (0.33\pm0.035)$ kg·mol^{-1} [95CAP/VIT]. Uncert[ainty] estimated in [2001LEM/FUG] (see Appendix A). In the first book of this series [92GRE/F[U]] $\varepsilon(Pu^{3+},ClO_4^-) = (1.03\pm0.05)$ kg·mol^{-1} was tabulated based on references [89ROB], [89RIG/ROB], [90R[IG]]. Capdevila and Vitorge's data [92CAP], [94CAP/VIT] and [95CAP/VIT] were unavailable at that time.

(I) Derived from $\Delta\varepsilon = \varepsilon(Np^{4+},ClO_4^-) - \varepsilon(Np^{3+},ClO_4^-) = (0.35\pm0.03)$ kg·mol^{-1} [89ROB], [89RIG/R[OB]], [90RIG].

(J) Using the measured value of $\Delta\varepsilon = \varepsilon(U^{4+},ClO_4^-) - \varepsilon(U^{3+},ClO_4^-) = (0.35\pm0.06)$ kg·mol^{-1} p.89 [90R[IG]] where the uncertainty is recalculated in [2001LEM/FUG] from the data given in this thesis, $\varepsilon(U^{3+},ClO_4^-) = (0.49\pm0.05)$ kg·mol^{-1} (see footnote (y)), a value for $\varepsilon(U^{4+},ClO_4^-)$ can be calculate[d] the same way as is done for $\varepsilon(Np^{4+},ClO_4^-)$ and $\varepsilon(Pu^{4+},ClO_4^-)$. This value, $\varepsilon(U^{4+},ClO_4^-) = (0.84\pm 0$ kg·mol^{-1} is consistent with that tabulated $\varepsilon(U^{4+},ClO_4^-) = (0.76\pm 0.06)$ kg·mol^{-1}, since the uncertai[nties] overlap. The authors of the present work do not believe that a change in the previously selected valu[e of] $\varepsilon(U^{4+},ClO_4^-)$ is justified at present.

B.3 Tables of ion interaction coefficients

Table B-5: Ion interaction coefficients, $\varepsilon(j,k)$ kg · mol^{-1}, for anions j with k = Li, Na and K, taken from Ciavatta [80CIA], [88CIA] unless indicated otherwise. The uncertainties represent the 95% confidence level. The ion interaction coefficients marked with † can be described more accurately with an ionic strength dependent function, listed in Table B-6.

$j \quad k \rightarrow$ $\downarrow$	Li$^+$	Na$^+$	K$^+$
OH$^-$	$-(0.02 \pm 0.03)$†	(0.04 ± 0.01)	(0.09 ± 0.01)
F$^-$		$(0.02 \pm 0.02)^{(a)}$	(0.03 ± 0.02)
HF$_2^-$		$-(0.11 \pm 0.06)^{(a)}$	
Cl$^-$	(0.10 ± 0.01)	(0.03 ± 0.01)	(0.00 ± 0.01)
ClO$_3^-$		$-(0.01 \pm 0.02)$	
ClO$_4^-$	(0.15 ± 0.01)	(0.01 ± 0.01)	
Br$^-$	(0.13 ± 0.02)	(0.05 ± 0.01)	(0.01 ± 0.02)
BrO$_3^-$		$-(0.06 \pm 0.02)$	
I$^-$ (p)	(0.16 ± 0.01)	(0.08 ± 0.02)	(0.02 ± 0.01)
IO$_3^-$		$-(0.06 \pm 0.02)^{(b)}$	
HSO$_4^-$		$-(0.01 \pm 0.02)$	
N$_3^-$		$(0.0 \pm 0.1)^{(b)}$	
NO$_2^-$	(0.06 ± 0.04)†	(0.00 ± 0.02)	$-(0.04 \pm 0.02)$
NO$_3^-$	(0.08 ± 0.01)	$-(0.04 \pm 0.03)$†	$-(0.11 \pm 0.04)$†
H$_2$PO$_4^-$		$-(0.08 \pm 0.04)$†	$-(0.14 \pm 0.04)$†
HCO$_3^-$		$(0.00 \pm 0.02)^{(d)}$	$-(0.06 \pm 0.05)^{(i)}$
SCN$^-$		(0.05 ± 0.01)	$-(0.01 \pm 0.01)$
HCOO$^-$		(0.03 ± 0.01)	
CH$_3$COO$^-$	(0.05 ± 0.01)	(0.08 ± 0.01)	(0.09 ± 0.01)
SiO(OH)$_3^-$		$-(0.08 \pm 0.03)^{(a)}$	
Si$_2$O$_2$(OH)$_5^-$		$-(0.08 \pm 0.04)^{(b)}$	
B(OH)$_4^-$		$-(0.07 \pm 0.05)$†	
Am(SO$_4$)$_2^-$		$-(0.05 \pm 0.05)^{(c)}$	
Am(CO$_3$)$_2^-$		$-(0.14 \pm 0.06)^{(r)}$	
PuO$_2$CO$_3^-$		$-(0.18 \pm 0.18)^{(o)}$	
NpO$_2$(OH)$_2^-$		$-(0.01 \pm 0.07)^{(q)}$	
NpO$_2$CO$_3^-$		$-(0.18 \pm 0.15)^{(f)}$	
(NpO$_2$)$_2$CO$_3$(OH)$_3^-$		$(0.00 \pm 0.05)^{(k)}$	
UO$_2$(OH)$_3^-$		$-(0.09 \pm 0.05)^{(b)}$	
UO$_2$F$_3^-$		$-(0.14 \pm 0.05)^{(b)}$	
UO$_2$(N$_3$)$_3^-$		$(0.0 \pm 0.1)^{(b)}$	
(UO$_2$)$_2$CO$_3$(OH)$_3^-$		$(0.00 \pm 0.05)^{(b)}$	

Table B-5 (continued)

j $k \rightarrow$ ↓	Li$^+$	Na$^+$	K$^+$
SO_3^{2-}		$-(0.08 \pm 0.05)$†	
SO_4^{2-}	$-(0.03 \pm 0.04)$	$-(0.12 \pm 0.06)$	$-(0.06 \pm 0.02)$
$S_2O_3^{2-}$		$-(0.08 \pm 0.05)$†	
HPO_4^{2-}		$-(0.15 \pm 0.06)$†	$-(0.10 \pm 0.06)$†
CO_3^{2-}		$-(0.08 \pm 0.03)^{(d)}$	(0.02 ± 0.01)
$SiO_2(OH)_2^{2-}$		$-(0.10 \pm 0.07)^{(a)}$	
$Si_2O_3(OH)_4^{2-}$		$-(0.15 \pm 0.06)^{(b)}$	
CrO_4^{2-}		$-(0.06 \pm 0.04)$†	$-(0.08 \pm 0.04)$†
$NpO_2(HPO_4)_2^{2-}$		$-(0.1 \pm 0.1)$	
$NpO_2(CO_3)_2^{2-}$		$-(0.02 \pm 0.14)^{(k)}$	
$UO_2F_4^{2-}$		$-(0.30 \pm 0.06)^{(b)}$	
$UO_2(SO_4)_2^{2-}$		$-(0.12 \pm 0.06)^{(b)}$	
$UO_2(N_3)_4^{2-}$		$-(0.1 \pm 0.1)^{(b)}$	
$UO_2(CO_3)_2^{2-}$		$-(0.02 \pm 0.09)^{(d)}$	
$(UO_2)_2(OH)_2(SO_4)_2^{2-}$		$-(0.14 \pm 0.22)$	
PO_4^{3-}		$-(0.25 \pm 0.03)$†	$-(0.09 \pm 0.02)$
$Si_3O_6(OH)_3^{3-}$		$-(0.25 \pm 0.03)^{(b)}$	
$Si_3O_5(OH)_5^{3-}$		$-(0.25 \pm 0.03)^{(b)}$	
$Si_4O_7(OH)_5^{3-}$		$-(0.25 \pm 0.03)^{(b)}$	
$Am(CO_3)_3^{3-}$		$-(0.23 \pm 0.07)^{(r)}$	
$Np(CO_3)_3^{3-}$			$-(0.15 \pm 0.07)^{(n)}$
$NpO_2(CO_3)_2^{3-}$		$-(0.33 \pm 0.17)^{(f)}$	
$P_2O_7^{4-}$		$-(0.26 \pm 0.05)$	$-(0.15 \pm 0.05)$
$Fe(CN)_6^{4-}$			$-(0.17 \pm 0.03)$
$NpO_2(CO_3)_3^{4-}$		$-(0.40 \pm 0.19)^{(e)}$	$-(0.62 \pm 0.42)^{(g)}$
$NpO_2(CO_3)_2OH^{4-}$		$-(0.40 \pm 0.19)^{(m)}$	
$U(CO_3)_4^{4-}$		$-(0.09 \pm 0.10)^{(b)(d)}$	
$UO_2(CO_3)_3^{4-}$		$-(0.01 \pm 0.11)^{(d)}$	
$(UO_2)_3(OH)_4(SO_4)_3^{4-}$		(0.6 ± 0.6)	
$NpO_2(CO_3)_3^{5-}$		$-(0.53 \pm 0.19)^{(f)}$	$-(0.22 \pm 0.03)^{(s)}$
$UO_2(CO_3)_3^{5-}$		$-(0.62 \pm 0.15)^{(d)}$	

(continued on next page)

Table B-5 (continued)

$j \; k \rightarrow$ ↓	Li$^+$	Na$^+$	K$^+$
$Np(CO_3)_5^{6-}$			$-(0.73 \pm 0.68)^{(j)}$
$NpO_2)_3(CO_3)_6^{6-}$		$-(0.46 \pm 0.73)^{(e)}$	
$J(CO_3)_5^{6-}$		$-(0.30 \pm 0.15)^{(d)}$	$-(0.70 \pm 0.31)^{(i)}$
$UO_2)_3(CO_3)_6^{6-}$		$(0.37 \pm 0.11)^{(d)}$	
$UO_2)_2NpO_2(CO_3)_6^{6-}$		$(0.09 \pm 0.71)^{(l)}$	
$UO_2)_5(OH)_8(SO_4)_4^{6-}$		(1.10 ± 0.5)	
$UO_2)_4(OH)_7(SO_4)_4^{7-}$		(2.80 ± 0.7)	

Evaluated in the NEA–TDB uranium review [92GRE/FUG].

Estimated in the NEA–TDB uranium review [92GRE/FUG].

Estimated in the NEA–TDB americium review [95SIL/BID].

These values differ from those reported in the NEA–TDB uranium review. See the discussion in [95GRE/PUI]. Values for CO_3^{2-} and HCO_3^- are based on [80CIA].

Calculated in [2001LEM/FUG] (Section 12.1.2.1.2)

Calculated in [2001LEM/FUG] (Section 12.1.2.1.3)

Calculated in [2001LEM/FUG] (Section 12.1.2.2.1)

$\varepsilon(NpO_2(CO_3)_3^{4-}, NH_4^+) = -(0.78 \pm 0.25)$ kg·mol^{-1} is calculated in [2001LEM/FUG] (Section 12.1.2.2.1)

Calculated in [2001LEM/FUG] from Pitzer coefficients [98RAI/FEL]

Calculated in [2001LEM/FUG] (Section 12.1.2.1.4)

Estimated by analogy in [2001LEM/FUG] (Section 12.1.2.1.2)

Estimated by analogy in [2001LEM/FUG] (Section 12.1.2.2.1)

Estimated in [2001LEM/FUG] by analogy with $NpO_2(CO_3)_3^{4-}$

Estimated by analogy in [2001LEM/FUG] (Section 12.1.2.1.5)

Estimated in [2001LEM/FUG] by analogy with $\varepsilon(NpO_2CO_3^-, Na^+)$

$\varepsilon(I^-, NH_4^+) \approx \varepsilon(SCN^-, Na^+) = (0.05 \pm 0.01)$ kg·mol^{-1}

Estimated in [2001LEM/FUG] (Section 8.1.3)

Evaluated in section 12.6.1.1.1 from $\Delta\varepsilon_n$ (in NaCl solution) for the reactions $An^{3+} + nCO_3^{2-} \rightleftharpoons An(CO_3)_n^{(3-n)}$ (based on $\varepsilon(Am^{3+}, Cl^-) = (0.23 \pm 0.02)$ kg·mol^{-1} and $\varepsilon(Na^+, CO_3^{2-}) = -(0.08 \pm 0.03)$ kg·mol^{-1}

Evaluated in Appendix A, discussion of [98ALM/NOV] from $\Delta\varepsilon$ for the reactions $KNpO_2CO_3(s) + 2CO_3^{2-} \rightleftharpoons NpO_2(CO_3)_3^{5-} + K^+$ (in K$_2$CO$_3$–KCl solution) and $K_3NpO_2(CO_3)_2(s) + CO_3^{2-} \rightleftharpoons NpO_2(CO_3)_3^{5-} + 3K^+$ (in K$_2$CO$_3$ solution) (based on $\varepsilon(K^+, CO_3^{2-}) = (0.02 \pm 0.01)$ kg·mol^{-1}).

Table B-6: Ion interaction coefficients, $\varepsilon(1,j,k)$ and $\varepsilon(2,j,k)$ kg · mol^{-1}, for cations j with Cl$^-$, ClO$_4^-$ and NO$_3^-$ (first part), and for anions j with k = Li$^+$, Na$^+$ and K$^+$ (second part), according to the relationship $\varepsilon = \varepsilon_1 + \varepsilon_2 \log_{10} I_m$. The data are taken from Ciavatta [80CIA], [88CIA]. The uncertainties represent the 95% confidence level.

$j\,k \rightarrow$ $\downarrow$	Cl$^-$		ClO$_4^-$		NO$_3^-$	
	ε_1	ε_2	ε_1	ε_2	ε_1	ε_2
NH$_4^+$			$-(0.088 \pm 0.002)$	(0.095 ± 0.012)	$-(0.075 \pm 0.001)$	$(0.057 \pm 0.00$
Tl$^+$			$-(0.18 \pm 0.02)$	(0.09 ± 0.02)		
Ag$^+$					$-(0.1432 \pm 0.0002)$	$-(0.0971 \pm 0.0$
Pb^{2+}					$-(0.329 \pm 0.007)$	$-(0.288 \pm 0.01$
Hg^{2+}					$-(0.145 \pm 0.001)$	$-(0.194 \pm 0.00$
Pb$_2^{2+}$					$-(0.2300 \pm 0.0004)$	$-(0.194 \pm 0.00$

$j\,k \rightarrow$ $\downarrow$	Li$^+$		Na$^+$		K$^+$	
	ε_1	ε_2	ε_1	ε_2	ε_1	ε_2
OH$^-$	$-(0.039 \pm 0.002)$	(0.072 ± 0.006)				
NO$_2^-$	(0.02 ± 0.01)	(0.11 ± 0.01)				
NO$_3^-$			$-(0.049 \pm 0.001)$	(0.044 ± 0.002)	$-(0.131 \pm 0.002)$	$(0.082 \pm 0.0$
H$_2$PO$_4^-$			$-(0.109 \pm 0.001)$	(0.095 ± 0.003)	$-(0.1473 \pm 0.0008)$	$(0.121 \pm 0.0$
B(OH)$_4^-$			$-(0.092 \pm 0.002)$	(0.103 ± 0.005)		
SO$_3^{2-}$			$-(0.125 \pm 0.008)$	(0.106 ± 0.009)		
SO$_4^{2-}$	$-(0.068 \pm 0.003)$	(0.093 ± 0.007)	$-(0.184 \pm 0.002)$	(0.139 ± 0.006)		
S$_2$O$_3^{2-}$			$-(0.125 \pm 0.008)$	(0.106 ± 0.009)		
HPO$_4^{2-}$			$-(0.19 \pm 0.01)$	(0.11 ± 0.03)	$-(0.152 \pm 0.007)$	$(0.123 \pm 0.0$
CrO$_4^{2-}$			$-(0.090 \pm 0.005)$	(0.07 ± 0.01)	$-(0.123 \pm 0.003)$	$(0.106 \pm 0.0$
PO$_4^{3-}$			$-(0.29 \pm 0.02)$	(0.10 ± 0.01)		

Appendix C
Assigned uncertainties[1]

This Appendix describes the origin of the uncertainty estimates that are given in the TDB tables of selected data. The original text in [92GRE/FUG] has been retained in [95SIL/BID], [99RAR/RAN] and [2001LEM/FUG], except for some minor changes. Because of the importance of the uncertainty estimates, the present review offers a more comprehensive description of the procedures used.

C.1 The general problem.

The focus of this section is on the uncertainty estimates of equilibria in solution, where the main problem is analytical, *i.e.*, the determination of the stoichiometric composition and equilibrium constants of complexes that are in rapid equilibrium with one another. We can formulate analyses of the experimental data in the following way: From n measurements, y_i, of the variable y we would like to determine a set of N equilibrium constants k_r, $r = 1, 2,..., N$, assuming that we know the functional relationship:

$$y = f(k_1, k_2, \ldots k_r \ldots k_N; a_1, a_2, \ldots.) \tag{C.1}$$

where a_1, a_2, *etc.* are quantities that can be varied but whose values (a_{1i}; a_{2i}; *etc.*) are assumed to be known accurately in each experiment from the data sets (y_i, a_{1i}, a_{2i},...), $i = 1, 2$, ... The functional relationship (C.1) is obtained from the chemical model proposed and in general several different models have to be tested before the "best" one is selected. Details of the procedures are given in Rossotti and Rossotti [61ROS/ROS].

When selecting the functional relationship (C.1) and determining the set of equilibrium constants that best describes the experiments one often uses a least-squares method.

---

[1] This Appendix essentially contains the text of the TDB-3 Guideline, [99WAN/OST], earlier versions of which have been printed in the previous NEA TDB reviews [92GRE/FUG], [95SIL/BID], [99RAR/RAN] and [2001LEM/FUG]. Because of its importance in the selection of data and to guide the users of the values in Chapters 3, 4, 5, 6, 7 and 8, the text is reproduced here with minor revisions.

C. Assigned uncertainties

Within this method, the "best" description is the one that will minimise the residual sum squares, U:

$$U = \sum_i w_i \left[y_i - F(k_1...k_N; a_{1i}, a_{2i}...) \right]^2 \qquad \text{(C}$$

where w_i is the weight of each experimental measurement y_i.

The minimum of the function (C.2) is obtained by solving a set of normal eq tions:

$$\frac{\partial U}{\partial k_r} = 0, r = 1,......N \qquad \text{(C}$$

A "true" minimum is only obtained if:

- the functional relationship (C.1) is correct, *i.e.*, if the chemical model is correct.
- all errors are random errors in the variable y, in particular there are no system errors.
- the random errors in y follow a Gaussian (normal) distribution.
- the weight $w_i(y_i, a_{1i}, a_{2i}, ...)$ of an experimental determination is an exact meas of its inherent accuracy.

To ascertain that the first condition is fulfilled requires chemical insight, such as formation of the coordination geometry, relative affinity between metal ions and vari donor atoms, *etc.* It is particularly important to test if the chemical equilibrium constant complexes that occur in small amounts are chemically reasonable. Too many experimen ists seem to look upon the least-squares refinement of experimental data more as an exer in applied mathematics than as a chemical venture. One of the tasks in the review of literature is to check this point. An erroneous chemical model is one of the more seri type of systematic error.

The experimentalist usually selects the variable that he/she finds most appropr to fulfill the second condition. If the estimated errors in a_{1i}, a_{2i} ... are smaller than the e in y_i, the second condition is reasonably well fulfilled. The choice of the error-carrying v able is a matter of choice based on experience, but one must be aware that it has impl tions, especially in the estimated uncertainty.

The presence of systematic errors is, potentially, the most important source of certainty. There is no possibility to handle systematic errors using statistics; statistical me ods may indicate their presence, no more. Systematic errors in the chemical model h been mentioned. In addition there may be systematic errors in the methods used. By c paring experimental data obtained with different experimental methods one can obtain ication of the presence and magnitude of such errors. The systematic errors of this type accounted for both in the review of the literature and when taking the average of data ained with different experimental methods. This type of systematic error does not seem affect the selected data very much, as judged by the usually very good agreement between equilibrium data obtained using spectroscopic, potentiometric and solubility methods.

The electrode calibration, especially the conversion between measured pH and $\log_{10}[H^+]$ is an important source of systematic error. The reviewers have when possible rected this error, as seen in many instances in Appendix A.

The assumption of a normal distribution of the random errors is a choice made in absence of better alternatives.

Finally, a comment on the weights used in least-squares refinements; this is important because it influences the uncertainty estimate of the equilibrium constants. The weights individual experimental points can be obtained by repeating the experiment several times and then calculating the average and standard deviation of these data. This procedure is ely used, instead most experimentalists seem to use unit weight when making a least-ares analysis of their data. However, also in this case there is a weighting of the data by number of experimental determinations in the parameter range where the different complexes are formed. In order to have comparable uncertainty estimates for the different complexes, one should try to have the same number of experimental data points in the centration ranges where each of these complexes is predominant; a procedure very rarely d.

As indicated above, the assignment of uncertainties to equilibrium constants is not traightforward procedure and it is complicated further when there is lack of primary experimental data. The uncertainty estimates given for the individual equilibrium constants orted by the authors and for some cases re-estimated by this review are given in the tas of this and previous reviews. The procedure used to obtain these estimates is given in original publications and in the Appendix A discussions. However, this uncertainty is a subjective estimate and to a large extent based on "expert judgment".

.2 Uncertainty estimates in the selected thermodynamic data.

e uncertainty estimate in the selected thermodynamic data is based on the uncertainty of individual equilibrium constants or other thermodynamic data, calculated as described in following sections. A weighted average of the individual $\log_{10}K$ values is calculated using the estimated uncertainty of the individual experimental values to assign its weight. The certainty in this average is then calculated using the formulae given in the following text. is uncertainty depends on the number of experimental data points – for N data point with

the same estimated uncertainty, σ, the uncertainty in the average is $\sigma/\sqrt{N}$. The aver
and the associated uncertainty reported in the tables of selected data are reported with ma
more digits than justified only in order to allow the users to back-track the calculations.
reported uncertainty is much smaller than the estimated experimental uncertainty and
users of the tables should look at the discussion of the selected constants in order to ge
better estimate of the uncertainty in an experimental determination using a specific metho

One of the objectives of the NEA Thermochemical Data Base (TDB) project is
provide an idea of the uncertainties associated with the data selected in this review. A
rule, the uncertainties define the range within which the corresponding data can be rep
duced with a probability of 95% at any place and by any appropriate method. In many cas
the statistical treatment is limited or impossible due to the availability of only one or
data points. A particular problem has to be solved when significant discrepancies occur
tween different source data. This appendix outlines the statistical procedures, which w
used for fundamentally different problems, and explains the philosophy used in this revi
when statistics were inapplicable. These rules are followed consistently throughout the
ries of reviews within the TDB Project. Four fundamentally different cases are considere

1. One source datum available

2. Two or more independent source data available

3. Several data available at different ionic strengths

4. Data at non-standard conditions: Procedures for data correction
   recalculation.

C.3 One source datum

The assignment of an uncertainty to a selected value that is based on only one experimer
source is a highly subjective procedure. In some cases, the number of data points, on wh
the selected value is based, allows the use of the "root mean square" [82TAY] deviation
the data points, X_i, to describe the standard deviation, s_X, associated with the average, $\overline{X}$

$$s_X = \sqrt{\frac{1}{N-1} \sum_{i=1}^{N}(X_i - \overline{X})^2}$$ (C

The standard deviation, s_X, is thus calculated from the dispersion of the equally weigh
data points, X_i, around the average $\overline{X}$, and the probability is 95% that an X_i is wit
$\overline{X} \pm 1.96\ s_X$, see Taylor [82TAY] (pp. 244-245). The standard deviation, s_X, is a measure
the precision of the experiment and does not include any systematic errors.

Many authors report standard deviations, s_X, calculated with Eq. (C.4) (but often multiplied by 1.96), but these do not represent the quality of the reported values in absolute terms. Therefore, it is thus important not to confuse the standard deviation, s_X, with the uncertainty, σ. The latter reflects the reliability and reproducibility of an experimental value and also includes all kinds of systematic errors, s_j, that may be involved. The uncertainty, σ, can be calculated with Eq. (C.5), assuming that the systematic errors are independent.

$$\sigma_X = \sqrt{s_X^2 + \sum_j (s_j^2)} \tag{C.5}$$

The estimation of the systematic errors s_j (which, of course, have to relate to $\overline{X}$ and be expressed in the same units), can only be made by a person who is familiar with the experimental method. The uncertainty, σ, has to correspond to the 95% confidence level preferred in this review. It should be noted that for all the corrections and recalculations made (e.g., temperature or ionic strength corrections) the rules of the propagation of errors have to be followed, as outlined in Section C.6.2.

More often, the determination of s_X is impossible because either only one or two data points are available, or the authors did not report the individual values. The uncertainty in the resulting value can still be estimated using Eq. (C.5) assuming that s_X^2 is much smaller than $\sum_j (s_j^2)$, which is usually the case anyway.

4 Two or more independent source data

Frequently, two or more experimental data sources are available, reporting experimental determinations of the desired thermodynamic data. In general, the quality of these determinations varies widely, and the data have to be weighted accordingly for the calculation of the mean. Instead of assigning weight factors, the individual source data, X_i, are provided with an uncertainty, σ_i, that also includes all systematic errors and represents the 95% confidence level, as described in Section C.1. The weighted mean $\overline{X}$ and its uncertainty, $\sigma_{\overline{X}}$, are then calculated according to Eqs. (C.6) and (C.7).

$$\overline{X} = \frac{\sum_{i=1}^{N}\left(\dfrac{X_i}{\sigma_i^2}\right)}{\sum_{i=1}^{N}\left(\dfrac{1}{\sigma_i^2}\right)} \tag{C.6}$$

$$\sigma_{\overline{X}} = \sqrt{\dfrac{1}{\sum_{i=1}^{N}\left(\dfrac{1}{\sigma_i^2}\right)}} \tag{C.7}$$

C. Assigned uncertainties

Eqs. (C.6) and (C.7) may only be used if all the X_i belong to the same parent distribution there are serious discrepancies among the X_i, one proceeds as described below under Sect C.2.1. It can be seen from Eq. (C.7) that $\sigma_{\bar{X}}$ is directly dependent on the absolute magtude of the σ_i values, and not on the dispersion of the data points around the mean. Thi reasonable because there are no discrepancies among the X_i, and because the σ_i values ready represent the 95% confidence level. The selected uncertainty, $\sigma_{\bar{X}}$, will therefore a represent the 95% confidence level.

In cases where all the uncertainties are equal, $\sigma_i = \sigma$, Eqs. (C.6) and (C.7) reduce Eqs. (C.8) and (C.9).

$$\bar{X} = \frac{1}{N}\sum_{i=1}^{N} X_i \qquad (C$$

$$\sigma_{\bar{X}} = \frac{\sigma}{\sqrt{N}} \qquad (C$$

Example C.1:

Five data sources report values for the thermodynamic quantity, X. The reviewer has signed uncertainties that represent the 95% confidence level as described in Section C.3.

i	X_i	σ_i
1	25.3	0.5
2	26.1	0.4
3	26.0	0.5
4	24.85	0.25
5	25.0	0.6

According to Eqs.(C.6) and (C.7), the following result is obtained:

$$\bar{X} = (25.3 \pm 0.2).$$

The calculated uncertainty, $\sigma_{\bar{X}} = 0.2$, appears relatively small, but is statistica correct, as the values are assumed to follow a Gaussian distribution. As a consequence Eq. (C.7), $\sigma_{\bar{X}}$ will always come out smaller than the smallest σ_i. Assuming $\sigma_4 = 0.10$ stead of 0.25 would yield $\bar{X} = (25.0 \pm 0.1)$ and $\sigma_4 = 0.60$ would result in $\bar{X} = (25.6 \pm 0$ In fact, the values $(X_i \pm \sigma_i)$ in this example are at the limit of consistency, i.e., the ra $(X_4 \pm \sigma_4)$ does not overlap with the ranges $(X_2 \pm \sigma_2)$ and $(X_3 \pm \sigma_3)$. There might be a be way to solve this problem. Three possible choices seem more reasonable:

i. The uncertainties, σ_i, are reassigned because they appear too optimistic after further consideration. Some assessments may have to be reconsidered and the uncertainties reassigned. For example, multiplying all the σ_i by 2 would yield $\overline{X} = (25.3 \pm 0.3)$.

i. If reconsideration of the previous assessments gives no evidence for reassigning the X_i and σ_i (95% confidence level) values listed above, the statistical conclusion will be that all the X_i do not belong to the same parent distribution and cannot therefore be treated in the same group (cf. item iii below for a non–statistical explanation). The values for $i = 1$, 4 and 5 might be considered as belonging to Group A and the values for $i = 2$ and 3 to Group B. The weighted average of the values in Group A is X_A ($i = 1, 4, 5) = (24.95 \pm 0.21)$ and of those in Group B, X_B ($i = 2, 3) = (26.06 \pm 0.31)$, the second digit after the decimal point being carried over to avoid loss of information. The selected value is now determined as described below under "Discrepancies" (Section C.4.1, Case I). X_A and X_B are averaged (straight average, there is no reason for giving X_A a larger weight than X_B), and $\sigma_{\overline{X}}$ is chosen in such a way that it covers the complete ranges of expectancy of X_A and X_B. The selected value is then $\overline{X} = (25.5 \pm 0.9)$.

i. Another explanation could be that unidentified systematic errors are associated with some values. If this seems likely to be the case, there is no reason for splitting the values up into two groups. The correct way of proceeding would be to calculate the unweighted average of all the five points and assign an uncertainty that covers the whole range of expectancy of the five values. The resulting value is then $\overline{X} = (25.45 \pm 1.05)$, which is rounded according to the rules in Section C.6.3 to $\overline{X} = (25.4 \pm 1.1)$.

4.1 Discrepancies

Two data are called discrepant if they differ significantly, *i.e.*, their uncertainty ranges do not overlap. In this context, two cases of discrepancies are considered. Case I: Two significantly different source data are available. Case II: Several, mostly consistent source data are available, one of them being significantly different, *i.e.*, an "outlier".

Case I. Two discrepant data: This is a particularly difficult case because the number of data points is obviously insufficient to allow the preference of one of the two values. If there is absolutely no way of discarding one of the two values and selecting the other, the only solution is to average the two source data in order to obtain the selected value, because the underlying reason for the discrepancy must be unrecognised systematic errors. There is no point in calculating a weighted average, even if the two source data have been given different uncertainties, because there is obviously too little information to give even only limited preference to one of the values. The uncertainty, $\sigma_{\bar{X}}$, assigned to the lected mean, $\bar{X}$, has to cover the range of expectation of both source data, X_1, X_2, as sho in Eq.(C.10),

$$\sigma_{\bar{X}} = |X_i - \bar{X}| + \sigma_{max} \qquad (C.1$$

where $i = 1, 2,$ and σ_{max} is the larger of the two uncertainties σ_i, see Example C.1.ii and I ample C.2.

Example C.2:

The following credible source data are given:

$$X_1 = (4.5 \pm 0.3)$$
$$X_2 = (5.9 \pm 0.5).$$

The uncertainties have been assigned by the reviewer. Both experimental meth are satisfactory and there is no justification to discard one of the data. The selected value then:

$$\bar{X} = (5.2 \pm 1.2).$$

Figure C-1: Illustration for Example C.2

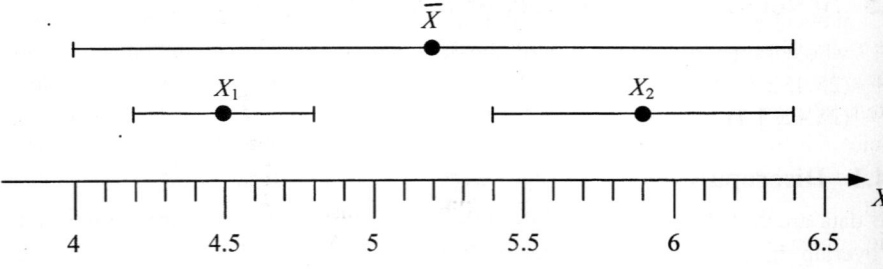

Case II. Outliers: This problem can often be solved by either discarding the outlying d point, or by providing it with a large uncertainty to lower its weight. If, however, the ou ing value is considered to be of high quality and there is no reason to discard all the ot data, this case is treated in a way similar to Case I. Example C.3 illustrates the procedure.

Example C.3:

The following data points are available. The reviewer has assigned the uncertainties and gives no justification for any change.

i	X_i	σ_i
1	4.45	0.35
2	5.9	0.5
3	5.7	0.4
4	6.0	0.6
5	5.2	0.4

There are two data sets that, statistically, belong to different parent distributions, A and B. According to Eqs. (C.6) and (C.7), the following average values are found for the two groups: X_A ($i = 1$) = (4.45 ± 0.35) and X_B (i = 2, 3, 4, 5) = (5.62 ± 0.23). The selected value will be the straight average of X_A and X_B, analogous to Example C.1:

$$\overline{X} = (5.0 \pm 0.9).$$

C.5 Several data at different ionic strengths

The extrapolation procedure for aqueous equilibria used in this review is the specific ion interaction model outlined in Appendix B. The objective of this review is to provide selected data sets at standard conditions, *i.e.*, among others, at infinite dilution for aqueous species. Equilibrium constants determined at different ionic strengths can, according to the specific ion interaction equations, be extrapolated to $I = 0$ with a linear regression model, yielding as the intercept the desired equilibrium constant at $I = 0$, and as the slope the stoichiometric sum of the ion interaction coefficients, $\Delta\varepsilon$. The ion interaction coefficient of the target species can usually be extracted from $\Delta\varepsilon$ and would be listed in the corresponding table of Appendix C.

The available source data may sometimes be sparse or may not cover a sufficient range of ionic strengths to allow a proper linear regression. In this case, the correction to $I = 0$ should be carried out according to the procedure described in Section C.6.1.

If sufficient data are available at different ionic strengths and in the same inert salt medium, a weighted linear regression will be the appropriate way to obtain both the constant at $I = 0$, $\overline{X}°$, and $\Delta\varepsilon$. The first step is the conversion of the ionic strength from the frequently used molar (mol·dm^{-3}, M) to the molal (mol·kg^{-1}, m) scale, as described in Section

2.2. The second step is the assignment of an uncertainty, σ_i, to each data point X_i at the m lality, $m_{k,i}$, according to the rules described in Section C.3. A large number of commerc and public domain computer programs and routines exist for weighted linear regression The subroutine published by Bevington [69BEV] (pp. 104 – 105) has been used for the c culations in the examples of this appendix. Eqs. (C.11) through (C.15) present the equation that are used for the calculation of the intercept $\bar{X}^\circ$ and the slope $-\Delta\varepsilon$:

$$\bar{X}^\circ = \frac{1}{\Delta}\left(\sum_{i=1}^{N}\frac{m_{k,i}^2}{\sigma_i^2}\sum_{i=1}^{N}\frac{X_i}{\sigma_i^2} - \sum_{i=1}^{N}\frac{m_{k,i}}{\sigma_i^2}\sum_{i=1}^{N}\frac{m_{k,i}X_i}{\sigma_i^2}\right) \quad (C.1)$$

$$-\Delta\varepsilon = \frac{1}{\Delta}\left(\sum_{i=1}^{N}\frac{1}{\sigma_i^2}\sum_{i=1}^{N}\frac{m_{k,i}X_i}{\sigma_i^2} - \sum_{i=1}^{N}\frac{m_{k,i}}{\sigma_i^2}\sum_{i=1}^{N}\frac{X_i}{\sigma_i^2}\right) \quad (C.1\text{:})$$

$$\sigma_{\bar{X}^\circ} = \sqrt{\frac{1}{\Delta}\sum_{i=1}^{N}\frac{m_{k,i}^2}{\sigma_i^2}} \quad (C.1\text{:})$$

$$\sigma_{\Delta\varepsilon} = \sqrt{\frac{1}{\Delta}\sum_{i=1}^{N}\frac{1}{\sigma_i^2}} \quad (C.1\text{·})$$

where $\quad \Delta = \sum_{i=1}^{N}\frac{1}{\sigma_i^2}\sum_{i=1}^{N}\frac{m_{k,i}^2}{\sigma_i^2} - \left(\sum_{i=1}^{N}\frac{m_{k,i}}{\sigma_i^2}\right)^2.$ (C.1

In this way, the uncertainties, σ_i, are not only used for the weighting of the data Eqs. (C.11) and (C.12), but also for the calculation of the uncertainties, $\sigma_{\bar{X}^\circ}$ and $\sigma_{\Delta\varepsilon}$, in E (C.13) and (C.14). If the σ_i represents the 95% confidence level, $\sigma_{\bar{X}^\circ}$ and $\sigma_{\Delta\varepsilon}$ will also so. In other words, the uncertainties of the intercept and the slope do not depend on the d persion of the data points around the straight line, but rather directly on their absolute unc tainties, σ_i.

Example C.4:

Ten independent determinations of the equilibrium constant, $\log_{10}{^*\beta}$, for the reaction:

$$UO_2^{2+} + HF(aq) \rightleftharpoons UO_2F^+ + H^+ \quad (C.16)$$

are available in $HClO_4/NaClO_4$ media at different ionic strengths. Uncertainties that rep sent the 95% confidence level have been assigned by the reviewer. A weighted linear gression, $(\log_{10}{^*\beta} + 2D)$ vs. m_k, according to the formula, $\log_{10}{^*\beta}$ (C.16) $+ 2D$ $\log_{10}{^*\beta^\circ}$ (C.16) $- \Delta\varepsilon\, m_k$, will yield the correct values for the intercept, $\log_{10}{^*\beta^\circ}$ (C.1

C.5 Several data at different ionic strengths

nd the slope, $\Delta\varepsilon$. In this case, m_k corresponds to the molality of ClO_4^-. D is the Debye–Hückel term, cf. Appendix B.

i	$m_{ClO_4^-, i}$	$\log_{10} {}^*\beta + 2D$	σ_i
1	0.05	1.88	0.10
2	0.25	1.86	0.10
3	0.51	1.73	0.10
4	1.05	1.84	0.10
5	2.21	1.88	0.10
6	0.52	1.89	0.11
7	1.09	1.93	0.11
8	2.32	1.78	0.11
9	2.21	2.03	0.10
10	4.95	2.00	0.32

The results of the linear regression are:

$$\text{intercept} = (1.837 \pm 0.054) = \log_{10} {}^*\beta° \quad (C.16)$$

$$\text{slope} = (0.029 \pm 0.036) = -\Delta\varepsilon$$

Calculation of the ion interaction coefficient $\varepsilon(UO_2F^+, ClO_4^-) = \Delta\varepsilon + \varepsilon(UO_2^{2+}, ClO_4^-) - \varepsilon(H^+, ClO_4^-)$: from $\varepsilon(UO_2^{2+}, ClO_4^-) = (0.46 \pm 0.03)$ kg·mol^{-1}, $\varepsilon(H^+, ClO_4^-)$ (0.14 ± 0.02) kg·mol^{-1} (see Appendix B) and the slope of the linear regression, $\Delta\varepsilon = (0.03 \pm 0.04)$ kg·mol^{-1}, it follows that $\varepsilon(UO_2F^+, ClO_4^-) = (0.29 \pm 0\ 05)$ kg·mol^{-1}. Note that the uncertainty ($\pm 0\ 05$) kg·mol^{-1} is obtained based on the rules of error propagation as described in Section C.6.2:

$$\sigma = \sqrt{(0.04)^2 + (0.03)^2 + (0.02)^2}$$

The resulting selected values are thus:

$$\log_{10} {}^*\beta° (C.16) = (1.84 \pm 0.05)$$

$$\varepsilon(UO_2F^+, ClO_4^-) = (0.29 \pm 0.05) \text{ kg·mol}^{-1}.$$

C.5.1 Discrepancies or insufficient number of data points

Discrepancies are principally treated as described in Section C.4. Again, two cases can [be] defined. Case I: Only two data points are available. Case II: An "outlier" cannot be discarded. If only one data point is available, the procedure for correction to zero ionic strength outlined in Section C.4 should be followed.

Case I. Too few molalities: If only two source data are available, there will be [no] straightforward way to decide whether or not these two data points belong to the same parent distribution unless either the slope of the straight line is known or the two data refer [to] the same ionic strength. Drawing a straight line right through the two data points is an inappropriate procedure because all the errors associated with the two source data would accumulate and may lead to highly erroneous values of $\log_{10} K°$ and $\Delta\varepsilon$. In this case, an ion interaction coefficient for the key species in the reaction in question may be selected [by] analogy (charge is the most important parameter), and a straight line with the slope $\Delta\varepsilon$ calculated may then be drawn through each data point. If there is no reason to discard one [of] the two data points based on the quality of the underlying experiment, the selected value will be the unweighted average of the two standard state data point obtained by this procedure, and its uncertainty must cover the entire range of expectancy of the two values, analogous to Case I in Section C.4. It should be mentioned that the ranges of expectancy of t[he] corrected values at $I = 0$ are given by their uncertainties, which are based on the uncertainties of the source data at $I \neq 0$ and the uncertainty in the slope of the straight line. The lat[ter] uncertainty is not an estimate, but is calculated from the uncertainties in the ion interaction coefficients involved, according to the rules of error propagation outlined in Section C.6. The ion interaction coefficients estimated by analogy are listed in the table of selected i[on] interaction coefficients (Appendix B), but they are flagged as estimates.

Case II. Outliers and inconsistent data sets: This case includes situations whe[re] it is difficult to decide whether or not a large number of points belong to the same pare[nt] distribution. There is no general rule on how to solve this problem, and decisions are left [to] the judgment of the reviewer. For example, if eight data points follow a straight line re[a]sonably well and two lie way out, it may be justified to discard the "outliers". If, howev[er] the eight points are scattered considerably and two points are just a bit further out, one c[ould] probably not consider them as "outliers". It depends on the particular case and on the jud[g]ment of the reviewer whether it is reasonable to increase the uncertainties of the data [to] reach consistency, or whether the slope, $\Delta\varepsilon$, of the straight line should be estimated [by] analogy.

Example C.5:

Six reliable determinations of the equilibrium constant, $\log_{10} \beta$, of the reaction:

C.5 Several data at different ionic strengths

$$UO_2^{2+} + SCN^- \rightleftharpoons UO_2SCN^+ \qquad (C.17)$$

available in different electrolyte media:

$I_c = 0.1$ M (KNO$_3$):	$\log_{10} \beta$ (C.17) = (1.19 ± 0.03)
$I_c = 0.33$ M (KNO$_3$):	$\log_{10} \beta$ (C.17) = (0.90 ± 0.10)
$I_c = 1.0$ M (NaClO$_4$):	$\log_{10} \beta$ (C.17) = (0.75 ± 0.03)
$I_c = 1.0$ M (NaClO$_4$):	$\log_{10} \beta$ (C.17) = (0.76 ± 0.03)
$I_c = 1.0$ M (NaClO$_4$):	$\log_{10} \beta$ (C.17) = (0.93 ± 0.03)
$I_c = 2.5$ M (NaNO$_3$):	$\log_{10} \beta$ (C.17) = (0.72 ± 0.03)

The uncertainties are assumed to represent the 95% confidence level. From the values at $I_c = 1$ M, it can be seen that there is a lack of consistency in the data, and that a linear regression similar to that shown in Example C.4 would be inappropriate. Instead, the use of $\Delta\varepsilon$ values from reactions of the same charge type is encouraged. Analogies with $\Delta\varepsilon$ are more reliable than analogies with single ε values due to canceling effects. For the same reason, the dependency of $\Delta\varepsilon$ on the type of electrolyte is often smaller than for single ε values.

A reaction of the same charge type as Reaction (C.17), and for which $\Delta\varepsilon$ is well known, is:

$$UO_2^{2+} + Cl^- \rightleftharpoons UO_2Cl^+. \qquad (C.18)$$

The value of $\Delta\varepsilon$ (C.18) = − (0.25 ± 0.02) kg · mol^{-1} was obtained from a linear regression using 16 experimental values between $I_c = 0.1$ M and $I_c = 3$ M Na(Cl,ClO$_4$) [92GRE/FUG]. It is thus assumed that:

$$\Delta\varepsilon \, (C.17) = \Delta\varepsilon \, (C.18) = - (0.25 \pm 0.02) \text{ kg} \cdot \text{mol}^{-1}.$$

The correction of $\log_{10} \beta$ (C.17) to $I_c = 0$ is done using the specific ion interaction equation, cf. TDB–2, which uses molal units:

$$\log_{10} \beta + 4D = \log_{10} \beta^\circ - \Delta\varepsilon \, I_m. \qquad (C.19)$$

D is the Debye-Hückel term in molal units and I_m the ionic strength converted to molal units by using the conversion factors listed in Table 2-5. The following list gives the details of this calculation. The resulting uncertainties in $\log_{10} \beta$ are obtained based on the rules of error propagation as described in Section C.6.2.

Table C-1: Details of the calculation of equilibrium constant corrected to $I = 0$, using (C.1

I_m	electrolyte	$\log_{10} \beta$	$4D$	$\Delta\varepsilon\, I_m$	$\log_{10} \beta°$
0.101	KNO$_3$	(1.19 ± 0.03)	0.438	− 0.025	(1.68 ± 0.03)[a]
0.335	KNO$_3$	(0.90 ± 0.10)	0.617	− 0.084	(1.65 ± 0.10)[a]
1.050	NaClO$_4$	(0.75 ± 0.03)	0.822	− 0.263	(1.31 ± 0.04)
1.050	NaClO$_4$	(0.76 ± 0.03)	0.822	− 0.263	(1.32 ± 0.04)
1.050	NaClO$_4$	(0.93 ± 0.03)	0.822	− 0.263	(1.49 ± 0.04)
2.714	NaNO$_3$	(0.72 ± 0.03)	0.968	− 0.679	(1.82 ± 0.13)[a]

(a) These values were corrected for the formation of the nitrate complex, $UO_2NO_3^+$, by using $\log_{10} K(UO_2NO_3^+) = (0.30 \pm 0.15)$ [92GRE/FUG].

As was expected, the resulting values, $\log_{10} \beta°$, are inconsistent and have the fore to be treated as described in Case I of Section C.4. That is, the selected value will be t unweighted average of $\log_{10} \beta°$, and its uncertainty will cover the entire range of expe tancy of the six values. A weighted average would only be justified if the six values $\log_{10} \beta°$ were consistent. The result is:

$$\log_{10} \beta° = (1.56 \pm 0.39).$$

C.6 Procedures for data handling

C.6.1 Correction to zero ionic strength

The correction of experimental data to zero ionic strength is necessary in all cases where linear regression is impossible or appears inappropriate. The method used throughout t review is the specific ion interaction equations described in detail in Appendix B. Two va ables are needed for this correction, and both have to be provided with an uncertainty at 95% confidence level: the experimental source value, $\log_{10} K$ or $\log_{10} \beta$, and t stoichiometric sum of the ion interaction coefficients, $\Delta\varepsilon$. The ion interaction coefficie (see Tables B.3 and B.4 of Appendix B) required to calculate $\Delta\varepsilon$ may not all be knov Missing values therefore need to be estimated. It is recalled that the electric charge has t most significant influence on the magnitude of the ion interaction coefficients, and that i in general more reliable to estimate $\Delta\varepsilon$ from known reactions of the same charge ty rather than to estimate single $\Delta\varepsilon$ values. The uncertainty of the corrected value at $I = 0$ calculated by taking into account the propagation of errors, as described below. It should noted that the ionic strength is frequently given in moles per dm^3 of solution (molar, M) a

C.6 Procedures for data handling

...s to be converted to moles per kg H_2O (molal, m), as the model requires. Conversion factors for the most common inert salts are given in Table 2.5.

Example C.6:

For the equilibrium constant of the reaction:

$$M^{3+} + 2\,H_2O(l) \rightleftharpoons M(OH)_2^+ + 2\,H^+, \qquad (C.20)$$

only one credible determination in 3 M $NaClO_4$ solution is known to be, $\log_{10}{}^*\!\beta$ (C.20) = -6.31, to which an uncertainty of ± 0.12 has been assigned. The ion interaction coefficients are as follows:

$$\varepsilon(M^{3+}, ClO_4^-) = (0.56 \pm 0.03)\ \text{kg·mol}^{-1},$$
$$\varepsilon(M(OH)_2^+, ClO_4^-) = (0.26 \pm 0.11)\ \text{kg·mol}^{-1},$$
$$\varepsilon(H^+, ClO_4^-) = (0.14 \pm 0.02)\ \text{kg·mol}^{-1}.$$

The values of $\Delta\varepsilon$ and $\sigma_{\Delta\varepsilon}$ can be obtained readily (cf. Eq. (C.22)):

$$\Delta\varepsilon = \varepsilon(M(OH)_2^+, ClO_4^-) + 2\varepsilon(H^+, ClO_4^-) - \varepsilon(M^{3+}, ClO_4^-) = -0.02\ \text{kg·mol}^{-1},$$

$$\sigma_{\Delta\varepsilon} = \sqrt{(0.11)^2 + (2\times 0.02)^2 + (0.03)^2} = 0.12\ \text{kg·mol}^{-1}.$$

The two variables are thus:

$$\log_{10}{}^*\!\beta\ (C.20) = -(6.31 \pm 0.12),$$
$$\Delta\varepsilon = -(0.02 \pm 0.12)\ \text{kg·mol}^{-1}.$$

According to the specific ion interaction model the following equation is used to correct for ionic strength for the reaction considered here:

$$\log_{10}{}^*\!\beta\ (C.20) + 6D = \log_{10}{}^*\!\beta^\circ\ (C.20) - \Delta\varepsilon\, m_{ClO_4^-}$$

D is the Debye-Hückel term:

$$D = \frac{0.509\sqrt{I_m}}{(1+1.5\sqrt{I_m})}.$$

The ionic strength, I_m, and the molality, $m_{ClO_4^-}$ ($I_m \approx m_{ClO_4^-}$), have to be expressed in molal units, 3 M $NaClO_4$ corresponding to 3.5 m $NaClO_4$ (see Section 2.2), giving D = 0.25. This results in:

$$\log_{10}{}^*\!\beta^\circ\ (C.20) = -4.88.$$

The uncertainty in $\log_{10} {}^*\beta°$ is calculated from the uncertainties in $\log_{10} {}^*\beta$ a $\Delta\varepsilon$ (cf. Eq. (C.22)):

$$\sigma_{\log_{10} {}^*\beta°} = \sqrt{\sigma^2_{\log_{10} {}^*\beta} + (m_{ClO_4^-} \sigma_{\Delta\varepsilon})^2} = \sqrt{(0.12)^2 + (3.5 \times 0.12)^2} = 0.44.$$

The selected, rounded value is:

$$\log_{10} {}^*\beta° \text{(C.20)} = -(4.9 \pm 0.4).$$

C.6.2 Propagation of errors

Whenever data are converted or recalculated, or other algebraic manipulations are performed that involve uncertainties, the propagation of these uncertainties has to be taken in account in a correct way. A clear outline of the propagation of errors is given by Bevingt [69BEV]. A simplified form of the general formula for error propagation is given Eq.(C.21), supposing that X is a function of $Y_1, Y_2, \ldots, Y_N$.

$$\sigma_X^2 = \sum_{i=1}^N \left(\frac{\partial X}{\partial Y_i} \sigma_{Y_i} \right)^2 \tag{C.2}$$

Eq. (C.21) can be used only if the variables, $Y_1, Y_2, \ldots, Y_N$, are independent or their uncertainties are small, i.e., the covariances can be disregarded. One of these two assumptions can almost always be made in chemical thermodynamics, and Eq. (C.21) can th almost universally be used in this review. Eqs. (C.22) through (C.26) present explicit form las for a number of frequently encountered algebraic expressions, where c, c_1, c_2 are constants.

$X = c_1 Y_1 \pm c_2 Y_2$: $\sigma_X^2 = (c_1 \sigma_{Y_1})^2 + (c_2 \sigma_{Y_2})^2$ (C.2

$X = \pm c Y_1 Y_2$ and $X = \pm \dfrac{c Y_1}{Y_2}$: $\left(\dfrac{\sigma_X}{X}\right)^2 = \left(\dfrac{\sigma_{Y_1}}{Y_1}\right)^2 + \left(\dfrac{\sigma_{Y_2}}{Y_2}\right)^2$ (C.2

$X = c_1 Y^{\pm c_2}$: $\dfrac{\sigma_X}{X} = c_2 \dfrac{\sigma_Y}{Y}$ (C.2

$X = c_1 e^{\pm c_2 Y}$: $\dfrac{\sigma_X}{X} = c_2 \sigma_Y$ (C.2

$X = c_1 \ln(\pm c_2 Y)$: $\sigma_X = c_1 \dfrac{\sigma_Y}{Y}$ (C.2

Example C.7:

A few simple calculations illustrate how these formulas are used. The values have not be rounded.

Eq. (C.22): $\Delta_r G_m = 2 \cdot [-(277.4 \pm 4.9)] \text{ kJ·mol}^{-1} - [-(467.3 \pm 6.2)] \text{ kJ·mol}^{-1}$
$= -(87.5 \pm 11.6) \text{ kJ·mol}^{-1}$.

Eq. (C.23): $K = \dfrac{(0.038 \pm 0.002)}{(0.0047 \pm 0.0005)} = (8.09 \pm 0.92)$

Eq. (C.24): $K = 4 \cdot (3.75 \pm 0.12)^3 = (210.9 \pm 20.3)$

Eq. (C.25): $K° = e^{\frac{-\Delta_r G_m^\circ}{RT}}$; $\quad \Delta_r G_m^\circ = -(2.7 \pm 0.3) \text{ kJ·mol}^{-1}$

$R = 8.3145 \text{ J·K}^{-1}\text{·mol}^{-1}$

$T = 298.15 \text{ K}$

$K° = (2.97 \pm 0.36)$.

Note that powers of 10 have to be reduced to powers of e, i.e., the variable has to multiplied by $\ln(10)$, e.g.,

$\log_{10} K = (2.45 \pm 0.10); \ K = 10^{\log_{10} K} = e^{(\ln(10) \cdot \log_{10} K)} = (282 \pm 65)$.

Eq. (C.26): $\Delta_r G_m^\circ = -RT \ln K°$; $\quad K° = (8.2 \pm 1.2) \times 10^6$

$R = 8.3145 \text{ J·K}^{-1}\text{·mol}^{-1}$

$T = 298.15 \text{ K}$

$\Delta_r G_m^\circ = -(39.46 \pm 0.36) \text{ kJ·mol}^{-1}$

$\ln K° = (15.92 \pm 0.15)$

$\log_{10} K° = \ln K° / \ln(10) = (6.91 \pm 0.06)$.

Again, it can be seen that the uncertainty in $\log_{10} K°$ cannot be the same as in $K°$. The constant conversion factor of $\ln(10) = 2.303$ is also to be applied to the uncertainty.

6.3 Rounding

e standard rules to be used for rounding are:

1. When the digit following the last digit to be retained is less than 5, the last digit retained is kept unchanged.

2. When the digit following the last digit to be retained is greater than 5, the last digit retained is increased by 1.

3. When the digit following the last digit to be retained is 5 and

   a) there are no digits (or only zeroes) beyond the 5, an odd digit in the last place to be retained is increased by 1 while an even digit is kept unchanged.

   b) other non–zero digits follow, the last digit to be retained is increased 1, whether odd or even.

This procedure avoids introducing a systematic error from always dropping or n dropping a 5 after the last digit retained.

When adding or subtracting, the result is rounded to the number of decimal plac (not significant digits) in the term with the least number of places. In multiplication a division, the results are rounded to the number of significant digits in the term with the lea number of significant digits.

In general, all operations are carried out in full, and only the final results a rounded, in order to avoid the loss of information from repeated rounding. For this reaso several additional digits are carried in all calculations until the final selected data set is d veloped, and only then are data rounded.

C.6.4 Significant digits

The uncertainty of a value basically defines the number of significant digits a value shou be given.

$$\text{Example: } (3.478 \pm 0.008)$$

$$(3.48 \pm 0.01)$$

$$(2.8 \pm 0.4)$$

$$(10 \pm 1)$$

$$(105 \pm 20).$$

In the case of auxiliary data or values that are used for later calculations, it is oft inconvenient to round to the last significant digit. In the value (4.85 ± 0.26), for examp the "5" is close to being significant and should be carried along a recalculation path in ord to avoid loss of information. In particular cases, where the rounding to significant dig could lead to slight internal inconsistencies, digits with no significant meaning in absolu terms are nevertheless retained. The uncertainty of a selected value always contains t same number of digits after the decimal point as the value itself.

Appendix D

Some limitations encountered in the use of the ionic strength correction procedures

the first book in the NEA-TDB series, [92GRE/FUG], the SIT approach was adopted as procedure for ionic strength corrections needed to compare experimental data obtained different ionic media and to reduce these to the standard state chosen for the NEA-TDB abase. The publication of the subsequent NEA-TDB reviews, [95SIL/BID], RAR/RAN] and [2001LEM/FUG] has established that the SIT approach provides an quate basis for the description of a very large category of aqueous systems.

In the OECD/NEA publication "Modelling in Aquatic Chemistry" [97ALL/BAN], re is an extensive discussion of ionic strength corrections showing that in most cases the cific ion interaction theory (SIT) [80CIA] is a good approximation, in particular when mating ion interaction parameters from thermodynamic equilibrium constants deter- ied in a single ionic medium; this is the most common situation encountered when evalu- g published data. However, in some particular cases, the simplifications introduced in SIT approach (cf. Appendix B) can lead to inaccuracies in the extrapolation to evaluate ilibrium constants at $I = 0$ (when using experimental data at low and high ionic strengths for equilibria involving ions with charge $|z| > 3$), and in the calculation of equilibrium stants in other ionic media. Some of these limitations, not discussed in the previous re- ws of the NEA-TDB series, are presented in the following sections. For a more extensive cussion the readers are referred to [97ALL/BAN]. The following discussion makes fre- nt reference to the Pitzer ion interaction model [91PIT], which can be formulated with- some of the underlying simplifications of the SIT approach. The reader should be aware, vever, that the correct use of Pitzer equations relies on the existence of very precise ex- imental data, such as are available for the ions in mixtures of strong electrolytes, but are vailable for chemical systems containing complexes. For this reason, in practice, the zer method has only been used on chemical systems where the speciation is well known. ere are no studies where the stoichiometry, equilibrium constants and the interaction parameters have been determined by use of the Pitzer model in a unique manner; the reas being the very strong correlation between the parameters in the Pitzer model. Some of the problems are discussed with examples in [97ALL/BAN]. In order to use the Pitzer model describe experimental equilibrium constants it is necessary to have data at low ion strength, but if such data are unavailable, then the equilibrium constant at zero ionic streng can be estimated more expediently by the SIT approach as was done throughout the NE TDB series.

D.1 Implications of neglecting ternary ion interactions and the use of constant values for $\varepsilon(j,k,I_m)$

Within the SIT approach (Eq.(B.1) in section. B.1.1), activity coefficients are described b Debye-Hückel term and an expansion accounting for ion-ion interactions between pairs ions of opposite charge. The ion interaction coefficients, $\varepsilon(j,k,I_m)$, are not strictly const but may vary with the ionic strength, particularly at very low and high ionic strengths. Ho ever, in practice, the ionic strength dependence of the ion interaction coefficients can o be determined for strong electrolyte systems and not for complexes, and this is the reas why in the SIT approach this dependence is most frequently neglected (see Appendix B. Consequently, the specific ion interaction term of Eq. (B.1) is a linear function of the mo electrolyte concentration. With increasing ionic strength, in particular at $I_m > 4$ mol · kg experimental data often deviate from this linear relationship. At these high ionic strengt the Pitzer approach is *a priori* more accurate because it takes more ion-ion interactions i account, such as triple ion interactions (reflected in the terms containing the $C_{i,j}^{\phi}$ and ψ_i parameters). The possible inaccuracies when using the SIT at $I_m > 4$ mol · kg^{-1} are we known and already mentioned in section B.1.1.

At low ionic strength ($I_m < 0.1$ m), the use of constant values for $\varepsilon(j,k)$ in the S approach may also impose a limitation. This is circumvented in Pitzer's model by the int duction of semi-empirical functions including the binary parameters, $\beta_{i,j}^{(0)}$, $\beta_{i,j}^{(1)}$ and β which enable accurate fits to be made to experimental osmotic and activity coefficients the cases of 3:1 and 2:2 electrolytes.

[97ALL/BAN] contains an extensive discussion of the magnitude of ion interact coefficients and the relation between the SIT coefficients, $\varepsilon(i,j)$, and the correspond Pitzer parameters, $\beta_{i,j}^{(0)}$ and $\beta_{i,j}^{(1)}$. The concentration dependence of the ion interaction coe cients in the SIT equation is of minor importance at low ionic strengths as the effect of interaction coefficients, $\varepsilon(i,j)$, on $\log_{10}\gamma_i$ is limited. Unfortunately, the Pitzer equations often used incorrectly as a tool to describe experimental data as accurately as possible by arbitrary fit, which does not necessarily constrain the equilibrium constants and activity efficients to realistic values. Numerous recent publications use the ion interaction equati

D.1: Implications of neglecting ternary ion interactions and the use of constant values for ε(j,k,Im)

Pitzer for ionic strength corrections, once a suitable chemical speciation model has been established. In these cases, if highly charged ions are involved in the reaction, the values obtained for equilibrium constants at $I = 0$ may differ from those obtained with the SIT approach (as in this and other reviews of the NEA-TDB series). The magnitude of the observed differences is illustrated below for some cases taken from the present review.

Examples for differences in equilibrium constants at $I = 0$ obtained with the SIT and Pitzer equations

The following examples illustrate that it is inadvisable to "mix" ionic interaction parameters and equilibrium constants at zero ionic strength obtained by the SIT and Pitzer methods. The present review selects in all cases the values extrapolated to $I = 0$ with the SIT approach. Therefore, these constants should *not* be combined with Pitzer parameters for the back-extrapolation to high ionic strength for reasons that are apparent in the following case studies.

a) Fanghänel et al. [98FAN/WEG2], [99FAN/KON] reported spectroscopic data in 0 – 6 m NaCl for the stepwise formation of the tetracarbonate complex of Cm(III) according to the reaction:

$$Cm(CO_3)_3^{3-} + CO_3^{2-} \rightleftharpoons Cm(CO_3)_4^{5-}. \tag{D.1}$$

The linear SIT extrapolation to $I = 0$ (Figure 12-9, section 12.6.1.1.1) appears to be straightforward and leads to $\log_{10} K_4^\circ = -(1.6 \pm 0.2)$. However, using a fixed value of $\log_{10} K_4^\circ = -1.6$ it is impossible to fit the experimental data in a reasonable way with the Pitzer equations. Conversely, using the Pitzer method throughout for the extrapolation to $I = 0$, Fanghänel et al. [99FAN/KON] calculated a significantly different value of $\log_{10} K_4^\circ = -2.16$ (the uncertainty estimation is lacking in the original publication, but it is deemed to be smaller than 0.6 logarithmic units).

b) In general the discrepancies between the two methods are much smaller than in the first example. As observed by [95FAN/NEC] and [97PAS/CZE] for the formation constants of $NpO_2(CO_3)_3^{5-}$ and $UO_2(CO_3)_3^{4-}$, the Pitzer model [95FAN/NEC] calculated $\log_{10} \beta_3^\circ$ ($NpO_2(CO_3)_3^{5-}$, 298.15 K) = (5.37 ± 0.36), while the SIT gave (5.54 ± 0.09), close to (5.50 ± 0.15) selected by [2001LEM/FUG]. Based on the experimental data accepted in [92GRE/FUG], Pashalidis et al. [97PAS/CZE] calculated $\log_{10} \beta_3^\circ$ ($UO_2(CO_3)_3^{4-}$, 298.15 K) = 21.3 (no uncertainty reported) with the Pitzer equations, which is somewhat lower than (21.60 ± 0.05) obtained in [92GRE/FUG] with the SIT.

c) In Appendix A, (*cf.* discussion of [97NOV/ALM] and [98ALM/NOV]), the SIT extrapolation yielded $\log_{10} K_{s,3}^\circ$ (D.2) = $-(7.65 \pm 0.11)$ and $\log_{10} K_{s,3}^\circ$ (D.3) = $-(9.96 \pm 0.06)$ for the reactions:

$$\text{KNpO}_2\text{CO}_3(s) + 2\,\text{CO}_3^{2-} \rightleftharpoons \text{NpO}_2(\text{CO}_3)_3^{5-} + \text{K}^+ \quad \text{(D.2)}$$

$$\text{K}_3\text{NpO}_2(\text{CO}_3)_2(s) + \text{CO}_3^{2-} \rightleftharpoons \text{NpO}_2(\text{CO}_3)_3^{5-} + 3\,\text{K}^+. \quad \text{(D.3)}$$

while the [97NOV/ALM] calculated $\log_{10} K^\circ_{s,3}$ (D.2) = $- (8.2 \pm 0.1)$ and $\log_{10} K^\circ_{s,3}$ (D.3) $- (10.5 \pm 0.1)$ with the Pitzer equations.

D.2 Implications of neglecting anion-anion and cation-cation interactions

In section B.1.1 the simplification is made that ion interaction coefficients, $\varepsilon(j,k)$, may set equal to zero for ions j and k of the same charge sign (assumption 2). This simplification is an extension of the convention, $\varepsilon(i,i) = 0$, between the same ions, which is generally applied in the calculation of ionic activity coefficients, γ_i, with the SIT. Ionic activity coefficients calculated with the Pitzer equations are usually based on an analogous convention ($\lambda_{(i,i)} = 0$, $\mu_{(i,i,i)} = 0$). In the simple case of a binary electrolyte solution MX this leads to:

$$\gamma_M = \gamma_X = \gamma_{\pm(MX)} \quad \text{(D.4)}$$

(splitting convention for ionic activity coefficients).

The use of conventional activity coefficients follows a general principle in the field of aqueous thermodynamics (e.g., the standard Gibbs energies of formation are based on the convention, $\Delta_f G^\circ_m (\text{H}^+, 298.15\,\text{K}) = 0$) because the absolute single ion values cannot be determined directly by experiments. Their evaluation or estimation requires the application extra-thermodynamic assumptions (cf. [85MAR2]). For example: the electrochemical studies of Rabinovich et al. [74RAB/ALE] and Schwabe et al. [74SCH/KEL] have shown th absolute activity coefficients, $^*\gamma_i$, of cation M and anion X in binary electrolyte solution (MX = HCl, NaCl, NaClO$_4$, etc.) are quite different. The ratio ($^*\gamma_M/^*\gamma_X$) increases as a function of the electrolyte concentration, e.g., ($^*\gamma_{Na}/^*\gamma_{Cl}$) increases from 1.1 in 0.1 m NaCl to 2 in 3 m NaCl [74RAB/ALE]. The effect of neglecting ion-ion interactions between ions the same charge is illustrated in the following:

(1) The interactions between like ions, e.g., Na$^+$-Na$^+$, Cl$^-$-Cl$^-$ or ClO$_4^-$-ClO$_4^-$ pure NaCl or NaClO$_4$ solutions, respectively, may be neither negligible nor equal to o other. However, the convention, $\varepsilon(i,i) = 0$, has no impact on the modelling of equilibriu reactions or on their ionic strength dependencies, because the reactions are electroneutr ($\Delta z = 0$). Therefore, single-ion activity coefficients can always be combined into mean ior activity coefficients, which are thermodynamically well defined. In the same way, the d crepancies between the conventional and absolute $\Delta_f G^\circ_m$ values of aqueous ions cancel c in the standard Gibbs energies of reaction $\Delta_r G^\circ_m$.

D.2 Implications of neglecting anion-anion and cation-cation interactions

(2) The SIT interaction coefficients, $\varepsilon(A, X)$ and $\varepsilon(C, M)$ of anion A and cation C a solution of electrolyte, MX, may differ considerably from $\varepsilon(A, X')$ and $\varepsilon(C, M')$ in an ... er electrolyte solution, M'X' (e.g., for an anion A in NaCl or NaClO$_4$ solution: $\varepsilon(A, Cl^-) \neq \varepsilon(A, ClO_4^-) \neq 0$). The consequences of neglecting these interactions are dis... ssed in the following sections. Anion-anion and cation-cation interactions are included in ... Pitzer equations and are allowed for by the introduction of the mixing parameters, θ_{ij}.

2.1 On the magnitude of anion-anion and cation-cation interactions

... shown above, the binary cation-anion interaction coefficients used in the NEA-TDB to ... culate the trace activity coefficients of cation C or anion A in an electrolyte solution, MX ... noted here by "$\varepsilon(C, X)$" and "$\varepsilon(A, M)$"), include the non-zero anion-anion and cation-... ion interactions, i.e., they are used for the sums:

$$\text{"}\varepsilon(C, X)\text{"} = \varepsilon(C, X) + \varepsilon(C, M) \tag{D.5}$$

...

$$\text{"}\varepsilon(A, M)\text{"} = \varepsilon(A, M) + \varepsilon(A, X), \tag{D.6}$$

pectively. In most cases, the inaccuracies arising from setting $\varepsilon(C, M)$ and $\varepsilon(A, X)$ equal ... zero are negligible compared to other uncertainties. This holds for most univalent cations ... anions. However, there are some cases where the inaccuracies become sufficiently large ... merit attention. The examples below illustrate the magnitude of the possible inaccuracies ... their chemical implications. For ease of understanding the interaction coefficients, "$\varepsilon(C,$ ..." and "$\varepsilon(A, M)$", are given with indices denoting the medium, MX, in which the interac-... n coefficients are derived:

$$\varepsilon(C, X)_{MX} = \varepsilon(C, X) + \varepsilon(C, M), \tag{D.7}$$

$$\varepsilon(A, M)_{MX} = \varepsilon(A, M) + \varepsilon(A, X). \tag{D.8}$$

...ample 1:
...ace activity coefficients of the H$^+$ ion in NaCl and CsCl solution

... e compilation of Pitzer [91PIT] includes interaction coefficients for the ternary systems, ... l-NaCl and HCl-CsCl, with mixing parameters derived from emf data for traces of HCl ... the two chloride media. The trace activity coefficients of the H$^+$ ion in NaCl and CsCl ... utions become increasingly different with increasing ionic strength (Figure D-1). These ... ferences are neglected by the SIT approach which, in contrast to the Pitzer equations, ... es not account for the interactions between H$^+$ and Na$^+$ or Cs$^+$ ions.

Figure D-1: Trace activity coefficients of the H$^+$ ion in NaCl and CsCl solution at 25°C.

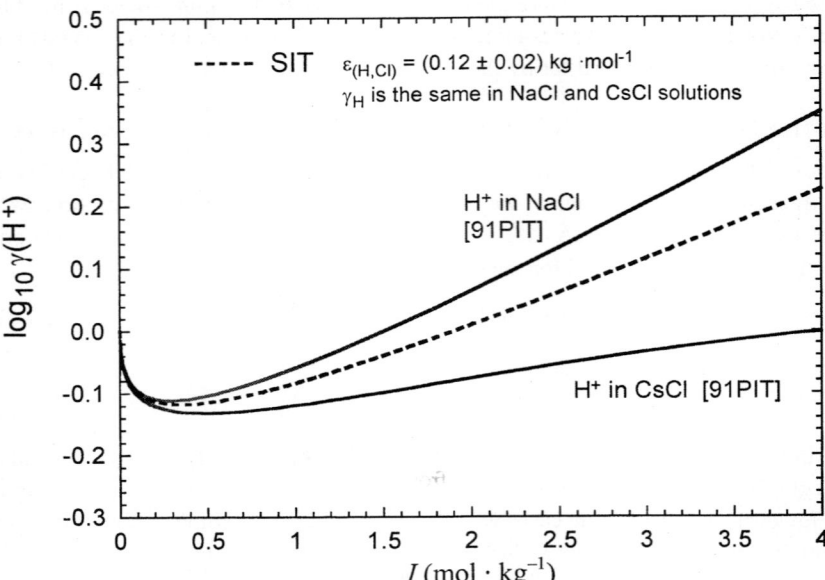

Example 2:
Ion interaction coefficients for the limiting Np(V) carbonate complex

Numerous solubility and spectroscopic data for the limiting Np(V) carbonate complex $NpO_2(CO_3)_3^{5-}$, have led to a well-ascertained formation constant of $\log_{10} \beta_3$ (5.50 ± 0.15) selected by [2001LEM/FUG]. The ion interaction coefficient selected [2001LEM/FUG] is based on experimental data in 0.1 – 3 M NaClO$_4$ solution, in combination with $\varepsilon(NpO_2^+, ClO_4^-) = $ (0.25 ± 0.05) kg · mol^{-1} and $\varepsilon(Na^+, CO_3^{2-}) = -$ (0.08 ± 0 kg · mol^{-1}:

$$\varepsilon(NpO_2(CO_3)_3^{5-}, Na^+)_{NaClO_4} = -(0.53 \pm 0.19) \text{ kg} \cdot \text{mol}^{-1}.$$

Evaluation of the solubility and spectroscopic data in NaCl solution from [94RUN/KIM], [94NEC/KIM], [96RUN/NEU], in combination with fixed values $\log_{10} \beta_3^\circ$, $\varepsilon(Na^+, CO_3^{2-})$ and $\varepsilon(NpO_2^+, Cl^-) = $ (0.09 ± 0.05) kg · mol^{-1} fr

01LEM/FUG], result in a considerably different interaction coefficient of:

$$\varepsilon(NpO_2(CO_3)_3^{5-}, Na^+)_{NaCl} = -(0.29 \pm 0.11) \text{ kg} \cdot \text{mol}^{-1}.$$

This apparent inconsistency was noted in [2001LEM/FUG] and was ascribed to ~~pos~~sible shortcomings in the experimental data or in the data treatment with the Pitzer approach [95FAN/NEC], [96RUN/NEU].

In the present review, the solubility data reported by [97NOV/ALM] and [ALM/NOV] for $KNpO_2CO_3$(s) in KCl-K_2CO_3 solution and for $K_3NpO_2(CO_3)_2$(s) in pure CO_3 solution are evaluated with the SIT (cf. Appendix A). In combination with $\log_{10}\beta_3^\circ$, $\varepsilon(K^+, CO_3^{2-})$ and $\varepsilon(NpO_2^+, Cl^-)$ from [2001LEM/FUG], the experimental data yield two consistent values of:

$$\varepsilon(K^+, NpO_2(CO_3)_3^{5-})_{KCl} = -(0.21 \pm 0.07) \text{ kg} \cdot \text{mol}^{-1}$$

$$\varepsilon(K^+, NpO_2(CO_3)_3^{5-})_{K_2CO_3} = -(0.23 \pm 0.02) \text{ kg} \cdot \text{mol}^{-1},$$

respectively. As the SIT coefficient for the interaction of $NpO_2(CO_3)_3^{5-}$ with K^+ may be expected to differ slightly, but not dramatically from that with Na^+, these values are compatible with $\varepsilon(Na^+, NpO_2(CO_3)_3^{5-})_{NaCl} = -(0.29 \pm 0.11) \text{ kg} \cdot \text{mol}^{-1}$ derived from the available data in NaCl media. However, they appear incompatible with $\varepsilon(Na^+, NpO_2(CO_3)_3^{5-})_{NaClO_4} = -(0.53 \pm 0.19) \text{ kg} \cdot \text{mol}^{-1}$ derived in [2001LEM/FUG] from data in $NaClO_4$ solution. A plausible explanation would be to consider that the anion-anion interaction coefficients, $\varepsilon(NpO_2(CO_3)_3^{5-}, Cl^-)$ and $\varepsilon(NpO_2(CO_3)_3^{5-}, ClO_4^-)$ or those for the carbonate ion involved in these evaluations, are quite different and not negligible.

Example 3:
Trace activity coefficients of CO_3^{2-} and SO_4^{2-} in NaCl and $NaClO_4$ solutions

Faghänel et al. [96FAN/NEC] reported experimental H_2CO_3 dissociation constants in 0.1, 3 and 5 M $NaClO_4$ in comparison with the corresponding value in NaCl solution from [RUN/KIM] and used the ion interaction model of Pitzer to treat the data. The equilibrium constants in NaCl solution are consistent with the well-defined parameter set reported in the literature [84HAR/MOL], [91PIT], whereas the results in $NaClO_4$ required the evaluation of mixing parameters for $\theta_{i,j}$ and $\psi_{i,j,k}$ for CO_3^{2-} and HCO_3^- in $NaClO_4$ solution. According to the results of [96FAN/NEC], the trace activity coefficients of the CO_3^{2-} ion in NaCl and $NaClO_4$ solutions above one molal are considerably different (Figure D-2).

A similar behaviour is expected for the trace activity coefficients of the SO_4^{2-} ion in NaCl and $NaClO_4$ solution. The Pitzer parameters for the system, Na-SO$_4$-Cl, are well known [84HAR/MOL], [91PIT]. There are no data for the ternary system Na-SO$_4$-ClO$_4$, such data do exist for the analogous system, Na-SO$_4$-TcO$_4$. The physical and chemical properties of the TcO$_4^-$ ion are very similar to those of the ClO$_4^-$ ion [97KON/NEC] and Pitzer parameters determined by isopiestic measurements in the system, Na-SO$_4$-Tc [98NEC/KON2] have been also confirmed by solubility experiments [98NEC/KO [98NEC/KON2]. The corresponding trace activity coefficients of the SO$_4^{2-}$ ion in NaCl a NaTcO$_4$ solutions are shown in Figure D-3. Indeed, the differences in the trace activity co ficients of CO$_3^{2-}$ and SO$_4^{2-}$ in chloride and perchlorate media are comparable, not o qualitatively but also quantitatively.

These differences cannot be described with the simplified SIT approach applied the NEA-TDB reviews (*cf.* calculated dashed lines in Figure D-2 and Figure D-3), where interaction coefficients between ions of the same charge sign are generally set equal to ze The consequences of these findings, in particular for the important actinide carbonate s tems, are discussed in the following section.

D.2 Implications of neglecting anion-anion and cation-cation interactions

Figure D-2: Trace activity coefficients of the CO_3^{2-} ion in NaCl and NaClO$_4$ solutions at 25°C.

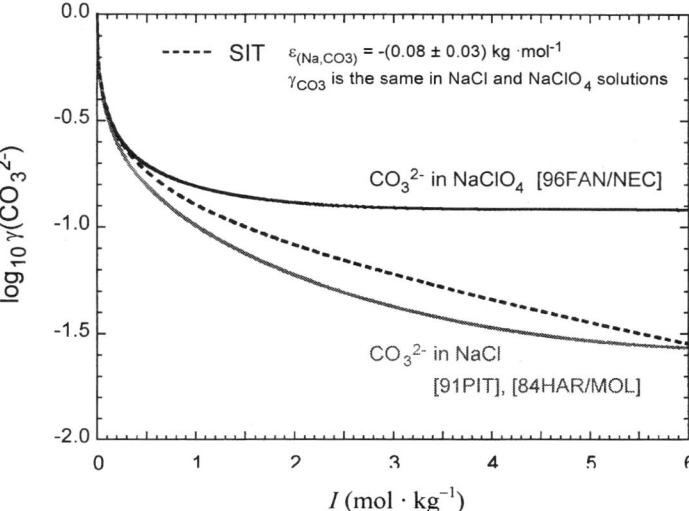

Figure D-3: Trace activity coefficients of the SO_4^{2-} ion in NaCl and NaTcO$_4$ solutions at 25°C.

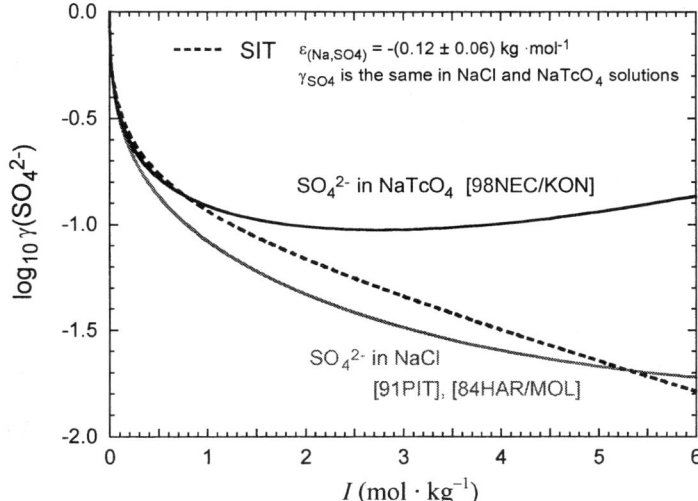

D.2.2 Discussion of ion interaction coefficients for the carbonate ion and the consequences for selected data

In Appendix D.4 of [95SIL/BID] the interaction coefficient selected in [92GRE/FU $\varepsilon(Na^+, CO_3^{2-}) = -(0.05 \pm 0.03)$ kg · mol^{-1}, which originated from activity or osmotic coe cient data in binary Na$_2$CO$_3$ solutions, was slightly changed to $\varepsilon(Na^+, CO_3^{2-}) = -(0.08 \pm 0.03)$ kg · mol^{-1} [80CIA] in order to reproduce better the experimental data in aqueous uranium carbonate systems. This value has been used in the subsequent NEA-T reviews, [99RAR/RAN], [2001LEM/FUG] and also in the present review to calculate c bonate trace activity coefficients in sodium salt solutions.

As shown in Figure D-2, the carbonate trace activity coefficients calculated w the SIT approach are reasonably consistent with those calculated for NaCl media with Pitzer equations. The corresponding Pitzer ion interaction coefficients [91PIT] are based a large number of widely accepted data coming from different experimental methods. small deviations in $\gamma_{(CO_3^{2-})}$ are probably due to triple ion interactions or higher-order ter in the Pitzer equations. By combining, $\varepsilon(Na^+, CO_3^{2-}) = -(0.08 \pm 0.03)$ kg · mo $\varepsilon(H^+, Cl^-) = (0.12 \pm 0.02)$ kg · mol^{-1} and $\log_{10} K°((D.9), 298.15\ K) = -(18.15 \pm 0.$ [95SIL/BID] for the reaction:

$$CO_2(g) + H_2O(l) \rightleftharpoons 2H^+ + CO_3^{2-}, \qquad (D$$

the available equilibrium constants, $\log_{10} K$ ((D.9), 0 – 3.2 m NaCl), are reproduced sa factorily [96FAN/NEC]. Accordingly there is no reason to cast doubt on the value:

$$\varepsilon(Na^+, CO_3^{2-})_{NaCl} = -(0.08 \pm 0.03)\ kg · mol^{-1}.$$

However, the carbonate trace activity coefficients calculated for NaClO$_4$ me with the Pitzer parameters of Fanghänel *et al.* [96FAN/NEC] differ significantly from th calculated with the SIT assuming $\varepsilon(Na^+, CO_3^{2-}) = -(0.08 \pm 0.03)$ kg · mol^{-1}. This discr ancy arises from differences in the experimental values for the dissociation constants of bonic acid. At low ionic strength (0 – 0.1 m NaClO$_4$) the results of [96FAN/NEC] ag well with the auxiliary data selected in the NEA TDB, but at NaClO$_4$ concentrations mol · kg^{-1} they deviate considerably. The equilibrium constants determined [96FAN/NEC] for reaction (D.9) lead to $\Delta\varepsilon_{NaClO_4}$ (D.9) = (0.32 ± 0.03) kg · mol^{-1} a combined with $\varepsilon(H^+, ClO_4^-) = (0.14 \pm 0.02)$ kg · mol^{-1}, to:

$$\varepsilon(Na^+, CO_3^{2-})_{NaClO_4} = +(0.04 \pm 0.05)\ kg · mol^{-1}.$$

The carbonic acid dissociation constants determined by [96FAN/NEC] in 1.0, and 5.0 M NaClO$_4$ were not confirmed by independent measurements. On the contrary already discussed in [2001LEM/FUG], the values in 1.0 and 3.0 M NaClO$_4$ contradict th ported by other authors, *e.g.*, by the group of Sillén *et al.* [58NIL/REN], [58FRY/NIL], and are therefore not accepted. However, as the actinide carbonates are of great importance in natural systems, the possible consequences of different ion interaction coefficients in NaCl and NaClO$_4$ media will have to be discussed in light of additional experimental data.

If there are sufficient experimental data at different ionic strengths in a given medium to perform a linear SIT regression as shown in Figure B-1, Appendix B.1.3.2, the values of $\Delta\varepsilon$ and $\log_{10} K_{s,n}^\circ$ for carbonate solids are not affected. The same holds for $\Delta\varepsilon$ and $\log_{10} \beta^\circ$ for aqueous carbonate complexes (*e.g.*, for the formation constants selected for the Np(V) carbonate complexes in [2001LEM/FUG]). Of course, different individual $\varepsilon(j,k)$, values would be calculated from $\Delta\varepsilon$ if $\varepsilon(Na^+, CO_3^{2-})_{NaClO_4} = (0.04 \pm 0.05)$ kg · mol^{-1} is used instead of $-(0.08 \pm 0.03)$ kg · mol^{-1}.

On the other hand, if the experimental data set is more restricted, particularly if determined at higher ionic strengths, the interaction coefficients used for the conversion to I = 0 may have a significant effect on the selected equilibrium constant. This problem arises for instance in the calculated solubility constants of solid Np(V) carbonates, NaNpO$_2$CO$_3$ · 3.5H$_2$O(s) and Na$_3$NpO$_2$(CO$_3$)$_2$(s), and solid Am(III) carbonate, Am$_2$(CO$_3$)$_3$ · xH$_2$O(s). The possible consequences are discussed in detail in sections 12.6.2.1.1 and 12.6.1.1.3.2, respectively.

Bibliography

[00PIS] Pissarjewsky, L., Die Überuran-, Übermolybdän- und Überwolframsäuren und entsprechende Säuren, *Z. Anorg. Chem.*, **24**, (1900), 108-122, in German. Cited on pages: 192, 397.

[05OEC/CHA] Oechsner de Coninck, W.,Chauvenet, Sur un sulfate double de lithium et d'uranyle, *Bull. Cl. Sci., Acad. R. Belg.*, (1905). Cited on page: 417.

[22BRO] Brønsted, J. N., Studies of solubility: IV. The principle of specific interaction of ions, *J. Am. Chem. Soc.*, **44**, (1922), 877-898. Cited on page: 709.

[22BRO2] Brønsted, J. N., Calculation of the osmotic and activity functions in solutions of uni-univalent salts, *J. Am. Chem. Soc.*, **44**, (1922), 938-948. Cited on page: 709.

[35GUG] Guggenheim, E. A., The specific thermodynamic properties of aqueous solutions of strong electrolytes, *Philos. Mag.*, **19**, (1935), 588-643. Cited on page: 709.

[36SCA] Scatchard, G., Concentrated solutions of strong electrolytes, *Chem. Rev.*, **19**, (1936), 309-327. Cited on page: 709.

[37KIE] Kielland, J., Individual activity coefficients of ions in aqueous solutions, *J. Am. Chem. Soc.*, **59**, (1937), 1675-1678. Cited on page: 404.

[40COU/PIT] Coulter, L. V., Pitzer, K. S., Latimer, W. M., Entropies of large ions: The heat capacity, entropy and heat of solution of K_2PtCl_6, $(CH_3)_4NI$ and $UO_2(NO_3)_2 \cdot 6H_2O$, *J. Am. Chem. Soc.*, **62**, (1940), 2845-2851. Cited on page: 485.

[42JEN/AND] Jenkins, F. A., Anderson, O. E., Reports RL 4.6.3 and 4.6.4, (1942), (from the citation in [48MUE]). Cited on page: 216.

[42SCH/THO] Schelberg, A., Thompson, R. W., (1942), Report A-809, (from the citation in [51KAT/RAB]). Cited on page: 229.

[42THO/SCH] Thompson, R. W., Schelberg, A., (1942), Report A-179, (from the citation in [51KAT/RAB]). Cited on page: 227.

[43ALT/LIP] Altman, D., Lipkin, D., Weissman, S., (1943), Report RL-4.6.22, (from the citation in [51KAT/RAB]). Cited on pages: 218, 219.

[43ALT] Altman, D., (1943), Report RL-4.6.156, (from the citation in [51KAT/RAB]). Cited on pages: 201, 214, 225.

[43WEB] Webster, R. A., (1943), Report CK-873, (from the citation in [51KAT/RAB]). Cited on page: 225.

[44FER/PRA] Ferguson, W. J., Prather, J., (1944), Report A-3143, (from the citation in [58MAC]). Cited on page: 222.

[44FER/PRA2] Ferguson, W. J., Prather, J., Scott, R. B., (1944), Report A-1920, (from the citation in [51KAT/RAB]). Cited on page: 223.

[44NOT/POW] Nottorf, R., Powell, J., The vapor pressure of UBr_4, Ames Project, Chemical Research 10 May to 10 June, 1944, Spedding, F. H., Wilhelm, H. A., Eds., Report CC-1504 (A-2087), pp. 9-17, (1944). Cited on page: 227.

[45DAV/STR] Davidson, P. H., Streeter, I., The equilibrium vapor pressure of UCl_4 above $UOCl_2$, (1945), Report RL-4.6.920, (from the citation in [48MUE]). Cited on page: 216.

[45DAV] Davidson, P. H., The flow of UCl_4 vapor through tubes of cylindrical cross-section, (1945), Report RL-4.6.931, (from the citation in [48MUE]). Cited on page: 216.

[45FER/RAN] Ferguson, W. J., Rand, R. D., U.S. Natl. Bureau of Standards, Report A-3357, (1945). Cited on page: 224.

[45WAG/GRA] Wagner, E. L., Grady, H. F., Miller, A. J., The vapor pressure curves of uranium tetrachloride, Report C.2350.5, (1945), (from the citation in [48MUE]). Cited on page: 216.

[46GRE] Gregory, N. W., The vapor pressure of uranium tetraiodide, US Atomic Energy Commission, Report AECD-3342, (1946). Cited on page: 229.

[46GRE2] Gregory, N. W., Applications of a diaphragm gauge for the investigation of vapor pressures and chemical equilibria at high temperatures: I. The vapor pressure and gas density of UCl_4, Report BC-16, (1946), (from the citation in [48MUE]). Cited on page: 216.

[46GRE3] Gregory, N. W., The vapour pressure of UCl_3Br, UCl_2Br_2, and UBr_4, Report BC-3, (1946), (from the citation in [48MUE]). Cited on page: 226.

[47GIN/COR] Ginnings, D. C., Corruccini, R. J., Heat capacities at high temperatures of uranium, uranium trichloride and uranium tetrachloride, *J. Res. Natl. Bur. Stand.*, **39**, (1947), 309-316. Cited on pages: 222, 223.

[47JOH] Johnsson, K. O., The vapor pressure of uranium tetrafluoride, Carbon and Carbon Chemicals Corporation, Report Y-42, (1947), 19 pp. Cited on page: 201.

[47ROB/LEV] Robinson, R. A., Levien, B. J., The vapor pressures of potassium ferricyanide and thorium nitrater solutions at 25°C *Trans. R. Soc. N.Z.*, **76**, (1947), 295. Cited on page: 403.

[47RYO/TWI] Ryon, A. D., Twichell, L. P., Carbide and chemicals corporation, Oak Ridge, Tenesse, USA, Report, (1947). Cited on page: 201.

[48MUE] Mueller, M. E., The vapor pressure of uranium halides, U.S. Atomic Energy Commission, Report AECD-2029, (1948), 118 pp. Cited on pages: 216, 226.

[48THO/SCH] Thompson, R. W., Schelberg, A., OSRD Proj. SSRC-5 Princeton Report 24 A-809, (1948), (from the citation in [48MUE]). Cited on page: 216.

[49KAS] Kasha, M., Reactions between plutonium ions in perchloric acid solution: rates, mechanisms and equilibria, *The transuranium elements, research papers*, Seaborg, G. T., Katz, J. J., Manning, W. M., Eds., pp. 295-334, McGraw-Hill, New York, USA, (1949). Cited on pages: 316, 642.

[49SUT] Sutton, J., The hydrolysis of the uranyl ion Part, I., *J. Chem. Soc.*, **S57**, (1949), 275-286. Cited on page: 177.

[50KRA/NEL] Kraus, K. A., Nelson, F., Hydrolytic behavior of metal ions: I. The acid constants of uranium(IV) and plutonium(IV), *J. Am. Chem. Soc.*, **72**, (1950), 3901-3906. Cited on page: 560.

[51KAT/RAB] Katz, J. J., Rabinowitch, E., *The chemistry of uranium*, Dover Publications, New York, (1951), 609 pp. Cited on pages: 214, 223, 225, 226.

[51ROB/LIM] Robinson, R. A., Lim, C. K., The osmotic and activity coefficients of uranyl nitrate, chloride, and perchlorate at 25°C, *J. Chem. Soc.*, (1951), 1840-1843. Cited on pages: 239, 574.

[52COH/HIN] Cohen, D., Hindman, J. C., The neptunium(IV)-neptunium(V) couple in perchloric acid. The partial molal heats and free energies of formation of neptunium ions, *J. Am. Chem. Soc.*, **74**, (1952), 4682-4685. Cited on page: 639.

[52ROS/WAG] Rossini, F. D., Wagman, D. D., Evans, W. H., Levine, S., Jaffe, I., Selected values of chemical thermodynamic properties, NBS Circular, Report 500, (1952). Cited on page: 192.

[53AGA/AGA] Agalonova, A. L., Agafonov, I. L., Ionisation constants of inorganic acids and bases in aqueous solution, *Zh. Fiz. Khim.*, **27**, (1953), 1137. Cited on page: 657.

[53JON] Jones, M. M., A study of plutonium trifluoride precipitated from aqueous solution, Hanford Atomic Products Operation, Richland Washington, USA, Report HW-30384, (1953). Cited on page: 359.

[54FEA] Feay, D. C., Some chemical properties of curium, Ph. D. Thesis, Radiation Laboratory, University of California, (1954), Berkeley, 50 pp., UCRL-2547. Cited on pages: 354, 355.

[54KOC/CUN] Koch, C. W., Cunningham, B. B., The vapor phase hydrolysis of the actinide halides. I. Heat and free energy of the reaction $AmCl_3(s) + H_2O(g) \rightleftarrows AmOCl(s) + 2HCl(g)$, *J. Am. Chem. Soc.*, **76**, (1954), 1470-1471. Cited on page: 362.

[54MIL] Milkey, R. G., Stability of dilute solutions of uranium, lead and thorium ions., *Anal. Chem.*, **26**, (1954), 1800-1803. Cited on page: 400.

[54SCH/BAE]	Schreyer, J. M., Baes, Jr., C. F., The solubility of uranium(VI) orthophosphates in phosphoric acid solutions, *J. Am. Chem. Soc.*, **76**, (1954), 354-357. Cited on pages: 242, 292.

[55CAR/CUN]	Carniglia, S. C., Cunningham, B. B., Vapor pressures of americium trifluoride and plutonium trifluoride, heats and free energies of sublimation, *J. Am. Chem. Soc.*, **77**, (1955), 1451-1453. Cited on page: 360.

[55GAY/LEI]	Gayer, K. H., Leider, H., The solubility of uranium trioxide, $UO_3 \cdot H_2O$, in solutions of sodium hydroxide and perchloric acid at 25°C, *J. Am. Chem. Soc.*, **77**, (1955), 1448-1550. Cited on pages: 399, 400.

[55RAB/COW]	Rabideau, S. W., Cowan, H. D., Chloride complexing and disproportionation of Pu(IV) in hydrochloric acid, *J. Am. Chem. Soc.*, **77**, (1955), 6145-6148. Cited on page: 322.

[55SCH]	Schreyer, J. M., The solubility of uranium(IV) orthophosphates in phosphoric acid solutions, *J. Am. Chem. Soc.*, **77**, (1955), 2972-2974. Cited on page: 288.

[55WAI]	Wait, E., A cubic form of uranium trioxide, *Inorg. Nucl. Chem. Lett.*, **1**, (1955), 309-312. Cited on page: 419.

[56BAE]	Baes, Jr., C. F., A spectrophotometric investigation of uranyl phosphate complex formation in perchloric acid solution, *J. Phys. Chem.*, **60**, (1956), 878-883. Cited on page: 241.

[56CHU/SHA]	Chukhlantsev, V. G., Sharova, A. K., Solubility products of uranyl arsenates, *Russ. J. Inorg. Chem.*, **1**, (1956), 39-44. Cited on page: 290.

[56CHU/STE]	Chukhlantsev, V. G., Stepanov, S. I., Solubility of uranyl and thorium phosphates, *Russ. J. Inorg. Chem.*, **1**, (1956), 135-141. Cited on pages: 288, 289.

[56HIE]	Hietanen, S., Studies on the hydrolysis of metal ions: 17. The hydrolysis of the uranium(IV) ion, U^{4+}, *Acta Chem. Scand.*, **10**, (1956), 1531-1546. Cited on page: 184.

[56SHC/VAS2] Shchukarev, S. A., Vasil'kova, I. V., Efimov, A. I., Kirdyashev, V. P., Determination of the saturated vapor pressure of UCl_4 and the disproportionation pressure of $UOCl_2$, *Zh. Neorg. Khim.*, **1**, (1956), 2272-2277, in Russian. Cited on page: 216.

[56STU/SIN] Stull, D. R., Sinke, G. C., *Thermodynamic properties of the elements*, American Chemical Society, Washington, D.C., (1956). Cited on page: 448.

[56WAR/WEL] Ward, M., Welch, G. A., The chloride complexes of trivalent plutonium, americium and curium, *J. Inorg. Nucl. Chem.*, **2**, (1956), 395-402. Cited on page: 322.

[57HEA/WHI] Hearne, J. A., White, A. G., Hydrolysis of the uranyl ion, *J. Chem. Soc.*, (1957), 2168-2174. Cited on page: 580.

[57RAB] Rabideau, S. W., The hydrolysis of plutonium(IV), *J. Am. Chem. Soc.*, **79**, (1957), 3675-3677. Cited on page: 315.

[58BRU] Brusilovskii, S. A., Investigation of the precipitation of hexavalent uranium hydroxide, *Proc. Acad. Sci. USSR*, **120**, (1958), 343-347. Cited on pages: 399, 400.

[58FRY/NIL] Frydman, M., Nilsson, G., Rengemo, T., Sillén, L. G., Some solution equilibria involving calcium sulfite and carbonate: III. The acidity constants of H_2CO_3 and H_2SO_3, and $CaCO_3 + CaSO_3$ equilibria in $NaClO_4$ medium at 25°C, *Acta Chem. Scand.*, **12**, (1958), 868-872. Cited on pages: 568, 763.

[58JOH/BUT] Johnson, O., Butler, T., Newton, A. S., Preparation, purification, and properties of anhydrous uranium chlorides, *Chemistry of uranium*, Katz, J. J., Rabinowitch, E., Eds., vol. 1, pp. 1-28, USA Energy Commission, Oak Ridge, Tennessee, USA, (1958). Cited on pages: 216, 217, 218, 219.

[58MAC] MacWood, G. E., Thermodynamic properties of uranium compounds, Chemistry of uranium: Collected papers, Katz, J. J., Rabinowitch, E., Eds., pp. 543-609, US Atomic Energy Commission, Oak Ridge, Tennessee, USA, (1958). Cited on page: 222.

[58NIL/REN] Nilsson, G., Rengemo, T., Sillén, L. G., Some solution equilibria involving calcium sulfite and carbonate: I. Simple solubility equilibria of CO_2, SO_2, $CaCO_3$, and $CaSO_4$, *Acta Chem. Scand.*, **12**, (1958), 868-872. Cited on page: 763.

[58YOU/GRA] Young, H. S., Grady, H. F., Physical constants of uranium tetrachloride, *Chemistry of uranium*, Katz, J. J., Rabinowitch, E., Eds., vol. 2, pp. 749-756, USA Energy Commission, Oak Ridge, Tennessee, USA, (1958). Cited on page: 216.

[59POP/GAL] Popov, M. M., Gal'chenko, G. L., Senin, M. D., Specific heats and heats of fusion of UCl_4 and UI_4 and heat of transformation of UI_4, *Russ. J. Inorg. Chem.*, **4**, (1959), 560-562. Cited on pages: 217, 223.

[59POP/KOS] Popov, M. M., Kostylev, F. A., Zubova, N. V., Vapour pressure of uranium tetrafluoride, *Russ. J. Inorg. Chem.*, **4**, (1959), 770-771. Cited on page: 201.

[59ROB/STO] Robinson, R. A., Stokes, R. H., *Electrolyte solutions*, 2nd. Edition, Butterworths, London, (1959), 559 pp. Cited on pages: 574, 714.

[60BAB/KOD] Babko, A. K., Kodenskaya, V. S., Equilibria in solutions of uranyl carbonate complexes, *Russ. J. Inorg. Chem.*, **5**, (1960), 1241-1244. Cited on pages: 399, 400.

[60LAN/BLA] Langer, S., Blankenship, F. F., The vapour pressure of uranium tetrafluoride, *J. Inorg. Nucl. Chem.*, **14**, (1960), 26-31. Cited on page: 201.

[60OOS] Oosting, M., Quantitative extraction equilibriums. IV, *Recl. Trav. Chim. Pays-Bas*, **79**, (1960), 627-634. Cited on page: 400.

[60RAB/KLI] Rabideau, S. W., Kline, R. J., A spectrophotometric study of the hydrolysis of plutonium(IV), *J. Phys. Chem.*, **64**, (1960), 680-682. Cited on page: 315.

[60SAV/BRO] Savage, Jr., A. W., Browne, J. C., The solubility of uranium(IV) fluoride in aqueous fluoride solutions, *J. Am. Chem. Soc.*, **82**, (1960), 4817-4821. Cited on pages: 211, 212.

[60STE/GAL] Stepanov, M. A., Galkin, N. P., The solubility product of the hydroxide of tetravalent uranium, *Sov. At. Energy*, **9**, (1960), 817-821. Cited on page: 184.

[61AKI/KHO] Akishin, P. A., Khodeev, Yu. S., Determination of the heat of sublimation of uranium tetrafluoride by the mass-spectroscopic method, *Russ. J. Phys. Chem.*, **35**, (1961), 574-575. Cited on pages: 201, 202.

[61KAR] Karpov, V. I., The solubility of triuranyl phosphate, *Russ. J. Inorg. Chem.*, **6**, (1961), 271-272. Cited on pages: 288, 289.

[61LEW/RAN] Lewis, G. N., Randall, M., *Thermodynamics, as revised by Pitzer, K. S. and Brewer, L.*, 2nd. Edition, McGraw-Hill, New York, (1961), 723 pp. Cited on pages: 16, 708, 710, 714, 716.

[61PET] Peterson, A., Studies on the hydrolysis of metal ions: 32. The uranyl ion, UO_2^{2+}, in Na_2SO_4 medium, *Acta Chem. Scand.*, **15**, (1961), 101-120. Cited on pages: 232, 452, 664, 665, 672, 673.

[61ROS/ROS] Rossotti, F. J. C., Rossotti, H., *The determination of stability constants and other equilibrium constants in solution*, McGraw-Hill, New York, (1961). Cited on pages: 25, 735.

[61WAT/DOU] Waterbury, G. R., Douglass, R. M., Metz, Thermogravimetric behavior of plutonium metal, nitrate, sulfate and oxalate, *Anal. Chem.*, **33**, (1961), 1018-1023. Cited on pages: 315, 668.

[62BAE/MEY] Baes, Jr., C. F., Meyer, N. J., Acidity measurements at elevated temperatures: I. Uranium(VI) hydrolysis at 25 and 94°C, *Inorg. Chem.*, **1**, (1962), 780-789. Cited on pages: 478, 580.

[62DAV] Davies, C. W., *Ion association*, Butterworths, London, (1962), 190 pp. Cited on page: 708.

[62GEL/MOS] Gel'man, A. D., Moskvin, A. I., Zaitseva, V. P., Carbonate compounds of plutonyl, *Sov. Radiochem.*, **4**, (1962), 138-145. Cited on page: 325.

[62HOS/GAR]	Hostetler, P. B., Garrels, R. M., Transportation and precipitation of U and V at low temperature with special reference to sandstone-type U deposits, *Econ. Geol.*, **57**, (1962), 137-167. Cited on pages: 291, 549.
[62LIE/STO]	Lietzke, M.H., Stoughton, R.W., The calculation of activity coefficients from osmotic coefficient data, *J. Phys. Chem.*, **66**, (1962), 508-509. Cited on page: 418.
[62PER/BER]	Perez-Bustamante, J. A., Bermudez Polonio, J., Fernandez Cellini, R., Contribution to the study of the carbonate complexes of uranium, *An. Fís. Quím.*, **58B**, (1962), 677-704, in Spanish. Cited on page: 400.
[62RAN/FOX]	Rand, M. H., Fox, A. C., Street, R. S., Radiation self-damage in plutonium compounds, *Nature (London)*, **195**, (1962), 567-568. Cited on pages: 315, 668.
[62VAS]	Vasil'ev, V. P., Influence of ionic strength on the instability constants of complexes, *Russ. J. Inorg. Chem.*, **7**, (1962), 924-927. Cited on page: 709.
[62WIL]	Wilcox, D. E., A method for estimating the heat of formation and free energy of formation of inorganic compounds, Univ. of California, UCRL report, (1962). Cited on pages: 5, 432.
[63BAR/SOM]	Bartušek, M., Sommer, L., Ph. D. Thesis, Institut für Analytische Chemie der J.E. Purkyne Universität, (1963), Brno, Czechoslovakia. Cited on page: 456.
[63COR/ALI]	Cordfunke, E. H. P., Aling, P., Thermal decomposition of hydrated uranium peroxides, *Recl. Trav. Chim. Pays-Bas*, **82**, (1963), 257-263. Cited on pages: 192, 398, 464.
[63DUN/SIL]	Dunsmore, H. S., Sillén, L. G., Studies on the hydrolysis of metal ions: 47. Uranyl ion in 3 M (Na)Cl medium, *Acta Chem. Scand.*, **17**, (1963), 2657-2663. Cited on page: 584.
[63JAC/RAN]	Jackson, E. E., Rand, M. H., The oxidation behaviour of plutonium dioxide and solid solutions containing plutonium dioxide, UK Atomic Energy Authorithy, Report AERE R 3636, (1963). Cited on pages: 315, 668.

[63KAN]   Kangro, W., Recovery of heavy metals, especially uranium, from poor ores with the aid of chlorine gas. I, *Z. Erzbergbau Metallhüttenwes.*, **16**, (1963), 107-112, in German. Cited on page: 220.

[63LEF]   Lefort, M., Oxydo-réduction du couple TcO_2-TcO_4^- en solutions diluées sous l'effet du rayonnement γ, *Bull. Soc. Chim. Fr.*, (1963), 882-884, in French. Cited on pages: 391, 685.

[63RUS/JOH] Rush, R. M., Johnson, J. S., Hydrolysis of uranium(VI): Absorption spectra of chloride and perchlorate solutions, *J. Phys. Chem.*, **67**, (1963), 821-825. Cited on pages: 456, 457.

[64BAN/PAT] Bansal, B. M. L., Patil, S. K., Sharma, H. D., Chloride, nitrate and sulphate complexes of europium (III) and americium (III), *J. Inorg. Nucl. Chem.*, **26**, (1964), 993-1000. Cited on page: 367.

[64COR]   Cordfunke, E. H. P., Heats of formation of some hexavalent uranium compounds, *J. Phys. Chem.*, **68**, (1964), 3353-3356. Cited on pages: 188, 190, 262, 412, 539, 633.

[64CRI/COB] Criss, C. M., Cobble, J. W., The thermodynamic properties of high temperature aqueous solutions. IV. Entropies of the ions up to 200°C and the correspondence principle, *J. Am. Chem. Soc.*, **86**, (1964), 5385-5390. Cited on page: 158.

[64CRI/COB2] Criss, C. M., Cobble, J. W., The thermodynamic properties of high temperature aqueous solutions. V. The calculation of ionic heat capacities up to 200°C. Entropies and heat capacities above 200°C, *J. Am. Chem. Soc.*, **86**, (1964), 5390-5393. Cited on pages: 158, 541.

[64KEE/KRU] Keenan, T. K., Kruse, F. H., Potassium double carbonates of pentavalent neptunium, plutonium and americium, *Inorg. Chem.*, **3**, (1964), 1231-1232. Cited on pages: 311, 569.

[64LEV]   Levinson, L. S., Heat content of molten uranium, *J. Chem. Phys.*, **40**, (1964), 3584-3585. Cited on pages: 157, 430.

[64SEK]   Sekine, T., Complex formation of La(III), Eu(III), Lu(III) and Am(III) with oxalate, sulphate, chloride and thiocyanate ions, *J. Inorg. Nucl. Chem.*, **26**, (1964), 1463-1465. Cited on pages: 365, 367.

[64SHI/MAR] Shiloh, M., Marcus, Y., The chemistry of trivalent neptunium plutonium and americium in halide solutions, Israel Atomic Energy Commission, Report IA-924, (1964), 26 pp. Cited on page: 356.

[64SIL/MAR] Sillén, L. G., Martell, A. E., *Stability constants of metal ion complexes*, *Special Publication*, No. 17, Chemical Society, London, (1964), 754 pp. Cited on pages: 17, 283.

[65BAE/MEY] Baes, Jr., C. F., Meyer, N. J., Roberts, C. E., The hydrolysis of thorium(IV) at 0 and 95°C, *Inorg. Chem.*, **4**, (1965), 518-527. Cited on page: 403.

[65MUT/HIR] Muto, T., Hirono, S., Kurata, H., Some aspects of fixation of uranium from natural waters, Japanese Atomic Energy Research Institute, Report NSJ-Tr 91, (1965), 27 pp. Cited on pages: 288, 289, 292.

[65MUT] Muto, T., Thermochemical stability of ningyoite, *Mineral. J.*, **4**, (1965), 245-274. Cited on pages: 287, 289, 292, 549.

[65PAR] Parker, V. B., *Thermal properties of aqueous uni-univalent electrolytes, National Standard Reference Data Series*, National Bureau of Standards, Ed., No. Rep NSRDS-NBS-2, US Government printing office, Washington, D.C., (1965), 66 pp. Cited on page: 398.

[65PER] Perez-Bustamante, J. A., Solubility product of tetravalent plutonium hydroxide and study of the amphoteric character of hexavalent plutonium hydroxide, *Radiochim. Acta*, **4**, (1965), 67-75. Cited on pages: 316, 642.

[65SEK2] Sekine, T., Solvent extraction study of trivalent actinide and lanthanide complexes in aqueous solutions. II. Sulfate complexes of La(III), Eu(III), Lu(III), and Am(III) in 1 M Na(ClO$_4$), *Acta Chem. Scand.*, **19**, (1965), 1469-1475. Cited on pages: 365, 367.

[65STA] Stary, J., Study of complexes of americium and promethium by using solvent extraction, *Nucl. Sci. Abstr.*, **19**, (1965), 275. Cited on page: 683.

[65VES/PEK] Veselý,V., Pekárek, V., Abbrent, M., A study of uranyl phosphates: III. Solubility products of uranyl hydrogen phosphate, uranyl orthophosphate and some alkali uranyl phosphates, *J. Inorg. Nucl. Chem.*, **27**, (1965), 1159-1166. Cited on pages: 288, 289, 292.

[65WOL/POS] Wolf, A. S., Posey, J. C., Rapp, K. E., α-uranium pentafluoride: I. Characterization, *Inorg. Chem.*, **4**, (1965), 751-757. Cited on pages: 203, 207.

[66COR] Cordfunke, E. H. P., Thermodynamic properties of hexavalent uranium compounds, Thermodynamics, Proc. Symp. held 22-27 July, Vienna, Austria, vol. 2, pp. 483-495, International Atomic Energy Agency, Vienna, (1966). Cited on pages: 192, 398.

[66DRO/KOL] Drobnic, M., Kolar, D., Calorimetric determination of the enthalpy of hydration of UO_3, *J. Inorg. Nucl. Chem.*, **28**, (1966), 2833-2835. Cited on page: 412.

[66DRO/PAT] Drowart, J., Pattoret, A., Smoes, S., Mass spectrometric studies of the vaporization of refractory compounds, *Proc. Br. Ceram. Soc.*, (1966), 67-89. Cited on page: 161.

[66GUG] Guggenheim, E. A., *Applications of Statistical Mechanics*, Clarendon Press, Oxford, (1966). Cited on pages: 709, 710.

[66LOO/COR] Loopstra, B. O., Cordfunke, E. H. P., On the structure of α-UO_3, *Recl. Trav. Chim. Pays-Bas*, **85**, (1966), 135-142. Cited on pages: 188, 407.

[66OLS/MUL2] Olson, W. M., Mulford, R. N. R., The melting point and decomposition pressure of neptunium mononitride, *J. Phys. Chem.*, **70**, (1966), 2932-2934. Cited on page: 300.

[66ROB] Robins, R. G., Hydrolysis of uranyl nitrate solutions at elevated temperatures, *J. Inorg. Nucl. Chem.*, **28**, (1966), 119-123. Cited on page: 538.

[67CAR/CHO] de Carvalho, R. G., Choppin, G. R., Lanthanide and actinide sulfate complexes: I. Determination of stability constants, *J. Inorg. Nucl. Chem.*, **29**, (1967), 725-735. Cited on pages: 365, 366, 367.

[67COR/LOO]	Cordfunke, E. H. P., Loopstra, B. O., Preparation and properties of the uranates of calcium and strontium, *J. Inorg. Nucl. Chem.*, **29**, (1967), 51-57. Cited on pages: 261, 262, 263, 425, 426, 467, 633, 634, 635.

[67GEL/MOS]	Gel'man, A. D., Moskvin, A. I., Zaitsev, L. M., Mefod'eva, M. P., *Complex compounds of transuranides*, Davey, New York, (1967), 152 pp. Cited on page: 325.

[67GRY/KOR]	Gryzin, Yu. I., Koryttsev, K. Z., A study of the behaviour of UO_3 and its hydrates in solutions with an electrode of the third kind, *Russ. J. Inorg. Chem.*, **12**, (1967), 50-53. Cited on page: 400.

[67MER/SKO]	Merkusheva, S. A., Skorik, N. A., Kumok, V. N., Serebrennikov, V. V., Thorium and uranium(IV) pyrophosphates, *Sov. Radiochem.*, **9**, (1967), 683-685. Cited on pages: 289, 292.

[67PER/KRY]	Perkovec, V. D., Kryukov, P. A., Ionisation constants for water at temperature up to 150°C, VINITI, Report 189-68, (1967). Cited on page: 400.

[68AZI/LYL]	Aziz, A., Lyle, S. J., Naqvi, S. J., Chemical equilibria in americium and curium sulphate and oxalate systems and an application of a liquid scintillation counting method, *J. Inorg. Nucl. Chem.*, **30**, (1968), 1013-1018. Cited on pages: 365, 366, 367.

[68CHO/CHU]	Choporov, D. Y., Chudinov, E. T., Melting point and saturated vapor pressure of neptunium tetrachloride, *Sov. Radiochem.*, **10**, (1968), 208-213. Cited on page: 216.

[68NAI]	Nair, G. M., Americium(III)-sulphate complexes, *Radiochim. Acta*, **10**, (1968), 116-119. Cited on pages: 365, 367.

[68PAT/DRO]	Pattoret, A., Drowart, J., Smoes, S., Etudes thermodynamiques par spectrométrie de masse sur le système uranium-oxygène, Thermodynamics of nuclear materials 1967, pp. 613-636, IAEA, Vienna, (1968). Cited on page: 161.

[69ABE]	Aberg, M., Crystal structure of $[(UO_2)_2(OH)_2Cl_2(H_2O)_4]$, *Acta Chem. Scand.*, **23**, (1969), 791. Cited on page: 597.

[69ACK/RAU]   Ackermann, R. J., Rauh, E. G., Chandrasekharaiah, M. S., A themodynamic study of the urania-uranium system, *J. Phys. Chem.*, **73**, (1969), 762-769. Cited on page: 161.

[69AZI/LYL]   Aziz, A., Lyle, S. J., Equilibrium constants for aqueous fluoro complexes of scandium, yttrium, americium(III) and curium(III) by extraction into di-2-ethylhexyl phosphoric acid, *J. Inorg. Nucl. Chem.*, **31**, (1969), 3471-3480. Cited on pages: 354, 355.

[69BAR/MIK]   Barbanel', Yu. A., Mikhailova, N. K., Study of the complex formation of Am(III) with the Cl^- ion in aqueous solutions by the method of spectrophotometry, *Sov. Radiochem.*, **11**, (1969), 576-579. Cited on page: 356.

[69BEV]   Bevington, P. R., *Data reduction and error analysis for the physical sciences*, McGraw-Hill, New York, (1969), 336 pp. Cited on pages: 744, 750.

[69CAJ/PRA]   Caja, J., Pravdic, V., Contribution to the electrochemistry of uranium(V) in carbonate solutions, *Croat. Chem. Acta*, **41**, (1969), 213-222. Cited on page: 250.

[69COM]   Comité International des Poids et des Mesures, The International Practical Temperature Scale of 1968, *Metrologia*, **5**, (1969), 35-47. Cited on page: 35.

[69GRE/VAR]   Grenthe, I., Varfeldt, J., A potentiometric study of fluoride complexes of uranium(IV) and uranium(VI) using the U(VI)/U(IV) redox couple, *Acta Chem. Scand.*, **23**, (1969), 988-998. Cited on pages: 210, 211.

[69GRU/MCB]   Gruen, D. M., McBeth, R. L., Vapor complexes of uranium pentachloride and uranium tetrachloride with aluminium chloride: The nature of gaseous uranium pentachloride, *Inorg. Chem.*, **8**, (1969), 2625-2633. Cited on page: 219.

[69KEL/FAN]   Keller, C., Fang, D., Über Karbonatokomplexe des dreiwertigen Americiums sowie des vier- und sechswertigen Urans und Plutoniums, *Radiochim. Acta*, **11**, (1969), 123-127, in German. Cited on pages: 381, 382, 430.

[69KNA/LOS]   Knacke, O., Lossmann, G., Müller, F., Zur thermischen Dissoziation und Sublimation von UO_2F_2, Z. Anorg. Allg. Chem., **371**, (1969), 32-39, in German. Cited on pages: 207, 208.

[69MAR/SHI]   Marcus, Y., Shiloh, M., A spectrophotometric study of trivalent actinide complexes in solution. IV. Americium with chloride ligands, Isr. J. Chem., **7**, (1969), 31-43. Cited on page: 356.

[69NOR]   Norén, B., A solvent extraction and potentiometric study of fluoride complexes of thorium(IV) and uranium(IV), Acta Chem. Scand., **23**, (1969), 931-942. Cited on pages: 210, 211.

[69ROS]   Rossotti, H., *Chemical applications of potentiometry*, D. Van Nostrand, Princeton N. J., (1969), 229 pp. Cited on pages: 24, 25.

[69SAL/HAK]   Salomaa, P., Hakala, R., Vesala, S., Aalto, T., Solvent deuterium isotope effects on acid-base reactions: Part III. Relative acidity constants of inorganic oxyacids in light and heavy water. Kinetic applications, Acta Chem. Scand., **23**, (1969), 2116-2126. Cited on page: 657.

[69SCH/KRA]   Schwochau, K., Krasser, W., Schwingungsspektren und Kraftkonstanten der Hexahalogeno-Komplexe des Technetium(IV) und Rhenium(IV), Z. Naturforsch., **24a**, (1969), 403-407, in German. Cited on page: 663.

[69SHI/GIV]   Shiloh, M., Givon, M., Marcus, Y., A spectrophotometric study of trivalent actinide complexes in solutions. III. Americium with bromide, iodide, nitrate and carbonate ligands, J. Inorg. Nucl. Chem., **31**, (1969), 1807-1814. Cited on page: 375.

[69ZMB]   Zmbov, K. F., Heats of formation of lower actinide fluorides from mass spectrometric studies, Proc. 1st International Conf. on Calorimetry and Thermodynamics, Polish Scientific Publisher, Warsaw, Poland, (1969). Cited on pages: 199, 200.

[70ABE]   Aberg, M., On the structures of the predominant hydrolysis products of uranyl(VI) in solution, Acta Chem. Scand., **24**, (1970), 2901-2915. Cited on page: 597.

[70BRA/COB]   Brand, J. R., Cobble, J. W., The thermodynamic functions of neptunium(V) and neptunium(VI), *Inorg. Chem.*, **9**, (1970), 912-917. Cited on page: 639.

[70CHU/CHO4] Chudinov, E. G., Choporov, D. Y., Vapor pressure of solid uranium tetrafluoride, *Russ. J. Phys. Chem.*, **44**, (1970), 1106-1109. Cited on page: 201.

[70GOO/LON]   Goodenough, J. B., Longo, J. M., Crystallographic and magnetic properties of perovskite and perovskite-related compounds, Landolt-Börnstein Tables, III, 4A, pp. 126-314, Springer Berlin, (1970). Cited on pages: 260, 543, 617.

[70MAN]   Mann, J. B., Ionization cross sections of the elements, *Recent developments in Mass Spectroscopy*, Ogata, K., Ed., pp. 814-819, Univ. Park Press, Baltimore, (1970). Cited on page: 446.

[70MOR/COB]   Morss, L. R., Cobble, J. W., 160th National meeting of the American Chemical Society, Chicago, (1970). Cited on page: 404.

[71AHR/KUL]   Ahrland, S., Kullberg, L., Thermodynamics of metal complex formation in aqueous solution: I. A potentiometric study of fluoride complexes of hydrogen, uranium(VI) and vanadium(IV), *Acta Chem. Scand.*, **25**, (1971), 3457-3470. Cited on pages: 209, 451.

[71COR/LOO]   Cordfunke, E. H. P., Loopstra, B. O., Sodium uranates: Preparation and thermochemical properties, *J. Inorg. Nucl. Chem.*, **33**, (1971), 2427-2436. Cited on pages: 271, 274, 287, 398, 399.

[71JEN]   Jensen, K. A., *Nomenclature of inorganic chemistry, IUPAC Commission on Nomenclature of Inorganic Compounds*, Pergamon Press, Oxford, (1971), 110 pp. Cited on page: 13.

[71MOS]   Moskvin, A. I., Correlation of the solubility products of actinide compounds with the properties of the metal ions and acid anions forming them, *Sov. Radiochem.*, **13**, (1971), 299-300. Cited on page: 288.

[71NAU/RYZ] Naumov, G. B., Ryzhenko, B. N., Khodakovsky, I. L., *Handbook of thermodynamic data*, Atomizdat, Moscow, (1971), in Russian; Engl. transl.: Report USGS-WRD-74-001 (Soleimani, G. J., translator; Barnes I., Speltz, V., *eds.*), U.S. Geological Survey, Menlo Park, California, USA, 1974, 328p. Cited on pages: 286, 288, 289, 290.

[71NIK/PIR] Nikolaeva, N. M., Pirozhkov, A. V., Determination of the solubility product of uranyl hydroxide at elevated temperatures, *Izv. Sib. Otd. Akad. Nauk SSSR, Ser. Khim. Nauk*, **4**, (1971), 73-81, in Russian. Cited on pages: 191, 399, 538.

[71NIK] Nikolaeva, N. M., The study of hydrolysis and complexing of uranyl ions in sulphate solutions at elevated temperatures, *Izv. Sib. Otd. Akad. Nauk SSSR, Ser. Khim. Nauk*, **7**, (1971), 61-67. Cited on page: 400.

[71POR/WEB] Porter, R. A., Weber, Jr., W. J., The interaction of silicic acid with iron(III) and uranyl ions in dilute aqueous solutions, *J. Inorg. Nucl. Chem.*, **33**, (1971), 2443-2449. Cited on pages: 252, 253, 254, 593, 612.

[71SAN/VID] Santalova, N. A., Vidavskii, L. M., Dunaeva, K. M., Ippolitova, E. A., Enthalpy of formation of uranium trioxide dihydrate, *Sov. Radiochem.*, **13**, (1971), 608-612. Cited on page: 412.

[71SIL/MAR] Sillén, L. G., Martell, A. E., *Stability constants of metal ion complexes, Suppl. No. 1, Special Publication*, No. 25, Chemical Society, London, (1971), 865 pp. Cited on pages: 17, 422, 438.

[71TAY/HUR] Taylor, J. C., Hurst, H. J., The hydrogen atom location in α and β forms of uranyl hydroxide, *Acta Crystallogr.*, **B27**, (1971), 2018-2022. Cited on page: 505.

[72DWO] Dworkin, A. S., Enthalpy of uranium tetrafluoride from 298 to 1400 K: Enthalpy and entropy of fusion, *J. Inorg. Nucl. Chem.*, **34**, (1972), 135-138. Cited on pages: 202, 213.

[72FUG] Fuger, J., Thermodynamic properties of simple actinide compounds, *Lanthanides and actinides*, Bagnall, K. W., Ed., vol. 7, pp. 157-210, Butterworths, London, (1972). Cited on page: 648.

[72GUI/GIG] Guido, M., Gigli, G., Balducci, G., Dissociation energy of CuCl and Cu_2Cl_2(g) gaseous molecules, *J. Chem. Phys.*, **57**, (1972), 3731-3735. Cited on page: 393.

[72MCD/COL] McDowell, W. J., Coleman, C. F., The sulfate complexes of some trivalent transplutonium actinides and europium, *J. Inorg. Nucl. Chem.*, **34**, (1972), 2837-2850. Cited on pages: 365, 366, 367.

[72MET/GUI] Metivier, H., Guillaumont, R., Hydrolyse du plutonium tétravalent, *Radiochem. Radioanal. Lett.*, **10**, (1972), 27-35, in French. Cited on pages: 314, 316, 318, 642, 643, 648, 687, 692.

[72MUS] Musikas, C., Formation d'uranates solubles par hydrolyse des ions uranyle(VI), *Radiochem. Radioanal. Lett.*, **11**, (1972), 307-316, in French. Cited on page: 177.

[72NIK/SER] Nikitin, A. A., Sergeyeva, E. I., Khodakovsky, I. L., Naumov, G. B., Hydrolysis of uranyl in the hydrothermal region, *Geokhimiya*, (1972), 297-307, in Russian; Engl. transl.: AECL translation Nr. 3554, Atomic Energy of Canada Ltd., Pinawa, Manitoba, Canada, 21pp. Cited on pages: 538, 566.

[72SAN/VID] Santalova, N. A., Vidavskii, L. M., Dunaeva, K. M., Ippolitova, E. A., Enthalpy of formation of uranium trioxide hemihydrate, *Sov. Radiochem.*, **14**, (1972), 741-745. Cited on pages: 190, 539.

[72TAY/KEL] Taylor, J. C., Kelly, J. W., Downer, B., A study of the $\beta \rightarrow \alpha$ phase transition in $UO_2(OH)_2$ by dilatometric, microcalorimetric and X-ray diffraction techniques, *J. Solid State Chem.*, **5**, (1972), 291-299. Cited on page: 191.

[73BAT] Bates, R. G., *Determination of pH, theory and practice*, John Wiley & Sons, New York, (1973), 479 pp. Cited on page: 24.

[73GRE/CHE] Greaves, C., Cheetham, A. K., Fender, B. E. F., Sodium uranium bronze and related phases, *Inorg. Chem.*, **12**, (1973), 3003-3007. Cited on page: 419.

[73GRU/HEC] Gruber, J. B., Hecht, H. G., Interpretation of the vapor spectrum of UCl_4, *J. Chem. Phys.*, **59**, (1973), 1713-1720. Cited on page: 216.

[73HOE/SIE]   Hoekstra, H. R., Siegel, S., The uranium trioxide-water system, *J. Inorg. Nucl. Chem.*, **35**, (1973), 761-779. Cited on page: 190.

[73HUL/DES]   Hultgren, R., Desai, P. D., Hawkins, D. T., Gleiser, M., Kelley, K. K., Wagman, D. D., *Selected values of the thermodynamic properties of the elements*, American Society for Metals, Metals Park, Ohio, USA, (1973), 636 pp. Cited on pages: 148, 447.

[73MET]   Metivier, H., Contribution à l'étude de l'hydrolyse du plutonium tétravalent et de sa complexation par des acides d'intérêt biologique, Ph. D. Thesis, Université Paris VI, (1973), Paris, France, in French. Also published as CEA-R-4477. Cited on page: 314.

[73MOS]   Moskvin, A. I., Some thermodynamic characteristics of the processes of formation of actinide compounds in a solid form: I. Energy and entropy of the crystal lattice, heats of formation and heats of solution, *Sov. Radiochem.*, **15**, (1973), 356-363. Cited on page: 288.

[73PIT]   Pitzer, K. S., Thermodynamics of electrolytes: I. Theoretical basis and general equations, *J. Phys. Chem.*, **77**, (1973), 268-277. Cited on pages: 708, 710.

[73PIT/MAY]   Pitzer, K. S., Mayorga, G., Thermodynamics of electrolytes. II. Activity and osmotic coefficients for strong electrolytes with one or both ions univalent, *J. Phys. Chem.*, **77**, (1973), 2300-2308. Cited on page: 710.

[74BIN/SCH]   Binnewies, M., Schäfer, H., Gasförmige Halogenidkomplexe und ihre Stabilität, *Z. Anorg. Allg. Chem.*, **407**, (1974), 327-344, in German. Cited on page: 219.

[74COR/PRI]   Cordfunke, E. H. P., Prins, G., Equilibria involving volatile UO_2Cl_2, *J. Inorg. Nucl. Chem.*, **36**, (1974), 1291-1293. Cited on page: 220.

[74DHA/TRI] Dharwadkar, S. R., Tripathi, S. N., Karkhanavala, M. D., Chandrasekharaiah, M. S., Thermodynamic properties of gaseous uranium hydroxide, 4th Thermodynamic of Nuclear Materials Symp., held 21-25 October 1974, in Vienna, vol. 2, pp. 455-465, IAEA, Vienna, Austria, (1974). Cited on pages: 162, 163, 401, 454, 455, 591.

[74HEL/KIR] Helgeson, H. C., Kirkham, D. H., Theoretical prediction of the thermodynamic behavior of aqueous electrolytes at high pressures and temperatures: I. Summary of the thermodynamic/ electrostatic properties of the solvent, *Am. J. Sci.*, **274**, (1974), 1089-1198. Cited on page: 715.

[74KAK/ISH] Kakihana, H., Ishiguro, S., Potentiometric and spectrophotometric studies of fluoride complexes of uranium(IV), *Bull. Chem. Soc. Jpn.*, **47**, (1974), 1665-1668. Cited on pages: 210, 211.

[74MOC/NAG] Mochizuki, A., Nagashima, K., Wakita, H., The synthesis of crystalline hydrated double carbonates of rare earth elements and sodium, *Bull. Chem. Soc. Jpn.*, **47**, (1974), 755-756. Cited on pages: 382, 430.

[74OHA/HOE3] O'Hare, P. A. G., Hoekstra, H. R., Thermochemistry of uranium compounds: IV. Standard enthalpy of formation of sodium uranium(V) trioxide (NaUO$_3$), *J. Chem. Thermodyn.*, **6**, (1974), 965-972. Cited on pages: 272, 474.

[74PAR] Parsons, R., Manual of Symbols and Terminology for Physicochemical. Quantities and Units. Appendix III. Electrochemical nomenclature, *Pure Appl. Chem.*, **37**, (1974), 503-516. Cited on page: 24.

[74PIT/KIM] Pitzer, K. S., Kim, J. J., Thermodynamics of electrolytes. IV. Activity and osmotic coefficients for mixed electrolytes, *J. Am. Chem. Soc.*, **96**, (1974), 5701-5707. Cited on page: 710.

[74PIT/MAY] Pitzer, K. S., Mayorga, G., Thermodynamics of electrolytes. III. Activity and osmotic coefficients for 2-2 electrolytes, *J. Solution Chem.*, **3**, (1974), 539-546. Cited on page: 710.

[74RAB/ALE]   Rabinovich, V. A., Alekseeva, T. E., Real activity coefficients of the individual ions in solutions of electrolytes. II. Comparison of the real and chemical activity coefficients of the ions in solutions of hydrogen chloride and alkali metal chlorides at 25°C, *Sov. Electrochem.*, **10**, (1974), 502-506. Cited on page: 756.

[74SCH/KEL]   Schwabe, K., Kelm, H., Queck, C., Zum Problem der Einzelionenaktivitäten in konzentrierten Elektrolytlösungen, *Z. Phys. Chem. (Leipzig)*, **255**, (1974), 1149-1156, in German. Cited on page: 756.

[74STE]   Stephens, H. P., Determination of the enthalpy of liquid copper and uranium with a liquid argon calorimeter, *High Temp. Sci.*, **6**, (1974), 156-166. Cited on pages: 157, 430.

[74TAR/GAR]   Tardy, Y., Garrels, R. M., A method of estimating the Gibbs energies of formation of layer silicates, *Geochim. Cosmochim. Acta*, **38**, (1974), 1101-1116. Cited on page: 254.

[74VIS/VOL]   Visyashcheva, G. I., Volkov, Y. F., Simakin, G. A., Kapshukov, I. I., Bevz, A. S., Yakovlev, G. N., Carbonate compounds of pentavalent actinides with alkali metal cations: I. Composition and some properties of solid carbonates of pentavalent neptunium with potassium obtained from K_2CO_3 solutions, *Sov. Radiochem.*, **16**, (1974), 832-837. Cited on pages: 311, 569.

[74VOL/KAP2]   Volkov, Y. F., Kapshukov, I. I., Visyashcheva, G. I., Yakovlev, G. N., Carbonate compounds of pentavalent actinides with alkali metals cations: IV. X-Ray investigation of dicarbonates of neptunium(V), plutonium(V) and americium(V) with potassium, *Sov. Radiochem.*, **16**, (1974), 846-850. Cited on pages: 311, 569.

[75APE/SAH]   Apelblat, A., Sahar, A., Properties of aqueous thorium nitrate solutions. 3. Partial molal heat capacities and heats of dilution at 30°C, *J. Chem. Soc. Faraday Trans.*, **1**, (1975), 71. Cited on pages: 159, 541.

[75BIE]   Biedermann, G., Ionic media, Dahlem workshop on the nature of seawater, pp. 339-362, Dahlem Konferenzen, Berlin, (1975). Cited on page: 709.

[75BRA/KEM] von Braun, R., Kemmler-Sack, S., Roller, H., Seemann, I., Wall, I., Uber Perowskitphasen im System BaO-UO$_{2+x}$ mit x<1, *Z. Anorg. Allg. Chem.*, **415**, (1975), 133-155, in German. Cited on pages: 266, 534.

[75COR/EGM] Cordfunke, E. H. P., van Egmond, A. B., van Voorst, G., Investigations on cesium uranates-I. Characterization of the phases in the Cs-U-O system, *J. Inorg. Nucl. Chem.*, **37**, (1975), 1433-1436. Cited on page: 405.

[75COR/OUW] Cordfunke, E. H. P., Ouweltjes, W., Prins, G., Standard enthalpies of formation of uranium compounds: I. β-UO$_3$ and γ-UO$_3$, *J. Chem. Thermodyn.*, **7**, (1975), 1137-1142. Cited on pages: 394, 398.

[75COR] Cordfunke, E. H. P., (1975), Personal communication, (from the citation in [76FUG/OET]). Cited on pages: 262, 633.

[75DEG/CHO] Degischer, G., Choppin, G. R., Complex chemistry in aqueous solutions, *Gmelin Handbuch der anorganischen Chemie: Transurane, Band 20,Teil **D1**: Chemie in Lösung*, pp. 129-176, Springer-Verlag, Berlin, (1975). Cited on pages: 354, 355.

[75FRE] Fred, M., (1975), Argonne National Laboratory, private communication, (from the citation in [76OET/RAN]). Cited on page: 335.

[75HIL/CUB] Hildenbrand, D. L., Cubicciotti, D. D., *Mass spectrometric study of the molecular species formed during vaporization of UCl$_4$ in Cl$_2$ and AlCl$_3$*, No. UCRL-13657, Lawrence Livermore National Laboratory, Stanford, Res. Inst., Menlo Prak, California, USA, (1975), 22 pp. Cited on page: 216.

[75OHA/HOE2] O'Hare, P. A. G., Hoekstra, H. R., Thermochemistry of uranium compounds: VI. Standard enthalpy of formation of Cs$_2$U$_2$O$_7$, thermodynamics of formation of cesium and rubidium uranates at elevated temperatures, *J. Chem. Thermodyn.*, **7**, (1975), 831-838. Cited on page: 287.

[75OLO/HEP] Olofsson, G., Hepler, L. G., Thermodynamics of ionization of water over ranges of temperaure and pressure, *J. Solution Chem.*, **4**, (1975), 127-143. Cited on page: 401.

[75PIT] Pitzer, K. S., Thermodynamics of electrolytes. V. Effects of higher-order electrostatic terms, *J. Solution Chem.*, **4**, (1975), 249-265. Cited on page: 710.

[75SPI/ZIN] Spitsyn, V. I., Zinov'ev, V. E., Gel'd, P. V., Balakhovskii, O. A., Thermophysical and magnetic properties of technetium over a wide range of temperatures, *Proc. Acad. Sci. USSR Phys. Chem. Sect.*, **221**, (1975), 225-228. Cited on pages: 389, 530.

[75STO/KHA] Stozhenko, T. P., Khanaev, E., Afanas'ev, Yu. A., The enthalpies of additio of water to certain lanthanide trifluorides, *Russ. J. Phys. Chem.*, **49**, (1975), 1241-1242. Cited on page: 359.

[75UEN/SAI] Ueno, K., Saito, A., Solubility and absorption spectra of a carbonato complex of pentavalent neptunium, *Radiochem. Radioanal. Lett.*, **22**, (1975), 127-133. Cited on page: 570.

[76BAE/MES] Baes, Jr., C. F., Mesmer, R. E., *The hydrolysis of cations*, Wiley & Sons, New York, (1976), 489 pp. Cited on pages: 28, 184, 427, 710.

[76BOU/BON] Bousquet, J., Bonnetot, B., Claudy, P., Mathurin, D., Turck, G., Enthalpies de formation standard des formiates d'uranyle anhydre et hydrate par calorimétrie de réaction, *Thermochim. Acta*, **14**, (1976), 357-367, in French. Cited on page: 290.

[76CHO/UNR] Choppin, G. R., Unrein, P. J., Thermodynamic study of actinide fluoride complexation, 4th International Transplutonium Element Symposium, held in Baden-Baden, 13-17 September, 1975, pp. 97-107, North-Holland Publ., Amsterdam, Holland, (1976). Cited on pages: 210, 211, 354, 355.

[76FUG/OET] Fuger, J., Oetting, F. L., *The chemical thermodynamics of actinide elements and compounds: Part 2. The actinide aqueous ions*, International Atomic Energy Agency, Vienna, (1976), 65 pp. Cited on pages: 297, 318, 319, 321, 358, 639, 648, 690.

[76MOR/MCC] Morss, L. R., McCue, M. C., Partial molal entropy and heat capacity of the aqueous thorium(IV) ion. Thermochemistry of thorium nitrate pentahydrate, *J. Chem. Eng. Data*, **21**, (1976), 337-341. Cited on pages: 158, 159, 403, 404, 541.

[76NGU/POT] Nguyen-Trung, C., Poty, B., Solubilité de UO_2 en milieu aqueux à 400-500°C à 1 kbar, (1976). Annual Report (CREGU, Nancy) 37-39. Cited on page: 561.

[76OET/RAN] Oetting, F. L., Rand, M. H., Ackermann, R. J., *The chemical thermodynamics of actinide elements and compounds: Part 1. The actinide elements*, International Atomic Energy Agency, Vienna, (1976), 111 pp. Cited on pages: 157, 335, 448, 455, 648.

[76PIT/SIL] Pitzer, K. S., Silvester, L. F., Thermodynamics of electrolytes. VI. Weak electrolytes including H_3PO_4, *J. Solution Chem.*, **5**, (1976), 269-278. Cited on page: 710.

[76REA/LAN] Reardon, E. J., Langmuir, D., Activity coefficients of $MgCO_3^0$ and $CaSO_4^0$ ion pairs as a function of ionic strength, *Geochim. Cosmochim. Acta*, **40**, (1976), 549-554. Cited on page: 404.

[76SCA] Scatchard, G., *Equilibrium in solutions: Surface and colloid chemistry*, Harvard University Press, Cambridge, Massachusetts, (1976), 306 pp. Cited on page: 709.

[76SHA] Shannon, R. D., Revised effective ionic radii and systematic studies of interatomic distances in halides and chalcogenides, *Acta Crystallogr.*, **A32**, (1976), 751-767. Cited on page: 334.

[76TAR/GAR] Tardy, Y., Garrels, R. M., Prediction of Gibbs energies of formation: I. Relationships among Gibbs energies of formation of hydroxides, oxides and aqueous ions, *Geochim. Cosmochim. Acta*, **40**, (1976), 1051-1056. Cited on page: 534.

[76TOS] Tossidis, I. A., Dissociation of phosphoric and arsenic acid in aqueous organic solvents, *Inorg. Nucl. Chem. Lett.*, **12**, (1976), 609. Cited on page: 657.

[76WEI/WIS] Weigel, F., Wishnevsky, V., Hauske, H., The vapor phase hydrolysis of $^{241}AmCl_3$ and $^{243}AmCl_3$, heats of formation of $^{241}AmOCl$ and $^{243}AmOCl$, *Transplutonium 1975*, Muller, W., Lindner, R., Eds., pp. 217-226, North-Holland, Amsterdam, (1976). Cited on page: 362.

[77COR/OUW2] Cordfunke, E. H. P., Ouweltjes, W., Standard enthalpies of formation of uranium compounds: IV. α-UO_2SO_4, β-UO_2SeO_4, and UO_2SeO_3, *J. Chem. Thermodyn.*, **9**, (1977), 1057-1062. Cited on pages: 188, 193, 269, 272, 407, 413, 419.

[77DUP/GUI] Duplessis, J., Guillaumont, R., Hydrolyse du neptunium tétravalent, *Radiochem. Radioanal. Lett.*, **31**, (1977), 293-302, in French. Cited on pages: 295, 296, 298, 314, 648, 693, 694.

[77FER] Fernelius, W. C., *How to name an inorganic substance, IUPAC Commission on Nomenclature of Inorganic Compounds*, Pergamon Press, Oxford, (1977), 36 pp. Cited on page: 13.

[77HIL] Hildenbrand, D. L., Thermochemistry of gaseous uranium pentafluoride and uranium tetrafluoride, *J. Chem. Phys.*, **66**, (1977), 4788-4794. Cited on pages: 201, 205.

[77SIM] Simakin, G. A., Real oxidation potentials of the couples AmO_2^{2+}-AmO_2^+, NpO_2^{2+}-NpO_2^+ in solutions of potassium and sodium carbonates, *Sov. Radiochem.*, **19**, (1977), 424-426. Cited on pages: 309, 310.

[77VOL/VIS] Volkov, Y. F., Visyashcheva, G. I., Kapshukov, I. I., Study of carbonate compounds of pentavalent actinides with alkali metal cations: V. Production and identification of hydrate forms of sodium monocarbonatoneptunylate, *Sov. Radiochem.*, **19**, (1977), 263-266. Cited on pages: 308, 309, 482.

[78ALL/BEA] Allard, B., Beall, G. W., Predictions of actinide species in the groundwater, Workshop on the environmental chemistry and research of the actinide elements, held 8-12 October, 1978, in Warrenton, Virginia, USA, (1978). Cited on page: 288.

[78BOY] Boyd, G. E., Osmotic and activity coefficients of aqueous $NaTcO_4$ and $NaReO_4$ solutions at 25°C, *J. Solution Chem.*, **7**, (1978), 229-238. Cited on page: 547.

[78BRA] Brandenburg, N. P., Methods for estimating the enthalpy of formation of inorganic compounds; Thermochemical and crystallographic investigations of uranyl salts of group VI elements, Ph. D. Thesis, University of Amsterdam, (1978), Amsterdam, The Netherlands, 141 pp. Cited on pages: 234, 235.

[78COR/OHA]   Cordfunke, E. H. P., O'Hare, P. A. G., *The chemical thermodynamics of actinide elements and compounds: Part 3. Miscellaneous actinide compounds*, International Atomic Energy Agency, Vienna, (1978), 83 pp. Cited on pages: 192, 262, 287, 290, 633.

[78KOB/KOL]   Kobets, L. V., Kolevich, T. A., Umreiko, D. S., Crystalline hydrated forms of triuranyl diorthophosphate, *Russ. J. Inorg. Chem.*, **23**, (1978), 501-505. Cited on page: 242.

[78LAK]   Lakner, J. F., Isotherms for the $U-UH_3-H_2$ system at temperatures of 700-1065°C and pressures to 137.9 MPa, Lawrence Livermore National Laboratory, Report UCRL-52518, (1978). Cited on page: 417.

[78LAN]   Langmuir, D., Uranium solution - mineral equilibria at low temperatures with applications to sedimentary ore deposits, *Geochim. Cosmochim. Acta*, **42**, (1978), 547-569. Cited on pages: 288, 289, 291, 292.

[78NIK/PIR]   Nikolaeva, N. M., Pirozhkov, A. V., The solubility product of U(IV) hydroxide at the elevated temperatures, *Izv. Sib. Otd. Akad. Nauk SSSR, Ser. Khim. Nauk*, **5**, (1978), 82-88, in Russian. Cited on page: 184.

[78PER/THO]   Perachon, G., Thourey, J., Etude par calorimétrie de la solvatation des halogénures alcalino-terreux dans les solutions aqueuses d'acides halogénés correspondants: I. Enthalpies de dissolution et de dilution des halogénures alcalino-terreux, *Thermochim. Acta*, **27**, (1978), 111-124, in French. Cited on pages: 261, 466.

[78PIT/PET]   Pitzer, K. S., Peterson, J. R., Silvester, L. F., Thermodynamics of electrolytes: IX. Rare earth chlorides, nitrates, and perchlorates, *J. Solution Chem.*, **7**, (1978), 45-56. Cited on page: 710.

[78SIN/PRA]   Singh, Z., Prasad, R., Venugopal, V., Sood, D. D., The vaporization thermodynamics of uranium tetrachloride, *J. Chem. Thermodyn.*, **10**, (1978), 129-134. Cited on page: 216.

[79BRA/PIT]   Bradley, D. J., Pitzer, K. S., Thermodynamics of electrolytes. 12. Dielectric properties of water and Debye-Hückel parameters to 350°C and 1 kbar, *J. Phys. Chem.*, **83**, (1979), 1599-1603. Cited on page: 715.

[79CIA/FER]   Ciavatta, L., Ferri, D., Grimaldi, M., Palombari, R., Salvatore, F., Dioxouranium(VI) carbonate complexes in acid solution, *J. Inorg. Nucl. Chem.*, **41**, (1979), 1175-1182. Cited on page: 245.

[79GOL]   Goldberg, R. N., Evaluated activity coefficients and osmotic coefficients for aqueous solutions: Bi-univalent compounds of lead, copper, manganese and uranium, *J. Phys. Chem. Ref. Data*, **8**, (1979), 1005-1050. Cited on page: 574.

[79HAA/WIL]   Haacke, D. F., Williams, P. A., The aqueous chemistry of uranium minerals: Part I. Divalent cation zippeite, *Mineral. Mag.*, **43**, (1979), 539-541. Cited on pages: 282, 283, 287, 292.

[79JOH/PYT]   Johnson, K. S., Pytkowicz, R. M., Ion association and activity coefficients in multicomponent solutions, *Activity coefficients in electrolyte solutions*, Pytkowicz, R. M., Ed., II, pp. 1-62, CRC Press, Boca Raton, Florida, (1979). Cited on page: 708.

[79KLE/HIL]   Kleinschmidt, P. D., Hildenbrand, D. L., Thermodynamics of the dimerization of gaseous UF_5, *J. Chem. Phys.*, **71**, (1979), 196-201. Cited on pages: 204, 206, 207.

[79LIN/KON]   Lingerak, W. A., Konijn, P. C., Slanina, J., The high precision determination of uranium by computer controlled titrimetry, Proceedings of the first annual symposium on safeguards and nuclear material management, pp. 157-159, Brussels, Belgium, (1979). Cited on page: 632.

[79MIL]   Millero, F. J., Effects of pressure and temperature on activity coefficients, *Activity coefficients in electrolyte solutions*, Pytkowicz, R. M., Ed., II, pp. 63-151, CRC Press, Boca Raton, Florida, (1979). Cited on pages: 708, 716.

[79PIT]   Pitzer, K. S., Theory: Ion Interaction Approach, *Activity coefficients in electrolyte solutions*, Pytkowicz, R. M., Ed., vol. 1, pp. 157-208, CRC Press, Boca Raton, Florida, (1979). Cited on pages: 708, 710.

[79PRA/NAG] Prasad, R., Nagarajan, K., Singh, Z., Bhupathy, M., Venugopal, V., Sood, D. D., Thermodynamics of the vaporization of thorium and uranium halides, Thermodynamics of nuclear materials 1979, Proc. Symp. held 29 January to 2 February, 1979, in Jülich, Federal Republic of Germany, vol. 1, p. 45, International Atomic Energy Agency, Vienna, Austria, (1979). Cited on pages: 226, 227.

[79PYT] Pytkowicz, R. M., Activity coefficients, ionic media, and equilibria in solutions, *Activity coefficients in electrolyte solutions*, Pytkowicz, R. M., Ed., II, pp. 301-305, CRC Press, Boca Raton, Florida, (1979). Cited on page: 708.

[79SYL/DAV] Sylva, R. N., Davidson, M. R., The hydrolysis of metal ions: Part 1. Copper(II), *J. Chem. Soc. Dalton Trans.*, (1979), 232-235. Cited on page: 169.

[79TAG/FUJ] Tagawa, H., Fujino, T., Yamashita, T., Formation and some chemical properties of alkaline-earth metal monouranates, *J. Inorg. Nucl. Chem.*, **41**, (1979), 1729-1735. Cited on page: 286.

[79VOL/VIS2] Volkov, Yu. F., Visyashcheva, G. I., Tomilin, S. V., Kapshukov, I. I., Rykov, A. G., X-ray diffraction analysis of composition and crystal structure of some pentavalent actinide carbonates, Nauchno-Issledovatel'skij Inst. Atomnykh Reaktorov, Report NIIAR-16(375), (1979), 29 pp. Cited on page: 309.

[79WHI] Whitfield, M., Activity coefficients in natural waters, *Activity coefficients in electrolyte solutions*, Pytkowicz, R. M., Ed., II, pp. 153-299, CRC Press, Boca Raton, Florida, (1979). Cited on pages: 11, 12.

[79WHI2] Whiffen, D. H., *Manual of symbols and terminology for physicochemical quantities and units, IUPAC Commission on Physicochemical Symbols, Terminology and Units*, Pergamon Press, Oxford, (1979), 41 pp. Cited on page: 708.

[80ALW/WIL] Alwan, A. K., Williams, P. A., The aqueous chemistry of uranium minerals: Part II. Minerals of the liebigite group, *Mineral. Mag.*, **43**, (1980), 665-667. Cited on pages: 282, 290, 291.

[80ALW]   Alwan, A. K., The chemistry of formation of some mineral species, Ph. D. Thesis, University of Wales, UK, (1980). Cited on page: 404.

[80BEN/TEA]   Benson, L. V., Teague, L. S., A tabulation of thermodynamic data for chemical reactions involving 58 elements common to radioactive waste package systems, Lawrence Berkeley Laboratory, Report LBL-11448, (1980), 97 pp. Cited on pages: 290, 291.

[80CHI/WES]   Chirico, R. D., Westrum, Jr., E. F., Thermophysics of the lanthanide hydroxides: I. Heat capacities of $La(OH)_3$, $Gd(OH)_3$, and $Eu(OH)_3$ from near 5 to 350 K. Lattice and Schottky contributions, *J. Chem. Thermodyn.*, **12**, (1980), 71-85. Cited on page: 350.

[80CIA]   Ciavatta, L., The specific interaction theory in evaluating ionic equilibria, *Ann. Chim. (Rome)*, **70**, (1980), 551-567. Cited on pages: 709, 710, 721, 722, 723, 724, 729, 731, 733, 735, 753, 762.

[80COR/WES]   Cordfunke, E. H. P., Westrum, Jr., E. F., Investigations on caesium uranates: VII. Thermochemical properties of $Cs_2U_4O_{12}$, Thermodynamics of nuclear materials 1979, Proc. Symp. held 29 January to 2 February, 1979, in Jülich, Federal Republic of Germany, pp. 125-141, International Atomic Energy Agency, Vienna, Austria, (1980). Cited on pages: 269, 278, 405.

[80COT/WIL]   Cotton, F. A., Wilkinson, G., *Advanced inorganic chemistry. A comprehensive Text*, 4th. Edition, Wiley-Interscience, New York, (1980). Cited on page: 334.

[80KHO/MAT]   Khopkar, P. K., Mathur, J. N., Complexing of californium(III) and other trivalent actinides by inorganic ligands, *J. Inorg. Nucl. Chem.*, **42**, (1980), 109-113. Cited on pages: 365, 367.

[80LEI]   Leitnaker, J. M., Re-interpretation of the vapor pressure measurements over UF_5, *High Temp. Sci.*, **12**, (1980), 289-296. Cited on pages: 203, 205, 207.

[80LEM/TRE] Lemire, R. J., Tremaine, P. R., Uranium and plutonium equilibria in aqueous solutions to 200°C, *J. Chem. Eng. Data*, **25**, (1980), 361-370. Cited on pages: 158, 409, 542, 564, 612.

[80NAG/BHU] Nagarajan, K., Bhupathy, M., Prasad, R., Singh, Z., Venugopal, V., Sood, D. D., Vaporization behaviour of uranium tetrafluoride, *J. Chem. Thermodyn.*, **12**, (1980), 329-333. Cited on page: 201.

[81BRO/NYV] Broul, M., Nyvlt, J., Soehnel, O., *Solubility in inorganic two-component systems*, Elsevier Sci. Publ., Amsterdam, (1981). Cited on page: 574.

[81CHI/AKH] Chiotti, P., Akhachinskij, V. V., Ansara, I., Rand, M. H., *The chemical thermodynamics of actinide elements and compounds: Part 5. The actinide binary alloys*, International Atomic Energy Agency, Vienna, (1981), 275 pp. Cited on page: 447.

[81CIA/FER] Ciavatta, L., Ferri, D., Grenthe, I., Salvatore, F., Spahiu, K., Studies on metal carbonate equilibria: 3. The lanthanum(III) carbonate complexes in aqueous perchlorate media, *Acta Chem. Scand.*, **A35**, (1981), 403-413. Cited on page: 586.

[81COR/OUW] Cordfunke, E. H. P., Ouweltjes, W., Standard enthalpies of formation of uranium compounds: VI. MUO_3 (M = Li, Na, K, and Rb), *J. Chem. Thermodyn.*, **13**, (1981), 187-192. Cited on pages: 193, 269, 272, 287, 413, 419, 420.

[81COR/OUW2] Cordfunke, E. H. P., Ouweltjes, W., Standard enthalpies of formation of uranium compounds: VII. UF_3 and UF_4 (by solution calorimetry), *J. Chem. Thermodyn.*, **13**, (1981), 193-197. Cited on pages: 194, 258, 464, 492.

[81FIN/CHA] Fink, J. K., Chasanov, M. G., Leibowitz, L., Thermophysical properties of uranium dioxide, *J. Nucl. Mater.*, **102**, (1981), 17-25. Cited on pages: 193, 666.

[81GLU/GUR] Glushko, V. P., Gurvich, L. V., Bergman, G. A., Veits, I. V., Medvedev, V. A., Khachkuruzov, G. A., Yungman, V. S., *Thermodynamic properties of individual substances*, vol. 3, Hemisphere Publ., New York, (1981). Cited on pages: 199, 394.

[81GOL/TRE] Gol'tsev, V. P., Tretyakov, A. A., Kerko, P. F., Barinov, V. I., Thermodynamic characteristics of calcium uranates, *Vestsi Akad. Navuk BSSR, Ser. Fiz.-Ener. Navuk*, (1981), 24-29, in Russian. Cited on page: 286.

[81GOL/TRE2] Gol'tsev, V. P., Tretyakov, A. A., Kerko, P. F., Barinov, V. I., Malevich, V. M., Thermodynamic characteristics of magnesium uranates, *Vestsi Akad. Navuk BSSR, Ser. Fiz.-Ener. Navuk*, (1981), 18-23, in Russian. Cited on page: 286.

[81HEL/KIR] Helgeson, H. C., Kirkham, D. H., Flowers, G. C., Theoretical prediction of the thermodynamic behavior of aqueous electrolytes at high pressures and temperatures: IV. Calculation of activity coefficients, osmotic coefficients, and apparent molal and standard and relative partial molal properties to 600°C and 5 kb, *Am. J. Sci.*, **281**, (1981), 1249-1516. Cited on pages: 715, 716.

[81LIN/BES] Lindemer, T. B., Besmann, T. M., Johnson, C. E., Thermodynamic review and calculations: Alkali-metal oxide systems with nuclear fuels, fission products, and structural materials, *J. Nucl. Mater.*, **100**, (1981), 178-226. Cited on pages: 274, 276, 287, 399, 527, 638.

[81OBR/WIL] O'Brien, T. J., Williams, P. A., The aqueous chemistry of uranium minerals: Part 3. Monovalent cation zippeites, *Inorg. Nucl. Chem. Lett.*, **17**, (1981), 105-107. Cited on pages: 282, 287.

[81OHA/FLO] O'Hare, P. A. G., Flotow, H. E., Hoekstra, H. R., Cesium diuranate($Cs_2U_2O_7$): Heat capacity (5 to 350 K) and thermodynamic functions to 350 K; A re-evaluation of the standard enthalpy of formation and the thermodynamics of (cesium + uranium + oxygen), *J. Chem. Thermodyn.*, **13**, (1981), 1075-1080. Cited on pages: 269, 277, 546.

[81STO/SMI] Stohl, F. V., Smith, D. K., The crystal chemistry of the uranyl silicates minerals, *Am. Mineral.*, **66**, (1981), 610-625. Cited on page: 292.

[81STU/MOR] Stumm, W., Morgan, J. J., *Aquatic Chemistry. An introduction emphasizing chemical equilibria in natural waters*, 2nd. Edition, John Wiley and Sons, New York, (1981), 780 pp. Cited on page: 23.

[81VOC/PIR]   Vochten, R., Piret, P., Goeminne, A., Synthesis, crystallographic data, solubility and electrokinetic properties of copper-, nickel- and cobalt-uranylphosphate, *Bull. Minéral.*, **104**, (1981), 457-467. Cited on pages: 282, 283, 284, 288.

[82BAR/JAC]   Barrett, S. A., Jacobson, A. J., Tofield, B. C., Fender, B. E. F., The preparation and structure of barium uranium oxide $BaUO_{3+x}$, *Acta Crystallogr.*, **B38**, (1982), 2775-2781. Cited on pages: 266, 534.

[82BID]   Bidoglio, G., Characterization of Am(III) complexes with bicarbonate and carbonate ions at groundwater concentration levels, *Radiochem. Radioanal. Lett.*, **53**, (1982), 45-60. Cited on pages: 375, 586.

[82COR/MUI]   Cordfunke, E. H. P., Muis, R. P., Ouweltjes, W., Flotow, H. E., O'Hare, P. A. G., The thermodynamic properties of Na_2UO_4, $Na_2U_2O_7$, and $NaUO_3$, *J. Chem. Thermodyn.*, **14**, (1982), 313-322. Cited on pages: 472, 473.

[82DRE]   Drever, J. I., *The geochemistry of natural waters*, Prentice-Hall, Englewood Cliffs, N. J., (1982), 388 pp. Cited on page: 23.

[82GLU/GUR]   Glushko, V. P., Gurvich, L. V., Bergman, G. A., Veits, I. V., Medvedev, V. A., Khachkuruzov, G. A., Yungman, V. S., *Thermodynamic properties of individual substances*, Glushko, V. P., Ed., Vol. IV, Nauka, Moscow, USSR, (1982), 623 pp., in Russian. Cited on pages: 148, 157, 158, 161, 162, 195, 196, 198, 201, 203, 204, 207, 208, 213, 402, 430, 454, 455, 494, 649.

[82HAM]   Hamann, S. D., The influence of pressure on ionization equilibria in aqueous solutions, *J. Solution Chem.*, **11**, (1982), 63-68. Cited on page: 37.

[82HEM]   Hemingway, B. S., Thermodynamic properties of selected uranium compounds and aqueous species at 298.15 K and 1 bar and at higher temperatures. Preliminary models for the origin of coffinite deposits, US Geological Survey, Open File Report 82-619, (1982), 89 pp. Cited on pages: 287, 290, 291, 292, 549.

[82HOG]   Högfeldt, E., Inorganic ligands, Stability constants of metal-ion complexes, vol. 21, *IUPAC Chemical Data Series*, A, Pergamon Press, (1982). Cited on page: 422.

[82LAF]   Laffitte, M., A report of IUPAC commission I.2 on thermodynamics: Notation for states and processes, significance of the word "standard"' in chemical thermodynamics, and remarks on commonly tabulated forms of thermodynamic functions, *J. Chem. Thermodyn.*, **14**, (1982), 805-815. Cited on pages: 14, 15, 31, 32, 35, 707.

[82LAU/HIL] Lau, K. H., Hildenbrand, D. L., Thermochemical properties of the gaseous lower valent fluorides of uranium, *J. Chem. Phys.*, **76**, (1982), 2646-2652. Cited on pages: 199, 200.

[82LUN]   Lundqvist, R., Hydrophilic complexes of the actinides: I. Carbonates of trivalent americium and europium, *Acta Chem. Scand.*, **A36**, (1982), 741-750. Cited on pages: 337, 338, 339, 368, 369, 370, 372.

[82MAY]   Maya, L., Hydrolysis and carbonate complexation of dioxouranium(VI) in the neutral-pH range at 25°C, *Inorg. Chem.*, **21**, (1982), 2895-2898. Cited on page: 506.

[82MOR]   Morss, L. R., Complex oxide systems of the actinides, *Actinides in perspective*, pp. 381-407, Pergamon Press, Oxford, (1982). Cited on pages: 286, 287.

[82NAI/CHA] Nair, G. M., Chander, K., Joshi, J. K., Hydrolysis constants of plutonium(III) and americium(III), *Radiochim. Acta*, **30**, (1982), 37-40. Cited on pages: 337, 338, 339.

[82RAR/MIL] Rard, J. A., Miller, D. G., Isopiestic determination of osmotic and activity coefficients of aqueous CsCl, $SrCl_2$ and mixtures of NaCl and CsCl at 25°C, *J. Chem. Eng. Data*, **27**, (1982), 169-173. Cited on page: 615.

[82RAR/SPE] Rard, J. A., Spedding, F. H., Isopiestic determination of the activity coefficients of some aqueous rare-earth electrolyte solutions at 25°C, *J. Chem. Eng. Data*, **27**, (1982), 454-461. Cited on page: 574.

[82ROY/PRA] Roy, K. N., Prasad, R., Venugopal, V., Singh, Z., Sood, D. D., Studies on (2 UF_4 + H_2 ⇌ 2 UF_3 + 2 HF) and vapour pressure of UF_3, *J. Chem. Thermodyn.*, **14**, (1982), 389-394. Cited on page: 199.

[82SIL] Silva, R. J., The solubilities of crystalline neodymium and americium trihydroxides, Lawrence Berkeley Laboratory, Report LBL-15055, (1982), 57 pp. Cited on pages: 337, 338, 344, 345, 346, 349, 350, 351, 411, 477, 554, 636.

[82SUL/WOO] Sullivan, J. C., Woods, M., Thermodynamics of plutonium(VI) interaction with bicarbonate, *Radiochim. Acta*, **31**, (1982), 45-50. Cited on pages: 325, 558.

[82TAY] Taylor, J. R., *An introduction to error analysis: The study of uncertainties in physical measurements*, University Science Books, Mill Valley, CA, USA, (1982), 270 pp. Cited on page: 738.

[82WAG/EVA] Wagman, D. D., Evans, W. H., Parker, V. B., Schumm, R. H., Halow, I., Bailey, S. M., Churney, K. L., Nuttall, R. L., The NBS tables of chemical thermodynamic properties: Selected values for inorganic and C_1 and C_2 organic substances in SI units, *J. Phys. Chem. Ref. Data*, **11**, Suppl. 2, (1982), 1-392. Cited on pages: 26, 31, 32, 148, 192, 287, 289, 290, 432.

[82WEI/WIS] Weigel, F., Wishnevsky, V., Güldner, R., The vapor phase hydrolysis of $PuBr_3$ and $AmBr_3$: Heats of formation of PuOBr and AmOBr, *J. Less-Common Met.*, **84**, (1982), 147-155. Cited on page: 364.

[83CAC/CHO] Caceci, M. S., Choppin, G. R., The determination of the first hydrolysis constant of Eu(III) and Am(III), *Radiochim. Acta*, **33**, (1983), 101-104. Cited on pages: 337, 338, 339.

[83CAC/CHO2] Caceci, M. S., Choppin, G. R., The first hydrolysis constant of uranium(VI), *Radiochim. Acta*, **33**, (1983), 207-212. Cited on page: 475.

[83CHO2] Choppin, G. R., Solution chemistry of the actinides, *Radiochim. Acta*, **32**, (1983), 43-45. Cited on page: 334.

[83EDE/BUC] Edelstein, N. M., Bucher, J. J., Silva, R. J., Nitsche, H., Thermodynamic properties of chemical species in nuclear waste, Lawrence Berkeley Laboratory, Report ONWI-399 and LBL-14325, (1983), 115 pp. Cited on pages: 338, 339, 344, 345, 347.

[83FER/GRE] Ferri, D., Grenthe, I., Salvatore, F., Studies on metal carbonate equilibria. 7. Reduction of the tris(carbonato)dioxouranate(VI) ion, $UO_2(CO_3)_3^{4-}$, in carbonate solutions, *Inorg. Chem.*, **22**, (1983), 3162-3165. Cited on pages: 373, 721.

[83FUG/PAR] Fuger, J., Parker, V. B., Hubbard, W. N., Oetting, F. L., *The chemical thermodynamics of actinide elements and compounds: Part 8. The actinide halides*, International Atomic Energy Agency, Vienna, (1983), 267 pp. Cited on pages: 214, 222, 223, 224, 230, 358, 427, 466, 544.

[83FUG] Fuger, J., Uranium: chemical thermodynamic properties-selected values, *Gmelin Handbook of Inorganic Chemistry*, Suppl. A6, pp. 165-192, Springer-Verlag, Berlin, (1983). Cited on pages: 286, 287, 290, 633.

[83KAG/KYS] Kaganyuk, D. S., Kyskin, V. I., Kazin, I. V., Calculation of enthalpies of formation for radioelement compounds, *Sov. Radiochem.*, **25**, (1983), 65-69. Cited on page: 287.

[83KOH] Kohli, R., Heat capacity and thermodynamic properties of alkali metal compounds: II. Estimation of the thermodynamic properties of cesium and rubidium zirconates, *Thermochim. Acta*, **65**, (1983), 285-293. Cited on page: 286.

[83MAY] Maya, L., Hydrolysis and carbonate complexation of dioxoneptunium(V) in 1.0 M $NaClO_4$ at 25°C, *Inorg. Chem.*, **22**, (1983), 2093-2095. Cited on pages: 305, 307, 308, 309, 482.

[83MOR/WIL] Morss, L. R., Williams, C. W., Enthalpies of formation of strontium dichloride and of the strontium ion (Sr^{2+}) in water and in 1 mol · dm^{-1} HCl, and an assessment of the enthalpies of formation of alkaline-earth dichlorides, *J. Chem. Thermodyn.*, **15**, (1983), 279-285. Cited on pages: 261, 466.

[83MOR/WIL2] Morss, L. R., Williams, C. W., Choi, I. K., Gens, R., Fuger, J., Thermodynamics of actinide perkovskite-type oxides: II. Enthalpy of formation of Ca_3UO_6, Sr_3UO_6, Ba_3UO_6, Sr_3NpO_6, and Ba_3NpO_6, *J. Chem. Thermodyn.*, **15**, (1983), 1093-1102. Cited on pages: 263, 634.

[83OBR/WIL] O'Brien, T. J., Williams, P. A., The aqueous chemistry of uranium minerals: 4. Schröckingerite, grimselite, and related alkali uranyl carbonates, *Mineral. Mag.*, **47**, (1983), 69-73. Cited on pages: 249, 282, 283, 287, 290, 291, 404.

[83RAI/STR] Rai, D., Strickert, R. G., Moore, D. A., Ryan, J. L., Am(III) hydrolysis constants and solubility of Am(III) hydroxide, *Radiochim. Acta*, **33**, (1983), 201-206. Cited on pages: 344, 345, 347, 348, 349, 636.

[83RYA/RAI] Ryan, J. L., Rai, D., The solubility of uranium(IV) hydrous oxide in sodium hydroxide solutions under reducing conditions, *Polyhedron*, **2**, (1983), 947-952. Cited on pages: 186, 646.

[83SPA] Spahiu, K., Carbonate complex formation in lanthanoid and actinoid systems, Ph. D. Thesis, The Royal Institute of Technology, (1983), Stockholm, Sweden. Cited on pages: 721, 728, 729.

[83VOC/PEL] Vochten, R., Pelsmaekers, J., Synthesis, solubility, electrokinetic properties and refined crystallographic data of sabugalite, *Phys. Chem. Miner.*, **9**, (1983), 23-29. Cited on pages: 282, 283, 288.

[84ALL/OLO] Allard, B., Olofson, U., Torstenfelt, B., Environmental Actinide Chemistry, *Inorg. Chim. Acta*, **94**, (1984), 205-221. Cited on page: 639.

[84ANA/ATK] Ananthaswamy, J., Atkinson, G., Thermodynamics of concentrated electrolyte mixtures. 4. Pitzer-Debye-Hückel limiting slopes for water from 0 to 100°C and from 1 atm to 1 kbar, *J. Chem. Eng. Data*, **29**, (1984), 81-87. Cited on page: 715.

[84BER/KIM] Bernkopf, M. F., Kim, J. I., Hydrolysereaktionen und Karbonatkomplexierung von dreiwertigem Americium im natürlichen aquatischen System, Inst. für Radiochemie der Tech. Univ. München, Report RCM-02884, (1984), 200 pp. Cited on pages: 344, 348, 349, 375, 378, 379, 469, 518, 586, 636.

[84BRE]	Brewer, L., The responsibility of high temperature scientists, *High Temp. Sci.*, **17**, (1984), 1-30. Cited on page: 335.
[84COR/KUB]	Cordfunke, E. H. P., Kubaschewski, O., The thermochemical properties of the system uranium-oxygen-chlorine, *Thermochim. Acta*, **74**, (1984), 235-245. Cited on page: 219.
[84FRE/LAN]	*Handbook on the physics and chemistry of the actinides*, Freeman, A. J., Lander, G. H., Eds., vol. 1, North-Holland, Amsterdam, (1984), 515 pp. Cited on page: 4.
[84FRE]	Freeman, R. D., Conversion of standard (1 atm) thermodynamic data to the new standard-state pressure, 1 bar (10^5 Pa), *J. Chem. Eng. Data*, **29**, (1984), 105-111. Cited on page: 32.
[84FUG/HAI]	Fuger, J., Haire, R. G., Peterson, J. R., The enthalpy of solution of californium metal and the standard enthalpy of formation of Cf^{3+}(aq), *J. Less-Common Met.*, **98**, (1984), 315-321. Cited on page: 358.
[84GEN/WEI]	van Genderen, A. C. G., van der Weijden, C. H., Prediction of Gibbs energies of formation and stability constants of some secondary uranium minerals containing the uranyl group, *Uranium*, **1**, (1984), 249-256. Cited on pages: 288, 289, 290, 291.
[84GOR/SMI]	Gorokhov, L. N., Smirnov, V. K., Khodeev, Yu. S., Thermochemical characterization of uranium UF_n molecules, *Russ. J. Phys. Chem.*, **58**, (1984), 980. Cited on pages: 199, 200, 204, 205, 225.
[84GRE/EAR]	Greenwood, N. N., Earnshaw, A., *Chemistry of the elements*, Pergamon Press, Oxford, (1984). Cited on page: 334.
[84GRO/DRO]	Grønvold, F., Drowart, J., Westrum, Jr., E. F., *The chemical thermodynamics of actinide elements and compounds: Part 4, The actinide chalcogenides (excluding oxides)*, International Atomic Energy Agency, Vienna, (1984), 265 pp. Cited on page: 234.

[84HAR/MOL]   Harvie, C. E., Moller, N., Weare, J. H., The prediction of mineral solubilities in natural waters: The Na-K-Mg-Ca-H-Cl-SO$_4$-OH-HCO$_3$-CO$_3$-CO$_2$-H$_2$O system to high ionic strengths at 25°C, *Geochim. Cosmochim. Acta*, **48**, (1984), 723-751. Cited on pages: 306, 479, 498, 570, 586, 615, 616, 759, 760.

[84HOS]   Hostettler, J. D., Electrode electrons, aqueous electrons, and redox potentials in natural waters, *Am. J. Sci.*, **284**, (1984), 734-759. Cited on page: 23.

[84KUB]   Kubaschewski, O., An empirical estimation of entropies and heat capacities of gaseous double molecules, *High Temp. High Pressures*, **16**, (1984), 197-198. Cited on page: 219.

[84LAU/HIL]   Lau, K. H., Hildenbrand, D. L., Thermochemical studies of the gaseous uranium chlorides, *J. Chem. Phys.*, **80**, (1984), 1312-1317. Cited on pages: 214, 215, 217, 229, 393.

[84MAR/MES]   Marshall, W. L., Mesmer, R. E., Pressure-density relationships and ionization equilibria in aqueous solutions, *J. Solution Chem.*, **13**, (1984), 383-391. Cited on page: 38.

[84NAS/CLE]   Nash, K. L., Cleveland, J. M., The thermodynamics of plutonium(IV) complexation by fluoride and its effect on plutonium(IV) speciation in natural waters, *Radiochim. Acta*, **36**, (1984), 129-134. Cited on pages: 354, 355.

[84NAS/CLE2]   Nash, K. L., Cleveland, J. M., Thermodynamics of the system: Americium(III)-fluoride. Stability constants, enthalpies, entropies and solubility products, *Radiochim. Acta*, **37**, (1984), 19-24. Cited on pages: 359, 360.

[84NRI2]   Nriagu, J. O., Formation and stability of base metal phosphates in soils and sediments, *Phosphate minerals*, Nriagu, J. O., Moore, P. B., Eds., pp. 318-329, Springer-Verlag, Berlin, (1984). Cited on pages: 288, 292.

[84RAI]   Rai, D., Solubility product of Pu(IV) hydrous oxide and equilibrium constants of Pu(IV)/Pu(V), Pu(IV)/Pu(VI) and Pu(V)/Pu(VI) couples, *Radiochim. Acta*, **35**, (1984), 97-106. Cited on pages: 316, 317, 642, 695, 706.

[84RAR] Rard, J. A., Solubility of Eu(NO$_3$)$_3$.6H$_2$O(c) in water at 298.15 K, *J. Chem. Thermodyn.*, **16**, (1984), 921-925. Cited on page: 574.

[84SIL/NIT] Silva, R. J., Nitsche, H., Thermodynamic properties of chemical species of waste radionuclides, NRC Nuclear waste geochemistry'83, Alexander, D. H., Birchard, G. F., Eds., NUREG/CP-0052, pp. 70-93, U.S. Nuclear Regulatory Commission, Washington, D.C., USA, (1984). Cited on pages: 377, 378, 379, 380, 469, 518, 519.

[84SMI/GOR] Smirnov, V. K., Gorokhov, L. N., Thermodynamics of the sublimation and decomposition of uranyl fluoride, *Russ. J. Phys. Chem.*, **58**, (1984), 346-348. Cited on pages: 204, 207, 208.

[84VIE/TAR] Vieillard, P., Tardy, Y., Thermochemical properties of phosphates, *Phosphate minerals*, Nriagu, J. O., Moore, P. B., Eds., pp. 171-198, Springer-Verlag, Berlin, (1984). Cited on pages: 288, 289.

[84VIT] Vitorge, P., Mesure de constantes thermodynamiques de composés trans-uraniens pour prévoir leur géochimie: complexation du Np(V) et de l'Am(III) par les carbonates, hydrolyse du Pu(VI) et de l'Am(III), Seminar held 6-10 Febr. in Sofia (Bulgaria), International Atomic Energy Agency, Report IAEA-SR-104/25, (1984), 14 pp. Cited on pages: 305, 382.

[84VOC/GOE] Vochten, R., Goeminne, A., Synthesis, crystallographic data, solubility and electrokinetic properties of meta-zeunerite, meta-kirchheimerite and nickel-uranylarsenate, *Phys. Chem. Miner.*, **11**, (1984), 95-100. Cited on pages: 282, 283, 290.

[84VOC/GRA] Vochten, R., de Grave, E., Pelsmaekers, J., Mineralogical study of bassetite in relation to its oxidation, *Am. Mineral.*, **69**, (1984), 967-978. Cited on pages: 282, 283, 288.

[84WIL/MOR] Williams, C. W., Morss, L. R., Choi, I. K., Stability of tetravalent actinides in perovskites, Geochemical behavior of disposed radioactive waste, vol. 246, *ACS Symp. Ser.*, pp. 323-334, American Chemical Society, (1984). Cited on pages: 266, 267, 534.

[85ALD/BRO] Aldridge, J. P., Brock, E. G., Filip, H., Flicker, H., Fox, K., Galbraith, H. W., Holland, R. F., Kim, K. C., Krohn, B. J., Magnuson, D. W., Maier II, W. B., McDowell, R. S., Patterson, C. W., Person, W. B., Smith, D. F., Werner, G. K., Measurement and analysis of the infrared-active stretching fundamental (v_3) of UF_6, *J. Chem. Phys.*, **83**, (1985), 34-48. Cited on pages: 206, 463.

[85BAR/PAR] Bard, A. J., Parsons, R., Jordan, J., *Standard Potentials in Aqueous Solution*, Bard, A. J., Parsons, R., Jordan, J., Eds., International Union of Pure and Applied Chemistry. Marcel Dekker, Inc., New York, (1985), 834 pp. Cited on pages: 24, 582.

[85BRA/LAG] Bratsch, S. G., Lagowski, J. J., Lanthanide thermodynamic predictions. 6. Thermodynamics of gas-phase ions and revised enthalpy equations for solids at 298.15 K, *J. Phys. Chem.*, **89**, (1985), 3310-3316. Cited on page: 5, 543.

[85BRU/GRE] Bruno, J., Grenthe, I., Potentiometric techniques applied to the modelling of actinide migration in natural water systems, *Toxicol. Environ. Chem.*, **10**, (1985), 257-264. Cited on pages: 160, 427.

[85DAV/FOU] David, F., Fourest, B., Duplessis, J., Hydration thermodynamics of plutonium and transplutonium ions, *J. Nucl. Mater.*, **130**, (1985), 273-279. Cited on page: 581.

[85DIC/LAW] Dickens, P. G., Lawrence, S. D., Weller, M. T., Lithium insertion into α-UO_3 and U_3O_8, *Mater. Res. Bull.*, **20**, (1986), 635-641. Cited on page: 407.

[85FER/GRE] Ferri, D., Grenthe, I., Hietanen, S., Nàer-Neumann, E., Salvatore, F., Studies on metal carbonate equilibria: 12. Zinc(II) carbonate complexes in acid solution, *Acta Chem. Scand.*, **A39**, (1985), 347-353. Cited on page: 728.

[85FRE/KEL] *Handbook on the physics and chemistry of the actinides*, Freeman, A. J., Keller, C., Eds., vol. 3, North-Holland, Amsterdam, (1985), 520 pp. Cited on page: 4.

[85FRE/LAN]   *Handbook on the physics and chemistry of the actinides*, Freeman, A. J., Lander, G. H., Eds., vol. 2, North-Holland, Amsterdam, (1985), 503 pp. Cited on page: 4.

[85FUG]   Fuger, J., Thermochemistry of the alkali metal and alkaline earth-actinide complex oxides, *J. Nucl. Mater.*, **130**, (1985), 253-265. Cited on page: 287.

[85HIL/GUR]   Hildenbrand, D. L., Gurvich, L. V., Yungman, V. S., *The chemical thermodynamics of actinide elements and compounds: Part 13. The gaseous actinide ions*, International Atomic Energy Agency, Vienna, (1985), 187 pp. Cited on pages: 195, 196, 206, 207, 214, 216, 217, 224, 226, 228, 229, 360.

[85KIM]   Kim, J. I., Basic actinide and fission products chemistry in the CEC-coordinated project: Migration of Radionuclides in the Geosphere, (MIRAGE), Inst. für Radiochemie, Tech. Univ. München, Germany, Report RCM-02085, (1985). Cited on page: 382.

[85KIM2]   Kim, J. I., Basic actinide and fission product chemistry, and first summary report covering work period January to December 1984, MIRAGE project, EUR-9543-EN, pp. 9-40, Inst. für Radiochemie, Tech. Univ. München, Germany, (1985). Cited on page: 382.

[85LAU/BRI]   Lau, K. H., Brittain, R. D., Hildenbrand, D. L., Complex sublimation/decomposition of uranyl fluoride: Thermodynamics of gaseous UO_2F_2 and UOF_4, *J. Phys. Chem.*, **89**, (1985), 4369-4373. Cited on pages: 204, 205, 207, 208.

[85MAG/CAR]   Magirius, S., Carnall, W. T., Kim, J. I., Radiolytic oxidation of Am(III) to Am(V) in NaCl solutions, *Radiochim. Acta*, **38**, (1985), 29-32. Cited on pages: 342, 352, 353.

[85MAR2]   Marcus, Y., *Ion solvation*, John Wiley and Sons limited, Chichester, (1985). Cited on page: 756.

[85MUL]   Muller, A. B., (1985), Private communication, OECD Nuclear Energy Agency, Paris. Cited on page: 4.

[85NIT/EDE] Nitsche, H., Edelstein, N. M., Solubility and speciation of actinide ions in near-neutral solutions, Lawrence Berkeley Laboratory, Report LBL-18900, (1985), 75 pp. Cited on pages: 344, 345, 347, 639.

[85PHI/PHI] Phillips, S. L., Phillips, C. A., Skeen, J., Hydrolysis, formation and ionization constants at 25°C, and at high temperature-high ionic strength, Lawrence Berkeley Laboratory, Report LBL-14996, (1985). Cited on pages: 286, 287, 288, 289, 291.

[85RAI/RYA] Rai, D., Ryan, J. L., Neptunium(IV) hydrous oxide solubility under reducing and carbonate conditions, *Inorg. Chem.*, **24**, (1985), 247-251. Cited on pages: 298, 302, 523, 648.

[85SAW/RIZ] Sawant, R. M., Rizvi, G. H., Chaudhuri, N. K., Patil, S. K., Determination of the stability constant of Np(V) fluoride complex using a fluoride ion selective electrode, *J. Radioanal. Nucl. Chem.*, **89**, (1985), 373-378. Cited on page: 420.

[85SOH/NOV] Söhnel, O., Novotný, P., *Densities of aqueous solutions of inorganic substances*, Elsevier, Amsterdam, (1985), 335 pp. Cited on pages: 29, 30.

[85SPA] Spahiu, K., Studies on metal carbonate equilibria: 11. Yttrium(III) carbonate complex formation in aqueous perchlorate media of various ionic strengths, *Acta Chem. Scand.*, **A39**, (1985), 33-45. Cited on pages: 374, 586.

[85STO2] Storms, E. K, Sublimation thermodynamics of UO_{2-x}, *J. Nucl. Mater.*, **132**, (1985), 231-243. Cited on page: 161.

[85TSO/BRO] Tso, T. C., Brown, D., Judge, A. I., Halloway, J. H., Fuger, J., Thermodynamics of the actinoid elements: Part 6. The preparation and heats of formation of some sodium uranates(VI), *J. Chem. Soc. Dalton Trans.*, (1985), 1853-1858. Cited on pages: 287, 399.

[85UNE] Une, K., Oxygen potential of $Cs_2U_4O_{12}/Cs_2U_4O_{13}$ equilibrium, *J. Nucl. Sci. Technol.*, **22**, (1985), 586-588. Cited on pages: 278, 279, 405.

[86AVO/BIL] Avogadro, A., Billon, A., Cremers, A., Henrion, P., Kim, J. I., Jensen, B. S., Hooker, P. J., The MIRAGE project: Actinide and fission product physico-chemical behaviour in geological environment, Radioactive Waste Management and Disposal, pp. 331-345, Cambridge University Press, Luxembourg, (1986). Cited on page: 382.

[86BRA/LAG] Bratsch, S. G., Lagowski, J. J., Actinide thermodynamic predictions: 3. Thermodynamics of compounds and aquo ions of the 2+, 3+, and 4+ oxidation states and standard electrode potentials at 298.15 K, *J. Phys. Chem.*, **90**, (1986), 307-312. Cited on page: 5, 543.

[86BRU/FER] Bruno, J., Ferri, D., Grenthe, I., Salvatore, F., Studies on metal carbonate equilibria: 13. On the solubility of uranium(IV) dioxide, $UO_2(s)$, *Acta Chem. Scand.*, **40**, (1986), 428-434. Cited on page: 184.

[86BRU] Bruno, J., Stoichiometric and structural studies on the Be^{2+}-H_2O-$CO_2(g)$ system, Ph. D. Thesis, The Royal Institute of Technology, (1986), Stockholm, Sweden. Cited on page: 729.

[86COD] CODATA, The 1986 adjustment of the fundamental physical constants. A report of the CODATA Task Group on Fundamental Constants, prepared by Cohen, E. R. and Taylor, B. N., Pergamon Journals, Oxford, CODATA Bulletin, Report 63, (1986), 36 pp. Cited on pages: 35, 36.

[86DAV] David, F., Thermodynamic properties of the lanthanide and actinide ions in aqueous solution, *J. Less-Common Met.*, **121**, (1986), 27-42. Cited on page: 581.

[86DIC/PEN] Dickens, P. G., Penny, D. J., Weller, M. T., Lithium insertion in uranium oxide phases, *Solid State Ionics*, **18-19**, (1986), 778-782. Cited on pages: 89DIC/LAW, 269, 405, 413.

[86EWA/HOW] Ewart, F. T., Howse, R. M., Thomason, H. P., Williams, S. J., Cross, J. E., The solubility of actinides in the near-field, Scientific Basis for Nuclear Waste Management IX, held 9-11 September 1985, in Stockholm, vol. 50, pp. 701-708, (1986). Cited on pages: 348, 349, 375.

[86FEL/WEA] Felmy, A. R., Weare, J. H., The prediction of borate mineral equilibria in natural waters: application to Searles Lake, California, *Geochim. Cosmochim. Acta*, **50**, (1986), 2771-2783. Cited on page: 498.

[86FRE/KEL] *Handbook on the physics and chemistry of the actinides*, Freeman, A. J., Keller, C., Eds., vol. 4, North-Holland, Amsterdam, (1986), 567 pp. Cited on page: 4.

[86GRE/ROB] Grenthe, I., Robouch, P., Vitorge, P., Chemical equilibria in actinide carbonate systems, *J. Less-Common Met.*, **122**, (1986), 225-231. Cited on pages: 305, 306, 307.

[86HOV/TRE] Hovey, J. K., Tremaine, P. R., Thermodynamics of aqueous aluminium : Standard partial molar heat capacities of aluminium(+3) from 10 to 55°C, *Geochim. Cosmochim. Acta*, **50**, (1986), 453-459. Cited on pages: 159, 541.

[86KAT/SEA] *The chemistry of the actinide elements*, 2nd. Edition, Katz, J. J., Seaborg, G. T., Morss, L. R., Eds., Chapman and Hall, London, (1986), 1674 pp. Cited on page: 4.

[86LIE/KIM] Lierse, Ch., Kim, J. I., Chemisches Verhalten von Plutonium in natürlichen aquatischen Systemen: Hydrolyse, Carbonatkomplexierung und Redoxreaktionen, Inst. Für Radiochemie, Technische Universität München, Report RCM-02286, (1986), 234 pp. Cited on pages: 316, 642, 652, 688.

[86MAL/SRE] Mallika, C., Sreedharan, O. M., Thermodynamic stabilities of TeO_2 and Sb_2Te_3 by a solid-oxide electrolyte e.m.f. technique, *J. Chem. Thermodyn.*, **18**, (1986), 727-734. Cited on page: 393.

[86MOR] Morss, L. R., Thermodynamic properties, *The chemistry of the actinide elements, 2nd ed.*, Katz, J. J., Seaborg, G. T., Morss, L. R., Eds., vol. 2, pp. 1278-1360, Chapman and Hall, London, (1986). Cited on pages: 286, 287, 290.

[86NOR/MUN] Nordstrom, D. K., Munoz, J. L., *Geochemical Thermodynamics*, Blackwell Sci. Publ., Palo Alto, (1986). Cited on page: 23.

[86VOC/GRA] Vochten, R., de Grave, E., Pelsmaekers, J., Synthesis, crystallographic and spectroscopic data, solubility and electrokinetic properties of metakahlerite and its Mn analogue, *Am. Mineral.*, **71**, (1986), 1037-1044. Cited on pages: 282, 283, 284, 290.

[86WAN] Wanner, H., Modelling interaction of deep groundwaters with bentonite and radionuclide speciation, National Co-operative for the Storage of Radioactive Waste (Nagra), Report EIR-Bericht Nr. 589 and Nagra NTB 86-21, (1986), 103 pp. Cited on pages: 288, 289, 291.

[87ALE/OGD] Alexander, C. A., Ogden, J. S., Real time mass spectrometric evaluation of fission product transport at temperature and pressure, Proceedings of the symposium on chemical phenomena associated with radioactivity releases during severe nuclear plant accidents, held at 9-12 September 1986, Anaheim, California, pp. 21-34, U. S. Nucl. Regul. Comm. NUREG CP-0078, (1987). Cited on pages: 163, 406, 591.

[87BON/KOR] Bondarenko, A. A., Korobov, M. V., Sidorov, L. N., Karasev, N. M., Enthalpy of formation of gaseous uranium pentafluoride, *Russ. J. Phys. Chem.*, **61**, (1987), 1367-1370. Cited on page: 205.

[87BRU/CAS] Bruno, J., Casas, I., Lagerman, B., Muñoz, M., The determination of the solubility of amorphous $UO_2(s)$ and the mononuclear hydrolysis constants of uranium(IV) at 25°C, Sci. Basis Nucl. Waste Management X, Symp. held 1-4 December, 1986 in Boston, Massachusetts, vol. 84, pp. 153-160, (1987). Cited on pages: 187, 559.

[87CIA/IUL] Ciavatta, L., Iuliano, M., Porto, R., The hydrolysis of the La(III) ion in aqueous perchlorate solution at 60°C, *Polyhedron*, **6**, (1987), 1283-1290. Cited on page: 40.

[87CRO/EWA] Cross, J. E., Ewart, F. T., Tweed, C. J., Thermochemical modelling with application to nuclear waste processing and disposal, UK Atomic Energy Authority, Report AERE-R12324, (1987), 45 pp. Cited on page: 375.

[87DUB/RAM] Dubessy, J., Ramboz, C., Nguyen-Trung, C., Cathelineau, M., Charoy, B., Cuney, M., Leroy, J., Poty, B., Weisbrod, A., Physical and chemical controls ($f(O_2)$, T, pH) of the opposite behaviour of U and Sn-W as exemplified by hydrothermal deposits in France and Great-Britain and solubility data, *Bull. Minéral.*, **110**, (1987), 261-281. Cited on pages: 185, 560.

[87FRE/LAN] *Handbook on the physics and chemistry of the actinides*, Freeman, A. J., Lander, G. H., Eds., vol. 5, North-Holland, Amsterdam, (1987), 375 pp. Cited on page: 4.

[87GAR/PAR] Garvin, D., Parker, V. B., White, Jr., H. J., CODATA *thermodynamic tables: Selection of some compounds of calcium and related mixtures: A prototype set of tables*, Springer-Verlag, Berlin, (1987), 356 pp. Cited on pages: 28, 214.

[87LAU/HIL] Lau, K. H., Hildenbrand, D. L., Thermochemistry of the gaseous uranium bromides UBr through UBr$_5$, *J. Phys. Chem.*, **86**, (1987), 2949-2954. Cited on pages: 225, 228.

[87LOU/BES] Louis, C., Bessiere, J., Diagrammes potentiel-niveau d'acidité dans les milieux H_2O-H_3PO_4 -II. Systèmes électrochimiques faisant intervenir le proton, *Talanta*, **34**, (1987), 771-777, in French. Cited on page: 241.

[87MAT/OHS] Matsui, T., Ohse, R. W., Thermodynamic properties of uranium nitride, plutonium nitride and uranium-plutonium mixed nitride, *High Temp. High Pressures*, **19**, (1987), 1-17. Cited on pages: 238, 447.

[87RAI/SWA] Rai, D., Swanson, J. L., Ryan, J. L., Solubility of $NpO_2 \cdot xH_2O$(am) in the presence of Cu(I)/Cu(II) redox buffer, *Radiochim. Acta*, **42**, (1987), 35-41. Cited on pages: 297, 298, 316, 640, 642, 648, 649, 663, 693.

[87RIG/VIT] Riglet, C., Vitorge, P., Grenthe, I., Standard potentials of the (MO_2^{2+}/MO_2^+) systems for uranium and other actinides, *Inorg. Chim. Acta*, **133**, (1987), 323-329. Cited on page: 728.

[87ROB/VIT] Robouch, P., Vitorge, P., Solubility of $PuO_2(CO_3)$, *Inorg. Chim. Acta*, **140**, (1987), 239-242. Cited on pages: 325, 330, 558, 679, 680, 681.

[87VEN/IYE]	Venugopal, V., Iyer, V. S., Sundaresh, V., Singh, Z., Prasad, R., Sood, D. D., Standard molar Gibbs free energy of formation of $NaCrO_2$ by e.m.f. measurements, *J. Chem. Thermodyn.*, **19**, (1987), 19-25. Cited on page: 510.
[88ATK/BEC]	Atkins, M., Beckley, A., Glasser, F. P., Influence of cement on the near field environment and its specific interaction with uranium and iodine, *Radiochim. Acta*, **44/45**, (1988), 255-261. Cited on pages: 282, 284.
[88CIA]	Ciavatta, L., (1988), Università de Napoli, Naples, Italy, Private Communication, (from the citation in [92GRE/FUG]). Cited on pages: 724, 731, 735.
[88COR/OUW]	Cordfunke, E. H. P., Ouweltjes, W., Standard enthalpies of formation of uranium compounds: XIV. BaU_2O_7 and $Ba_2U_2O_7$, *J. Chem. Thermodyn.*, **20**, (1988), 235-238. Cited on pages: 265, 532, 634.
[88DIC/POW]	Dickens, P. G., Powell, A. V., Chippindale, A. M., Alkali metal insertion of uranium oxides, *Solid State Ionics*, **28/30**, (1988), 1123-1127. Cited on pages: 188, 269, 272, 406, 413, 420.
[88HED]	Hedlund, T., Studies of complexation and precipitation equilibria in some aqueous aluminium(III) systems, Ph. D. Thesis, University of Umeå, (1988), Sweden. Cited on page: 729.
[88HOV/HEP3]	Hovey, J. K., Hepler, L. G., Tremaine, P. R., Apparent molar heat capacities and volumes of aqueous $HClO_4$, HNO_3, $(CH_3)_4NOH$, and K_2SO_4 at 298.15 K, *Thermochim. Acta*, **126**, (1988), 245-253. Cited on page: 541.
[88KEN/MIK]	Kennedy, C. M., Mikelsons, M. V., Lawson, B. L., Pinkerton, T. C., A formate based precursor for the preparation of technetium complexes, *Appl. Radiat. Isot.*, **39**, (1988), 213-225. Cited on page: 407.
[88LEM]	Lemire, R. J., Effects of high ionic strength groundwaters on calculated equilibrium concentrations in the uranium-water system, Atomic Energy of Canada Ltd., Report AECL-9549, (1988), 40 pp. Cited on page: 291.

[88MIL/CVI] Mills, I., Cvitaš,T., Homann, K., Kallay, N., Kuchitsu, K., *Quantities, units and symbols in physical chemistry, IUPAC*, Blackwell Scientific Publications, Oxford, (1988), 134 pp. Cited on pages: 11, 21, 25.

[88MUL/SHE] Mulford, R. N. R., Sheldon, R. I., Density and heat capacity of liquid uranium at high temperatures, *J. Nucl. Mater.*, **154**, (1988), 268-275. Cited on pages: 157, 430, 448.

[88OHA/LEW] O'Hare, P. A. G., Lewis, B. M., Nguyen, S. N., Thermochemistry of uranium compounds: XVII. Standard molar enthalpy of formation at 298.15 K of dehydrated schoepite $UO_3 \cdot 0.9H_2O$. Thermodynamics of (schoepite + dehydrated schoepite + water), *J. Chem. Thermodyn.*, **20**, (1988), 1287-1296. Cited on pages: 190, 191, 407, 409, 434, 535, 539.

[88PAR/POH] Parks, G. A., Pohl, D. C., Hydrothermal solubility of uraninite, *Geochim. Cosmochim. Acta*, **52**, (1988), 863-875. Cited on pages: 185, 560.

[88PHI/HAL] Phillips, S. L., Hale, F. V., Silvester, L. F., Siegel, M. D., Thermodynamic tables for nuclear waste isolation: Vol.1, Aqueous solutions database, Lawrence Berkeley Lab., Report LBL-22860, NUREG/CR-4864, SAND87-0323, (1988). Cited on pages: 286, 287, 288, 289, 291.

[88SHO/HEL] Shock, E. L., Helgeson, H. C., Calculation of the thermodynamic and transport properties of aqueous species at high pressures and temperatures: Correlation algorithms for ionic species and equation of state predictions to 5 kb and 1000°C, *Geochim. Cosmochim. Acta*, **52**, (1988), 2009-2036. Cited on pages: 38, 158, 562, 715.

[88STA/KIM] Stadler, S., Kim, J. I., Chemisches Verhalten von Americium in natürlichen wässrigen Lösungen: Hydrolyse, Radiolyse und Redox-reaktionen, Inst. für Radiochemie der Technischen Universität, München, Report RCM-01188, (1988), 141 pp. Cited on pages: 337, 338, 339, 340, 341, 342, 344, 345, 346, 348, 349, 352, 353, 409, 410, 411, 448.

[88STA/KIM2]   Stadler, S., Kim, J. I., Hydrolysis reactions of Am(III) and Am(V), *Radiochim. Acta*, **44/45**, (1988), 39-44. Cited on pages: 342, 344, 349, 352, 409.

[88STA/NIT]   Standifer, E. M., Nitsche, H., First evidence for hexagonal AmOHCO$_3$, *Lanthanide Actinide Res.*, **2**, (1988), 383-384. Cited on page: 378.

[88TAN/HEL]   Tanger, IV, J. C., Helgeson, H. C., Calculation of the thermodynamic and transport properties of aqueous species at high pressures and temperatures: Revised equations of state for the standard partial molal properties of ions and electrolytes, *Am. J. Sci.*, **288**, (1988), 19-98. Cited on pages: 38, 715.

[88TAS/OHA]   Tasker, I. R., O'Hare, P. A. G., Lewis, B. M., Johnson, G. K., Cordfunke, E. H. P., Thermochemistry of uranium compounds: XVI. Calorimetric determination of the standard molar enthalpy of formation at 298.15 K, low-temperature heat capacity, and high-temperature enthalpy increments of $UO_2(OH)_2 \cdot H_2O$ (schoepite), *Can. J. Chem.*, **66**, (1988), 620-625. Cited on pages: 189, 408, 412.

[88ULL/SCH]   Ullman, W. J., Schreiner, F., Calorimetric determination of the enthalpies of the carbonate complexes of U(VI), Np(VI), and Pu(VI) in aqueous solution at 25°C, *Radiochim. Acta*, **43**, (1988), 37-44. Cited on page: 325.

[88WAN]   Wanner, H., The NEA Thermochemical Data Base Project, *Radiochim. Acta*, **44/45**, (1988), 325-329. Cited on page: 4.

[89BRU/SAN]   Bruno, J., Sandino, M. C. A., The solubility of amorphous and crystalline schoepite in neutral to alkaline solutions, Sci. Basis Nucl. Waste Management XII, held 10-13 October, in Berlin, vol. 127, pp. 871-878, (1989). Cited on pages: 434, 443, 444, 512.

[89CHA/RAO]   Chatt, A., Rao, R. R., Complexation of europium(III) with carbonate ions in groundwater, *Mater. Res. Soc. Symp. Proc.*, **127**, (1989), 897-904. Cited on page: 373.

[89COR/KON] Cordfunke, E. H. P., Konings, R. J. M., Westrum, Jr., E. F., Recent thermochemical research on reactor materials and fission products, *J. Nucl. Mater.*, **167**, (1989), 205-212. Cited on pages: 222, 671, 690.

[89COX/WAG] Cox, J. D., Wagman, D. D., Medvedev, V. A., *CODATA Key Values for Thermodynamics*, Hemisphere Publ. Corp., New York, (1989), 271 pp. Cited on pages: 5, 31, 37, 39, 45, 133, 134, 135, 148, 158, 214, 230, 238, 239, 318, 321, 394, 398, 430, 485, 492, 574.

[89DIC/LAW] Dickens, P. G., Lawrence, S. D., Penny, D. J., Powell, A. V., Insertion compounds of uranium oxides, *Solid State Ionics*, **32-33**, (1989), 77-83. Cited on pages: 188, 269, 270, 271, 406, 413, 493.

[89FEL/RAI] Felmy, A. R., Rai, D., Schramke, J. A., Ryan, J. L., The solubility of plutonium hydroxide in dilute solution and in high-ionic-strength chloride brines, *Radiochim. Acta*, **48**, (1989), 29-35. Cited on page: 636.

[89GRE/BID] Grenthe, I., Bidoglio, G., Omenetto, N., Use of thermal lensing spectrophotometry (TLS) for the study of mononuclear hydrolysis of uranium(IV), *Inorg. Chem.*, **28**, (1989), 71-74. Cited on page: 187.

[89GUI/GRI] Guillermet, A. F., Grimvall, G., Thermodynamic properties of technetium, *J. Less-Common Met.*, **147**, (1989), 195-211. Cited on page: 671.

[89GUR/DEV] Gurevich, V. M., Devina, O. A., Sergeyeva, E. I., Gavrichev, K. S., Gorbunov, V. E., Efimov, M. E., Khodakovsky, I. L., Experimental study of thermodynamic properties of $UO_2(OH)_2$(c) and UO_2CO_3(c), Academy of Sciences of the USSR, Report 3-10, (1989), 1 pp. Cited on pages: 414, 537, 538.

[89HOV/HEP] Hovey, J. K., Hepler, L. G., Apparent and partial molar heat capacities and volumes of aqueous $HClO_4$ and HNO_3 from 10 to 55°C, *Can. J. Chem.*, **67**, (1989), 1489-1495. Cited on page: 541.

[89HOV/NGU] Hovey, J. K., Nguyen-Trung, C., Tremaine, P. R., Thermodynamics of aqueous uranyl ion: Apparent and partial molar heat capacities and volumes of aqueous uranyl perchlorate from 10 to 55°C, *Geochim. Cosmochim. Acta*, **53**, (1989), 1503-1509. Cited on pages: 159, 562, 563, 564.

[89KIM/KAN] Kim, J. I., Kanellakopulos, B., Solubility products of plutonium(IV) oxide and hydroxide, *Radiochim. Acta*, **48**, (1989), 145-150. Cited on pages: 316, 642, 649, 692.

[89LIE/GRE] Liebman, J. F., Greenberg, A., *From atoms to polymers: isoelectric analogies*, vol. 11, *Molecular structure and energetics*, VCH Publishers, New York, USA, (1989). Cited on pages: 5, 436.

[89MOR/ELL] Morss, L. R., Eller, P. G., Enthalpy of formation of $BaPuO_3$; stability of perovskite as a nuclear-waste matrix for Pu^{4+}, *Radiochim. Acta*, **47**, (1989), 51-54. Cited on pages: 331, 645.

[89NIT/STA] Nitsche, H., Standifer, E. M., Silva, R. J., Americium(III) carbonate complexation in aqueous perchlorate solution, *Radiochim. Acta*, **46**, (1989), 185-189. Cited on pages: 368, 369, 372.

[89PAZ/KOC] Pazukhin, E. M., Kochergin, S. M., Stability constants of hydrolyzed forms of americium(III) and solubility product of its hydroxide, *Sov. Radiochem.*, **31**, (1989), 430-436. Cited on pages: 344, 349, 522, 525.

[89PET/SEL] Petrov, V. G., Seleznev, V. P., Pozdnyakov, S. V., Synthesis methods and physico-chemical properties of uranium fluorophosphates, Mendeleyev Chemico-Technological Institute, Report 4-32, (1989), 2 pp. Cited on pages: 242, 415.

[89RED/SAV] Red'kin, A. F., Savelyeva, N. I., Sergeyeva, E. I., Omelyanenko, B. I., Ivanov, I. P., Khodakovsky, I. L., Investigation of uraninite ($UO_2(c)$) solubility under hydrothermal conditions, *Sci. Geol., Bull.*, **42**, (1989), 329-334. Cited on pages: 185, 560.

[89RIG/ROB] Riglet, C., Robouch, P., Vitorge, P., Standard potentials of the (MO_2^{2+}/MO_2^+) and (M^{4+}/M^{3+}) redox systems for neptunium and plutonium, *Radiochim. Acta*, **46**, (1989), 85-94. Cited on pages: 639, 728, 730.

[89ROB] Robouch, P., Contribution à la prévision du comportement de l'américium, du plutonium et du neptunium dans la géosphère; données chimiques, Commissariat à l'Energie Atomique, Report CEA-R-5473, (1989). Cited on pages: 368, 369, 372, 380, 381, 728, 730.

[89SER/SAV] Sergeyeva, E. I., Savelyeva, N. I., Red'kin, A. F., Zotov, A. V., Omelyanenko, B. I., Ivanov, I. P., Khodakovsky, I. L., Complex formation of U(IV) and U(VI) in aqueous solutions in the range 25-600°C and 1-1000 bars (experimental study), Proceedings of a Workshop, Academy of Sciences of the USSR, Report 3-11, (1989), 1 pp. Cited on pages: 165, 180, 245, 247, 415.

[89SHO/HEL] Shock, E. L., Helgeson, H. C., Corrections to Shock and Helgeson (1988) Geochimica et Cosmochimica Acta, 52, 2009-2036, *Geochim. Cosmochim. Acta*, **53**, (1989), 215. Cited on pages: 38, 715.

[89SHO/HEL2] Shock, E. L., Helgeson, H. C., Sverjensky, D. A., Calculation of the thermodynamic and transport properties of aqueous species at high pressures and temperatures: Standard partial molal properties of inorganic neutral species, *Geochim. Cosmochim. Acta*, **53**, (1989), 2157-2183. Cited on pages: 38, 715.

[89STA/MAK] Standritchuk, O. Z., Maksin, V. I., Zapol'skii, A. K., Thermodynamic aspects of the solubility of sulphamates, *Russ. J. Phys. Chem.*, **63**, (1989), 1282-1286. Cited on pages: 239, 416.

[89TAT/SER] Tatarinova, E. E., Serezhkina, L. B., Serezhkin, V. N., Solubility in the Li_2XO_4-UO_2XO_4-H_2O (X = S, Se) systems at 25°C, *Zh. Neorg. Khim.*, **34**, (1989), 2157-2159, in Russian, English translation in [89TAT/SER2]. Cited on pages: 233, 416.

[89TAT/SER2] Tatarinova, E. E., Serezhkina, L. B., Serezhkin, V. N., Solubility in the Li_2XO_4-UO_2XO_4-H_2O (X = S, Se) systems at 25°C, *Russ. J. Inorg. Chem.*, **34**, (1989), 1227-1228. From a citation in this Bibliography

[89YAM/FUJ] Yamashita, T., Fujino, T., Thermodynamics of quaternary strontium-yttrium-uranium oxides, Japan Atomic Energy Research Institute, Report 3-7, (1989), 2 pp. Cited on pages: 260, 417.

[90ARC/WAN]   Archer, D. G., Wang, P., The dielectric constant of water and Debye-Hückel limiting law slopes, *J. Phys. Chem. Ref. Data*, **19**, (1990), 371-411. Cited on page: 715.

[90BEC/NAG]   Beck, M. T., Nagypál, I., *Chemistry of complex equilibria*, Horwood Limited Publishers, New York, (1990), 402 pp. Cited on page: 25.

[90BRU/GRE]   Bruno, J., Grenthe, I., Lagerman, B., On the UO_2^{2+}/U^{4+} redox potential, *Acta Chem. Scand.*, **44**, (1990), 896-901. Cited on page: 548.

[90CAP/VIT]   Capdevila, H., Vitorge, P., Temperature and ionic strength influence on U(VI/V) and U(IV/III) redox potentials in aqueous acidic and carbonate solutions, *J. Radioanal. Nucl. Chem.*, **143**, (1990), 403-414. Cited on pages: 250, 629.

[90CIA]   Ciavatta, L., The specific interaction theory in equilibrium analysis: Some empirical rules for estimating interaction coefficients of metal ion complexes, *Ann. Chim. (Rome)*, **80**, (1990), 255-263. Cited on pages: 708, 722.

[90COR/KON]   Cordfunke, E. H. P., Konings, R. J. M., Potter, P. E., Prins, G., Rand, M. H., *Thermochemical data for reactor materials and fission products*, Cordfunke, E. H. P., Konings, R. J. M., Eds., North-Holland, Amsterdam, (1990), 695 pp. Cited on pages: 393, 528, 542, 610, 658.

[90COR/KON2]   Cordfunke, E. H. P., Konings, R. J. M., Ouweltjes, W., The standard enthalpies of formation of MO(s), $MCl_2(s)$, and M^{2+}(aq, ∞), (M = Ba, Sr), *J. Chem. Thermodyn.*, **22**, (1990), 991-996. Cited on pages: 4, 426.

[90COS/LAK]   Costantino, M. S., Lakner, J. F., Bastasz, R., Synthesis of monolithic uranium hydride and uranium deuteride, *J. Less-Common Met.*, **159**, (1990), 97-108. Cited on pages: 195, 417.

[90FEL/RAI]   Felmy, A. R., Rai, D., Fulton, R. W., The solubility of $AmOHCO_3(c)$ and the aqueous thermodynamics of the system Na^+-Am^{3+}-HCO_3^--CO_3^{2-}-OH^--H_2O, *Radiochim. Acta*, **50**, (1990), 193-204. Cited on pages: 335, 342, 368, 369, 370, 372, 378, 379, 469, 518, 636, 637.

[90FUG/HAI] Fuger, J., Haire, R. G., Wilmarth, W. R., Peterson, J. R., Molar Enthalpy of formation of californium tribromide, *J. Less-Common Met.*, **158**, (1990), 99-104. Cited on page: 358.

[90GOU/HAI] Goudiakas, J., Haire, R. G., Fuger, J., Thermodynamics of lanthanide and actinide perovskite-type oxides. IV. Molar enthalpies of formation of MM'O$_3$ (M = Ba or Sr, M' = Ce, Tb, or Am) compounds, *J. Chem. Thermodyn.*, **22**, (1990), 577-587. Cited on pages: 543, 617.

[90HAL/JEF2] Hall, R. O. A., Jeffery, A. J., Mortimer, M. J., Spirlet, J. C., Heat capacity of PuTe between 10 and 300 K, AEA Technology report, Report AERE-R-13490, (1990). Cited on page: 324.

[90HAY/THO] Hayes, S. L., Thomas, J. K., Peddicord, K. L., Material property correlations for uranium mononitride IV. Thermodynamic properties, *J. Nucl. Mater.*, **171**, (1990), 300-318. Cited on pages: 236, 237, 417.

[90IYE/VEN] Iyer, V. S., Venugopal, V., Sood, D. D., Thermodynamic studies on the Rb$_2$U$_4$O$_{12}$(s)-Rb$_2$U$_4$O$_{13}$(s) system, *J. Radioanal. Nucl. Chem.*, **143**, (1990), 157-165. Cited on pages: 276, 418.

[90KUM/BAT] Kumok, V. N., Batyreva, V. A., Interpretation of the solubility diagrams of selenate systems at 25°C, *Zh. Neorg. Khim.*, **35**, (1990), 2663-2667, in Russian, English translation in [90KUM/BAT2]. Cited on pages: 233, 418.

[90KUM/BAT2] Kumok, V. N., Batyreva, V. A., Interpretation of the solubility diagrams of selenate systems at 25°C, *Russ. J. Inorg. Chem.*, **35**, (1990), 1514-1517. From a citation in this Bibliography

[90LEI] Leigh, G. J., *Nomenclature of inorganic chemistry, recommendations 1990, issued by the Commission on Nomenclature of Inorganic Chemistry*, IUPAC, Ed., Blackwell Sci. Publ., Oxford, (1990), 289 pp. Cited on page: 13.

[90MON] Monnin, C., The influence of pressure on the activity coefficients of the solutes and on the solubility of minerals in the system Na-Ca-Cl-SO$_4$-H$_2$O to 200°C and 1 kbar, and to high NaCl concentration, *Geochim. Cosmochim. Acta*, **54**, (1990), 3265-3282. Cited on page: 38.

[90OEL/HEL]   Oelkers, E. H., Helgeson, H. C., Triple-ion anions and polynuclear complexing in supercritical electrolyte solutions, *Geochim. Cosmochim. Acta*, **54**, (1990), 727-738. Cited on pages: 715, 716.

[90PAR/SAK]   Park, Y. Y., Sakai, Y., Abe, R., Ishii, T., Harada, M., Kojima, T.,Tomiyasu, H., Desactivation mechanism of excited uranium(VI) complexes in aqueous solutions, *J. Chem. Soc. Faraday Trans.*, **86**, (1990), 55. Cited on page: 594.

[90PER/SAP]   Pershin, A. S., Sapozhnikova, T. V., Hydrolysis of Am(III), *J. Radioanal. Nucl. Chem.*, **143**, (1990), 455-462. Cited on page: 344.

[90PHI]   Phillips, S. L., Calculation of thermodynamic properties for monomeric U(VI) hydrolysis products at 298.15 K and zero ionic strength, Lawrence Berkeley Laboratory, Report LBL-28015, (1990), 20 pp. Cited on pages: 183, 418.

[90POW]   Powell, A. V., The preparation and characterisation of uranium oxide insertion compounds and related phases, Ph. D. Thesis, University of Oxford, (1990), UK. Cited on pages: 188, 193, 269, 272, 406, 413, 419, 420.

[90PRA/MOR]   Pratopo, M. I., Moriyama, H., Higashi, K., Carbonate complexation of neptunium(IV) and analogous complexation of groundwater uranium, *Radiochim. Acta*, **51**, (1990), 27-31. Cited on page: 688.

[90RAI/FEL]   Rai, D., Felmy, A. R., Ryan, J. L., Uranium(IV) hydrolysis constants and solubility product of $UO_2 \cdot xH_2O$(am), *Inorg. Chem.*, **29**, (1990), 260-264. Cited on pages: 185, 186, 435, 505, 559, 646, 647.

[90RIG]   Riglet, C., Chimie du neptunium et autres actinides en milieu carbonate, Commissariat à l'Energie Atomique, Report CEA-R-5535, (1990), 267 pp. Cited on pages: 305, 728, 730.

[90ROS/REI]   Rösch, F., Reimann, T., Buklanov, V., Milanov, M., Khalkin, V. A., Dreyer, R., Electromigration of carrier-free radionuclides. XIV. Complex formation of ^{241}Am-Am(III) with oxalate and sulphate in aqueous solution, *J. Radioanal. Nucl. Chem.*, **140**, (1990), 159-169. Cited on pages: 365, 367.

[90SAW/CHA2]   Sawant, R. M., Chaudhuri, N. K., Patil, S. K., Potentiometric studies on aqueous fluoride complexes of actinides: stability constants of Th(IV)-, U(IV)-, and Pu(IV)-fluorides, *J. Radioanal. Nucl. Chem.*, **143**, (1990), 295-306. Cited on pages: 210, 211, 420.

[90SEV/ALI]   Sevast'yanov, R. G., Alikhanyan, A. S., Krasovskaya, T. I., Kuznetsov, N. T., Synthesis and evaporation of uranium hexachloride, *Vys. Vesh.*, (1990), 103-105, Part 6, in Russian, English translation available. Cited on pages: 218, 219, 421.

[90TAN]   Tananaev, I. G., Hydroxides of pentavalent americium, *Sov. Radiochem.*, **32**, (1990), 305-307. Cited on pages: 352, 524.

[90TAY/DIN]   Taylor, J. R., Dinsdale, A. T., A thermodynamic assessment of the Ni-O, Cr-O and Cr-Ni-O systems using the ionic liquid and compound energy models, *Z. Metallkd.*, **81**, (1990), 354-366. Cited on pages: 472, 510, 511, 526, 658.

[90VOC/HAV]   Vochten, R., van Haverbeke, L., Transformation of schoepite into the uranyl oxide hydrates: becquerelite, billietite and wölsendorfite, *Mineral. Petrol.*, **43**, (1990), 65-72. Cited on pages: 180, 282, 284, 286, 287, 422, 484, 531, 705.

[91AGU/CAS]   Aguilar, M., Casas, I., de Pablo, J., Torrero, M. E., Effect of chloride concentration on the solubility of amorphous uranium dioxide at 25°C under reducing conditions, *Radiochim. Acta*, **52/53**, (1991), 13-15. Cited on pages: 221, 423.

[91AND/CAS]   Anderson, G. M., Castet, S., Schott, J., Mesmer, R. E., The density model for estimation of thermodynamic parameters of reactions at high temperatures and pressures, *Geochim. Cosmochim. Acta*, **55**, (1991), 1769-1779. Cited on page: 38.

[91BID/CAV]   Bidoglio, G., Cavalli, P., Grenthe, I., Omenetto, N., Qi, P., Tanet, G., Studies on metal carbonate equilibria, Part 21: Study of the U(VI)-H_2O-CO_2(g) system by thermal lensing spectrophotometry, *Talanta*, **38**, (1991), 433-437. Cited on pages: 245, 247, 423.

[91BRU/GLA] Brücher, E., Glaser, J., Toth, I., Carbonate exchange for the complex $UO_2(CO_3)_3^{4-}$ in aqueous solution as studied by ^{13}C NMR spectroscopy, *Inorg. Chem.*, **30**, (1991), 2239-2241. Cited on pages: 247, 424.

[91CAR/BRU] Carroll, S. A., Bruno, J., Mineral-solution interactions in the U(VI)-CO_2-H_2O system, *Radiochim. Acta*, **52**, (1991), 187-193. Cited on pages: 247, 424.

[91CAR/LIU] Carnall, W. T., Liu, G. K., Williams, C. W., Analysis of the crystal field spectra of the actinide tetrafluorides. I. UF_4, NpF_4 and PuF_4, *J. Chem. Phys.*, **95**, (1991), 7194-7203. Cited on page: 361.

[91CHO/MAT] Choppin, G. R., Mathur, J. N., Hydrolysis of actinyl(VI) cations, *Radiochim. Acta*, **52/53**, (1991), 25-28. Cited on pages: 166, 175, 176, 424, 549, 589, 704.

[91COR/KON] Cordfunke, E. H. P., Konings, R. J. M., The vapour pressure of UCl_4, *J. Chem. Thermodyn.*, **23**, (1991), 1121-1124. Cited on pages: 216, 425.

[91COR/VLA] Cordfunke, E. H. P., van Vlaanderen, P., Onink, M., Ijdo, D. J. W., $Sr_3U_{11}O_{36}$: crystal structure and thermal stability, *J. Solid State Chem.*, **94**, (1991), 12-18. Cited on pages: 263, 425, 426, 633.

[91DIN] Dinsdale, A. T., SGTE data for pure elements, *CALPHAD: Comput. Coupling Phase Diagrams Thermochem.*, **15**, (1991), 317-425. Cited on page: 31.

[91FAL/HOO] Falck, W. E., Hooker, P. J., Uranium solubility and solubility controls in selected Needle's Eye groundwaters, UK Department of the Environment, Report DoE/HMIP/RR/91/008, (1991), 24 pp. Cited on page: 290.

[91FEL/RAI] Felmy, A. R., Rai, D., Mason, M. J., The solubility of hydrous thorium(IV) oxide in chloride media: Development of an aqueous ion-interaction model, *Radiochim. Acta*, **55**, (1991), 177. Cited on page: 410.

[91FIN/EWI] Finch, R. J., Ewing, R. C., Uraninite alteration in an oxidizing environment and its relevance to the disposal of spent nuclear fuel, Swedish Nucl. Fuel Waste Managem. Co., Report SKB-TR-91-15, (1991), 137 pp. Cited on page: 292.

[91FRE/KEL] *Handbook on the physics and chemistry of the actinides*, Freeman, A. J., Keller, C., Eds., vol. 6, North-Holland, Amsterdam, (1991), 742 pp. Cited on page: 4.

[91FUJ/YAM] Fujino, T., Yamashita, T., Ouchi, K., Phase relation and thermodynamic properties of cubic fluorite-type solid solution, $Ba_{y/2}Y_{y/2}U_{1-y}O_{2+x}$ ($x < 0$ or $x > 0$), *J. Nucl. Mater.*, **183**, (1991), 46-56. Cited on pages: 267, 426.

[91GIR/LAN] Giridhar, J., Langmuir, D., Determination of E° for the UO_2^{2+}/U^{4+} couple from measurement of the equilibrium: UO_2^{2+} + Cu(s) + $4H^+ \rightleftharpoons U^{4+} + Cu^{2+} + 2H_2O$ at 25°C and some geochemical implications, *Radiochim. Acta*, **54**, (1991), 133-138. Cited on pages: 160, 427, 548.

[91GUI/BAL] Guido, M., Balducci, G., Identification and stability of U_2O_2, U_2O_3 and U_2O_4 gaseous oxides molecules, *J. Chem. Phys.*, **95**, (1991), 5373-5376. Cited on page: 162.

[91HAL/HAR] Hall, R. O. A., Harding, S. R., Mortimer, M. J., Spirlet, J. C., Heat capacity of NpSb between 2-300 K, AEA Technology report, Report AEA-FS-0049, (1991), 12 pp. Cited on page: 301.

[91HAL/MOR] Hall, R. O. A., Mortimer, M. J., Harding, S. R., Spirlet, J. C., Low temperature heat capacity of plutonium selenide over the temperature range 7-300 K, AEA Technology report, Report AEA-FS-0048H, (1991). Cited on page: 323.

[91HIL/LAU] Hildenbrand, D. L., Lau, K. H., Brittain, R. D., The entropies and probable symmetries of the gaseous thorium and uranium tetrahalides, *J. Chem. Phys.*, **94**, (1991), 8270-8275. Cited on pages: 199, 200, 216, 227, 427.

[91HIL/LAU2] Hildenbrand, D. L., Lau, K. H., Redetermination of the thermochemistry of gaseous UF_5, UF_2 and UF, *J. Chem. Phys.*, **94**, (1991), 1420-1425. Cited on pages: 205, 226, 428.

[91KIM/KLE] Kim, J. I., Klenze, R., Neck, V., Sekine, T., Kanellakopulos, B., Hydrolyse, Carbonat- und Humat-Komplexierung von Np(V), Institut für Radiochemie, Technische Universität München, Report RCM 01091, (1991), 61 pp. Cited on page: 305.

[91MEI/KIM] Meinrath, G., Kim, J. I., The carbonate complexation of the Am(III) ion, *Radiochim. Acta*, **52/53**, (1991), 29-34. Cited on pages: 368, 369, 370, 372, 380, 381, 428.

[91MEI/KIM2] Meinrath, G., Kim, J. I., Solubility products of different Am(III) and Nd(III) carbonates, *Eur. J. Solid State Inorg. Chem.*, **28**, (1991), 383-388. Cited on pages: 379, 380, 381, 428, 519.

[91MEI] Meinrath, G., Carbonate complexation of the trivalent americium under groundwater conditions, Ph. D. Thesis, Technische Universität München, (1991), Garching, Germany, 199 pp. , in German. Cited on pages: 382, 428, 429, 430, 480, 636.

[91PIT] Pitzer, K. S., Ion interaction approach: theory and data correlation, *Activity coefficients in electrolyte solutions*, Pitzer, K. S., Ed., pp. 75-153, CRC Press, Boca Raton, Florida, (1991). Cited on pages: 306, 382, 430, 479, 480, 712, 753, 757, 759, 760, 762.

[91RAO/CHA] Rao, R. R., Chatt, A., Studies on stability constants of europium(III) carbonate complexes and application of SIT and ion-pairing models, *Radiochim. Acta*, **54**, (1991), 181-188. Cited on page: 373.

[91RAR/MIL] Rard, J. A., Miller, D. G., Corrected values of osmotic and activity coefficients of aqueous $NaTcO_4$ and $HTcO_4$ at 25°C, *J. Solution Chem.*, **20**, (1991), 1139-1147. Cited on page: 547.

[91SAN] Sandino, M. C. A., Processes affecting the mobility of uranium in natural waters, Ph. D. Thesis, The Royal Institute of Technology, (1991), Stockholm, Sweden. Cited on pages: 176, 241, 443, 512, 584, 626.

[91SHE/MUL] Sheldon, R. I., Mulford, R. N. R., Correction to the uranium equation of state, *J. Nucl. Mater.*, **185**, (1991), 297-298. Cited on pages: 157, 430, 448.

[91SHI]	Shilov, V. P., Estimated stability of the U(V) complex with $P_2W_{17}O_{61}^{10-}$, Sov. Radiochem., **33**, (1991), 622-624. Cited on page: 281.

[91VIT/TRA]	Vitorge, P., Tran The, P., Solubility limits of radionuclides in interstitial water - Americium in cement. Task 3 - Characterization of radioactive waste forms. A series of final reports (1985-89) - No. 34, Commission of the European Communities Luxembourg, Report EUR 13664, (1991), 39 pp. Cited on pages: 348, 349.

[91VOC/HAV]	Vochten, R., van Haverbeke, L., Sobry, R., Transformation of schoepite into uranyl oxide hydrates of the bivalent cations Mg^{2+}, Mn^{2+} and Ni^{2+}, J. Mater. Chem., **1**, (1991), 637-642. Cited on page: 286.

[91WAN]	Wanner, H., On the problem of consistency of chemical thermodynamic data bases, Scientific basis for nuclear waste management XIV, vol. 212, pp. 815-822, Materials Research Society, (1991). Cited on page: 4.

[92ADN/MAD]	Adnet, J. M., Madic, C., Bourges, J., Redox and extraction chemistry of actinides U, Np, Pu, Am complexed with the phosphotungstate ligand: $P_2W_{17}O_{61}^{10-}$, 22ème Journées des Actinides, 22-25 April, 1992, Méribel, France, pp. 15-16, (1992). Cited on pages: 386, 431.

[92BLA/WYA]	Blaise, J., Wyart, J. F., *Selected constants: Energy levels and atomic spectra of actinides*, Delplanque, M., Dayet, J., Gasgnier, N., Pépin, I., Eds., *International Tables of Selected Constants*, No. 20, Tables Internationales des Constantes, Paris, (1992). Cited on pages: 157, 335, 431.

[92CAP]	Capdevila, H., Données thermodynamiques sur l'oxydoréduction du plutonium en milieux acide et carbonate. Stabilité de Pu(V), Ph. D. Thesis, Université de Paris-Sud, Orsay, France, 5 June, (1992), in French. Also published as CEA-R-5643 Commissariat à l'Energie Atomique, France, (1993). Cited on pages: 250, 326, 629, 728, 730.

[92CHO/DU]   Choppin, G. R., Du, M., f-Element complexation in brine solutions, *Radiochim. Acta*, **58/59**, (1992), 101-104. Cited on pages: 221, 239, 431.

[92DUC/SAN]   Ducros, M., Sannier, H., Method of estimation of the enthalpy of formation and free enthalpy of formation of inorganic compounds, *Thermochim. Acta*, **196**, (1992), 27-43. Cited on pages: 5, 432.

[92DUE/FLE]   Dueber, R. E., Fleetwood, J. M., Dickens, P. G., The insertion of magnesium into α-U_3O_8, *Solid State Ionics*, **50**, (1992), 329-337. Cited on pages: 257, 434.

[92FIN/MIL]   Finch, R. J., Miller, M. L., Ewing, R. C., Weathering of natural uranyl oxide hydrates: schoepite polytypes and dehydration effects, *Radiochim. Acta*, **58/59**, (1992), 433-443. Cited on pages: 189, 434, 505, 587.

[92FUG/KHO]   Fuger, J., Khodakovsky, I. L., Sergeyeva, E. I., Medvedev, V. A., Navratil, J. D., *The Chemical Thermodynamics of Actinide Elements and Compounds, Part 12. The Actinide Aqueous Inorganic Complexes*, International Atomic Energy Agency, Vienna, (1992), 224 pp. Cited on pages: 4, 175, 183, 435, 436, 485, 646.

[92FUG]   Fuger, J., Thermodynamic properties of actinide aqueous species relevant to geochemical problems, *Radiochim. Acta*, **58/59**, (1992), 81-91. Cited on pages: 175, 337, 338, 435.

[92GRE/FUG]   Grenthe, I., Fuger, J., Konings, R. J. M., Lemire, R. J., Muller, A. B., Nguyen-Trung, C., Wanner, H., *Chemical Thermodynamics of Uranium*, Wanner, H., Forest, I., Nuclear Energy Agency, Organisation for Economic Co-operation, Development, Eds., vol. 1, *Chemical Thermodynamics*, North Holland Elsevier Science Publishers B. V., Amsterdam, The Netherlands, (1992), 715 pp. Cited on pages: xi, 4, 5, 6, 26, 29, 39, 43, 44, 80, 98, 114, 133, 134, 135, 148, 149, 157, 158, 159, 160, 162, 164, 165, 168, 169, 170, 172, 173, 174, 175, 176, 177, 178, 179, 180, 181, 182, 183, 184, 185, 186, 187, 188, 189, 190, 191, 192, 193, 195, 196, 199, 200, 203, 205, 206, 207, 208, 209, 210, 211, 212, 213, 214,

[92GRE/FUG]   215, 218, 219, 220, 221, 222, 223, 224, 226, 229, 230, 231, 234, 236, 237, 238, 239, 240, 241, 242, 243, 245, 246, 247, 250, 251, 252, 254, 255, 257, 259, 260, 261, 262, 263, 266, 268, 271, 272, 274, 276, 277, 278, 279, 282, 284, 285, 286, 287, 288, 289, 290, 291, 292, 294, 297, 299, 315, 318, 319, 321, 323, 327, 363, 364, 366, 394, 397, 398, 399, 401, 402, 404, 407, 408, 409, 412, 416, 417, 418, 420, 421, 422, 423, 424, 425, 426, 427, 428, 431, 432, 433, 434, 435, 436, 438, 439, 441, 442, 443, 444, 446, 447, 449, 451, 452, 453, 454, 455, 456, 457, 458, 460, 461, 463, 467, 468, 469, 473, 476, 478, 485, 486, 487, 488, 489, 491, 492, 494, 495, 497, 505, 506, 509, 510, 512, 516, 517, 530, 531, 535, 536, 537, 538, 539, 540, 541, 542, 543, 548, 549, 551, 552, 553, 555, 556, 557, 558, 559, 560, 562, 563, 564, 565, 566, 567, 574, 578, 579, 582, 583, 584, 587, 589, 593, 594, 598, 599, 600, 603, 605, 608, 609, 611, 612, 619, 620, 621, 623, 624, 625, 626, 628, 630, 633, 635, 638, 639, 640, 642, 646, 647, 648, 649, 653, 654, 655, 656, 657, 658, 660, 663, 664, 665, 667, 669, 674, 675, 676, 687, 690, 698, 699, 704, 705, 708, 721, 723, 728, 729, 730, 733, 736, 748, 749, 754, 756, 763.

[92HAS2]   Hassan, Refat M., The oxidation of uranium(IV) by polyvalent metal ions. A linear free-energy correlation, *J. Coord. Chem.*, **27**, (1992), 255-266. Cited on pages: 160, 436.

[92HIS/BEN]   Hisham, M. W. M., Benson, S. W., Thermochemistry of inorganic solids. 10. Empirical relations between the enthalpies of formation of solid halides and the corresponding gas phase halide anions, *J. Chem. Eng. Data*, **37**, (1992), 194-199. Cited on pages: 5, 436.

[92IUP]   IUPAC Commission on Atomic Weights & Isotopic Abundances, Atomic weights of the elements 1991, *Pure Appl. Chem.*, **64**, (1992), 1519-1534. Cited on page: 195.

[92KHO]   Khodakovsky, I. L., About possible systematic error in thermodynamic properties of uranyl and U^{4+} aqueous ions, Abstracts of papers, IUPAC conference on Chem. Thermodyn., Snowbird, UT, US, 16-21 August 1992, p. 308, (1992). Cited on pages: 486, 540.

[92KIM/SER]   Kimura, T., Serrano, G. J., Nakayama, S., Takahashi, K., Takeishi, H., Speciation of uranium in aqueous solutions and in precipitates by photoacoustic spectroscopy, *Radiochim. Acta*, **58/59**, (1992), 173-178. Cited on pages: 247, 436.

[92KRA/BIS]   Kramer-Schnabel, U., Bischoff, H., Xi, R. H., Marx, G., Solubility products and complex formation equilibria in the systems uranyl hydroxide and uranyl carbonate at 25°C and $I = 0.1$ M, *Radiochim. Acta*, **56**, (1992), 183-188. Cited on pages: 165, 166, 168, 169, 172, 321, 438, 516, 584, 606, 607, 643.

[92LIE/HIL]   Lieser, K. H., Hill, R., Hydrolysis and colloid formation of thorium in water and consequences for its migration behaviour - comparison with uranium, *Radiochim. Acta*, **56**, (1992), 37-45. Cited on pages: 247, 438.

[92MOR/WIL]   Morss, L. R., Williams, C. W., Enthalpies of formation of rare earth and actinide(III) hydroxides; their acid-base relationships and estimation of their thermodynamic properties, Sci. Basis Nucl. Waste Management XV, symp. held 4-7 November 1991 in Strasbourg, vol. 257, pp. 283-288, (1992). Cited on pages: 439, 477.

[92NEC/KIM]   Neck, V., Kim, J. I., Kanellakopulos, B., Solubility and hydrolysis behaviour of neptunium(V), *Radiochim. Acta*, **56**, (1992), 25-30. Cited on pages: 296, 480, 640, 641.

[92NGU/BEG]   Nguyen-Trung, C., Begun, G. M., Palmer, D. A., Aqueous uranium complexes. 2. Raman spectroscopic study of the complex formation of the dioxouranium(VI) ion with a variety of inorganic and organic ligands, *Inorg. Chem.*, **31**, (1992), 5280-5287. Cited on pages: 439, 674.

[92NGU/SIL]   Nguyen, S. N., Silva, R. J., Weed, H. C., Andrews, Jr., J. E., Standard Gibbs free energies of formation at the temperature 303.15 K of four uranyl silicates: soddyite, uranophane, sodium boltwoodite and sodium weeksite, *J. Chem. Thermodyn.*, **24**, (1992), 359-376. Cited on pages: 255, 256, 291, 292, 440, 442, 498, 535, 549, 555, 558, 677, 678.

[92RUN/MEI] Runde, W., Meinrath, G., Kim, J. I., A study of solid-liquid phase equilibria of trivalent lanthanide and actinide ions in carbonate systems, *Radiochim. Acta*, **58/59**, (1992), 93-100. Cited on pages: 377, 378, 379, 380, 381, 428, 519.

[92SAN/BRU] Sandino, M. C. A., Bruno, J., The solubility of $(UO_2)_3(PO_4)_2 \cdot 4H_2O(s)$ and the formation of U(VI) phosphate complexes: Their influence in uranium speciation in natural waters, *Geochim. Cosmochim. Acta*, **56**, (1992), 4135-4145. Cited on pages: 165, 166, 167, 168, 174, 176, 177, 178, 240, 288, 321, 443, 444, 512, 626.

[92SAT/CHO] Satoh, I., Choppin, G. R., Interaction of uranyl(VI) with silicic acid, *Radiochim. Acta*, **56**, (1992), 85-87. Cited on pages: 252, 253, 254, 444, 593.

[92SHO/OEL] Shock, E. L., Oelkers, E. H., Johnson, J. W., Sverjensky, D. A., Helgeson, H. C., Calculation of the thermodynamic properties of aqueous species at high pressures and temperatures. Effective electrostatic radii, dissociation constants and standard partial molal properties to 1000°C and 5 kbar, *J. Chem. Soc. Faraday Trans.*, **88**, (1992), 803-826. Cited on page: 715.

[92SIL] Silva, R. J., Mechanisms for the retardation of uranium(VI) migration, Sci. Basis Nucl. Waste Management XV, vol. 257, pp. 323-330, (1992). Cited on pages: 488, 704.

[92VEN/IYE] Venugopal, V., Iyer, V. S., Jayanthi, K., Thermodynamic studies on $A_2U_4O_{12}$ and $A_2U_4O_{13}$ (A = Cs/Rb) system by emf and calorimetric measurements at high temperatures, *J. Nucl. Mater.*, **199**, (1992), 29-42. Cited on pages: 267, 269, 276, 278, 279, 405, 418, 444, 510, 546.

[92VEN/KUL] Venugopal, V., Kulkarni, S. G., Subbanna, C. S., Sood, D. D., Vapour pressures of uranium and uranium nitride over UN(s), *J. Nucl. Mater.*, **186**, (1992), 259-268. Cited on pages: 237, 446.

[92WIM/KLE] Wimmer, H., Klenze, R., Kim, J. I., A study of hydrolysis reaction of curium(III) by time resolved laser fluorescence spectroscopy, *Radiochim. Acta*, **56**, (1992), 79-83. Cited on pages: 338, 339, 340, 411, 447.

[93BOI/ARL]   Boivineau, M., Arlès, L., Vermeulen, J. M., Thévenin, T., High-pressure thermophysical properties of solid and liquid uranium, *Physica B (Amsterdam)*, **190**, (1993), 31-39. Cited on pages: 157, 448.

[93CAR]   Carroll, S. A., Precipitation of Nd-Ca carbonate solid solution at 25°C, *Geochim. Cosmochim. Acta*, **57**, (1993), 3383-3393. Cited on pages: 379, 519.

[93ERI/NDA]   Eriksen, T. E., Ndalamba, P., Cui, D., Bruno, J., Caceci, M. S., Spahiu, K., Solubility of the redox sensitive radionuclides ^{99}Tc and ^{237}Np under reducing conditions in neutral to alkaline solutions. Effect of carbonate, SKB, SKB Technical Report, Report TR-93-18, (1993), 32 pp. Cited on pages: 251, 298, 302, 303, 304, 448, 449, 523, 648, 689.

[93FER/SAL]   Ferri, D., Salvatore, F., Vasca, E., Glaser, J., Grenthe, I., Complex formation in the U(VI)-OH$^-$-F$^-$ system, *Acta Chem. Scand.*, **47**, (1993), 855-861. Cited on pages: 165, 166, 169, 209, 210, 451, 551, 587.

[93FUG/HAI]   Fuger, J., Haire, R. G., Peterson, J. R., Molar enthalpies of formation of BaCmO$_3$ and BaCfO$_3$, *J. Alloys Compd.*, **200**, (1993), 181-185. Cited on pages: 260, 543, 617.

[93GIF/VIT]   Giffaut, E., Vitorge, P., Evidence of radiolytic oxidation of ^{241}Am in Na$^+$ / Cl$^-$ /HCO$_3^-$ / CO$_3^{2-}$ media, Sci. Basis Nucl. Waste Management XVI, vol. 294, pp. 747-751, (1993). Cited on pages: 375, 383, 469.

[93GIF/VIT2]   Giffaut, E., Vitorge, P., Capdevila, H., Corrections de température sur les coefficients d'activité calculés selon la TIS, Commissariat à l'Energie Atomique, Report CEA-N-2737, (1993), 29 pp. Cited on page: 716.

[93GRE/LAG]   Grenthe, I., Lagerman, B., Ternary metal complexes: 2. The U(VI)-SO$_4^{2-}$-OH$^-$ system, *Radiochim. Acta*, **61**, (1993), 169-176. Cited on pages: 178, 231, 232, 233, 452, 453, 664, 665, 672, 673.

[93HIN]   Hinatsu, Y., The magnetic susceptibility and structure of BaUO$_3$, *J. Solid State Chem.* **102**, (1993), 566-569. Cited on page: 267.

[93JAY/IYE] Jayanthi, K., Iyer, V. S., Singh Mudher, K. D., Venugopal, V., High-temperature calorimetric studies on the thermal properties of $Rb_2U(SO_4)_3(s)$, *Thermochim. Acta*, **230**, (1993), 95-102. Cited on pages: 277, 453, 657.

[93KRI/EBB] Krikorian, O. H., Ebbinghaus, B. B., Adamson, M. G., Fontes, Jr., A. S., Fleming, D. L., Experimental studies and thermodynamic modeling of volatilities of uranium, plutonium, and americium from their oxides interacted with ash, University of California, Lawrence Livermore National Laboratory, Report UCRL-ID-114774, (1993). Cited on pages: 162, 163, 403, 454, 455, 649.

[93KUB/ALC] Kubaschewski, O., Alcock, C. B., Spencer, P. J., *Materials thermochemistry*, 6th. Edition, Pergamon Press Ltd., Oxford, (1993), 363 pp. Cited on page: 543.

[93LEE/BYR] Lee, J. H., Byrne, R. H., Complexation of trivalent rare earth elements (Ce, Eu, Gd, Tb, Yb) by carbonate ions, *Geochim. Cosmochim. Acta*, **57**, (1993), 295-302. Cited on page: 374.

[93LEM/BOY] Lemire, R. J., Boyer, G. D., Campbell, A. B., The solubilities of sodium and potassium dioxoneptunium(V) carbonate hydrates at 30, 50 and 75°C, *Radiochim. Acta*, **61**, (1993), 57-63. Cited on page: 305.

[93MCB/GOR] McBride, B.J., Gordon, S., Reno, M. A., Thermodynamic data for fifty reference elements, NASA, Report 3287, (1993), 7 pp. Cited on pages: 157, 455.

[93MEI/KAT] Meinrath, G., Kato, Y., Yoshida, Z., Spectroscopic study of the uranyl hydrolysis species $(UO_2)(OH)_2^{2+}$, *J. Radioanal. Nucl. Chem.*, **174**, (1993), 299-314. Cited on pages: 165, 166, 170, 171, 456, 458, 460, 475, 507, 514, 515, 517, 580, 594, 597, 609.

[93MEI/KIM] Meinrath, G., Kimura, T., Carbonate complexation of the uranyl(VI) ion, *J. Alloys Compd.*, **202**, (1993), 89-93. Cited on pages: 171, 172, 245, 321, 456, 457, 458, 475, 514, 515, 516, 584, 606, 607, 643.

[93MEI/KIM2]   Meinrath, G., Kimura, T., Behaviour of U(VI) solids under conditions of natural aquatic systems, *Inorg. Chim. Acta*, **204**, (1993), 79-85. Cited on pages: 168, 170, 171, 245, 458, 461, 512, 514, 515, 643.

[93MEI/TAK]   Meinrath, G., Takeishi, H., Solid-liquid equilibria of Nd^{3+} in carbonate solutions, *J. Alloys Compd.*, **194**, (1993), 93-99. Cited on pages: 379, 519.

[93MIZ/PAR]   Mizuguchi, K., Park, Y. Y., Tomiyasu, H., Electrochemical and spectroelectrochemical studies on uranyl carbonato and aqua complexes, *J. Nucl. Sci. Technol.*, **30**, (1993), 542-548. Cited on pages: 160, 250, 462, 635.

[93NAI/VEN]   Naik, J. S., Venugopal, V., Kulkarni, S. G., Subbanna, C. S., Sood, D. D., Vaporization studies over U(l) + UC, UC and $(U_{0.79}Ce_{0.21})C$ by Knudsen effusion mass spectrometry, *Met. Mater. Processes*, **5**, (1993), 23-32. Cited on pages: 244, 462.

[93NIK/TSI]   Nikitin, M. I., Tsirel'nikov, V. I., Determination of enthalpy of formation of gaseous uranium pentafluoride, *High Temp.*, **30**, (1993), 730-735. Cited on pages: 205, 463.

[93ODA]   Oda, T., Statistical thermodynamic properties of uranium hexafluoride, Dept. of Fuel Safety Research, Tokai Research Est., Japan Atomic Energy Research Institute, Report JAERI-M 93-052, (1993). Cited on pages: 206, 463.

[93OGA]   Ogawa, T., Thermodynamic properties of $(U, Pu)N_{1-x}$ with a sublattice formalism - Equilibria involving the nonstoichiometric nitrides, *J. Nucl. Mater.*, **201**, (1993), 284-292. Cited on pages: 237, 463.

[93OGO/ROG]   Ogorodnikov, V. V., Rogovoi, Yu J., Regularities in the changes of properties of cubic transition metal mononitrides, *Inorg. Mater.*, **29**, (1993), 591-595. Cited on pages: 237, 463.

[93PAS/RUN]   Pashalidis, I., Runde, W., Kim, J. I., A study of solid-liquid phase equilibria of Pu(VI) and U(VI) in aqueous carbonate systems, *Radiochim. Acta*, **61**, (1993), 141-146. Cited on pages: 245, 324, 325, 330, 464, 556, 557, 643, 679, 680, 681.

[93PAT/DUE]   Patat, S., Dueber, R. E., Dickens, P. G., Thermochemical and electrochemical study of magnesium insertion into α-UO_{3-x}, Solid State Ionics, **59**, (1993), 151-155. Cited on pages: 194, 257, 258, 271, 273, 280, 464, 492, 493, 494.

[93SAL/KUL]   Sali, S. K., Kulkarni, N. K., Sampath, S., Jayadevan, N. C., Thermochemical studies on Tl-U-O system, Proc. IX National Symposium of Thermal Analysis, Goa, India, pp. 325-328, Bombay, India, (1993). Cited on page: 279.

[93STO/CHO]   Stout, B. E., Choppin, G. R., Nectoux, F., Pagès, M., Cation-cation complexes of NpO_2^+, Radiochim. Acta, **61**, (1993), 65-67. Cited on pages: 257, 465, 681.

[93TAK/FUJ]   Takahashi, K., Fujino, T., Morss, L. R., Crystal chemical and thermodynamic study on $CaUO_{4-x}$, $(Ca_{0.5}Sr_{0.5})UO_{4-x}$ and α–$SrUO_{4-x}$ ($x = 0$ 0.5), J. Solid State Chem., **105**, (1993), 234-246. Cited on pages: 259, 261, 465.

[94AHO/ERV]   Ahonen, L., Ervanne, H., Kaakola, T., Blomqvist, R., Redox chemistry in uranium-rich groundwater of Palmottu uranium deposit, Finland, Radiochim. Acta, **66/67**, (1994), 115-121. Cited on pages: 160, 467.

[94CAP/VIT]   Capdevila, H., Vitorge, P., Potentiels redox des couples PuO_2^{2+}/PuO_2^+ et Pu^{4+}/Pu^{3+} à force ionique et température variables. Entropie et capacité calorifique, Commissariat à l'Energie Atomique, Report CEA-N-2762, (1994), 73 pp. Cited on pages: 728, 730.

[94CAS/BRU]   Casas, I., Bruno, J., Cera, E., Finch, R. J., Ewing, R. C., Kinetic and thermodynamic studies of uranium minerals. Assessment of the long-term evolution of spent nuclear fuel, Swedish Nucl. Fuel Waste Manag. Co., Report SKB-TR-94-16, (1994), 73 pp. Cited on pages: 286, 291.

[94CHO/RIZ]   Choppin, G. R., Rizkalla, E. N., Solution chemistry of actinides and lanthanides, Ch.128, Handbook on the physics and chemistry of rare earths, Gschneidner Jr., K. A., Eyring, L., Choppin, G. R., Lander, G. H., Eds., vol. 18, pp. 559-590, Elsevier Science B.V. Amsterdam, (1994). Cited on page: 334.

[94COR/IJD] Cordfunke, E. H. P., Ijdo, D. J. W., $Sr_2UO_{4.5}$: a new perovskite-type strontium uranate, *J. Solid State Chem.*, **109**, (1994), 272-276. Cited on pages: 260, 265, 467, 634.

[94CZE/BUC] Czerwinski, K. R., Buckau, G., Scherbaum, F., Kim, J. I., Complexation of the uranyl ion with aquatic humic acid, *Radiochim. Acta*, **65**, (1994), 111-119. Cited on page: 509.

[94FAN/KIM] Fanghänel, T., Kim, J. I., Paviet, P., Klenze, R., Hauser, W., Thermodynamics of radioactive trace elements in concentrated electrolyte solutions: hydrolysis of Cm^{3+} in NaCl-solutions, *Radiochim. Acta*, **66/67**, (1994), 81-87. Cited on pages: 337, 338, 339, 340, 344, 411, 468, 479, 480, 585, 614, 637.

[94GIF] Giffaut, E., Influence des ions chlorure sur la chimie des actinides, Ph. D. Thesis, Université de Paris-Sud, Orsay, France, (1994), 259 pp. , in French. Cited on pages: 323, 356, 368, 369, 370, 372, 375, 376, 377, 378, 379, 383, 468, 470, 471, 482, 483, 518, 636, 653, 654, 655, 656, 728.

[94HEN/MER] Henry, M., Merceron, T., An independent method for data selection of long-life radionuclides (actinides and fission products) in the geosphere, *Radiochim. Acta*, **66/67**, (1994), 57-61. Cited on pages: 390, 471.

[94JAY/IYE] Jayanthi, K., Iyer, V. S., Kulkarni, S. G., Rama Rao, G. A., Venugopal, V., Molar Gibbs energy of formation of $NaUO_3(s)$, *J. Nucl. Mater.*, **211**, (1994), 168-174. Cited on pages: 271, 274, 275, 472, 512, 638, 639.

[94KAL/MCC] Kaledin, L. A., McCord, J. E., Heaven, M. C., Laser spectroscopy of UO: characterization and assignment of states in the 0 to 3 eV range, with a comparison to the electronic structure of ThO, *J. Mol. Spectrosc.*, **164**, (1994), 27-65. Cited on pages: 161, 474.

[94KAT/MEI] Kato, Y., Meinrath, G., Kimura, T., Yoshida, Z., A study of U(VI) hydrolysis and carbonate complexation by time-resolved laser-induced fluorescence spectroscopy (TRLFS), *Radiochim. Acta*, **64**, (1994), 107-111. Cited on pages: 474, 507, 509, 515, 516, 597, 609, 613.

[94KIM/CHO]   Kim, W. H., Choi, K. C., Park, K. K., Eom, T. Y., Effects of hypochlorite on the solubility of amorphous schoepite at 25°C in neutral to alkaline solutions, *Radiochim. Acta*, **66/67**, (1994), 45-49. Cited on pages: 224, 475.

[94KIM/KLE]   Kim, J. I., Klenze, R., Wimmer, H., Runde, W., Hauser, W., A study of the carbonate complexation of Cm(III) and Eu(III) by time resolved laser fluorescence spectroscopy, *J. Alloys Compd.*, **213/214**, (1994), 333-340. Cited on pages: 369, 372, 475.

[94LIU/CAR]   Liu, G. K., Carnall, W. T., Jursich, G., Williams, C. W., Analysis of the crystal field spectra of the actinide tetrafluorides. II. AmF_4, CmF_4, $Cm^{4+}:CeF_4$, and $Bk^{4+}:CeF_4$, *J. Chem. Phys.*, **101**, (1994), 8277-8289. Cited on page: 361.

[94MEI]   Meinrath, G., Np(V) carbonates in solid state and aqueous solution, *J. Radioanal. Nucl. Chem.*, **186**, (1994), 257-272. Cited on pages: 305, 307, 308.

[94MER/FUG]   Merli, L., Fuger, J., Thermochemistry of a few neptunium and neodymium oxides and hydroxides, *Radiochim. Acta*, **66/67**, (1994), 109-113. Cited on page: 641.

[94MOR/WIL]   Morss, L. R., Williams, C. W., Synthesis of crystalline americium hydroxide, $Am(OH)_3$, and determination of its enthalpy of formation; estimation of the solubility-product constants of actinide(III) hydroxides, *Radiochim. Acta*, **66/67**, (1994), 89-93. Cited on pages: 349, 350, 476, 477, 554.

[94NEC/KIM]   Neck, V., Kim, J. I., Kanellakopulos, B., Thermodynamisches Verhalten von Neptunium(V) in konzentrierten NaCl- und NaClO$_4$-Lösungen, Kernforschungszentrum Karlsruhe, Report KfK 5301, (1994), 37 pp. Cited on pages: 305, 307, 308, 758.

[94NEC/RUN]   Neck, V., Runde, W., Kim, J. I., Kanellakopulos, B., Solid-liquid equilibrium reactions of neptunium(V) in carbonate solution at different ionic strength, *Radiochim. Acta*, **65**, (1994), 29-37. Cited on pages: 305, 306, 307, 308, 309, 640, 641.

[94OCH/SUZ] Ochiai, A., Suzuki, Y., Shikama, T., Hotta, E., Haga, Y., Suzuki, T., Transport properties and specific heat of UTe and USb, *Physica B (Amsterdam)*, **199-200**, (1994), 616-618. Cited on pages: 234, 243, 477.

[94OMO/MUR] Omori, T., Muraoka, Y., Suganuma, H., Solvent extraction mechanism of pertechnetate with tetraphenylarsonium chloride, *J. Radioanal. Nucl. Chem.*, **178**, (1994), 237-243. Cited on pages: 390, 477.

[94OST/BRU] Östhols, E., Bruno, J., Grenthe, I., On the influence of carbonate on mineral dissolution: III. The solubility of microcrystalline ThO_2 in CO_2-H_2O media, *Geochim. Cosmochim. Acta*, **58**, (1994), 613-623. Cited on pages: 251, 327, 503, 637, 640.

[94RIZ/RAO] Rizkalla, E. N., Rao, L. F., Choppin, G. R., Sullivan, J. C., Thermodynamics of uranium(VI) and plutonium(VI) hydrolysis, *Radiochim. Acta*, **65**, (1994), 23-27. Cited on pages: 174, 478, 580.

[94RUN/KIM] Runde, W., Kim, J. I., Chemical behaviour of trivalent and pentavalent americium in saline NaCl-solutions. Studies of transferability of laboratory data to natural conditions, Technische Universität München, Report RCM 01094, (1994), 236 pp. Cited on pages: 305, 306, 307, 308, 309, 310, 337, 338, 339, 340, 342, 343, 344, 345, 348, 349, 352, 353, 368, 369, 370, 372, 375, 376, 377, 382, 383, 428, 429, 430, 470, 471, 479, 480, 481, 482, 483, 636, 758, 759.

[94SAL/KUL] Sali, S. K., Kulkarni, N. K., Sampath, S., Jayadevan, N. C., Solid state reactions of uranium oxide and alkali metal chromates: characterisation of new uranates, *J. Nucl. Mater.*, **217**, (1994), 294-299. Cited on pages: 276, 484, 684.

[94SAM] Samant, M. S., Ph. D. Thesis, University of Bombay, (1994). Cited on page: 596.

[94SAN/GRA] Sandino, M. C. A., Grambow, B., Solubility equilibria in the U(VI)-Ca-K-Cl-H_2O system: transformation of schoepite into becquerelite and compreignacite, *Radiochim. Acta*, **66/67**, (1994), 37-43. Cited on pages: 180, 181, 182, 286, 484, 531, 704, 705.

[94SER/DEV] Sergeyeva, E. I., Devina, O. A., Khodakovsky, I. L., Thermodynamic database for actinide aqueous inorganic complexes, *J. Alloys Compd.*, **213/214**, (1994), 125-131. Cited on pages: 159, 174, 183, 485, 486, 538, 540, 646.

[94SER/SER] Serezhkina, L. B., Serezhkin, V. N., Solubility in the UO_2SeO_4-$(CH_3)_2NCONH_2$-H_2 system, *Radiochemistry (Moscow)*, **36**, (1994). Cited on pages: 233, 487.

[94SIE/PHI] Siekierski, S., Phillips, S. L., *Solubility data series: Actinide Nitrates*, vol. 55, IUPAC Publications, Oxford University Press, Oxford, UK, (1994). Cited on page: 491.

[94TOR/CAS] Torrero, M. E., Casas, I., de Pablo, J., Sandino, M. C. A., Grambow, B., A comparison between unirradiated $UO_2(s)$ and schoepite solubilities in 1 M NaCl medium, *Radiochim. Acta*, **66/67**, (1994), 29-35. Cited on pages: 168, 175, 179, 321, 487, 584.

[94TOU/PIA] Touzelin, B., Pialoux, A., U-Ba-O system study by high temperature X-ray diffraction under controlled atmosphere, *J. Nucl. Mater.*, **217**, (1994), 233. Cited on pages: 265, 488, 616, 650.

[94YAM/SAK] Yamaguchi, T., Sakamoto, Y., Ohnuki, T., Effect of the complexation on solubility of Pu(IV) in aqueous carbonate system, *Radiochim. Acta*, **66/67**, (1994), 9-14. Cited on pages: 327, 328, 329, 523, 652, 688.

[95ALL/BUC] Allen, P. G., Bucher, J. J., Clark, D. L., Edelstein, N. M., Ekberg, S. A., Godhes, J. W., Hudson, E. A., Kaltsoyannis, N., Lukens, W. W., Neu, M. P., Palmer, P. D., Reich, T., Shuh, D. K., Tait, C. D., Zwick, B. D., Multinuclear NMR, Raman, EXAFS, and X-ray structure of $[C(NH_2)_3]_6[(UO_2)_3(CO_3)_6]$ $\cdot 6.5H_2O$, *Inorg. Chem.*, **34**, (1995), 4797-4807. Cited on pages: 245, 246, 489.

[95ARB] Arblaster, J. W., The thermodynamic properties of ruthenium on ITS-90, *CALPHAD: Comput. Coupling Phase Diagrams Thermochem.*, **19**, (1995), 339-347. Cited on page: 671.

[95BAN/GLA] Banyai, I., Glaser, J., Micskei, K., Toth, I., Zékany, L., Kinetic behavior of carbonate ligands with different coordination modes: Equilibrium dynamics for uranyl(2+) carbonato complexes in aqueous solution. A ^{13}C and ^{17}O NMR study, *Inorg. Chem.*, **34**, (1995), 3785-3796. Cited on pages: 489, 567.

[95BIO/MOI] Bion, L., Moisy, P., Madic, C., Use of heteropolyanion ligand as analytical reagent for off-line analysis of uranium(IV) in the PUREX process, *Radiochim. Acta*, **69**, (1995), 251-257. Cited on pages: 281, 489.

[95CAP/VIT] Capdevila, H., Vitorge, P., Redox potentials of PuO_2^{2+} / PuO_2^+ and Pu^{4+} / Pu^{3+} at different ionic strengths and temperatures. Entropy and heat capacity, *Radiochim. Acta*, **68**, (1995), 51-62. Cited on pages: 577, 687, 728, 730.

[95CLA/HOB] Clark, D. L., Hobart, D. E., Neu, M. P., Actinide carbonate complexes and their importance for actinide environmental chemistry, Los Alamos National Laboratory, Report LA-UR-94-1718, (1995), 85 pp. Cited on pages: 491, 640.

[95COH/LOR] Cohen-Adad, R., Lorimer, J. W., Phillips, S. L., Salomon, M., A consistent approach to tabulation of evaluated data: Application to the binary systems $RbCl-H_2O$ and $UO_2(NO_3)_2-H_2O$, *J. Chem. Inf. Comput. Sci.*, **35**, (1995), 675-696. Cited on pages: 238, 491, 492, 574.

[95DUE/PAT] Dueber, R. E., Patat, S., Dickens, P. G., Thermochemical and electrochemical study of sodium and zinc insertion into $\alpha\text{-}UO_{3-y}$ and $\alpha\text{-}U_3O_8$, *Solid State Ionics*, **80**, (1995), 231-238. Cited on pages: 257, 270, 271, 272, 273, 280, 406, 492, 493, 494.

[95EBB] Ebbinghaus, B. B., Calculated thermodynamic functions for gas phase uranium, neptunium, plutonium, and americium oxides (AnO_3), oxyhydroxides ($AnO_2(OH)_2$), oxychlorides (AnO_2Cl_2) and oxyfluorides(AnO_2F_2), Lawrence Livermore National Laboratory, Report UCRL-JC-122278, (1995), 45 pp. Cited on pages: 162, 163, 207, 402, 403, 455, 494, 591.

[95ELI/BID]   Eliet, V., Bidoglio, G., Omenetto, N., Parma, L., Grenthe, I., Characterisation of hydroxide complexes of uranium(VI) by time-resolved fluorescence spectroscopy, *J. Chem. Soc. Faraday Trans.*, **91**, (1995), 2275-2285. Cited on pages: 165, 495, 509, 580, 594, 597, 611, 613.

[95FAN/KIM]   Fanghänel, T., Kim, J. I., Klenze, R., Kato, Y., Formation of Cm(III) chloride complexes in $CaCl_2$ solutions, *J. Alloys Compd.*, **225**, (1995), 308-311. Cited on pages: 337, 356, 357, 358, 369, 469, 496, 547, 585, 614, 662.

[95FAN/NEC]   Fanghänel, T., Neck, V., Kim, J. I., Thermodynamics of neptunium(V) in concentrated salt solutions: II. Ion interaction (Pitzer) parameters for Np(V) hydrolysis species and carbonate complexes, *Radiochim. Acta*, **69**, (1995), 169-176. Cited on pages: 311, 481, 570, 637, 755, 759.

[95FEL/RAI]   Felmy, A. R., Rai, D., Mason, M. J., Fulton, R. W., The aqueous complexation of Nd(III) with molybdate: The effects of both monomeric molybdate and polymolybdate species, *Radiochim. Acta*, **69**, (1995), 177-183. Cited on pages: 385, 636.

[95GRE/PUI]   Grenthe, I., Puigdomènech, I., Rand, M. H., Sandino, M. C. A., *Corrections to the Uranium NEA-TDB review, Appendix D, in: Chemical Thermodynamics of Americium (authored by Silva, R. J., Bidoglio, G., Rand, M. H. and Robouch, P. B., Wanner, H., Puigdomènech, I.)*, Nuclear Energy Agency, Organisation for Economic Co-operation, Development, Ed., pp. 347-374, North Holland Elsevier Science Publishers B. V., Amsterdam, The Netherlands, (1995). Cited on pages: 192, 222, 422, 733.

[95HAA/MAR]   Haaland, A., Martinsen, K.-G., Swang, O., Volden, H. V., Booij, A. S., Konings, R. J. M., Molecular structure of monomeric uranium tetrachloride determined by gas electron diffraction at 900 K, gas phase infrared spectroscopy and quantum-chemical density-functional calculations, *J. Chem. Soc. Dalton Trans.*, (1995), 185-190. Cited on pages: 216, 425, 494, 496, 595.

[95HOB/KAR]   Hobbs, D. T., Karraker, D. G., Recent results on the solubility of uranium and plutonium in Savannah river site waste supernate, *Nucl. Technol.*, **114**, (1995), 318-324. Cited on pages: 178, 496.

[95KAT/KIM]   Kato, Y., Kimura, T., Yoshida, Z., Nitani, N., Solid-liquid phase equilibria of Np(V) and of U(VI) under controlled CO_2 partial pressures, *Radiochim. Acta*, **74**, (1995), 21-25. Cited on page: 607.

[95MOL/MAT]   Moll, H., Matz, W., Schuster, G., Brendler, E., Bernhard, G., Nitsche, H., Synthesis and characterization of uranyl orthosilicate $(UO_2)_2SiO_4 \cdot 2H_2O$, *J. Nucl. Mater.*, **227**, (1995), 40-49. Cited on pages: 496, 519.

[95MOR/GLA]   Moroni, L. P., Glasser, F. P., Reaction between cement components and U(VI) oxide, *Waste Manage. (Oxford)*, **15**, (1995), 243-254. Cited on pages: 255, 257, 259, 497, 498.

[95MOU/DEC]   Moulin, C., Decambox, P., Moulin, V., Decaillon, J. G., Uranium speciation in solution by time-resolved laser-induced fluorescence, *Anal. Chem.*, **67**, (1995), 348-353. Cited on pages: 509, 516.

[95NEC/FAN]   Neck, V., Fanghänel, T., Rudolph, G., Kim, J. I., Thermodynamics of neptunium(V) in concentrated salt solutions: Chloride complexation and ion interaction (Pitzer) parameters for the NpO_2^+ ion, *Radiochim. Acta*, **69**, (1995), 39-47. Cited on pages: 296, 310, 526, 641, 642, 695.

[95NEC/RUN]   Neck, V., Runde, W., Kim, J. I., Solid-liquid equilibria of neptunium(V) in carbonate solutions of different ionic strengths: II. Stability of the solid phases, *J. Alloys Compd.*, **225**, (1995), 295-302. Cited on pages: 305, 306, 307, 308.

[95NOV/CRA]   Novak, C. F., Crafts, C. C., Dhooge, N. J., A data base for thermodynamic modeling of +III actinide solubility in concentrated Na-Cl-SO_4-CO_3-PO_4 electrolytes, Sandia National Lab., Report-SAND-95-2010C, (1995), 17 pp. Cited on pages: 322, 498.

[95NOV/ROB]   Novak, C. F., Roberts, K. E., Thermodynamic modeling of neptunium(V) solubility in concentrated Na-CO_3-HCO_3-Cl-ClO_4-H-OH-H_2O systems, Scientific Basis for Nuclear Waste Management XVIII, held 1994, in Kyoto, vol. 353, pp. 1119-1128, (1995). Cited on pages: 498, 568.

[95PAL/NGU]   Palmer, D. A., Nguyen-Trung, C., Aqueous uranyl complexes. 3. Potentiometric measurements of the hydrolysis of uranyl(VI) ion at 25°C, *J. Solution Chem.*, **24**, (1995), 1282-1291. Cited on pages: 165, 166, 169, 170, 174, 175, 176, 499, 500, 501, 502, 626, 661, 674, 675.

[95PAN/CAM]   Pan, P., Campbell, A. B., Dissolution of $Np_2O_5(c)$ in CO_2 free aqueous solutions at 25°C and pH 6-13: Final report, Atomic Energy of Canada Limited, Report RC-1407 and COG-I-95-203, (1995), 45 pp. Cited on pages: 585, 616.

[95PAS/KIM]   Pashalidis, I., Kim, J. I., Ashida, T., Grenthe, I., Spectroscopic study of the hydrolysis of PuO_2^{2+} in aqueous solution, *Radiochim. Acta*, **68**, (1995), 99-104. Cited on page: 175.

[95PER/SHI]   Peretrukhin, V. F., Shilov, V. P., Pikaev, A. K., Alkaline chemistry of transuranium elements and technetium elements and technetium and the treatment of alkaline radioactive wastes, Westinghouse Hanford Company, Report WHC-EP-0817, (1995), 171 pp. Cited on page: 686.

[95PUI/BRU]   Puigdomènech, I., Bruno, J., A thermodynamic data base for Tc to calculate equilibrium solubilities at temperatures up to 300°C, Swedish Nucl. Fuel Waste Manag. Co., Report SKB TR-95-09, (1995), 52 pp. Cited on pages: 390, 502.

[95RAI/FEL]   Rai, D., Felmy, A. R., Moore, D. A., Mason, M. J., The solubility of Th(IV) and U(IV) hydrous oxides in concentrated $NaHCO_3$ and Na_2CO_3 aqueous solutions, Scientific Basis for Nuclear Waste management XVIII, held 1994, in Kyoto, vol. 353, pp. 1143-1150, (1995). Cited on pages: 183, 251, 503, 617, 618, 650, 651.

[95RAI/FEL2]   Rai, D., Felmy, A. R., Fulton, R. W., Nd^{3+} and Am^{3+} ion interactions with sulfate ion and their influence on $NdPO_4(c)$ solubility, *J. Solution Chem.*, **24**, (1995), 879-895. Cited on pages: 504, 636.

[95SAL/JAY]   Sali, S. K., Jayanthi, K., Iyer, V. S., Sampath, S., Venugopal, V., High temperature chemistry of $Rb_2U_4O_{11}$-$Rb_2U_4O_{12}$ system, NUCAR-95: nuclear and radiochemistry symposium, held in Kalpakkam, India, 21-24 Feb. 1995, pp. 222-223, Bhabha Atomic Research Centre, Mumbai, India, (1995). Cited on pages: 276, 504.

[95SIL/BID]   Silva, R. J., Bidoglio, G., Rand, M. H., Robouch, P., Wanner, H., Puigdomènech, I., *Chemical Thermodynamics of Americium*, Nuclear Energy Agency, Organisation for Economic Co-operation, Development, Ed., vol. 2, *Chemical Thermodynamics*, North Holland Elsevier Science Publishers B. V., Amsterdam, The Netherlands, (1995), 374 pp. Cited on pages: xi, 4, 6, 44, 80, 98, 113, 114, 245, 250, 282, 284, 318, 319, 333, 334, 335, 336, 337, 338, 339, 342, 343, 344, 345, 346, 347, 349, 351, 352, 353, 354, 355, 356, 358, 359, 360, 362, 363, 364, 365, 366, 368, 369, 370, 373, 374, 375, 377, 378, 379, 380, 381, 383, 384, 385, 410, 411, 422, 423, 428, 431, 448, 468, 469, 476, 477, 487, 518, 519, 521, 525, 544, 547, 554, 557, 579, 582, 586, 620, 625, 627, 663, 683, 690, 728, 729, 733, 735, 753, 762.

[95YAJ/KAW]   Yajima, T., Kawamura, Y., Ueta, S., Uranium(IV) solubility and hydrolysis constants under reduced conditions, Scientific basis for nuclear waste management XVIII, held in Kyoto, 1994, vol. 353, pp. 1137-1142, (1995). Cited on pages: 183, 185, 186, 504, 579, 646, 647.

[96ALL/SHU]   Allen, P. G., Shuh, D. K., Bucher, J. J., Edelstein, N. M., Palmer, C. E. A., Silva, R. J., Nguyen, S. N., Marquez, L. N., Hudson, E. A., Determinations of uranium structures by EXAFS: Schoepite and other U(VI) oxide precipitates, *Radiochim. Acta*, **75**, (1996), 47-53. Cited on pages: 505, 672.

[96BER/GEI]   Bernhard, G., Geipel, G., Brendler, V., Nitsche, H., Speciation in uranium in seepage waters of a mine tailing pile studied by time-resolved laser-induced fluorescence spectroscopy, *Radiochim. Acta*, **74**, (1996), 87-91. Cited on pages: 248, 506, 516, 529, 530, 590, 670, 685.

[96BRE/GEI] Brendler, V., Geipel, G., Bernhard, G., Nitsche, H., Complexation in the system $UO_2^{2+}/PO_4^{3-}/OH^-$ (aq): potentiometric and spectroscopic investigations at very low ionic strengths, *Radiochim. Acta*, **74**, (1996), 75-80. Cited on pages: 239, 240, 507, 621.

[96BUR/MIL] Burns, P. C., Miller, M. L., Ewing, R. C., U^{6+} minerals and inorganic phases : a comparison and hierarchy of crystal structures, *Can. Mineral.*, **34**, (1996), 845-880. Cited on page: 254.

[96BYR/SHO] Byrne, R. H., Sholkovitz, E. R., Marine chemistry and geochemistry of the lanthanides, *Handbook on the physics and chemistry of rare earths*, Gschneidner Jr., K. A., Eyring, L., Eds., vol. 23, pp. 497-593, Elsevier Science B.V. Amsterdam, (1996). Cited on page: 334.

[96CAP/VIT] Capdevila, H., Vitorge, P., Giffaut, E., Delmau, L., Spectrophotometric study of the dissociation of the Pu(IV) carbonate limiting complex, *Radiochim. Acta*, **74**, (1996), 93-98. Cited on page: 326.

[96CLA/CON] Clark, D. L., Conradson, S. D., Ekberg, S. A., Hess, N. J., Neu, M. P., Palmer, P. D., Runde, W., Tait, C. D., EXAFS studies of pentavalent neptunium carbonato complexes. Structural elucidation of the principal constituents of neptunium in groundwater environments, *J. Am. Chem. Soc.*, **118**, (1996), 2089-2090. Cited on pages: 250, 686.

[96DIA/GAR] Díaz Arocas, P., Garcia-Serrano, J., Quiñones, J., Geckeis, H., Grambow, B., Coprecipation of mono-, di-, tri-, tetra- and hexavalent ions with Na-polyuranates, *Radiochim. Acta*, **74**, (1996), 51-58. Cited on pages: 165, 508.

[96FAL/REA] Falck, W. E., Read, D., Thomas, J. B., CHEMVAL2: Thermodynamic Database, Report, (1996). Cited on page: 578.

[96FAN/NEC] Fanghänel, T., Neck, V., Kim, J. I., The ion product of H_2O, dissociation constants of H_2CO_3 and Pitzer parameters in the system $Na^+/H^+/OH^-/HCO_3^-/CO_3^{2-}/ClO_4^-/H_2O$ at 25°C, *J. Solution Chem.*, **25**, (1996), 327-343. Cited on pages: 306, 308, 556, 759, 762.

[96FIN/COO] Finch, R. J., Cooper, M. A., Hawthorne, F. C., Ewing, R. C., The crystal structure of schoepite, $[(UO_2)_8O_2(OH)_{12}](H_2O)_{12}$, Can. Mineral., **34**, (1996), 1071-1088. Cited on pages: 189, 587.

[96GEI/BRA] Geipel, G., Brachmann, A., Brendler, V., Bernhard, G., Nitsche, H., Uranium(VI) sulfate complexation studied by time-resolved laser-induced fluorescence spectroscopy (TRLFS), Radiochim. Acta, **75**, (1996), 199-204. Cited on pages: 230, 231, 508, 672.

[96HIL/LAU] Hildenbrand, D. L., Lau, K. H., Thermochemistry of gaseous AgCl, Ag_3Cl_3 and CuCl, High Temp. Mater. Sci., **35**, (1996), 11-20. Cited on page: 393.

[96IYE/JAY] Iyer, V. S., Jayanthi, K., Sali, S. K., Sampath, S., Venugopal, V., Molar Gibbs free energy of formation of $Rb_2U_4O_{11}(s)$ by solid oxide electrolyte galvanic cell, J. Alloys Compd., **235**, (1996), 1-6. Cited on pages: 276, 504, 510, 684.

[96JAY/IYE] Jayanthi, K., Iyer, V. S., Rama Rao, G. A., Venugopal, V., Molar Gibbs energy formation of $RbUO_3(s)$, J. Nucl. Mater., **232**, (1996), 233-239. Cited on pages: 275, 511.

[96KAT/KIM] Kato, Y., Kimura, T., Yoshida, Z., Nitani, N., Solid-liquid phase equilibria of Np(VI) and of U(VI) under controlled CO_2 partial pressures, Radiochim. Acta, **74**, (1996), 21-25. Cited on pages: 168, 171, 172, 174, 245, 321, 512, 516, 597, 606.

[96KON/BOO] Konings, R. J. M., Booij, A. S., Kovács, A., Girichev, G. V., Giricheva, N. I., Krasnova, O. G., The infrared spectrum and molecular structure of gaseous UF_4, J. Mol. Struct., **378**, (1996), 121-131. Cited on pages: 201, 512, 595.

[96KOV/BOO] Kovács, A., Booij, A. S., Cordfunke, E. H. P., Kok-Scheele, A., Konings, R. J. M., On the fusion and vaporisation behaviour of UCl_3, J. Alloys Compd., **241**, (1996), 95-97. Cited on pages: 222, 513.

[96KUL/MAL] Kulyako, Y. M., Malikov, D. A., Trofimov, T. I., Myasoedov, B. F., Behaviour of transplutonium and rare earth elements in acidic and alkaline solutions of potassium ferricyanide, Mendeleev Commun., **5**, (1996), 173-174. Cited on pages: 352, 513.

[96MEI/KAT] Meinrath, G., Kato, Y., Kimura, T., Toshida, Z., Solid-aqueous phase equilibria of uranium(VI) under ambient conditions, *Radiochim. Acta*, **75**, (1996), 159-167. Cited on pages: 165, 166, 168, 170, 171, 172, 321, 514, 517, 597, 606, 609.

[96MEI/KLE] Meinrath, G., Klenze, R., Kim, J. I., Direct spectroscopic speciation of uranium(VI) in carbonate solutions, *Radiochim. Acta*, **74**, (1996), 81-86. Cited on pages: 245, 516.

[96MEI/SCH] Meinrath, G., Schweinberger, M., Hydrolysis of the uranyl(VI) ion - A chemometric approach, *Radiochim. Acta*, **75**, (1996), 205-210. Cited on pages: 166, 170, 171, 514, 517, 553, 580, 597, 606.

[96MEI] Meinrath, G., Coordination of uranyl(VI) carbonate species in aqueous solutions, *J. Radioanal. Nucl. Chem.*, **211**, (1996), 349-362. Cited on pages: 514, 607.

[96MER/FUG] Merli, L., Fuger, J., Thermochemistry of selected lanthanide and actinide hydroxycarbonates and carbonates, *Radiochim. Acta*, **74**, (1996), 37-43. Cited on pages: 378, 379, 380, 469, 518, 519.

[96MOL/GEI] Moll, H., Geipel, G., Matz, W., Bernhard, G., Nitsche, H., Solubility and speciation of $(UO_2)_2SiO_4 \cdot 2H_2O$ in aqueous systems, *Radiochim. Acta*, **74**, (1996), 3-7. Cited on pages: 253, 254, 255, 443, 496, 498, 519, 520, 558.

[96NAK/YAM] Nakayama, S., Yamaguchi, T., Sekine, K., Solubility of neptunium(IV) hydrous oxide in aqueous solutions, *Radiochim. Acta*, **74**, (1996), 15-19. Cited on pages: 298, 648.

[96PAR/PYO] Park, Y. J., Pyo, H. R., Kim, W. H., Chun, K. S., Solubility studies of uranyl hydrolysis precipitates, *J. Korea. Chem. Soc.*, **40**, (1996), 599-606. Cited on pages: 165, 520.

[96PAV/FAN] Paviet, P., Fanghänel, T., Klenze, R., Kim, J. I., Thermodynamics of curium(III) in concentrated electrolyte solutions: formation of sulfate complexes in $NaCl/Na_2SO_4$ solutions, *Radiochim. Acta*, **74**, (1996), 99-103. Cited on pages: 365, 366, 367, 368, 521, 585, 614.

[96PER/KRY] Peretrukhin, V. F., Kryutchov, S. V., Silin, V. I., Tananaev, I. G., *Determination of the solubility of Np(IV)-(VI), Pu(III)-(VI), Am(III)-(VI) and Tc(IV), (V) hydroxo compounds in 0.5-14 M NaOH solutions*, Westinghouse Hanford Company, P.O. Box 1970, Richland, Washington, (1996). Cited on pages: 294, 296, 315, 352, 521, 576.

[96RAK/TSY] Rakitskaya, E. M., Tsyplenkova, V. L., Panov, A. S., Interaction of uranium dioxide with water, *Russ. J. Inorg. Chem.*, **41**, (1996), 1893-1894. Cited on pages: 165, 183, 185, 525.

[96RAO/RAI] Rao, L., Rai, D., Felmy, A. R., Fulton, R. W., Novak, C. F., Solubility of $NaNd(CO_3) \cdot 6H_2O$(c) in concentrated Na_2CO_3 and $NaHCO_3$ solutions, *Radiochim. Acta*, **75**, (1996), 141-147. Cited on pages: 383, 637.

[96RAO/RAI2] Rao, L., Rai, D., Felmy, A. R., Solubility of $Nd(OH_3)$(c) in 0.1 M NaCl at 25°C and 90°C, *Radiochim. Acta*, **72**, (1996), 151-155. Cited on page: 636.

[96ROB/SIL] Roberts, K. E., Silber, H. B., Torretto, P. C., Prussin, T., Becraft, K., Hobart, D. E., Novak, C. F., The experimental determination of the solubility product for NpO_2OH in NaCl solutions, *Radiochim. Acta*, **74**, (1996), 27-30. Cited on pages: 352, 353, 499, 525.

[96RUN/NEU] Runde, W., Neu, M. P., Clark, D. L., Neptunium(V) hydrolysis and carbonate complexation: Experimental and predicted neptunyl solubility in concentrated NaCl using the Pitzer approach, *Geochim. Cosmochim. Acta*, **60**, (1996), 2065-2073. Cited on pages: 296, 305, 307, 308, 309, 310, 343, 375, 383, 470, 479, 481, 482, 483, 641, 642, 656, 758, 759.

[96SAL/IYE] Sali, S. K., Iyer, V. S., Jayanthi, K., Sampath, S., Venugopal, V., Thallium uranates and other (Tl,U,O) compounds: a structural and thermodynamic study, *J. Alloys Compd.*, **237**, (1996), 49-57. Cited on pages: 279, 526.

[96SHO/BAM] Shoup, S. S., Bamberger, C. E., Haire, R. G., Novel plutonium titanate compounds and solid solutions $Pu_2Ti_2O_7$-$Ln_2Ti_2O_7$: relevance to nuclear waste disposal, *J. Am. Ceram. Soc.*, **79**, (1996), 1489-1493. Cited on pages: 331, 527.

[96YAM/HUA] Yamawaki, M., Huang, J., Yamaguchi, K., Yasumoto, M., Sakurai, H., Suzuki, Y., Investigation of the vaporization of $BaUO_3$ by means of mass spectrometry, *J. Nucl. Mater.*, **231**, (1996), 199-203. Cited on pages: 267, 527.

[97ALL/BAN] Allard, B., Banwart, S. A., Bruno, J., Ephraim, J. H., Grauer, R., Grenthe, I., Hadermann, J., Hummel, W., Jakob, A., Karapiperis, T., Plyasunov, A. V., Puigdomènech, I., Rard, J. A., Saxena, S., Spahiu, K., *Modelling in Aquatic Chemistry*, Grenthe, I., Puigdomènech, I., Eds., Nuclear Energy Agency, Organisation for Economic Co-operation and Development, Paris, France, (1997), 724 pp. Cited on pages: 25, 285, 366, 478, 547, 617, 753, 754.

[97ALL/BUC] Allen, P. G., Bucher, J. J., Shuh, D. K., Edelstein, N. M., Reich, T., Investigation of aquo and chloro complexes of UO_2^{2+}, NpO_2^+, Np^{4+}, and Pu^{3+} by X-ray absorption fine structure spectroscopy, *Inorg. Chem.*, **36**, (1997), 4676-4683. Cited on pages: 221, 293, 301, 322, 528, 679.

[97ALL/SHU] Allen, P. G., Shuh, D. K., Bucher, J. J., Edelstein, N. M., Palmer, C. E. A., Marquez, L. N., EXAFS spectroscopic studies of uranium(VI) oxide precipitates, Aqueous chemistry and geochemistry of oxides and oxyhydroxides, vol. 432, pp. 139-143, (1997). Cited on page: 274.

[97AMA/GEI] Amayri, S., Geipel, G., Schuster, G., Baraniak, L., Bernhard, G., Nitsche, H., Synthesis and characterization of calcium uranyl carbonate: $Ca_2[UO_2(CO_3)_3] \cdot 10\ H_2O$, Forschungszentrum Rossendorf Institute of Radiochemistry, Annual report, (1997). Cited on pages: 248, 506, 670, 685.

[97ASA/SUG] Asahina, K., Suganuma, H., Omori, T., Solvent extraction of tetrachloronitridotechnetate(VI) ion with tetraphenylarsonium chloride, *J. Radioanal. Nucl. Chem.*, **222**, (1997), 25-28. Cited on pages: 392, 529.

[97BER/GEI] Bernhard, G., Geipel, G., Brendler, V., Reich, T., Nitsche, H., Validation of complex formation of Ca^{2+}, UO_2^{2+} and CO_3^{2-}, Forschungszentrum Rossendorf Institute of Radiochemistry, Annual report, (1997). Cited on pages: 248, 249, 529, 590, 670, 685.

[97BIO/MOI]   Bion, L., Moisy, P., Vaufrey, F., Méot-Reymond, S., Simoni, E., Madic, C., Coordination of U^{4+} in the complex in solid sate and in aqueous solution, *Radiochim. Acta*, **78**, (1997), 73-82. Cited on page: 489.

[97BOU]   Boucharat, N., Etude métallurgique du technétium en vue de sa transmutation en réacteur à neutrons rapides, Ph. D. Thesis, Université d'Aix-Marseille I, (1997), in French. Cited on pages: 389, 530, 671.

[97BRU/CAS]   Bruno, J., Casas, I., Cera, E., Duro, L., Development and application of a model for the long-term alteration of UO_2 spent nuclear fuel, test of equilibrium and kinetic mass transfer models in the Cigar Lake ore deposit, *J. Contam. Hydrol.*, **26**, (1997), 19-26. Cited on pages: 165, 531.

[97CAS/BRU]   Casas, I., Bruno, J., Cera, E., Finch, R. J., Ewing, R. C., Characterization and dissolution behavior of a becquerelite from Shinkolobwe, Zaire, *Geochim. Cosmochim. Acta*, **61**, (1997), 3879-3884. Cited on pages: 181, 422, 531, 704, 705.

[97CHA/SAW]   Chaudhuri, N. K., Sawant, R. M., Stability constants of the fluoride complexes of actinides in aqueous solution and their correlation with fundamental properties, Bangladesh Agricultural Research Council, Report E/022, (1997), 48 pp. Cited on page: 354.

[97CLA/CON]   Clark, D. L., Conradson, S. D., Neu, M. P., Palmer, P. D., Runde, W., Tait, C. D., XAFS structural determination of Np(VII). Evidence for trans dioxo cation under alkaline solution conditions, *J. Am. Chem. Soc.*, (1997), 5259-5260. Cited on pages: 293, 532, 697.

[97COR/BOO]   Cordfunke, E. H. P., Booij, A. S., Smit-Groen, V., van Vlaanderen, P., Structural and thermodynamic characterization of the perovskite-related $Ba_{1+y}UO_{3+x}$ and $(Ba, Sr)_{1+y}UO_{3+x}$ phases, *J. Solid State Chem.*, **131**, (1997), 341-349. Cited on pages: 265, 266, 528, 532.

[97DAV/FOU]   David, F., Fourest, B., Structure of trivalent lanthanide and actinide aquo ions, *New J. Chem.*, **21**, (1997), 167-176. Cited on page: 581.

[97DEM/SER] Demchenko, E. A., Serezhkina, L. B., Gar'kin, V. P., Serezhkin, V. N., Solubility in UO_2-$RCONH_2$-H_2O (R = $CHCl_2$ or CCl_3) systems, *Russ. J. Inorg. Chem.*, **42**, (1997), 726-728. Cited on pages: 231, 534.

[97ELL/ARM] Elless, M. P., Armstrong, A. Q., Lee, S. Y., Characterization and solubility measurements of uranium-contaminated soils to support risk assessment, *Health Physics*, **72**, (1997), 716-726. Cited on pages: 165, 534.

[97FEL/RAI] Felmy, A. R., Rai, D., Sterner, S. M., Mason, M. J., Hess, N. J., Conradson, S. D., Thermodynamic models for highly charged aqueous species: Solubility of Th(IV) hydrous oxide in concentrated $NaHCO_3$ and Na_2CO_3 aqueous solutions, *J. Solution Chem.*, **26**, (1997), 233-248. Cited on pages: 326, 327, 637, 650.

[97FIN] Finch, R. J., Thermodynamic stabilities of U(VI) minerals: Estimated and observed relationships, *Mater. Res. Soc. Symp. Proc.*, **465**, (1997), 1185-1192. Cited on pages: 282, 534, 535.

[97GEI/BER] Geipel, G., Bernhard, G., Brendler, V., Brachmann, A., Nitsche, H., Speciation of uranium in natural waters of an uranium mining area, NRC4 4th International Conference on Nuclear and Radiochemistry, held 8-13 September 1996 in St Malo, France, vol. 2, p. EP48, (1997). Cited on pages: 165, 536.

[97GEI/RUT] Geipel, G., Rutsch, M., Bernhard, G., Nitsche, H., Increase of the ionic strength of uranyl containing solutions, Forschungszentrum Rossendorf Institute of Radiochemistry, Annual report, (1997). Cited on pages: 160, 537.

[97GRE/PLY] Grenthe, I., Plyasunov, A., On the use of semiempirical electrolyte theories for the modeling of solution chemical data, *Pure Appl. Chem.*, **69**, (1997), 951-958. Cited on page: 716.

[97GUR/SER] Gurevich, V. M., Sergeyeva, E. I., Gavrichev, K. S., Gorbunov, V. E., Kuznetova, T. P., Khodakovskii, I. L., Thermodynamic properties of uranyl hydroxide $UO_2(OH)_2$(cr, α, rhomb), *Geochem. Int.*, **35**, (1997), 74-87. Cited on pages: 159, 189, 190, 191, 192, 414, 537, 538, 539, 587, 588.

[97HOV] Hovey, J. K., Thermodynamics of hydration of a 4+ aqueous ion: partial molar heat capacities and volumes of aqueous thorium(IV) from 10 to 55°C, *J. Phys. Chem. B*, **101**, (1997), 4321-4334. Cited on pages: 159, 404, 541.

[97HUA/YAM] Huang, J., Yamawaki, M., Yamaguchi, K., Yasumoto, M., Sakurai, H., Suzuki, Y., Investigation of vaporization thermodynamics of $SrUO_3$ by means of mass spectrometry, *J. Nucl. Mater.*, **247**, (1997), 17-20. Cited on pages: 260, 542, 543.

[97ION/MAD] Ionova, G., Madic, C., Guillaumont, R., Covalency effects in the standard enthalpies of formation of trivalent lanthanide and actinide halides, *Radiochim. Acta*, **78**, (1997), 83-90. Cited on pages: 5, 543.

[97IOU/KRU] Ioussov, A., Krupa, J. C., Luminescence properties and stability constants of curium(III) complexes with lacunary heteropolyanions $PW_{11}O_{39}^{7-}$ and $SiW_{11}O_{39}^{8-}$ in nitric acid solutions, *Radiochim. Acta*, **78**, (1997), 97-104. Cited on pages: 386, 387, 544.

[97IYE/JAY] Iyer, V. S., Jayanthi, K., Venugopal, V., Thermodynamic properties of $K_2U_4O_{12}(s)$ and $K_2U_4O_{13}(s)$ by emf and calorimetric measurements, *J. Solid State Chem.*, **132**, (1997), 342-348. Cited on pages: 267, 269, 275, 544.

[97JAY/IYE] Jayanthi, K., Iyer, V. S., Venugopal, V., Thermodynamic studies on $Cs_4U_5O_{17}(s)$ and $Cs_2U_2O_7(s)$ by emf and calorimetric measurements, *J. Nucl. Mater.*, **250**, (1997), 229-235. Cited on pages: 267, 269, 277, 278, 545.

[97KAL/HEA] Kaledin, L. A., Heaven, M. C., Electronic spectroscopy of UO, *J. Mol. Spectrosc.*, **185**, (1997), 1-7. Cited on pages: 161, 474.

[97KON/FAN] Könnecke, T., Fanghänel, T., Kim, J. I., Thermodynamics of trivalent actinides in concentrated electrolyte solutions. Modelling the chloride complexation of Cm(III), *Radiochim. Acta*, **76**, (1997), 131-135. Cited on pages: 335, 344, 356, 358, 479, 496, 547, 585, 614, 635, 729.

[97KON/NEC] Könnecke, T., Neck, V., Fanghänel, T., Kim, J. I., Activity coefficients and Pitzer parameters in the systems $Na^+/Cs^+/Cl^-/TcO_4^-$ or ClO_4^-/H_2O at 25°C, *J. Solution Chem.*, **26**, (1997), 561-577. Cited on pages: 547, 615, 760.

[97KRE/GAN] Kremer, C., Gancheff, J., Kremer, E., Mombrú,A. W., González, O., Mariezcurrena, R., Suescun, L., Cubas, M. L., Ventura, O. N., Structural and conformational analysis of Tc(V) and Re(V) dioxo complexes. X-ray crystal structure of $[TcO_2(tn)_2]I \cdot H_2O$, *Polyhedron*, **16**, (1997), 3311-3316. Cited on pages: 391, 548.

[97KRI/FON] Krikorian, O. H., Fontes, Jr. A. S., Ebbinghaus, B. B., Adamson, M. G., Transpiration studies on the volatilities of $PuO_3(g)$ and $PuO-2(OH)_2(g)$ from $PuO_2(s)$ in the presence of steam and oxygen and application to plutonium volality in mixed-waste thermal oxidation processes, *J. Nucl. Mater.*, **247**, (1997), 161-171. Cited on pages: 162, 455.

[97KRI/RAM] Krishnan, K., Rama Rao, G. A., Singh Mudher, K. D., Venugopal, V., Vapour pressure measurements of $TeO_2(g)$ over $UTeO_5(s)$, NUCAR 97: nuclear and radiochemistry symposium. Held in Calcutta, India, 21-24 Jan 1997, pp. 220-221, Bhabha Atomic Research Centre, Mumbai, India, (1997). Cited on pages: 234, 548.

[97LAN] Langmuir, D., *Aqueous environmental geochemistry*, Prentice Hall, New Jersey, USA, (1997). Cited on pages: 160, 183, 242, 254, 548, 639.

[97LUB/HAV] Lubal, P., Havel, J., The study of complex equilibria of uranium(VI) with selenate, *Talanta*, **44**, (1997), 457-466. Cited on pages: 233, 549, 550, 551.

[97LUB/HAV2] Lubal, P., Havel, J., Spectrophotometric and potentiometric study of uranyl hydrolysis in perchlorate medium. Is derivative spectrophotometry suitable for search of the chemical model ?, *Chem. Pap.*, **51**, (1997), 213-220. Cited on pages: 166, 175, 550, 551.

[97MAS/COU] Maslennikov, A. G., Courson, O., Peretrukhin, V. F., David, F., Masson, M., Technetium electrochemical reduction in nitric solutions at mercury and carbon electrodes, *Radiochim. Acta*, **78**, (1997), 123-129. Cited on pages: 391, 551.

[97MEI/SCH] Meinrath, G., Schweinberger, M., Merkel, B. J., Hydrolysis of the uranyl(VI) ion - A chemometric approach, NRC4 4th International Conference on Nuclear and Radiochemistry, held 8-13 September 1996 in St Malo, France, vol. 1, p. DP7, (1997). Cited on pages: 166, 170, 171, 517, 553.

[97MEI] Meinrath, G., Chemometric and statistical analysis of uranium(VI) hydrolysis at elevated U(VI) concentrations, *Radiochim. Acta*, **77**, (1997), 221-234. Cited on pages: 166, 170, 171, 172, 514, 552, 597, 598, 606.

[97MEI2] Meinrath, G., Uranium(VI) speciation by spectroscopy, *J. Radioanal. Nucl. Chem.*, **224**, (1997), 119-126. Cited on pages: 551, 553, 597, 606, 609, 643, 674.

[97MER/LAM] Merli, L., Lambert, B., Fuger, J., Thermochemistry of lanthanum, neodymium, samarium and americium trihydroxides and their relation to the corresponding hydroxycarbonates, *J. Nucl. Mater.*, **247**, (1997), 172-176. Cited on pages: 321, 350, 351, 439, 553, 554.

[97MOL/GEI] Moll, H., Geipel, G., Reich, T., Brendler, V., Bernhard, G., Nitsche, H., Interaction of uranium(VI) with silicon species in aqueous solutions, NRC4 4th International Conference on Nuclear and Radiochemistry, held 8-13 September 1996 in St Malo, France, vol. 1, p. DP8, (1997). Cited on pages: 252, 253, 554, 611.

[97MUR] Murphy, W. H., Retrograde solubility of source term phases, Scientific basis for nuclear waste management XX, vol. 465, pp. 713-720, (1997). Cited on pages: 165, 257, 555.

[97NAK/ARA] Nakajima, K., Arai, Y., Suzuki, Y., Vaporization behaviour of neptunium mononitride, *J. Nucl. Mater.*, **247**, (1997), 33-36. Cited on pages: 300, 644, 645.

[97SCA/ANS] Scapolan, S., Ansoborlo, E., Moulin, C., Madic, C., Uranium speciation in aqueous medium by means of capillary electrophoresis and time-resolved laser-induced fluorescence, NRC4 4th International Conference on Nuclear and Radiochemistry, held 8-13 September 1996 in St Malo, France, vol. 1, p. DP5, (1997). Cited on pages: 165, 562.

[97SHO/SAS] Shock, E. L., Sassani, D. C., Willis, M., Sverjensky, D. A., Inorganic species in geologic fluids: Correlations among standard molal thermodynamic properties of aqueous ions and hydroxide complexes, *Geochim. Cosmochim. Acta*, **61**, (1997), 907-950. Cited on pages: 165, 562, 563, 564.

[97SHO/SAS2] Shock, E. L., Sassani, D. C., Betz, H., Uranium in geologic fluids: Estimates of standard partial molal properties, oxidation, potentials, and hydrolysis constants at high temperatures and pressures, *Geochim. Cosmochim. Acta*, **61**, (1997), 4245-4266. Cited on pages: 158, 541, 563.

[97STE/FAN] Steinle, E., Fanghänel, T., Klenze, R., Complexing of Cm(III) with monosilicic acid, *Wiss. Ber. - Forschungszent. Karlsruhe*, (1997), 143-152, in German. Cited on pages: 384, 385, 565, 614, 683.

[97SUG/SAT] Suganuma, H., Satoh, I., Omori, T., Choppin, G. R., Am^{3+}-F^- interaction in mixed system of dimethyl sulfoxide and water, *Radiochim. Acta*, **77**, (1997), 211-214. Cited on pages: 354, 355.

[97SUG/SAT2] Suganuma, H., Satoh, I., Omori, T., Yagi, M., Am^{3+}-F^- interaction in mixed system of methanol and water, *Radiochim. Acta*, **77**, (1997), 207-209. Cited on pages: 354, 355.

[97SZA/AAS] Szabó, Z., Aas, W., Grenthe, I., Structure, isomerism, and ligand dynamics in dioxouranium(VI) complexes, *Inorg. Chem.*, **36**, (1997), 5369-5375. Cited on page: 567.

[97TAK] Takahashi, Y., High temperature heat-capacity measurement up to 1500 K by the triple-cell DSC, *Pure Appl. Chem.*, **69**, (1997), 2263-2269. Cited on pages: 193, 566.

[97TOR/BAR]   Torrero, M. E., Baraj, E., de Pablo, J., Giménez, J., Casas, I., Kinetics of corrosion and dissolution of uranium dioxide as a function of pH, *Int. J. Chem. Kinet.*, **29**, (1997), 261-267. Cited on pages: 183, 185, 566.

[97VAL/RAG]   Valsami-Jones, E., Ragnarsdottir, K. V., Solubility of uranium oxide and calcium uranate in water, and $Ca(OH)_2$-bearing solutions, *Radiochim. Acta*, **79**, (1997), 249-257. Cited on pages: 165, 259, 274, 566.

[97WRU/PAL]   Wruck, David A., Palmer, C. E. A., Silva, Robert J., Actinide speciation in elevated temperature aqueous solutions by laser photoacoustic spectroscopy, Book of abstracts, 212th ACS National Meeting, Orlando, FL, August 25-29, 1996, p. 3, American Chemical Society, Washington, D.C., Washington, D.C., USA, (1997). Cited on pages: 174, 368, 369, 372, 566, 567.

[98AAS/MOU]   Aas, W., Moukhamet-Galeev, A., Grenthe, I., Complex formation in the ternary U(VI)-F-L system (L= Carbonate, oxalate and picolinate), *Radiochim. Acta*, **82**, (1998), 77-82. Cited on pages: 248, 567.

[98ALM/NOV]   Al Mahamid, I., Novak, C. F., Becraft, K. A., Carpenter, S. A., Hakem, N., Solubility of Np(V) in $K-Cl-CO_3$ and $Na-K-Cl-CO_3$ solutions to high concentrations: measurements and thermodynamic model predictions, *Radiochim. Acta*, **81**, (1998), 93-101. Cited on pages: 310, 311, 312, 498, 568, 569, 570, 572, 639, 734, 755, 759.

[98AMA/BER]   Amayri, S., Bernhard, G., Nitsche, H., Solubility of $Ca_2[UO_2(CO_3)_3] \cdot 10H_2O$, liebigite, Forschungszentrum Rossendorf Institute of Radiochemistry, Annual report, (1998). Cited on page: 506.

[98APE/KOR]   Apelblat, A., Korin, E., Vapour pressures of saturated aqueous solutions of ammonium iodide, potassium iodide, potassium nitrate, strontium chloride, lithium sulphate, sodium thiosulphate, magnesium nitrate and uranyl nitrate from T = (278 to 323) K, *J. Chem. Thermodyn.*, **30**, (1998), 459-471. Cited on pages: 239, 492, 574.

[98BAL/HEA] Baldas, J., Heath, G. A., Macgregor, S. A., Moock, K. H., Nissen, S. C., Raptis, R. G., Spectroelectrochemical and computational studies of tetrachloro and tetrabromo oxo- and nitridotechnetium(V) and their Tc(VI) counterparts, *J. Chem. Soc. Dalton Trans.*, (1998), 2303-2314. Cited on pages: 392, 574.

[98BAN/SAL] Banerjee, A., Sali, S. K., Venugopal, V., Calorimetric studies on $Rb_2U_4O_{11}(s)$, THERMANS 98: Eleventh National Symposium on Thermal Analysis, held in Jammu, India, 2-5 Mar., pp. 94-95, Indian Thermal Analysis Society, Mumbai, India, (1998). Cited on pages: 276, 575.

[98BAR/RUB] Bardin, N., Rubini, P., Madic, C., Hydration of actinyl(VI), MO_2^{2+}, M = U, Np, Pu. An NMR study, *Radiochim. Acta*, **83**, (1998), 189-194. Cited on pages: 160, 575.

[98BEN/FAT] Ben Said, K., Fattahi, M., Delorme-Hiver, A., Abbé, J. C., A novel approach to the oxidation potential of the Tc(VII)/Tc(IV) couple in hydrochloric acid medium through the reduction of TcO_4^- by Fe^{2+} ion, *Radiochim. Acta*, **83**, (1998), 195-203. Cited on pages: 390, 575.

[98BUD/TAN] Budantseva, N. A., Tananaev, I. G., Fedoseev, A. M., Delegard, C. H., Behaviour of plutonium(V) in alkaline media, *J. Alloys Compd.*, **271-273**, (1998), 813-816. Cited on pages: 313, 576.

[98CAP/VIT] Capdevila, H., Vitorge, P., Solubility product of $Pu(OH)_4(am)$, *Radiochim. Acta*, **82**, (1998), 11-16. Cited on pages: 313, 316, 317, 577, 642, 649, 687, 695, 706.

[98CAS/DIX] Casteel, W. J., Dixon, D. A., LeBlond, N., Mercier, H. P. A., Schrobilgen, G. J., Lewis-acid properties of technetium(VII) dioxide trifluoride, TcO_2F_3: Characterization by ^{19}F, ^{17}O, and ^{99}Tc NMR spectroscopy and Raman spectroscopy, density functional theory calculations of TcO_2F_3, $M^+TcO_2F_4^-$ [M = Li, Cs, $N(CH_3)_4$], and $TcO_2F_3 \cdot CH_3CN$, and X-ray crystal structure of $Li^+TcO_2F_4^-$, *Inorg. Chem.*, **37**, (1998), 340-353. Cited on page: 577.

[98CAS/PAB] Casas, I., de Pablo, J., Giménez, J., Torrero, M. E., Bruno, J., Cera, E., Finch, R. J., Ewing, R. C., The role of pe, pH, and carbonate on the solubility of UO_2 and uraninite under nominally reducing conditions, *Geochim. Cosmochim. Acta*, **62**, (1998), 2223-2231. Cited on pages: 183, 187, 577.

[98CHA/DON] Chartier, D., Donnet, L., Adnet, J. M., Electrochemical oxidation of Am(III) with lacunary heteropolyanions and silver nitrate, *Radiochim. Acta*, **83**, (1998), 129-134. Cited on pages: 385, 386, 387, 578, 630.

[98CHA/TRI] Chandratillake, M., Trivedi, D. P., Randall, M. G., Humphreys, P. N., Kelly, E. J., Criteria for compilation of a site-specific thermodynamic database for geochemical speciation calculations, *J. Alloys Compd.*, **271-273**, (1998), 821-825. Cited on pages: 165, 313, 317, 578, 652, 688.

[98CHA] Chase, Jr., M. W., Monograph No. 9. NIST-JANAF Thermochemical Tables, *J. Phys. Chem. Ref. Data*, (1998). Cited on pages: 204, 206, 220.

[98CLA/CON] Clark, D. L., Conradson, S. D., Keogh, D. W., Palmer, P. D., Scott, B. L., Tait, C. D., Identification of the limiting species in the plutonium(IV) carbonate system. Solid state and solution molecular structure of the $[Pu(CO_3)_5]^{6-}$ ion, *Inorg. Chem.*, **37**, (1998), 2893-2899. Cited on pages: 326, 502, 579.

[98CLA/EWI] Clark, S. B., Ewing, R. C., Schaumloffel, J. C., A method to predict free energies of formation of mineral phases in the U(VI)-SiO_2-H_2O system, *J. Alloys Compd.*, **271-273**, (1998), 189-193. Cited on page: 254.

[98CON/ALM] Conradson, S. D., Al Mahamid, I., Clark, D. L., Hess, N. J., Hudson, E. A., Neu, M. P., Palmer, P. D., Runde, W., Tait, C. D., Oxidation state determination of plutonium aquo ions using X-ray absorption spectroscopy, *Polyhedron*, **17**, (1998), 599-602. Cited on pages: 313, 580, 659.

[98DAI/BUR]   Dai, S., Burleigh, M. C., Simonson, J. M., Mesmer, R. E., Xue, Z.-L., Application of chemometric methods in UV-vis absorption spectroscopic studies of uranyl ion dimerization reaction in aqueous solutions, *Radiochim. Acta*, **81**, (1998), 195-199. Cited on pages: 174, 180, 580.

[98DAV/FOU]   David, F., Fourest, B., Hubert, S., Le Du, J. F., Revel, R., Den Auwer, C., Madic, C., Morss, L. R., Ionova, G., Mikalko, V., Vokhmin, V., Nikonov, M., Berthet, J. C., Ephritikine, M., Aquo ions of some trivalent actinides, EXAFS data and thermodynamic consequences, OCDE-AEN Workshop Proceedings, Speciation, techniques and facilities for radioactive materials at synchrotron light sources, Grenoble, France, October 4-6, p. 95, (1998). Cited on pages: 159, 580, 661.

[98DIA/GRA]   Díaz Arocas, P., Grambow, B., Solid-liquid phase equilibria of U(VI) in NaCl solutions, *Geochim. Cosmochim. Acta*, **62**, (1998), 245-263. Cited on pages: 168, 174, 179, 181, 274, 321, 583.

[98DIA/RAG]   Diakonov, I. I., Ragnarsdottir, K. V., Tagirov, B. R., Standard thermodynamic properties and heat capacity equations of rare earth hydroxides: II. Ce(III)-, Pr-, Sm-, Eu(III)-, Gd-, Tb-, Dy-, Ho-, Er-, Tm-, Yb-, and Y-hydroxides. Comparison of thermochemical and solubility data, *Chem. Geol.*, **151**, (1998), 327-347. Cited on pages: 350, 351.

[98DIA/TAG]   Diakonov, I. I., Tagirov, B. R., Ragnarsdottir, K. V., Standard thermodynamic properties and heat capacity equations for rare earth element hydroxides. I. La(OH)$_3$(s) and Nd(OH)$_3$(s). Comparison of thermochemical and solubility data., *Radiochim. Acta*, **81**, (1998), 107-116. Cited on page: 350.

[98EFU/RUN]   Efurd, D. W., Runde, W., Banar, J. C., Janecky, D. R., Kaszuba, J. P., Palmer, P. D., Roensch, F. R., Tait, C. D., Neptunium and plutonium solubilities in a Yucca mountain groundwater, *Environ. Sci. Technol.*, **32**, (1998), 3893-3900. Cited on pages: 297, 313, 317, 321, 584, 616, 639, 641.

[98ERI/BAR] Erin, E. A., Baranov, A. A., Volkov, A. Y., Chistyakov, V. M., Timofeev, G. A., Thermodynamics of actinide redox reactions in potassium phosphotungstate solutions, *J. Alloys Compd.*, **271-273**, (1998), 782-785. Cited on page: 385.

[98FAN/KIM] Fanghänel, T., Kim, J. I., Spectroscopic evaluation of thermodynamics of trivalent actinides in brines, *J. Alloys Compd.*, **271-273**, (1998), 782-785. Cited on pages: 335, 585.

[98FAN/WEG] Fanghänel, T., Weger, H. T., Schubert, G., Kim, J. I., Bicarbonate complexes of trivalent actinides - stable or unstable?, *Radiochim. Acta*, **82**, (1998), 55-57. Cited on pages: 374, 375, 585, 586, 614.

[98FAN/WEG2] Fanghänel, T., Weger, H. T., Könnecke, T., Neck, V., Paviet-Hartmann, P., Steinle, E., Kim, J. I., Thermodynamics of Cm(III) in concentrated electrolyte solutions. Carbonate complexation at constant ionic strength (1 m NaCl), *Radiochim. Acta*, **82**, (1998), 47-53. Cited on pages: 369, 370, 372, 373, 375, 585, 586, 587, 614, 755.

[98FAZ/YAM] Fazekas, Z., Yamamura, T., Tomiyasu, H., Deactivation and luminescence lifetimes of excited uranyl ion and its fluoro complexes, *J. Alloys Compd.*, **271-273**, (1998), 756-759. Cited on pages: 210, 587.

[98FIN/HAW] Finch, R. J., Hawthorne, F. C., Ewing, R. C., Structural relations among schoepite, metaschoepite and "dehydrated schoepite", *Can. Mineral.*, **36**, (1998), 831-845. Cited on pages: 164, 179, 189, 536, 587, 588.

[98GEI/BER] Geipel, G., Bernhard, G., Brendler, V., Nitsche, H., Complex formation between UO_2^{2+} and CO_3^{2-}: studied by laser-induced photoacoustic spectroscopy (LIPAS), *Radiochim. Acta*, **82**, (1998), 59-62. Cited on pages: 245, 588, 589.

[98GEI/BER2] Geipel, G., Bernhard, G., Brendler, V., Nitsche, H., Complex formation between UO_2^{2+} and CO_3^{2-}: studied by laser-induced photoacoustic spectroscopy (LIPAS), *Forsch. Ross.*, (1998), 22. Cited on page: 589.

[98GEI/BER3] Geipel, G., Bernhard, G., Brendler, V., Nitsche, H., Uranyl hydroxo carbonate complexes by laser induced spectroscopy, *Forsch. Ross.*, (1998), 23. Cited on pages: 245, 248, 590, 685.

[98GEI/BER4] Geipel, G., Bernhard, G., Nitsche, H., Validation of the complex formation between Ca^{2+}, UO_2^{2+} and CO_3^{2-} using EDTA adjust Ca^{2+} concentration, *Forsch. Ross.*, (1998), 11. Cited on pages: 248, 590, 670, 685.

[98GOR/SID] Gorokhov, L. N., Sidorova, I. V., $UO_2(OH)_2$ molecules in the U-O-H system and water vapor effects on the rate of uranium dioxide vaporization, *Russ. J. Phys. Chem.*, **72**, (1998), 1038-1040. Cited on pages: 162, 163, 402, 403, 455, 495, 591.

[98HUA/YAM] Huang, J., Yamawaki, M., Yamaguchi, K., Ono, F., Yasumoto, M., Sakurai, H., Sugimoto, J., Thermodynamic study of the vaporization of Cs_2UO_4 by high temperature mass spectrometry, *J. Alloys Compd.*, **271-273**, (1998), 625-628. Cited on pages: 277, 591.

[98ITO/YAM] Ito, H., Yamaguchi, K., Yamamoto, T., Yamawaki, M., Hydrogen absorption properties of U_6Mn and U_6Ni, *J. Alloys Compd.*, **271-273**, (1998), 629-631. Cited on pages: 194, 592.

[98JAY/IYE] Jayanthi, K., Iyer, V. S., Venugopal, V., Thermal properties of $KUO_3(s)$ and $K_2U_2O_7$ by high temperature Calvet calorimeter, THERMANS 98: Eleventh National Symposium on Thermal Analysis, held in Jammu, India, 2-5 Mar., pp. 74-76, Indian Thermal Analysis Society, Mumbai, India, (1998). Cited on pages: 267, 269, 275, 592, 638.

[98JEN/CHO] Jensen, M. P., Choppin, G. R., Complexation of uranyl(VI) by aqueous orthosilicic acid, *Radiochim. Acta*, **82**, (1998), 83-88. Cited on pages: 252, 253, 444, 593, 612.

[98KAP/GER] Kaplan, D. I., Gervais, T. L., Krupka, K. M., Uranium(VI) sorption to sediments under high pH and ionic strength conditions, *Radiochim. Acta*, **80**, (1998), 201-211. Cited on pages: 165, 593.

[98KIT/YAM] Kitamura, A., Yamamura, T., Hase, H., Yamamoto, T., Moriyama, H., Measurement of hydrolysis species of U(VI) by time-resolved laser-induced fluorescence spectroscopy, *Radiochim. Acta*, **82**, (1998), 147-152. Cited on pages: 165, 594, 597.

[98KON/HIL] Konings, R. J. M., Hildenbrand, D. L., The vibrational frequencies, molecular geometry and thermodynamic properties of the actinide tetrahalides, *J. Alloys Compd.*, **271-273**, (1998), 583-586. Cited on pages: 201, 216, 226, 229, 595.

[98KRI/RAM] Krishnan, K., Rama Rao, G. A., Singh Mudher, K. D., Venugopal, V., Thermal stability and vapour pressure studies on UTe_3O_9 and $UTeO_5(s)$, *J. Nucl. Mater.*, **254**, (1998), 49-54. Cited on pages: 234, 235, 236, 548, 595.

[98MAR/SMI] Martell, A. E., Smith, R. M., Montecates R. J., NIST critically selected stability constants of metal complexes, Report Version V, (1998). Cited on page: 657.

[98MAS/MAS] Maslennikov, A. G., Masson, M., Peretrukhin, V. F., Lecomte, M., Technetium electrodeposition from aqueous formate solutions: electrolysis kinetics and material balance study, *Radiochim. Acta*, **83**, (1998), 31-37. Cited on pages: 391, 597.

[98MEI/FIS] Meinrath, G., Fischer, S., Köhncke, K., Voigt, W., Solubility behaviour of uranium(VI) in alkaline solution, Uranium mining and hydrology II, Proceedings of the international conference and workshop in Freiberg, Germany in September, pp. 246-255, Verlag Sven von Loga, Köln, Germany, (1998). Cited on pages: 168, 181, 608, 609.

[98MEI/KAT] Meinrath, G., Kato, Y., Kimura, T., Yoshida, Z., Stokes relationship in absorption and fluorescence spectra of U(VI) species, *Radiochim. Acta*, **82**, (1998), 115-120. Cited on pages: 597, 609.

[98MEI] Meinrath, G., Chemometric analysis: Uranium(VI) hydrolysis by UV-vis spectroscopy, *J. Alloys Compd.*, **275-277**, (1998), 777-781. Cited on pages: 166, 170, 172, 173, 517, 552, 553, 594, 597, 605.

[98MEI2] Meinrath, G., Direct spectroscopic speciation of schoepite-aqueous phase equilibria, *J. Radioanal. Nucl. Chem.*, **232**, (1998), 179-188. Cited on pages: 165, 166, 167, 168, 169, 170, 172, 321, 605.

[98MEI3] Meinrath, G., Aquatic Chemistry of Uranium. A review focusing on aspects of environmental chemistry, *Freiberg Online Geoscience*, **1**, (1998), 101. Cited on pages: 165, 166, 170, 172, 607.

[98MIS/NAM] Mishra, R., Namboodiri, P. N., Tripathi, S. N., Bharadwaj, S. R., Dharwadkar, S. R., Vaporization behaviour and Gibbs energy of formation of $UTeO_5$ and UTe_3O_9, *J. Nucl. Mater.*, **256**, (1998), 139-144. Cited on pages: 234, 235, 236, 596, 609, 610, 627.

[98MOL/GEI] Moll, H., Geipel, G., Brendler, V., Bernhard, G., Nitsche, H., Interaction of uranium(VI) with silicic acid in aqueous solutions studied by time-resolved laser-induced fluorescence spectroscopy (TRLFS), *J. Alloys Compd.*, **271-273**, (1998), 765-768. Cited on pages: 252, 253, 254, 520, 554, 593, 611.

[98MOU/LAS] Moulin, C., Laszak, I., Moulin, V., Tondre, C., Time-resolved laser-induced fluorescence as a unique tool for low-level uranium speciation, *Appl. Spectrosc.*, **52**, (1998), 528-535. Cited on pages: 165, 612, 613.

[98NAK/NIS] Nakagawa, T., Nishimaki, K., Urabe, T., Katsura, M., Thermodynamic study on α-U_2N_{3+x} using N-rich starting materials ($x \geq 0.6$), *J. Alloys Compd.*, **271-273**, (1998), 658-661. Cited on pages: 236, 613.

[98NEC/FAN] Neck, V., Fanghänel, T., Kim, J. I., Aquatic Chemistry and thermodynamic modeling of trivalent actinides, *Wiss. Ber. - Forschungszent. Karlsruhe*, (1998), 1-108, (in German). Cited on pages: 335, 365, 366, 367, 368, 369, 370, 375, 382, 480, 521, 614, 636, 637.

[98NEC/KON] Neck, V., Könnecke, T., Fanghänel, T., Kim, J. I., Activity coefficients and Pitzer parameters for the fission product ions Cs^+ and TcO_4^- in the system $Cs^+/$-$Na^+/$-$K^+/$-$Mg^{2+}/$-$Cl^-/$-$SO_4^{2-}/$-$TcO_4^-/$-H_2O at 25°C, *Radiochim. Acta*, **83**, (1998), 75-80. Cited on pages: 615, 760.

[98NEC/KON2]   Neck, V., Könnecke, T., Fanghänel, T., Kim, J. I., Pitzer parameters for the pertechnetate ion in the system Na^+/-K^+/-Mg^{2+}/-Ca^{2+}/-Cl^-/-SO_4^{2-}/-TcO_4^-/-H_2O at 25°C, *J. Solution Chem.*, **27**, (1998), 107-120. Cited on pages: 615, 760.

[98OGA/KOB]   Ogawa, T., Kobayashi, F., Sato, T., Haire, R. G., Actinide nitrides and nitride-halides in high-temperature systems, *J. Alloys Compd.*, **271-173**, (1998), 347-354. Cited on pages: 237, 324.

[98PAN/CAM]   Pan, P., Campbell, A. B., The characterization of $Np_2O_5(c)$ and its dissolution in CO_2-free aqueous solutions at pH 6-13 and 25°C, *Radiochim. Acta*, **81**, (1998), 73-82. Cited on pages: 321, 585, 616.

[98PIA/TOU]   Pialoux, A., Touzelin, B., Etude du système U-Ca-O par diffractométrie de rayons X à haute température, *J. Nucl. Mater.*, **255**, (1998), 14-25. Cited on pages: 259, 616, 617, 650.

[98RAI/FEL]   Rai, D., Felmy, A. R., Hess, N. J., Moore, D. A., Yui, M., A thermodynamic model for the solubility of UO_2(am) in the aqueous K^+-Na^+-HCO_3^--CO_3^{2-}-OH^--H_2O system, *Radiochim. Acta*, **82**, (1998), 17-25. Cited on pages: 184, 250, 251, 326, 327, 449, 504, 617, 637, 647, 650, 651, 733.

[98SAI/CHO]   Saito, A., Choppin, G. R., Complexation of uranyl(VI) with polyoxometalates in aqeous solutions, *J. Alloys Compd.*, **271-273**, (1998), 751-755. Cited on pages: 281, 619.

[98SAV]   Savenko, V. S., Stability constants of $M(OH)_3^0$ hydroxo complexes from the solubility products of the corresponding crystalline hydroxides, *Russ. J. Inorg. Chem.*, **43**, (1998), 461-463. Cited on pages: 188, 620.

[98SCA/ANS]   Scapolan, S., Ansoborlo, E., Moulin, C., Madic, C., Investigations by time-resolved laser-induced fluorescence and capillary electrophoresis of the uranyl-phosphate species: application to blood serum, *J. Alloys Compd.*, **271-273**, (1998), 106-111. Cited on pages: 239, 620.

[98SER/RON] Serrano, J. A., Rondinella, V. V., Glatz, J. P., Toscano, E. H., Quiñones, J., Díaz Arocas, P., García-Serrano, J., Comparison of the leaching behaviour of irradiated fuel, SIMFUEL, and non-irradiated UO_2 under oxic conditions, *Radiochim. Acta*, **82**, (1998), 33-37. Cited on pages: 165, 621.

[98SHI] Shilov, V. P., Probable forms of actinides in alkali solutions, *Radiochemistry (Moscow)*, **40**, (1998), 11-16. Cited on pages: 621, 622, 686.

[98SPA/PUI] Spahiu, K., Puigdomènech, I., On weak complex formation: re-interpretation of literature data on the Np and Pu nitrate complexation, *Radiochim. Acta*, **82**, (1998), 413-419. Cited on pages: 301, 623.

[98SUZ/ARA] Suzuki, Y., Arai, Y., Thermophysical and thermodynamic properties of actinide mononitrides and their solid solutions, *J. Alloys Compd.*, **271-273**, (1998), 577-582. Cited on page: 237.

[98WER/SPA] Werme, L. O., Spahiu, K., Direct disposal of spent nuclear fuel: comparison between experimental and modelled actinide solubilities in natural waters, *J. Alloys Compd.*, **271-273**, (1998), 194-200. Cited on pages: 165, 624.

[98YAM/KIT] Yamamura, T., Kitamura, A., Fukui, A., Nishikawa, S., Yamamoto, T., Moriyama, H., Solubility of U(VI) in highly basic solutions, *Radiochim. Acta*, **83**, (1998), 139-146. Cited on pages: 165, 167, 168, 177, 178, 181, 245, 247, 321, 594, 609, 624, 625, 626, 661, 662.

[98YOO/CYN] Yoo, C., Cynn, H., Söderlind, P., Phase diagram of uranium at high pressures and temperatures, *Phys. Rev. B*, **57**, (1998), 10359-10362. Cited on pages: 157, 626.

[99AAS/STE] Aas, W., Steinle, E., Fanghänel, T., Kim, J. I., Thermodynamics of Cm(III) in concentrated electrolyte solutions. Fluoride complexation in 0-5 m NaCl at 25°C, *Radiochim. Acta*, **84**, (1999), 85-88. Cited on pages: 354, 355, 614, 627, 637.

[99BAS/MIS]   Basu, M., Mishra, R., Bharadwaj, S. R., Namboodiri, P. N., Tripathi, S. N., Kerkar, A. S., Dharwadkar, S. R., The standard molar enthalpies of formation of $UTeO_5$ and UTe_3O_9, *J. Chem. Thermodyn.*, **31**, (1999), 1259-1263. Cited on pages: 234, 235, 597, 611, 627.

[99BOU/BIL]   Bouby, M., Billard, I., Bonnenfant, A., Klein, G., Are the changes in the lifetime of the excited uranyl ion of chemical or physical nature?, *Chem. Phys.*, **241**, (1999), 353-370. Cited on pages: 160, 537, 628, 695.

[99CAP/COL]   Capone, F., Colle, Y., Hiernaut, J. P., Ronchi, C., Mass spectrometric measurement of the ionization energies and cross sections of uranium and plutonium oxide vapors, *J. Phys. Chem. A*, **103**, (1999), 10899-10906. Cited on pages: 161, 629.

[99CAP/VIT]   Capdevila, H., Vitorge, P., Redox potentials of M(VI)/M(V) limiting carbonate complexes (M = U or Pu) at different ionic strengths and temperatures. Entropy and heat capacity, *Czech. J. Phys.*, **49**, (1999), 603-609. Cited on pages: 250, 629.

[99CHA/DON]   Chartier, D., Donnet, L., Adnet, J. M., Evidence for a new composition of Am(IV) complexes with tungstophosphate (α_2-$P_2W_{17}O_{61}^{10-}$) and tungstosilicate (α-$SiW_{11}O_{39}^{8-}$) ligands in nitric acid medium, *Radiochim. Acta*, **85**, (1999), 25-31. Cited on pages: 385, 386, 387, 578, 630, 631.

[99CHE/EWI]   Chen, F., Ewing, R. C., Clark, S. B., The Gibbs free energies and enthalpies of formation of U^{6+} phases: An empirical method of prediction, *Am. Mineral.*, **84**, (1999), 650-664. Cited on pages: 254, 285, 440, 442.

[99CHE/EWI2] Chen, F., Ewing, R. C., Clark, S. B., Erratum of The Gibbs free energies and enthalpies of formation of U^{6+} phases: An empirical method of prediction, *Am. Mineral.*, **84**, (1999), 1208. Cited on pages: 254, 285.

[99CHO]   Choppin, G. R., Utility of oxidation state analogs in the study of plutonium behavior, *Radiochim. Acta*, **85**, (1999), 89-95. Cited on pages: 322, 631.

[99CLA/CON]   Clark, D. L., Conradson, S. D., Donohoe, R. J., Keogh, D. W., Morris, D. E., Palmer, P. D., Rogers, R. D., Tait, C. D., Chemical specification of the uranyl ion under highly alkaline conditions. Synthesis, structures, and oxo ligand exchange dynamics, *Inorg. Chem.*, **38**, (1999), 1456-1466. Cited on pages: 177, 626, 631, 661, 672, 674, 676.

[99COR/BOO]   Cordfunke, E. H. P., Booij, A. S., Huntelaar, M. E., The standard enthalpies of formation of uranium compounds. XV. Strontium uranates, *J. Chem. Thermodyn.*, **31**, (1999), 1337-1345. Cited on pages: 261, 262, 263, 264, 265, 398, 426, 467, 468, 631, 634, 666.

[99DOC/MOS]   Docrat, T. I., Mosselmans, J. F. W., Charnock, J. M., Whiteley, M. W., Collison, D., Livens, F. R., Jones, C., Edmiston, M. J., X-ray absorption of tricarbonatodioxouranate(V), $UO_2(CO_3)_3^{5-}$ in aqueous solution, *Inorg. Chem.*, **38**, (1999), 1879-1882. Cited on pages: 160, 635.

[99FAN/KON]   Fanghänel, T., Könnecke, T., Weger, H., Paviet-Hartmann, P., Neck, V., Kim, J. I., Thermodynamics of Cm(III) in concentrated salt solutions: Carbonate complexation in 0-6 m NaCl at 25°C, *J. Solution Chem.*, **28**, (1999), 447-462. Cited on pages: 369, 370, 371, 372, 373, 374, 375, 480, 585, 587, 614, 635, 636, 755.

[99FEL/RAI]   Felmy, A. R., Rai, D., Application of Pitzer's Equation for modeling the aqueous thermodynamics of actinide species in natural waters: A review, *J. Solution Chem.*, **28**, (1999), 533-553. Cited on pages: 636, 637, 706.

[99FIR]   Firestone, R. B., *Table of isotopes, Eight edition, Update*, Baglin, C. M., Franck Chu, S. Y., Eds., John Wiley & Sons, (1999). Cited on pages: 195, 336.

[99GRA/MUL]   Grambow, B., Müller, R., Solids, colloids and hydrolysis of U(VI) in chloride media, INE Forschungszentrum Karlsruhe, (1999), private communication, (from the citation in [99NEC/KIM]). Cited on page: 646.

[99HAS/WAN] Hashizume, K., Wang, W. E., Olander, D. R., Volatilization of urania in steam at elevated temperatures, *J. Nucl. Mater.*, **275**, (1999), 277-286. Cited on pages: 163, 637.

[99HRN/IRL] Hrnecek, E., Irlweck, K., Formation of uranium(VI) complexes with monomeric and polymeric species of silicic acid, *Radiochim. Acta*, **87**, (1999), 29-35. Cited on pages: 252, 253, 254, 637.

[99JAY/IYE] Jayanthi, K., Iyer, V. S., Rama Rao, G. A., Venugopal, V., Molar Gibbs energy formation of $KUO_3(s)$, *J. Nucl. Mater.*, **264**, (1999), 263-270. Cited on pages: 274, 275, 512, 546, 592, 638.

[99KAS/RUN] Kaszuba, J. P., Runde, W., The aqueous geochemistry of neptunium: dynamic control of soluble concentrations with applications to nuclear waste disposal, *Environ. Sci. Technol.*, **33**, (1999), 4427-4433. Cited on pages: 294, 296, 298, 639.

[99KNO/NEC] Knopp, R., Neck, V., Kim, J. I., Solubility, hydrolysis and colloid formation of plutonium(IV), *Radiochim. Acta*, **435**, (1999), 1-8. Cited on pages: 313, 314, 315, 316, 317, 642, 687, 692, 695, 706.

[99MEI/KAT] Meinrath, G., Kato, Y., Kimura, T., Yoshida, Z., Comparative analysis of actinide(VI) carbonate complexation by Monte Carlo resampling methods, *Radiochim. Acta*, **84**, (1999), 21-29. Cited on pages: 247, 643.

[99MEI/VOL] Meinrath, G., Volke, P., Helling, C., Dudel, E. G., Merkel, B. J., Determination and interpretation of environmental water samples contaminated by uranium mining activities, *J. Anal. Chem.*, **364**, (1999), 191-202. Cited on pages: 165, 644.

[99MIK/RUM] Mikheev, N. B., Rumer, I. A., Stabilization of the divalent state for the lanthanides and actinides in solutions, melts and clusters, *Radiochim. Acta*, **85**, (1999), 49-55. Cited on pages: 159, 644.

[99MOL/DEN] Moll, H., Denecke, M. A., Jalilehvand, F., Sandström, M., Grenthe, I., Structure of the aqua ions and fluoride complexes of uranium(IV) and thorium(IV) in aqueous solution an EXAFS study, *Inorg. Chem.*, **38**, (1999), 1795-1799. Cited on page: 160.

[99NAK/ARA] Nakajima, K., Arai, Y., Suzuki, Y., Vaporization behavior of NpN coloaded with PuN, *J. Nucl. Mater.*, **275**, (1999), 332-335. Cited on pages: 300, 644, 645.

[99NAK/ARA2] Nakajima, K., Arai, Y., Suzuki, Y., Yamawaki, M., Vaporization behavior of $BaPuO_3$, *J. Mass Spectrom. Soc. Japan*, **47**, (1999), 46-48. Cited on pages: 331, 645.

[99NEC/KIM] Neck, V., Kim, J. I., Solubility and hydrolysis of tetravalent actinides, Forschungszentrum Karlsruhe, Report FZKA 6350, (1999), 40 pp. Cited on pages: 183, 186, 295, 297, 298, 313, 314, 316, 319, 334, 579, 640, 642, 646, 647, 648, 649, 691, 692.

[99OLA] Olander, D. R., Thermodynamics of urania volatilization in steam, *J. Nucl. Mater.*, **270**, (1999), 187-193. Cited on pages: 163, 637, 649, 650.

[99PIA/TOU] Pialoux, A., Touzelin, B., Etude du système U-Sr-O par diffraction X à haute température, *Can. J. Chem.*, **77**, (1999), 1384-1393, in French. Cited on pages: 259, 616, 650.

[99RAI/HES] Rai, D., Hess, N. J., Felmy, A. R., Moore, D. A., Yui, M., A thermodynamic model for the solubility of NpO_2(am) in the aqueous K^+-HCO_3^--CO_3^{2-}-OH^--H_2O system, *Radiochim. Acta*, **84**, (1999), 159-169. Cited on pages: 250, 302, 303, 304, 326, 327, 637, 640, 650, 688, 689.

[99RAI/HES2] Rai, D., Hess, N. J., Felmy, A. R., Moore, D. A., Yui, M., Vitorge, P., A thermodynamic model for the solubility of PuO_2(am) in the aqueous K^+-HCO_3-CO_3^{2-}-OH^--H_2O system, *Radiochim. Acta*, **86**, (1999), 89-99. Cited on pages: 313, 317, 326, 327, 328, 329, 585, 637, 651, 652, 688, 692.

[99RAO/RAI] Rao, L., Rai, D., Felmy, A. R., Novak, C. F., Solubility of $NaNd(CO_3)_2 \cdot 6H_2O$(c) in mixed electrolyte (Na-Cl-CO_3-HCO_3) and synthetic brine solutions, Proc. Am. Chem. Soc. Symp. Exp. Model Stud. Actinide Speciation Non-ideal Syst., pp. 153-169, Plenum Publisher, (1999). Cited on page: 383.

[99RAR/RAN]   Rard, J. A., Rand, M. H., Anderegg, G., Wanner, H., *Chemical Thermodynamics of Technetium*, Sandino, M. C. A., Östhols, E., Nuclear Energy Agency Data Bank, Organisation for Economic Co-operation, Development, Eds., vol. 3, *Chemical Thermodynamics*, North Holland Elsevier Science Publishers B. V., Amsterdam, The Netherlands, (1999). Cited on pages: xi, 4, 5, 6, 39, 125, 126, 128, 133, 134, 135, 389, 390, 391, 478, 503, 521, 576, 671, 735, 753, 762.

[99RUN/REI]   Runde, W., Reilly, S. D., Neu, M. P., Spectroscopic investigation of the formation of PuO_2Cl^+ and PuO_2Cl_2 in NaCl solutions and application for natural brine solutions, *Geochim. Cosmochim. Acta*, **63**, (1999), 3443-3449. Cited on pages: 323, 469, 652, 653, 654, 655, 656, 679.

[99RUT/GEI]   Rutsch, M., Geipel, G., Brendler, V., Bernhard, G., Nitsche, H., Interaction of uranium(VI) with arsenate(V) in aqueous solution studied by time-resolved laser-induced fluorescence spectroscopy (TRLFS), *Radiochim. Acta*, **86**, (1999), 135-141. Cited on pages: 243, 561, 656.

[99SAX/RAM]   Saxena, M. K., Ramakumar, K. L., Venugopal, V., Heat capacity measurements on rubidium uranium sulphate (RBS), $Rb_2U(SO_4)_3$, by differential scanning calorimetry (DSC), NUCAR 99, Mumbai, India, 19-22 Jan., pp. 393-394, Bhabha Atomic Research Centre, Mumbai, India, (1999). Cited on pages: 277, 454, 657.

[99SIN/DAS]   Singh, Z., Dash, S., Krishnan, K., Prasad, R., Venugopal, V., Standard molar Gibbs energy of formation of $UTeO_5(s)$ by the electrochemical method, *J. Chem. Thermodyn.*, **31**, (1999), 197-204. Cited on pages: 235, 595, 596, 610, 658, 666.

[99SOD/ANT]   Soderholm, L., Antonio, M. R., Williams, C., Wasserman, S. R., XANES spectroelectrochemistry: A new method for determining formal potentials, *Anal. Chem.*, **71**, (1999), 4622-4628. Cited on pages: 293, 659.

[99SOW/CLA]   Sowder, A. G., Clark, S. B., Field, R. A., The transformation of uranyl oxide hydrates: The effects of dehydration on synthetic metaschoepite and its alteration to becquerelite, *Environ. Sci. Technol.*, **33**, (1999), 2552-3557. Cited on pages: 189, 588.

[99SUZ/TAM]   Suzuki, S., Tamura, K., Tachimori, S., Usui, Y., Solvent extraction of technetium(VII) by cyclic amides, *J. Radioanal. Nucl. Chem.*, **239**, (1999), 377-380. Cited on pages: 390, 659.

[99VAL/MAR]   Vallet, V., Maron, L., Schimmelpfennig, B., Leininger, T., Teichteil, C., Gropen, O., Grenthe, I., Wahlgren, U., Reduction behavior of the early actinyl ions in aqueous solution, *J. Phys. Chem. B*, **103**, (1999), 9285-9289. Cited on pages: 159, 580, 660.

[99WAH/MOL]   Wahlgren, U., Moll, H., Grenthe, I., Schimmelpfennig, B., Maron, L., Vallet, V., Gropen, O., Structure of uranium(VI) in strong alkaline solutions: a combined theoretical and experimental investigation, *J. Phys. Chem. A*, **103**, (1999), 8257-8264. Cited on pages: 160, 177, 626, 631, 661, 662, 672.

[99WAN/OST]   Wanner, H., Östhols, E., TDB-3: Guidelines for the assignment of uncertainties, Nuclear Energy Agency Data Bank, Organisation for Economic Co-operation and Development, Report, (1999), 17 pp. Cited on page: 6.

[2000ALL/BUC]   Allen, P. G., Bucher, J. J., Shuh, D. K., Edelstein, N. M., Craig, I., Coordination chemistry of trivalent lanthanide and actinide ions in dilute and concentrated chloride solutions, *Inorg. Chem.*, **39**, (2000), 595-601. Cited on pages: 322, 357, 662.

[2000BEN/FAT]   Ben Said, K., Fattahi, M., Musikas, C., Revel, R., Abbé, J. C., The speciation of Tc(IV) in chloride solutions, *Radiochim. Acta*, **88**, (2000), 567-571. Cited on pages: 391, 662.

[2000BRU/CER]   Bruno, J., Cera, E., Grivé,M., de Pablo, J., Sillen, P., Duro, L., Determination and uncertainties of radioelement solubility limits to be used by SKB in the SR 97 performance assessment exercise, *Radiochim. Acta*, **88**, (2000), 823-828. Cited on pages: 165, 663.

[2000BUN/KNO]   Bundschuh, T., Knopp, R., Müller, R., Kim, J. I., Neck, V., Fanghänel, T., Application of LIBD to the determination of the solubility product of thorium(IV)-colloids, *Radiochim. Acta*, **88**, (2000), 625-629. Cited on pages: 320, 321, 663.

[2000BUR/OLS] Burns, P. C., Olson, R. A., Finch, R. J., Hanchar, J. M., Thibault, Y., $KNa_3(UO_2)_2(Si_4O_{10})_2(H_2O)_4$, a new compound formed during vapor hydration of an actinide-bearing borosilicate waste glass, J. Nucl. Mater., **278**, (2000), 290-300. Cited on pages: 165, 664.

[2000COM/BRO] Comarmond, M. J., Brown, P. L., The hydrolysis of uranium(VI) in sulphate media, Radiochim. Acta, **88**, (2000), 573-577. Cited on pages: 178, 230, 231, 232, 233, 453, 664, 698.

[2000DAS/SIN] Dash, S., Singh, Z., Prasad, R., Venugopal, V., Calorimetric studies on the strontium-uranium-oxygen system, J. Nucl. Mater., **279**, (2000), 84-90. Cited on pages: 263, 264, 665.

[2000DAV/FOU] David, F., Fourest, B., Hubert, S., Purans, J., Vokhmin, V., Madic, C., Thermodynamic properties of Pu^{3+} and Pu^{4+} aquo ions, Plutonium Futures - The Science, Conference Transactions, Santa Fe, NM, USA, 10-13 July, p. 388, (2000). Cited on page: 581.

[2000FIN] Fink, J. K., Thermophysical properties of uranium dioxide, J. Nucl. Mater., **279**, (2000), 1-18. Cited on pages: 193, 666, 667.

[2000GOR/LUM] Gorshkov, N. I., Lumpov, A. A., Miroslavov, A. E., Suglobov, D. N., Synthesis of $[Tc(CO_3)_3(H_2O)_3]^+$ ion and study of its reaction with hydroxyl ion in aqueous solutions, Radiochemistry (Moscow), **42**, (2000), 231. Cited on page: 668.

[2000GOR/MIR] Gorshkov, N. I., Miroslavov, A. E., Lumpov, A. A., Suglobov, D. N., Sukhov, V. Yu., Complexation of $Tc(CO)_3(H_2O)_3^+$ with simple ligands in aqueous solutions and some data on its pharmarokinetics, NRC5 5th International Conference on Nuclear and Radiochemistry, Pontresina, Switzerland, 3-8 September, pp. 765-767, (2000). Cited on pages: 391, 667.

[2000GRE/WAN] Grenthe, I., Wanner, H., Östhols, E., TDB-2: Guidelines for the extrapolation to zero ionic strength, OECD Nuclear Energy Agency, Data Bank, Report, (2000), 34 pp. Cited on pages: 6, 441, 708, 714.

[2000HAS/ALL] Haschke, J. M., Allen, T. H., Morales, L. A., Reaction of plutonium dioxide with water: formation and properties of PuO_{2+x}, Science, **287**, (2000), 285-287. Cited on pages: 315, 668, 669.

[2000HAY/MAR] Hay, P. J., Martin, R. L., Schreckenbach, G., Theoretical studies of the properties and solution chemistry of AnO_2^{2+} and AnO_2^+ aquo complexes for An = U, Np, and Pu, *J. Phys. Chem. A*, **104**, (2000), 6259-6270. Cited on pages: 159, 669.

[2000KAL/CHO] Kalmykov, S. N., Choppin, G. R., Mixed $Ca^{2+}/UO_2^{2+}/CO_3^{2-}$ complex formation at different ionic strengths, *Radiochim. Acta*, **88**, (2000), 603-606. Cited on pages: 248, 249, 669, 670.

[2000KON/CLA] Konze, W. V., Clark, D. L., Conradson, S. D., Donohoe, R. J., Gordon, J. C., Gordon, P., Keogh, D. W., Morris, D. E., Tait, C. D., Identification of oligomeric uranyl complexes under highly alkaline conditions, Plutonium Futures - The Science, Conference Transactions, Santa Fe, NM, USA, 10-13 July, p. 261, (2000). Cited on pages: 178, 670.

[2000LAA/KON] van der Laan, R. R., Konings, R. J. M., The heat capacity of $Tc_{0.85}Ru_{0.15}$ alloy, *J. Alloys Compd.*, **297**, (2000), 104-108. Cited on pages: 389, 530, 671.

[2000LAN] Landolt-Bornstein Tables, *Thermodynamic properties of inorganic materials*, vol. 19, *New series, Group IV: Physical chemistry*, No. 3, compounds from $CoCl_3$ to Ge_3N_4, Springer Verlag, Berlin, (2000). Cited on page: 204.

[2000MEI/EKB] Meinrath, G., Ekberg, C., Landgren, A., Liljenzin, J. O., Assessment of uncertainty in parameter evaluation and prediction, *Talanta*, **51**, (2000), 231-246. Cited on page: 644.

[2000MEI/HUR] Meinrath, G., Hurst, S., Gatzweiler, R., Aggravation of licensing procedures by doubtful thermodynamic data, *J. Anal. Chem.*, **368**, (2000), 561-566. Cited on page: 644.

[2000MEI] Meinrath, G., Robust spectral analysis by moving block bootstrap designs, *Radiochim. Acta*, **415**, (2000), 105-115. Cited on page: 644.

[2000MEI2] Meinrath, G., Computer-intensive methods for uncertainty estimation on complex situations, *Chemom. Intell. Lab. Syst.*, **51**, (2000), 175-187. Cited on page: 644.

[2000MOL/REI] Moll, H., Reich, T., Szabó,Z., The hydrolysis of dioxouranium(VI) investigated using EXAFS and ^{17}O-NMR, *Radiochim. Acta*, **88**, (2000), 411-415. Cited on pages: 177, 178, 453, 672, 698, 702.

[2000MOL/REI2] Moll, H., Reich, T., Henning, C., Rossberg, A., Szabó,Z., Grenthe, I., Solution coordination chemistry of uranium in the binary UO_2^{2+}-SO_4^{2-} and the ternary UO_2^{2+}-SO_4^{2-}-OH^- system, *Radiochim. Acta*, **88**, (2000), 559-566. Cited on pages: 231, 232, 233, 502, 626, 664, 665, 672, 676.

[2000NEC/KIM] Neck, V., Kim, J. I., An electrostatic approach for the prediction of actinide complexation constants with inorganic ligands-application to carbonate complexes, *Radiochim. Acta*, **88**, (2000), 815-822. Cited on page: 673.

[2000NGU/BUR] Nguyen-Trung, C., Burneau, A., Quilès, F., Palmer, D. A., Détermination des spectres d'absorption UV-visible des hydroxydes d'uranyle(VI) individuels en présence de deux électrolytes non complexants, $NaClO_4$ et $(CH_3)_4NCF_3SO_3$, dans un large domaine de pH (0-14) par chimiométrie à 25°C, Journées Practis 1999, Villeneuve-lès-Avignon, France, 17-18 février, (2000). Cited on pages: 178, 551, 673.

[2000NGU/PAL] Nguyen-Trung, C., Palmer, D. A., Begun, G. M., Peiffert, C., Mesmer, R. E., Aqueous uranyl complexes 1. Raman spectroscopic study of the hydrolysis of uranyl(VI) in solutions of trifluoromethanesulfonic acid and/or tetramethylammonium hydroxide at 25°C and 0.1 MPa, *J. Solution Chem.*, **29**, (2000), 101-129. Cited on pages: 166, 175, 177, 661, 674, 675, 676, 677, 701, 702, 703.

[2000OST/WAN] Östhols, E., Wanner, H., TDB-0: The NEA Thermochemical Data Base Project, Nuclear Energy Agency Data Bank, Organisation for Economic Co-operation and Development, Report, (2000), 21 pp. Cited on page: 6.

[2000PER/CAS] Pérez, I., Casas, I., Martín, M., Bruno, J., The thermodynamics and kinetics of uranophane dissolution in bicarbonate test solutions, *Geochim. Cosmochim. Acta*, **64**, (2000), 603-608. Cited on pages: 256, 677, 678.

[2001VAL/WAH] Vallet, V., Wahlgren, U., Schimmelpfennig, B., Moll, H., Szabó, Z., Grenthe, I., Solvent Effects on Uranium(VI) fluoride and hydroxide complexes studied by EXAFS and Quantum Chemistry, *Inorg. Chem.*, **40**, (2001), 3516-3525. Cited on pages: 177, 451, 631, 696.

[2001WIL/BLA] Williams, C. W., Blaudeau, J. P., Sullivan, J. C., Antonio, M. R., Bursten, B., Soderholm, L., The coordination geometry of Np(VII) in alkaline solution, *J. Am. Chem. Soc.*, **13**, (2001), 4346-4347. Cited on pages: 293, 532, 686, 697.

[2001YAM/KUR] Yamanaka, S., Kurosaki, K., Matsuda, T., Uno, M., Thermophysical properties of $BaUO_3$, *J. Nucl. Mater.*, **294**, (2001), 99-103. Cited on page: 267.

[2002BRO] Brown, P. L., The hydrolysis of uranium(VI), *Radiochim. Acta*, **90**, (2002), 589-593. Cited on pages: 165, 167, 173, 174, 176, 178, 589, 697.

[2002GER/GRI] German, K. E., Grigoriev, M. S., Synthesis, structure and properties of zirconium(IV) and uranyl pertechnetate and perrhenate, The third Russian-Japanese seminar on technetium, Dubna, Russia, (2002). Cited on pages: 390, 700.

[2002GER/KOD] German, K. G., Kodina, G. E., Sekine, T., *The third Russian-Japanese seminar on technetium*, Dubna, Russia, (2002). Cited on pages: 390, 700.

[2002GRI/GER] Grigoriev, M. S., German, K. E., Den Auwer, C., Dancausse, J. P., Masson, M., Simonoff, M., Structure and properties of tetrapropylammonium pertechnetate and perrhenate, The third Russian-Japanese seminar on technetium, Dubna, Russia, (2002). Cited on pages: 700, 701.

[2002HAS/ALL] Haschke, J. M., Allen, T. H., Equilibrium and thermodynamic properties of the PuO_{2+x} solid solution, *J. Alloys Compd.*, **336**, (2002), 124-131. Cited on page: 315.

[2002KIR/GER] Kirakosyan, G. A., German, K. E., Tarasov, V. P., Grigoriev, M. S., The structure and properties of pertechnetic acid in solid state and solutions as probed be NMR, The third Russian-Japanese seminar on technetium, Dubna, Russia, (2002). Cited on pages: 390, 700.

[2002MAS/PER] Maslennikov, A. G., Peretrukhin, V. F., Silin, V. I., Kareta, A. V., Sladkov, V. E., David, F., Fourest, B., Courson, O., Masson, M., Lecomte, M., Delegard, C., Electrochemical reactions for technetium recovery and analysis in radwaste solutions, The third Russian-Japanese seminar on technetium, Dubna, Russia, (2002). Cited on pages: 390, 701.

[2002NGU] Nguyen-Trung, C., Etude expérimentale de l'hydrolyse en solution aqueuse des ions uranyle(VI) en présence d'électrolytes non complexants par spectroscopie de vibration à 25°C, 0.1 MP, Ph. D. Thesis, Université Henri Poincaré, (2002), Nancy 1, France. Cited on pages: 178, 701, 703.

[2002RAI/FEL] Rai, D., Felmy, A. R., Hess, N. J., LeGore, V. L., McCready, D. E., Thermodynamics of the U(VI)-Ca^{2+}-Cl^--OH^--H_2O system: solubility product of becquerelite, *Radiochim. Acta*, **90**, (2002), 495-503. Cited on pages: 165, 176, 178, 180, 284, 422, 704, 705.

[2002RAI/GOR] Rai, D., Gorby, Y. A., Fredrickson, J. K., Moore, D. A., Yui, M., Reductive dissolution of PuO_2(am): The effect of Fe(II) and hydroquinone, *J. Solution Chem.*, **31**, (2002), 433-453. Cited on pages: 316, 317, 705, 706.

[2002SHI/MIN] Shirasu, Y., Minato, K., Heat capacities of technetium metal and technetium-ruthenium alloy, *J. Alloys Compd.*, **337**, (2002), 243-247. Cited on pages: 389, 390.

[2002VAL/WAH] Vallet, V., Wahlgren, U., Szabó, Z., Grenthe, I., Rates and mechanism of fluoride and water exchange in $UO_2F_5^{3-}$ and $[UO_2F_4(H_2O)]^{2-}$ studied by NMR spectroscopy and wave function based methods, *Inorg. Chem.*, **41**, (2002), 5626-5633. Cited on page: 209.

[2003RAI/YUI] Rai, D. and Yui, M. and Moore, D. A., Solubility and solubility product at 22°C of UO_2(c) precipitated from aqueous U(VI) solutions, *J. Solution Chem.*, **32**, (2003), 1-17. Cited on page: 186.

List of cited authors

This chapter contains an alphabetical list of the authors of the references cited in this book. The reference codes given with each name corresponds to the publications of which the person is the author or a co-author. Note that inconsistencies may occur due to variations in spelling between different publications. No attempt has been made to correct for such inconsistencies in this volume.

Authors list

Author	Reference
Aalto, T.	[69SAL/HAK]
Aas, W.	[97SZA/AAS], [98AAS/MOU], [99AAS/STE]
Abbé, J. C.	[98BEN/FAT], [2000BEN/FAT], [2001BEN/SEI]
Abbrent, M.	[65VES/PEK]
Abe, R.	[90PAR/SAK]
Aberg, M.	[69ABE], [70ABE]
Ackermann, R. J.	[69ACK/RAU], [76OET/RAN]
Adamson, M. G.	[93KRI/EBB], [97KRI/FON]
Adnet, J. M.	[92ADN/MAD], [98CHA/DON], [99CHA/DON]
Afanas'ev, Yu. A.	[75STO/KHA]
Agafonov, I. L.	[53AGA/AGA]
Agalonova, A. L.	[53AGA/AGA]
Aguilar, M.	[91AGU/CAS]
Ahonen, L.	[94AHO/ERV]
Ahrland, S.	[71AHR/KUL]
Akhachinskij, V. V.	[81CHI/AKH]
Akishin, P. A.	[61AKI/KHO]
Al Mahamid, I.	[97NOV/ALM], [98ALM/NOV], [98CON/ALM]
Alcock, C. B.	[93KUB/ALC]
Aldridge, J. P.	[85ALD/BRO]
Alekseeva, T. E.	[74RAB/ALE]
Alexander, C. A.	[87ALE/OGD]
Alikhanyan, A. S.	[90SEV/ALI]
Aling, P.	[63COR/ALI]
Allard, B.	[78ALL/BEA], [84ALL/OLO], [97ALL/BAN]

(Continued on next page)

Authors list (continued)

Author	Reference
Allen, P. G.	[95ALL/BUC], [96ALL/SHU], [97ALL/BUC], [97ALL/SHU], [2000ALL/BUC]
Allen, T. H.	[2000HAS/ALL], [2001HAS/ALL], [2002HAS/ALL]
Altman, D.	[43ALT/LIP], [43ALT]
Alwan, A. K.	[80ALW/WIL], [80ALW]
Amayri, S.	[97AMA/GEI], [98AMA/BER], [2001BER/GEI]
Ananthaswamy, J.	[84ANA/ATK]
Anderegg, G.	[99RAR/RAN]
Anderson, G. M.	[91AND/CAS]
Anderson, O. E.	[42JEN/AND]
Andrews, Jr., J. E.	[92NGU/SIL]
Ansara, I.	[81CHI/AKH]
Ansoborlo, E.	[97SCA/ANS], [98SCA/ANS]
Antonio, M. R.	[99SOD/ANT], [2001WIL/BLA]
Apelblat, A.	[75APE/SAH], [98APE/KOR]
Arai, Y.	[97NAK/ARA], [98SUZ/ARA], [99NAK/ARA], [99NAK/ARA2]
Arblaster, J. W.	[95ARB]
Archer, D. G.	[90ARC/WAN]
Arlès, L.	[93BOI/ARL]
Armstrong, A. Q.	[97ELL/ARM]
Asahina, K.	[97ASA/SUG]
Ashida, T.	[95PAS/KIM]
Atkins, M.	[88ATK/BEC]
Atkinson, G.	[84ANA/ATK]
Avogadro, A.	[86AVO/BIL]
Aziz, A.	[68AZI/LYL], [69AZI/LYL]
Babko, A. K.	[60BAB/KOD]
Baes, Jr., C. F.	[54SCH/BAE], [56BAE], [62BAE/MEY], [65BAE/MEY], [76BAE/MES]
Bailey, S. M.	[82WAG/EVA]
Balakhovskii, O. A.	[75SPI/ZIN]
Baldas, J.	[98BAL/HEA]
Balducci, G.	[72GUI/GIG], [91GUI/BAL]
Bamberger, C. E.	[96SHO/BAM]
Banar, J. C.	[98EFU/RUN]
Banerjee, A.	[98BAN/SAL], [2001BAN/PRA]
Bansal, B. M. L.	[64BAN/PAT]
Banwart, S. A.	[97ALL/BAN]
Banyai, I.	[95BAN/GLA]
Baraj, E.	[97TOR/BAR]
Barbanel', Yu. A.	[69BAR/MIK]
Bard, A. J.	[85BAR/PAR]

(Continued on next page)

Authors list (continued)

Author	Reference
Bardin, N.	[98BAR/RUB]
Barinov, V. I.	[81GOL/TRE], [81GOL/TRE2]
Barrett, S. A.	[82BAR/JAC]
Bartušek, M.	[63BAR/SOM]
Bastasz, R.	[90COS/LAK]
Basu, M.	[99BAS/MIS]
Bates, R. G.	[73BAT]
Batyreva, V. A.	[90KUM/BAT]
Beall, G. W.	[78ALL/BEA]
Beck, M. T.	[90BEC/NAG]
Beckley, A.	[88ATK/BEC]
Becraft, K.	[96ROB/SIL]
Becraft, K. A.	[97NOV/ALM], [98ALM/NOV]
Begun, G. M.	[92NGU/BEG], [2000NGU/PAL]
Ben Said, K.	[98BEN/FAT], [2000BEN/FAT], [2001BEN/SEI]
Benson, L. V.	[80BEN/TEA]
Benson, S. W.	[92HIS/BEN]
Bergman, G. A.	[81GLU/GUR], [82GLU/GUR]
Berhard, G.	[97RUT/GEI]
Bermudez Polonio, J.	[62PER/BER]
Bernhard, G.	[95MOL/MAT], [96BER/GEI], [96BRE/GEI], [96GEI/BRA], [96MOL/GEI], [97AMA/GEI], [97BER/GEI], [97GEI/BER], [97GEI/RUT], [97MOL/GEI], [98AMA/BER], [98GEI/BER], [98GEI/BER2], [98GEI/BER3], [98GEI/BER4], [98MOL/GEI], [99RUT/GEI], [2000REI/BER], [2001BER/GEI]
Bernkopf, M. F.	[84BER/KIM]
Berthet, J. C.	[98DAV/FOU]
Besmann, T. M.	[81LIN/BES]
Bessiere, J.	[87LOU/BES]
Betz, H.	[97SHO/SAS2]
Bevington, P. R.	[69BEV]
Bevz, A. S.	[74VIS/VOL]
Bharadwaj, S. R.	[98MIS/NAM], [99BAS/MIS]
Bhupathy, M.	[79PRA/NAG], [80NAG/BHU]
Bidoglio, G.	[82BID], [89GRE/BID], [91BID/CAV], [95ELI/BID], [95SIL/BID]
Biedermann, G.	[75BIE]
Billard, I.	[99BOU/BIL], [2001BIL/RUS], [2001SEM/BOE]
Billon, A.	[86AVO/BIL]
Binnewies, M.	[74BIN/SCH]
Bion, L.	[95BIO/MOI], [97BIO/MOI]
Bischoff, H.	[92KRA/BIS]

(Continued on next page)

Authors list (continued)

Author	Reference
Blaise, J.	[92BLA/WYA]
Blankenship, F. F.	[60LAN/BLA]
Blaudeau, J. P.	[2001WIL/BLA]
Blomqvist, R.	[94AHO/ERV]
Boehme, C.	[2001SEM/BOE]
Boivineau, M.	[93BOI/ARL]
Bolvin, H.	[2001BOL/WAH], [2001BOL/WAH2]
Bondarenko, A. A.	[87BON/KOR]
Bonnenfant, A.	[99BOU/BIL]
Bonnetot, B.	[76BOU/BON]
Booij, A. S.	[95HAA/MAR], [96KON/BOO], [96KOV/BOO], [97COR/BOO], [99COR/BOO]
Bouby, M.	[99BOU/BIL]
Boucharat, N.	[97BOU]
Bourges, J.	[92ADN/MAD]
Bousquet, J.	[76BOU/BON]
Boyd, G. E.	[78BOY]
Boyer, G. D.	[93LEM/BOY]
Brücher, E.	[91BRU/GLA]
Brachmann, A.	[96GEI/BRA], [97GEI/BER]
Bradley, D. J.	[79BRA/PIT]
Brand, J. R.	[70BRA/COB]
Brandenburg, N. P.	[78BRA]
Bratsch, S. G.	[85BRA/LAG], [86BRA/LAG]
Brendler, E.	[95MOL/MAT]
Brendler, V.	[96BER/GEI], [96BRE/GEI], [96GEI/BRA], [97BER/GEI], [97GEI/BER], [97MOL/GEI], [97RUT/GEI], [98GEI/BER], [98GEI/BER2], [98GEI/BER3], [98MOL/GEI], [99RUT/GEI], [2001BER/GEI]
Brewer, L.	[84BRE]
Brittain, R. D.	[85LAU/BRI], [91HIL/LAU]
Brock, E. G.	[85ALD/BRO]
Broul, M.	[81BRO/NYV]
Brown, D.	[85TSO/BRO]
Brown, P. L.	[2000COM/BRO], [2002BRO]
Browne, J. C.	[60SAV/BRO]
Bruno, J.	[85BRU/GRE], [86BRU/FER], [86BRU], [87BRU/CAS], [89BRU/SAN], [90BRU/GRE], [91CAR/BRU], [92SAN/BRU], [93ERI/NDA], [94CAS/BRU], [94OST/BRU], [95PUI/BRU], [97ALL/BAN], [97BRU/CAS], [97CAS/BRU], [97PER/CAS], [98CAS/PAB], [2000BRU/CER], [2000PER/CAS]

(Continued on next page)

Authors list (continued)

Author	Reference
Brusilovskii, S. A.	[58BRU]
Brønsted, J. N.	[22BRO], [22BRO2]
Bucher, J. J.	[83EDE/BUC], [95ALL/BUC], [96ALL/SHU], [97ALL/BUC], [97ALL/SHU], [2000ALL/BUC]
Buckau, G.	[94CZE/BUC]
Budantseva, N. A.	[98BUD/TAN]
Buklanov, V.	[90ROS/REI]
Bundschuh, T.	[2000BUN/KNO]
Burleigh, M. C.	[98DAI/BUR]
Burneau, A.	[2000NGU/BUR], [2000QUI/BUR]
Burns, P. C.	[96BUR/MIL], [2000BUR/OLS]
Bursten, B.	[2001WIL/BLA]
Butler, T.	[58JOH/BUT]
Byrne, R. H.	[93LEE/BYR], [96BYR/SHO]
Caceci, M. S.	[83CAC/CHO], [83CAC/CHO2], [93ERI/NDA]
Caja, J.	[69CAJ/PRA]
Campbell, A. B.	[93LEM/BOY], [95PAN/CAM], [98PAN/CAM]
Capdevila, H.	[90CAP/VIT], [92CAP], [93GIF/VIT2], [94CAP/VIT], [95CAP/VIT], [96CAP/VIT], [98CAP/VIT], [99CAP/VIT]
Capone, F.	[99CAP/COL]
Carnall, W. T.	[85MAG/CAR], [91CAR/LIU], [94LIU/CAR]
Carniglia, S. C.	[55CAR/CUN]
Carpenter, S. A.	[97NOV/ALM], [98ALM/NOV]
Carroll, S. A.	[91CAR/BRU], [93CAR]
Casas, I.	[87BRU/CAS], [91AGU/CAS], [94CAS/BRU], [94TOR/CAS], [97BRU/CAS], [97CAS/BRU], [97PER/CAS], [97TOR/BAR], [98CAS/PAB], [2000PER/CAS]
Casteel, W. J.	[98CAS/DIX]
Castet, S.	[91AND/CAS]
Cathelineau, M.	[87DUB/RAM]
Cavalli, P.	[91BID/CAV]
Cera, E.	[94CAS/BRU], [97BRU/CAS], [97CAS/BRU], [97PER/CAS], [98CAS/PAB], [2000BRU/CER]
Chander, K.	[82NAI/CHA]
Chandrasekharaiah, M. S.	[69ACK/RAU], [74DHA/TRI]
Chandratillake, M.	[98CHA/TRI]
Charnock, J. M.	[99DOC/MOS]
Charoy, B.	[87DUB/RAM]
Chartier, D.	[98CHA/DON], [99CHA/DON]
Chasanov, M. G.	[81FIN/CHA]
Chase, Jr., M. W.	[98CHA]
Chatt, A.	[89CHA/RAO], [91RAO/CHA]

(Continued on next page)

Authors list (continued)

Author	Reference
De Carvalho, R. G.	[67CAR/CHO]
De Grave, E.	[84VOC/GRA], [86VOC/GRA]
De Pablo, J.	[91AGU/CAS], [94TOR/CAS], [97TOR/BAR], [98CAS/PAB], [2000BRU/CER]
Decaillon, J. G.	[95MOU/DEC]
Decambox, P.	[95MOU/DEC]
Degischer, G.	[75DEG/CHO]
Delegard, C.	[2002MAS/PER]
Delegard, C. H.	[98BUD/TAN]
Delmau, L.	[96CAP/VIT]
Delorme-Hiver, A.	[98BEN/FAT]
Demchenko, E. A.	[97DEM/SER]
Den Auwer, C.	[98DAV/FOU], [2002GRI/GER]
Desai, P. D.	[73HUL/DES]
Devina, O. A.	[89GUR/DEV], [94SER/DEV]
Dharwadkar, S. R.	[74DHA/TRI], [98MIS/NAM], [99BAS/MIS]
Dhooge, N. J.	[95NOV/CRA]
Diakonov, I. I.	[98DIA/RAG], [98DIA/TAG]
Dickens, P. G.	[85DIC/LAW], [86DIC/PEN], [88DIC/POW], [89DIC/LAW], [92DUE/FLE], [93PAT/DUE], [95DUE/PAT]
Dinsdale, A. T.	[90TAY/DIN], [91DIN]
Dixon, D. A.	[98CAS/DIX]
Docrat, T. I.	[99DOC/MOS]
Donnet, L.	[98CHA/DON], [99CHA/DON]
Donohoe, R. J.	[99CLA/CON], [2000KON/CLA]
Douglass, R. M.	[61WAT/DOU]
Downer, B.	[72TAY/KEL]
Drever, J. I.	[82DRE]
Dreyer, R.	[90ROS/REI]
Drobnic, M.	[66DRO/KOL]
Drowart, J.	[66DRO/PAT], [68PAT/DRO], [84GRO/DRO]
Du, M.	[92CHO/DU]
Dubessy, J.	[87DUB/RAM]
Ducros, M.	[92DUC/SAN]
Dudel, E. G.	[99MEI/VOL]
Dueber, R. E.	[92DUE/FLE], [93PAT/DUE], [95DUE/PAT]
Dunaeva, K. M.	[71SAN/VID], [72SAN/VID]
Dunsmore, H. S.	[63DUN/SIL]
Duplessis, J.	[77DUP/GUI], [85DAV/FOU]
Duro, L.	[97BRU/CAS], [97PER/CAS], [2000BRU/CER]
Dworkin, A. S.	[72DWO]
Díaz Arocas, P.	[96DIA/GAR], [98DIA/GRA]
Denecke, M. A.	[99MOL/DEN]

(Continued on next page)

Authors list (continued)

Author	Reference
Earnshaw, A.	[84GRE/EAR]
Ebbinghaus, B. B.	[93KRI/EBB], [95EBB], [97KRI/FON]
Edelstein, N. M.	[83EDE/BUC], [85NIT/EDE], [95ALL/BUC], [96ALL/SHU], [97ALL/BUC], [97ALL/SHU], [2000ALL/BUC]
Edmiston, M. J.	[99DOC/MOS]
Efimov, A. I.	[56SHC/VAS2]
Efimov, M. E.	[89GUR/DEV]
Efurd, D. W.	[98EFU/RUN]
Ekberg, C.	[2000MEI/EKB]
Ekberg, S. A.	[95ALL/BUC], [96CLA/CON]
Eliet, V.	[95ELI/BID]
Eller, P. G.	[89MOR/ELL]
Elless, M. P.	[97ELL/ARM]
Eom, T. Y.	[94KIM/CHO]
Ephraim, J. H.	[97ALL/BAN]
Ephritikine, M.	[98DAV/FOU]
Eriksen, T. E.	[93ERI/NDA]
Erin, E. A.	[98ERI/BAR]
Ervanne, H.	[94AHO/ERV]
Evans, W. H.	[52ROS/WAG], [82WAG/EVA]
Ewart, F. T.	[86EWA/HOW], [87CRO/EWA]
Ewing, R. C.	[91FIN/EWI], [92FIN/MIL], [94CAS/BRU], [96BUR/MIL], [96FIN/COO], [97CAS/BRU], [98CAS/PAB], [98CLA/EWI], [98FIN/HAW], [99CHE/EWI], [99CHE/EWI2]
Falck, W. E.	[91FAL/HOO], [96FAL/REA]
Fang, D.	[69KEL/FAN]
Fanghänel, T.	[94FAN/KIM], [95FAN/KIM], [95FAN/NEC], [95NEC/FAN], [96FAN/NEC], [96PAV/FAN], [97KON/FAN], [97KON/NEC], [97NEC/FAN], [97PAS/CZE], [97STE/FAN], [98FAN/KIM], [98FAN/WEG], [98FAN/WEG2], [98NEC/FAN], [98NEC/KON], [98NEC/KON2], [99AAS/STE], [99FAN/KON], [2000BUN/KNO], [2001BOL/WAH]
Fattahi, M.	[98BEN/FAT], [2000BEN/FAT], [2000VIC/FAT], [2001BEN/SEI]
Fazekas, Z.	[98FAZ/YAM]
Feay, D. C.	[54FEA]
Fedoseev, A. M.	[98BUD/TAN]

(Continued on next page)

Authors list (continued)

Author	Reference
Felmy, A. R.	[86FEL/WEA], [89FEL/RAI], [90FEL/RAI], [90RAI/FEL], [91FEL/RAI], [95FEL/RAI], [95RAI/FEL], [95RAI/FEL2], [96RAO/RAI], [96RAO/RAI2], [97FEL/RAI], [97RAI/FEL], [98RAI/FEL], [99FEL/RAI], [99RAI/HES], [99RAI/HES2], [99RAO/RAI], [2001RAI/MOO], [2002RAI/FEL]
Fender, B. E. F.	[73GRE/CHE], [82BAR/JAC]
Ferguson, W. J.	[44FER/PRA], [44FER/PRA2], [45FER/RAN]
Fernandez Cellini, R.	[62PER/BER]
Fernelius, W. C.	[77FER]
Ferri, D.	[79CIA/FER], [81CIA/FER], [83FER/GRE], [85FER/GRE], [86BRU/FER], [93FER/SAL]
Field, R. A.	[99SOW/CLA]
Filip, H.	[85ALD/BRO]
Finch, R. J.	[91FIN/EWI], [92FIN/MIL], [94CAS/BRU], [96FIN/COO], [97CAS/BRU], [97FIN], [98CAS/PAB], [98FIN/HAW], [2000BUR/OLS]
Fink, J. K.	[81FIN/CHA], [2000FIN]
Firestone, R. B.	[99FIR]
Fischer, S.	[98MEI/FIS]
Fleetwood, J. M.	[92DUE/FLE]
Fleming, D. L.	[93KRI/EBB]
Flicker, H.	[85ALD/BRO]
Flotow, H. E.	[81OHA/FLO], [82COR/MUI]
Flowers, G. C.	[81HEL/KIR]
Fontes, Jr. A. S.	[93KRI/EBB], [97KRI/FON]
Fourest, B.	[85DAV/FOU], [97DAV/FOU], [98DAV/FOU], [2000DAV/FOU], [2002MAS/PER]
Fox, A. C.	[62RAN/FOX]
Fox, K.	[85ALD/BRO]
Fred, M.	[75FRE]
Fredrickson, J. K.	[2002RAI/GOR]
Freeman, R. D.	[84FRE]
Frydman, M.	[58FRY/NIL]
Fuger, J.	[72FUG], [76FUG/OET], [83FUG/PAR], [83FUG], [83MOR/WIL2], [84FUG/HAI], [85FUG], [85TSO/BRO], [90FUG/HAI], [90GOU/HAI], [92FUG/KHO], [92FUG], [92GRE/FUG], [93FUG/HAI], [94MER/FUG], [96MER/FUG], [97MER/LAM], [2000RAN/FUG], [2001LEM/FUG]

(Continued on next page)

Authors list (continued)

Author	Reference
Fujii, T.	[2001FUJ/YAM]
Fujino, T.	[79TAG/FUJ], [89YAM/FUJ], [91FUJ/YAM], [93TAK/FUJ], [2001FUJ/PAR], [2001FUJ/SAT]
Fujiwara, K.	[2001FUJ/YAM]
Fukuda, K.	[2001FUJ/SAT]
Fukui, A.	[98YAM/KIT]
Fulton, R. W.	[90FEL/RAI], [95FEL/RAI], [95RAI/FEL2], [96RAO/RAI]
Funke, H.	[2000REI/BER]
Güldner, R.	[82WEI/WIS]
Gal'chenko, G. L.	[59POP/GAL]
Galbraith, H. W.	[85ALD/BRO]
Galkin, N. P.	[60STE/GAL]
Gancheff, J.	[97KRE/GAN]
Gar'kin, V. P.	[97DEM/SER]
García-Serrano, J.	[96DIA/GAR], [98SER/RON]
Garrels, R. M.	[62HOS/GAR], [74TAR/GAR], [76TAR/GAR]
Garvin, D.	[87GAR/PAR]
Gatzweiler, R.	[2000MEI/HUR]
Gavrichev, K. S.	[89GUR/DEV], [97GUR/SER]
Gayer, K. H.	[55GAY/LEI]
Geckeis, H.	[96DIA/GAR]
Geipel, G.	[96BER/GEI], [96BRE/GEI], [96GEI/BRA], [96MOL/GEI], [97AMA/GEI], [97BER/GEI], [97GEI/BER], [97GEI/RUT], [97MOL/GEI], [97RUT/GEI], [98GEI/BER], [98GEI/BER2], [98GEI/BER3], [98GEI/BER4], [98MOL/GEI], [99RUT/GEI], [2000REI/BER], [2001BER/GEI], [2001BOL/WAH]
Gel'd, P. V.	[75SPI/ZIN]
Gel'man, A. D.	[62GEL/MOS], [67GEL/MOS]
Gens, R.	[83MOR/WIL2]
German, K. E.	[2002GER/GRI], [2002GRI/GER], [2002KIR/GER]
German, K. G.	[2002GER/KOD]
Gervais, T. L.	[98KAP/GER]
Giffaut, E.	[93GIF/VIT], [93GIF/VIT2], [94GIF], [96CAP/VIT]
Gigli, G.	[72GUI/GIG]
Giménez, J.	[97TOR/BAR], [98CAS/PAB]
Ginnings, D. C.	[47GIN/COR]
Girichev, G. V.	[96KON/BOO]
Giricheva, N. I.	[96KON/BOO]
Giridhar, J.	[91GIR/LAN]
Givon, M.	[69SHI/GIV]

(Continued on next page)

Authors list (continued)

Author	Reference
Glaser, J.	[91BRU/GLA], [93FER/SAL], [95BAN/GLA]
Glasser, F. P.	[88ATK/BEC], [95MOR/GLA]
Glatz, J. P.	[98SER/RON]
Gleiser, M.	[73HUL/DES]
Glushko, V. P.	[81GLU/GUR], [82GLU/GUR]
Godhes, J. W.	[95ALL/BUC]
Goeminne, A.	[81VOC/PIR], [84VOC/GOE]
Gol'tsev, V. P.	[81GOL/TRE], [81GOL/TRE2]
Goldberg, R. N.	[79GOL]
González, O.	[97KRE/GAN]
Goodenough, J. B.	[70GOO/LON]
Gorbunov, V. E.	[89GUR/DEV], [97GUR/SER]
Gorby, Y. A.	[2002RAI/GOR]
Gordon, J. C.	[2000KON/CLA]
Gordon, P.	[2000KON/CLA]
Gordon, S.	[93MCB/GOR]
Gorokhov, L. N.	[84GOR/SMI], [84SMI/GOR], [98GOR/SID]
Gorshkov, N. I.	[2000GOR/LUM], [2000GOR/MIR]
Goudiakas, J.	[90GOU/HAI]
Grady, H. F.	[45WAG/GRA], [58YOU/GRA]
Grambow, B.	[94SAN/GRA], [94TOR/CAS], [96DIA/GAR], [98DIA/GRA], [99GRA/MUL], [2000VIC/FAT]
Grauer, R.	[97ALL/BAN]
Greaves, C.	[73GRE/CHE]
Greenberg, A.	[89LIE/GRE]
Greenwood, N. N.	[84GRE/EAR]
Gregory, N. W.	[46GRE], [46GRE2], [46GRE3]
Grenthe, I.	[69GRE/VAR], [81CIA/FER], [83FER/GRE], [85BRU/GRE], [85FER/GRE], [86BRU/FER], [86GRE/ROB], [87RIG/VIT], [89GRE/BID], [90BRU/GRE], [91BID/CAV], [92GRE/FUG], [93FER/SAL], [93GRE/LAG], [94OST/BRU], [95ELI/BID], [95GRE/PUI], [95PAS/KIM], [97ALL/BAN], [97GRE/PLY], [97PUI/RAR], [97SZA/AAS], [98AAS/MOU], [99MOL/DEN], [99VAL/MAR], [99WAH/MOL], [2000GRE/WAN], [2000MOL/REI2], [2000SZA/MOL], [2001BOL/WAH], [2001VAL/WAH], [2002VAL/WAH]
Grigoriev, M. S.	[2002GER/GRI], [2002GRI/GER], [2002KIR/GER]
Grimaldi, M.	[79CIA/FER]
Grimvall, G.	[89GUI/GRI]
Grivé, M.	[2000BRU/CER]
Gropen, O.	[99VAL/MAR], [99WAH/MOL], [2001BOL/WAH2]
Gruber, J. B.	[73GRU/HEC]

(Continued on next page)

Authors list (continued)

Author	Reference
Gruen, D. M.	[69GRU/MCB]
Gryzin, Yu. I.	[67GRY/KOR]
Grønvold, F.	[84GRO/DRO]
Guggenheim, E. A.	[35GUG], [66GUG]
Guido, M.	[72GUI/GIG], [91GUI/BAL]
Guillaumont, R.	[72MET/GUI], [77DUP/GUI], [97ION/MAD]
Guillermet, A. F.	[89GUI/GRI]
Gurevich, V. M.	[89GUR/DEV], [97GUR/SER]
Gurvich, L. V.	[81GLU/GUR], [82GLU/GUR], [85HIL/GUR]
Haacke, D. F.	[79HAA/WIL]
Haaland, A.	[95HAA/MAR]
Hadermann, J.	[97ALL/BAN]
Haga, Y.	[94OCH/SUZ]
Haire, R. G.	[84FUG/HAI], [90FUG/HAI], [90GOU/HAI], [93FUG/HAI], [96SHO/BAM], [98OGA/KOB]
Hakala, R.	[69SAL/HAK]
Hakem, N.	[97NOV/ALM], [98ALM/NOV]
Hale, F. V.	[88PHI/HAL]
Hall, R. O. A.	[90HAL/JEF2], [91HAL/HAR], [91HAL/MOR]
Halloway, J. H.	[85TSO/BRO]
Halow, I.	[82WAG/EVA]
Hamann, S. D.	[82HAM]
Hanchar, J. M.	[2000BUR/OLS]
Harada, M.	[90PAR/SAK]
Harding, S. R.	[91HAL/HAR], [91HAL/MOR]
Harvie, C. E.	[84HAR/MOL]
Haschke, J. M.	[2000HAS/ALL], [2001HAS/ALL], [2002HAS/ALL]
Hase, H.	[98KIT/YAM]
Hashizume, K.	[99HAS/WAN]
Hassan, Refat M.	[92HAS2]
Hauser, W.	[94FAN/KIM], [94KIM/KLE], [2001NEC/KIM2]
Hauske, H.	[76WEI/WIS]
Havel, J.	[97LUB/HAV], [97LUB/HAV2]
Hawkins, D. T.	[73HUL/DES]
Hawthorne, F. C.	[96FIN/COO], [98FIN/HAW]
Hay, P. J.	[2000HAY/MAR]
Hayes, S. L.	[90HAY/THO]
Hearne, J. A.	[57HEA/WHI]
Heath, G. A.	[98BAL/HEA]
Heaven, M. C.	[94KAL/MCC], [97KAL/HEA]
Hecht, H. G.	[73GRU/HEC]
Hedlund, T.	[88HED]

(Continued on next page)

Authors list (continued)

Author	Reference
Helgeson, H. C.	[74HEL/KIR], [81HEL/KIR], [88SHO/HEL], [88TAN/HEL], [89SHO/HEL], [89SHO/HEL2], [90OEL/HEL], [92SHO/OEL]
Helling, C.	[99MEI/VOL]
Hemingway, B. S.	[82HEM]
Henning, C.	[2000MOL/REI2], [2000REI/BER], [2001SEM/BOE]
Henrion, P.	[86AVO/BIL]
Henry, M.	[94HEN/MER]
Hepler, L. G.	[75OLO/HEP], [88HOV/HEP3], [89HOV/HEP]
Hess, N. J.	[96CLA/CON], [97FEL/RAI], [98CON/ALM], [98RAI/FEL], [99RAI/HES], [99RAI/HES2], [2002RAI/FEL]
Hiernaut, J. P.	[99CAP/COL]
Hietanen, S.	[56HIE], [85FER/GRE]
Higashi, K.	[90PRA/MOR]
Hildenbrand, D. L.	[75HIL/CUB], [77HIL], [79KLE/HIL], [82LAU/HIL], [84LAU/HIL], [85HIL/GUR], [85LAU/BRI], [87LAU/HIL], [91HIL/LAU], [91HIL/LAU2], [96HIL/LAU], [98KON/HIL]
Hill, R.	[92LIE/HIL]
Hinatsu, Y.	[93HIN]
Hindman, J. C.	[52COH/HIN]
Hirono, S.	[65MUT/HIR]
Hisham, M. W. M.	[92HIS/BEN]
Hobart, D. E.	[95CLA/HOB], [96ROB/SIL]
Hobbs, D. T.	[95HOB/KAR]
Hoekstra, H. R.	[73HOE/SIE], [74OHA/HOE3], [75OHA/HOE2], [81OHA/FLO]
Hoffman, D. C.	[2000STO/HOF]
Holland, R. F.	[85ALD/BRO]
Homann, K.	[88MIL/CVI]
Hooker, P. J.	[86AVO/BIL], [91FAL/HOO]
Hostetler, P. B.	[62HOS/GAR]
Hostettler, J. D.	[84HOS]
Hotta, E.	[94OCH/SUZ]
Houee-Levin, C.	[2001BEN/SEI]
Hovey, J. K.	[86HOV/TRE], [88HOV/HEP3], [89HOV/HEP], [89HOV/NGU], [97HOV]
Howse, R. M.	[86EWA/HOW]
Hrnecek, E.	[99HRN/IRL], [2000WAD/HRN]
Huang, J.	[96YAM/HUA], [97HUA/YAM], [98HUA/YAM]
Hubbard, W. N.	[83FUG/PAR]

Authors list (continued)

Author	Reference
Hubert, S.	[98DAV/FOU], [2000DAV/FOU]
Hudson, E. A.	[95ALL/BUC], [96ALL/SHU], [98CON/ALM]
Hultgren, R.	[73HUL/DES]
Hummel, W.	[97ALL/BAN]
Humphreys, P. N.	[98CHA/TRI]
Huntelaar, M. E.	[99COR/BOO]
Hurst, H. J.	[71TAY/HUR]
Hurst, S.	[2000MEI/HUR]
Högfeldt, E.	[82HOG]
Ijdo, D. J. W.	[91COR/VLA], [94COR/IJD]
Ionova, G.	[97ION/MAD], [98DAV/FOU], [2001DAV/VOK]
Ioussov, A.	[97IOU/KRU]
Ippolitova, E. A.	[71SAN/VID], [72SAN/VID]
Irlweck, K.	[99HRN/IRL], [2000WAD/HRN]
Ishiguro, S.	[74KAK/ISH]
Ishii, T.	[90PAR/SAK]
Ito, H.	[98ITO/YAM]
Iuliano, M.	[87CIA/IUL]
Ivanov, I. P.	[89RED/SAV], [89SER/SAV]
Iyer, V. S.	[87VEN/IYE], [90IYE/VEN], [92VEN/IYE], [93JAY/IYE], [94JAY/IYE], [95SAL/JAY], [96IYE/JAY], [96JAY/IYE], [96SAL/IYE], [97IYE/JAY], [97JAY/IYE], [98JAY/IYE], [99JAY/IYE]
Jackson, E. E.	[63JAC/RAN]
Jacobson, A. J.	[82BAR/JAC]
Jaffe, I.	[52ROS/WAG]
Jakob, A.	[97ALL/BAN]
Jalilehvand, F.	[99MOL/DEN]
Janecky, D. R.	[98EFU/RUN]
Jayadevan, N. C.	[93SAL/KUL], [94SAL/KUL]
Jayanthi, K.	[92VEN/IYE], [93JAY/IYE], [94JAY/IYE], [95SAL/JAY], [96IYE/JAY], [96JAY/IYE], [96SAL/IYE], [97IYE/JAY], [97JAY/IYE], [98JAY/IYE], [99JAY/IYE]
Jeffery, A. J.	[90HAL/JEF2]
Jenkins, F. A.	[42JEN/AND]
Jensen, B. S.	[86AVO/BIL]
Jensen, K. A.	[71JEN]
Jensen, M. P.	[98JEN/CHO], [2001NEC/KIM2]
Johnson, C. E.	[81LIN/BES]
Johnson, G. K.	[88TAS/OHA]
Johnson, J. S.	[63RUS/JOH]

(Continued on next page)

Authors list (continued)

Author	Reference
Johnson, J. W.	[92SHO/OEL]
Johnson, K. S.	[79JOH/PYT]
Johnson, O.	[58JOH/BUT]
Johnsson, K. O.	[47JOH]
Jones, C.	[99DOC/MOS]
Jones, M. M.	[53JON]
Jordan, J.	[85BAR/PAR]
Joshi, J. K.	[82NAI/CHA]
Joubert, L.	[2001JOU/MAL]
Judge, A. I.	[85TSO/BRO]
Jursich, G.	[94LIU/CAR]
Köhncke, K.	[98MEI/FIS]
Könnecke, T.	[97KON/FAN], [97KON/NEC], [98FAN/WEG2], [98NEC/KON], [98NEC/KON2], [99FAN/KON]
Kaakola, T.	[94AHO/ERV]
Kaganyuk, D. S.	[83KAG/KYS]
Kakihana, H.	[74KAK/ISH]
Kaledin, L. A.	[94KAL/MCC], [97KAL/HEA]
Kallay, N.	[88MIL/CVI]
Kalmykov, S. N.	[2000KAL/CHO]
Kaltsoyannis, N.	[95ALL/BUC]
Kanellakopulos, B.	[89KIM/KAN], [91KIM/KLE], [92NEC/KIM], [94NEC/KIM], [94NEC/RUN]
Kangro, W.	[63KAN]
Kaplan, D. I.	[98KAP/GER]
Kapshukov, I. I.	[74VIS/VOL], [74VOL/KAP2], [77VOL/VIS], [79VOL/VIS2]
Karapiperis, T.	[97ALL/BAN]
Karasev, N. M.	[87BON/KOR]
Kareta, A. V.	[2002MAS/PER]
Karkhanavala, M. D.	[74DHA/TRI]
Karpov, V. I.	[61KAR]
Karraker, D. G.	[95HOB/KAR]
Kasha, M.	[49KAS]
Kaszuba, J. P.	[98EFU/RUN], [99KAS/RUN]
Kato, Y.	[93MEI/KAT], [94KAT/MEI], [95FAN/KIM], [95KAT/KIM], [96KAT/KIM], [96MEI/KAT], [98MEI/KAT], [99MEI/KAT]
Katsura, M.	[98NAK/NIS]
Katz, J. J.	[51KAT/RAB]
Kawamura, Y.	[95YAJ/KAW]
Kazin, I. V.	[83KAG/KYS]
Keenan, T. K.	[64KEE/KRU]
Keller, C.	[69KEL/FAN]

(Continued on next page)

Authors list (continued)

Author	Reference
Kelley, K. K.	[73HUL/DES]
Kelly, E. J.	[98CHA/TRI]
Kelly, J. W.	[72TAY/KEL]
Kelm, H.	[74SCH/KEL]
Kemmler-Sack, S.	[75BRA/KEM]
Kennedy, C. M.	[88KEN/MIK]
Keogh, D. W.	[98CLA/CON], [99CLA/CON], [2000KON/CLA]
Kerkar, A. S.	[99BAS/MIS]
Kerko, P. F.	[81GOL/TRE], [81GOL/TRE2]
Khachkuruzov, G. A.	[81GLU/GUR], [82GLU/GUR]
Khalkin, V. A.	[90ROS/REI]
Khanaev, E.	[75STO/KHA]
Khodakovsky, I. L.	[71NAU/RYZ], [72NIK/SER], [89GUR/DEV], [89RED/SAV], [89SER/SAV], [92FUG/KHO], [92KHO], [94SER/DEV], [97GUR/SER]
Khodeev, Yu. S.	[61AKI/KHO], [84GOR/SMI]
Khopkar, P. K.	[80KHO/MAT]
Kielland, J.	[37KIE]
Kim, J. I.	[84BER/KIM], [85KIM], [85KIM2], [85MAG/CAR], [86AVO/BIL], [86LIE/KIM], [88STA/KIM], [88STA/KIM2], [89KIM/KAN], [91KIM/KLE], [91MEI/KIM], [91MEI/KIM2], [92NEC/KIM], [92RUN/MEI], [92WIM/KLE], [93PAS/RUN], [94CZE/BUC], [94FAN/KIM], [94KIM/KLE], [94NEC/KIM], [94NEC/RUN], [94RUN/KIM], [95FAN/KIM], [95FAN/NEC], [95NEC/FAN], [95NEC/RUN], [95PAS/KIM], [96FAN/NEC], [96MEI/KLE], [96PAV/FAN], [97KON/FAN], [97KON/NEC], [97NEC/FAN], [97PAS/CZE], [98FAN/KIM], [98FAN/WEG], [98FAN/WEG2], [98NEC/FAN], [98NEC/KON], [98NEC/KON2], [99AAS/STE], [99FAN/KON], [99KNO/NEC], [99NEC/KIM], [2000BUN/KNO], [2000NEC/KIM], [2001NEC/KIM], [2001NEC/KIM2]
Kim, J. J.	[74PIT/KIM]
Kim, K. C.	[85ALD/BRO]
Kim, W. H.	[94KIM/CHO], [96PAR/PYO]

(Continued on next page)

Authors list (continued)

Author	Reference
Kimura, T.	[92KIM/SER], [93MEI/KIM], [93MEI/KIM2], [94KAT/MEI], [95KAT/KIM], [96KAT/KIM], [96MEI/KAT], [98MEI/KAT], [99MEI/KAT]
Kirakosyan, G. A.	[2002KIR/GER]
Kirdyashev, V. P.	[56SHC/VAS2]
Kirkham, D. H.	[74HEL/KIR], [81HEL/KIR]
Kitamura, A.	[98KIT/YAM], [98YAM/KIT], [2001KIT/KOH]
Klein, G.	[99BOU/BIL]
Kleinschmidt, P. D.	[79KLE/HIL]
Klenze, R.	[91KIM/KLE], [92WIM/KLE], [94FAN/KIM], [94KIM/KLE], [95FAN/KIM], [96MEI/KLE], [96PAV/FAN], [97STE/FAN]
Kline, R. J.	[60RAB/KLI]
Knacke, O.	[69KNA/LOS]
Knopp, R.	[99KNO/NEC], [2000BUN/KNO]
Kobayashi, F.	[98OGA/KOB]
Kobets, L. V.	[78KOB/KOL]
Koch, C. W.	[54KOC/CUN]
Kochergin, S. M.	[89PAZ/KOC]
Kodenskaya, V. S.	[60BAB/KOD]
Kodina, G. E.	[2002GER/KOD]
Kohara, Y.	[2001KIT/KOH]
Kohli, R.	[83KOH]
Kojima, T.	[90PAR/SAK]
Kok-Scheele, A.	[96KOV/BOO]
Kolar, D.	[66DRO/KOL]
Kolevich, T. A.	[78KOB/KOL]
Konijn, P. C.	[79LIN/KON]
Konings, R. J. M.	[89COR/KON], [90COR/KON], [90COR/KON2], [91COR/KON], [92GRE/FUG], [95HAA/MAR], [96KON/BOO], [96KOV/BOO], [98KON/HIL], [2000LAA/KON], [2001KON], [2001KON2]
Konze, W. V.	[2000KON/CLA]
Korin, E.	[98APE/KOR]
Korobov, M. V.	[87BON/KOR]
Koryttsev, K. Z.	[67GRY/KOR]
Kostylev, F. A.	[59POP/KOS]
Kovács, A.	[96KON/BOO], [96KOV/BOO]
Kramer-Schnabel, U.	[92KRA/BIS]
Krasnova, O. G.	[96KON/BOO]
Krasovskaya, T. I.	[90SEV/ALI]
Krasser, W.	[69SCH/KRA]
Kraus, K. A.	[50KRA/NEL]

(Continued on next page)

Authors list (continued)

Author	Reference
Kremer, C.	[97KRE/GAN]
Kremer, E.	[97KRE/GAN]
Krikorian, O. H.	[93KRI/EBB], [97KRI/FON]
Krishnan, K.	[97KRI/RAM], [97SAL/KRI], [98KRI/RAM], [99SIN/DAS]
Krohn, B. J.	[85ALD/BRO]
Krupa, J. C.	[97IOU/KRU]
Krupka, K. M.	[98KAP/GER]
Kruse, F. H.	[64KEE/KRU]
Kryukov, P. A.	[67PER/KRY]
Kryutchov, S. V.	[96PER/KRY]
Kubaschewski, O.	[84COR/KUB], [84KUB], [93KUB/ALC]
Kuchitsu, K.	[88MIL/CVI]
Kulkarni, N. K.	[93SAL/KUL], [94SAL/KUL]
Kulkarni, S. G.	[92VEN/KUL], [93NAI/VEN], [94JAY/IYE]
Kullberg, L.	[71AHR/KUL]
Kulyako, Y. M.	[96KUL/MAL]
Kumok, V. N.	[67MER/SKO], [90KUM/BAT]
Kurata, H.	[65MUT/HIR]
Kurosaki, K.	[2001YAM/KUR]
Kuznetova, T. P.	[97GUR/SER]
Kuznetsov, N. T.	[90SEV/ALI]
Kyskin, V. I.	[83KAG/KYS]
Lützenkirchen, K.	[2001BIL/RUS], [2001SEM/BOE]
Laffitte, M.	[82LAF]
Lagerman, B.	[87BRU/CAS], [90BRU/GRE], [93GRE/LAG]
Lagowski, J. J.	[85BRA/LAG], [86BRA/LAG]
Lakner, J. F.	[78LAK], [90COS/LAK]
Lambert, B.	[97MER/LAM]
Landgren, A.	[2000MEI/EKB]
Landolt-Bornstein	[2000LAN]
Langer, S.	[60LAN/BLA]
Langmuir, D.	[76REA/LAN], [78LAN], [91GIR/LAN], [97LAN]
Laszak, I.	[98MOU/LAS]
Latimer, W. M.	[40COU/PIT]
Lau, K. H.	[82LAU/HIL], [84LAU/HIL], [85LAU/BRI], [87LAU/HIL], [91HIL/LAU], [91HIL/LAU2], [96HIL/LAU]
Lawrence, S. D.	[85DIC/LAW], [89DIC/LAW]
Lawson, B. L.	[88KEN/MIK]
Le Du, J. F.	[98DAV/FOU]
LeBlond, N.	[98CAS/DIX]
LeGore, V. L.	[2002RAI/FEL]
Lecomte, M.	[98MAS/MAS], [2002MAS/PER]

(Continued on next page)

Authors list (continued)

Author	Reference
Lee, J. H.	[93LEE/BYR]
Lee, S. Y.	[97ELL/ARM]
Lefort, M.	[63LEF]
Leibowitz, L.	[81FIN/CHA]
Leider, H.	[55GAY/LEI]
Leigh, G. J.	[90LEI]
Leininger, T.	[99VAL/MAR]
Leitnaker, J. M.	[80LEI]
Lemire, R. J.	[80LEM/TRE], [88LEM], [92GRE/FUG], [93LEM/BOY], [2001LEM/FUG]
Leroy, J.	[87DUB/RAM]
Levien, B. J.	[47ROB/LEV]
Levine, S.	[52ROS/WAG]
Levinson, L. S.	[64LEV]
Lewis, B. M.	[88OHA/LEW], [88TAS/OHA]
Lewis, G. N.	[61LEW/RAN]
Liebman, J. F.	[89LIE/GRE]
Lierse, Ch.	[86LIE/KIM]
Lieser, K. H.	[92LIE/HIL]
Lietzke, M.H.	[62LIE/STO]
Liljenzin, J. O.	[2000MEI/EKB]
Lim, C. K.	[51ROB/LIM]
Lindemer, T. B.	[81LIN/BES]
Lingerak, W. A.	[79LIN/KON]
Lipkin, D.	[43ALT/LIP]
Liu, G. K.	[91CAR/LIU], [94LIU/CAR]
Livens, F. R.	[99DOC/MOS]
Longo, J. M.	[70GOO/LON]
Loopstra, B. O.	[66LOO/COR], [67COR/LOO], [71COR/LOO]
Lorimer, J. W.	[95COH/LOR]
Lossmann, G.	[69KNA/LOS]
Louis, C.	[87LOU/BES]
Lubal, P.	[97LUB/HAV], [97LUB/HAV2]
Lukens, W. W.	[95ALL/BUC]
Lumpov, A. A.	[2000GOR/LUM], [2000GOR/MIR]
Lundqvist, R.	[82LUN]
Lyle, S. J.	[68AZI/LYL], [69AZI/LYL]
Müller, F.	[69KNA/LOS]
Müller, R.	[99GRA/MUL], [2000BUN/KNO]
Méot-Reymond, S.	[97BIO/MOI]
MacWood, G. E.	[58MAC]

(Continued on next page)

Authors list (continued)

Author	Reference
Macgregor, S. A.	[98BAL/HEA]
Madic, C.	[92ADN/MAD], [95BIO/MOI], [97BIO/MOI], [97ION/MAD], [97SCA/ANS], [98BAR/RUB], [98DAV/FOU], [98SCA/ANS], [2000DAV/FOU]
Madisson, A. P.	[2000YEH/MAD]
Magirius, S.	[85MAG/CAR]
Magnuson, D. W.	[85ALD/BRO]
Maier II, W. B.	[85ALD/BRO]
Maksin, V. I.	[89STA/MAK]
Maldivi, P.	[2001JOU/MAL]
Malevich, V. M.	[81GOL/TRE2]
Malikov, D. A.	[96KUL/MAL]
Mallika, C.	[86MAL/SRE]
Mann, J. B.	[70MAN]
Marcus, Y.	[64SHI/MAR], [69MAR/SHI], [69SHI/GIV], [85MAR2]
Mariezcurrena, R.	[97KRE/GAN]
Maron, L.	[99VAL/MAR], [99WAH/MOL]
Marquardt, C. M.	[2001NEC/KIM2]
Marquez, L. N.	[96ALL/SHU], [97ALL/SHU]
Marsden, C.	[2001BOL/WAH2]
Marshall, W. L.	[84MAR/MES]
Martín, M.	[2000PER/CAS]
Martell, A. E.	[64SIL/MAR], [71SIL/MAR], [98MAR/SMI]
Martin, R. L.	[2000HAY/MAR]
Martinsen, K.-G.	[95HAA/MAR]
Marx, G.	[92KRA/BIS]
Maslennikov, A. G.	[97MAS/COU], [98MAS/MAS], [2002MAS/PER]
Mason, M. J.	[91FEL/RAI], [95FEL/RAI], [95RAI/FEL], [97FEL/RAI], [97RAI/FEL]
Masson, M.	[97MAS/COU], [98MAS/MAS], [2002GRI/GER], [2002MAS/PER]
Mathur, J. N.	[80KHO/MAT], [91CHO/MAT]
Mathurin, D.	[76BOU/BON]
Matsuda, T.	[2001YAM/KUR]
Matsui, T.	[87MAT/OHS]
Matz, W.	[95MOL/MAT], [96MOL/GEI], [2000REI/BER]
Maya, L.	[82MAY], [83MAY]
Mayorga, G.	[73PIT/MAY], [74PIT/MAY]
McBeth, R. L.	[69GRU/MCB]
McBride, B.J.	[93MCB/GOR]
McCord, J. E.	[94KAL/MCC]
McCready, D. E.	[2002RAI/FEL]
McCue, M. C.	[76MOR/MCC]

(Continued on next page)

Authors list (continued)

Author	Reference
McDowell, R. S.	[85ALD/BRO]
McDowell, W. J.	[72MCD/COL]
Medvedev, V. A.	[81GLU/GUR], [82GLU/GUR], [89COX/WAG], [92FUG/KHO]
Mefod'eva, M. P.	[67GEL/MOS]
Meinrath, G.	[91MEI/KIM], [91MEI/KIM2], [91MEI], [92RUN/MEI], [93MEI/KAT], [93MEI/KIM], [93MEI/KIM2], [93MEI/TAK], [94KAT/MEI], [94MEI], [96MEI/KAT], [96MEI/KLE], [96MEI/SCH], [96MEI], [97MEI/SCH], [97MEI], [97MEI2], [98MEI/FIS], [98MEI/KAT], [98MEI], [98MEI2], [98MEI3], [99MEI/KAT], [99MEI/VOL], [2000MEI/EKB], [2000MEI/HUR], [2000MEI], [2000MEI2]
Merceron, T.	[94HEN/MER]
Mercier, H. P. A.	[98CAS/DIX]
Merkel, B. J.	[97MEI/SCH], [99MEI/VOL]
Merkusheva, S. A.	[67MER/SKO]
Merli, L.	[94MER/FUG], [96MER/FUG], [97MER/LAM]
Mesmer, R. E.	[76BAE/MES], [84MAR/MES], [91AND/CAS], [98DAI/BUR], [2000NGU/PAL]
Metivier, H.	[72MET/GUI], [73MET]
Metz	[61WAT/DOU]
Meyer, N. J.	[62BAE/MEY], [65BAE/MEY]
Micskei, K.	[95BAN/GLA]
Mikalko, V.	[98DAV/FOU]
Mikelsons, M. V.	[88KEN/MIK]
Mikhailova, N. K.	[69BAR/MIK]
Mikheev, N. B.	[99MIK/RUM]
Milanov, M.	[90ROS/REI]
Milkey, R. G.	[54MIL]
Miller, A. J.	[45WAG/GRA]
Miller, D. G.	[82RAR/MIL], [91RAR/MIL]
Miller, M. L.	[92FIN/MIL], [96BUR/MIL]
Millero, F. J.	[79MIL]
Mills, I.	[88MIL/CVI]
Minato, K.	[2002SHI/MIN]
Miroslavov, A. E.	[2000GOR/LUM], [2000GOR/MIR]
Mishra, R.	[98MIS/NAM], [99BAS/MIS]
Miyairi, M.	[97OMO/MIY]
Mizuguchi, K.	[93MIZ/PAR]
Mochizuki, A.	[74MOC/NAG]

(Continued on next page)

Authors list (continued)

Author	Reference
Moisy, P.	[95BIO/MOI], [97BIO/MOI]
Moll, H.	[95MOL/MAT], [96MOL/GEI], [97MOL/GEI], [98MOL/GEI], [99MOL/DEN], [99WAH/MOL], [2000MOL/REI], [2000MOL/REI2], [2000SZA/MOL], [2001BOL/WAH], [2001VAL/WAH]
Moller, N.	[84HAR/MOL]
Mombrú, A. W.	[97KRE/GAN]
Monnin, C.	[90MON]
Montecates R. J.	[98MAR/SMI]
Moock, K. H.	[98BAL/HEA]
Moore, D. A.	[83RAI/STR], [95RAI/FEL], [97RAI/FEL], [98RAI/FEL], [99RAI/HES], [99RAI/HES2], [2000RAI/MOO], [2001RAI/MOO], [2002RAI/GOR] [2003RAI/YUI]
Moore, R. C.	[2001RAI/MOO]
Morales, L. A.	[2000HAS/ALL], [2001HAS/ALL]
Morgan, J. J.	[81STU/MOR]
Moriyama, H.	[90PRA/MOR], [98KIT/YAM], [98YAM/KIT], [2001FUJ/YAM]
Moroni, L. P.	[95MOR/GLA]
Morris, D. E.	[99CLA/CON], [2000KON/CLA]
Morss, L. R.	[70MOR/COB], [76MOR/MCC], [82MOR], [83MOR/WIL], [83MOR/WIL2], [84WIL/MOR], [86MOR], [89MOR/ELL], [92MOR/WIL], [93TAK/FUJ], [94MOR/WIL], [98DAV/FOU]
Mortimer, M. J.	[90HAL/JEF2], [91HAL/HAR], [91HAL/MOR]
Moskvin, A. I.	[62GEL/MOS], [67GEL/MOS], [71MOS], [73MOS]
Mosselmans, J. F. W.	[99DOC/MOS]
Moukhamet-Galeev, A.	[98AAS/MOU]
Moulin, C.	[95MOU/DEC], [97SCA/ANS], [98MOU/LAS], [98SCA/ANS]
Moulin, V.	[95MOU/DEC], [98MOU/LAS]
Muñoz, M.	[87BRU/CAS]
Mueller, M. E.	[48MUE]
Muis, R. P.	[82COR/MUI]
Mulford, R. N. R.	[66OLS/MUL2], [88MUL/SHE], [91SHE/MUL]
Muller, A. B.	[85MUL], [92GRE/FUG]
Munoz, J. L.	[86NOR/MUN]
Muraoka, Y.	[94OMO/MUR]
Murphy, W. H.	[97MUR]
Musikas, C.	[72MUS], [2000BEN/FAT], [2000VIC/FAT]
Muto, T.	[65MUT/HIR], [65MUT]
Myasoedov, B. F.	[96KUL/MAL]
Nàer-Neumann, E.	[85FER/GRE]

(Continued on next page)

Authors list (continued)

Author	Reference
Nagarajan, K.	[79PRA/NAG], [80NAG/BHU]
Nagashima, K.	[74MOC/NAG]
Nagypál, I.	[90BEC/NAG]
Naik, J. S.	[93NAI/VEN]
Nair, G. M.	[68NAI], [82NAI/CHA]
Nakagawa, T.	[98NAK/NIS]
Nakajima, K.	[97NAK/ARA], [99NAK/ARA], [99NAK/ARA2]
Nakama, S.	[2001FUJ/SAT]
Nakayama, S.	[92KIM/SER], [96NAK/YAM]
Namboodiri, P. N.	[98MIS/NAM], [99BAS/MIS]
Naqvi, S. J.	[68AZI/LYL]
Nash, K. L.	[84NAS/CLE], [84NAS/CLE2]
Naumov, G. B.	[71NAU/RYZ], [72NIK/SER]
Navratil, J. D.	[92FUG/KHO]
Ndalamba, P.	[93ERI/NDA]
Neck, V.	[91KIM/KLE], [92NEC/KIM], [94NEC/KIM], [94NEC/RUN], [95FAN/NEC], [95NEC/FAN], [95NEC/RUN], [96FAN/NEC], [97KON/NEC], [97NEC/FAN], [98FAN/WEG2], [98NEC/FAN], [98NEC/KON], [98NEC/KON2], [99FAN/KON], [99KNO/NEC], [99NEC/KIM], [2000BUN/KNO], [2000NEC/KIM], [2001NEC/KIM], [2001NEC/KIM2]
Nectoux, F.	[93STO/CHO]
Nelson, F.	[50KRA/NEL]
Neu, M. P.	[95ALL/BUC], [95CLA/HOB], [96CLA/CON], [96RUN/NEU], [97CLA/CON], [97NEU/REI], [97RUN/NEU], [98CON/ALM], [99RUN/REI], [2000REI/NEU]
Newton, A. S.	[58JOH/BUT]
Nguyen, S. N.	[88OHA/LEW], [92NGU/SIL], [96ALL/SHU]
Nguyen-Trung, C.	[76NGU/POT], [87DUB/RAM], [89HOV/NGU], [92GRE/FUG], [92NGU/BEG], [95PAL/NGU], [2000NGU/BUR], [2000NGU/PAL] , [2002NGU]
Nikitin, A. A.	[72NIK/SER]
Nikitin, M. I.	[93NIK/TSI]
Nikolaeva, N. M.	[71NIK/PIR], [71NIK], [78NIK/PIR]
Nikonov, M.	[98DAV/FOU]
Nilsson, G.	[58FRY/NIL], [58NIL/REN]
Nishikawa, S.	[98YAM/KIT]
Nishimaki, K.	[98NAK/NIS]
Nissen, S. C.	[98BAL/HEA]
Nitani, N.	[95KAT/KIM], [96KAT/KIM]

(Continued on next page)

Authors list (continued)

Author	Reference
Nitsche, H.	[83EDE/BUC], [84SIL/NIT], [85NIT/EDE], [88STA/NIT], [89NIT/STA], [95MOL/MAT], [96BER/GEI], [96BRE/GEI], [96GEI/BRA], [96MOL/GEI], [97AMA/GEI], [97BER/GEI], [97GEI/BER], [97GEI/RUT], [97MOL/GEI], [97RUT/GEI], [98AMA/BER], [98GEI/BER], [98GEI/BER2], [98GEI/BER3], [98GEI/BER4], [98MOL/GEI], [99RUT/GEI], [2000REI/BER], [2001BER/GEI], [2001LEM/FUG]
Norén, B.	[69NOR]
Nordstrom, D. K.	[86NOR/MUN]
Nottorf, R.	[44NOT/POW]
Novak, C. F.	[95NOV/CRA], [95NOV/ROB], [96RAO/RAI], [96ROB/SIL], [97NOV/ALM], [98ALM/NOV], [99RAO/RAI]
Novotný, P.	[85SOH/NOV]
Nriagu, J. O.	[84NRI2]
Nuttall, R. L.	[82WAG/EVA]
Nyvlt, J.	[81BRO/NYV]
O'Brien, T. J.	[81OBR/WIL], [83OBR/WIL]
O'Hare, P. A. G.	[74OHA/HOE3], [75OHA/HOE2], [78COR/OHA], [81OHA/FLO], [82COR/MUI], [88OHA/LEW], [88TAS/OHA]
Oakes, C. S.	[2000RAI/MOO]
Ochiai, A.	[94OCH/SUZ]
Oda, T.	[93ODA]
Oechsner de Coninck, W.	[05OEC/CHA]
Oelkers, E. H.	[90OEL/HEL], [92SHO/OEL]
Oetting, F. L.	[76FUG/OET], [76OET/RAN], [83FUG/PAR]
Ogawa, T.	[93OGA], [98OGA/KOB]
Ogden, J. S.	[87ALE/OGD]
Ogorodnikov, V. V.	[93OGO/ROG]
Ohnuki, T.	[94YAM/SAK]
Ohse, R. W.	[87MAT/OHS]
Olander, D. R.	[99HAS/WAN], [99OLA]
Olofson, U.	[84ALL/OLO]
Olofsson, G.	[75OLO/HEP]
Olson, R. A.	[2000BUR/OLS]
Olson, W. M.	[66OLS/MUL2]
Omelyanenko, B. I.	[89RED/SAV], [89SER/SAV]
Omenetto, N.	[89GRE/BID], [91BID/CAV], [95ELI/BID]
Omori, T.	[94OMO/MUR], [97ASA/SUG], [97OMO/MIY], [97SUG/SAT], [97SUG/SAT2]
Onink, M.	[91COR/VLA]

(Continued on next page)

Authors list (continued)

Author	Reference
Ono, F.	[98HUA/YAM]
Oosting, M.	[60OOS]
Östhols, E.	[94OST/BRU], [99WAN/OST], [2000GRE/WAN], [2000OST/WAN], [2000WAN/OST]
Ouchi, K.	[91FUJ/YAM]
Ouweltjes, W.	[75COR/OUW], [77COR/OUW2], [81COR/OUW], [81COR/OUW2], [82COR/MUI], [88COR/OUW], [90COR/KON2]
Pérez, I.	[97PER/CAS], [2000PER/CAS]
Pagès, M.	[93STO/CHO]
Palmer, C. E. A.	[96ALL/SHU], [97ALL/SHU], [97WRU/PAL]
Palmer, D. A.	[92NGU/BEG], [95PAL/NGU], [2000NGU/BUR], [2000NGU/PAL]
Palmer, P. D.	[95ALL/BUC], [96CLA/CON], [97CLA/CON], [97RUN/NEU], [98CLA/CON], [98CON/ALM], [98EFU/RUN], [99CLA/CON]
Palombari, R.	[79CIA/FER]
Pan, P.	[95PAN/CAM], [98PAN/CAM]
Panov, A. S.	[96RAK/TSY]
Park, K.	[2001FUJ/PAR]
Park, K. K.	[94KIM/CHO]
Park, Y. J.	[96PAR/PYO]
Park, Y. Y.	[90PAR/SAK], [93MIZ/PAR]
Parker, V. B.	[65PAR], [82WAG/EVA], [83FUG/PAR], [87GAR/PAR]
Parks, G. A.	[88PAR/POH]
Parma, L.	[95ELI/BID]
Parsons, R.	[74PAR], [85BAR/PAR]
Pashalidis, I.	[93PAS/RUN], [95PAS/KIM], [97PAS/CZE]
Patat, S.	[93PAT/DUE], [95DUE/PAT]
Patil, S. K.	[64BAN/PAT], [85SAW/RIZ], [90SAW/CHA2]
Patterson, C. W.	[85ALD/BRO]
Pattoret, A.	[66DRO/PAT], [68PAT/DRO]
Paviet, P.	[94FAN/KIM], [96PAV/FAN]
Paviet-Hartmann, P.	[98FAN/WEG2], [99FAN/KON]
Pazukhin, E. M.	[89PAZ/KOC]
Peddicord, K. L.	[90HAY/THO]
Peiffert, C.	[2000NGU/PAL]
Pekárek, V.	[65VES/PEK]
Pelsmaekers, J.	[83VOC/PEL], [84VOC/GRA], [86VOC/GRA]
Penny, D. J.	[86DIC/PEN], [89DIC/LAW]
Perachon, G.	[78PER/THO]
Peretrukhin, V. F.	[95PER/SHI], [96PER/KRY], [97MAS/COU], [98MAS/MAS], [2002MAS/PER]

(Continued on next page)

Authors list (continued)

Author	Reference
Perez-Bustamante, J. A.	[62PER/BER], [65PER]
Perkovec, V. D.	[67PER/KRY]
Pershin, A. S.	[90PER/SAP]
Person, W. B.	[85ALD/BRO]
Peterson, A.	[61PET]
Peterson, J. R.	[78PIT/PET], [84FUG/HAI], [90FUG/HAI], [93FUG/HAI]
Petrov, V. G.	[89PET/SEL]
Phillips, C. A.	[85PHI/PHI]
Phillips, S. L.	[85PHI/PHI], [88PHI/HAL], [90PHI], [94SIE/PHI], [95COH/LOR]
Pialoux, A.	[94TOU/PIA], [98PIA/TOU], [99PIA/TOU]
Pikaev, A. K.	[95PER/SHI]
Pinkerton, T. C.	[88KEN/MIK]
Piret, P.	[81VOC/PIR]
Pirozhkov, A. V.	[71NIK/PIR], [78NIK/PIR]
Pissarjewsky, L.	[00PIS]
Pitzer, K. S.	[40COU/PIT], [73PIT], [73PIT/MAY], [74PIT/KIM], [74PIT/MAY], [75PIT], [76PIT/SIL], [78PIT/PET], [79BRA/PIT], [79PIT], [91PIT]
Plyasunov, A. V.	[97ALL/BAN], [97GRE/PLY], [97PUI/RAR]
Pohl, D. C.	[88PAR/POH]
Popov, M. M.	[59POP/GAL], [59POP/KOS]
Porter, R. A.	[71POR/WEB]
Porto, R.	[87CIA/IUL]
Posey, J. C.	[65WOL/POS]
Potter, P. E.	[90COR/KON], [2001LEM/FUG]
Poty, B.	[76NGU/POT], [87DUB/RAM]
Powell, A. V.	[88DIC/POW], [89DIC/LAW], [90POW]
Powell, J.	[44NOT/POW]
Pozdnyakov, S. V.	[89PET/SEL]
Prasad, R.	[78SIN/PRA], [79PRA/NAG], [80NAG/BHU], [82ROY/PRA], [87VEN/IYE], [99SIN/DAS], [2000DAS/SIN], [2001BAN/PRA]
Prather, J.	[44FER/PRA], [44FER/PRA2]
Pratopo, M. I.	[90PRA/MOR]
Pravdic, V.	[69CAJ/PRA]
Prins, G.	[74COR/PRI], [75COR/OUW], [90COR/KON]
Prussin, T.	[96ROB/SIL], [97NOV/ALM]
Puigdomènech, I.	[95GRE/PUI], [95PUI/BRU], [95SIL/BID], [97ALL/BAN], [97PUI/RAR], [98SPA/PUI]

(Continued on next page)

Authors list (continued)

Author	Reference
Streeter, I.	[45DAV/STR]
Strickert, R. G.	[83RAI/STR]
Stull, D. R.	[56STU/SIN]
Stumm, W.	[81STU/MOR]
Subbanna, C. S.	[92VEN/KUL], [93NAI/VEN]
Suescun, L.	[97KRE/GAN]
Suganuma, H.	[94OMO/MUR], [97ASA/SUG], [97OMO/MIY], [97SUG/SAT], [97SUG/SAT2]
Sugimoto, J.	[98HUA/YAM]
Suglobov, D. N.	[2000GOR/LUM], [2000GOR/MIR]
Sukhov, V. Yu.	[2000GOR/MIR]
Sullivan, J. C.	[82SUL/WOO], [94RIZ/RAO], [2001LEM/FUG], [2001WIL/BLA]
Sundaresh, V.	[87VEN/IYE]
Sutton, J.	[49SUT]
Suzuki, S.	[99SUZ/TAM]
Suzuki, T.	[94OCH/SUZ]
Suzuki, Y.	[94OCH/SUZ], [96YAM/HUA], [97HUA/YAM], [97NAK/ARA], [98SUZ/ARA], [99NAK/ARA], [99NAK/ARA2]
Sverjensky, D. A.	[89SHO/HEL2], [92SHO/OEL], [97SHO/SAS]
Swang, O.	[95HAA/MAR]
Swanson, J. L.	[87RAI/SWA]
Sylva, R. N.	[79SYL/DAV]
Szabó, Z.	[97SZA/AAS], [2000MOL/REI], [2000MOL/REI2], [2000SZA/MOL], [2001VAL/WAH], [2002VAL/WAH]
Söhnel, O.	[85SOH/NOV]
Tachimori, S.	[99SUZ/TAM]
Tagawa, H.	[79TAG/FUJ]
Tagirov, B. R.	[98DIA/RAG], [98DIA/TAG]
Tait, C. D.	[95ALL/BUC], [96CLA/CON], [97CLA/CON], [97RUN/NEU], [98CLA/CON], [98CON/ALM], [98EFU/RUN], [99CLA/CON], [2000KON/CLA]
Takahashi, K.	[92KIM/SER], [93TAK/FUJ]
Takahashi, Y.	[97TAK]
Takeishi, H.	[92KIM/SER], [93MEI/TAK]
Tamura, K.	[99SUZ/TAM]
Tananaev, I. G.	[90TAN], [96PER/KRY], [98BUD/TAN]
Tanet, G.	[91BID/CAV]
Tanger, IV, J. C.	[88TAN/HEL]
Tarasov, V. P.	[2002KIR/GER]
Tardy, Y.	[74TAR/GAR], [76TAR/GAR], [84VIE/TAR]
Tasker, I. R.	[88TAS/OHA]

(Continued on next page)

Authors list (continued)

Author	Reference
Tatarinova, E. E.	[89TAT/SER]
Taylor, J. C.	[71TAY/HUR], [72TAY/KEL]
Taylor, J. R.	[82TAY], [90TAY/DIN]
Teague, L. S.	[80BEN/TEA]
Teichteil, C.	[99VAL/MAR]
Thévenin, T.	[93BOI/ARL]
Thibault, Y.	[2000BUR/OLS]
Thomas, J. B.	[96FAL/REA]
Thomas, J. K.	[90HAY/THO]
Thomason, H. P.	[86EWA/HOW]
Thompson, R. W.	[42SCH/THO], [42THO/SCH], [48THO/SCH]
Thourey, J.	[78PER/THO]
Timofeev, G. A.	[98ERI/BAR]
Tofield, B. C.	[82BAR/JAC]
Tomilin, S. V.	[79VOL/VIS2]
Tomiyasu, H.	[90PAR/SAK], [93MIZ/PAR], [98FAZ/YAM]
Tondre, C.	[98MOU/LAS]
Torrero, M. E.	[91AGU/CAS], [94TOR/CAS], [97PER/CAS], [97TOR/BAR], [98CAS/PAB]
Torretto, P. C.	[96ROB/SIL]
Torstenfelt, B.	[84ALL/OLO]
Toscano, E. H.	[98SER/RON]
Toshida, Z.	[96MEI/KAT]
Tossidis, I. A.	[76TOS]
Toth, I.	[91BRU/GLA], [95BAN/GLA]
Touzelin, B.	[94TOU/PIA], [98PIA/TOU], [99PIA/TOU]
Tran The, P.	[91VIT/TRA]
Tremaine, P. R.	[80LEM/TRE], [86HOV/TRE], [88HOV/HEP3], [89HOV/NGU]
Tretyakov, A. A.	[81GOL/TRE], [81GOL/TRE2]
Tripathi, S. N.	[74DHA/TRI], [98MIS/NAM], [99BAS/MIS]
Trivedi, D. P.	[98CHA/TRI]
Trofimov, T. I.	[96KUL/MAL]
Tsirel'nikov, V. I.	[93NIK/TSI]
Tso, T. C.	[85TSO/BRO]
Tsyplenkova, V. L.	[96RAK/TSY]
Turck, G.	[76BOU/BON]
Tweed, C. J.	[87CRO/EWA]
Twichell, L. P.	[47RYO/TWI]
Ueno, K.	[75UEN/SAI]
Ueta, S.	[95YAJ/KAW]
Ullman, W. J.	[88ULL/SCH], [2001LEM/FUG]
Umreiko, D. S.	[78KOB/KOL]

(Continued on next page)

Authors list (continued)

Author	Reference
Une, K.	[85UNE]
Uno, M.	[2001YAM/KUR]
Unrein, P. J.	[76CHO/UNR]
Urabe, T.	[98NAK/NIS]
Usui, Y.	[99SUZ/TAM]
Vallet, V.	[99VAL/MAR], [99WAH/MOL], [2001VAL/WAH], [2002VAL/WAH]
Valsami-Jones, E.	[97VAL/RAG]
Van Egmond, A. B.	[75COR/EGM]
Van Genderen, A. C. G.	[84GEN/WEI]
Van Haverbeke, L.	[90VOC/HAV], [91VOC/HAV]
Van Vlaanderen, P.	[91COR/VLA], [97COR/BOO]
Van Voorst, G.	[75COR/EGM]
Van der Laan, R. R.	[2000LAA/KON]
Van der Weijden, C. H.	[84GEN/WEI]
Varfeldt, J.	[69GRE/VAR]
Vasca, E.	[93FER/SAL]
Vasil'ev, V. P.	[62VAS]
Vasil'kova, I. V.	[56SHC/VAS2]
Vaufrey, F.	[97BIO/MOI]
Veits, I. V.	[81GLU/GUR], [82GLU/GUR]
Ventura, O. N.	[97KRE/GAN]
Venugopal, V.	[78SIN/PRA], [79PRA/NAG], [80NAG/BHU], [82ROY/PRA], [87VEN/IYE], [90IYE/VEN], [92VEN/IYE], [92VEN/KUL], [93JAY/IYE], [93NAI/VEN], [94JAY/IYE], [95SAL/JAY], [96IYE/JAY], [96JAY/IYE], [96SAL/IYE], [97IYE/JAY], [97JAY/IYE], [97KRI/RAM], [97SAL/KRI], [98BAN/SAL], [98JAY/IYE], [98KRI/RAM], [99JAY/IYE], [99SAX/RAM], [99SIN/DAS], [2000DAS/SIN], [2001BAN/PRA]
Vermeulen, J. M.	[93BOI/ARL]
Vesala, S.	[69SAL/HAK]
Veselý, V.	[65VES/PEK]
Vichot, L.	[2000VIC/FAT]
Vidavskii, L. M.	[71SAN/VID], [72SAN/VID]
Vieillard, P.	[84VIE/TAR]
Visyashcheva, G. I.	[74VIS/VOL], [74VOL/KAP2], [77VOL/VIS], [79VOL/VIS2]
Vitorge, P.	[84VIT], [86GRE/ROB], [87RIG/VIT], [87ROB/VIT], [89RIG/ROB], [90CAP/VIT], [91VIT/TRA], [93GIF/VIT], [93GIF/VIT2], [94CAP/VIT], [95CAP/VIT], [96CAP/VIT], [98CAP/VIT], [99CAP/VIT], [99RAI/HES2], [2001LEM/FUG]

Authors list (continued)

Author	Reference
Vochten, R.	[81VOC/PIR], [83VOC/PEL], [84VOC/GOE], [84VOC/GRA], [86VOC/GRA], [90VOC/HAV], [91VOC/HAV]
Voigt, W.	[98MEI/FIS]
Vokhmin, V.	[98DAV/FOU], [2000DAV/FOU], [2001DAV/VOK]
Volden, H. V.	[95HAA/MAR]
Volke, P.	[99MEI/VOL]
Volkov, A. Y.	[98ERI/BAR]
Volkov, Y. F.	[74VIS/VOL], [74VOL/KAP2], [77VOL/VIS], [79VOL/VIS2]
Von Braun, R.	[75BRA/KEM]
Wadsak, W.	[2000WAD/HRN]
Wagman, D. D.	[52ROS/WAG], [73HUL/DES], [82WAG/EVA], [89COX/WAG]
Wagner, E. L.	[45WAG/GRA]
Wahlgren, U.	[99VAL/MAR], [99WAH/MOL], [2001BOL/WAH], [2001BOL/WAH2], [2001VAL/WAH], [2002VAL/WAH]
Wait, E.	[55WAI]
Wakita, H.	[74MOC/NAG]
Wall, I.	[75BRA/KEM]
Wang, P.	[90ARC/WAN]
Wang, W. E.	[99HAS/WAN]
Wanner, H.	[86WAN], [88WAN], [91WAN], [92GRE/FUG], [95SIL/BID], [99RAR/RAN], [99WAN/OST], [2000GRE/WAN], [2000OST/WAN], [2000WAN/OST], [2001LEM/FUG]
Ward, M.	[56WAR/WEL]
Wasserman, S. R.	[99SOD/ANT]
Waterbury, G. R.	[61WAT/DOU]
Weare, J. H.	[84HAR/MOL], [86FEL/WEA]
Weber, Jr., W. J.	[71POR/WEB]
Webster, R. A.	[43WEB]
Weed, H. C.	[92NGU/SIL]
Weger, H. T.	[98FAN/WEG], [98FAN/WEG2], [99FAN/KON]
Weigel, F.	[76WEI/WIS], [82WEI/WIS]
Weisbrod, A.	[87DUB/RAM]
Weissman, S.	[43ALT/LIP]
Welch, G. A.	[56WAR/WEL]
Weller, M. T.	[85DIC/LAW], [86DIC/PEN]
Werme, L. O.	[98WER/SPA]
Werner, G. K.	[85ALD/BRO]

(Continued on next page)

Authors list (continued)

Author	Reference
Westrum, Jr., E. F.	[80CHI/WES], [80COR/WES], [84GRO/DRO], [89COR/KON]
White, A. G.	[57HEA/WHI]
White, Jr., H. J.	[87GAR/PAR]
Whiteley, M. W.	[99DOC/MOS]
Whitfield, M.	[79WHI]
Wilcox, D. E.	[62WIL]
Wilkinson, G.	[80COT/WIL]
Williams, C.	[99SOD/ANT]
Williams, C. W.	[83MOR/WIL], [83MOR/WIL2], [84WIL/MOR], [91CAR/LIU], [92MOR/WIL], [94LIU/CAR], [94MOR/WIL], [2001WIL/BLA]
Williams, P. A.	[79HAA/WIL], [80ALW/WIL], [81OBR/WIL], [83OBR/WIL]
Williams, S. J.	[86EWA/HOW]
Willis, M.	[97SHO/SAS]
Wilmarth, W. R.	[90FUG/HAI]
Wimmer, H.	[92WIM/KLE], [94KIM/KLE]
Wipff, G.	[2001SEM/BOE]
Wishnevsky, V.	[76WEI/WIS], [82WEI/WIS]
Wolf, A. S.	[65WOL/POS]
Woods, M.	[82SUL/WOO]
Wruck, D. A.	[97WRU/PAL]
Wyart, J. F.	[92BLA/WYA]
Xi, R. H.	[92KRA/BIS]
Xue, Z.-L.	[98DAI/BUR]
Yagi, M.	[97SUG/SAT2]
Yajima, T.	[95YAJ/KAW]
Yakovlev, G. N.	[74VIS/VOL], [74VOL/KAP2]
Yamada, K.	[2001FUJ/SAT]
Yamada, M.	[2001FUJ/PAR]
Yamaguchi, K.	[96YAM/HUA], [97HUA/YAM], [98HUA/YAM], [98ITO/YAM]
Yamaguchi, T.	[94YAM/SAK], [96NAK/YAM]
Yamamoto, T.	[98ITO/YAM], [98KIT/YAM], [98YAM/KIT]
Yamamura, T.	[98FAZ/YAM], [98KIT/YAM], [98YAM/KIT]
Yamana, H.	[2001FUJ/YAM]
Yamanaka, S.	[2001YAM/KUR]
Yamashita, T.	[79TAG/FUJ], [89YAM/FUJ], [91FUJ/YAM]
Yamawaki, M.	[96YAM/HUA], [97HUA/YAM], [98HUA/YAM], [98ITO/YAM], [99NAK/ARA2]
Yasumoto, M.	[96YAM/HUA], [97HUA/YAM], [98HUA/YAM]
Yeh, M.	[2000YEH/MAD]
Yoo, C.	[98YOO/CYN]

(Continued on next page)

Authors list (continued)

Author	Reference
Yoshida, Z.	[93MEI/KAT], [94KAT/MEI], [95KAT/KIM], [96KAT/KIM], [98MEI/KAT], [99MEI/KAT]
Young, H. S.	[58YOU/GRA]
Yui, M.	[98RAI/FEL], [99RAI/HES], [99RAI/HES2], [2000RAI/MOO], [2002RAI/GOR], [2003RAI/YUI]
Yungman, V. S.	[81GLU/GUR], [82GLU/GUR], [85HIL/GUR]
Zékany, L.	[95BAN/GLA]
Zaitsev, L. M.	[67GEL/MOS]
Zaitseva, V. P.	[62GEL/MOS]
Zapol'skii, A. K.	[89STA/MAK]
Zinov'ev, V. E.	[75SPI/ZIN]
Zmbov, K. F.	[69ZMB]
Zotov, A. V.	[89SER/SAV]
Zubova, N. V.	[59POP/KOS]
Zwick, B. D.	[95ALL/BUC]
Whiffen, D. H.	[79WHI2]